Solar technology for building: Volume 1

Volume 1: Sessions 1 to 4

Solar technology for building

Proceedings of the first International Conference on Solar building technology
25–29 July 1977
At the Royal Institute of British Architects, London
Organised by UNESCO and NELP

Edited by Costis Stambolis

RIBA Publications Limited

Organised by:

United Nations Educational and Scientific Organisation and

North East London Polytechnic, Faculty of Environmental Studies,

Department of General Surveying and Construction

Sponsored by:

International Solar Energy Society

Intermediate Technology Development Group

New Mexico Solar Energy Association

UK Solar Trade Association

ISBN: 0 900630 66 3

Published for UNESCO and NELP

by RIBA Publications Limited, Finsbury Mission,

Moreland Street, London EC1V 8VB

Printed in Great Britain by Unwin Brothers Ltd,

The Gresham Press, Old Woking, Surrey

Contents

Volume 1

Volume 2

Opening ceremony and Solar Show

The opening ceremony

Left to right:
C. Stambolis, Conference Co-ordinator
Alderman Edwards, Chairman of the Board of Governors, NELP
J. Overall, Chairman of the Organising Committee
Alex Eadie MP, Under-Secretary of State for Energy, UK
Dr B. Boiko UNESCO representative

A view of the Solar Show at Regents Park, organised in parallel with the Conference.

Conference opening

1

1.1 Editorial introduction

C. Stambolis (Conference co-ordinator)
Department of General Surveying and Construction
North East London Polytechnic

Since the so called 'energy crisis' of 1973 the world has become more aware of the need to lessen its dependence on mineral oil resources; their proven and un-proven deposits are finite and cannot possibly support an ever increasing demand for energy consumption. If we stop considering nuclear energy as being the only viable alternative to power generation for the future, in view of the mounting controversy, two types of energy strategy appear to be feasible in the coming years of energy scarcity. The first strategy implies the establishment of national and international policies whose aim shall be to achieve sizeable reductions of current levels of energy demand. The second strategy should aim at utilizing at a maximum level every conceivable form of inexhaustible natural energy sources. These sources include solar and wind energy, hydrogen, bio gas, ocean thermal gradients, hydro, wave and geothermal energy. One of the most promising of these alternative energy sources is solar energy. Today, one of the most successful applications of solar energy is its use for servicing the energy needs of buildings i.e. provision of domestic hot water, space heating and cooling and even electricity.

Hundreds of thousands of solar water heating units have been installed and are successfully operating in many countries, situated over a wide range of latitudes from 55° North - 55° South. Thousands of solar heated and cooled buildings have already been built in different places all over the world and many more thousands are under construction. A new industry is being born and many new areas of work have been created for architects and engineers, builders and installers. The prospects for utilising the sun's energy for the benefit of mankind are now better than they have ever been before. However, if solar energy is to make world wide impact, in lessening our dependence on mineral oil and contributing to the creation of an unpolluted environment, international co-operation is vital at all levels, especially between industrialised and less industrialised countries.

Another important factor of solar energy is that it can provide a basic energy source in many poor areas of the world where no other energy, in whatever form, is available because of the high cost which is often required. It is appropriate to quote here Farrington Daniels, who in the introduction of his pioneering work, 'Direct Use of the Sun's Energy', wrote the following: "The rapidly developing countries are eagerly looking for new sources of industrial power. Some are expecting too much too soon, and it is important to emphasize that education and research and the spirit of public service are more important than kilowatts. If however it is a question of 'water or no water' or 'power or no power', no price is too high to pay". Solar energy could therefore play a leading role in the development of many sunny but poor areas of the world. This fact was clearly demonstrated during the conference when many new projects and ideas were announced by scientists, architects and engineers who work in developing regions.

The importance of the 'International Conference on Solar Building Technology' was that for the first time a major international conference of its kind backed by United Nations dealt exclusively with the subject of utilising the sun's energy for servicing buildings and discussed ways in which solar energy could be equally used throughout the world for providing essential energy needs in human settlements. The conference was organised jointly by UNESCO and North East London Polytechnic. It took place at the Royal Institute of British Architects from July 25-29, 1977 with the participation of more than 300 delegates from 40 different countries. From 87 scheduled papers, a total of 71 were actually presented, and the provisional conference proceedings, containing all the abstracts and some of the invited papers in full, were published by NELP and UNESCO in two volumes and were available during the conference.

These two volumes of conference proceedings contain in full all the papers which were presented during the conference, and those papers whose authors were unable to come in person at the conference, but which had been accepted for presentation by the conference's scientific committee. During the conference a special evening session was organised to which a number of authorities had been invited to discuss some UK based projects. The papers resulting from these presentations have not been published under a separate section but have been included under the most appropriate section of the proceedings.

Some editorial liberties were taken during the preparation of this publication. Although some papers were published virtually untouched, a significant number had to be edited, mainly from a language point of view, in order to express clearly important points made by the authors. As editor I assume responsibility for the modifications made but do not necessarily agree with all the ideas expressed in the papers. However, everything possible was done to enhance the value of each paper which was accepted for publication in both style and clarity of ideas, in both the languages of the conference i.e. English and French.

These proceedings are divided into nine sections following the exact sequence of the nine sessions of the conference. The first four sections are published in the first volume and the remaining five are published in the second. The subjects covered in the first four sections of the first volume are: Architectural, Technical and Socio-economic aspects; Flat plate collectors; Design methodology and Space Heating and Cooling. In the second volume the subjects included are: Space Heating and Cooling Projects; Solar Energy Utilisation in Developing Countries with special reference to buildings; Solar Water Heating; General Solar Energy applications related to buildings. Finally the conclusions and recommendations of the conference's scientific committee have been reproduced in section nine. In the first volume the remarks and speeches made during the opening session of the conference have been published, in an edited form, including the statement made by Mr Alex Eadie, M.P., Under Secretary of State for Energy of H.M. Government.

The high technical standard of the conference was guaranteed by the conference's scientific committee which met for the first time in London in May 1977, and selected the most suitable papers. Chairman of the Scientific Committee was Professor Erich Farber, of the University of Florida, who chaired the committee during the conference and co-ordinated the work of the various working groups. A number of world authorities were also invited to present papers on specific topics in order to introduce the various themes of the conference programme. This proved very successfull since every invited speaker also reviewed the state of the art in his field. Also the published versions of the invited papers (as these were included in the provisional conference proceedings) provided a very good ground for the discussions which took place during the conference.

The greatest value of a conference like this is that the participants have an opportunity to meet, listen to each other and exchange ideas informally outside the various sessions. However, the effectiveness of any conference is much greater if following it the full proceedings are published. In this way two main goals can be achieved. First, the amount of effort which has gone into the organisation of the conference is not wasted and the many different ideas presented become unified and constitute important reference material. Secondly, the message of the conference can reach a far wider audience than the few participants who managed to make it to the conference room. Although many solar conferences are held every year very few proceedings become available. It is hoped that with the publication of these volumes one further step has been taken in making knowledge in this fast growing field widely available.

I wish to thank all who have assisted in the preparation of these volumes, too numerous to name here. My appreciation goes also to Professor Erich Farber who by chairing the executive committee of the scientific committee was largely responsible for summarising the proposals put forward by the four working groups during the conference. Finally I would like to thank all contributing authors for their assistance and co-operation.

Dr. Erich A. Farber
Professor & Research Professor
Director, Solar Energy & Energy Conversion Laboratory
University of Florida, USA

It was indeed a pleasure and a privilege to act as chairman for the International Conference on Solar Building Technology, and the United Nations Educational and Scientific Organization (UNESCO) and the North East London Polytechnic (NELP) have to be congratulated for organizing such a meeting under the sponsorship of the International Solar Energy Society, the Intermediate Technology Development Group, the UK Solar Trade Association and the New Mexico Solar Energy Association.

This meeting demonstrated how international cooperation can be used to solve problems and how different nations with different backgrounds can work together. I counted forty nations from all parts of the world representing their people and participating in the events.

It was heartwarming to see that nations who in the past followed the lead of some of the so-called 'technologically advanced' nations, use their own initiative and develop energy conversion methods particularly applicable to their own needs and not developed by someone else. This is the method by which progress toward the improvement of mankind is made.

Four working groups on:

(a) Educational aspects
(b) Passive systems
(c) Active systems
(d) Economic, social, legal, and political aspects

worked hard and produced concrete suggestions and specific recommendations as presented in this report of the Conference Proceedings.

Topics for Short Courses and Workshops were suggested, new devices and patents presented and discussed, experts pinpointed as well as subjects to be vigorously supported.

I heard many complimentary remarks about the conference but what made me proud to be associated with this group was that throughout the meeting all political differences were set aside resulting in a true spirit of international cooperation.

<u>Scientific Committee:</u>

Professor Erich Farber, Chairman
University of Florida, FL, USA

John Eberhard,
AIA Research Group, Washington DC, USA

Dr Douglas Balcomb,
Los Alamos Scientific Laboratories, NM, USA

Dr Boris Boiko,
UNESCO

Professor Bryan Brinkworth,
University College, Cardiff, UK

George Chouleur,
Architect, Nimes, France

Professor C. V. Chinnapa,
University of Technology, Lae, Papua, New Guinea

Dr Morris Davies,
Liverpool University, UK

Demos Daskalakis,
Solar Engineer, Athens, Greece

Professor Peter Dunn,
Reading University, UK

Professor Tom Scott-Dean,
Kansas University, USA

Peter van Dresser,
Sundwellings Project, Santa Fe, New Mexico, USA

Professor E. Hahne,
Stuttgart University, FGR

Ian Hogan,
Architect, London, UK

Dominic Michaelis,
Architect, Solar Energy Development, London, UK

Dr Cleveland McVeigh,
Brighton Polytechnic, UK

Tielman Nikolopoulos,
Architect, Sir William Hallcrow & Partners, London, UK

Professor John Page,
Sheffield University, UK

Dr Argyris Spyridonos,
NRC, Demokritos, Athens, Greece

Costis Stambolis,
NELP, London, UK

T. Steemers,
EEC, Belgium

Dr Harry Tabor,
Scientific Research Foundation, Jerusalem, Israel

Bill Yanda
Solar Sustenance Project, New Mexico, USA

Dr Mehdi Bahadori,
Puhleri University, Shirare, Iran

Ahmet Vatansever,
Engineering Consultant, Istanbul, Turkey

John Overall
Head of the Department of General Surveying and Construction
North East London Polytechnic, Forest Road, London E17 4JB

Ladies and Gentlemen a very good morning to you. As Chairman of the organising committee it is with very great pleasure that I welcome you here this morning. You may care to know that we have approximately 280 delegates either with us or about to join us and that they are drawn from 40 different nations. Now I know that my colleagues wish to comment upon our geographical spread and possibly upon the national mix that we have but I too would like to say something. What I would like to say is thank you, thank you for coming because without you there wouldn't be a conference. Nevertheless now that you are here we have a number of things to do. It is for us through our efforts to make this conference a conference to remember. Let it be said that the 1977 International Conference on Solar Building Technology maintained the spirit of scientific cooperation and through its recommendations added something to man's store of knowledge and furthermore that it helped to lessen his burden of hardship. Let us insure and let us be able to recall to ourselves that we were there.

So much for my welcome, if it is a welcome, it was intended to be, what of my duty, well I have the honour to be your chairman for this first session and thereby introduce to you our honoured guests and speakers this morning but I can't do this until we are properly constituted, therefore without further delay I really have great pleasure in declaring the 1977 International Conference on Solar Building Technology formally in session. Thank you.

The first of my colleagues that I would like to introduce you to this morning is the conference co-ordinator and one of the persons who has been instrumental in making sure that the conference in fact takes place. He is Mr Costis Stambolis who would like to address you now and make a few formal announcements. Thank you.

<u>Organising Committee:</u>

John Overall, Chairman,
Head of the Department of General Surveying and Construction,
North East London Polytechnic (NELP)

W. Foy,
Dean, Faculty of Environmental Studies,
NELP

Ellis Hillman,
Principal Lecturer, Faculty of Environmental Studies, NELP

Dr Boris Boiko,
Department of Higher Education and Technology,
UNESCO

Costis Stambolis, Conference Co-ordinator,
Department of General Surveying and Construction, NELP

Doreen Royds, Conference Secretary,
Department of General Surveying and Construction, NELP

David Gerassi, Exhibition Organiser,
UK Solar Trade Association, London, UK

1.4 Opening remarks

Costis Stambolis
Conference Co-ordinator
Research Associate
North East London Polytechnic

Allow me to make a few introductory remarks concerning the organisation of this conference.

Representatives from more than 40 different countries, from a wide spectrum of socio-economic and climatic backgrounds, have assembled here today to exchange ideas and report progress on solar energy utilisation.

This conference is the first major international gathering to focus its attention on the use of solar energy for buildings.

Solar energy applications in the built environment have intensified greatly over the last few years. This is logical since the energy consumed in the domestic sector is a sizeable portion of any country's national energy budget. It must also be noted that solar energy systems are relatively simple and that their technology is well understood.

The scientific content of this conference consists of, first, a number of invited papers, which will introduce and amplify themes in the various scheduled sessions, and secondly, of papers submitted to the conference, and accepted for presentation. It is hoped that the papers to be presented and the ensuing discussion will provide a base for further action by government, industry and academic institutions. Solar energy use in human settlements is at present the most economically justifiable application.

Let us hope that by the end of the day we will have all benefitted by becoming a little more knowledgeable and more eager to apply what we have learned.

Alderman Edwards,
Chairman of the Board of Governors
North East London Polytechnic

As the chairman has said I am here to welcome you on behalf of the North East London Polytechnic. Sometime ago I was asked if I would be willing to play some nominal part in an international conference on solar building technology. Rather more recently I was asked if I would give the welcoming address and very much more recently I found I was completely lacking in background knowledge about the true purpose of the conference and my part in it. However to go through the formalities I would most sincerely welcome each and everyone of you to this conference especially those delegates from overseas, especially those who are contributing to the conference and a very special welcome to Alex Eadie, Member of Parliament, Her Majesty's Under Secretary of State for Energy.

For the benefit of those of you from overseas without an intimate knowledge of the British system of education I thought I might quickly put you into the picture as far as polytechnics are concerned. Obviously education in the United Kingdom has been developing over many hundreds of years but over the last century that which we call higher education has been developing in two distinct and separate streams, the university stream, with which you are almost certainly all very familiar, and the local authority stream. Now historically for something like a hundred years public institutions, institutions financed out of public funds under the control of local authorities, have prepared students for degrees and higher degrees. About 8 years ago the Secretary of State of the day decided that the highest of higher education in the United Kingdom should be concentrated in about 30 institutions to be known as Polytechnics and first and foremost amongst these is the North East London Polytechnic. It is as its name suggests located over a wide area of North East London. It is built around three major predecessor institutions, the Barking Regional College of Technology, formally the South East Essex Technical College, the South West Essex Technical College at Walthamstow and the West Ham College of Technology based in the London Borough of New Ham. I think it is worthy to note that although the institutions or type of institution may be unfamiliar to you in size and breadth of studies it exceeds most of the universities in the United Kingdom.

Being a major institution of higher education it is right and proper that we should be involved in research and development and obviously coming from North East London we should be concerned with applied research, research which has day to day applications. Your conference is I hope one of many that will be looking not only at new forms of energy but new ways of utilising forms of energy that have existed from the year dot. Obviously solar energy has been with us a long time and I hope it is going to be with us in the future but it is only in recent years that some thought has been given to utilising it on any real scale particularly in temperate climates. We were unable to arrange for this conference to be held inside our own buildings, as they are not situated in accessible parts of London, so we have come to the Royal Institute of British Architects.

I come here to welcome you on behalf of the North East London Polytechnic but although I greatly regret that my time with you both this morning and throughout the week has to be strictly limited I'm sure that each and everyone of you will find the occasion worthwhile. I hope collectively the world knowledge of ways and means of solar energy building technology will be furthered and that at the end of it a publication containing a comprehensive fund of knowledge on this subject will be produced.

1.6 UNESCO's greetings

Boris Boiko
Department of Higher Education and Technology
United Nations Educational and Scientific Organisation

Mr Chairman, Ladies and Gentlemen on behalf of the Director General and Assistant Director General for science of UNESCO I have the honour and pleasure to welcome you at the International Conference on Solar Building Technology. The announcement of this international conference has found a worldwide response and we are glad to welcome about 300 participants from about 40 nations and international institutions. The aim of this conference by the active contribution of participants is to review the present state of science and technology of solar buildings and to formulate conclusions, suggestions and recommendations for future action in such a modern field as solar energy. Recent events in many countries make it apparent that a take off point for large scale solar energy development may be at hand. There is widespread appreciation of the limited lifetime of fossil fuel supplies and in many countries there is also concern about the pollution caused by the use of fossil fuels. Up to now rather than learning how its current solar income evolved the world is living off its rapidly shrinking solar savings, i.e. the coal, oil and gas that were produced by the action of the sun millions of years ago. If we assume that the world will reduce the energy consumption to such an extent that after the year 2,000 no future increase in energy consumption will be necessary even then the remaining resources of fossil fuels will be completely insufficient for the future of mankind. There is therefore wide recognition that solar energy as a renewable and non-polluting source may play a marginal role along with other forms of energy in the next few years.

Solar energy is practical for almost all countries. There is also a remarkable absence of negative factors which would tend to inhibit international co-operation. International co-operation would help to reduce the time for developing technical equipment and to overcome the gap between industrialised and developing countries by initiating common projects and solar energy applications. We are happy that many well known scientists from all over the world will participate in the working sessions and discussions. Besides the exchange of information during this international conference recommendations for future planning and co-ordination resulting from it will represent the highlight of this conference. I should like to wish to all of you that you have an efficient and successful meeting and the results of which will help the advancement to solar building technology in all countries. Thank you.

Alex Eadie MP
Under-Secretary of State for Energy
United Kingdom

It gives me great pleasure to have been invited and to be here today, and to open this International Conference on Solar Building Technology. I should like, on behalf of you all, to thank the North East London Polytechnic, in association with UNESCO, for organising this meeting, which is so relevant and timely to our energy situation.

It is, I think, appropriate that on this occasion, I should outline for you the work which the British Government is supporting in your field, solar energy, and on other alternative energy technologies.

The shock of the oil price rises which followed the Yom Kippur war in 1973 forced Britain and the whole of the industrialised world to look afresh at the energy available to use and how we use it. It made us look at the short term - the need to reduce our dependence on imported oil by developing our indigenous supplies of energy and by trying to use energy less wastefully. But it also made us look further ahead, to the days when Britain's - and the world's - fossil fuels will be running out.

In Britain we have coal reserves to last us 300 years at present rates of consumption, and our known North Sea oil and gas will last into the next century without taking into account any new discoveries. But for the sake of the generations to come we have to look now for the possible fuels of the future.

For the past 20 years nuclear power has provided us in the UK with a steadily growing proportion of our electricity - we now get 13 per cent of our electricity from nuclear power stations. But there is active debate at the moment about the implications of large-scale development of nuclear fission - we do not yet know if the use of fast reactors will be publicly acceptable - and nuclear fusion is unlikely to be available on a large-scale until well into the next century. So we need to examine other alternatives as the renewable sources of energy to give ourselves the widest possible choice for the future. Vast quantities of energy are in principle available in nature - from the sun, the tides, the heat of the earth itself, and from the winds and the waves. The problem with all natural sources of energy, though, is that they are difficult or expensive to tap for an industrial society which needs a concentrated and continuous supply. In Britain we have embarked on an extensive programme of research and development into these renewable, or alternative, sources of energy, to determine what contribution they would make to our energy economy.

Following a series of feasibility studies, the Energy Technology Support Unit, which was set up at Harwell by the Government in 1974, has estimated that renewable sources might possibly be supplying us by early next century with up to the equivalent of about 40 million tonnes of coal a year - or about 12 per cent of all the energy we consumed in the UK last year - provided they are vigorously exploited and no limiting technical difficulties are found.

Solar energy

While wave energy may provide our largest potential source amongst these alternative sources, solar energy would appear to offer a large contribution, and may be the first to have a substantial impact, in providing low grade heat. My department has established a £3.6M four year programme on solar energy research and development, which brings the total expenditure of the British Government on solar research to approximately £6M over that period. In drawing up the programme the department found it very helpful to compare its thinking with that put forward in the ISES (UK) solar report. I am pleased to say that there is a good deal of common ground in our respective views and I hope that working relationships already established will continue. The UK is also participating in the solar energy programme of the European Economic Community and of the International Energy Agency. We consider these international programmes to be very important, and to provide most useful information for participants.

The total solar energy imput to the United Kingdom amounts to 80 times the country's present demand for primary energy. On average over the year this imput is only slightly lower than that for Australia, India or South West USA. However, the variations during the year are large, and our demand for energy peaks in winter, and not in the summer as, for example, in USA and Japan. Further, the amount of direct sunlight we receive is limited, so the use of focussing devices to concentrate solar radiation, boil water and hence operate large power production systems, is not attractive in the UK.

The application of solar energy which has been identified as the area of greatest potential return is its use for water and space heating, and this is a major subject area in the Department of Energy programme. This complements work being carried out by the Department of the Environment, and the Building Research Establishment of that Department will undertake a large share of the work for the Department of Energy. There is considerable research and development effort in industry in this area, and a major part of my department's programme is being carried out through contracts in support of existing industrial initiatives. The response of industry to this part of the programme - and it was only introduced in February last - has been most encouraging.

The Meteorological Office has for many years recorded and published data relevant to the use of solar energy in building. We are supporting additional work to ensure that more information is collected in urban areas, where the majority of the potential applications for solar energy are located, and that it is analysed and presented in the forms most useful to the emerging solar energy industry.

In addition, the United Kingdom's solar energy programme also includes work on research into biological systems, aimed at the assessment of economic production of liquid and gaseous fuels from this source, and through the Department of Industry's programme, on the development of pholovoltaic devices.

Wave energy

Wave energy is potentially most attractive for the UK, which is not surprising, given our long coastline and rough seas. It has been estimated that if usable energy can be extracted from waves with an efficiency of about 25%, an installation of wave devices situated in the most favourable seas off North West Britain and about 600 miles long, could provide about half the present UK requirement for electricity.

So we are undertaking a 2½ year research programme to study a number of specific floating devices and to carry out generic studies of common interest to all devices. This programme, which has recently been increased from £1M to £2.5M aims to establish the feasibility - technically and economically - of the scale extraction of power from waves. If successful, a prototype device could be operating at sea at some time in the 1980's.

Geothermal energy

We are also looking at geothermal energy, in a three-year programme costing £840,000. Geothermal energy is at present being used mainly in those parts of the world where there are naturally occurring geysers and hot springs. The prospect of locating a concealed steam field in the UK can largely be discounted. But heat may exist either in hot dry rocks or in hot water held in porous sedimentary basins. There is not enough information available at present to allow a realistic assessment and the Government's research programme is aimed at getting more data.

Wind energy

The present position on the use of wind energy differs from that of most of the other alternative sources, in that the major part of the technology required to generate electricity using wind machines is established. Large machines must be used in positions with high average wind speeds, and as there are limits to the machine size and the number of sites, the estimated contribution of wind energy in the United Kingdom is not large. The present uncertainties are the economics and the public acceptibility of the use of large machines on hill tops or coasts.

To establish a firm basis for economic assessment the Government has initiated a design study for a reference design of aero-generator. This is being jointly funded with an industrial consortium and two electricity utilities are contributing to the work. The study should enable progress to be made to the construction of a prototype machine, if, on detailed consideration of its findings, that is considered to be desirable.

Tidal energy

The Severn Estuary provides the United Kingdom with potentially one of the most suitable tidal power sites in the world, since the tidal range is high, the geography of the estuary is favourable and there is a large integrated electricity supply network close by.

It might be attractive to harness this power, but studies carried out for the Government and recently published indicate that the costs of a tidal barrage would be large, and that there is considerable uncertainty about the effects on the tidal regime in the Estuary. In addition there are no sites of comparable attractiveness in the UK to which the expensive technology could later be applied. We are assessing these factors before deciding what future studies to initiate.

The total Government expenditure on the research and development programme on the alternative energy sources already announced totals nearly £10M. These programmes are for the first phase of the research and development required for exploitation. The Government will not hesitate to build up these research programmes once the technical and economic feasibility of a renewable energy source has been proved and its potential contribution to the UK's energy needs appear to warrant it.

We shall proceed step by step, collaborating with our international partners and with other interested bodies in the UK, to ensure that all our energy options are fully and efficiently explored. Solar energy may well provide a valuable input to our future energy supplies. And I can assure you that its potential will not be neglected.

I am sure this conference added significantly to the advance of solar energy applications in one of the most promising areas.

1.8 Delegates

Complete list of delegates attending the conference.
An asterisk * preceeding a name indicates the author of a paper.

* M.S. Abd Elefiz	Alexandria University, Faculty of Engineering, Alexandria, Egypt
M.S. Abou-Hussein	Faculty of Engineering, Cairo University, Giza, Egypt
E.A. Adeyemi	Department of Architecture, Ahmadu Bello University
B.A. Ajakaiye	University of Ife, Ile-Ife, Nigeria
* G. Alexandroff	Ecole National Superience des Beaux Arts de Paris, Paris, France
J.M. Alexandroff	Ecole National Superience des Beaux Arts de Paris, Paris, France
Y. Alpan	c/o Ketteneh Bros, John Carpenters House, John Carpenters Street, London EC4, UK
K.Y. Amefia	Embassy of Republic of Togo, Avenue de Tervevren, Brussels, Belgium
D.C. Anderson	Department of Engineering, University of Lancaster, Bailrigg, Lancaster, UK
M. Arditi	Paris University, Paris, France
Dr. Arnaout	Beirut Arab University, Beirut, Lebanon
M.G. Aston	Scott Tallon Walker Architects, 3 Petersham Place, London SW7, UK
D. Ataman	State Academy of Fine Arts, Findikli, Istanbul, Turkey
M. Attard	206 High Street, Hamrun, Malta
* N.M. Bahadori	Director, Solar Energy Center, Pahlevi University, Shiraz, Iran
G.R. Bainbridge	University of Newcastle upon Tyne, 37 Jesmond Road, Newcastle upon Tyne, UK
* D. Balcomb	Solar Energy Group, Los Alamos Scientific Laboratory of University of California, Los Alamos, New Mexico 87545, USA
R.A. Beardmore	Nottingham CC, County Hall, West Bridgford, Nottingham, UK
S.H. Berg	Anders Berg Arkitalthkontor AB, Kommendorsgaten 30 Stockholm, Sweden
D. Behrman	UNESCO, 7 Place de Fontenoy, 75700 Paris, France
C.R. Bell	Int. Inst. for Applied Systems Analysis, Laxenburg, Austria
G. Bernard	Digby Stuart College of the Sacred Heart, Roehampton Lane, London, UK
* R. Bernard	Universite de Lyon 1, 69261 Villewibamme, France
* R. Best	Centro de Investigacion de Materiales, National University of Mexico, Mexico 20, D.F.
S. Bielikoff	Elf Aquitaine Soc., 72 rue Jean Kicot, 75007 Paris, France
P.T. Blacker	Atomic Energy Authority, Winfrith, Dorchester, Dorset, UK
D. Blackhouse	David Parry Quiggin & Gee Associates, 57/59 Seel Street, Liverpool L1 4AZ, UK
* B. Boiko	UNESCO, Department of Higher Education and Technology, 7 Place de Fontenoy, 75700 Paris, France
F. Boileau	Parc de Brotonne, 2nd Marbec, 76580 ie Trait
T. Bolander	84 Avenue Montjoie, 1180 Brussels, Belgium
* B. Brinkworth	Solar Energy Unit, University College, Cardiff, UK
L.V. Brattle	University of Surrey, Home Economics Department, Sussex, UK
F. Brune	Compagnie Francaise des Petroles, 5 rue Michel Ange, Paris 16e, France
R. Burton	Ahrends Burton & Koralek, 1 Spencer Court, 7 Chalcot Road, London NW1, UK
J. Bussy	Williams & Winkley, 24 Townsend Road, London NW8, UK
B.E. Butt	Heating & Ventilating Engineer, High Road, Finchley, London N12, UK
R.J. Capdet	Honeywell Europe s.a., 14 Avenue Henri Matisse, 1140 Brussels, Belgium
Hon. Mrs B. Carter	'Brencar', Wakeham, Rogate, Nr. Petersfield, Hants, UK
* P.F.M. Casalonga	Ingénieur des Travaux Publics, Etudiant en Architecture Groupe A.B.C., Cabinet d'Etudes, 6 Stretta San Lazaru (et Laboratoire) 20000 Ajaccio, France
M. Cetincelik	Turkish Solar Energy Research Foundation, PO Box 37, Bakanliklar, Ankara, Turkey
N. Charlesworth	Ergotechnics, 14 Sandhills Way, Calcot, Reading, Berks, UK
M. Chavlaiguet	Secretariat D'Etat a la Culture, 11 Quai Malaguais, Paris, France
C. Chervilini	CRAIES, c/o Agsm Lungadige Galtarossa, 8-37100 Verona, Italy
* C.V. Chinappa	P.N.G. University of Technology, Lae, Papua, New Guinea
K. Chinciner	45 Norman Grove, Bow, London E3, UK

P. Chouard	Electricite de France les Renardieres, F77250 Moret/Loing, France
* G. Chouleur	6 rue Fresque, Nimes, France
J.R. Claydon	Gateshead MBC, Windmill Hill, Gateshead, Tyne & Wear, UK
R.A. Cockle	Percy Thomas Partnership, Ridgefield House, Manchester, UK
M. Combarnous	University de Bordeaux, Tolence, France
M. Consalvi	Tecnocasa s.p.a., 43 Viale Lombardia, 67100 L'Angulla, Italy
J. Considine	15 Haverstock Street, London N1, UK
C.J. Cooper	Rhys-Davies Services Co, 66 Hagley Road, Edgbaston, Birmingham, UK
* J.M. Cottier	Polisolar A.G., SA, Ltd., Aastrasse 102, Postfach, CH-3005 Bern, Switzerland
B. Cristoff	Ege University, Muhendislik Bilimleri Fakultesi, Izmir, Turkey
W. Cruickshanks	Department of General Surveying & Construction, Newcastle Polytechnic, Newcastle upon Tyne, UK
F. Cucchua	Instituto Technico Industriale Statale, Perugia, Italy
D.M. Curtis	Oscar Faber & Partners, 18 Upper Marlborough Road, St. Albans, Herts, UK

O. Damerdji	Universite des Sciences et de la Technologie d'Alger, Algerie
* D. Daskalakis	Thermasol Ltd, B Mela 44, Holargos, Athens, Greece
* M. Davies	Department of Building Engineering, The University, Liverpool, UK
A. de Herde	Faculte des Science Appliques Unite Architecture, Batiment Vinci Place du Levant 1, B-1348 Louvain-le-Neuve, Belgium
L. Delvaux	University of Liege, Liege, Belgium
D.M. Dennis	Western New Mexico University, Silver City, New Mexico, USA
* M. Devanathan	Tea Research Institute of Ceylon, St. Coombs, Telawukela, Sri Lanka
* P. Dunn	Department of Engineering & Cybernetics, University of Reading, Whiteknights, Reading, Berkshire, UK

* J. Eberhard	AIA Research Corporation, 1735 New York Avenue, Washington DC 20006, USA
* G. Echaniz	Institute of Optica, CSIC, Madrid, Spain
Prof. El Abd	Faculty of Engineering, Beirut Arab University, Beirut, Lebanon
J. Evans	5/10 Bannerman Street, Cremorne, NSW 2090, Australia

* A. Faist	Ecole Polytechnique Federale de Lausanne Groupe Energie Solaive, 14 Avenue de l'Eglise Anglaise, 1007 Lausanne, Switzerland
* E. Farber	Head of Solar Energy Laboratory, Department of Mechanical Engineering, University of Florida, Gainsville, Florida 32611, USA
P.L. Fraenkel	Intermediate Technology, Development Group Ltd, King Street, London, UK
M. Farge	Choire de Thermique, Conservatoire National Des Arts et Metiers, 292 Rue St Martin, 75141 Paris, CEDEX 03, France
W.S. Faville	Atkins Research & Development, Woodcote Grove, Epsom, Surrey, UK
M. Feruglio	Ind. A Zanussi s.p.a. 33170, Pordenone, Italy
D. Feuermann	Seetru Ltd, 43 Corn Street, Bristol BS1, UK
A. Findeli	Universite de Montreal, 3680 Darlington, Montreal, Quebec, Canada
P. Fineman	Tudor Safety Glass Company, Queensborough, Kent, UK
F. Fingado	29 The Avenue, Flat 1, Hitchin, Herts, UK
W.W. Foy	Dean of Faculty of Environmental Studies, North East London Polytechnic, Forest Road, Walthamstow, London E17, UK

H. Gad	Mansura University, Egypt
P.G. Galli	Department of Liberal Studies in Science, Manchester University, Manchester, UK
P. Gandenzi	via Mortelli Untiche, Pordenone, Italy
J.B. Gay	Laboratoire de Physique Theorique, Lausanne, Switzerland
D. Gerassi	Distrimex Ltd, 88 The Avenue, London NW6, UK
A.G. Guy	Pilkington Bros Ltd, Latham, Ormskirk, Lancs, UK
* W.B. Gillet	Solar Energy Unit, University College, Cardiff, UK
J.P. Gleizes	Ministere de l'Equipement, SCHE, Paris, France
D. Gough	Pilkington Bros Ltd, St. Helens, Merseyside, UK
R. Griffiths	Polytechnic of Central London, 35 Marylebone Road, London NW1, UK

* A. Grosfils	ATRAC, s.a., Avenue de Tervveren 98, B-1040, Brussels, Belgium
H.M. Guilmoto	Guernsey Gas Light Co Ltd, St. Peter Port, Guernsey, Channel Islands
E. Hahne	Stuttgart University, Stuttgart, Federal Republic of Germany
D.O. Hall	University of London, King's College, Department of Plant Sciences, 68 Half Moon Lane, London SE24 9JF, UK
P.J. Hallman	Building Research Station, Garston, Watford, Herts, UK
* Y.H. Hamid	Inst. of Solar Energy & Related Env. Research, Council for Scientific & Technological Research, PO Box 6094, Khartoum, Sudan
A.G. Hardacre	Harwell Atomic Energy Centre, Harwell, Oxfordshire, UK
H. Harding	Under Secretary HFE. (1), Department of Education & Science, London, UK
R. Harten	S.G. Leiden ZW, Jacob Catslaan, 65 Leiden, Nederland
* G. Hassan	Chelmsford Institute of Higher Education, Chelmsford, Essex, UK
M.S. Hassan	Department of Architectural Engineering, Helwan Mutrih, Cairo, Egypt
* O St. C Headley	University of West Indies, Department of Chemistry, St. Augustine, Trinidad and Tobago, West Indies
* A.S. Hegazy	Faculty of Engineering, Menefia University, Shebin El-ben, Egypt
W.F. Heshuysen	Netherlands Energy Research Foundation (ECN), Westerduinweg e, Petten (NH), Netherlands
G.J.E. Hextall	Int. Copper Research Association Inc, Orchard House, Mutton Lane, Potters Bar, Herts, UK
E.R. Hitchin	British Gas Corporation, Watson House, Peterborough Road, London SW6, UK
* N.M. Hoda	Appropriate Technology Development Association, Gandi Bhawan, Lucknow 226001, India
* H.C.A. Hoekstra	Eindhoven University of Technology, Eindhoven, Holland
A. Holder	Altec Solar Products, Northfield Road, Ealing, London W13, UK
G. Hug	Etablissement Enseignement Superieur Polytechnique, Universite de Madagascar, Madagascar
* I.M. Hussain	Department of Aeronautical and Mechanical Engineering, University of Salford, Salford, Gr. Manchester, UK
K. Jones	University of Newcastle upon Tyne, 37 Jesmond Road, Newcastle upon Tyne NE1, UK
P. Joyce	Claritude Ltd, 19 Dunraven Street, London W1, UK
J. Kapus	Bled, Ribensica 16, Jugoslavia
* M.P. Keath	Department of Development & Technical Services, London Borough of Harrow, Station Road, Harrow, Middlesex, UK
* D. Kelbaugh	70 Pine Street, Princeton, New Jersey 08540, USA
M.S. Khady	Mechanical Power Engineering Department, Cairo University, Cairo, Egypt
H. Khalil	Donald Smith Seymour & Rooney, 82/89 Uxbridge Road, Ealing, London, UK
* K.H. Khalil	Mechanical Power Engineering Department, Cairo University, Cairo, Egypt
F. Kialos	CEAER, BP 117, Butare, Rwanda, South Africa
J.C. Kiminas	Olympic Plating Inc. 208-15th Street, SW. Canton, Ohio 44707, USA
R. Kriesi	Polisolar, A.G. Postfach/Guterstr, 24, 3000 Bern, Switzerland
G. Lanuza	Apartado 2, Benidorm, Spain
R. Latran	Cabinet d'Ingenieurs-Conseils, rue des Espagnds, 53 Bis, Arlon, Belgium
H.W. Lende	Trinity University, 715 Stadium Drive, San Antonio, Texas, USA
* P. Lessieur	Compagnie Francaise des Petroles, Total, 5 rue Michel Ange, Paris 16, France
G. Lewis	Architect, Pound House, St. Leonards Road, Nazeing, Essex, UK
J. Owen Lewis	Solar Energy Society of Ireland, University College, Dublin 2, Eire
R. Linnell	Department of Environment, Lunar House, 40 Wellesley Road, Croydon, Surrey, UK
K. Lodge	CUBE Ltd, Cheshire, UK
G. Long	ETSU, Harwell Atomic Energy Station, Harwell, Oxfordshire, UK
S. Los	Istituto Universitario di Architettura Venezia, Calle Amai 197, Tolentini, Venice, Italy

N. Pulitzer Los	Istituto Universitario di Architettura Venezia, Calle Amai 197, Tolentini, Venice, Italy
J.N. Lowe	ISES, 21 Albermarle Street, London W1, UK
N. Lowe	ISES, 21 Albermarle Street, London W1, UK
W. Lu	City of Dallas, Department of Urban Planning, 500 S. Ervay, Dallas, Texas, USA
* C.R. Lundahl	Office of Research, Western New Mexico University, Silver City, New Mexico 88061, USA
Mathieu	Ministry of Industry, SDIEN, 28 rue de la Source, 75016 Paris, France
L. Marko	K-Konsult, S-11780 Stockholm, Sweden
* R. Marshall	Solar Energy Unit, University College, Cardiff, UK
M. Martin	Messerschmitt-Bolkow-Blohm, 8 Munchen 80, Postfach 80 11 69, Federal German Republic
* E. Mayer	Centro de Investigation de Materiales, UNAM, Ap. Postal 70-360, Mexico 20, D.F.
* K.E. McCartney	Architectural Association, School of Architecture, Bedford Square, London WC1B, UK
D.S. McIlhagger	Williamson, Brown, Crozier & Wyatt, 15 Wellington Park, Belfast, UK
F.B. McKee	McKee Solaronics Ltd, 12 Queensborough Road, Southminster, Essex, UK
* C. McVeigh	Department of Mechanical and Production Engineering, Brighton Polytechnic, Moulsecoomb, Brighton, Sussex, UK
* H.J. McQueen	Department of Liberal Studies in Science, University of Manchester, Manchester, UK
H.N. Michell	John Laing Research & Development Ltd, Manor Way, Boreham Wood, Herts, UK
* J. Mitchel	14 rue des Poissonniers, 92204 Neuilly sur Seine, Cedex BP 32, France
P. Miller	Survey Design & Construction Ltd, 19 Grove Place, Bedford, UK
G.M. Mills	School of Sciences & Society, University of Bradford, Bradford, UK
Miyamoto	Solar Heater Project, Central Research Lab, Japan
D. Molard	14 Grande Rue, 77 790 Jarennes sur Seine, France
* P. Monaghan	Department of Statistics, Trinity College, Dublin University, Dublin 2, Eire
S.R. Montgomery	University College, Torrington Place, London WC1E, UK
B. Mooney	North East London Polytechnic, Department of Chemical Engineering, West Ham Precinct, Romford Road, London E15 4LZ, UK
P.M. Moreton	McLellan & Partners, 1 Warwick Row, London SW1, UK
R.G. Morgan	Solar Energy Unit, University College, Cardiff, UK
J. Mornekar	Eindhoven University of Technology, Eindhoven, Holland
J.A. Morton	UK-ISES, 21 Albermarle Street, London W1, UK
O.O. Mojola	Faculty of Technology, University of Ife, Ile-Ife, Nigeria
P. Myers	PSA, Lunar House, Wellesley Road, Croydon, Surrey, UK
D.M. Nelson	Architect, London, UK
F. Neubauer	Dornier System G.m.b.H, PO Box 1360, D-7990 Friedrichshafen, Federal Republic of Germany
* N.I. Ngoka	Towning Hill & Partners, 18 Great George Street, Bristol BS1, UK
F. Nicolle	47 rue Daguere, Paris, France
T. Nikolopoulos	Sir William Hallcrow & Partners, 364 Kensington High Street, London W14, UK
I. Nissan	Centro de Investigacion de Materiales, UNAM, Ap. Postal 70-360, Mexico
A.S. Norman	Scarletts Cottage, Recreation Road, Colchester, Essex, UK
J. Norton	Prime Minister's Office, 132 Avenue Raissi, Tehran, Iran
J.C. O'Connell	Air Distribution Equipment (M & W) Ltd, 64 Whitebarn Road, Llanishen, Cardiff, UK
T. Oka	20 Arakiche Shinju ku-ku, Tokyo, Japan
D.J. Oppenheim	2 Chapman Street, North Melbourne, Victoria, Australia 3051
H. Ottosson	Linkoping University, Department of Physics, S 581 83 Linkoping, Sweden
J. Overall	Head of Department of General Surveying & Construction, North East London Polytechnic, Forest Road, Walthamstow, London E17, UK

M. Pacetti	Facolta di Ingegneria Universita di Ancona, Via Della Montagnola 30, 1-60100, Ancona, Italy
* J. Page	Head of Department of Building Science, Sheffield University, Sheffield, S. Yorkshire, UK
S. Pallis	Hilltop House, Warren Road, Chatham, Kent, UK
E. Panzhauser	13 Karlsplass A-1040 Vienna, Austria
* J.K. Parich	International Institute for Applied Systems Analysis, Schloss Laxenburg, 2361 Laxenburg, Austria
A. Pelecanos	BP, Syntagma Square, Athens, Greece
M. Perkins	Hampshire CC, Architect's Department, Three Ministers House, Winchester, Hants, UK
* R.D. Piacentini	Universidad Nacional de Rosario, Avenue Pellagrini 250, 2000 Rosario, Argentina
A. Player	East Sussex CC, Architect's Department, County Hall, Lewes, Sussex, UK
P. Proust	Rhone-Poulenc Batiment, 25 quai Paul Doumer, Courbevoie, France
M.T. Regnard	Loreto Convent, Hatfield Road, St Albans, Herts, UK
R.W. Richardson	ICI Plastics Division, Bessemer Road, Welwyn Garden City, Herts, UK
* R. Rigopoulos	University of Patras, Physics Laboratory II, Patras, Greece
R.J. Riordan	AIA Research Corporation, 1735 New York Avenue, Washington DC 20006, USA
P. Robertson	Scott Sutherland School of Architecture, Garthdee, Aberdeen, UK
* G.G. Rogers	Department of Building Science, University of Sheffield, Sheffield, UK
C. Rolz	Central American Research Inst, for Industry, PO Box 1552, Guatemala, Central America
F. Romas-Bergeno	Institute de Optica, CSIC, Madrid, Spain
F.A. Romig	Int. Inst. for Environment & Dev, Energy Project Office, 3 Tanza Road, Hampstead, London NW3, UK
* A. Russell-Cowan	Solar Centre Ltd, 176 Ifield Road, Chelsea, London SW10, UK
D. Russell-Cowan	Solar Centre Ltd, 176 Ifield Road, Chelsea, London SW10, UK
M. Ruyffelaere	Electricite de France, Direction de la Distribution, Service Commercial - Division TAE, Cedex 8, 92080 Paris, France
M. Ryba	144 rue de Floudre, 75019 Paris, France
* I.A. Sakr	Solar Laboratory, National Research Centre, Doki, Cairo, Egypt
* S. Satcuanathan	Department of Mechanical Engineering, University of West Indies, St. Augustine, Trinidad and Tobago, West Indies
V. Schaefer	Im Schilf 6, CH-8044 Zurich, Switzerland
F. Scharf	Robert Rosch GmbH, Abt. FMGS, Postfach 50, 7000 Stuttgart 1, Federal Republic of Germany
* J.C. Scott	Western New Mexico University, Silver City, New Mexico 88061, USA
* T. Scott-Dean	School of Architecture and Urban Design, University of Kansas, Lawrence, Kansas 66045, USA
G. Scudo	Centro di Documentaziome, Facolta di Architettura, Via Bonardiz, Milano, Italy
* D. Sellmeyer	Department of Management Science, Imperial College of Science and Technology, London, UK
D.E. Sharpley	Don Engineering (SW) Ltd, Wellington Trading Estate, Wellington, Somerset, UK
I. Sherif	Faculty of Engineering, Alexandria University, Alexandria, Egypt
F. Sigurdson	908-72 Donald Street, Winnipeg, Manitoba, Canada
D. Simpson	HGS, 22 Carlisle Place, London SW1P, UK
A.R. Smith	St. Osyth's Farm, 80 Jaywick Lane, Clacton-on-Sea, Essex, UK
* G.C. So'Brien	Department of Mechanical Engineering, University of West Indies, St. Augustine, Trinidad and Tobago, West Indies
N.A. Spelman	Robinson Developments, Winnell Industrial Estate, Winchester, Hants, UK
* A. Spyridonos	Nuclear Research Centre, Demokritos, Agia Paraskevi, Athens, Greece
* N.C. Srivastava	Department of Mechanical Engineering, University of Nairobi, PO Box 30197, Nairobi, Kenya
* C. Stambolis	Intermediate Technology Development Group, 9 King Street, London WC2E 8HN, UK

* T. Steemers	European Economic Community (EEC), rue de la Loi, Brussels, Belgium
* G. Stevens	Airworks, 55 Colebrooke Row, London N1 8AF, UK
T. Stevens	Solar Energy Dev., 16 South Wharf Road, London W2, UK
* R. Stewart	Atmospheric Sciences Research Centre, State University of New York at Albany, Albany, 1400 Washington Avenue, New York 12222, USA
O. Stubbs	Admin. Assistant to Director of North East London Polytechnic, Forest Road, Walthamstow, London E17, UK
* D.A. Svendsen	Solar Energy Unit, University College, Cardiff, UK
R. Swaine	The Calor Group Ltd, Calor House, Windsor Road, Slough, Bucks, UK
* H. Tabor	Scientific Research Foundation, Hebrew University Campus, Jerusalem, Israel
L. Taub	Building Systems Design, 86 Virginia Street, Valley Stream, New York 11580, USA
D. Taylor	Department of the Environment, PSA, Room A222, Whitgift Centre, Wellesley Road, Croydon, Surrey, UK
G. Taylor	Polytechnic of the South Bank, London, UK
S.R.A. Taylor	19 The Vale, Stock, Ingatestone, Essex, UK
* W.P. Teagan	Senior Staff, Arthur D. Little Inc., 20 Acorn Park, Cambridge, Mass, 02140, USA
* E. Tedeschi	IADIZA, Casilla Correo 507, 5500 Mendoza, Argentina
* D. Tinaut	Instituto de Optica, Daza de Valdés, Serrano 121, Madrid 6, Spain
* R. Tomkins	Department of Management Science, Imperial College, Exhibition Road, London SW7, UK
Traisnel	CNRS, 282 Boulevard Saint-Germain, 75700 Paris, France
J. Tugby	Crowlon Ltd, 38 Lyndhurst Rise, Chigwell, Essex, UK
S.G. Turner	Department of the Environment, PSA, Room B802, Whitgift Centre, Wellesley Road, Croydon, Surrey, UK
S. Ulkay	States Engineering & Architectural Academy, Istanbul, Turkey
I. Ulvanos	National Swedish Board for Tech. Dev, S-10072 Stockholm 43, Sweden
S. Ulvanos	National Swedish Board for Tech. Dev, S-10072 Stockholm 43, Sweden
F.B.J. Underhill	Carminowe Farm, Pendeen, Nr. Penzance, Cornwall, UK
* P. van Dresser	Sundwellings Project, 634 Garcia Street, Santa Fe, New Mexico 87501, USA
* C.W.J. van Koppen	Eindhoven University of Technology, Department of Mechanical Engineering, Eindhoven, Holland
* J.T.T. van Wolde	Eindhoven University of Technology, Department of Mechanical Engineering, Eindhoven, Holland
J.M. Vardon	South Australian Institute of Technology, North Ice Adelaide, c/o University College of London, London, UK
* A. Vatansever	General Secretary, Turkish Solar Energy Research Foundation, Findikzade, Sair M. Emin Sokak, 417 Capa, Istanbul, Turkey
J. Vleggaan	AK20 Research Laboratories, Department CRT, Ve Lperweg 76, Arnhem, Netherlands
L. Vretblad	Anders Berg Arkitektkontor Ab. Kommendorsgatan 30, S11448, Stockholm, Sweden
S. Wajc	Vrije Universiteit Brussel, 2 Pleinlaan, 1050 Brussels, Belgium
* I. Wallner	Arbeitsgemeinschaft Solarenergie e.v., Krupp Str. 5, 43 Essen, Federal Republic of Germany
* K. Wasserman	Solar Vision Inc., Church Hill, Harrisville, NH 03450, USA
I.C. Wiles	Department of Architecture & Building Science, Strathclyde University, Glasgow, UK
B. Williams	Tudor Safety Glass Co, Queensborough, Kent, UK
W.N. Wilson	North East London Polytechnic, Department of Engineering, Barking Precinct, Longbridge Road, Dagenham, Essex, UK
A. Winlow	80 Killyon Road, London SW8, UK
W.H. Winning	British Industrial Plastics, PO Box 18, Tat Bank Road, Oldbury, Lancs, UK

Woollaston	Copper Development Association, Orchard House, Mutton Lane, Potters Bar, Herts, UK
* J.B. Wright	Department of Env. Sciences, Open University, Walton Hall, Milton Keynes, Bucks, UK
* B. Yanda	Solar Sustenance Project, RT1, Box 107 AA, Santa Fe, New Mexico 87501, USA
S. Yannas	Architectural Association, Graduate School, 34/36 Bedford Square, London WC1, UK
N. Zarb-Adami	Malta Development Corporation, Development House, Floriana, Malta

Solar energy utilisation in buildings

Erich A. Farber, Professor & Research Professor
Director, Solar Energy & Energy Conversion Laboratory
University of Florida
Gainesville, Florida

Introduction

It is almost impossible to pick up a newspaper or magazine today without being reminded about our energy situation. This is not new; the writer, when studying history in high school over four decades ago, noticed that the nations and civilizations who had most energy and used most of it advanced most rapidly.

When he looked where the energy came from it was mostly from coal at that time, some from oil, and some from gas. These energy sources seemed like a savings account to him, which in time would be used up (Fig. 1). When looking around for what could be used then, the only energy source found large enough and permanent enough was solar energy, so whether we like it or not we have to learn to live off this energy income.

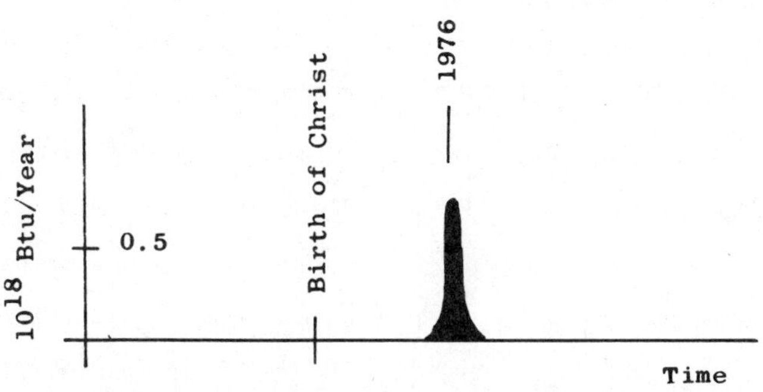

Fig. 1
Conventional world Energy resources utilization

At the present time practically all the energy used comes from fossil fuels (coal, oil and gas) and these resources are the only sources for our medicines, preservatives, pesticides, fertilizers, all our plastics, clothing, etc., things which are just as important for our survival as energy and therefore we should really reserve the fossil fuels for those purposes and get our energy elsewhere.

To solve our problems we have to use all energy sources which we can find, such as wind, geothermal, ocean thermal, waves, tides, etc., but none of these except solar energy is widespread enough to be able to help materially with the solution of our problems.

We will have to use nuclear energy to some extent but it is hoped that it be used wisely. Having worked in the nuclear field the writer feels that we can solve its technical problems but he is concerned with the possible missuse and abuse through wars, riots, sabotage - in other words the human factor, which is much more difficult to control.

Conservation

So what can be done? First of all, we have to use our energy resources more wisely. We build buildings which are glass boxes or perfect solar traps and then have to put big air-conditioning systems into them to throw the collected energy back out so that we can live and work in them. Or, we build buildings without windows and have to put lighting into them. An incandescent light bulb gives about 7% light, 93% heat; fluorescent is somewhat better - 15% light and 85% heat, and even up North we have to run the air-conditioning in these buildings in the winter to keep them cool enough. When travelling around the world one finds the same building designs near the equator as in the Northern inhabited latitudes, a practice that does not make much sense.

Buildings can be built to require a minimum of energy, often through energy storage. The ancient people took advantage of this. Building partially or completely underground, where the temperature is constant. wine cellars, thick walls in buildings, "Hogans" (mud igloos) of the American Indians, etc. are examples. Trees can be used for shading.

Conservation measures should always be used since supplying the needed energy is thereby made considerably easier.

Fig. 2
Schematic of an energy conserving building

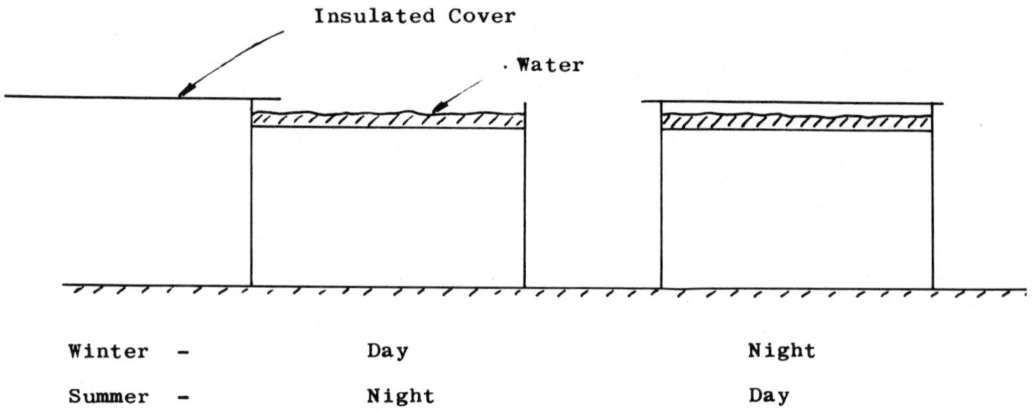

Storage can also be supplied in the form of a shallow pool of water on the roof, provided with a cover. By heating the water through solar irradiation during the day in the winter and cooling it by night radiation and evaporation during the summer can be used for temperature control in the building (Fig. 2).

Solar energy

As was pointed out earlier, solar energy is the largest and most permanent income of energy we have and eventually we will have to learn to live off it. Solar energy has different basic characteristics from fossil and nuclear fuels and they should be considered when converting and utilizing solar energy.

Fossil fuels and nuclear fuels can be converted efficiently and effectively only in large power plants. These plants are some distance from where the energy is being used so that distribution becomes a necessity and since electricity can easily be distributed we logically developed an electric economy. Solar energy, on the other hand, is already well distributed and therefore conversion on location and to the form of energy which is needed seems a much better match than to force solar energy into the pattern which was developed for fossil fuels and which we are used to.

We receive from the sun about one horsepower per square yard, one kilowatt per square meter, or about 32 times the energy every year which we believe to have in our savings account.

Most often solar radiation maps are drawn reporting the solar energy falling on a horizontal surface. The numbers for Northern latitudes are much smaller but mostly because of the geometry, since the sun is much lower in the sky. The values change as soon as the collecting surfaces are properly oriented.

Solar energy arrives both as direct and diffuse radiation. If flat plate or non-concentrating devices are used they can collect both kinds. Concentrating devices can convert only the direct portion and are unable to work on cloudy days.

Since solar energy is intermittent (the sun does not shine during the night) and continuous operation is needed, storage must be supplied. The amount of energy needed must be collected but not necessarily at the same time. When more is collected than is needed the excess is deposited into the storage, and later (such as at night) when more energy is needed than is collected, the extra amount is drawn out of storage.

An instrument used to measure the solar radiation is shown in Fig. 3.

Fig. 3
Solar radiation measuring instrument

Fig. 4
University of Florida Solar Calorimeter

Solar properties of materials

When one intends to build devices for the conversion of solar energy it is necessary to know the solar properties of the materials to be used - how much of the energy is reflected, how much is absorbed and how much is transmitted. The Solar Calorimeter shown in Fig. 4, the only instrument of its kind, measures these quantities and the results are published in the Guide of the American Society of Heating, Refrigeration and Air Conditioning Engineers.

Many misconceptions exist with regard to these properties. For instance, over 20 years ago, we developed in our laboratory a green paint which was a better absorber than most black paints. The idea originated when we observed that leaves absorb more solar energy than graphite in a good portion of the solar spectrum. Later we found that there are about two dozen parameters such as grain size, grain material, grain shape, pigment to binder ratio, thickness of paint, surface condition, substrata material, substrata finish, etc. which determine and control the solar properties.

Having knowledge of the characteristics of solar energy, how much is available at a particular location, and the solar properties of material it is possible to design and build solar conversion devices which will perform as desired.

Water heating

Fig. 5
Solar water heater

Probably the oldest and most widespread application of solar energy is for water heating. It is only required to have a solar collector and a storage tank. Thousands of such heaters (Fig. 5) exist around the world in many countries.

The collector, usually of the flat plate design, consists of a box (wood, metal, etc.) with about 5 cm of insulation in the bottom, above which rests a metal sheet to which tubes are well connected (very important), the plate and tubes are finished with a good absorbing coating, and the box is covered with one or two sheets of glazing, two sheets for colder climates to reduce the heat losses. The sun shines through the glazing, is absorbed by the plate, converted into heat which flows to the tubes and through the tubes into the water circulating through the tubes. The water heated in this manner is then stored in an insulated tank until it is needed.

If the bottom of the tank is placed above (usually 2/3 m or more) the top of the collector, no circulating pump is required. If it is desirable to have the storage tank lower than the collector then a small circulating pump must be used which is controlled to run when the collector is warmer than the water in the storage tank. Such solar water heaters perform very satisfactorily in the United States, Australia, Israel, etc.

Fig. 6
University of Florida Solar House

Fig. 7
Baseboard convectors in UF solar house

Space heating

Space heating can be provided by, essentially, a large solar water heater system. The hot water from the storage tank can be circulated by a small circulating pump through heating coils as needed. The size of collector must be determined and matched to the heating requirements of the building. With a system of reasonable design an average collection of about 2500Kcal/m^2day can be expected and when coupled with storage of about 160 liters water per square meter of collector satisfactory performance can be expected. Houses have been built requiring from 20 to 40% of their roof area as collectors (Fig. 6, 7).

Fig. 8
Air heater

If in addition to heating active cooling is also required, a water system is almost essential. If only heating is needed an air collector and rock storage can be used (Fig. 8). The air collector consists of a box containing overlapping metal sheets (similar to venetian blinds) with a glass cover. The sun heats the metal plates and the air streaming between them is heated. The storage bin containing crushed rock (all about the same size), 5 to 8cm in diameter, will store any excess energy collected for later use.

In an air system more energy is required to circulate the air than the water in a hot water system. Phase change material etc. can be used for thermal storage but they are found generally to be considerably more expensive systems.

Heating (and hot water) should whenever possible be coupled with cooling since solar systems used both summer and winter will pay for themselves much faster than if used only for half of the year.

Space cooling

The most expensive part of solar systems are the collectors, therefore any increase in cost of them makes the total system less attractive economically and as said above they should be used as much of the time as possible and for as many purposes as possible. This requires at least heating, cooling and hot water.

Space cooling can be accomplished with concentrating solar collectors producing steam or vapor, operating steam jet systems, or engines driving compression systems, or generators producing electricity used in conventional air-conditioning systems. Concentrating collectors have the drawback that they can only utilize the direct portion of solar radiation, wasting the diffuse portion. They cannot work on a cloudy day. They usually require tracking devices making them more complicated and expensive.

One can use flat plate collectors and couple them directly or through storage to conventional absorption air-conditioning systems. Unfortunately, the temperatures required by the conventional absorption systems (designed for waste steam, internal combustion exhausts, or gas flames as the driving source) are higher than the conventional flat plate collector can easily produce for prolonged periods.

The third, and we feel most promising method, is to use flat plate collectors which are satisfactory for heating water and heating and design absorption air-conditioning systems capable of operating under these conditions. We have designed and built twelve systems which can operate with input temperatures as low as 60°C.

The absorption systems can be of the intermittent or the continuous type, essentially the same as the conventional systems. To be able to operate at lower temperatures, however, they have to have different concentrations and different size components, heat transfer surfaces, etc. (Fig. 9, 10, 11).

The detail of these systems is covered in the references but because of lack of space cannot be presented here.

In typical American homes about 85% of the total energy used goes for water heating, space heating and cooling. Even if only those 'Big Three' services are provided by solar energy the greatest part of the energy required in the home can be supplied by the sun.

Swimming pool heating

Swimming pool heating is widely used in the United States. To heat such large quantities of water requires large amounts of energy. This energy can be provided by the sun. If one has a house heating system, the available collectors can be used and the swimming pool can be heated at essentially no cost.

Fig. 9
5 Ton solar absorption air conditioner (Low T)

Fig. 10
Ice machine (Low T)

Fig. 11
Intermittent solar air conditioner in UF solar house

Fig. 12
Solarly heated swimming pool and standard

When the collectors are cooler than the water in the storage tank but warmer than the swimming pool (early morning and late afternoon) they can be used to heat the swimming pool rather than sitting there doing nothing. For a typical swimming pool it takes a collector area about equal to the surface area of the pool to raise the temperature a minimum of 6°C above what it would otherwise be. This, in Florida, will extend the swimming season by about three months.

If no heating system and the associated collectors exist, the most economical method of heating swimming pools is by floating a transparent plastic cover on the surface of the pool with as few air bubbles under the plastic as possible. This will raise the pool temperature (if the pool is in the sun) at least 6°C above what it would be otherwise. In addition, the cover will keep out dirt, prevent evaporation losses, and reduce drastically the chlorine requirements (Fig. 12).

Thus, a cover of the right type will give at least 6°C and each additional collecting surface equal to the surface area of the pool another 6°C or more. This pool cover solar heating method saved the swimming programs in the US during the 1973 energy crisis.

Liquid waste recycling

Potable or drinking water is a commodity in short supply in many parts of the world. Big cities in coastal areas are having to drill wells at greater and greater distances in order to get their fresh water since salt water intrusion into their closer wells has made many of them useless. One method of solving or at least reducing this problem is to reclaim and recycle some of the liquid waste from buildings.

Fig. 13
Liquid waste recycling unit

Fig. 14
Outdoor solar cooker

This can be done with solar distillation units both on large and small scale. Fig. 13 shows such a still which can recycle part of the liquid waste from a house, kitchen sink and the bathroom which is designed to also collect the rain water falling on it. It has a basin, which is filled with the waste water. The sun shining through the glass cover heats the water, it steams, and the steam is then condensed on the cooler glass, runs down and is collected as distilled water. This still of 2.2m^2 produces about 12 liters of fresh water a day from solar energy and in Gainesville collects about the same amount from rain.

Cooking

Three different cooking systems have been used in our Solar House and Laboratory. The first is an outdoor cooker made from an umbrella as shown in Fig. 14. It is of the concentrating type. The other two are for indoor cooking.

A flat plate collector is coupled to an insulated steam double boiler which is always hot all day, and when used for cooking the food is placed inside the hot double boiler. The system with little storage reaches boiling temperatures about three hours after sunup and stays hot for about two hours after sundown.

The third and most versatile system consists of a concentrating collector heating oil to about 450°C which is then stored in a storage tank. Inside the building is a stove and oven which looks just like an electric range but has hot oil circulating in the coils instead of electricity. The user throttles the oil flow to get the desired burner temperatures. In this manner, 24-hour cooking and baking is achieved.

Refrigeration

Solar refrigeration can be accomplished by the methods described under air-conditioning except that the temperatures produced are below freezing. The refrigeration-absorption system can be operated from the hot water in the storage tank when needed or it can produce ice during the day and the ice can be used for refrigeration during the night. The latter has the advantage that the ice produced during the day at only one station can provide refrigeration at several places.

Clothes drying

Clothes drying can be accomplished with the hot water in the large storage tank by having a heat exchanger in a conventional dryer and only a small motor for tumbling of the clothes. The wife of the student couple living in our Solar House at this time is a nurse, and she said that her uniforms come out much better with the moderate solar heat than with the harsh electric heater.

Electricity production (Mechanical energy)

Electricity or mechanical energy can be produced with concentrating solar collectors as described before or even with flat plate collectors and vapor engines but the thermal efficiency is very low (less than 10%) and so the conversion to heat is in general more desirable.

The generation of electricity by using engines, driving generators, is one method but the low efficiency requires large collector surfaces. The generation of electricity through the use of solid state devices also low in conversion efficiency, is still very expensive at this time, even though technically it is impossible as shown in Fig. 15.

Fig. 15
Solar to electricity conversion system

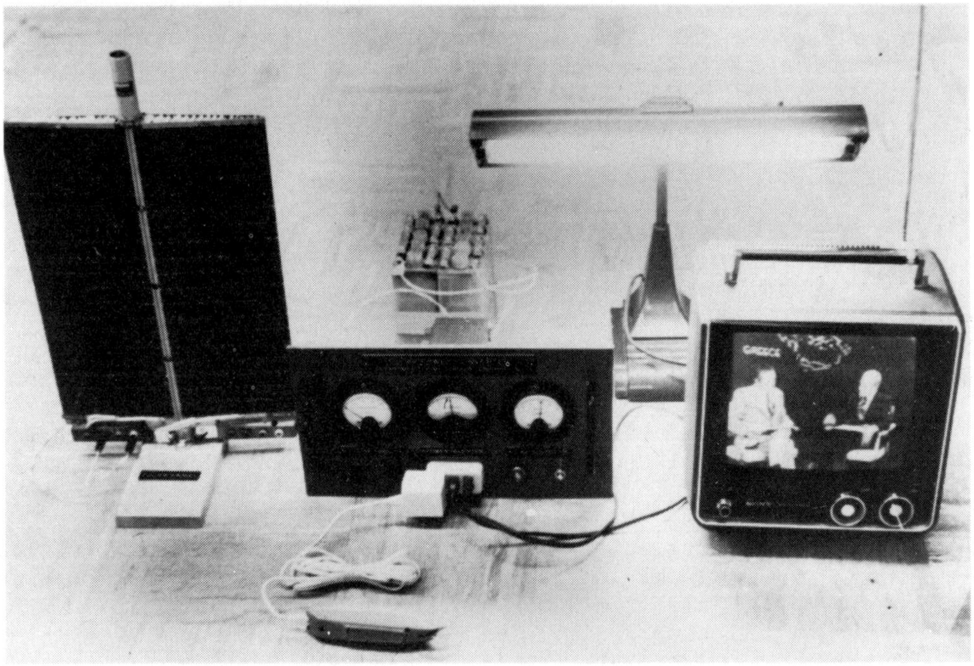

Transportation

Our laboratory has built two solar electric cars and two solar electric buses. One of the cars is shown in Fig. 16, 17. It has a top speed of over 100 kph, over 100 kilometer range and accelerates faster than most cars. One of the buses is shown in Fig. 18.

It is possible to develop a solar electric, energy free, and pollution free transportation system by having solar battery charging stations instead of gasoline stations. In driving when batteries run down one stops at one of three solar battery changing stations to have the

Fig. 16
Solar electric car

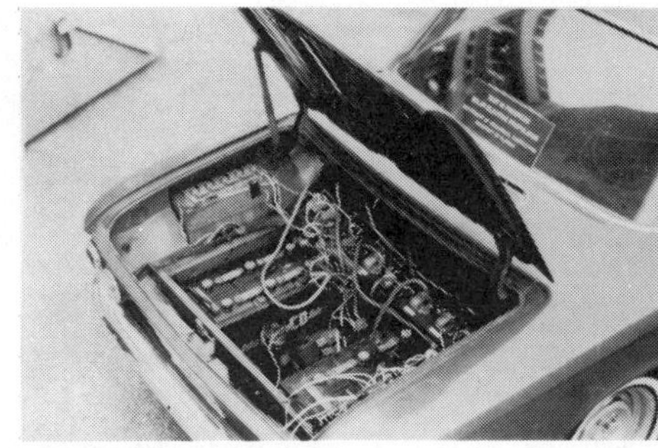

Fig. 17
Motor-generator and battery system

Fig. 18
Solar electric bus

batteries exchanged for charged ones. This can be done in the time it takes to fill a gasoline tank. Then, the batteries are charged for the next user.

Buildings with solar systems can be made to look like conventional buildings with the collectors added, or built into the roof or walls, the collectors can be hidden or made to stand out for everyone to see. Personal preference has to be the guide. Three examples, a home, chain grocery store, and Federal office building indicating the above mentioned approaches are shown in Figs. 19, 20, 21, 22.

Fig. 19
University of Florida solar house

Fig. 20
Winn Dixie grocery chain store

Fig. 21
Solar collectors for store

Fig. 22
Federal office building (Saginaw, Michigan)

Closure

Through this paper it has been shown that it is technically possible to provide all the energy needs in a typical home or building from solar energy. Some of the uses are at present economically competitive with conventional methods such as water heating, space heating and cooling, swimming pool heating, and liquid waste recycling. The production of mechanical energy and electrical energy from solar energy is at this time still more expensive than the conventional methods and can therefore in general not be recommended.

Sometimes the question comes up why solar energy is not used more since the technical capability is here. The answers are that:

1. Most people do not know that all these things can be done.
2. The equipment is relatively unavailable.
3. The initial cost is higher than that of conventional systems.

The above points can be corrected by education, having more manufacturers going into the solar business (several chain stores, such as Sears, now carry solar water heaters), and looking at the life cycle cost of a system (the initial cost and the operating cost over the life of the system). Solar systems, although more expensive to start with, cost nothing or very little to operate.

We really do not have a choice, since just like a family we have to learn to live off an income and the sun is the only energy income large enough and permanent enough to sustain us, and the sooner we learn to use it the better we are off and the easier will be the changeover. The writer's generation does not need solar energy, since it will not run out of fossil fuels as long as it lives but the writer has devoted a considerable part of his life to solar energy since he would like to leave a world behind him which is better than the world he is living in.

References

The main references used were the over 400 publications of the University of Florida Solar Energy and Energy Conversion Laboratory and the over 100 thesis and dissertations produced by this group.

A few selected references are listed below:

Farber, E.A., Solar energy, Its Conversion and Utilization, Building Systems Design, March 1974

Farber, E.A., et al, Clear day design values, Applications of solar energy for heating and cooling of buildings, ASHRAE GRP 170, 1977

Zarem, A.M., et al, Introduction to the utilization of solar energy, McGraw-Hill 1963

Farber, E.A., Selective surfaces and solar absorbers, Journal of Solar Energy Science and Engineering, Vol. III, No. 2, April 1959

Pennington, C.W., University of Florida, ASHRAE Solar Calorimeter, ASHRAE Journal, Vol. 18, No. 3, March 1966

Farber, E.A., Sonnenenergie, UMSCHAU (in Wissenschaft und Technik) D, 6792 D, 1975, Jahrgang, 3.

Farber, E.A., et al, An experimental determination of shading coefficients for selected insulating reflective glasses and draperies, ASHRAE Journal 1977 - This paper received the Crosby Field Award, for Best Technical Paper and Best Paper of 1976.

Farber, E.A., et al, Formulation of a data base for the analysis, evaluation and selection of a low temperature solar air conditioning system, NSF/RANN/SE/GI-39323/FR/74/2, July 1974

Farber, E.A., et al, Utilization of solar energy at the University of Florida Solar Research Residence, Proceedings, International Symposium on lower-cost housing problems, Atlanta, May 1976

Many more references from NASA, Erda, ASME, ASHRAE, ISES, etc. could be listed but they can be found in any good library.

1.10

Opening lecture
Geographical variations in the climatic factors influencing solar building design

J.K. Page,
Department of Building Science,
University of Sheffield, Sheffield, U.K.

Abstract

This paper discusses geographical variations in the climatological factors to be considered in the systematic design of solar houses. Special attention is given to the short wave radiation climatology of vertical and inclined surfaces in different parts of the world under different conditions of cloudiness and atmospheric turbidity. As diffuse radiation is dominant in high latitudes and in Equatorial regions the need to estimate diffuse radiation exchanges accurately is stressed. The various meteorological factors affecting heat losses are also considered, including the effects of wind, long wave radiation exchanges and temperature. The need to give greater attention to micrometeorological aspects of site assessment is also stressed. In addition to drawing from published meteorological data, extensive use has been made of the Department of Building Science's new range of radiation computer programs to prepare diagrams for average days for vertical slopes as well as clear day conditions to illustrate the precise effects of latitude on solar energy availability.

Introduction

The meteorological influences that need to be considered in the systematic design of solar houses are extremely complex. It is difficult at our current stage of knowledge to produce simple global climatological classifications to assist architects in solar house design. Weather not only affects the day to day supply of solar energy, but also the magnitude and the temporal patterns of the different demands for the use of that energy, say for space heating or for cooling. Weather in many parts of the world is highly variable from day to day, and in all parts of the world, outside the equatorial regions, the climate also varies strongly from season to season. The need to match the supply of the solar energy with diurnally varying energy demands usually creates a need for energy storage but the establishment of appropriate energy storage design strategies is entirely dependent on the availability of adequate statistical descriptions of the probable relationships between energy demands and solar energy supplies. Figure 1 taken from the UK-ISES Report (1) attempts to describe the complex inter-relationships that exist between component design for solar houses and the multi-dimensional external meteorological environment. Figure 1 distinguishes between the passive collection of solar energy through windows, etc. and the active external collection of solar energy using external solar collectors on the roof, etc. It indicates the problem of solar controls and energy dumping. It also distinguishes between energy storage in the building fabric itself and energy storage in separate specially constructed energy stores. It assumes the potential for energy collaboration with other types of energy supply system, ambiently based, fossil fuel based, and where appropriate nuclear based.

Fig. 1

The systematic design of solar building involves understanding the interactions between the energy demand system and the different energy supply systems, no less than three of which are used in a typical solar building. The solar systems interact with the wider energy supply system. Many of the factors are weather sensitive. These factors are shown in double boxes.

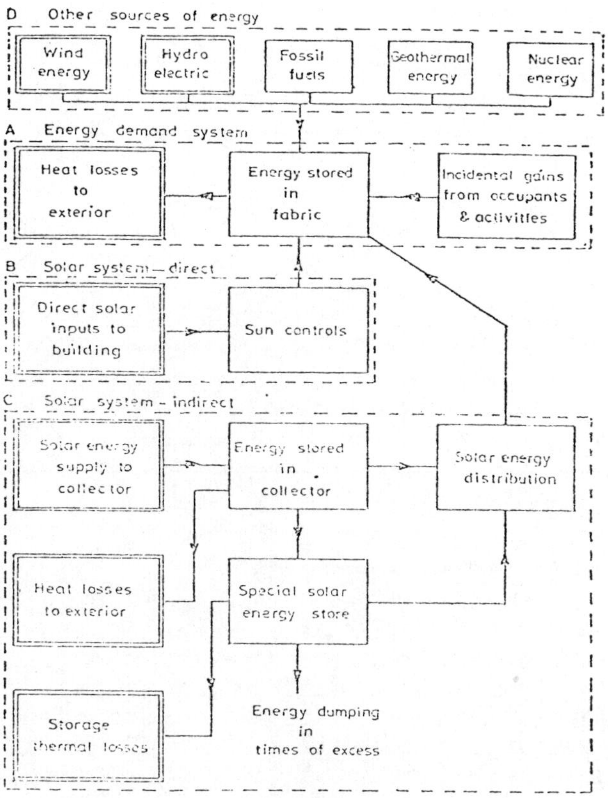

The numerous meteorologically sensitive aspects of system design are all ringed in double boxes. The large number of climatic interactions is the clearest feature of the diagram. It is clear that solar system design is difficult.

The design of solar houses needs to be strongly conditioned by the precise geographical location of the proposed project and its associated climate analysed in detail. This paper attempts to briefly outline some of the variations found across the world in the climatological variables relevant to solar house design. Schemes for particular areas must be seen to rest within a particular climatological context, and their success must be judged within that geographical context. We must seek an appropriate solar technology for each geographical location. The logical form of solar buildings is strongly conditioned by latitude and cloudiness characteristics. Architects must put aside any belief there can be any universal international solar style.

Accurate climatological analysis is the foundation of successful solar house design and borrowing designs out of their climatological context can lead to disaster. For example Figure 2 compares the measured monthly mean daily irradiation on a vertical south facing wall at Odeillo in the French Pyrenees with that measured at Bracknell, just west of London, England. The difference of latitude is just over 10 degrees, but the energy availability in winter

Fig. 2 Measured monthly mean daily irradiation on vertical south walls at Odeillo and Kew. Note large winter differences.

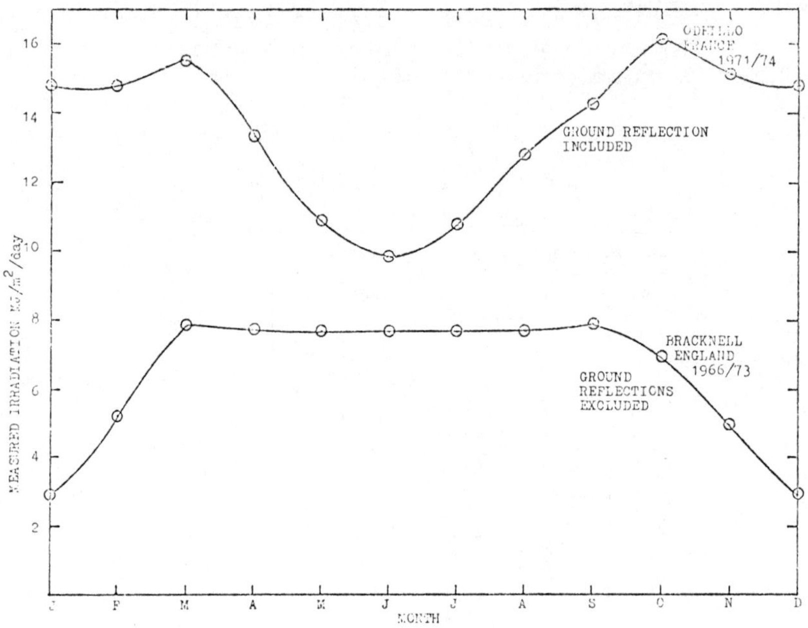

is vastly different. Odeillo has a clear sunny winter climate with an average in January of about five and a half hours of bright sunshine. Bracknell has a typical UK winter climate with a monthly mean bright sunshine level of only about one and a half hours a day. The high level mountain climate of Odeillo is also much clearer than the low level winter climate of the UK and it was for a very good reason that Felix Trombe took the French CRNS Solar Energy Laboratory to Odeillo. The Trombe passive solar wall system may have proved itself successful in Odeillo (2), but Figure 2 shows that one must not assume that it would necessarily be successful in mid-winter in the UK, however attractive the concept might seem at a superficial glance. We need, therefore, to clarify the climatological design principles right at the start of this conference.

Global patterns of solar irradiation

Figure 3 maps the annual mean global irradiance on a horizontal plane. The range across the main inhabited parts of the world is from 100 W/m^2 in high latitude climates to 300 W/m^2 in the Red Sea area. The great desert areas of the world typically located between 20° and 30°N and 20 - 25°S have mean irradiances of 250 W/m^2, over 2$\frac{1}{2}$ times as great as countries like the UK. The low mean irradiance values at the Equator are also evident. The Congo Basin and Paris have very similar mean annual irradiances.

Figure 4 shows recorded monthly mean daily totals of irradiation $\overline{H}_h$ for six stations of widely differing latitude, some north of the Equator and some south. The diffuse irradiation $\overline{H}_{dh}$ is also shown. The variation in mean daily global irradiation with season increases with latitude. The even supply of solar energy in the equatorial region from month to month is very obvious at Kinshasa. The diffuse irradiation forms a big proportion

Fig. 3 Annual mean global irradiance on a horizontal plane at the surface of the earth (W/m^2 averaged over 24 hours).

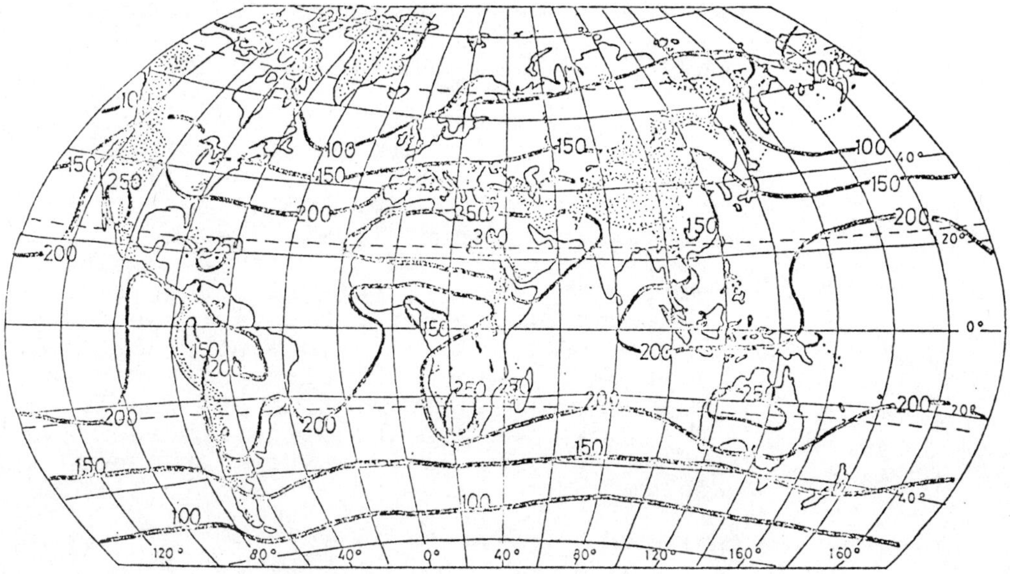

Fig. 4 Measured mean monthly global and diffuse irradiation on a horizontal surface for 6 stations of widely differing latitude. Note high proportion of diffuse irradiation at high latitudes and close to Equator.

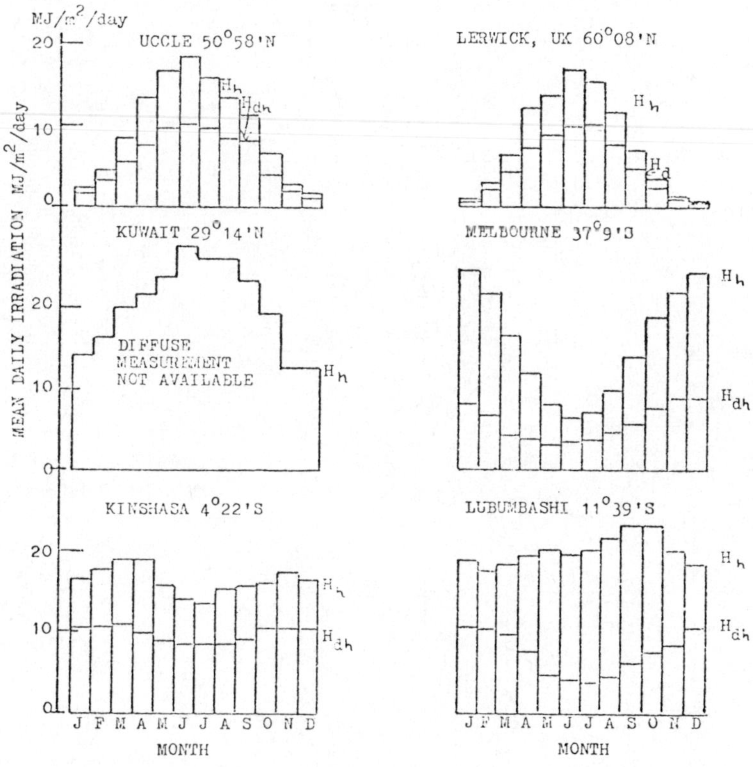

of the short wave irradiation at Kinshasa in the Congo Basin. Lubumbashi a few degrees in latitude further to the south from the Equator has a two season climate, sometimes wet and humid and cloudy, sometimes dry and sunny according to the position of the inter-tropical convergence zone. The proportion of diffuse varies accordingly. Kuwait, a desert type area, has a high insolation throughout the year, but, being situated 29°N of the Equator, the horizontal monthly mean irradiation levels during the winter half of the year fall back as the sun moves to its southward position. Melbourne, Australia at 37°S has the minimum global irradiation level in June. The diffuse proportion at Melbourne is relatively high during the southern hemisphere winter, but this proportion decreases in the summer period. Uccle outside Brussels in Belgium shows a big range from summer to winter with a large proportion of diffuse radiation through the year. The annual range is even greater at Lerwick in the Shetlands to the north of the Scottish mainland at latitude 60°N. The winter solar energy availability is very low indeed. The prospects for solar houses close to the Arctic circle are poor.

It is useful to start by looking at the monthly changes of horizontal surface irradiance G_h. Figure 5 presents variations in the measured mean monthly irradiance with time of day at Lubumbashi in Central Africa, for four selected months together with information about the associated percentage possible sunshine. In January with the

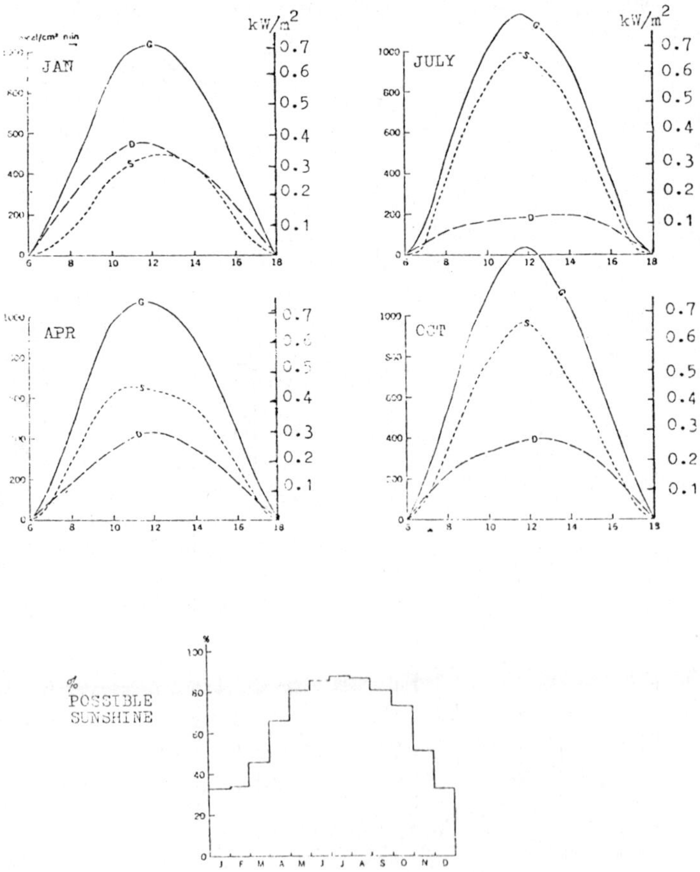

Fig. 5 Variations in mean global, direct, and diffuse irradiance on a horizontal surface in a low latitude climate with a clear and a cloudy season - Lumbumbashi, 11°39'S 27°28'E, 1298m 1955-60

percentage possible sunshine as low as 34%, the diffuse irradiation exceeds the direct. In July the percentage possible sunshine is about 87%, and every day is virtually cloudless. The direct horizontal surface irradiance dominates, and the proportion of diffuse is very small. In October the sun is overhead, so the mean global irradiance reaches its peak, but there is slightly more cloud and the diffuse has risen compared with July.

Solar building design involves understanding such radiation patterns in order to make best use of them. The knowledge of irradiation of sloping and vertical surfaces is critical for proper design, and it is essential to have methods for developing architectural design data from basic meteorological information. However understanding the transmission properties of the atmosphere in relation to the astronomical movements of the earth about the sun is complex. I want to devote most of my time to considering the problem of modelling these important properties for architectural design purposes.

The absorption and scattering processes in the atmosphere (3), (4)

The principal extra-terrestrial variations in the solar energy supply arise from the variations in the distance between the earth and sun that are the consequence of the earth's elliptical orbit. The extraterrestrial solar irradiance normal to the beam at mean solar distance G_{sc} known as the solar constant is about 1353 W/m^2. The extra-terrestrial intensity varies by ± 3.3% being greatest on January 1st and least on July 5th. The direct beam intensity is substantially reduced as it passes through the atmosphere. The lower the sun the greater the path length and the greater the reduction due to absorption and scattering. Some radiation is absorbed by gases in the atmosphere, mainly ozone, oxygen and carbon dioxide, some by water (vapour and cloud droplets) and some by aerosol, man-made and natural. In terms of absorption/unit air mass ozone absorbs about 3% of the solar constant G_{sc}, oxygen and CO_2 absorb less than 2% and water vapour some 10-15% depending on its concentration. Variations in the surface irradiance due to absorption depend mainly on the water vapour distribution. Bannon (5) has mapped global data on the amount of precipitabl water vapour in the atmosphere. The precipitable water held gaseously in the atmosphere varies considerably from area to area and is greatest in the equatorial regions where it may reach 100 mm. In high latitudes in winter about 5 mm is typical. The water vapour absorption removes energy mainly from the infra red part of the spectrum. Air molecules scatter the blue light preferentially. Scattering by cloud droplets and aerosols depends strongly on particle size and on wavelength. Scattering by aerosol removes from 5% to 40% of G_{sc} in unit air mass, but a significant part of the scattered radiation still reaches the earth's surface as diffuse radiation. Figure 6 shows the main spectral features of the direct solar beam. The theoretical direct solar irradiance G^*_{bn} expected at the surface depends on (i) the solar constant G_{sc}, (ii) the sun-earth radius vector, (iii) the total water vapour content of the atmosphere expressed as precipitable water vapour, and (iv)

Fig. 6 Spectral irradiance curves for direct sunlight extraterrestrially and at sea level; shaded areas indicate absorption due to atmospheric constituents, mainly H_2O, CO_2 and O_3. Wavelengths potentially utilised in different solar energy applications are indicated at the top.

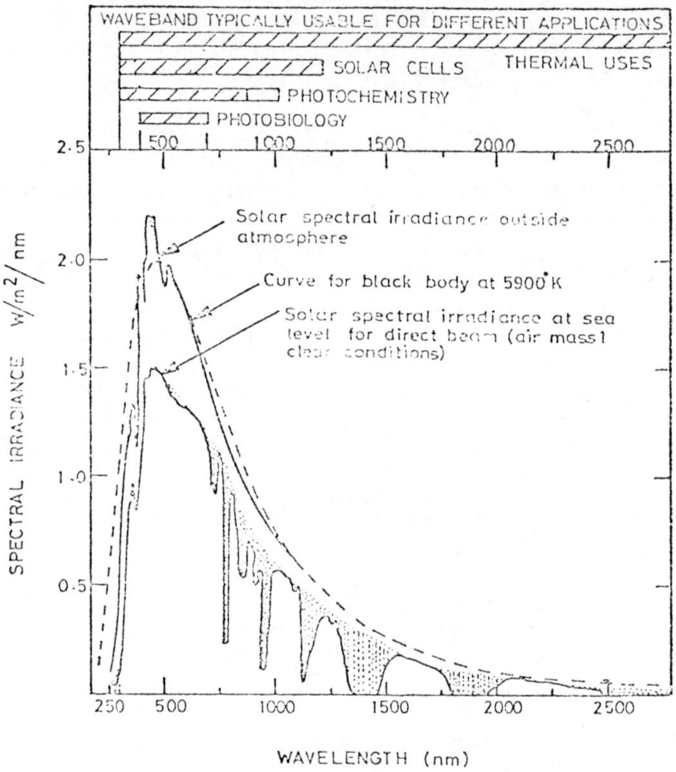

the air mass m which is related to solar elevation α by the formula $m = 1/\sin\alpha$ for $\alpha > 10^{\circ}$ for stations at or close to sea level. For lower altitudes the Smithsonian Tables (6) provided suitable values. The major variations of G^*_{bn} are due to water vapour and air mass. In practice the measured beam irradiance G_{bn} is almost always less than G^*_{bn} and the additional attenuation is ascribed to aerosol and described by a turbidity coefficient.

A variety of turbidity coefficients have been proposed, many of them depending on radiation measurements in a narrow waveband (for discussion refer Robinson (4)). For solar energy applications there are advantages in using the Monteith and Unsworth turbidity coefficient τ_a which is based on measurements of G_{bn} over the whole solar spectrum 0.3 - 3 μm.

τ_a is defined by the relationship

$$\tau_a = -\frac{1}{m}\log_e\frac{G_{bn}}{G^*_{bn}} \qquad (1)$$

We have used this turbidity coefficient in Sheffield extensively in the development of our computer models for the study of solar radiation.

Geographical variations in atmospheric turbidity

There are very large geographical variations in mean annual atmospheric turbidity. Large seasonal differences are found as well. The clearest region is Antarctic. The Arctic is also very clear, and air masses originating from these polar regions having a low precipitable moisture content and also a low dust burden have low turbidities. The solar beam intensity in such high latitude regions may be surprisingly high in spite of the low solar altitude. In the great desert areas of the world, a considerable amount of dust enters the atmosphere, and the turbidity may be relatively high throughout the year. The high midday solar altitudes associated with lower latitudes mean the midday air mass is low, so that the energy that penetrates is still very considerable. In the humid equatorial regions the turbidity is often lower than in the desert regions on either side, even though there is a lot of cloud which substantially reduces the average irradiation by intercepting the direct beam for a high proportion of the day. In these areas there are a lot of small water droplets which produce substantial forward scattering and a characteristically milky sky. In two season low latitude climates where a rainy season intervenes into a clear season, the rainy season typically reveals lower atmospheric turbidities. The high rainfall tends to wash the dust particles out of the sky. Further the fast growing vegetation and the moist ground surfaces suppress surface dust generation due to turbulence. In temperate mid-latitude climates the turbidity is characteristically lowest in the middle of winter and increases to a maximum around July. The atmospheric water vapour contents also increase from winter to summer so the absorption in the infra red absorption bands increases so reducing summer direct beam irradiance.

Figure 7 shows typical month to month variations in the mean monthly maximum pseudo turbidities for several N. European sites. The pseudo turbidity was derived from the daily totals of direct and diffuse radiation using our computer solar radiation model, by dividing the observed mean monthly daily direct radiation on a horizontal surface by the percentage possible sunshine for the month in question, fixing the daylengths as the length of time during which the direct solar irradiance is above 200 W/m^2 (which is the intensity at which burning of the sunshine recorder first starts) assuming a Monteith and Unsworth turbidity of $\tau_a = 0.2$. Figure 7 shows the highest turbidities are associated with urban situations, Kew and Hamburg. The remote seashore sites have substantially lower turbidities than the urban sites. The clearest conditions at sea level being found at Valentia in the far west of Ireland, which is a distance from the main continental sources of pollution. Further the air masses come mainly off the clean Atlantic Ocean. One high level French mountain station, Odeillo, is included. The low turbidities associated with this particular site are evident. At Odeillo some turbidity increase from winter to summer is present, but is more subdued than at Kew. Published data for the USA (7) bears out very much the same annual pattern. The global computer solar radiation models which we have developed in the University of Sheffield allow these turbidity variations and the monthly variations in precipitable water vapour to be properly taken account of by the user.

Fig. 7 Annual variations in maximum pseudo turbidity - Northern Europe

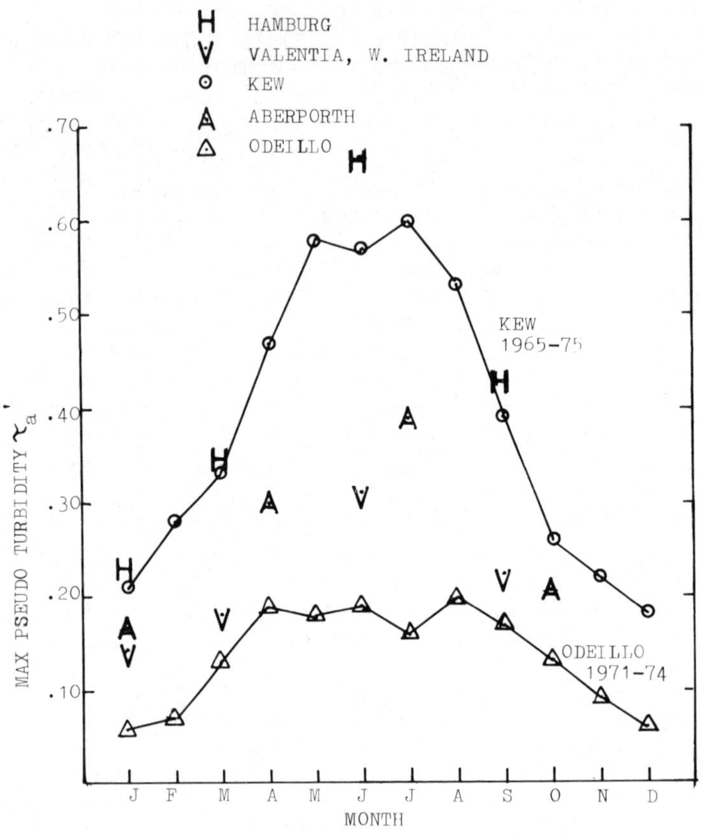

Fig. 8

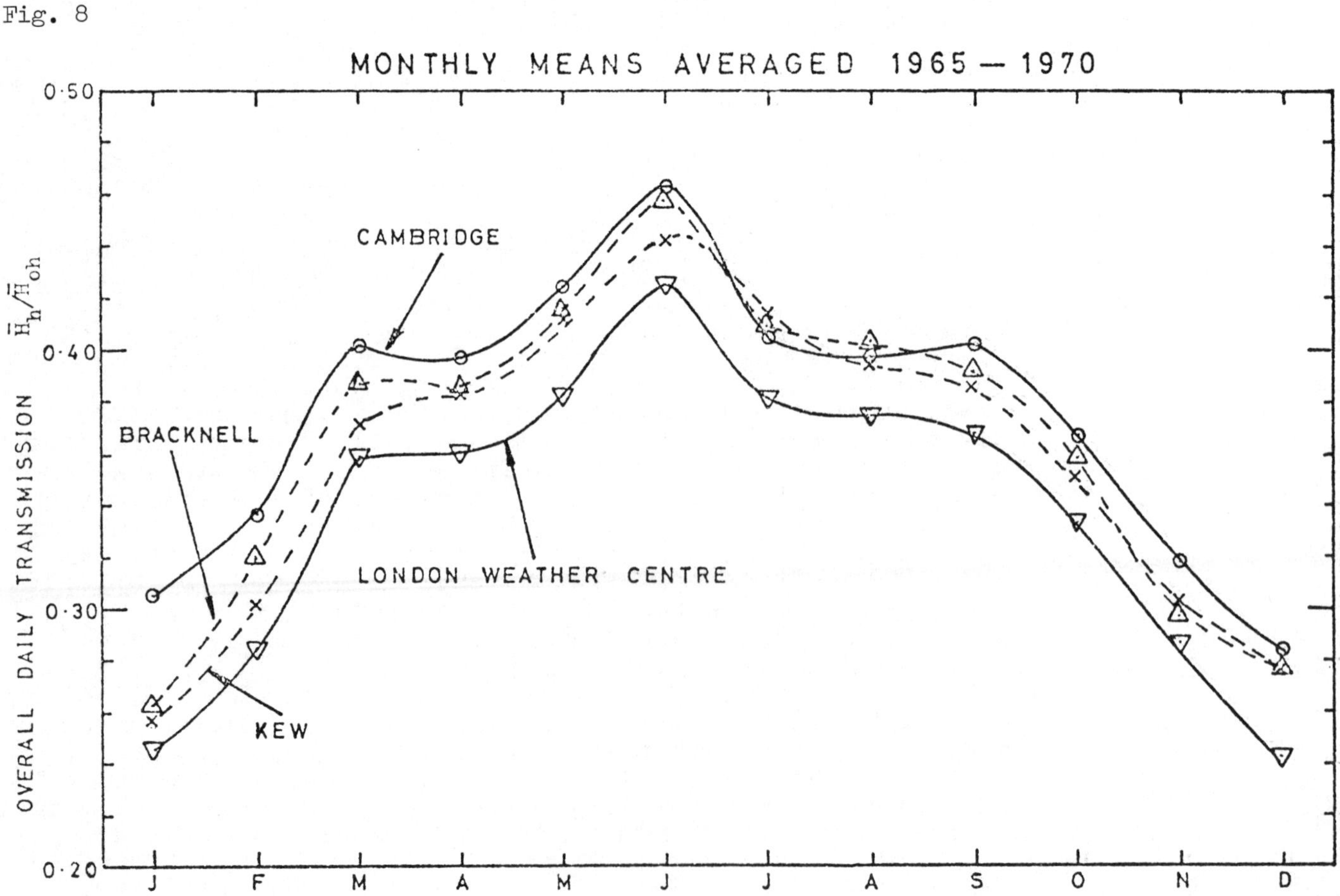

Urban influences on turbidity

In addition to the general geographical factors discussed above, one has also to consider urban pollution influences, which may extend over very wide areas. Figure 8 gives data for the mean atmospheric transmission $\bar{H}_h/\bar{H}_{oh}$ around London where $\bar{H}_h$ is the global irradiation and $\bar{H}_{oh}$ is the corresponding extraterrestrial irradiation. The fall off from the centre of London to Cambridge some 100 km away is considerable. Fortunately, under the influence of clean air legislation, the smoke pollution burden of the UK atmosphere has been falling, so over the last decade the turbidities in many of the large towns have decreased.

Table 1

CHARACTERISTIC VALUES OF THE MONTEITH AND UNSWORTH TURBIDITY COEFFICIENTS τ_a FOR THE U.K. SHOWING THE INFLUENCE OF AIR MASS TYPE AND LOCATION IN RELATION TO POLLUTION SOURCES

	Air mass type	τ_a
Northerly island site, minimum	polar	0.05
pollution from land sources	average	0.20
	continental	0.35
Rural or coastal site exposed	polar	0.10
to natural aerosol pollution and	average	0.25
small amounts of smoke	continental	0.40
Urban site within or close to a	polar	0.25
large town (say population	average	0.40
exceeding 100,000)	continental	0.55

Typical values of τ_a for different types of air mass from Unsworth (3) with corrections for urban and other polluting influences in high latitudes are given in Table 1. Figure 9 shows the effect of turbidity on vertical and horizontal surface irradiance on a south facing vertical surface at latitude 50°N for March 21st for cloudless days. The range of values of τ_a selected were chosen to match the actual range found across the UK from the clearest conditions in the far west to the inland urban polluted situations found in towns with poor control of smoke emissions, fortunately now a minority in the UK. The halving of available energy on cloudless days is a serious loss and careful attention needs to be given to urban pollution effects in assessing the feasibility of solar houses in high latitudes. The effects of ground albedo are also important, and a snow ground cover will substantially increase vertical surface irradiation. One point has been plotted in Figure 9 to demonstrate the influence of increasing the ground albedo from 0.25 to 0.70. As the turbidity increases so the

Fig. 9 The effect of changes in turbidity on horizontal surface and vertical surface irradiation on cloudless days. Latitude 50°N March 21st. Ground albedo = 0.25 and precipitable water vapour 10mm.

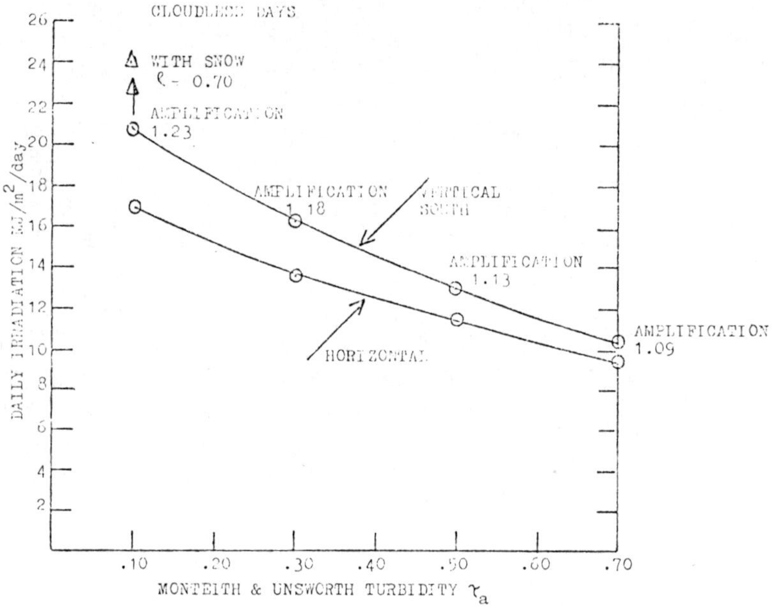

amplification factor for vertical and tilted surfaces will decrease. The appropriate vertical surface amplication figures are included on Figure 9.

The diffuse radiation climate - cloudless days

The amount of diffuse solar radiation on cloudless days also varies considerably from one part of the world to another because dust particles especially from deserts and general atmospheric pollution from man's activities add substantially to the atmospheric scattering. The fall off in the direct beam irradiance due to scattering by dust and other pollution is partly compensated by the associated increases in the scattered diffuse irradiation reaching the earth's surface. Approximately half the scattered energy moves towards the surface, the precise proportion depending on the nature of the scattering particles. Increase in station altitude tends to produce a fall off in both Rayleigh and atmospheric scattering due to dust so the diffuse energy tends to be relatively low on cloudless days at high altitude stations, as the clear days blue skies typically found clearly demonstrate. In contrast in the low level humid tropics, water droplets are present in large numbers and a great deal of forward scattering takes place. The diffuse horizontal surface cloudless day irradiance levels are high in such areas and on clear days the sky tends to have a bright milky appearance.

It is possible to find systematic links between the observed turbidity and the associated diffuse radiation on cloudless days. In our clear sky work at Sheffield we have made extensive use of the clear sky US observations of Parmelee (8) linking his original US observations with observations in other countries including our own country and have devised ways of integrating Parmelee's clearness index approach to our own overall approach based on the extensive use of the Monteith and Unsworth turbidity coefficient. Unsworth and Monteith (9) themselves report, that for cloudless conditions, when the solar altitude is above 30°, the ratio of the diffuse irradiance on a horizontal surface, G_{dh}, to global irradiance on a horizontal surface, G_h, in Central England may be expressed as a function of turbidity τ_a as $G_{dh}/G_h = c + d(\tau_a)$ where $c = 0.097 \pm 0.009$ and $d = 0.68 \pm 0.04$. Page has extended this work to lower altitudes. Page (10) shows that values obtained from Parmelee compare favourably with Unsworth's values for solar altitudes of more than 30° and with observations made by Blackwell (11) at Kew. Similar data obtained by Doigniaux in Belgium also compares well with that derived by Page from Parmelee.

Parmelee's data were reduced as follows. Values of G_{bh}, the direct irradiance on the horizontal plane, were first calculated for different air masses using Unsworth's tables for values of τ_a between 0.05 and 0.4, assuming 20 mm of precipitable water vapour. Parmelee's graph giving values of G_{dh} plotted against G_{bh} for different values of α was then used to find G_{dh} for different values of τ_a, hence G_{dh}/G_H was calculated as a function of τ_a and solar altitude α. No corrections could be made for solar distance, as Parmelee's observations were made over the whole year.

Blackwell also provided values of G_{dh} and G_h for cloudless days at Kew for March, June, September and December. G_{bh} was first found by subtraction, and G_{bn} the normal incidence irradiance calculated using the solar altitude for the middle of each hour (L.A.T.). Then, assuming water vapour contents on cloudless days of 7.5 mm in December, 10 mm in March, 20 mm in June and 15 mm in September and reducing to mean solar distance, values of τ_a were obtained for each hour for cloudless days, for each of the four months for which the data was available. The corresponding values of G_{dh}/G_h were then calculated and plotted against α, the data being divided into different transmission categories according to the derived values of τ_a.

Reasonable agreement was obtained between the results derived from Parmelee and the Blackwell observations at Kew, though the Kew values were typically a little lower for a

Table 2

VALUES OF $\bar{H}_{dh}/\bar{H}_h$ FOR DIFFERENT TURBIDITY VALUES AND SOLAR ALTITUDE

Air mass m	1.5	2	3	4	5	6	1.15-2.0
Solar altitude h	42°	30°	20°	14.5	11.5	9.6	Unsworth
Turbidity τ_a							
0.05	.092	.13	.17	.22	.25	.28	.13
0.10	.12	.17	.24	.29	.34	.39	.17
0.20	.19	.24	.33	.42	.50	.57	.23
0.30	.24	.32	.45	.56	.65	.72	.30
0.40	.30	.41	.55	.67	.77	.83	.37

given turbidity value. The mean Monteith and Unsworth turbidity values for Blackwell's cloudless days between 1947-52 at Kew were found to be 0.15 in September, 0.13 in December, 0.23 in June and 0.24 in March. Values of G_{dh}/G_h are tabulated for various values of τ_a in Table 2. Unsworth's values which are independent of solar altitude for air mass 1.15-2.0 in summer are also given in Table 2. They correspond reasonably well with values derived from Parmelee. This ratio however should fall with increasing solar altitude, a phenomenon invariably observed in practice. The errors by using a fixed ratio above 30° are minimal however as the slope of the graph above 30° solar altitude is small. This is not true below 30°.

Parmelee (8) has presented results of diffuse solar irradiance both on horizontal and vertical surfaces. His results were derived from measurements made at Cleveland, Ohio, on days which, although cloudless differed in atmospheric clarity. A study of his results for a horizontal surface shows that, for a fixed solar altitude, α, a linear relationship exists between the diffuse horizontal irradiance and the direct horizontal irradiance. The relationship takes the following form:

$$G_{dh} = a_o - a_1 G_{bh} \quad (Wm^{-2}) \qquad (2)$$

where G_{dh} is the diffuse irradiance on a horizontal surface, (Wm^{-2}), G_{bh} is the direct irradiance on a horizontal surface, (Wm^{-2}) and a_o and a_1 are constants for a particular solar altitude α and are given in Table 3.

Equation 2 has been incorporated in our Sheffield clear sky computer programs, the constants a_o and a_1 being determined for a particular solar altitude by interpolation on the values given in Table 3. Thus knowing the solar altitude one can derive the direct irradiance on the horizontal surface G_{bh} from the Monteith and Unsworth turbidity calculation and estimate the corresponding associated diffuse clear sky irradiance G_{dh} on the horizontal surface.

Table 3

The figures in parentheses in Table 3 were obtained by extrapolation.

Solar altitude (degrees)	a_o	a_1
0	(0)	(0.290)
10	(63.1)	(0.295)
20	134.9	0.314
30	222.1	0.360
40	284.3	0.362
50	383.0	0.424
60	484.6	0.492
70	(552.1)	(0.520)
80	(604.3)	(0.545)
90	(624.7)	(0.560)

Diffuse radiation on vertical and inclined surfaces on clear days

There is not space to discuss in detail the relationships between vertical surface diffuse irradiation, and horizontal surface irradiation under clear sky conditions. Figure 10 which is derived from Parmelee's observations shows the importance of the precise orientation of the surface receiving the diffuse radiation relative to solar position. The ratio of the diffuse vertical surface irradiance to the horizontal surface diffuse irradiance is expressed in Figure 10 as a function of the cosine of the angle of incidence of the direct sun on the surface for different angles of the sun. The large variations are immediately apparent. This figure demonstrates forcibly that the isotropic approximation is very inaccurate for the estimation of incident diffuse cloudless sky radiation on vertical surfaces. Fuller discussion may be found in Page (12). However, these non isotropic features of the clear sky are modelled in our Sheffield clear sky computer programs.

Fig. 10 **This diagram based on Parmelee's work demonstrates the large errors which may result from using the isotropic approximation for diffuse radiation**

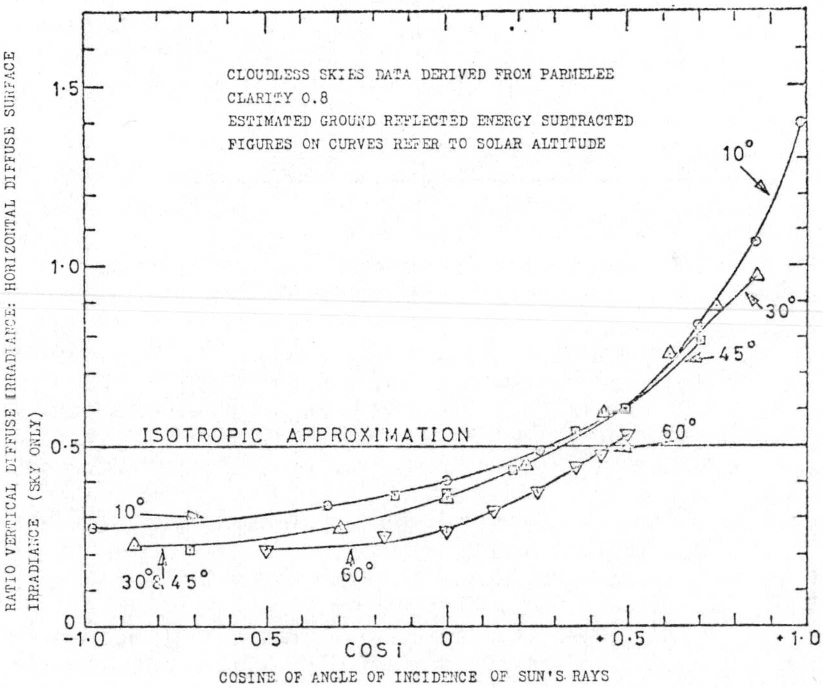

Figure 11 shows the effects of change of latitude on the cloudless irradiation of a horizontal surface and a vertical south surface calculated for a fixed turbidity of τ_a and ground albedo, using our program SUN3 for Dec 21st. It shows in December the most favoured latitudes for passive solar houses lie between 30 and 45°N. Further north a lot of the low sun's energy is lost in the atmosphere. The clear sky south vertical amplification factor however becomes quite large at high latitudes due to the favourable angle of incidence of the sun on the south wall, but the energy available close to the Arctic Circle on Dec 21st must necessarily be very low due to the long path length.

Fig. 11 Variations in the irradiation on a horizontal surface and on a vertical surface facing south with change of latitude $\tau_a = 0.150$, precipitable $H_2O = 5.0mm$, ground albedo = 0.25

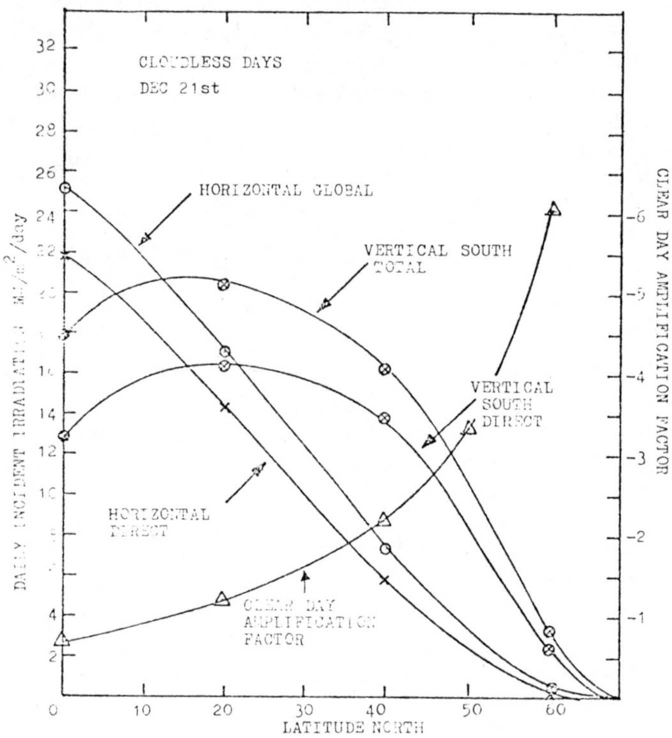

The climatology of diffuse irradiation under average conditions

In cloudy climates diffuse irradiation dominates. The study of the climatology of diffuse radiation is therefore very important in solar building applications, especially at high latitudes and in the equatorial region where there is also a lot of cloud. Well over half the short wave energy income may be diffuse in such areas and it is important to have reasonably accurate methods for estimating the diffuse radiation available on building surfaces.

It is well known that the maximum amounts of diffuse radiation are received on days with partially clouded skies. Broken cumulus, in particular, can give very high values of diffuse radiation, and when the sun is shining through scattered broken cloud instantaneous values of global radiation can be recorded that are higher than the solar constant.

The basic problem is how, in the absence of local measurements, to separate the diffuse horizontal surface radiation, $\bar{H}_{dh}$, from the global radiation, $\bar{H}_h$. Page (13) found it was fortunately possible, using data from stations where both diffuse and direct irradiation are observed to set up reasonably reliable regression equations of the form:

$$\bar{H}_{dh}/\bar{H}_h = c + d(\bar{H}_h/H_{oh}) \qquad (3)$$

where $\bar{H}_{dh}$ = mean monthly values of daily diffuse irradiation on a horizontal plane,

$\bar{H}_h$ = mean monthly value of daily global irradiation on a horizontal plane,

$\bar{H}_{oh}$ = mean monthly daily irradiation on a horizontal plane in the absence of any atmosphere,

c & d are climatically determined regression constants which vary significantly from area to area in the world.

Liu and Jordan (14) in contrast put forward a single formula based on Blue Hill Observatory.
Typical values of c and d for N. Europe and the US are given in Table 4, while Table 5 gives some values for the African continent calculated by the author.
Corresponding data is available for Canada from Tuller (15)

The mean regression equation for the ten stations scattered across the world studied by the author in 1961 was found to be:

$$\bar{H}_{dh}/\bar{H}_h = 1.00 - 1.13\ \bar{H}_h/\bar{H}_{oh}$$

An important feature of this formula for estimating diffuse radiation is that it mathematically predicts the occurrence of a high level diffuse irradiation with relatively modest amounts of sunshine.

Rearranging equation 3, we get:

$$\bar{H}_{dh} = c\bar{H}_h + d(\bar{H}_h/\bar{H}_{oh})^2$$

This is a parabolic curve, and we may find the maximum value by simple differentiation, thus it is easy to show that:

$$\bar{H}_{dh(max)} = - c^2\bar{H}_{oh}/4d$$

and that this value occurs when $\bar{H}_h/\bar{H}_{oh} = -c/2d$

Observed relationships between daily irradiation and percentage possible sunshine for two stations close to the Equator in Africa are given in Figure 12. Both global and diffuse daily irradiation on a horizontal surface are plotted against percentage possible sunshine. In the equatorial climate of Kinshasa the maximum diffuse occurs with % possible sunshine is around 30%. Lubumbashi has a very sunny climate for part of the year. During this period the diffuse radiation is very low compared with Kinshasa at the same percentage possible sunshine where humid conditions prevail throughout the year. In the wet season at Lubumbashi the characteristics are similar to Kinshasa. Figure 12 shows that the diffuse radiation cannot be obtained by linear interpolation between clear and overcast day observations. Kasten (16) has comparable data for Hamburg given in terms of cloud cover and solar altitude for different seasons. The maximum diffuse typically occurs with cloud cover between 4 and 6 oktas when the diffuse irradiance is typically about 40% greater than the diffuse irradiance for cloudless sky conditions.

Table 4

REGRESSION EQUATIONS FOR $\bar{H}_{dh}/\bar{H}_h$ AGAINST $\bar{H}_h/\bar{H}_{oh}$ USING MONTHLY MEAN DAILY VALUES OF $\bar{H}_h$ AND $\bar{H}_{dh}$ OVER PERIOD, AND DIFFUSE MULTIPLIERS FROM KEW

Station	Values of c & d	Corr. Coeff.	$\frac{-c^2}{4d}$	Diffuse multiplier from Kew
London Weather Centre 51°31'N 77m 1965–70	c .990 d −1.103	0.98	.222	.949
Kew 51°28'N 6m 1965–70	c .980 d −1.026	0.97	.234	1.000
Bracknell 51°21'N 73m 1965–70	c .995 d −0.990	0.95	.250	1.068
Cambridge 52°12' 13m 1965–70	c .937 d −0.841	0.91	.261	1.115
Aberporth 52° 8' 115m 1965–70	c 1.064 d −1.140	0.97	.248	1.061
Lerwick 60°08' 82m 1965–70	c 1.078 d −1.14	0.96	.254	1.089
Hamburg 53°30' 1964–73	c 1.043 d −1.0386	.961	.262	1.119
Valentia 51°56' 20m 1964–74	c .958 d − .8508	.944	.270	1.153
Uccle 50°49' 120m 1951–65	c .9715 d − .9369	.968	.252	1.076
Blue Hill USA 42°13'N 205m Nov 1945 – Oct 1949	c .72 d − .67	.15	.193	.827

Table 5

VALUES OF CONSTANTS c & d IN REGRESSION EQUATION $\bar{H}_{dh}/\bar{H}_h = c + d\ (\bar{H}_h/\bar{H}_{oh})$ FOR STATIONS IN THE AFRICAN CONTINENT

STATION LATITUDE & HEIGHT (M)	VALUES OF c & d	CORR. COEFF.	RANGE OF MONTHLY MEAN VALUES $\bar{H}_h/\bar{H}_{oh}$	$\frac{-c^2}{4d}$	DIFFUSE MULTIPLIER	COMMENTS
KISANGANI 0°31'N 437 Sept 1952– Aug 1953	c 1.07 d −1.16	.93	39%–53%	.247	1.055	Hot humid climate
KINSHASA 4°22'S 450 Feb 1951–Dec 1952	c 1.08 d −1.21	.96	32%–52%	.241	1.032	Hot humid climate
WINDHOEK 22°34'S 1728 Aug 1951–Feb 1954	c 0.88 d −0.95	.95	57%–78%	.204	.871	Hot dry climate – high
PRETORIA 25°45'S 1369 Jan 1951–Feb 1954	c 0.98 d −1.16	.93	55%–69%	.207	.885	Hot dry climate – high
TANANARIVE 13°53IS 1310 Feb 1953–Feb 1954	c 1.20 d −1.39	.89	48%–65%	.259	1.107	Hot humid climate – high
DURBAN 29°50'S 5 July 1951–Feb 1954	c 1.10 d −1.43	.97	46%–61%	.212	.904	Hot humid climate
CAPETOWN 33°54'S 17 Sept 1951–Feb 1954	c 1.07 d −1.26	.93	56%–68%	.227	.971	Mediterranean type climate

Fig. 12 Relationship between percentage possible sunshine and global and diffuse irradiance on a horizontal surface for two low latitude stations

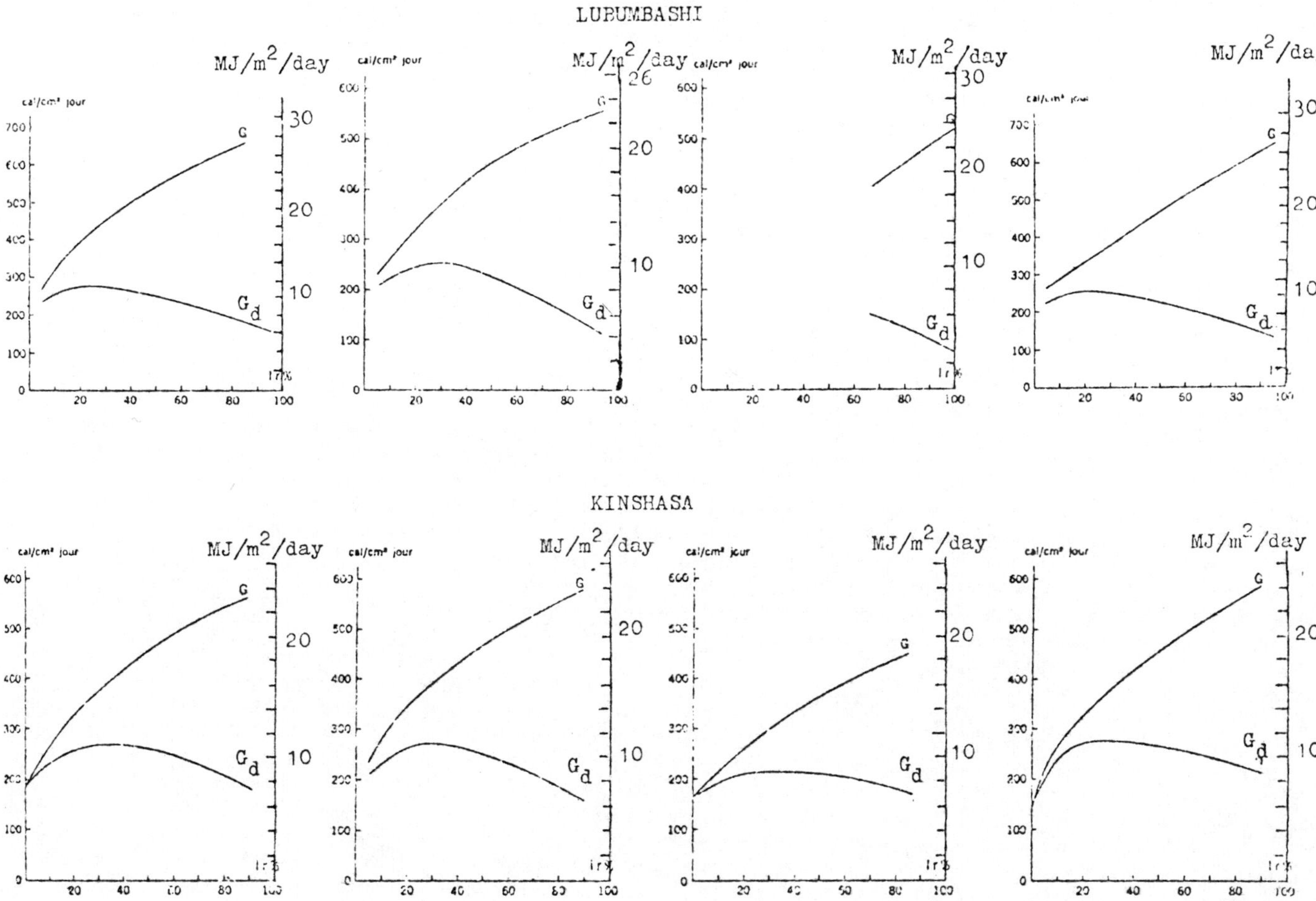

Tables 4 and 5 imply the relative amount of diffuse energy for a given value of the monthly mean atmospheric transmission coefficient $\overline{H}_h/\overline{H}_{oh}$ varies from climate to climate. Some measure of the proportional variation from place to place may be obtained from values of $-c^2/4d$, which as I have shown define the maximum values of $\overline{H}_{dh}$ as a ratio to $\overline{H}_{oh}$. Values of $-c^2/4d$ are included in Tables 4 and 5. These values are also expressed as a ratio to the corresponding value of $-c^2/4d$ for Kew, England. We use this figure in our average day programs in Sheffield as a diffuse multiplier to correct our mean diffuse irradiance figures on the computer based on Kew to match actual observations at other stations more accurately. The multiplier may be greater than one, for example in clean high latitude relatively overcast climates, like Valentia, Ireland, or less than one for example in clear sunny climates like those found in the inland parts of Southern Africa. These corrections appear to hold over quite larger regions provided proper account is taken of local urban influences. The urban influence appears to depress the diffuse multiplier within a specific region by some 5 - 10%. Our experience in the UK is that the reduction in pollution which we are achieving in our cities has had a bigger impact on diffuse energy availability than on direct energy availability.

Hourly relationships for diffuse irradiance

We have additionally studied climatologically for several north western European stations the observed relationships between monthly mean hourly diffuse horizontal surface irradiance and the mean monthly solar altitudes in the middle of each observational hour recorded. We have found, provided long term means are taken, there is a very high correlation between monthly mean hourly diffuse horizontal surface irradiance and solar altitude for a number of stations in Western Europe. The relationship function is remarkably simple:

$$G_{dh} = a' + b'\alpha \quad (4)$$

where G_{dh} is the monthly mean hourly diffuse irradiance for a particular hour when the altitude of the sun at the mid-point of that hour is α. This is the function we use in our average day computer model to predict diffuse radiation under average sky conditions in the first stage of the computation. Table 6 sets out some of the values of a' and b' we have recently determined. If the daily irradiation data are studied on a monthly annual basis for a particular station for a run of individual years, it is found that, as the average monthly global irradiation increases, the average diffuse irradiation also simultaneously increases in low sunshine climates. If however one is dealing with a very sunny climate the opposite is true. This observational result is, of course, implied by the parabolic nature of the relationship function already discussed:

$$\bar{H}_{dh} = c\bar{H}_h + d(\bar{H}^2_h/\bar{H}_{oh})$$

(This relationship is, of course, derived from the study of whole day data rather than from hourly values.) The rising or falling characteristic simply depends on whether $\bar{H}_h/\bar{H}_{oh}$ is above or below the peak value of $-c/2d$. By incorporating this additional relationship function into our computer program we are able to predict annual variations of annual monthly mean diffuse radiation from annual monthly mean sunshine data. We have not had time to study the monthly mean hourly diffuse relationship functions for the Southern Hemisphere in detail. The relationship discussed certainly holds approximately for the Australian data reported by Paltridge (17) and is more accurate than Paltridge's method of estimation in spite of its far greater simplicity. Nevertheless the relationship appears to hold somewhat less accurately than in the Northern Hemisphere and it is possible that the remarkably simple result found for temperate areas of the Northern Hemisphere depends on the fact that the influence of the summer increase in turbidity on diffuse irradiation coincides with and precisely balances the summer decrease due to changes in the mean solar distance, while just the opposite situation will exist in the Southern Hemisphere. Pending further studies equation 4 should be used with caution in the Southern Hemisphere.

Having explored the diffuse radiation problem in considerable detail, we can now consider the problem of predicting the actual irradiation on slopes under average conditions.

Table 6

CONSTANTS IN MONTHLY MEAN HOURLY DIFFUSE HORIZONTAL SURFACE IRRADIANCE FORMULA

$$G_{dh} = a' + b'\alpha \text{ watts/m}^2$$

WHERE α IS THE SOLAR ALTITUDE IN THE MIDDLE OF THE HOURLY PERIOD CONSIDERED

DATA DERIVED FROM HOURLY OBSERVATIONS

Station	Comment	a'	b'	Correlation Coefficient
Kew 1959-68	Suburban London	2	4.532	0.996
Eskdalemuir 1959-68	Inland site in hills - possibly some pollution influence from Glasgow	2	4.798	0.997
Lerwick 1959-68	Northern exposed coastal position	2	5.068	0.997
Aberporth 1959-68	Coastal site West Wales	2	5.176	0.994
Hamburg 1964-73	Airport site	2	5.36	Fitted by eye. Preliminary figure based on four months - Jan, Mar, June, Sept.
Valentia 1964-74	Extreme westerly of coast Ireland Pollution levels v. low	2	5.60	Fitted by eye. Preliminary figure based on four months - Jan, Mar, June, Sept.

DATA DERIVED FROM DAILY OBSERVATIONS OF $\bar{H}_h$ & $\bar{H}_{dh}$

Station	Comment	a'	b'	Correlation Coefficient
Kew 1965-75	Suburban London with reduced pollution	2	4.804	Used computer radiation model to determine correction to 1959-68 period
Bracknell 1965-75	New town in countryside outside London	2	5.068	Used computer radiation model to determine correction to 1959-68 period

Climatological variations in the mean irradiation of sloping surfaces

At high latitudes the levels of irradiation on horizontal surfaces even on cloudless days are very low in the winter half of the year. In summer the longer day compensates to a considerable extent for the lower mid-day sun. The low winter horizontal surface values are due partly to the unfavourable angles of incidence. Fortunately the solar energy density falling on a collector may be substantially increased by tilting the collector at an appropriate angle towards the Equator. Table 7 presents data on observed values of the mean irradiation incident on vertical surfaces at different stations as a ratio to the mean irradiation simultaneously incident on a horizontal surface. If the mornings are clearer than the afternoons an appropriate offset towards the easterly direction will give higher incident energy, and vice versa, if the afternoons are clearer. Figure 13 for Kinshasa shows in some climates very strong asymmetry may occur. The high humidities in the equatorial zone produce heavy cloud obstruction during the morning which lifts to some extent as the day progresses. Providing one can estimate the direct beam irradiance, which one can do if the hourly variations of turbidity and

Table 7

OBSERVED RATIOS OF MEAN MONTHLY VERTICAL SURFACE IRRADIATION TO MEAN MONTHLY HORIZONTAL SURFACE IRRADIATION

MONTH	VERT SOUTH KINSHASA 4°22'S 1954–60	VERT SOUTH BLUE HILL 42° 13' 1952–56	VERT SOUTH ODEILLO 42°29'N 1971–74	VERT SOUTH BRACKNELL* 51°21'N 1967–73	VERT SOUTH HAMBURG 53°30' 1952
J	0.46	1.80	1.76	1.38	1.40
F	0.37	1.38	1.33	1.26	1.3
M	0.27	0.93	0.90	1.00	1.15
A	0.26	0.61	0.67	0.73	0.80
M	0.30	0.44	0.49	0.60	0.62
J	0.29	0.39	0.43	0.52	0.59
J	0.31	0.42	0.44	0.55	0.60
A	0.31	0.54	0.59	0.66	0.75
S	0.44	0.79	0.85	0.87	1.00
O	0.39	1.23	1.24	1.20	0.96
N	0.44	1.60	1.69	1.59	1.30
D	0.49	1.94	2.09	1.65	1.50

* Observations exclude ground reflections, observations adjusted to a ground albedo of 0.20.

Fig. 13 Variations in mean global, direct and diffuse irradiance on a horizontal surface in a cloudy equatorial climate. Note a symmetry of direct irradiance about solar noon, Kinshasa 4°22'S, 15°15'E, 438m, 1954-1960.

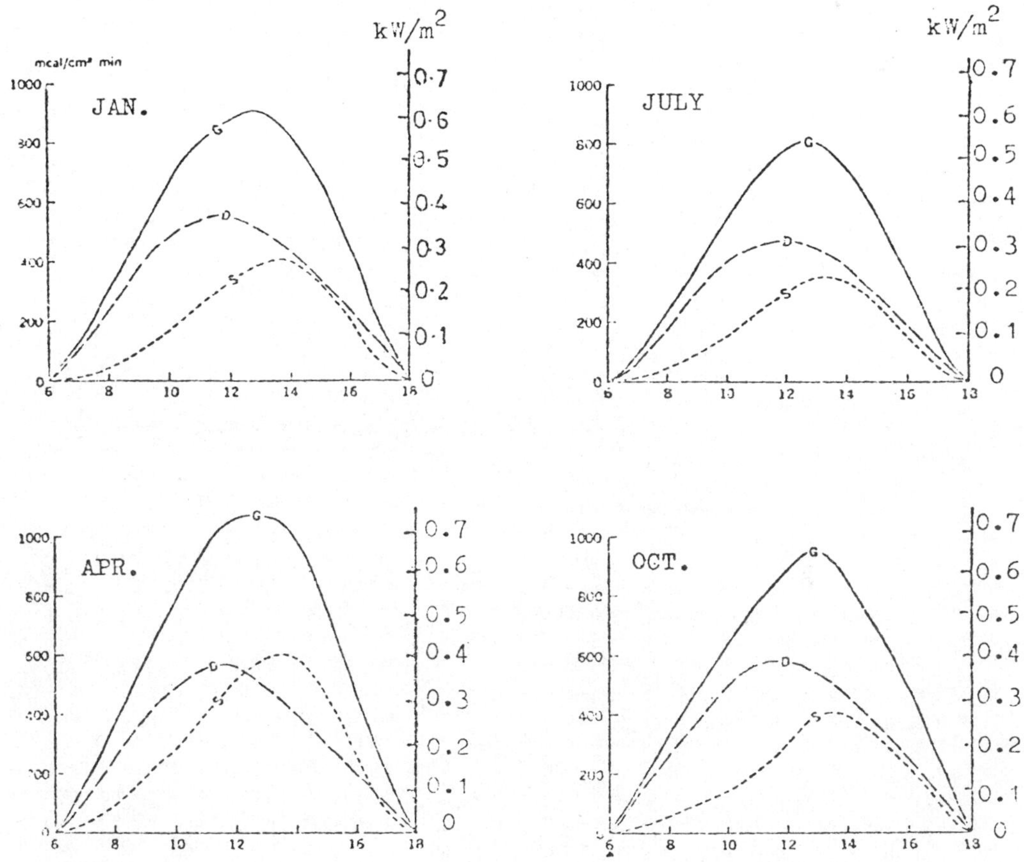

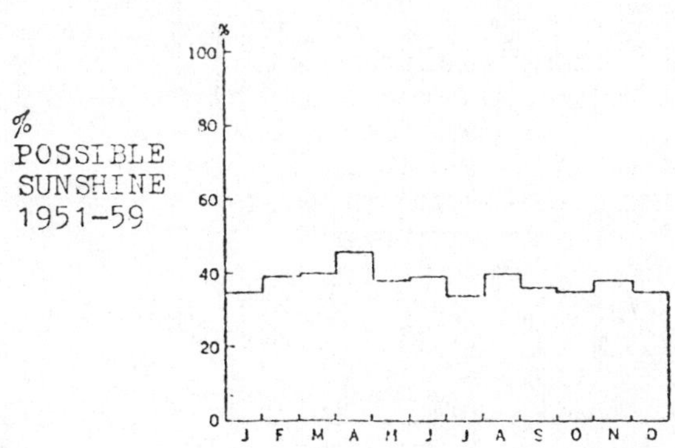

sunshine are known, it is fairly easy to calculate the tilt which will collect the greatest amount of direct solar energy on clear and partially clouded days, using standard trignometrical methods.

However the estimation of the mean diffuse irradiation on slopes is more difficult because it is not isotropic. One must also consider the effects of ground albedo, which are greatest for the steepest slopes. Furthermore the ground albedo may vary strongly with the actual angle of incidence of the direct beam on that surface (18). The author presented a fairly simple approach for estimating the available energy on slopes to the Rome conference (13) and an updated version has been recently issued (12).

In Sheffield we now prefer to handle this problem in more detail on the digital computer and our average day solar radiation AVDY program generates hourly data for slopes of any inclination and any orientation from mean monthly values of the sunshine recorded. At present the accuracy is fully tested for the UK, but the program appears reliable between latitudes 40° and 60°N at this stage. Further checks are in progress. If appropriate values of the turbidity are available, it is certainly possible to estimate the mean slope irradiation ithin 10%. Our current climatological model considers the diffuse sky irradiation to be made up of two parts, one part coming from the clear blue sky and the other part from the overcast sky. The division of the diffuse into two parts is done on the basis of possible sunshine amount. The clear sky part is distributed in such a way that the area of high radiance around the sun is correctly allowed for as described by Rodgers et al (19). The radiance of diffuse overcast sky part is assumed to follow the Moon and Spencer distribution adopted by the CIE (20). The radiance of the overcast sky L_θ at an angle θ from the horizontal plane is related to the zenith radiance L_z by the relationship function:

$$L_\theta = L_z\left(\frac{1}{3} + \frac{2}{3}\sin\theta\right) \qquad (5)$$

This formula indicates the radiance at the zenith of the overcast sky is three times the radiance at the horizon. The diffuse sky irradiance on a vertical surface under overcast sky conditions is .3955 of the diffuse irradiance on a horizontal surface (21). As a horizontal surface receives the greatest overcast diffuse sky irradiation, so as the proportion of overcast diffuse radiation rises, the optimal tilt for maximum surface irradiation decreases. However, solar collectors are not linear systems, so it may be better to collect energy efficiently on the few clear days using a steeper tilt. The answer to such complex questions however can only be accurately determined by detailed simulation.

We have used our average day program to carry out a preliminary systematic study of the effects of sunshine amount, and diffuse multiplier on the irradiance multiplier for vertical south facing surfaces. Our preliminary results for Dec 15th are presented in Figure 14 for latitudes between 30°N and 60°N. It will be seen that as the sunshine and atmospheric transmission decrease, so the energy multiplier falls, and the benefits of southerly steep tilts are very small in situations where there is little sunlight. December, of course, is the most extreme month in radiation terms. Figure 14 brings out the enormous differences between sunny and cloudy climates and I must stress the dangers about being too optimistic about the actual amounts of solar energy available. Clear day analysis can lead to excessively optimistic estimates of solar energy prospects. Even in low latitudes considerable problems may be experienced in climates where there is a lot of cloud. Figure 15 compares the mean monthly variations in the vertical surface irradiation for the two low latitude

Fig. 14

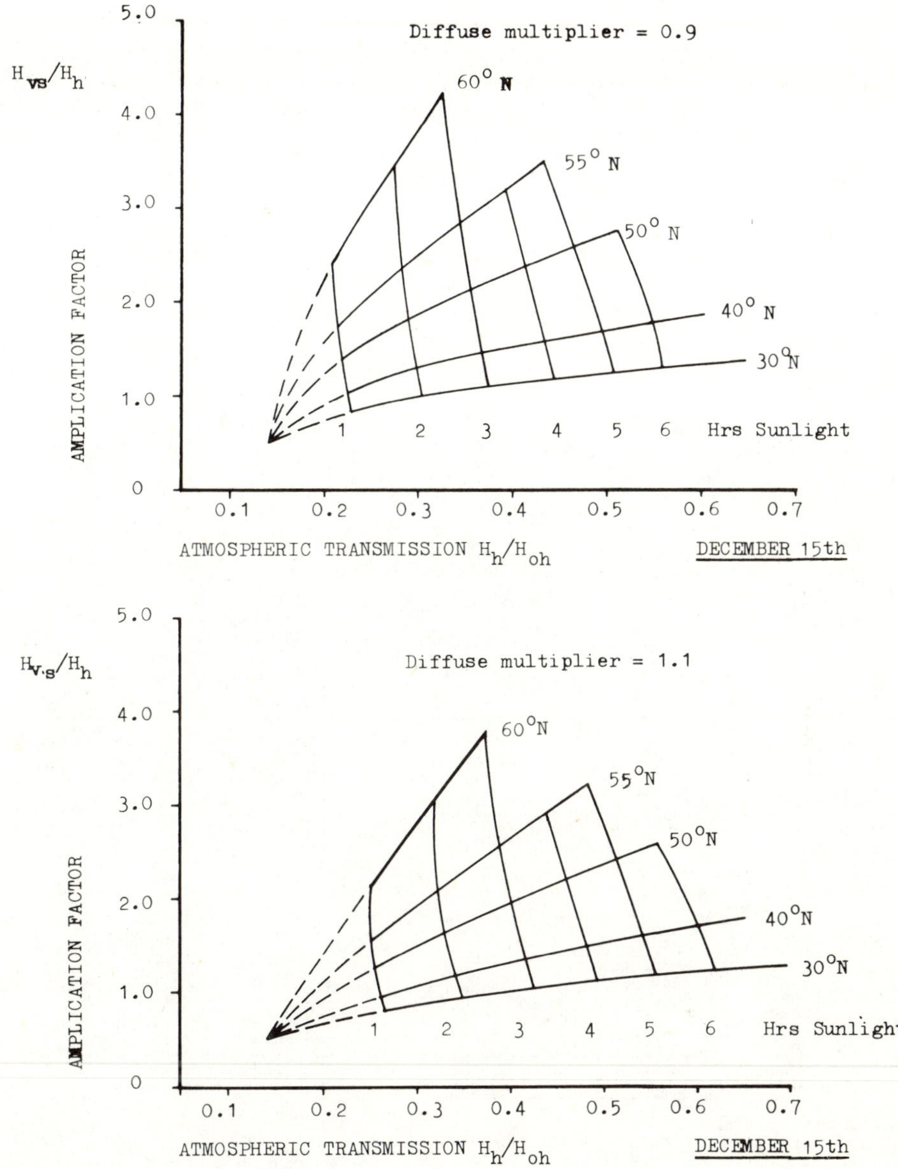

African sites already discussed. The lower the latitude the less the advantages of using steeply tilted surfaces, as a comparison between Figure 15 and Figure 4 shows.

The broad climatological implications of tilt are clear. Tilting towards the Equator reduces the winter-summer energy density range, especially at high latitudes. Steep equatorial tilts increase winter energy availability at high latitudes and decrease summer energy availability. The optimum tilt depends on the purpose for which the solar energy is to be used. The corrections to horizontal surface data for increases in available energy density due to favourable tilt are greatest at high latitudes, and are least in equatorial regions. The more diffuse energy there is the shallower the tilt required for maximum incident energy. There may be big differences between mean insolation in the morning and the afternoon, which may make it desirable to twist the collector in the more favoured direction to take account of differences in cloudiness between morning and afternoon.

Fig. 15 Mean monthly irradiation falling on vertical surfaces facing two cardinal points at Lubumbashi and Kinshasa. Note roughly equal irradiation of all surfaces in the humid equatorial climate of Kinshasa, and the marked differences which appear in the sunny season at Lubumbashi with the sun standing north of the elevation

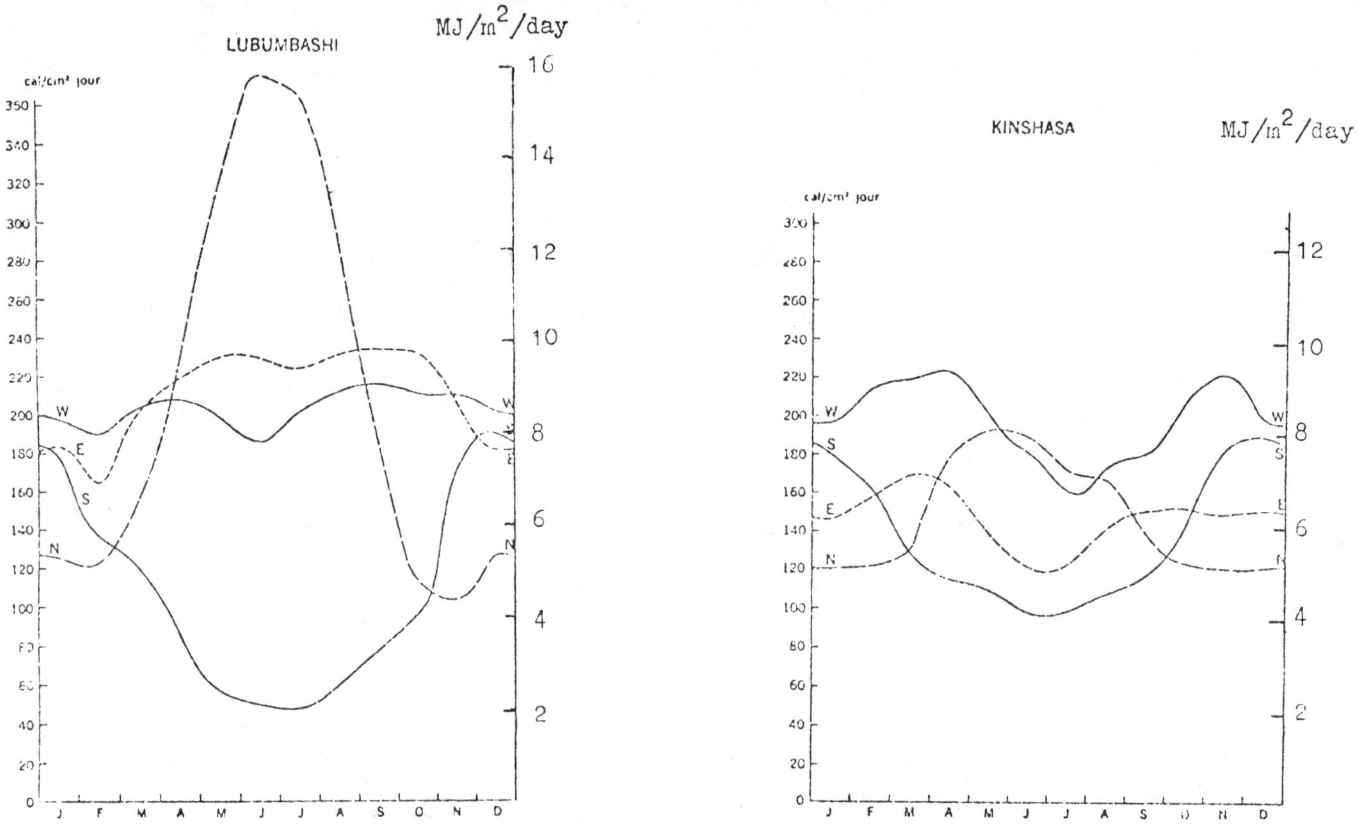

Climatological factors affecting heat losses from flat plate solar collection systems

The overall thermal efficiency of collection of solar energy depends on the balance between the energy gains and the thermal energy losses. If the collection system is properly insulated, most of the heat loss will occur from the front face. The long wave radiative losses will depend on the equivalent radiative temperature of the sky, which is strongly influenced by the amount of cloud cover.

Unsworth and Monteith (22) have shown that measurements of ϕ_t, the long wave flux density on a horizontal surface under cloudless conditions in the Sudan and also at Sutton Bonington, UK is well represented by the straight line:

$$\phi_t = c' + d' T_a^4 \qquad (6)$$

where T_a is the screen air temperature in degrees Kelvin

is Stefan Boltzman constant.

The constants c' and d' were found to have values of – – 119 ± 16 Wm^{-2} and 1.06 ± 0.04 respectively. When the screen air temperature is expressed in °C the cloudless sky relationship reduced to:

$$\phi_t = 213 + 5.5\, t_a \quad W/m^2$$

The downward radiation is increased by the presence of cloud, and the flux of downward radiation below different types of cloud can be estimated from Kondratyev (23). The difference between the outgoing long wave radiation and the incoming short wave radiation is known as the net long wave radiation. It normally has a negative value. When the radiation temperature of the ground is equal to the screen air temperature, the ratio of the net long wave radiation under cloudy skies to the corresponding net flux under clear skies is equal to 1 - 0.84c, where c is the fraction of sky covered by cloud. Thus the net long wave radiation exchange under overcast conditions typically falls to 16% of the clear sky value. The long wave radiance of the sky is not uniform and Unsworth (24) has dealt in detail with the geometry of interception of long wave radiation by slopes. Long wave energy exchanges with the ground become increasingly important as the slope angle gets greater. We have used Unsworth's paper to prepare an interactive program for predicting long wave radiation exchanges and have used it to produce Figure 16. Figure 16 shows that steeply sloping collectors have a somewhat more favourable long wave energy balance than collectors with small tilts, especially in clear weather, and more account needs to be taken of this climatological factor in assessing collector performance.

The convection losses are influenced by external wind speeds and air temperatures. The wind speed, of course, increases substantially with height. The aerodynamic factors that influence the detailed flow over the collection surface are important, but are poorly considered at present, and there is considerable scope for further study of how to reduce the energy losses from collectors due to wind flow. Our own wind tunnel and full scale work on energy transfer in Sheffield has so far been confined to the study of the actual energy losses from the windows of a 19 storey building. Our experience to date indicates that the actual heat transfer coefficients are different from those used in conventional engineering calculations. Wind is, of course, particularly important in the case of unglazed collectors. Low mounting positions on sites well sheltered by low trees are likely to be best for solar collection when using unglazed collectors for swimming pools. Exposed sites on the tops of poles are favourable for the efficient operation of solar cells whose

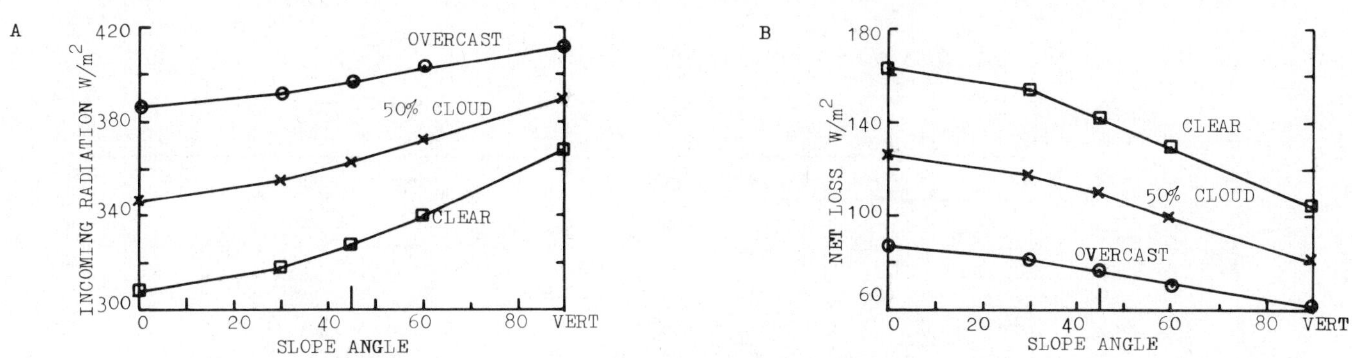

Fig. 16 Effect of slope angle and sky clarity on (A) total incoming long wave radiation (sky + ground + ground reflectance) (B) net long wave radiation loss from surface to sky and ground with air temp at 17°C, collector surface temp 30°C, collector surface emittance 0.95, ground emittance 0.90, and ground surface temperature 20°C.

Fig. 17 Effect of height and insolation on air temperature regimes on sunny days and on overcast days - S.E. England summer.

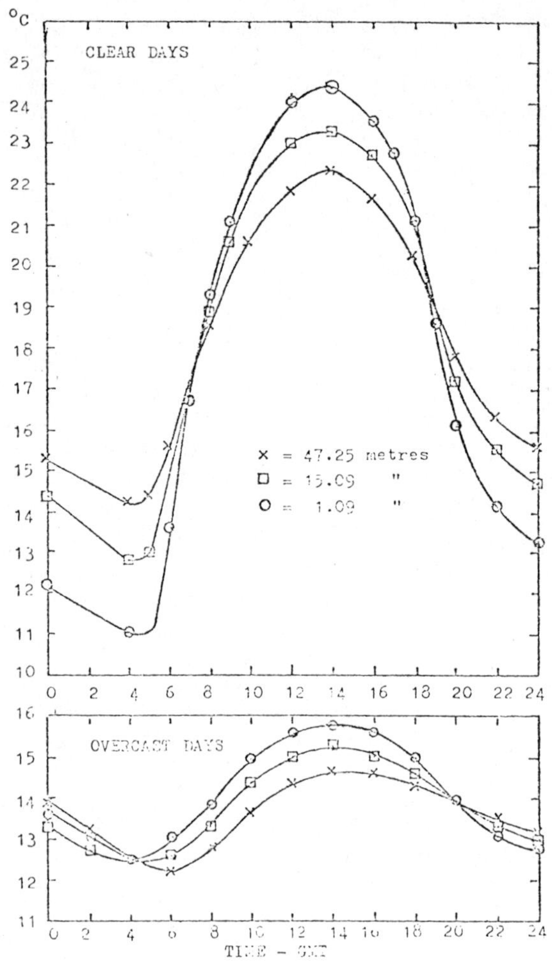

performance is adversely affected by significant rises of temperature. Increasing attention is being paid to microclimatic factors in building design (for example for general review refer Page (25)), but so far little systematic attention seems to have been given to the importance of microclimatic considerations in the solar house design field. For example Figure 17 shows typical relationships found between air temperature and height in the UK for clear and overcast weather.

Climatological factors affecting space heating and space cooling demands

Finally in assessing the climatic feasibility of using solar energy for space heating and space cooling, it is obviously important to simultaneously study the impact of climate on the actual energy demands in buildings. In space cooling, especially in situations where the radiant heat load through the windows and/or the rest of the structure dominates over the latent heat load, the solar radiation supply will be closely linked to the energy demand patterns, and the problems of solar energy storage will be minimal. A few hours storage may suffice. In space heating, especially in very high latitudes, the opposite situation will often

Fig. 18 Diurnal variations in shade air temperature at Kew for three classes of daily irradiation, smooth curve - days of high radiation, dotted curve - average days, and smooth curve with circles - days of low irradiation, for the months of March, June, September and December.

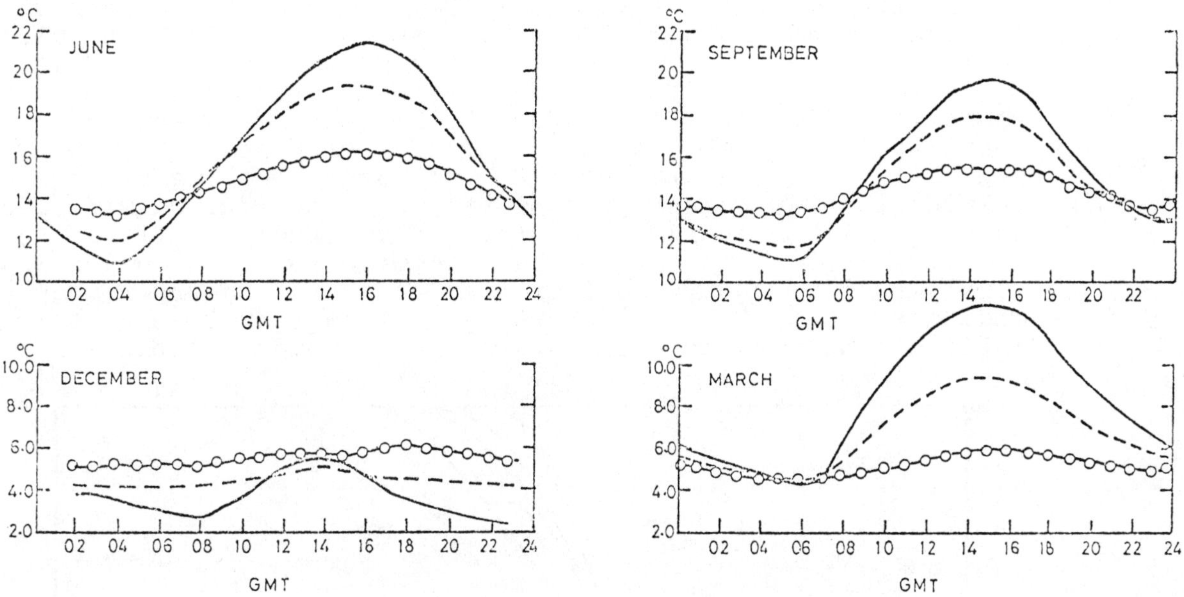

prevail, and the solar supply will be six months out of phase with the heating energy demand.

The solar heat gain variables and the heat loss variables are, of course, statistically interlinked, and the temperature regimes on sunny days are quite different from overcast days. The nights may be much colder after sunny days than when the weather is overcast, as the clouds associated with overcast days substantially reduce the outgoing radiation and so stabilise the night-day temperature differences. Figure 18 illustrates the screen air temperature regimes found for days of different radiation classes at Kew, England. Statistically overcast days in winter are warmer throughout the twenty-four hours than clear days. In the UK it is normally windier on overcast days than on clear days. In other parts of the world this relationship could well be reversed as high insolation in dry desert areas tends to stir up the daytime surface winds due to strong vertical convection. In the humid tropics the wind speeds are often very low on overcast days and the weather may become very oppressive indeed.

Such complex climatological interactions can only be properly demonstrated by appropriate local meteorological studies and observations. I wish to stress the need for meteorological services to observe and prepare meteorological summaries for solar house design which respect a design logic that needs the simultaneous assessment of solar heat gains and collection heat losses. Guidance is also needed on the effects of height and of shelter.

In high latitudes economic solar energy utilisation may only become feasible when and if we can develop cheap long term solar energy stores to store summer energy. In mid-latitudes between 25° and 40°, there is usually a far better balance between winter energy supplies on south facing surfaces, and space heating demands. In such areas successful space heating solutions can be built in less stringent energy storage criteria. Some measure of the relative scale of the problems to be solved can be compiled by comparing the winter degree days (to some suitable base) with the available radiation on a favourably inclined surface (say tilt equal to the latitude, or the latitude + 10°). The centre of the US located around latitude 35° - 40°N is for example much colder in winter than the south west of the UK warmed by the Gulf Stream at latitude 51°N, but has a much better solar radiation climate. The length of the heating season in North Western Europe is very long and solar space heating can play an alternative role to full space heating by helping to keep buildings comfortable in spring, summer, and autumn, while the mid-winter load is met from fossil fuels, or possibly, in future, from wind and wave power which are in very plentiful supply at that time of year. UK experience indicates that with only short term storage available it is difficult to meet as much as 40% of the domestic space

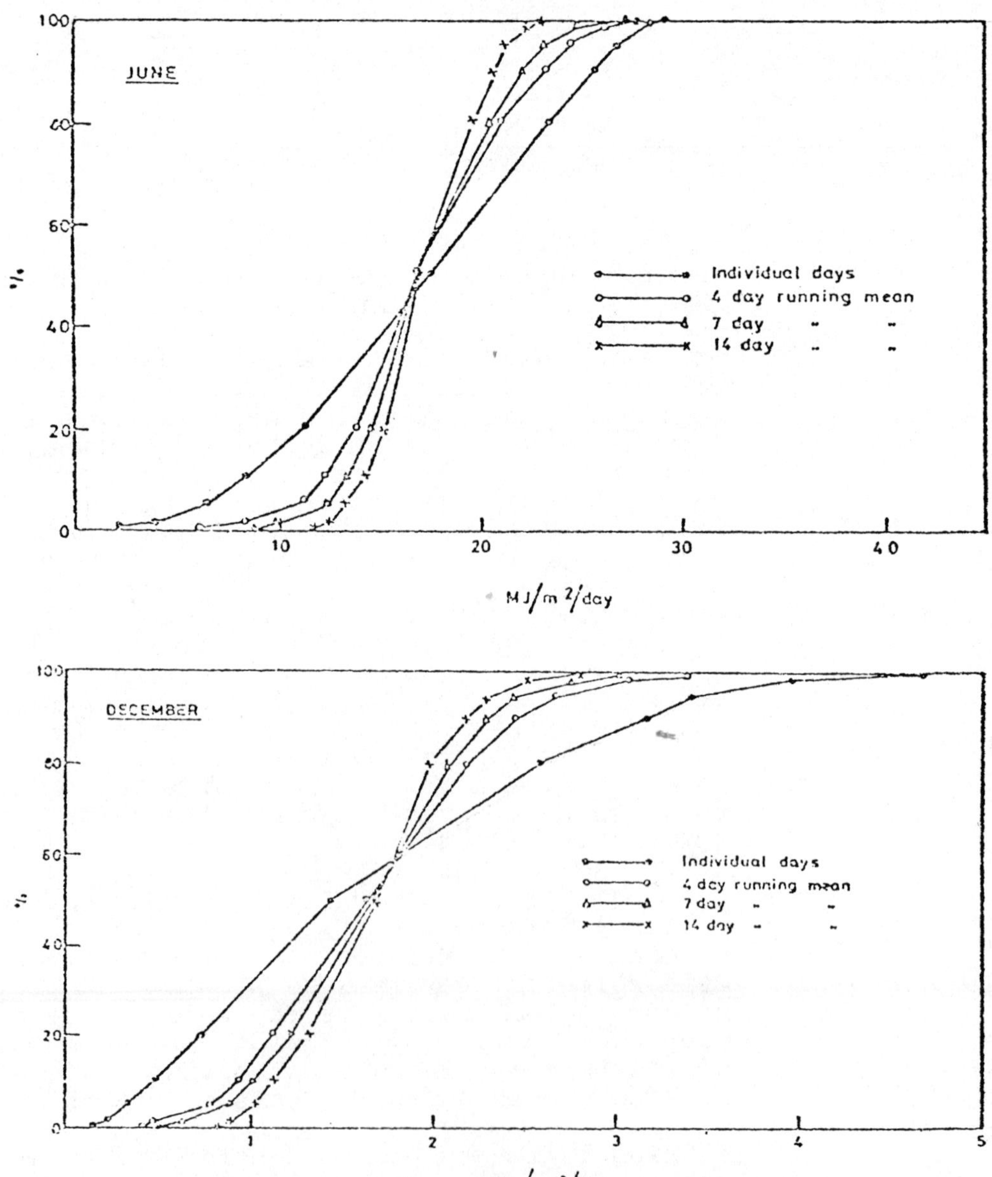

Fig. 19 Cumulative frequency distribution of average values of total daily global solar irradiation based on running means over different periods for a) June and b) December at Kew, 1951-1970.

Fig. 20 Measured mean monthly daily irradiation on vertical surfaces facing south, north, east and west at Bracknell, England, latitude 51°21'N. Ground reflections are excluded. Period 1966-73.

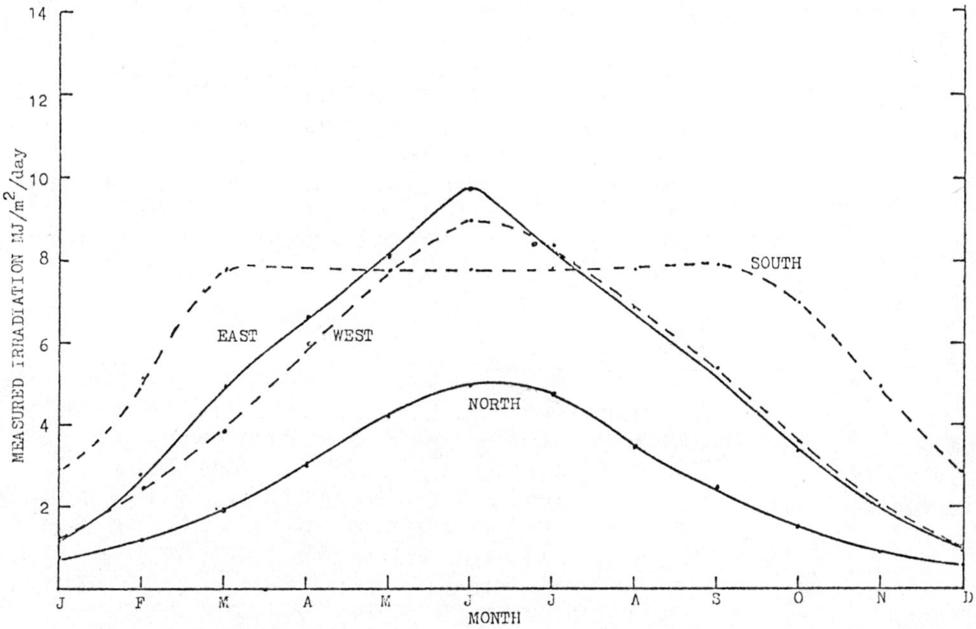

heating load from the sun in our latitudes, and this contribution comes mainly in the period March to May, and in October.

The pattern of day to day weather is also important as it affects storage needs. There is no space to discuss here the best ways of describing the temporal patterns for design purposes. Simple statistical radiation data for the US may be found in Jordan and Liu's recent review (26). The UK Meteorological Office have compiled statistics on the variations in solar energy on a horizontal surface using running means for different periods from one day to fourteen days. Data presented from the UK-ISES Report (1) shown in Figure 19 demonstrates the very large variability in the irradiation for single days, especially in December. The variability is substantially reduced if a four day period is considered, but there is little further change for the fourteen day running means. If short term storage is to be used in such a climate,the benefits of fourteen day storage capacity will be modest compared with four day storage over one day storage. If really long term storage is available, as in the Danish Zero Energy House (27), then very long term running means over months can be considered, and the principles of storage design are fundamentally changed in nature.

Conclusion

Further collaboration between meteorologists, building scientists and architects will be needed to produce a sounder climatological foundation for solar house design. The computer can play an important role in assisting developments in this area because many of the calculations are relatively long and tedious to perform. Our computing approach in Sheffield enables appropriate design aids be rapidly produced for different centres but considerable further detailed meteorological study is still needed to make design forecasts more reliable in various parts of the world. More study for example is needed of the links between meteorological factors influencing the demands for energy, with the factors that influence the supply of solar energy. The microclimatological considerations need greater attention. As solar house design is a very difficult systems engineering problem, collaboration in systems analysis will be vital to success, and the will to achieve effective interdisciplinary communication essential. We must take pride in what we achieve as a team rather than simply in the professional activities through which we make our personal contribution to the wider human good.

Acknowledgements

The computer based research project on solar radiation estimation reported in this paper is partly financed by the Science Research Council. The help of my collaborators in this project, G.G. Rodgers and C.G. Souster, has been invaluable in the task of preparing the tables and diagrams. Figures 1,3,6, 18 and 19 have been drawn from the UK Section of the International Solar Energy Society's comprehensive report, "Solar Energy - a UK Assessment", and I am grateful to the Society for letting me reproduce them. Figures 5, 12, 13, and 15 have been taken from the excellent Atlas Climatique du Bassin Congolais which provides a fund of good information on how to present radiation and other meteorological data for a whole territory (28). Time unfortunately did not allow me to seek their formal permission, but I hope they will accept the need for the more experienced to help the less experienced in this vital field. At present few tropical countries can provide such excellent data.

References

1. UK-ISES, Solar Energy: A UK Assessment, UK-ISES, c/o The Royal Institution, 21 Albemarle Street, London, W1X 4BS, pp. 375, (1976).

2. Trombe, F., Maisons solaires, Techniques de l'ingénieur, C 777, 3, 1-5, (1974).

3. Robinson, N., (Ed.), Solar radiation, Elsevier, pp. 346, (1966). (This is a standard reference that explains these factors in great detail.)

4. Unsworth, M.H., Variations in the short wave radiation climate of the UK, Proc. Conf. UK Meteorological data and solar energy applications, Feb. 1975. UK-ISES c/o The Royal Institution, 21 Albemarle Street, London, W1X 4BS, 18-36.

5. Bannon, J.H. & Steele, L.P., Average water vapour content of the air, Meteorological Office, Geographical Memoir 13, No. 102, H.M.S.O., London, (1960).

6. List, R.J., (Ed.), Smithsonian Meteorological Tables, Smithsonian Miscellaneous Collections, 114, 6th Ed., pp. 527, (1951).

7. Bilton, T., Flowers, E.C., McCormick, R.A. & Kurfis, K.R., Atmospheric turbidity with the dual-wavelength sunphotometer, Report Solar Energy Data Workshop, Nov. 1973, National Oceanic and Atmospheric Admin., Environmental Research Laboratories, Air Resources Laboratories, NSF-RA-N-74-062, US Govt. Printing Office, Washington, 61-67 (1974). (See also J. Appl. Meteor., 8, 955-962 (1969).

8. Parmelee, G.V., Irradiation of vertical and horizontal surfaces by diffuse solar radiation from cloudless skies, Heating, Piping and Air Conditioning, August 1954, 129-135, (1954).

9. Unsworth, M.H. & Monteith, J.L., Aerosol and solar radiation in Britain, Quart. J.R. Met. Soc., 98, 778-797, (1972). (Errata to key table, 99, 598).

10. Page, J.K., Supplementary note on values of $\bar{D}/\bar{K}$ on cloudless days for solar altitudes below 30°, Proc. Conf. UK Meteorological data and solar energy applications, 1975, 37-39, UK-ISES, c/o The Royal Institution, 21 Albemarle Street, London, W1X 4BS, (1975).

11. Blackwell, M.J., Five years continuous recording of total and diffuse solar radiation at Kew Observatory, Met. Res. Pub. 985, Met. Office, London, (1954).

12. Page, J.K., The estimation of monthly mean values of daily short wave irradiation on vertical and inclined surfaces from sunshine records for latitudes 60°N to 40°S, Dept. of Building Science, Univ. of Sheffield, Internal Note No. 32, pp. 33 (1976).

13. Page, J.K., The estimation of monthly mean values of daily total short wave radiation on vertical and inclined surfaces from sunshine records for latitudes 40°N - 40°S, Proc. of U.N. Conf. on New Sources of Energy, Rome, 1961, Conf. Paper No. 35/5/98, (1961).

14. Liu, B.Y.K. & Jordan, R.C., The inter-relationship and characteristic distribution of direct, diffuse and total solar radiation, Solar Energy, 4, 1-19, (1960).

15. Tuller, S.E., The relationship between diffuse, total and extra terrestrial solar radiation, Solar Energy, 18, 259-264, (1976).

16. Kasten, F., Der einfluss der bewölkung auf die kurz- und langwelligen strahlungsflüsse am boden, Annalen der Meteorologie, Neue Folge, 12, 65-68, Offenbach, (1977).

17. Paltridge, G.W. & Proctor, D., Monthly mean solar radiation statistics for Australia, Solar Energy, 18, 235-244, (1976).

18. Tomps, R.C. & Coulson, K.L., Solar radiation incident upon slopes of different orientations, Solar Energy, 19, 179-184, (1977).

19. Page, J.K., Rodgers, G.G. & Souster, C.G., An interactive computer design methodology for the design of solar houses, Proc. NELP/UNESCO Conf., Solar Building Technology, July 1977, Paper 4.2, (1977).

20. Hopkinson, R.G., Petherbridge, P. & Longmore, J., Daylighting, Chapter 2, 29-58, Heinemann, (1966).

21. Walsh, J.W.T., The science of daylight, McDonald, London, 128-129, (1961).

22. Unsworth, M.H. & Monteith, J.L., Long wave radiation at the ground, Angular distribution of incoming radiation, Quart. J.R. Met. Soc., 101, 13-29, (1975).

23. Kondratyev, K.Ya, Radiation in the atmosphere, Academic Press, London,(1969).

24. Unsworth, M.H., Long wave radiation at the ground, (ii) Geometry of interception by slopes, solids and obstructed planes, Quart. J.R. Met. Soc., 101, 25-34, (1975).

25. Page, J.K., Application of building climatology to the problems of housing and building for human settlements, WMO Tech. Note No. 150, WMO, Geneva, pp. 64, (1976).

26. Jordan, R.C. & Liu, B.Y.K., Applications of solar energy for heating and cooling buildings, ASHRAE, New York, pp. 206 (1977). (Refer especially to Chapter 5)

27. Esbensen, T.V. & Korsgaard, V., Dimensioning of the solar heating system in the Zero Energy House in Denmark, Proc. Conf. European Solar Houses, April 1976, UK-ISES, c/o The Royal Institution, 21 Albemarle Street, London W1X 4BS.

28. Bultot, F., Atlas Climatique du Bassin Congolais, Institut National pour l'Étude Agronomique du Congo, Democratic Republic of the Congo, (I.N.E.A.C.), (1971).

Architectural, technical and socio-economic aspects

2

2.1 Active and passive solar heating of buildings

J. Douglas Balcomb
Solar Energy Group
Los Alamos Scientific Laboratory
Los Alamos, New Mexico, USA

Abstract

An overview of both active and passive solar heating approaches for buildings is presented. Passive solar heating concepts--in which the thermal energy flow is by natural means--are described according to five classifications: direct gain, thermal storage wall, solar greenhouses, roof ponds, and convective loops. Results of simulation analyses are presented for a variety of climates. Active systems utilizing both liquid-heating collectors and air-heating collectors are described. Trends in the recent development of solar heating are discussed.

Introduction

Within the very short time of approximately three years, solar heating has developed from the status of a curiosity being explored by a few pioneers scattered around the world to a vigorously developing technology with a bright hope for the future. One notable aspect of this development is its diversity and this will be highlighted in this paper. This diversity is especially evident in space heating applications which range from the relatively simple and yet sophisticated passive systems approaches to some very complex equipment-oriented active system applications such as the combined use of solar collectors and heat pumps. Active solar cooling can add yet another level of complexity.

As an energy system, solar heating of buildings is markedly different from other alternative energy systems. For one thing, the technology is generally available to implement the system. For another, it is a distributed rather than a centralized system. It tends to make the individual in his home less dependent rather than more dependent on sophisticated technology and on the intricate economic, social, and industrial networks that distribute conventional energy forms. Furthermore it is an approach wherein a person can contribute to alleviation of the energy problem in an individual way, without sacrifice of the way of life that abundant energy has made possible--that is, without energy privation.

Solar heating systems are generally divided into two categories: *passive* systems in which the energy flow is by natural means and *active* systems which utilize external energy such as electrically-driven pumps or fans. The paper illustrates these different approaches by the use of examples of various houses and buildings which have been constructed. Unfortunately it is not convenient to reproduce these slides in the printed paper.

The burgeoning solar energy industry presents a fast changing scene. A large number of companies, mostly small firms, have entered the market anticipating a vigorous business. There are now a wide variety of viable and reliable systems available. Until the winter of 1976-77, solar energy in the U. S. could be characterized as an industry in search of a market. Confronted with high initial costs, buyers were cautious. The brutal and sobering winter in the eastern U. S. has stimulated a much more serious interest than has ever existed before. There are now approximately 350 firms in the U. S. alone marketing solar collectors, with a current annual average sales volume of over 600 square meters per firm and many other firms marketing other components and whole systems.

It is generally conceded that solar heating is now cost-effective throughout most of the U. S. against the cost of electric heating on a life-cycle cost basis. Generally speaking however, because of tax and financing considerations, it is far more favorable for individuals to install solar heating systems on buildings than for commercial enterprises to do so.

The prognosis for the future is bright. A multi-billion dollar industry is predicted by 1985. President Carter has called for 2.5 million solar homes by that date. While this may be optimistic it is certainly achievable. Future building construction will certainly be better built, both from an energy conservation and a passive solar heating standpoint, and a growing fraction will employ active solar heating systems.

Passive Solar Space Heating

Solar gains through windows, walls, modified walls, sky lights, clerestory windows, and roof sections provide an opportunity to dramatically reduce the total heating energy requirements of a building. When the thermal energy flow is wholly by natural means, such as radiation, conduction, and natural convection, and when solar energy contributes more than half of the total thermal requirements from the building, then it is properly referred to as a passive solar heated structure.

Passive solar heating works very well. This has been demonstrated time and again in a wide variety of buildings located in a wide variety of climates. The occupants of these buildings testify to their comfort, to the ease of their operation, and especially to their low fuel bills.

I have just moved in to a passive solar house in Santa Fe, New Mexico. It is an elegant and relatively sophisticated design, conceived and built by the Nichols in Santa Fe who are now constructing their fifth solar house in a small subdivision of sixteen units. It is a solar greenhouse type of design and is 80 to 90% solar heated in our 3600°C-day climate.

Passive solar heating systems can be cataloged into five basic types. The first and simplest of these is the *direct gain* approach in which one simply has an expanse of glass, usually double glass, facing south. The building should have a considerable thermal mass in the flow and massive masonry construction with insulation on the outside. The building becomes a live-in solar collector. The characteristic sun angles result in a good situation since the south face is exposed to a maximum amount of solar energy in the cold winter months when the sun angles are low and a minimum amount of solar energy in the summer when the sun angles are high. Thus the basic seasonal characteristic of control is automatic.

An elegant example of the direct gain approach is the Wallasey school in Liverpool, England. This is certainly the largest passive solar heated structure in the world and one of the first. The building was designed by the architect, A. E. Morgan, who unfortunately died soon after its completion in 1962. The building has been carefully monitored by M. G. Davies, of the University at Liverpool, who has determined that the solar heating fraction is 50 to 70%.[1,2] The remainder of the heat energy required by the building is from the lighting and the students.

The building is a large two-story structure with 17 to 25 centimeters of concrete forming the roof, the back wall, the floor and side walls, and with 12 centimeters of expanded polystyrene insulation on the outside. The solar wall is an expanse of two sheets of glass, 8 meters tall and 70 meters long, facing south. The figured inner-layer of glass refracts the sun's rays so as to irradiate the roof and the floor fairly uniformly, providing natural daylighting. It is interesting to note that the auxiliary system installed in the Wallasey school has never been needed.

Another direct gain system is the David Wright house in Santa Fe, New Mexico which uses an adobe earth brick construction with insulation on the outside.[3] The auxiliary heating system in this house is a wood burning stove. Temperature fluctuations during sunny winter days are approximately 10°C.

The direct gain approach can take more complicated forms with the use of clerestory windows or skylights to provide energy in back rooms.

The second type of system is the *thermal storage wall* in which the thermal storage is in a wall which blocks the sun after it comes through the glazing and this wall then stores the heat energy. The wall is usually painted black or a dark color to be a good absorber. The wall can be made of water in containers or masonry. A small section of thermal storage wall was used in the Wallasey school discussed earlier.

The thermal storage wall approach overcomes two disadvantages of the direct gain approach. The building interior is shielded from the brightness and the fabric-degrading effects of the direct sun and the interior of the building is also somewhat buffered from the temperature variations normally associated with direct gain structures.

This approach has been popularized by structures built in Odeillo, France by F. Trombe and J. Michel.[4] The masonry walls in the house built in 1967 are 60 centimeters thick. The primary mechanism for heating the house is by radiation and convection from the face of the wall facing the room. Thermal energy diffuses through the wall from the outside to the inside. In addition, about 30% of the energy is from a thermocirculation path which operates only during the day by natural convection through vents at the top and bottom of the wall.

Published data taken over a period of one year indicate that about 70% of the total thermal energy required by the building (which is thermostatically controlled to a temperature of 20°C) was provided by solar energy.[5]

Additional examples of thermal storage wall houses have been built in Princeton, New Jersey by the architect, Doug Kelbaugh,[6] and in Los Alamos, New Mexico by Bruce Hunn.[7] Each of these buildings utilizes a two-story masonry wall with some novel approaches. Kelbaugh cut windows through the concrete wall providing a view to the south and some direct gain heating to the building. Hunn utilizes a fan to force air taken from the space between the glazing and the wall into a rock-bed located underneath the ground floor. Energy from the rock-bed is then used to heat a portion of the house which does not have a thermal storage wall.

The thermal storage wall can also be constructed of water in containers rather than masonry. An example is the Steven Baer house in Albuquerque, New Mexico in which the thermal storage wall consists of surplus oil drums laid on their side providing a massive thermal storage.[8] A system of movable insulation is used in which a door forming the south wall lowers during the winter day to allow the sun in and actually reflect some sun onto the storage. In this case only a single layer of glass is used. The door can then be raised at night to reduce heat loss.

A mix of the preceding two concepts is the *solar greenhouse* in which one builds a greenhouse onto the south side of a building with some kind of thermal storage wall between the greenhouse and the house. The temperature in the greenhouse does not require very good control as long as the plants do not freeze. Solar energy provides, typically, all of the heat required for the greenhouse as well as providing substantial energy for heating the house.[9] Solar greenhouses are an excellent means of providing solar heat to a house which has already been constructed.

My own house, discussed earlier, is of the solar greenhouse type.[10] The wall separating the house and the greenhouse is a 36 cm thermal storage wall of adobe earth-brick. The greenhouse glazing is 37 m^2 of sealed double glass. The remainder of the house is of wood frame construction employing approximately 20 cm of rockwool and fiberglass insulation for a net conductance of approximately 0.17 W/°C m^2. Proper passive solar heating design requires a proper combination of building and glazing orientation, glazing area, thermal storage, and thermal insulation.

The fourth type of passive design is the *roof pond* in which the thermal storage is in the ceiling of the building. In this case movable insulation is needed because the sun angles are altogether wrong. The sun provides large inputs in the summer and small inputs in the winter. However it is a good natural cooling system because by using movable insulation one can take advantage of night time radiation. This concept has been implemented in the Atascadero, California house which uses the Harold Hay "Skytherm" concept.[11] A system of insulating panels on the roof is used which slide back and forth on tracks. The water bags are left exposed to the sky during the day in the winter and during the night in the summer. This provides heat input in the winter and heat loss on a summer night. The insulating panels are put in place during a winter night to conserve heat or during a summer day in order to exclude the sun which is reflected from the top of the white surface. This system has worked very well and has operated without any backup in the rather mild climate of Atascadero providing good thermal comfort in a small building.[12]

The fifth type of passive system is a natural *convective loop*. The classic thermosiphon water heater fits into this category. Natural convective loops which use air as the heat transport fluid have also been built and work well. A good example is the Paul Davis house in Corrales, New Mexico.[13] In these designs the thermal storage is necessarily located above the solar collector and the heat transport fluid then flows by natural circulation.

Passive test rooms have been set up at Los Alamos to study the behavior of passive solar heating elements under carefully controlled conditions.[14] One year of test data have been obtained on a pair of test rooms which utilize thermal storage wall concepts, one with cylindrical water storage tubes and the other with a thick masonry wall. The storage walls are located behind a vertical double-glazed wall. These totally passive buildings are so climatically adaptive that their average temperatures during the mid-winter months are actually higher than during the summer months. During January the average temperature inside the rooms was approximately 35°C above the outside ambient temperature but during the summer the room temperature is only slightly above the average ambient temperature. The primary purpose for the construction of these test rooms was to validate mathematical analysis techniques developed by the Los Alamos Scientific Laboratory. The temperature histories in these two rooms are very accurately predicted by the computer analysis. Once validated, these computer models can be used to predict the performance of similar passive designs in any climate where weather data and solar data are available on an hourly basis.

Some results from these analyses are given in the following table.

PREDICTED ANNUAL SOLAR HEATING RESULTS FOR A REFERENCE THERMAL STORAGE WALL PASSIVE SOLAR HEATING SYSTEM IN VARIOUS CLIMATES

City	Year Starting	Heating °C-days	Latitude	Solar Heating,* $kWhr/m^2$	Solar Heating Fraction, Percent
Los Alamos, NM	9/1/72	4080	35.8	190	56.5
El Paso, TX	7/1/54	1390	31.8	158	97.5
Ft. Worth, TX	7/1/60	1370	32.8	120	80.8
Madison, WI	7/1/61	4350	43.0	141	41.6
Albuquerque, NM	7/1/62	2360	35.0	200	84.1
Medford, OR	7/1/61	2930	42.3	150	56.1
Bismarck, ND	7/1/54	4580	46.8	170	46.4
New York, NY	6/1/58	2920	40.6	151	60.2
Dodge City, KS	7/1/55	7880	37.8	186	71.8
Nashville, TN	7/1/55	2110	36.1	124	65.2
Boston, MA	7/1/57	3070	42.3	148	56.8
Seattle, WA	7/1/63	2890	47.5	133	52.2
Lincoln, NE	7/1/58	3330	40.8	169	59.1
Boulder, CO	1/1/56	3150	40.0	197	70.0
Vancouver, BC	1/1/70	3280	49.1	145	52.7
Edmonton, ALB	1/1/70	6490	53.5	119	24.7
Winnipeg, Man	1/1/70	6380	49.8	106	22.6
Ottawa, Ont	1/1/70	4910	45.3	119	31.9
Fredericton, NB	1/1/70	4910	45.8	126	33.9
Hamburg, Germany	1/1/73	3620	53.2	78	27.5
Denmark	?	3800	56	136	43.8
Tokyo, Japan	?	1830	34.6	158	85.8

*The values in the solar heating column are the net energy flow through the inner face of the wall into the building.

These calculations are made for a passive system with a thermal storage wall constructed of 45 cm of masonry having the following properties (typical of dense concrete):

Thermal conductivity = 1.73 W/m°C

Heat capacity = 2.01×10^6 $J/m^3{}^\circ C$

The building thermal load, exclusive of the south wall, is equivalent to 2.84 W/°C per square meter of south glazing. Auxiliary energy is supplied as required to maintain the building room temperature within a band of 18.3°C to 23.9°C. In each case the analysis is based on hourly data for the year specified.

The table dramatically illustrates the wide climatic viability of the thermal storage wall approach.

Active Solar Heating Systems

In an active system external energy is utilized to pump a working fluid, generally either liquid or air, to transfer heat from a separate solar collector into either the building or into thermal storage.[16] By utilizing a separate collector heat losses at night are greatly reduced compared to a passive system approach in which heat losses generally extend throughout the 24 hour period. However since the collectors typically operate at much higher temperatures than passive systems, the day time efficiency is much lower.

There are two main categories of active solar heating systems. The most widely used utilize collectors designed to heat a liquid pumped through passages in thermal contact with the collecting surface. These liquid-heating collectors are especially appropriate in cases where the thermal storage medium is water. Water storage is normally associated with domestic hot water applications and also building heating systems which utilize water as the working fluid.

Although a few installations also utilize water as the heat transport fluid from the collector, most utilize a non-freezing primary working fluid. In this case the normal practice is to utilize a heat exchanger to transfer heat from the primary collector coolant to a water thermal storage tank. Those installations which do utilize water as the primary coolant are either located in climates where freezing is not a problem or utilize a drain-down system during nights when freezing might occur.

The second basic type of active system utilizes air-heating collectors. These collectors are generally less expensive than liquid-heating type, are far less subject to corrosion degradation, and are totally free of the possibility of liquid leaks. For these systems, the convenient form of thermal storage is a bin filled with rocks or gravel.

In many parts of the U. S., home heating systems which utilize forced air are common. There is a growing popularity of the air-heating collector rock-bed storage option in many of these locations. While there is some penalty in collector performance, due to the poor heat transfer properties of air, most of this performance disadvantage is regained in the system operation due to the heat-transfer effectiveness and the thermal stratification of the rock bed. Thus the overall system performance is comparable to that of systems which utilize liquid-heating collectors.

Another attractive design approach utilizes a heat pump in conjunction with solar collectors.[17] The heat pump can operate effectively between a moderate temperature source which is supplied by solar energy and the higher temperatures generally required by the building heat distribution system. At the same time the collectors can operate in a very efficient regime replenishing the moderate temperature thermal storage. This is a particularly attractive approach either in severely cold climates where electric heat pumps cannot work effectively or in climates where active cooling is required in the summer and the heat pump can be used in a cooling mode during these periods to reject heat from the building thus doing double duty.

Although flat-plate collectors continue to dominate the market a variety of concentrating and other advanced collectors are becoming relatively common in the marketplace. The concentrating collectors either use reflection or refraction in order to achieve concentration ratios from four to ten. Single-axis tracking of the sun is usually required for these collectors. They operate at relatively high efficiency but collect only the direct beam component of the solar radiation. A large number of collectors are now available which utilize some advanced techniques to minimize heat loss from the collector.[18] Some utilize evacuated spaces within glass tubes to minimize loss. Others utilize special "selective" surfaces to reduce the infrared re-radiation losses from the absorber surface. In some instances the extra performance achieved from such advanced concepts warrants the extra cost associated with the construction.

Solar heat can also be used to provide active cooling for a building. The present preferred approach is to utilize collectors capable of higher temperature performance with a storage system capable of storing water at a higher temperature, and an absorption chiller which has been derated to run at the temperatures anticipated from the collector-storage combination. For most installations the add-on cost associated with the extra equipment exceeds the savings.

The problem with solar cooling is basic thermodynamics. The only way to achieve a substantial performance increase in solar cooling is to increase the temperature of the solar heat supply. This necessarily increases the cost associated with thermal storage and thermal energy transport and one wonders whether solar cooling can ever be economically practical. Nonetheless there are many installations which demonstrate the technical feasibility of the concept. Installations which demonstrate the use of other techniques are also being installed such as Rankine-cycle chillers and several varieties which use desiccants in combination with air-heating collectors.

Trends

The solar heating of buildings has become a boom phenomena in the U. S. At the end of 1976 there were between 1000 and 1500 solar heated buildings and the growth rate over the last few years corresponds to a doubling in number every eight months.[19] If this trend continues through 1985 then we can expect that President Carter's goal of 2.5 million solar homes will have been met or slightly exceeded.

There are major concerns continually voiced about both the durability and the reliability of solar equipment being installed, especially for the active system approaches. To be economic solar collectors must survive for the lifetime of the building in a very hostile thermal, humidity, ultraviolet, and physical environment. Based on the materials of their construction, one can expect that many collectors presently marketed are unlikely to meet their requirements. Many people are now talking about certification of solar energy equipment and systems, however it remains a difficult matter to cope with the bewildering variety of collectors and the large number of small firms which have entered the marketplace. There are also concerns that standardization would inhibit the development of innovative concepts.

Although passive system approaches are relatively old in concept they have received little attention. There is a recent burgeoning of interest in these approaches and a growing realization that passive system approaches can be used in a wide variety of climates without sacrifices in lifestyle or comfort. There are many more architectural options open with passive systems and the designs tend to be aesthetically more pleasing. The principle problem has been the lack of a quantitative basis for incorporation of the basic concepts into architectural design.[20] Work done recently tends to alleviate these concerns.[21] Another problem is that these systems are not as amenable to control of building temperature as active systems. The challenge confronting passive solar energy design is not so much one of heat collection, since a window is an efficient solar collector, but one of storage and control of the heat so as to maintain suitable comfort standards within the building. New materials and new design approaches may help substantially in alleviating this problem. Systems of movable insulation such as hinged doors, sliding panels, rotating louvers, and air-blown polystyrene beads are being employed to control solar gains and thermal losses. These techniques can provide the desired degree of control in the passive design approach. However it also increases the complexity of the design and the cost of the installation.

Improved glazing materials which have lower heat loss coefficients by virtue of infrared reflection or low surface emittance are now being investigated. The potential for these materials is enormous, especially if some degree of control over their optical properties can be obtained.

The trends predicted for the U. S. can be extended to other countries. Solar heating is particularly appropriate to other areas which have a temperate climate such as Southern Europe, Australia, Argentina, Japan, etc. In these locations there is sufficient heating load to make solar heating cost-effective and sufficient sun to make the system work effectively.

Historically, solar energy has been viewed with particular interest in Israel and Japan, where solar water heaters have been plentiful, and in Australia where several air-handling systems have been developed. At the present time predominant interest in the development of solar energy for heating and cooling is within these countries and also within the U. S. and Northern Europe. Major research and development projects have been initiated in Germany, Denmark, and Japan as well as the continuation of the long standing activities in Australia. To some extent the character of development of solar systems will depend on both the climate and the traditional methods of home heating. Thus, for example, in Northern Europe one can expect the development of systems which utilize liquid heating collectors in preference to those which utilize air heating collectors, simply because nearly all centrally heated heated homes utilize hydronic systems. The development of passive systems can be expected to occur throughout the world because they have been shown to work well in quite a variety of climates. Excellent examples have been built at Odeillo in France and as far north as Liverpool, England as I have illustrated. One can expect the development of passive systems to be particularly favored in those areas where traditional construction practices utilize massive construction such as stone, brick and mortar or adobe earth bricks. Throughout developing countries one can expect solar heating and cooling to emerge primarily along the lines of passive system approaches. These tend to be more simple and tend to utilize native materials more effectively. They should therefore be lower in price. This is particularly true since the construction tends to be more labor intensive and labor costs are generally lower in the developing countries. The development of solar heating and cooling systems would seem an appropriate technology for all developing countries since it is generally simple in concept and execution, inexpensive, and suitable for small scale application.[22]

References:

1. M. G. Davies,"The Contribution of Solar Gain to Space Heating," Solar Energy, Vol. 18, pp 361-367 (1976).

2. J. D. Perry, Jr., "The Wallasey School," Proceedings of the Conference on Passive Solar Heating and Cooling, Albuquerque, NM, May 18-19, 1976, pp 223-237 (1976). (Available from the National Technical Information Service, U. S. Dept. of Commerce, Springfield, VA, 22161, Stock # LA-6637-C, $13.00.)

3. B. T. Rogers, "Some Performance Estimates for the Wright House," ibid Ref. 2, pp 189-195.

4. F. Trombe, Maisons Solaires, Techniques de l"Ingenieur (3), 1974, C777.

5. F. Trombe, J. F. Robert, M. Cabanet, and B. Sesolis, "Some Performance Characteristics of the CNRS Solar House Collectors," ibid Ref. 2, pp 201-222.

6. D. Kelbaugh, "Kelbaugh House," ibid Ref. 2., pp 119-128.

7. B. Hunn, "A Hybrid Passive/Active Solar House," Proceedings of the 1977 Annual Meeting, American Section, ISES, Vol. 1, Section 1, pp 11-16 to 11-20 (1977).

8. B. Anderson, "The Solar Home Book," (1977).

9. R. Fisher and W. F. Yanda, "Solar Greenhouses," John Muir Press, Santa Fe, NM (1976).

10. W. D. Nichols, "Unit 1, First Village," ibid Ref. 2, pp 137-149.

11. J. Yellott, "Early Tests of the 'Skytherm' System," ibid Ref. 2, pp 54-62.

12. "Research Evaluation of a System of Natural Air Conditioning," Cal. Poly. State Univ., Jan., 1975.

13. P. Davis, "To Air is Human: Some Humanistic Principles in the Design of Thermosiphon Air Heaters," ibid Ref. 2, pp 40-45.

14. J. D. Balcomb, R. D. McFarland and S. W. Moore, "Simulation Analysis of Passive Solar Heated Buildings--Comparison with Test Room Results," ibid Ref. 7, pp 11-5 to 11-19.

15. ibid Ref. 14, pp 11-1 to 11-3; also J. D. Balcomb, J. C. Hedstrom, and R. D. McFarland, "Simulation Analysis of Solar Heated Buildings-Preliminary Results," Solar Energy, Vol. 19, pp 277-282 (1977); and also (same authors) "Passive Solar Heating of Buildings," LA-UR-77-1162, available from Los Alamos Scientific Laboratory, MS-571, Los Alamos, NM 87545.

16. "ERDA's Pacific Regional Solar Heating Handbook," available from Superintendent of Documents, Washington, D. C. 20402. Stock # 060-000-0024-7, (1976).

17. "Solar Energy Heat Pump Systems for Heating and Cooling Buildings," ERDA Doc. COO-2560-1, Con 7506130, Workshop Proceedings June 12-14, 1975, Penn. State Univ. (1976).

18. J. D. Balcomb and J. E. Perry, Jr., "Assessment of Solar Heating and Cooling Technology," LA-6379-MS, available from NTIS (see Ref. 2), (1977).

19. W. A. Shurcliff, "Solar Heated Buildings, A Brief Survey," 13th Edition, Cambridge, MA.

20. J. D. Balcomb, "Simulation Analysis as a Design Tool," ibid Ref. 2, pp 238-246.

21. ibid Ref. 15.

22. E. F. Schumacher, "Small is Beautiful: Economics as if People Mattered," Harper and Row, New York (1973).

Solar and regional architecture

M. Gerber
School of Architecture
Toulouse, France

Abstract

A direct relationship between passive solar energy buildings and traditional regional architecture exists: a passive solar heated house takes advantage of climatology and of topography. Local building materials often possess a good thermic inertia. These conditions result necessarily in a good local and regional integration of the building. For example in the South-West Mediterranean part of France, the walls of an ancient 3 level house will be re-used for the living spaces. The North extension (storage space) and the South extension (greenhouse) are acting as buffer spaces, protecting the central areas from the North wind and adding heat from the South. The shape of the building deviates wind by a traditional tile roof slope and particular wall angles. Its compact shape and the conventional materials relates the structure to its natural environment. Solar devices are: glassed south elevation, fixed and flexible shades, storage in the building masses and a water tank.

Introduction

We are surrounded by an architecture of international style, tall glass-walled buildings or surburban houses standardized on the same pattern, using the same materials everywhere in the world. These buildings either high rise or individual suburban houses subserve fashion, marketing trends, technological break-throughs. They are conceived for standardized sites and do not take into account specific characteristics of climate, topography, etc. They even accentuate the negative factors of the climate, like whirlwinds between the high rise buildings and solar reflexion on their elevations. Such buildings ignore the environment, they do not take advantage of it. Mechanical equipment is designed to keep a constant 20°C inside without concern for the energy costs. They even ignore the human aptitudes. Physicians and biologists observe that nyctemeral rhythms in light and warmth, as well as seasonal variations are necessary for the balance of human beings. Therefore not only must we avoid these artificial conditions, but it is even better to live within light variations. Harold Hay is one of the first to say that 'genetic adaptation of life to sun and sky radiation has resulted in a human scale of temperature (18 to 28°C) from which architectural passive systems need not depart'.

On the other hand we are confronted with a great variety of structures, forms and building material as we travel through different regions and countries, looking at ancient village dwellings. We are surprised to see how well they fit in the landscape and how efficiently they flatten the curve of the variations of outside temperature differences to obtain inside an ambient atmosphere, cool in summer and tempered in winter.

These considerations demonstrate that the problems of energy must become an integrated part of architectural analysis. As a matter of fact, every structure acts and reacts in a certain way with the environment and particularly with the sun. If well conceived it becomes a collector which regulates favorably human environment. There is a direct relationship between the morphology and the choice of the material of a building and how it may take advantage or not of the elements of the climate in order to succeed in the achievement of comfortable and pleasant life.

Architectural analysis

Therefore when we plan to design a low profile energy building using solar energy, we have first to study climatic conditions of each site: insolation, dominant winds, rainfalls, temperature extremes. The topography is then to be considered since it directs the implantation of the building on the ground together with the climatic specifications: or may take advantage of a hill for wind protection, or of a well orientated slope for maximal insolation. Low profile energy building requires also an investigation on the energy costs of the building materials. The use of local resources avoids transport and transformation costs. The technical solutions of the building will also be considered from an energy saving point of view. Audacious structures requesting a technological 'tour de force' should not be retained as requiring too much energy and money. Therefore the often incongruous results of technological freedom will be avoided. On the other hand local materials like stone, adobe or even concrete allow storage and establish a sufficient time lag to provide most of the heat when needed (thermic inertia).

At this point, one notes that the gathered data from these various searches meet the characteristics of regional architecture which is mainly the results of energy saving: local building materials are principally used, avoiding infrastructure and transport costs. Climatic, geographical elements are implicated in building morphology for ensuring an optimal thermic comfort at the lowest cost.

Examples to prove these statements can be found all over the world: The American Indian cliff dwellers took advantage of a natural microclimate (caves) which protected them from the sun in summer but largely exposed their buildings to its heat in winter (Montezuma castle).

The Swiss Bern House creates its own microclimate against cold, snow and rain by using protecting roof and balconies and it reduces the loss of heat by realizing extremely compact buildings.

The Pyrenean village exposes each of its houses to the sun on south-east or south-west slopes.

Of course regional architecture is not the sole result of physical determinations. Cultural and sociological data get into the concept and morphology of regional architecture. Generally, the more extreme the climatic characteristics are, the more important is their influence on the morphology of the building (e.g. the igloo of the Eskimos). On the contrary, when climatic conditions are less imperative, more room is left for other factors to be expressed (culture, economy). In that case morphology may reflect more directly social organisation inside the house like special religious or ritual areas, reserved rooms for men or women.

Thus in order to be an integrated part of a real regional architecture, a low profile energy building using passive solar energy in addition to the physical data must include cultural, sociological and economic features. Therefore a passive solar house takes advantage of the climate (insolation, dominant wind) and of topography. It uses local building materials preferentially since they often possess a good thermic inertia (adobe, stone, bricks). These conditions result necessarily in good local and regional integration of the building. But also sociological and economic data which render this solar house convenient and satisfactory for our actual way of life ought to be integrated in the design.

Obviously we cannot simply reproduce the ancient regional architecture which was made for people living outside of the house most of the time.

The indoor life was mainly restricted to basic functions. Besides the ancient house was built essentially for energy conservation without real attempts for energy trapping. Nowadays, we spend much more time in the house than outside for leisure and various creative activities. The habitability of the house is most important, large living-rooms with large openings, extensions such as terraces or greenhouses contribute to the quality of the life in the house. Noteworthy, most of these improvements participate in the completion of a passive solar system.

Thus the type of building realized according to the concepts of low profile energy and solar passive system may be quite adapted to our way of life. Moreover it can use soft technical devices and be proposed as a real sample of human ingeniousness and progress (e.g. Steve Baer's sky-lid, beadwall).

It is the connection of all these elements which provides an authentic type of architecture, quite remote from what is generally proposed as traditional architecture: these commercial products hook isolated characteristics of ancient country architecture (tile or slate roofs, arches, stuccos) on cheap standardized structures. There is no longer any relationship between these elements and their original function.

On the contrary, the intrication of climatic, geographical cultural specifications together with actual economic and sociological concerns will result in a true living regional architecture.

The project

We like to exemplify this direct relationship between regional architecture and the use of solar energy passive system by describing our own approach in Languedoc (South-West Mediterranean part of France) in an area located a few kilometers from the Mediterranean sea, at 43° of North latitude.

Climate: generally speaking the Mediterranean sea determines the climate and thus influences the vegetation and the natural landscape. Summer droughts are interrupted by occasional heavy, short rains in fall. This rain persists less frequently in winter and spring. Although the sea influences the climate, the area is also affected by perturbations from the Atlantic Ocean and the nearby mountains.

The climate is characterized by its luminosity. The yearly average is 2700 hours of sunshine. The monthly averages account for (1) horizontal surfaces 6.8KWh/m2 in July and 1.6KWh/m2 in January, (2) vertical surfaces to the south, 3.5KWh/m2 in July and 2.5KWh/m2 in January, for average conditions of normal sky (sources Meteorologie Nationale). Strong, sometimes violent winds from the north-west accentuate the summer drought. They are responsible for temperature drops in winter.

The yearly average wind frequency is: 236 days $v > 5m/s$, 116 days $v > 16m/s$, 17 days $v > 28m/s$. Summers are hot, winters are mild, frost occurs sporadically. The daily average in summer is $22^{\circ}C$ and in winter $6^{\circ}C$. Rainfall is quite irregular from year to year and may vary between 250mm and 500mm with an average around 400mm.

Site: a cluster of ancient ruins tops the edge of a plateau (90m above sea level) overlooking the plains and the sea. The worn out landscape provides no more protection from the main elements of the climate. Trees and/or forests ought to have existed when the ruins were dwellings, yet misuse has depleted these natural resources. Also during the years of vacancy, doors, windows and parts of the roof have been taken away for re-use on other buildings. Therefore only the outside walls of the ancient 9 x 8 meters 3 level house are re-usable. They are built in limestone rubble with lime putty and sand, 60cm thick on the base.

Sociological context

The existing ruins have been abandonned approximately one century ago, most probably for economic reasons.

Today beside the heavy pressure of tourism for housing, there is a small demand from young farmers to get installed in the area. Therefore few acres of land have been brought into cultivation with the governmental aid for new plantations, predominantly wine, but also almond and olive trees.

One of these ruins will now be rehabilitated for agricultural use. The program requires space for a farmer's family, parents and two children, plus space for storage of materials and tools as well as a small workshop. Animals are sheltered in one of the ruins next to the house.

The design

The analysis has shown that there are two main climatic factors to be taken into consideration. They are the sun and the wind. They both will influence the design of the future structure.

Their strong impact have already determined the ancient building: south windows only, no openings to the north and north-west.

Whatever the physical restrictions may be, the design approach has to integrate satisfying organization and pleasant living.

The projection locates the living spaces in the existing structure (Fig. 1 and Fig. 2). Kitchen, dining and living-room with a big open fireplace completed by an aerocirculation for heat recuperation are on the ground level. Parent's bedroom and working area are on the second level and communicate with the living-room. Children's bedroom and a small guest room are on the third floor. Situated to the north part in addition to the existing walls are storage spaces and the workshop. These areas are not heated and function as passive buffer spaces protecting living areas from dominant northwinds. Situated to the south, a greenhouse provides extension for the living and dining-room areas. It acts as an active buffer space, adding heat in winter into the house by natural convection. The staircase is used as return air duct. In summer hot air is directly ventilated towards the exterior (Fig. 2); it is to be noticed that summer gardening in this area is as profitable as winter gardening, since growing vegetables in a greenhouse requires much less watering than in open field, mainly because of the protection from the dry north wind. That is a notable advantage in the semi-arid area.

Fig. 1
Groundlevel

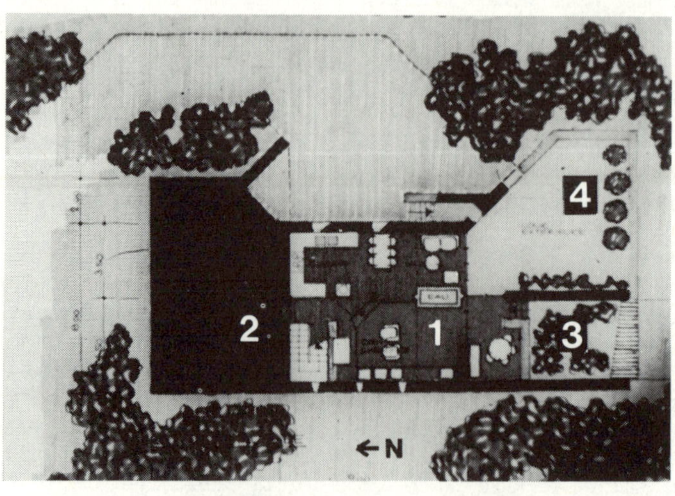

1. Kitchen, dining and living-room area (existing structure)
2. Storage and workshop area (passive bufferspace)
3. Greenhouse (active bufferspace)
4. Outside court

Flat plate collectors integrated in the greenhouse structure (Fig. 2 and 4) provide heat for domestic warm water. Warmed water gets to the watertank through the primary fluid loop by thermosyphon. The watertank with heat exchanger is placed above bathroom and kitchen.

Fig. 2

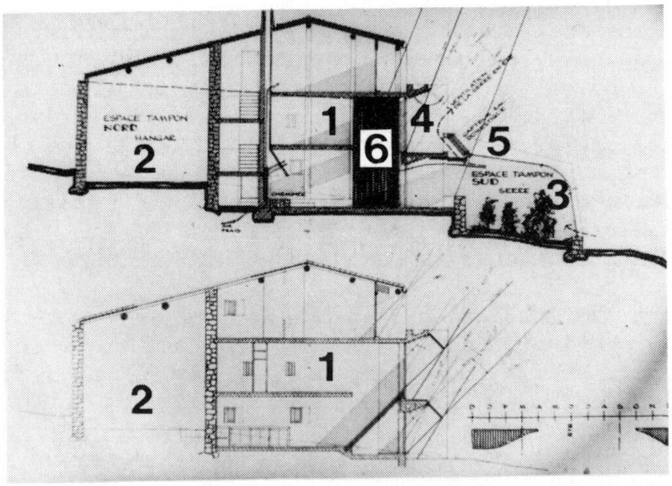

1. 3 level areas - living spaces
2. Storage and workshop area
3. Greenhouse
4. Greenhouse summer ventilation
5. Flat plate collectors for domestic water
6. Water tank

Fig. 3

Model view from North-East, roof sloping to the North and wind deviating wall angles on the East

Fig. 4

Model view from South-West, glassed south wall and greenhouse

The solar devices are as follows: the south elevation is almost entirely glassed on the 3 levels. The ratio of the glassing area to the heatable volume is 0.07 m2/m3. Fixed shading (concrete slabs balconies) protects the glass from sun in summer and exposes it completely in winter. Direct storage is allowed (1) in the building masses, stone walls, concrete slabs and columns which accounts for an equivalent of 2 metric tons of concrete for 1m2 of glassing, (2) in a 6000 litre galvanized steel, 2 stories high, cylindrical water tank, which increases substantially the total thermal mass. The purpose is to use it as a variable thermal stockage capacity. Tests and measurements will show how much water is needed and if it is necessary to modify the thermal mass during the heating season. As a possibility we study the addition of reflectors to improve the direct insolation of the cylindrical surface of the water tank. Insulated curtains on the inside of the glassing protects the building from energy losses at night.

Though the insolation of a building may be conveniently regulated with fixed shading in summer, spring and fall are much more critical in that respect. One may need some heat in March, but very seldom in September. We are designing flexible shadings (Fig. 6) for controlling insolation at these periods of the year. They are light, rigid 100cm panels, hinged on one side to the concrete slab and they may be moved manually with a winch.

Theoretically, these adjustments are only needed twice a year. However they may also be used in winter as reflectors if conveniently coated.

An architectural model has been built and solar incidences throughout the year have been simulated to complete the study (Fig. 5 and Fig. 6).

A considerable amount of wind energy estimated at 3.000KWh/m2 a year at 40m above the ground is available for the production of electricity by means of a wind generator.

However for everyday living, wind may be considered as a nuisance and the building must be protected from excessive losses of energy and create wind protected outside areas.

The building shape tends to deviate wind without turbulences by the appropriate slope of a traditional tile roof and by particular angles of the walls. It is shown in Fig. 3 how the mud is modified by the main roof and by the 45° angles of the east wall. These wall angles create areas largely protected from wind where the entries of the building are located.

Figs. 5, 6

Insolation on South elevation, April 21 (noon) flexible shadings pulled in

Insolation on South elevation, August 21 (noon) flexible shadings pulled out

These efforts to ensure wind protection confer to the building a compact shape, which, in addition to the use of local, traditional materials, relates at most the structure to its natural environment.

A system of measurements has been worked out and will be integrated in the structure as construction goes along.

It consists of: (a) 16 thermocouples placed in carefully chosen locations throughout the building, (b) one electronic control unit, and (c) one small lineprinter unit.

At regular intervals, which may vary between one and ten hours, the 16 measurements will be transcripted automatically on a paper ribbon and may then be expressed in graphs indicating daily and seasonal temperature fluctuations on the 16 different locations.

Building construction has started and we hope to be able to begin measurements by the end of the year.

We are actually in the process of testing an integrated solar greenhouse under the same climatic conditions as the projected building.

Fig. 7
Greenhouse

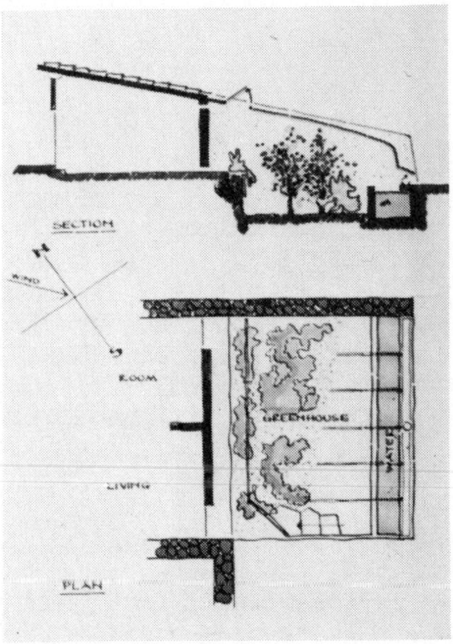

The thermal masses consist of rocks, stone walls and a 5.000 litre open water tank (Fig. 7). By July 15 we have not measured excessive temperatures, which would indicate that lots of thermal masses and good ventilation may protect it from overheating without any auxiliary energy for cooling, even in hot arid climate. But it is of course too early to draw valid conclusions which should be based on extended studies. We should be able soon to come up with some experienced figures which may help to define more specific parameters for designing low energy profile buildings.

2.3 United States solar energy programs: Their potential for developing countries

J.P. Eberhard
AIA Research Corporation
Washington, D.C., U S A

Abstract

Passive systems for capturing the energy of the sun to heat, cool and ventilate houses have been a part of the long history of architecture. This paper explores the concept of passive systems and the relationship of such concepts to vernacular architecture. The next section reviews an earlier study by the AIA Research Corporation of energy use in the residential sector of 132 nations. This is followed by more information on the base of solar energy research being developed in the United States - a base which has seen federal funding expand from $500,000 in 1970 to $240 million in 1977. This base can now be used to assist developing countries in technology utilization (the process of effective use of existing resources and techniques) and in technology transfer (the process of transfering the use of resources or techniques to new areas of application). Some recommendations are made for how UNESCO and other UN agencies might assist in this effort.

Introduction

For the last 50 years, modern building has applied mechanical technologies in expansionist and additive ways, with minimum regard for land variations, cultural differences, and climatic givens. Sophisticated building techniques allow a building to disregard context, generating instead an 'international', anonymous style. The reasons that generated this architecture belong to the advancement and mechanization of a society which possessed unlimited resources. More recently, these 'successful' urban and suburban models have been directly applied to the growth of developing countries with little thought directed toward the regional differences that land, climate, and culture represent to architecture.

In most of the world we should now be exploring ways in which the architectural solutions to new buildings and the redesign potential for existing buildings can reduce their dependence on fossil fuel. There are two fundamental strategies for doing this - both of which are challenges to the imagination and skills of the architectural, engineering and manufacturing community. One is to reduce the demand for energy and the other is to replace or substitute non-fossil fuel solutions for those which depend on petroleum or natural gas for their energy supplies (including the use of electricity generated by such fuels).

The use of energy in buildings can be reduced drastically by exploring first the design criteria and program requirements usually associated with client needs. Since it is the human requirement for comfort associated with the buildings functional use that forms the basis for energy demand, these design requirements become the basis for classifying buildings energy demand patterns. We can further reduce our demand for energy use in buildings by a judicious balance between isolating the interiors of buildings from an alien climate (a condition we tend to exaggerate in our glass-enclosed, hermetically-sealed office buildings) and opening the interiors to enable the building to respond in a positive way to a friendly climate (practice long followed by many countries with mild climates).

Once we have determined that some form of supplementary heating, cooling or ventilation will be needed in addition to all of the solutions provided by the natural design of the building, then we move to a stage of selecting passive solar solutions. These would include the potential for passive solar heating methods, as well as cooling by induced ventilation, thermal cold storage, dessication, evaporative cooling, etc.

If we find that passive systems will still be inadequate to provide the level of comfort desired, then we will want to explore the active systems of solar energy which include: solar plate collectors, concentrating collectors, photovoltaic cells for converting sunlight directly to electrical energy, wind energy and bio-mass conversions.

If after all of this we find that we still require additional supplementary support we will turn to more conventional heating, cooling and ventilation equipment. But we will have lowered our demand for the size and complexity of such fossil fueled systems to a minimum.

Although we are always searching for 'new' approaches, especially technologically advanced ones, many energy conserving solutions are to be found in existing architecture. Vernacular architecture, being inherently responsive to regional climatic forces, can effectively serve as a foundation of building vocabularies through which passive solar technologies can be applied. An operational definition of passive design follows as the next step in energy conserving design.

> PASSIVE DESIGN is not a style of building, or a packaged technological system that can be plugged into any building type. It is determined by how a building works to store and give off heat from the sun, how it cools and provides ventilation ... It is an attitude toward building that can employ various conceptual and physical principles of thermal dynamics as applied to material and form as they react to land and climate. These principles can be translated through a range of building vocabularies, and become qualities inherent in the building construction and operation. It is a sophistication to challenge the design profession-response to the environment (Wright's organic architecture, Olmstead's city beautiful, Le Corbusier's use of Greek vernacular architecture, etc.).

World Energy Use in Housing

In 1975, the American Institute of Architects Research Corporation (AIA/RC) was asked to prepare an estimate of energy requirements in the residential sectors of various countries and to project this to a possible total requirement for energy use in the year 2,000 A.D. While professional economists balked at the complexity of the task, a team of young architects on our staff proceeded with a logical analysis based on the best available data. The results are less important in their absolute values than in their revelation of sharp contrasts in energy use between developed nations (especially the United States) and developing nations. (See Table I.1.)

The method used for estimating energy use by the world for residential purposes in the year 2,000 is as follows:

- We first classified housing into twelve types according to their size, type of materials, extent of heating and cooling, use of appliances, etc. (See Table I.2.)

- Using United Nations sources we determined estimated populations for most countries in the world by the year 2,000.

- We next developed an estimate of the number of housing units needed to house this population depending on the number of family units housed per unit. This was done for each of 120 nations for whom we could discover data. (Summarized in Tables II.1 to II.6.)

- We then developed as much data as possible regarding the type of energy used in houses in each of 60 countries. (See examples by their stage of economic development, Tables III.1 to III.7.) We extrapolated from this data when we could not find information for a particular country.

- We then sought information on the likely economic progress of each nation between now and the year 2,000 and upgraded the probable energy use and levels of housing in those countries expected to show economic progress.

- With all of this accumulated data, we then calculated the high, low, and reasonably likely energy demand for the housing sector in the year 2,000. (See Table IV.1.) This amounts to 9×10^{16} BTU's per year (or 9.5×10^{9} giga joules) or the equivalent of about 39 million barrels of oil equivalent per day, more than twice the present U.S. energy consumption per day.

What is most revealing is that the twenty or so most developed nations of the world would demand about 90% of the energy used in the residential sector, with the United States demand representing about one-third of the total.

It would seem likely, therefore, that the most pressure from the market place and the largest economic advantages for solar use in housing will be in the 20+ most developed countries, with the U.S. being again the most interested of all.

TABLE I.1
Energy Use Per Capita

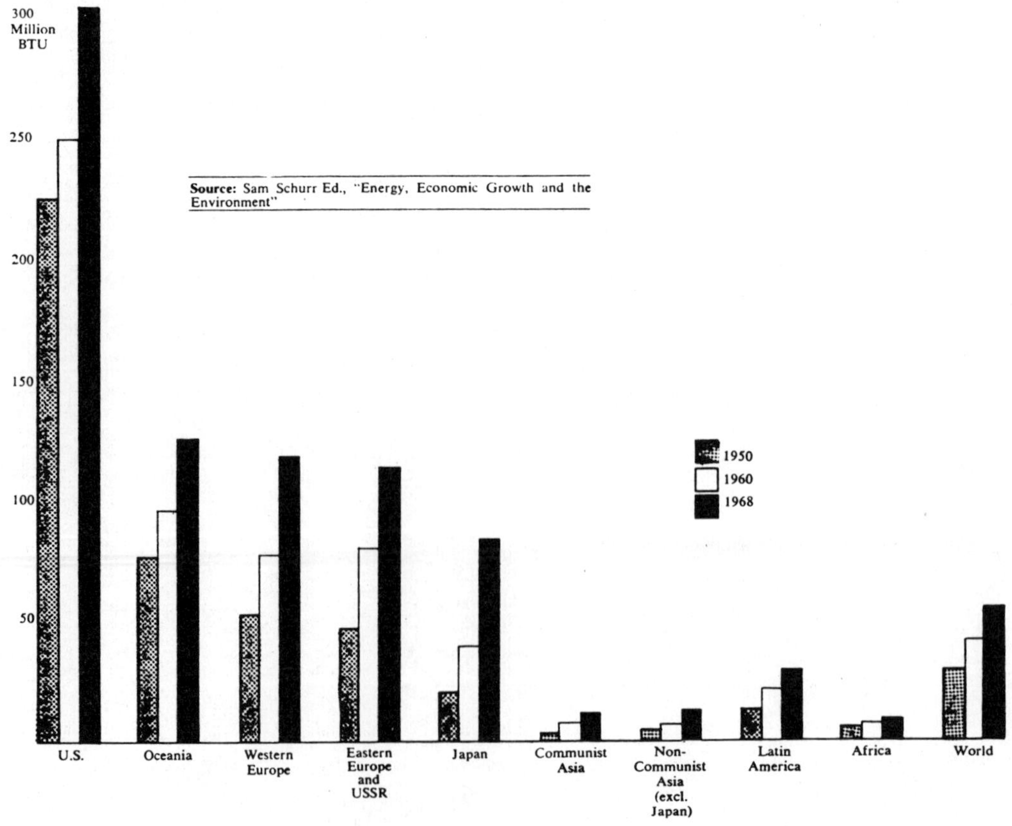

TABLE I.2
Housing Types

TYPE	Central Heat	Central Air	Local Heat	Single Family	Multi-family	1-3 rooms	3+ rooms	Stone, Concrete, Brick or Wood	Regional Materials	Major Appliances	Electric or Gas Cooking	Electricity	Kerosene Cooking	Kerosene Lighting	Noncommercial Fuels - Cooking	Noncommercial Fuels - Lighting
1	●	●		●			●	●		●	●	●				
2	●			●			●	●		●	●	●				
3	●	●			●		●	●		●	●	●				
4	●				●		●	●		●	●	●				
5			●	●			●	●		●	●	●				
6				●	●	●		●		●	●	●				
7			*	●			●	●	●		●	●	●			
8				●	●	●	●	●			●	●				
9				●	●	●			●			●	●			
10				●	●	●			●				●	●		
11				●	●	●			●					●	●	
12				●	●	●			●						●	●

*Using noncommercial fuels.

TABLE II.1

Housing Requirements in Stage I in 2000 (in thousands)

Country	K=1	K=1.5	K=2
Senegal	1,814	1,209	907
Guinea	1,757	1,171	878
Sierra Leone	1,144	763	572
Mali	2,194	1,463	1,097
Upper Volta	2,330	1,554	1,165
Ghana	4,549	3,034	2,274
Togo	1,092	728	546
Dahomey	1,441	961	720
Niger	2,097	1,398	1,048
Chad	1,717	1,145	858
Central African Republic	702	468	351
Sudan	6,243	4,831	3,621
Ethiopia	10,666	7,114	5,333
Uganda	4,024	2,684	2,012
Rwanda	1,913	1,275	956
Burundi	1,645	1,097	822
Tanzania	10,049	6,702	5,024
Somalia	1,314	876	657
Cyprus	275	183	138
Sri Lanka	4,141	2,762	2,070
Malaguasy Republic	3,530	2,354	1,765
	65,637	43,772	32,814

Source: AIA/RC calculations based on the U.N. useful housing solution.

TABLE II.2

Housing Requirements in Stage II in 2000 (in thousands)

Country	K=1	K=1.5	K=2
Liberia	479	319	239
Ivory Coast	1,998	1,332	999
Nigeria	26,274	17,524	13,137
Zaire	8,162	5,444	4,081
Kenya	4,981	3,322	2,490
Egypt	15,415	10,281	7,707
Zambia	2,620	1,773	1,310
Burma	10,506	7,007	5,253
Jordan	1,264	843	632
Mongolia	641	427	320
Yemen, People's Democratic Republic of	693	462	346
Afghanistan	6,628	4,420	3,314
Nepal	3,785	2,524	1,892
Bangladesh	33,777	22,529	16,888
Khmer Republic	3,423	2,283	1,711
Viet-Nam, Democratic Republic of	6,385	4,258	3,192
Viet-Nam, Republic of	12,092	8,065	6,046
	139,123	92,813	69,557

Source: AIA/RC calculations based on the U.N. useful housing solution.

TABLE II.3

Housing Requirements in Stage III in 2000 (in thousands)

Country	K=1	K=1.5	K=2
Mauritania	540	360	270
Morocco	8,270	5,516	4,135
Algeria	7,534	5,025	3,767
Tunisia	2,470	1,647	1,235
Libyan Arab Republic	947	631	473
Congo, People's Republic of	434	289	217
Angola	22,853	15,242	11,426
Malawi	2,143	1,424	1,071
Mozambique	3,332	2,222	1,666
Rhodesia	2,860	1,907	1,430
Thailand	15,126	9,983	7,563
Malaysia	4,520	3,014	2,260
Iraq	5,608	3,704	2,804
Syrian Arab Republic	3,267	2,179	1,633
Iran	13,114	8,747	6,557
Indonesia	59,361	39,593	29,680
Lebanon	--------not available--------		
Saudi Arabia	3,625	2,417	1,812
Yemen Arab Republic	2,954	1,970	1,477
Pakistan	30,485	20,333	15,242
Laos	1,248	832	624
Columbia	10,517	7,014	5,258
Bolivia	1,938	1,292	969
Paraguay	--------not available--------		
India	204,583	136,456	102,291
Philippines	16,625	11,088	8,312
Guatemala	2,295	1,530	1,147
Honduras	1,300	867	650
El Salvador	1,892	1,261	946
Nicaragua	880	586	440
Cuba	3,260	2,174	1,630
Haiti	2,871	1,914	1,435
	436,852	291,222	218,420

Source: AIA/RC calculations based on the U.N. useful housing solution

TABLE II.4

Housing Requirements in Stage IV in 2000 (in thousands)

Country	K=1	K=1.5	K=2
Equador	3,329	2,220	1,664
Turkey	21,958	14,645	10,979
Singapore	670	446	335
Korea Republic	12,092	8,065	6,046
Hong Kong	1,891	1,261	945
Korea, Democratic Republic	12,092	8,065	6,046
China, Republic of	7,032	4,690	3,516
Costa Rica	920	613	460
Panama	866	577	433
Dominican Republic	2,526	1,684	1,263
Puerto Rico	880	586	440
Trinidad & Tobago	338	225	169
Jamaica	795	524	397
Papua New Guinea	1,303	1,085	869
South Africa	8,964	5,978	4,487
Czechoslovakia	5,795	3,865	2,897
Poland	13,740	9,164	6,870
Hungary	4,071	2,715	2,035
Mexico	23,287	15,532	11,643
Peru	7,126	4,753	3,563
Uruguay	--------not available--------		
Argentina	--------not available--------		
Chile	--------not available--------		
Venezuala	5,117	3,413	2,558
	134,792	90,106	67,633

Source: AIA/RC calculations based on the U.N. useful housing solution.

TABLE II.5

Housing Requirements in Stage V in 2000 (in thousands)

Country	K=1	K=1.2	K=1.5
Portugal	3,363	2,801	2,243
Malta	88	73	58
Bulgaria	2,951	2,458	1,968
Greece	2,764	2,302	1,843
Romania	8,882	7,398	5,924
Yugoslavia	8,845	7,367	5,899
Albania	--------not available--------		
Brazil	45,747	38,107	30,513
United Kingdom	22,426	18,680	14,958
New Zealand	1,303	1,085	869
	96,369	80,271	64,275

Source: AIA/RC caluclations based on the U.N. useful housing solution

TABLE II.6

World Housing Requirements, Under Varying L's, By Stage in 2000 (in thousands)

Stage	K=1	K=1.2	K=1.5	K=2
I	65,637	54,675	43,772	32,814
II	139,123	115,889	92,813	69,557
III	436,852	363,897	291,222	218,420
IV	134,792	112,281	90,106	67,633
V	96,369	80,271	64,275	48,184
VI	165,142	137,555	110,142	82,571
United States	85,300	71,054	56,895	46,895
Canada	10,197	8,494	6,801	5,098
Soviet Union	103,614	86,310	69,110	51,807
China	261,809	218,086	174,626	130,904
Total	1,498,835	1,248,512	999,762	763,883

Source: AIA/RC calculations based on United Nations' demographic projections and the useful housing solution.

TABLE III.1

Housing Data, By Country

Stage I	Nigeria	Chad
Environmental Modification Systems		
Fuel: Commercial v.s. Non-commercial		oil or keresone for lighting
Type of Structure		nomadic tents in north
Building Size	average of 1.4 rooms per unit	
Building Material	adobe construction	98% of Fort Lamy is of traditional mud construction
Electric Lighting	81% of Lagos has electric lighting	very few homes
Appliances		

TABLE III.2

Housing Data, By Country

Stage I	Senegal	Sierra Leone
Environmental Modification Systems		
Fuel: Commercial v.s. Non-commercial	kerosene, wood for cooking and lighting	urban (cooking) 94.6% wood 2.5% kerosene .6% charcoal 1.2% bottled gas .1% electricity
Type of Structure	primarily single family, detached, althouth some apartments	66% single family, detached 15% row housing 19% apartments
Building Size		
Building Material	vegetable fibers, thatch, adobe brick with iron roofs	47% masonry 9% wattle and daub 35% mud and wattle
Electric Lighting	for urban wealthy only	
Appliances		

TABLE III.3

Housing Data, By Country

Stage II	Iran	Malaysia
Environmental Modification Systems		
Fuel: Commercial v.s. Non-commercial	cooking: bottled gas used in urban areas	
Type of Structure		
Building Size	number of rooms: (1) 23.3% (5) 6.5% (2) 29.5% (6) 4.4% (3) 17.4% (7+) 6.1% (4) 12.8%	
Building Material	urban: 33 % concrete 23.4% wood, iron 43.9% straw, mud brick rural: 1.1% concrete 1.6% wood 97.3% mud	50% bamboo 35% timber and bamboo 14% timber . 1% brick
Electric Lighting	25%	
Appliances		

TABLE III.4
Housing Data, By Country

Stage III	Columbia, 1964	Ecuador, 1962
Environmental Modification Systems	heating needed only in higher elevations, often wood fueled little air conditioning	heating in higher elevations little air conditioning
Fuel: Commercial v.s. Non-commercial	mostly non-commercial fuel	approximately 71% non-commercial fuels
Type of Structure	mostly single family	mostly single family
Building Size	number of rooms per unit: (1) 14% (4-5) 19% (2) 33% (6+) 10% (3) 19%	number of rooms: (1) 44% (4) 6% (2) 31% (5+) 7% (3) 12%
Building Material		45% vegetable fibers, thatch 31% adobe 13% wood 11% masonry
Electric Lighting		32% electric lighting
Appliances		

TABLE III.5
Housing Data, By Country

Stage IV	Uruguay, 1963	Argentina
Environmental Modification Systems	mostly local heat little or no air conditioning	10% with central heat 80% local heat little air conditioning
Fuel: Commercial v.s. Non-commercial	mostly commercial little wood	mostly commercial (oil, gas, electricity)
Type of Structure	row houses, single family	80-90% single family 10-20% apartments
Building Size	number of rooms per unit: (1) 4% (4) 30% (2) 10% (5) 18% (3) 19% (6+) 10%	number of rooms per unit: (1) 14% (4) 16% (2) 25% (5) 8% (3) 27% (6+) 13%
Building Material	concrete, masonry walls, tile roofs, manufactured panels	75% brick walls 13% adobe walls 41% cement floors 32% wood floors 39% metal roofs 31% cement roofs
Electric Lighting	79% electric lighting 11% kerosene lighting 10% other, or none	69% electric lighting 29% kerosene lighting
Appliances		

TABLE III.6
Housing Data, By Country

Stage V	Denmark	Japan
Environmental Modification Systems	38.7% central heating	5.9% room air conditioning 68.9% heating stoves
Fuel: Commercial v.s. Non-commercial	cooking: 90% electric or gas	
Type of Structure	45% single family detached housing 10% duplex	generally single family housing
Building Size	number of rooms per unit: (1) 3.5% (4) 23.5% (2) 22.9% (5) 10.2% (3) 30.8% (6) 8.9%	number of rooms per unit: (1) 9.1% (5) 12.9% (2) 20.4% (6) 8.3% (3) 21.2% (7+) 9.4% (4) 18.7%
Building Material		traditional wood for single family housing
Electric Lighting	99%	
Appliances	31% water heating 90% electric or gas stoves	44% water heating 84.6% refrigerator 28.6% electric range 88.3% washing machine

TABLE III.7
Housing Data, By Country

Stage VI	Canada	United States
Environmental Modification Systems	7% room air conditioning 81% central heating 10% local heating	36% room air conditioning 63% central heating 36% local heating
Fuel: Commercial v.s. Non-commercial	for heating: 55% oil central 34% gas for cooking: 3% oil 11% gas, piped 2% gas, bottled 82% electric	for cooking: 49% gas 40% electricity 8% bottled gas
Type of Structure	61% single family detached 7% row housing 32% apartments	69% single family detached housing 13% 2-4 units per structure 14% 5 units per structure
Building Size	number of rooms per unit: (1) 1% (5) 24% (2) 3% (6) 20% (3) 10% (7+) 23% (4) 18%	18% single family 69% 2 or 3 bedroom 13% 4+ bedrooms 5% 4+ floors
Building Material		
Electric Lighting	100%	100%
Appliances	94% water heating 99% refridgeration 77% washing machine 48% dryers 69% television	95.5% television 28% freezers 96% water heating 42% clothes dryer 71% washing machines 19% dishwashers

TABLE IV.1

World Residential Energy Requirements in 2000 A.D. (in million tce's)*

High		Low		Reasonable	
1) I	10.5	1) I	1.3	1) I	2.2
2) II	22.3	2) II	11.1	2) II	11.1
III	87.3	III	43.7	III	58.2
IV	94.4	IV	47.3	IV	63.0
V	378.7	V	189.4	V	315.5
VI	1203.9	VI	601.9	VI	1203.9
3) U.S.A.	1107.7	3)	622.2	3)	706.4
4) Canada	261.0	4)	174.0	4)	178.5
5) U.S.S.R.	1309.6	5)	530.7	5)	861.0
6) China	77.0	6)	77.0	6)	77.0
Totals	4552.4		2298.6		3476.8

Source: AIA/RC calculations

* These figures express the effective demand for energy in the year 2000 in "ton coal equivalents" per day. They represent the energy needed to operate the required number of accommodations. In other words, if all the housing accommodations needed to shelter the world's population in the year 2000 exist, this would be the amount of energy needed to operate them.

U.S. Experience In Active Solar Energy

The United States has the largest number of actual installations of solar energy systems in the world. As well as the widest range of types, sizes and technological alternatives. These range from the small passive methods of heating hot water for residential use in bathing and washing to multi-million dollar installations of solar systems to generate electricity through very high temperature generation in solar collection towers.

It is estimated that there are currently 3,000 to 5,000 solar installations in residential units and 800 to 1,000 systems installed in commercial and institutional buildings. Many more installations are on the way because of large demonstration programs being funded by the federal government. Most optimistic projections are forecasting a $1 billion market for solar equipment by the end of this century. Because of the level of interest in solar energy evidenced by the Congress of the United States and the Carter administration, it seems certain that many more concepts and devices will be invented in the next few years and virtually certain that the cost of solar heating and cooling systems for buildings will gradually decrease over the long term (but probably not in the near term).

The major elements and key activities of the Solar Heating and Cooling Program of the U.S. Energy Research and Development Administration (as of November 1976) are as Fig.1.

The AIA Research Corporation (established by the American Institute of Architects) has been active in solar energy research since 1973. Our work has included:

- The development of guidelines for the design of residential uses of solar energy in the various climatic regions of the United States

- An analysis of the barriers and constraints to the widespread utilization of solar energy

- A research report on the use of passive approaches to the use of solar energy

- The management of a national data collection process by architects and engineers for the solar demonstration program of the Department of Housing and Urban Development

- The organization of a national competition for more than 1600 students in schools of architecture to explore more advanced design concepts of solar energy use in homes

Fig. 1

MAJOR ELEMENTS AND KEY ACTIVITIES OF THE PROGRAM

MAJOR ELEMENTS	KEY ACTIVITIES
Demonstrations: Residential Commercial	– Primary emphasis on space heating and domestic hot water only in cycles 1 and 2, with increasing emphasis on combined heating and cooling systems in subsequent cycles. – Existing solar energy systems in cycle 1 with improved available systems, in subsequent cycles, employing generally combined heating and cooling systems.
Development in Support of Demonstrations	– Develop prototype systems – Improve existing systems – Integrate subsystems – Component/subsystem development – Test and evaluation
Research and Development	– Component/subsystem development – Materials and processes – Systems analysis and performance
Data Collection, Evaluation, and Dissemination	– Data collection system development – Central data processing facility – Information Center – Solar demonstration systems evaluation – Data bank operation – Solar energy system performance simulations
Standards and Criteria	– Interim performance criteria – Definitive performance standards – Laboratory accreditation
Environmental and Resource Assessment	– Environmental consequences – Technology assessment – Resource data base analysis
Market Development	– Federal agency involvement – State and local program development – Barrier assessment – Incentives/cost effectiveness studies

The recommendations made for how UNESCO and other UN agencies might assist in the effort include:

- Provide technical information briefs to government agencies and other influential groups in developing countries. This should be done by a direct link with specific institutions, not just a 'broadcast' of technical information to any and all agencies and institutions

- Identification of business organizations that could be 'catalytic' in introducing solar heating and cooling into a developing country, and the exploration of ways to underwrite the risks in such ventures

- Provide program materials to mass-media to assist in educating the general public to the possibilities of solar energy utilization, especially low-cost passive systems that can be built by individual owners. This program would obviously have to be 'packaged' to climatic and technological conditions of each developing country

- Support areas of research needed for low-technology use of solar energy for developing countries; including:

(a) The development of a Climate Framework which would provide for each country the tradeoffs between solutions that are defensive against an 'alien' climate and solutions that enhance the use of a 'friendly' climate.

(b) Restudy and implement new explorations of indigenous, traditional and modern building design solutions responsive to regional climatic condition. A catalog of ideas could be produced that would be useful to people everywhere.

(c) Specific applications of the passive use of solar heating and cooling to buildings of relatively modest construction. To a limited extent active systems of non-complex nature could also be explored.

(d) Specific possibilities of storing and preparing food with solar applications, especially in countries where the major use of energy is presently for this purpose.

Summary

There are some policy makers who believe the energy situation in the world is only one of those passing concerns of the moment and that in a few years our concerns will find a new subject to address. I believe this is a mistaken notion, and any country which fails to deal with the seriousness of the problem is going to face massive difficulties within ten years.

There are some policy makers who believe that while solar energy is an interesting idea it has little or no importance compared to finding ways to increase the supply of conventional energy sources. I believe they may be correct if they are from a developing country, but I believe they are badly mistaken if they are from one of the twenty nations that consume most of the fossil fuels.

There are some policy makers who believe that solar energy applied to heating and cooling is an opportunity for technologically sophisticated companies who market their products in technologically sophisticated countries. I hope that I have helped to make the case for the use of low-technology solutions to be designed into the traditional dwelling designs of countries everywhere.

2.4 An architect's approach to solar building design

G. Chouleur, Architect
6 Rue Fresque
Nimes 30000, France

Vous tous, avez conscience des inquiétudes énergétiques, particulièrement en ce qui concerne les énergies fossilés. Qu'on soit d'un pays producteur, ou d'un pays non producteur, on peut apprécier l'épuisement des ressources et connaître, déjà, l'échéance. Ces inquiétudes sont l'inquiétude de l'humanité toute entière. Mais, on le sait il n'y a pas d'inquiétude sans espérance, ni d'espérance sans inquiétude. L'espérance, c'est le soleil. Les inquiétudes mineures sont les moyens d'utiliser cette énergie. L'inquiétude majeure est que, même en utilisant toute l'énergie solaire déversée sur la terre, en l'an 2160 la terre sera à court d'énergie et risquera de mourir de disette énergétique. C'est l'inquiétude majeure. Il restera le proche lointain des planètes, c'est l'espérance majeure. Parmi les inquiétudes mineures, une à laquelle beaucoup de personnes sont sensibles, c'est l'intégration de l'énergie solaire à l'environnement et, en particulier à l'architecture. C'est celà que je vais discuter.

L'expérience que j'ai acquise, dans différentes réalisations, m'a permis de mieux situer le problème et d'y apporter ce que je crois être des solutions. En effet, j'ai réalisé en 1947 l'eau chaude sanitaire solaire de la maison Chatelain, puis, les années eu phoriques arrêtant tout désir de recherche, j'ai attendu 1972 pour adapter au solaire la maison de Monsieur Cherence. Enfin, en 1974-75, je réalisais cinq maisons à apport solaire dans la Cité d'Aramon, petite commune du sudde la France et, enfin, trois maisons à chauffage solaire intégral pour les besoins du personnel de la Centrale Thermique d'Aramon (Fig.1).

Figs. 1, 2

Les cinq maisons, dans la Cité d'Aramon, ont été construites avant que la décision de faire du chauffage solaire ait été prise. Elles sont à apport solaire. Les capteurs ont été ajoutés sur cinq des 122 villas réalisées. Ils sont en aluminium, protégés par un Rôhm, c'est à dire un métacrylate dont on aurait pensé qu'il avait toutes les vertus (Fig.2).

Les capteurs ont une surface de 38 m2 pour les villas de 7 pièces et de 32 m2 pour celles de 5 pièces. La distribution des calories collectées se fait par plancher chauffant. L'intégration n'est pas particulièrement heureuse (Fig.3).

L'économie en énergie, contrôlée par les Etudes et Recherches d'Electricité de France est de 48,7% (Fig.4).

Je dirai plus loin les erreurs de cette réalisation.

Figs. 3, 4

A l'occasion de la construction de ces cinq villas, j'ai été chargé, par Electricité de France, de construire trois villas à chauffage solaire intégral. Le projet a été choisi pour obtenir un meilleur coefficient de forme. L'isolation a été soignée, la cuve de stockage enterrée et, pour plus de diversité dans notre recherche, la distribution des calories collectées a été faite, dans une des villas par plancher chauffant, dans l'autre par des radiateurs en tôle d'acier, dans la troisième par des radiateurs en fonte (Fig.5).

La ventilation mécanique rejette l'air extrait et laisse entrer de l'air froid par des ouies aux fenêtres. Les économies d'énergie sont beaucoup plus conséquentes dans la villa qui comporte la distribution par plancher chauffant. Elles ont été cette année de 100% (Fig.6).

Figs. 5, 6

Pour les autres villas, à radiateurs, les rendements sont moins importants. J'attends le résultat des contrôles. L'intégration architecturale, sur les trois villas, n'a pas donné lieu à des remarques désagréables. Le fait d'avoir utilisé de la glace, d'avoir traité les volumes avec des cadres horizontaux transversaux, le fait d'avoir peint sur les capteurs une décoration, rendent cette intégration valable (Fig.7).

Fig. 7

Les dispositions que je vous présente vous permettent une opinion. Même défavorable, il me sera intéressant de la connaître. Mais, vous vous êtes rendu compte que les villas construites pour le solaire sont plus agréables d'aspect que celles où le système absorbant a été rajouté à un projet existant.

Les erreurs:

D'abord sur les cinq villas dans la Cité:

- L'utilisation du métacrylate est, pour l'instant, à déconseiller. Son aspect grisâtre est déplaisant. Les cadres posés verticalement rehaussent l'immeuble au lieu de l'apaiser et la forte dilatation du Rôhm apporte des accroissements de dimensions, pui des rétractions, qui sont causes de cassure.

- La cuve étant en tôle d'acier et les capteurs en aluminium, des courants de couple ont percé ces capteurs.

- Le pas du plancher chauffant est trop grand pour un bon rendement.

- Le système de régulation est trop sophistiqué et trop fragile.

Puis des villas sur le site:

- Il ne faut pas enterrer la cuve, elle devient hors contrôle,

- la régulation est trop sophistiquée, les thermistances tiennent mal. Il faut simplifier;

- le chauffage par radiateur est trop encombrant. Le rendement est moins bon. Seul le chauffage par plancher chauffant, à faible pas, est à recommander.

Figs. 8, 9

J'ai réalisé également à Aramon l'eau chaude sanitaire solaire à une maison de personnes âgées. En voici quelques particularités, la présentation n'est pas achevée, mais le résultat sera une économie de 100% (Figs.8, 9).

Au cours de ces expériences, je me suis trouvé confronté au problème de l'intégration, tant à l'intérieur qu'à l'extérieur. L'aspect des cinq villas est peu agréable, mais celui des trois villas est assez bon.

Il faut donc construire pour le chauffage solaire des maisons conçues, des l'origine, pour cette utilisation. Il est évident que, dans des maisons à toit-terrasse, l'on peut utiliser cette surface pour y implanter des capteurs. Les toitures-terrasses ne conviennent pas à toutes les architectures, surtout dans la façade méditerranéenne de la France. De plus, ces surfaces sont justes suffisantes pour des villas à simple rez de chaussée. Sur le site d'Aramon, le toit-terrasse, aux trois villas à chauffage solaire intégral, n'aurait pu supporter que 36 m2 de capteurs. Il en faut 45, et même en accroissant l'albedo jusqu'à disposer un miroir sur terrasse, nous ne pourrions pas

Fig. 10

satisfaire les besoins. Dès que l'immeuble comporte deux ou plusieurs niveaux le toit est insuffisant. Il faut alors utiliser les abords ou les façades. Pour utiliser les façades, j'ai été amené à établir un projet de capteurs plans, à inclination fixe ou variable, dont la face intérieure peut porter quelle couleur ou décoration que ce soit. Ces capteurs à redans présentent cet aspect. La gendarmerie de Bagnols sur Cèze, qui comporte 32 logements, un petit casernement et des bureaux, utilise, pour la première fois, ces capteurs décoratifs (Fig.10).

Comme l'on voir dans les diapositives présentées, le capteur, au lieu d'être une contrainte pour l'architecte, devient un élément de décoration. L'art retrouve sa place. Pour faciliter la décoration les sous-faces peuvent être démontables, permettant le travail en atelier et la pose de l'ensemble après coup. A la gendarmerie de Bagnols sur Cèze, les capteurs ne sont pas la seule originalité (Fig.11).

La cuve de stockage, d'un volume de 20 m3, est une à grand allongement, c'est à dire que le rapport hauteur utile sur diamètre est trés élevé. Cette disposition permet une meilleure stratification des températures du fluide caloporteur à l'intérieur de la cuve. L'eau provenant des absorbeurs est introduite dans la cuve dans la zone de température la plus proche de la température de la sortie, au moyen de sondes, de comparateurs et d'électro-vanne trois voies. Par le même dispositif, l'eau alimentant le plancher chauffant est sortie de la zone la plus proche en température de la température affichée d'utilisation (Fig.12).

Figs. 11, 12

L'eau chaude sanitaire passe au travers de deux échangeurs, l'un bas, l'autre haut, et est distribuée par un bouclage dans lequel le liquide est eu continuel mouvement de façon que dès qu'un robinet est ouvert l'eau qui en sort est à la température désirée. Pour la Gendarmerie de Bagnols encore, l'utilisation de l'énergie solaire n'a été décidée hélas qu'après exécution des plans. Une intention solaire, à l'origine, eut permis un meilleur résultat.

Figs. 13, 14

Ces possibilités d'intégration agréable ressortent sur le projet d'immeuble collectif à 4 niveaux que je vous présente (Fig.13).

L'une des diapositives montre un immeuble à pignons aveugles, permettant leur juxtaposition. L'autre, un immeuble à pignons éclairés pour des implantations isolées. Les deux systèmes peuvent s'associer (Fig.14).

Vous voyez là que l'effet décoratif peut aller jusqu'aux plus grandes audaces. Les maisons retrouvent un aspect éclatant. Pour ceux qui veulent conserver un classicisme d'aspect, la sous-face peut être simplement blanche ou grise. Mais rêvons, rêvons d'une ville ainsi décorée par les plus grands peintres et l'art retournant au bâtiment retrouve son emploi et sa raison d'être. La rue et la ville prennent une vie lumineuse. Dans cet ensemble, autres originalités: le toit est un miroir à différentes inclinaisons, pour mieux alimenter les capteurs de toiture en flux solaire, tandis que la cuve en position centrale à l'intérieur du bâtiment est du type 'Bagnols sur Cèze'. Avec ce photo, voici une ville prototype qui comporte à la fois des capteurs à redans et des capteurs plans. Les uns et les autres portent leur décoration (Figs.15, 16). Il apparaît, à la vue des photos, que l'aspect de la maison reste agréable et riche en couleurs. Plusieurs maisons meuvent être accolées ou disposées en grappes, constituant un quartier ou un groupe d'habitations aux chatoyants aspects.

Figs. 15, 16

Continuant les études par l'intégration des capteurs à l'architecture, voici un projet de six villas comportant, sur la façade Sud des capteurs sur portiques et sur la façade Nord, d'autres capteurs (Fig.17).

L'ensemble de ces capteurs à redans ajoute à ces villas regroupées sur un patio une décoration dont le choix reste à l'architecte et au décorateur. L'apport gaz n'intervient que par une seule chaudière de ménage pour l'ensemble des six villas. Les eaux provenant de ces absorbeurs sont envoyés dans une cuve verticale de grande hauteur, comme celle de la gendarmerie de Bagnols. Dans ces cuves à grand allongement - et je ne l'ai pas dit à propos de Bagnols - des dipositifs anti-convection, constitués par une série de cônes selon le dessin qui

Fig. 17

apparaît sur l'écran, laissent un espace vide entre les flancs de la cuve, et dirigent le fluide réchauffé vers la partie centrale. Ces cônes aident la stratification et l'apaisent en même temps.

L'Electricite de France, qui m'a donné les moyens de réaliser Aramon et que je remercie là, ainsi que les Etudes et Recherches des Renardieres qui ont suivi et suivent ces réalisations, m'a demandé encore de construire en Corse à Vazzio, près de l'aéroport d'Ajaccio, dix villas à chauffage solaire intégral. Ces villas procèderont du prototype que je vous ai montré. Les capteurs sont portés par les garages et la cuve de grande hauteur est dans un petit bâtiment accolé.

Il m'appartiendra avec Electricite de France de définir une campagne de contrôles 'intelligents' qui diront la valeur du système et ses erreurs. Les erreurs m'intéressent davantage que le succés. L'erreur apprend. Le succés sclérose. Ce n'est pas parce que l'on n'est pas sûr que l'on ne doit pas entreprendre. La vérité n'est qu'une suite d'erreurs corrigées.

Je ne parlerai pas ici d'un Collège pour 600 élèves, que je dois réaliser à Aramon, et d'un Collège de 900 élèves pour Saint-Gilles. Je dois revoir totalement les façades en tenant compte d'une meilleure intégration à l'architecture avec le procédé exposé.

Il est intéressant, dans le cadre de mon propos, de vous parler des capto-réservoirs. Ils ont également la face noire, vers le soleil, et la sous-face colorée. Ils peuvent contenir de 20 à 100 litres au m2. Ils sont obturés par différents systèmes isolants tel qu'un rideau tournant sur deux axes, l'un fou et l'autre moteur, ou par des panneaux mobiles se plaçant alternativement devanant et derrière l'absorbeur. Ces volets peuvent porter la décoration.

Ces systèmes sont commandés par un comparateur de températures permettant la circulation d'eau et obturant l'absorbeur lorsque la température d'impact solaire est inférieure à la température de sortie du capteur ou lorsque cette même température est inférieure à la température de l'eau dans la cuve. Les déperditions sont ainsi arrêtées.

Ce dispositif permet, au matin, de remettre en exposition l'absorbeur seulement lorsque la température à la sortie de celui-ci est inférieure à la température d'impact solaire. En conséquence, les surfaces des insolateurs peuvent être diminuées et cette diminution de surface permet une meilleure intégration. Les cuves à grand allongement facilitent le stockage de moyenne durée. Des études sur le stockage moyenne durée et sur le fluide caloporteur sont en cours. La mon utilisation directe du chauffage solaire de mi novembre à début mars, permet d'incliner la surface absorbante de 117° à 135°. La sous-face devient plus importante et avec elle la partie décorée. Le stockage moyenne durée est donc un des paramètres de l'intégration. Les cuves à grand allongement, à condition d'avoir un liquide caloporteur approprié, permettent des températures plus élevées en tête de la cuve et, par là, une température plus haute dans les planchers ou les radiateurs. L'intégration architecturale intérieure en sera considérablement améliorée. Les prix seront abaissés puisque l'on pourra alors utiliser les appareils d'émission du commerce. Le plancher chauffant reste cependant la meilleure distribution possible pour utiliser des fluides caloporteurs à moyenne ou basse température. Il évite les convections désagréables et les poussières; il évite les bruits des ventilo-convecteurs. Le soin apporté au problème de ventilation, avec traitement de l'air neuf au travers de l'air extrait, diminue de façon considérable la surface des capteurs et de là leur intégration. En effet, la déperdition globale des villas d'Aramon est de 6 kw500 environ, dont 2 kw200 pour le renouvellement d'air. Si sur cet air extrait l'on récupère seulement 50% des calories perdues, la surface des capteurs passe de 45 m2 à 38 et le prix diminue.

Dans un collège pour 600 élèves, où la déperdition globale est de l'ordre de 600 kwh, en raison des taux de renouvellement d'air (2, 3, 4, 5, ou 25 volumes heure) la déperdition par l'air extrait est de 350 kwh. Avec une hypothèse de récupération de 50% des calories rejetées, l'économie serait de près de 30% en surface de capteur et de 20% en coût. Une V.M.C. intelligente, avec récupération, est donc ainsi un paramètre important de l'intégration.

Pour terminer avec l'intégration architecturale et ses paramètres j'ajoute que je vais réaliser un gymnase à apport solaire, dont les capteurs à la partie haute sont un élément décoratif (Fig.18).

Fig. 18

Enfin, pour les piscines, et je sais que je touche là un des problèmes qui intéresse et inquiète mon excellente collègue et amie, l'Honorable Mrs Carter, il faut considérer les piscines couvertes et les piscines ouvertes. Le chauffage des piscines couvertes passe obligatoirement par la récupération des énergies humides perdues. Pour les piscines ouvertes, le mur porte-capteur est à la fois un abri aux vents du Nord dominants et un magnifique ensemble décoratif. Pour l'instant, l'énergie solaire peut allonger la durée d'utilisation de la piscine d'un mois avant l'hutilisation habituelle, et d'un mois après cette même date. Mais le coût d'exploitation des piscines est si élevé en France, et encore plus dans le Royaume Uni, que sans l'utilisation de l'énergie solaire la moitié des piscines fermeront sous deux ans.

Enfin, j'étudie un village solaire de 600 villas et logements qui utilisera à la fois les capteurs à redans et les concentrateurs à miroir pour augmenter la température de l'eau. L'intégration et l'originalité en sont trés étudiées.

Le plus grand projet soumis à mon étude est une mouvelle cité de Nimes. Nimes est une ville de 140.000 habitants, au sud de la France. Elle possède deux des plus beaux monuments romains d'Europe. Un joyau: la Maison Carrée; une autorité: les arènes, et, à quelques kilomètres: une áudace, le Pont du Gard (Figs.19, 20, 21).

Figs. 19, 20, 21

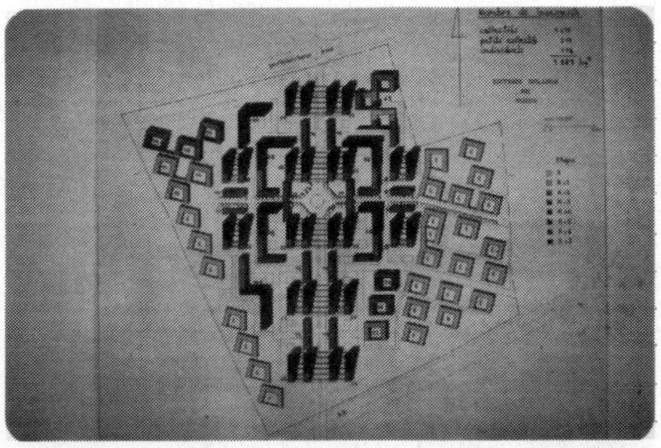

Cependant, je dois dire que d'aucuns pensent que la terre se réchauffe, d'autres qu'elle se refroidit. Mes collègues soviétiques croient au réchauffement et veulent placer la population totale du monde dans un satellite artificiel. Je ne suis pas d'accord avec eux, malgré toute l'amitié que je leur porte. Ce sont les usines et les activités polluantes qui doivent être mises sur orbite à 200, 500 ou 1.000 km de la terre. Ils seront nourris d'énergie en prise directe sur le soleil et le problème de l'énergie sera résolu pour des siècles. La terre sera restituée aux hommes chauffés à l'énergie solaire. A cette époque, il n'y aura plus ou presque plus de pétrole. Nos amis des Etats producteurs de pétrole se seront reconvertis. Ils en ont l'intelligence et les moyens. Ils y pensent. Le peu de pétrole qui restera servira aux soins de la chevelure.

La terre redeviendra un paradis... heureusement, Dieu me pardonne, nous succomberons encore au péché originel.

2.5 Solar construction in a biotechnic economy

P. van Dresser
Sundwellings Project
634 Garcia Street, Santa Fe, New Mexico 87501

Abstract

This paper deals with the application of simple solar energy solutions in Northern New Mexico. The impact of solar energy utilisation is examined in relation to the particular economic conditions of this region and certain generalised conclusions are drawn.

Introduction

It has long been a truism of speculative thought that many of the problems peculiar to modern society stem from our tendency to specialization, fragmentation, or "reductionism" in our approach to the world.

A result of this tendency has been a mechanistic ordering of man's activities, often hostile to the interplay of natural processes and to the inner needs of man himself.

The recent wave of interest in solar energy utilization is but one component of a general revulsion from the excesses of exploitive industrialism and life-denying "megatechnism" which has been the physical embodiment of this tendency. This revulsion has found expression under the banners of the ecology and environmental protection movement, of the various trends towards a counter-cultural or alternative lifestyle, and of the strivings of minority groups and exploited peoples for cultural and economic self-determination. Most recently it has penetrated into the sanctums of technology itself, where it manifests as a drive for the development of "intermediate", "appropriate", "soft", or "low impact" technologies.

A common element in most of these movements is the general proposition that man must learn to live within the life balance of the biosphere, without exploiting destructively either nature or his fellows. This obviously implies a marked de-emphasis in our massively indiscriminate use of the elemental energies locked in fossil fuels and atomic nuclei, and a corresponding adjustment to the gentler forces of plant growth, the movement of winds and waters, and finally sunshine - of which all these others are, of course, transformations.

While the great bulk of our material needs have, through most of history been met by utilization of these energies indirectly through agriculture, in recent years - as a result of the new orientation - increasing attention is being paid to the direct use of solar radiation to satisfy some of the more demanding requirements of contemporary society. Of the various techniques under experiment, as for electric power generation, industrial process heat, etc., that for the space heating of our habitations and buildings is one of the most promising, and the one with which we are concerned here.

The potential benefits of this application are very great, extending over large sectors of the world where fossil fuels for heating are unobtainable or prohibitively expensive, and where vegetative fuels have been so depleted as to endanger the fertility of the soil or the permanence of the plant cover. A good deal has been written about these benefits, and also about the desirability of research in the field of solar space heating, and of demonstration and assistance programs to speed up the applications of the techniques as widely as possible. And, of course, within the past decade or less, efforts in this direction in several countries have gained considerable momentum.

Speaking from my own experience in the United States, which is considered to be the leader in this field, I find a serious descrepancy in the nature of these efforts, The development of sunwarmed homes and other buildings, although on the whole a beneficial and life-affirming objective, is not on this account automatically freed of all the negative aspects of our commitment to big technology and large-scale industrialism. We have a marked tendency to "attack" the "problem" of solar construction in the same spirit we set up the Manhattan Project or the mass manufacture of Sherman tanks. Solar homes are viewed as engineered products and we naturally apply to their design and fabrication the well-tried techniques of the heating and air conditioning industry, of hydronic engineering and assembly-line metal fabrication. We think of solar heating installations as fluid-handling and heat-transfer systems which may be "retrofitted" into existing structures or built into new ones. And these systems, of course, tend to be complex and costly.

It may be safely said that the bulk of research and demonstration in solar space heating now being funded in the United States, especially those through scientific and academic institutions, is of this nature. Besides the funding of specific buildings, there is, of course, a considerable amount of supporting research into rather sophisticated matters such as the molecular metallurgy of selective surfaces, the design of computer simulations of collector-storage systems, etc.

The rationalization of this type of research program is that, with the application of mass production techniques to the systems and devices so perfected, costs will drop and they will become available to the average citizen and will thus eventually solve the problem of energy for habitations and for domestic use throughout much of our nation.

There is much justification for this rationalization in the short view and the immediate commercial-industrial context. But a growing minority of us feel it is inadequate in the longer view and in the larger context of ecological balance and viability.

The gist of the exceptions which are taken to the adequacy of the "high technology" approach to solar construction may be stated fairly simply. The lavish use of high-energy materials - steel, copper, aluminum, cement, plastics, etc. - which this approach necessitates and the sophisticated devices which it requires - pumps, blowers, servo-valves, heat exchangers, electronic controls, etc. - involves us even more completely in the industrial labyrinth than we are at present. If this version of solar utilization became really widespread, it is possible that our per capita energy consumption might actually be higher than it is now, due to the heavy requirements of extracting, refining, fabricating, and transporting such materials. The high energy costs will inevitably be reflected in high monetary costs which will limit the application of this type of technology to affluent sectors of society, which will further increase the have/have-not tensions which are disrupting civilization.

As an alternative to this unsatisfactory prospect, people are exploring, in various parts of our country, the possibilities of solar dwelling construction of available low energy material, using sunheating or "suntempering" techniques based on the application of simple physical and geometrical principles rather than on engineering virtuosity. Since the results of this grassroots "solution of the housing and energy problem" are so promising in the larger frame of reference and the longer view, and since New Mexico is one of the focal areas for this type of folk-technical development, I should like to tell you something about it here.

Fig. 1
Taos Pueblo

Fig. 2
van Dresser solar adobe, Santa Fe.

First to be noted is the fact that New Mexico - in particular the mountainous northcentral portion of it - is one of the few areas of the United States in which a folk-and-land-based economy and lifestyle is still partially operative. This is partly due to historical accident; the fact that its settlement by Europeans was achieved by land from the south in a non-commercial period and context. As a result of this circumstance, and its long isolation from the vigorously industrializing East Coast of the continent, there developed in this region a relatively stable and self-sufficient pastoral and agrarian economy, mainly at the village level of urbanization, with a simple but adequate low-energy craft technology based on local natural resources (an "eotecnic economy", to use Lewis Mumfords terminology).

A highly distinctive and functional architecture evolved in this milieu, combining both Spanish and Pueblo Indian traditions and utilizing the universally available material, adobe, for its main ingredient. See Figure 1. This architecture still pretty much dominates the region. A high proportion of the dwellings in Santa Fe are of this material and most village and farm homes. Additionally this technique culminated in effective public buildings, notably churches and schools. In the earlier decades of this century, some handsome government buildings were built of this medium.

From the 1920s and 1930s on, eastern immigrants into the region began to pay serious attention to the traditional architecture and many homes, ranging from the modest to the quite pretentious, were built of adobe in the native vernacular or in sophisticated elaborations of it.

Recently - mostly within the past decade - the new settlers in the region, many of them ecologically concerned emigrants from metropolitan centers have been experimenting with the introduction of solar heating principles into the conformation of adobe dwellings. (At this point I should probably mention that, contrary to popular impression, northern New Mexico is not a hot desert country, but experiences fairly severe winters with many zero and below-zero fahrenheit nights, and generous snowfall at higher elevations.)

I may lay some claim to pioneering in this movement of solar adaptation, having in 1958 rebuilt a small native house into a solar dwelling with roof-mounted air collectors and underfloor rock storage. This is now considered to be the second oldest continuously operated solar residence in the nation. (See Fig.2.)

The movement gathered momentum, very slowly at first, but accelerating in recent years until a current census by the New Mexico Solar Energy Association indicates that within a 100-mile radius of Santa Fe there are at least 200 such structures and probably 75 will be added next year. These are homes built largely by private individuals, many with their own hands, and without benefit of government or foundation subsidies. Quite recently some commercial builders have entered the field and some handsome adobe homes have been built speculatively and sold at very substantial prices to affluent purchasers.

Naturally the variation in type and degree of elaboration in this efflorescence of home building was considerable, ranging from almost primitive jacals (the New Mexican term for a small pole-and-mud-plaster cabin) with plastic-covered solar gain frameworks on their south walls, to highly finished near-mansions with engineered collectors, storage banks, and electronic control systems.

After several years of this grassroots experimentation, it may be said that there is emerging a marked consensus towards the "low-tech" or passive versions of solar dwellings, depending for their functions on simple processes of absorption, conduction, and convection channeled by the orientation, geometry, and massing of the structure.

Roughly four main versions of this approach may be identified. At present the two most favoured systems are the "attached greenhouse" arrangement, and the "direct-gain" configuration. The latter is the simplest and perhaps the most thermally efficient, consisting as it does essentially of the opening up of the south of the building with large windows permitting liberal entrance of sunlight into the interior.

The recipe for a successful building of this kind is boiling down to a few precepts, such as:

Proportion the structure, in plan and section, so that during the winter months low-angle sunlight will penetrate well into the depth of the lived-in portion of the house. This usually results in a longish and narrowish building on an approximately east-west axis, with a typical shed-roof profile, high side to the south. However, the plan may be liberalized by the use of clerestory windows giving into north rooms or by "stepping up" a hillside on a north-south axis, as was done in the original Wright-Terry house in Santa Fe. Skylights may also be used, although these introduce complications as regards leaking, night-sky heat loss, and closures.

Double-glaze all solar gain windows and provide effective insulating closures, either in the form of exterior shutters or hatches, or heavy and tightly fitted interior drapes.

Provide mass walls, partitions, and floors of moderate conductivity masonry or masonry-like materials (usually adobe). Insulate these walls against heat leakage to the exterior, especially north walls and foundations. See that mass floor receiving sunlight are dark in colour and are not covered with carpeting. The tonnage of walls and partitions in a fair-sized adobe house may approach 150 which, when warmed only a few degrees by interior convection and infra-red re-radiation, will keep the house comfortable with very little auxiliary heat for several days of cloudy weather. Thus the mass of the house itself acts as heat storage bank, eliminating the need for water tanks or even rock bins.

Insulate roofs thoroughly and reduce to a minimum non-solar-gain windows, especially to the north. Double-glaze those that are necessary and provide storm sash as well if budget permits.

Provide overhangs for solar windows calculated to begin shading out most sunlight at the start of the summer months. A refinement is to provide a means of varying the extent of this overhang so that around the fall equinox in warm weather shading is maximized, but in cold springs, when the solar angle is the same, more sunlight may be admitted.

Calculations made by Dr. F.C. Wessling of the University of New Mexico's Department of Mechanical Engineering suggest that a well-designed direct-gain house may experience a net gain of about 400 BTUs per square foot of solar window on a typical January "design day" and that from 72-80% of the total heating needs may be met by this input. Several houses built on these principles are coming close to this prediction. (See Fig.3.)

The next most popular system, the attached greenhouse, consists essentially in replacing the south solar windows with a relatively narrow conservatory, integrated into the south facade. This heat trap is vented into the main body of the house at the ceiling and floor level, as its principal contribution to the energy budget will be in the form of warm air. The convective circulation of this air may be enhanced by providing underfloor return ducts intaking from the north side of the house and low-wattage booster fans may also be used. But the vents should be closed at night to discourage reverse circulation. Under these conditions the greenhouse may drop to 45° F. or so on cold winter nights, which does not endanger most temperate zone plants. Several schematics of this type of installation are shown in Figures 8 -17. Note the typical profile of such greenhouses. Contrary to traditional practice, most of the transparent surface is in the south wall which is vertical or only slightly slanted to the north. This provides ample solar intake from the low-angle winter sun. The roof is only partially transparent, thus providing needed shade in summer and minimizing heat loss to the night sky in winter. Containers of water are often placed along the rear wall to provide some thermal storage. The charm of this system is the "side benefits" it yields in the form of a green and lush space intimate with the house and, of course, the salads, vegetables, and flowers it produces. It should be mentioned that the transparent surfaces are usually double glazed and in especially cold situations closures of some sort are provided, preferably outside. In other respects, the prescription for such houses is similar to that for direct-gain systems.

B.T. Rogers, an engineer consultant to the solar program of the Los Alamos Scientific Laboratories, has calculated that in the vicinity of Name, New Mexico (elevation about 6,500 feet), an attached double-glazed greenhouse of the type described will deliver a surplus of from 308 to 479 BTUs per square foot per day during a typical December.

The Trombe wall is a runner-up to these other systems in terms of practicality for low cost, simple solar construction. I am sure you are all familiar with the Trombe wall principle which has achieved international usage in the vocabulary of solar buffs. In New Mexico the system has been subjected to some interesting computer simulation analysis by the solar division of Los Alamos Scientific Laboratories. (See Fig.4.)

Fig. 3
Terry Wright solar adobe, Santa Fe

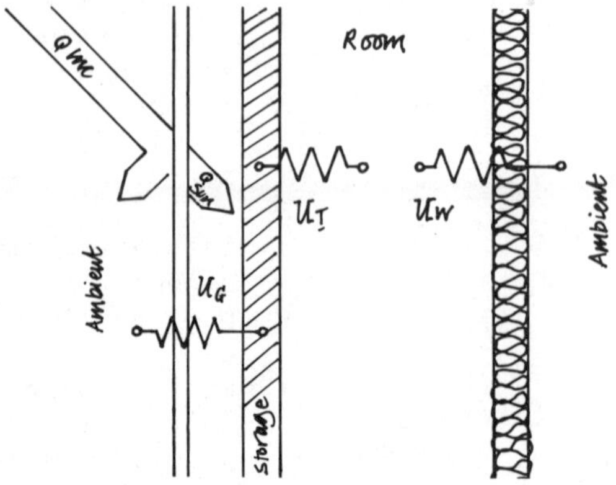

Fig. 4
Trombe wall heat-flow diagram

Here we have a schematic of the essential heat flow relationships involved in a space heated by a Trombe wall. Assigning realistic values to the various constants (mainly conductivity, transmissivity, the mass of storage per square foot of window and the proportion of window to floor area), and utilizing the weather and insolation date for 1972-73 at Los Alamos, the heat balance for the space was calculated on an hourly basis for the entire year. The results of this calculation show this simple solar trap arrangement providing 73.25% of the space-heating requirements over a year in which 7,350 degree-days of heating were required. During the worst month, December, the system indicated a capacity to meet 61% of the heating load.

In January when average ambient temperatures hovered in the 20°F. range, this analysis indicated the system holding a uniform 65°F., with a few slight rises above this during days of strong sunshine. These figures and the summary of results are from a preliminary paper by Drs. Balcomb and Hedstrom in the winter of 1976 issue of the Journal of the New Mexico Solar Energy Association.

A particular virtue of the Trombe system is its suitability for applying to existing houses which have south masonry walls. Frequently a fair measure of "solarization" can be achieved in this way without great expense.

The "waterwall" or "drumwall" we tend to consider a variant of the Trombe wall with improved performance due to the greater specific heat of the storage mass and the more rapid absorption of solar energy due to internal convective circulation. The original and best known dwelling using this system, the Baer house near Albuquerque, has been much publicized and discussed. A less well-known example, a 7,000 square foot warehouse at the Benedictine Monastery near Pecos, New Mexico, is impressive in its successful performance. This is a more conventional masonry and timber structure with a half-height drumwall occupying its south facade, 2nd the building is said to be operating over 90% on solar input.

Not many fully evolved drumwall installations are to be found in our region, possibly due to the increasing cost and difficulty of procuring used steel 55-gallon barrels. However, large plastic water containers for this purpose are entering the market and numerous designs are in the making for buildings utilizing them. A good deal of somewhat haphazard use is made of water containers of various kinds, stacked in greenhouses or even built into adobe walls or benches, simply to add storage mass.

The fourth passive system being experimented with in New Mexico is the convective-air-loop rock storage system. In this arrangement, an air collector panel is placed below and to the south of the building to be heated and warm air is ducted from it into a rock storage bin, ideally under the floor of the building. The space above is warmed either by venting warm air from the rock bin or allowing its stored heat to radiate upward through the floor. Not too many working examples of this system have been built, probably because of the site requirements (a pronounced southward slope) and also because of the rather large amount of underground excavation and ductwork that are required, not to speak of the hauling and placing of tons of rocks which should be clean and of fairly uniform size. The two best examples I know of in New Mexico are the Davis house outside of Albuquerque and the Masterson pottery studio at Cienega, south of Santa Fe. This last building is of particular interest because of its great simplicity and low cost. The collector panel is incorporated in the south foundation wall of the building and no ductwork is required since the warm air from the collector moves under the floor, sinks down through the shallow rock bed, and returns to the lower edge of the collector along the bottom of the excavation. The solar effect is reinforced by a large direct-gain window closed by barn-

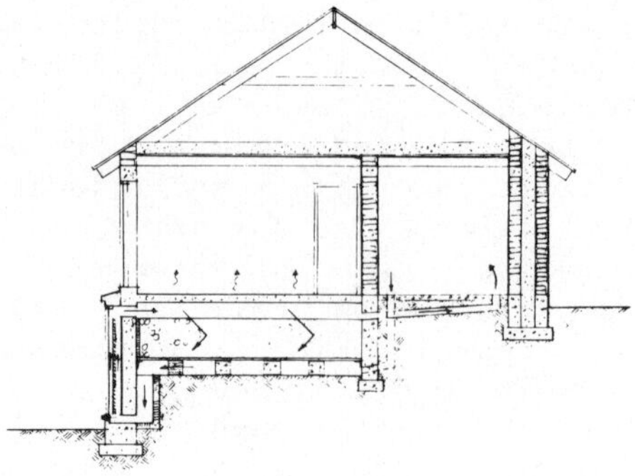

Fig. 5
Convective rock storage schematic

Fig. 6
Martinez forced-air storage

like sliding doors. I collaborated with the owner in designing this system. Figure 5 schematicizes one installation of this type.

Theoretical analysis of convective rock storage systems is rather complicated, but such studies as have been made suggest that their thermal efficiency tends to be lower than other passive systems, due mainly to the difficulty of achieving sufficiently high air velocity from natural draft to adequately absorb the collector output. With air-loop-rock-storage systems using forced circulation, this difficulty need not occur and calculations presented in the LASL "Solar Heating Handbook for Los Alamos" show that a properly proportioned system of this kind can achieve solar heating effectiveness very close to those of the more sophisticated liquid type - that is, approaching 80% of the total load for comparable installation parameters. Blower power consumption can theoretically be as small as 0.01 watts per square foot of collector. Thus the system, although not strictly passive, is minimally power consumptive.

More installation of this type can be found in our region than the strictly convective-loop type. Figure 6 shows the Martinez house in Albaquerque.

It need hardly be said that many, perhaps most, passive or sem-passive solar houses combine two or more of these systems. The Masterson studio already mentioned combines direct-gain with convective-loop-rock storage, while the Benedictine monastery reinforces its drumwall system with direct gain, both from a bank of windows above the drumwall and clerestory windows giving into the warehouse section. The Martinez house combines direct gain through a south-facing loggia or conservatory with underfloor forced-air rock storage, and even adds an air-to-air heat pump for good measure

Other examples may be cited.
Figure 7 shows a house by Wayne Nichols of Santa Fe which incorporates a rather elegant solarium or atrium concept with excess warm air channeled through rock bins for energy storage.

Fig. 7
Nichols atrium house

Unfortunately many of these homes which you have just seen, even though essentially passive or semi-passive in design, are by no means low in cost. A number of them under 2,000 square feet in floor area have been sold within the past year for $100,000 or more; hence they do not seem to point the way towards low or even moderate-cost energy conserving housing.

The popular contemporary (and characteristically North American) answer to this problem is, as I pointed out at the start of this discussion, mass production. The reasoning is well established: handicrafted buildings (which most of those I have shown you are) inevitably become expensive because of the high cost of labour, particularly skilled labour. The obvious remedy is to mechanize their production, developing standardized parts or modules.

Without disputing the validity of this reasoning for immediate conditions in certain parts of the industrialized and urbanized world, I believe a strong case can be made for intensive promulgation of the handicrafted, low-tech, or folk-indigenous approach to solar housing in the longer view and broader geographic context. The urban-industrial core of society has larely achieved its status by exploiting the energy and metallic resources of the rest of the globe and this pattern of development simply cannot be projected indefinitely. As the accessibility of the resources necessary for megatechnic industry decreases and their cost of extraction increases, the areas in which this pattern of civilization can be maintained will inevitably shrink.

If we are intelligent we will seek to replace this pattern with one based on the skilled and scientific use of the readily available materials of the earth and of the renewable energies of plant growth and all the other manifestations of the sun's largesse. The technology resulting from this form of economic evolution will be manpower, skill, and science-intensive, rather than machine and mechanical-power-intensive. For centralization, the economies of scale, and massive transportation it will substitute decentralization (or human-scaled recentralization) and the internal logistic efficiencies resulting from the close coupling of production and consumption in organic communities deriving their support largely from the life processes of the regions in which they are situated. For brevity I refer to such an economy as "biotechnic", borrowing the phrase from Lewis Mumford whom I believe, in turn, adapted it from the great pioneering regionalist Patrick Geddes.

In such an economy, the handcrafted home of low-energy materials, adapted sensitively to the climatic realities of its site, makes ecologic, economic, and technologic sense. Unlike most of the experimental solar homes we are now building, its construction costs will not be inflated by the excessive living costs engendered in an overcentralized, overcomplex, overmetropolitanized and power-gluttonous society. Nor will it be unduly dependent on the products of remote industrial complexes with their entrainment of energy and resources depletion and biospheric pollution.

For such reasons I have become involved, both personally and professionally, in the effort to help nurture the growth of an indigenous solar technology in the receptive soil of New Mexico. One form this effort took was the launching by a few of us of a modest project called the "Sundwellings Program", which I would like to tell you about. This was a two-year investigation funded by several agencies of the State of New Mexico of the characteristics of the do-it-yourself solar home building technique that is evolving in this region, with particular reference to its relationship with the traditional adobe construction methods and to simple means for improving its thermal performance. The investigation was carried out by a team of architects, engineer-physicists, and practical homebuilders and tinkerers, all familiar with solar concepts and

Figs. 8-13
Sundwellings test houses

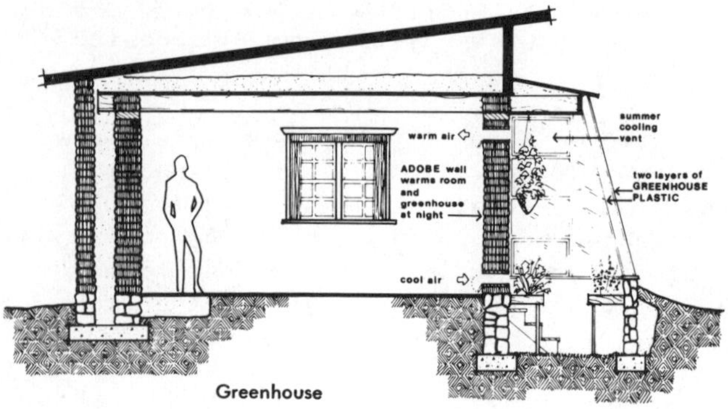

Fig. 8

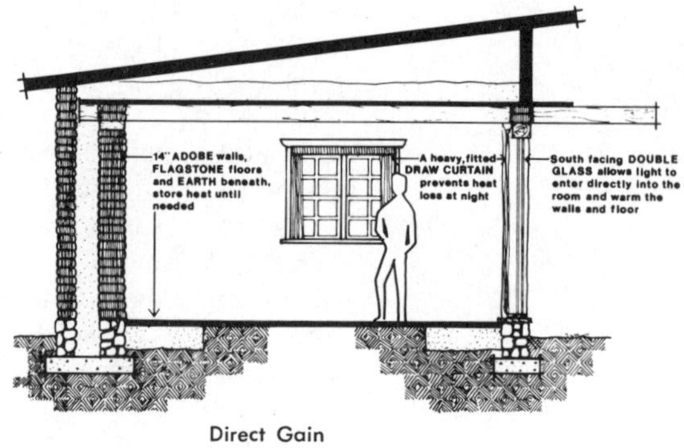

Fig. 9

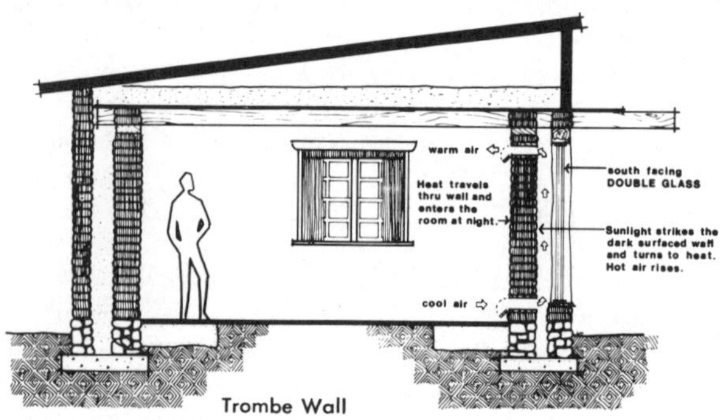

Fig. 10

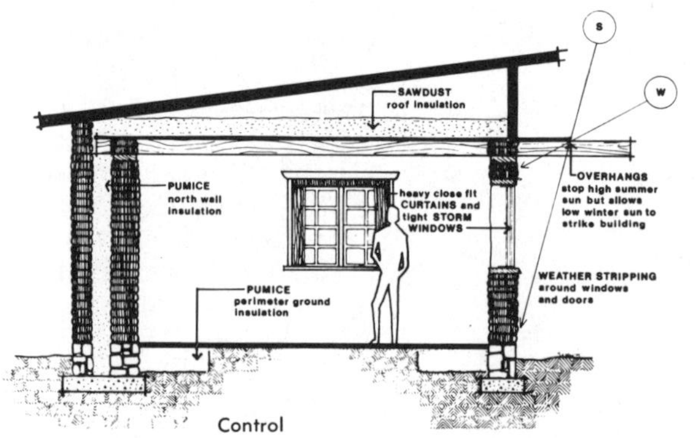

Fig. 11

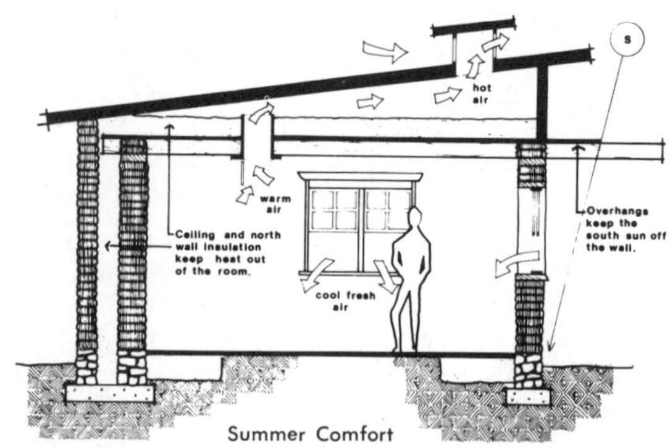

Fig. 12

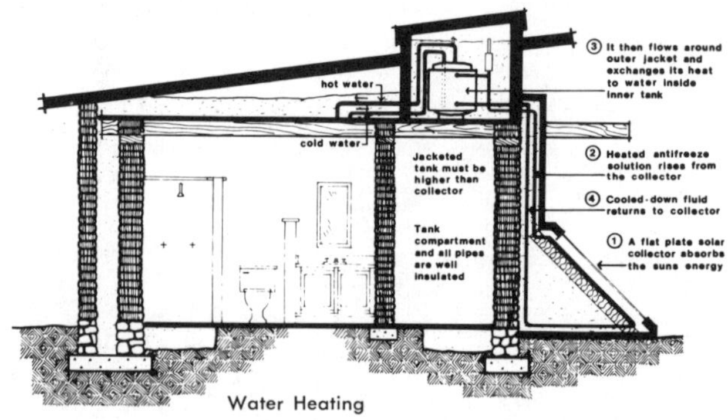

Fig. 13

with the lifestyle of the region. The project culminated with the actual construction by a group of local youths of Spanish and Indian descent of four adobe-solar buildings on a site in the heart of the northern New Mexican uplands region.

An interesting point about the project design is that three basic passive solar systems were built into structures otherwise exactly similar in size, geometry, and location. The systems were: direct-gain, greenhouse, and Trombe wall. A fourth building of the same geometry was given no special solar feature and was intended to act as a "control". The illustrations on figure 8-13 were prepared by Mark Chalom, field architect on the construction project, show schematics of the four buildings. The point to emphasize is that the construction materials are 90% obtained within a few miles of the site - adobe brick, of course, made immediately on the site; rafter logs or vigas cut in the adjoining forest; flagstone quarried locally; and rough lumber from a local sawmill. Roof unsulation was coarse sawdust from the same mill (gratis) and wall and perimeter insulation was locally mined pumice. The principal manufactured materials required were glass, hardware, and slate-coated roofing felt (this last because no wood suitable for shingles or shakes grows in our region).

We were fortunate in securing the cooperation of the solar section of the Los Alamos Scientific Laboratories in arranging for very thorough thermal monitoring of the performance of these buildings. Over fifty thermocouples were embedded in floor, walls, attic space, etc., of each structure, and a small weather station including vertical and horizontal pyranometers was installed. Readings from these various sensors are being recorded in a Hewlett-Packard Data Acquisition center and calculator-computer. When this information is gathered and analyzed through one or more heating seasons, absolute and comparative performance data for differing passive systems will be available. We believe this combination of sophisticated analytic techniques in a folk building situation is unique.

It is obvious that buildings of this kind, while low in commercial energy and manufactured materials costs, are high in labour and "sweat equity". In a metropolitan or suburban fiscal environment, they would not be considered "economic". But I must reiterate that we are addressing ourselves to large sectors of the world in which labour is abundant and manufactured materials scarce and expensive. It may come as a surprise to some of us here that even in the affluent United States of North America there are many such sectors of "underdevelopment" in which building techniques of this kind are most appropriate and would substantially help in the attainment of a high level of domestic amenity without benefit of mass production and deficit financing, with their attendant social burden. I believe, as I have indicated earlier, that this approach to the shelter needs of mankind will become more and more relevant as energy and materials shortages mount and the real costs of supersophisticated power sources become visible.

As part of this particular research and demonstration effort we undertook to suggest a prototype design for a modest suntempered building which could be built, in stages, by a family of modest means willing to do a good deal of the work themselves. (There are many such in New Mexico.) These designs are not intended to be taken literally, but hopefully will open up to the average person in our region alternative concepts of house design which will fit into this whole trend towards energy conservation through climatic adaptation and use of simple materials, while still harmonizing reasonably well with traditional taste and feelings of propriety. These feelings should rightfully be considered as part of the human ecology of the regional community. Figures 14-17, prepared by Santa Fe architect William Lumpkins from the Sundwellings Team studies show the various stages of this prototype and alternative cross-sections.

I have attempted to sum up the conclusions of this study in a book, "Homegrown Sundwellings", which has just been published. I said that I was involved personally as well as professionally in this sort of solar endeavour. We are at present constructing a suntempered "life support" home in the mountain village which is our permanent home base in northern New Mexico, incorporating a number of interesting biotechnic features. It is built, however, principally of native materials and is described in some detail in my book.

I am afraid that I have not been able to do justice to the many facts of the movement towards a solar-oriented economy that is taking place in New Mexico, but it is exciting and gratifying to feel one is contributing to it.

Figs. 14-17 Sundwellings prototype house

Fig. 14

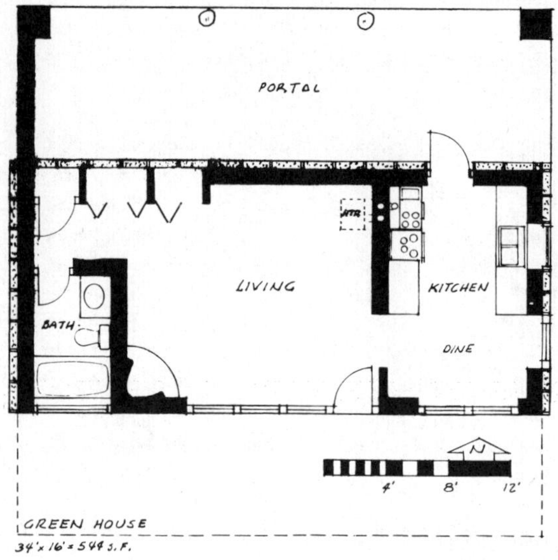

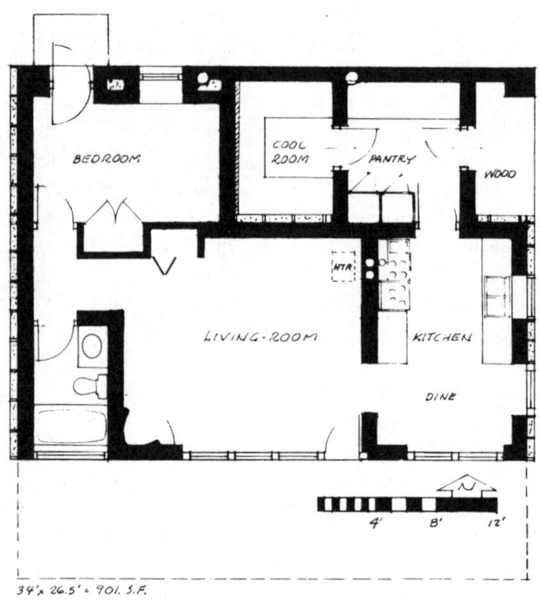

Fig. 15

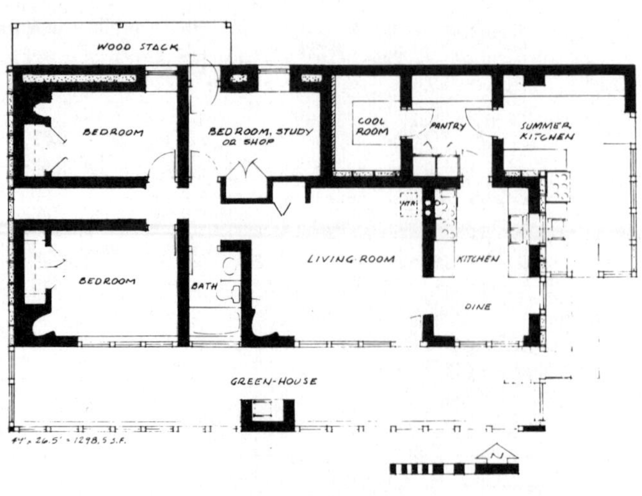

Fig. 16

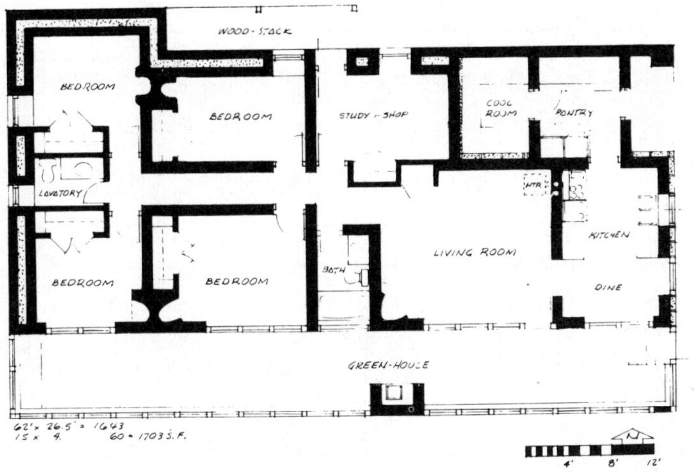
WOOD-STACK
BEDROOM
BEDROOM
STUDY + SHOP
COOL ROOM
PANTRY
LAVATORY
HTR
KITCHEN
LIVING ROOM
BATH
BEDROOM
BEDROOM
DINE
GREEN-HOUSE
62' x 26.5' = 1643
15 x 4. 60 = 1703 S.F.
4' 8' 12'
N

Fig. 17

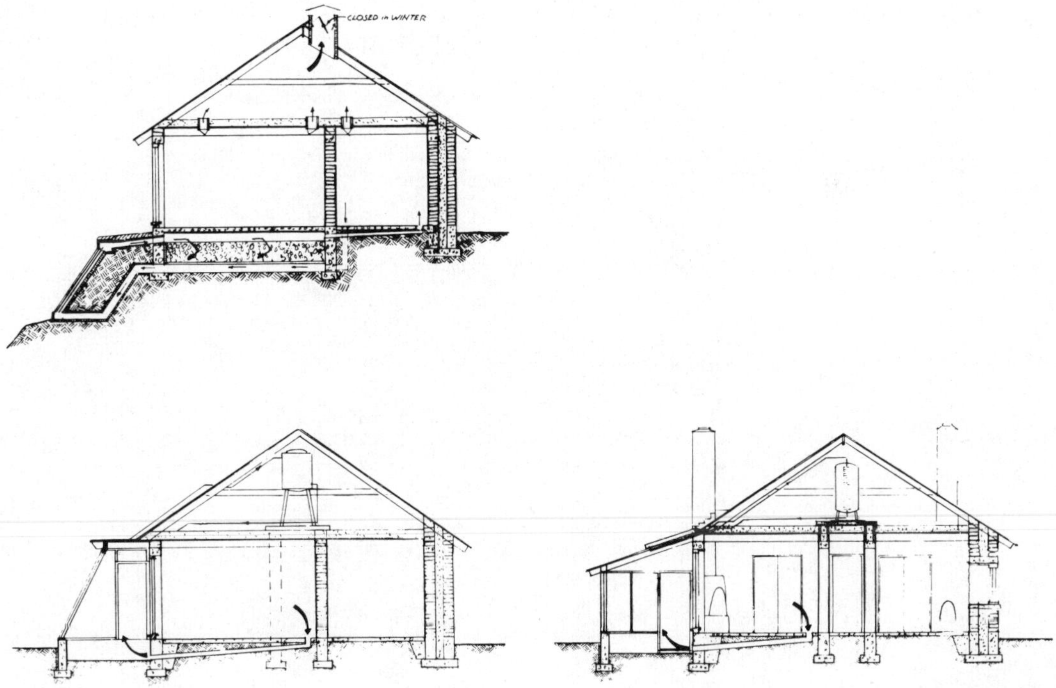
CLOSED in WINTER

2.6 Individual and collective output and profit earning capacity in solar energy applications

P.F.M. Casalonga
Ingénieur des Travaux Publics
Etudiant en Architecture groupe A.B.C.
Cabinet d'Etudes 6 Stretta San Lazaru
(et Laboratoire)
20000 Ajaccio, France

Abstract

The purpose of this paper is to introduce our approach to a methodology which is systemic rather than analytic, owing to the many imperfections of an analytic approach.

This paper is to be presented and discussed within the framework of: 'Socio-economic aspects of solar energy applications in the built environment'.

This paper also aims to take advantage of a preceding study which covered the following study areas:

1) Output/Profit - earning capacity
2) Individual Cost/Collective Cost
3) Global or Localized Energy output
4) Instant or long-term profit-earning capacity
5) The Question of the Fittings Size
6) Social Factors
7) Optimization

Condensé

Il s'agit d'exposer ici l'approche d'une méthodologie qui se veut systémique plus qu'analytique, en raison des nombreuses imperfections de l'approche analytique.

Communication présentée dans le cadre de la commission: 'Aspects socio-économiques des applications de l'Energie Solaire dans l'environnement bâti'.

Il a pour but de faire le point d'une étude en cours comprenant:

1) Rendement/Rentabilité
2) Coût individuel et coût collectif
3) Rendement énergétique global ou localisé
4) Rentabilité instantanée ou à long terme
5) Problème de la dimension des installations
6) Facteurs sociaux
7) Optimisation

Introduction

Il s'agit d'exposer ici l'approche d'une méthodologie qui se veut systémique plus qu'analytique, en raison des nombreuses imperfections de l'approche analytique.

Rendement/Rentabilité

Trop souvent la recherche du rendement maximum semble être l'objectif unique des chercheurs en matière d'héliotechnique.

Fig. 1.1

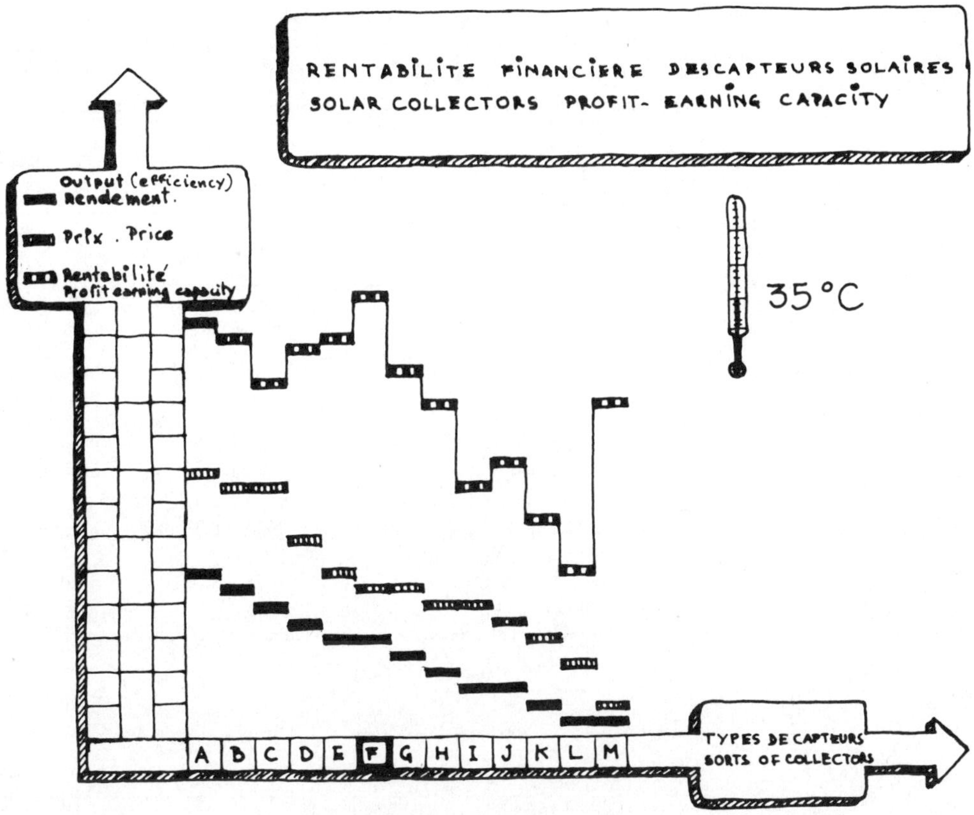
RENTABILITE FINANCIERE DESCAPTEURS SOLAIRES
SOLAR COLLECTORS PROFIT- EARNING CAPACITY
Output (efficiency)
Rendement.
Prix . Price
Rentabilité
Profit earning capacity
35 °C
A B C D E F G H I J K L M
TYPES DE CAPTEURS
SORTS OF COLLECTORS

Fig. 1.2

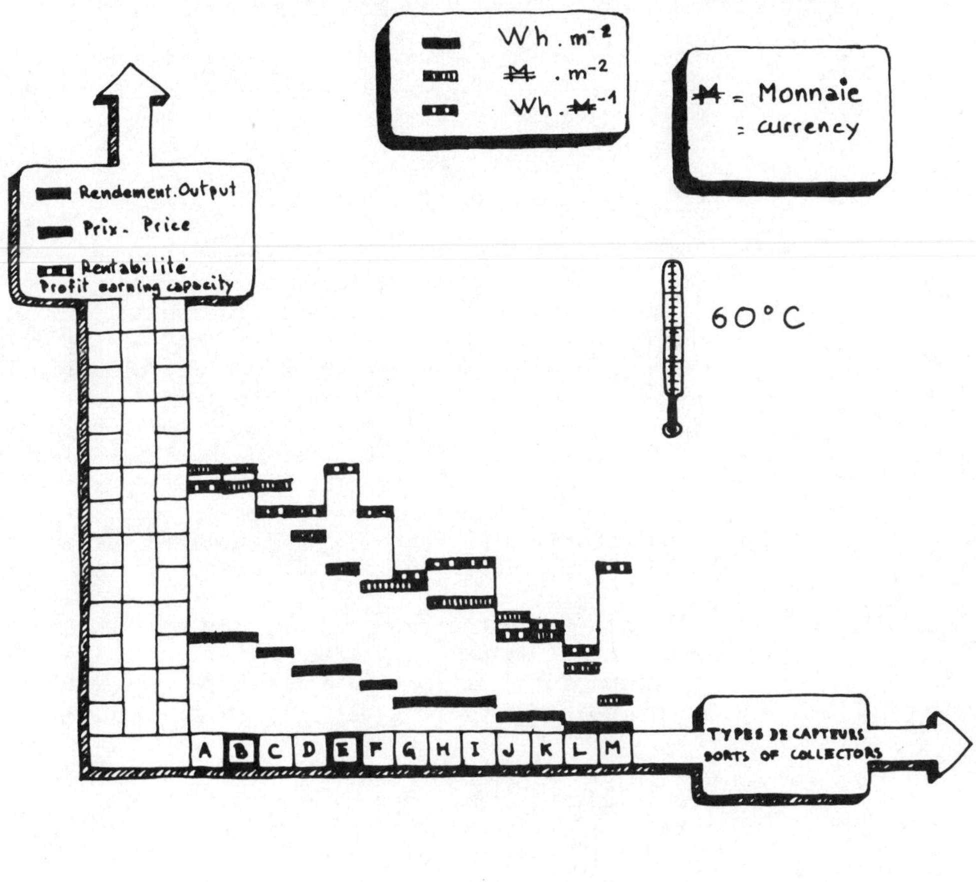
Wh . m⁻²
₼ . m⁻²
Wh . ₼⁻¹
₼ = Monnaie
= currency
Rendement. Output
Prix . Price
Rentabilité
Profit earning capacity
60 °C
A B C D E F G H I J K L M
TYPES DE CAPTEURS
SORTS OF COLLECTORS

En fait la comparaison entre la variation du rendement et celle de la variation de la rentabilité, doit conduire à rechercher un optimum qui peut varier dans le temps et suivant les conditions d'utilisation. (Figures 1.1. et 1.2.). En comparant différents types de capteurs en fonction de leur rendement, mais aussi en fonction de leur prix par unité de surface, on s'aperçoit que le capteur qui produit le plus d'énergie par unité monétaire n'est pas forcément celui qu'a le meilleur rendement thermique.

Les variations de cette Rentabilité financière des capteurs solaires montrent qu'il est peut-être intéressant de s'orienter vers la fabrication de capteurs à bas rendement mais à très bas coût de fabrication.

De plus cela montre que les essais de capteurs solaires doivent être effectués à plusieurs niveaux de température afin de pouvoir choisir le capteur le mieux adapté au problème à résoudre (très basse température pour les piscines, basse pour le chauffage par le sol, moyenne pour le chauffage classique, haute pour la réfrigération, très haute pour la production d'énergie).

Dans les figures le rendement exprime la quantité d'énergie fournie par chaque capteur en fonction d'une même quantité d'énergie solaire reçue, il est exprimé en Wh/m2.

Le prix est exprimé en unité monétaire quelconque, symbolisée par M̶/m2.

La rentabilité financière exprime donc l'énergie fournie par le capteur par unité monétaire investie, symbolisée par Wh/M̶.

Dans les cas de figure exposés le capteur (F) est le plus rentable financièrement pour les utilisations à basse température, et les capteurs (B) et (E) pour les utilisations à moyenne température.

Coût individuel et coût collectif

L'optimum du rapport rendement/rentabilité ne sera pas forcément le même suivant que l'on se place du point de vue de l'intérêt de l'individu ou de l'intérêt de la collectivité. Si, et cela est souhaitable, c'est ce dernier point de vue qui l'emporte, une des conséquences logiques est que c'est la collectivité qui doit prendre en charge le supplément de coût actuellement supporté par l'individu. Plusieurs techniques d'intervention sont possibles: par des subventions, par des prêts bonifiés, par des déductions fiscales aux acheteurs ou au fabricants etc. De toutes façons, quelle que soit la méthode choisie, l'intervention de la collectivité doit couvrir la totalité du surcoût d'investissement solaire, que l'investisseur soit un particulier ou une collectivité, qu'il soit public ou privé.

En effet, soit l'énergie solaire est rentable pour la Société (état, région etc.) et dans ce cas elle fait ce qu'il faut pour assurer son développement, soit elle n'est pas rentable, et dans ce cas il n'y a aucune raison de financer ni 50% ni même 10% ou 1% du surcoût solaire.

Dans l'état actuel des choses, à l'exception de l'eau chaude sanitaire, les investissements pour l'utilisation de l'énergie solaire ont une durée d'amortissement trop longue et trop mal définie (de 8 à 100 ans suivant le pessimisme, les hypothèses de base, les moyens de calcul ou les motivations de celui qui fait le calcul) pour qu'un individu puisse envisager de les prendre à sa charge (en excluant le cas des motivations de satisfaction intellectuelle, philosophique, patriotique, ou la trop grande crédulité dans les affirmations des fabricants).

Fig. 2.1

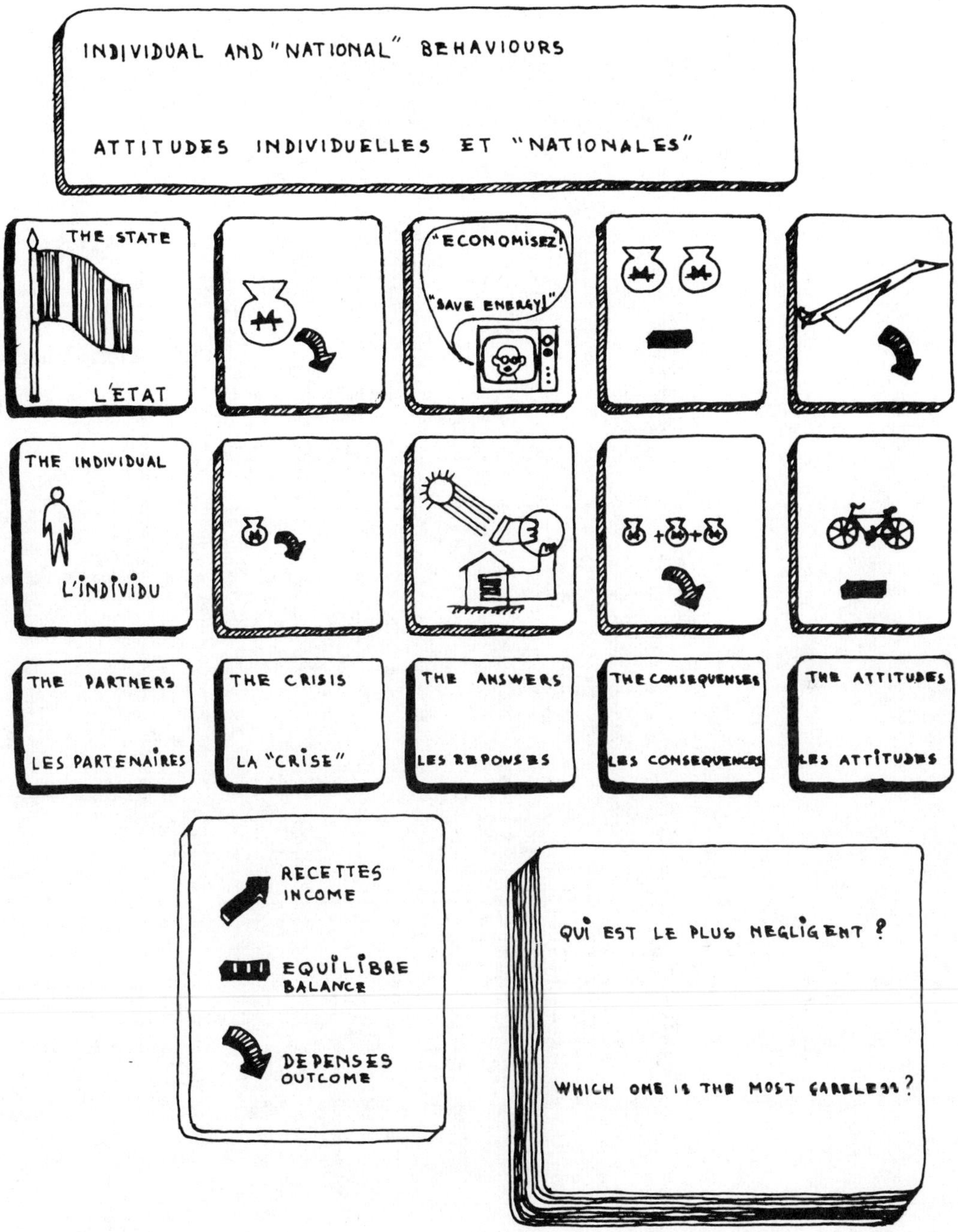

De plus, si l'individu fait des économies d'énergie qui se traduisent pour lui par une perte financière immédiate, la communauté continue à dépenser les devises ainsi épargnées dans un gaspillage énergétique continu.

Un simple exemple: En France, avec le déficit d'exploitation du Concorde, de 220 millions de francs par an, on pourrait financer le surcoût solaire de 10.000 à 20.000 installations individuelles par an (chauffage et eau sanitaire), sans parler de l'économie de kérosène (et d'ozone) qui serait entraînée par la mise à la réforme du supersonique. (Figure 2.1) mais (hélas) cet exemple serait applicable à la plupart des pays industrialisés qui n'ont pas compris que la crise de l'énergie devait les conduire à remettre en cause leurs modèles de croissance, et à établir enfin une politique énergétique qui soit, non pas à l'échelle des ménages mais à l'échelle du globe (en n'oubliant personne, s'il vous plaît!).

Fig. 2.2

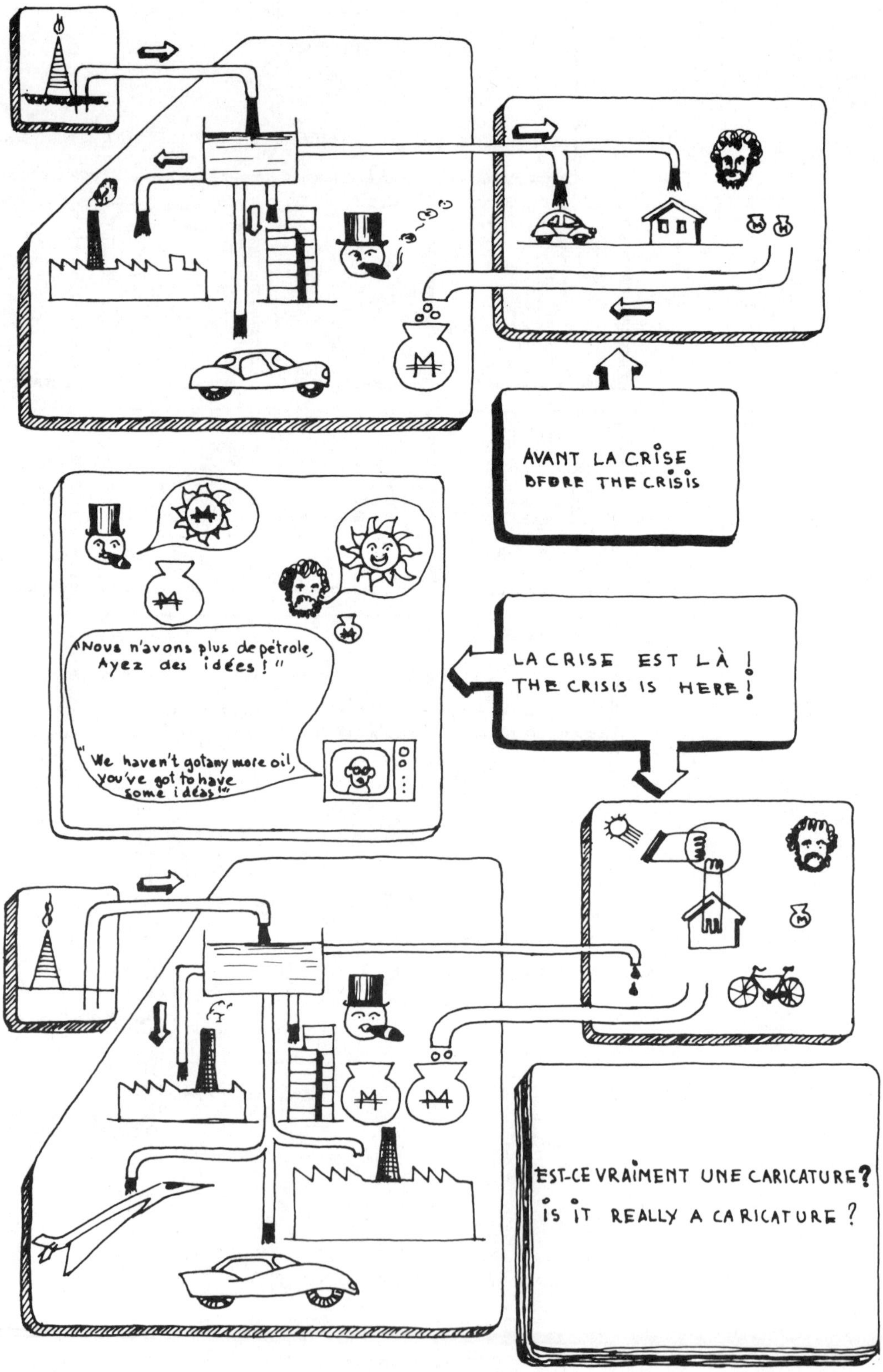

D'autant plus que l'individu, une fois passée l'euphorie de 'l'énergie solaire, c'est beau, c'est écologique, c'est l'utopie réalisée, etc.', se rend compte que le système est encore capable d'utiliser cette technique (qui devrait être au maximum décentralisée et autogérée dans une optique de développement global) au profit soit d'un capitalisme privé, soit d'un capitalisme d'état, en tout cas au profit d'une organisation centralisée qui fait son profit au détriment du citoyen qui ne se sent plus concerné. (Figure 2.2)

Rendement énergétique global ou localisé

Très souvent des unités de production énergétique sont évaluées sans tenir compte ni de leur environnement, ni de ce qui se passe en amont ou en aval, ce qui fausse considérablement les résultats vus sous l'angle collectif.

Le rendement énergétique doit tenir compte non seulement de la comparaison de la production d'énergie, mais aussi de la valeur énergétique des matériaux constituant les appareils utilisés.

Si on reprend les capteurs examinés sur les Figures 1.1 et 1.2 et que l'on tient compte de l'énergie qu'il a fallu utiliser pour les fabriquer, on s'aperçoit que doit exister, à côté de la notion de rentabilité financière, la notion de rentabilité énergétique (Figures 3.1 et 3.2). Cette rentabilité peut-être exprimée en Wh fourni par Wh investi.

Cette approche par la rentabilité énergétique à l'avantage de celle par la rentabilité financière, d'être beaucoup moins soumise aux fluctuations diverses et de permettre d'utiliser pour les comparaisons un étalon plus significatif et plus fiable.

Cela permet de traiter le problème de l'énergie en considérant réellement un point de départ et un point d'arrivée, avec une entropie. A la limite, il serait vain, en effet, de fabriquer des capteurs solaires qui nécessiteraient plus d'énergie pour leur construction qu'ils n'en produiraient pendant leur durée de vie. Dans les cas de figure exposés les capteurs (A) et (G) sont les plus rentables énergétiquement pour les basses températures et le capteur (B) pour les moyennes températures.

Cette approche doit également être effectuée pour les différents matériaux et pour les différentes opérations nécessitées par la construction d'une habitation par exemple (Figure 3.3) en distinguant la consommation d'énergie fossile et celle d'énergie renouvelable.

De plus, pour comparer la rentabilité énergétique des différents choix d'aménagement du cadre bâti, il convient de comparer systématiquement toutes les phases de consommation d'énergie qui interviennent. (Figure 3.4)

Enfin, la comparaison de l'énergie solaire avec les autres sources d'énergie n'est possible que si l'on tient compte de l'ensemble des phases successives qui entraînent une consommation d'énergie non renouvelable.

Rentabilité instantanée ou à long terme

Les calculs de rentabilité, doivent tenir compte des investissements instantanés mais aussi des coûts d'exploitation. Bien que ceci soit évident, cette pratique est loin d'être entrée dans les moeurs, tant des utilisateurs que des financiers publics ou privés.

La Figure (4.1) donne l'allure de la variation, en fonction de l'époque, des dépenses d'investissement pour une installation utilisant une énergie renouvelable. On se rend compte que jusqu'à l'année où les deux courbes se rencontrent, le surcoût doit être pris en charge par la collectivité.

La Figure 4.2 donne l'allure de la variation (en fonction de l'époque), de l'actualisation des dépenses d'exploitation correspondant à la durée de vie de l'installation.

Fig. 3.1

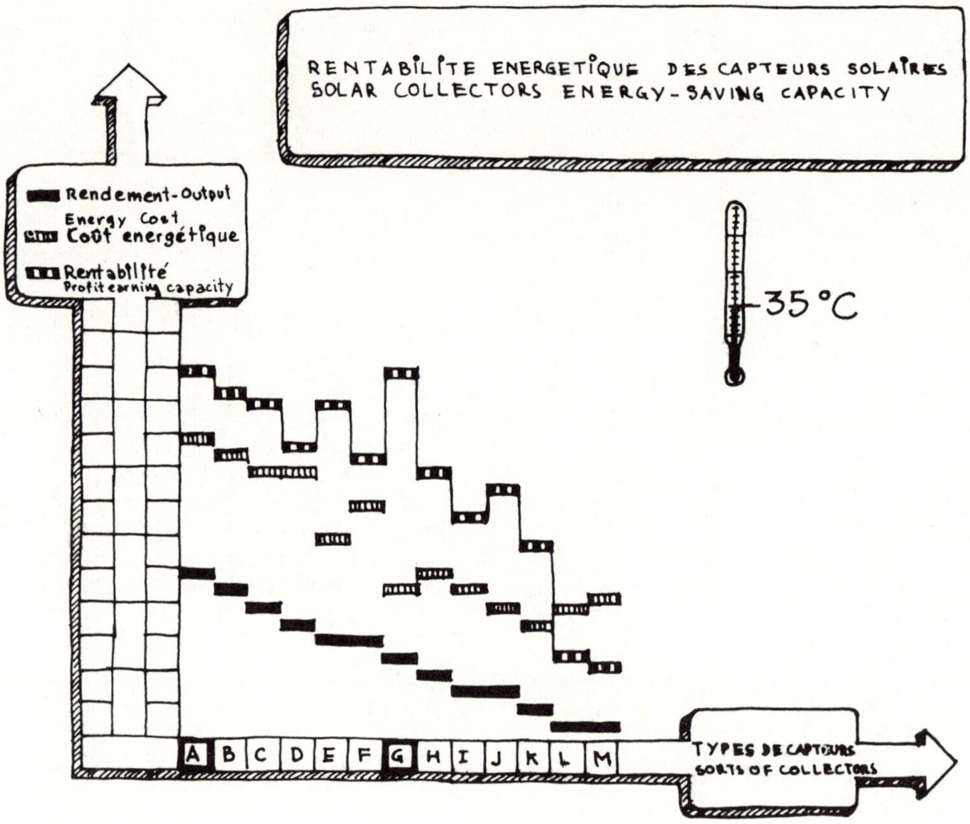
RENTABILITE ENERGETIQUE DES CAPTEURS SOLAIRES
SOLAR COLLECTORS ENERGY-SAVING CAPACITY
Rendement-Output
Energy Cost
Coût energétique
Rentabilité
Profit earning capacity
35 °C
A B C D E F G H I J K L M
TYPES DE CAPTEURS
SORTS OF COLLECTORS

Fig. 3.2

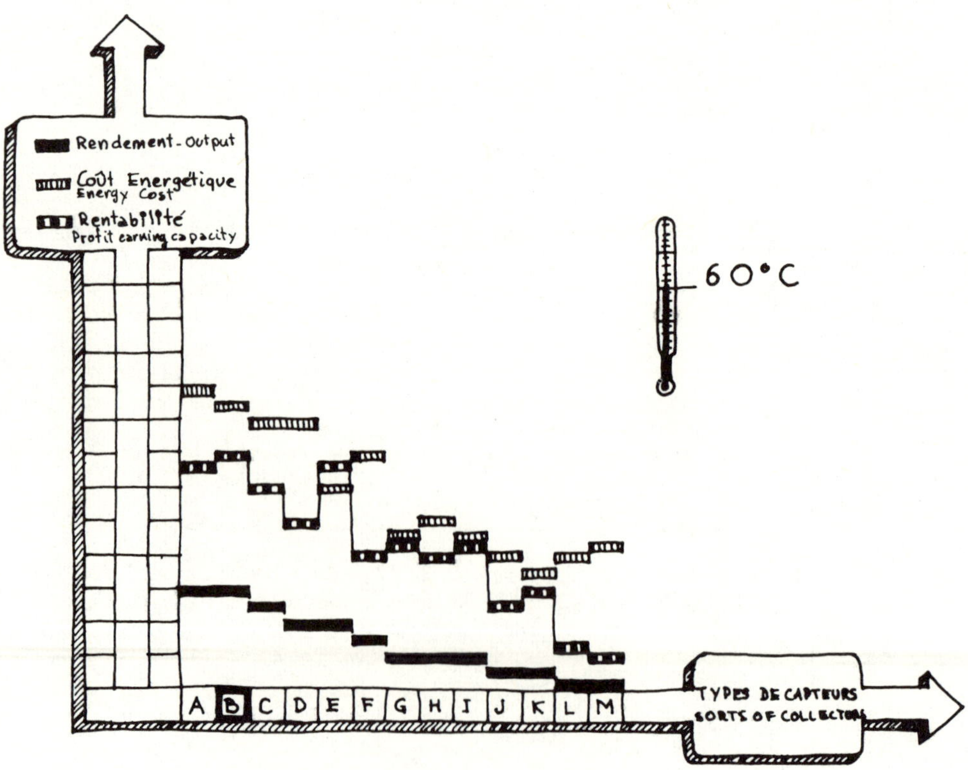
Rendement-Output
Coût Energétique
Energy Cost
Rentabilité
Profit earning capacity
60 °C
A B C D E F G H I J K L M
TYPES DE CAPTEURS
SORTS OF COLLECTORS

Fig. 3.3

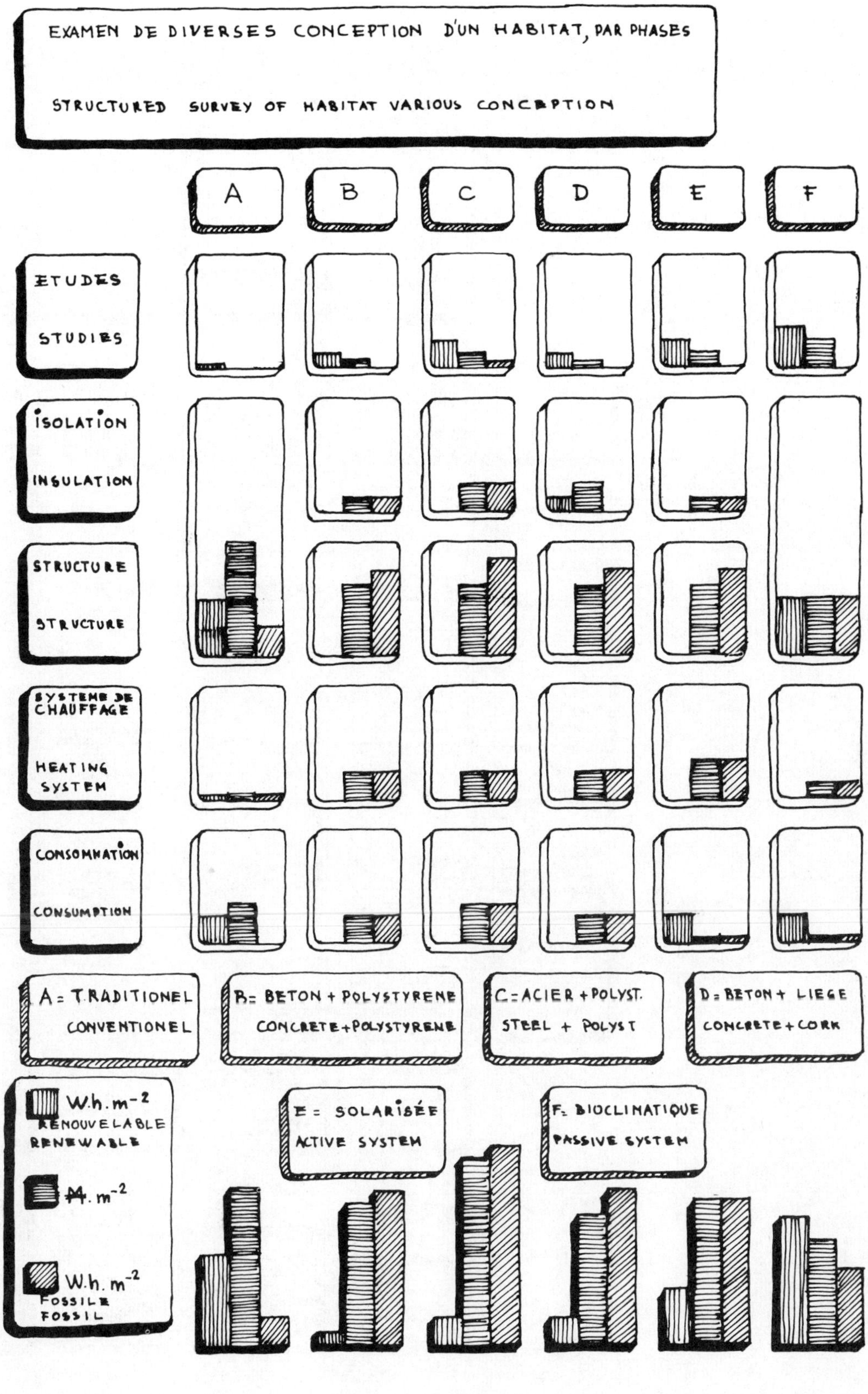
EXAMEN DE DIVERSES CONCEPTION D'UN HABITAT, PAR PHASES
STRUCTURED SURVEY OF HABITAT VARIOUS CONCEPTION
A
B
C
D
E
F
ETUDES
STUDIES
ISOLATION
INSULATION
STRUCTURE
STRUCTURE
SYSTEME DE CHAUFFAGE
HEATING SYSTEM
CONSOMMATION
CONSUMPTION
A = TRADITIONEL
CONVENTIONEL
B = BETON + POLYSTYRENE
CONCRETE + POLYSTYRENE
C = ACIER + POLYST.
STEEL + POLYST
D = BETON + LIEGE
CONCRETE + CORK
W.h.m-2
RENOUVELABLE
RENEWABLE
M.m-2
W.h.m-2
FOSSILE
FOSSIL
E = SOLARISÉE
ACTIVE SYSTEM
F = BIOCLIMATIQUE
PASSIVE SYSTEM

Fig. 3.4

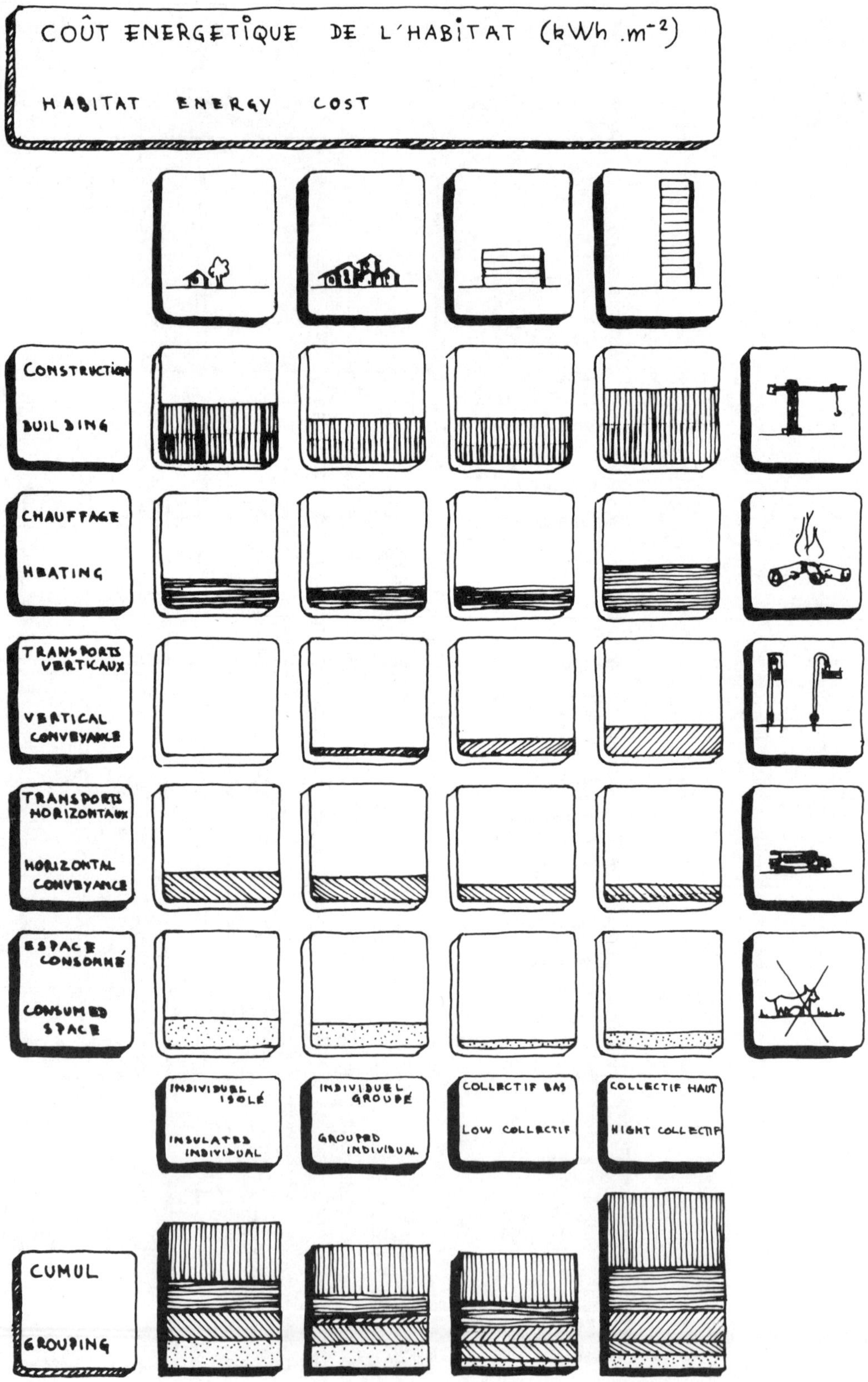
COÛT ENERGETIQUE DE L'HABITAT (kWh .m-2)
HABITAT ENERGY COST
CONSTRUCTION
BUILDING
CHAUFFAGE
HEATING
TRANSPORTS VERTICAUX
VERTICAL CONVEYANCE
TRANSPORTS HORIZONTAUX
HORIZONTAL CONVEYANCE
ESPACE CONSOMMÉ
CONSUMED SPACE
INDIVIDUEL ISOLÉ
INSULATED INDIVIDUAL
INDIVIDUEL GROUPÉ
GROUPED INDIVIDUAL
COLLECTIF BAS
LOW COLLECTIF
COLLECTIF HAUT
HIGHT COLLECTIF
CUMUL
GROUPING

Fig. 4.1

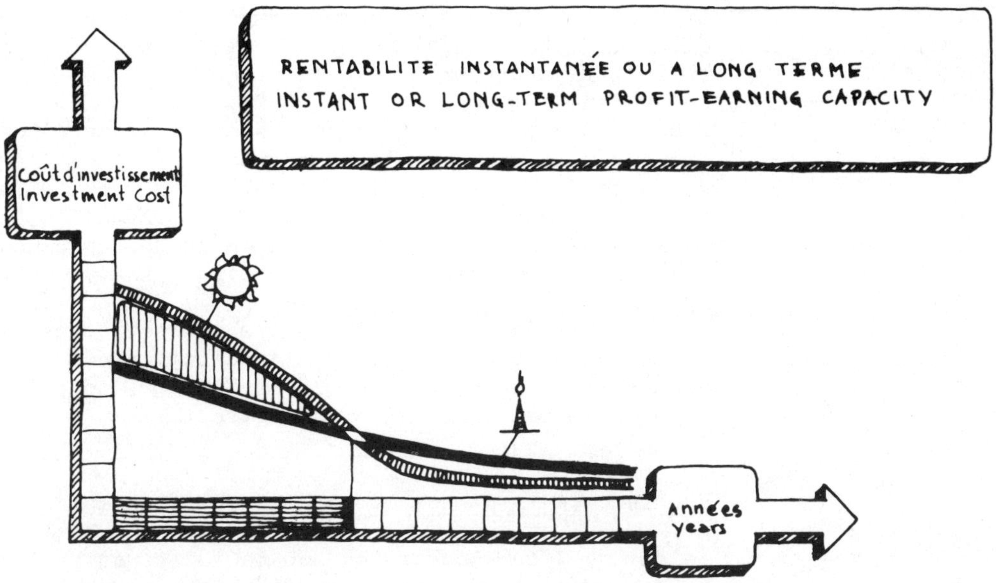

Fig. 4.2

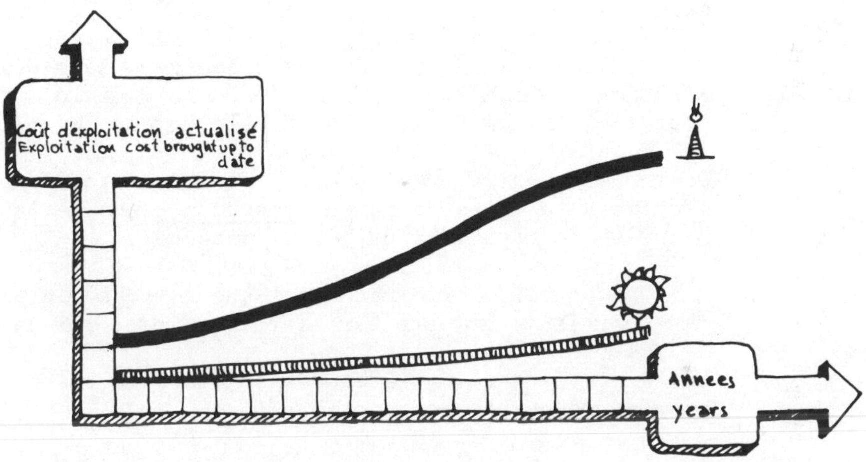

Fig. 4.3

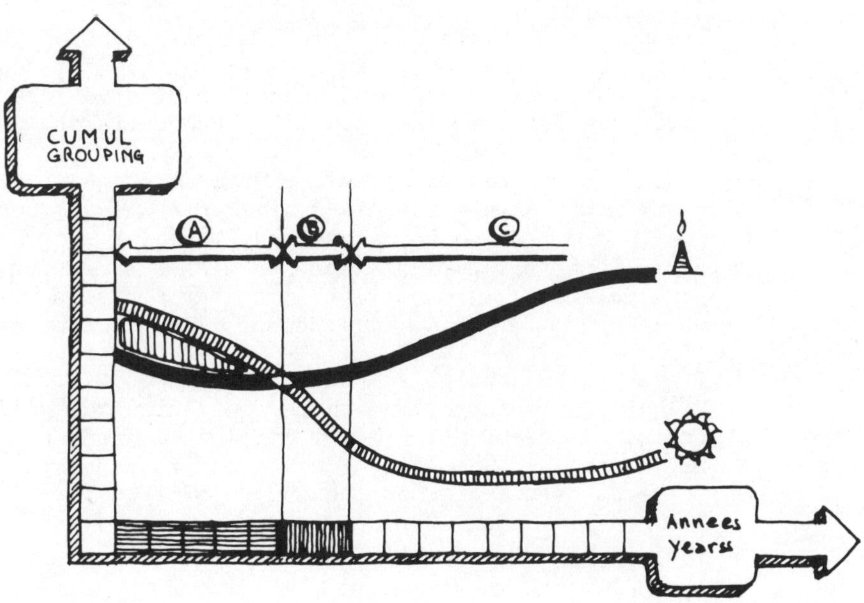

La Figure 4.3 donne le cumul des deux figures précédentes. On s'aperçoit que l'on atteint beaucoup plus tôt la date à laquelle l'énergie solaire devient financièrement rentable.

Une politique énergétique cohérente consisterait à ce que la collectivité prenne en charge jusqu'à l'année d'équivalence correspondant au cumul la totalité du surcoût solaire, et entre cette année et l'année d'équivalence correspondant au seul coût d'investissement une partie seulement de ce surcoût (phase B), les aides étant supprimées ensuite (phase C).

Cette politique aurait d'ailleurs pour effet de rapprocher de la date présente ces étapes car, par l'incitation qu'elle créerait, elle augmenterait le nombre des installations, ce qui tendrait probablement à en faire baisser le prix de revient.

Problème de la dimension des installations

Selon que l'on s'occupe ou non des différents éléments de rendement et de rentabilité, localisés ou globaux, individuels ou collectifs, instantanés ou à long terme, on arrive à des conclusions qui peuvent être tres différentes du point de vue de la dimension des installations.

Le premier réflexe d'un scientifique sera sans doute de parler de l'économie d'échelle qui justifiera le choix d'unités de grande taille plutôt que celui d'unités de taille moyenne ou petite.

Certes une centrale hélio-électrique de type THEM de 3 mégawatt par exemple aura un rendement de production énergétique supérieur a celui de 6 unités de type THEK de 500 kW chacune.

Mais si on calcule le rendement énergétique global, en tenant compte de tous les facteurs, arrivera-t-on aux mêmes résultats?

La Figure 5 donne une idée de la méthodologie qu'il faudrait employer pour approcher cette question.

Facteurs sociaux

Les données sociales, culturelles politiques et philosophiques influent fortement sur les choix et l'optimisation doit tenir compte d'un mouvement général qui tend actuellement plus vers les structures autonomes que vers les structures centralisées.

Si on reprend les exemples des Figures 3.3 - 3.4 et 5 et que l'on y ajoute une perte d'énergie équivalente au coût social et culturel, tel que le désir de décentralisation, le refus des dégradations esthétiques opérées par des implantations de grande dimension, la recherche d'un équilibre etc.... on arrivera certainement à des résultats qui modifieront des choix strictement financiers ou énergétiques.

Il est certain qu'il y a une trés grande difficulté à évaluer l'équivalent énergétique ou le coefficient de pondération de ces facteurs sociaux mais il est impossible de ne pas en tenir compte, et il faudra faire des optimisations en retenant plusieurs hypothèses, et faire en dernier ressort un choix politique.

Optimisation

En examinant en détail les critères d'optimisation, on s'aperçoit que l'avenir de l'utilisation de l'énergie solaire est plus dans les projets d'architecture bioclimatique (passive system), ou dans les projets d'architecture solarisée (active system) combinés avec des petites unités hélio-électriques, que dans l'alimentation de l'habitat en énergie à partir de grosses centrales hélio-électriques, et à plus forte raison que dans l'alimentation à partir de stations solaires satellisées (projet Glaser).

Fig. 5

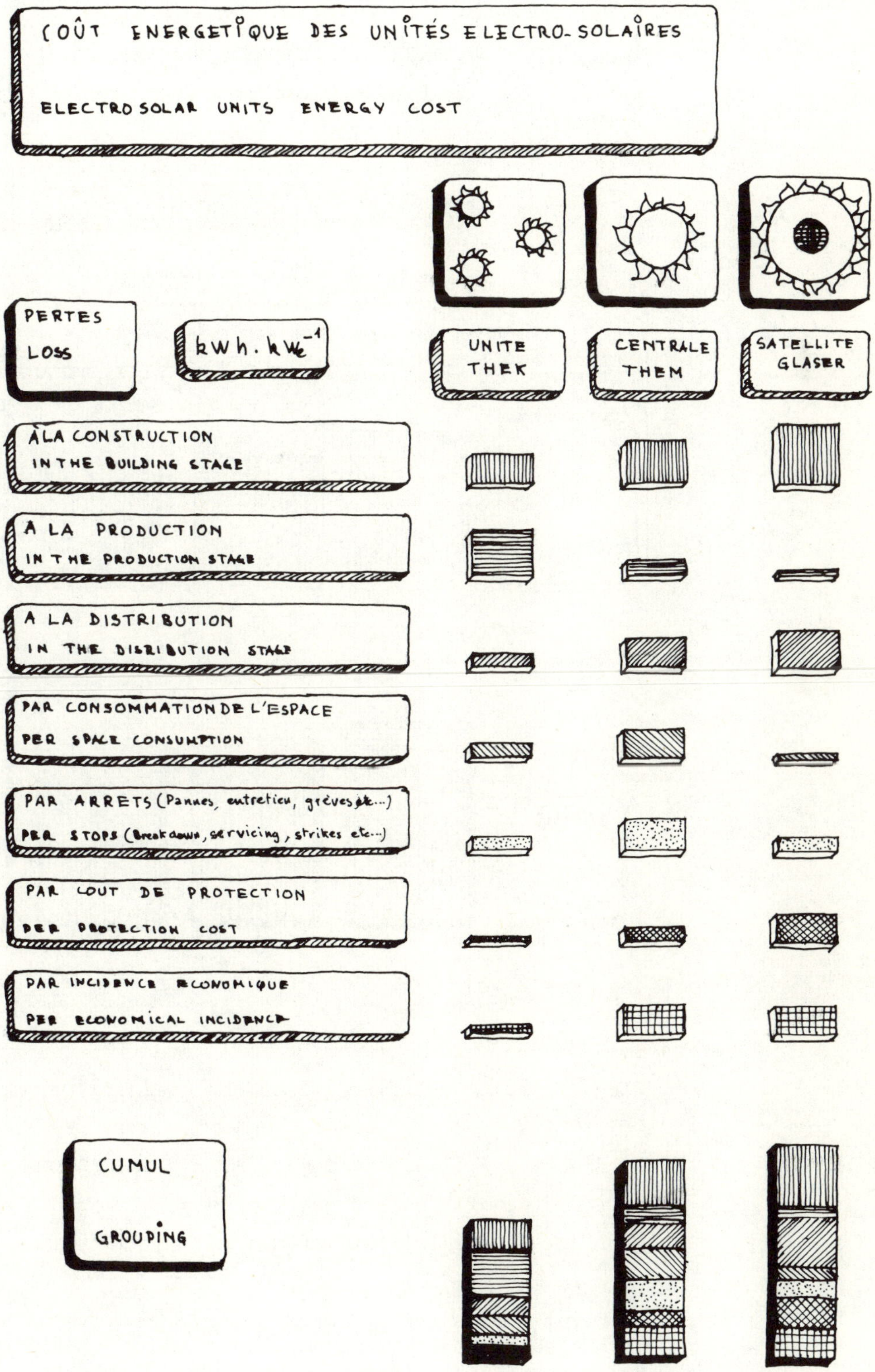

Fig. 6

Fig. 6.1

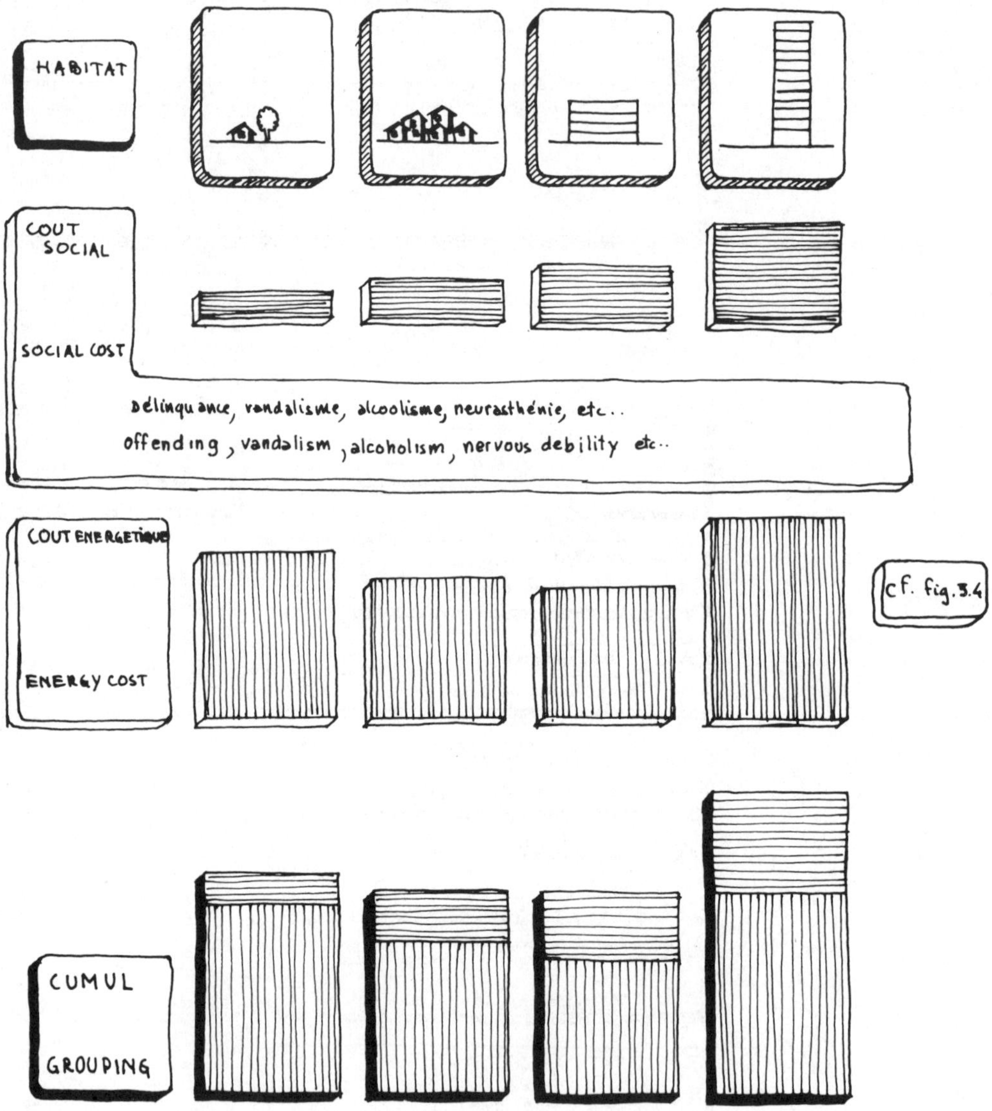

Bibliographie

Clegg Peter: 'New Low-cost sources of Energy for the House' Garden Way Publishing

De Rosnay Joël: 'Le Macroscopé - Vers une vision globale' - Seuil, Paris 1975

Farrington Daniels: 'Direct use of the sun's energy', Yale University, 1964

Institut Economique et Juridique de l'Energie de Grenoble: 'Alternatives au nucléaire, réflexions sur les choix énergétiques de la France' Presses Universitaires de Grenoble, 1975

De Chapellier, Breuzard, Billet: 'Energie Solaire et Habitat' L'Affranchi, Paris 1976

Leckie, Masters, Whitehouse, Young: 'Other homes and garbage, designs for self-sufficient living', Sierra Club Books, San Francisco 1975

Molo Molo: 'L'Energie, c'est vous!', Vivre, Stock Paris 1974

Mihajlo-Mesarovic-Eduard-Pestel: 'Stratégie pour demain - 2ème rapport au Club de Rome' Seuil, 1974

Nicolas et Vaye: 'La face cachée du Soleil', Bricolo lézardeur, Paris 1974

Nicolas et Vaye: 'Recherches sur les enveloppes bioclimatiques' La face cachée du Soleil, Paris, Mars 1977

Passat Jacques: 'Energies pour aujourd'hui et domain' Eyrolles, Paris 1976

Percebois Jacques: 'L'Energie Solaire, perspectives économiques' - Energie et Société, éditions du CNRS, 1975

Reiser: 'Chronique de l'Energie Solaire' - La Gueule Ouverte, Paris 1974 à 1977

Szokolay: 'Solar Energy and Building' The Architectural Press, London

Vaillant J.R.: 'Utilisations et promesses de l'énergie solaire', Eyrolles 1976

2.7 Factors influencing solar energy diffusion in the housing industry

C.R. Lundahl
Director of Research and Assistant Professor of Sociology

J.C. Scott
Assistant Professor of Business Administration

D.M. Dennis
Associate Professor of Education and Psychology

Western New Mexico University
Silver City, New Mexico, U.S.A.

Abstract

This study sought to explore and to evaluate a number of factors expressed in terms of action statements which comprise the multi-dimensional acceptability construct for the New Mexican housing industry.

Six key groups were selected for in-depth study because of their impact on the housing industry.

Five action statements were found to have very high and consistent rankings not only from the key group interviewees but also from larger segments of each key group responding to mailed questionnaires. Ten additional action statements were found to have a moderate influence on the diffusion of solar-energy technology and eight action statements were found to have minimal effect.

The study suggests that any program seeking to expedite the diffusion of solar-energy technology in the housing industry must be based on a thorough understanding of a multidimensional construct of solar-energy acceptability and on the opinions of all of the key groups involved.

Introduction

Energy resources affect all aspects of human life—our economic, social, cultural, and political lives. No facet of contemporary society is free from the influence of energy-related matters.

Since the 1973 Arab oil embargo and the subsequent energy shortages, one thing has become clear: Energy from fossil fuel sources will be available for a limited period of time and at rapidly escalating prices. Alternative sources of energy must be found to supplement and later to replace the dwindling supply of fossil fuels. Each increase in the price of scarce fossil fuels spurs more interest in the abundant solar energy.

Solar energy is one underutilized source for the heating and cooling of houses, a major drain on energy resources. The rate of solar-energy diffusion in the housing industry is a major concern if solar energy is to supplement or to replace fossil fuel sources for heating and/or cooling purposes. This article discusses the multidimensional construct encompassing those factors influencing the acceptability and adoption of solar heating and/or cooling technology in the New Mexican housing industry.

Acceptability Construct

The vast majority of American solar energy effort has been spent in the areas of development and refinement of solar technology. But little is known about the socioeconomic factors which influence the use of solar energy in the housing industry, and interest in these socioeconomic factors is a recent phenomena.

A search of the literature suggests that the factors influencing the acceptability of solar heating and cooling systmes are intertwined in a multidimensional construct consisting of technology-related considerations, economic-related considerations, and culture-related considerations.

A number of technology-related factors influence the acceptability of solar-energy technology. Among the factors creating potential obstacles for the rapid adoption of solar systems are building codes (1, 2, 3, 4, 5) and modifications of construction techniques which may necessitate changes in building codes (2, 6, 7). Another prime concern of the groups involved in the housing industry is the integration of solar systems with aesthetic considerations (5, 7). Concern has also been expressed about the dependability of solar systems (5, 6, 7, 8) and the necessity for adequate solar-system warranties (5, 7). Repair and maintenance (3, 5, 6, 7) as well as the availability of adequately trained solar technicians (2, 5, 7) are also important considerations. Excessive temperature variations in solar dwellings may be a problem area that can be overcome by backup systems (5, 7).

Economic-related factors also influence the acceptability of solar-energy technology. The energy crunch may create the need for solar systems to supply a larger portion of heating and/or cooling energy than they presently do (1, 2, 3, 8, 9, 11, 12). Furthermore, as more solar units are demanded and become available, prices for solar units will decrease (2, 3, 8, 9). Other factors influencing the acceptability of solar heating and cooling systems include tax subsidies (2, 3, 5, 9), guaranteed mortgages (2, 3, 5), and hook-up charges for solar-energy backup systems (5, 7).

When examining the acceptability of solar heating and/or cooling systems for residences, culture-related factors must be considered. Attitudes toward traditional heating and/or cooling systems (8, 13) can not be overlooked. The environmental acceptance of solar-energy systems (2, 7, 8, 9, 10, 13) is a concern of many individuals. The availability of adequate information for potential users of solar energy (2, 3, 8) is also an important factor.

Interview Procedures and Findings

In order to gain insight into the multidimensional acceptability construct of solar-energy technology, the investigators chose to rely to a large degree on the responses from six highly select key groups of people. These key groups were selected for in-depth study because of their impact on the New Mexican housing industry: architects, consumers, contractors, energy suppliers, financiers, and governmental officials. After careful sample selection, five representatives from each key group were painstakingly interviewed by two well-trained interviewers using a modified-delphi technique. A modified-delphi technique attempts to formulate group judgements of experts unidentified to each other through the use of interviews in an iterative process.

Each interviewer was provided with an interview guide which standardized the interview procedures. After a brief explanation of the research project and its goals, the interviewer marked the demographic items. Next, the interviewer presented and explained a matrix board which was used to classify the action statements from both the personal and statewide perspectives as evaluated by the respondent. These action statements were developed from the important factors identified through the literature review, and they dealt with possible developmental strategies relating to solar-energy technology in the housing industry. After each action statement card was evaluated and placed on the matrix board by the respondent, his reactions and reasonings were probed and recorded by the interviewer. Once all action statements has been placed on the matrix board and discussed, the respondent was asked for other possible action statements which might influence the acceptability of solar-energy technology in the New Mexican housing industry and yet were not included in the deck of action statement cards. After discussion of these items, a summary statement was elicited from each respondent by the interviewer.

Data from the interviews were analyzed group-by-group and by totals utilizing a set of master matrices and coded responses. Comments accompanying each matrix placement were also evaluated. A character code was assigned to each respondent so that the analyst could easily recognize the matrix placement of an individual and of his key group. All responses were utilized in determining the rank assigned to each action statement. The ranking was based upon the arithmetic means of response values on a five-point scale in terms of both the statewide and personal perspectives.

According to Table 1, five action statements had the highest composite arithmetic mean ratings—all above 4.500. These five action statements comprise the upper range items. The ranking order of these action statements was exactly the same from the personal perspective as it was from the composite perspective. These five action statements were also the five highest in statewide importance although the ranking order was not the same as that found in the composite rankings.

The succeeding two action statements in Table 1 appear to be influenced by divergent nonsolar attitudes toward aesthetics and pollution. In addition, it must be emphasized that there is a significant difference between the fifth ranked action statement and the sixth ranked action statement. This difference of .320 signifies that two out of three respondents ranked the sixth action statement one category lower on either the statewide or personal perspective than they did on the fifth action statement. These two action statements and the next eight action statements comprise the middle range category. The latter eight action statements have arithmetic mean composite scores from 4.105 to 3.925, and, thus, they have a narrow range of .180.

The remaining eight action statements in Table 1 comprise the lower range action statements. These action statements have a broad arithmetic mean composite range of .860. The significance of this range is that nine out of ten respondents rated the lowest ranking action statement one level lower on both perspectives than they rated the highest ranking action statement in the lower range category.

The interview comments, with their response patterns and similarities, gave depth and meaning to the rankings. The interview comments were divided into four main categories: (1) state of the art, (2) need for more research, (3) attitudes toward solar-energy factors, and (4) miscellaneous comments.

The state of the art comments were often diametrically opposed to each other. Thus, there were those who believed that solar-energy technology was "already to go" while others believed that solar-energy

TABLE 1 Number Ranking and Range Placement of the Interviews Action Statement According to Composite Arithmetic Means.

Ranking	Action Statement	Composite Arithmetic Mean	Range
1	Make the initial cost of solar heating and/or cooling systems comparable to conventional systems	4.695	Upper
2	Provide more information on solar energy	4.645	Upper
3	Design solar systems with long-term dependability	4.615	Upper
4	Make energy consumption a critical factor in house design	4.600	Upper
5	Make solar heating and/or cooling units as available as conventional units	4.535	Upper
6	Make houses with solar heating and/or cooling systems as attractive as nonsolar houses	4.215	Middle
7	Use solar energy systems to reduce pollution in New Mexico	4.150	Middle
8	Emphasize the possibility of an energy shortage to stimulate the use of solar energy	4.105	Middle
9	Develop solar systems with lower maintenance costs than for conventional systems	4.100	Middle
10	Make the long-term cost (payback period) of solar heating and/or cooling systems lower than for conventional systems	4.100	Middle
11	Design solar heating and/or cooling systems with reasonable temperatures and variations	4.050	Middle
12	Develop solar heating and/or cooling systems which do not require major modifications in house constructions	4.025	Middle
13	Issue strong warranties for solar heating and/or cooling systems	4.000	Middle
14	Develop solar systems with lower repair costs than for conventional systems	3.970	Middle
15	Change the building codes to accommodate solar heating and/or cooling systems in homes	3.925	Middle
16	Emphasize that solar heating and/or cooling systems are socially acceptable to New Mexicans	3.880	Lower
17	Emphasize the pride in owning a solar heating and/or cooling system	3.820	Lower
18	Award a tax subsidy for the use of solar heating and/or cooling systems in houses	3.765	Lower
19	Make solar heating systems that require no backup systems	3.735	Lower
20	Establish the same utility hookup charge for conventional and solar backup systems	3.695	Lower
21	Establish a fund to guarantee mortgage loans for houses with solar heating and/or cooling systems	3.500	Lower
22	Implement manpower training programs for solar-related jobs	3.425	Lower
23	Emphasize that traditional heating and cooling systems are outmoded	3.025	Lower

heating and/or cooling systems would never be on a "par" with conventional systems.

Respondents indicated that more study is needed on matters in addition to technology. Some specific items mentioned frequently by the respondents as needing additional research were differing hook-up charges, building codes, and shortening payback periods.

Attitudes toward solar-energy factors influenced many interview comments relating to economic factors. A number of similar remarks were noted, including that an energy shortage might drive up the price of fuels and that long-term payback periods are directly related to initial costs. Other comments were somewhat unique but nevertheless quite important. Remarks in this category included that lower maintenance and repair costs are beneficial to low income groups and that social acceptability has economic connotations. Some of the comments about action statements reflected negative attitudes toward solar energy. Typical of this type of comments was this statement ". . . pride of [solar] ownership is snob appeal." The comments of some respondents expressed disbelief: ". . . I do not believe in retrofitting" or ". . . long-term warranties are already the acceptable [accepted] practice."

Miscellaneous interview comments included some which demonstrated keen insight. For example, one person remarked that when more requirements are placed on a solar-energy heating and/or cooling system, more delay in the diffusion of solar-energy systems can be expected.

The comments of the interviewees revealed that each respondent brought to an interview a set of attitudes and values, many of which influenced his responses. Members of each key group tended to view each action statement in a similar manner from the statewide perspective, but they often disagreed with each other from their personal perspective. This fact suggests that solar-energy diffusion can not be successfully accomplished without a thorough analysis of not only the characteristics of the key groups but also the individual characteristics of the members comprising the key groups.

Interview Validation

Questionnaire instruments were developed and administered to larger segments of each key group in order to validate the interview responses. Tabulation statistics and descriptive statistics were computed from the questionnaire data to verify the placement of the action statements. Arithmetic means and standard deviations for each key group from both the statewide and personal perspectives were calculated for each action statement, along with composite arithmetic means. These composite scores were used in preparing the questionnaire ranking of the action statements.

A comparison of the interview and questionnaire rankings shown in Table 2 reveals that the five highest ranking interview action statements were also included in the six highest ranking questionnaire action statements and that the eight lowest ranking action statements in the interview data were found to be the eight lowest ranking action statements in the questionnaire data. There was an average difference of two places between the interview and the questionnaire data for the eight lowest ranking action statements.

Rank correlation between the interview and the questionnaire item rankings was found to be .88. Correlations were calculated for each individual questionnaire key group versus the total interview group, and these varied from .69 to .88. An interesting aspect of these

TABLE 2

Comparison of the Highest and Lowest Ranking Interview and Questionnaire Action Statements

Interview Ranking*	Action Statement	Questionnaire Ranking
1	Provide more information on solar energy	2
2	Design solar systems with long-term dependability	1
3	Make energy consumption a critical factor in house design	3
4	Make solar heating and/or cooling units as available as conventional units	4
5	Make houses with solar heating and/or cooling systems as attractive as nonsolar houses	6
14	Emphasize that solar heating and/or cooling systems are socially acceptable to New Mexicans	15
15	Emphasize the pride in owning a solar heating and/or cooling system	18
16	Award a tax subsidy for the use of solar heating and/or cooling systems in houses	19
17	Make solar heating systems that require no backup systems	14
18	Establish the same utility hookup charge for conventional and solar backup systems	16
19	Establish a fund to guarantee mortgage loans for houses with solar heating and/or cooling systems	20
20	Implement manpower training programs for solar-related jobs	17
21	Emphasize that traditional heating and cooling systems are outmoded	21

*These rankings were adjusted to accomodate for the differences in the interview and questionnaire action statements.

TABLE 3

Interview and Questionnaire Rankings and Ranges of Action Statements with 95+% Significance (Chi Square Significance $\leq$ .05)

Chi Square Significance	Action Statement	Interview Ranking	Questionnaire Ranking	Range
.005 (S) .04 (P)	Make the long-term cost (payback period) of solar heating and/or cooling systems lower than for conventional systems	10	12	Middle
— .02 (P)	Develop solar systems with lower maintenance and repair costs than for conventional systems	11½*	8	Middle
— .02 (P)	Emphasize that solar heating and/or cooling systems are socially acceptable to New Mexicans	14	15	Lower
— .04 (P)	Award a tax subsidy for the use of solar heating and/or cooling systems in houses	16	19	Lower
— .05 (P)	Establish a fund to guarantee mortgage loans for houses with solar heating and/or cooling systems	19	20	Lower
— .05 (P)	Establish the same utility hookup charge for conventional and solar back-up systems	18	16	Lower

(S) Refers to statewide importance
(P) Refers to personal desirability
*This ranking is an average of the two separate action statements—maintenance and repair—which appeared in the interview format.

correlations is that the variability of one or two action statements account for 25 to 58 percent of the differences in correlation.

Cross tabulation analysis was computed. This computation yielded chi-square values for each action-statement perspective. The chi-square values were then matched to degrees of freedom to yield the chi-square significance values. The six action statements shown in Table 3 had chi-square significance values of .05 or less, thus demonstrating a significance at or above the 95 percent level from either the statewide perspective or the personal perspective. Two of these action statements were in the middle range on both the interview and questionnaire rankings. The remaining four action statements had rankings in the lower range, indicating their minimal influence on the total diffusion of solar-energy technology. However, the high significance values do indicate a consistent value placement for all key groups.

An analysis of variance was also computed. An F-ratio value resulted from this calculation, and when matched to the degree of freedom, an F-probability value was derived. Six action statements were identified where variances from both perspectives were found to exist at or above the 95 percent significance level (F probability $\leq$ .05). These action statements are listed in Table 4. Four of these six action statements also had 95 percent significance levels in cross tabulation analysis. These are the first, second, fourth, and fifth action statements listed in Table 3. Note that two of these action statements were low-ranking items. The remaining action statements were in the middle range. This fact is important because the natural effect of diverse opinion is to pull the ranking of an action statement toward mid-value. For this reason, those middle-range action statements listed in Table 4 may have a highly significant influence on the acceptability of solar-energy technology for some of the key groups.

To quantify the differences in response evaluations existing between pairwise groups, a measure of remoteness was derived from the analysis of variance computation and the results of Duncan's multiple range test. Duncan's test notes the pair-wise groups between which a significant difference in opinion exists on each action statement perspective at a given level of significance—in this case, 95 percent level of significance. The arithmetic mean averages on a given action statement for each differing group were then obtained from the descriptive statistics. The difference between the arithmetic mean averages then was multiplied by a scaling factor of 15, and the resulting value was labeled as the measure of remoteness.

The measure of remoteness indicates how much the response of one key group differs from that of another key group. Thus, the larger the measure of remoteness is, the greater the difference in opinion between the two key groups is. All measures of remoteness for the action statements on which pair-wise groups differed were added together to yield the total value of remoteness. The total value of remoteness indicates how far apart one key group is idealogically from another key group.

The total measures of remoteness were then linked together to form a relationship construct. Figure 1 shows both the individual and total group values of remoteness. The figure further illustrates that architects had points of disagreement with three other key groups while energy suppliers had points of disagreement with all other key groups. On the other hand, contractors had the lowest total value of remoteness—64.0. Energy suppliers had the highest total value of remoteness—216.7—and therefore have the greatest ideologically differences with the other key groups. Approximately half of this idealogical difference is accounted for by the differences between energy suppliers and consumers. All six key groups are arranged in a

TABLE 4

Differences in Perspectives at 95+% Significance (F Probability ≤ .05) on Questionnaire Action Statements Between Pair-wise Key Groups as Revealed by Analysis of Variance using Duncan's Multiple Range Test

Action Statement	F Probability	Key Group Pairs
Award a tax subsidy for the use of solar heating and/or cooling systems in houses (S)	0.002	Energy Suppliers--Architects Energy Suppliers--Consumers Energy Suppliers--Governmental Officials
Make the long-term cost (payback period) of solar heating and/or cooling systems lower than for conventional systems	0.003	Consumers--Contractors Governmental Officials--Contractors Governmental Officials--Energy Suppliers Governmental Officials--Financiers Consumers--Financiers
Establish a fund to guarantee mortgage loans for houses with solar heating and/or cooling systems (S)	0.004	Consumers--Financiers
Develop solar systems with lower maintenance and repair costs than for conventional systems (P)	0.005	Consumers--Energy Suppliers Consumers--Financiers
Develop solar systems with lower maintenance and repair costs than for conventional systems (S)	0.012	Energy Suppliers--Consumers
Design solar heating and/or cooling systems with reasonable temperatures and variations (S)	0.019	Architects--Consumers
Award a tax subsidy for the use of solar heating and/or cooling systems in houses (P)	0.024	Energy Suppliers--Architects Energy Suppliers--Consumers Energy Suppliers--Governmental Officials
Emphasize the possibility of an energy shortage to stimulate the use of solar energy (P)	0.029	Energy Suppliers--Architects Energy Suppliers--Contractors Energy Suppliers--Consumers Energy Suppliers--Governmental Officials
Make the long-term cost (payback period) of solar heating and/or cooling systems lower than for conventional systems (P)	0.041	Energy Suppliers--Consumers Financiers--Consumers
Design solar heating and/or cooling systems with reasonable temperatures and variations (P)	0.041	Architects--Consumers
Establish a fund to guarantee mortgage loans for houses with solar heating and/or cooling systems (P)	0.041	Consumers--Financiers
Emphasize the possibility of an energy shortage to stimulate the use of solar energy (S)	0.055	Energy Suppliers--Consumers Energy Suppliers--Governmental Officials

(S) Refers to statewide importance
(P) Refers to personal desirability

Figure 1.

Legend for Figure 1

"A" represents architects, "Cs" represents consumers, "Ct" represents contractors, "E" represents energy suppliers, "F" represents financiers, and "G" represents governmental officials.

The number on the vector lines are the value of units of remoteness and the numbers under the key group names are the total values of remoteness for those key groups.

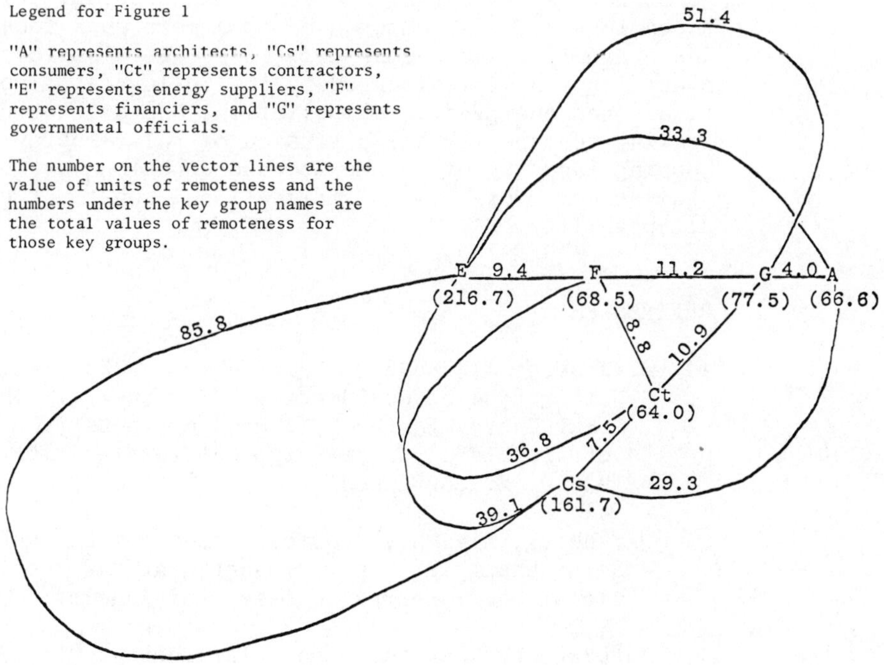

A Relationship Construct Between Pair-wise Key Groups Based on the Values of Remoteness Expressed as Line Vectors.

linear relationship which forms the nucleus for the construct. This arrangement contains half of the individual measures of remoteness. The other half of the measures of remoteness display curvilinear relationships and, thus, illustrate the complexity of the differences existing between nonadjacent key groups. Three of the key groups were within a range of 4.5 units of remoteness, suggesting a rather close agreement among the key groups of contractors, architects, and financiers.

Conclusions

Five action statements were found to have very high and consistent rankings according to the key groups studied. These actions should have a high priority in any solar-energy diffusion program:

1. Make the initial cost of solar heating and/or cooling systems comparable to conventional systems.
2. Provide more information on solar energy.
3. Design solar systems with long-term dependability.
4. Make energy consumption a critical factor in house design.
5. Make solar heating and/or cooling units as available as conventional units.

Ten action statements were found to have moderate influence on the diffusion of solar-energy technology; the remaining eight action statements were found to have minimal effect on the acceptability of solar-energy technology.

Each of the six key groups of architects, consumers, contractors, energy suppliers, financiers, and governmental officials demonstrated characteristic response patterns toward the various action statements. This suggests that any successful program for the diffusion of solar-energy technology must consider the opinions of all of the key groups.

Data from this study suggest that all of the action statements studied have at least some influence on the acceptability of solar-energy technology. These action statements and possible other unexplored factors intertwine to form a complex multidimensional construct. This construct, as an entity, governs the acceptability of solar-energy technology. It is composed of much more than technological-related and economic-related considerations. It also includes culture-related considerations as evidenced by the influence of nonsolar attitudes toward the energy crisis, pollution, and aesthetics. Any program seeking to expedite the diffusion of solar-energy technology in the housing industry of New Mexico and a nation must be based on a thorough understanding of this multidimensional construct of solar-energy acceptability.

REFERENCES

1. Hirshberg, Alan and Richard Schoen. "Barriers to the Widespread Utilization of Residential Solar Energy: The Prospects for Solar Energy in the U.S. Housing Industry," Pasadena, California: Jet Propulsion Laboratory, California Institute of Technology, 1973. (Mimeographed)

2. Hirshberg, Alan S., "Policy Issues for the Rapid Implementation of Solar Energy." Paper presented at the meeting of the International Solar Energy Society, Los Angeles, California, July, 1975.

3. Interagency Task Force on Solar Energy, "Federal Energy Administration Project Independence Blueprint, Final Task Report," November, 1974. (Washington: Government Printing Office, 1974).

4. McCue, Gerald, William R. Ewald, and the Midwest Research Institute. Creating the Human Environment. Chicago: University of Illinois Press, 1970. 339 pp.

5. TRW Systems Group, "Solar Heating and Cooling of Buildings, Phase O. Volume 2: Final Report," TRW 25168-002, May, 1974. 534 pp.

6. Scott, Jerome E., Ronald W. Melicher, and Donald M. Sciglimpaglia, "Demand Analysis, Solar Heating and Cooling of Buildings, Phase I Report, Section 2, Lending Institution Attitudes Toward Solar Heating and Cooling of Residences," NSF-RA-N-74-190, December, 1974, (Washington: Government Printing Office, 1974). 169 pp.

7. Westinghouse Electric Corporation, "Solar Heating and Cooling of Buildings, Phase O: Final Report, Volume I," NSF-RA-N-74-023B, May, 1974. 361 pp.

8. NSF/NASA Solar Energy Panel, "An Assessment of Solar Energy as a National Energy Source," NSF/RA/N-73-001, December, 1972. (Washington: Government Printing Office, 1972.) 85 pp.

9. Duffie, John A. and William A. Beckman, "Solar Heating and Cooling," Science, CXCI, No. 4223, January 16, 1976, pp. 143-149.

10. Lundahl, Craig R. and James Calvert Scott. "Solar Rays—An Abundant Source of Energy for the Southwest," Opinion Papers—What Are the Best Ways to Prevent Shortages of Clean Water, Public Land Uses, and Energy Resources, Las Cruces, New Mexico: Regional Environmental, Education, Research, and Improvement Organization, 1974.

11. San Martin, Robert L. "The Solar Energy Alternative," Opinion Papers—What are the Best Ways to Prevent Shortages of Clean Water, Public Land Uses, and Energy Resources, Las Cruces New Mexico: Regional Environmental, Education, Research, and Improvement Organization, 1974.

12. Schoen, Richard, Alan Hirshberg, and Jerome Weingart. New Energy Technologies for Buildings: Institutional Problems and Solutions. Cambridge, Massachusetts: Ballinger Publishing Company, 1975. 244 pp.

13. Miller, R.J. and J.A. Duffie, "Thoughts on Economic-Social Implications of Solar Energy Use." Paper presented at the 1970 International Solar Energy Society Conference, Melbourne, Australia, March, 1970.

2.8 Technical and economic considerations of solar applications for space and water heating in Greece

D.P. Daskalakis
Mechanical and Electrical Engineer (T.U. Athens)
Thermosol Ltd
Athens, Greece

Abstract

Solar systems for water and space heating are more efficient in Greece than in other countries due to the high insolation, low manufacturing and installation costs, and the high cost of energy. Such systems properly selected and engineered can be economically efficient for the users and they may substantially contribute in lowering the energy dependence of the country on imported oil, today covering 85% of its energy needs.

In this paper various solar systems solutions and applications are considered, costs analysed and classified in table 2, together with the cost of oil or electricity to be saved, and the pay-back periods resulting from this analysis.

Introduction

Following the energy crisis and the increasing interest for solar heating of buildings and domestic hot water, it was realised in the industrially developed countries where solar heating was first studied and experienced, that the investment cost of solar systems is substantial, compared to the cost of energy saved, especially in areas with low insolation.

The high cost of solar energy systems in developed countries is due to two main factors. (a) Collectors are expensive due to low production and high labour costs and overheads, and (b) installation is expensive due again to high labour costs.

In developing countries where labour is not so expensive, collectors, (being not highly engineered products), could be manufactured under low production rates in substantially lower costs than in developed countries. The installation costs are for the same reason far lower.

In addition the cost of energy in such countries (excluding the oil producers) is higher due to the taxation policy of these countries for imported goods affecting their foreign exchange balance.

This paper is concerned with the situation in Greece, but the results quoted could be applied, I think, as a yardstick for similar developing countries.

Insolation and climatic conditions in Greece

Greece is situated between 35° and 41° north latitude, with Athens in the middle (38°) for which insolation data is available, and on which the considerations of this paper are based.

This insolation and climatic data is as shown in Table 1:

Table 1

Month	Total solar radiation (horizontal) Kwh/m2	Degree-Days
January	67	278
February	86	231
March	123	217
April	160	92
May	172	---
June	206	---
July	225	---
August	200	---
September	163	---
October	103	35
November	70	107
December	60	222
Total	1635	1182

Solar radiation incident on collectors of various tilt was calculated (1) as follows:

(a) For tilt 45°, yearly total 1800 Kwh/m2
(b) For tilt L + 15° = 53°, November-April 800 Kwh/m2
(c) For tilt L - 15° = 23°, April-November 1500 Kwh/m2
(d) For tilt L - 15° = 23°, May- September 1000 Kwh/m2

Case (b) is for space heating, (a) for year round domestic hot water (DHW), and (c), (d) for summer hotels DHW and swimming pool heating.

Collectors and systems efficiencies

Several solar collection systems were considered for space heating, year round DHW and for summer hotel and swimming pool water heating, with single glazed and unglazed collectors with an efficiency as shown in Figure 1. We also calculated the collected solar energy, according to reference (2), which in turn is based on reference (1), for the climatic conditions of the Athens area, for the considered systems.

Fig. 1

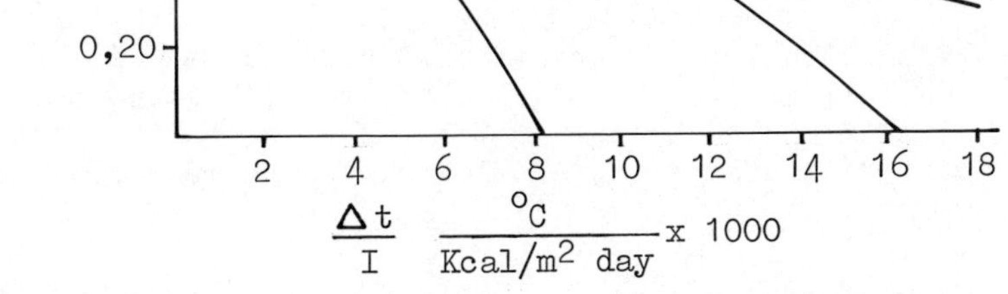

The results were approximately as follows for the following collection efficiencies:

(a) For space heating 0,50. We assume 0,40 for Thomason collectors, and 0,60 for vertical collectors with horizontal reflectors increasing the incident radiation by 25%, as in the Mathew solar house of Coos Bay, Oregon (3).

(b) For domestic water heating or preheating, and swimming pool heating, collection efficiencies are ranging from 0,40 to 0,90, depending on collector type, season of application and water and temperature, as shown in table 2.

Table 2

	Unit	Cost per unit $	Swimming Pool heating		Water heating							
			1	2	3	4	5	6	7	8	9	10
1 Collectors	m2											
- Single glaze collectors												
Prefabricated		55					55	55			55	55
Prefabricated as architectural elements		35								35		
Trickle type		17				17			17			
- No glaze collectors												
Plastic		15	15	-								
Roof		5	-	5	5							
2 Thermal storage	m3	120-250	-	-	5	18	10	18	18	18	18	18
3 Collectors installation	m2	4- 14	5	-	-	-	10	10	-	-	10	14
4 Connection tubing, pump, thermostat	m2	10- 20	10	15	10	15	20	20	20	20	20	20
5 Heat exchanger	m2	5- 10	-	-	10	10	10	10	-	10	10	10
Total			30	20	30	60	105	113	55	83	113	117
Plus 20% contractors profit	$/m2		6	4	6	12	21	23	11	17	23	23
Total cost of system	$/m2		36	24	36	72	126	136	66	100	136	140
Incident solar radiation	KWH/m2		1000	1000	1000	1800	1800	1500	800	800	800	1000
Mean collection efficiency			0,90	0,70	0,70	0,40	0,50	0,60	0,40	0,50	0,50	0,60
Solar energy collected	KWH/m2		900	700	700	720	900	900	320	400	400	600
Cost of above energy produced produced from												
- a Electricity	$/KWH	0,060	-	-	-	43	54	-	-	-	-	-
- b Oil	$/KWH	0,023	21	16	16	16,5	21	21	7,5	9,2	9,2	14
Pay-back period for electricity saving	Years		1,7	1,5	2	4	5	5,5	6,5	7,5	9,5	7,5

Systems costs

For analysing systems costs we consider the cost of the components and the installation expenses.

(a) Collectors
Prefabricated solar heat collectors (single glazed, flat black) are now offered on the Greek market at prices starting from $55/m2 although the market is not expended and they are not yet mass produced. It is expected that in the near future with mass production and cheaper designs this price will be lowered.

If such collectors are installed to substitute architectural elements of the building, such as walls, facades, balcony protecting walls etc, the cost of the collector is then shown to be the initial cost minus the cost of the substituted element, which we assume to be of the order of $20/m2.

Finally we consider the costs required to convert an aluminum roof of a building to a trickle type collector by adding back insulation if it does not exist, black painting glassfibre glazing, water distribution and collection tubing.

Such costs range from $14 to $20/m2.

For swimming pool heating and domestic water preheating during the summer, unglazed collectors were considered. Such collectors may be the prefabricated plastic type, or the trickle type when sloping roofs are used.

If collection temperatures range from 26° - 32°C, for similar ambient summer day temperatures, collection efficiencies could be as high as 0,90.

The cost of such collectors is $15/m2 for the plastic prefabricated typ type, and may drop to $5/m2 for the arrangement of an existing roof to a trickle type unglazed, and without insulation, collector.

Double glazed collectors were not considered, because as shown in Figure 1 the climatic conditions of Greece do not justify it. They are only justified in the colder northern areas of the country.

(b) Thermal storage
For Thermal storage, water and rocks were considered in the following sizes for the relevant applications:

- Swimming pool heating. No particular storage, the pool itself being the storage.
- Summer hotel water heating, 20-40 lit/m2 sufficient only for the morning loads, practically each day being sunny.
- Year round water heating, 50-70 lit/m2 and 20-30 lit/m2 for rock storage.
- Space heating, 40-75 lit/m2 for water storage of 0,080,0,160m3 for rock storage.

The cost of water storage tanks, insulated and installed ranges from $120/m3 to $250/m3 for large tanks, corresponding to non-pressure and pressure tanks. Cost for rocks is $40-80/m3.

(c) Other components
The other components of a solar heating installation with the relevant costs considered are as follows:

- Collector installation and frame for prefabricated collectors $4-14/m2. For collectors being elements of the building or used as such this cost is included within the building cost.
- Connection tubing with insulation, $10-20/m2 of collector.
- Heat exchanger if necessary costs $5-10/m2 of collector.
- Pump and thermostat costs $5-10/m2 of collector.

Finally, for the complete installation a 20% contractors profit and overheads was considered which is common practice for such jobs in the country.

Economic considerations

The above cost elements of solar heating installations for various cases of application, collectors and systems construction, and their total costs are shown in table 2.

The same table shows the solar radiation incident on each square meter of collector of various tilt together with the relevant collection efficiency of the system considered above and the solar energy collected.

The cost of the respective energy saved is based on $0,16/kg of oil, and an efficiency of 0,60 when burnt in conventional systems.

This efficiency is very optimistic, especially for water heating in hotels where the consumption in Greece is 1 kg of oil per client per day, during the summer season.

If the solar energy collected is used for domestic hot water for dwellings, then electricity is saved costing $0,06/kWh. Electric water heaters are used exclusively in Greece for domestic hot water, even in the winter season when oil central heating is used.

We considered the efficiency of water heaters as being 1,0 although their insulation is very poor (3cm of cork chips).

On considering the data shown in table 1, the cost of energy saved and the cost of the various relevant systems, we calculated and tabulated their pay-back periods in years, considering a 10% per year rate of increase of the energy cost.

The cases considered in the table are various examples from many possible solutions and are as follows:

For swimming pool heating
1. System with plastic collector.
2. System with the roof used as unglazed collector.

For water heating
3. Water preheating system with glazed collector for summer hotels (May -September).
4. Water heating system with a trickle type collector (factories or dwellings).
5. Water heating system for year round use with prefabricated collectors (hotels or dwellings).
6. Water heating systems for summer hotels (April-November).

For space heating
7. System with trickle type collectors.
8. System with collectors used as architectural elements.
9. System with prefabricated collectors.
10. System with prefabricated vertical collectors and aluminium foil reflectors on the roof.

Conclusions

From the above considerations and the pay-back periods of solar heating systems in Greece the results indicate that such systems are economically justifiable and can be proved to be substantial energy savers in the future.

Therefore they should be considered for a nationwide programme of experimental application in order to be tested and included in the future energy programme of the Country.

References

(1) S.A. Klein, W.A. Beckman, and J.A. Duffie. 'A design procedure for solar heating systems', University of Wisconsin, Madison, Wisconsin, U S A

(2) E.J. Beck, Jr and R.L. Field. 'Solar heating of buildings and domestic hot water', U S Department of the Navy, Civil engineering laboratory Port Hueneme CA 93043 U S A (Report R 835)

(3) John S. Reynolds et alii. 'The Atypical Mathew solar house at Coos Bay, Oregon. Solar Energy Vol.19 No.3 1977.

The status of solar technology development in the Federal Republic of Germany

I. Wallner
Arbeitsgemeinschaft Solarenergie e.V., ASE
Kruppstrasse 5, 4300 Essen 1, F.R.G.

Abstract

Based on scientific and technological know-how, components and systems have been developed during the last 3 to 4 years in the Federal Republic in all areas of solar technology. In addition to the research work carried out at universities, and support efforts by the Federal Ministry of Research and Technology, industry has been particularly active in furthering research work.

Numerous firms active in the area of solar technology have joined in the 'Arbeitsgemeinschaft Solarenergie e.V.' (ASE, a joint working group for solar energy); goals are to work jointly on technical projects of mutual interest, to inform the public about the possibilities for utilizing solar energy, and to further international cooperation.

Since the climatic conditions in the Federal Republic require that all scientific and technical possibilities for utilizing solar energy be explored, in certain cases quite sophisticated systems have been developed. The main emphasis of solar technology lies in the following areas:

Heating and cooling systems
Collectors
Heat pumps
Energy storage
Glazings and coatings
Solar electric energy
Solar power stations
Construction technology and materials
Desalination
Consulting

Based upon the fact that 55% of the primary energy demand of the Federal Republic is covered by mineral oil, and the fact that about 60% to 70% of the energy demand in industry, trade, agriculture, domestic use, and transportation is heating energy, one can say that the utilization of solar energy in the Federal Republic in some of the above mentioned areas is one way of reducing the Federal Republic's dependence on oil. In this context one must not only consider the direct utilisation of solar energy by means of collectors, but also indirect use, e.g. by using air/water heat pumps.

Very promising results have been achieved in numerous experimental facilities, so that cooperative agreements have been concluded between companies active in the area of solar technology and companies constructing pre-fabricated homes.

As regarding the discussions relevant to the construction of power stations, one can say that despite some positive results, there would be no discussion of the question 'electricity or solar energy,' 'oil or solar energy'. On the contrary, multi-valent systems will be proposed; for instance, 'oil as a storable energy source to cover peak-period demand' and 'electricity as auxiliary energy for the operation of solar systems'.

Introduction

As recently as five years ago, the utilization of solar energy in our country could not even be discussed. Only companies involved in the space program were conducting developmental research, especially in the area of photovoltaic conversion. However, three years ago, research on the use of solar energy for water and space heating was intensified even on a broad basis.

The reason for the sharply increasing interest in the use of solar energy was and remains the reduction of energy costs. Perhaps we should be grateful - and certainly future generations - that the tripling of oil prices in October 1973 created a shock effect which led to increased efforts towards becoming more independent of energy imports. At present 55% of the primary energy demand in the Federal Republic is covered by oil, 96% of which must be imported.

The threatening depletion of fossil fuels and the still unknown effects of other energy sources - for instance, nuclear energy - also lead to intensified research and development efforts, among other areas, also in the field of solar energy.

When considering the utilization of solar energy, in the Federal Republic we must take into consideration that solar energy is an unreliable energy source; this fact has, of course, implications for the techniques and application possibilities which have been developed, and those which will be developed.

Compared with an overcast January day, a sunny day in June receives 50 times more radiation (600W/m2 as compared with 20W/m2 or 8kWh/m2 compared with 0.16kWh/m2). In Figure 1 one can see the solar energy distribution in FRG as compared with the situation in other countries.

Figure 2 shows the average number of sunshine hours in our country stated in hours per year. From this figure, one can also draw comparisons with other countries. The maximum values are 1900-2000 hours per year.

Figs. 1, 2

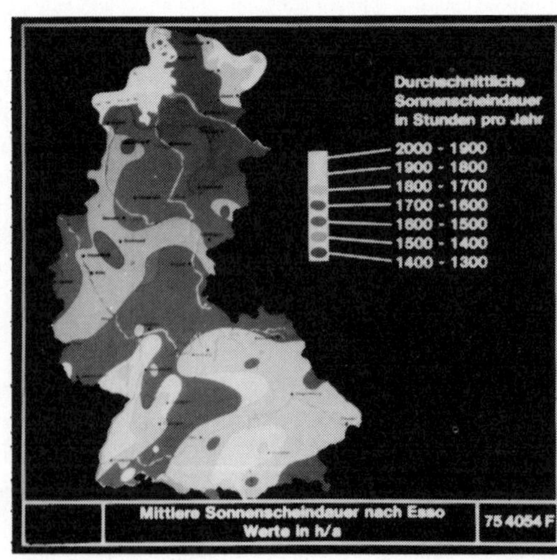

In the next figure (Fig. 3), which shows the values for the USA, one can see that the best values reached in the Federal Republic are only as high as the lowest values in the USA.

Furthermore, we must take into consideration the fact that the time period in which solar energy is available does not correspond to that in which a large quantity of energy is needed (Fig. 4).

Figs. 3, 4

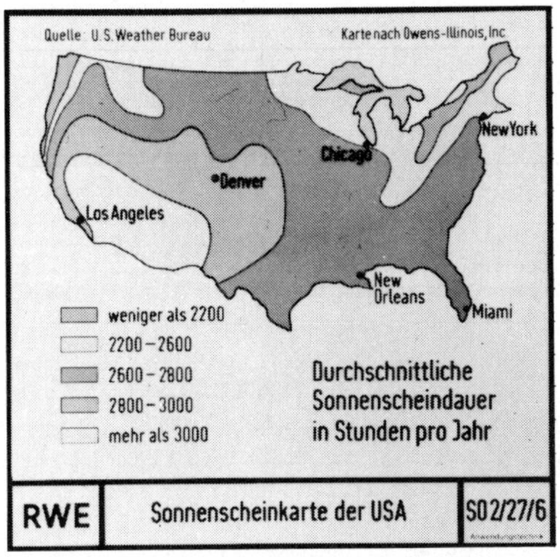

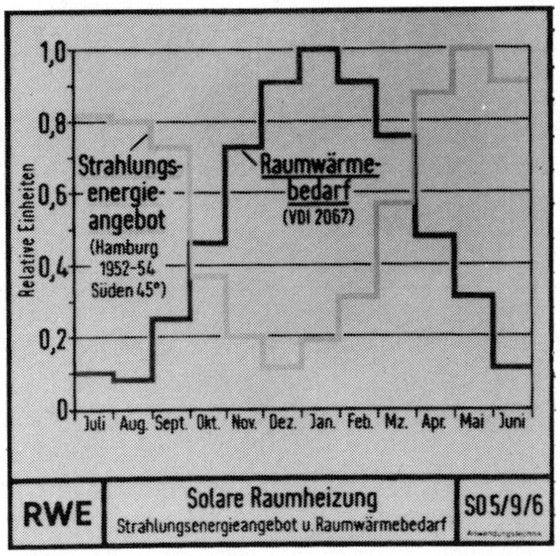

For instance, solar generated electricity could only be produced in our country during the noonday hours on a few summer days. On a few further reasonably favorable days, this would also be theoretically possible. Thus this form of solar energy utilization is not a feasible alternative for the Federal Republic.

Despite these rather unfavorable climatic conditions, in our country solar energy can play a role in water and space heating when certain technical innovations are used. The climatic conditions thus require that all possibilities for the utilization of solar energy be carefully evaluated from the scientific perspective. This is already being implimented.

Many facilities designed for water and space heating, but also for air-conditioning and generation of electricity in sunny countries, have been constructed. Furthermore numerous peripheral components have been developed. Also, questions of particular interest to structural engineers and architects are being investigated at the same time.

Within certain limitations, solar energy can pose an alternative to heating with oil. With respect to the Federal Republic, this represents one of the major opportunities for solar facilities, since 60 to 70% of the total energy requirement for industry, trade, agriculture, domestic use, and transportation is heating energy.

If we consider only the domestic sector, which will demand the highest proportion of energy in the next decade, we realise that this sector uses 90% of its energy as heating energy, including that for hot water heating.

For some industrial companies, for the Federal Ministry of Research, and for a steadily growing number of home owners or people interested in building a home, a certain fact is clear: heating with the sun has a future.

The primary initiatives for research and development in the Federal Republic come from the following three groups:
1. The Federal Ministry of Research and Technology;
2. A group of institutes located in various universities which have formed a cooperative association ('Institutsgemeinschaft') based at the University of Stuttgart; and, of course,
3. Industry.

With reference to the participation of the Federal Ministry of Research and Technology, I would like to mention that R & D projects are generally only partially subsidized, in most cases by 50%. Governmental expenditures for solar energy have risen from 1.5 million DM in 1974 to 6 million DM in 1975 and 12 million DM in 1976. The financial ceiling foreseen in the program is 14 million DM per year.

As of August 1976, a total of 24 solar energy projects were being subsidized by the Federal Government:

- 3 system studies
- 15 projects on solar-thermal conversion
  solar houses and demonstration projects
  complete systems
  components
- 3 projects on solar-mechanical conversion for electricity generation and water pumping
- 2 on photovoltaic conversion, and
- 1 on photochemical conversion.

Furthermore a solar data collection network will be established consisting of a number of small stations set up in meteorologically typical regions. The instrumentation of the stations will be standardized by international cooperation within the International Energy Agency. This data collection project will also be funded by the Federal Government.

Of course, many companies have been responsible for a large portion of the research and development.

Nineteen leading companies have united their efforts in the 'Arbeitsgemeinschaft Solarenergie e.V.' (ASE, a joint working group for Solar energy); the goal of this group is to promote the investigation, testing and development of possibilities for using solar energy. These goals also include joint work on mutually interesting technical projects, dissemination to the public of information on the possibliities for utilizing solar energy, and promotion of international collaboration.

In the meantime, these companies have begun manufacturing and delivering components for solar facilities, as well as complete systems.

The spectrum of their activities within the scope of solar energy utilization basically encompasses the following areas:

Space heating	Solar electric energy
Space cooling	Solar power stations
Collectors	Construction technology and materials
Heat pumps	Desalination
Energy storage	
Glazings and coatings	

At this point I would especially like to mention the heat pump, since these companies believe that solar energy utilization includes not only the direct utilization of solar energy with collectors, but also the indirect use of solar energy, e.g. by the use of heat pumps.

Following this general survey, I would now like to report on a few developments, although I will only be able to go into detail in a few cases. However, should you desire further information, I would suggest that you contact individually the companies participating in ASE.

Collectors

With respect to the development of collectors, one can say that a wide variety of types of flat-plate collectors has been developed and marketed in the Federal Republic, as they also have in other countries.

Figs. 5, 6

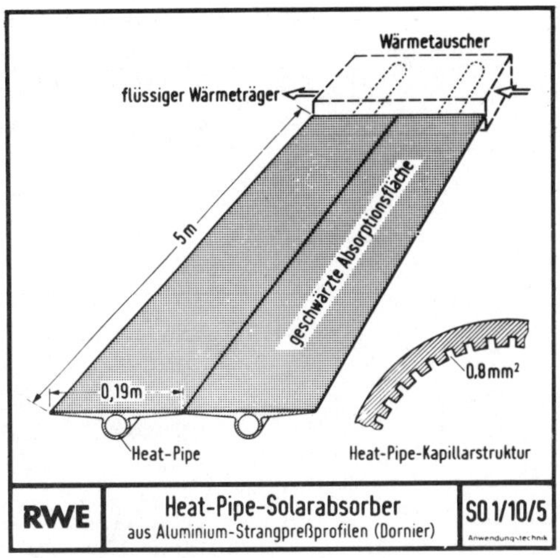

Among the leading manufacturers who are members of ASE for instance are, BBC, MBB, Brötje and Esser (Fig. 5). However, collectors not belonging to the 'classic' type have also been developed, e.g., the Dornier collector (Fig. 6), which operates on the heatpipe principle and which is, incidentally, also used in desalination plants. In the vacuum-type collector category, I would like to mention ERNO and Philips. The following figure (Fig. 7) shows the construction of the Philips collector: Radiation passes through the evacuated glass tubes; the metal profile conducts the heat absorbed by the cobalt-sulfide selective absorber surface to the heat transfer medium in the tubes (Fig. 8).

When considering the amortization and political economics associated with the use of solar water heating, it is important to consider the fact that conventional oil-fired boilers only operate at a very low efficiency level during the summer. A relatively large quantity of heating oil can be saved when the oil-fired boilers are inactivated and replaced by a few collectors. If oil prices continue to rise, this method of heating water during the summer months will thus become particularly interesting from the economic perspective.

Figs. 7, 8

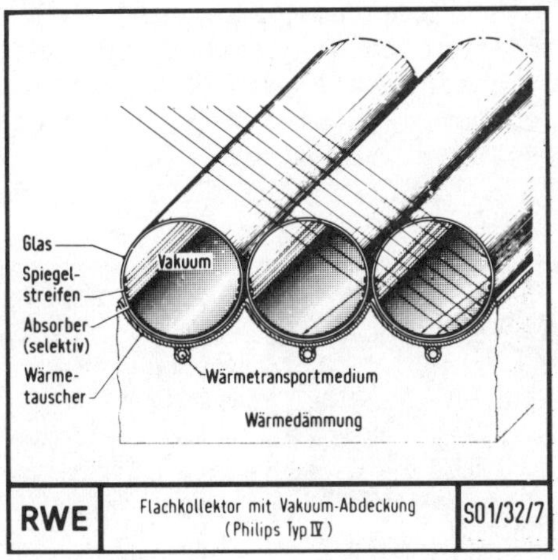

Based upon the practical experience gained, I would like to give a few details:

The total collector surface area for a 4-person household with higher-than-average hot water consumption is about 6 to 8m2; the hot water storage tank should hold between 300 and 600 l. During longer overcast periods - even during the summer - supplementary heating of the water will be necessary. Utilization of the oil-fired boiler for this purpose is not reasonable, since the heat loss from the boiler during the warm-up phase would largely cancel out the useful solar energy contained in the pre-heated water. During the summer months, electrical energy represents the most suitable form of auxiliary energy. This type of auxiliary heating can most easily be accomplished by equipping the storage tank with an electric heating coil.

During the heating season, the oil-fired boiler represents the most suitable form of auxiliary heating, since electricity is a form of line-bound energy and should not be used to take up the energy slack during winter months which have low insolation levels.

The following figure (Fig. 9) shows a facility equipped with 3m2 BBC-collectors and a 150 l storage tank; 140 l of hot water are taken daily from the storage tank, a figure which corresponds to the normal daily hot water requirement for two persons.

The next figure (Fig. 10) gives the energy balance for this experimental facility. When the hot water is to be supplied a temperature of 45°C, an average of about 80% of the hot water requirement can be supplied by solar energy between May and September; taken on a yearly average, 50% of the hot water requirement can be covered by solar energy. These values depend on the overall design of the system, the annual variation in insolation conditions, and above all, on the habits of the consumers.

Depending on the efficiency of the oil-fired boiler, on the design of the facility, and on the consumption patterns, one can expect an annual saving of heating oil of 100 to 200 l per square meter collector surface area. These relatively high savings are possible because of the low summertime efficiency of conventional oil-fired boilers. The amount of oil saved during the heating season is only small - about 10 to 15% - since the conventional hot water preparation system operates with a high degree of efficiency during this period of low insolation. Viewed from the perspective of economic feasibility, solar water heating is particularly interesting in those areas in which fossil-fired heating systems operate with a low degree of efficiency during the summer months.

Figs. 9, 10

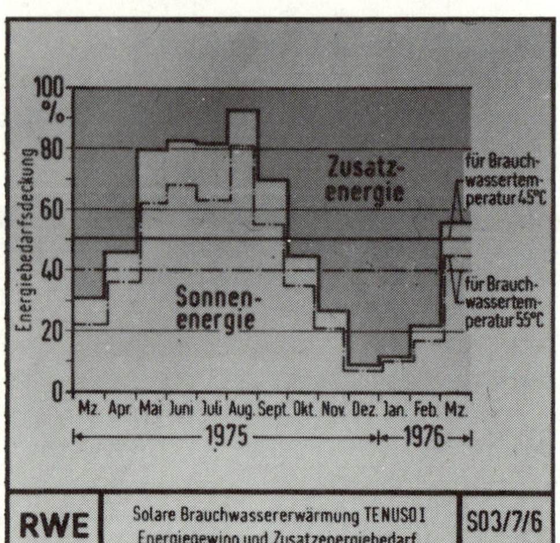

With respect to the hardware available, one can clearly see that the companies do not market only their collectors, but rather concentrate their efforts on selling complete systems.

Space heating

As far as space heating in our country is concerned, we must devote particular attention to developing effective systems appropriate to our climatic conditions.

As one can see in this figure (Fig. 11), the conditions for using solar energy for space heating are not very good in our country; the period in which solar energy is abundantly available does not correspond to the period in which space heating is most required. The insolation during the months from November to February is very low; but about 60% of the heat demand for space heating occurs during this same period. Long-term storage of excess heat collected by the solar collectors during the summer months for use during the peak heating period will not be feasible for some time to come, since this type of storage is very expensive. Therefore, space heating with collectors will - for the present - be essentially limited to covering the space heat demand during the transitional months. During the peak heating period, for most part, space heating will be taken over by a conventional heating system.

Figs. 11, 12

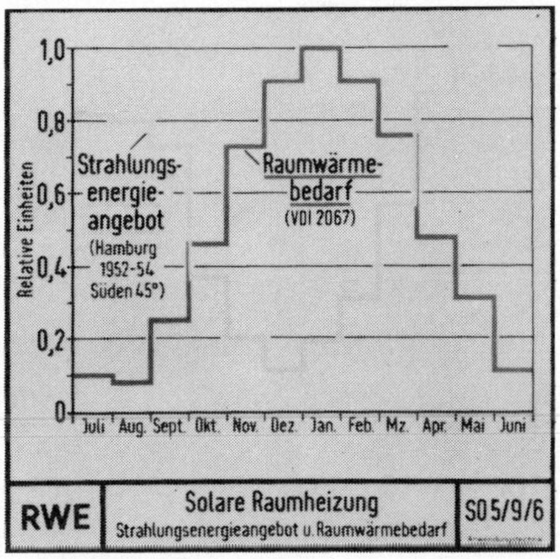

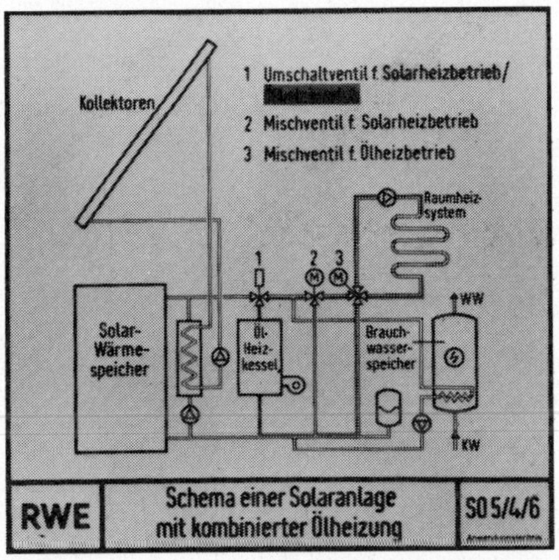

When solar collectors are used for space heating, special provision must be made for extra insulation. The utilization of low-temperature heating systems is necessary if one wants to utilize the highest proportion of the 'collected' solar energy, unless highly efficient collectors are used. Experience to date has shown that a collector surface area of about 0.3 to 0.4 times the actual living surface area which is to be heated would be required to cover the heating demand during the transitional heating period. In order to guarantee a reserve capacity for a few days, we also provide for a storage tank with a water capacity of 5 to 15m3, in addition to the solar collectors.

For the period during which solar energy cannot contribute effectively in satisfying the heat demand, a conventional heating system - such as an oil-fired heating system - is the most reasonable alternative. Figure 12 shows such a system.

Solar energy can be further exploited when heat pumps are used:

When the feed-in temperature drops below the required level, the heat pump comes into operation. In this case the solar storage tank is cooled down to 5°C. If the insolation level is so low that even the

Figs. 13, 14

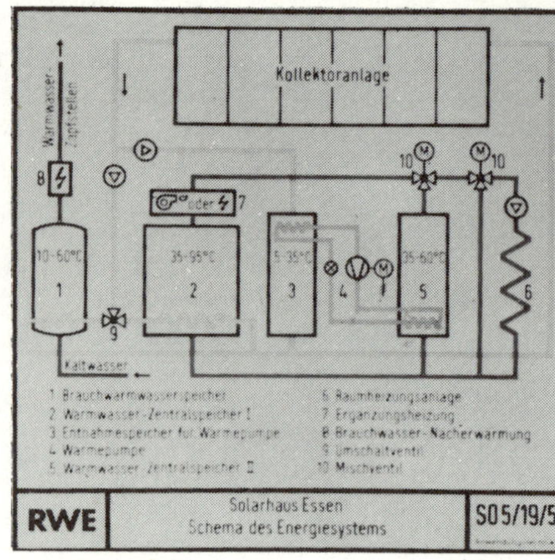

Fig. 15

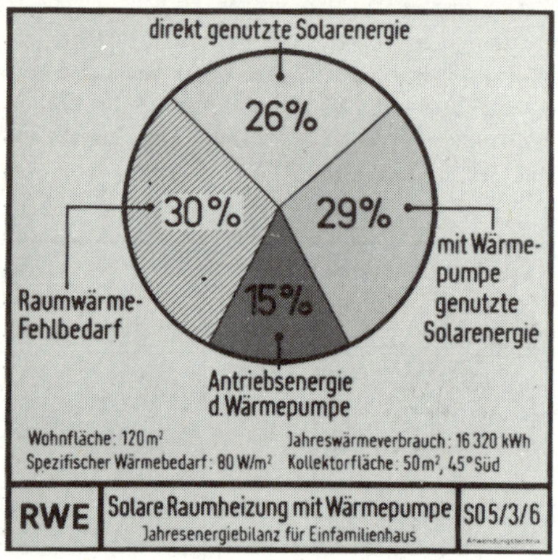

heat pump can no longer cover the heat demand, the oil-fired boiler comes into operation.

This system is used in the solar house developed by the Dornier company and the electric utility company RWE (Fig. 13, 14).

To date, the following annual energy balance has been achieved during operation (Fig. 15): this annual energy balance shows clearly that with this system 55% of the energy requirement can be satisfied by solar energy and another 15% by the electrically operated heat pump. The remaining requirement must be provided by an oil-fired boiler.

One can summarize the results for solar assisted space heating in the Federal Republic as follows: If only collectors are employed, about 40% of the heat demand can be covered by solar energy, and the remaining 60% must be provided by additional energy sources. If one employs collectors and a heat pump, 55% of the heat demand can be covered by solar energy; 15% of the energy demand will be covered by electrical energy, which is required to operate the heat pump. The remaining 30% is provided by conventional energy sources.

You will probably be particularly interested to know that, for instance, the companies MBB, BBC and Philips have 'solar' agreements with companies producing pre-fabricated houses. From some of these companies one can already buy houses which are equipped with solar systems for heating water; other systems will be on the market soon. The following figures (Fig. 16-19) give an impression of some of our solar houses.

Figs. 16, 17

Figs. 18, 19

The utilization of solar energy in open-air swimming pools

In our country, the use of solar energy for swimming pools is particularly favorable, since the heat demand occurs at the same time as the availability of solar energy. In August 1976, the largest solar facility of this type constructed to date was put into operation. This facility is in Wiehl, a small town in the vicinity of Cologne (Fig. 20). Not only does this facility utilize solar radiation through the use of collectors, but it also makes use of a number of measures which significantly reduce the energy demand.

The facility in Wiehl consists of an open-air swimming pool and a multi-purpose hall. The large BBC solar collector array for heating the swimming pool water is located on the roof of the multi-purpose hall.

The multi-purpose hall is used as an ice skating rink during the winter. The refrigeration unit which was installed for producing ice and providing heat can also be used for heating the water of the outdoor swimming pool during the summer. This would be necessary when the solar collector system cannot supply sufficient heat due to unfavorable weather conditions. In order to heat the pool water, the refrigeration unit - which now operates as a heat pump - extracts heat from the ground surrounding the multi-purpose hall.

The open-air swimming pool has a total water surface area of 1460m2. The three large basins are equipped with a covering system which will reduce the heat requirement for these basins by 40 to 50%. In addition, this open-air swimming pool utilizes for the first time a heat recovery system. A 30% reduction in the energy demand is expected. Finally, the non-swimmers' pool was lined with dark-blue tiles, because these tiles show a high degree of absorptance for global radiation. This, too, is utilization of solar energy. The heat demand for the non-swimmers' pool is thereby reduced by 5%. Thus, one can expect a heat demand for the swimming pool at Wiehl which is about one-third the normal level.

If one takes into consideration that the heat stored in the ground is solar energy, the heat supplied to the open-air swimming pool in Wiehl is largely solar energy. Solar energy in this sense accounts for 90% of

Figs. 20, 21

the supplied heat. Electric energy continues to be used for operating the motors, the regulation equipment, etc.

The large collector array on the roof of the multi-purpose hall consists of 1100 collectors with about 1500m2 of absorber surface area (Fig. 21).

The roof sheating of the multi-purpose hall is made from aluminium. Since the underside of the collectors is also made of aluminium, global radiation is reflected from these surfaces. Part of this reflected radiation strikes the collectors and thus increases its absorption.

During the five-month swimming season, the efficiency of the collector system is about 55%. This means that during this period, for each square meter absorber surface, 2kWh of useful energy are collected each day. This high efficiency is due to the fact that the radiant energy is converted into heat in the temperature range of only 26 to 30°C.

At the end of this section on collectors, I would like to present a survey of the market potential of collectors. There are about 2800 open-air swimming pools in the Federal Republic; they have an average water surface area of about 1300m2. If one takes the specific value of 1 to 1.5m2 collector surface area per m2 water surface area, then one reaches a sum of about 3 million m2 collector surface area. In addition, the growth rate for open-air swimming pools lies at about 10%. Calculations of the potential in the area of domestic warm water, specifically for houses built after 1960, showed that 15 million collectors would be required if one assumes 1m2 collector per person. One would require 35 million collectors for solar space heating systems which, however, simultaneously supply domestic warm water. In these calculations only houses constructed after 1965 were considered, since for a variety of reasons the installation of solar systems is impossible in many houses. Of course, these figures are only rough calculations which should be considered with this in mind. Nevertheless, they show that there is a considerable market for solar collectors and complete systems if they become economically feasible.

In conclusion, I would still like to mention one additional aspect which is also receiving intensive attention and which represents an important concomitant. This is the question of collector capability and international agreement in this sector.

In order to enable users of collectors to compare the different collector types, a common basis for evaluation is necessary. An important aspect is the objective measurement of thermal output. Based upon criteria developed in the USA by NBS, the companies within the ASE have developed a test procedure; this test procedure is presently being tested in many laboratories throughout the world under the auspices of the International Energy Agency and the research program of the European Community. This method provides for indoor-outdoor testing.

In addition to these criteria, criteria which will enable evaluation of the thermal and mechanical characteristics of collectors are currently also being established. Among other reasons, such criteria are necessary so that the user can have confidence in this new technology. This work should be completed by Spring 1978.

Generation of electricity

I would also like to present a short survey of the state-of-the-art in the Federal Republic for developmental work on solar generation of electricity:

Solar generators were able to demonstrate their capabilities in numerous satellite programs. AEG, for instance, has built reliable, modern energy systems for use in space.

In addition to the development of space applications, projects for terrestrial application have either been realized or just started.

Promising research results are already available today. For instance, AEG has developed a modular generator concept which would allow construction of generators from the watt to the multi-kilowatt range, depending on the number of generator building blocks used.

The following figure (Fig. 22) shows numerous modules for operating a water pump.

Although photovoltaic generators in the power range up to 10kW are currently considered to be the most economical, solar-thermal generators appear to be more economical in the high power range. Naturally this depends in each case on the application.

However, before large-scale application can be realized in the area of photovoltaic solar generators, many technological and economic problems must be solved.

A preliminary step in this direction was already taken towards the end of last year. Photovoltaic solar generators with large area solar cells could be manufactured of an unconventional silicon base material; the efficiency level of these cells is 10%!

Figs. 22, 23

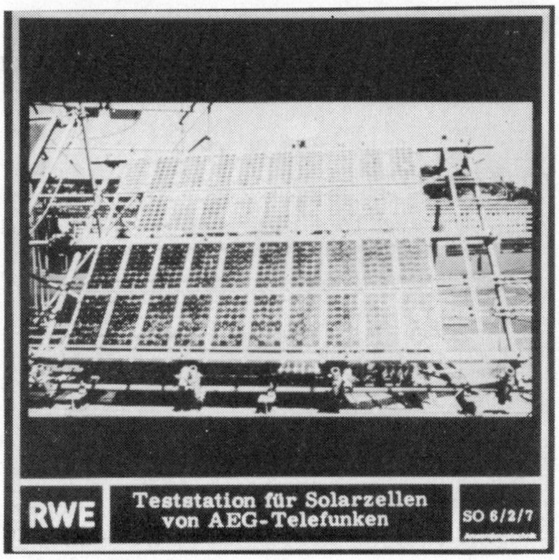

Solar generators of this type offer the possibility that reasonably priced systems can be developed through concentrated developmental research.

Intensive work is also being carried out in the area of solar-thermal generators. An example is the AEG project for a 10kW solar power plant based on a cold-steam screw-piston engine as the heat engine.

In addition to development in this direction, MBB and MAN, for instance, are working on solar-thermal power stations, MAN has devoted its attention to the development of concentrating collectors for use in solar power stations primarily located in sunny countries.

In this pilot facility (Fig. 23), cylindrical parabolic mirrors are used. The constant realignment of the mirrors in the direction of the sun is carried out electrically by simple solar sensors and mechanical adjustment motors which use only a small percentage of the energy produced. The steam which is produced is stored in an insulated energy storage unit from which the required energy can be extracted at the desired power level, either when it is needed or during the night.

Two different types are being tested as steam engines: a conventional 10 to 100kW linear-piston steam engine and a new and economical 10-30kW rotary-piston steam engine based on the screw expansion principle. In the experimental facility, the waste heat of the steam engines is alternatively passed to an absorption unit which provides cooling.

Fig. 24

This small solar-thermal power plant is being developed for different sizes. Depending on the number of collectors, an output of 15-1000kW will be produced. This figure (Fig. 24) shows a 50-100kW facility with whose output pumps are used to irrigate a desert area and the adjoining work and living quarters are provided with electricity and air-conditioning. Depending on the size of the facility and based upon series production, it is estimated that the cost of electricity from such facilities will be about 10-30 German Pfennige/kWh; thus, at least for remote areas, one can hope for the achievement of economic feasibility.

Figure 24 also gives an impression of how such a facility could look in its environment. A desalination plant which uses the surplus steam energy could also be linked up to the facility.

The question of energy preparation in our country is not an 'either-or' question. The question 'oil or solar energy' or 'electricity or solar energy' will not be the subject under discussion. On the contrary, multi-valent systems will be proposed; for instance, 'oil as a storable energy source to cover peak-period demand' and

'electricity as auxiliary energy for the operation of solar systems.'

In my presentation I could only give a very general survey of the status of solar energy technology in the Federal Republic. Furthermore, all figures were with reference to our country. Utilization of present systems will thus yield substantially better results in areas with better climatic conditions.

2.10

Development of a solar heating system with an auxiliary heater for an urban building

Y.N. Malevsky, S.I. Smirnov, B.V. Tarnizhevsky.
The Krzhizhanovsky Power Engineering Institute,
Moscow, U.S.S.R.

Y.A. Konstantinovsky,
Objedinenie Crymteploset,
Simferopol, U.S.S.R.

Abstract

This paper presents the conceptual design and performance of a solar service water heating system for an hotel in the Crimea. Data is given on the system design and the three modes of operation are explained. Further information is also provided on the solar collector to be used in the above project with its design characteristics and thermal performance described in detail. Finally the main objectives of the planned programme of experiments and tests which will take place at the above installation are described.

In 1975 the U.S.S.R. consumed about 500 mln. tonnes of fuel equivalent, that is, about a third of the total energy consumption in the country, to produce heat, including low quality thermal energy at temperatures below 100°C [1]. Moreover, about 30% of total heat requirements [2] were met by small-scale boiler plants of 0.5-3.0 G. cal/hr output. The fuel consumption per unit of thermal energy delivered by these plants to end users is about 1.5 or 2 times that of large centralized residential or industrial hot water and power stations. Therefore, in regions with favourable solar radiation conditions, the essential fuel saving may be achieved by developing and operating integrated solar heating and/or service hot water systems which could be incorporated in existing small boiler plants.

The paper presents the conceptual design and performance of an integrated solar system with an auxiliary gas-fired water heater for supplying a 215 person hotel in Simferopol (44°50' N), the Crimea, with service hot water. The system now under construction is to be completed this year by incorporating solar components into the existing conventional installation supplying space heating and service hot water exclusively for the hotel. The operating gas-fired plant has two NIISTU-5 boilers totalling I.I2 G.cal/hr (th). Service hot water exclusively for the hotel. The operating gas-fired plant has two NIISTU-5 boilers totalling I.I2 G.cal/hr (th). Service hot water load increases noticeably in summer reaching 0.5-0.6 G.cal/day.

The system design has been jointly developed by the Krzhizhanovsky Power Engineering Institute and an operating and maintenance enterprise, Crymteploset.

Service Hot Water

The schematic diagram of the integrated service hot water system is shown in Fig.1. Primary system components include an array of flat-plate solar collectors, heat exchanger, storage tanks, auxiliary heater and associated pumps and piping.

Since tap water gives rise to rapid corrosion of aluminum absorbers, the looped heat transfer scheme with an intermediate heat exchanger has been accepted. In summer, tap water without inhibitors will be employed as a heat transfer fluid in collectors. Calculations based on the chemical analysis of water to be used show that after a relatively short period, pH in a closed loop will reduce to 5.6 due to HCO_3^- decomposition. This value of pH corresponds to the minimum

Fig. 1 Flow diagram of integrated service hot water system

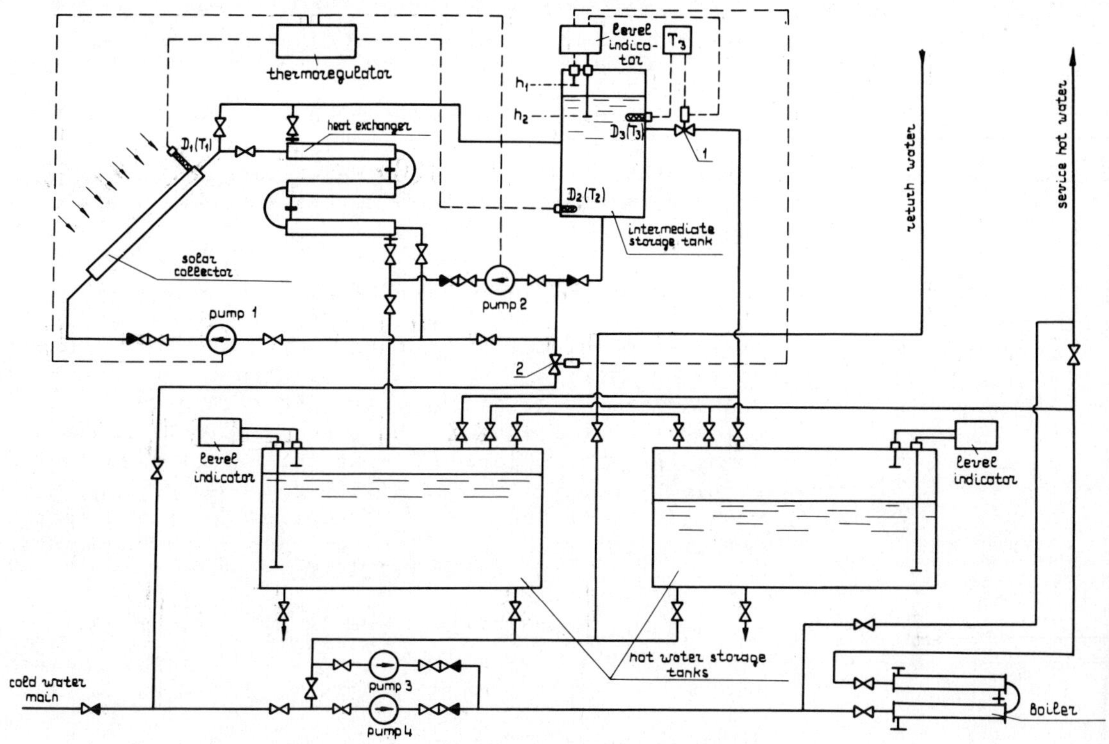

corrosion rate for aluminium. In winter, a mixture of water and ethylene glycol (automotive antifreeze) is supposed to be used in the collector loop.

Solar heat is transferred to hot water in a single-pass counter flow heat exchanger. The heated water flows from the exchanger into intermediate storage. The cooler water accumulated at the bottom of the intermediate tank returns to the exchanger by circulating pump 2. When storage temperature reaches a present value, T^*_3, the hot water drains down to main storage. The automatic control system operates as follows. As the exit collector temperature monitored by sensors D_1 (T_1) exceeds the temperature at the intermediate storage outlet where the D_2 (T_2) sensor is located, the differential thermoregulator PTRD-2 switches pumps 1 and 2 on. The pumps initiate the water flow until T_1 remains higher than T_2. When T_1 becomes lower than T_2, the pumps shut off automatically.

The intermediate tank has a two-position level indicator ESU-2M and temperature sensor D_3 (T_3), which control electric drives of valves 1 and 2.

Valve 1, directing hot water into main storage, opens as intermediate storage temperature T_3 reaches preset amount, T^*_3, and the water level in the intermediate tanks remains above the lower level h_2. If one of these conditions is not satisfied, that is, h is less than h_2, or T_3 below T^*_3, valve 1 shuts down.

Make-up water from cold main is supplied via the line connected to the intake of pump 2. The line contains electrically-driven value 2 which opens when the water level in the intermediate tanks drops below h_1, that is, below the upper contacts of the level indicator. Conversely, the valve shuts down if $h=h_1$.

There are three alternative modes of hot water supply:

- directly from the storage tanks;
- from the storage tanks via the auxiliary heater;
- from the cold water main via the auxiliary heater.

In all three modes, the hot water flow is initiated by pumps 3 or 4. The return hot water flows either into the storage tanks or to the suction sides of the pumps.

The 40 m^3 total capacity storage tank is estimated to be sufficient to meet 48-hour hot water requirements. It should be noted that the utilization of thermal storage may prove to be beneficial not only with solar hot water systems but also with conventional boilers. In summer a fuel-fired heating/hot water plant is usually operated considerably below its rated output. Besides, at night such a plant should be shut down. As a consequence, there is an increased fuel consumption and accelerated wearing of facility components. The storage tanks make it possible to reduce manpower required for plant operation in summer, cut down the frequency of plant shut-downs and operate the plant under the design output while refilling storage.

The storage tank filling time is 3.5 hr with boiler operating at the rated output. The storage filling may take place simultaneously with hot water supply to the users.

Solar Collection System

<u>Heat absorbing surface.</u> The solar installation contains 320 fixed absorbers with 230.4 m^2 overall area. The solar collection system is sized to meet load of hot water system in summer provided monthly insolations approach their long-term average values. In practice, however, the boiler start-ups would be necessary from time to time due to unfavourable sky conditions. The further increase in absorbing area would give rise to excessive production of hot water, initially in summer, then in spring and fall, and, at last, in winter. As a result, the annual increment in fuel saving associated with additional collection area would gradually decrease to zero. Therefore, the increase of solar installation investments might pay back only if fuel costs are sufficiently high.

<u>Solar collector design and thermal performance.</u> The main component of the solar collector (Fig.2) is its flat-plate "roll-bond" type aluminium panel with internal tubes. The manufacturing technique for the panels was developed by the Institute of Light-Metal Alloys. The collector exposed surface is coated with flat black paint whose spectral absorptivity is above 0.9. The panel mounted in a steel housing has a

Fig. 2
Absorber panel

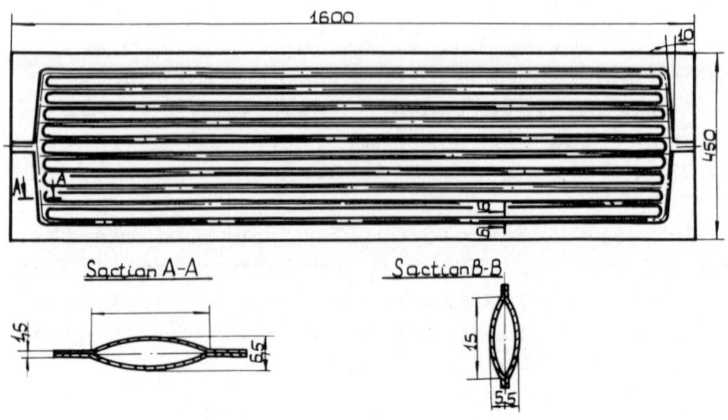

30 mm layer of polystyrene foam attached to the panel backside. The collector is glazed with a window pane sealed at its edges by a non-setting cement.

The thermal performance of collectors were established during tests conducted in 1975-1976 in Eupatoria, 65 km from Simferopol. The test installation contained four solar collectors, head tank, control valves and instrumentation. Collectors were mounted on a common supporting structure of variable tilt. Solar energy is transferred from the absorbing panels by water flowing from the head tank where the constant preset water level is maintained by a float valve.

During the tests the following parameters were recorded:

- water temperatures at the collector inlet and outlet;
- overall water flow rate across the installation;
- ambient temperature;
- total solar radiation incident on the receiving surface;
- temperature profile in the absorber panel along the fluid paths, etc.

Two types of tests were conducted: (1) at constant temperature, T_{out}, at the panel outlet and (2) at constant flow rate, g, through the installation during a test day. The results obtained were used for computing various thermal characteristics, including:

- diurnal average values of the flow rate (at constant T_{out}, and of hot water temperature (at constant g);
- total diurnal values of insolation and usable energy;
- diurnal average values of the collector efficiency.

Table 1 and Figs. 3,4 give some results obtained in summer season. These results were used in sizing heat absorbing area required.

Table 1

Experimental results for tests at Eupatoria 1976,
(α = $15^{o}15'$; ψ_n = O)

Test No	Date	Time	Test duration, hrs	Average values over the test period: Direct and diffused solar radiation incident on absorbing surface, $W.m^{-2}$	Inlet collector temperature T_{in}, °C	Outlet collector temperature T_{out}, °C	Ambient temperature T_o, °C	Water flow rate through the installation g, $l.m^{-2}.hr^{-1}$	Total values over the period: Total direct and diffused radiation incident on the absorbing surface H, W. hr. m^{-2}	Collector output, $l.m^{-2}$	Solar system thermal output Q, W. hr. m^{-2}	Diurnal average efficiency, %	Parameter kept constant
1.	23.06	10.00 17.30	7.30	683.0	29.0	64.1	32.0	8.34	5122	62.5	2555	49.9	g = const 8.3 l/m^2.hr
2.	01.07	9.35 16.55	7.20	841.7	26.2	59.5	26.0	11.6	6169	85.0	3297	53.4	t_{out} = const 60°C
3.	04.07	9.10 17.12	8.02	758.3	25.7	50.5	26.5	15.6	6091	125.2	3611	59.3	t_{out} = const 50°C
4.	06.07	8.50 17.15	8.25	645.4	25.0	49.5	24.0	11.7	5432	98.7	2809	51.7	g = const 11.7l/m^2.hr

Fig. 3
Experimental results for solar collector test, July 5, 1976

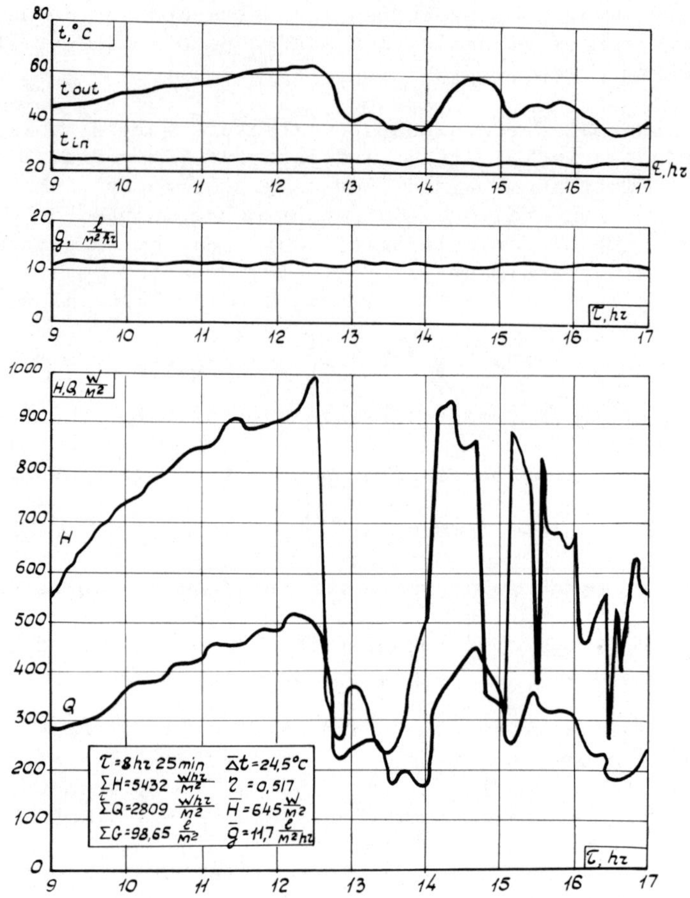

Fig. 4
Experimental results for solar collector test, July 1, 1976

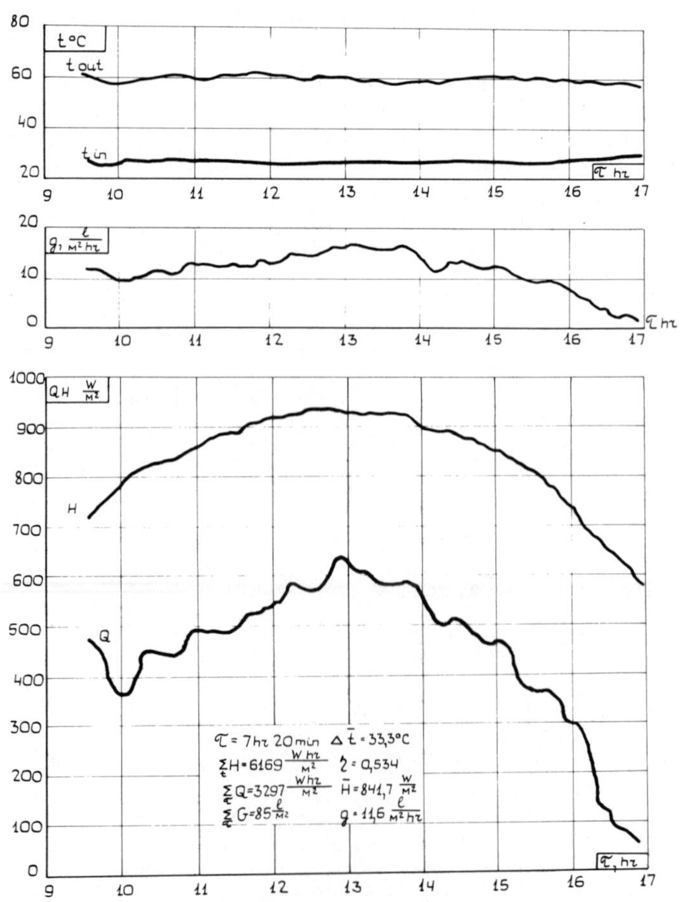

Fluid circuits. The heat transfer fluid is fed from the heat exchanger via an inlet manifold to eight identical parallel arrays. Each array comprises ten rows of absorbing panels, each row consisting of four panels connected in series. An array has two (upper and lower) headers. From the outlet array headers the heated fluid flows into the heat exchanger via an outlet manifold.

The uniform distribution of heat transfer fluid over arrays is controlled by means of valves in accordance with the temperature at the outlets of lower array manifolds.

Solar collector orientation. Parameters determining the collector orientation are the tilt, α , and the azimuth of the horizontal projection of the normal to the collector surface measured from the north-south direction, ψ_n . These angles are accepted to be of 20° and 30° respectively. Space restrictions in the vicinity of the existing boiler plant were one of the main factors that dictated the selection of the orientation which was less advantageous than the south orientation (ψ_n =0). It may be expected, that these restrictions would be frequently encountered when reconstructing boiler plants situated in residential regions where space limitations exist.

According to calculations, the relative decrease in total insolation on the absorbing surface with 30° azimuth and 20° tilt angles is insignificant as compared to south orientation. This decrease does not exceed 1 per cent at sun declination of 23.5^o (June 22) and is about 4 per cent during the second half of September.

The solar collectors are situated near the hotel boiler plant on two similar trusses (four absorber arrays on a truss). To minimize the area needed for the installation, the trusses are elevated over the ground on the height of 3.8-6.7m.

Program objectives

The primary objectives of the project are to:

- demonstrate the practicality of integrated solar hot water systems added to existing boiler plants;
- accumulate operational experience and appraise the economical viability of the reconstruction of operating boiler plants;
- conduct experimental studies and prepare recommendations for designing similar systems.

The main objective of the program of experiments and tests are to:

- identify possible modes of system operation;
- modify conceptual thermal and fluid flow schemes to improve efficiency of solar and conventional production of thermal energy in a single integrated installation;
- acquire data necessary for the comparison between test and simulation results;
- determine efficiencies of converting and transmitting solar energy in separate components and in the system as a whole.

The system adjustment and start-up are to take place in 1977; the tests are planned for 1978.

REFERENCES.

1. Melentiev, L.A. The Influence of Heat Energy Production on the U.S.S.R. Energy Balance. The presentation to Scientific and Technological Conference devoted to Fiftieths Anniversary of the Introduction of district heating plants into the National Economy of the U.S.S.R. December 1975, Leningrad.

2. Chernukhin, A.A., and Flakserman, Yu.N., Economic structure in the U.S.S.R. power industry, The Second Edition, Energia, Moscow, 1975 (Russ).

J. C. McVeigh
Brighton Polytechnic
United Kingdom

The UK government budget for the year 1976 was estimated to be about £300,000 but in February 1977 a very substantial increase in solar research and development funding was announced by the government. A sum of £3.6 million has been allocated over 4 years, with the main emphasis on solar water heating applications for both domestic and space heating in buildings. The UK is also actively participating in several programmes of the European Commission.

Solar space heating

The importance of the building sector in our national energy pattern can be demonstrated from available statistics on energy usage which show that some 40 to 50% of the total usage of primary energy can be attributed to building services (3). The demand on our conventional energy supplies can be reduced by increasing the use of solar energy and by adopting energy conservation techniques. An indication of the many possibilities for reducing the energy consumption in this sector was given by the Building Research Establishment (4) who predicted that over 15% of the total usage of primary energy could be saved by measures which would not impair environmental standards.

Completed space heating schemes

The first three buildings listed in Table 1 below have been extensively reported and appropriate reference sources are given. Their location is indicated by the reference letter in Figure 1, which also gives the mean annual values of wind and solar energy in MJ/m2.

The Anglesey Solar Bungalow has not been systematically monitored but is claimed to be unique in two respects:

(a) it is believed to be the first privately built house in Britain to have an active solar space heating system and
(b) the first house to have solar space heating by underfloor circulation.

Full details of its performance were given in the 4th number of Sun at Work in Britain (12). The Granada House is in many respects the most significant solar house in Europe. The stages in the conversion of an old coach house into a solar heater energy-conserving modern dwelling were first shown early in 1976 in a Granada TV series and two accounts of the work have been published (13, 14). Its significance lies in the relative simplicity and elegance of the approach to every aspect of the conversion and the fact that virtually unskilled labour was used at every stage. A modified Thomason-type trickle collector occupies the entire south-westerly roof area.

The Wates house at the National Centre for Alternative Technology, Machynlleth, was designed more to have a substantially reduced energy demand rather than to have its energy supply from ambient sources (15). Its energy demand has been calculated to be about one fifth that of a similar sized conventional sized house, with a U-value of 0.075W/m2 °C for the walls. A feature which attracted general interest was the quadruple glazing on the windows. Space heating is provided by an air-to-air heat pump and another heat pump is used for the domestic hot water system, operating between the 40 gallon hot water tank and a 60 gallon waste water tank below the ground floor.

Table I
Completed space-heating schemes

Title	*Date*	*Location*	*Description of special features and references*
The Curtiss House	1956	Rickmansworth, Herts. (A)	Passive space heating system with large south-facing double-glazed windows, heat pump. [5, 6]
St. Georges School, Wallasey	1960	Wallasey, Merseyside. (B)	Passive space heating system with large south-facing double-glazed windows. High thermal capacity and well insulated. [7, 8, 9]
The Milton Keynes House	1975	Milton Keynes, Bucks. (C)	Combined solar water and space heating system in a standard terrace ouse. Closed water circuit. [10, 11]
The Anglesey Solar Bungalow	1975	Malltraeth, Anglesey. (D)	Combined solar water heating with underfloor space heating in a new bungalow. Closed water circuit. [12]
The Granada House	1976	Macclesfield, Cheshire. (E)	Combined solar water and space heating in a major conversion of an existing building. Trickle open water circuit. [13, 14]
The Wates House	1976	National Centre for Alternative Technology, Machynlleth, Wales. (F)	Very high insulation, heat reclaim systems, heat pump. Independent of mains and powered by wind generator. [15]
NCAT Exhibition Hall and Offices	1976	National Centre for Alternative Technology, Machynlleth, Wales. (F)	Conversion from old workshop with 200 m^2 floor area, trickle collector, underfloor space heating, three interconnecting heat storage tanks. [15]
The Blyth House	1976	Blyth, Notts. (G)	The first design with integrated solar roof panels for water and space heating. [16]
The Washington Low Energy Project	1977	Washington New Town, Co. Durham. (H)	A low energy housing scheme with 33 houses, high insulation and mechanical ventilation. [17, 18]

Fig. 1

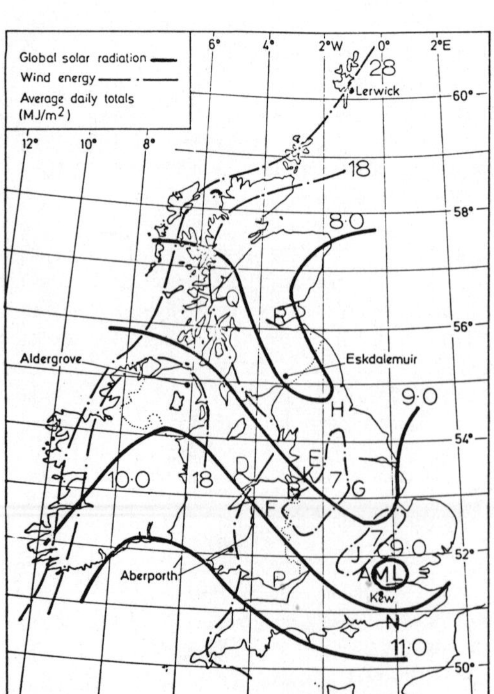

Annual mean global daily irradiation (MJ/m^2) and daily wind energy available at the rotor (MJ/m^2)

The Blyth house represents an important step towards the mass-produced solar house, as it was built speculatively as a commercial venture. The solar panels were specially designed to be set down between rafters at 600mm centres and are fitted into insulated compartments formed by boarding below the rafters. The tops of the compartments are glazed directly on to the rafters, so that the glass surface was below the level of the tiles, facilitating traditional flashing from the tiles to the glass surface. Standard lists of equipment based on this principle are now available (16). The Washington New Town low energy project began early in 1973 and its main aim was to apply existing knowledge to the design of housing to reduce the annual cost of energy to the occupier and at the same time either maintain existing capital costs or not increase them by more than 10% (17). Thirty-three houses are involved in the project and all are highly insulated with controlled ventilating systems. The solar radiation incident on the roof, together with heat transferred through the first floor ceilings, preheats the input air to the space-heating system. Monitoring is being carried out by the Building Science Section of the School of Architecture, University of Newcastle upon Tyne (18).

Space heating schemes under construction

These schemes range from the highly sophisticated, fully instrumented approach of a major Government research establishment to a modest conversion of an existing pre-war terraced house, and are listed in Table II.

Table II
Space heating schemes under construction

Title	*Estimated Completion Date*	*Location*	*Description of special features and references*
BRE low energy houses	1977 (J)	Building Research Establishment, Garston, Herts.	Three wooden-framed, 5 person terrace houses, covering solar energy, heat pump and heat reclaim systems. [19, 20, 21]
Bebington Environmental Housing Scheme	1977 (K)	Higher Bebington, Merseyside.	Nine houses with double-glazed south-facing solar wall, up to 60% energy savings predicted. [22]
Islington Solar Conversion	1977 (L)	London Borough of Islington	A pre-war terraced house being fitted with solar-assisted water and space heating. [23]
North London Solar House	1977 (M)	Outskirts of London (Barnet).	A private house with an integrated solar roof (45 m^2) and high insulation. [24]
Integrated Solar Dwelling	1977 (N)	Brighton Polytechnic	An experimental autarchic single-storey house designed to be independent of all mains services. [25]

The three experimental houses being built at the Building Research Establishment have been designed to demonstrate various applications of energy saving techniques outlined in their 1975 report (2). One house will be primarily devoted to solar energy studies and the system includes a 20m2 conventional solar water heater and three heat pumps (19, 20, 21). The Higher Bebington houses were designed by Peter Greenwood, Director of Building Science at the University of York and Dr. H. Ward. They are the first houses in the UK to incorporate a double-glazed solar wall to absorb solar radiation which will be re-radiated into the house. The performance of the nine solar heated houses will be compared with five conventionally built houses on the same site. Estimated energy savings of 30-60% are predicted. One solar house and one traditional house will also have a flat plate solar collector for domestic water heating (22).

The Islington solar conversion focusses attention on the existing housing stock and is the first Science Research Council project on an older house. A Georgian terraced house has been selected from Islington Council's rehabilitation programme and will have both solar water and space heating (23). The North London solar house at Barnet has a completely integrated solar water heating system for both domestic and space heating. The storage capacity is relatively small - approximately 4500 l (24). The Brighton Integrated Solar Dwelling is in a class of its own, as the architect, John Shore, has designed it to be completely independent of all services (25).

Monitored solar water heating systems

There are several commercial companies active in this field, but the main support for monitored solar water heating activities comes from the Universities and Polytechnics. The University of Cardiff (26) has completed the installation of solar heaters in a new block of student appartments (P). The Department of Architecture and Building Science at the University of Strathclyde are investigating several solar water heating and low energy projects in Scotland (27) including a low cost solar system at a youth summer camp near Aberfoyle (Q). The Napier College, Edinburgh (28) are involved in a project with a commercial company in which 70m2 of a new type of collector (vacuum formed aluminium sheet fitted round a serpentine stainless steel pipe) will form part of the roof of a new student hostel in Fife (R).

Acknowledgements

The author would like to thank all those who contributed either directly or indirectly to this paper, including M. Armor (Design and Materials Ltd), Professor B. J. Brinkworth (Cardiff), D. Gerassi (Distrimex Ltd), P. Greenwood (York), Dr. B. Justin (Pilkington Bros. Ltd), Kevin McCartney (AA School), A. W. K. McGregor (Napier College), Dr. R. W. Todd (NCAT), Dr. S. J. Wozniak (BRE) and Dr. P. Yaneske (Strathclyde).

References

(1) J.C. McVeigh, 'Developments in Solar Energy Utilization in the United Kingdom', conference 'Le Chauffage Solaire de Base', Charleroi (1975).

(2) UK Section, ISES, 'Solar Energy': A UK Assessment (1976).

(3) J. Dick, 'Ambient Energy in the Context of Buildings', Ambient Energy and Building Design, Construction Industry Conference Centre (1977).

(4) Energy Conservation, 'A Study of Consumption in Buildings and Possible Means of Saving Energy in Housing', Building Research Establishment Working Party Report CP 56/75 (June 1975).

(5) E.J.W. Curtis and M. Komedere, 'The Heat Pump', Architectural Design (June 1956).

(6) E.J.W. Curtis, 'Solar Energy Applications in Architecture', Department of Environmental Design, Polytechnic of North London, (February 1974)

(7) M.G. Davies, 'Model Studies of St. George's School, Wallasey'. JIHVE, 39, 77 (July 1971).

(8) M.G. Davies, N.S. Sturrock and A.C. Benson, 'Some Results of Measurements in St. George's School, Wallasey', JIHVE, 39, 80 (July 1971).

(9) M.G. Davies, 'The Contribution of Solar Gain to Space Heating', Sun at Work in Britain, 3, 11, (June 1976).

(10) S.V. Szokolay, 'Design of an Experimental Solar Heated House at Milton Keynes', UK ISES Conf. on Low Temperature Collection of Solar Energy, (April 1974).

(11) 'Solar Heated House in Milton Keynes', Milton Keynes Development Corporation, Wavendon Tower, Wavendon, Milton Keynes, MK17 8LX, UK.

(12) J.B. Wright, 'The Anglesey Solar Bungalow - Energy Costs', Sun at Work in Britain, 4, 11, (December 1976).

(13) T.P. McLoughlin, 'The House of the Future', TV Times Family Series (1976).

(14) B. Trueman and D.R. Wilson, 'Experience from "House of the Future" Project of Granada TV', Ambient Energy and Building Design, Construction Industry Conference Centre (1977).

(15) R.W. Todd, 'Low Energy Housing at the National Centre for Alternative Technology', ibid.

(16) Information from Design and Materials Ltd., Carlton in Lindrick Industrial Estate, Worksop, Notts., S81 9LB.

(17) A.C. Hardy, 'Washington Housing', Building Science Section, School of Architecture, University of Newcastle upon Tyne (1975).

(18) 'Research in Domestic Heating', Building Services Engineer, 45, No. 1, A8, (April 1977).

(19) K.J. Seymour-Walker, 'Low Energy Experimental Houses', BRE News, No. 34, 12, (1975).

(20) K.J. Seymour-Walker, 'BRE Proposals for Experimental Low Energy Houses', Conf. European Solar Houses, UK ISES (April 1976).

(21) K.J. Seymour-Walker, 'Experimental Low Energy Houses at the Building Research Establishment, Ambient Energy and Building Design', Construction Industry Conference Centre (1977).

(22) Information from Peter Greenwood, Director of Building Science, Institute of Advanced Architectural Studies, University of York, The King's Manor, York, YO1 2EP.

(23) Information from Kevin McCartney, The Architectural Association, School of Architecture, 34-36 Bedford Square, London, WC1B 3ES.

(24) Information from Distrimex Ltd., 88 The Avenue, London, NW6 7NN.

(25) Information from John Shore, Department of Civil Engineering and Building, Brighton Polytechnic, Brighton, BN2 4GJ.

(26) Information from Solar Energy Unit, Department of Mechanical Engineering and Energy Studies, University of Cardiff, P.O. Box 97, Cardiff, CF1 1XP.

(27) Information from Dr. P. Yaneske, Department of Architecture and Building Science, 131 Rotten Row, Glasgow, G4 ONG.

(28) Information from A.W.K. McGregor, Napier College, Collington Road, Edinburgh, EH10 5DT.

Flat plate collectors

3

3.1 Flat plate collectors for use in buildings

A.V. Spyridonos
Nuclear Research Center "Demokritos"
Aghia Paraskevi Attikis, Greece

Abstract

In the present work a general review is presented of the state of the art of the flat-plate collector.

The flat plate collector is the main part of an installation for the exploitation of the solar energy.

On the other hand a suitable auxiliary system was developed for the transfer of heat into the interior of the building.

This review is submitted after an invitation from the International Conference on Solar Building Technology/North East London Polytechnic/ UNESCO.
In a brief introduction the increasing interest in the conversion of the solar energy is underlined.

The solar energy conversion can be accomplished by the collectors. There are two types of collectors. The flat-plate collector and the concentrator.

By means of the flat-plate collector thermal electrical and physico-chemical conversions can be accomplished.

In the present review we examine the thermal conversions.

In general, the flat plate collector of thermal conversion is a black or gray body and its application is much more extensive than that of the other collectors.

The different parts of this type of collector, namely the absorber, the heat transfer fluid, the cover, the selective surfaces etc. will be described.

On the other hand flat-plate collectors which concentrate the solar radiation were presented.

We think that the characteristics of these collectors make them resemble more to the flat plate collectors than to the concentrators.

A flat plate collector in which the absorber consists of lines of parallel tubes is analyzed.

A flat-plate collector in which the absorber consists of a serpentine tube will also be described.

The flat-plate collector in which the absorber consists of a series of heat pipes has created a significant interest. This type of collectors presents certain particularities with respect to its function. In this type the vacuum technique was also applied for reducing the heat losses.

In fact, during the last decade the development of the flat-plate collector can be characterized quite significantly.

There is a tendency to increase the yield of the flat plate collector by reducing the heat losses from the cover.

Another tendency is to increase the yield by improving the heat transfer from the absorber.

Resumé

Par ce travail nous présentons une revue de l'aspect technologique de l'insolateur plan.

L'insolateur plan est la partie principale d'une installation solaire.

L'insolateur est lié avec un système auxiliaire approprié, capable de transporter la chaleur à l'intérieure d'un local.

Dans cette revue nous nous occuperons seulement de l'insolateur plan, et nous laisserons de côté le système auxiliaire.

Cette revue, est réalisée, d'après l'invitation du "International Conference on Solar Building Technology/North East London Polytechnic/UNESCO.

Nous fairons une introduction courte en soulignant que l'interêt sur la conversion du rayonemment solaire connaît dernièrement une vive évolution.

La conversion du rayonnement solaire peut être realisée dans le sens le plus général par des collecteurs c'est-à-dire par des insolateurs plans et par des concentrateurs.

L'insolateur peut être à conversion thermique, à conversion électrique et à conversion physicochimique, dans le sens la plus général du mot.

Dans cette revue nous nous occuperons de l'insolateur plan à conversion thermique.

L'insolateur à conversion thermique est un corps noir en général ou gris qui connaît dernièrement une application beaucoup plus vaste que tout autre collecteur.

Nous ferons la description de ce type d'insolateur. Plus particulièrement nous examinerons à part l'absorbeur, le fluide caloporteur, la couverture, la surface sélective, le coffrage etc.

Par ailleurs des insolateurs plans à concentration qui sont déjà presentés sont inclus avec l'insolateur plan parce que cet insolateur qui porte des séries de miroirs au dessus de son absorbeur a des caractéristiques plus proches à l'insolateur plan qu'aux concentrateurs proprement-dits.

Nous ferons l'analyse de l'insolateur plan qui a un absorbeur composé par des tubes en parallèle.

Nous croyons aussi necessaire de présenter l'insolateur qui a un absorbeur composé par un tube en serpentine.

L'insolateur dont l'absorbeur est composé par des caloducs connaît dernièrement un trés vif interêt.

Ce type d'insolateur présente certaine particularité quant à son fonctionnement.

L'application du vide pour réduire les pertes de la chaleur a été déjà appliquée à ce type d'insolateur.

Nous remarquons donc que dans la dernière décennie l'évolution de l' insolateur plan est considérable.

Il y a une tendence à augmenter le rendement de l'insolateur au moyen de la diminution des pertes thermiques par la couverture. Une autre tendance est d'augmenter le rendement de l'insolateur au moyen de l'augmentation du transfert thermique à l'intérieur de l'absorbeur.

Introduction

Par ce travail nous présentons une revue de l'aspect technologique de l'insolateur plan.

L'insolateur plan est la partie principale d'une installation solaire.

L'insolateur plan est lié à un système de chauffage auxiliaire qui sert à la distribution de la chaleur à l'intérieur d'un local.

Un travail considérable a déjà été réalisé pour l'évolution de l'insolateur et du système auxiliaire.

Il y a cinq raison qui ont causé cette évolution:

1) La crise mondiale de l'énergie
2) L'énergie solaire est inépuisable
3) L'énergie solaire est gratuite
4) L'énergie solaire est non polluante
5) L'énergie solaire peut être soumise à une exploitation économique

L'énergie solaire est fournie sur place.

Par ailleurs il faut noter que:

1) L'énergie solaire est intermittente
2) L'énergie solaire est influencée par des condition météorologiques
3) L'énergie solaire doit être stockée afin que son utilisation soit rationnelle
4) L'énergie solaire n'est pas disponible au même degré sur toute la terre
5) L'énergie solaire a un flux, qui est considéré bas, par conséquent il est nécessaire d'avoir des grandes surfaces quant à son captage

Considérant aujourd'hui les avantages et les inconvenients de l'énergie solaire, un certain progrès à été accompli.

Par ailleurs il est vrai qu'environ les 30% de l'énergie consommée dans les pays industrialisés est utilisée pour le chauffage de locaux et l'eau sanitaire.

C'est à cette catégorie de chauffage que nous pouvons intervenir efficacement actuellement par l'énergie solaire.

L'insolateur plan est connue depuis longtemps. Il est appuru d'abord avec un absorbeur qui avait une nappe de tubes ou un absorbeur avec un tube en serpentine ou simplement un absorbeur plat. Cette configuration a été conservés jusqu'à nos jours. Par ailleurs des multiples variantes ont été développées entre-temps.

L'insolateur qui a une nappe de tubes a été étudié par Hottel et Woerzt en 1942 [1]. Un pas important a été realisé quelques années plus tard par les travaux de Bliss Jr. en 1959 [2] et ensuite par Whillier, [3] [4], [6]. L'insolateur qui a un tube, en serpentine, a été étudié tout récemment par Abdel-Khalik en 1975 [30].

L'introduction de la technique du vide et l'adaptation du caloduc à l'absorbeur ont été réalisées tout récemment.

L'insolateur à concentration de conversion thermique a été appliqué récemment par Sevcik et Winston [14], par Bannerot [19] et d'autres. Cet insolateur port des miroirs avec une génératrice linéaire. Une ou plusieurs réflexions du rayonnement solaire incident sont effectuées avant d'être concentrés et absorbées par l'absorbeur.

Un grand effort a été réalisé pour l'isolation de l'insolateur par la face avant et la face arrière. En ce qui concerne la face arrière de l'absorbeur, l'isolation évolué au fur et à mesure des isolants disponibles dans le marché soit: la laine de verre, le liège, le polystyrène etc.

En ce qui concerne la face avant le situation est beaucoup plus compliquée. La face avant de l'insolateur reçoit d'une part le rayonnement solaire et d'autre part en reemet une partie, l'insolateur perd une quantité de chaleur par la face avant par conduction et par convection. C'est pour cela qu'on couvre l'absorbeur d'une matière transparente qui est capable en même temp de réduire les pertes par rayonnement et par convection.

Le verre est une matière déjà appliquée, d'autres encore, matières d'origine plastique, sont en application ou en étude. Les matières plastiques présentont une déterioration quand elles sont frapées par le rayonnement ultraviolet, d'autres plastiques qui résistent au rayonnement ultraviolet sont actuellement chers.

L'absorption du rayonnement solaire par l'absorbeur dépend de l'état de sa surface. Pour augmenter l'absorption de l'absorbeur des traitements physicochimiques ou des pigments sont appliquées à la surface de l'absorbeur.

D'ailleurs des traitments mécaniques sont appliquées comme la rugosité de la surface de l'absorbeur ou l'ajutage de surface antirayonnantes. De très intéressants travaux ont été réalisées par Tabor [18], Sparrow [31], Francia [5] et le groupe de l'Université d'Aix-Marseille [20].

Des nombreuses études ont été réalisées avec des differents matériaux de construction.

D'intéressantes analyses sont en évolution quant au comportement hydraulique, thermique et mécanique de l'insolateur.

En définitive l'insolateur, tel qu'il existe actuellement est en pleine expansion et son application industrielle est très prochaine.

L'insolateur à conversion thermique

L'insolateur peut être à conversion thermique, à conversion électrique et à conversion physicochimique, dans le sens le plus général du mot.

Dans cette revue nous nous occuperons de l'insolateur plan à conversion thermique.

L'insolateur à conversion thermique est un corps noir en général ou gris qui connaît dernièrement une application beaucoup plus vaste que tout autre collecteur.

Nous férons la description de ce type d'insolateur. Nous avons aussi pensé à inclure à cette revue l'insolateur de conversion thermique et à concentration.

Un insolateur à conversion thermique expose au rayonnement solaire une surface qui en absorbe une partie maximum et reemet à la température qui découle de cette exposition une énergie minimum. La chaleur obtenue est transmise à un fluide à chauffer puis éventuellement à un élèment de stockage de manière à rendre son utilisation plus indépendante de sa production (Brenniere, Perrot) [23] [20].

Par ailleurs l'insolateur à concentration de conversion thermique se compose de miroirs et de surfaces sélectives de dimensions réduites. Ces miroirs ont une génératrice, qui est une ligne droite. Les dimensions de ces miroirs sont telles qu'elles peuvent être contenues dans un coffrage de dimensions similaires avec un coffrage de l'insolateur à corps gris. C'est alors que l'insolateur à conversion thermique et l' insolateur à concentration de conversion thermique ont des caractéristiques similaires en comparaison du concentrateur soit:

1) Il y a la possibilité d'augmenter la surface de l'insolateur avec une précision réduite.
2) Il y a la possibilité d'intégrer l'insolateur dans un toit d'un local.
3) L'orientation de l'insolateur n'est pas tellement critique.
4) L'entretien de l'insolateur est plus aisé.

L'insolateur à conversion thermique se compose essentiellement des parties suivantes:

1. La couverture
2. L'absorbeur
3. Le liquide caloporteur
4. L'isolation
5. Le coffrage

La couverture

L'insolateur est couvert par une matière, en général transparente. Comme matière transparente sont utilisées le verre et les films plastiques.

La couverture peut avoir une, deux, trois ou aucun vitrage. Celui-ci dépend de la température moyenne de l'absorbeur et de la sélectivité de la surface de l'absorbeur.

Le verre au faible pourcentage en oxyde de fer est très satisfaisant du point de vue des proprietés optiques. Par ailleurs des plastiques Tedlar, Flexxan sont déjà appliqués.

Il y a la possibilité d'utiliser seulement des vitrages espacés de quelques mm ou d'utiliser des vitrages et des plastiques ensemble. La distance entre chaque vitrage ou chaque matière transparente joue un rôle dans le rendement de l'insolateur. On prévoit, en général, une certaine distance entre chaque plaque transparente de l'ordre de 25 mm.

Dans la pratique le verre est surtout utilisé à cause de sa grande transmission (90%) des courtes longueurs d'onde que forme la rayonnement solaire et à cause de sa très faible transmission des grandes longueurs d'onde formant l'émission de la plaque absorbante.

La transmission du verre diminue quand le contenue en oxyde de fer augmente. La couleur verte du verre devient alors plus foncées.

La transmission au rayonnement diffus est sensiblement constante, celle du rayonnement direct est fonction de l'angle d'incidence.

Par des traitements de la couverture il est possible d'augmenter la transmission de celle-ci: en diminuant la perte par réflexion par des traitement anti-reflets.

Des expériences concluantes ont été réalisées avec des dépots de fluorure de magnésium MgF_2 sur du verre ordinaire. Pour une incidence normale, le coefficient de réflexion augmente pour les grandes longueurs d'onde, c'est-à-dire pour les radiations thermiques. Des expériences similaires ont été réalisées avec l'oxyde d'étain (SnO_2) et l'oxyde d'indium (In_3O_2).

Les films ou pellicules en plastique possèdent également une transmission importante pour les courtes longueurs d'ondes, mais la plupart des variétés utilisées ont une bande de transmission au milieux du spectre de radiation thermique et elles peuvent avoir des transmissions pour les grandes longueurs d'onde de l'ordre de 40% [18]. Les plastiques posent également le problème de leur dégradation aux ultraviolets mais présentent d'avantage de résister à la grêle et autre chutes: en films très minces ils sont complètement flexibles.

La transmission du vitrage ou des films plastiques peut être réduite quand il y a de la poussière.

Le recouvrement de la couverture par la poussière ne pose pas, en général, de grands problèmes pour le rendement de l'insolateur, aux moyens et aux grands angles de latitude. Ceci n'est pas absolument vrais car il dépand des conditions locales. Par contre dans un climat aride ou semi-aride on doit tenir compte de la poussière. Le facteur dont on doit tenir compte, dû à la poussière, est variable avec le temps; mais on peut considérer avec une certaine assurance qu'après 30 jours le facteur reste constant (Garg) [21]. La transmission est alors fonctions de l'inclinaison de l'insolateur et peut varier de 95% à 30% ce qui est trop important dans les pays de la zone de $\pm 40^\circ$ de latitude.

Il faut aussi mentionner que la facteur de la transmission dû à la poussière est plus petit pour les plastiques que pour les vitrages.

L'absorbeur

L'absorbeur est la partie principale de l'insolateur. C'est sur lui que la conversion thermique a lieu.

Nous distinguons deux catégories d'absorbeur:

a) L'absorbeur corps-gris
b) L'absorbeur à concentration

L'absorbeur crops-gris

La conversion du rayonnement solaire en chaleur est réalisée au niveau de la surface de l'absorbeur.

L'absorbeur absorbe le maximum de radiation transmise au travers du vitrage, en réemetre le minimum et transferer le maximum de l'énergie retenue au fluide caloporteur (Breniere) [23].

L'absorption du rayonnement solaire par la surface de l'absorbeur peut être augmentée par les méthodes suivantes:

1. Par un traitement spécial de la surface de l'absorbeur [18].
2. Par des pigments qui couvrent la surface de l'absorbeur.

3. Par la rugosité de la surface de l'absorbeur Fig. 1 [40] [31] [14]
4. Par l'usage de structures antirayonnantes Fig. 2 [5] [22] [14] [24] [26].

Fig. 1

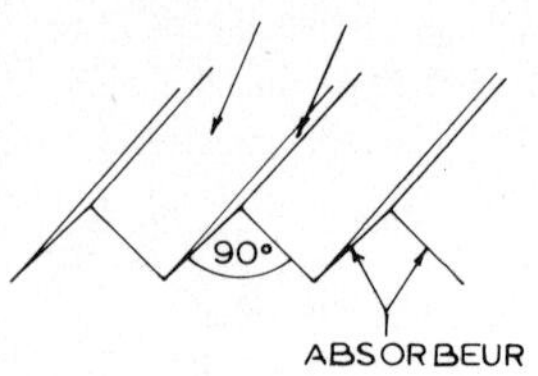

Fig. 2

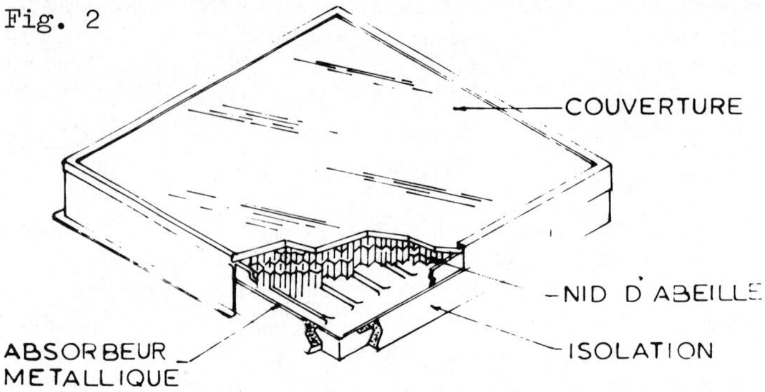

De nombreuses études ont été faites dans le monde pour améliorer les propriétés de la plaque absorbante, particulièrement pour diminuer les pertes par rayonnement qui deviennent importante lorsque la température de la plaque augmente au-delà de 80°C.

Des surfaces sélectives absorbantes dans le spectre visible et faiblement émissives dans l'infrarouge permettent de réduire les pertes.

Par un traitement special de la surface absorbante on peut avoir un absorbeur résultant de l'interférence de plusieurs couches déposées sur une base peu émissive. Le dépôt rend l'ensemble absorbant pour le rayonnement solaire et transparent pour les rayonnements de grande longueur d'onde; aussi, on peut avoir un absorbeur résultant du dépôt d'un matériau transparent dans le visible et réflecteur dans l'infrarouge sur une base absorbante dans tout le spectre. Par la déposition d'un pigment sur une surface on peut avoir un absorbeur ayant un état de surface particulier donnant lieu à des réflexions multiples et dont le bilan correspond à un coefficient d'absorption α très grand et à une émissivité ε faible.

C'est surtout le premier type de surface qui a été le plus developpé jusqu'à present.

Signalons que le traitement del la tôle d'acier peut être realisé avec le procédé de "Nickel-Black", la tôle d'aluminium par la déposition d'oxyde de cuivre et la tôle de cuivre par la déposition d'oxyde de cuivre.

Il faut pourtant protéger ces traitements contre la condensation et l'excés de la température de l'absorbeur.

Par la rugosité de la surface de l'absorbeur peut augmenter son absorption. Des travaux ont été réalisé par Sparrow et alii [45]. En général la rugosité est effectuée en séries de V d'angle 60° et 90° (Fig.1). On a aussi des rugosités en forme de V-arc gothique, et V-arc parabolique.

Par la rugosité on peut avoir des réflexions multiples du rayonnement solaire incident et par cette manière augmenter l'absorption.

La sélectivité de la surface de l'absorbeur diminue son pouvoir émissif, mais ne permet pas de diminuer les pertes par convection à la surface de l'absorbeur. Il est possible de réduire les pertes par rayonnement et les pertes par convection en recouvrant l'absorbeur d'une structure antirayonnante. Des nids d'abeille, des tubes de verre placés perpendiculairement à la surface de l'absorbeur peuvent former une structure antirayonnante (Francia [5], Perrot [20], Group Université Aix-Marseille [20], Hollands [26]). Ces tubes sont transparents pour le rayonnement solaire mais absorbent presque intégralement le rayonnement IR émis par l'absorbeur. Il vont donc s'échauffer et rayonner mais leur structure est telle qu'ils rayonneront sur eux-mêmes. En fait ces cellules réalisent un effet de serre meilleur qu'une simple feuille de verre. En ce qui concerne le matériaux à utiliser pour la fabrication de ces cellules, le verre paraît être un matériau de choix de par ses proprietés, mais on se heurte à de grandes difficultés technologiques lors de la réalisation. Pour le moment le matériau utilisé est le polyterephtalate d'éthylene (Mylar).

Les matériaux utilisés, pour la construction de l'absorbeur sont surtout les métaux: acier, cuivre, aluminium, mais il est également possible d'utiliser les plastiques et le verre.

Les principales configurations de l'absorbeur corps-gris avec fluide caloporteur d'eau ou un liquide sont: [11] [14] [22] [23] [29].

1) Tube in strip, integré à la plaque (Fig. 3).
2) Tubes après gonflement (Roll Bond) (Fig. 4).
3) Tubes attachés au-dessus ou en-dessous (Fig. 5 et Fig. 6).
4) Les plaques de l'absorbeur sont rivetées ou soudées pour former des conduites.

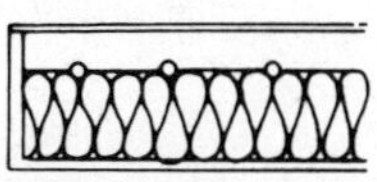
Fig. 3

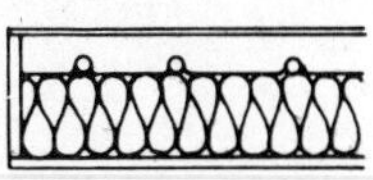
Fig. 5

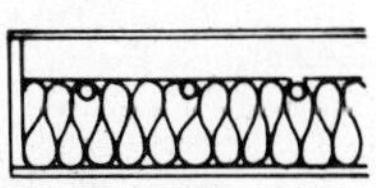
Fig. 6

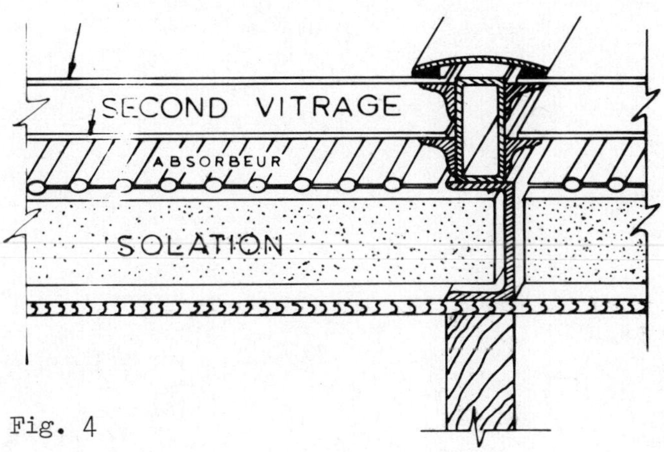

Fig. 4

Les principales configurations de l'absorbeur corps gris avec fluide caloporteur de l'air sont: [29] [27] [23].

1) L'absorbeur est une plaque plane
2) L'absorbeur est une plaque plane, l'air passe d'abord par le conduit supérieur et ensuite par le conduit inférieur. Ces deux conduits sont séparés par l'absorbeur.
3) L'absorbeur est une plaque à ailettes (Fig. 7) vers la face du conduit.
4) L'absorbeur est de type toil-Metrix, l'air passe à travers (Fig. 8).
5) L'absorbeur a des rugosités en forme de V (Fig. 9).
6) L'absorbeur est en plaque de verre clair et noir (Fig. 9a).

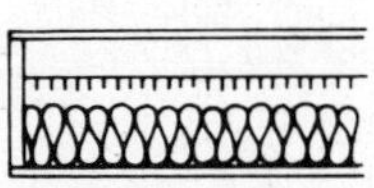
Fig. 7

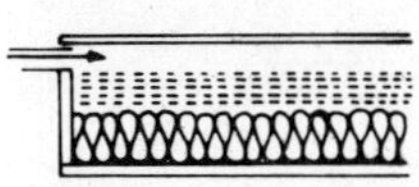
Fig. 8

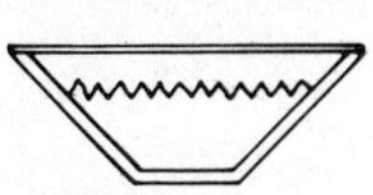
Fig. 9

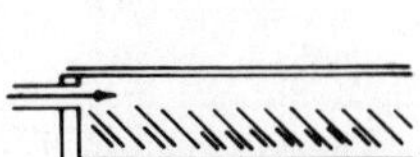
Fig. 9a

a) Insolateur avec une nappe des tubes

L'énergie absorbée par l'insolateur dépend de l'insolation I du facteur de transmission de la couverture τ du facteur d'absorption de l'absorbeur α, du facteur de poussière D, du facteur d'ombre S [4] [7] [11] [13] |22|.

Les pértes de chaleur q_L de l'insolateur dépendent de la température moyenne de l'absorbeur t_c, du facteur d'émission ε de l'absorbeur et des conditions ambiantes: la température de l'air ambiant t_a, la température effective du ciel t_s, de la vitesse du vent V, du nombre des plaques transparents et de leur distance entre eux, du matériel des plaques transparentes et des pertes de chaleur de la face arrière de l'absorbeur.

L'isolation de la face arrière peut être augmentée de façon que les pértes par elle couvrent seulement quelques pourcentages des pertes totales.

La transmission du rayonnement solaire à travers une matière transparente dépend: [4]

$$\tau = \left[\frac{1-r}{1+r} \right] e^{-KL} \qquad (1)$$

1. de l'épaisseur L de la matière transparente que le rayonnement traverse.
2. Le facteur d'extinction $K=0{,}078\ cm^{-1}$ pour un bon verre et $K=0{,}314\ cm^{-1}$ pour un verre de mauvaise qualité (existence d'oxyde de fer).
3. Du facteur r de réflexion.

$$r = \frac{1}{2} \left[\frac{\sin^2(i-i')}{\sin^2(i+i')} + \frac{tg^2(i-i')}{tg^2(i+i')} \right] \qquad (2)$$

ou i et i' sont l'angle d'incidence et l'angle de réfraction.

Si nous avons plusieurs plaques transparentes nous aurons:

$$\tau_{1,2,3,4 \ldots n} = \left[\frac{1-r}{1+(2n-1)r} \right] e^{-[K_1L_1+K_2L_2 \ldots K_nL_n]} \qquad (3)$$

Dans les calculs il est plus convenable d'utiliser le produit transmissibilité-absorptivité Fe car le verre s'échauffe par l'énergie qui en absorbe. Pour un absorbeur $\alpha=0{,}95$ et $\varepsilon=0{,}95$ ayant un vitrage d'épaisseur 3,11 mm et de $K=0{,}314$ on aura

$$a_1 = 0{,}23, \quad F_e = 1{,}008\ \tau\alpha, \quad F_c = 0{,}795, \quad F_e = 0{,}817$$

Pour un absorbeur avec deux vitrage aux mêmes proprietés :

$$a_1 = 0{,}17 \qquad F_c = 1{,}012\ \tau_1\tau_2 a \quad \text{et} \quad F_c = 0{,}663$$

$$a_2 = 0{,}63 \qquad F_e = 0{,}724 \qquad (3b)$$

On aura en général

$$F_e = F_c + a_1(1-e^{-K_1L_1})+a_2\tau_1(1-e^{-K_2L_2}) \qquad (4)$$

Par ailleurs le facteur d'absorption d'une plaque peinte en noir varie avec l'angle d'incidence

angle d'incidence	0	10	20	30	40	50	60	70	80	90
α	0,96	0,96	0,96	0,95	0,94	0,92	0,88	0,82	0,67	0,00

Fig. 10

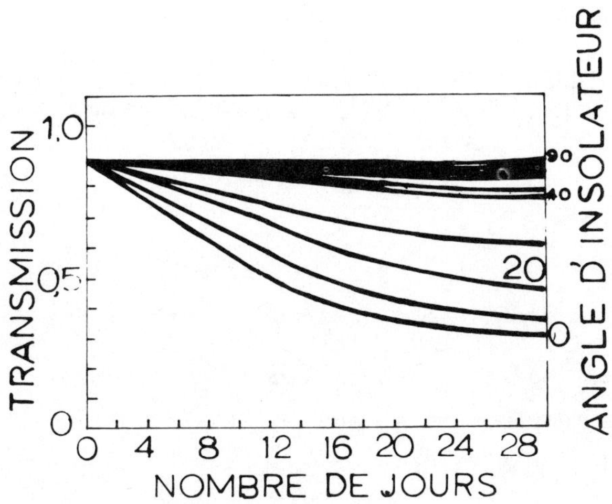

L'influence du nombre de plaques transparentes sur la transmission est:

Température	Nombre de plaques	transparentes
$t_c - t_a$	$\varepsilon = 0,9$	$\varepsilon = 0,2$
-5 ∿ 5°C	0	0
5 ∿ 30°C	1	1
30 ∿ 50°C	2	1
50 ∿ 75°C	3	2

Les pertes dues à l'accumulation de la poussière sur la couverture ont été étudiées par Garg [21] . Le résultat de son expérience est représenté à la fig. 10. On remarque que la transmission est fonction de l'angle de l'insolateur et du nombre des jours d'exposition. La transmission varie entre 30% à 90%.

Le facteur de l'ombre varie avec l'angle d'incidence. On peut considérer, avec une bonne approximation, un facteur d'ombre de l'ordre s=97% pris à midi.

Le rayonnement solaire global se compose de deux parties:

1. Le rayonnement direct I_D

2. Le rayonnement diffus I_d

Le rayonnement diffus n'est pas uniforme dans le ciel, sa brillance varie et il est plus fort an voisinage du soleil et à l'horizon.

Le rayonnement diffus est en général de l'ordre de 8-10% du rayonnement global mais dans des pays humides et industriels il peut être de l'ordre de 16-20% et dans certains cas il est encore plus.

Le rayonnement diffus est considéré pour un angle de incidence de 50°. La quantité d'énergie absorbée pour une surface est alors:

$$q_a = (F_e \cdot D \cdot S)_i I_D + (F_e D \cdot S)_{i=50} I_d \qquad (5)$$

et on peut en définitive calculer la quantité d'énergie absorbée pour l'équation (Whillier): [4][7]

$$q_a = 0.98\{F_e \cdot D \cdot S\}_i I \qquad (6)$$

où I est l'insolation, globale pour unité de surface et de temps corrigée avec le facteur d'angle, soit le facteur astronomique du soleil.

Les pértes de chaleur par la face avant sont définies par la relation suivante:

$$q_{up} = U_{up}(t_c - t_a)A_p \qquad (7)$$

où U_{up} est le coefficient du transfert thermique de l'absorbeur à l'air

t_c la température moyenne de l'absorbeur

t_a la température moyenne de l'air ambiant

A_p la surface de l'absorbeur

Les pértes par derrière sont définies par la relation suivante:

$$q_{re} = U_{re}(t_c - t_a)A_p \qquad (8)$$

où U_{re} est la coefficient du transfert thermique de l'absorbeur à l'air. U_{re} doit avoir la forme de

$$U_{re} = \frac{1}{\frac{\delta_{in}}{k_{in}} + \frac{\delta_{co}}{k_{co}} + \frac{1}{h_w}} \qquad (9)$$

où δ_{in}, δ_{co} sont les épaisseurs de l'isolation et du coffrage, k_{in}, k_{co} et h_w la conductivité thermique de l'isolation, la conductivité thermique de la matière avec laquelle est construit de coffrage et le coefficient de convection. Ce coefficient de convection est égal à:

$$h_w = 5{,}7 + 3{,}8V \qquad (10)$$

quand le vent refroidit la face arrière de l'insolateur, et V est la vitesse du vent en m/sec. Dans le cas où l'insolateur est protégé dans le toit on aura

$$h_w = 1.45(t_{co} - t_a)^{0,25} \qquad (11)$$

où t_{co} est la température moyenne du coffrage.

Les pértes de chaleur de l'insolateur par la circonférence sont éstimées par la relation suivante:

$$q_{ed} = 0{,}454(D_o P)\cdot(t_c - t_a) = U_{ed} A_c (t_c - t_a) \qquad (12)$$

où D_o est la profondeur du coffrage

et P son périmètre.

Les pértes de chaleur totales sont alors définies par la relation suivante:

$$U_L = U_{up} + U_{re} + U_{ed} \frac{A_c}{A_p} \qquad (13)$$

où A_p est la surface de l'absorbeur
et $A_c = D_o P$

Pour un insolateur bien dessiné on doit avoir 0,1 $U_{up}=U_{ed}$ et U_L 10 à 20% plus grand que U_{up}.

Nous revenons maintenant à la définition du coefficient de transfert de la chaleur U_{up}.

Nous distinguons deux cas:

a) La couverture comporte un vitrage

b) La couverture comporte deux vitrages:

Nous nous limitons ici car nous ne voulons par élargir beaucoup le têxte du cet article.

Couverture avec un vitrage

$$U_{up} = \frac{1}{\dfrac{1}{h_{c1}+E_{c1}h_{rc1}} + \dfrac{1}{h_w+\varepsilon_1 h_{r1s}\left(\dfrac{t_1-t_s}{t_1-t_a}\right)}} \qquad (14)$$

Couverture avec deux vitrages

$$U_{up} = \frac{1}{\dfrac{1}{h_w+\varepsilon_1 h_{r1s}\dfrac{(t_1-t_s)}{(t_1-t_a)}} + \dfrac{1}{h_{c2}+E_{c2}h_{rc2}} + \dfrac{1}{h_{21}+E_{21}h_{r21}}} \qquad (15)$$

où t_1 t_2 est la température du premier et du second vitrage

t_c est la température moyenne de 1 absorbeur

t_s est la température équivalante de l'air

t_a est la température de l'air ambiant

h_{c1}, h_{c2}, et h_{21} sont les coefficients de convection de la chaleur entre l'absorbeur et le vitrage, entre l'absorbeur et le second vitrage et entre les deux vitrages.

Ce coefficient est défini pour une température de 10°C pour la relation [11]

$$h_{10} = 1,14 \; \frac{\Delta T^{0,310}}{\ell^{0,070}} \qquad (16)$$

pour une température t on aura

$$\frac{h_t}{h_{10}} = 1-0,0018(\bar{t}-10) \qquad (17)$$

où $\bar{t}$ est la température moyenne entre les plaques planes, ΔT est la différence de température entre elles et ℓ leur distance.

La température équivalente t_s est définie par la relation suivante

$$t_s = t_a - 6 \qquad (18)$$

où h_{r1s}, h_{rc1}, h_{rc2} et h_{r21} sont les facteurs de rayonnement entre le vitrage et l'air, l'absorbeur et le vitrage, l'absorbeur et le second vitrage et entre les deux vitrages, h_{rxy} est défini par la relation suivante:

$$h_{rxy} = C \frac{t_x^4 - t_y^4}{t_x - t_y} \qquad (19)$$

où C est la constante Stefan-Boltzmann

où ε_1 est l'émissivité du vitrage

où E_{c1}, E_{c2} et E_{21} sont les facteurs d'émission entre l'absorbeur et le vitrage, l'absorbeur et le second vitrage et entre les deux vitrages, E_{xy} est défini par la relation suivante:

$$E_{xy} = \frac{1}{\frac{1}{\varepsilon_x} + \frac{1}{\varepsilon_y} - 1} \qquad (20)$$

x et y sont des indices.

Nous considérons l'absorbeur avec une nappe des tubes. C'est le modèle de Hottel, Bliss et Whillier; nous nous limiterons seulement à ce modèle et nous laisserons à part tous les autres. La théorie par ailleurs est la même et il suffit de quelques adaptations pour arriver aux autres modèles [1] [2] [3] [4] [7].

On définit d'abord un facteur d'efficacité des ailettes par la relation suivante voir fig. 11

$$F = \frac{tgh(aL')}{aL'} \qquad (21)$$

ou

$$a^2 = \frac{U_L}{kM} \qquad (22)$$

k est la conductivité thermique de l'absorbeur et M l'épaisseur de l'ailette.

Fig. 11

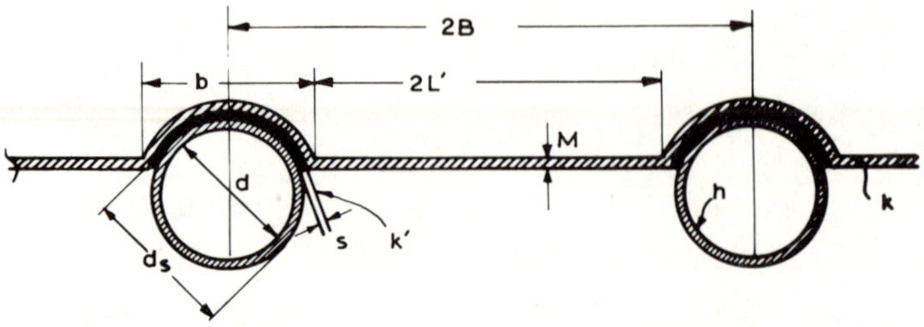

On définit le facteur d'efficacité de l'insolateur par la relation suivante:

$$F' = \frac{U_o}{U_L} \tag{23}$$

où U_o est le coefficient global de transfert de la chaleur, égal à:

$$\frac{1}{U_o} = \frac{2B}{\pi dh} + \frac{2B}{\pi dk/s_o} + \frac{2B}{R} + \frac{2B}{U_L(d_s+2L'F)} \tag{24}$$

où 2B est la distance des axes entre deux tubes
2L' est la distance entre deux tubes
d, d_s sont le diamètres intérieur et extérieur du tube et d_s-d/2=50
R est la résistance thermique de soudage
s est l'épaisseur du soudage
k' est la conductivité thermique du soudage
h est le coefficient de convection de chaleur à l'intérieur des tube

La résistance R est définie par la relation suivante:

$$R = \frac{k'b}{s} \tag{25}$$

On définit le facteur du flux par la relation

$$F'' = \frac{1-e^{-\frac{F'U_L}{GC_p}}}{\frac{F'U_L}{GC_p}} \tag{26}$$

où C_p est la chaleur spécifique du fluide caloporteur
et G le débit du fluide par unité de surface.

La relation (26) peut être exprimée aussi par la relation

$$F'' = F_R/F' \tag{28}$$

La fonction F" en fonction de la fonction $\frac{G \cdot C_p}{U_L F'}$ est presentée dans la figure 12.

Fig. 12

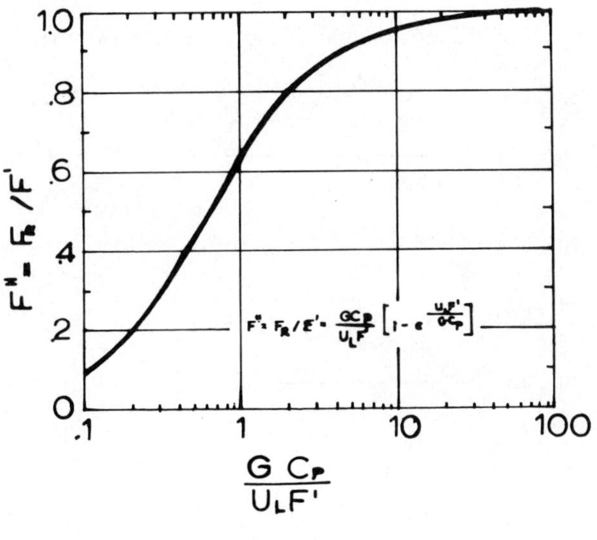

La chaleur qui est absorbée par un insolateur est définie par la relation suivante:

$$q_u = F''F'\left[q_a - U_L(t_1' - t_a)\right] \qquad (28)$$

où t_1' est la température d'entrée du fluide à la nappe des tubes.

Il faut noter que pour un écoulement laminaire d'un liquide à l'intérieur d'un tube de l'absorbeur nous avons [11]

$$Nu = \frac{RePrd}{4L_o} \ln\left[1 - \frac{2,654}{Pr^{0,167}\left(\frac{RePrd}{L_o}\right)^{0,5}}\right]^{-1} \qquad (29)$$

où Nu est le nombre de Nusselt

Re est le nombre de Reynolds

Pr est le nombre de Prandtl

et L_o la longueur du tube

Si l'écoulement est turbulent on aura:

pour les liquides

$$Nu = 0,0155Re^{0,83}Pr^{0,5} \qquad (30)$$

pour les gaz

$$Nu = 0,022R_e^{0,8}Pr^{0,6} \qquad (31)$$

Pour un insolateur refroidi par l'air, où une plaque est chauffée l'autre est froide on aura:

$$Nu = 0,0158Re^{0,8} \qquad (32)$$

Un autre aspect du transfert de la chaleur à l'intérieur de l'insolateur est la conductivité, de contact entre le tube et la plaque de l'absorbeur.

En général pour des tubes bien soudés ou bloqués contre la plaque de l'absorbeur nous avons C=1,73-8,65 w/mC pour le blocage et C=3,46-34,6 w/mC pour la soudure.

Pour un instant donné le rendement de l'insolateur est défini par la relation:

$$\eta = \frac{F''F'\left[q_a - U_L(t_1' - t_a)\right]}{I} \qquad (33)$$

Le rendement moyen est définit par le relation:

$$\eta_m = \frac{\int_{t_1}^{t_2} q_u dt}{\int_{t_1}^{t_2} I dt} \qquad (34)$$

b) Insolateur avec un tube en forme de serpentine

L'absorbeur avec un tube en serpentine était connu et appliqué depuis longtemps. C'est ainsi que Abdel-Khalik [30] définit un facteur de flux de l'insolateur F"

$$F'' = \frac{F_R}{F'} \qquad (35)$$

On définit d'abord un nombre sans dimensions, (Fig. 13)

Fig. 13

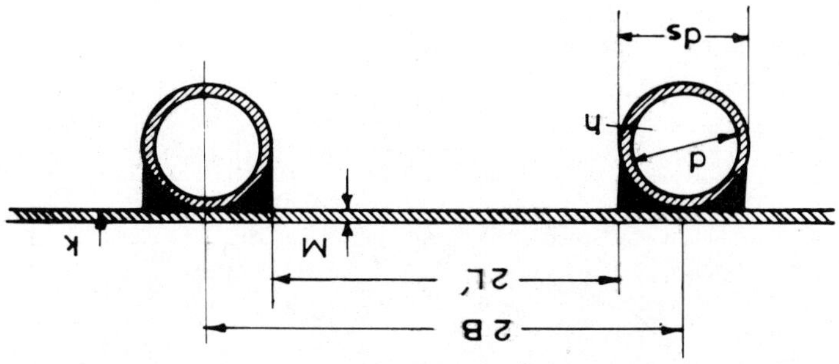

$$\eta^2 = \frac{U_L[2L']^2}{kM} \qquad (36)$$

Ensuite on définit un nombre

$$\kappa = \frac{kM\eta}{2L'\sinh\eta} \qquad (37)$$

et un nombre

$$\gamma = -2\cosh\eta - \frac{d_s U_L}{\kappa} \qquad (38)$$

et ensuite

$$R = \frac{1}{\pi d h} \qquad (39)$$

Par la définition encore de

$$F' = \frac{N\kappa L''}{U_L A_p} \frac{\kappa R(1+\gamma)^2 - 1 - \gamma - \kappa R}{(\kappa R(1+\gamma)-1)^2 - (\kappa R)^2} \qquad (40)$$

et

$$F_1 = \frac{1}{\kappa R(1+\gamma)^2 - 1 - \gamma - \kappa R} \qquad (41)$$

on aura en définitive

$$F'' = \frac{F_R}{F'} = \Phi_1\{F_1, GC_p/F'U_L\} \qquad (42)$$

celui-ci est valable pour N=2 où N est le nombre de segments de la serpentine et L" sa longueur. Quand nous avons plusieurs segments nous pouvons faire l'usage du diagramme de la figure 14.

Fig. 14

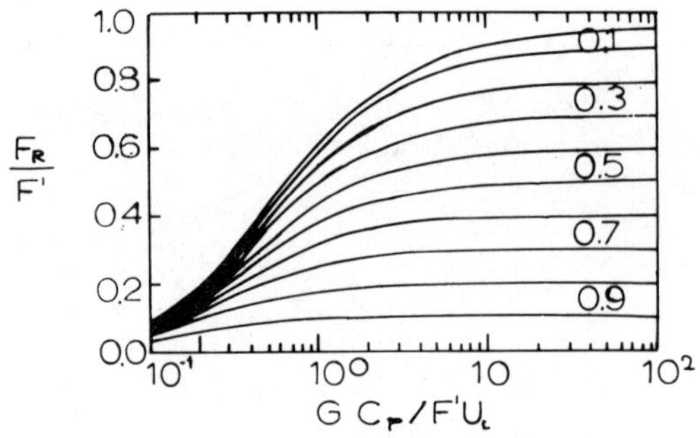

La quantité de chaleur absorbée par cet insolateur est:

$$q_u = GC_p(t_1''-t_1') = F_R|q_a-U_L(t_1'-t_a)| \qquad (43)$$

où t_1'' est la température de sortie du liquide. Tous les symboles ont la même signification qu'à l'insolateur de Hottel.

c) Insolateur composé par des caloducs

Pour un absorbeur qui se compose de caloducs, les pértes de chaleur par la face avant et la face arrière de l'insolateur peuvent être calculées avec les formules pour un absorbeur avec une nappe de tubes.

L'absorption peut être réalisée de la même manière, soit on peut avoir des surfaces sélectives ou ajouter des nids d'abeille.

En ce qui concerne le transfert de la chaleur à l'intérieur du caloduc la situation est complètement différente, de l'insolateur où un liquide à de faibles vitesses circule à l'intérieur des tubes. A l'intérieur au caloduc le liquide qui est véhicule change de phase en s'évaporant et en se condensant ensuite au condensateur. Il y a des livres spéciaux qui traitent cette matière comme ceux Dunn [32] et de Chisholm [33].

Pour ne pas élargir beaucoup le têxte du présent ouvrage nous nous limitons à certains aspects généraux du caloduc (Fig. 15).

Fig. 15

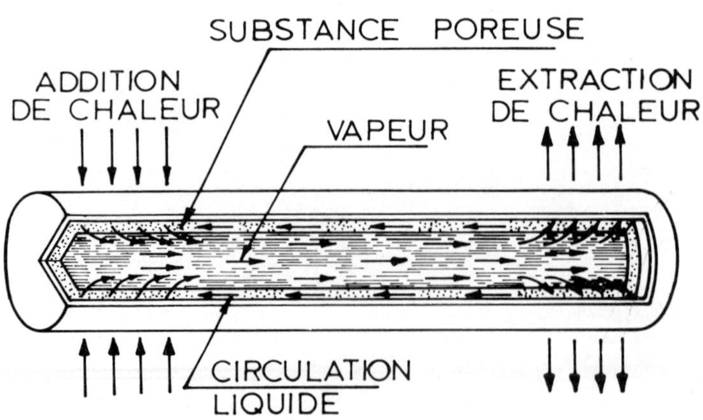

Chaque caloduc est défini quant à son fonctionnement dans une région de température. Dans cette région de température il peut absorber une quantité de chaleur maximum. Les propriétés du caloduc dépendent des propriétés physiques du liquide qui circule à l'intérieur du caloduc, des propriétés physiques de la substance poreuse, de son dimensionnement, et de son inclinaison.

Un caloduc à haute température est limité par quatre paramètres. Pour un caloduc à moyenne et basse température, certains de ces paramètres s'annulent.

Ces paramètres sont [32].

1) La vitesse de la vapeur ne doit pas dépesser la vitesse du son.
2) La vapeur ne doit pas entraîner de liquide par la surface poreuse.
3) La perte de charge du liquide et de la vapeur lors de sa circulation doit être plus petite que la pression due à la hauteur de capilarité.
4) L'ébullition du liquide ne doit pas conduire au séchage du caloduc.

Les avantages qu'offre l'application du caloduc à l'insolateur plan sont:

1) Il fonctionne comme une diode.
2) Il est un système compact et étanche. Il présente la moindre corrosion.
3) Il est un système qui peut être appliqué n'importe où.
4) Il n'a pas besoin d'antigel.
5) Il est capable de chauffer, l'air, l'eau, les fluides.
6) Il absorbe moins d'énergie pour son refroidissement, car il présente de pertes de charges diminuées.
7) Il est capable d'augmenter considérablement le rendement de l'insolateur.

Les incovénients sont:

1) L'application du caloduc à l'insolateur peut augmentér les frais de construction de celui-ci.
2) L'addition des résistances thermiques à l'insolateur peut conduire à la diminution du rendement de celui-ci.

Le caloduc donc s'offre à l'insolateur. Des applications sont déjà présentées dans certains cas notons celles de Mazarrasa |8| de Francken |35|, de Ortabasi|34| et de la Société Dornier. Il faut noter l'application qui a été présentée par Ortabasi.

Caloduc en cuivre, matière poreuse en cuivre, fluide solution glycol-eau. Le caloduc est posé à l'intérieur d'un tube de verre sous vide. La chaleur est fournie par concentration, le rendement du caloduc est de 60%.

L'absorbeur à concentration

L'absorbeur à concentration peut être considéré comme une autre catégorie de l'insolateur plan, parce qu'il peut être installée et intégré essentiellement sur un toit. Quant à son fonctionnement cet insolateur comporte pour une surface d'absorption, une ou deux surface catoptriques.

Le rayonnement solaire incident tombe sur une surface refléchissante et subit une ou plusieurs reflexions avant d'arriver à l'absorbeur. L'absorbeur peut être une surface selective ou un corps-gris. Quand le concentrateur a une surface de réflexion, il est partialement concentrateur et quand il a deux surfaces, il est concentrateur. Ceci dépend évidement des dimensions entre l'absorbeur et les surfaces reflechissantes. En général, quand les dimensions de l'absorbeur sont très réduites en comparaison des surfaces reflechissantes c'est alors que nous avons un concentrateur. Cet insolateur a des dimensions réduites de façons que ces dimensions soient similaires aux dimensions de l'insolateur plan à conversion thermique, corps gris.

Les principales variantes de cet insolateur sont:

1. Deux surfaces réflechissantes paraboliques avec une génératrice linéaire, c'est le cas de Sevcik et Winston [14].
2. Deux surfaces réflechissantes plates, et une surface absorbante; c'est le cas de Bannerot et Howell [14].
3. Une surface réfléchissante plate et une ou deux surfaces absorbantes.
4. Des surfaces réfléchissants, cylindrique ou parabolique à l'intérieur d'un tube de verre, sous vide. C'est le procédé Philips (Fig. 16) et Ortabasi [34].

Fig. 16

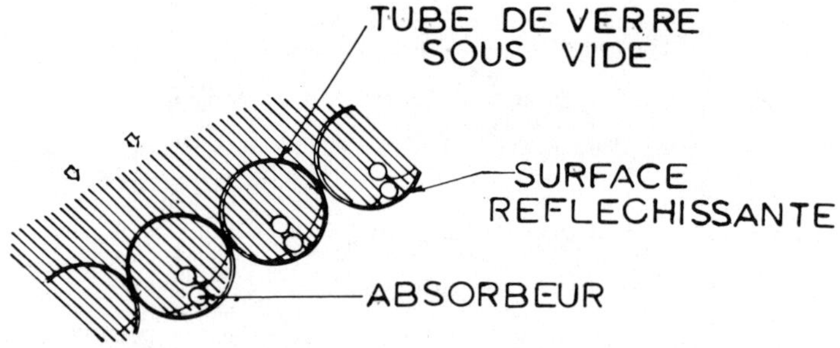

Le besoin d'avoir des températures élevées en comparaison de celles qu' on peut avoir par insolateur plan est la principale raison de cette évolution. On peut monter par un insolateur à concentration à des températures de l'ordre de 100∿150°C et on peut chauffer l'eau et l'air. En général les surfaces réflechissantes ont la direction est-ouest et leur facteur de concentration varie entre 1∿10.

Le facteur de concentration de l'insolateur à concentration de Secvik est

$$c = \frac{1}{\sin\Theta_{max}} \quad (44)$$

la hauteur théorique des miroirs est

$$L = \frac{1}{2}\,(d_1+d_2)\cot\cdot\Theta_{max} \quad (45)$$

où d_1, d_2 est la distance du front d'entrée de deux surfaces réflechissantes et la distance de sortie de deux surface réfléchissantes, Θ_{max} est l'angle lié avec l'ouverture des miroirs (Fig. 17).

Fig. 17

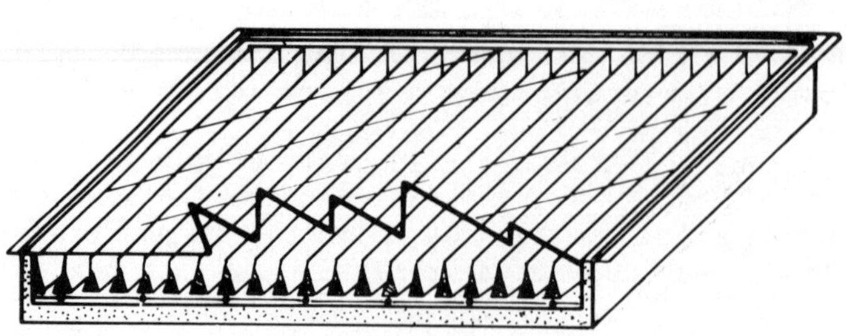

La concentration de l'insolateur que propose Bannerot est: |14|

$$c = 1 + \frac{2tg^{n}/2}{tg(\gamma_{20}+n)+tg\ ^{n}/2} \qquad (46)$$

où n/2 est l'angle que forment les deux surfaces catoptriques plates (Fig. 18), γ_{20} est l'angle du rayonnement incident avec la normale à la surface absorbante.

Fig. 18

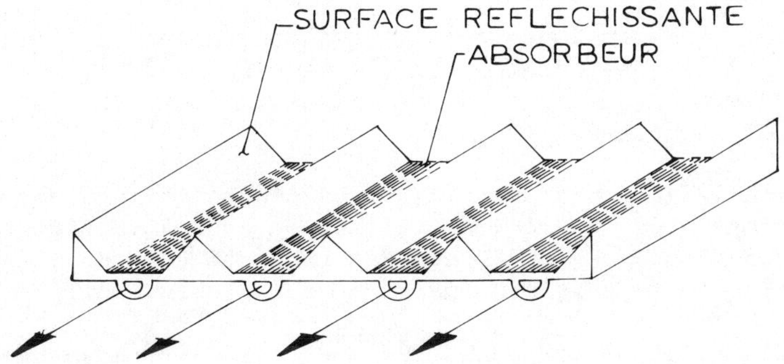

Ces insolateurs sont couverts par un vitrage et ils sont bien isolés à leur face arrière. Au-dessous de l'absorbeur peut circuler un fluide, en général l'eau, ou l'air. L'usage de l'air offre à ces installations certains avantages:

Ces avantages sont: [27] [14] [22].

1) Le rendement d'un absorbeur refroidi par l'air peut approcher le rendement d'un absorbeur refroidi par l'eau. Car l'air ne change pas de phase et on peut monter en température.
2) Par l'air nous n'avons pas besoin de l'antigel.
3) Par l'air la maintenance est réduite.
4) Par l'air les problèmes de corrosion sont réduits.
5) Par l'air les problèmes de l'étancheité n'existent pas.
6) Par l'air la vie de l'installation et de l'insolateur devient plus grande.
7) Par l'air nous pouvons chauffer à presque tout les climats.

L'insolateur avec surfaces réfléchissantes auxiliaires.

L'augmentation du rendement et de l'éfficacité de l'insolateur peut être réalisée par l'ajutage de surfaces réfléchissantes auxiliaires |17||15||16|. Ces surfaces sont posées à côté de l'insolateur qui est orienté vers le sud. On peut ajuter deux surfaces réfléchissantes à l'insolateur une du côté est de l'insolateur et l'autre du côté ouest. On peut aussi ajuter deux surfaces réfléchissantes du côté sud de l'insolateur et du côté nord de l'insolateur. On peut aussi ajuter seulement une surface du côté sud ou du côté nord.

Le rayonnement réfléchissante solaire incident, réfléchi par ces surfaces, est absorbé par l'insolateur.

La concentration du rayonnement qu'on peut entendre dans ce cas est de l'ordre 1∿3. C'est le cas d'un insolateur à hautes températures.

D'après Tabor |17| l'addition de deux miroirs à côtés, c'est-à-dire à l'est et à l'ouest de l'insolateur, augmentera la température de sortie de l'insolateur. Quand les miroirs auxiliaires sont posés du côté sud et di côtés nord de l'insolateur l'énergie produite par l'insolateur a une courbure aiguë et ceci n'est pas satisfaisant pour la production de l'énergie. Quand on ajoute un miroir du côté est ou du côté ouest et qu'on le transpose pendant le midi on peut reçevoir une énergie quasi uniforme.

Le fluide caloporteur

Le liquide caloporteur utilisé couramment est l'eau additionnée, lorsque ceci s'avère nécessaire, d'un antigel, par exemple glycol. Ceci peut provoquer quelques oxydations et il faut contrôler et recharger d'installation de temps en temp.

L'utilisation d'autre liquides peut être intéressante dans le cas où la température en régime statique ou dynamique peut dépasser 100°C. Ce liquide sera véhiculé par des tubes fixés sous l'absorbeur ou formera un film de très faible épaisseur sur l'absorbeur.

L'air, dont nous avons déjà mentionné les avantages, est le seul fluide gazeux utilisé jusqu'à present dans les capteur à gaz.

L'isolation

Afin d'éviter les pértes à l'arrière et sur les côtés de l'absorbeur il est indispensable d'assurer une bonne isolation de ces parties. Pour cela, on tapisse le fond et les côtés du capteur de couches d'isolant.

On peut caractériser un isolant par sa conductivité thermique qui est d'autant plus faible que le caractère isolant est meilleur.

Les matériaux utilisés sont surtout la laine de verre 0,057 W/mC à 50°C pour une densité de 123 Kg/m^3.

la polyuréthane 0,026 W/mc à 50°C pour une densité de 39 Kg/m^3.

le polystyrène expansé 0,031 W/mc à 50°C pour une densité de 36 Kg/m^3.

le liège 0,042 W/mc à 50°C pour une densité de 95 Kg/m^3.

L'isolant idéal étant le vide.

Citons à ce propos les promesses intéressantes des capteurs fonctionnant sous vide, comme les capteurs de Blum, d'Ortabasi, et de Philips.

Le coffrage

Le coffrage de l'insolateur peut être fabriqué par des différents materiaux: soit: La tôle ordinaire en acier ou en aluminium, la fibre de verre, la mousse de polyméthane, et la mousse de polyvinychlorure. Le matériau choisi devra régister aux variations de température, aux chocs et à l'attaque des agents corrosifs.

L'étanchéité du coffrage joue un certain rôle quant au fonctionnement et la durée de vie de l'insolateur. Des courants d'air peuvent exister entre le vitrage et le coffrage. Par conséquent le rendement de l'insolateur peut être diminué par le mauvais contact et la pénétration de précipitations.

En général l'étanchéité de l'insolateur est nécessaire soit à l'endroit de contact du vitrages avec le coffrage soit sur le coffrage lui même (soudage, joints mécaniques).

Des différents matériaux sont proposés pour ameliorer l'étanchéité du coffrage. Un des premiers materiaux utilisés était le caoutchouc. Mais le caoutchouc se détériore par le rayonnement solaire et les conditions climatologiques. Aujourd'hui, pour rendre l'insolateur étanche une gamme de matières à base de silicone ou à base de caoutchouc synthétique est disponible.

D'après Clark [19] on propose quelques matières résistant au rayonnement solaire, à la chaleur et aux conditions météorologiques:

1. Silastic 655 Rubber
2. Silastic 250 Rubber
3. D.C. Building Sealant
4. Silicone Resin X9-63100 on Silicon.

Mode d'emploie de l'insolateur

Un autre aspect de l'utilisation de l'insolateur en général, est son comportement thermique et hydraulique quand il est lié avec un autre insolateur.

L'insolateur peut être lié avec un autre insolateur en série ou en parallèle. D'intéressants travaux sont réalisées sur ce sujet par Garg |9|, Dunkle et Davey |38|. Si nous avons plusieurs insolateurs nous pouvons concevoir aussi un couplage mixte. C'est-à-dire des groupes, d'insolateurs liés en parallèle, sont liés en série ou le contraire (Spyridonos)|36|.

Le rendement de l'insolateur dépend de sa température d'entrée il est évident que des insolateur liés en série n'auront pas le même rendement. Le rendement de chaque insolateur,à un débit donné, va diminuer au fur et à mesure que la température d'entrée va augmenter. Pour conservér le rendement il faut une isolation supplémentaire à chaque insolateur consécutif. Morse et alii ont étudié ce problème |37|.

Quant à la perte de charge et le comportament thermique de l'insolateur Dunkle et Davey |38| proposent que le meilleur résultat du point de vue extraction de chaleur est:

1) Le nombre maximum des insolateurs liés en parallèle est 8.
2) Les insolateurs peuvent avoir une liaison mixte soit: trois insolateurs sont liées en série et forment un groupe. Chaque groupe est reliée en parallele avec l'autre.

On remarque une chute de température du fluide après le second insolateur de chaque groupe. La température du fluide monte de nouveau lors de son entrée au groupe consécutif.

Il faut prévoir un débit de l'ordre de 37,17∿49,5 lt/hm^2 dans chaque insolateur. Ce débit peut donner des résultats satisfaisants et une augmentation de la température du fluide (de l'eau) raisonnable. Quant à la pérte de charge, elle dépand aussi de la méthode de construction de l'absorbeur. Il faut éliminer autant que possible les obstructions à l'intérieur de l'absorbeur.

Conclusions

De remarquables progrès sont réalisés quant à l'évolution de l'insolateur pandent la dernière décennie.

Pourtant, pour que l'insolateur soit établi à l'échelle industrielle, il faut accomplir encore des efforts.

Par ailleurs à cause de cela sa normalisation n'est pas encore établie.

Nous constatons alors deux catégories d'insolateurs:

a) L'insolateur corps-gris refroidi à l'eau ou à l'air.

b) L'insolateur à concentration refroidi à l'eau ou à l'air.

L'insolateur corps-gris se divise en quatre sections:

1. Absorbeur plat avec une nappe de tubes soudés ou vissés à deux tubes horizontaux (admission de l'eau froide en bas, sortie de l'eau chaude en haut).
2. Absorbeur plat avec un tube en serpentine soudé on attaché ou serré avec joints.
3. Absorbeur plat avec une nappe des caloducs à ailettes.
4. Absorbeur plat, ou en ailettes ou en rugosités en V avec une ou deux conduites, refroidi à l'air.

L'insolateur à concentration se divise en trois sections:

1. Absorbeur avec miroirs plans concentrateur ou partiallement concentrateur, refroidi à l'air.
2. Absorbeur en miroirs plans ou paraboliques, concentrateur refroidi par l'eau, ou l'air ou au moyen des caloducs.
3. Absorbeur avec une nappe de tubes de verre sous vide, concentrateur, refroidi au moyen de l'eau par des caloducs ou des tubes.

Il est très dificcile de comparer tous ces procédés de captage de l'énergie solaire sans savoir les paramètres qui interviennent. Le rendement de chaque insolateur dépand de plusieurs paramètres. Par ailleurs chaque auteur n'a pas exprimé le rendement de l'insolateur avec le même abscisse. Certains ont étudié l'insolateur en fonction de la différence de la température moyenne absorbeur-air d'autres en fonction de la température moyenne de l'absorbeur etc. Pour comparer les insolateurs il faut exprimer le rendement en fonction de la différence de la température d'entrée du fluide et de l'air divisée par l'insolation par unité de surface ou sous une forme sans dimensions ce qui n'est pas toujours le cas. Ensuite on peut introduire comme paramètres la sélectivité de la surface de l'absorbeur, le nombre des vitrages et leurs distances, le debit fluide, les pértes de chaleur en général et les conditions climatologiques locales. C'est pour cela que nous irons présenter certains résultats, où leur facteur commun est seulement le rendement.

La conclusion générale est que le rendement, de tous ces insolateurs pour $\Delta T=0$ ou t petit, est de l'ordre de 70∿85%. Ceci est valable, actuellement, pour toutes les catégories d'insolateurs qui ont une bonne isolation une construction soignée et une surface sélective. Pour les insolateurs qui sont définis comme crops gris le rendement peut varier de 50% à 80%.

D'après Brenecke et Justi |39| Fig. 19 le rendement d'un insolateur dans les meilleures conditions du point du vue isolation ne peut pas dépasser 90% (Fig. 19). C'est le cas où l'insolateur est complétement isolé et les pértes par rayonnement et conduction sont devennues nulles. Celui-ci est difficile à appliquer. Au contraire des situations où il n'existe pas de pértes par rayonnement et convection peuvent être envisagées. Dans cette voie a travaillé Ortabasi |34| Fig. 20 on compare un insolateur sous vide avec un caloduc chauffé par concentration dans un tube sous vide où les pértes de convection peuvent devenir nulles et celles du rayonnement approcher le zéro (Fig. 20).

Fig. 19

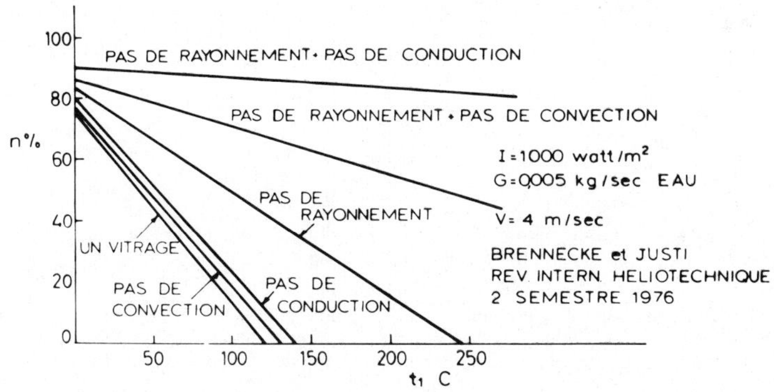

Fig. 20

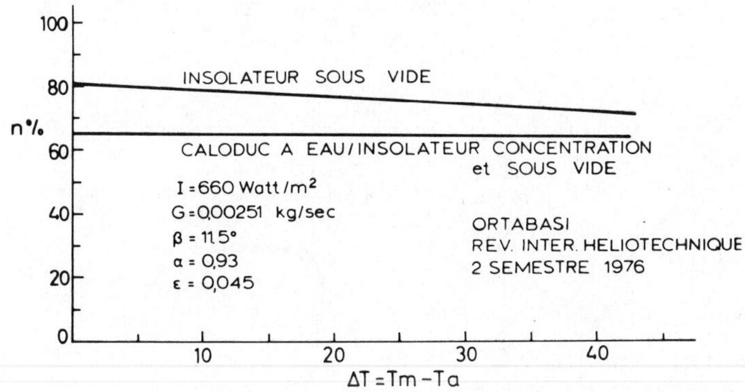

A la fig. 21 un insolateur classique a été expérimenté par Simon et Harlement |13|. On voit bien que le rendement pour $\Delta T/I = 0$ est 75% ce qui est raisonable. Des résultats meilleurs sont donnés par Bruno et alii |12|. Il y a une augmentation de la pente de la droite du rendement.

Fig. 21

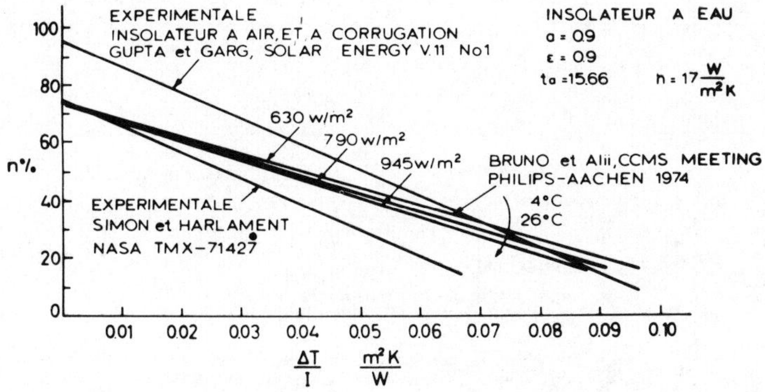

Sur le même diagramme est indiquée d'expérience realisée par Gupta et Garg |40| avec un insolateur refroidi à l'air. Cet insolateur peint avec une peinture noire à un absorbeur à rugosités. Son rendement est remarquable car pour $\Delta T/I = 0$ le rendement est 95%.

Au contraire un travail réalisé par Close |27| et antérieurement par Tabor, (Fig. 22) n'est indiqué que pour insolateurs peints en noire, α=0,9 et ε=0,9 le rendement ne peut pas dépasser 65%. (Fig. 22).

Si les surfaces étaient sélectives Close indique que le rendement peut monter jusqu'à 85% environ.

Par ailleurs, d'après Lorsch |29|, Fig. 23 le rendement augmente et la pente de la droite augmente aussi en fonction de l'augmentation des distances entre les vitrages, l'abscisse t est la température moyenne de l'absorbeur (Fig. 23).

Fig. 22

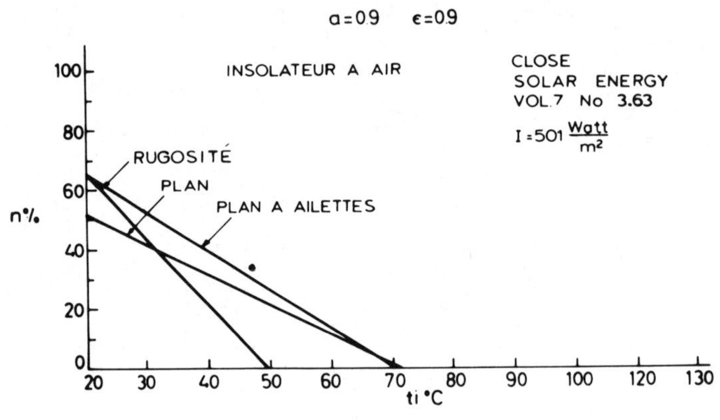

Fig. 23

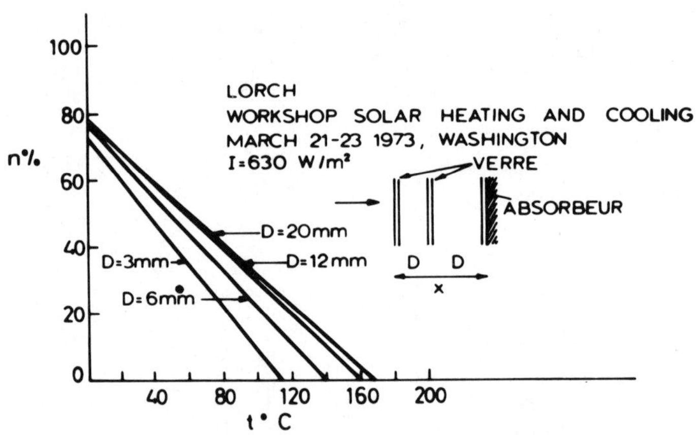

Dans la fig. 24 Bannerot |14| présente un insolateur à concentration en comparaison d'un insolateur plan. Le rendement d'un insolateur à concentration (α=0,9, ε=0,9) est plus grand qu'un insolateur plan corps gris (α=0,9, ε=0,9) (Fig. 24).

Fig. 24

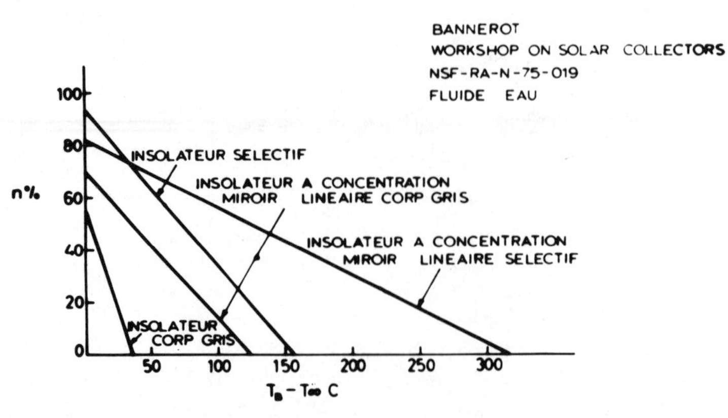

Approximativement la même figure est présentée par Sevcik et Winston |14| (Fig. 25) où le rendement d'un insolateur à concentration est supérieure à un insolateur sans concentration.

Fig. 25

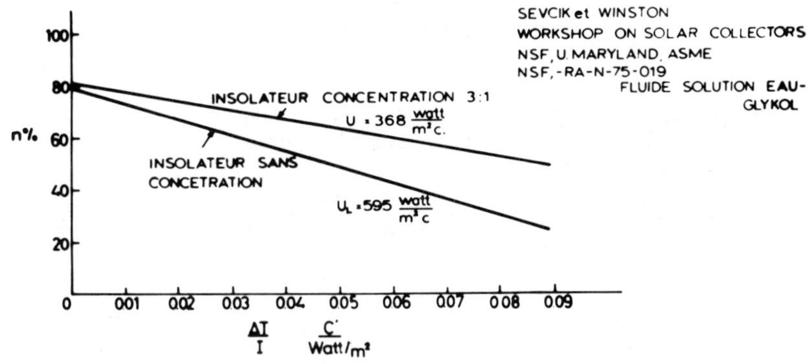

Dans les pages antérieures il est mentionné que l'effort principal pour améliorer le rendement d'un insolateur, est consacré à la sélectivité et l'isolation de la face que reçoit le rayonnement solaire incident, tandis que l'isolation arrière reste quasi identique. On remarque par ailleurs tous les insolateurs sont presque équivalents du point de vue rendement, nous évoquons pourtant l'influence de la pente de la droite car l'insolateur fonctionnera rarement à $\Delta T/I \simeq 0$. C'est alors plus important d'avoir un rendement meilleur pour $\Delta T/I >> 0$. Ce qui rend l' insolateur plus utilisable.

Il faut mentionner avant de terminer que le rendement peut augmenter par l'augmentation de la vitesse de l'écoulement du fluide à l'intérieur de l'absorbeur. Duncle signale pourtant une certaine gamme des debits, il y a certainement une limitation; par contre par le caloduc la nature du coefficient de transfert de chaleur est complètement différente de celle d'un absorbeur où se vêhicule un liquide.

En effet il faut compter certaines résistances dues à la conductivité de la substance poreuse du caloduc mais d'autre part il faut compter l'importance du coefficient de transfert de chaleur lors de l'ébullition du liquide à l'intérieur du caloduc. En choisissant convenablement ces coefficients on peut avoir une augmentation importante du coefficient global de transfert de la chaleur et ensuite une augmentation du rendement de l'insolateur.

Bibliographie

1. H.C. Hottel and B.B. Woertz. The performance of flat plate solar heat collector, Transactions of ASME Fef. 1942, Vol. 64.
2. Ray.W. Bliss Jr. The derivation of several plate efficiency factors useful in the design of flat-plate solar heat collectors. Solar Energy, Vol. III No. 4. December 1959.
3. Aus Whillier, Plastic Covers for solar collectors. Solar Energy, Vol. 7, No. 3, 1963.
4. Aust. Whillier, Design factors influencing solar collectors performance.
5. Giov. Francia, Un nouveau collecteur de l'énergie rayonnante solaire, theorie et verification experimentale, Congrès UNESCO 5/71 Rome 1964.
6. Aust. Whillier, Thermal reistance of the tube-plate bond in solar heat collectors, Solar Energy, Vol. 8, No. 3, 1964.
7. Argyr. Spyridonos. The solar water heater,Technika Chronika, Oct. 1971, vol. 10, Athens.

8. R. Mazarrasa, Nouveau systême d'utilisation de l'énergie solaire, Bul. COMPLES Juin 1972, p. 9-14.
9. H. P. Garg, Design and performance of large size solar water heater, Solar Energy, Vol. 14, No. 3, Feb. 1973.
10. M. Touchais, Methode simplifier d'estimation de la chaleur fournie par un insolateur. Rencontre Intern. de la COMPLES, Madrid 23-28 Sept.1974.
11. J.D. Duffie and W.A. Beckman, Solar energy thermal processes. John Wiley and Sons, New York 1974.
12. R. Bruno, W. Hermann, H. Hörster, R. Kersten and F. Mahdjuri. On collector testing methods, Phylips Forschungslaboratorium, Aachen, CCMS Meeting NATO, Odeillo, 4 Oct. 1974.
13. St. L. Sargent, Method of testing for rating solar collectors, Center for Building technology, Nat.Bur.of Standard, Washington, D.C. 20234.
14. Workshop on Solar Collectors for Heating and Cooling of Buildings. Report NSF-RA-N-75-019 N. York, Nov. 21-23 1974.

15. D.K. McDaniels, D.H. Lawndes, H. Methew, S. Reynolds and R. Gray, Enhanced solar energy collectors using reflector solar thermal collector combinations, Solar Energy, Vol. 17, No. 5, 1975.
16. Stev.C. Seitel. Collector performance enhancement with flat collectors. Solar Energy, Vol. 17, N. 5. 1975.
17. H. Tabor, Mirror boosters for solar collectors, Solar Energy, Vol.10 No. 3, 1966.
18. H. Tabor, Selective surfaces for solar collectors, ASHRAE.
19. A.A. Clark, Environmental resistance data on silicone polymers of utility in solar collectors. ISES International Congrès, Los Angeles 1975, Paper 30/13.
20. M. Perrot, Index analytique COMPLES 1976.
21. H.P. Garg, Effect of dirt on transparent covers in flat plate solar energy collectors. Solar Energy, Vol. 15, No. 4 April 1974.
22. A. Eggers-Lura, Flat plate solar collectors and their application to dwelling. A/S Intern. Solar Power Co. Ltd. and Com.Eur.Communities DK-2860 Söborg,Denmark.
23. F. Breniere, Energie Solaire et climatisation des locaux. Editions Européenes Thérmique et Industrie, 3, rue Henri Heine Paris 1976.
24. H. Buchberg and D.K. Edwards. Design consideration for solar collectors with cylindrical glass honeycombs. Solar Energy, Vol. 18, No. 3, 1976.
25. Hans Rau, Heliotechnik, Udo Pfriemer Verlag, München, 1976.
26. K.G.T. Hollands, Honeycomb devices in flat plate solar collectors. Solar Energy, Vol. 9, No. 3, 1965.
27. D.J. Close, Solar air heaters. Solar Energy, Vol. 7, No. 3, 1963.
28. D. Balcomb. J.C. Hedstrom, St.W.Morse and Ken.C. Herz. Research on integrated solar collector roof structures, ISES Intern.Congres Los Angeles 1975, LA-UR-1335, Los Alamos Scient.Lab.
29. J.I. Yellot, Utilisation of sun and sky radiation for heating and cooling of building ASHRAE.
30. S.I. Abdel-Khalik, Heat removal factor for a flat plate collector with a serpentine tube. Solar Energy Vol. 18, No. 1, 1976.
31. E.M. Sparrow, Radiant absorption characteristics of concave cylindrical surfaces, J. of Heat Transfer ASME, Nov. 1962, p. 283.
32. P.D. Dunn and D.A. Rey, Heat Pipes, Perhamon Press 1976.
33. D. Chisholm, The Heat Pipe. Mills and Boon Ltd. London 1971.
34. U. Ortabasi, A tubular evacuated solar collector utilizing a heat pipe as absorber. Revue Intern. Heliotechnique 2. Semestre 1976, p. 14.
35. J.C. Francken, The heat pipe fin, novel design of a planar collector. Int.Sol.Energy Society Conference, Los Angeles 1975 paper 34/8.

36. A Spyridonos, Solar Heat Technology, Synchroni Technologia, Periodos B. 11/24, Sept.-Oct. 1975.
37. R.N. Morse et alii, High temperature solar water heating, Intern. Congrès "Le Soleil au Service de l'Homme", Paris 1-6 Juillet 1973, EH60.
38. R.V. Dunkle and E.T. Davey, Flow distribution in solar absorber banks. ISES Conference Melbourne 2-6 March, 1970, paper 4/35.
39. P. Breneke and E. Justi, Studies into reduction of radiative heat losses of flat plate solar collectors. Revue Intern. Heliotechnique 2 Semestre 1976, p. 18.
40. C.I. Gupta and H.P. Garg, Performance studies on solar collector performance evaluation. Solar Energy, vol. 11, No. 1, 1967.

3.2 Practical selective surfaces

H. Tabor
Scientific Director
The Scientific Research Foundation, Jerusalem, Israel

Abstract

Interest in selective surfaces produced by electro-deposition - such as nickel-black and chrome-black - induced the American Electroplaters' Society to hold a symposium on these and other coatings for solar collectors.

This report summarises the thirty odd papers presented at the symposium, where the accent was on practical aspects.

Introduction

Although selective black surfaces, i.e. surfaces of low thermal emittance and high solar absorbtance - were first demonstrated at the First World Symposium on Solar Energy in Phoenix, Arizona, in 1955 - and solar collectors using such surfaces have been commercially available - in particular in Israel and Australia from the late fifties - it is only in the last few years that the subject has become suddenly "popular" with a flurry of activity, particularly in the U.S. Particular interest was created in electro-deposited layers by recent work conducted at NASA-LEWIS Space Centre based upon the work of this reporter and his colleagues in the fifties and sixties. As a result of this interest, Glen McDonald of NASA-LEWIS persuaded the American Electroplaters' Society to host a special symposium on surfaces for solar collectors: in addition to nine papers dealing with electro-plated nickel-black and chrome-black, there were thirty additional papers covering other forms of selective surfaces, paints and some application studies.

The writer attended the symposium at the request of the MIROMIT SUN-HEATER COMPANY, who have graciously agreed to the publication of this report, which tries to summarise the material presented and to draw some conclusions concerning the present state of the art.

1. Program

Thirty-three papers were listed in the program (with some additional presentations added).

These papers were divided as follows:
1 introductory status paper by Dr. F.H. Morse of ERDA
1 paper on collector installations
3 papers on instrumentations (α and ϵ)
3 " on coatings on aluminium
2 " on coatings on copper
1 paper on conversion coatings
1 " on sputtered coatings
1 " on interference coatings
1 " on lead-oxide layers
1 " on thin films
1 " on general requirements
2 papers on dendritic coatings
6 " on paints, enamels etc.
9 " on nickel black and chrome-black

MORSE's introductory paper dealt with such questions as: testing, high cost, corrosion resistance and durability, inadequate performance data on existing surfaces: durability had not yet been established. Dealing with applications, he covered the following points: for cooling: (1) passive systems, i.e. good design, nocturnal cooling etc; (2) active systems, i.e. moderately high temperatures required; (3) the need to match the collector to the machine; (4) use of dessicant systems. For heating: (5) passive aspects - storage; (6) solar-assisted heat pumps; (7) air heating; (8) liquid heating. On the problems related to coatings he referred to: durability and failure modes; effect of high temperature, UV and humidity; lower-cost, high-volume techniques; better instrumentation for production, i.e. quality control; need to compare transparent heat mirrors with selective absorbers; problems of heat rejection ("selective whites"); coatings for plastics; etching of glass i.e. anti-reflecting coatings also IR reflecting coatings.

The discussion on the paper by J.R. Williams (Georgia Tech) [1]* describing a collector installation on the Shenandoah Community Center indicated the following:

Collector costs $\$12/ft^2$
Total system $\$40/ft^2$

This revealed some confusion since the system cost, for determining solar viability, should not include the equipment that would be used anyway, but only the extra equipment required with the collectors.

The author said that selective coatings should last 30 years but no one would give a guarantee for more than one year!

The papers on instrumentation discussed some very expensive sophisticated instruments and some somewhat simpler instruments for measuring α and ε. The discussion revealed - what was stressed particularly by the practical "plating" men in the audience - that there was a need for good quality-control instrumentation for use in the shop, particularly as electroplaters+ could not judge what was a "good" coating in view of the special optical characteristics required.

Selective Coatings on Aluminium

R.K. Williams (Birchwood Casey) [17] described a proprietary dipping process that produced a black oxide coating. But the surface had poor corrosion resistance and needed to be protected with an organic sealant, otherwise a white corrosion product appeared. The sealant resulted in increasing the emittance - to that of the order of 0.4 - so the result coating (with $\alpha = 0.96 - 0.97$) was only partially selective.

* Numbers in square brackets are page numbers in the Symposium volume.

+ This assumed that the collector manufacturer did not do his own coating: indeed,the plant required to coat full-sized plates was so expensive that only a collector manufacturer producing a large volume could afford to do his own plating. Incidentally, in the early development of the industry, this factor would encourage the use of simpler coating methods even if the performance were inferior.

Powers and Cochran (ALCOA) [41] described a chemical system which also resulted in partial selectivity i.e. $\varepsilon \sim 0.35, \alpha \sim 0.93$. The system worked on alloys but not on pure aluminium and the impression gained was that it was a tricky process - with a tendency to fading with time. Costs were estimated at \$1.54/m^2 for a single-tank process and \$0.87/m^2 for a double-tank process.

The third Al process was based upon anodising (J. L. Woods of Permaloy [117] again resulted in ε values of the order of 0.4 or higher and some risk of fading of the dyes used.

In summary, the situation as regards conversion coatings on aluminium did not appear to have changed much in the past twenty years: consistant results and good selectivity were not easily achieved. Of course, added coatings, such as electro-plated nickel-black or chrome-black after some substrate had been put on - or paints - were applicable to aluminium as to any other metal.

Coatings on Copper

The earliest selective coatings on copper had been produced by chemical oxidation (Ebonol process) and the paper by Hajdu of Enthone [29] discussed this. They had worked to get better stability and corrosion resistance. Their best results were $\alpha = 0.85 - 0.95$; $\varepsilon = 0.1 - 0.4$. Temperature stability: 500 hours at 375°F caused defoliation: 500 hours at 100°F and 100% RH caused degradation. They had had some collectors out in the field at Long Island Sound (a marine atmosphere) for 2½ years and had not visually observed any corrosion products.

The paper from A. D. Little [111] indicated that the Enthone-type coating was better (in durability) than a chrome-black process.

Conversion Coatings

(a) Bishop of R.O. Hull Co. [109] was not impressed with conversion coatings on copper (or on other material).

(b) Tin. Bishop's paper discussed the blackening of tin by both chemical and electro-chemical means. No performance or corrosion data were given. (Tin thickness 0.0002" = 5 microns).

(c) Zinc. Horner and Greene of Allied-Kelit [125]. This metal could be blackened with a molybdenum dipping process. The minimum dip of 0.5 minutes gave $\alpha = 0.88$ and $\varepsilon = 0.14$ for non-cyanide zinc and $\varepsilon = 0.12$ (α not stated) for 0.25 minutes on the low-cyanide zinc. However, corrosion protection was poor - the zinc corroded.

Zinc could also be blackened with a chromate dip. The best result seemed to be with the ½ strength bath i.e. $\alpha = 0.93$ and $\varepsilon = 0.28$ for non-cyanide zinc and $\alpha = 0.91$ and $\varepsilon = 0.18$ for the low cyanide zinc. The corrosion resistance of chromate coatings on zinc was good (though no details for the actual coatings used).

(d) Nickel. This material could be darkened by heating in air but the α values were not very high. Molybdate colouring also did not produce very high values. Sulfide coatings (obtained by first oxidising the nickel and then dipping in sodium sulfide solution) gave $\alpha = 0.93$, $\varepsilon = 0.15$ but no data was given on stability.

(e) Iron. Blackening of iron was discussed by Borzoni of Honeywell [89.] One method, the Ebonol S dip method produced similar optical results but poor rust resistance. "Presto Black" did not produce a satisfactory surface. The Ebonol S treatment could be improved using organic overcoats which improved humidity resistance. Best α values were about 0.89, still on the low side. (There was no indication that the organics chosen were temperature stable to 200°C, needed for no-flow conditions).

Isaacs [139] described the colouring of stainless steels using a patented process developed by International Nickel. This involved immersion first in a solution of chromic and nitric acids, then in chromic and phosphoric acids (anodically) to get a better surface. Interference films produced: on matt finish, a "blue" gave $\alpha = 0.92$, $\varepsilon = 0.14$ and a "black" gave $\alpha = .90$, $\varepsilon = 0.18$. Costs were 20c/ft^2 not including the substrate.

Sputtered Coatings

The paper by Thornton [63] presented what appeared to this reviewer to be a rather optimistic picture of a process for depositing metals by electrical discharge in vacuum: i.e. if enough machinery were built and a large enough output assured, production costs of 10-15 cents/ft^2 for the coating process (on foil) were predicted. (Note that chrome-black on copper foil is being offered commercially at this price).

Interference Coatings

The paper by Weingarner of OCLI [37] dealt with multilayer interference coatings similar to those developed in Israel (Weinberger, Harris, Doron, Tabor), except that he considered vacuum coating (as against the electro-plating method). He argued that high-vacuum coating gave a far wider choice: it was also more expensive: the author predicted \$1/ft^2.

Lead Oxide Layers

Lead oxide is a semi-conductor and substantially transparent to long-wave radiation. It has therefore been proposed at various times as a base for selective paints.

Bowen of Ametek [121] reported on electro-plated lead oxide where the ratio of lead to oxygen could be varied over quite a range. Very high α values (c. 0.99) were claimed with ε values around 0.25 to 0.4. No data on corrosion resistance or temperature stability was available, but the company was offering collectors using this coating.

Thin Films

Libbey-Owen-Ford [19] discussed heat mirrors on glass as an alternative to selective coatings.

Here it is worth returning to theory. The effective emissivity of a surface 1 close to a surface 2 is given by:

$$E_e = \frac{1}{\frac{1}{E_1} + \frac{1}{E_2} - 1}$$

the equation resulting from inter-reflections between the surfaces. Since the equation is symmetrical in E_1 and E_2, it is immaterial which surface has the low emittance. So putting a low emittance (i.e. IR reflecting) surface on the glass cover of a collector is an interesting alternative to a selective absorber surface with the advantage that it does not have to be stable to the elevated absorber temperatures. But it does have to be highly transparent to solar radiation, a condition not too easy to achieve. Since the yield q of a collector is given by:

$$q = \alpha\beta Q - U\Delta T$$

where $U\Delta T$ represents the heat-loss and where β is the solar transmissivity of the window, any percentage reduction in β has exactly the same effect as a similar reduction in absorbtivity α of the absorber. Writing $\beta = \beta_g \beta_f$ where β_g is the transmission of the untreated window and β_f the additional effect of the IR reflecting film, β_f is found to be just under 0.9 for the best films: it may easily be considerably less.

Dendrite Coatings

It was known in the literature that, using CVD (Chemical Vapour Deposition), tungsten surfaces had been made in dendritic form i.e. needles that gave strong absorbtance in the visible with low emittance values. This was fine for high temperature absorbers, but tungsten was expensive, so only suitable for high-concentration systems. A paper from IBM [133] discussed such coatings. Grimmer of LASL [79] discussed work on nickel dendrites by CVD methods on such bases as aluminium. High α values were obtained, but the ϵ values were not very low. Costs were high - of the order of $\$3/m^2$ for materials only!

Paints and Enamels

All these have high emittance. So the interest is in high α, good stability and low cost. (Radiation loss might be reduced by IR reflecting windows).
(a) Porcelain enamels [35] could give α up to 0.99 for direct beam and up to .97 for diffuse radiation. High stability to heat and weather. Costs said to be low but no figures were given in the paper, though Smith [47] indicated 1972 cost of $\$4$-$10/m^2$ for 26 gauge enamelled steel (including the cost of the steel). (b) Black paint $\alpha = 0.9$, $\epsilon = 0.6$ was described by Ytterhaus of Chamberlain [43] who claimed good stability even on steel substrates.

The 3M Company [131] described their 101-$C10_2$ flat black coatings (Nexel) apparently usable to 300°F. Cost of 10-20c/ft^2 materials cost only! Wolf, Ray and Lavan [103] discussed a number of paints stable up to as high as 490°F.

One paper from the Nuclear Division of Union Carbide, 101 discussed a paint comprising micro-spheroid carbon particles* in a suitable binder. Whilst good results were reported, no α or ϵ values were given: nor was there any mention of stability of the binder.

* Some forms of carbon show considerable IR transparency.

Nickel Black and Chrome Black

As already indicated, this subject took a major place in the symposium primarily as the result of the work of Glen McDonald - the organiser of the symposium - who, together with his colleagues revived interest in nickel-black and chrome-black. Chrome-black had been sponsored as a more stable surface than nickel-black, though it was more expensive. Most of the reported work by McDonald and those who followed him used a nickel layer as the substrate. The total cost, because of this, was high and this had stimulated interest in alternative selective blacks - such as the conversion coatings - that might be cheaper.

Sandia Labs[21] appeared to have done very good work on black-chrome. They showed black-chrome on nickel to be stable to 400°C in vacuum applications.

Lindsay of Oxy Metals Corp.[25] showed that different quality of nickel substrate had no effect on the α and ε values but did affect the chrome-black plating time for optimum results: i. e. the morphology of the nickel affected the black-chrome deposition process. Benning of Harshaw [31] also described work in this field. He pointed out good humidity resistance (if the base nickel were 12μ thick on iron) and $\alpha \sim 0.95$, $\varepsilon \sim 0.12$.

Barber of PPG [51] described purchaser's requirements for a selective coating. Keeling, Asher and Gurtler of Motorala [53] indicated chrome black as more stable than nickel black (though the latter required much higher plating currents).

They found the substrate nickel to affect the results i. e. $\alpha = 0.88$ on bright nickel and $\alpha = 0.95$ on sulfamate nickel. They claimed a result for black chrome on bright aluminium foil of $\alpha = 0.964$ with $\varepsilon = 0.023$ (a figure this reviewer finds hard to accept). Moore of LASL [57] reported on commercial steel collectors with Harshaw "Perflow" nickel and "Chrome Onyx" black with $\alpha \sim 0.94$, $\varepsilon = 0.08$. Borzoni of Honeywell [89] compared black-nickel and black-chrome, finding the former to give better optical properties ($\alpha = 0.96$, $\varepsilon = 0.07$) cf. $\alpha = 0.95$, $\varepsilon = 0.10$ for black-chrome but that the nickel-black was less stable. Thus some failed after 24 hours in the humidity test (i. e. the black washed away) and this was with nickel substrate. This nickel substrate, 25u i. e. 1 mil thick, represented more than half the cost of the total coating. With chrome-black, the nickel only needed to be $\frac{1}{4}$ mil thick, thus bringing down the cost of the chrome-black process. Black-chrome on bare steel was also tried but rusted unless covered with an organic overcoating. This gave $\alpha = 0.94$, $\varepsilon = 0.24$ and good humidity resistance. This offered a cheaper selective coating, since the nickel was eliminated.

Black-chrome on galvanised steel gave $\alpha = 0.95$, $\varepsilon = 0.25$ but the humidity resistance was poor.

Oleson of Corillium Corp. [107] suggested that their Solarchrome process (directly onto the base metal) was better and cheaper than their earlier Chromonyx as licensed to Harshaw. The base metal could be copper, aluminium (zincated), stainless steel or carbon steel. The stainless steel results were particularly good on all aspects (though no α or ε figures were actually given), but even the carbon steel showed increased corrosion resistance provided it first got the Gemini II chromium process prior to the Solarchrome. Indicated total costs of less than $1 per ft^2 were suggested.

Conclusions:

The systems of selective surfaces discussed can be divided into two groups - where the absorber surface is provided with low emissivity whilst, at the same time, a high absorbtivity - and where the cover window is made IR reflective to achieve a similarly low effective emittance. The second group has the advantage of not requiring to operate at elevated temperatures. But the method discussed (nor other methods proposed for producing such coatings on large glass sheets) are unlikely to be cheap unless and until the process can be incorporated in the glass-making process itself, i.e. some bold glass producer, anticipating a very large market, is prepared to introduce a large-scale coating process as part of his overall process as part of his overall production process.

For the absorber selective surfaces, not all the ideal answers have yet been found. Processes for direct treatment of oxidisable metals such as aluminium and iron are not too successful and, in the present state of the art, it seems necessary to first coat these absorbers with an intermediate substrate layer such as nickel. This intermediate material seems to be the popular substrate for the electro-plated nickel-black and chrome-black processes. Both these processes, on a bright base, appear to give reasonable selectivity, i.e. 0.9, 0.1: some claims are made that chrome-black is more stable than nickel-black, especially at higher temperatures, though it is a more expensive process. In both cases the stability is greatly influenced by the stability of the substrate metal.

Copper is converted into a practical selective surface by chemical oxidation, with α and ε figures similar to those for the nickel-black and chrome-black layers. Not all copper oxides are equally stable against atmospheric corrosion and none of them appears suitable for elevated temperatures.

There appears little progress in selective paints comprising a semi-conductor layer in a suitable binder, primarily because of the limitations of the binder. Lead oxide, that has been proposed in the past as a suitable semi-conductor for this purpose, has been successfully deposited electrolytically with very high values: the values are moderate.

Cost data appears unreliable and it is difficult, at this stage, to compare cost estimates made by different researchers. The range covers from a few cents to a dollar per sqr. foot, the latter figure probably being unacceptable for collectors without optical concentration. Clearly, for concentrating systems, the permissible cost per unit area of absorber surface can be much higher, thus opening the door to more sophisticated surfaces, such as the dendrite surfaces, heat mirrors, etc.

Measurement of selective surface properties for solar collectors

3.3

R. G. Morgan and W. B. Gillett
Solar Energy Unit,
University College, Cardiff, UK.

Abstract

The concepts of spectral and directional emissivity are described and their influence on three approaches to emissivity measurement is discussed. A simple calorimetric method of measurement is described and also an arrangement for direct radiation measurement which can produce results quickly at low cost. Results obtained by the different measurement techniques are compared for a sample of commercial black chrome selective surface. A cheap approximate method for estimating the solar absorptivity of surfaces is also presented.

Introduction

When discussing equipment for the conversion of solar radiation directly into heat it is common to refer to the absorptivity (α) of the absorbing surface and to define this for the absorption integrated over the solar spectrum. The spectral distribution of the solar radiation incident upon a solar absorber is dependent on the local atmospheric conditions and on the nature of the reflector or transparent cover used, but can generally be approximated by an air mass 2 distribution between 0.3 and 2.5 μm (1).

Although it is known from Kirchoff's Law that the absorptivity and emissivity of a surface for any given wavelength and direction are equal, it is common to use the term emissivity (ε) when referring to the thermal radiation emitted by a solar absorber. The absorber exchanges thermal radiation with its transparent cover or local environment at wavelengths corresponding to its operating temperature. Flat plate collectors generally operate at temperatures between 280 and 400 K and emit thermal radiation with wavelengths between 2.5 and 30.0μm.

A good selective surface is one which has a high absorptivity for solar radiation whilst emitting little thermal radiation. In a conventional flat plate collector, a surface with a low emissivity produces a lower cover glass temperature and hence a lower overall loss coefficient than a black surface. The magnitude of the performance improvement varies with the operating conditions but is generally small for low temperature heating (eg swimming pools) and greater for higher temperature applications. A reduction in radiation heat transfer between the absorber and cover glass leads to a greater temperature difference and hence a greater convective transfer between the two. The improvement in instantaneous performance is therefore usually calculated by an iterative technique and a system model is required to establish the real long term benefits of using selective surfaces (2).

The predicted instantaneous performance improvements have been substantiated by collector performance testing and it can be seen from the effect on instantaneous collector performance that reductions in emissivity below 0.1 will produce a negligible improvement in energy collection (3). Computer models suggest that improvements in annual energy collection of the order of 13% may be achieved by the use of selective surfaces in domestic water heating systems in the UK (4). There is however insufficient practical experience in the operation of systems with good selective surfaces in the UK to confirm these predictions.

There is a strong incentive to develop selective surfaces, and there has been a widespread effort over the last few years to develop a low cost selective surface with absorptivity near to 0.95 and emissivity near to 0.1. This has generated a requirement for a cheap and rapid means of evaluating the surfaces, and equipment has been developed at Cardiff to fulfil this need.

Measurement of Emissivity

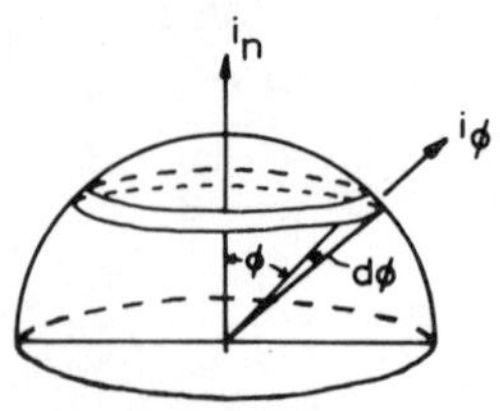

Fig. 1 Spacial distribution of radiation

The spacial distribution of the radiation emitted by a black body is defined by Lambert's Cosine Law.

$$i_{b\phi} = i_{bn} \mathrm{Cos}\ \phi \qquad \ldots.(1)$$

where i_ϕ is the intensity of radiation or the energy flow per unit solid angle per unit area, at an angle ϕ to the surface normal (Fig.1). But real surfaces may not always have such a convenient distribution and it is therefore important to define precisely what is being measured and to ensure that the correct parameters are subsequently used in heat transfer calculations.

The power radiated by a black surface may be obtained by integrating the intensity over the hemisphere, and is known to be related to its absolute temperature by the Boltzman Law.

$$P_b = \int_{\phi=0}^{\pi/2} 2\ \pi i_{bn}\ \mathrm{Cos}\phi.\mathrm{Sin}\phi.d\phi = \pi i_{bn} = \sigma T^4 \qquad \ldots.(2)$$

But for a real surface the intensity may not vary sinusoidally with incidence angle and is usually defined in terms of an angularly dependent emissivity compared with a black body at the same temperature.

$$i_\phi = \varepsilon'(\phi,T).\frac{\sigma T^4 \mathrm{Cos}\phi}{\pi} \qquad \ldots.(3)$$

Hence the total power radiated by a real surface may be obtained by integrating over the hemisphere.

$$P = 2\sigma T^4 \int_{\phi=0}^{\pi/2} \varepsilon'(\phi,T)\ \mathrm{Sin}\ \phi.\mathrm{Cos}\ \phi.d\phi \qquad \ldots.(4)$$

The radiation emitted by a black body at any given temperature T will have a spectral distribution given by Planck's Law.

$$P_b(\lambda) = \frac{C_1.\lambda^{-5}}{e^{C_2/\lambda T}-1} \qquad \ldots.(5)$$

Where C_1 and C_2 are constants.

And the wavelength of maximum emission is given by Wien's Displacement Law

$$\lambda\ \mathrm{max}\ T = 0.0029\ \mathrm{M\ K} \qquad \ldots.(6)$$

Now the power radiated from a real surface may be related to the power emitted by a black surface at the same temperature by using an overall effective emissivity even though in general their spectral distributions may be different.

$$P = \int_{\lambda=0}^{\infty} \varepsilon(\lambda,T).\ P_b(\lambda,T).d\lambda = \varepsilon\sigma T^4 \qquad \ldots.(7)$$

In general ε will be a function of temperature and although most materials considered for use as selective surfaces exhibit a fairly constant emissivity over the wavelengths 2.5 to 30 m, it is important to note that real surfaces do not necessarily conform to an emission spectrum of the simple Planckian form, and the emitted spectral distribution may vary with temperature. Any method used to determine ε must therefore be employed over the complete emission spectrum.

In most designs of solar collector the radiation exchanges occur in configurations with large aspect ratios and hence hemispherical emissivity values are required for engineering calculations. It is known that metallic reflectors tend to exhibit values of hemispherical emissivity (ε) which are slightly higher than their normal emissivity (ε_n) and typical values are given by Rogers and Mayhew (5).

$$1.1 \leqslant \varepsilon/\varepsilon_n \leqslant 1.3 \qquad \ldots.(8)$$

Hence errors may be introduced by the approximation which is commonly used of taking one value of emissivity at near normal incidence. In order to determine the effective hemispherical emissivity ε experimentally, equ.4 should be evaluated not only over the complete emission spectrum but also at a number of angles to the normal.

According to Kirchoff's Law for radiation of a given wavelength and angle of incidence, the absorptivity and emissivity of a surface are equal, and for an opaque surface each with the reflectance sum to unity.

$$\alpha(\lambda,\phi) = \varepsilon(\lambda,\phi) = 1 - \rho(\lambda,\phi) \qquad \ldots.(9)$$

Hence there are three potential ways of obtaining the emissivity; from measurements of absorption, emission or reflection.

Reflectance Techniques

Commercially available spectrophotometers can be used to measure monochromatic specular reflectance for wavelengths from 0.2 to 50μm, but in order to determine the emissivity from equation 9 the total reflectance including both specular and diffuse components must be measured. Integrating spheres are available for the measurement of total reflectance in the visible and near infra-red regions but are not widely available for longer wavelengths (6). If integrating spheres are used then a number of ports are required to permit the evaluation of emissivity over a range of incidence angles.

Bearing in mind the restrictions explained above, it is possible to obtain an approximation for the emissivity by a less thorough technique.
(a) For example, if a spectrophotometer is available with a specular reflectance attachment for the appropriate range of wavelengths then the specular reflectance $\rho_s(\lambda)$ may be determined for a range of incidence angles.

Now $$\rho(\lambda,\phi) = \rho_s(\lambda,\phi) + \rho_d(\lambda,\phi) \qquad \ldots.(10)$$

hence using equation (9) it follows that

$$\varepsilon'(\lambda,\phi) \leqslant 1 - \rho_s(\lambda,\phi) \qquad \ldots.(11)$$

This might be used as an indication that a surface has selective properties in order to justify a complete investigation.
(b) Similarly if it is thought that the spectral emissivity of a particular surface is approximately constant over the appropriate emission spectrum, the radiation from a monochromatic laser beam reflected by the surface may be measured for a range of incidence angles with a goniometrically mounted detector to estimate both

specular and diffuse reflectance. A sample of commercial black chromium selective black was measured using a laser at 9.6μm wavelength and the reflectance was found to be predominantly specular and to vary between 0.75 and 0.85 over the surface. A sample of matt black paint showed a reflectance of greater than 0.99 by this method.

Either of these approximate methods might be useful as a guidance to suggest that a surface is selective, but neither is a suitable method for surface development purposes.

In conclusion, reflectance techniques are complicated by the need for both wavelength and angular integrals and equipment costs are usually high. Despite the high accuracy possible for each monochromatic measurement with reflectance measuring equipment, the finally derived emissivity must contain an accumulation of errors. The accuracy of reflectance measurements depends on the difference between two signals, and for good selective surfaces with a high value of reflectance this difference is small and the accuracy is correspondingly reduced.

Calorimetric Techniques

The emissivity of a surface may be deduced from measured heat flows and temperatures when it is participating in a heat transfer process which includes a radiation exchange.

A simple arrangement is shown in Fig.2 where a black surface is electrically heated and exchanges heat with an air cooled specimen surface. The dimensions of the enclosure may be selected to minimise convective exchange and the hot components are well insulated from atmosphere such that rear and edge losses from the heated region are negligible. The heat exchange process may be represented by a heat balance using notation detailed in Fig.2.

Fig. 2 Calorimetric emissivity measurement

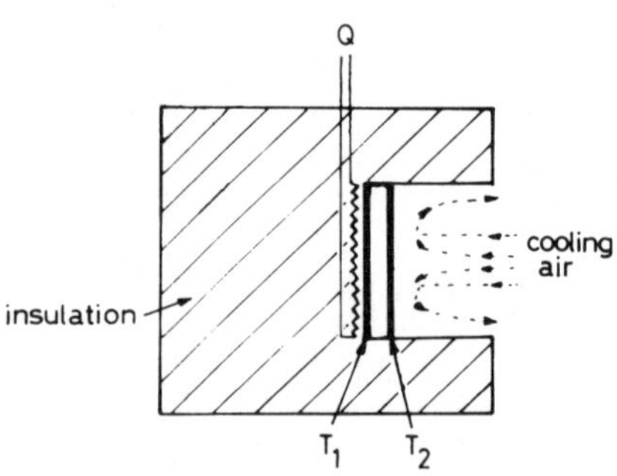

$$Q = A\sigma\varepsilon_e(T_1^4 - T_2^4) + AU(T_1 - T_2) + h(\bar{T} - Ta) \qquad ....(12)$$

$$\text{where } \varepsilon_e = \frac{1}{(\frac{1}{\varepsilon_1} + \frac{1}{\varepsilon_2} - 1)} \qquad ....(13)$$

For a well insulated enclosure $h \to 0$ and for small values of $(T_1 - T_2)$ equ.12 may be simplified using the approximation for small differences in 4th powers to

$$Q = (4A\sigma\varepsilon_e\bar{T}^3 + AU)\ (T_1 - T_2) \qquad ....(14)$$

$$\text{where } \bar{T} = \frac{T_1 + T_2}{2}$$

If air is the gas used between the two radiating surfaces then the value of U will vary with $\bar{T}$, and for small plate sizes edge effects may be significant. The arrangement must therefore be calibrated for known values of ε_1 and ε_2 to determine the variation of $U(\bar{T})$ over the temperature range to be used for testing. Once calibrated the device may be operated at a number of values of electrical power to provide points where ε_2 can be determined using equations 13 and 14.

The advantages of a method of this type are:

1. It can be operated at typical collector operating temperatures and hence the emission spectrum is inherently correct.
2. The enclosure is similar to that of a solar absorber with a cover glass and hence the emissivity measured has the appropriate spacial distribution.
3. No sophisticated or expensive equipment is required to perform the measurements.

There are however, a number of significant disadvantages of this simple method.

1. The system must be permitted to reach a steady state before a reading is valid and this is likely to take around 12 hours. Hence it may take a week to obtain a suitable set of data points from which to determine the emissivity.
2. The radiative component is approximately 30% of the total heat transfer for a black specimen surface, but only 1% for a good selective surface. Hence the accuracy of the measurement is very much poorer for good selective surfaces.

A typical set of results showing measured values of emissivity over a range of mean operating temperatures for a sample of commercial black chromium is shown in Fig.3.

The accuracy may be improved by performing a similar heat transfer process in a vacuum where the convective heat transfer can be reduced to a negligible value. There is also a possibility of using a cyclic temperature fluctuation in a vacuum if the surfaces can be mounted on specimen plates with a sufficiently low thermal capacity. The advantages of a cyclic system are that only the cyclic changes of temperature need be measured rather than absolute values, and control of the temperature of the surroundings is less important. However, for a suitable reduction in convective heat transfer, a vacuum of the order of 10^{-4} Torr would be required and hence the development of a suitable piece of equipment is likely to be expensive.

Direct Radiation Measurement

If a surface is heated to a known temperature and quantitative measurements are made of its emitted radiation flux, then its emissivity may be directly calculated. There are two primary classes of infra-red detector which might be used for this purpose.

1. Photon Detector: The incident photons interact with the detector surface to give a change in electrical properties and hence such a device must by nature have a non-linear spectral response. This means that the output cannot be directly related to the input radiation without spectral compensation and hence photon detectors are unsuitable for simple measurement purposes.

2. Thermal Detectors: The heating effect of the incident flux on the surface of a thermal detector causes a change in electrical properties. The surface can be made to absorb a wide band of wavelengths and hence to exhibit a flat spectral response over the range of interest.

A thermopile could be used as a thermal detector but it is experimentally easier to utilise a pyro-electric detector which is sensitive only to changes in incident radiation level. A technique has been developed at Cardiff where the emission from a heated test surface is continually compared with that from a reference surface at a known temperature, and the response of the system is such that it is insensitive to slow fluctuations in the environment temperature. (Fig.3).

Fig. 3
Direct radiation measurement

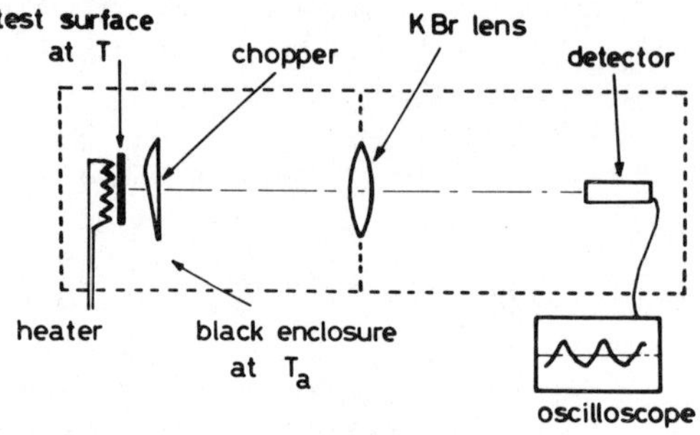

The equipment is calibrated using a black reference surface and the temperature differential of the detector output is determined over a small range of absolute temperatures because the measurement of the temperature difference seen by the detector contains the largest source of error. The slope for a specimen surface is compared with that of the reference surface under the same conditions to determine the emissivity (Fig.4).

Fig. 4
Typical detector output data

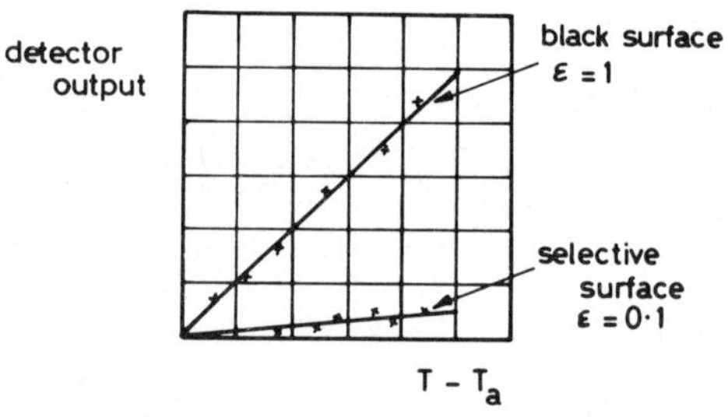

In the Cardiff device the optics are arranged to measure the emissivity over an area of $4 \times 10^{-6} M^2$ using a potassium bromide lens with a flat spectral response over the range 2 to 30μm. The continual comparison is provided by a chopper located between the heated specimen surface and the detector. The front surface of the chopper blade is highly polished and reflects the background radiation from the black enclosure into the detector. The chopping frequency used is 10Hz and this permits steady heating of the test specimen through 10-20°C with manual recording of the detector output and temperatures. The detector receives radiation from a small solid angle (approx 1/65 of the hemisphere) and by rotating the specimen surface, measurements can be made to permit evaluation of the angular integral (Equ.4).

Once the device has been calibrated a set of readings to determine the emissivity of a test surface at any given incidence angle can be performed in 20 - 30 minutes.

Measurement of Solar Absorptivity

For most solar collectors a value of hemispherical absorptivity greater than 95% is desirable and this is readily achievable with matt black paints. However, in the search for selectivity it is common to find lower values of solar absorptivity and a measurement is required to determine the absorptivity of surfaces over the solar spectrum from 0.3 to 2.5μm.

Integrating spheres are commercially available for spectrophotometers over these wavelengths and hence monochromatic total reflectance can be readily measured and used to compute an effective absorptivity for solar radiation.

For rapid measurements to determine an approximate value for the absorptivity of selective surfaces a photo-electric reflectometer has been used at Cardiff. This device contains a tungsten filament lamp which provides a near normal incidence beam of radiation in the visible and near infra-red wavelength regions. This is a poor approximation for solar radiation (1). An annular detector receives both the direct and diffusely reflected radiation and this may be calibrated using surfaces of known reflectance.

Although the results obtained with this device are in good agreement with those published for standard surfaces, it is not possible to use it for other than near normal incidence, and it could give erroneous results for a surface which did not exhibit approximately grey body properties for wavelengths between 0.3 and 2.5μm. The advantages of the device are that it is cheap and gives results quickly. Two calibration check points can be established and the absorptivity of a specimen surface measured within 5 minutes.

Discussion

It has been shown that the radiative behaviour of real surfaces is complex and that there are a number of ways of measuring the properties required to determine the benefits of selective surfaces for solar utilization equipment.

The required thermal emissivity value has been measured by three methods for a commercial sample of black chromium and it can be seen that good agreement has been achieved between the methods (Table 1). The most rapid method of assessment which also has the greatest accuracy uses a pyro-electric detector for direct radiation measurement.

A photo-electric reflectometer has been shown to be a useful tool for rapid assessment of solar absorptivity although a spectrophotometer with an integrating sphere would be more accurate.

Table 1 EMISSIVITY VALUES FOR COMMERCIAL SAMPLE OF BLACK CHROME SELECTIVE SURFACE

Method of Measurement	ε	Comment
1. Total Reflectance (7)	0.04-0.17	depending on plating time
2. Laser spectral reflectance at 9.6μm	0.15-0.25	surface variation
3. Calorimetric method	0.15±0.05	experimental error
4. Direct radiation method	0.12±0.02	experimental error

Nomenclature

A	Area of specimen surface
α	absorptivity
ε	hemispherical emissivity
$\varepsilon'(\phi,T)$	directional total emissivity at temperature T
$\varepsilon'(\phi,\lambda)$	directional spectral emissivity
$\varepsilon\ (\lambda,T)$	hemispherical spectral emissivity at temperature T
ε_e	effective emissivity for exchange between surfaces 1&2
h	overall heat transfer coefficient between heated region and ambient
i	radiation intensity
λ	wavelength of radiation
P	hemispherical radiative emissive power
ϕ	angle from surface normal
ρ	hemispherical total reflectance
ρ_s	specular reflectance
ρ_d	diffuse reflectance
Q	Electrical power input to heater
σ	Boltzmán constant
T	Temperature of surface
Ta	Ambient air temperature
$\bar{T}$	Arithmetic mean temperature between surfaces 1&2
U	Convective heat transfer coefficient between surfaces 1&2

Suffices

b	black body
n	normal to surface

References

(1) W.B. Gillett, 'Solar simulators and indoor testing'. ISES-UK Conference (C11) April, 1977.

(2) J.A. Duffie and W.A. Beckman, 'Solar energy thermal processes' Wiley 1974.

(3) NBS Technical Note 899 'Development of proposed standards for testing solar collectors and thermal storage devices' Feb. 1976.

(4) B.J. Brinkworth, 'Factors affecting the performance of solar water heating systems'. International Conference on Solar Building Technology NELP/UNESCO July, 1977.

(5) G.F.C. Rogers and Y.R. Mayhew, 'Engineering thermodynamics, work and heat transfer' Longmans 1967.

(6) W.G. Egan and T. Hilgeman, 'Integrating spheres for measurements between 0.185 m and 12 m' Applied Optics Vol. 14, No. 5, May 1975.

(7) G. MacDonald et al., 'Fundamental studies of black chrome for solar collector use'. NASA TMX - 73461 Aug. 1976.

Numerical heat analysis of solar heat absorbers using digital computers

3.4

I. A. Sakr
Solar Energy Laboratory, National Research Center, Cairo, A.R.E.

M. S. Abd Elhafiez
Faculty of Engineering, Alexandria University, Alexandria, A.R.E.

A. S. Hegazy
Faculty of Engineering, Monoufia University, Shebin, A.R.E.

ABSTRACT

The present research offers a numerical method for analyzing the solar heat absorbers of concentrating collectors without neglecting radiation loss. The analysis is based on the classification of the solar absorbers into two groups. The first group includes flat plate, circular flat plate, and cylindrical heat absorbers. The second group contains spherical and pancake heat absorbers. The first group of absorbers may be directly analyzed by applying the proposed method given below. The absorbers in the second group must be modified to equivalent longitudinal cylindrical absorbers with the same area.

This proposed method was applied to some cases of solar absorbers to obtain the distribution of the absorber surface temperature and the fluid bulk temperature along the absorber length. The maximum length of a heat absorber, at which there is no temperature gradient between the absorber outer surface and the fluid, is computed also for each case.

The results show that the maximum length of a heat absorber is affected by the intensity of solar radiation input to the absorber. The maximum length is long for low values of this radiation flux. It decreases rapidly with the increase of the value of the radiation flux input to the absorber, until a minimum length where the rate of heat transfer to the fluid, and the rate of heat losses are maximum. With an increase in the value of the radiation flux, the maximum length increases slowly due to a decrease in the rates of heat transfer. Also, the results show that the rate of increase of the fluid bulk temperature is high at small distances from the inlet point, and decreases for long distances. The circular flat plate heat absorber appears to be more efficient than the other types of absorbers. It requires less length (smaller radius) than other absorbers to heat a fluid to the same temperature at the same value of solar radiation flux input to the absorber.

INTRODUCTION

Solar energy is presently being economically collected and used at low temperatures (about 100°C above ambient temperature) with flat plate collectors. The limitation of a flat plate collector is its low energy level. This energy can be mechanically concentrated to raise its temperature to the desired level. This has been done using different designs of reflecting surfaces.

In spite of the existence of many papers analyzing flat plate collectors, little has been published about heat analysis of solar concentrators' absorbers [1-3]. The majority of researchers in their studies considered the heat transfer equations of the heat absorbers and their solutions neglecting the heat loss by radiation or assuming it linear. In both cases, a big error results in evaluating the system temperatures, especially if the intensity of input solar radiation to the absorber, and the temperature of the heat absorber surface are very high.

THE PROPOSED NUMERICAL METHOD

To simplify the analysis, the following assumptions were made:

1. Performance is steady state;
2. Thermal properties are independent of temperature and the incident angle of the solar radiation beam;

3. One dimensional heat flow through pipes;
4. The thermal conductivity of insulation was neglected if the outer surface temperature of the insulation was considered to be equal to that of the outer surface of the heat absorber;
5. Constant rate of heat energy per unit area received by the absorber.

The general equations governing the heat transfer of flat plate, cylindrical, pancake, and spherical heat absorbers at a distance x from inlet may be written in the form:

$$I_N C_a \eta_c = \frac{A_i}{A_a} h_i (T_x - T_{fx}) + \frac{A_l}{A_a} h_c (T_x - T_a) + \frac{A_l}{A_a} \epsilon\sigma (T_x{}^4 - T_a{}^4); \qquad (1)$$

$$\dot{m} C_p \frac{dT_{fx}}{dx} = A_i h_i (T_x - T_{fx}); \qquad (2)$$

Where:

A_i Area of heat transfer between the outer surface of the heat absorber and the fluid per unit length of the absorber;
A_l Area of heat transfer loss per unit length of the absorber;
A_a Absorber (target) surface area per unit length of the absorber;
h_i Overall coefficient of heat transfer between the outer surface of the absorber and the fluid;
h_c Convective coefficient of heat transfer between the outer surface of the absorber and the surroundings;
T_x Outer surface temperature of the absorber at a distance x from the inlet;
T_{fx} Fluid bulk temperature at a distance x from the inlet;
T_a Ambient temperature;
η_c Concentrator reflection efficiency;
$\dot{m}$ Fluid mass flow rate;
C_p Average specific heat of the fluid;
ϵ Emissivity of the outer surface of the absorber;
σ Stefan Boltzman constant.

The solution of equations (1), and (2) is not easy because of the existing non-linear term expressing the rate of heat loss by radiation. But these two equations can be solved by numerical methods. A proposed method for solving these two equations is summarized below:

Equation (2) can be rewritten in the form:

$$\frac{dT_{fx}}{dx} = K_1 (T_x - T_{fx}); \qquad (3)$$

Where:

$$K_1 = \frac{A_i h_i}{\dot{m} C_p}$$

The total length of the absorber is divided to equal small segments, each one having a length δ then T_x, and T_{fx} of each segment may be evaluated as follows:

From equation (1), T_x for a segment at a distance x from inlet can be found by using the Newton-Raphson method of iteration, and considering T_{fx} for this segment is constant i.e.

$$T_{x_{n+1}} = \frac{3\frac{A_l}{A_a}\epsilon\sigma T_{x_n}^4 + \frac{A_i}{A_a} h_i T_{fx} + \frac{A_l}{A_a} h_c T_a + \frac{A_l}{A_a}\epsilon\sigma T_a + I_N C_a \eta_c}{\frac{A_l}{A_a} h_c + \frac{A_i}{A_a} h_i + 4\frac{A_l}{A_a}\epsilon\sigma T_{x_n}^3}; \qquad (4)$$

Where:

n Denotes the number of iteration

Equation (4) is repeated for each obtained T_x till two successive values of T_x are equal or nearly equal. T_{fx} for a segment at the same distance can be found from equation (3) by using the fourth-order Runge-Kutta method, and considering T_x for this segment is constant as follows:

$$G_1 = \delta K_1 \left[T_{x-\delta} - T_{fx-\delta}\right]$$

$$G_2 = \delta K_1 \left[T_{x-\delta} - (T_{fx-\delta} + \frac{G_1}{2})\right]$$

$$G_3 = \delta K_1 \left[T_{x-\delta} - (T_{fx-\delta} + \frac{G_2}{2})\right]$$

$$G_4 = \delta K_1 \left[T_{x-\delta} - (T_{fx-\delta} + G_3)\right]$$

$$G = \frac{1}{6}(G_1 + 2G_2 + 2G_3 + G_4) \qquad (5)$$

$$T_{fx} = T_{fx-\delta} + G \qquad (6)$$

But for circular flat plate absorber equation (2) is modified to the following form:

$$\dot{m}\, C_p \frac{dT_{fr}}{dr} = \pi r h_i (T_r - T_{fr}); \qquad (7)$$

Where:

r Radius of the absorber from inlet;
T_r Absorber outer surface temperature at a radius r from inlet;
T_{fr} Fluid bulk temperature at a radius r from inlet.

The above equation can be rewritten as follows:

$$\frac{dT_{fr}}{dr} = K_3 r (T_r - T_{fr}); \qquad (8)$$

Where:

$$K_3 = \frac{\pi h_i}{\dot{m} C_p}$$

Replacing the subscript x by r in equation (1), the solution of it and equation (8) is similar to that of (1), and (3). But T_{fr} is evaluated using the fourth-order Runge-Kutta method as given below:

$$G_1' = \delta K_3 r \left[T_{r-\delta} - T_{fr-\delta}\right]$$

$$G_2' = \delta K_3 r \left[T_{r-\delta} - (T_{fr-\delta} + \frac{G_1'}{2})\right]$$

$$G_3' = \delta K_3 r \left[T_{r-\delta} - (T_{fr-\delta} + \frac{G_2'}{2})\right]$$

$$G_4' = \delta K_3 r \left[T_{r-\delta} - (T_{fr-\delta} + G_3')\right]$$

$$G' = \frac{1}{6}(G_1' + 2G_2' + 2G_3' + G_4') \qquad (9)$$

$$T_{fr} = T_{fr-\delta} + G' \tag{10}$$

In the case of considering thermal conductivity of the absorber insulation, the above proposed numerical method can be applied as explained in Appendix A.

PROCEDURE FOR USING THE PREVIOUS PROPOSED NUMERICAL METHOD

1. The total length of the absorber is divided into small equal segments, each one having a length δ.
2. Substituting in equation (4) with $T_{fx} = T_{fo}$ (T_{fo} is the inlet fluid temperature), and beginning the iteration process with $T_x = T_{fo}$. Continuing the iteration T_o can be determined (T_o is the absorber surface temperature at inlet).
3. Substituting in equations (5) and (6) with $T_{x-\delta} = T_o$, and $T_{fx-\delta} = T_{fo}$, to obtain $T_{f\delta}$ in case of flat plate, cylindrical, pancake, and spherical heat absorbers. However, in case of circular flat plate heat absorber, substitute with the same values in equations (9) and (10).
4. Repeat (2) with $T_{fx} = T_{f\delta}$, to get T_δ. Then repeat (3) with $T_{x-\delta} = T$, to obtain $T_{f2\delta}$. This work is continued for the total length of the absorber.
5. A maximum length of a heat absorber for a certain system is that length at which there is no temperature gradient between the absorber outer surface and the fluid (i.e. there is no heat transfer between the absorber and the fluid). This maximum length can be found by comparing T_x and T_{fx} obtained for each segment.

This work is very difficult to be carried out manually but it can be easily done with a digital computer. For this purpose, the authors devised two computer programs. One for solving flat plate, cylindrical, spherical and pancake heat absorbers, the other for solving circular flat plate. The block diagram for both is given in Appendix B.

The procedure explained above was widely applied to different absorbers with different configurations and working conditions.

EXAMPLE OF USING THE PROPOSED NUMERICAL METHOD

The method described above was applied to different absorbers having the following initial data:

$h_i = 10$ W/sq. m°K $\quad$ $h_c = 5$ W/sq. m°K

$T_{fo} = T_a = 20$°C $\quad$ $\dot{m}C_p = 15$W/sec. °K

$I_N C_a \eta_c$ varies from 5.0 to 500 KW/sq. m

$\epsilon = 0.2$

The digital computer IBM 1620 was used for running the two computer programmes.

The black body temperature for different values of solar radiation flux ($I_N C_a \eta_c$) were computed by substituting in equation (4) with $\epsilon = 1.0$, and $T_{fx} = 0.0$. The results are presented in Figure 1.

Figure 2 shows the absorbers maximum lengths of different types versus the intensity of solar radiation input to the absorber ($I_N C_a \eta_c$).

A dimensionless fluid temperature ($T_{fx}/T_b 5000$), ($T_b 5000$ is the black body temperature at input flux to the absorber 5000 W/sq. m), distribution versus a dimensionless distance ($x/x_{max.}$) is plotted in Figure 3 for flat plate, cylindrical, spherical and pancake types. A dimensionless fluid temperature ($T_{fr}/T_b 5000$) distribution versus a dimensionless radius ($r/r_{max.}$) is given also in Figure 3.

Figure 4 shows the collection efficiencies of different types of solar heat absorbers.

Fig. 1

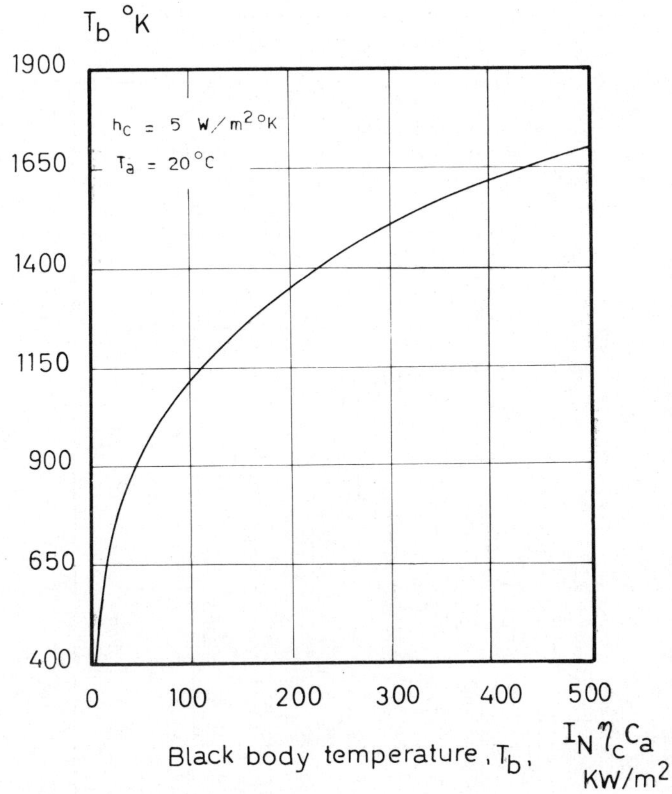
T_b °K
1900
1650
1400
1150
900
650
400
h_c = 5 W/m²°K
T_a = 20°C
0
100
200
300
400
500
Black body temperature, T_b,
$I_N \eta_c C_a$
KW/m²

Fig. 2

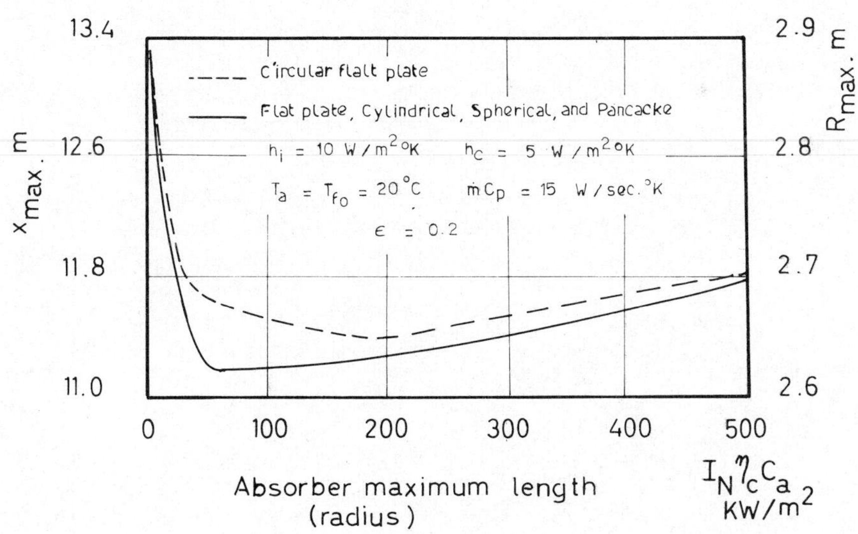
13.4
12.6
11.8
11.0
x_{max}, m
2.9
2.8
2.7
2.6
R_{max}, m
Circular flalt plate
Flat plate, Cylindrical, Spherical, and Pancacke
h_i = 10 W/m²°K
h_c = 5 W/m²°K
T_a = T_{fo} = 20°C
$\dot{m}C_p$ = 15 W/sec.°K
ε = 0.2
0
100
200
300
400
500
Absorber maximum length
(radius)
$I_N \eta_c C_a$
KW/m²

Fig. 3

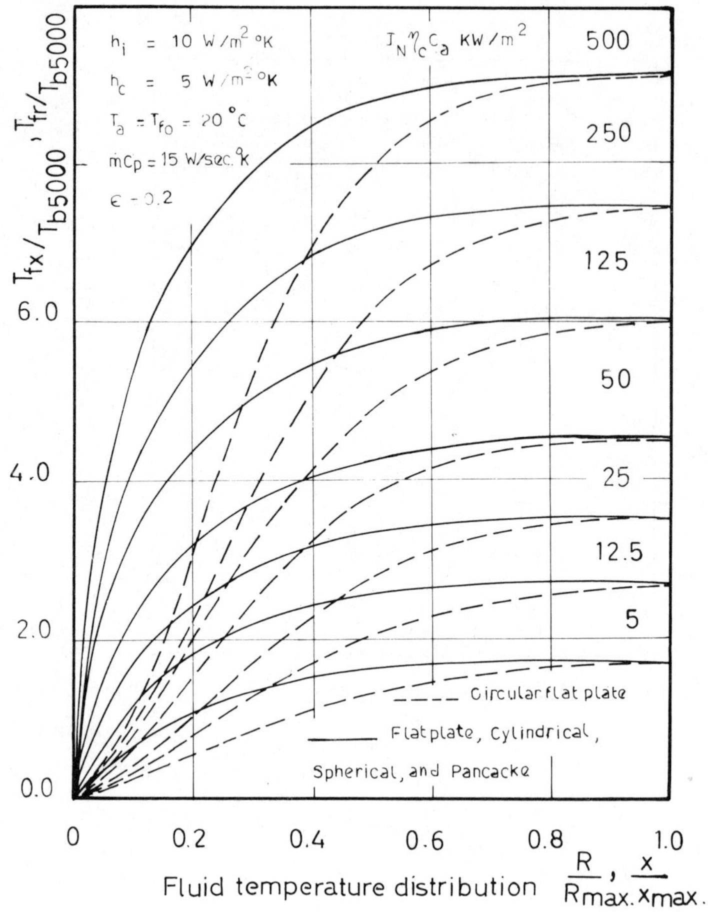
h_i = 10 W/m^2 °K
h_c = 5 W/m^2 °K
$T_a = T_{fo}$ = 20 °C
$\dot{m}C_p$ = 15 W/sec.°K
ε = 0.2
$I_N \eta_c C_a$ KW/m^2
500
250
125
50
25
12.5
5
Circular flat plate
Flatplate, Cylindrical, Spherical, and Pancacke
T_{fx}/T_{b5000}, T_{fr}/T_{b5000}
6.0
4.0
2.0
0.0
0
0.2
0.4
0.6
0.8
1.0
Fluid temperature distribution $\frac{R}{R_{max.}}, \frac{x}{x_{max.}}$

Fig. 4

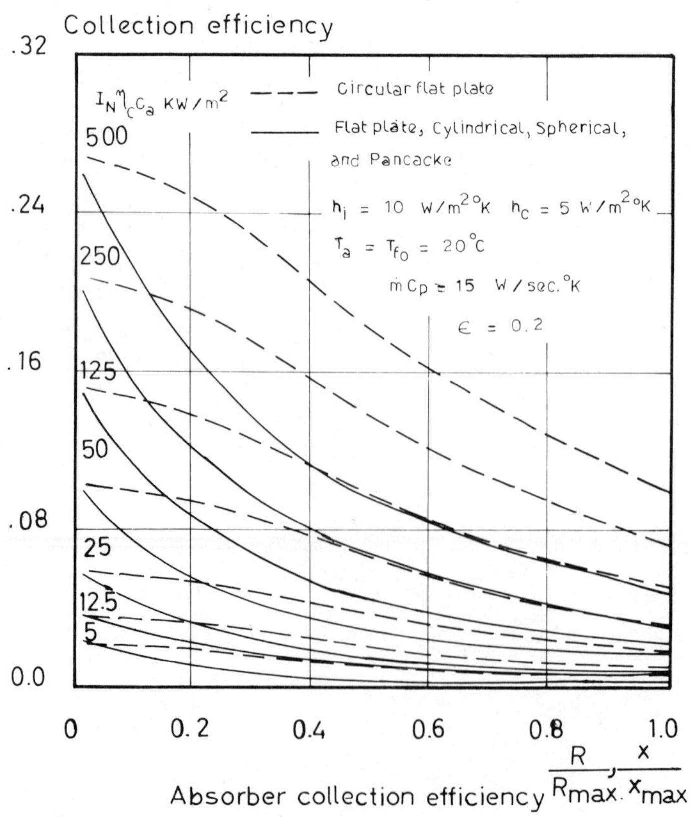
Collection efficiency
.32
.24
.16
.08
0.0
$I_N \eta_c C_a$ KW/m^2
Circular flat plate
Flat plate, Cylindrical, Spherical, and Pancacke
h_i = 10 W/m^2°K h_c = 5 W/m^2°K
$T_a = T_{fo}$ = 20°C
$\dot{m}C_p$ = 15 W/sec.°K
ε = 0.2
500
250
125
50
25
12.5
5
0
0.2
0.4
0.6
0.8
1.0
Absorber collection efficiency $\frac{R}{R_{max.}}, \frac{x}{x_{max}}$

Appendix A

Use of the proposed numerical method for analyzing solar absorbers considering the insulation; thermal conductivity.

The heat transfer equations representing the performance of solar heat absorbers, except circular flat plate one, considering the thermal conductivity of insulation may be written as follows:

$$\frac{K}{s}(T_{px} - T_x) = h_c (T_x - T_a) + \epsilon\sigma (T_x^4 - T_a^4); \qquad (A1)$$

Where:

K Thermal conductivity of insulation material;
T_x Outer surface temperature of the insulation;
T_{px} Outer surface temperature of the absorber;
s Insulation thickness.

$$I_N C_a \eta_c = \frac{KA_1}{sA_a}(T_{px} - T_x) + \frac{A_i h_i}{A_s}(T_{px} - T_{fx})$$

$$I_N C_a \eta_c = F_1 (T_{px} - T_x) + F_2 (T_{px} - T_{fx}); \qquad (A2)$$

Where:

$$F_1 = \frac{KA_1}{sA_a}, \quad F_2 = \frac{A_i h_i}{A_a}$$

$$\frac{d\,T_{fx}}{dx} = K_1 (T_{px} - T_{fx}); \qquad (A3)$$

K_1 is defined as before.

From equation (A2)

$$T_{px} = \frac{I_N C_a \eta_c + F_1 T_x + F_2 T_{fx}}{F_1 + F_2} \qquad (A3)$$

Substituting T_{px} from equation (A4) in equation (A1)

$$\frac{K}{s}\left[\frac{I_N C_a \eta_c + F_1 T_x + F_2 T_{fx}}{F_1 + F_2} - T_x\right] = h_c (T_x - T_a) + \epsilon\sigma (T_x^4 - T_a^4) \qquad (A5)$$

Equations (A3), A4), and (A5) can be solved numerically as in the case before;

$$T_{xn+1} = \frac{3\,T_{x-\delta}^4 + h_c T_a + \epsilon\sigma T_a^4 + \frac{K}{s}\,\frac{I_N C_a \eta_c + F_2 T_{fx-\delta}}{F_1 + F_2}}{4\epsilon\sigma\,T_x^3 + h_c + \frac{KF_2}{s\,(F_1 + F_2)}} \qquad (A6)$$

Substituting T_x obtained by iteration from equation (A6) in equation (A4) to obtain T_{px}. Equation (A3) can be solved using the fourth-order Runge-Kutta method by using equations (5), and (6), and replacing T_x by T_{px}.

Replacing the subscript x by r in equation (A1), and (A2), the proposed method can be applied to the circular flat plate absorber considering the insulation; thermal conductivity to solve these two equations simultaneously with equation (8). T_r is found by using equation (A3) and putting the subscript r instead of x. T_{fr} is obtained by using equations (9), and (10).

Appendix B

Block diagram for using the proposed numerical method on digital computer

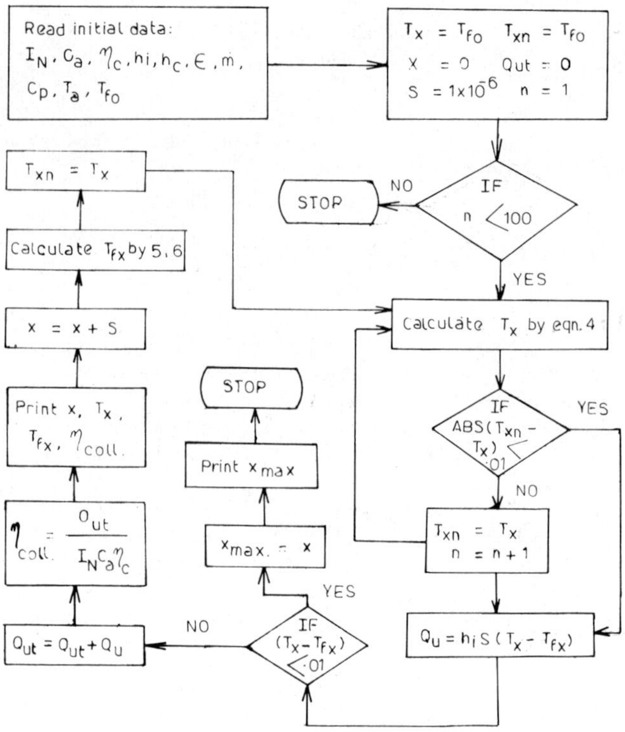

CONCLUSION

In spite of the simplicity of the given analysis the following results are observed:

1. The maximum length, or the maximum radius of an absorber is long for the low values of the input radiation flux to that absorber ($I_N C_a \eta_c$). That is due to the low temperature of the absorber surface, so that the rate of heat transfer to the fluid, and the rate of heat loss by convection and radiation are small;
2. Increasing the value of $I_N C_a \eta_c$, the maximum length or radius of a heat absorber decreases rapidly until a minimum length or radius, where the rates of heat transfer to the fluid and the losses are maximum;
3. More increase in the value of ($I_N C_a \eta_c$) results in the maximum gradual increase of the length or radius due to a decrease in rates of heat transfer;
4. The rate of increase of the fluid temperature is high at small distances from inlet, and decreases for long distances. Also, the collection efficiency is high at inlet and gradually decreases till a minimum at the maximum length or radius;
5. It is obvious that the circular flat plate heat absorber is more efficient than the other types. It requires less length (radius) to heat a fluid to the same temperature at the same value of solar radiation input to the absorber ($I_N C_a \eta_c$). That is due to the increase of the area of heat transfer with the increase of the radius.

REFERENCES

(1) M. H. Cobble, 'Heat exchangers for solar concentrators' Solar Energy J. Vol. 7, No. 1, 1963.

(2) M. H. Cobble, 'Temperature fields of solids heated by solar concentrators' Solar Energy J. Vol. 7, No. 3, 1963.

(3) A. I. Hegazy, 'Solar refrigeration systems' M.Sc. Thesis, Cairo University, 1975

(4) D. G. Moursund, C. S. Duris 'Elementary theory and application of numerical analysis' McGraw-Hill Book Company.

Factors effecting the optimum design of solar energy absorbing surfaces

K. H. Khalil, M. S. Khader and H. Gad
Mechanical Engineering Department,
Faculty of Engineering, Cairo University, Egypt

I. A. Sakr
Solar Energy Laboratory, National Research Centre, Egypt

ABSTRACT

This paper studies the factors affecting the design of solar energy absorbing surfaces, in order to improve the amount of energy absorbed per square meter. Therefore the study examines firstly the data of the available solar energy and then analyses the geometries of the currently used solar heaters. Estimates are made for heating capacity per person/day, required heating area per person and optimum orientation angle for each month. Based on this data comparison of different shapes of solar heaters is then made in order to select the optimum shape.

The paper also presents the results of experimental performance measurements of the heaters studied. As a result of these experiments the velocity profile and temperature distribution inside these heaters have also been determined.

These studies show that (a) the average hot water consumption per person in Egypt is about 27 Lit/day which requires a surface area of about 0.28m^2, (b) flat heaters with flow rate of about 26 Lit/sq. mt and orientation angle of 30° from horizontal are best for such locations. Water circulation inside the solar water heater is affected by an important factor which influences heat collection and C.O.E. This factor was studied experimentally and results were discussed with a view to minimize losses due to insufficient circulation.

INTRODUCTION

Many investigations have been carried out [1 to 10] in order to study the effect of the different factors on the performance of solar water heaters. In this paper some of these factors are reinvestigated with special emphasis on water circulation and temperature distribution inside flat plate solar collectors.

A solar water heater unit consists of a collector fitted to an insulated storage tank. The collector consists of the absorbing surface with a glass or plastic cover and is insulated at the base and sides. Water flows into the collector from below, carrying out the heat from the absorbing plate to the storage tank.

Circulation of water may be natural or forced. Natural circulation is usually used as it is more economic and satisfactory.

The capacity and size of the solar water heater may be indicated by the hot water required per day. This capacity is influenced by (a) family size, (b) time of the year, and (c) nature of the use, and (d) winter temperature level.

The area of the collector can be calculated from the relation:

$$\text{C.O.E} = \frac{\dot{m}\ C_p\ \Delta T}{H} \qquad (1)$$

Where C.O.E. is the coefficient of effectiveness and m is the mass flow rate, Kg/hr. m^2

ΔT is the temperature rise, C°
H is the total solar flux on the tilted surface, kcal/hr. m^2

The results of such calculations are given in Fig. 1.

Fig. 1
Area of collector and heat gain in different months of the year

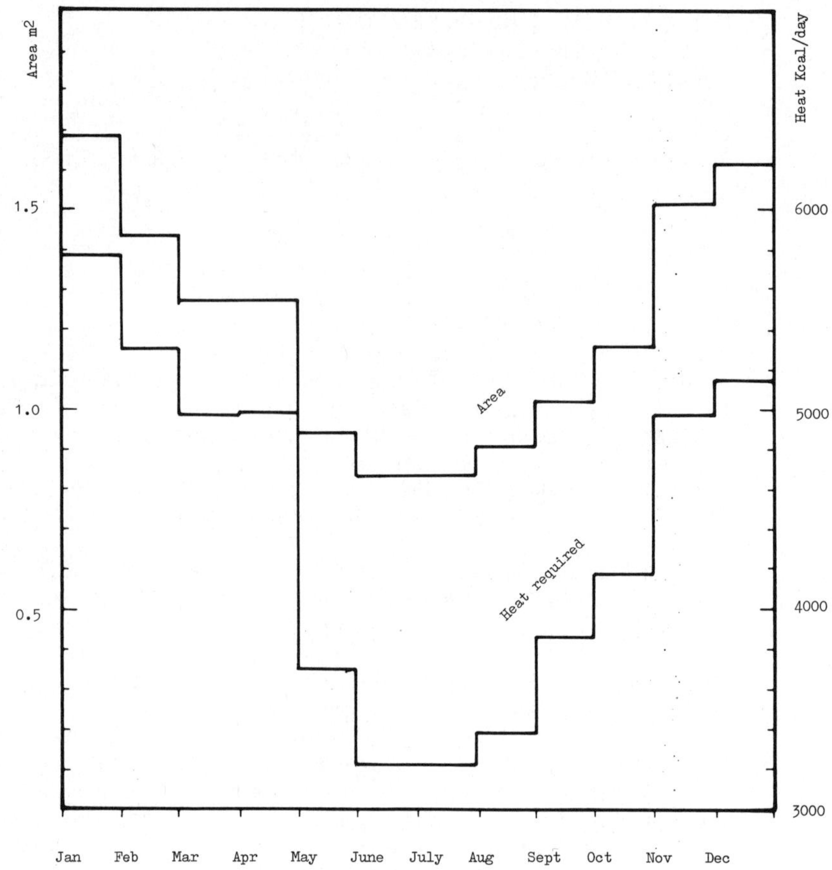

An electric heater may be used as an auxiliary means of heating in winter, during night-time and also for overload and cloudy days, which reduces the size of the solar water heater and increases its reliability, also the consumer benefits from regular hot water.

Angle of inclination

The collector absorbs both the direct and diffuse components of solar radiation. The angle of incidence of the sun rays on the collector plate must be minimized to obtain a maximum solar radiation intensity on the plate of the absorber.

The suns' altitude changes during the day throughout the year, so that there is an optimum angle of inclination of the collector plate with the horizontal, which gives a maximum average intensity of solar radiation throughout the year.

Solving the problem mathematically by computer programming, it was shown that the optimum angle of inclination of the collector plate with the horizontal changes during the year, assuming the values of 15° in summer and 45° in winter with an average of 30° over the whole year.

EXPERIMENTAL WORK

The experimental work here is divided into two parts, the first concerning the performance of the solar collector and the second dealing with the circulation of hot water through the collector.

Performance of the solar collector

Three collectors were used under different operating conditions in order to compare and predict the optimum operating conditions.

All collectors used are corrugated iron sheets (0.8mm thickness with 8mm between plates and an area of $0.3m^2$ for each collector (50 × 60cm). The upper plate being corrugated so that there are a number of channels and fins in between, the plates are painted with a blackened selective coating. The water inlet and outlet are located at the middle of the lower and upper headers respectively. Each collector is covered with an ordinary glass window (3mm thickness) and insulated at the bottom and sides.

Collector (I)

Wide channels (10cm) and narrow fins (1cm) angle of inclination 30°.

Collector (II)

Similar to that above but inclined at 45° with horizontal.

Collector (III)

Narrow channels and fins with angle of inclination 45°.

All experiments were carried out on clear sunny days.
The mass flow rate water is kept constant during each experiment.

After reaching the steady state, the following parameters were measured; inlet and outlet temperature, ambient air temperature, water flow rate and solar intensity, heat collected by the water Qu; and the C.O.E. were calculated from:

$$Q_u = \dot{m}\, C_p\, (T_o - T_i) \qquad (2)$$

$$\text{C.O.E.} = \frac{Q_u}{H} \qquad (3)$$

Figure 2 shows a sample of these results.

Fig. 2

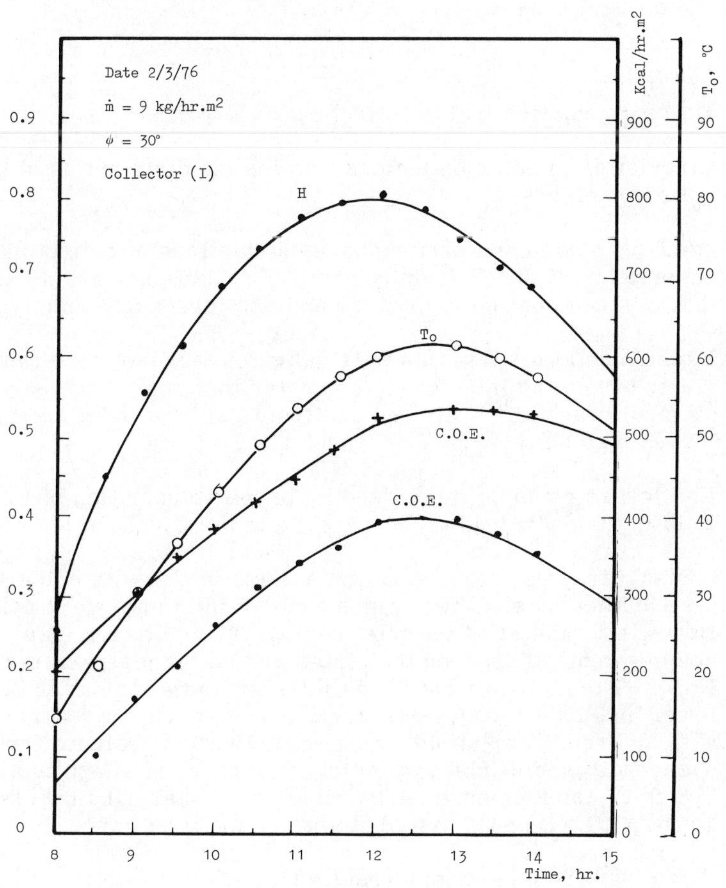

RESULTS

It can be seen from the previous measurements that:

—H, To. Qu and C.O.E. increase from morning up to a maximum value at noon, with a time lag between H and Qu.

—Combining all results for different mass flow rates, it is noticed that the C.O.E. has a maximum value at a mass flow rate of about 26 litres/hour m^2 see Fig 3. On the same figure a comparison of results with some previous workings [4] and [6] are made.

Fig. 3
Comparison of results with the previous work

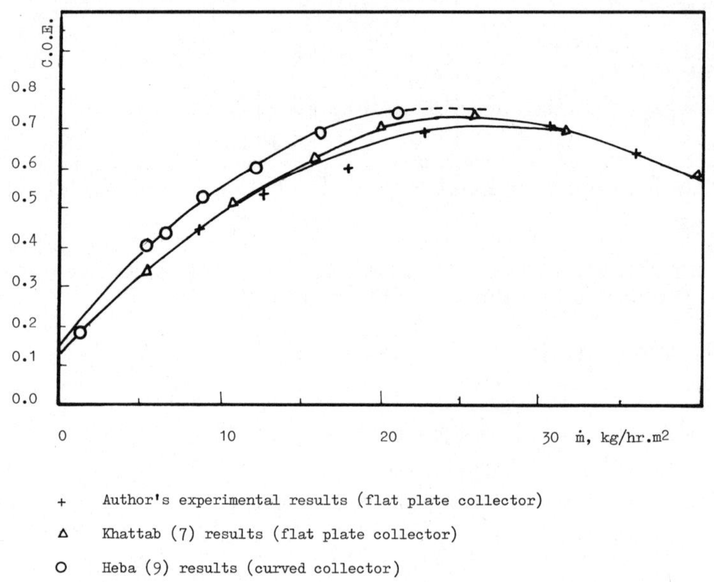

<u>Water circulation inside collectors</u>

Circulation in solar collectors implies the distribution of hot water inside the collector risers.

Standard design and performance calculations of collectors are based on the assumption of the uniformity of flow. Actual measurements on collectors showed that this assumption is not true and end losses are mainly due to these reasons.

Experimental work in this part indicates that the mean temperature of fluid at the far end from the inlet point, is greater than that of the part nearer to it, so that the upward losses at the far end and from the inlet point are greater. These excessive losses may be termed 'End losses'.

End losses could be minimized by proper circulation of water through the collector. This in turn will lead to an increase of the C.O.E. of the collector.

A visual test was carried out on a glass model with coloured water to show flow distribution under different positions of inlet and outlet points. This showed that flow will be faster at the inlet region. The collector (I) was fitted with five thermocouples centred between the plates and on the plates at the top of each channel, Fig. 4. The collector had three inlet and outlet points at the middle and ends of the lower and upper headers respectively. Experiments were carried out on a collector with an area $0.3m^2$ at 30° angle of inclination designed for a flow rate of 26kg/hr. The conditions of inlet and outlet were changed. Results are shown in Figures 5, 6, 7, 8, 9, 10, for temperature distribution, isothermal flow distribution, velocity distribution, C.O.E. and Q_u distributions for the first case.

It is clear from the above results that:

Fig. 4

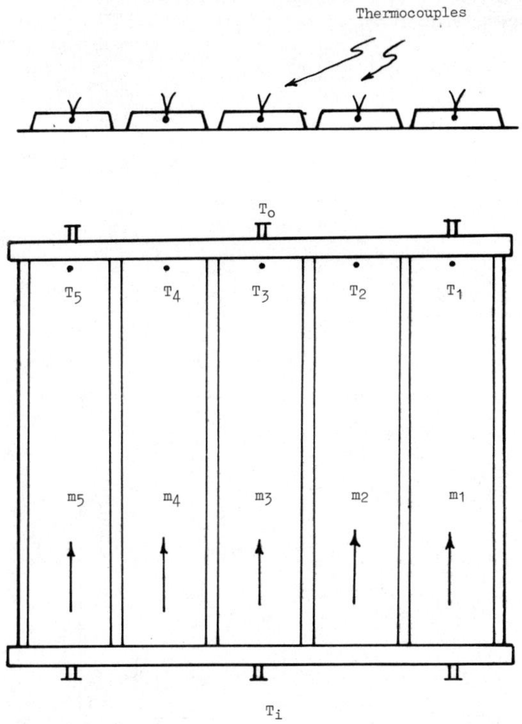

Fig. 5
Temperature distribution across the collector

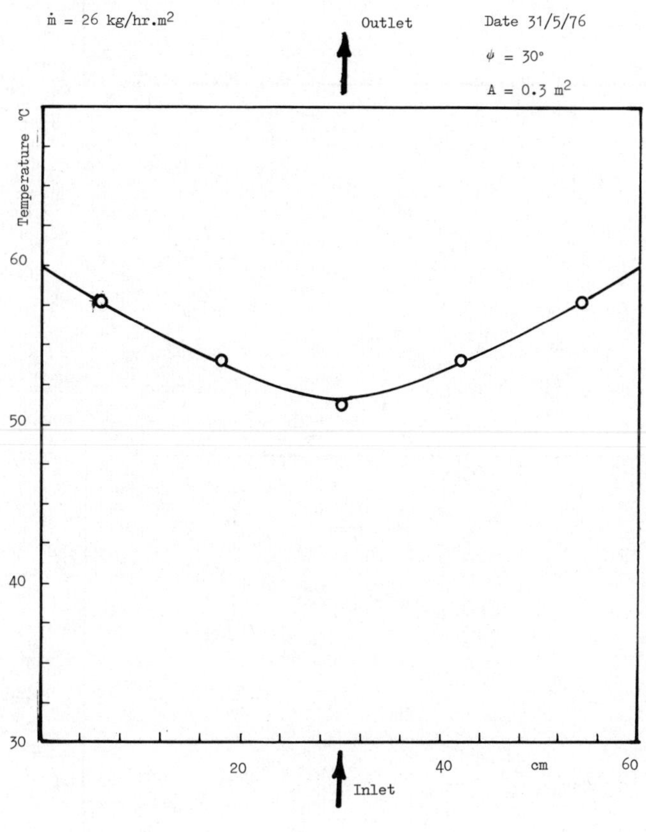

The mass flow rate through the collector decreases as the distance from the inlet point increases. The temperature and hence the temperature difference increase in the same direction, but not the heat output ($\dot{m} - C_p \, \Delta T$) and C.O.E. So the heat output and C.O.E. at the far end from the inlet point is lower than the average. Moreover the isothermal lines are not perpendicular to the direction of flow. If the flow distribution is adjusted to become uniform all over the collector and have the value corresponding to a maximum C.O.E., the temperature difference and hence the heat output and C.O.E. will be uniform and have the maximum value.

Fig. 6
Isothermal lines across the collector

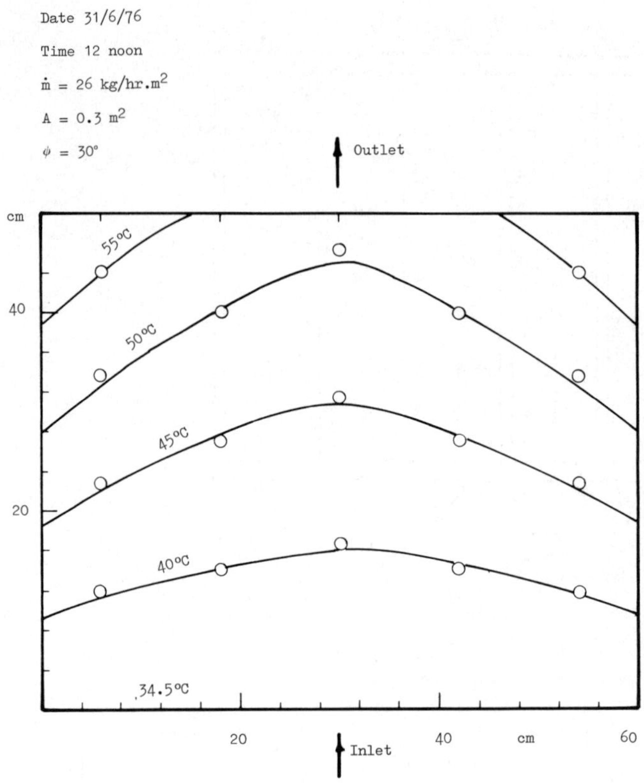

Fig. 7
Velocity distribution across the collector

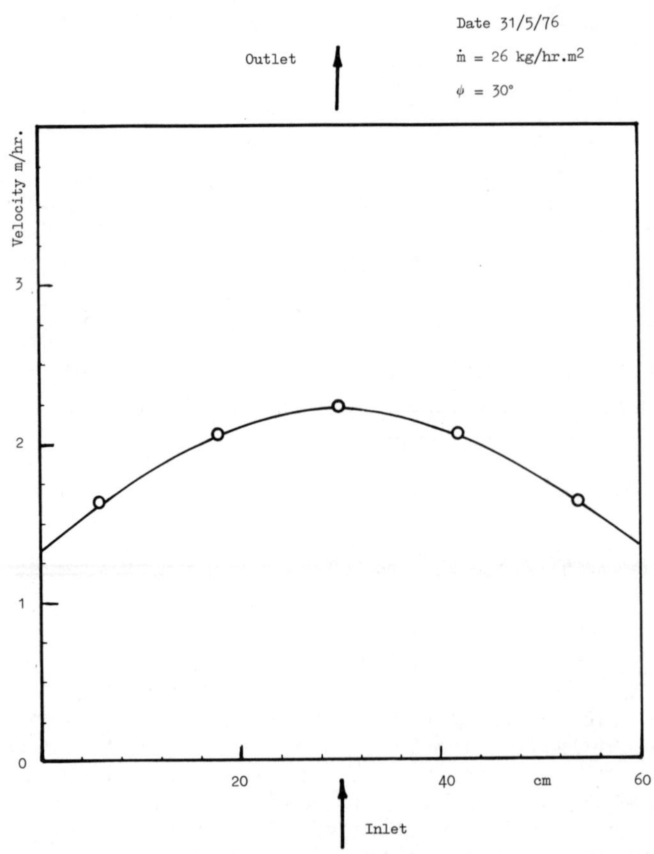

Fig. 8
Output and C.O.E.
distrubution

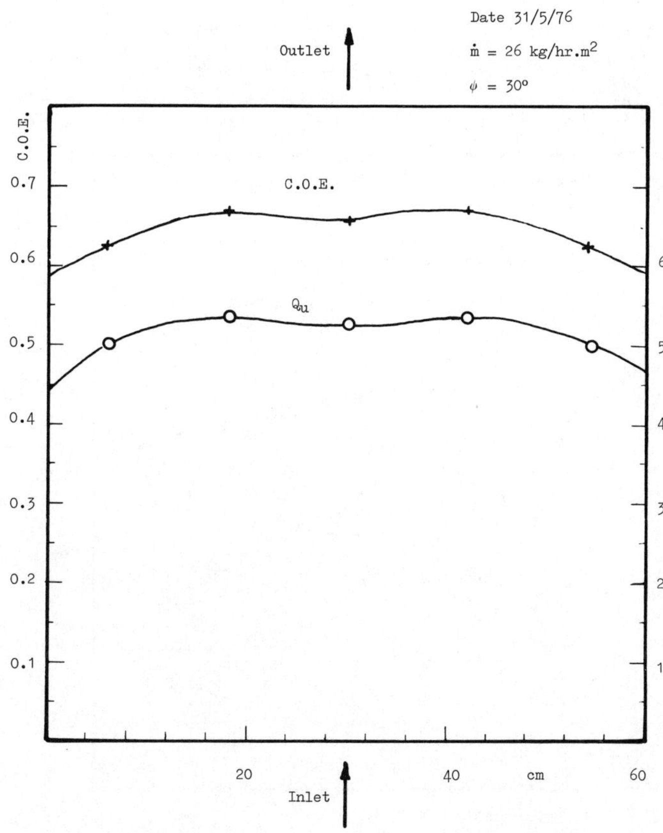

Fig. 9
Heat output and
C.O.E. distribution

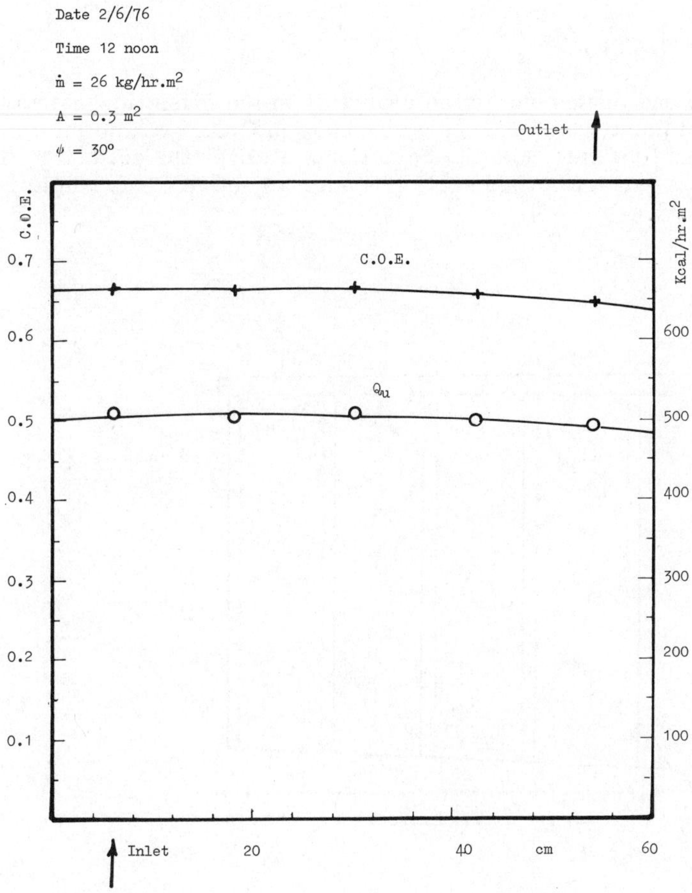

Fig. 10
Heat output and C.O.E. distribution

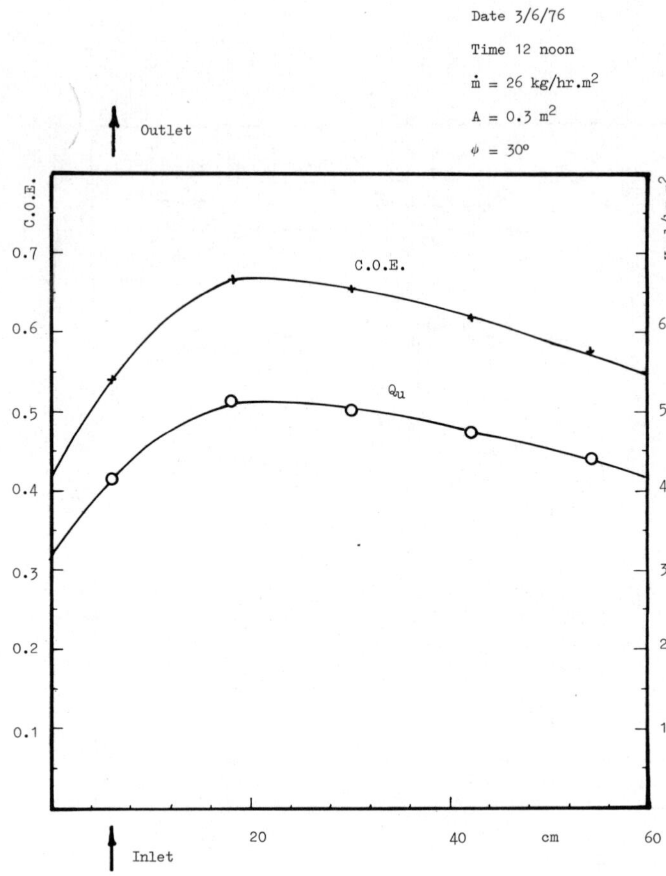

CONCLUSION

The end losses due to the reduction in the mass flow rate at the far end from the inlet point, which causes a decrease in C.O.E. can be minimized by adjusting the mass flow rate, so as to be uniform all over the collector. This could be done by using variable cross section header as shown for example in Fig. 11.

Fig. 11
Collector with uniform flow rate

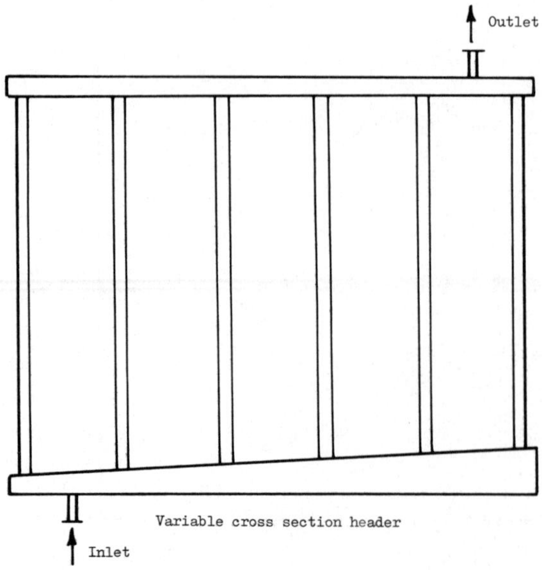

REFERENCES

(1) Hottel, H.C. and Weatz, B. B., 'Performance of flat solar collectors' Trans. ASME. pp. 91-104, February 1942.

(2) Whillier, A., 'Solar energy collection and its utilization for house heating' Sc.D. Thesis, M.I.T. 1953.

(3) Hottel, H.C., 'The performance of flat plate solar energy collectors' Proceedings of a course symposium held at M.I.T. August 1950, pp. 58-79. published M.I.T. 1954.

(4) N.M. Khattab M.Sc., Thesis, Faculty of Engineering Cairo University, 1976.

(5) 'Utilization of solar energy air-conditioning', M.Sc. Thesis, SAFWAT H.H. Cairo 1966.

(6) M.M.Y. Heba, 'Performance of curved solar water heater in domestic use' M.Sc. Thesis, University of Alexandria 1974.

(7) Whillier, A., 'Plastic covers for solar collectors' Solar Energy Vol. 7, No. 3. pp. 148-151, 1954.

(8) 'Natural circulation from a flat plate collector to hot storage tank' A.S.M.E. Paper. August 1975.

(9) Phillips, N. F. and Pate, R. A., 'A hot liquid energy storage system utilizing natural circulation' Publication 74-WA/HT-16, 1973.

(10) Dunkle, R. V. and Davey, E. T., 'Flow distribution in absorben banks' Melbourne International Solar Energy Society Conference 1970.

3.6 A flat plate solar collector for operating cooling systems

K. H. Khalil
Faculty of Engineering, Cairo University

I. A. Sakr and A. I. Hegazi
Solar Energy Laboratory, National Research Centre

ABSTRACT

Operation of an absorption refrigerator by solar energy needs temperatures higher than those of flat plate solar collectors. To avoid the use of a tracking solar concentrator, this paper presents a design study of a flat reflecting solar collector suited for solar refrigeration and intermittent temperature applications. The collector is an E-W oriented trapezoidal enclosure with flat reflecting mirrors and tubular absorber at the bottom. Collector geometry, concentration ratios and temperature boosting factors over flat plate collectors are deduced. The obtained temperatures with such collectors range from 100° to 200°C for different opening angles θ.

INTRODUCTION

One of the most appropriate areas for solar energy application is refrigeration and air conditioning. A solar operated refrigerator Fig. 1 consists mainly of a solar collection unit, an absorption refrigeration machine and a storage system. The design of a solar operated refrigerator is controlled by different factors, among which is the available solar radiation, see Fig. 2, and the design of a suitable

Fig. 1
Schematic diagram of solar refrigerator

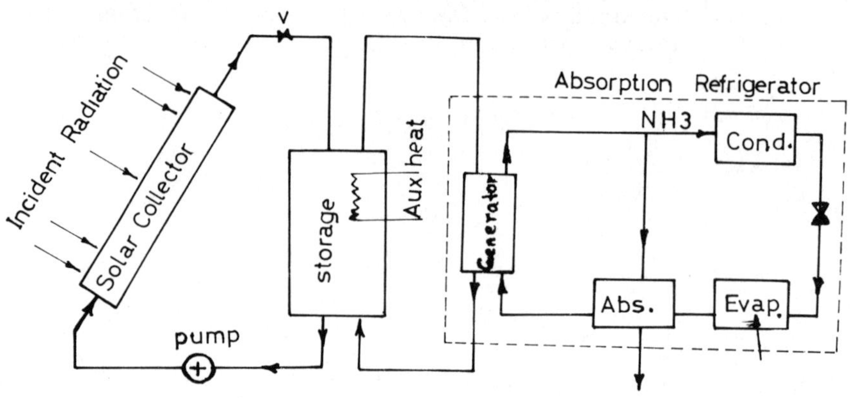

Fig. 2
Daily radiation
A: Total on horizontal
B: Direct on normal

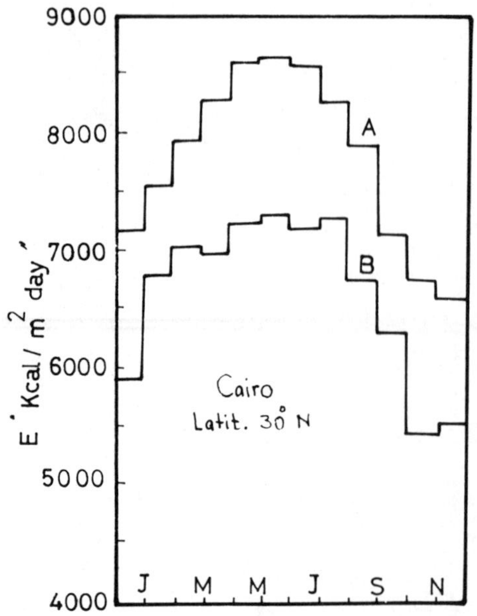

collector capable of feeding the systems thermal requirements. As the necessary temperature range for this application, which lies between 90° and 150°C, could not be attained with usual flat plate collectors, this work investigates a developed collector capable of producing such temperature levels.

THE TRAPEZOIDAL SOLAR–FLAT BOOSTER

Fig. 3 shows the new collector which has the flat plate characteristics besides concentrating the direct beam radiation. It is composed of an array of trapezoidal enclosures, see Fig. 4, with side reflecting walls, tubular absorbers at the bottom and a glass cover at the top. It has the advantage of being capable of collecting diffused radiation as well as concentrating beam radiation. Also the collector construction eliminates convection heat losses.

Fig. 3
Trapezoidal channels solar flat booster

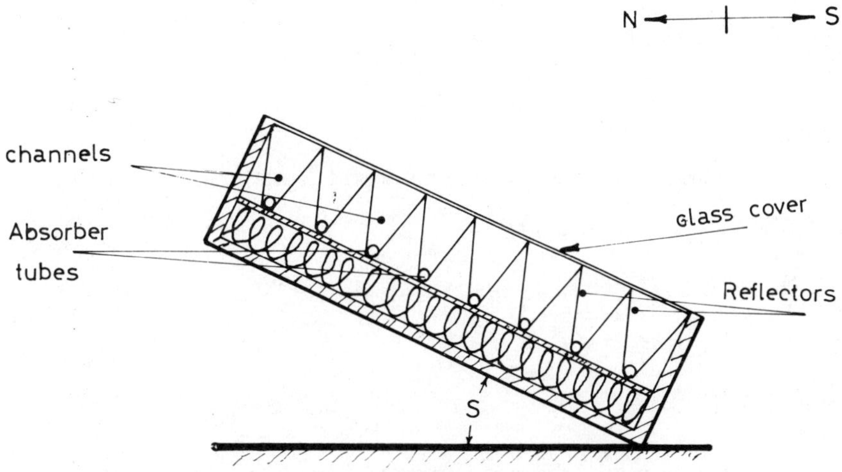

Fig. 4
Geometry of the trapezoidal channels

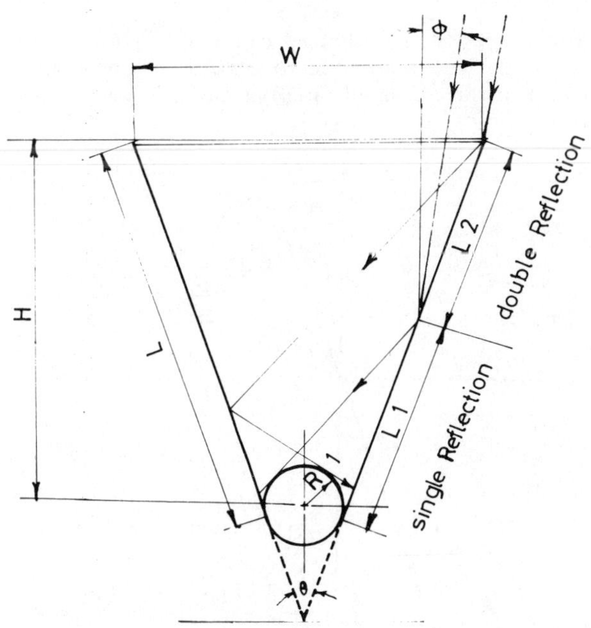

COLLECTOR GEOMETRY

The geometrical parameters of the collector L.W.H. are deduced, based on two main assumptions. Firstly, it assumes that the reflected radiation is uniformly distributed over the absorber surface, secondly the analysis is restricted to the case that normal solar beam does not suffer more than two reflections to reach the absorber surface.

Due to the reflection sine law the two reflection channels are confined to collectors with opening angles θ less than 45°, while channels with $\theta \geqslant 45°$ are treated so that the normal beam suffers only single reflection to reach the absorber. If L_1, L_2, are the lengths of single and double reflection zones, the side reflector length of single and double reflection channels as deduced from the analysis will be:

$$L_1 = (1 + \cos\theta)/2 \sin\theta/2 \quad (1)$$

$$L = L_1 + L_2$$

$$= \frac{2\cos\theta}{\sin\theta/2}(\cos^2\theta/2) \quad (2)$$

consequently the width and height of both channels will be:

$$W_1 = (2 + \cos\theta) \quad (3)$$

$$W = (4\cos\theta/2.\ \cos\theta + 1) \quad (4)$$

$$H_1 = L_1 \cos\theta/2$$

$$H = L\cos\theta/2 \quad (5)$$

DIRECT BEAM CONCENTRATION FACTOR

The ratio of the intensity of reflected radiation on absorber surface to the direct solar intensity on collector surface is denoted as the concentration factor C. The analysis considered the effect of opening angle θ, projected deviation angle ϕ and

Fig. 5
Critical angles of deviation

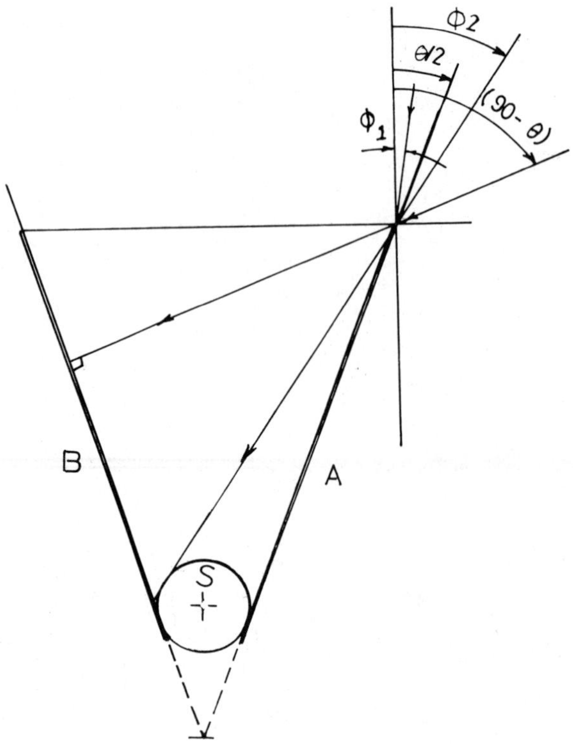

surface reflectivity ζ. The projected deviation angle is defined as the angle between the normal to the surface and the projection of the solar beam. The collector can concentrate solar beams in the range of $O < \phi \leq (90 - \theta)$ which is divided into four zones shown in Fig. 5, each with specified characteristics.

(1) $O \leq \phi < \phi_1$

In this zone the solar beam reaches all the surfaces and the single reflection zone of surface A is increased gradually so that it reaches the whole length at $\phi = \phi_1$ where:

$$\phi_1 = (\tan^{-1}(2H/(W + 1)) + (\theta - 90))$$

(2) $\phi_1 \leq \phi < \theta/2$

The solar beam deviated by angles in this range is received by the three surfaces A, B and S but it is only subjected to single reflection on surface A. At $\phi = \theta/2$ surface A is completely shaded.

(3) $\theta/2 \leq \phi < \phi_2$

Only absorber surface and surface B are subjected to solar rays, while A is completely shaded. As ϕ exceeds $\theta/2$ the absorber surface is also gradually shaded up to the end at $\phi = \phi_2$ where:

$$\phi_2 = \tan^{-1}((W + 1)/2H)$$
$$= \theta = \phi_2$$

(4) $\phi_2 \leq \phi < (90 - \theta)$

When ϕ is more than ϕ_2 only surface B is radiated. This irradiation is decreased gradually and the useful reflection completely diminishes at $(90 - \theta)$ where the beam is normal to the surface.

Tables 1, 2 show the variation of the previous critical angles with collector opening angle, θ, both for double and single reflection collectors.

The concentration factor was deduced for each zone separately for the different opening angles. Figs 6, 7 show the computer results. It can be seen that the deviation angles ϕ have great influence on concentration ratios. The new collector should be tilted to receive the maximum possible solar radiation as well as the minimum beam deviation angles. A new tilting routine has been proposed to achieve these goals, by which the maximum deviation angle all over the year does not exceed 10°. This routine proposes a monthly adjustment according to the following relation:

$$S = (\theta_s + Z_n)/2$$

Where

θ_s is the angle confined between the projected beam on an N-S surface and vertical at starting

z_n is the zenith angle at noon.

Also the beam deviation at any time (t) is

$$\phi_t = \pm (\theta t - s)$$

Where

$$\theta_t = \tan^{-1}(\cos \alpha \tan Z)_t$$

Table 1 Limiting critical angles of double reflection channels

Apex angle θ	Tubular absorber ϕ_1	ϕ_2	$\theta/2$	$(90 - 2\theta)$	$(90 - \theta)$
5	1.246	3.75	2.5	80	85
10	2.467	7.537	5	70	80
15	3.637	11.369	7.5	60	75
20	4.729	15.281	10	50	70
25	5.71	19.30	12.5	40	65
30	6.55	23.46	15	30	60
35	7.219	27.79	17.5	20	55
40	7.669	32.35	20	10	50

Table 2 Limiting critical angles of double reflection channels

θ	$\theta/2$	$(90 - \theta)$
45	22.5	45
50	25	40
55	27.5	35
60	30	30
65	32.5	25
70	35	20
75	37.5	15
80	40	10
85	42.5	5

Fig. 6 Concentration ratio vs deviation angle for double reflection channels

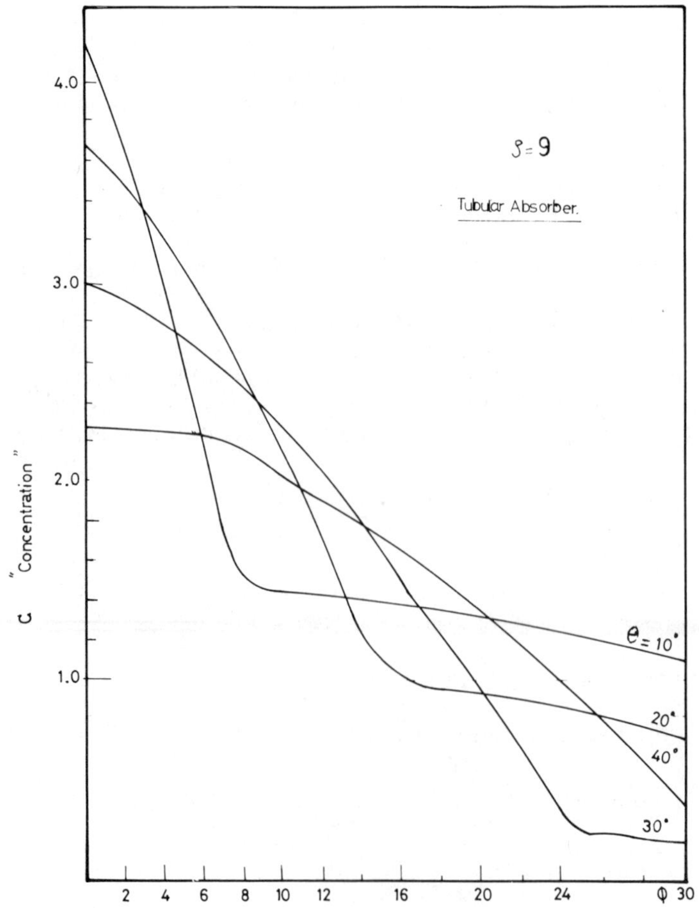

Fig. 7
Concentration ratio for single reflection channel

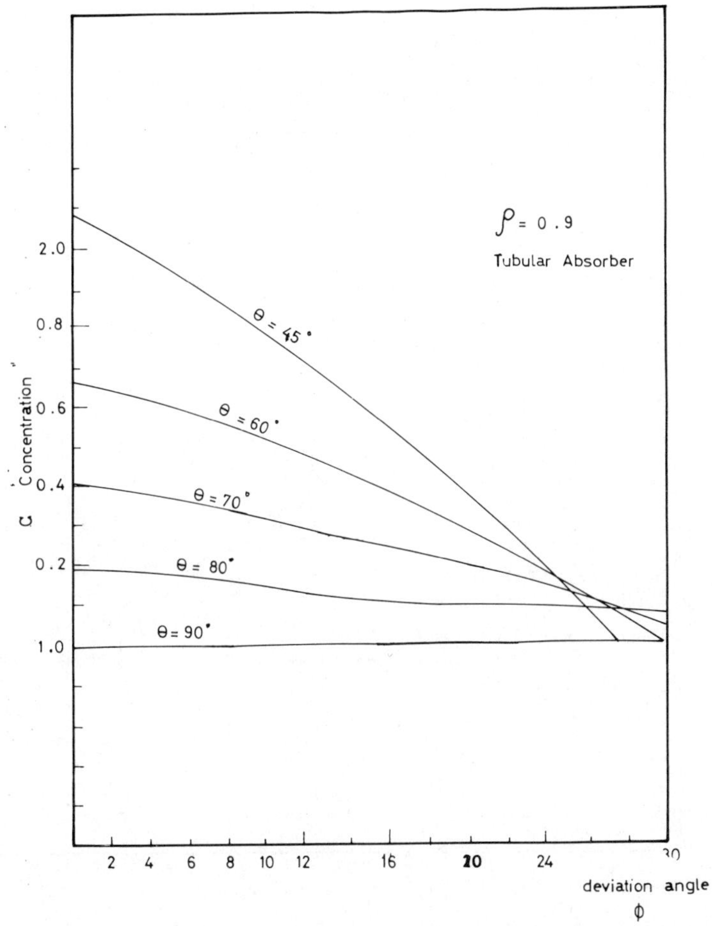

ATTAINABLE TEMPERATURE AND BOOSTING FACTOR

The attainable temperature in a solar collector is a function of different factors. These factors are the intensity of the available radiation I, the tilting angle or deviation angle in our case, concentration level and different heat losses. Beside increasing the concentration level, the new tilting routine, helps in stabilizing as much as possible the output temperature of the new collector i.e. minimum concentration is obtained with maximum solar intensity. Fig 8 is the temperature distribution over a day in January for different collectors. It can be seen that a collector with $30 \leqslant \theta \leqslant 60°$ is suitable for solar refrigeration systems, as the temperatures obtained are in the range of $100° \leqslant t°c < 150°C$.

The gain in temperature level achieved by the new collector construction is measured by the boosting factor. It is defined by the percentage increase in output temperature over that obtained by an ordinary flat plate collector. Table 3 shows the maximum and minimum temperatures attained, with the corresponding flat plate temperature and the boosting factor in each case. Fig 9 shows the dependence of the temperature boosting factor on collector opening angle. It is noted that this factor is almost constant in the range of single reflection channels.

CONCLUSION

Absorption cooling systems can be operated using the trapezoidal solar flat-booster collectors, with opening angles between 30° and 60°. The choice of the collector geometry depends on the required attainable temperature.

Fig. 8
Temperature variation vs time for different apex angle for January

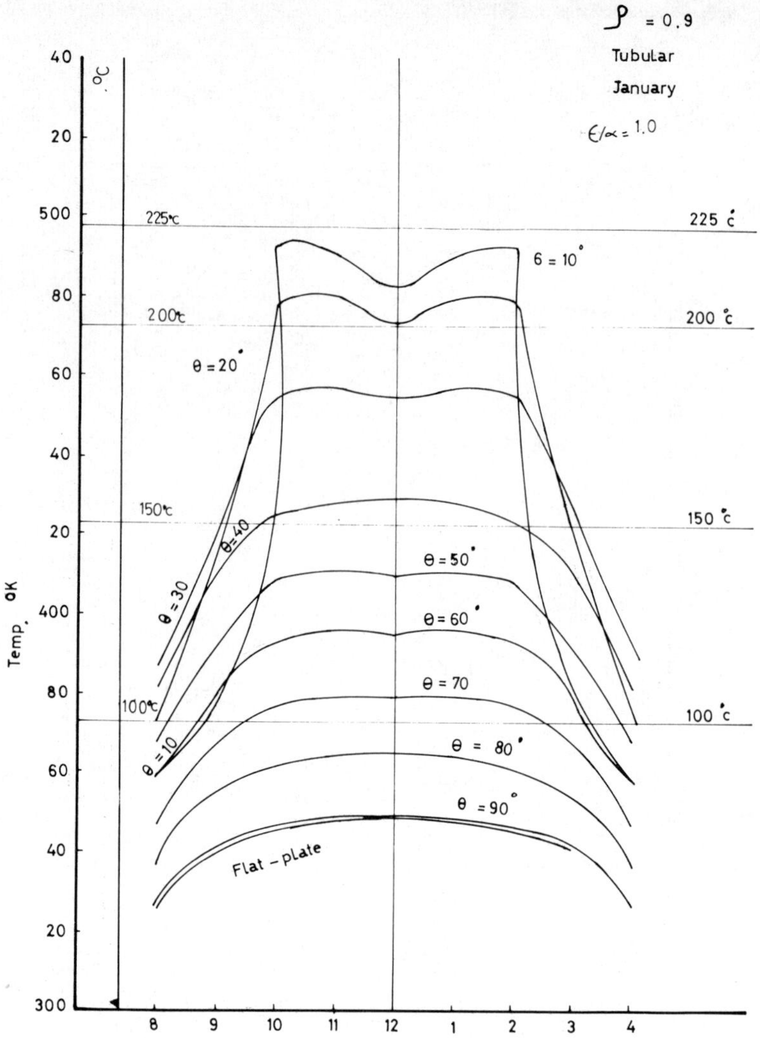

Fig. 9
The change of boosting factor vs apex angle

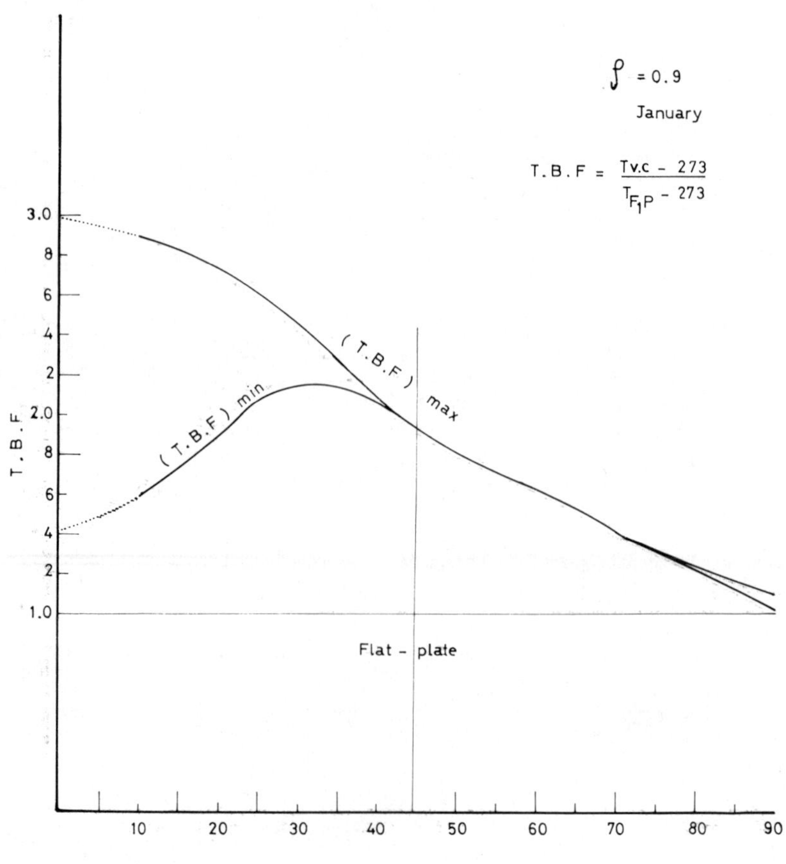

Table 3 Temperature boosting factors for trapezoidal collectors

θ	Max temp	$T_{F.P}$ = 349 K	Min Temp	$T_{F.P}$ = 325 C
degree	Tmax	(B.F) max	Tmin	(B.F) min
10	492.5	2.89	357.2	1.595
20	479.8	2.725	371.4	1.863
30	457.5	2.43	388.0	2.183
40	428.7	2.05	383	2.079
50	411.0	1.82	368.4	1.806
60	396.0	1.62	359.0	1.626
70	380.0	1.408	347.6	1.413
80	364.0	1.194	337.2	1.216
90	350.0	1.015	327.1	1.026

This table is calculated for a collector with tubular absorber, surface reflectivity of 0.9 and a tilting angle S of 54° in January.

NOMENCLATURE

C	Concentration ratio 'dimensionless'
H	Height of the collector channel/'Absorber' diameter
H_1	length of single reflection channel/'Absorber' diameter
L	length of side reflectors/absorber diameter
L_1	length of single reflection zone/'Absorber' diameter
L_2	length of double reflection zone/'Absorber' diameter
W	Opening width/absorber diameter
W_1	Width of single reflection channel/'Absorber' diameter
T	Attainable black body temperature °K
T.B.F.	Temperature boosting factor
S	Tilting

GREEK SYMBOLS

α	azimuth angle of the sun.	degrees
θ	Opening angle	degrees
ρ	Reflectivity of the mirror	
ϕ	Beam deviation angle	degrees
ϕ_1	The first critical angle	degrees
ϕ_2	The second critical angle	degrees

REFERENCES

(1) H. Tabor, 'Stationary mirror systems for solar collectors' Solar Energy 2, 27. 1958

(2) K. G. T. Hollands, 'A concentrator for thin film solar cells' Solar Energy 13, 163. 1971

(3) R. Winston, 'Principles of solar concentrators of novel design' Solar Energy 2. 1974

(4) Ari Rabi, 'Comparison of solar concentrators' Solar Energy, 2. 1976

All lengths given are dimensionless

3.7 Selective coatings for solar water heaters

Ashok Chandra
Physics Division
Central Building Research Institute
Roorkee (U.P.) India

Abstract

Selective coatings provide reasonably high efficiency. Two types of coatings are discussed here, they are NiS-ZnS and copper oxide layers on the aluminium plates, the solar absorptance (α) of both are 0.94 and 0.88 whereas the thermal emittance (ϵ) are 0.15 and 0.20 respectively. The maximum exit water temperature for NiS-ZnS coated and copper oxide coated solar water heaters, are 85°C and 77°C respectively.

Introduction

Flat plate absorbers coated with ordinary dull black paint, used as heat exchangers of solar water heaters, have low efficiency. For an efficient converter, sufficient absorption (α) of solar wavelengths (2-2μ) is one requirement whereas its emittance (ϵ) in infra red region (beyond 2μ) plays an important role in determining the re-radiation losses at a given temperature of operation of the absorber. If ϵ is reduced much below that of a black body, the absorber operates at a reasonably high efficiency. Sun emits 95% of its energy at wavelengths smaller than 2μ. The spectral ranges of the solar input and the re-radiation losses overlaps very little on either side of 2μ(1). The point at which the curve depicting solar input as a function of wavelength intersects the curve showing re-radiation losses as a function of wavelength determines the characteristic wavelength. An ideal selective surface should be black below and poor in emitting above this characteristic wavelength.

Though the metals provide high infra red reflectance (low emittance) they have low solar absorption. The selectivity can be achieved by overcoating the metallic surface with a thin layer of semi conductor/ dielectric which should have high, solar absorption and infra red transparency (2). The figures of merit of such a selective surface will be quite high.

This paper deals with the optical and thermal performances of two selective coatings deposited on commercially available aluminium sheets. The coatings which have been chosen in our case are NiS-ZnS and copper oxide whose α/ϵ ratios have been calculated from spectral reflectance curves. Plate temperatures and water exit temperatures have also been given for both the coatings.

Deposition of selective coatings

Selective coatings were deposited by using ion-exchange method. Before the application of these coatings Al plates are thoroughly cleaned in an alkaline solution of 5% sodium hydroxide maintained at 70°C. The plate was dipped in this solution for about 60 seconds, then water washed and again dipped in 50% nitric acid solution for 30 seconds. The plates are then free from grease, dust etc. Two types of coatings were used, (a) NiS-ZnS coating and (b) Copper oxide coating.

(a) NiS-ZnS coating
The cleaned Al plate was dipped as a cathode in an electrolytic bath containing 60g/l nickel ammonium sulphate, 10g/l zinc sulphate, 16g/l

sodium thiyocynate and 2g/l citric acid. A carbon rod was used as an annode. The current density fixed in the bath was $2mA/cm^2$ for 3 minutes. After this electrolytic process the Al plate has a black coating of NiS-ZnS. The plate was then water washed and dried.

(b) Copper oxide coating
After proper cleaning the Al plate was immersed for 20 seconds in a bath containing 450g/l sodium hydroxide and 90g/l zinc oxide maintained at $25^{\circ}C$. Zinc is deposited on the plate. The plate with this deposit was washed in water and again dipped for 20 seconds in the bath. After the above zincating process Zn film is covered with a thin copper layer by immersing it for 2 minutes in an electroplating bath of 2.25% copper cyanide, 3.4% sodium cyanide, 1.5% sodium carbonate. Current density was adjusted to $0.003A/cm^2$. Now the Al plate coated with zinc and copper films was dipped for 5 minutes in the solution of 500g/l sodium hydroxide and 250g/l sodium chlorite maintained at $90^{\circ}C$. This gave a black copper oxide layer. The coated plate is taken out, water washed and then dried.

The above two coatings were deposited separately on the corrugated Al sheet, dimensions 34m x 34m. The corrugations had a hydraulic diameter equal to 1.2cm and centre to centre distance of 4cm behind this the G.I. pipes were tied with the sheet. These absorbers were housed in an M.S. box, dimensions 50 x 41 x 10cm. The box had a 5cm insulation of loose glass wool around its perimeter and 2cm on the back. The box was covered with a single sheet of colourless window glass of extinction coefficient, 0.077/cm. The orientation of the solar water heater was kept at 45°, facing south.

Reflectance and temperature profiles

The reflectances of the above two selective coatings systems were measured at various wavelengths (Fig. 1) at the National Physical Laboratory, New Delhi, India and α and ϵ were calculated for the above two coatings from the reflectance spectra by the method of weighted ordinate system (3). The α and ϵ for NiS-ZnS were 0.94 and 0.15 respectively and the values of α and ϵ for copper oxide coatings were

Fig.1
Spectral reflectance

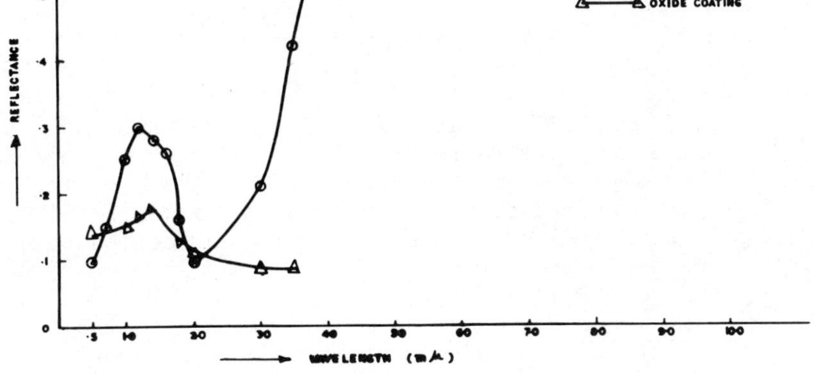

Fig. 2
Plate temperatures

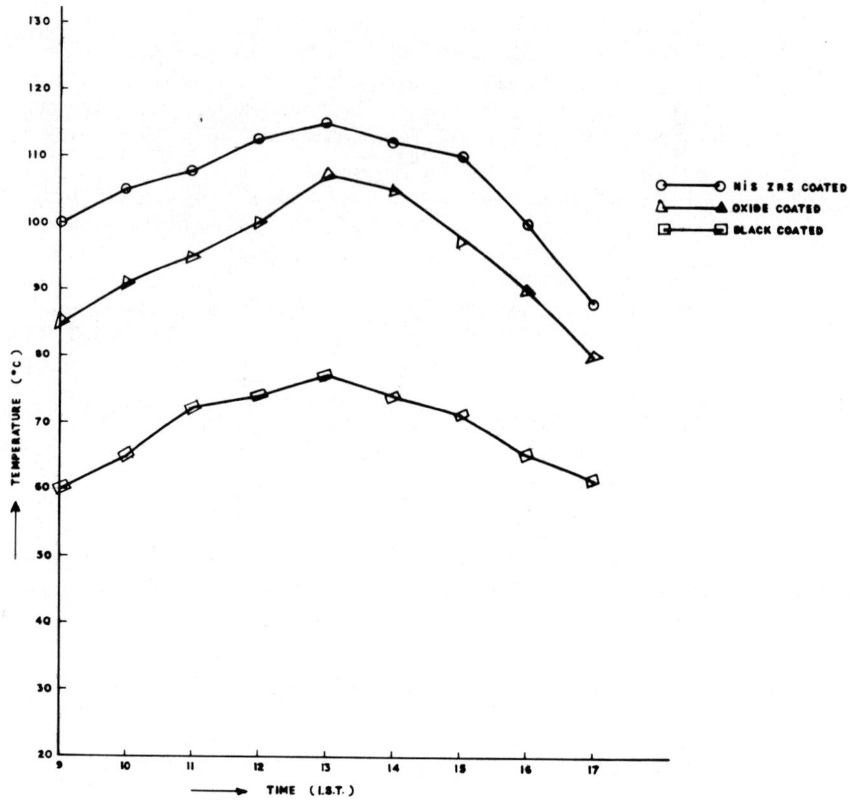

Fig. 3
Outlet water temperature from solar water heaters

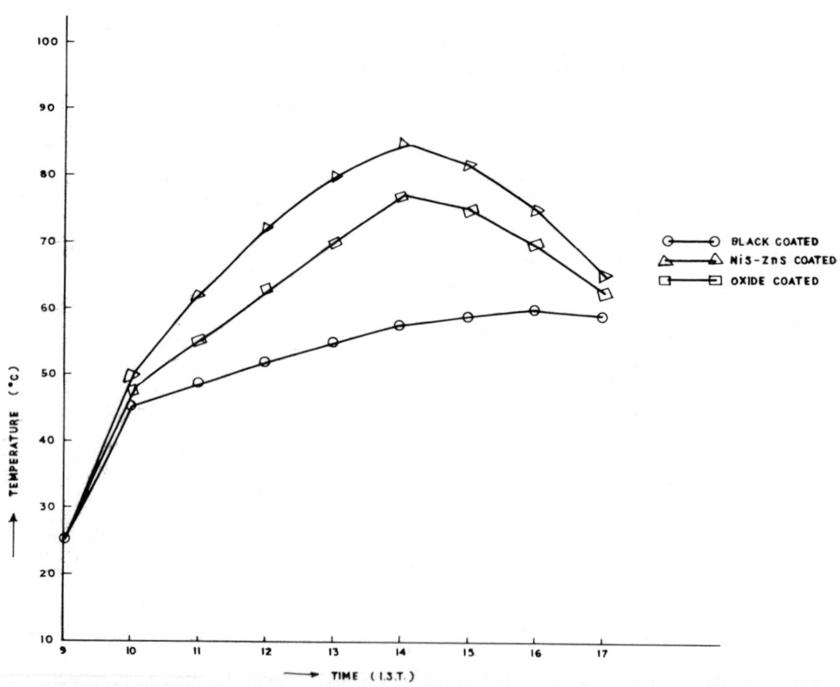

0.88 and 0.20 respectively. The plate temperatures and water exit temperatures were recorded with the help of thermocouples. (Fig. 2, 3). The temperatures were compared with those of the lamp black painted solar water heater. The maximum water temperatures were 85°C, 77°C and 59°C of NiS-ZnS, oxide and lamp black coated solar water heaters, respectively.

From the above observations we found that the NiS-ZnS coating performs better in comparison to the copper oxide coating for use in a solar water heater.

Acknowledgements

The author wishes to thank Professor S. K. Joshi, Head, Physics Department University of Roorkee, Roorkee, India for his valuable comments in the preparation of this paper. The author is also thankful to Mr. Suresh Chandra Ganguly for his help in the experimental work.

References

(1) Meinel, A.B. et al., 'Physics Looks at Solar Energy'. Physics Today February 1972.

(2) Seraphin, B.O. et al., 'Photothermal Solar Energy Conversion and Optical Properties of Solids', Optical Properties of Solids - New Development, North Holland Publishing Co. Amsterdam, 1976, page 927.

(3) Olson, O.H. 'Selected Ordinates for Solar Absorptivity Calculations' Appl. optics Vol.2 no.1 January 1963, page 109.

3.8

Development of a new low cost flat plate collector for tropical countries

E. R. Mayer, I. Nissan, R. Martinez
Material Research Center
National University of Mexico
Apdo. Postal 70-360 Mexico 20 D.F.

Abstract

In order to increase the efficiency and reduce the cost of flat plate collectors used in domestic and industrial applications, a solar panel has been developed using a construction process allowing the absorber plate to be integrated to the insulated case. The design uses only one corrugated metallic sheet permanently sealed to the sides of the case and allowing the heat carrying fluid to flow in the intermediate space between the plate and the bottom of the case. Such a collector presents a good thermal and structural resistance and can be inserted as a roof panel in areas where roof insulation is necessary. The collector efficiency was calculated for different construction designs using a new computer model.

Introduction

Extensive use of solar energy for residential and industrial water heating systems requires the development of economically viable solar collectors.

Actually a flat plate collector is an heat exchanger in which the output heat is being transferred by a carrying fluid circulating in metallic tubes in good thermal contact with a black metallic sheet receiving the solar radiation. In order to reduce the heat losses this exchanger is placed in a box and insulated on five sides, while on the upper part one or more transparent covers eliminate part of the convection and radiation losses when temperatures above 40°C are needed (1).

Design

This paper reports a collector in which the case is being used not only as a container but also as an active part of the fluid carrying system.

As it can be seen in figures 1 and 2 the basic component is the case prepared by moulding, using fibreglass reinforced plastic. During the moulding process, reinforcement structures are being incorporated to the bottom plate, on the outside, as well as two longitudinally cut tubes which are located on two opposite edges. These tubes are sealed to the base plate and in communication with the interior through a set of holes.

Once the case is finished, the absorber plate - a corrugated galvanised steel sheet - is placed on the bottom of the box and screwed at its edges to the plastic case and also at some contact points with the reinforcement structures lying underneath. In order to provide a permanent seal an epoxy resin is applied between the fibreglass reinforced plastic material and the metallic sheet. Once the system is waterproof, a polyurethane foam and a polyester film are sprayed on the back. The cover glazing is adjusted and sealed on the upper part.

The use of fibreglass reinforced material in the construction of the box is justified by its high tensile strength, resistance to weathering and microbiological activity, good chemical resistance, low thermal expansion and conductivity. The uniform and light-weight body can be

Fig.1

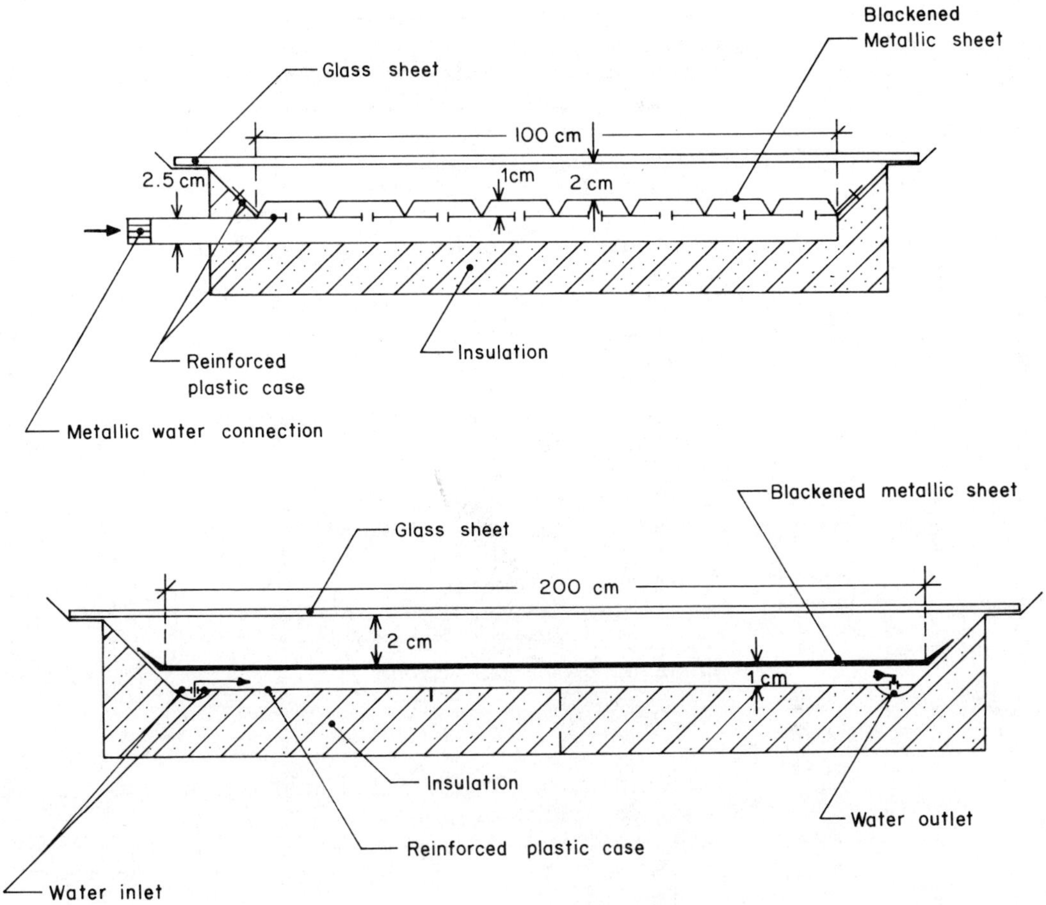

Fig.2

Schematic design of a solar-heated collector using an integrated plastic reinforced case.

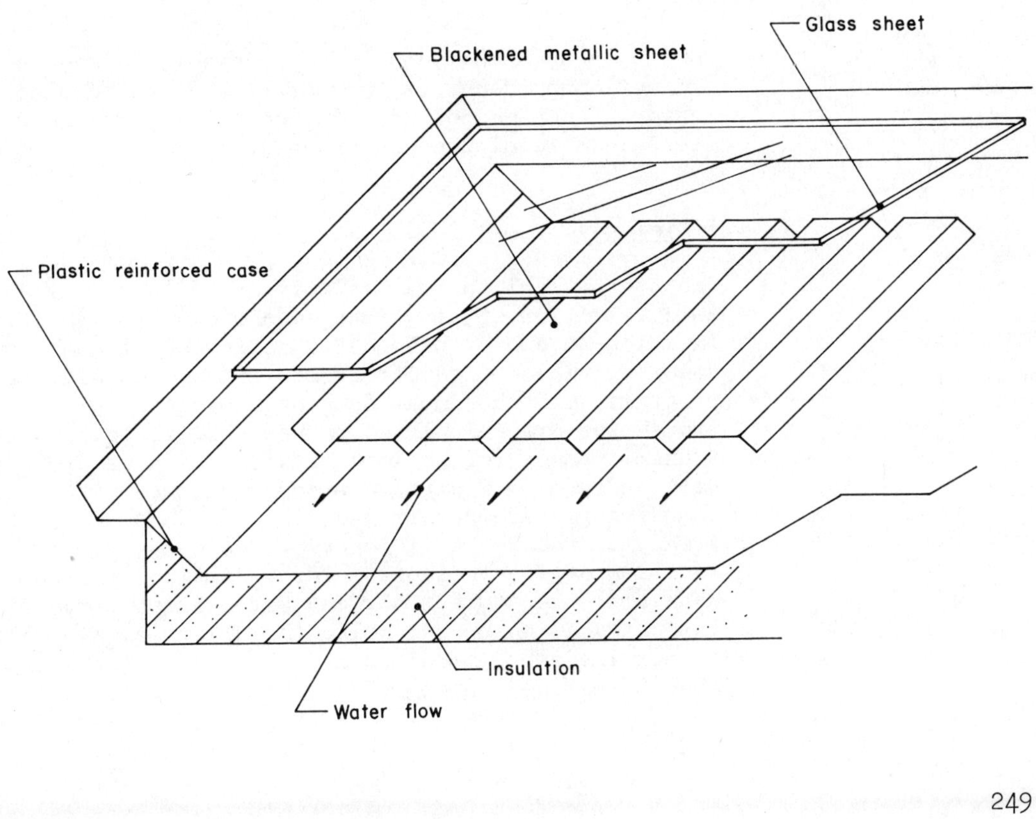

perfectly integrated preserving its good strength as a panel. During manufacturing of reinforced plastics, additional stabilizers and fillers can be added in order to improve the material performance and to protect it from thermal and ultraviolet radiation damage.

Characteristics and costs

The equipment built and reported in this paper has the following characteristics:

absorbing area	$2m^2$
opening	$2.31m^2$
glass cover	$2.51m^2$
fluid flow channels	$0.01m^2$
air layer thickness	0.02m
case total area	$3.00m^2$
weight with structures	20.5kg
thickness of composite material	0.003m
composition of case material	30% fibreglass 70% polyester
insulation thickness	0.05m
insulation density	$24kg/m^3$
insulation thermal conductivity	$0.025Watt/m^oC$
absorber plate thickness	0.001m

A breakdown of construction costs gives the following values for handmade collectors:

case including structures	US$	20.00
absorber plate		10.00
insulation		5.00
glass cover (5mm)		28.00
production cost including overhead		70.00
Total	US$	133.00
Cost per m^2	US$	66.50

Obviously this cost can be considerably reduced using a semi-automatic production procedure to prepare the case and also if agricultural residues such as polymerised coconut fibres are used instead of the fibreglass reinforced resin. (2).

Efficiency

The theoretical thermal efficiency of these collectors has been calculated for steady state conditions (3). An iterative process is used in which an energy balance is performed at each cover and at the absorber plate. Figure 3 presents the flow diagram of the computer program. With the input data of the collector, the cover's optical properties are calculated as well as its total solar absorptance and transmittance; reflections between surfaces are also being considered. Next through an iterative process all heat transfer coefficients and temperatures are determined. This is done until the system reaches a steady state. The efficiency of the collector is then obtained as a function of the incident solar radiation. The computer program input data are: the ambient air and water inlet temperatures, solar radiation intensity, wind speed, collector orientation, collector's thermal properties and dimensions and the relevant optical properties of the absorber plate coating. The output consists of the solar reflection and transmission of the system, component temperatures and collector efficiency. The program allows comparison of collectors with and without Honeycombe structures between the absorber plate and the first cover.

Fig.3
Flow diagram

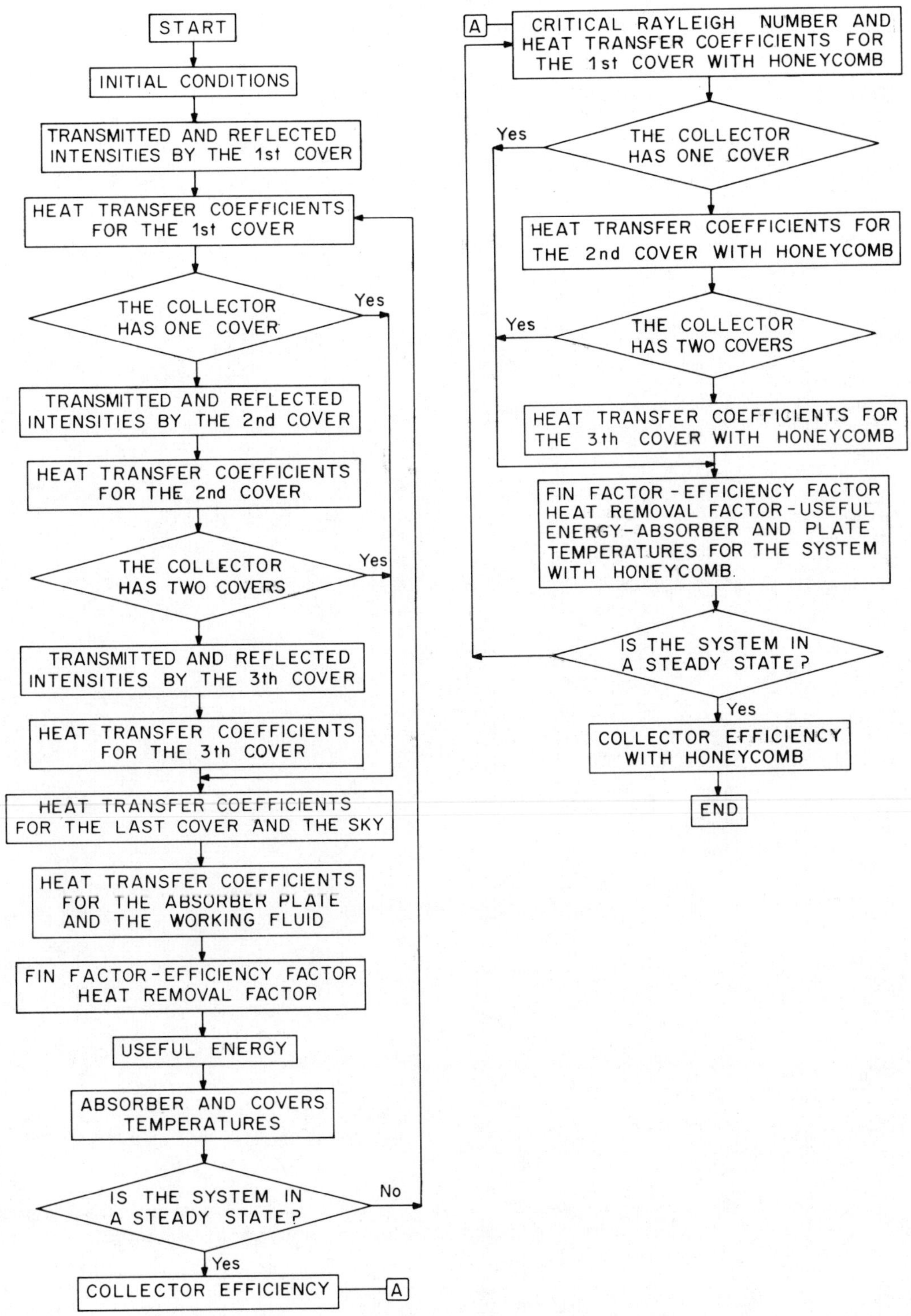

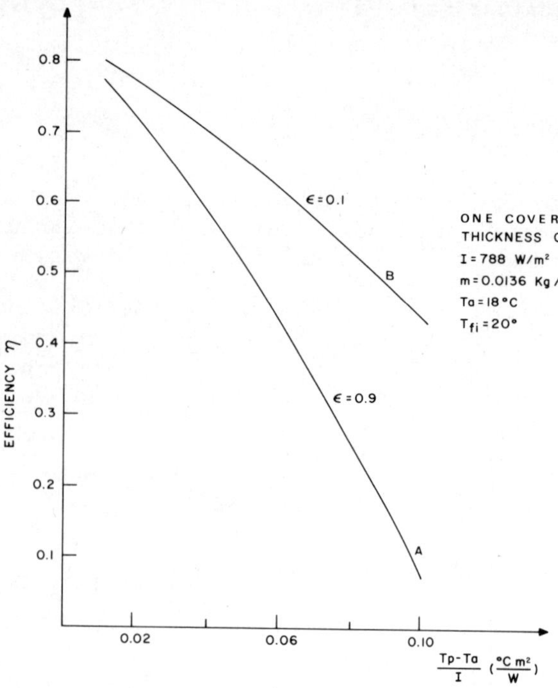

Fig.4

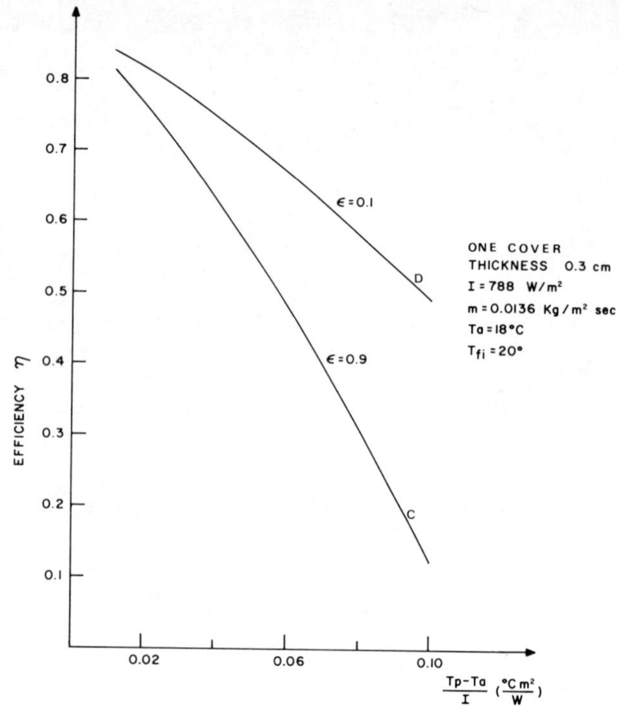

Fig.5

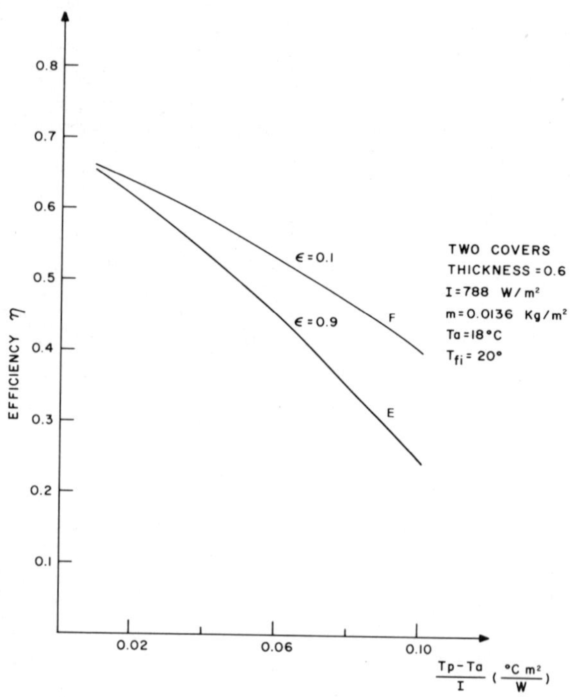

Fig.6

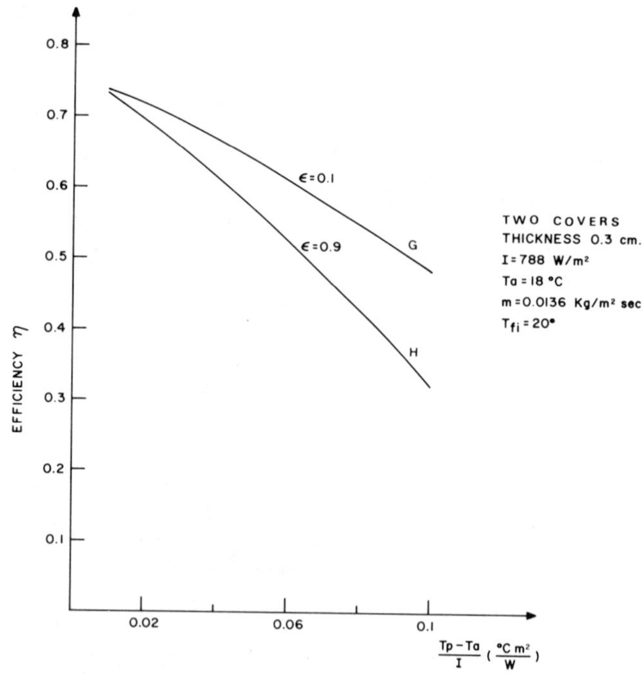

Fig.7

In figures 4, 5, 6 and 7, efficiency as function of $T_p - T_a/I$, is reported for different cases in which water was used as circulating fluid:

a. one 6mm glass cover and non selective absorber coating ($\alpha = \epsilon = 0.9$)
b. one 6mm glass cover and selective absorber coating ($\alpha = 0.9$, $\epsilon = 0.1$)
c. one 3mm glass cover and non selective absorber coating ($\alpha = \epsilon = 0.9$)
d. one 3mm glass cover and selective absorber coating ($\alpha = 0.9$, $\epsilon = 0.1$)
e. two 6mm glass covers and non selective absorber coating ($\alpha = \epsilon = 0.9$)
f. two 6mm glass covers and selective absorber coating ($\alpha = 0.9$, $\epsilon = 0.1$)
g. two 3mm glass covers and non selective absorber coating ($\alpha = \epsilon = 0.9$)
h. two 3mm glass covers and selective absorber coating ($\alpha = 0.9$, $\epsilon = 0.1$)

As was previously expected, the maximum efficiency was obtained by using those covers of minimum thickness. As it is well known, the use of selective coating improves the efficiency, hence the low price of the polymerized materials makes the cost of the collector structure so cheap that selective coatings - often expensive - can be introduced without increasing the overall cost of the whole collector.

There are other systems that have been reported in literature. They improve the collector efficiency by means of honeycombs cells. These systems can be evaluated also by using our computer program. Preliminary results have shown a noticeable efficiency increase at medium and high temperature ranges. Extensive experimental tests are underway at the Material Research Center (4).

Conclusions

The construction procedure reported has proved that the use of the collector case as an active part of the fluid carrying system is very effective with regard to cost reduction and efficiency increase. Within reasonable limits any module size can be achieved and mass produced. Due to the fact that the insulation is being applied on the reverse of the case, thus in the interior of the building using such modules as roof panels, weathering and maintenance problems are considerably reduced, while installation is much easier and adapted to tropical conditions.

Nomenclature

T_p average absorber plate temperature
T_a ambient temperature
I incident solar radiation
T_{fi} temperature at which the fluid enters the solar collector
α absorptivity of the absorbing plate
ϵ emissivity of the absorbing plate

Acknowledgement

The authors wish to thank Dr. Ignacio Alvarez and Fis. Raoul Herrera for the computer work reported in this paper, M.Sc. Everado Hernandez for his help and comments, Mr H Cortez for drawing the figures and Miss Mercedes Spindola for the translation of the paper.

References

1. Meinel A., Meinel M., Applied Solar Energy, An Introduction. Addison-Wesley Publishing Company (1976).
2. Ballado D., Torres M., 'Obtencion de polimeros reforzados a partir de la lignina contenida en el coco' Professional Thesis, UNAM Mexico (1976).
3. Herrera R., 'Simulacion digital de un sistema anticonvectivo y antirradiativo en colectores solares planos' Professional Thesis, UNAM Mexico (1977).
4. Nissan I., 'Analisis de la eficiencia termica de colectores solares planos' Professional Thesis, UIA Mexico (1977).

3.9 Some new ideas on solar collectors

R. Rigopoulos
Physics Laboratory II,
University of Patras, Patras, Greece

Our solar energy group (J. Caouris, R. Rigopoulos, J. Trypanagnostopoulos and P. Yianoulis), has proposed and investigated a collector which, under certain conditions, can have improved collection efficiency.

In a common flat plate collector, when the absorber plate temperature is increased, the cover temperature is also increased as a result of radiative and convective heat transfer from the absorber plate to the cover. The cover temperature determines the heat losses of the collector, assuming perfect back and side insulation.

Several methods have been used to reduce the cover plate temperature. Selective coatings are used to reduce the radiative heat transfer, honeycomb structures or vacuum can be used to reduce convective transfer; two cover plates are used to reduce the front cover temperature.

We propose a collector in which a working fluid flows in several layers above the absorbing plate (Fig. 1). The fluid enters the collector at a temperature T_i which is close to the ambient temperature T_a and as it flows successively through the layers 1,2,....,n, is heated up. The fluid must have high transmittance in the visible and near i.r. part of the spectrum, and nearly zero transmittance at longer wevelenghts. The layers are constructed by using n transparent plates.

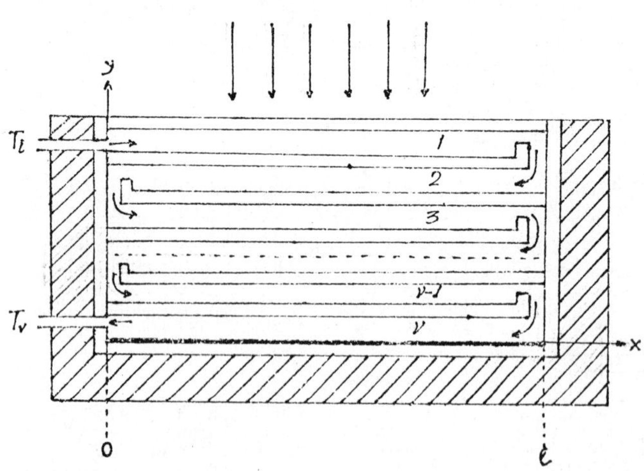

Fig. 1 A vertical section of the collector is shown. The numbers 1, 2,..., ν refer to the fluid layers from top to bottom.

We should note that the reflective losses of this system are minimal coming only from the first plate-air interface. Since the other interfaces are between the plates and the working fluid, which have nearly equal indices of refraction, the reflection losses from these interfaces are negligible. We also have improved heat transfer between the absorber and the working fluid, since the fluid is flowing over all the heated side of the absorber.

A theoretical analysis for this system shows that the important parameters for the behaviour of the system are: the number of layers n, and the quantities K_j/D_j where D_j is the thickness of the jth plate K_j is its coefficient of thermal conductivity. Actually we should take into account the film heat transfer coefficients $h_{j,j}$ and $h_{j,j-1}$ between the jth plate and the working fluid in the layers j and j-1 correspondingly. We can use then an effective conductivity $(K/D)_{j,eff}$ such that:

$$\left(\frac{K}{D}\right)_{j,eff} = \left[\left(\frac{K_j}{D_j}\right)^{-1} + h_{j,j}^{-1} + h_{j,j-1}^{-1}\right]^{-1} \qquad (1)$$

The efficiency of the collector is improved when the value of $(K/D)_{eff}$ becomes smaller, for all transparent plates. This has important implications since $(K/D)_{eff}$ depends not only on the conductivity of the plates, but also on the physical properties of the fluid and on its velocity, through (1). It is therefore obvious that one should select plates with K as small as possible, and the fluid in such a way, that the film coefficients h are as small as possible. The number of layers n should generally be larger in order to have higher efficiencies at relatively high temperatures, but there is a limit which depends on the absorption of the solar radiation by the plates.

Fig. 2

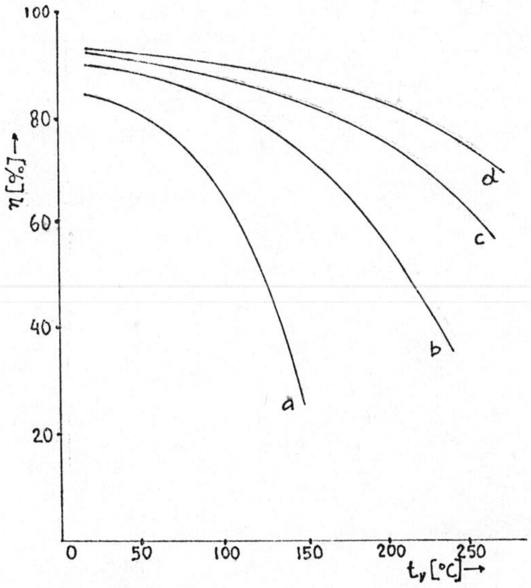

Efficiency vs. outlet fluid temperature of a four layer collector. The parameters which have been used are described in the text. The four curves correspond to:

a: 0.1 W cm^{-2}
b: 0.2 W cm^{-2}
c: 0.3 W cm^{-2}
d: 0.4 W cm^{-2}

In Fig. 2 we give the calculated efficiency of such a collector, consisting of four layers as a function of the outlet fluid temperature t_f[°C]. The transparent plates are assumed to be plastic so that $(K/D)_{eff} \approx 0.001 \cdot [W\ cm^{-2}K^{-1}]$ for all plates; $T_a = T_i = 290$ [°K]; the wind heat transfer coefficient is assumed to be $h_o = 2\times10^{-3} [W\ cm^{-2}K^{-1}]$ and corresponds to a wind velocity of about 4[m sec^{-1}]. The four curves of the figure correspond to solar energy fluxes of: 0.1,0.2,0.3, and 0.4 [W/cm^2]. We can achieve small concentration ratios by using flat mirrors along the sides of the collector. The same parameters are used to calculate the efficiency of a five layer collector (Fig. 3).

To find the efficiency of the collector under variable meteorogical conditions, we assume a concentration equal to four and we vary the insolation rate. The outlet fluid

Fig. 3

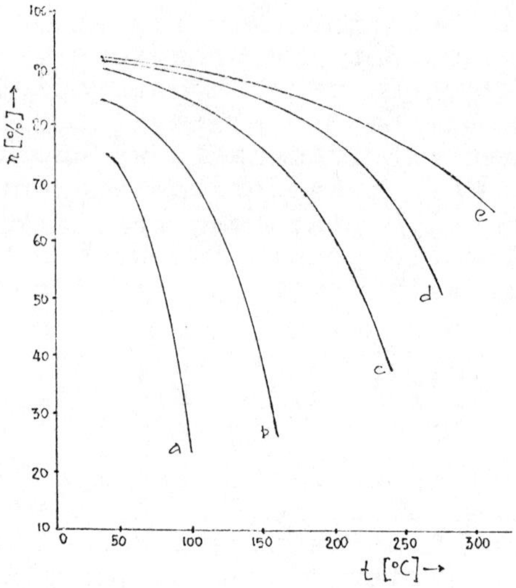

Five layer collector efficiency, for:

a: 0.05 W cm^{-2}
b: 0.1 W cm^{-2}
c: 0.2 W cm^{-2}
d: 0.3 W cm^{-2}
e: 0.4 W cm^{-2}

temperature is considered as a parameter. The rest of the parameters are the same as those of Fig. 3. The results are shown in Fig. 4 for a five layer collector (solid lines). For comparison purposes we have calculated also the corresponding efficiencies for a conventional collector with a selective black paint (a=1, ε=0.3) and one glass cover (dashed lines).

Fig. 4

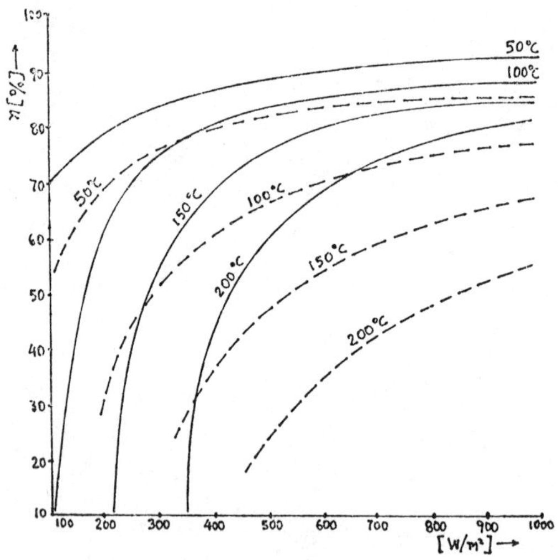

Comparison of the efficiency of a five layer collector (solid lines) with the efficiency of a conventional collector (dashed lines), for various outlet fluid temperatures.

The high efficiency of the above type collector as well as the high final temperature of the heated liquid, permits the production, except of hot water, and of electricity with a conversion factor better than 15% for final liquid temperature of 200°C.

One unit of 5 m^2 effective collection area could be enough for the needs of a home. The combination of a number of such units could provide the electricity and the hot water for a block of flats, a village or even for a town, if the hot water can be distributed through an appropriate distribution network.

The testing of flat plate solar heat collectors

G. Hassan
Principal Lecturer
Chelmer Institute of Higher Education
Chelmsford, Essex, UK

Abstract

Details are given of the author's method of analysing the performance of flat plate solar heat collectors. The method does not require the measurement of mass flow rate of the working fluid merely the measurement of temperatures and the intensity of radiation from the sun.

The testing procedure may be automated using microprocessing techniques.

The results are compared favourably with the results obtained from equations given by Hottell and Woertz.

Purpose of paper

The purpose of this paper is to make public a simple method of analysing the performance of flat plate solar heat collectors.

Plan of paper

The plan is to give some information about the way in which such a method has been devised so that others may judge its soundness and usefulness.

Fig. 1 shows the energy streams incident upon a collector and for this system the basic energy equation may be written as:

$$q_u = E_i - \text{LOSSES}$$

where q_u = Useful energy

and E_i = Incident energy

The losses may be classified under External and Internal losses and these are set out below.

EXTERNAL LOSSES

(1) Loss due to reflection*

(2) Loss from the sides

(3) Loss from the back*

(4) Loss due to shading

(5) Loss from the top*

(6) Loss due to dirt on cover

INTERNAL LOSSES

(7) Loss due to reflection at the inner glass surface

(8) Loss due to internal convection currents

Fig.1

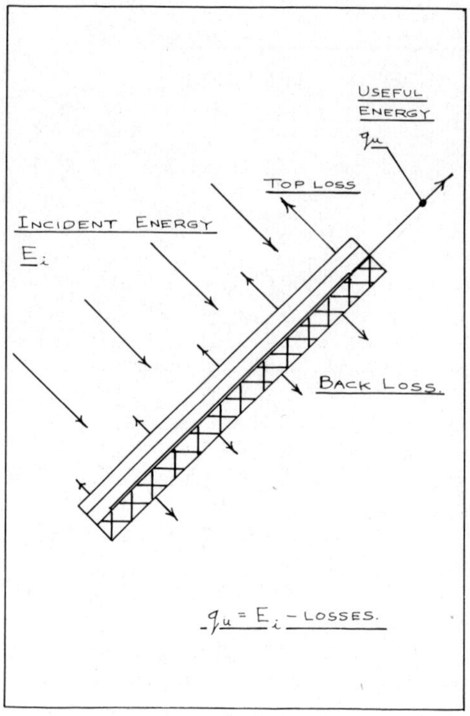

(9) Loss due to the angle of the plate affecting the absorption coefficients

(10) Loss due to fin efficiency

(11) Loss due to bonding of tubes to plate

(12) Loss due to absorption by cover

(13) Loss due to thermal capacity of collector.

Some of these losses are small when compared with the other energy quantities and may be neglected. Some of the losses affect the others, and some may be easily calculated or estimated. For a full discussion on each of the losses please see reference (1). Asterisks have been placed against three losses which may be calculated and estimated, and these are listed below in their general order of magnitude.

Loss number

(5) Loss from the top of the collector

(1) Loss due to reflection

(3) Loss from the back of the collector.

EVALUATION OF LOSSES

(1) Loss due to reflection at the outer glass surface

For angles of incidence up to and including 50°

Reflection Loss $\simeq 0.15 \times E_i$ (reference (1))

(3) Loss from the back of the collector

This may be calculated from the equation:

$$q_B = U_B (t_m - t_a)$$

or $$q_B = U_B t_v$$

where q_B = sum of convection and radiation losses

U_B = overall thermal transmittance of the materials used on the underside of the plate. This may be calculated from tables and information given in reference (2)

t_m = mean temperature of the plate = $\frac{t_o + t_i}{2}$

t_o = outlet temperature

t_i = inlet temperature

t_a = air temperature

t_v = a temperature index = $t_m - t_a$

(5) Loss from the top

This may be calculated from the equation:

$$q_G = U(t_{IG} - t_a)$$

or $$q_G = U \Delta t$$

where q_G = sum of convection and radiation losses from the glass

U = overall thermal transmittance for the outer glass. This may be calculated by referring to reference (2)

t_{IG} = temperature between inner and outer glass covers (or between single glass cover and the plate)

t_a = air temperature taken inside a Stephenson screen

Δt = a temperature index = $t_{IG} - t_a$

THE BASIC ENERGY EQUATION which is approximate may now be stated in the following form:

$$q_u = E_i - (0.15\, E_i + U_B t_v + U \Delta t)$$

or $$q_u = 0.85\, E_i - U_B t_v - U \Delta t$$

The particular equation including approximate losses for the author's collector is given below in SI units

$$q_u = 0.85\, E_i - 0.74\, t_v - 6.53 \Delta t \quad W/m^2$$

The performance of the collector will be described when information is obtained on the variation of t_v and Δt with E_i.

Measurements needed to test a collector using the author's method

t_o = outlet temperature

t_i = inlet temperature

t_a = air temperature taken inside a Stephenson screen

t_{IG} = air temperature between inner and outer covers

E_i = incident energy using a solarimeter

THE ELEGANCE OF THIS METHOD OF TESTING WILL BE APPRECIATED WHEN IT IS REALISED THAT IT IS NOT NECESSARY TO MEASURE THE MASS FLOW RATE OF THE WORKING SUBSTANCE PASSING THROUGH THE COLLECTOR.

A known drawback to this method of inferential testing rests with the positioning of the temperature sensing element for the measurement of t_{IG}. The preferred position and that used for compiling the information given in this paper is as shown in Fig.2. Clearly, if other investigators wish to use this method then in order that their results may be compared with those of the author and with one another then it is important that they should place their temperature sensing element in the position shown in Fig.2.

Fig.2

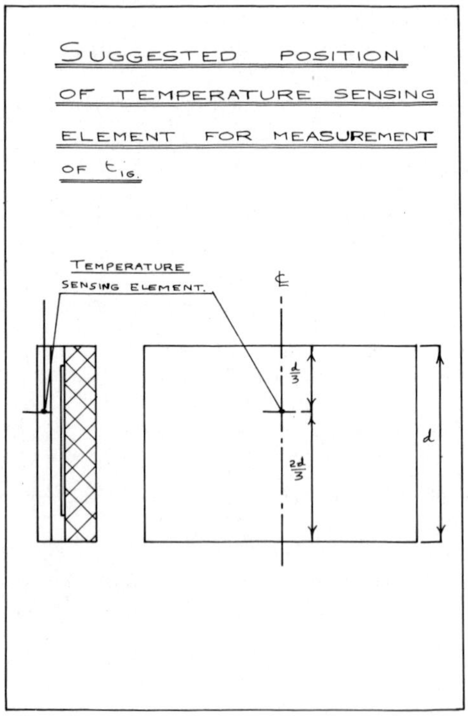

RESULTS

Fig.3 is made up of a family of curves depicting the plot of Δt against t_v for seven particular values of E_i. The values of E_i given are the exact equivalent (in W/m^2) of the incident energy measured in imperial units (Btu/ft^2h) and this is because the original work was done in imperial units. It is possible to produce the graph shown in Fig.4 using the particular equation of approximate losses for the collector tested and these lines may then be described as derived lines.

Fig.3

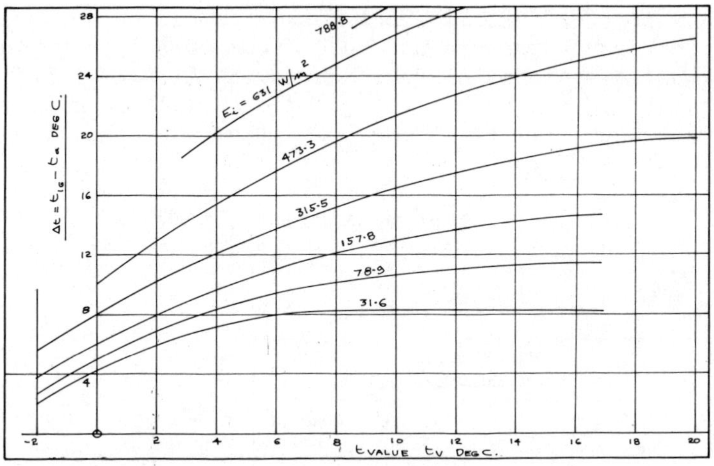

Fig.4

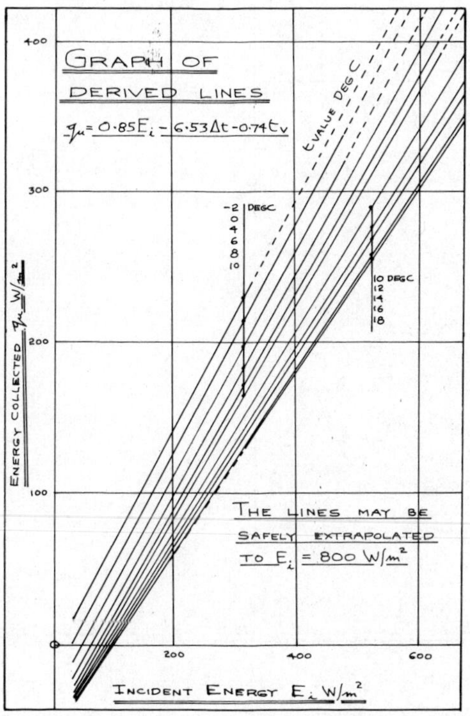

Fig.5

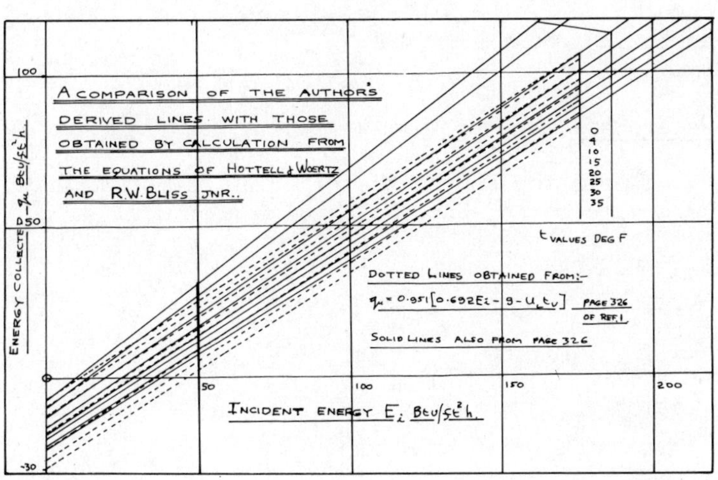

Any proposed experimental method should be compared with the results of other investigators and such a comparison is given in Fig.5. The equations of Hottell and Woertz and of R.W. Bliss, Junior have been solved by the author using series solution methods for his own collector reference (1). It is concluded that the comparison is sufficiently close to commend the method of testing outlined.

An indication of the magnitude of the various losses for a particular set of operating values is shown in Fig.6 in the form of an energy flow diagram.

Fig.6

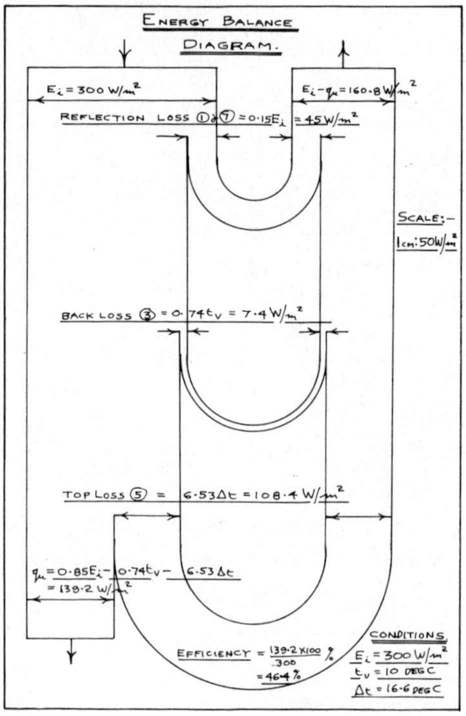

Future experimental work at the Chelmer Institute

A collector having one glass cover is to be tested using the foregoing method and the data is to be obtained and processed using microprocessing techniques. This work has been made possible by a grant made to the author by the Science Research Council, to whom thanks and acknowledgements are due.

References

(1) Thesis for PhD on the 'Thermal performance of flat plate solar heat collectors', University of London, July 1969.

(2) IHVE Guide, 1970.

Energetic efficiency of some particular types of flat plate collectors

3.11

J. Brodin, Professor, Van Vui Pham, Chargé de recherches au CNRS,
C. Sinnasse, Head of the Physics-Electricity Laboratory
Ecole Nationale Supérieure de l'Aéronautique et de l'Espace
10 Avenue Edouard Bélin, Boîte Postale 4032
31055 - Toulouse Cedex

ABSTRACT

In this paper, we have tried to outline the main characteristics of flat plate collectors.

Having studied the greenhouse effect and formulated the relationship between the energetic efficiency and the solar flux, we treat the experimental results by the segment synthesis method to take into account the cumulative effect of the energy. We can thus show the evolution of the quasi static parameter and the energetic efficiency. The results obtained agree with the theoretical analysis.

During the period of July-September 1976, we have obtained for the Toulouse region a heat gain of about 2000 Kcal/sq.m per day at a temperature of 50°C.

INTRODUCTION

The solar collector is the basic element to convert solar energy into thermal energy. It differs totally from classical heat exchangers because it transfers radiated energy from a powerful and remote source to a fluid.

It is known that the radiation is characterized by a solar constant: the quantity of energy that the sun sends normally per square centimeter and per minute, through an area situated beyond our atmosphere and at the mean distance between earth and sun. It is equal to 1.93 calg or 8.07×10^7 ergs. On earth, the solar constant's variation depends on the transparency of our atmosphere (clouds, humidity, fog etc.) and fluctuates due to the surface modification of the sun. Its spectrum lies between 0.3 and 3 microns with a peak at 0.435 micron. The problem is therefore to collect this energy with or without radiation concentration.

The flat plate collectors, which are studied in this paper, are the devices which utilize the direct radiation as well as the diffuse one without any orientation requirements towards the sun. Flat plate collectors require only easy maintenance thus reducing costs. However the temperatures obtained are low (theoretically up to 100°C).

The focused collectors or concentrators have concave reflectors to concentrate the radiation. The energy flux is thus increased on a small area. They have the advantage of supplying a higher temperature to the fluid but they require complicated equipment which set us orientation devices for the concentrating reflectors.

PRINCIPLE

The principle for the operation of a flat plate collector is based on the greenhouse effect.

Greenhouse effect

A black plate, exposed to the incident solar flux, absorbs the energy, heats up and emits back outwards an energy flux R.

At the equilibrium temperature T_e, $R = \Sigma$, where Σ is the incident solar flux.

According to Stefan (1)

$$\Sigma = \sigma T_e^4$$

$$= 1.357 \times 10^{-12}\, T_e^4 \text{ (cal. s}^{-1}\text{ cm}^{-2}\,°\text{K}^{-4})$$

$$= 5.675 \times 10^{-8}\, T_e^4 \text{ (w m}^{-2}\,°\text{K}^{-4})$$

For $\Sigma = 1100$ w/sq.m, $T_e = 64.8\sqrt[4]{1100} = 371°K$

The plate re-emits then, after Wien, a wavelength $\lambda = 7.7\ \mu$, where Wien $\lambda_m T = 0.2884$ cm°K.

As the solar spectrum is situated between 0.3 and 3 microns, it is enough that a pane whose transparency becomes limited at 3 microns is placed. The solar radiation is let in and the radiation emitted by the plate is prevented from going out.

Poynting (2) has established that (Appendix 1)

$$R = \frac{t_i + \frac{a}{2}}{1 - z_1 - \frac{a_1}{2}}\Sigma$$

The greenhouse effect will be so much more efficient that $\dfrac{t_i + \frac{a}{2}}{1 - z_1 - \frac{a_1}{2}} > 1$

The denominator $1 - z_1 - \frac{a_1}{2}$ is reduced to a maximum, on superimposing another glass, thus reducing $t_i + \frac{a}{2}$ also.

The greenhouse effect can be improved by putting thin layers of filter coatings on the glass, reducing $1 - z_1 - \frac{a_1}{2}$.

Design of the flat plate collector

Figure 1 gives the schematic representation of the flat plate collector design.

Fig. 1
Design of a flat plate collector

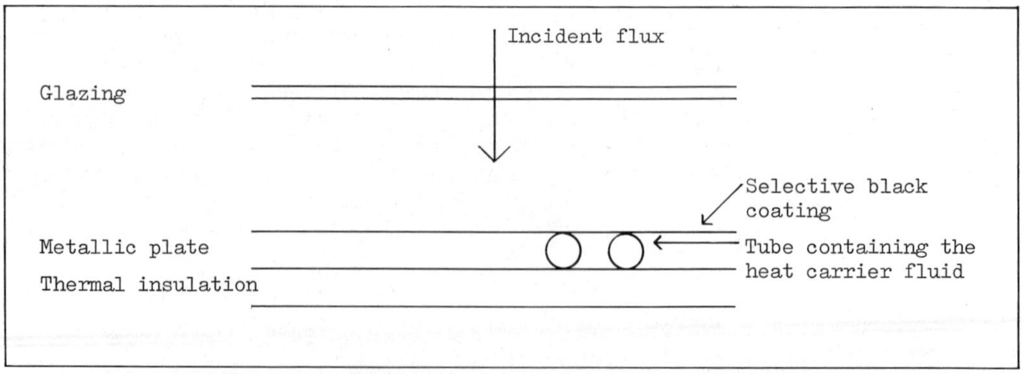

The solar flux which penetrates across the glass is absorbed by a blackened plate. The heat collected by the latter is transferred towards the device in which the heat is utilized by a heat carrier liquid. To reduce the heat losses, the plate is placed on a thermally insulated support.

FUNCTIONING

Functioning at the thermal equilibrium

This functioning enables us to determine the different parameters of the collector: quasi static parameter σ_1, time constant τ.

The plate is heated up by the solar flux, with the heat unextracted by the heat carrier fluid. The maximum temperature is reached, when the incident solar power is exactly balanced by the collector losses.

According to Stefan's law, $T_e = \sqrt[4]{\frac{\Sigma}{\sigma}}$ where T_e is the equilibrium temperature.

Let us introduce the greenhouse effect. We have (Appendix 1)

$$R = \frac{t_i + \frac{a}{2}}{1 - z_1 - \frac{a_1}{2}} \Sigma$$

and

$$\frac{R}{T_e^4} = \frac{t_i + \frac{a}{2}}{1 - z_1 - \frac{a_1}{2}} \frac{\Sigma}{T_e^4} = \sigma \text{ or } \frac{\Sigma}{T_e^4} = \sigma \frac{1 - z_1 - \frac{a_1}{2}}{t_i + \frac{a}{2}} = \sigma_1$$

σ_1 characterizes the greenhouse effect of the collector.

To determine the time constant τ which characterizes the response time of the collector, let us admit that the heating speed of the collector is proportional to $T_e - T$ and in inverse ratio to the equivalent thermal mass M, then we have (Appendix 2)

$$\tau = \frac{M}{k} \text{ and } T = T_e(I - e^{-t/\tau}) + T_a e^{-t/\tau}$$

We note that, at $t = 0, T = T_a$: ambient temperature
and at $t \gg \tau, T = T_e$: equilibrium temperature

If we solve with respect to the coefficient of proportionality k, we have (Appendix 3)

$$k = 4\, T_e^3 \sigma_1 (t + \frac{a}{2})$$

Consequently, $\tau = \dfrac{M}{4T_e^3\sigma_1(t_i + \frac{a}{2})}$

Consequences

(a) τ is proportional to the equivalent thermal mass. Therefore more precautions should be taken concerning the thermal insulation of the collector.

(b) To show the influence of the equilibrium temperature, we use Stefan's law

$$\Sigma = \sigma_1 T_e^4$$

$$\tau = \frac{T_e M}{4\, \Sigma(t_i + \frac{a}{2})} = \frac{T_e M}{4R(1 - z_1 - \frac{a_1}{2})}$$

We find again the utility of the greenhouse effect.

(c) The efficiency, is, by definition, the ratio between the received energy and the extracted heat energy. We have in the case of functioning at thermal equilibrium, where no heat is extracted by the heat carrier fluid, $\eta = 0$.

Functioning at maximum heat extraction

In this case, the efficiency is maximum. When a quantity of heat, at a temperature near the ambient one, is extracted, the conduction and emission losses become negligible; the rise in the fluid temperature is low (1°C), but the whole incident flux is absorbed in the form of heat.

The time constant of the collector is neglected, because, we operate it with a high flow rate. This heat regime cannot be utilised effectively. So we search for an intermediary regime where we can extract a quantity of heat at usable temperatures.

INTERMEDIARY REGIME

The two above regimes are extreme cases. In practice, the collector is used in a regime such that the temperature reached is sufficient so that the fluid can be used without much loss of efficiency. The latter is (Appendix 4).

$$\eta = \frac{M(T_e - T_o)(1 - e^{-t/\tau})}{t\Sigma} \; e^{\overline{\Delta x}\left(\frac{2 + x_o}{1 - x_o}\right)}$$

where $\overline{\Delta x} = \dfrac{\overline{T - T_o}}{T_e}$

As $e^{\overline{\Delta x}\left(\frac{2 + x_o}{1 - x_o}\right)}$ is in the order of 1 to 1.4, we can write

$$\eta = \frac{M(T_e - T_o)(1 - e^{-t/\tau})}{t\Sigma}$$

Consequences

(a) The efficiency varies according to the incident flux Σ. So the study is focused on the precise knowledge of Σ and its daily and monthly variations. The flux determination is conducted experimentally by different apparatus, especially by using the pyrheliometer (4). In our case, we measure the voltage which indicates the sunshine, multiplied by a factor 1.7921 Kcal per sq.m per minute.

(b) The efficiency varies depending on the equilibrium temperature T_e. We then suppose that the collector at each instant is at a steady state, that is to say $T = T_e$.

(c) The efficiency varies depending on the equivalent thermal mass M. In fact, what is interesting to us is the fluid temperature at the outlet of the collector and the amount of usable fluid; the extracted thermal energy is then $E_{coll} = M(T_\Delta - T_{en})$ at instants $t \ll \tau$ where T_{en} is the fluid temperature at the inlet of the collector, T_Δ its temperature at the outlet.

If we consider that the fluid is water and that there is a circulation, we can write

$$E_{coll} = \frac{D(T_\Delta - T_{en})}{S} \text{ Kcal. m}^{-2} \text{ mn}^{-1}$$

where D is the flow rate (1/mn), S the collector area (square meter).

Fig. 2
Spiral type flat plate collector

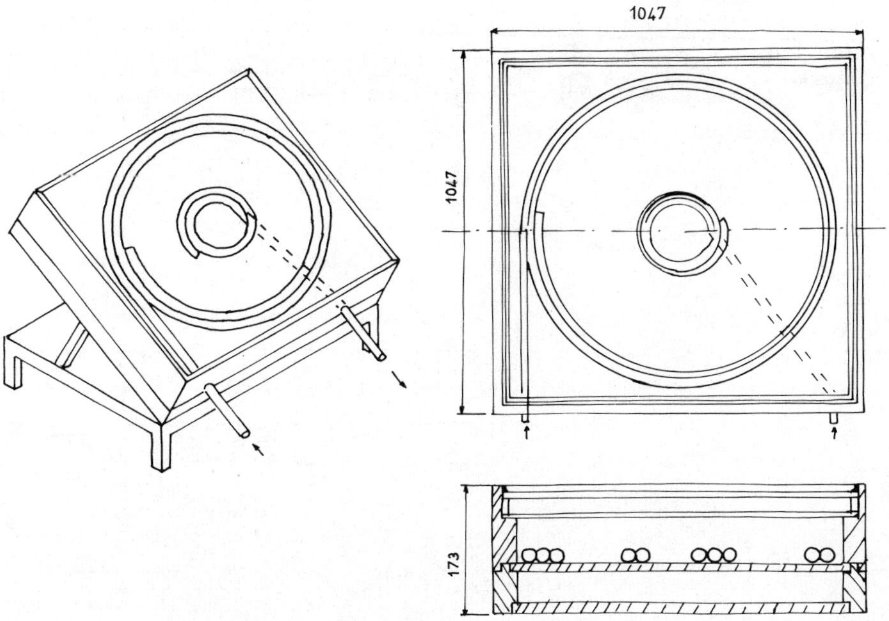

APPLICATIONS

We want to utilize the plate as solar radiation absorber as well as the channel for the heat transfer fluid. We had a choice of three materials: the butaprene (commercial name: Nitril) is chosen for its black colour and for its temperature resistance; the silicone rubber (commercial name: Versilic) which has an exceptional thermal stability (−80°C to 220°C); the glass which is a traditional material resists at high temperatures, the disadvantages being its brittleness and its complicated realization in the serpentine form.

Figures 2, 3, 4 show the three different designs for a flat plate collector and the layout of the fluid circulation.

Fig. 3
Quincunx type flat plate collector

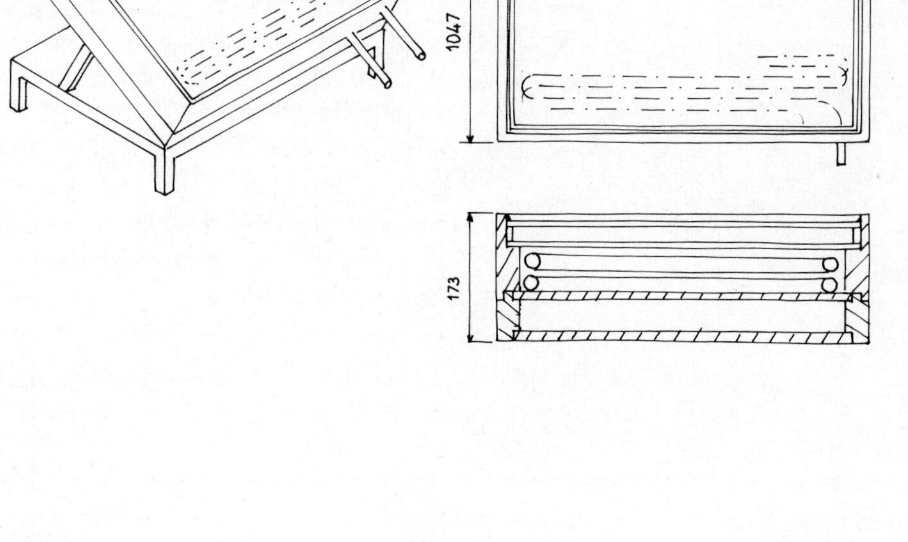

Fig. 4
Water circulation and measurement block diagram

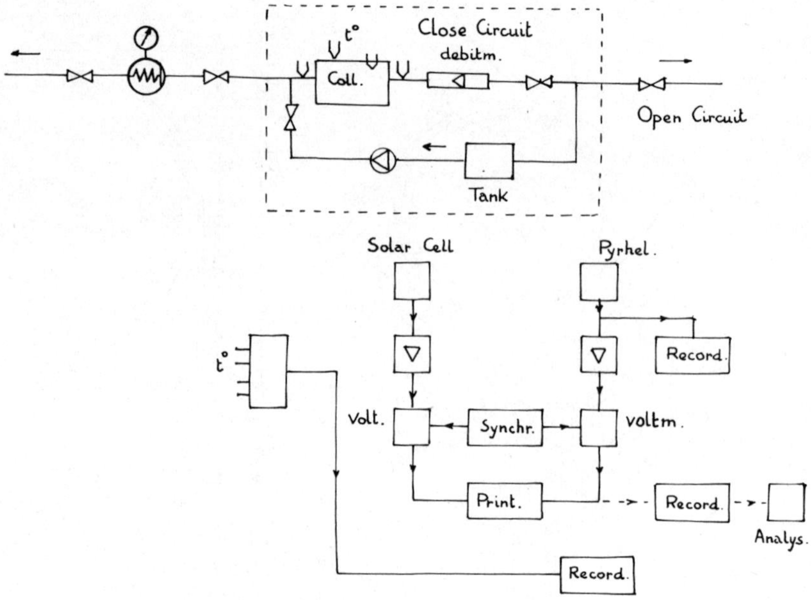

EXPERIMENTAL RESULTS AND INTERPRETATIONS

Greenhouse effect

Figure 5 shows the results related to the greenhouse effect with single and double glazing. The x coordinate is graduated in solar hour and the y coordinate is in Kcal per sq.m per minute. The flat plate collector is a spiral in silicone rubber. At first glance, it seems that glazing is necessary. The collector without glazing collects lesser energy than that with double glazing, for an equivalent amount of solar radiation. However, it is not fair to compare the results of the single glazing with those of the double glazing because the incident solar radiation in each of the two abovesaid cases is totally different. In order to compare them, we calculate the energetic efficiency in terms of cumulated energy (Appendix 4). The result (Fig. 6) shows that there is the greenhouse effect and that the single glazing is enough. It seems hazardous to come to a hasty conclusion that the single glazing is better than the double one, because the space between the two glazings may have a significant effect in insulating the air layer between the collector and the first glass.

Fig. 5
Energy vs time (solar hour)—parameter : number of glasses

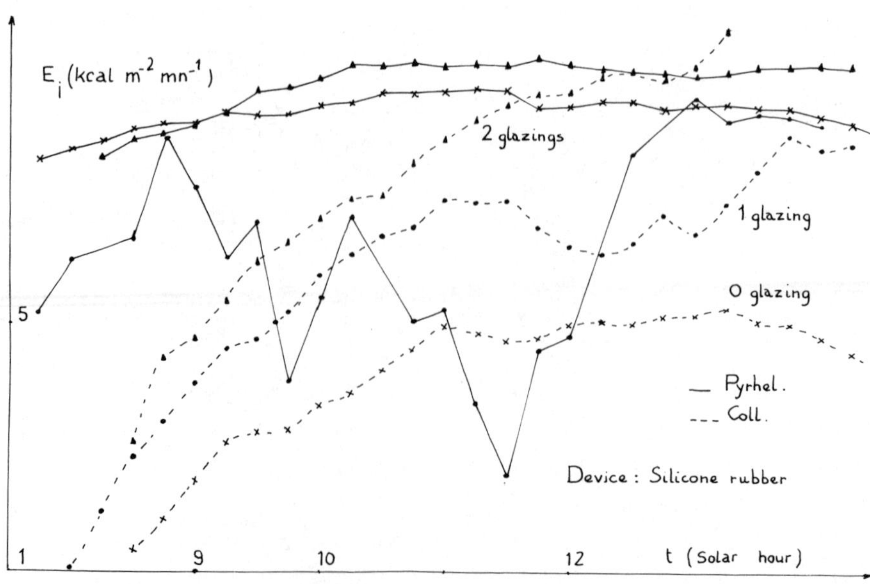

Fig. 6
Efficiency vs cumulated energy—parameter : number of glasses

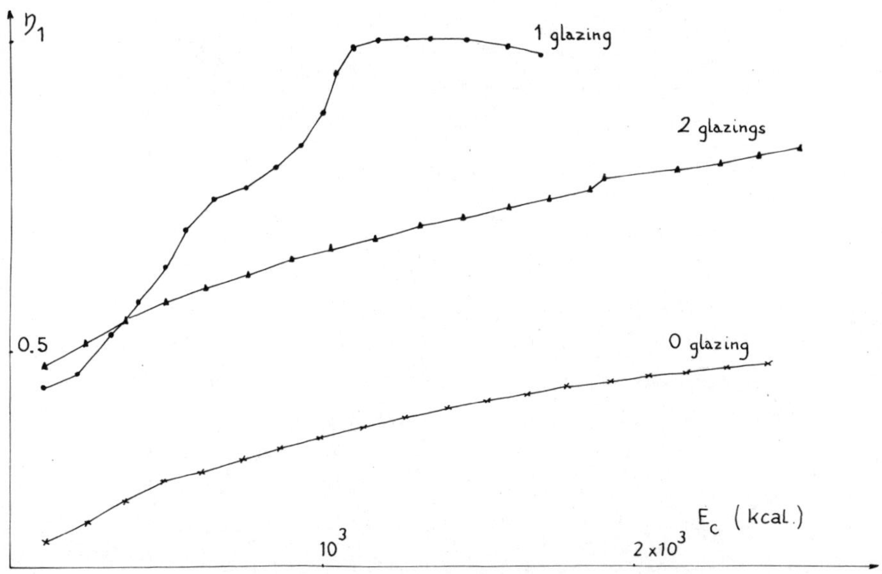

Fig. 7
$\frac{\sigma_1}{\sigma}$ vs cululated energy—parameter : number of glasses

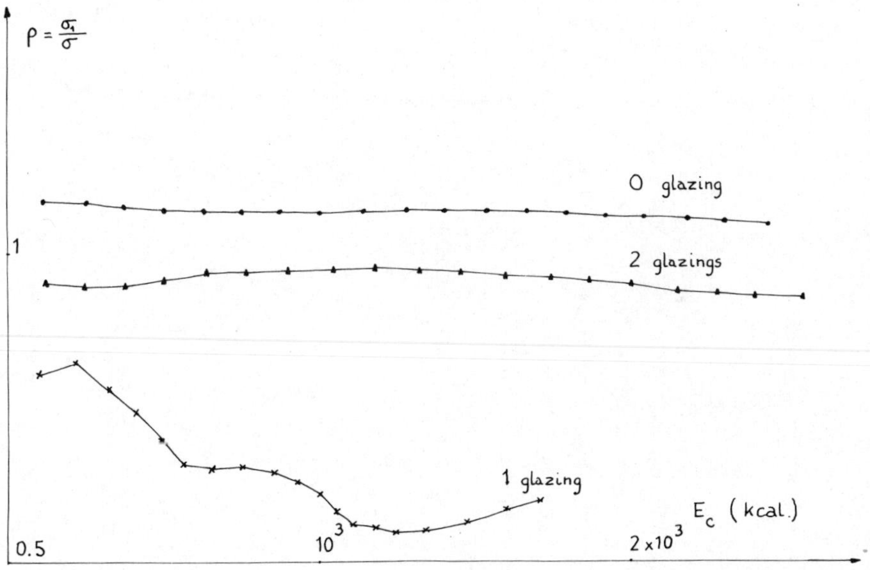

We proceed in the same way to draw the graph of ρ, the ratio between the quasi static parameter of the collector and Stefan's constant, in function with the incident cumulated energy (Fig. 7) (Appendix 3). The result supports the idea that there is a greenhouse effect present for a single glazed flat plate solar collector.

Flow rate influence

The same procedure (Fig. 9, 10) is good enough to treat the results, presented in Figure 8. The relationship is thus confirmed. However, we have to take into consideration the ratio ρ for adjusting the fluid temperature.

$$E_{coll} = \frac{D(T_\Delta - T_{en})}{S}$$

Fig. 8
Energy vs time—parameter: water flow rate

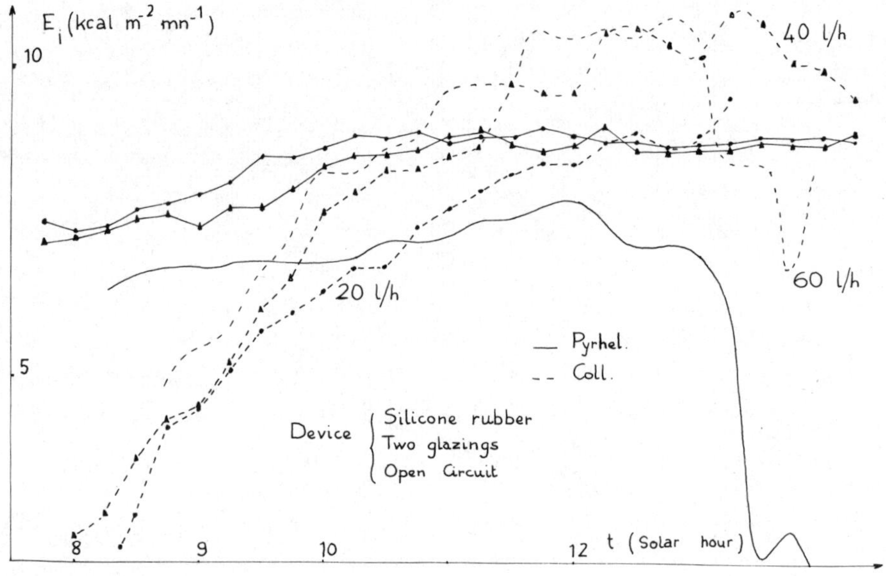

Fig. 9
Energetic efficiency—parameter: water flow rate

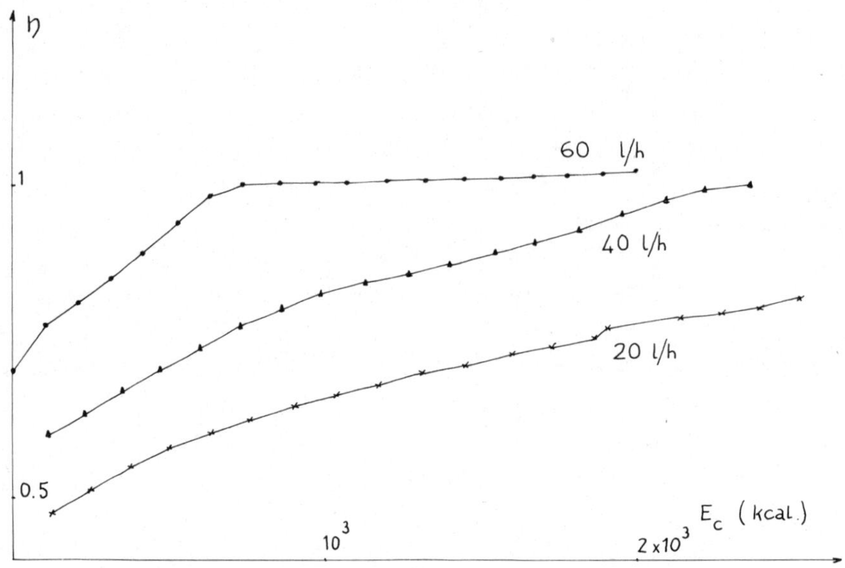

Fig. 10
$\frac{\sigma_1}{\sigma} = \rho$—parameter: water flow rate

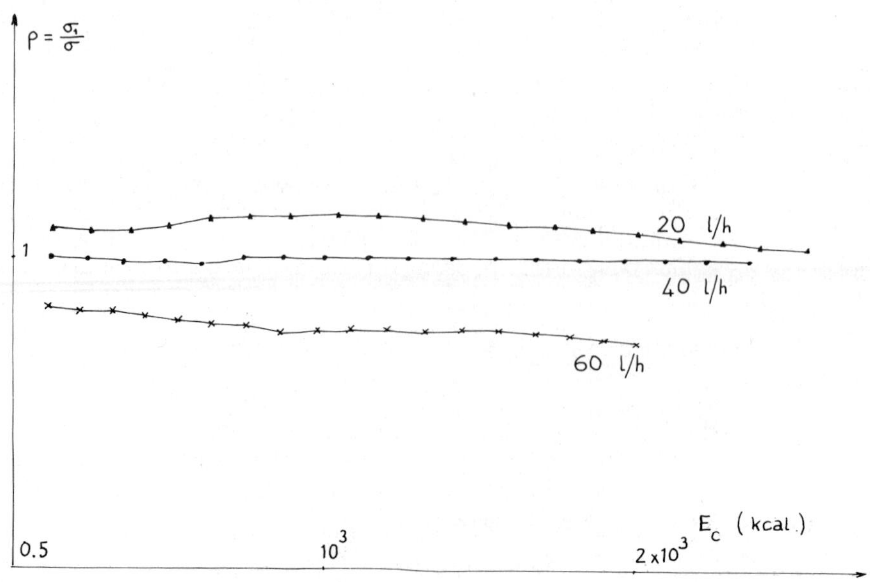

Open circuit and close circuit

In order to have a sufficiently high fluid temperature, the fluid in the collector is recycled many times. Then its calories are transferred to a classical heat exchanger. We have realized this for a flow rate of 60 l/h. The experimental results are presented in Fig. 11. Figure 12 shows that the open circuit has a better efficiency but the temperature is less high (Fig. 13).

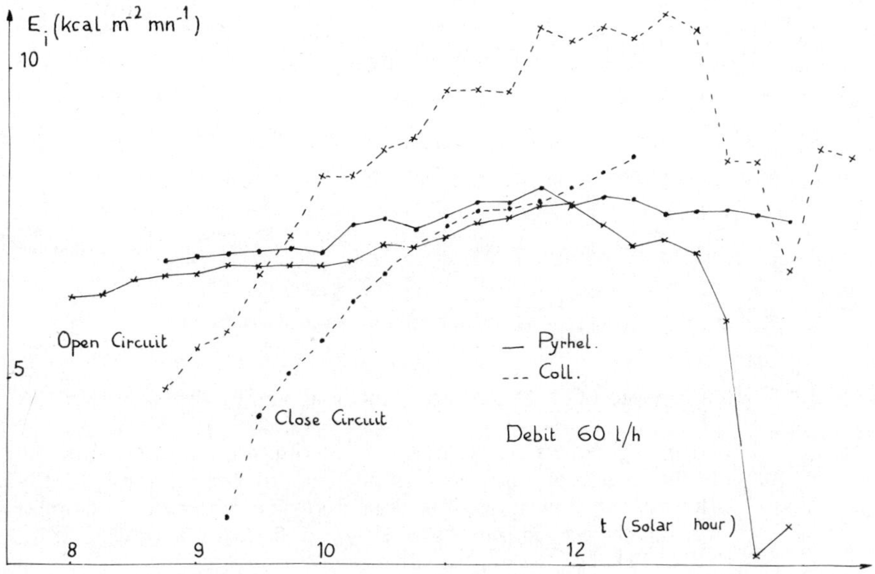

Fig. 11
Energy vs time—
parameter : circuit
mode

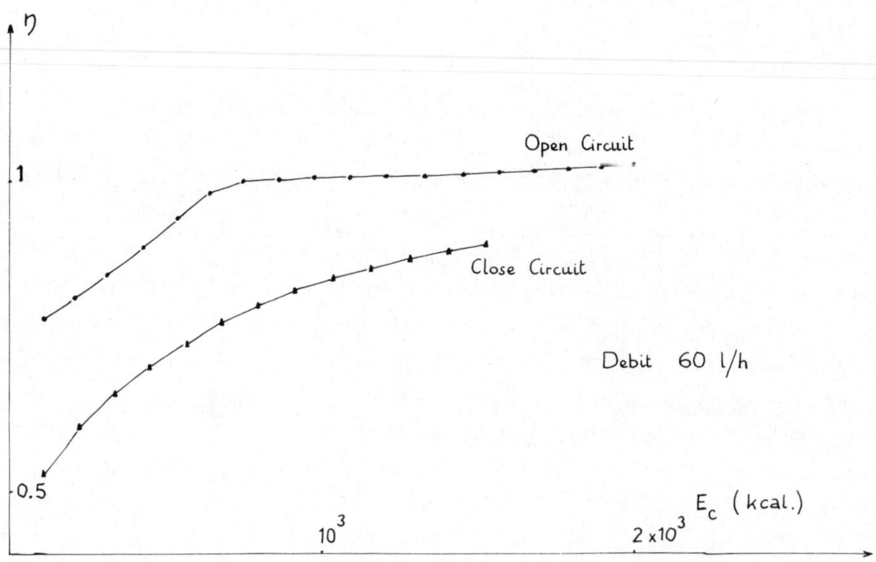

Fig. 12
Energetic efficiency—
parameter : circuit
mode

Fig. 13
$\frac{\sigma_1}{\sigma} = \rho$–parameter:
circuit mode

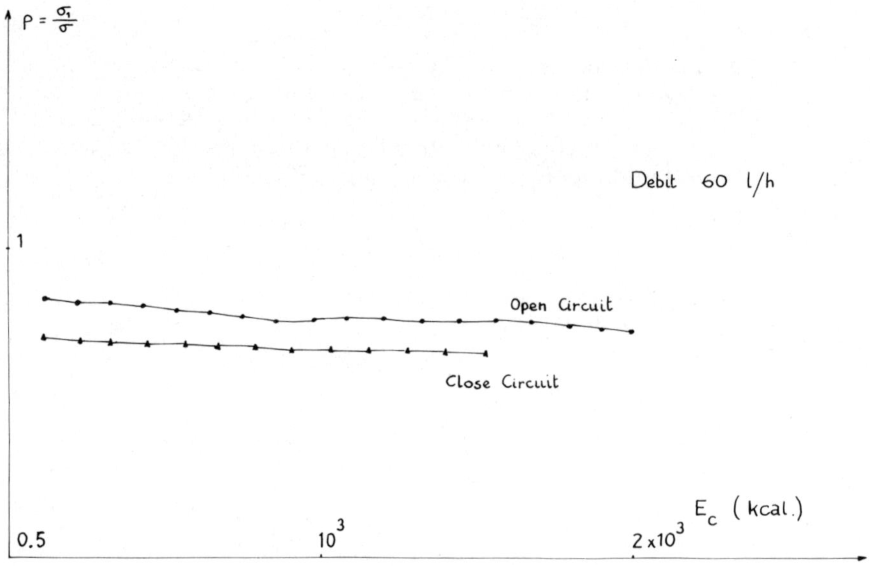

Influence of the collector material and of the nature of the fluid

Although not many tests were conducted, they are quite sufficient to foresee that the butaprene tube and the silicone rubber tube are better than the glass tube as far as the heat transfer and the heat rise are concerned. Concerning the nature of the fluid, we have carried out tests with tinted water and with tinted water mixed with glycol (Figs. 14, 15, 16, 17, 18, 19).

Fig. 14 Energy vs time–parameter: material

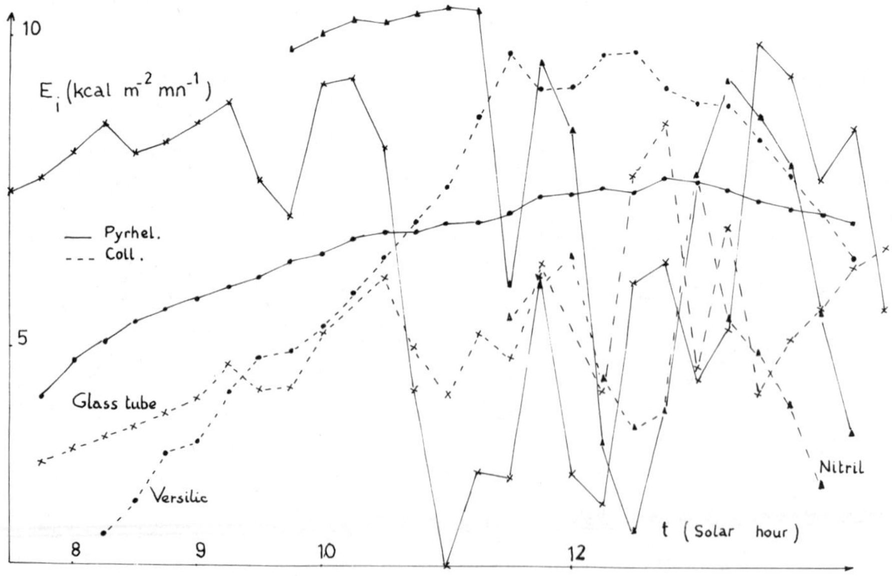

Fig. 15
Energetic efficiency—parameter: material

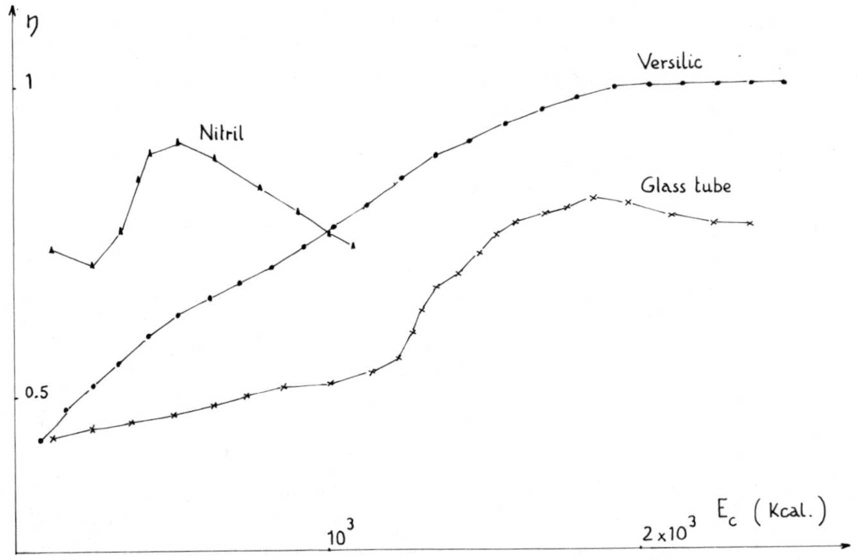

Fig. 16
$\frac{\sigma_1}{\sigma} = \rho$—parameter: fluid

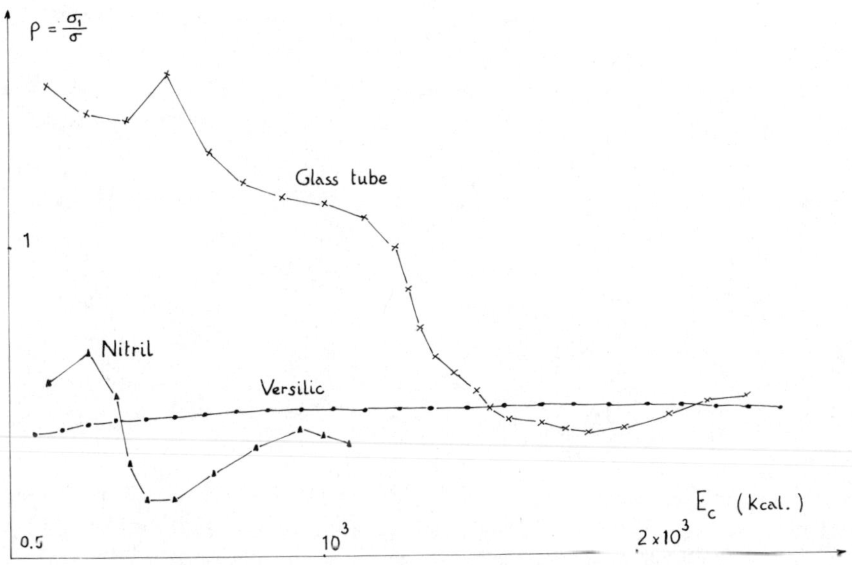

Fig. 17
Energy vs time—parameter: fluid

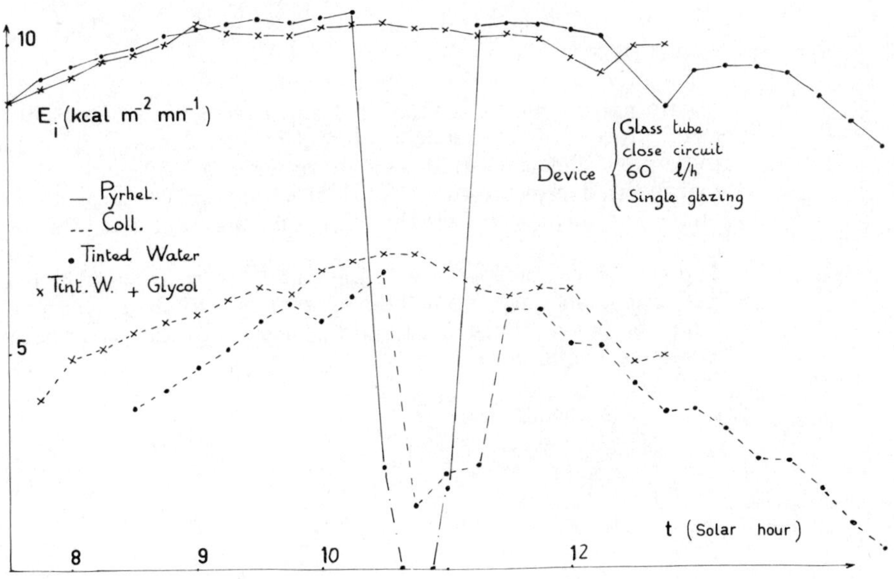

Fig. 18
Energetic vs time–
parameter : fluid

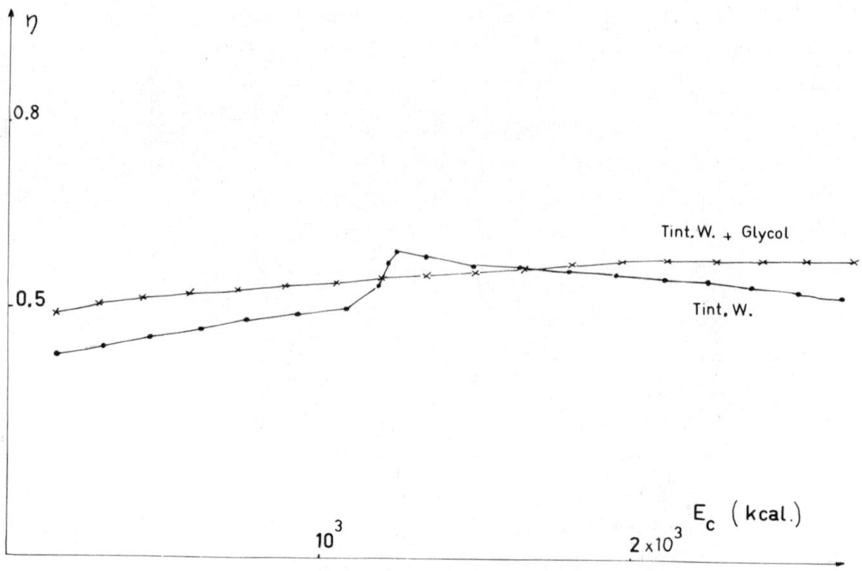

Fig. 19
$\frac{\sigma_1}{\sigma} = \rho$–parameter :
fluid

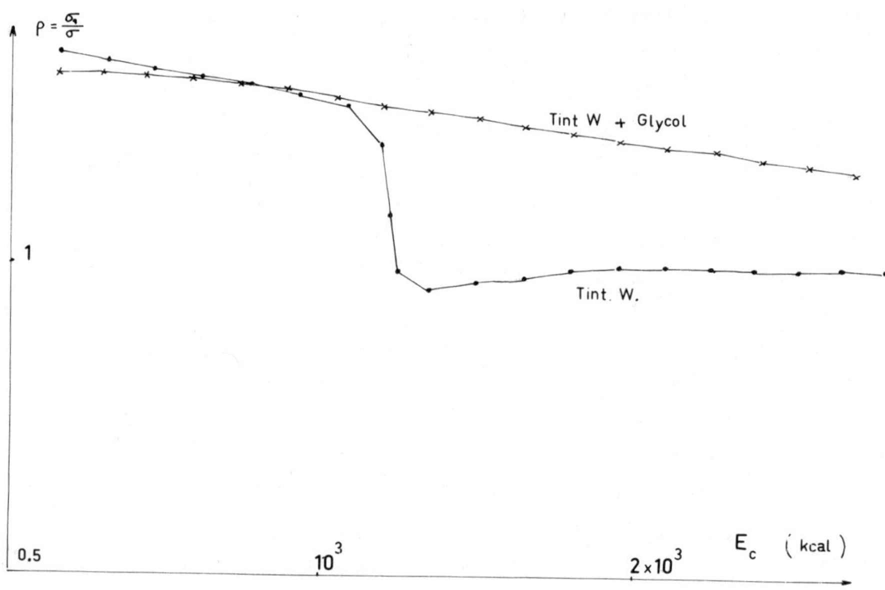

CONCLUSION

In this report, we have studied the different parameters characterizing the collector: quasi static parameter, heat collection efficiency. The calculated efficiency seems to be very high and at times; this is undoubtedly due to the way for which heat is extracted, due to the functioning of calory extraction and due to the cumulated energy effect and the way by which efficiency is calculated (5).

The non-linear properties of the flat plate collector is obvious. It seems very interesting to elaborate, from the abovesaid parameters, an analogical model, enabling future improvements and the optimisation of the structure of the collector for the Toulouse region.

APPENDIX 1

Poynting calculation

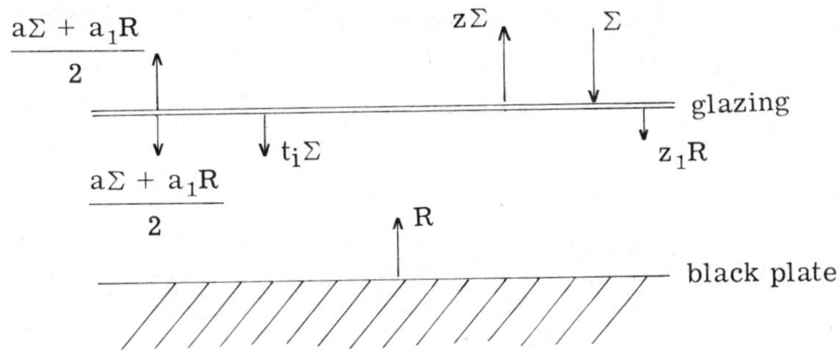

Let a, r, t_i be the absorption, reflection and transmission coefficients of the glazing with respect to the solar radiation Σ.

We have $a + r + t_i = I$

Let a_1, r_1, t_1 be the absorption, reflection and transmission coefficients of the glazing with respect to the radiation reflected by the black plate.

We have $a_1 + r_1 + t_1 = I$

Assuming the width of the pane to be very small so that the two sides are at the same temperature, we can then say that the pane emits from each side half of what it absorbs totally.

$$\frac{a\Sigma + a_1 R}{2}$$

As the plate emits all that it receives, we have

$$R = t_i\Sigma + z_1 R + \frac{a\Sigma + a_1 R}{2} = \Sigma\left(t_i + \frac{a}{2}\right) + \left(z_1 + \frac{a_1}{2}\right)R$$

$$R = \frac{\left(t_i + \frac{a}{2}\right)\Sigma}{1 - z_1 - \frac{a_1}{2}}$$

APPENDIX 2

Relationship between the plate temperature and the equilibrium temperature

Let ϑ_{coll} be the heating speed of the collector, T the plate temperature, T_e the equilibrium temperature of the plate, M the equivalent thermal mass, we admit that the heating speed is directly proportional to $T_e - T$ and in inverse ratio to the equivalent thermal mass M.

$$\vartheta_{coll} = \frac{dT}{dt} = -\frac{k}{M}(T - T_e)$$

$$\frac{dT}{T - T_e} = -\frac{k}{M}dt$$

But $\mathrm{Log}(T - T_e) = -\frac{k}{M}t + c^{te}$ $T - T_e = Ae^{-k/M\, t}$

At $t = t_o$ $T = T_a$ $T_a - T_e = A$

$$T = T_e(1 - e^{-t/\tau}) + T_a e^{-t/\tau}$$

with $\tau = \frac{M}{k}$ time constant of the collector.

APPENDIX 3

Evaluation of the coefficient of proportionality k

In the case of a plate absorbing certain energy, we have

$$R = (t_i + \frac{a}{2}) + (z_1 + \frac{a_1}{2})\sigma(T_1^4 - T^4)$$

where T_1 is the glass temperature.

We suppose that the plate at each instant is at steady state; that is to say that T is the temperature corresponding to the solar flux S.

$$S = \sigma_1 T^4$$

where σ_1 is the quasi static parameter of the collector.

So, during the evolution of the collector towards a steady state, the quantity of heat absorbed by the collector is

$$R' = (\Sigma - S)(t_i + \frac{a}{2})$$

This quantity of heat is equal to the quantity of heat absorbed $M\frac{dT}{dt}$

$$M\frac{dT}{dt} = (\Sigma - S)(t_i + \frac{a}{2}) = (t_i + \frac{a}{2})\sigma(T_e^4 - T^4)$$

$$= \sigma_1' (T_e^4 - T^4)$$

where $\sigma_1' = \sigma_1(t_i + \frac{a}{2})$ kinetic constant of the collector.

$$\frac{dT}{dt} = \frac{\sigma_1}{M}(T_e^4 - T^4)$$

$$\frac{dT}{T_e^4 - T^4} = \frac{\sigma_1}{M} dt$$

Consequently

$$\frac{4T_e^3\sigma_1'}{M} t = \log \frac{T_e + T}{T_e - T} + 2\ A_2 ctg \frac{T}{T_e} - \log \frac{T_e + T_o}{T_e - T_o} - 2\ A_2 ctg \frac{T_o}{T_e}$$

where T_o is the initial temperature of the collector.

Let us put $\frac{T}{T_e} = x, \frac{T_o}{T_e} = x_o$

We have $\frac{4T_e^3\sigma_1'}{M} t = \log \frac{1 + x}{1 - x} + 2\ A_2 ctg\ x - 2\ A_2 ctg\ x_o - \log \frac{1 + x_o}{1 - x_o}$

But $\log \frac{1 + x}{1 + x_o} \# \frac{\Delta x}{1 + x_o}$

$2\ A_2 ctg\ x - 2\ A_2 ctg\ x_o \# \Delta x$

We have

$$\frac{4\,T_e^3\sigma'_1}{M}t = \log(1+x) - \log(1-x) - \log(1+x_o) + \log(1-x_o) + \Delta x$$

$$= \log\frac{1+x}{1+x_o} - \log\frac{1-x}{1+x_o} + \Delta x = \frac{\Delta x}{1+x_o} + \Delta x - \log\frac{1-x}{1-x_o}$$

$$= \Delta x\left(\frac{2+x_o}{1+x_o}\right) - \log\frac{1-x}{1-x_o}$$

But $$\log\frac{1-x}{1-x_o} = \log\frac{T_e - T}{T_e - T_o}$$

Neglecting $\Delta x\left(\frac{2+x_o}{1+x_o}\right)$, we can identify with (Appendix 2)

$$\log(T - T_e) = -\frac{k}{M}t$$

to have $k = 4\,T_e^3\,\sigma'_1$

APPENDIX 4

Energetic efficiency of the collector

By definition, the efficiency η is the ratio between the received power and the extracted heat power.

$$\eta = \frac{\text{extracted heat power}}{\text{received power}}$$

Similarly, the energetic efficiency

$$\eta = \frac{\text{extracted heat energy}}{\text{incident energy}} = \frac{E_{coll}}{E_i}$$

But $E_i = t\Sigma$

$$E_{coll} = M\int_o^t \left(\frac{dT}{dt}\right)dt$$

We have seen (Appendix 3)

$$\text{Log}\,\frac{1-x}{1-x_0} = -\frac{k}{M}t + \Delta x\left(\frac{2+x_0}{1+x_0}\right)$$

$$\text{Log}\,(1-x) = \log(1-x_0) - \frac{k}{M}t + \Delta x\left(\frac{2+x_0}{1+x_0}\right)$$

$$(1-x) = (1-x_0)e - \frac{k}{M}t + \Delta x\left(\frac{2+x_0}{1-x_0}\right)$$

$$\left(1-\frac{T}{T_e}\right) = \left(1-\frac{T_0}{T_e}\right)e - \frac{k}{M}t + \Delta x\left(\frac{2+x_0}{1-x_0}\right)$$

$$(T_e - T) = (T_e - T_0)e^{-\frac{k}{M}t + \overline{\Delta x}\left(\frac{2 + x_0}{1 - x_0}\right)}$$

$$T = \left(1 - e^{-\frac{k}{M}t + \overline{\Delta x}\left(\frac{2 + x_0}{1 - x_0}\right)}\right) T_e + T_0 e^{-\frac{k}{M}t + \overline{\Delta x}\left(\frac{2 + x_0}{1 - x_0}\right)}$$

$$\frac{dT}{dt} = T_e \frac{k}{M} t\, e^{-\frac{k}{M}t + \overline{\Delta x}\left(\frac{2 + x_0}{1 - x_0}\right)} - T_0 \frac{k}{M} t\, e^{-\frac{k}{M}t + \overline{\Delta x}\left(\frac{2 + x_0}{1 - x_0}\right)}$$

$$E_{coll} = \int_0^t k(T_e - T_0)\varphi\, e^{-\frac{k}{M}\varphi + \overline{\Delta x}\left(\frac{2 + x_0}{1 - x_0}\right)}$$

$$= M(T_e - T_0)(1 - e^{-t/\tau})e^{\overline{\Delta x}\left(\frac{2 + x_0}{1 - x_0}\right)}$$

$$\text{so } \eta\ \frac{M(T_e - T_0)(1 - e^{-t/\tau})e^{\overline{\Delta x}\left(\frac{2 + x_0}{1 - x_0}\right)}}{t\Sigma}$$

In practice, it is interesting to consider the collector as a quadripole where the input is the incident solar energy and the output the extracted energy. It is obvious that the cumulation effect exists. In order to take into account this effect, we proceed by a segment synthesis: we then divide the x axis in regular intervals (15 minutes), we calculate the area subtended by the curves. We divide then the area subtended by the curve indicating the solar radiation by that of the calories extracted by the collector to obtain the energetic efficiency value. To obtain the following points, we proceed in the same manner, but the new area is the sum of the areas (each one taken at 15 minutes intervals). Thus we draw the graph of the energetic efficiency as a function of the incident cumulated energy.

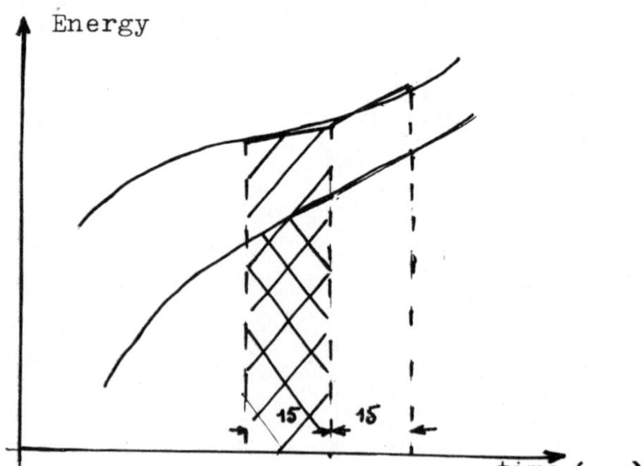

REFERENCES

(1) J. Brodin, 'Thermodynamique' Ed. Ensae, 1966 (Chapter 7)

(2) H. Bousse, 'Emission, chaleur solaire, eclairage' Ed. Delagrave, 1925 (Chapter 9)

(3) Haltiner and Martin, 'Dynamical and physical meteorology' McGraw Hill, 1957 (Chapter 6)

(4) 'Bulletin de Physiciens', 1976

(5) Beckmann and Duffie, 'Solar energy thermal processes'

J.E.H. Mustoe,
Department of Architecture,
North East London Polytechnic,
London, England.

Abstract

The principles upon which the operation of flat plate solar collectors are based are described and the main limitations upon their thermal performance are enumerated. In particular, it emerges that large absorbers like flat plate collectors cannot be raised to useful temperatures without their rate of heat loss becoming cripplingly high. This point is reached at a lower temperature than is required by economical building heating systems. However, the new designs of concentrating solar collectors, which operate with smaller and hotter absorbers, promise to overcome these limitations and to open the way to efficient solar central heating and hot water installations. The properties of five types of concentrating solar collectors are discussed, and the cusp collector is recommended for further development.

Introduction

The type of solar collector device most commonly in use today is the flat plate collector. The plate usually consists of a thin metal envelope resembling a central heating panel-type radiator through which water is circulated. The plate is covered by glass on its sunward side and is insulated against heat loss behind. The function of the glass is to prevent evaporative and conductive heat loss by protecting the panel from wind and from rain. It also enables the designer to exploit the so-called greenhouse effect. Glass is transparent to the infa-red radiation emitted by the sun in the 0.3 to 1.0 μm wavelength band which reaches the panel unimpeded where, by being absorbed, it raises its temperature. Radiation emitted by the panel at a temperature of less than 100°C is of longer wavelength, between 5.0 and 20.0 μm, to which glass is opaque. As a result of this one way traffic of infa-red radiation the glass, the air between it and the panel, and the panel itself are all further raised in temperature. Nevertheless, experience has shown that flat plate collector panels produce flow temperatures above 40°C only on days which in the British climate are exceptionally sunny.

The reasons for this disappointing level of performance are

(i) Flat plate collectors have a large surface area. For any unit area rate of heat loss, therefore, the total quantity of heat lost will be high. Radiative and conductive heat losses are multiplied by this factor.

(ii) The air gap between the collector panel and its glass cover forms a convection cell whose action transfers heat from the panel to the glass. The warmer glass loses heat at a greater rate to the outside air. The operation of the convection cell can be suppressed but not eliminated by baffles in the air space.

(iii) Flat plate collectors, because of their size, have a high thermal capacity. In our maritime climate the intensity of solar radiation fluctuates rapidly and through a large range during the course of a typical day's weather. A high thermal capacity panel will have warmed only slightly during the course of a transitory period of sunshine, and its flow temperature will consequently be little effected. We need in this country solar collectors that are thermally nimble.

(iv) The solar flux in our climate rarely exceeds $1kW/m^2$ and then only for short periods. Any type of collector which functions in such a relatively weak flux of energy will reach an equilibrium temperature at a low level. A low temperature equilibrium is imposed primarily by the awkward fact that an object radiates heat at a rate proportional to the fourth power of its absolute temperature. A slight rise in collector temperature causes a large increase in the rate of heat loss, and consequently a prompt return to equilibrium.

The performance of a flat plate collector can be improved in a number of ways. Two improvements calling for notice are,

(i) The panel can be provided with a selectively absorbing finish. A variety of metallic finishes have the property of absorbing nearly all the incident solar radiation and inhibiting the escape of all but a small proportion of the infa-red radiated by the panel. As with the greenhouse effect of a glass cover, the result is to raise the temperature of the panel. However, depositing a selective finish onto a large object like a panel collector is expensive, and these finishes are subject to various forms of degradation. Furthermore, a hotter panel loses heat more rapidly by conduction through its insulation, by radiation, and because the convection cell under the glass is accelerated.

(ii) Some of the potential benefits of selective coatings can be realised by evacuating the air space between panel and glass, and thus destroying the convection cell. Apart from the cost of this operation, it is open to the objection that it is impossible to maintain a permanent vacuum in a solar collector over a long period unless elaborate engineering details are resorted to.

Flat plate collector panels incorporating selective coatings, vacuum environments, and very thick and elaborate insulation have been designed and tested. There is no doubt that flow temperatures well above 40°C can be reached with these designs, although it should be noted that the thermal inertia of these highly developed panels is as great as the simpler types. However, they are very expensive and are consequently unlikely to make the difficult transition from laboratory test rig to building product in everyday use.

Solar collectors as building components

The limitations upon the performance of flat plate collectors we have discussed are inherent, and cannot be circumvented by ingenuity or clever design. It follows inescapably, therefore, that building heating systems whose energy is supplied by flat plate collectors will obtain only moderately hot water for distribution and use. Only in summer, when the building heat load is lightest, will the heating system operate with a flow temperature as high as 40°C. We must face two unwelcome consequences of this fact.

(i) Heat transfer by conduction and convection as well as by radiation takes place more efficiently the higher the temperature difference. For example, a central heating radiator operating at 40°C will emit only as much heat as another half its size at 60°C. Heat exchangers, radiators, fan convectors and pipework will all be large and expensive if the system must operate at the very moderate temperatures to be expected from flat plate collectors.

(ii) The quantity of heat that can be stored in a tank of water or in a rock chamber is increased when the temperature of the store is raised. Moderate operating temperatures result in large and heavy heat stores which are expensive and awkward to plan for.

The large size, low operating temperature, and high thermal inertia of the flat plate collector may therefore be seen to stem from its combined function of interrupting the solar flux and of absorbing it. Its properties are the reverse of those needed in solar collectors which are to provide energy for the purpose of servicing buildings. We need building solar collectors which are small, hot, and which respond quickly to changes in the intensity of sunlight.

Concentrating collectors

Solar collectors with the required characteristics can be made by separating the function of interrupting the solar flux from the task of absorbing it. Devices of this type are usually known as concentrating solar collectors. In this section of the paper the term collector will be used to describe the part whose function is to increase the density of the solar flux, and the thermally active part which converts the solar flux into useful heat is referred to as the absorber. A number of designs are described below.

Boosted Flat Plate (Figure 1)

A set of flat mirrors on either side of or surrounding a flat plate absorber form the simplest type of concentrating collector. Very large unweildy mirrors are required if the concentration factor is to exceed 2, and shadows cast by the mirrors reduce the performance of the boosted design below that of a plain flat plate collector when sunshine reaches it at an angle greater than 30° from the centreline, although this difficulty can be mitigated by the use of moveable mirrors. Boosted flat plate collectors have found few applications because the mirrors are expensive and cumbersome, and need frequent cleaning since the large space bounded by the plate and mirrors is difficult to seal hermetically. Furthermore, the thermal capacity of the flat plate absorber remains large, limiting its ability to respond to weather changes.

Fig. 1
Boosted Flat Plate

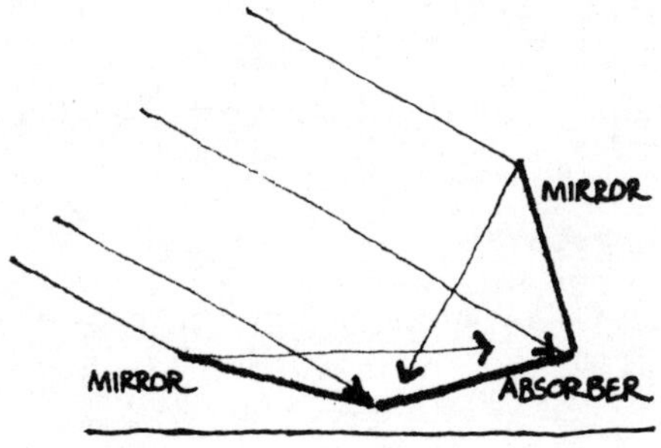

Parabola (Figure 2)

Very large concentrations of solar flux can be achieved by exploiting the geometry of the parabola. Solar radiation arriving on the axis of the collector is reflected to the focus of the parabola. Devices of this type usually take the form of a trough shaped parabolic collector with the absorber a water filled pipe fixed along the line of the focus. Although the theoretically infinite flux concentration cannot be reached because the sun is not a point source and because of mirror imperfections, parabolic troughs can raise steam in clear weather in the British climate.

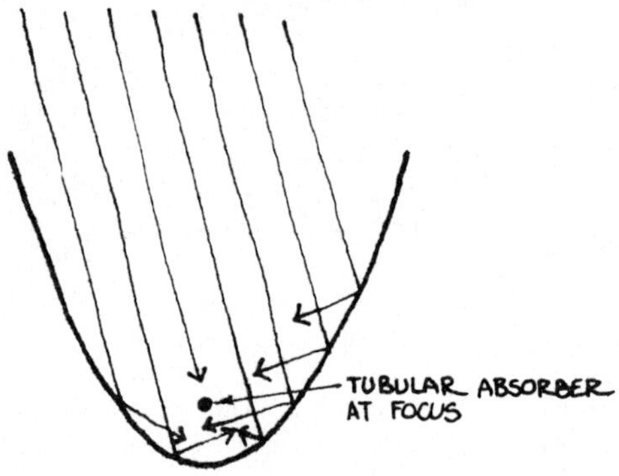

Fig. 2
Parabola

The potential performance of parabolic collectors can only be reached if expensive mechanisms are available to keep the sun accurately on the axis of the parabola. Since more than 50% of the solar radiation available in our climate arrises in diffused form, after scattering by dust and water vapour in the atmosphere, a parabolic collector can utilise less than half the available energy. Diffuse sunshine is reflected away from the focus as shown in figure 2. Although parabolic concentrators find applications where high temperatures must be produced, it should be noted that for the purpose of heating any but very large buildings water at more than 90°C is unnecessary. They are in a sense too effective, as well as too complicated, for general application in the field of building.

Winston Compound Parabolic Collector (Figure 3)

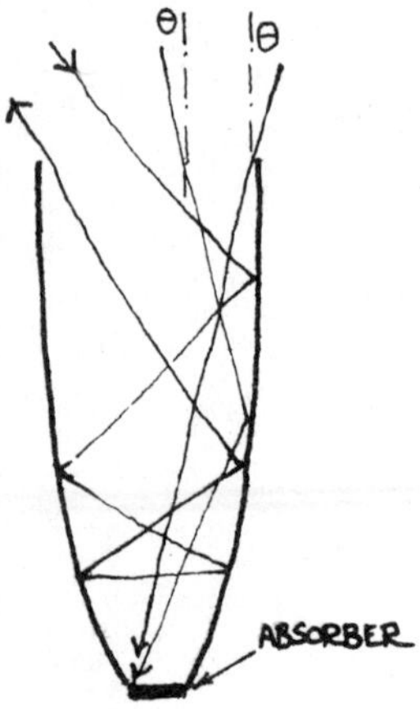

Fig. 3
Winston CPC

The invention by Baranov (1) in 1965 and the development by Hinterberger and Winston (2) in 1966 of compound parabolic collectors were crucial steps in the development of solar optics and have moved the subject into a new and promising phase. By employing these geometries it is possible to concentrate the solar flux, direct and diffuse, in an accurately determined manner and to place in the concentrated flux small efficient absorbers of low thermal capacity.

The essential features of what has come to be known as the Winston CPC are

(i) The focus of one parabola lies on the toe of the other.

(ii) The two parabola segments are aligned so that their outer edges lie parallel to the axis of the system.

(iii) The solar flux is concentrated in the ratio of the size of the entrance aperature to that of the exit aperature. A detailed treatment of the design is given by Winston (3).

The effect of this geometry is that any light ray entering the collector within the acceptance angle θ will reach the absorber. Rays entering at an angle greater than θ are reflected back out of the collector, as illustrated in figure 3. If the collector is designed to concentrate by a factor of 3 then the absorber will be only one third the area of a flat plate collector receiving the same quantity of radiation and in consequence it will have a much lower thermal inertia. Because it operates in a concentrated solar flux its equilibrium temperature will be higher and the absorber will be able to deliver heat at a higher temperature to the building. It should be noted that a Winston CPC of trough configuration will not need to track the sun diurnally, and at concentrations of 3 or less, with acceptance angle θ of about 20°, the collector will not need to be moved with the rotation of the seasons. This design exhibits all the characteristics required of a solar collector in our climate. The absorber is hot and thermally nimble, and the collector is mechanically simple.

Fin Compound Parabolic Collector (Figure 4)

The absorber of the Winston CPC is irradiated on one side only. Heat losses from the back, even with heavy insulation, are serious in view of the relative high operating temperature. An alternative configuration is available in the form of the fin CPC. Here both sides of the absorber are exposed to the concentrated solar flux, and since only one edge of the absorber is in contact with the collector the path for conductive heat loss is minimised. However, the fin shaped absorber poses heat removal difficulties, for water can be pumped through a thin wafer shaped pipe only against great frictional resistance unless the fin and consequently the whole collector is very large. It should be noted that the absorber of a fin CPC need be only half the size of the Winston type for a given concentration ratio because it receives energy on both its sides. Its thermal performance will therefore be better.

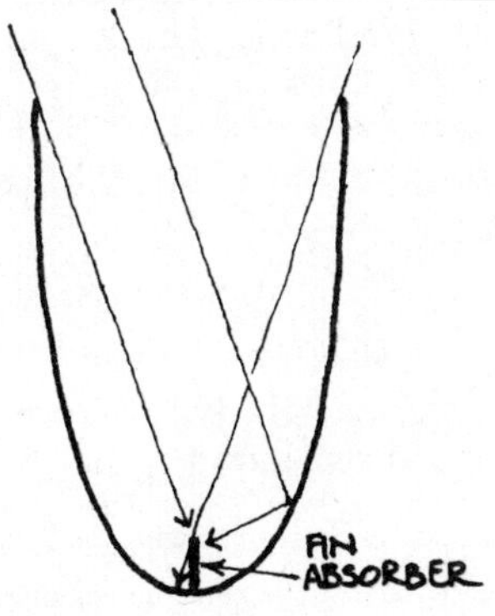

Fig. 4 Fin CPC

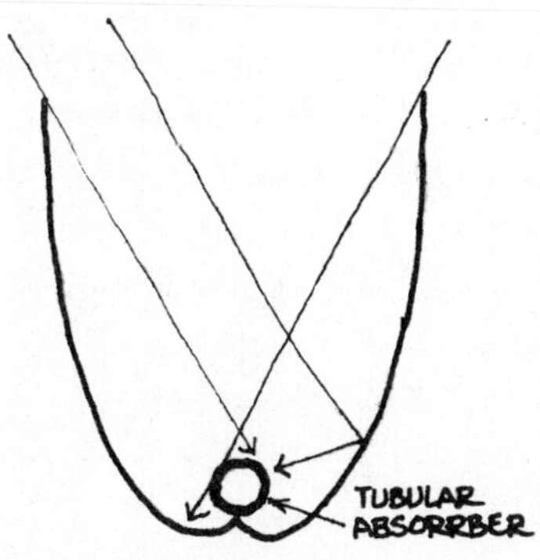

Fig. 5 Cusp CPC

Cusp Compound Parabolic Collector (Figure 5)

When the absorber is of circular cross section, enabling normal plumbing and fabricating techniques to be employed, the shape of the collector must be adapted to ensure that all radiation entering within the acceptance angle reaches the absorber. This is done by forming the inner section of the collector into symetrical cusps whose shape is the evolute of the circular absorber. The optical properties are similar to the Winston and fin CPC's with the concentration factor being the ratio of the width of the entrance aperture to the circumference of the absorber.

Comparison of Performance (Figure 6)

It will be seen that the Winston CPC provides the highest level of optical performance, achieving a concentration in the case illustrated of 3 with an acceptance angle of 20° from its optical axis. Neither the fin or cusp CPC can match this because parts of their collectors lie in the shade of the absorber. The effect is slightly more marked in the case of the more bulby absorber of the cusp collector, with the result that its optically effective area is confined to the shaded part of the diagram.

Fig. 6
Comparative
Performance

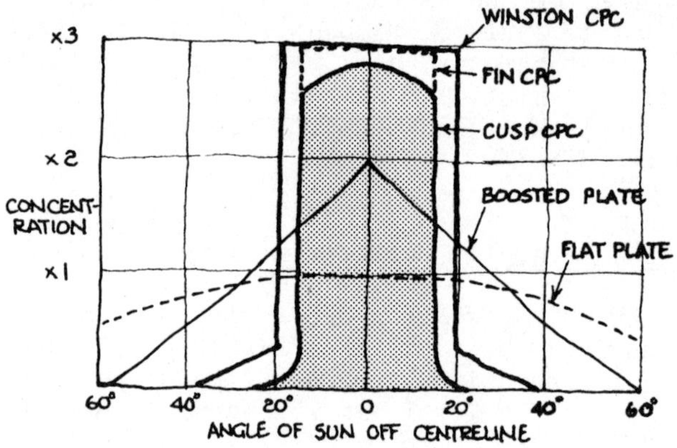

Nonetheless, the thermal properties of the cusp CPC are so much better than the Winston that future development work should, I believe, be directed towards realising its great potential for solar powered building heating systems.

Footnote

A full discussion of compound parabolic collectors, and of some additional configurations, will be found in Comparison of Solar Concentrators by Ari Rable, Solar Energy, Vol. 18, pp 93-111 (1976). For further reading on solar optics see chapter 6 of Applied Solar Energy by Meinal and Meinel, Addison-Wesley Publishing Company 1976.

References

1. V.K. Baranov, Properties of paraboloidotorial phocons. Optiko-mechanicheskaya promishlennost, no 6, 1965.

2. H. Hinterberger and R. Winston. Efficient light coupler for threashold Cerenkov Counters. Review of Scientific Instruments. Vol. 37, p 1094 (1966).

3. R. Winston. Principles of solar concentrators of a novel design. Solar Energy, Vol. 16, pp 89-95 (1974).

Design methodology and space heating and cooling

4

4.1

A design methodology for the design of solar space heated and cooled buildings

C. Stambolis*
Research Associate
Department of General Surveying and Construction
North East London Polytechnic
Forest Road, London E17 4JB

Abstract

General design criteria for solar buildings are outlined; and the various parts in the whole architectural and engineering design process, which are traditionally fragmented, are identified. A generalised solar building design method is described. The above design method can be applied for the design of any type of solar building, provided that the necessary assumptions are made and parameters identified, which are relevant for the particular type of the building concerned.

Introduction

As solar heating and cooling methods for buildings become more diversified, it is becoming increasingly important that the designer is given ample guidance in selecting and designing a system which best suits the needs of his building.

When the architect or the engineer or even the house owner first start considering the options of a solar solution it is paramount that they become aware of the many different ways in which solar energy can be used for servicing the energy needs of buildings. We often come across the case where the influence of a particular commercial solar product has seriously affected the design of a building's solar system. The professional designer is usually misinformed, or not at all informed, about solar energy's utilisation in the built environment. Even worse, when the designer begins to acquire some information on solar energy, he or she is at a loss to see which is the best way to proceed and design a solar system for the building concerned. Usually valuable time is lost in finding out what is already known and in designing systems which are unsuitable for a particular climatic and socio-economic situation. It is therefore important that a methodology is developed whereby designers can follow a step by step approach in selecting and designing suitable solar building systems.

Solar building types

There are two broad categories of solar building types. The first one includes buildings, which by virtue of their design enhance solar radiation through their structure, in which case, depending on the type of solar system used, a wall or a roof may become a solar collector. In this case the thermal energy flow is by natural means and buildings of this type are called 'passive solar buildings'. The second category of solar buildings are based on active systems utilising both liquid-heating and air-heating collectors. Solar collectors, which can be either flat-plate or focusing, can be incorporated in the building's envelope by becoming part of the roof's or wall's construction or be added on at a later stage. In either case the use of collector components in a building implies that the heating or cooling system is an active one and the term 'active solar system or building' is appropriate.

The different types of solar buildings, together with the relevant performance figures, have been described by the author in a paper which was presented to the WMO/UNESCO 'Solar Energy Symposium' (1). In that paper a classification was proposed according to which the various

*Present address:
Intermediate Technology
Development Group,
9 King's Street,
London WC2E 8HN

types of solar buildings were divided depending on the way in which the heat transfer fluid of the collector circulates (i.e. forced or natural circulation). In Table 1 a summary of this classification is reproduced with a small re-arrangement so as to classify building types under the active/passive terms also.

Table 1 Solar building classification

Types of solar system	Human interference	Examples
1. Passive (Natural fluid circulation) 1.1 Auto-kinetic Direct gain, Thermal storage wall Solar greenhouse Convective loop	Very little or none	Odeillo houses (Trombe/Michel). David Wrights house, Santa Fe, New Mexico. Paul Davis house, Corrales, New Mexico.
1.2 Assisted natural circulation Roof pond (water) Thermal storage wall	Minimal (Human activated or automatic support system on a daily basis)	H. Hay's, Atascapero 'Skytherm'. Steve Baer's house Corrales, New Mexico
2. Active Forced fluid circulation	None (continuous automated support system)	Colorado State University Solar houses, Milton Keynes, UK. Thomason houses, Timonium scool, Solar One, Delaware.

In an excellent publication prepared by the American Institute of Architect's Research Corporation (2) several solar dwelling design concepts are fully explored. Each solar design concept is well described by means of clear cut diagrams and drawings and information is provided too on its adaptability and performance under different climatic conditions. Design data is given on sun angles, solar controls, collectors and storage media, building orientation, wind effect etc. In another important publication by Bruce Anderson (3) additional information is provided on many existing and operating solar buildings. A number of operational details are also presented which help the reader to identify the pros and cons of the different solar systems.

Design methods

Several design methods have been proposed for the design of solar heated/cooled buildings but almost invariably all of them deal a priori with a particular solar system. For example Field (4) in his a 'Design Manual for Solar Heating of Buildings and Domestic Hot Water', has developed a comprehensive method, based on worksheets, where ample and easy to understand information is given on how to calculate heat losses, incident solar radiation, expected collector performance etc. in order to arrive at the optimum number of collector area, tilt and storage volume. However, implicit in this approach is the assumption that the system basically consists of liquid or air flat-plate collectors. Similarly G. Löf (5) has outlined the basic principles involved in the design of systems for space heating with solar energy but again this is limited to active systems. Klein, Duffie and Beckman (6) have carried out extensive computer simulation studies and have devised a fairly accurate design procedure for solar heating systems. The main features of this procedure, which is again limited to flat-plate liquid collector collectors and a water storage tank, is a simple graphical method requiring monthly average meteorological data which architects and heating engineers can use to design economical solar heating systems. In a further study the above team have devised a similar design procedure, based on computer simulation, for the design of solar air heating systems. (7)

In an effort to popularise design procedures ERDA prepared a handbook on solar heating for the Pacific Coast Region of the USA (8). This is the first major effort by any government so far, to study the design requirements for a particular geographical area of the country. The climate of the region is discussed by selected cities in terms of the effect of climate on solar heating requirements. The four states covered are California, Oregon, Washington and Arizona. Again the study is limited in presenting design parameters and performance characteristics of active systems (liquid and air type flat-plate collectors). One of the most important contributions of this study is the introduction of a simple, empirical, monthly solar-load ratio method for determining the collector area.

The Solar Energy Group of the Los Alamos Scientific Laboratory has for sometime now been studying both in a theoretical and experimental way the performance of passive solar heated buildings (9). Solar gains through windows, walls, modified walls, skylights, clerestory windows, and roof sections provide an opportunity to dramatically reduce the total heating energy requirements of a building. Many such passive solar heating elements are currently available to a designer presenting a large number of possible system designs. The Los Alamos Group have employed computer simulation analysis in order to devise ways to aid in the selection of components. The results indicate that a performance comparable to that of a conventional active solar heating system should be achievable in an optimized design passive solar heating system.

A generalised design method

As we have already seen several design procedures have been proposed for the design of particular types of solar buildings. However, all these procedures imply that the designer has somehow made up his mind to choose a particular active or passive solar system. How a designer is supposed to arrive at a decision as to what system to choose is an area where very little work has been carried out. This is so because many times the solution is obvious right from the beginning; or because the architect, engineer and owner want to try out a particular system without meaning that this is necessarily the most suitable solution.

I have been largely concerned with what is happening at the very beginning of the design process and for this reason I have examined the possibility of devising a generalised design method for the design of solar buildings. The outcome of my efforts is presented in Fig. 1 where in a diagramatic form I have outlined the design process. This approach implies that the designer is informed beforehand of the possible solar systems and has good accessibility to local data. Computer simulation is not necessary since in most cases where good climatic data exists, a hand calculator may be used.

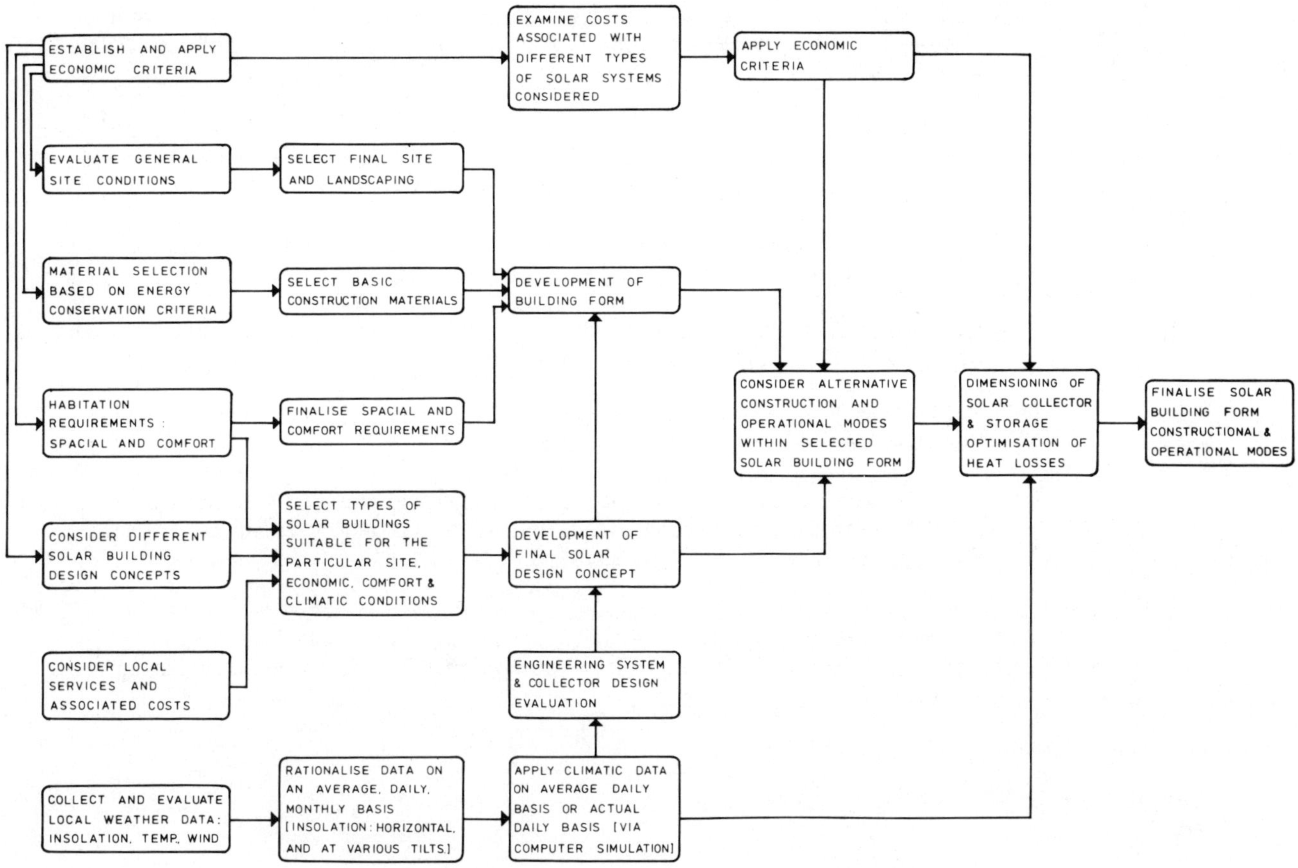

Fig. 1

Discussion

The proposed generalised design method is not a final statement. At this stage it can be used more as a checklist and general design guide. My intent is to refine this method further and produce a series of worksheets which will be adjustable to the variable socio-economic and climatic conditions of the globe. A number of case studies too are at present being analysed so that the weak points of this method, and possible additions, can be identified and the necessary changes be introduced. Nevertheless, I feel that even at this stage the proposed method points to the general direction which designers ought to look at when considering the design of a solar building. Simply, there is an infinite number of solutions and we need some kind of discipline in order to arrive at some sensible solution.

References

(1) C. Stambolis, 'Solar Houses: A General Appraisal of Solar Building Design and Technology', UNESCO/WMO Solar Energy Symposium, Geneva September 1976, (published in the seminar proceedings of 'Solar Energy Applications in the Built Environment', NELP, May 1977).

(2) AIA Research Corporation, 'Solar Dwelling Design Concepts', (prepared for the US Department of Housing and Urban Development,) US Government Printing Office, Washington DC, May 1976.

(3) B. Anderson, 'The Solar Home Book', Cheshire Books, Harrisville, N. Hampshire 1976.

(4) R.L. Field, 'Design Manual for Solar Heating of Buildings and Domestic Hot Water', Solpue Co., PO Box 2351, Gaithersbury, MD 20760 April 1977.

(5) G.O.G. Löf, 'Systems for Space Heating with Solar Energy', Chapter XII in 'Applications of Solar Energy for Heating and Cooling of Buildings', ASHRAE, GRP 170, New York, 1977.

(6) S.A. Klein, W.A. Beckman and J.A. Duffie, 'A Design Procedure for Solar Heating Systems', Solar Energy, Vol. 18, pp 113-127.

(7) S.A. Klein, W.A. Beckman and J.A. Duffie, 'A Design Procedure for Solar Air Heating Systems', Solar Energy, Vol. 19, pp 509-512.

(8) ERDA, San Francisco Operations Office, 'Pacific Regional Solar Heating Handbook', San Francisco, November 1976.

(9) J.D. Balcomb, J.C. Hedstrom and R.D. McFarland, 'Simulation Analysis of Passive Solar Heated Buildings - Preliminary Results,' Solar Energy, Vol. 19, pp 277-282.

All communications should be addressed to: C. Stambolis
Intermediate Technology Development Group, 9 King's Street London
WC2E 8HN, UK

An interactive computer design methodology for the design of solar houses

G.G. Rodgers, J.K. Page and C.G. Souster
Department of Building Science
University of Sheffield, Sheffield, U.K.

Abstract

A research group in the Department of Building Science, University of Sheffield is carrying out a systematic programme of study for the purpose of providing an improved design methodology for solar houses based on interactive computing methods. The methodology being developed covers all latitudes in the world and is designed to be both flexible and very easy to use by those who are not skilled in computer techniques. A modular approach has been adopted towards the production of the computer programs. A number of the modules are fully operational, others are in an advanced stage of development. The final aim is to combine the modules into a thermal energy balance model which will allow the thermal performance of a solar house design to be examined systematically very quickly.

The modules which are already operational are described in this paper.

Introduction

Page (1), in his paper to this conference, discusses the dynamic interactions which exist between the different parts of the overall design system of a solar building. Figure 1 of Page's paper summarises these interactions. The work carried out at Sheffield so far has concentrated on providing reliable solar radiation data for the purpose of predicting accurately the solar inputs to both the direct and indirect solar systems identified by Page. In addition, as Page discusses in his paper, the long wave radiation losses and convection losses from building surfaces are being studied and detailed work is at present being carried out with the aim of providing climatological data linking the radiation inputs with associated external air temperatures and wind speeds so that both heat gains and simultaneous heat losses can be predicted accurately.

A range of interactive computer programs has been written for predicting the solar energy available on a surface of any given slope and orientation under different conditions at any time in the year, anywhere in the world. The programs deal with clear, average and overcast days and provide information on both the direct and diffuse radiation. It was decided that our programs should not be based on conventional standard radiation curves but should flexibly allow for local variations in the radiation climate. The clear sky programs allow for variations in atmospheric water content and sky clarity and the programs dealing with the average amounts of radiation received allow in addition for differing hours of sunshine during the day. The average day programs have been checked systematically against U.K. and European meteorological observations and good agreement with actual observations is indeed achieved. We intend to explore wider applications later.

A solar collector must be covered with a transparent material such as glass. The windows of a room also collect energy. Solar energy transmission through this part of the system must be considered along with purpose built collectors on the roof. The transmission of glazing is strongly dependent on angle of incidence. Any systematic design procedure must allow for the directional transmission properties of the actual glazing materials used. Our existing clear sky programs allow the user to examine the effects of inserting glasses of different properties into the model. Single or double glazing can be studied. The average and overcast sky programs are currently being modified to deal with glazing in a similar way.

The mathematical theory behind our radiation models is somewhat involved. However we realise that as research workers we have an obligation to provide solar radiation data to the architects and engineers who are direct participants in the design process. It was for this reason that we decided to incorporate the theory in interactive computer programs which can be used with ease by any member of the design team. The programs are run from a remote terminal which is connected to the University's I.C.L. 1907 computer. The terminal which we use may be either a teletype terminal, which produces numerical tabular results or a visual display unit which provides graphs or diagrams in addition to tabular results. The way in which the programs are used is illustrated in a film prepared to accompany this paper. The film shows the simple method of supplying data to the computer and shows examples of the production of both tabular and graphical results.

The programs currently available will now be described.

Solar geometry

Many of our programs provide information on the position of the sun with respect to some specified surface in terms of either the solar altitude and azimuth or the horizontal and vertical shadow angles for plotting on the plans and sections of building drawings. Figure 1 shows a diagram produced on the visual display unit by our interactive program SHAN. The diagram shows the position of the sun on 25th July with respect to a rectangular building in Sheffield with its major facade facing south-west. The sun's position is plotted at each hour from sunrise to sunset in terms of horizontal shadow angles on plan and vertical shadow angles on the two perpendicular sections A-B and C-D. These angles can be transferred directly to an architects full scale drawings so that decisions on building orientation or shading, for example, can be made.

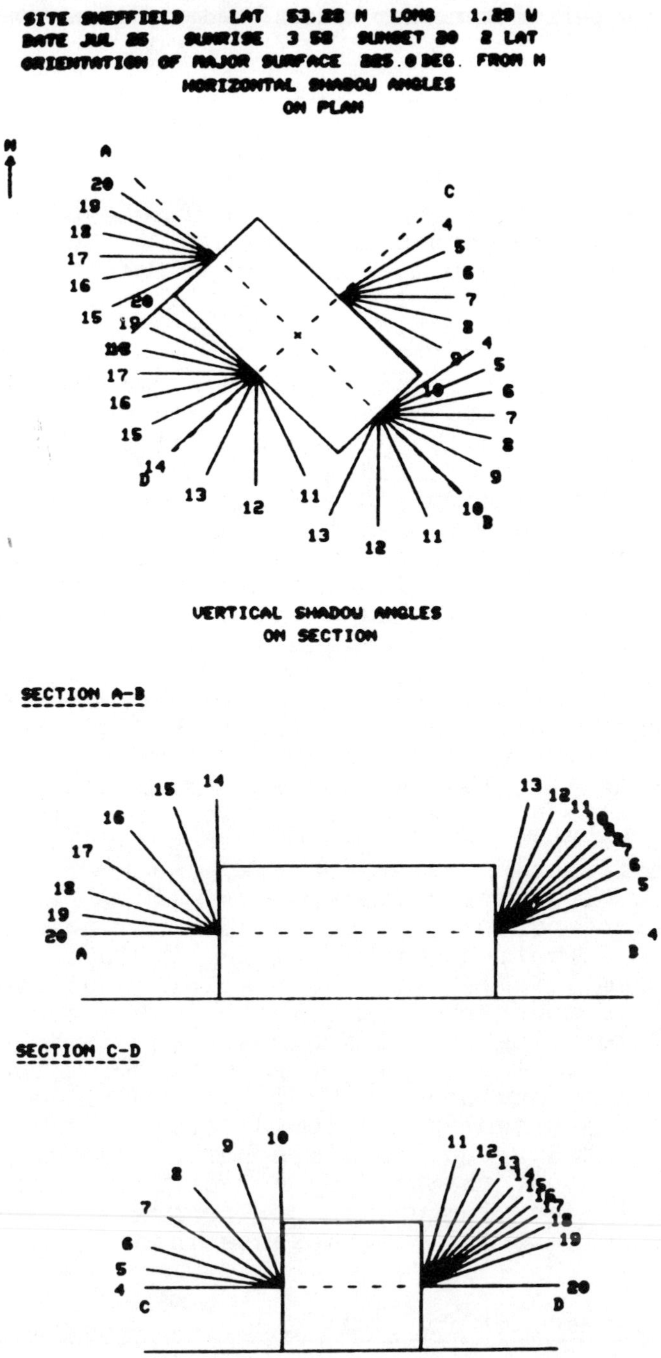

Fig. 1
Sample output of our shadow angles program SHAN for a building in Sheffield with its major surface facing south-west on July 25th.

Solar radiation in clear sky conditions

The first step in providing a clear sky short wave radiation model which could be used for surfaces of any slope and orientation anywhere on the earth's surface was to adopt a suitable meteorologically validated radiation transmission model of the atmosphere which allowed for local variations in both atmospheric water content and sky clarity. The model proposed by Unsworth and Monteith (2) was adopted. This model had been rigorously checked out for different parts of the U.K. and similar checks for the Sudan had shown it to be reasonably applicable in lower latitudes also.

The model is based on a turbidity coefficient, τ_a, which relates the measured direct normal solar irradiance G_{bn} to the clean air direct normal irradiance G^*_{bn} by the formula

$$G_{bn} = G^*_{bn} \exp(-\tau_a m) \qquad (1)$$

where m is the optical air mass and G^*_{bn} is the irradiance normal to the beam for a clean atmosphere containing known amounts of absorbing and scattering gasses and after correcting to mean solar distance. The chief factors affecting G^{*}_{bn} are the water vapour content of the atmosphere and the path length. The model closely follows the theory described by Unsworth (3).

The clear sky diffuse irradiance received by a surface is non-isotropic and depends on the atmospheric turbidity, increasing with increasing turbidity. Page (1) has shown that the diffuse irradiance measurements of Parmelee (4) in the U.S.A. compare favourably with measurements made in the U.K. and in Belgium. The Parmelee data gives comprehensive diffuse irradiances on horizontal surfaces and on vertical surfaces of numerous orientations and so this data was taken as a basis for our turbidity dependent clear sky diffuse irradiance model.

Two major meteorological inputs are required by our clear sky programs. These are the turbidity and the precipitable water vapour content of the atmosphere. Unsworth and Monteith (2) provide some information on the variation of turbidity over the summer months at different sites in the U.K. We have used our computer programs to analyse sets of radiation data for the U.K. and have derived statistical values of τ_a for different classes of radiation day. This procedure can be carried out for any area in the world, from observations of global and diffuse irradiance, or of the beam irradiance. The precipitable water vapour content of the atmosphere is available from meteorological records. The Meteorological Office have recommended appropriate values for clear days for each month of the year in the U.K. and Bannon and Steele (5) provide world-wide data on monthly average values for several months in the year. Recommended values of atmospheric turbidity and precipitable water content in the U.K. for different classes of radiation day for use in our programs are given in a Department of Building Science internal note (6).

Typical results from the clear sky programs

We have a number of clear sky programs which predict the direct and diffuse irradiances on fixed horizontal or sloping surfaces or on surfaces which follow the sun. User Guides have been prepared for each program and these explain the operation of the program in terms of the data which the user must provide and with examples of the results which are produced. Detailed descriptions of the theoretical bases of the programs are written up initially as internal reports but with a view to wider publication in the usual scientific and technological journals.

Program SUN3 is our most comprehensive clear sky program. It deals with the irradiation of sloping surfaces of any orientation at any place in the world on any given day in the year. Figure 2 shows typical results for a south facing surface with a slope of 40° located at Sheffield on 25th July. Between sunrise and sunset the program tabulates the position of the sun in terms of hourly values of the horizontal and vertical shadow angles for use on plan and section and hourly values of the direct, diffuse, ground reflected and total irradiances (Wm^{-2}). The irradiances are integrated at 15 minute intervals over the day to give the corresponding daily irradiations (kJm^{-2}, Whm^{-2}). The tabulated data is then drawn up in the form of a graph. The graph in Figure 2 has been drawn on a standard set of axes in which the time scale runs from 0 to 24 hours and the irradiance scale runs from 0 to 1200 Wm^{-2}. The user could have chosen to see the graph drawn to a larger scale if required.

Fig. 2
Results from program SUN3 for a south facing surface with a 40° slope in Sheffield on July 25th.

UNIVERSITY OF SHEFFIELD DEPARTMENT OF BUILDING SCIENCE

PROGRAM SUN3 (VERSION JAN 1977)

SITE SHEFFIELD

LAT 53.38 N LONG 1.30 W HT 140.0 M

SURFACE BEARING 180.0 FROM N. SURFACE TILT 40.0

JUL.25 (DEC.= 19.64 DEG. E.T.= -6.27 MIN. C.F.=0.980)

TA=0.380 HBO=20.0 MM ALBEDO=0.20

SUNRISE 3 58 LAT SUNSET 20 2 LAT

SURFACE IRRADIANCE RESULTS FOR CLOUDLESS SKIES

TIME HOURS	V.SHADOW (DEGREES)	H.SHADOW (DEGREES)	DIRECT (W/M^2)	DIFFUSE (W/M^2)	GRD RFLCTD (W/M^2)	TOTAL (W/M^2)
4.00	180	-125	0	0	0	0
5.00	162	-114	0	25	1	26
6.00	127	-102	18	51	3	72
7.00	91	-90	121	95	6	222
8.00	72	-78	268	139	10	417
9.00	63	-63	419	168	13	600
10.00	59	-46	546	184	15	745
11.00	57	-25	630	191	17	838
12.00	56	0	659	193	17	869
13.00	57	25	630	191	17	838
14.00	59	46	546	184	15	745
15.00	63	63	419	168	13	600
16.00	72	78	268	139	10	417
17.00	91	90	121	95	6	222
18.00	127	102	18	51	3	72
19.00	162	114	0	25	1	26
20.00	180	125	0	0	0	0

DAILY G.RFLCTD DIFFUSE IRRADIATN = 532.0 KJ/M^2 (148.0 WH/M^2)

DAILY SKY DIFFUSE IRRADIATION = 6849.0 KJ/M^2 (1903.0 WH/M^2)

DAILY DIRECT IRRADIATION = 16778.0 KJ/M^2 (4661.0 WH/M^2)

DAILY TOTAL IRRADIATION = 24158.0 KJ/M^2 (6711.0 WH/M^2)

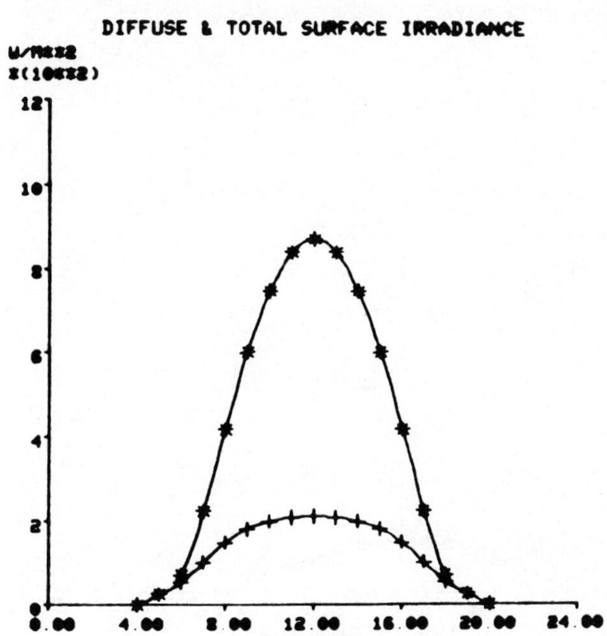

In order to demonstrate the versatility of the program, Figures 3-5 show the results obtained for a north facing vertical surface on 25th July located at Kew (London), Melbourne (Australia) and at a site with snow cover within the Arctic Circle. The differences arising from site locations in the northern and southern hemispheres are illustrated in the Kew and Melbourne results. At the site within the Arctic Circle the sun does not set on the day in question. All the graphs are drawn to the standard scale for comparison.

Fig. 3
Results from program SUN3 for a north facing vertical surface in London on 25th July.

UNIVERSITY OF SHEFFIELD DEPARTMENT OF BUILDING SCIENCE

PROGRAM SUN3 (VERSION JAN 1977)

SITE LONDON(KEW)

LAT 51.28 N LONG 0.19 W HT 6.0 M

SURFACE BEARING 0.0 FROM N. SURFACE TILT 90.0

JUL.25 (DEC.= 19.64 DEG. E.T.= -6.27 MIN. C.F.=0.969)

TA=0.330 H2O=20.0 MM ALBEDO=0.20

SUNRISE 4 7 LAT SUNSET 19 53 LAT

SURFACE IRRADIANCE RESULTS FOR CLOUDLESS SKIES

TIME HOURS	V.SHADOW (DEGREES)	H.SHADOW (DEGREES)	DIRECT (W/M^2)	DIFFUSE (W/M^2)	GRD RFLCTD (W/M^2)	TOTAL (W/M^2)
5.00	16	66	18	17	3	38
6.00	51	77	50	37	13	100
7.00	87	89	8	38	28	74
8.00	106	101	0	40	44	84
9.00	115	115	0	43	58	101
10.00	119	132	0	44	68	112
11.00	121	154	0	45	74	119
12.00	122	-180	0	45	77	122
13.00	121	-154	0	45	74	119
14.00	119	-132	0	44	68	112
15.00	115	-115	0	43	58	101
16.00	106	-101	0	40	44	84
17.00	87	-89	8	38	28	74
18.00	51	-77	50	37	13	100
19.00	16	-66	18	17	3	38

DAILY G.RFLCTD DIFFUSE IRRADIATN = 2348.0 KJ/M^2 (652.0 WH/M^2)

DAILY SKY DIFFUSE IRRADIATION = 2067.0 KJ/M^2 (574.0 WH/M^2)

DAILY DIRECT IRRADIATION = 559.0 KJ/M^2 (155.0 WH/M^2)

DAILY TOTAL IRRADIATION = 4974.0 KJ/M^2 (1382.0 WH/M^2)

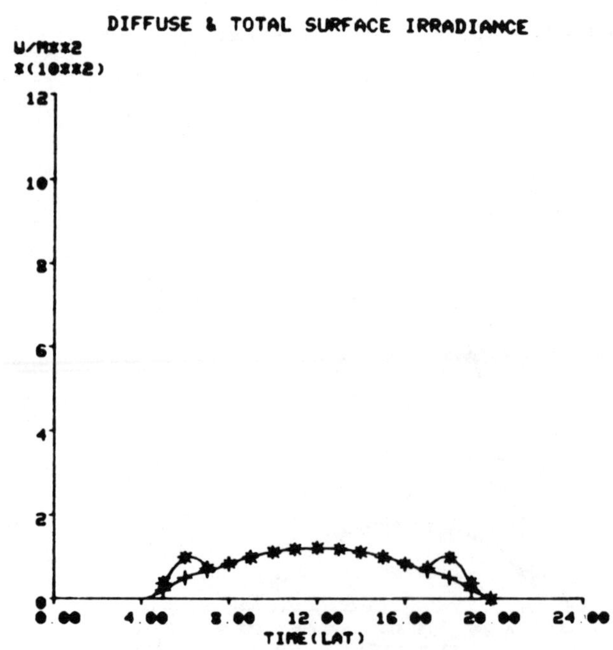

Fig. 4 Results from program SUN3 for a north facing vertical surface in Melbourne, Australia on 25th July.

UNIVERSITY OF SHEFFIELD DEPARTMENT OF BUILDING SCIENCE

PROGRAM SUN3 (VERSION JAN 1977)

SITE MELBOURNE

LAT 38. 0 S LONG 145. 0 E HT 150.0 M

SURFACE BEARING 0.0 FROM N. SURFACE TILT 90.0

JUL.25 (DEC.= 19.64 DEG. E.T.= -6.27 MIN. C.F.=0.969)

TA=0.340 HBO= 5.0 MM ALBEDO=0.20

SUNRISE 7 0 LAT SUNSET 17 0 LAT

SURFACE IRRADIANCE RESULTS FOR CLOUDLESS SKIES

TIME HOURS	V.SHADOW (DEGREES)	H.SHADOW (DEGREES)	DIRECT (W/M^2)	DIFFUSE (W/M^2)	GRD RFLCTD (W/M^2)	TOTAL (W/M^2)
8.00	16	56	60	30	6	96
9.00	25	45	215	78	18	311
10.00	30	32	342	114	31	487
11.00	32	16	422	131	39	592
12.00	32	0	449	136	42	626
13.00	32	-16	422	131	39	592
14.00	30	-32	342	114	31	487
15.00	25	-45	215	78	18	311
16.00	16	-56	60	30	6	96

DAILY G.RFLCTD DIFFUSE IRRADIATN = 835.0 KJ/M^2 (232.0 WH/M^2)

DAILY SKY DIFFUSE IRRADIATION = 3037.0 KJ/M^2 (844.0 WH/M^2)

DAILY DIRECT IRRADIATION = 9091.0 KJ/M^2 (2525.0 WH/M^2)

DAILY TOTAL IRRADIATION = 12962.0 KJ/M^2 (3601.0 WH/M^2)

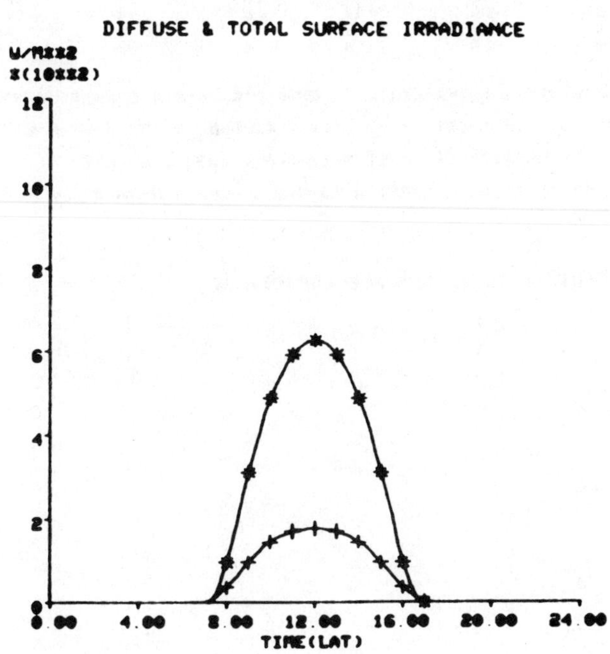

Fig. 5 Results from program SUN3 for a north facing vertical surface at a site within the Arctic Circle on 25th July.

UNIVERSITY OF SHEFFIELD DEPARTMENT OF BUILDING SCIENCE

PROGRAM SUN3 (VERSION JAN 1977)

SITE ARCTIC

LAT 70. 0 N LONG 0. 0 W HT 100.0 M

SURFACE BEARING 0.0 FROM N. SURFACE TILT 90.0

JUL.25 (DEC.= 19.64 DEG. E.T.= -6.27 MIN. C.F.=0.969)

TA=0.080 HBO= 8.0 MM ALBEDO=0.70

YOUR SITE LIES AT OR WITHIN THE ARCTIC CIRCLE AND FOR THE DATE WHICH YOU HAVE SPECIFIED THE SUN NEVER SETS

SURFACE IRRADIANCE RESULTS FOR CLOUDLESS SKIES

TIME HOURS	V SHADOW (DEGREES)	H.SHADOW (DEGREES)	DIRECT (W/M^2)	DIFFUSE (W/M^2)	GRD RFLCTD (W/M^2)	TOTAL (W/M^2)
1.00	0	14	45	1	0	46
2.00	2	28	147	8	4	159
3.00	7	42	250	16	14	280
4.00	16	56	259	23	33	315
5.00	34	69	205	26	61	292
6.00	70	83	80	23	94	197
7.00	106	97	0	21	129	150
8.00	124	112	0	24	163	187
9.00	133	127	0	26	191	217
10.00	138	144	0	27	213	240
11.00	140	162	0	28	227	255
12.00	140	-180	0	28	231	259
13.00	140	-162	0	28	227	255
14.00	138	-144	0	27	213	240
15.00	133	-127	0	26	191	217
16.00	124	-112	0	24	163	187
17.00	106	-97	0	21	129	150
18.00	70	-83	80	23	94	197
19.00	34	-69	205	26	61	292
20.00	16	-56	259	23	33	315
21.00	7	-42	250	16	14	280
22.00	2	-28	147	8	4	159
23.00	0	-14	45	1	0	46
24.00	0	0	0	0	0	0

DAILY G.RFLCTD DIFFUSE IRRADIATN = 8959.0 KJ/M^2 (2489.0 WH/M^2)

DAILY SKY DIFFUSE IRRADIATION = 1704.0 KJ/M^2 (473.0 WH/M^2)

DAILY DIRECT IRRADIATION = 6957.0 KJ/M^2 (1933.0 WH/M^2)

DAILY TOTAL IRRADIATION = 17620.0 KJ/M^2 (4894.0 WH/M^2)

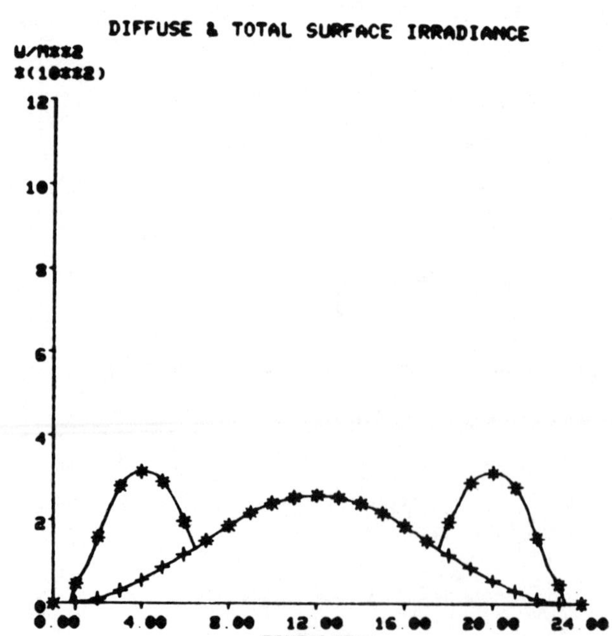

Solar heat gains through glazing on clear days

The clear sky solar radiation programs have been used as the basis of programs which determine solar heat gains through glazing materials.

The program GLAS determines the short wave radiation gains through one or two sheets of glass in a surface of any slope and orientation. The ability of the glass to transmit short wave radiation is specified in terms of the KL value which is the product of the absorption coefficient K and the glass thickness L. The KL value may be determined from the normal incidence solar transmittance values quoted by the glass manufacturers. Petherbridge (7) has shown that the glass spectral transmission curves for monochromatic radiation can be used with a constant KL value for the glass to provide accurate solar transmission and absorption data for any angle of incidence of the solar beam in the case of clear and lightly heat absorbing glasses. In order to deal with the transmission of the non-isotropic short wave diffuse radiation the total diffuse irradiance has been split up into a circumsolar diffuse component which comes from a bright patch of sky in the vicinity of the sun and an isotropic background diffuse component. The circumsolar component then augments the direct beam. Justification for splitting up the beam in this way is given by Louden (8). In addition, we have used this diffuse model to predict the diffuse irradiances on sloping surfaces and the results are in reasonable agreement with those based on measurements made by Kondratyev and Fedorova (9). When vertical glazing is specified the program also works out the heat gains due to the proportion of the energy absorbed in the glass which is released inwards by convection and by long wave radiation, based on the standard procedures reported in the Institution of Heating and Ventilating Engineers Guide (10). Research is being carried out in the Department with the aim of refining these procedures to take account of the effect of the local climate on the convective and long wave radiative heat exchanges from building surfaces.

Our program SOLWIN provides information on the solar heat gain through vertical windows on clear days. The program operates in a similar way to GLAS but in this case the properties of the window system are specified in terms of long wave and short wave shading coefficients which are standard values quoted by the British glass manufacturer, Pilkington Brothers Limited, for their range of glasses and for various combinations of glasses and blinds (11). The program allows glazing to be inserted in one wall or into several walls with different orientations and the program provides information at hourly intervals between sunrise and sunset on the position of the sun and on the solar gain due to the short wave transmitted radiation, the solar gain due to the energy absorbed in the glass which is released into the building and finally the total solar gain. These instantaneous solar gains (W), are integrated over the day to produce the corresponding daily totals (kJ, kWh).

Figure 6 shows results of the solar gain on 25th July for a room in a building located in London. The room has 14.7m^2 of 6mm clear float glass in each of two walls, one of which faces due east, the other due south. The graph

Fig. 6 Results from program SOLWIN for the solar gain through windows of standard clear glass in the east and south walls of a room in London on 25th July.

```
UNIVERSITY OF SHEFFIELD  DEPARTMENT OF BUILDING SCIENCE

INTERACTIVE PROGRAM SOLWIN (FORTRAN) VERSION FEB 1977

SITE TITLE:LONDON
LAT   51.28 N     LONG    0.19 W     HT     6.0 M      ALBEDO 0.20

WINDOW DESC.:6 MM CLEAR FLOAT
AREA OF GLAZING         29.40 M^2
WINDOW DETAILS
ASPECT AZIMUTH   AREA    SHADING COEFS
        (DEG)    (M^2)        SW    LW
   1     90.0     14.7    0.92  0.06
   2    180.0     14.7    0.92  0.05

JUL.25      TA=0.330     H20=20.0 MM
------
SUNRISE  4  7 LAT     SUNSET 19 53 LAT
```

SOLAR TRANSMISSION THROUGH WINDOWS ON CLOUDLESS DAYS

TIME (HRS)	ALTITUDE (DEGREES)	AZIMUTH (DEGREES)	TRANSMIT-TED SW (W)	CONVCTD/ RERAD LW (W)	TOT. SOLAR GAIN (W)
5.00	6	66	919	50	969
6.00	15	77	4015	218	4234
7.00	25	89	6739	366	7105
8.00	34	101	8427	458	8885
9.00	43	115	9542	519	10060
10.00	50	132	9457	514	9971
11.00	56	154	8169	444	8613
12.00	58	180	7105	386	7491
13.00	56	206	6745	367	7112
14.00	50	228	5696	310	6005
15.00	43	245	4085	222	4307
16.00	34	259	2329	127	2456
17.00	25	271	1370	74	1444
18.00	15	283	828	45	873
19.00	6	294	337	18	356

```
DAILY TRANSMITTED S.W.      =    272567 KJ (     75.713 KWH)
DAILY CONVECTED/L.W.        =     14813 KJ (      4.115 KWH)
DAILY TOTAL GAINS           =    287381 KJ (     79.828 KWH)
24 HOUR MEAN TOTAL GAIN     =     3.326 KW
```

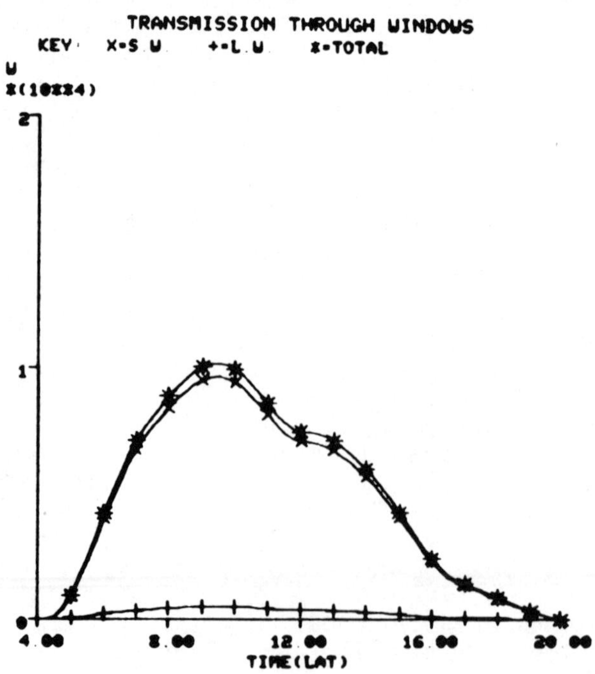

Fig. 7 Results from program SOLWIN for the solar gain through windows of heat absorbing glass in the east and south walls of a room in London on 25th July.

```
UNIVERSITY OF SHEFFIELD  DEPARTMENT OF BUILDING SCIENCE
INTERACTIVE PROGRAM SOLWIN (FORTRAN) VERSION FEB 1977

SITE TITLE·LONDON
LAT  51.28 N    LONG   0.19 W     HT    6.0 M     ALBEDO 0.20

WINDOW DESC.·6 MM ANTISUN 50/50 BRONZE
AREA OF GLAZING        29.40 M^2
WINDOW DETAILS
ASPECT AZIMUTH  AREA   SHADING COEFS
        (DEG)   (M^2)      SW    LW
   1     90.0    14.7    0.51  0.18
   2    180.0    14.7    0.51  0.18

JUL.25      TA=0.330     H20=20.0 MM
------
SUNRISE  4  7 LAT     SUNSET 19 53 LAT

      SOLAR TRANSMISSION THROUGH WINDOWS ON CLOUDLESS DAYS

TIME    ALTITUDE   AZIMUTH     TRANSMIT     CONVCTD/    TOT. SOLAR
                               -TED SW      RERAD LW       GAIN
(HRS)  (DEGREES)  (DEGREES)      (W)          (W)          (W)

 5.00       6          66          509          180          689
 6.00      15          77         2226          786         3012
 7.00      25          89         3736         1318         5054
 8.00      34         101         4672         1649         6320
 9.00      43         115         5289         1867         7156
10.00      50         132         5243         1850         7093
11.00      56         154         4529         1598         6127
12.00      58         180         3939         1390         5329
13.00      56         206         3739         1320         5059
14.00      50         228         3157         1114         4272
15.00      43         245         2264          799         3064
16.00      34         259         1291          456         1747
17.00      25         271          759          268         1027
18.00      15         283          459          162          621
19.00       6         294          187           66          253

DAILY TRANSMITTED S.W.      =     151097 KJ (      41.971 KWH)
DAILY CONVECTED/L.W.        =      53328 KJ (      14.813 KWH)
DAILY TOTAL GAINS           =     204425 KJ (      56.785 KWH)
24 HOUR MEAN TOTAL GAIN     =      2.366 KW
```

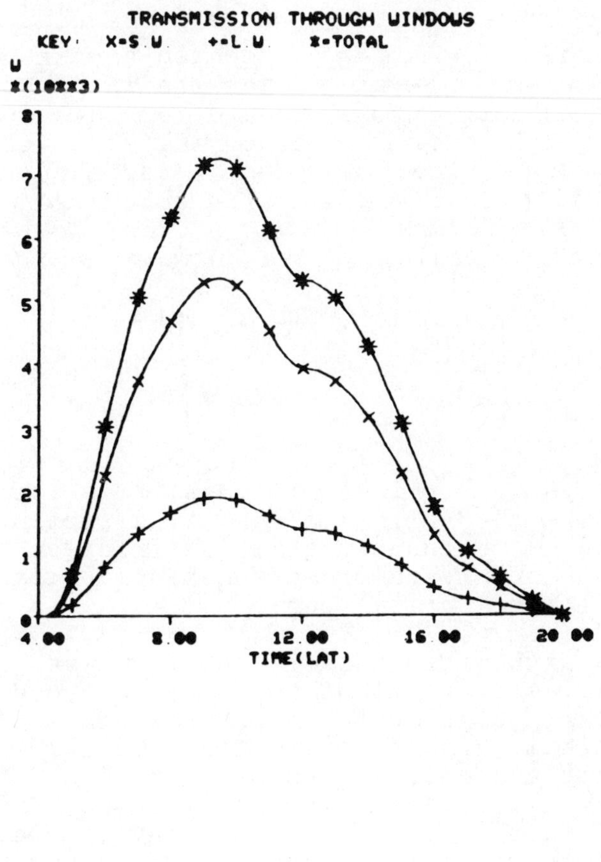

produced by the computer shows that the peak gain occurs in the morning and is due to the east facing windows. For clear glass the proportion of the energy absorbed in the glass which is subsequently released indoors is a small proportion of the short wave gain. Figure 7 shows the effect of replacing the clear glass by a solar control heat absorbing glass. The overall solar gain is reduced but the contribution from the energy absorbed by the glass released indoors is far larger than with clear glass.

Solar radiation on average days

When designing a solar house it is essential to know not simply what solar radiation can be expected on clear days but, more importantly, what solar radiation can be expected at a particular site on average during a particular month of the year. We have devoted a great deal of effort during the past year to the development of an average day model and we now have one which checks out extremely well with actual radiation measurements on both horizontal and vertical surfaces in the U.K. We are also checking out the model for Valentia (W. Ireland), Hamburg, Uccle (Nr. Brussels), Odeillo, and the initial results indicate that the model covers sites in N.W. Europe reasonably accurately.

Our original average day model was based on the Meteorological Office climatological summaries of the mean monthly values of the horizontal hourly mean global irradiance $\bar{G}_h$ and the horizontal hourly mean diffuse irradiance $\bar{G}_{dh}$ at Kew over the ten year period 1959-1968. An estimate of the mean horizontal direct irradiance $\bar{G}_{bh}$ was made at each hour in a particular month by subtracting $\bar{G}_{dh}$ from $\bar{G}_h$. Because each hourly value of $\bar{G}_{bh}$ is obtained by averaging over the whole hour when in fact bright sunshine was on average recorded for only a fraction f, say, of that hour, a more realistic value of the instantaneous direct horizontal surface irradiance during the fraction f of the hour for which the sun is shining is $\bar{G}_{bh}/f$. This figure allows a pseudo-turbidity, τ'_a, to be determined according to the Unsworth and Monteith definition as follows:

$$\tau'_a = -\frac{1}{m} \ln \left[\frac{\bar{G}_{bh}}{f} \times \frac{1}{G^*_{bn} \sin \alpha} \right] \qquad (2)$$

where m is the optical air mass, α is the solar altitude and G^*_{bn} is the theoretical clean air direct normal irradiance. This technique assumes that the fraction f which represents the mean monthly recorded sunshine during each hour of a particular month is known.

In most situations this is not the case. Frequently only the mean daily total of hours of recorded bright sunshine for a particular month, $\bar{n}_o$, is all that is available. We assume that this period is the time for which the measured direct irradiance, G_{bn} is greater than 200 Wm^{-2} and the program calculates an average fraction of sunshine f, which is applied to each hour of the day by considering the maximum possible number of hours of bright sunshine which would be recorded on a totally clear day in the middle of the month with a sunshine recorder that starts to record at 200 Wm^2. This value is obtained by running our clear sky program with a fixed reference turbidity of 0.2 and recording the number of hours for which the predicted value of G_{bn} is greater than 200 Wm^{-2}. If this value is $\bar{n}_r$ then the average fraction f is obtained as $f = \bar{n}_o/\bar{n}_r$. Using this monthly fraction the diurnal variation in the pseudo-turbidity at Kew has been determined for each month of the year. The daily variations for four selected months are shown in Figure 8. These observed curves have been fitted by an appropriate cosine function, the arguments of which vary with month and are stored in our average day program. The hourly values of $\bar{G}_{bh}$ obtained from these pseudo-turbidities are finally integrated over the day, for each month, to give the mean daily total direct irradiation on the horizontal, $\bar{H}_{bh}$.

Fig. 8

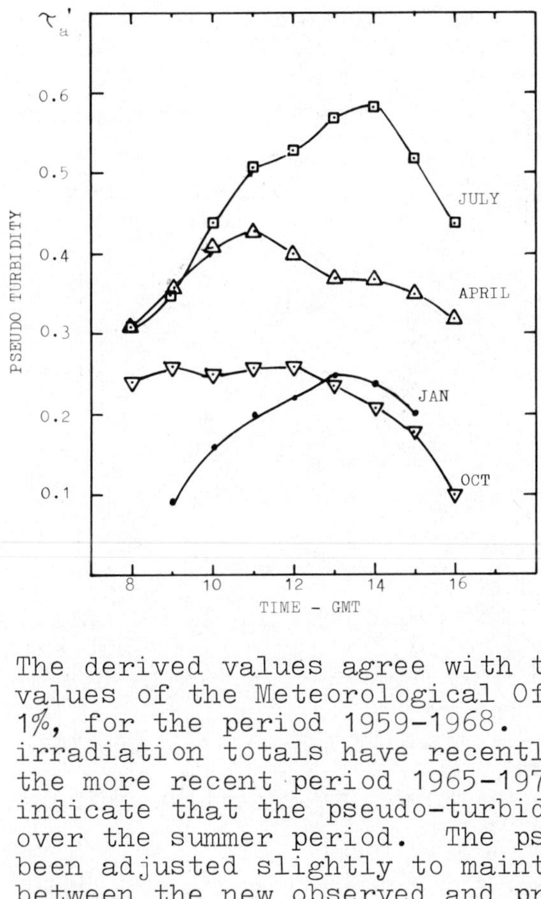

Variation of pseudo turbidity with time of day - Kew, England

The derived values agree with those obtained from mean values of the Meteorological Office observations to within 1%, for the period 1959-1968. New observed mean daily irradiation totals have recently been acquired for Kew for the more recent period 1965-1975 (excluding 1973) which indicate that the pseudo-turbidity has increased slightly over the summer period. The pseudo-turbidity curves have been adjusted slightly to maintain the maximum error between the new observed and predicted values of $\bar{H}_{bh}$ below 1%.

The mean monthly hourly diffuse irradiances on a horizontal surface have been estimated from the Kew average day data for the period 1959-1968. It was found that when the mean monthly horizontal diffuse irradiances were plotted hour by hour against solar altitude over the whole year the points could be fitted by the following straight line formula:

$$\bar{G}_{dh} = 2.18 + 4.532\,\alpha \qquad (3)$$

where α is the solar altitude in degrees. When this equation was used to predict the monthly mean daily total diffuse irradiations the values agreed with the observed values for Kew to within 5% for the period 1959-1968.

However, when compared with corresponding observed values for the later period 1965-1975 (excluding 1973) the predicted values were too low, particularly in winter when the error was approximately 11%. A comparison of the observed values of the monthly mean daily diffuse irradiations over the two periods shows that on average the values for the later period are some 6% higher than for the period 1959-1968. Therefore the coefficients of Equation (3) have been adjusted upwards so as to bring the predicted monthly mean daily total diffuse irradiations to within 5% of the observed values for the later period. Equation (3) thus becomes:-

$$\bar{G}_{dh} = 2 + 4.804\,\alpha \tag{4}$$

Having produced a model which agreed reasonably well with the most recent observed long term average diffuse irradiation data the next step was to modify the model to make the daily total diffuse irradiation vary in line with the observations with the average number of hours of recorded sunshine in any given month in a particular year.

Page (12) has shown that the following Angstrom type regression equations can be used to represent mean monthly irradiations:

$$\frac{\bar{H}_h}{\bar{H}_{oh}} = a + b\,\frac{\bar{n}}{N_o} \tag{5}$$

$$\frac{\bar{H}_{dh}}{\bar{H}_h} = c + d\,\frac{\bar{H}_h}{\bar{H}_{oh}} \tag{6}$$

where $\bar{H}_h$ is the mean monthly daily global irradiation on a horizontal plane.

$\bar{H}_{oh}$ is the mean monthly daily irradiation on a horizontal plane in the absence of any atmosphere.

$\bar{H}_{dh}$ is the mean monthly daily diffuse irradiation on a horizontal plane.

$\bar{n}$ is the monthly mean daily hours of observed bright sunshine.

$\bar{N}_o$ is the mean daily number of hours of daylight in a particular month.

where a, b, c and d are climatically determined regression constants which vary with locality. Equations (5) and (6) may be combined to give:

$$\frac{\bar{H}_{dh}}{\bar{H}_{oh}} = e + f\,\frac{\bar{n}}{\bar{N}_o} + g\left(\frac{\bar{n}}{\bar{N}_o}\right)^2 \tag{7}$$

where e, f and g are constants which can be expressed in terms of a, b, c and d.

Observations show that the curve passes through a maximum when $\bar{n}/\bar{N}_o \approx 0.5$. Following this idea the method of least squares was used to fit a quadratic function to the values of $\bar{H}_{dh}/\bar{H}_{oh}$ and $\bar{n}/\bar{N}_o$ which were obtained for each month of each year of the 1965-1975 Kew data. The following curve was obtained:

$$\frac{\bar{H}_{dh}}{\bar{H}_{oh}} = 0.1534 + 0.2891\frac{\bar{n}}{\bar{N}_o} - 0.2442\left(\frac{\bar{n}}{\bar{N}_o}\right)^2 \qquad (8)$$

The maximum value of $\bar{H}_{dh}/\bar{H}_{oh}$ is 0.24 and this occurs at $\bar{n}/\bar{N}_o = 0.59$.

Equation (8) can be used to give the mean monthly daily diffuse irradiation as a function of sunshine. Appropriate values of $\bar{H}_{oh}$ and $\bar{N}_o$ for a particular month can be found from our programs. When the mean daily hours of bright sunshine in a month is taken as the mean value for each month over the 1965-1975 period for Kew, $\bar{n}_m$, Equation (8) gives the mean monthly daily diffuse horizontal irradiation for that period, $\bar{H}_{dh}(\bar{n}_m)$. When average sunshine hours are used for a month in a particular year in the period 1965-1975, $\bar{n}_y$, the corresponding mean monthly daily diffuse horizontal irradiation for that year, $\bar{H}_{dh}(\bar{n}_y)$, is produced. Thus the diffuse irradiations can be predicted in terms of sunshine hours. In order to determine the mean monthly values of the hourly diffuse horizontal irradiances as a function of sunshine in a particular year it was assumed that the mean monthly values given by Equation (4) could be modified by multiplying by the ratio $\bar{H}_{dh}(\bar{n}_y)/\bar{H}_{dh}(\bar{n}_m)$. Equation (4) thus becomes:

$$\bar{G}_{dh}(\bar{n}_y) = (2 + 4.804\,\alpha).\frac{\bar{H}_{dh}(\bar{n}_y)}{\bar{H}_{dh}(\bar{n}_m)} \qquad (9)$$

Having modified the irradiances in this way the average day program was run for each month of each year in the 1965-1975 period for Kew, using the mean daily hours of sunshine recorded for each individual month in the ten year period. The predicted results were in excellent agreement with the observations. Figure 9 shows the observed and predicted values of the monthly mean global and diffuse horizontal surface irradiations at Kew for two months, June and December for the period 1965-1975.

Page (1) has explained how we correct the Kew horizontal surface model to allow for local climatical variations and how we have translated the horizontal direct and diffuse irradiance on to surfaces of any slope and orientation. The program has been checked out against mean monthly global irradiations on both horizontal and vertical surfaces at Bracknell over the period 1966-1973. Figure 10 shows the predicted and observed global irradiations on surfaces facing north, south, east and west. The agreement between the observed and predicted values is excellent.

Figs. 9, 10

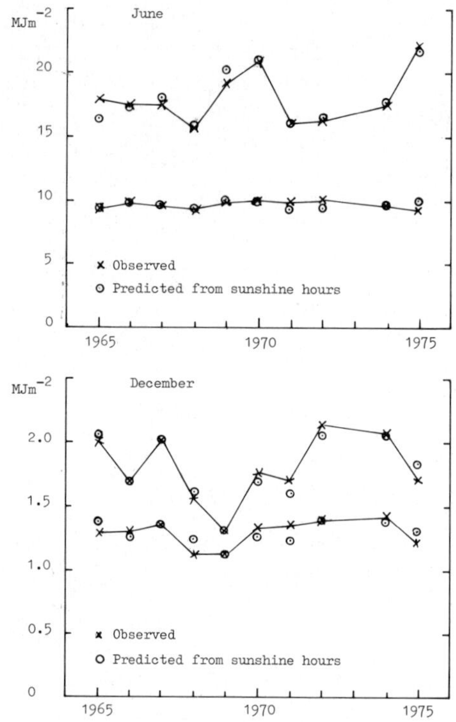

Monthly Mean Global and Diffuse Horizontal Surface Irradiation at Kew for the period 1965-1975.

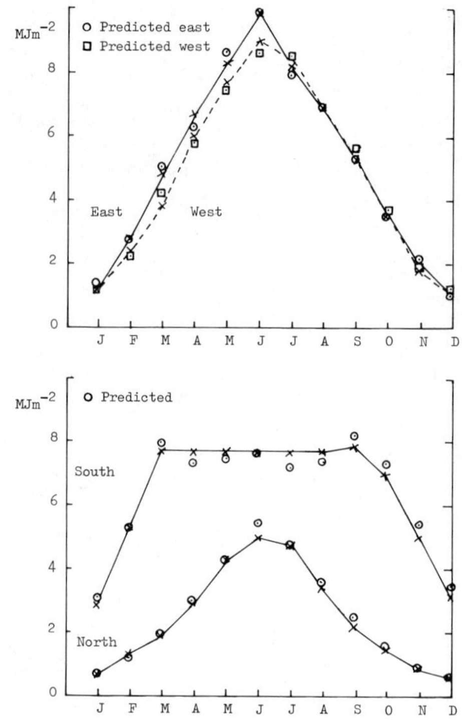

Monthly Global Vertical Surface Irradiation at Bracknell averaged over the period 1966-73.

Conclusions

The work described in this paper forms the basis of an improved design methodology for solar houses. Much yet remains to be done but we are confident that our interactive computer solar radiation modules, when combined with our continuing climatological and heat transfer studies will enable the speedy and accurate analysis of the thermal performance of solar house designs anywhere in the world. Moreover our system will be an open one, capable of being used by all those who have an interest in solar building design.

Acknowledgements

The research project described in this paper is being partly supported by a research grant from the Science Research Council. The grant supports one of the authors (C.G.Souster) as a postdoctoral research assistant and a computer programmer, Miss J. L. Thompson, who is responsible for much of the routine programming effort. We are particularly grateful to the U.K. Meteorological Office at Bracknell for the provision of data and for much helpful advice.

References

1. J. K. Page, Geographical variations in the climatic factors influencing solar building design, Proc. NELP/UNESCO Conf., London, Paper 1/2.

2. M. H. Unsworth and J. L. Monteith, Aerosol and solar radiation in Britain, Quart. J.R. Met. Soc. 98, 778-797, (1972).

3. M. H. Unsworth, Variations in the short wave radiation climate of the U.K., Proc. Conf. on U.K. Meteorological data and Solar Energy Applications, U.K. Section, International Solar Energy Society, London, (1975), pp 18-36.

4. G. V. Parmelee, Irradiation of vertical and horizontal surfaces by diffuse solar radiation from cloudless skies, Trans. Am. Soc. Heat. Vent. Engrs. 60, 341-356, (1954).

5. J. K. Bannon and L.P. Steele, Average water vapour content of the air, Meteorological Office, Geophysical Memoirs, 13, No. 102, H.M.S.O., London, (1960).

6. C.G. Souster, J.K. Page and G.G. Rodgers, A guide to appropriate values of atmospheric turbidity and precipitable water content for different months and different classes of radiation day in the U.K., Internal Note, Department of Building Science, University of Sheffield, (1976).

7. P. Petherbridge, Transmission characteristics of window glasses and sun controls, C.I.E. Conf. Sunlight in Buildings, Newcastle, (1965), Proc. Boucentrum, Rotterdam, pp 183-198, (1967).

8. A. G. Louden, The interpretation of solar radiation measurements for building problems, ibid, pp 111-118.

9. K.Y. Kondratyev and M.P. Fedorova, Radiation regime of inclined surfaces, Paper presented to the UNESCO/WMO Solar Energy Symposium, Geneva, (1976), Document Eng. S/Doc. 2. (Proceedings to be published by WMO).

10. I.H.V.E. Guide, Book A, Section A6, Institution of Heating and Ventilating Engineers, London, (1970).

11. Solar heat gain through windows, Report by Environmental Advisory Service, Pilkington Brothers Limited, 1974.

12. J.K. Page, The estimation of monthly mean values of daily short wave irradiation on vertical and inclined surfaces from sunshine records for latitudes 60°N to 40°S. Internal Research Report No. BS32. Department of Building Science, University of Sheffield, (1976).

4.3 The semi-transparent solar wall collector concept

A. Faist and J-B. Gay
Laboratoire de Physique Théorique
Ecole Polytechnique Fédérale
1007 Lausanne, Switzerland

Abstract

A new concept of solar wall collector is proposed. It consists of three sheets of glass with proper transmittance properties. Air is blown between the two inner glasses along a correctly choosen path and act as collect solar energy.The thermal properties of the system have been analysed and optimised with the aid of a computer model. Working characteristics have been determined, taking the solar intensity and the outdoor temperature as parameters. A full scale model has been built and tested on an open air testing facility. First working characteristics are presented.

Introduction

South facing wall collectors have already been proposed in different forms (1,2,3). In general, they are not transparent and thus give rise to rather low window percentage on the equipped wall. In conceiving a wall collector, one should try to achieve at least the following characteristics :

- good thermal insulation
- non-vanishing collection efficiency at low solar intensity
- possibility of monitoring the incoming solar energy
- visual contact with the outer space
- easy maintenance.

Principle

The semi-transparent wall collector is made of three glass panes. The first and the third ones (outdoor and indoor glasses, see Fig.1) are made of standard glass. The second one is an absorbing glass. A fan gives the possibility of circulating air between the two inner glasses, the air path being chosen so as to optimise heat exchange (Fig.2).

At low solar intensity (below 500 [W/m^2], 160[Btu/h.ft^2]) the air fan is not activated. Solar energy is partially converted into heat in the absorbing glass (at a rate of about 50%) raising its temperature. Thus the indoor glass is warmed up, and the heat losses through the wall collector are lowered, cancelled out, or even reversed, giving rise to a heat gain. The remaining solar energy is absorbed by the indoor walls, floor and furniture.

At higher intensities, the air fan is activated and warm air is available for air renewal, of for warming up a heat storage consisting of a rock bed or an integrated floor storage.

Fig. 1 Schematic diagram of the semi-transparent solar wall collector. Identification of the different temperatures and heat transfers.

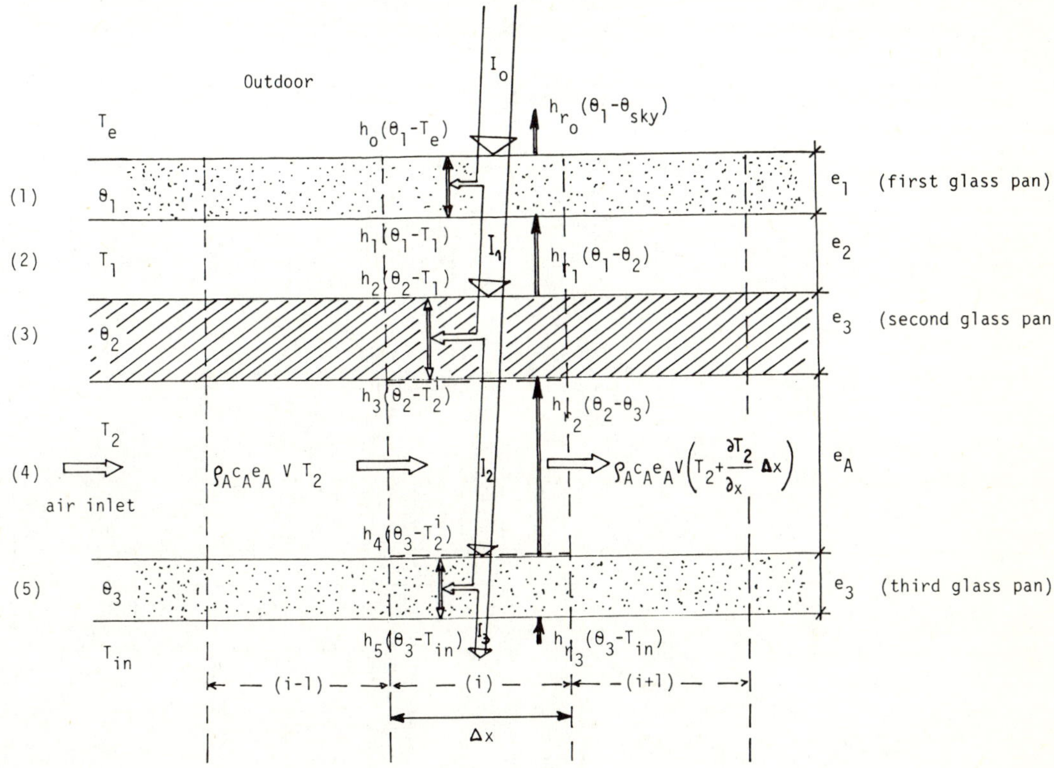

Fig. 2
Diagram of the air flow through the collector

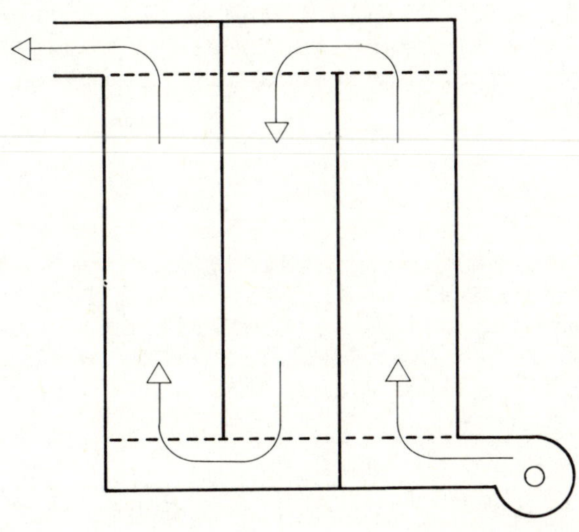

WORKING CHARACTERISTICS

1. Air fan off

When the air fan is not activated, the temperatures of the different glasses can easily be calculated as functions of the incoming solar intensity and of outdoor temperature (the indoor temperature is supposed to be constant : $20^oC, 68^oF$).

Fig. 3 Temperatures of glasses as functions of solar intensity and outdoor temperature.

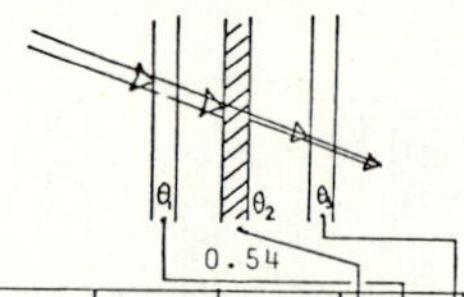

T_{out} [°C] \ [W/m²]	50	100	150	200	250	300	350	400	450	500	550	600
25	25.3 26.0 22.6	26.1 29.1 23.9	26.9 32.1 25.3	27.6 35.2 26.6	28.4 38.3 28.0	29.1 41.3 29.3	29.9 44.4 30.7	30.7 47.4 32.0	31.4 50.5 33.4	32.2 53.5 34.7	32.9 56.6 36.1	33.7 59.7 37.4
20	20.8 23.1 21.4	21.5 26.1 22.7	22.3 29.2 24.1	23 32.2 25.4	23.8 35.3 26.8	24.6 38.4 28.1	25.3 41.4 29.5	26.1 44.5 30.8	26.8 47.5 32.2	27.6 50.6 33.5	28.4 53.7 34.9	29.1 56.7 36.2
15	16.2 20.1 20.1	16.9 23.2 21.5	17.7 26.2 22.8	18.5 29.3 24.2	19.2 32.3 25.5	20.0 35.4 26.9	20.7 38.5 28.3	21.5 41.5 29.6	22.3 44.6 31	23 47.6 32.3	23.8 50.7 33.7	24.5 53.8 35
10	11.6 17.1 18.9	12.4 20.2 20.3	13.1 23.3 21.6	13.9 26.3 23	14.6 23.4 24.3	15.4 32.4 25.7	16.2 35.5 27	16.9 38.6 28.4	17.7 41.6 29.7	18.4 44.7 31.1	19.2 47.7 32.4	20.0 50.8 33.8
5	7 14.2 17.7	7.8 17.3 19.1	8.5 20.3 20.4	9.3 23.4 21.8	10.1 26.4 23.1	10.8 29.5 24.5	11.6 32.6 25.8	12.3 35.6 27.2	13.1 38.7 28.5	13.9 41.7 29.9	14.6 44.8 31.2	15.4 47.8 32.6
0	2.5 11.2 16.5	3.2 14.3 17.8	4 17.4 19.2	4.7 20.4 20.5	5.5 23.5 21.9	6.2 26.5 23.2	7 29.6 24.6	7.8 32.7 26.0	8.5 35.7 27.3	9.3 38.8 28.7	10 41.8 30.0	10.8 44.9 31.4
-5	-2.1 8.3 15.3	-1.4 11.3 16.6	-0.6 14.4 18	0.2 17.5 19.3	0.9 20.5 20.7	1.7 23.6 22.0	2.4 26.6 23.4	3.2 29.7 24.7	3.9 32.8 26.1	4.7 35.8 27.4	5.5 38.9 29.8	6.2 41.9 30.1
-10	-6.7 5.3 14.1	-5.9 8.4 15.4	-5.2 11.4 16.8	-4.4 14.5 18.1	-3.7 17.6 19.5	-2.9 20.6 20.8	-2.1 23.7 22.2	-1.4 26.7 23.5	-0.6 29.8 24.9	0.1 32.9 26.2	0.9 35.9 27.5	1.7 39 28.9

Fig.3 gives a summary of the glass temperatures. We would like to point out that a low solar intensity is sufficient to cancel out the heat losses. As an example, at an outdoor temperature of 0°C (32°F) and a solar intensity of 200 [W/m²] (64 [Btu/h ft²]) heat losses are already cancelled out. Such a radiative intensity is easily achieved, even with a partially clouded sky.

2. Air fan on

When the air fan is activated, the temperatures of the glasses are varying along the air path. The local temperatures can be computed after having written down the heat transfer equations. If we consider figure 1, we can write for the successive glass panes and air spaces :

$$-\frac{\partial \theta_1}{\partial t}\rho_v c_v e_1 + I_1^{abs} - h_o(\theta_1 - T_{out}) - h_1(\theta_1 - T_1) - h_{r_o}(\theta_1 - \theta_{sky}) - h_{r_1}(\theta_1 - \theta_2) = 0 \qquad (1)$$

$$-\frac{\partial T_1}{\partial t}\rho_a c_a e_i + h_1(\theta_1 - T_1) + h_2(\theta_2 - T_1) = 0 \qquad (2)$$

$$-\frac{\partial \theta_2}{\partial t}\rho_v c_v e_2 + I_2^{abs} - h_2(\theta_2 - T_1) - h_3(\theta_2 - T_2) - h_{r_1}(\theta_2 - \theta_1) - h_{r_2}(\theta_2 - \theta_3) = 0 \qquad (3)$$

$$-\left[\frac{\partial T_2}{\partial t} - V\frac{\partial T_2}{\partial x}\right]\rho_a c_a e_a + h_3(\theta_2 - T_2) + h_4(\theta_3 - T_2) = 0 \qquad (4)$$

$$-\frac{\partial \theta_3}{\partial t}\rho_v c_v e_3 + I_3^{abs} - h_4(\theta_3 - T_2) - h_5(\theta_3 - T_{in}) - h_{r_2}(\theta_3 - \theta_2) - h_{r_3}(\theta_3 - T_{in}) = 0 \qquad (5)$$

where :

ρ_v, ρ_a = density of glass (resp. of air)

C_v, C_a = specific heat of glass (ref. of air)

e_i = thickness of glass pane (resp.of air gap)

$I_1^{abs} \ldots I_3^{abs}$ = amount of solar intensity absorbed by each glass pane

$h_o \ldots h_5$ = convection heat exchange coefficient

$h_{r_o} \ldots h_{r_3}$ = radiative heat exchange coefficient

As temperatures of glasses vary slowly with time, one can omit the derivatives and rewrite the equations in a more convenient matrix form :

$$(h_o+h_1+h_{r_o}+h_{r_1})\,\theta_1^i - h_1T_1^i - h_{r_2}\theta_2^i = I_i^{abs} + h_oT_{out} + h_{r_o}\theta_{sky} \tag{6}$$

$$h_1\theta_1^i - (h_1+h_2)T_1^i + h_2\theta_2^i = 0 \tag{7}$$

$$-h_{r_1}\theta_1^i - h_2T_1^i + (h_2+h_3+h_{r_1}+h_{r_2})\,\theta_2^i - h_3T_2^i - h_{r_3}\theta_3^i = I_2^{abs} \tag{8}$$

$$h_3\theta_2^i - (h_3+h_4+\rho_A c_A e_A \frac{V}{\Delta x})\,T_2^i + h_4\theta_3^i = -\rho_A c_A e_A \frac{V}{\Delta x}T_2^{i-1} \tag{9}$$

$$-h_{r_2}\theta_2^i - h_4T_2^i + (h_4+h_5+h_{r_2}+h_{r_3})\theta_3^i = I_3^{abs} + (h_5+h_{r_3})T_{in} \tag{10}$$

The air temperature (T_2^i) and the local glass temperatures can then be computed by using a step calculation. After a certain number of trials, we found out that for a step (ΔX) smaller than 0.1 (m), the results are insensitive to the step size.

As it is interesting to have a heat transfer coefficient as big as possible between the circulating air and the warmed up glass, we must prevent laminar flow : the air speed has to be greater than a minimal value given by :

$$v_{air} > v_{critical} = \frac{Re_{critical} \cdot \nu}{D_h} \tag{11}$$

where :

ν = kinematic viscosity of air

$D_h = \frac{2\ e.b}{e + b}$ (hydraulic diameter)

e : distance between glass plates
b : width of air channel

In our case $v_{critical}$ = 0.6 (m/s) (2[ft/s]). If this condition is satisfied, the air flow is turbulent and the heat exchange coefficient is given by the following expression :

$$h_{3,4} = 0.0288 \frac{k}{D_h} Pr^{1/3} \left(\frac{v_\infty \cdot x}{\nu} \right)^{0.8}$$

with :

$$k = 2.68 \times 10^{-2} \; [W/m^{o}C] \text{ (thermal conductivity of air)}$$

Using the above described formalism, we have computed the temperature of the air at the collector outlet, assuming that the inlet temperature is equal to the outdoor one.

Figure 4 summarizes some results : it gives the amount of air at 20°C (68°F) that comes out of the collector as a function of the outdoor temperature and of the solar intensity.

Fig. 4 Volume of air at 20°C which can be extracted from the collector (area: 5m²) as function of solar intensity and outdoor temperature.

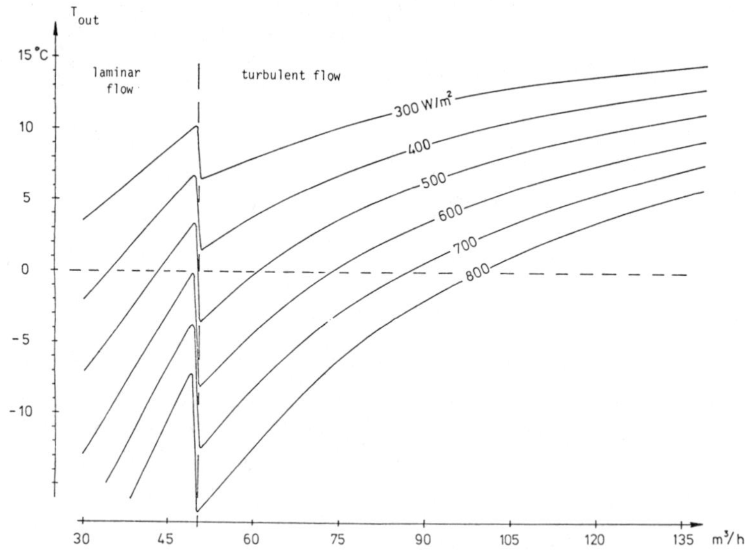

EXPERIMENTAL SETUP

Experimental measurements are achieved in real climate on a test facility located at Ecublens near Lausanne (alt. 410m, lat. 46°30'). The solar wall is about 5 (m^2) ($54ft^2$) of surface and is backed by a strongly isolated cell of about 12 (m^3) (425 ft^3) volume. Using an auxiliary electric heater, the inner temperature of the cell never drops below 20 (°C), (68°F) but can strongly exceed this value by solar overheating because of the low thermal capacitance of the system. Meteorological data as well as temperatures and energy flow data are operated every half hour and automatically stored on a paper strip. Data are analysed and processed on a CYBER 7326 computer.

FIRST RESULTS

Figure 5 exhibits for one day the hourly values of the following temperatures :

- outdoor temperature : T_{out}
- temperature of the air at the collector outlet : T_{air}
- indoor temperature : T_{in}

The solar energy measured on a vertical south facing surface is also plotted as a function of time, the dashed curve gives the values measured in the same conditions behind the wall collector. For the presented day, the fraction of solar energy collected by the air is 19%.

Fig. 5
Values of the solar intensity

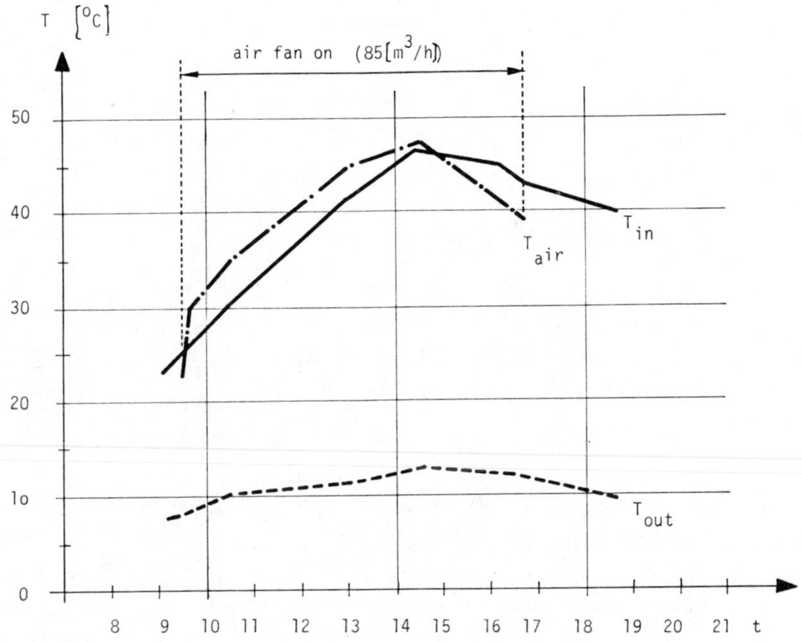

Main temperatures for a typical clear day (16th october 1976)

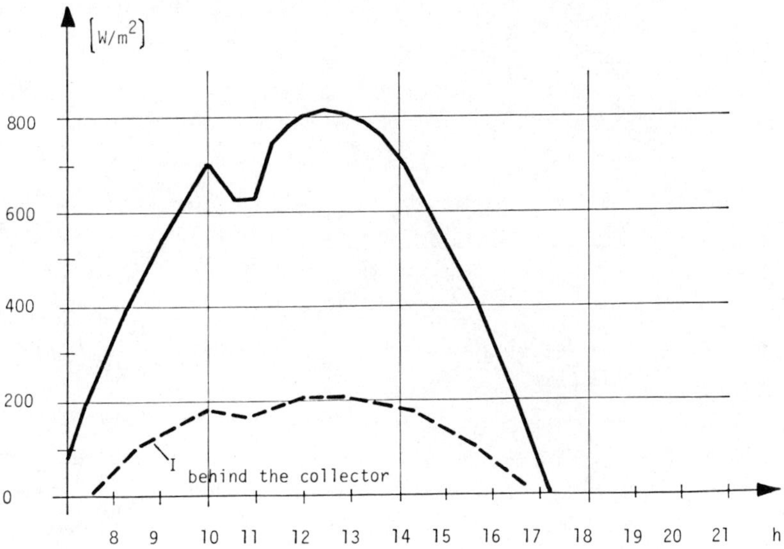

Fig. 6

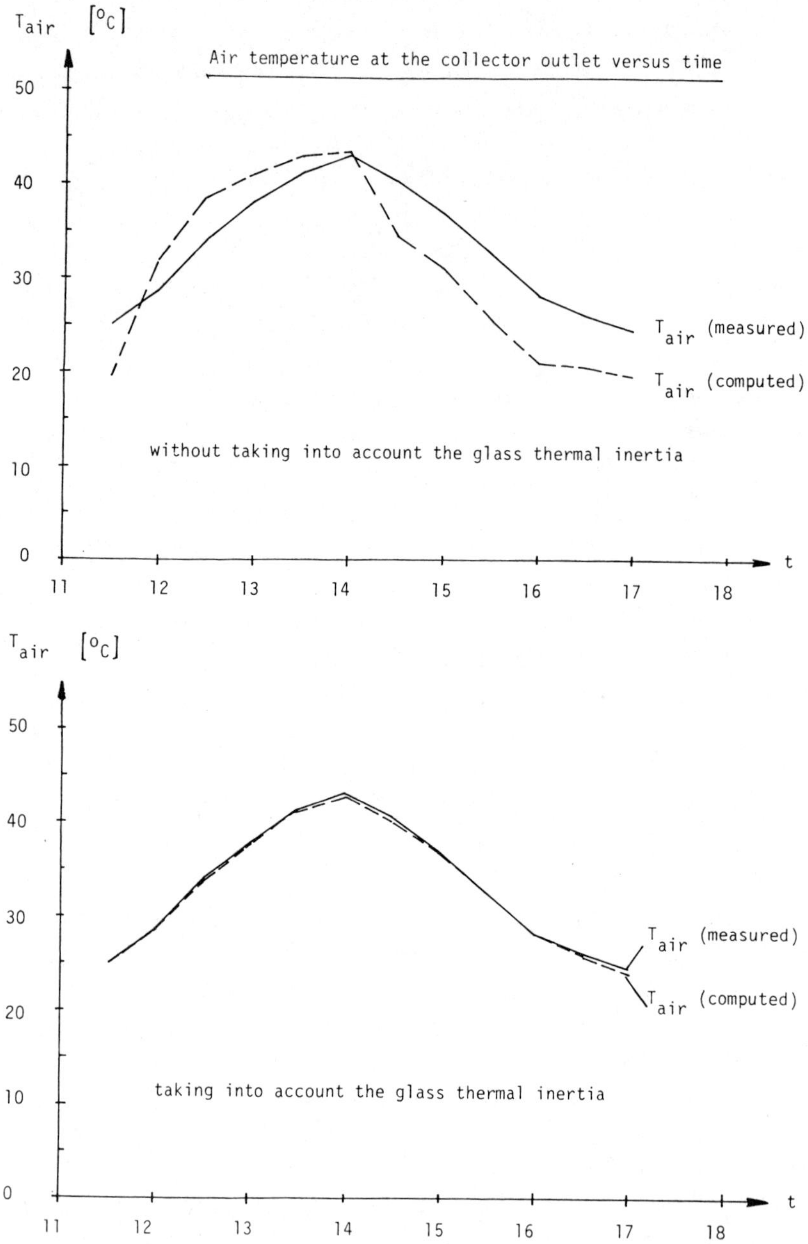

Fig. 7

Indoor temperature of the measuring cell versus time (measured ϕ and computed-values)

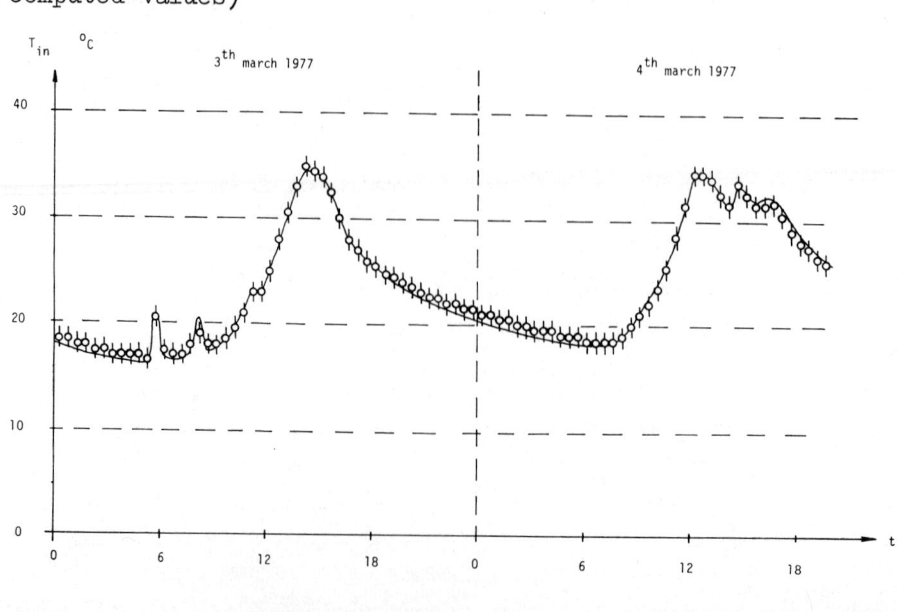

A preliminary analysis of the current results give rise to the following conclusions :

1. The fraction of solar energy, collected by the air, fluctuates between 17 and 20%, depending on the solar intensity variations during the day.
2. The amount of solar radiation transmitted through the wall collector is 23% of the total incident energy. However the fraction of that energy, which will be converted into heat, will depend on the size of the room situated behind the collector and on the furniture.

CONCLUSION

The concept of a semi-transparent wall collector has been tested mathematically and experimentally. First results are in complete agreement with the model. A more detailed analysis of the results accumulated during the last heating period is under completion and will permit the evaluation of the efficiency of the system over a full heating season.

REFERENCES

(1) E.R. Hitchin et al. St.-George's County Secondary School, Wallasey. Journal of the Institution of Heating and Ventilating Engenieers 33 (1966) 325.

(2) F. Trombe et al. Etude sur le chauffage des habitations par utilisation du rayonnement solaire. Revue générale de thermique 48 (1965) 1345.

(3) N. Fuchillo. Semi-transparent solar collector window systems. Solar Energy 17 (1975) 159.

Jacques Michel
14 Rue des Poissoniers
92204 Neuilly sur Seine Cedex, B.P.32

Introduction

Les maisons à chauffage solaire d'Odeillo et de Chauvency-le-Chateau utilisent l'effet de serre; le rayonnement solaire traverse un vitrage et sa réception par la surface noire du mur récepteur, permet de localiser l'apport thermique. On obtient ainsi un piège efficace des calories solaires.

Les habitations réalisées d'après le brevet Anvar/Trombe/Michel s'inspirent des quatre principes suivants:

1 Emploi de l'effet de serre

2 Emploi de façades Sud, Sud-Est, Sud-Ouest

3 Stockage à court terme, déphasage des calories à travers le mur

4 Transport des calories par thermocirculation de l'air pour le chauffage ou la ventilation d'été.

Parti architectural

Les trois maisons solaires d'Odeillo, réalisées en 1974, expriment les besoins du plan intérieur des programmes de trois familles, la crête rocheuse sur laquelle les maisons sont implantées et les orientations préférentielles en ont déterminé la morphologie.

L'utilisation de plusieurs niveaux de planchers qui épousent les courbes du terrain et le groupement des maisons, préfigurent en quelque sorte le collectif solaire éloigné de l'esthétique gratuite du modernisme actuel.

En 1972, j'avais également proposé sur le site d'Odeillo, l'implantation de 31 logements collectifs à chauffage solaire, utilisant des composants industrialisés réalisables à 1.000 Frs du m2 habitable; les formes exprimaient l'environnement, le terrain, les orientations et les paramètres solaires et techniques.

L'Administration timorée a refusé en bloc ce projet, préoccupée uniquement à cette époque, par l'architecture d'esthétique passeiste.

En 1977, on refuse toujours l'architecture exacte en synthèse avec les apports naturels, mais on impose dorénavant un faux modernisme, exemple: textes des 2 refus de permis de construire la maison Ancet à Annecy.

Figs. 1, 2

3 maisons, d'Odeillo

31 logements d'Odeillo

La maison solaire de Chauvency-le-Chateau, édifiée en Lorraine en 1969 était étudiée pour le plus grand nombre, et realisable à l'époque en série, dans les prix plafonds H.L.M., incluant les récupérateurs solaires dans le prix plafond.

L'objectif était de construire des habitations individuelles, en bande ou en collectif, avec un module d'espace habitable et juxtaposable équipé chacun de la serre verticale dans le rapport adequat surface-volume.

Fig. 3

Le maison solaire de Chauvency-le-Chateau

La maison était construite en ossature tubulaire en acier bardée extérieurement de différents composants modulés, dont les capteurs.

Un prototype industrialisé plus élaboré a été exposé à la Foire de Paris 1973 et visité par 200.000 visiteurs.

Figs. 4a, 4b

La maison solaire - Un prototype industrialise plus elaboré

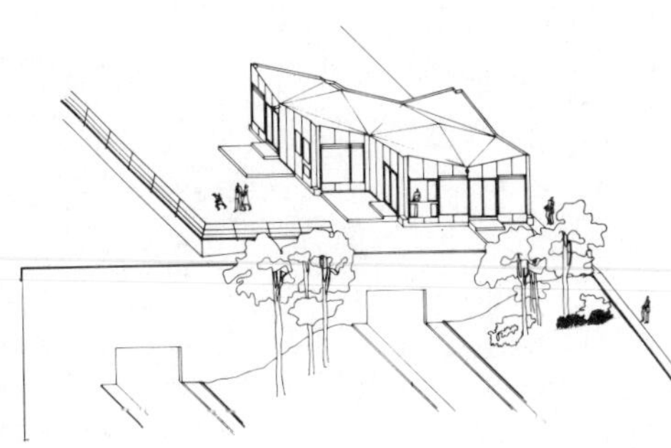

Foire de Paris 1973

Bilans de performance

Les bilans d'exploitation des maisons d'Odeillo situent les économies réalisées suivant les années, entre 60 à 70% de la consommation du chauffage, les radiateurs d'appoint étant thermostatés à 19 ou 20°C, entre le 1er Octobre et le 20 Mai.

Les études entreprises depuis la fin de 1974 ont permis d'établir les bilans thermiques de la thermocirculation d'air et de la restitution du mur accumulateur et de déterminer un rendement global moyen du dispositif de chauffage solaire vis-à-vis de l'énergie solaire incidente. Par exemple:

L'analyse des gradians de température des 23, 24 et 25 Décembre 1974 montre que le 24 Décembre, malgré l'absence d'énergie solaire reçue dans le plan de l'insolateur, la température de surface intérieure des murs est restée (comme le montrent les mesures des deux journées intermédiaires) entre 20 et 30°C.

Les estimations effectuées pendant l'hiver 1973/1974, dans la maison de Chauvency-le-Chạteau, donnent un gain de consommation de 46,2% apporté par le chauffage solaire sur le bilan de chauffage 'tout électrique', soit:

- Volume chauffé: 275 m3

- Coefficient de déperdition volumique: G = 1,22 W/m3°C

- Surface de capteurs: 45 m2

- Consommation théorique d'électricité en solution 'tout électrique': 16.800 kW/h.

- Consommation du chauffage d'appoint: 9035 kW/h.

  16.800 - 9.035 = 7.765 kW/h, soit:

$$\frac{7.765}{16.800} = 46,2\%$$

De ce qui précède, il ressort qu'avec un minimum d'amélioration de la conception et de la réalisation de cette maison située en Lorraine, il est raisonnable d'escompter un gain de consommation de 50% pour ce type d'habitat solaire dans l'Est de la France.

Possibilites de realisation actuelle

Etude de l'intégration et application de la fonction héliothermique de captage solaire à l'Architecture en concevant les façades captrices pour former un seul plan vitré comprenant:

- Les surfaces captantes spécifiques

- Les éléments vitrages, portes et fenêtres dont les bâtis et les volets intérieurs réalisent des enceintes thermiques supplémentaires renforçant les apports naturels.

- Ils seront à position verticale formant façade, ou oblique, formant toiture.

- Leur conception visera une réalisation par éléments composés ou démontés (KIT).

- L'objet précis est la mise au point de matérial héliothermique permettant de créer des 'paquets solaires' industrialisés pour des contenants habitables, concernant l'habitation et les bâtiments tertiaires en partant de volumes de 200 m3 jusqu'à 6.000 m3 et au-delà, ceaux-ci comprenant la mise au point de procédés et de composants dans les conditions de marché, d'insolateurs verticaux et masse thermique, d'insolateurs toiture sans inertie, de stockages thermiques différés, d'automatisation régulation et circuit aéraulique, d'émetteurs d'appoint et de récupérateurs de calories.

L'influence prévisible sur les structures du bâtiment, amènera une extension de l'activité du bâtiment à des produits élaborés, l'optimalisation du service façade et toiture (lumière, calories, frigories, isolation).

Elargissement de l'ensemble à une fonction globale d'enveloppe.

Interet economique

Dans le cas de traitement de façade ou de toiture continue, le fait d'intégrer les fonctions capteurs, portes et baies, constitue un avantage économique très appréciable.

Raisonnement identique en ce qui concerne la conception globalisée de l'asservissement et du chauffage d'appoint pour des volumes donnés.

Coût du produit mis en oeuvre actuellement pour une maison à chauffage solaire de 105 m2 habitable, volume intérieur 280 m3.

- Surface de capteurs verticaux 30 m2 x 400 F = 12.000 Frs. H.T.

- Coût du stockage (compris dans murs).

- Coût de l'installation de distribution d'air = 7.000 Frs H.T.

L'installation héliothermique peut se passer de l'asservissement; dans ce cas, l'investissement capteur de 12.000 Frs est plus intéressant par rapport à l'énergie solaire utilisée pendant la saison de chauffe (6000 kW/h dans la région parisienne) et en relation avec la T.E.P. économisée actuellement, qui est de 1, 3 T.

Ce type de maison à chauffage solaire a été exposé en Avril 1976, à la Foire de Paris, et proposé avec possibilité de réalisation immédiate de 10.000 maisons, au Plan Construction qui a refusé, jugeant le principe trop simple et sans interet.

Conclusions

A mon sens, l'application de l'énergie solaire au domaine architectural ne pourra commencer si l'on en voit seulement l'aspect technique sur le plan des économies d'énergie; les études actuelles et les conclusions cartésiennes sous forme de bilans ou rapportées en pourcentage des besoins énergétiques du pays sont basées sur l'esprit de 'consommation' de la dernière décennie.

Pour ma part, pour reprendre une expression de Le Corbusier, je partirais de la 'tringle à rideaux' et dessinerais la maison ou l'appartement solaire dans la proportion et les besoins des habitants. J'utiliserais le soleil pour les chauffer ou climatiser en été, sa lumière pour sculpter l'édifice et l'éclairer, des brise-soleil, loggias et parasols pour les protéger pendant la saison chaude; ceci, par exemple, dans la manière de penser du 'Capitol' de Chandigarh aux Indes, dessiné entre 1952 et 1956, où les aspects positifs et négatifs du climat, du lieu et des orientations ont produit une architecture de balance, de réponses en contrepoints très riches, pulsée en outre par le nombre d'or et basée sur l'unité de 6 pieds = 1,83m.

Celà aurait l'avantage de recréer une architecture très fondamentale et de parer au déclin des formes gratuites du 'Design' et de l'architecture dite moderne et de pallier au passeisme de refuge.

Bien entendu, la sagesse voudra que la transition architecturale issue du mariage avec les formes actuelles tienne compte dans l'immédiat, des habitudes, des archétypes de la maison, pour ne pas heurter et amener progressivement la sensibilité générale aux formes nouvelles en symbiose avec l'habitant, le soleil, les climats, les techniques, les matériaux, le lieu.

Appendix 1

Bilan Thermique et economique
Maison solaire type Foire de Paris 1976
Maison solaire type Saint Cheron
Centre AFPA Beziers

Je décrirai les projets en cours d'études ou de réalisation, mon action se situant également dans la mise au point de matériel héliothermique permettant de créer des 'Paquets solaires' industrialisés pour des contenants habitables concernant l'habitation et les bâtiments tertiaires, en partant de volumes de 200 m3, jusqu'à 15.000 m3 et au-delà, en recherchant les conditions de marché.

1. Maison solaire type Foire de Paris 1976

L'habitation fait appel au procédé naturel de chauffage solaire (en façade Sud-Est) utilisant le brevet Anvar, F. Trombe, J. Michel 'Eté-Hiver' avec une variante à ventilation asservie, qui permet de contrôler et de réguler le dispositif (en façade S.W.).

L'originalité, par rapport aux réalisations antérieures, sera l'emploi d'une serre standard, équipée d'un vitrage sélectif qui comporte un rideau réfléchissant l'énergie solaire en été et sert d'écran thermique la nuit pour limiter les déperditions nocturnes.

Fig. 5
Maison solaire -
Foire de Paris
1976

Les façades captrices formant une seule peau vitrée englobant sur le même plan les vitrages doubles des baies coulissantes dont les bâtis câdres et les volets intérieurs, réalisent des enceintes thermiques supplémentaires, renforçant les apports naturels.

Une autre amélioration consiste dans l'emploi d'un capteur vertical en agrégat lourd de béton étuvé, avec à l'arrière, un circuit secondaire vertical isolé thermiquement de l'intérieur de la maison et récupérant les calories qui parviennent en déphasage de fin de nuit.

Pour la ventilation asservie, l'ensemble des ces modules de captation est relié au caisson du ventilateur par une gaine collectrice. L'air est ensuite distribué à partir du ventilateur vers les pièces, au moyen d'un réseau de gaines et de bouches de soufflage.

Il sera fait appel à des composants industrialisés pour l'équipement et la réalisation de cet habitat solaire de 5 pièces principales, extrapolable aux types 4 et 6 pièces en partant des mêmes éléments.

Le prix de réalisation du premier prototype (Caisse Nationale de Crédit Agricole Foire de Paris 1976), le place dans les conditions de marché des habitations économiques.

Table 1

Bilans thermiques

Types de maison	Sh	Vh	G	Dju	C	Sc	Es	Cs
IA	105	280	0,99	2800	13971	35	5113	36,6
IB	105	280	1,06	2500	13356	35	5412	40,5
IC	105	280	1,15	2000	11592	28	5969	51,5

Sh: Surface habitable (m2)
Vh: Volume habitable (m3)
G: Coefficient de déperdition volumique (W/m3°C) calculé compte tenu des règles d'isolation E.D.F.
Dju: Degrés-jour unifiés, donnés par le règlement du HOT (page 14. de l'édition du 06/02/75).
C: Consommation en chauffage (kWh). C = 18 CVD ju
Sc: Surface de capteurs (m2)
Es: Energie solaire utilisée (kWh), d'après les bilans établis par M. Robert - CNRS à Odeillo
Cs: Contribution solaire maximale (%)

2. Maisons solaires utilisant le Brevet Anvar-Trombe-Michel 'Climatisation jour-nuit'

Les habitations sont équipées de caissons isolants simples en toiture et de caissons capteurs à assemblage identique, utilisant la climatisation naturelle avec asservissement et régulation. La toiture Sud capte le rayonnement solaire dont la longueur d'onde est comprise entre 0,25 et 4 microns. La toiture Nord émet un rayonnement du type rayonnement infra-rouge, dont la longueur d'onde est comprise entre 4 et 30 microns.

Pendant la période de captation du rayonnement solaire, il est possible de créer une circulation d'air chaud et pendant la période d'émission du rayonnement infra-rouge, on peut créer une circulation d'air frais.

Le dispositif d'accumulation des calories ou des frigories peut être installé en-dessus ou en dessous de la partie habitable de l'habitation.

Une réalisation expérimentale de ce type est en cours de construction dans la région parisienne, à St-Cheron.

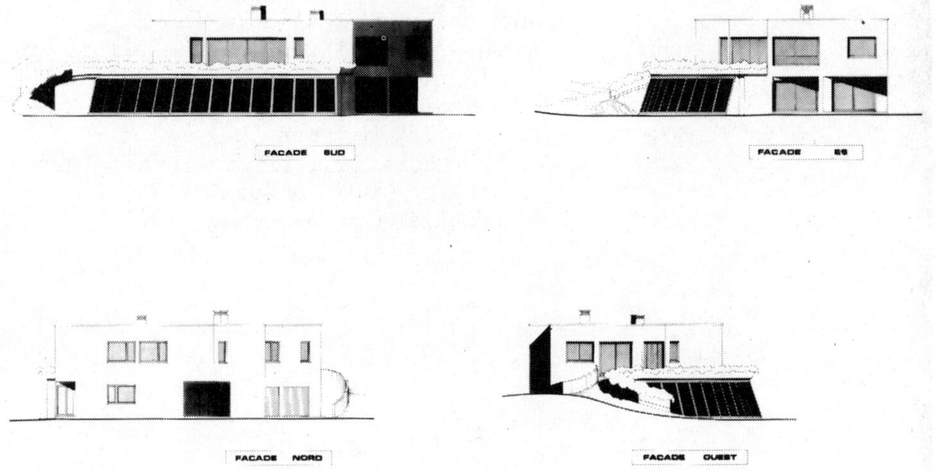

Fig. 6
Maison solaire à St. Cheron

Ces habitations sont conçues avec les mêmes composants que l'habitation précédente et reçoivent une toiture dont l'inclinaison Sud est située à 45° sur l'horizontale ou 60° dans le cas d'un comble habité. Avec ce dispositif, on peut choisir les meilleurs angles de captage pour le soleil et le froid et utiliser pour ce dernier, les surfaces sélectives (peintures blanches à l'oxyde de titane, par exemple).

Table 2 Depenses d'investissement

	1A	1B	1C
Coût du stockage (F.H.T.)		Compris dans mur	
Coût des capteurs (F.H.T.) et des registres	14.000	14.000	11.200
Coût de l'installation de distribution d'air (F.H.T.) (Facultatif)	8.000	8.000	8.000
Coût de l'energie sans chauffage solaire (F.H.T.)	3.493	3.339	2.898
Coût de l'énergie d'appoint (F.H.T.)	2.215	1.986	1.406
Economie (F.H.T.)	1.278	1.353	1.492
Nb de TEP économisées annuellement	1,1	1,2	1,3

On aura ainsi une toiture noire ou de valeur sombre côté soleil et blanche à l'opposé, avec un angle de 30° sur l'horizontale.

Table 3 Bilans thermiques

Types de maison	Sh	Vh	G	Dj	C	Sc	Es	Cs
IIA	105	280	0,99	2800	13971	45	6857	49,1
IIB	105	280	1,06	2500	13356	42	6828	51,1
IIC	105	280	1,15	2000	11592	28	6746	58,2

Sh: Surface habitable (m2)
Vh: Volume habitable (m3)
G: Coefficient de déperdition volumique (W/m3°C) calculé compte tenu des règles d'isolation E.D.F.
Dju: Degrés-jour unifiés, donnés par le règlement du HOT (page 14. de l'édition du 06/02/75).
C: Consommation en chauffage (kWh). C = 18 CVD ju
Sc: Surface de capteurs (m2)
Es: Energie solaire utilisée (kWh).d'après les bilans établis par M. Robert - CNRS à Odeillo
Cs: Contribution solaire maximale (%)

Table 4 Depenses d'investissement

	1A	1B	1C
Coût du stockage (F.H.T.)	5.145	5.035	3.840
Coût des capteurs	11.700	11.000	7.280
Coût de l'installation de distribution de l'air (F.H.T.)	12.000	12.000	12.000
Coût de l'énergie sans chauffage solaire (F.H.T.)	3.493	3.339	2.898
Coût de l'énergie d'appoint (F.H.T.)	1.779	1.632	1.212
Economie (F.H.T.)	1.714	1.707	1.686
Nb de TEP économisées annuellement	1,5	1,5	1,5

3. Centre AFPA Beziers
(Association Nationale pour la Formation Professionnelle des Adultes)

Une première tranche de 11 ateliers de 4.300 m2 de surface sera totalement 'solarisée' grâce aux serres verticales assurant la température de base pendant l'hiver et la ventilation d'été par thermocirculation naturelle d'air frais pris au Nord.

Le prix plafond des ateliers, de 1200 Frs le m2 hors-oeuvre comprend l'équipement héliothermique: 1.544 m2 de façades Sud avec serres solaires et incorporation des portes et fenêtres à 400 Frs le m2 posé.

Le logement de fonction de l'AFPA est également construit avec des serres verticales.

Le bloc administratif, restauration, foyer, est alimenté par préchauffage solaire, grâce à un champ de collecteurs laminaires à eau, disposés en terrasse et reliés à un stockage avec échangeurs sur circuit chaufferie.

- Economie annuelle d'énergie: 127 TEP.

Fig. 7
Perspective d'un des bâtiments AFPA à Beziers

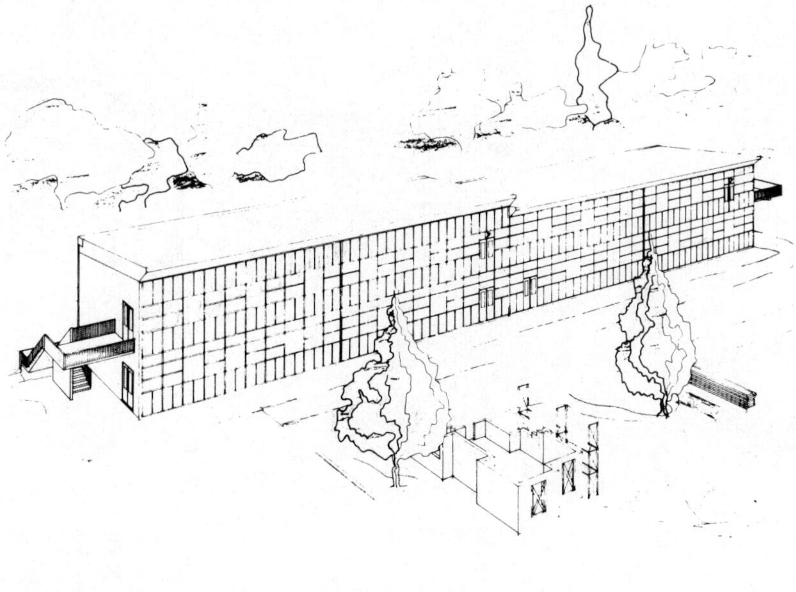

Appendix 2

Ensemble sportif a chauffage solaire

Gymnase type C

Architectes: J. Michel et F. Vardaguer
Ingénieur: W. Diamant
Ingénieurie générale et construction: Sté de Construction Durafour.

Description

- Halle de 40m x 20m x 7m sous entrait, volume 6.000 m3 les annexes sont positionnées dans le prolongement des lon pans sur les pignons Est ou Ouest.

- La conception d'ensemble du gymnase tient compte d'une isolation globale, située en épiderme extérieur.

- Le long pan Sud et le pignon opposé aux annexes, réalisent une surface verticale de captation solaire de 322 m2, comportant une paroi transparente extérieure parallèle à un élément de mur délimitant une enceinte thermique; sa structure alvéolaire contribue à la diminution des pertes thermiques (convection et rayonnement) - Brevet J. Michel, W. Diamant, Sté Durafour.

Les bandeaux d'acrotère de la Halle recevront une surface de captation de collecteurs laminaires à eau de 96 m2, suffisante pour assurer les besoins d'eau sanitaire des annexes, ainsi que le chauffage de base de celle-ci.

La température hivernale de base de 12°C de la halle est assurée par l'énergie solaire, l'appoint de chauffage éventuel pourra être réalisé à l'aide d'aérothermes électriques ou sera effectué directement sur le stockage d'eau solaire avec l'électricité ou le gaz liquéfié.

-L'orientation préférentielle de cet ensemble sportif à chauffage solaire est située dans un dièdre Sud-Est, Sud-Ouest.

-Le prix d'un gymnase solaire de série étant fixé à 1.800.000 Frs F.TTC, un gymnase classique de même confort coûtant 1.600.000 Frs F.TTC, la différence de 200.000 Frs. F.TTC serait amortie sur la seule économie de fuel en six années environ, ceci, sans tenir compte de l'augmentation inéluctable du prix du fuel, ni de la suppression des frais d'entretien chaufferie et du personnel chargé de l'exploitation.

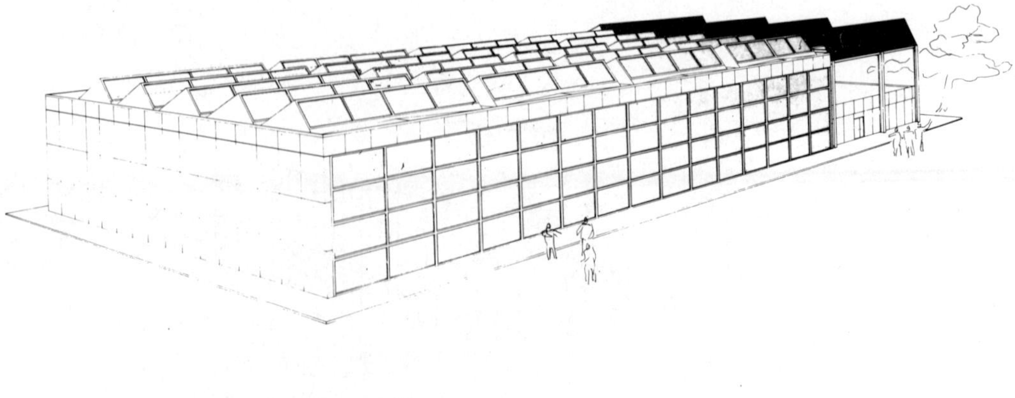

Fig. 8
Ensemble Sportif à chauffage solaire: Gymnase Type C

Appendix 3

Bloc Fenêtre a fonction climatique et phonique

Application des brevets Anvar-Trombe-Michel

ETE-HIVER no 71 23 778 et additif no 72 22 209
JOUR-NUIT no 75 32 921 et additif no 76 19 960

Le développement des architectures utilisant les façades captrices formant une seule peau vitrée englobant sur le même plan, le vitrage des baies (dont les bâtis cadres et les volets intérieurs) qui réalisent des enceintes thermiques supplémentaires renforçant les apports naturels, nous amène tout naturellement à traiter le bloc fenêtre à fonctions diverses et pouvant s'intégrer dans la peau isolante ou vitrée des bâtiments neufs ou à rénover.

Description du bloc fenêtre

Ce bloc est constitué de 2 chassis coulissants, séparés par un espace, réalisant les fonctions d'isolation phonique et thermique, dans lequel se développe un store à lamelles orientables. L'une des faces est traitée en corps noir absorbant le rayonnement solaire de longueur d'ondes comprise entre 0,25 et 4 microns. L'autre face est traitée avec un colorant sélectif émettant dans l'infra-rouge de longeur d'ondes comprise entre 4 et 40 microns.

On réalise 4 fonctions:

- En hiver, une enceinte thermique chaude qui distribue les calories solaires à la pièce arrière par la thermocirculation naturelle de l'air à l'aide de registres réglables.
- En été, la ventilation par effet de serre inversé
- Pendant la nuit en hiver, l'écran isolant émissif inversé évite les pertes nocturnes internes
- En été on pallie à l'ensoleillement excessif par l'inversion de l'écran.

Le bloc fenêtre conserve évidemment sa fonction d'élément vitré, libéré de toutes sujétions de stores.

4.5 Efficient analysis of long term performance of solar heating systems

P. F. Monaghan*
Statistics and Operations Research Lab. (SORL),
Trinity College, Dublin University,
Ireland

Abstract

This paper addresses the problem of computer run-time in the computer analysis of long-term performance of solar heating systems. The approach taken is to model each day's radiation by a parabola, and other climatic conditions by their daily averages. Daily performance is calculated by algebra, rather than by the iterative, numerical integration techniques used in conventional hour-by-hour simulations. Results given by these alternative simulation methods are compared for a variety of solar heating systems using Canadian and Irish climatic data. Over a wide range of parameters these estimates differ by -8% to +2%. The day-by-day simulation program produces results 8 to 60 times faster than hour-by-hour programs of varying degrees of complexity.

Introduction

To design a solar heating system which is most cost-effective in a given location, analysis of long-term performance is required.

Two features of solar system analysis have led to the wide use of highly detailed computer simulation methods:-

1. The dependence of solar energy collected on short-term climatic conditions, leads to a large data input and an immense amount of calculation.
2. The large number of coupled non-linear equations which model a system often require iterative solutions.

For this reason many computer simulation programs have been developed, which use hour-by-hour meteorological data and solve the system equations by numerical integration. TRNSYS(1) is perhaps the best known. Although the accuracy and adaptability of these programs is not in question, many research groups have stated that the cost of computer facilities rules out their general use in the design of solar heating systems.

In particular, Oonk, Duffie and Beckman(2) state that these simulations are too costly for residential buildings and suggest that the data requirement for simulation needs to be reduced. To alleviate this problem three approaches have been taken:-

1. Graphs and handbooks suitable for use by designers and architects have been produced from computer simulation results (3).
2. Non-simulation methods of analysing solar heating systems by stochastic analysis are currently being developed (4,5).
3. Intermediate methods, using computer simulation together with simple algebraic summaries of the climatic data, are being developed with a view to reduction of computer run-time (6,7,8).

The problem with the first method is that performance charts can only be produced with a certain amount of restriction. Additionally, these are based on past results of specific

---

* Work initiated at Queen's University at Kingston, Canada.

computer simulations and can quickly become obsolete in the changing world of energy economics and solar technology. The stochastic techniques, though still in early stages of development, display great potential. Lameiro's method (5), however, has run into computational difficulties due to the use of unwieldy numerical methods. The method of Haslett and Monaghan(4) uses the simplest algebraic approximations to system components and employs algebraic integration. Although the method gives very useful information in solar system design, it is unlikely that this method can match the accuracy of simulation techniques, particularily where a high degree of sophisticated control is critical.

For these reasons, this paper deals with a method of reducing computer run-time in simulation design methods. The approach taken is to reduce the amount of climatic data required and to solve the equations algebraically on a day-by-day basis rather than by the hour-by-hour numerical methods at present employed.

2. Analysis

The premise upon which this work is based is that most components in a solar heating system do not require analysis on the basis of a small time-step. Previously, analyses have been based on hour-by-hour climatic data only because solar collector performance is very strongly dependent on transient conditions, and, in particular, on solar radiation. The building heating load and the secondary components of a solar heating system (e.g. auxiliary heater, heat exchangers, controls, heat pumps etc.) do not require transient analysis. Transient climatic conditions are effectively damped from these components by the thermal inertia of the building and the solar store. For this reason this work concentrates on modifications to solar radiation and collector modelling, on the assumption that other components may be modelled conventionally.

2.1 Collector and Radiation Modelling

The best known model of solar collector performance is the Hottel-Whillier-Bliss (HWB) model, where the instantaneous rate of energy collection is given by,

$$Q_u = A_c F_r (H_T(\tau\alpha) - U_L(T_{f,i} - T_a)) \quad (1)$$

or alternatively,

$$Q_u = A_c F_r (H_a - LOSS) \quad (2)$$

H_a = rate of energy absorption by the collector plate

LOSS = rate of total connective and radiative losses to the environment from the collector plate.

It should be noted that H_a is dependent only on input climatic data and collector constants. However, LOSS is dependent on the state of the system being modelled, because of the influence of the storage temperature. The HWB equation is normally solved by inputting hour-by-hour values of climatic data and numerically integrating Q_u in small time-steps. The solution of this equation over a typical day is shown graphically in Figure 1. The system concerned is shown in Figure 2. The radiation profile, H_a, is determined by climatic conditions while the LOSS level increases during the day due to increases in the store temperature.

Figs. 1, 2

Variation of absorbed radiation, H_a, and collector losses, LOSS, on a clear winter day at Ottawa

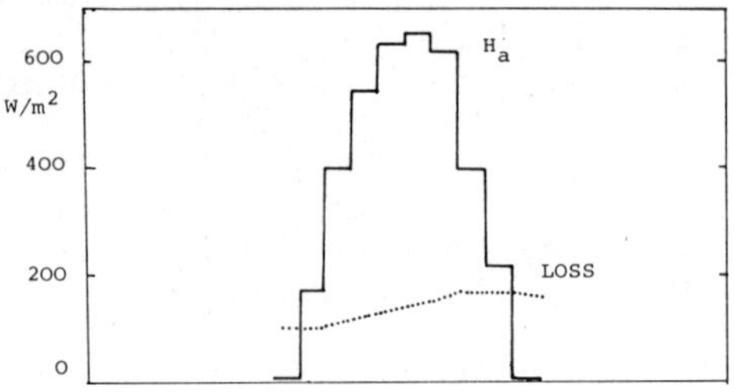

A solar space heating system

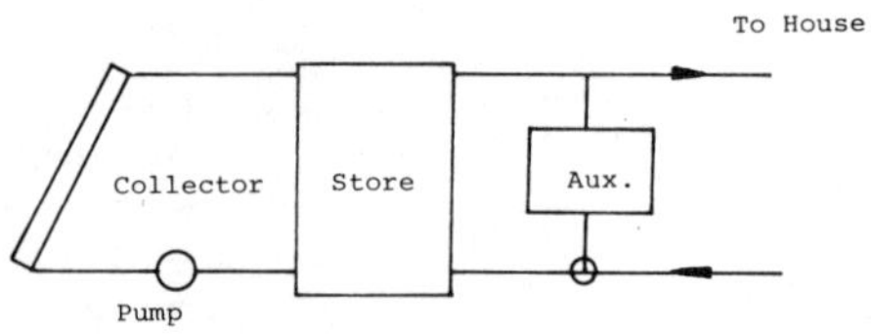

Because a collector is typically controlled to operate only when H_a is greater than LOSS, it should be noted that the <u>total</u> energy collected over any day, DQ_u, could be calculated by multiplying the <u>area</u> between H_a and LOSS by A_cF_r .

This graphical representation of the HWB equation leads to the proposal to model H_a and LOSS by functions which will describe their respective variations over the day. If this can be achieved then an algebraic calculation of the area between the functions will represent the daily total energy collected. To facilitate the following discussion of a function to represent H_a let us assume, in the first instance, that LOSS is constant over a day i.e. that the store is infinitely large.

2.2 Modelling of the Radiation Profile

The simplest representation of the distribution of daily radiation is a square wave. An example, where radiation is averaged from sunrise to sunset, is shown in Figure 3.2 alongside a typical radiation profile. This particular square wave will always underpredict energy collected, as demonstrated by the addition of constant LOSS levels. Although the total areas under the two H_a profiles are equal, the areas between H_a and any LOSS line is not. Nevertheless, Haslett and Monaghan (4) have shown that square waves can be used to give upper and lower estimates of system performance in terms of closed form expressions.

The simplest non-linear representation of the radiation profile is a parabola, the area under which is equal to the total daily radiation. This functional distribution is the basis for the methods presented in this paper. Two other groups, working independently, have both used cosine waves to represent the radiation profile. Swanson and Boehm (7) used cosine waves together with correction factors in a non-simulation technique for long-term solar collector performance

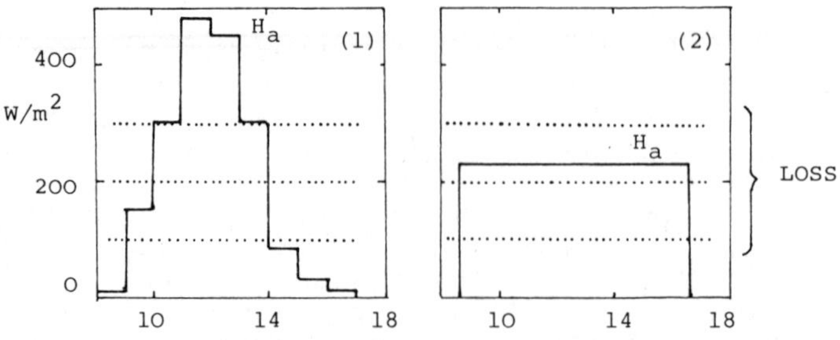

Fig. 3

Comparison of typical values of absorbed energy, H_a, to a daily averaged model of H_a, with fixed LOSS levels.

predictions. Bruno et alia (8) have studied water heating systems using cosine waves and "day length reduction factors" in simulation work.

Although cosine functions may be more realistic in some ways, parabolae are used in this work, principally for algebraic and computational simplicity. Thus H_T is represented as,

$$H_T = A_r - C_r(12 - HR)^2 \quad (3)$$

and an entire day's radiation is represented by two parameters, A_r and C_r. These may be determined from two conditions: (i) the total area under the curve, $4\sqrt{A_r^3/C_r}/3$, must equal the total daily radiation, DH_T, and (ii) the "effective day length", D_L $(= 2\sqrt{A_r/C_r})$, must give a true representation of the width of the parabola base at various times of the year. In this work D_L is regarded as constant for each month and it is estimated by fitting a two-parameter parabola to the average daily radiation profile for each month. Thus, for each day, A_r and C_r are calculated from (6,pp 30-40),

$$A_r = 1.5(DH_T/D_L) \quad (4)$$

$$C_r = A_r/(D_L/2)^2 \quad (5)$$

The results of this procedure for a clear day and a cloudy day are shown in Figure 4. It is obvious that, on clear days, a parabola is an excellent model of daily radiation while on cloudy days the parabola will underpredict DQ_u as previously described in the case of the square wave. However, on cloudy days less energy is collected and underpredictions caused by these are less important in the long-term.

Having formed a function for H_T a value of $(\tau\alpha)$ is required. In this work $(\tau\alpha)$ is calculated by the expression (6,pp40-46)

$$(\tau\alpha) = DP_{bT}\,(\tau\alpha)_{bmax} + (1 - DP_{bT})\,(\tau\alpha)_d \quad (6)$$

and so the absorbed energy is represented by,

$$H_a = H_T(\tau\alpha) = A_r(\tau\alpha) - C_r(\tau\alpha)(12 - HR)^2 \quad (7)$$

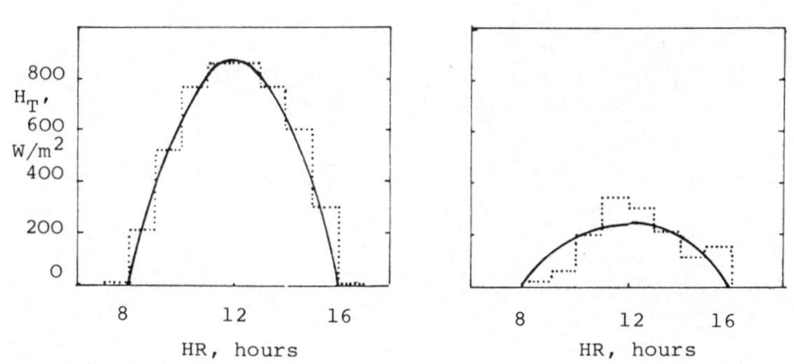

Fig. 4 Radiation, H_T, incident of a tilt of 60° for clear and cloudy winter days at Ottawa.
...... Measured radiation
______ Parabolas for given D_L, DH_T

2.3 Modelling of the LOSS level

The total radiative and convective heat losses from the collector plate are given by,

$$LOSS = U_L(T_{f,i} - T_a) \tag{8}$$

where U_L is a function of T_p, T_a and W_v. In this work T_a and W_v are represented by their daily average values. T_p is a function of actual operating conditions and, for simplicity of computation, the previous day's value is always used on the assumption that this will cause no bias to results in the long term. The variation of $T_{f,i}$ (and its effect on LOSS) throughout a day is taken into account in the integration procedure which is now described. To simplify the discussion let us assume that there is no heat exchanger between the collector and the store so that $T_{f,i} = T_s$.

Consider Figure 5. If the times of switching on and off the collector pump, t_{on} and t_{off}, can be found then the area, BAL, can be expressed as,

$$BAL = (A_r - 144C_r)(t_{off} - t_{on}) + 12C_r(\tau\alpha)(t_{off}^2 - t_{on}^2) - C_r(\tau\alpha)(t_{off}^3 - t_{on}^3)/3 - (LOSS_e + LOSS_m)(t_{off} - t_{on})/2 \tag{9}$$

and once again,

$$DQ_u = A_c F_R BAL \tag{10}$$

The expression assumes, for simplicity and adaptability, that the response of the LOSS level is linear. Although, in reality, the exact response of LOSS takes the form of an exponential curve superimposed on a quadratic curve, the linear assumption is found to be adequate. The main problem is to find t_{off}, which is the point of intersection of the rising LOSS line and H_a. To find t_{off}, iteration is required as LOSS and BAL are mutually dependent while the collector is operating. In this way DQ_u is calculated for each day of the year. In the periods t_{off} to midnight and midnight to t_{on} the temperature of load withdrawal is assumed to be the value of T_s at the beginning of those periods, i.e. the rate of change of T_s is linear.

In Figure 6, the daily results of an hour-by-hour simulation are compared to those of the parabolic method for a highly responsive collector-store system.

Figs. 5, 6

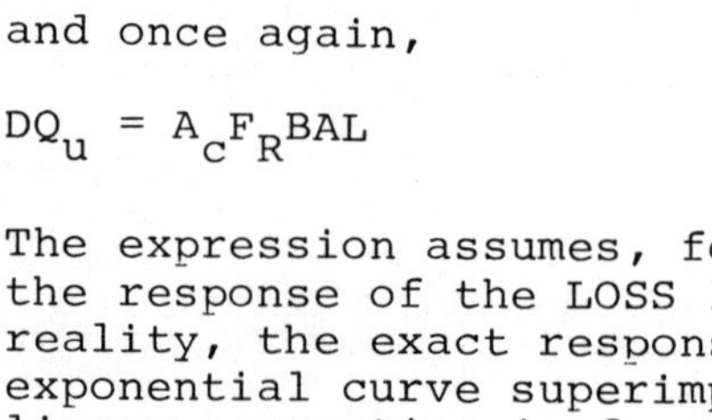

The area, BAL, when multiplied by $A_c F_r$ is the total daily energy collected.

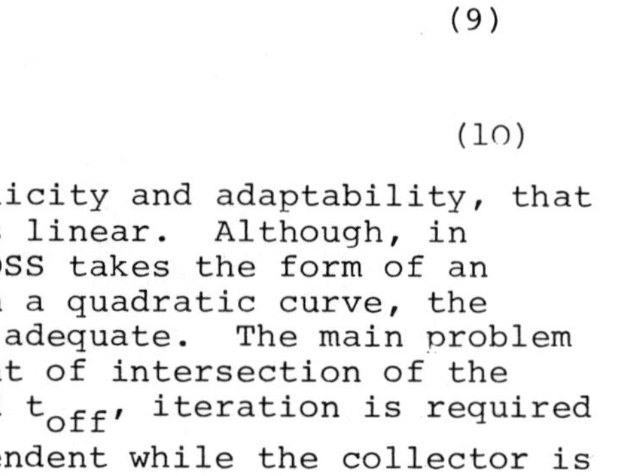

Comparison of one day's simulation results for a highly responsive solar heating system.
...... Hour-by-hour
______ Day-by-day

3 Testing the Method

To test the proposed simulation method two computer programs were developed, PARABOLA and HOURSTAT. PARABOLA is a solar system simulation program which uses the modelling techniques presented in the previous section. HOURSTAT uses hour-by-hour climatic data, the standard HWB collector model and essentially represents the TRNSYS (1) approach to simulation. It was developed because the TRNSYS program itself placed very severe demands on the university computer system.

The validation of PARABOLA is carried out in two parts which are presented and discussed separately. However, the common features of all the following work are:-

1. Collector and storage fluid is water
2. Building heating demand is represented by a degree-day model.
3. The stores are unstratified and the maximum allowable T_S is 95°C.
4. The collector pump switches on if energy can be collected.

3.1 Phase 1

Purely for preliminary testing of this approach to radiation and collector modelling a simple yet realistic space heating system was selected. The chosen system is shown in Figure 2 and was programed in both PARABOLA and HOURSTAT. To assist in identification of errors in the collector model, a major assumption is made for energy withdrawal from the store. It is assumed that, in any time period, the store temperature, T_S, can be reduced to 26°C and that the remaining required energy can be supplied by auxiliary.

Thirteen test runs were carried out for different data years at Ottawa, Toronto and Edmonton to test the generality of PARABOLA. To some extent storage mass was also varied. The constant parameters are listed in Appendix 1.

The results for these runs are presented in Figure 7. The last column (% DIFF) is the percent difference between PARABOLA predictions and HOURSTAT predictions, for annual solar energy gained.

Location	Year	Storage Mass Kg	Special Comment	Solar Energy Supplied, KWH		% DIFF	% Demand Supplied by Solar
				PARABOLA	HOURSTAT		
Ottawa	1970	22,000		20,286	20,209	+ 0.38	69.4
Ottawa	1970	11,000		19,532	19,452	+ 0.41	66.8
Ottawa	1971	22,000		19,202	19,231	- 0.15	66.2
Ottawa	1971	4,400		17,654	17,738	- 0.48	60.8
Ottawa	1972	4,400		17,609	17,617	- 0.05	58.4
Ottawa	1972	11,000		18,791	18,661	+ 0.69	62.3
Ottawa	1972	22,000		19,546	19,405	+ 0.72	64.8
Ottawa	1972	110,000		25,212	25,149	+ 0.25	83.6
Ottawa	1972	22,000	$\dot{m}_c$ = 5.055 kg/s	19,804	19,673	+ 0.71	65.7
Toronto	1970	22,000		18,539	18,617	- 0.42	71.4
Toronto	1970	110,000		24,102	24,203	- 0.42	92.9
Edmonton	1971	22,000	Tilt = 70°	22,508	23,228	- 3.2	60.2
Edmonton	1971	11,000	Tilt = 70°	21,837	22,492	- 3.0	58.4

Unless otherwise stated: collector mass flow rate, $\dot{m}_c$ = 1.264 kg/s; Tilt = 60°.

Fig. 7 Annual performance predictions for the space heating system (Fig. 2) with 50m² of collector area.

3.2 Discussion 1

The results presented in Figure 7 certainly indicate that PARABOLA is capable of duplicating resutls produced by more complex hour-by-hour simulations.

A full discussion of the source of errors is given in (6,pp91-95) but a summary is given here:-

1. The higher errors at Edmonton are caused by a local distortion of the radiation profile.

2. All errors are caused by limitations in the parabolic representation of the radiation profile.

3. Amount of underprediction by parabola is greatest on days when the LOSS level is close to the peak of the parabola.

4. The amount of underestimation is also great when actual radiation profiles are widely dispersed away from a parabolic profile by variable cloud cover.

5. It is not apparent that any error was caused by the linear assumption for storage response.

6. The results do not always show underprediction because the approximate estimate of $(\tau\alpha)$ by equation (6) is positively biased. This, in turn, causes an upward bias in annual energy collected.

Therefore, parabolic representations of radiation will generally perform worst when LOSS levels are high and daily radiation is low or highly variable. LOSS levels will be high in systems where:

1. One cover non-selective collectors are used.

2. Storage temperature is, in general, high due to high demand temperatures.

3. T_S can rise dramatically in a single day's operation due to highly responsive collector-store systems.

In Phase 2 of the validation the day-to-day simulation method is tested with some of these more rigorous conditions and with more realistic load withdrawal modelling.

3.3 Phase 2

The water heating system and space heating system, Figures 8 and 9 respectively, are modelled both in PARABOLA and HOURSTAT. All runs were carried out using data from Valentia, Ireland in 1970.

Figs. 8, 9

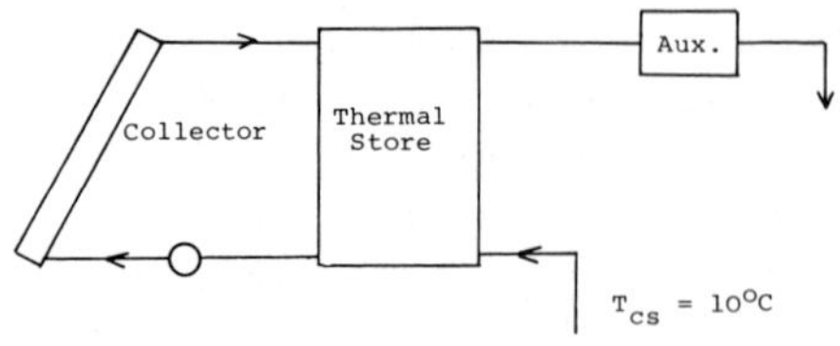

Water heating system

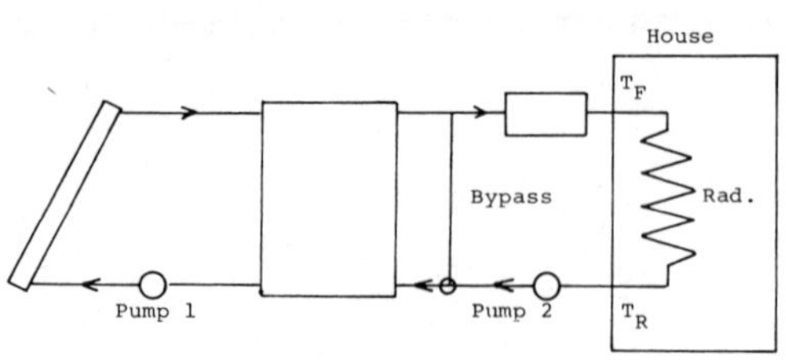

Space heating system with auxiliary in series and storage bypass

Fig. 10 Estimates of annual solar energy supplied, SOLAR, by a solar water heating system at Valentia in 1970.

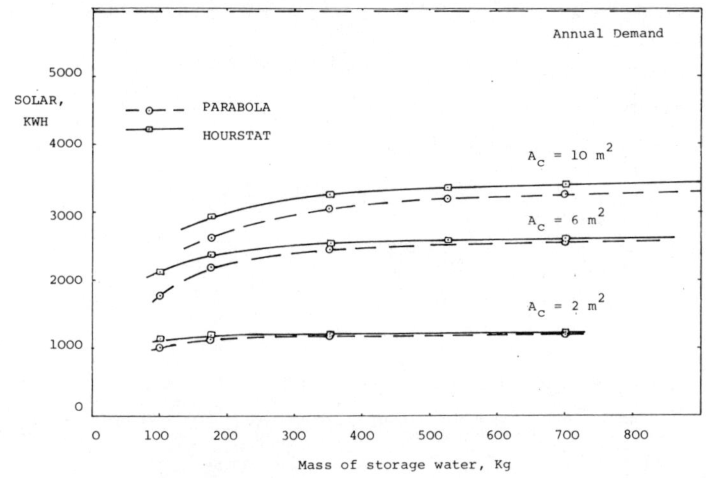

3.3.1 Water Heating System

A conventional water heating system is shown in Figure 8. The hot water demand profile is that suggested by T.I.L. Denmark (9). The simulation results are shown in Figure 10 and constant system parameters are given in Appendix 1.

3.3.2 Space Heating with Auxiliary in Series with the Store

In the space heating system featured in Figure 9, the auxiliary has a fixed output temperature, T_F, and room temperature is controlled by switching Pump 2 on and off. Depending on T_S, the heat distribution system can have three modes of operation:

1. Auxiliary supplies all the demand.
2. The store supplies all the demand.
3. The load is shared if T_S is between T_F and T_R, the radiator return temperature.

The annual results of the two simulations are presented in Figures 11 and 12.

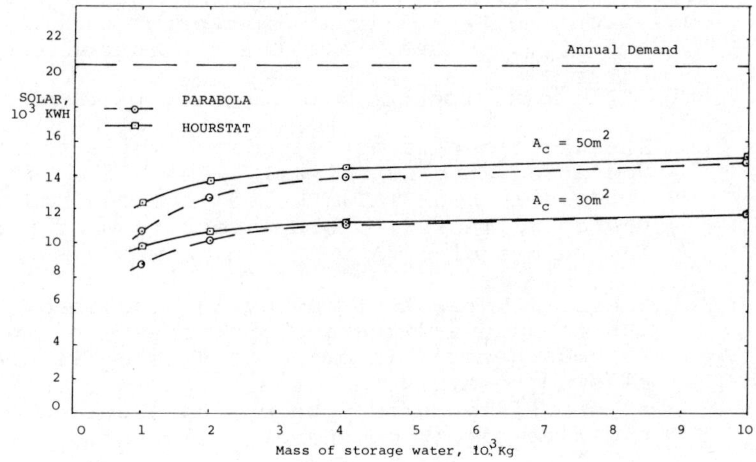

Fig. 11 Estimates of annual solar energy supplied, SOLAR, by a solar space heating syatem at Valentia in 1970.

Fig. 12 Estimates of annual solar energy supplied, SOLAR, by a solar space heating syatem at Valentia in 1970.

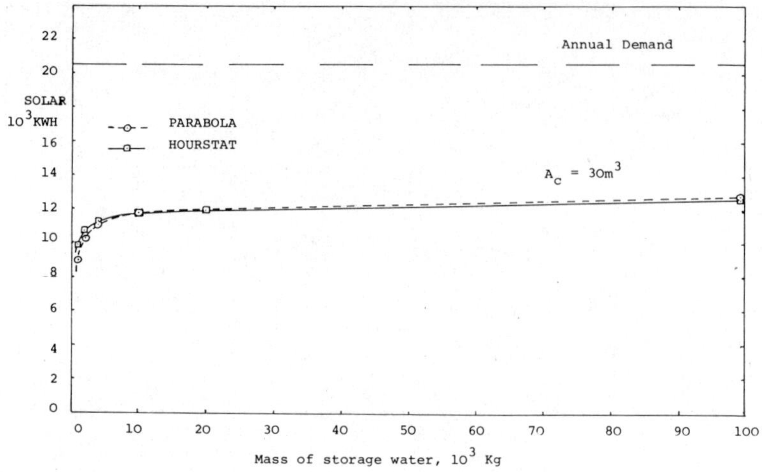

3.4 Discussion 2

The results in Figures 10 and 11 relate to systems expected to exhibit many of the features which will cause day-by-day modelling to be most inaccurate. The general trends may be summarised:

1. In the range of parameters of general interest the differences between PARABOLA and HOURSTAT are between -8% and +2%.

2. At very low ratios of m_s/A_c PARABOLA can exhibit errors greater than these but these systems would generally be ruled out by economic considerations.

3. In general, PARABOLA underpredicts for highly responsive stores and for systems where the store is much smaller than the daily demand. For much larger stores PARABOLA can overpredict the hour-by-hour program (Figure 12).

The two reasons for this trend are: (i) the limitations in the parabola model at high LOSS levels and (ii) the coarseness of the load withdrawal pattern in PARABOLA. For highly responsive systems the LOSS level rises rapidly during the day (Figure 6) causing high average LOSS levels and hence greatest underprediction. The PARABOLA withdrawal pattern is equivalent to tri-daily instantaneous withdrawals at midnight, t_{on} and t_{off}. Compared to the continuous withdrawal policy in HOURSTAT, this produces two opposing effects:

1. Because water is not withdrawn while the collector is operating, the LOSS level rises higher in PARABOLA than it does in HOURSTAT, thereby causing underprediction in energy collected. This effect is most pronounced for low values of m_s/A_c.

2. Where withdrawal of energy is concerned, however, instantaneous withdrawal policies are generally advantageous because incoming cold water is not continuously allowed to mix with hot storage water. For larger stores this positive effect on energy supplied out-weighs the negative effect on energy collected.

It is clear that if continuous load demand were more accurately modelled, the PARABOLA results would move closer to HOURSTAT results for both large and small stores. This can be done by allowing the store temperature to rise and fall exponentially rather than linearly. This causes clumsier algebra and further computational complexity but does not change the fundamental approach.

However, it is felt that the results, as they stand, are acceptable for use in economic analyses particularily when one considers the degree of error involved in predictions of individual building heating demands (10) and future fuel costs. In particular, the distinctive "knee" of the performance curve (Figure 12) is very accurately defined by PARABOLA and estimates above the "knee" are within ±2% of HOURSTAT results.

The objective of this work has been to make computer design of solar systems cheaper and more convenient by reducing the demand on computer facilities. PARABOLA ran from 8 to 16 times faster than hour-by-hour programs developed by this group, depending on their degree of complexity. Compared to the TRNSYS program however, PARABOLA ran 60 times faster and needed only one quarter of the computer core requirement.

4 Conclusion

The method of day-by-day simulation outlined in this paper has been tested for three different solar heating systems, in a variety of climates. Compared to conventional hour-by-hour simulation techniques, annual solar gain predictions differ by -8% to +2% over a wide range of parameters. Errors in the method can exceed these limits when some or all of the following conditions are combined: (i) very highly responsive collector-store combination, (ii) stores much smaller than daily hot water demand, (iii) most inefficient solar collectors and (iv) a required high storage temperature.

However, the errors are acceptable over the range of parameters of economic interest. The method may be improved by more careful modelling of the demand withdrawal pattern, without fundamental changes to the approach. The day-by-day simulation method is 8 to 60 times faster than conventional hour-by-hour simulation methods.

5 Acknowledgements

The advice given on the project by Professor C K. Rush, Department of Mechanical Engineering, Queen's University at Kingston and J. Haslett, SORL, is gratefully acknowledged. The computing assistance of R. Huckle is acknowledged.

This paper describes work on which the author is engaged in connection with a project being carried out by SORL on contract to the Institute of Industrial Research and Standards, Dublin and supported by the Energy R and D program of the European Communities. Acknowledgement is made to the Director of SORL for permission to publish.

6 Nomenclature

A_r, C_r	parabola parameters for daily radiation profile on a tilted surface, W/m^2
A_c	collector area, m^2
BAL	area enclosed by H_a and LOSS curves, Wh/m^2
DH_T	total daily radiation on a tilted surface, Wh
D_L	effective day length, hours
DP_{bT}	proportion of daily radiation on a tilted surface which is beam radiation
DQ_u	total daily energy collected, Wh
F_r	collector heat removal factor
H_a	radiation absorbed by the collector plate, W/m^2
HR	the time of day, hours
H_T	radiation incident on a tilted surface , W/m^2
$LOSS(LOSS_m, LOSS_e)$	the instantaneous rate of convective and radiative losses from the collector plate, W/m^2 (at t_{on}, t_{off})
m_s	mass of store, kg
Q_u	rate of energy collection, W
T_a	ambient temperature, ^oC
T_F, T_R	fluid temperatures to and from the heat distribution system, ^oC
$T_{f,i}$	temperature of fluid at inlet to collector, ^oC
t_{on}, t_{off}	the times of switching on and off the collector pump, hours
T_p	collector plate temperature, ^oC
T_s	storage temperature, ^oC
U_L	overall loss coefficient for collector, $W/m^2\,^oC$
W_v	wind speed, m/s
$(\tau\alpha)((\tau\alpha)_{bmax}(\tau\alpha)_d)$	the overall transmittance-absorptance product of the collector (for beam radiation at noon, diffuse radiation)

7 References

1. Klein, S. A., Cooper, P. I., Beckman, W. A. and Duffie, J. A.: University of Wisconsin Engineering Expt. Station. Report No. 38 (1974)."TRNSYS, A Transient Simulation Program".

2. Oonk, R. L., Beckman, W. A. and Duffie, J. A.: J. Solar Energy, 17, 21 (1975). "Modelling of the CSU Heating/ Cooling Systems".

3. Beckman, W. A., Klein, S. A, and Duffie, J. A.: J. Solar Energy, 18 , 113 (1976). "A Design Procedure for Solar Heating System".

4. Haslett, J. and Monaghan, P. F.: A paper to be presented at the International Conference on Residental Solar Heating, Liege, Belgium (September 1977). "Direct Modelling of Solar Energy Systems for Space and Water Heating".

5. Lameiro, G. F.: Ph.D. Thesis, Colorado State University (1977). "Stochastic Models of Solar Energy Systems".

6. Monaghan, P. F.: M.Sc. Thesis, Dept. of Mechanical Engineering, Queen's University at Kingston, Canada (1976). "Reduction of Computer Run-time in Design of Solar Heating Systems by Use of Processed Climatic Data".

7. Swanson, S. R. and Boehm, R. F.: J. Solar Energy, 19 ,129 (1977). "Calculation of Long Term Solar Collector Heating Performance".

8. Bruno, R., Hermann, W., Horster, H., Kersten and Mahdjuri, F.: To be published Vol. 1 No. 4 of the International J. of Energy Research (1977). "Analysis and optimisation of Solar Hot Water Systems".

9. Thermal Insulation Laboratory, Technical University of Denmark: Private Communication March 1st, 1977

10. Hardy, A. C.: To be published in Institute of Mathematics and its Applications, Symposium Proceedings, No. 15 (1977). "Energy Consumption in Buildings".

Appendix 1. Constant Parameters

	Phase 1	Phase 2	
Collector Parameters:		Water	Space
Tilt, degrees	60,70	62	62
Orientation, degrees	0	0	0
Ground reflectance	0.2,0.7	0.1	0.1
No. of glass covers	2	1	2
Absorptance of covers (total)	0.025	0.037	0.037
Glass emittance	0.9	0.88	0.88
Plate absorptance	0.95	0.95	0.95
Plate emittance	0.95	0.9	0.9
Fluid mass flow rate, Kg/s.m^2	0.025	0.0167	0.0167
Back and edge losses, W/^{o}C	0.63	0.4	0.4
Collector efficiency factor, F'	0.98	0.95	0.95
System Parameters:			
Hot water demand, Kg/day	-	350	-
Req. supply temp., ^{o}C	-	50	-
Cold water temp., ^{o}C	-	10	-
Max. allowable T_S, ^{o}C	95	95	95
House Temp., ^{o}C	18	20	20
Heating demand, KWh/^{o}C-day	6	-	6
T_F ,^{o}C	-	-	34.27
T_R, ^{o}C	-	-	29.5
Mass flow rate to radiators, Kg/s	-	-	0.25

4.6

The solar house of the Eindhoven University of Technology

J. Hamaker, Professor of Architectural Physics
H.C.A. Hoekstra, Senior Scientific Officer
Department of Building Technology

C.W.J. van Koppen, Professor of Heat Technology
J.T.T. van Wolde, Junior Research Officer
Department of Mechanical Engineering

Sectional Physical Engineering in relation to Building Design
Eindhoven University of Technology
Eindhoven, Holland

INTRODUCTION

The recently completed Solar House of the Eindhoven University of Technology (Fig. 1) is the result of more than three years of research and development work. At the beginning of 1973 an exploratory study was made (1), which led to the then rather unexpected conclusion that, even in the Dutch climate, the sun could be considered as a viable source of thermal energy. Particularly when the prices of the conventional energy sources, namely oil and natural gas, were bound to rise in the future. As such a rise was to be expected, due to the gradual depletion of the gas and oil fields, the study ended with the recommendation that research on the application of solar energy for heating purposes should be started on a modest scale.

Fig. 1
South-east view of the Solar House

It was clear from the start that in the follow-up of this recommendation both architectural and mechanical engineering problems would have to be solved. The already existing co-operation between the Departments of Mechanical Engineering and Architecture was therefore taken as the basis for the further research and development work. The idea to build a solar house, with the aim of carrying out the scientific research in the practical dimensions which are essential for all technology, grew up out of this co-operation.

Three years is a rather short period of preparation for a field experiment in a new area of research. Such a fast realization of the solar house would have been impossible without the data on solar heating installations already existing in other countries (Australia, USA, Israel).

A strong effort of our own however was indispensable for the 'translation' of these data to meet conditions in The Netherlands. In this connection the customary styles of living and building play their role, but the different and generally unfavourable climatic conditions in The Netherlands play an even greater role. The fact that the current data on solar radiation are ill-suited for the design of solar heating installations actually proved to be the biggest handicap in the 'translation', and even now this handicap has only partly been overcome. Therefore a separate investigation has been started recently at the EUT, with the aim of obtaining more suitable insolation data.

THE SOLAR HEATING SYSTEM

Figure 2 is a sketch of the solar heating system. Important parts of this system are:

- the solar collector, which transforms solar radiant energy into heat,
- the watertank storage in which momentarily surplus heat can be stored for low radiation periods,
- the gas-fired auxiliary boiler, which supplies heat during prolonged low radiation periods, and
- the heating system, which distributes the heat in the dwelling house.

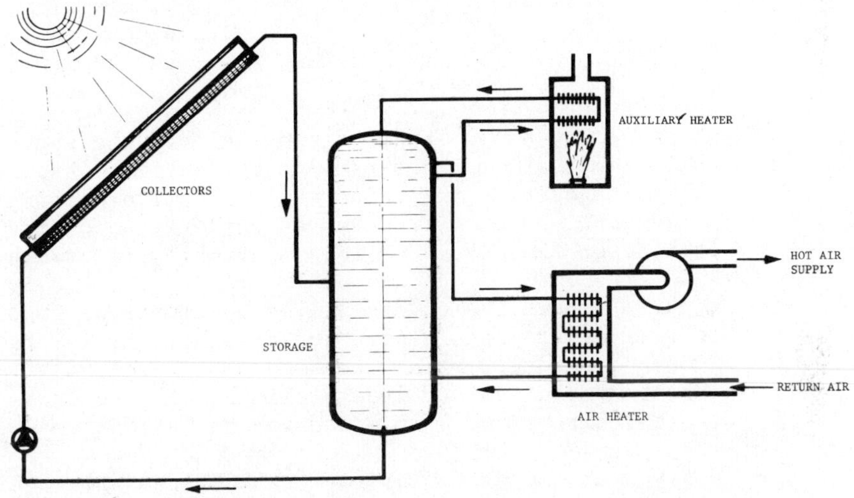

Fig. 2
Simplified scheme of the solar heating system

The solar collectors

The construction of the collectors is shown in Figure 3. The most important parts of the collector are the absorber plates; for this solar house, these plates are designed as one-piece radial finned aluminium tubes (Fig. 6).

The outside of the finned-tube collector plates is provided with a selective absorber coating. This layer consists of chrome-oxide and has the property of absorbing solar radiation (to about 92%; the collector looks black) and decreases the heat radiation from the plate to the glass cover (plate emittance for heat radiation is less than 15%). The width of the finned-tubes is 150mm. The whole solar-roof surface consists of forty parallel finned-tubes with lengths of 9 metres and 7.5 metres (Fig. 1).

The net collector surface amounts to 51m2. Water is pumped through the tubes from the bottom to the top of the collector and transports the absorbed energy to the watertank.

The pump is controlled automatically by the difference in temperature between sensor elements applied to the finned-tube and to the bottom of the storage tank.

Figs. 3, 4

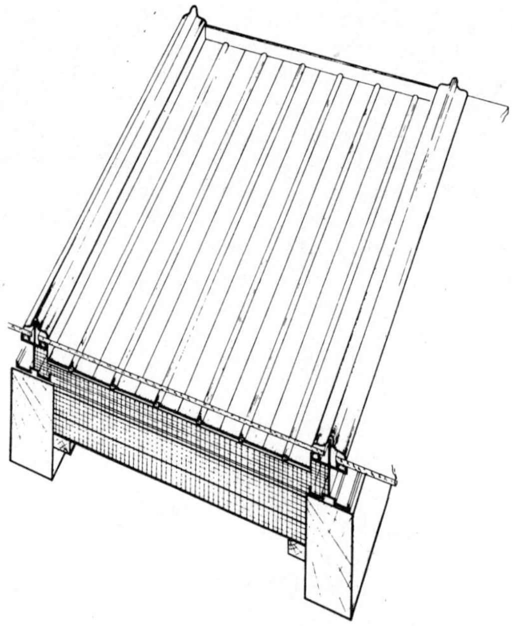

Construction of the collectors on the solar roof.

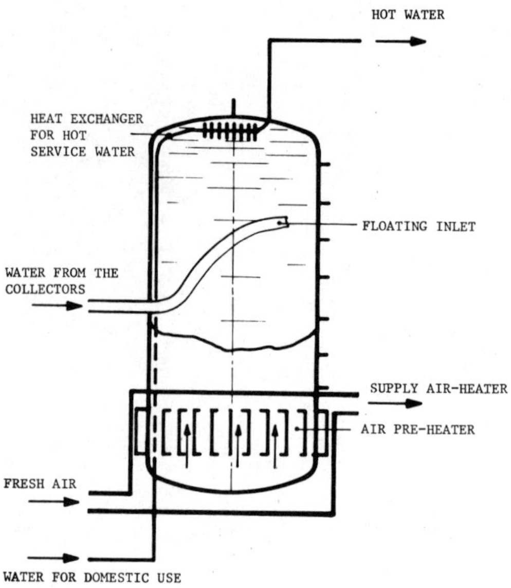

Scheme of the watertank storage with a floating inlet, a heat exchanger for hot service water and an air pre-heater.

The glass cover in front of the finned-tubes serves to prevent the collected heat from escaping to the environment. It also protects the tubes, especially the vulnerable selective coating against weather influences and dust from the atmosphere. According to experience, dust falling on the glass covers, is normally removed by the rain.

Figure 3 shows the covers fitted on aluminium support rails. This type of rail is frequently used for greenhouses. The rails are carried by wooden beams supporting the solar roof. The rails are fixed on the beam in the middle and are freely sliding at both ends to provide room for thermal expansion. This expansion may amount to 2cm.

Noteworthy is the 'solar roof' which does not consist of separate collector boxes with a water supply and exhaust each, but forms an integral part of the roof structure in this solar house. It is to be expected that the cost of a solar roof will be reduced if the roof is designed as an integral whole instead of a large number of relatively small collectors assembled on the roof.

Aluminium protecting sheets are mounted behind the finned-tubes, and behind these sheets there is an insulation 12cm thick. The insulation prevents nuisance to the occupants, because the tubes reach high temperatures (sometimes over 150°C). For further technical details concerning this solar roof we refer to (2).

The watertank storage

The storage of temporarily surplus heat takes place in a watertank of 4m3 content. This watertank is shown in Figure 4. In order to utilize the heat as well as possible, special attention has been given to the design of the storage tank, and particularly to the thermal stratification of the water.

Thermal stratification means the highest temperature at the top and the lowest at the bottom of the tank. The temperature increasing gradually from the bottom to the top.

The water from the collector has an alternating temperature because of the fluctuating intensity of the solar radiation. The water coming from the collectors is always supplied to the level in the tank where the temperature is equal to that of the water from the collectors by the use of a floating inlet (a flexible thin walled plastic hose). The temperatures and the use to which the water can be put at these temperatures are preserved in this way. The floating inlet makes 5% more energy available to the whole solar heating system. Besides there is a possibility of reducing the dimensions and costs of some parts of the solar heating system.

The hotwater supply of the dwelling can be provided with a coil shown in Figure 4. The water that is to be heated flows through the coil from the bottom to the top and reaches a temperature close to the top tank temperature. The auxiliary heater does not allow the latter temperature to drop below 50°C.

A casing is mounted around the lower part of the watertank; fresh air for the dwelling flows through it. The air is heated in it and the water in the bottom is cooled simultaneously. The temperature of the water flowing to the collectors is kept as low as possible in this way, which increases the efficiency of the collectors, see (11). To improve the heatflow from the tank to the air, thin fins are fixed on the bottom part of the tank. This yields an additional 5% of energy for the whole solar heating system.

The capacity of the heat storage is enough to heat the house for two days with average winter weather. It would be technically possible to use a larger watertank with a greater storage capacity. Particularly in the transition seasons the output of the solar heating system would be increased; however the extra cost of greater storage (about Dfl. 500,-- for each m3 extra) does not bring a commensurate increase in heat output.

The auxiliary heater

The auxiliary heater does not allow the water temperature at the top of the tank to drop below 50°C. Figure 5 shows how the auxiliary heater is connected to the storage tank.

The temperature at the top of the tank is controlled by means of a thermostat fitted 0.5m from the top of the tank, which switches the auxiliary heater (a normal gas-fired boiler) on or off at switching temperatures of 50° and 60°C respectively. This hysteresis is necessary to give the boiler longer burning periods (about 3 min.). The start and stop losses can be reduced in this way.

Fig. 5

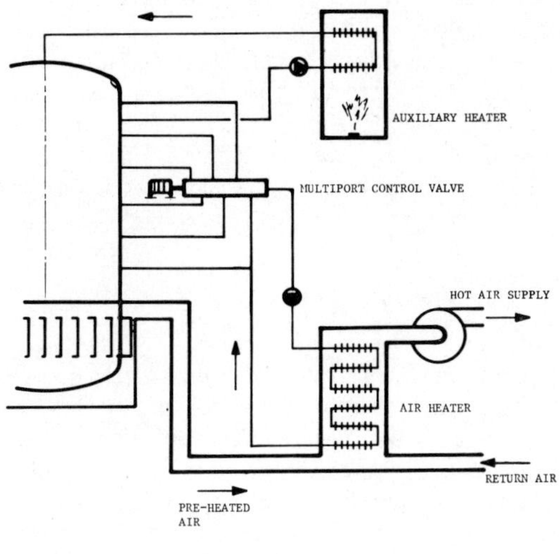

Auxiliary heater, heat supply system and air heater (schematically).

Frost protection

With frosty weather the water in the finned-tubes will freeze in dark cloudy weather and by night. Precautions must be taken to prevent the finned-tubes from being damaged by the expansion caused by the freezing. The most common methods of frost protection are:
- draining the collector,
- adding glycol to the water in the collector.

The first method could not be easily applied because the lowest part of the collector is lower than the top of the storage tank. The disadvantage of the second method is the necessity of a partition heat exchanger, which involves extra costs and decreases the heat gain of the solar heating system (with 5 to 10%). Therefore a completely new method of frost protection is applied, which is shown in Figure 6.

Fig. 6
Frost protection of the finned-tubes.

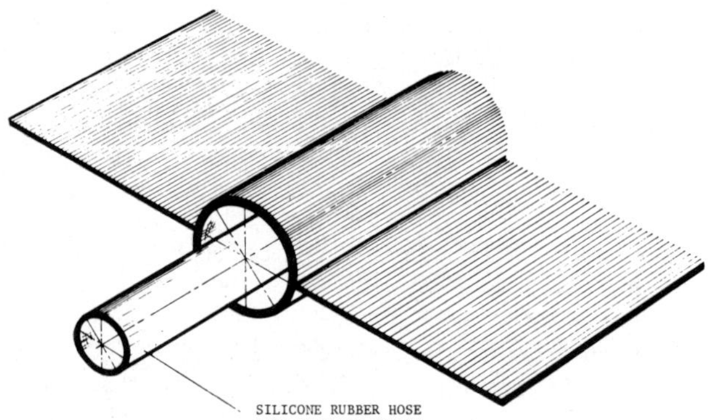

A silicon rubber hose, filled with glycol, is placed inside the finned-tubes. The hoses are all connected with a small expansion vessel outside the water system. The hose inside the tube is squeezed flat when the water freezes and the glycol flows to the expansion vessel. The experience gained with this system in the winter 76/77 was positive, and no damage to or leakage from the finned-tubes did occur.

The described frost protection system is not a universal solution to the freezing problem. It is only applicable in the finned-tubes or in a similar type of absorbing plate, and involves rather much mounting work. In the given situation it was the most suitable solution. General application of this system requires further development.

Thermal insulation

It is obvious that all the pipes of the solar heating system and the watertank containing hot water are well insulated. The insulation of the watertank consists of rockwool with a thickness of 10cm. The suspension of the pipes and the support of the watertank are specially constructed to diminish heat leakage via the connection elements. To reduce these losses the number of suspension points is reduced as much as possible and insulation is fixed between pipes and connection elements.

The heat supply system

A conventional hot air system distributes the warm air over the rooms in the dwelling. An unusual feature is the design of the airducts to prevent air leakage, which adds to the accuracy of the measurements. Airheating was preferred to radiatorheating or floorheating, because the dwelling is made draught-free. This building method was essential to reduce the total energy demand of the dwelling; however it may make the ventilation inadequate in a calm wind. So for both hygienic and safety reasons it is very desirable to apply forced ventilation in a draught-free house. An airheating system is readily combined with forced ventilation.

The air distribution system of a solar heating system must function at the lowest possible temperature level (11). The average of the supply and exhaust heating medium temperature being the decisive quantity. In this solar house heating system this average temperature is about 30°C, as the air supply temperature does not normally exceed 40°C. The number of revolutions of the fan supplying the air to the dwelling can be increased when the outside temperature falls below zero.

With a hot air system it is possible to respond to rapid fluctuations in the heat demand of the house by controlling the air supply temperature. This also was a reason to prefer airheating. It is generally known that well-insulated houses respond acutely to e.q. alternating insolation.

The heat from the watertank storage is transferred to the air in an air-water heat exchanger. Figure 5 shows how this transfer occurs. The water supply temperature from the watertank is restricted to the lowest temperature necessary for maintaining the room temperature. The control system makes use of the thermal stratification of the watertank storage.

Five supply pipes for the airheater are connected to the watertank at various levels. Further the pipes are connected to a control cylinder. Any difference between the room temperature and the set point causes the servomotor to turn the control cylinder to a higher or lower outlet port in the watertank. The adjustment of the control cylinder stops when the difference between setpoint and room temperature equals zero.

The performance of this control system in the winter 76/77 has been excellent; the room temperature was almost constant and furthermore the use of heat from the watertank was very efficient. For further detailed information about this system see (4), (5) and (6).

THE HOUSE

The house is fully adapted to the solar heating and the measuring system. This in order to make clear observations and a correct judgement about the potential of solar heating possible. An experimental solar house differs from a conventional house in the following respects:

- Provisions must be made to accommodate the piping and the measuring cables and space must be provided for the watertank storage and measuring instruments;
- The house must be designed and constructed with an open mind towards energy conservation; i.e. it has to be built draught-free and well insulated in order to approximate the ratio between the heat gain of the solar installation and the heat demand of future houses as close as possible;
- The solar roof must be integrated in the architectural design without loss of functional use of the inside space; the existing building codes have to be observed as well.

The first-mentioned provisions are described in the previous section. Additionally it is worth noting that the beams are made out of laminated wood to prevent deformations by the relatively high temperatures caused by the collector. On the ground floor a floorspace of about 7m2 and on the first floor an attic space of about 18m2 is reserved for the watertank storage and the measuring units. More than 50% of the attic area serves the measuring units and can be saved in the future solar heating systems.

Fig. 7

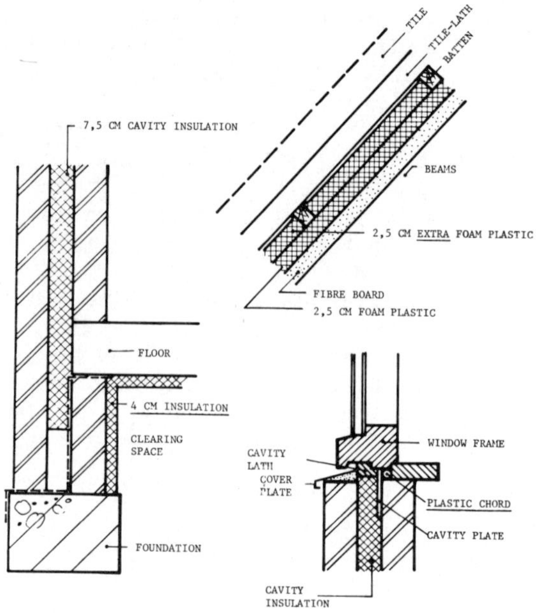

Some architectural provisions to reduce the heat demand.

Figs. 8a, 8b

South-west view of the Solar House

North-west view of the Solar House

Particular attention has been given in the design and construction to good thermal insulation; the reasons for this have already been outlined before. To reduce the heat transmission through the outside walls, the width of the cavities has been increased from the usual 5cm to 8cm and glass-wool with a thickness of 7.5cm has been inserted. The heat transmission coefficient obtained in this way equals 0.4 Watt/m2 ^{o}C *.

The roof surfaces also have a heat transmission coefficient of 0.4 W/m2 ^{o}C; specially manufactured roof boards are used for this purpose (see Figure 7). The underside of the groundfloor is insulated with 4cm of glass-wool (see Figure 4). It is self-evident that double pane windows are used throughout the house.

The window areas are relatively small and mainly concentrated on the south face, as can be seen in Figures 1 and 8a, b. The window area on the west, east and north face is restricted to the area required for inside space illumination.

In this context it may be interesting to note that in the Dutch climate double pane windows on the south face absorb more light energy during

* For comparison: The recently improved building codes require a heat transmission coefficient of 0.78 W/m2 ^{o}C; the value for the conventional Dutch cavity wall is 1.4 W/m2 ^{o}C.

the heating season than the thermal energy losses effected by them. From an energy point of view it is therefore recommendable to install spacious (double pane) windows on the south face.

As remarked before, much attention has been given to a draught-free construction of the house. For this purpose a specially designed draught-free construction is applied to all the window-frames. The construction is shown in Figure 7. All doors and other moving elements in the outer walls are provided with anti-draught strips. The connection between roof and brickwork has been filled with draught-tight insulation material. It is intended to check all these provisions to prevent draught by means of measurements.

According to calculations the provisions mentioned above reduce the heat demand of the house to 11 kW at an outside temperature of -10°C and an inside temperature of 20°C. This is a very low demand compared with conventional Dutch standards, especially for a house with a volume of about 800m3 (living area 220m2).

The integration of the solar roof in the architectural design of the solar house was the third aspect mentioned in the beginning of this section. Since the solar house was the first one in The Netherlands with an integral design, an entirely new architectural concept has been developed. The dimensions, inclination, material and the function of the solar roof together with the conventional requirements as to residential design form the elements of the new synthesis. The quality of this synthesis is subject to personal judgement in the last resort.

RESEARCH PROGRAMME AND MEASURING SYSTEM

For the next few years the research will be concentrated on the performance of the solar heating system. As this performance is related to the heat demand of the house, this demand will also be measured. However, these measurements will be of a general character. A number of measurements will be carried out in order to check the insulation. The measuring points in the solar heating system and the house total up to 50. In this paragraph the measuring system is described further, starting from the objectives of the research programme. Further details are given in (8).

The most important meteorological input datum for the performance of the solar heating system is the intensity of the solar radiation. This is measured with two solarimeters (see Figure 9a), both mounted in the plane of the solar roof surface and one of them behind a glass cover, in order to check the transparancy thereof. Since the solarimeters must be cleaned frequently they have been mounted within easy reach from the ground.

Figs. 9a, 9b

Solarimeter

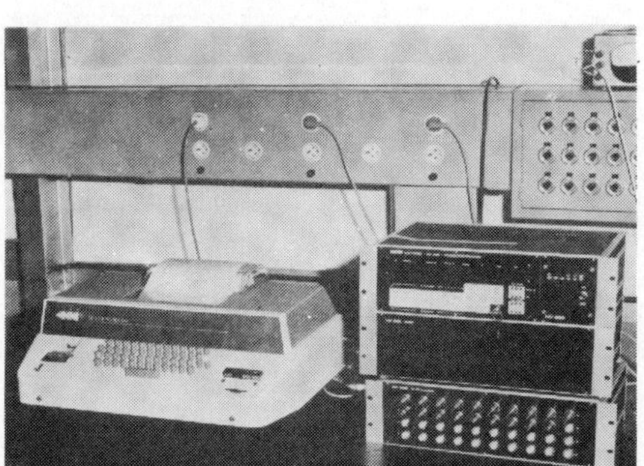

Micro-processor

A thermo sensor (PT 100) is suitably affixed outside the dwelling for measuring the ambient temperature. The velocity and the direction of the wind are measured by means of an anemometer, placed 1.5m above the highest point of the roof.

The heat gain of the collector is calculated from the difference between inlet and exhaust medium temperatures and the amount of water flowing through the collector.

The pump of the collector-circuit has a constant flow and the amount of flowing water can be calculated from the duty times of the pump. Six PT 100 temperature sensors are fixed on the finned-tubes in order to check the performance of the collector and the theoretical models. In the watertank storage, the inlet and outlet flow and the temperatures at eight different levels are measured. The latter to follow the thermal stratification changes in time.

The temperature rise of the fresh air when flowing along the bottom of the watertank storage is measured as well.

Because a large amount of service water flows incidentally through the heat exchanger in the watertank storage, it will be necessary to measure continuously the amount of hot service water, and its temperature.

The duty time, inlet and outlet temperatures of the auxiliary heater are continuously measured to calculate the energy supply. The gas consumption of the gas-fired boiler is noted daily to check the last calculations.

The amount of heat supplied with the heating system to the dwelling is proportioned to the amount of water flowing through the air-water heat exchanger and the water temperature difference between the inlet and outlet of the air-water heat exchanger. This data is all gathered for the further required calculations.

The heating for the house is partially obtained from the electricity and gas used in lighting and cooling, etc. It is necessary to estimate this amount of heat. The gas and electricity consumption therefore is noted daily.

In addition the duty time of the exhaust ventilation of the kitchen is noted since the associated heat loss from the house may be considerable.

Since the climatic conditions in The Netherlands are subject to rapid change, a large part of the foregoing measurements must be frequently repeated. Otherwise it will be impossible to follow the dynamic behaviour of the solar heating system.

The measuring system produced 72.000 measuring data every day when 50 measuring points have to be read per minute. It is impossible to work this data all out by hand and impracticable to punch it on tape.

Therefore a micro-computer has been applied to reduce this data to practical limits and to make simultaneous heat balance calculations in addition.

This micro-computer gives a read-out every half hour on a teletype and is connected by telephone to the university. The teletype prints the main data and also gives punched-tape information.

The object is to follow carefully the behaviour of the solar heating system for three heating seasons. The solar heating system can be analysed with this measuring system for all weather conditions to be expected in this country. The following weather conditions are for example possible: glazed frost, hoar frost, snow, rain, strong wind from various directions, frost, etc. A reliable basis for the design and development of solar house heating systems can be obtained in this way.

An additional important research objective is the behaviour of the system in practice (reliability, corrosion, lifetime, etc.), which demands continuous observation.

THE PERFORMANCE OF THE SOLAR HEATING SYSTEM

The performance of the solar heating system is presented in this section, as deduced from calculations and experiments. The results are preliminiary as further experiments are still going on to determine the performance of all the units involved accurately. However the properties of solar heating systems have become well enough known by now to allow a good first impression to be obtained from the given results.

The calculations are based on hourly meteorological data of the KNMI De Bilt. This data was already available at the start of this research project because there already was a research programme at the EUT into the influences of insolation, ambient temperature, wind velocity, etc. on the climate in houses and buildings. The average meteorological readings covering the last ten years were available on tape. In addition a computer programme which had been adapted to calculate the yield of the solar house heating system and the heat needed for this house was also available. The total amount of heat over a month, heating season or whole year, could thus be determined.

Not all the influences are covered by this programme, the calculations would in that case have been too time-consuming. For instance the reflection of radiation from the ground, small heat losses that can occur in parts of the solar heating system, shadow thrown by other buildings and trees in the neighbourhood of the house; effect of snow on the solar roof, melting of hoar frost in the morning, etc. are not included. Moreover a number of other influences are described by rough estimates; such influences are the heat production from the use of electricity, human bodies, cooking, the heat losses by natural ventilation, the heat needed for the domestic hot water.

Approximations and omissions will cause some differences between calculations and measurements. It appears from comparison that these differences remain within acceptable limits.

Figure 10 shows the calculated performance of the solar heating system over a year in a cold winter and again in a mild winter. For the cold winter the winter of 62/63 was taken and 67/68 for the mild winter. The non-shaded surfaces in Figure 10 show the amount of heat delivered by the solar heating system. The shaded surface in the bell-shaped diagram is the amount of heat delivered by the auxiliary unit.

The bell-shaped diagram itself gives the required amount of heat for the dwelling including the hot water for domestic use and corrected for the amount of heat from electricity, cooking, human bodies and insolation via the windows. In (9) more detailed information is given about the proportions of these influences.

It is to be expected that the heat needed for the house in a cold winter is greater than in a mild one as Figure 10 shows. The performance of the solar heating system is, however, greater in a cold winter, because there is a greater number of insolation house. The results of the calculations give the amount of heat needed for the house as 23.400 kWh for the cold winter of which the solar heating yields 10.700 kWh.

Fig. 10

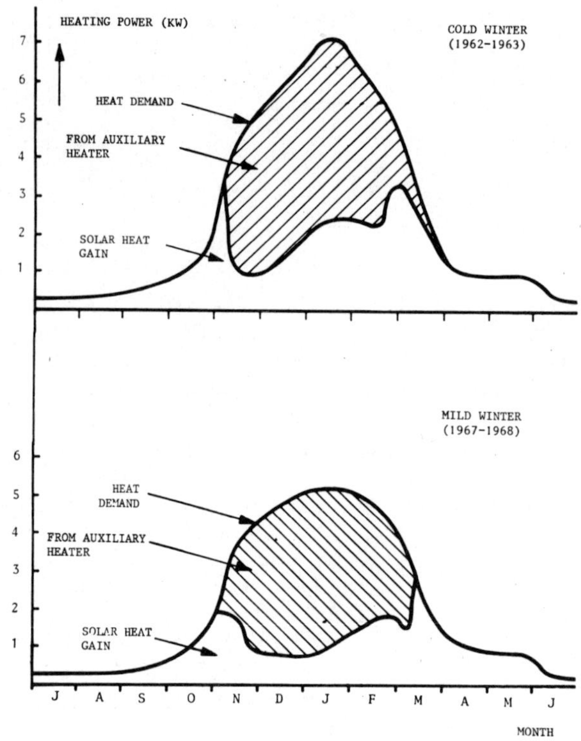

Calculated performance of the solar heating system in a year with a cold and a year with a mild winter.

In the year with the mild winter these results are respectively 19.500 kWh and 8.750 kWh. The saving in natural gas in both years is considerable, namely 53%; 7% of savings are achieved by switching off the auxiliary unit during the long summertime. The idle time losses of a gas-fired boiler for hot water can be eliminated in this way. Calculated for an average year, the latter savings amount to about 2.000m3 of natural gas.

The curves of Figure 10 that show the heat needed for the house are smooth. This is due to the fact that the curves are plotted on the monthly averages. Actually, however, there are day-to-day variations in the heat requirements for the house and in the heat delivered by the solar heating system. These variations are shown in Figure 11. This diagram shows the heat needed for the house and the heat delivered by the auxiliary unit from day-to-day and the heat content of the watertank

Fig. 11
Day to day heat economy of the Solar House (Calculated for the Dutch reference year).

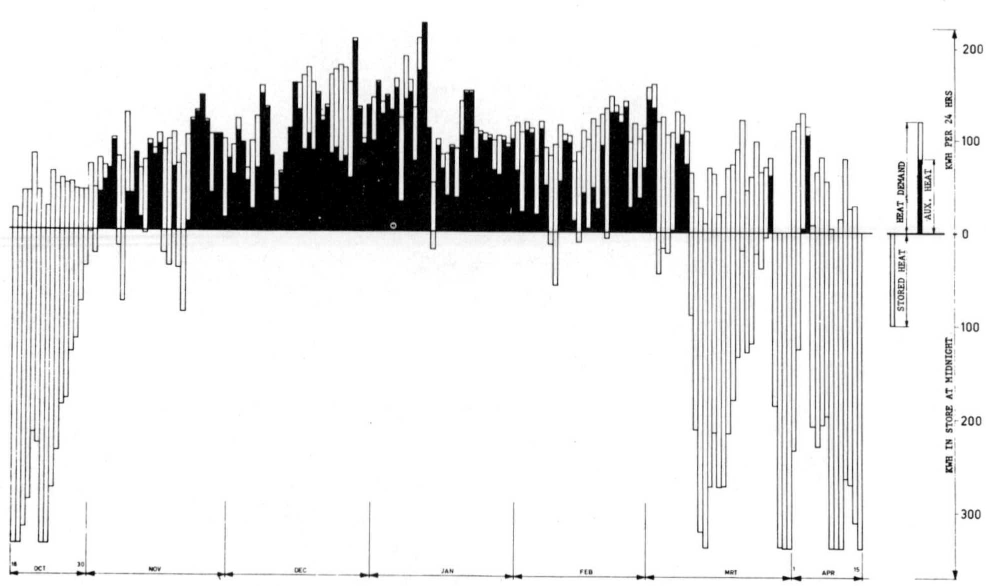

storage at the end of the day. This is calculated on hourly data for the period from mid-October to mid-April of an 'average year' worked out by the EUT (10).

In addition to the great variation from day-to-day, Figure 11 also shows the heat content of the watertank storage dropping to zero at the beginning of November. In November the winter situation gradually sets in and here the auxiliary heater has the larger part to play and a heat surplus is only incidentally supplied to the watertank storage. From mid-February to the end of March there is a period of heat surplus giving a full watertank storage and then the auxiliary heater is hardly even necessary.

REFERENCES

(1) C.W.J. van Koppen, Toepassingsmogelijkheden van zonne-energie voor huisverwarming en warmwatervoorziening onder Nederlandse omstandigheden (The feasibility of solar heating in The Netherlands) Eindhoven University of Technology, Dept. Mech. Eng., Report WPS3-73.08.R183. See also: Klimaatbeheersing, 3, nr. 5 (May 1974), pp 194-203.

(2) J.T.T. van Wolde, De constructie en de fysische eigenschappen van de collectoren van het zonnehuis van de Technische Hogeschool Eindhoven (The design and the physical characteristics of the collectors of the solar house of the Eindhoven University of Technology), Eindhoven University of Technology, Dept. Mech. Eng., Report WPS3-76.11.R259 (in preparation).

(3) J.T.T. van Wolde, De thermisch gelaagde warmteopslag van het zonnehuis van de Technische Hogeschool Eindhoven (The thermally stratified heat storage in the solar house of the Eindhoven University of Technology), Eindhoven University of Technology, Dept. Mech. Eng., Report WPS3-76.11.R260 (in preparation).

(4) J.T.T. van Wolde, De uitvoering en de regeltechnische eigenschappen van het warmte-distributiesysteem van het zonnehuis van de Technische Hogeschool Eindhoven (The design and dynamic characteristics of the heat distribution system in the solar house of the Eindhoven University of Technology) Eindhoven University of Technology, Dept. Mech. Eng., Report WPS3-76.11.R261 (in preparation).

(5) J.T.T. van Wolde, De optimalisatie en regeling van het verbruikscircuit van een zonneverwarmingsinstallatie (The optimalization and control of the heating circuit of a solar heating system), Eindhoven University of Technology, Dept. Mech. Eng., Report WPS3-76.11.R262 (in preparation).

(6) P.H.C. Voorter, Ontwerp en berekening van een lineaire mengkraan (Calculation and design of a linear miltiport mixing valve), Eindhoven University of Technology, Dept. Mech. Eng., Section of Heat Technology, 4th grade exercise.

(7) J. Hamaker en R.C. van de Bruggen, Eindhoven University of Technology, Dept. Architect., Information from current research.

(8) J.T.T. van Wolde, Het meetsysteem van het zonnehuis van de Technische Hogeschool Eindhoven (The measuring system in the solar house of the Eindhoven University of Technology), Eindhoven University of Technology, Dept. Mech. Eng., Report WPS3-76.11.R263 (in preparation).

(9) R.J.A. van Bruggen, Digitale berekening van de dynamische warmtehuishousing van gebouwen (Digital calculation of the dynamic heat economy of buildings), Verwarming en Ventilatie, 33, nr. 6 (June 1976), pp 335-346.

(10) A.A.J. v.d. Hoogen, Referentiejaar voor Nederland (A reference meteorological year for The Netherlands) Verwarming en Ventilatie, 33. nr. 2 (February 1976), pp 81-83.

(11) C.H. Hoogendoorn en C. den Ouden, Het rendement van zonnewarmtecollectoren bij Nederlandse meteorologische condities (The efficiency of solar collectors in the Dutch climate) Klimaatbeheersing, 4, nr. 4 (April 1975), pp 172-175.

Solar heating of a sports centre in Belgium

A. Grosfils, G. Descy, M. van Gysel, Ch. Heusquin
Engineering office ATRAC
Brussels, Belgium

Abstract

The sports centre which is to be built at Chevetogne (Province of Namur-Belgium) will be heated by a thermal system based on solar energy with use of heat pumps. It will include 2,000m2 of solar collectors. The present communication describes this application combining solar energy, heat pumps and electricity for heating of buildings, shower water and swimming pools.

1 GENERAL DESCRIPTION

1.1 Introduction

The engineering office ATRAC has been assigned by the Province of Namur to study the heating with help of solar energy of the sports centre of the 'Domaine de Chevetogne'. The objective of the present study is to determine a general concept and an estimate of energy and financial balance of the chosen system.

1.2 Description of the sports centre and its operation

The sports centre includes mainly:
- an olympic swimming pool, a diving pool and children's pools.

These outdoor swimming pools have a free surface of 1740m2 for a volume of 2400m3.
- The 'north' building includes:
  an indoor swimming pool (167m2)
  a sports hall
  two bars
  changing rooms for swimming pool, sports hall and outdoor sports
  a congress hall
  housing (not heated by the solar heating system).

- The 'south' building includes:
  restaurants and kitchen
  bars
  shops.

Each of these elements is considered as a thermal load.

We have also three boilers to heat water for:
- showers of the changing rooms of the north building,
- kitchen of the south building,
- showers of the outdoor swimming pool.

The specificity of a solar thermal system derives mainly from two elements:
- the input is not continuous,
- the input is not function of the needs.

As a consequence the usual dimensioning methods are inapplicable. It is therefore necessary to simulate the thermal loads under actual meteorological data in order to determine the instantaneous power balance. For this detailed simulation, we needed to specify the utilisation factor for each thermal load. We also distinguished two seasons called 'summer' and 'winter'.

Fig. 1
centre sportif
de Chevetogne:
Implantation

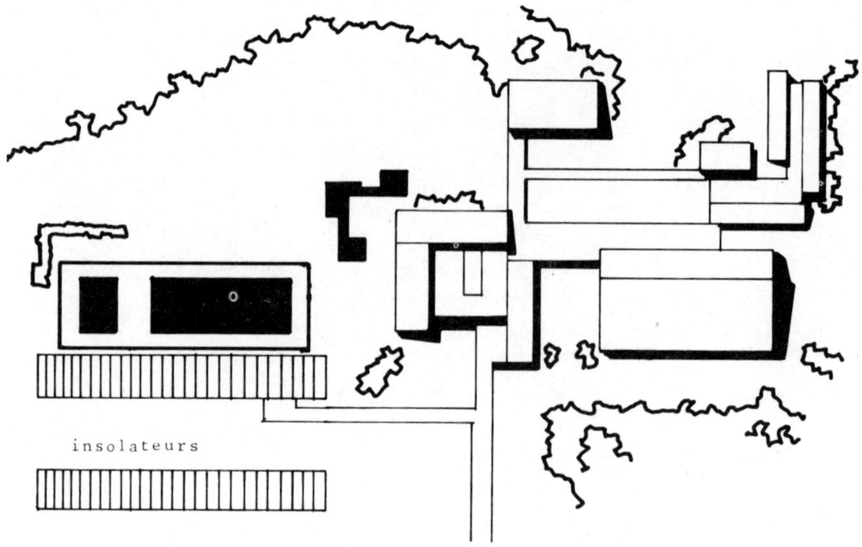

Fig. 2
centre sportif
de Chevetogne:
Schema de
principe

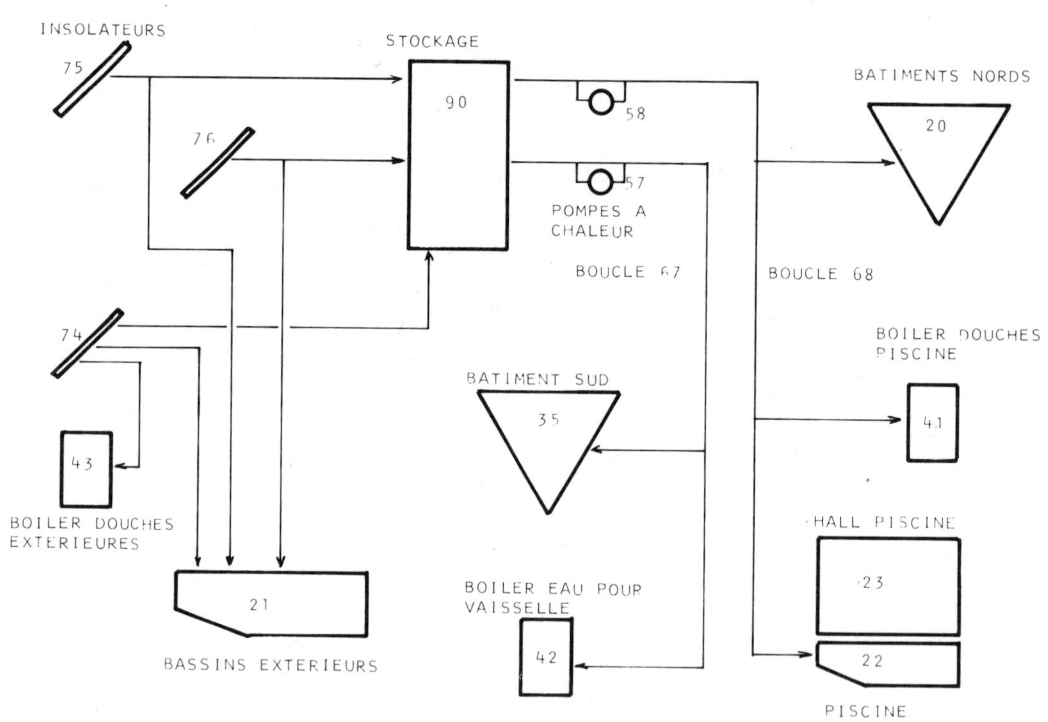

1.3 Solar thermal system description

All the solar collectors surfaces have been grouped south of the sports centre for architectural reasons.

We can distinguish two zones:

- one set of collectors heats in priority the boiler for outdoor showers and then the swimming pool or the heat storage.
- one set of collectors heats the outdoor swimming pool or the storage according to the season.

Heat pumps are introduced in the system to upgrade the energy at low temperature.

The other energy input will come from electricity as fuel oil and gas have been excluded from the project. All the thermal loads (except the outdoor swimming pool and its showers) are fed from the termal storage 90, either directly, or through a heat pump when the thermal level is insufficient.

The collectors operation varies with the season. In summer, they supply prioritively the outdoor swimming pool 21 and if the thermal needs of the swimming pool are satisfied, the collectors heat the storage 90.

We specified heat pumps between air and swimming pool water for the drying of the air of the swimming pool hall and the recovery of heat from drying.

2 MODEL AND SIMULATION

It has been necessary to build a mathematical model of each thermal load and input for a digital computer simulation. Such a simulation is only possible if complete meteorological data is available. In Belgium, we are in a lucky position to have a Meteorological Institute (IRM) that has done a very good and complete job for years.

We have, on magnetic tapes, the complete hourly or half-hourly data for temperature and radiation for 19 years (1958 to 1976) measured in Brussels. It is not very easy nor cheap to perform the simulations over 19 years. Therefore we have reduced this data by statistical methods to three 'standard-years': a good, a middle and a bad year. Each standard-year is composed by real months chosen on base of the temperature and radiation averages and standard deviations of this data.

For this application and others, we have built a complete programme for the simulation of thermal systems including solar energy, heat pumps, and auxiliary. The structure of this programme is fully modulerised. This means that it is based on a core programme that does the general management, input, output (of hourly, monthly and yearly data), etc... On this core programme we can add some sets of water or air collectors with different orientations, tilts, performances. We can also add storages subroutines describing sensible or latent heat storage types.

The different thermal loads already simulated are:
- outdoor swimming pool including the different losses and gains. The outdoor swimming pool can be covered by night or by non-swimming days to reduce the losses (with choice of night/day hours or swimming conditions).
- indoor swimming pool.
- water boilers.
- sport or swimming hall with complete balance of energy and enthalpy. (including drying heat pumps).
- building with different types of use.

The subroutines simulating the buildings take into account the gain from the sun through the windows and the thermal inertia of the building.

All the models follow utilization factors that change the nominal values of temperature, hygrometry. The utilization factors are subordinate to the season (summer/winter), the hour, the weather (outdoor swimming conditions).

We can also add to the programme subroutines of auxiliary heating or heat pumps. The heat pumps can take the energy on outside air or upgrade energy at low temperature level (water/water heat pump).

This programme called SOLSYSTEM allows us to simulate every system, from a simple swimming pool, to the sports centre of Chevetogne or a solar heated house.

Fig. 3

BILAN ANNUEL

ELEMENT	SOLAIRE KW.H	FRIG PAC KW.H	TOTAL SF KW.H	EL N EXC KW.H	EL N BIH KW.H	EL JOUR KW.H	TOTAL EL KW.H	APP TOT KW.H
BASSIN								
COL74	119154.	0.	119154.	0.	0.	0.	0.	119154.
COL75	320387.	0.	320387.	0.	0.	0.	0.	320387.
COL76	474302.	0.	474302.	0.	0.	0.	0.	474302.
PAC59	0.	0.	0.	0.	0.	0.	0.	0.
TOTAUX	913843.	0.	913843.	0.	0.	0.	0.	913843.
STOCKAGE								
COL74	18723.	0.	18723.	0.	0.	0.	0.	18723.
COL75	57377.	0.	57377.	0.	0.	0.	0.	57377.
COL76	122540.	0.	122540.	0.	0.	0.	0.	122540.
PAC59	0.	0.	0.	0.	0.	0.	0.	0.
BCL67	-41186.	0.	-41186.	0.	0.	0.	0.	-41186.
BCL68	-151647.	0.	-151647.	0.	0.	0.	0.	-151647.
TOTAUX	5806.	0.	5806.	0.	0.	0.	0.	5806.
BOILER 43								
COL74	34537.	0.	34537.	0.	0.	0.	0.	34537.
APPNT	0.	0.	0.	0.	0.	1513.	1513.	1513.
TOTAUX	34537.	0.	34537.	0.	0.	1513.	1513.	36051.
BOILER 41								
STOCK	12075.	0.	12075.	0.	0.	0.	0.	12075.
PAC58	0.	45102.	45102.	0.	4457.	10623.	15081.	60183.
APPNT	0.	0.	0.	0.	0.	116999.	116999.	116999.
TOTAUX	12075.	45102.	57177.	0.	4457.	127623.	132080.	189257.
BOILER 42								
STOCK	2836.	0.	2836.	0.	0.	0.	0.	2836.
PAC57	0.	4679.	4679.	0.	377.	850.	1227.	5906.
APPNT	0.	0.	0.	0.	0.	26368.	26368.	26368.
TOTAUX	2836.	4679.	7515.	0.	377.	27218.	27595.	35110.
BATIMENT NORD								
STOCK	11287.	0.	11287.	0.	0.	0.	0.	11287.
PAC58	0.	55126.	55126.	0.	4262.	15143.	19405.	74531.
APPNT	0.	0.	0.	187364.	0.	0.	187364.	187364.
TOTAUX	11287.	55126.	66413.	187364.	4262.	15143.	206769.	273182.
BATIMENT SUD								
STOCK	19621.	0.	19621.	0.	0.	0.	0.	19621.
PAC57	0.	14050.	14050.	0.	707.	3155.	3862.	17912.
APPNT	0.	0.	0.	94653.	0.	0.	94653.	94653.
TOTAUX	19621.	14050.	33672.	94653.	707.	3155.	98515.	132187.
PISCINE								
STOCK	4060.	0.	4060.	0.	0.	0.	0.	4060.
PAC58	0.	23996.	23996.	0.	3475.	4279.	7754.	31750.
APEAJ	0.	0.	0.	0.	109647.	75146.	184793.	184793.
APAIR	0.	0.	0.	0.	0.	1872.	1872.	1872.
PAC51	0.	0.	0.	0.	3972.	2172.	6144.	6144.
PAC52	0.	0.	0.	0.	13569.	27409.	40978.	40978.
PAC53	0.	0.	0.	0.	0.	12948.	12948.	12948.
TOTAUX	4060.	23996.	28056.	0.	130663.	123825.	254488.	282545.
TOTAUX	1004066.	142954.	1147019.	282017.	140466.	298478.	720961.	1867981.

3 RESULTS

Thanks to the SOLSYSTEM programme, we could have the evolutions of the energy balance of the system over one year. The annual energy balance (see table) presents the thermal needs of each load and gives the sources that cover these needs.

In the table 'solaire' means energy coming from the collector, 'Frig. PAC' means energy coming from the cold side of the heat pumps (Pompe A chaleur), 'E1' means electricity with the different tarifications (night exclusive, night, day). As there is, in this simulation, no heat pump on outside air, the 'Total SF' represents the energy coming from the solar collectors. The figures in the table show that the annual thermal needs of 1800MWh is covered for 62% by solar energy and 38% from auxiliary source.

Thanks to the energy consumption balance, we could evaluate the annual consumption costs (about 1.100.000bF/year or 30.500 US $\not{c}$). We also established the investment of the different solutions: solar thermal system or fully electric heating. The fully electric system has been evaluated at 27.000.000bF (860.000 $\not{c}$).

Knowing the investments, we calculated the net annual cost of the two systems including investments and consumption costs. The investment pay off is calculated on 10 or 20 years depending on the equipment. The net annual cost was 4.940.000bF for the initial system and 3.300.000bF for the solar thermal system. This gives an annual saving of about 1.700.000bF (47.000 $\not{c}$).

4 CONCLUSIONS

We can summarise the major results of the simulation of the solar thermal system of the sports centre of Chevetogne in three points:

1. The collected solar energy cover more than 60% of the thermal needs.
2. The solar thermal system leads to an annual saving of about 1.700.000bF (47.000 $\not{c}$) on net cost (consumption cost plus investment pay off).
3. Taking its energy from the environment, this solar thermal system saves, at a global level, more than 300 metric ton Petrol equivalent per year. The realisation of the solar system of Chevetogne is planned for 1978. The outdoor swimming pool will be open to the public before summer 1978 and the rest of the complex will follow during 1978 and 1979.

4.8 Simplified solar reheat for slab perimeter warm and cool air distribution systems

G.M. Clark
Professor and Research Supervisor
Department of Architecture
The Ohio State University, USA

INTRODUCTION

Given the current need for heat energy conservation and solar generation, a dominant criterion for the prioritizing of design strategies for new and retrofit construction is dual use or dual benefit of the construction task chosen. Integration of the components of sub-systems into other sub-systems requires careful design, scheduling and jurisdictional management and is often next to impossible because of these constraints. Certain materials and operations, however, are on the fringes of several trade practices and may possess the potential for further use optimization, through further design and testing. One such area is the proper use of warm air distribution under poured concrete slab-on-fill floor construction in comfort-sensitive occupancies such as residences, schools and churches. Recognized as a significant development in early organized housing research, this system offered the advantage of some gain of radiant heat from the warmed slab due to the "loop" and "spider" distribution patterns of the ductwork, particularly in the building perimeter zone.

PERFORMANCE PROBLEMS

Problems inherent to this relatively inexpensive substitute for joists over crawl space were found to be:

1. Conducted heat loss through exposed slab edges, resulting in excessive waste and discomfort.

   At its worst, this condition allowed such rapid conduction to the winter exterior air temperature that both slab and interior air mass were robbed of heat before it could be usefully transmitted, resulting in extreme discomfort and frost formation in the perimeter zone. Various available insulation materials of the rigid and non-permeable types were employed in the attempt to isolate the slab, ductwork and gravel mass from the foundation wall.

2. Proper engineering and construction of the supporting fill to insure against slab settlement, cracking and disruption of below-slab ductwork.

   Separation of the slab from its foundation support in order to prevent conduction introduced the need for new structural criteria and procedures for the slab support by the compacted gravel alone. While such techniques were familiar enough in highway and road base design, building designers and producers were apparently uncomfortable with this concept and made many mistakes resulting in either slab settlement or disruption of under-slab ducts and pipes. Expensive corrective measures and energy waste were the consequence to the degree that increased sensitivity to the best procedures became known and respected.

3. Choice of duct material for durability and prevention of "floating" during construction.

   Clay or concrete pipe as duct material came into common usage due to their compatibility with surrounding conditions and wall strengths to withstand live loads due to both construction and occupancy. Sheet metal register boxes which tied in to the duct system were specified on the basis of lateral pressures of the concrete during casting, and to the corrosive effect of the concrete.

4. Control of ground moisture migration from sub-strata.

   One function of the gravel fill interstices was to discourage capillary migration of the in-site moisture in the excavated sub-strata which could effectively negate the heat sink effect of the dry gravel mass. In extreme conditions of this type, ground water had penetrated the sub-floor tunnels and ducts to the extent that heat distribution and humidity conditions were disrupted.

Choice, therefore, of supporting fill material was dictated by its non-shrinking load-bearing capacity. For this reason, sub-grade of broken limestone or bank-run gravel was specified and installed with suitably minor compaction.

DUAL-USE STONE SLAB SUPPORT/SOLAR STORAGE

In active solar heated air collector and distribution systems, there is a general need for a heat storage volume containing an appropriate size mass of rock used as a heat sink and containing sufficient interstices for forced passage of the solar-heated air.

This paper proposes selective dual-use of the stone fill support for slab floor construction in suitable occupancy types. New structural constraints are implied by the probable larger aggregate size due to less compaction for air passage, and excessively conducted heat from this mass must be properly insulated from the slab itself.

There is presently some disagreement in this country among scientists and producers concerning the volume of rock storage required for optimum performance of an air system. Calculations which follow deal with these varying recommendations as bounds on a preliminary sensitivity analysis. The proportions of the storage mass are also an important consideration in the dual use concept due to the economic necessity of limiting the depth of supporting gravel fill for the slab. Depth may be varied, of course, from place to place under the floor depending on the supply and return duct locations in the plan. Another consideration in depth control is the air flow through the sub-slab stone mass to insure control of excessive temperature stratification.

The system hypothesized allows conductive heat transfer from the fill to, particularly, the loop ducts at the most remote locations from the bonnet and provides, therefore, terminal reheating of a "booster" nature. This system potentially simplifies the thermostatic control mechanism by continuing the sensitivity of the zone or central thermostats to the rebalanced closed loop heat supply from the conventional air distribution system. It also eliminates passage of any of the heated "living" air volume - as filtered and humidified on its route through the furnace - through the stone storage mass, thus eliminating dust or moisture pickup from that location.

The sophistication of the construction may vary with the need for the cooling mode in particular cases. Low cost housing in climates requiring heat may employ, for instance, a simple vertical supply and return duct connecting collector to sub-slab stone mass through which perimeter duct loops pass, in turn, connecting fireplace or basic furnace. The radiant effect from the heated slab may be generated by forced warm air through the ducts, or by the solar generated fill-storage combination, or both, depending on conditions.

Accompanying plans and detail illustrations show elements of the simple system just described. This application is a self-contained new room addition to an existing residence, and illustrates low-cost and low-energy use adjustments which can be made for improved selective use of the residence during the most severe winter conditions. Continuous or manually switched blower operation are implied in this mode.

SELECTED EXAMPLE

Given a house in Columbus, Ohio, calculate the south facing collector area, and approximate duct and storage sizes for a room addition measuring 12' x 14', which will be designed to accommodate a family of four during 0°F weather. It is desired that the new solar equipment will provide the necessary heat for the addition, which constitutes 7% of the volume - therefore 7% of the fossil fuel needed for heating - of the enlarged residence.

Using established methodology (and some extrapolation) for determining collector area for this location, a conversion factor of 333 BTUH/FT^2 OF COLLECTOR is employed.

The calculated heat load of the enlarged (and reinsulated) residence is 40,000 BTU/Hour

$$\text{Then: } \frac{40{,}000 \text{ BTUH}}{333\text{BTUH/FT}^2} = 120 \text{ SF}$$

Fig. 1

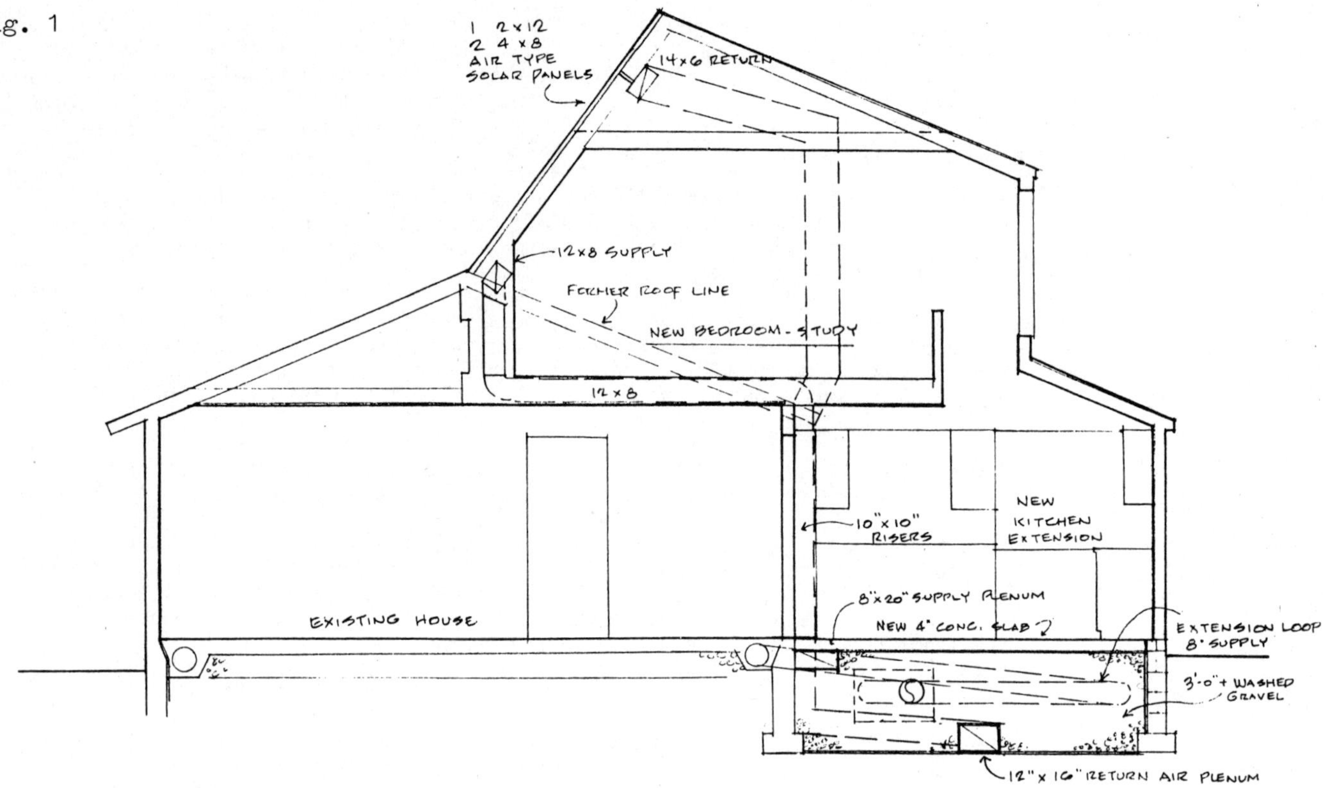

SECTION W
1/4" = 1'-0"

Fig. 2

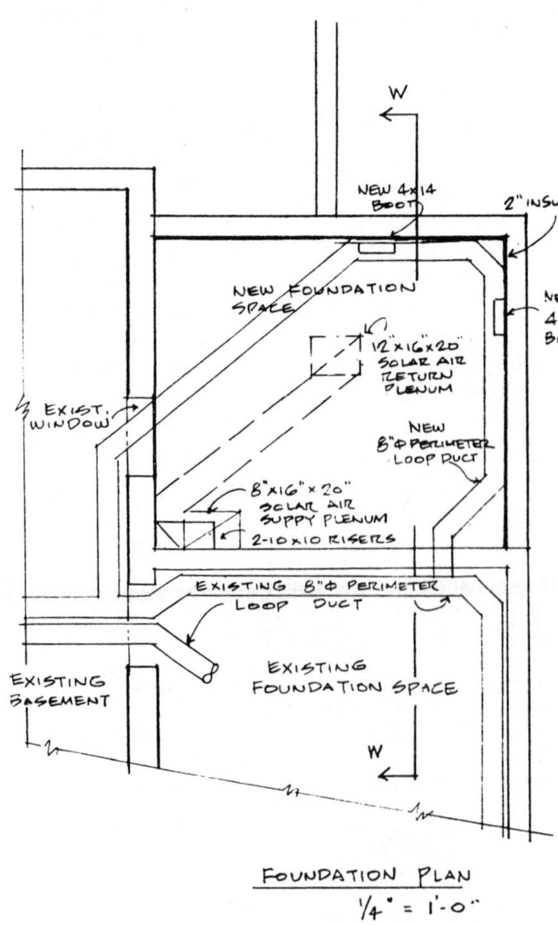

FOUNDATION PLAN
1/4" = 1'-0"

In the selected collector design, the standard flow rate across the collector is 2 CFM per ft.^{2} of collector. And the flow rate required is 240 CFM.

The pressure drop through the 4 panels in series is .34 W.G.

The sizing criteria for the rock bed heat storage varies from 1/2 ft.^{3} for 1 ft.^{2} collector to 1 ft.^{3} for 1 ft.^{2} collector. Therefore, the project will require

$$\left(\frac{1/2\ \text{ft.}^3}{\text{ft.}^2}\right)\left(120\right) \quad = \quad 60\ \text{ft.}^3 \text{ to } 120\ \text{ft.}^3 \text{ of storage.}$$

In the design, approximately 100 sq. ft. of new area within the foundation is indicated, due to the 2-story configuration of the addition. Storage depth required (other factors not yet considered) is 1'-0".

Tons of rock required:

$$120\ \text{ft.}^3 \times 100\left(\frac{\text{lbs.}}{\text{ft.}^3}\right) \times \frac{1\ \text{TON}}{2000\ \text{lbs.}} = 6\ \text{TONS}$$

Heat storage available:

$$120\ \text{ft.}^3 \times \frac{100\ \text{lbs.}}{\text{ft.}^3} \times \frac{.2\ \text{BTU}}{\text{lb.}\ ^\circ\text{F}} \times (135^\circ\text{F} = 65^\circ\text{F}) = 168000\ \text{BTU}$$

As measured against 2800 BTU/HR generated by the equipment for the addition when the sun is incident.

If the mean temperature could be maintained for the storage, these figures mean that the storage can supply $\frac{84000}{2800} = 30$ hours of heating without further sun generation. In fact, early in the "cloudy" period, there will be relatively high conduction from storage to the conventional duct loop such that less fossil fuel energy will be called for during this period at the central thermostat.

Let us examine the configuration of the storage from the standpoint of rock size and static pressure through the under-slab space. Usual recommendations for 5 feet of storage depth cannot be met with the dual use aspects of the gravel due to the usual 3 ft. sill to footing dimension of this particular area. The cross sectional area is ideally square in cross-section, and if not so allowed, air flow and rock size must be adjusted. In the design approximately 6' is practical horizontally between the air plenums of the supply and return ducts. If the required 120 sq. ft. is thus divided by 6', 20 sq. ft. of cross-sectional area is required. Since this may be only 3' in actual depth, a reduced cross sectional area is determined by comparison of the air flow area contained within the two rectangles of equal perimeter. 4.4/' x 4.47', and 3'x 5.94'. In the former case, the inscribed circle contains $\pi \times 2.235^2 = 15.9$ sq.ft. and in the latter, the ellipse contains 14 sq.ft.

Then: 240 CFM ÷ 14 = 17.14 FPM

Rock size should be 1 1/2" diameter and the static pressure drop for the 6' traverse and 17.14 FPM would be about .14" W.G.

The air inlet and outlet openings in the storage should be sized as follows:

Inlet: recommended velocities 700 to 900 FPM

This requires an inlet size of approximately 8" x 16". This accomplished in the design by forming 24" x 16" x 8" sheet metal boxes at the base of the 10" x 10" downdraft duct which has one side open to the gravel fill and over which coarse metal screen is placed (assumed to absorb 1/5 of the clear opening).

Outlet: at the base of 3' foundation box where weight of the gravel is a factor.

Build an 12" x 16" x 20" box of Wollmanized lumber to receive the 12" round plastic, clay or concrete pipe with 1" insulated duct liner.

Note that the perimeter loop extension from the adjacent house is dropped to 1/3 the depth of the storage, or about 1'-0", and new register boots extended from that level to floor.

CONCLUSIONS

Implications of this system include:

(1) Application to low cost housing situations, particularly in addition of rooms to existing houses which are thermally up-graded to meet the currently escalating costs of all types of fuel (including wood) for central heating. Financing for such "add-on" projects could include specific instructions for using the new addition only during future periods of future excessive fuel use, whether generated in-house or at a central (electrical) generating station.

(2) Possible summertime application as terminal reheat for cooling applications where humidity problems due to ambient air refrigeration is characteristic. Again, the system should be considered as a booster to energy consumptive summer heating necessarily directed to this purpose.

A one dimensional model of a solar heat store using a change of phase material

R.H.Marshall,
Solar Energy Unit, University College,
Newport Road, Cardiff, U.K.

Abstract

Several numerical schemes have been investigated for the calculation of the temperature distribution in a thermal store utilising a phase change material. These models include the temperature excess model, the equivalent specific heat model and a hybrid model. A two dimensional store undergoing natural convection can be reduced to a one dimensional model by substituting an "effective" conductivity as a function of instantaneous Rayleigh number for the conductivity in the liquid phase. By considering the process as quasi steady state a very simple model can be derived which gives only the fusion front as a function of position and time. This model can be readily programmed for a variable time step Runge Kutte scheme. Such a model is thus fast enough to be incorporated in a store optimisation program. The numerical models are discussed and some typical computer results are given.

Introduction

Basically, heat transfer at the containing walls of a thermal store is given by the equation

$$Q = U_{tot} A\Delta T$$

$$\text{where} \quad \frac{1}{U_{tot}} = \frac{1}{h_i} + \frac{t}{k} + \frac{1}{h_o} \qquad (1)$$

where h_o and h_i are the outside and inside heat transfer coefficients respectively, and k is the conductivity of the wall while t is its thickness.

If the heat transfer mechanism within the store is governed by conduction alone, then h_i is replaced by

$$h_i = \frac{k_s}{t_s} \qquad (2)$$

where the subscript s refers to the store medium.

The time optimisation cost function for the heat flow in or out of the store depends, then on U_{tot}. For a minimum time to store or time to extract heat U_{tot} must be maximised which means maximising h_i, h_o, and k/t.

The last two parameters are well known as a function of material and geometry but the first h_i, on which U_{tot} is most dependent, is only known for pure conduction or for a few cases of convection.

If a store utilizes convection as a means by which h_i may be increased, then h_i as a function of the fluid and geometry must be ascertained. Further, if a latent heat store is considered then the effect of the heat input on the fusion front velocity must be known.

Comparison of Salt Stores vs Non-Salt Stores

A popular storage medium has been found in the hydrated salts because of their very high storage energy density. Table 1 shows the comparison of a few materials including paraffin wax.

Table 1

Properties for storage of one million Btu heat using water or sodium sulphate decahydrate or paraffin

Materials	Density lb/ft^3	Sp. ht BTU / lb ^oF	Heat of reaction BTU/lb^oF $Na_2SO_4.10H_2O$	Heat storage of 10^6 BTU with 20^oF temperature limit: Mass lbs	Volume ft^3	Energy Density BTU/ft^3
Water	62.5	1.0	-	50,000	800	1250
$Na_2SO_4.10H_2O$	91.0	0.46	108	8,530	94	10683
$Na_2SO_4.10H_2O$ and water	79.5	-	105	12,800*	161*	6211
Paraffin	48.5	-	105	9,524	196	5102

If the cost function is based on energy density alone then the hydrated salts are very attractive. However, if the cost function considers the rates of heat transfer into and out of the store, the picture changes. Specifically, the salt store designs rely on heat transfer by conduction alone(call it k-limited) because of the presence of nucleating sites and "fillers". Other materials, such as the paraffin waxes, or water are not k-limited and therefore may be operated in the convection mode for which very large values of h_i are possible.

An effective comparison between the hydrated salts(highest energy density), water(the cheapest), and paraffin waxes(both high density and cheap) will only be possible when the phenomenom of convection in a phase change store has been studied.

This paper intends to describe the development of computer models suitable for the inclusion in an optimal design program for a thermal store utilizing a phase change material in the convection mode.

Since, in general, the definition of the cost function itself is a formidable task, in this paper attention will be restricted to modelling store designs which consider the heat transfer coefficient, h, as a function of position, time, paraffin phase change materials and different boundary conditions.

Computer Modelling

A separate report, A Review of Natural Convection in Enclosed Rectangular Cavities, Marshall (1), summarises the parameters and equations governing convection of a homogeneous single phase fluid in inclined cavities. The heat transfer coefficient h_i, appears non dimensionally in the Nusselt number, $Nu = \frac{hL}{k}$ and is related to the other parameters via an equation of the form

$$Nu = C\ Ra^m\ Ar^n\ Pr^\ell \qquad (3)$$

The Nusselt number is simply the effective conductivity of the liquid undergoing natural convection.

This equation can be rearranged in terms of fluid and geometric properties to yield

$$\bar{h} = C\frac{k}{L}\left(g\beta\Delta T\frac{L^3}{\nu^2}\right)^{m}\left(\frac{H}{L}\right)^{n}\left(\frac{\mu C_p}{k}\right)^{1+m} \qquad (4)$$

The fluid properties of an idea phase change material for natural convection should therefore be

1. a high value, k, (since 1 + m is <1)
2. a large Cp
3. a large value β
4. a low value μ
5. a large value ρ
6. a large ΔT
7. a large value λ

The last is not apparent from equation 4 but comes from the energy storage density defined as

$$E = \frac{\rho\lambda}{Vol} \qquad (5)$$

The geometric properties H, length of heated surface, and L, gap length, are variables whose choice relies on the exponents m and n. The exponents are known only for the three cases of the vertical slot, the Benard convection problem and the cross heated fuel tank. On the basis of their values a "zeroith" level store can be made.

Thus, a computer model backed up by experimental results is desired which will yield ultimately the exponents m and n, and the average Nusselt number Nu as a function of geometry, fluid properties and boundary conditions. Such a program is to be as rapid in execution time so that incorporation of the model into a system optimisation program is possible.

The next sections detail the numerical approaches considered and the current state of development of a code.

The One Dimensional 3 Equation Model

The motion of the fusion front undergoing a change of phase is described by the 3 equation model. Refer to Fig. 1.

$$\frac{\partial T_s}{\partial t} = \alpha_s \frac{\partial}{\partial x}\frac{\partial T_s}{\partial x} \qquad 6a$$

$$\frac{\partial T_L}{\partial t} = \alpha_L \frac{\partial}{\partial x}\frac{\partial T_L}{\partial x} \qquad 6b$$

$$\rho\lambda\frac{d\xi}{dt} = k_s \frac{\partial T_s}{\partial x} \; k_L - \frac{\partial T_L}{\partial x} \qquad 6c$$

Fig. 1

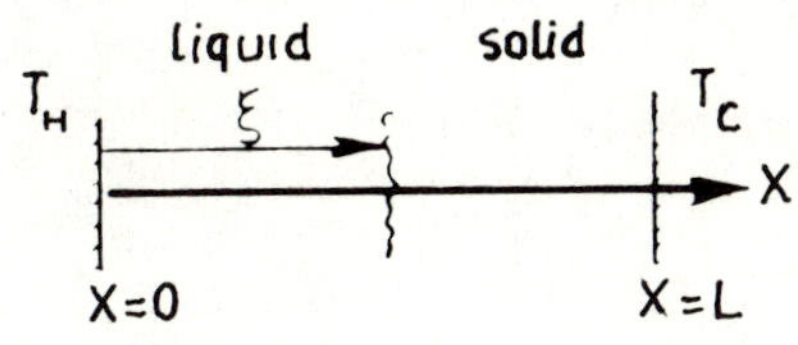

The solution for constant thermal properties ρ, α, k can be given analytically for only a few specific initial and boundary conditions. Özisik (1968) lists two solutions for the special case.

$$T(X,0) = T_m \quad t=o, \quad 0 < X < \infty \qquad 7$$

$$T(0,t) = T_H$$

as (approximately)

$$\xi = (K\ t)^{\frac{1}{2}} \qquad 8a$$

$$K = \left\{12\,\alpha \frac{1-(1+\mu)^{\frac{1}{2}}+\mu}{5+(1+\mu)^{\frac{1}{2}}+\mu}\right\} \qquad 8b$$

$$\mu = \frac{2(T_H - T_m)}{\lambda\ /Cp} \qquad 8c$$

and analytically

$$\xi = 2\,\gamma\,(\alpha t)^{\frac{1}{2}} \qquad 9a$$

with

$$\gamma\, e^{\gamma^2}\, \mathrm{erf}(\gamma) = \frac{T_H - T_m}{\lambda\ \pi/Cp} \qquad 9b$$

These solutions lack sufficient generality to be of great use but are useful for the testing of computer codes.

One may solve the three equation model given by equation 6 using standard finite difference techniques, Murray et al(2), Thomas et al(3), Lazaridis(4), Mastanalah(5), Martinek(6), Bain(7), all give examples of such solutions. However the drawbacks to these procedures are

1) there is a "starting error" when the fusion front, ξ, lies within the first $2\frac{1}{2}$ mesh points of a boundary.
2) the solution technique is explicit thereby requiring a large number of time steps and many iterations as the fusion front crosses a mesh point.

Fig. 2

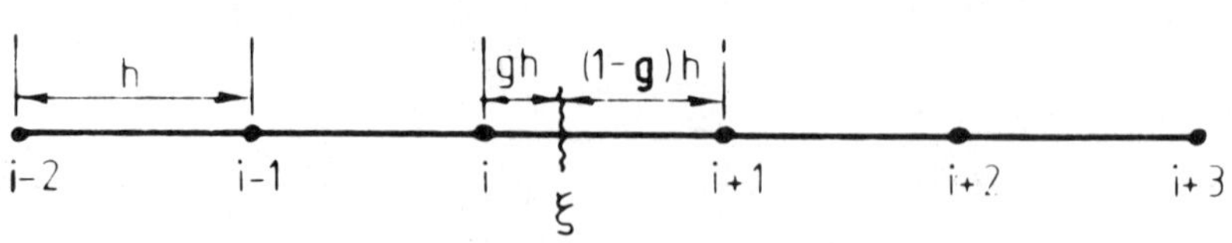

To clarify the first point a little consider the grid seen in Fig.2. At position i-1 the equation 6b may be written using the i-2 and i values of T for the $\frac{\partial}{\partial x}\, k\, \frac{\partial T}{\partial x}$ terms. Clearly then, the phase front must be at

$$\xi > 2h \qquad 10.$$

to permit a three point differencing scheme $0|h^2|$. Special techniques must be used whenever ξ lies within 2 mesh spacings of a wall.

Next, consider the equation for T_i. The explicit technique leads to

$$\frac{T_i^{n+1} - T_i^{n}}{\Delta t} = \alpha\left\{\frac{Tm + gT_{i-1} - (1 + g)T_i}{g(1+g)\ h^2/2}\right\} -(g-1)\frac{h}{3}\frac{\partial^3 T}{\partial x^3} \qquad 11.$$

This **equation** is singular when $g \to 0$. Further the truncation error, $(g-1)\frac{h}{3}$ Txx, is only $0|h|$ accurate. And still further and most importantly, the stability of eqn. 10 is governed by the restriction similar to

$$\Delta t < \tfrac{1}{2}\ gh^2/\alpha \qquad 12.$$

This means the time step size must go to zero as the fusion front goes through a mesh point!

These difficulties can be overcome by a change to a variable moving grid system or by using special techniques, see for example. Mastanalah(5) Lazaridis (4) or Murray et al(2). These procedures become expensive in terms of computer time especially when fins are involved as the explicit solution is governed by a time step size derived from the largest diffusivity, α, in the problem.

The explicit-implicit "Txcess" model(2 - equation)

The problem of singularities and an exceptionally small time step size when fins are involved may be avoided by using the Temperature Excess model. The temperature at any interior node is "pinned" at the melt temperature until such time that the time integral of the heat supplied to the node exceeds the heat of fusion divided by the specific heat. Mathematically this is given by ,

$$TE^{o} = 0 \qquad 13.$$
$$Txcess = \lambda/c_p.$$

If $T^n < Tm$ and $T^{n+1} > T_m$,

set $TE^{n+1} = TE^{n} + T^{n+1} - Tm$ 14.

Then only if $TE^{n+1} > Txcess$, 15.

set $T^{n+1} = TE - Txcess + Tm$; 16.

otherwise $T^{n+1} = Tm$.

The equation for $\frac{d\xi}{dt}$, eqn. 6c., is replaced by the above logic decisions given by eqns. 13-16. The solution requires only eqns. 6a and 6b where variable temperature properties are used.

If the type of node (or cell) is entirely solid(denote this by LTYPE = 1) then set

$$\rho = \rho_s \qquad 17a.$$

$$k = ks \qquad 17b.$$

$$cp = cp_s \qquad 17c.$$

If the type of cell is entirely molten(denote this by LTYPE = 3) then set

$$\rho = \rho_L \quad 18a.$$

$$k = k_L \quad 18b.$$

$$cp = cp_L \quad 18c.$$

If the type of cell is melting(denote this by LTYPE = 2) then one must use the weighted average between the solid and liquid phases which depends on the fraction, f, of the cell volume which has melted. With the definition,

$$f = TE/Txcess, \quad 19.$$

the property values become

$$\rho = f\ \rho_L + (1-f)\ \rho_s \quad 20a.$$

$$k = f\ k_L + (1-f)\ k_s \quad 20b.$$

$$C_p = f\ Cp_L + (1-f)\ Cp_s \quad 20c.$$

In the majority of cases, the value of Cp is constant and equal both the solid and liquid phase. Thus equation 20a and c can be further simplified to

$$\rho C_p = f\ \rho_L\ Cp_L + (1-f)\ \rho_s\ Cp_s \quad 21.$$

The fusion front is defined as the contour,

$$0 < f < 1 \quad 22.$$

It can be expected that the fusion front will smear over several mesh intervals. This is a disadvantage if the exact fusion front profile is required. A further disadvantage is that, in this form, the equations 6a and 6c must be solved explicitly as the non-linear effect of "pinning" a nodal temperature to the fusion temperature must be resolved before moving on to the next grid point.

However, the advantage of the method is in the reduction by 1/3 of the number of equations being solved.

Note that the heat stored is therefore

$$\frac{Q}{A}\ \text{stored} = \sum_{i=1}^{N} \rho i\ (Cp_i\ T_i + \lambda f_i)\ dx_i \quad 23.$$

This method is easily extended to two and three dimensions. Griggs et al(8) give an example of an implicit fin implicit phase change problem. Because of the smearing of the fusion front with this method they required a fine mesh (150 x 150 points) to achieve adequate resolution.

An alternative to the Griggs implicit solution would be to employ an implicit Runge Kutte scheme developed by the author as his PhD thesis Marshall(9). The method can adequately resolve the fusion front with a much courser mesh. The method also permits an automatically controlled variable time step. The method, however, is costly in terms of computer time.

As a further remark, any explicit solution for a case utilizing fins is subject to a severe stability restriction which is based on the diffusivity of the fin. Therefore, fins ought to be treated implicitly.

The 1 Dimensional Equivalent Cp(Spiked Cp)Model

The case of the PCM undergoing melting is treated adequately by the Txcess model if the boundary conditions are non-cyclic. If,however, one boundary condition is lowered while the other is raised in temperature or if the boundary conditions vary cyclically an additional set of logic conditions similar to eqns.13-16 are necessary to treat the equivalent Excess temperature generated by solidification. The resulting program then contains a larger number of time consuming If statements.

An imaginative alternative to the Txcess model is presented in Bonacina et al(10). They replace the full set of equations to T_s, T_L, ξ by the single variable property equation

$$\frac{\partial T}{\partial t} = \frac{1}{\rho Cp(T)} \nabla k(T) \nabla T. \qquad 24.$$

They show that a Dirac delta function for Cp can be used to treat the change of phase. Further they show that the Dirac dunction can be modelled by a rectangular "spike" of width 2 δ T. Mathematically, the variable property equations become

$$\rho Cp = (\rho Cp)_s \qquad T < Tm - \delta T \qquad 25a.$$

$$\rho Cp = (\rho Cp)_L \qquad T > Tm + \delta T \qquad 25b.$$

$$\rho Cp = \frac{\lambda}{2\delta T} + \frac{(\rho Cp_s) + (\rho cp)_L}{2} \qquad Tm-\delta T < T = < Tm + \delta T \qquad 25c.$$

$$k = k_s \qquad T < Tm - \delta T \qquad 25d.$$

$$k = k_L \qquad T > Tm \quad \delta T \qquad 25e.$$

$$k = k_s + \frac{k_L - k_s}{2\delta T} (T - (Tm - \delta T)) \qquad 25f$$

A schematic plot of the temperature property variations is seen in Fig 3.

Fig. 3

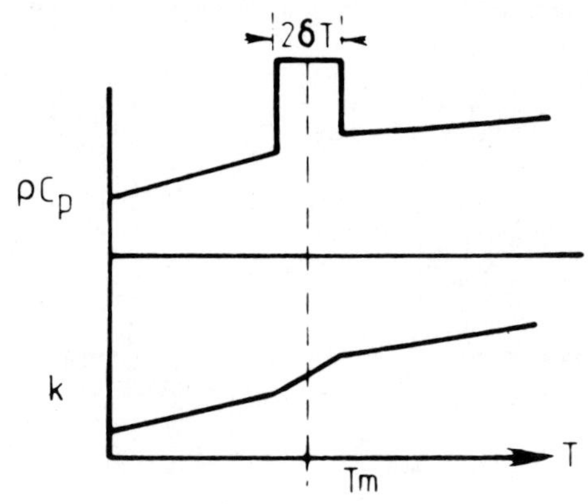

The authors use $2\delta T = 0.5^\circ C$ when the maximum ΔT was $30^\circ C$, i.e. $2\delta\ T$ of the order of 1.7% of total temperature drop.

This approach is applicable to both melting and solidification and to problems in 1, 2, or 3 dimensions.

There will be problems, however, with either the explicit of implicit solution if, the temperature "jumps" the spike i.e. $T^n < Tm - \delta T$ and $T^{n+1} > Tm + \delta T$. Basically this arises from too large a time step size, too large a grid spacing(i.e. a course mesh) or too small a δT! The remedy is dependent on experience and some degree of numerical experimentation. As an alternative, the implicit Runge Kutta scheme can be employed which with its associated intermediate evaluations will detect the jump.
(Note: the implicit Runge Kutta scheme was developed originally for shock front propagation problems analagous in fact, to the fusion front propagation).

With this method the heat stored is very simply given as the sum

$$Q\ \text{stored} = \Sigma\ \rho_i\ Cp_i\ (T)_i\ *\ T_i \qquad 26.$$

and the fraction of any mode undergoing melting is

$$f = \frac{T - (Tm - \delta T)}{2\delta T} \qquad 27.$$

A simplification of equation 24 results in a vast decrease in the number of averaging operations and IF statements. For example, expanding equation 24 for the 1-D case yield

$$\frac{\partial T}{\partial t} = \frac{1}{\rho Cp(T)_i} \quad \frac{1}{h}\ k(T)_{i+\frac{1}{2}} \quad \frac{(T_{i+1} - T_i)}{h} - k(T)_{i-\frac{1}{2}} \frac{(T_i - T_{i-1})}{h} \qquad 28.$$

Ideally one calculates

$$T_{i\pm\frac{1}{2}} = \frac{T_i + T_{i\pm1}}{2} \qquad 29.$$

and then computes

$$k_{i\pm\frac{1}{2}} = k(T_{i\pm\frac{1}{2}}) \qquad 30.$$

Note there are two averaging operation and two functional evaluation (or one calculation and 1 extra storage array $k_{i+\frac{1}{2}}$).

Since $k_S \simeq k_L$ for a PCM, the time saving or storage saving expedient is to compute

$$k_{i\pm\frac{1}{2}} = \frac{k_i\ (T_i) \pm k_{i\pm1}\ (T_{i\pm1})}{2} \quad + \text{T.E.} \qquad 31.$$

T.E. is easily shown to be

$$\text{T.E.} = \frac{1}{4}\ \frac{\partial^2 k}{\partial T^2}\ dT^2 \qquad 32.$$

so that for a "smooth" transition from k_S to k_L the error is second order, i.e. negligible.

Two Dimensional Models

The extension of this method to two dimensional stores is easily done if the two dimensional case is a "regular" case, i.e. either the Benard heated from below problem or the vertical slot problem and perhaps the cross heated tank problem. The effect of convection can be considered to lead to an "effective conductivity" of the liquid, k_L, which is a function of the instantaneous Ra and Ar based on the instantaneous gap length ξ.
In this case, the effective k is given by

$$\frac{k_{eff}}{k_L} = C \; Ra^m \; A^n \; Pr^{\ell} \qquad 33.$$

where the appropriate constant, C, and the exponents m,n,ℓ are given in Appendix A.

For other namely 14 cases, the full set of two dimensional Navier Stokes equations will have to be solved. These rely on the rapid solution of a Poisson pressure equation using primative variables of the form

$$\nabla^2 P = f(U, V, \Theta) \qquad 34.$$

The solution of the full two dimensional equations represents a formidable challenge and will require a sizeable computing expense. In fact, the London CDC 7600 is considered the only available computer up to the job. However, once a code is developed, the change to any of 17 temperature boundary conditions or the full set of 75 heat flux/temperature boundary conditions is quite straightforward.

Quasi Steady State Model

Although the above implicit transient methods are adequate for description of the melting process they are still too extravagant in terms of computer time if they are to be incorporated in an optimisation program. Hence, a simple model has been derived based on the quasi steady state idea. Basically any change at the hot side of the store is thought to give rise to an instantaneous linear temperature distribution in the store. The alalytic solution for a bar subject to a step change in temperature at one end is given (non dimensionally) by

$$T = \sum_{m=1}^{\infty} Cm \; Xm \; (Bm, X) \; e^{-Bmt} \qquad 35.$$

where Cm and Xm are Fourier series coefficients based on the initial temperature distribution. The decay terms, e^{-Bmt}, are the eigen value solutions to

$$\text{Sin } (Bm) = 0 \qquad 36.$$

$$\text{i.e. } Bm = \pm \; (0, \pi, 2\pi, \ldots.) \qquad 37.$$

Thus if boundary conditions change more slowly than e^{-t} (the slowest decay term) a quasi steady state can be assumed. The time dependent terms for the liquid and solid phases are dropped, then these equations are integrated spatially leaving only the time dependent equation for the fusion front

$$\rho\lambda \frac{d\xi}{dt} = k\ \text{solid}\ \frac{T_c - T_m}{1 - \xi} - k_{l_{eff}} \frac{T_m - T_H}{\xi} . \qquad 38.$$

With the non dimensional parameters,

$$\Theta = \frac{T - T_m}{T_H - T_m} \qquad 39a.$$

$$t^1 = \frac{t}{L^2 \alpha} \qquad 39b.$$

$$\rho_{ml} = \rho_m / \rho_l$$

$$k_{sl} = k_s / k_l$$

$$T_{xcess} = \lambda / C_{pl},$$

$$\xi = \xi / L$$

equation 38 becomes(after dropping primes)

$$\frac{d\xi}{dt} = \frac{k_{sl}}{\rho_{ml}} \frac{\Delta T_H}{T_{xcess}} \frac{\Theta_c}{1-\xi} + \frac{Nu(\xi)}{\rho_{ml}} \frac{\Delta T_H}{T_{xcess}} \frac{1}{\xi} . \qquad 40.$$

The term $Nu(\xi)$ is the effective conductivity of the liquid which accounts for natural convection occurring in the liquid phase. It is this term which makes eqn 40 non linear. The assumed temperature distribution is depicted in Fig 4 .

Fig. 4

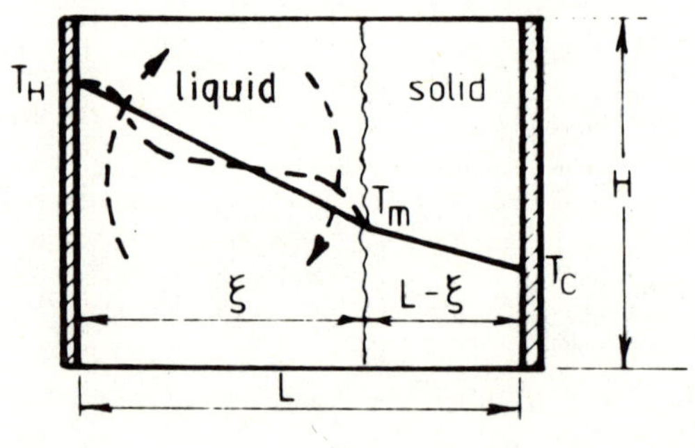

For the special case $T_c = T_m$ and $Nu(\xi) = 1$(i.e. pure conduction) eqn 40 can be integrated to yield

$$\xi = (\mu t)^{\frac{1}{2}} + \xi_0 \qquad 41.$$

$$\text{where} \quad \mu = 2 \frac{\Delta T_H}{T_{xcess}} \qquad 42.$$

For all practical cases of interest μ is less than 1.0. Özisik(14) gives approximate and analytic solutions already seen as eqns. 8 & 9 respectively.

For $\mu < .5$ eqn. 41 is within .1% of eqn. 8.

Eqn. 40 is readily integrated if $Nu(\xi)$ is constant(i.e. for conduction) to yield

$$\xi - 1 - \frac{C_1}{C_2} \ln\left(\frac{C_1}{C_2} + \xi - 1\right) = C_2 t + C_3 \qquad 43.$$

with $C_1 = \dfrac{\kappa_{sl}\ \Delta T_H\ \theta c}{\rho_{ml}\ Txcess}$

$C_2 = \dfrac{k_1\ \Delta T_H}{\rho_{ml}\ Txcess}$

$C_3 =$ the constant of integration.

For variable $Nu(\xi)$ (i.e. with convection) the Runge Kutte Merson method, see Merson(15), is advantageous. This method is fourth order accurate and provides an error bound estimate which thus enables a variable time step size. Therefore sharp transients merging to slow steady state solutions(and vice versa) are treated with ease. As with most Runge Kutte schemes, the method is"self starting".

Results

The three models, namely the equivalent (or "spiked Cp), the "TXCESS" and the quasi steady state models may now be compared. In Fig 5. is indicated the results for

$T_H = 1$

$Tc = Tm = 0$ (or $Tc = Tm - \delta T$ for spiked Cp)

$ksl = 1.25$

$\dfrac{\lambda}{Cp\ \Delta T_H} = 5$

$\xi(0) = 0$

This is the classical melting problem. The heavy straight line is the quasi steady state results which is within 0.2% of the analytic. The upper transient indicates the course of the spiked Cp model and is seen to overpredict the fusion front position. This is expected because although the equivalent ρCp is large at a melting node, the nodal temperature during melting can still become larger than the melting temperature. By definition, the fusion front is the thermocline where $T = Tm + \delta T$, see Bonacina et al(10).

Similarly, the lowest curve at the left of Fig.5 is the transient for the TXCESS model and it is seen to underpredict the fusion front position initially then to overpredict it in the later stages. This behaviour is expected since with this model, a nodal temperature is "pinned" at the fusion temperature until all the node has melted. In the early stages this means conduction to the second cell(and second to third etc) is delayed while the first melts. The later stages overpredict the fusion front position because the implicit solution scheme carries or "smears" the excess temperature to the solid region. Therefore, less excess temperature is required to melt the node. This is a drawback of the TXCESS model.

In reality melting of a paraffin wax does take place over a finite ΔT (i.e. the spiked Cp model) and yet the node does require a finite

Fig. 5

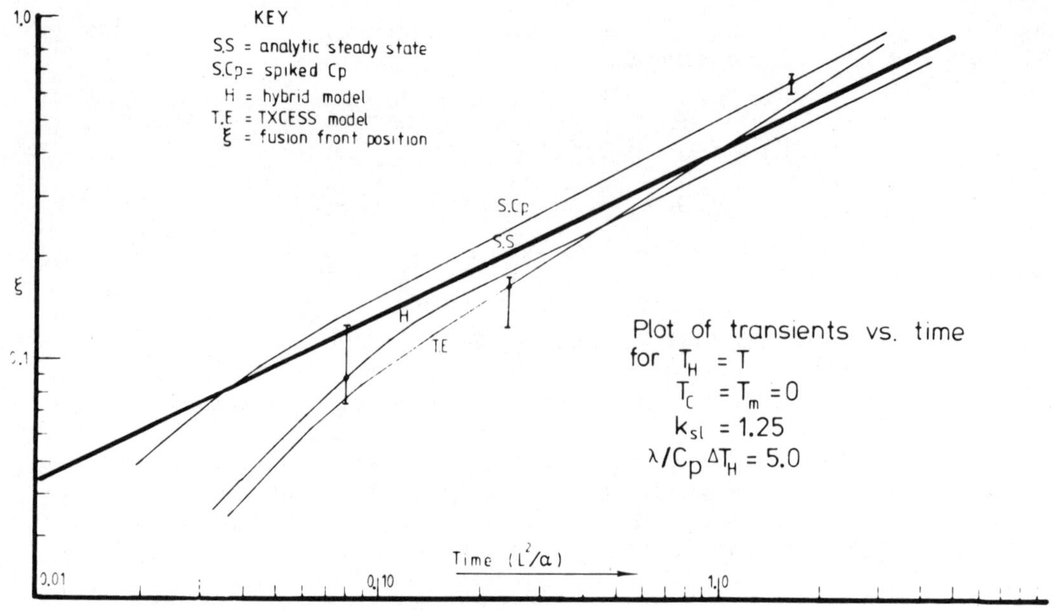

supply of heat in order to melt (the TXCESS model). Therefore a hybrid model, denoted by H in Fig. 5, was tried which is basically the TXCESS model together with the equivalent Cp model whereby

$$\text{TXCESS} = \frac{\lambda}{\text{Cpeq.}} \qquad 44.$$

and $$\rho_m \, \text{Cpeq} = \left(\frac{\rho m \lambda}{2\delta T} + \frac{\rho l Cp_l + \rho s \, Cp_s}{2} \right) - \delta T < T < \delta T, \qquad 45.$$

This model underpredicts the analytic but is better than either of the other two for the whole transient.

The vertical lines indicate the maximum resolution error of the fusion front position due to a finite grid. For the spiked Cp and TXCESS models the grid size was 1/20 while for the hybrid model a grid size of 1/10 was used. Thus, the error in the fusion front position can be quite large for the time when the fusion front is less than the grid size. This "starting error" problem may be avoided by using a non linear mesh and modifications are in progress in this direction.

In Fig. 6 are seen some results for a case with heat addition and removal. The same observation apply to these cases as in the above except that the hybrid model overpredicts the steady state model at all but the last stages. These results are significant in that for the case of simultaneous heat input and extraction the store can never be fully melted. The final position depends on the ratio ΔT_H to ΔTc and ksl and is given by

$$\xi_{max} = k_{sl} \frac{\Delta T_H}{\Delta T_c} \frac{1}{1 + k_{sl}} \qquad 46.$$

Thus, in designing a simultaneous heat in/out store k_{sl} should be greater than 1 and ΔT_c (i.e. $T_m - T_c$) should be small. The latter, however, conflicts with desired store requirement that the stored temperature should be as high as possible.

Fig. 6

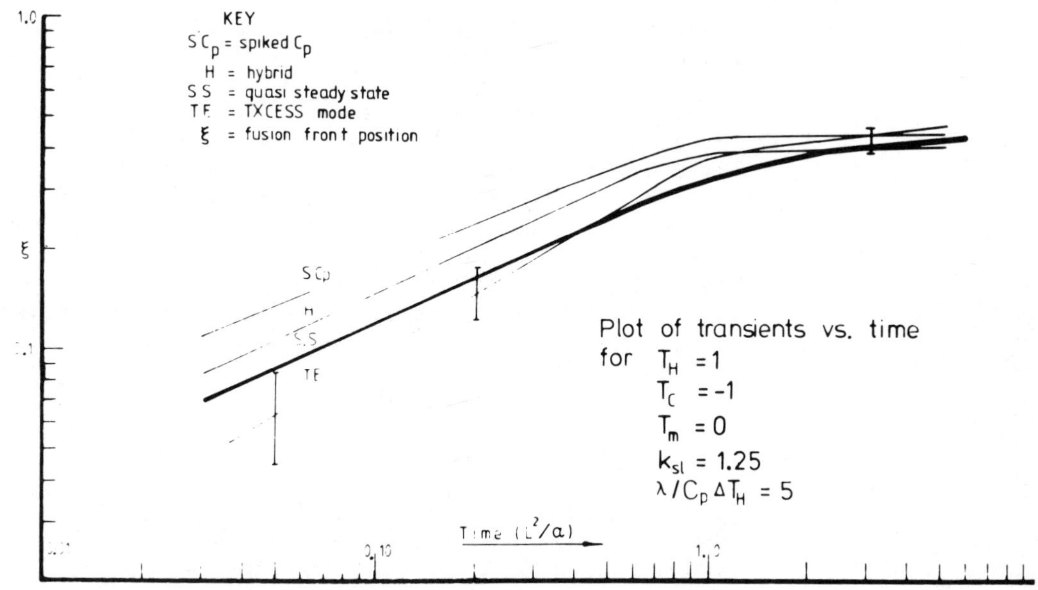

Fig. 7

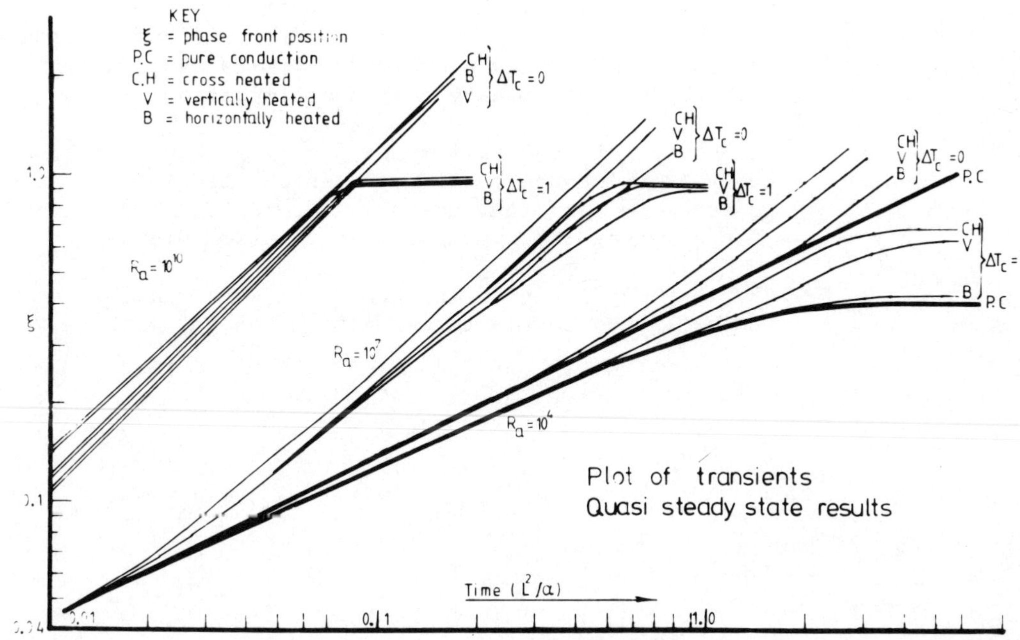

If the mechanism of heat transfer is by natural convection the quasi steady state analysis yields the results seen in Fig. 7 . Three types of store designs for two types of boundary conditions have been considered; the store heated from below cooled from above, heated/ cooled from vertical sides, and cross heated i.e. heated from below and cooled from vertical sides. For these cases the effective conductivity, Nu(ξ), is known as a function of instantaneous Rayleigh number.

For comparison the heavy solid lines indicate the result for pure conduction, Nu(ξ) =1. The results indicate that for a Rayleigh number of 10^4 based on the fully melted condition, ξ=1, the effects of convection on the transient are already considerable.

As the Rayleigh number is increased to 10^7 the transients occur about 1 order of magnitude faster than for pure conduction and at a Rayleigh number of 10^{10} they are two orders of magnitude greater. This implies that natural convection can significantly reduce the time to store.

This is crucial if solar energy is available only for short periods of time.

Of the designs the cross heated/cooled configuration appears to be the best followed by the vertically and horizontally heated cases respectively. However, the effective conductivity for the cross heated case and vertical case for a homogeneous fluid exhibits a strong aspect ratio effect, Ar=H/L. The experimental and numerical results for the former are scarce(see Mallinson (13)). For the latter (see Hollands et al(11)) the effective conductivity decreases as Ar to the power -0.25. The above results, then, for aspect ratio other than 1 may alter the choice of a design. These cases have yet to be explored.

Conclusions

Two numerical methods have been described for the modelling of the transient behaviour of a thermal store which uses a paraffin phase change material. A hybrid model was then developed to reduce the errors found with the separate models. However, these transient models are too expensive in computer time to be used in an optimisation package. Therefore, a quasi steady state one dimensional model was derived which is rapid enough and accurate enough for preliminary design optimisation.

The quasi steady state model was extended to include the effects of convection on the melting process. This showed that natural convection can significantly increase the fusion front velocity and hence reduce the time required to store heat.

Further experimental and numerical work remains to verify the effective conductivity assumption for a two phase medium.

R E F E R E N C E S

1. Marshall R.H. — A Review of Natural Convection in Enclosed Cavities, U.C. Cardiff, Dept. of Mech.Engr. Report 352, S.E.U. 57, 1977

2. Murray, W.D. & Landis, F. — Jnl. of Heat Transf., Vol.89, p.106, May, 1959

3. Thomas, ,L.J. & Westwater, J.W. — Chem. Eng. Prog. Symp., Series 59, part 41, p.155, 1963

4. Lazarodis, A. — Jnl. of Heat & Mass Transf., Vol. 13, p. 1459, 1970

5. Mastanalah, K. — Int. Jnl. for Num. Methods in Eng., Vol. 10, p.833, 1976

6. Martinek, F. — Jnl. of Spacecraft, Vol.7, No.9, p. 1032, 1970

7. Bain, R.L. & Stermole, F.J. & Golden, J.O. — Annul. Sum. Report No.1 NAS 8 30511 Nov. 1968 - Dec. 1969, Colerado School of Mines, Golden Colerado, 1969

8. Griggs, E.I. & Pitts, D.R. & Humphries, W.R. — ASME Paper 74/WA/HI/21 1974

9. Marshall, R.H. — A Hybrid Integration Scheme for the Solution of Viscous Compressible Flows, PhD Thesis, U. of New South Wales, Kensington, Australia, 1975

10. Bonacina, C. & Comine, G. & Fasano, A. & Primecero, M. — Int. Jnl. Heat & Mass Transf. Vol. 16, p. 1825, 1973

11. Hollands, K.G. & Unny, T.E. & Raithby, G.D. — Jnl. Heat Transf., p.p184-193, May, 1976

12. Raithby, G.D. & Hollands, K.G. & Unny, T.E. — ASME Publication 76-HT-37, 1976

13. Mallinson, G. — Natural Convection in Enclosed Cavities, PhD Thesis, U. of New South Wales, Kensington, Australia, 1973

14. Özisik, M.N. — Boundary Value Problems of Heat Conductors, Int. Textbook Co. Haddon Craftsman, Inc, 1968

15. Merson, R.H. — Proc. Conf. in Data Process & Auto-Comp. Machines, WRE, Salisbury, S.A. 3-8 June, 1957, Session 1, paper 110, pp. 19 - 22

NOMENCLATURE

Subscript

C	cold
H	hot or heated
i	inside
L	gap length
o	outside

Superscript

l	an exponent
m	an exponent
n	time level or an exponent

Roman

Ar	aspect ratio H/L of a two dimensional store
C_p	specific heat
E	energy storage density
f	fraction of node melted
h	heat transfer coefficient or the finite difference spatial step size
H	length of heated surface
k	conductivity
L	reference gap length
Pr	Prandtl number
q	heat flow rate
Q	heat
Ra	Raleigh number based on a fully melted store
t	wall thickness,or time variable,or turbulent
T	temperature
U	overall heat transfer coefficient

Greek

α	diffusivity
β	coefficient of thermal expansion
γ	intermediate variable
δ	small change
ξ	fusion front position
θ	non-dimensional temperature
μ	absolute viscosity, or a variable
ν	dynamic viscosity
ρ	density

APPENDIX

Summary of the 1-D, and 1-D equivalent conductivity models.

1-Dimensional equations

$$\frac{\partial T_s}{\partial t} = \frac{1}{\rho_s \; Cp_s(T)} \; k_s(T) \; \frac{\partial^2 T_s}{\partial x^2} \qquad \text{A1}$$

$$\frac{\partial T_L}{\partial t} = \frac{1}{\rho_L \; Cp_L(T)} \; k_L(T) \; \frac{\partial^2 T_s}{\partial x^2} \qquad \text{A2}$$

$$\rho_L \lambda \; \frac{\partial \xi}{\partial t} = k_s \frac{\partial T_s}{\partial x} - k_L \frac{\partial T_s}{\partial x} \qquad \text{A3}$$

$\lambda > 0$ for melting

$\lambda < 0$ for solidification

For Bottom Heated see Hollands et al(11)

$$Nu = fn(Ra\ Pr\ Ar) = 1 + 1.44\left\{\left(1 - \frac{1708}{Ra(\xi)}\right)\right\} + \left\{\left(\frac{Ra(\xi)}{5830}\right)^{1/3} - 1\right\} \qquad \text{A13}$$

where $Ra(\xi) = \frac{g\beta}{\nu^2}(TH - Tm)\ Xref^3 * \left(\frac{\xi^3}{X_{ref}^{\ 3}}\right)$

and ξ is the fusion front position determined by equation A3, or the point where f in equation 8b is > 0, or the point where $Tm - \delta T < T < Tm + \delta T$ in equation.

For Side Heated See Raithby et al (12)

$$Nu = fn(Ra\ Pr\ Ar) = Max\left\{1 \quad .375\left\{\frac{Ra(\xi)^{1/4}}{Ar(\xi)}\right\} .044\ Ra(\xi)^{1/3}\right\} \qquad \text{A14}$$

where Ra (ξ) is as above and

$$Ar(\xi) = \frac{Y\ ref}{\xi} \qquad \text{A15}$$

For Cross Heated see Mallinson (13)

$$Nu = .34 \quad Ra(\xi)^{.24} \quad Ar(\xi)^{-.28} \qquad \text{A16}$$

4.10 Long range solar heat storage

J.M. and G. Alexandroff
Ecole Nationale Supérieure
des Beaux-Arts de Paris,
France

Abstract

Today, long range heat storage, using water at a low temperature level from flat plate roof-collectors, is not economically feasible, at least for single house heating. This paper, however, explains a new and lower priced concept for heating small groups of houses (about 10 to 100). The heat storage is composed of trenches batteries, containing the water in Butyl rubber bags, at different temperature levels; the soil may then be reused for any urban purpose when the project is completed.

Introduction

Today, in different parts of the world, many studies and computations exist, as projects or full-scale tests, in order to solve the difficult economic problem presented by the need for long range storage of solar thermal energy. In many countries, because of the cloudy winters, an effective use of solar energy as an alternative installed capacity depends on solving that problem. Failing a solution to the problems, solar energy would be just a mid-season aid for heating buildings and satisfying domestic water requirements.

From another point of view, the existing solar installations tend to become overheated in summer, thus in certain cases creating thermal stresses for the inhabitants; also some installations can be damaged, especially during the holiday period, when they are not functioning.

This paper explains our work, effected in collaboration with A. Liebard, architect and R. Torrenti of the Commissariat à l'Energie Atomique, within the framework of the 'Ministère de l'Equipement' grant.

Paper

For a long time, the idea of long range energy storage was not accepted in responsible quarters, because of its economic non-feasibility, due to the required enormous capacity of concrete or metal containers and because of the insulation cost; however, the point is that the economic studies conducted so far dealt with single houses, and we became convinced that in this case the problem could not be economically solved. Therefore we have now directed our research towards collective dwelling units, or individual groups of houses and other kinds of building groups with individual collectors and collective storage; given that a big storage offers a better volume/surface ratio, and thus lower heat-losses.

Our research and calculations are based on water storage, because water is cheap, renewable and non-polluting. Further calculations have shown that the volume required needs to be only double, with water, than with molten salts which require expensive heat-exchangers; as for the bins, or rock-bins, the required volume is double that for water.

We soon renounced the use of containers made of metal, reinforced concrete or rigid plastic, and moved towards considering excavated basins with natural sloping sides and floating upper insulation. This answer has been adopted in France by petroleum technicians for solar or nuclear energy storage. The great failure of such a system lies at the important upper surface, and because of the fact that this surface cannot be used, makes the system not feasible in dwelling areas.

The above considerations led us to conceive a divided, or sectional storage, made up of a battery of trenches, in each of which lays a rubber bag (Butyl) containing the water. On the top of which are placed prefabricated or poured concrete slabs, then insulation. The excavated earth is finally restored to the complete area.

Fig. 1
Low-grade thermal storage

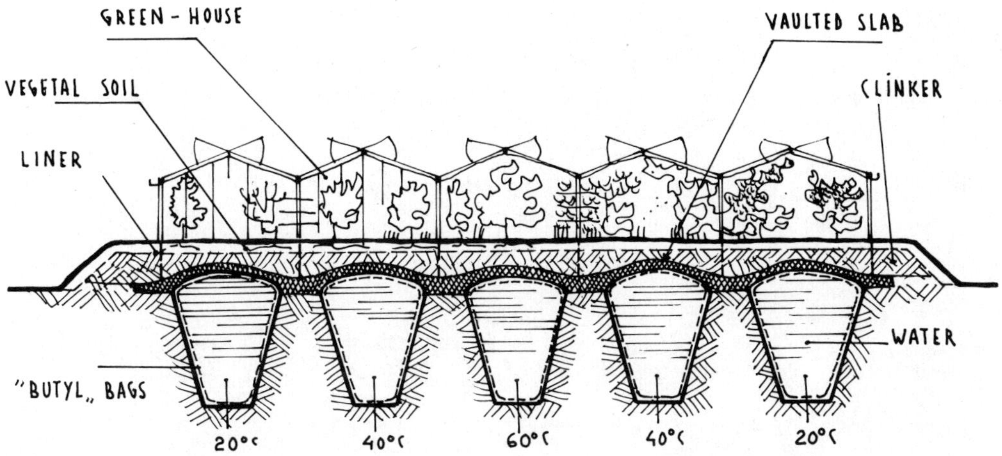

The interest of such a storage partition is, first, to avoid the huge excavation of a single pit, that is very expensive and difficult to make in dwelling areas, to save the accessible soil and, at the thermic point of view, to admit different temperature levels in every bag, the hotest being located in the center of the storage area.

An important fact is that the water bags are not insulated, so that the surrounding soil between each trench becomes an extension of the storage area.

The system requires a cohesive, homogeneous and load-bearing soil cover over the ground water-table.

Costs are lower than in any existing storage, as it includes no lost ground costs, no important earth carriage; no unusual equipment: work may be operated by any normal Trenche Shovel-Forming equipment. Our price calculations for the Cadarache Test Operation are estimated at £10 to £20/per cubic meter of water.

At an architecture level, the following results may be expected:

1. Easier integration for solar components into the new or existing buildings.
2. No storage in the houses.
3. A lower quantity of Solar Collectors are required for the same efficiency.
4. Moreover, a greater freedom in their exposure and tilt may be maintained, as south walls may be kept free for lighting and passive heat gains, by means of large windows, bow-windows, conservatories.

The integration of this kind of storage in the urban planning makes semi-autonomous dwelling units possible; in conjunction with electricity generated by wind; and furthermore, with food production by means of greenhouses located over the Storage area, and heating themselves from its caloric losses.

We found, that auxiliary heating was an important problem. In northern countries, even if complete solar heating is theoretically feasible, especially with a whole year of collected solar energy, computation shows that this may be economically unsound. In a location as, for instance Nancy in N.E. France, it would require a very large collector surface area (600m2, with a 4800m3 of water storage for ten well-insulated houses). This example proves the need for an auxiliary heating system.

Two main ways then appear:

1. The location of auxiliary heating units in the houses themselves (classical solar system)
2. A central auxiliary heating built by means of a small continuous-work boiler, burning gas or charcoal for example, in order to heat the long range storage central water bags if they fail to become hot enough. As a very attractive variant of this concept, increasing the central bags temperature may be obtained by using a heat pump, operating with the colder external bags of water as a heat reservoir.

The above examples show the wide range of improvements to heating problems which may be obtained in different types of northern dwellings by use of long range storage.

References

'Le Transport de chaleur par réseaux d'eau chaude à 90°C', par MM. Bourguet, Fischer et Lancel (Technip) - Techniques de l'énergie No 1 et 2 - 1977.

'Stockage de calories - Première approche des coûts des réservoirs' CEA - Département des Programmes - Getes Stockage - 1975.

'Installation de stockage thermique mixte de type modulaire' Brevet d'invention déposé le 1er Avril 77 au nom du CEA et de MM. Alexandroff et Liebard sous le No EN 77-9908.

'Bassins de stockage de longue durée de l'énergie solaire' CEA - SEEN-RT.77.081 - Saclay - Communication aux Journées Solaires de Nice - R. Torrenti.

'Intérêt économique d'un stockage de chaleur associé au chauffage à distance' Y. Coiba - Ed F. Chatou - DSE - 1976.

'Documents préparatoires à l'implantation d'un bassin de stockage à CADARACHE, à l'expérimentation dans la cuve Minerve, à l'accord Rhône-Poulenc, Elf, Technip', CEA. (Devis de Technip pour Diré, Mirail, ... Etudes et estimations du CEA, de Rhône-Poulenc, ...).

'Documentation de fabricants de matières plastiques: Latim, Fipec' (Esso-Chimie).

'Documents relatifs au projet de chauffage solaire d'un semi-collectif à Blagnac équipé d'un bassin de stockage de longue durée' (Action Gaz de France - CEA - Architectes Alexandroff et Liebard).

'Annual collection and storage of solar energy for the heating of buildings - Report No 1' J.T. Beard - University of Virginia (Contrat ERDA) - Janvier 1977.

St. George's School, Wallasey: A solar heated building

4.11

M.G. Davies,
Department of Building Engineering
The University, Liverpool L69 3B6, U.K.

Abstract

St. George's School, Wallasey, was completed in 1961 and appears to be the largest building so far constructed which makes use of solar gain for space heating by passive means. The article discusses the physical principles which underlie this process, the way they are realised in the school, and provides some information on the way the building has been found to function.

Buildings of the 1950's and 60's with their large glazed areas have made many of us uncomfortably aware of the heating power of the sun. The temperature inside a building with large south or near south facing areas of glass can reach unpleasantly high values during a summer heat wave. Relatively little attention seems to have been given to using this heat as part of the heating necessary in winter. However, such a scheme was devised and built by the late Emslie A. Morgan, Principal Assistant Borough Architect of Wallasey, Cheshire. His St. George's School, completed in 1961, (fig.1) has a large south facing wall and the radiant energy so admitted, together with heat from the electric lights and the body heat of the occupants, has at all times been sufficient to preserve an acceptable thermal environment. No conventional heating is used.

It is of interest to recount the physical principles which underlie a building of this sort. Glass transmits a large fraction, more than 70 per cent, of the radiation incident upon it. Daylight coming through a window falls on the surfaces within a room where it is absorbed after multiple reflections and the fabric warms up.

The surfaces themselves, however, are emitters of radiation, the average wavelength being much longer than the wavelengths of visible light. Since glass is opaque to this long wave thermal radiation, it acts as a one way valve and traps the radiation so that the temperature inside the room rises above the outside air temperature. Heat will of course be lost by conduction through the glass and other parts of the building fabric and by ventilation.

In clear sky conditions at noon, even in mid-December, enough sunshine will enter a room through a south facing window to cause a substantial rise in temperature. If the south facing wall were fully glazed, and all other surfaces were lined internally with some highly insulating material, conditions would become intolerably hot. Such clear days are of course infrequent in winter but the analysis suggests that solar gain might at any rate make some contribution to space heating.

In spring and autumn, the days are longer and the sun rises higher in the sky than in winter; the total radiation which can fall from a clear sky upon a horizontal surface is accordingly much greater than in December. The amount of radiation falling upon a south facing vertical surface varies much less, since the area of a wall seen by the sun when high in the sky is much less than when the sun is on the horizon. In British latitudes the maximum solar gain through a south facing window from a clear sky occurs in March and September, not June. Hence solar heating is best obtained, at the time of year when wanted, through vertical not horizontal, glazing.

A detailed heat balance study indicates that this form of solar wall construction on this site should lead to savings in the supplementary heat load.

A fairly even temperature by day and night is normally required within a building. As solar radiation can make no contribution to heat requirments by night, some form of thermal storage is necessary. Walls and floors of concrete and brick or other material of sufficient thickness and density will so serve. The internal furnishings will also contribute if massive enough. The material absorbs heat during the day and gives it back during the night and so stabilises the air temperature within the building.

A further requirement of a thermally confortable building is that the temperature of the internal surfaces should not be too cool and so the building fabric should be well insulated against heat flow to the outside. The insulation must be placed on the outside of the building and in fact serves two purposes: it reduces the overall heat loss, thereby reducing the running cost of whatever forms of supplementary heating is used, and it ensures that the structure is kept warm.

Solar heat is in short supply at the time it is needed most and if it is to contribute any significant part to the heating requirement during winter, heat losses and in particular, ventilation losses, must be carefully regulated. Special attention has to be given to the design of windows, ventilators and doorways to reduce air infiltration.

The only building in the U.K. so far constructed along these lines is St. George's School. (The dwellings built by David Wright in New Mexico are closely similar.) The latitude is 53°25N. The mean temperature in winter is about $4\frac{1}{2}$°C, with a high proportion of sunless days. The site is flat and unobstructed to the south. The main section of the solar wall, Figure 1, is 230f (70m) and 27ft (8.2m) high and faces 7° west of south. It is mostly double glazed, the separation between the leaves

Fig. 1 The solar wall from the south west. The Assembly Hall at the western end extends the full height of the building. The photograph shows the openable windows, a shutter-covered section of solar wall brickwork, the ventilator louvres, a small additional area of solar wall on the west elevation, and the marked slope of the roof.

being 24in. (610mm), a space large enough to provide access for maintenance. Walkways at different height and vertical partitions split the space into compartments. Seen from inside, parts of the solar wall are obscured by pinboards and most visible parts of the inner leaf are of diffusing glass.

Most walls, both external and internal, are built of 9in. (230mm) brick. The ground floor consists of a 4in. (100mm) concrete screed laid upon 6in. (150mm) of dense concrete. The intermediate floor is of 9in. (23mm), and the roof of 7in. (180mm) of concrete. This mass provides thermal storage and restricts the daily swings of temperature. A thickness of 5in. (125mm) of expanded polystyrene, with vapour barriers and weather protection, is fixed above the roof and outside much of the outside walls.

Fig. 2

An openable window in the open position. The clasp easily secures the window in the closed position. The lower and upper grilles make the window burglar proof. The aluminium panels, referred to in this article, are concealed behind the pinboards. The central heating radiators were installed as a precaution but have never been used.

To control ventilation, the solar wall is provided with openable windows, Figure 2; they are designed so that they can be secured in either the closed or fully open positions, and in either position are burglar-proof. The school has two storeys. The rooms on the upper floor at the westerly end extend the full depth, (38ft. 11.6m) of the building. The lower class rooms have a depth of about 23ft (7.0m), part of the remaining space being taken up by the east-west corridor. The entrances to this corridor form points of uncontrolled ventilation and to mitigate the consequent cooling, the architect designed units consisting of a pair of doors, (Figure 3), which swing outward, coupled so as to move together, well insulated and with padded adjoining edges to reduce air infiltration. All entrances to the building, including those into the Assembly Hall on the west end and into the Gymnasium at the east end, are of this pattern.

Fig. 3

These carefully constructed doors minimise draughts and consequent heat loss.

Some other features associated with thermal design should be mentioned. Aluminium panels are placed between the pinboards and the inside of the inner leaf of the solar wall so as to reflect to the outside part of the solar radiation during the summer season. The reverse sides of the sheets are painted black and when this side is presented outward during winter, solar radiation is absorbed and this heat is given mainly by convection into the room. Again, over certain lengths of the solar wall, notably at the Assembly Hall end and in the Gymnasium, portions of thick brickwork replace the inner leaf of the solar wall. The surface is painted black but can be obscured by hinged shutters painted with reflecting paint; the shutters should be moved in spring and autumn.

The upstairs rooms are more than 16ft. (4.9m) high at the solar wall end about 7ft (2.4m) at the north wall. Presumably this variation was felt to be necessary in order to ensure sufficient solar gain without an excessive cubic capacity. The angle of slope of the roof is 14°, coincidentally, it is believed, equal to the sun's elevation on the shortest day of the year at Wallasey.

The sun may be obscured for prolonged periods in winter and an additional source of heat must be provided. In St. George's School, this is achieved at no capital cost by regularly leaving the lights, (typically 1200W per downstairs classroom), on for perhaps 15 hours a day; a time switch turns them off between say 2200 and 0700 hours, and during the period of occupation the staff may turn them off if they wish. Furthermore, each child is a source of heat, equivalent, in round terms, to a 70W lamp. Thus the heat input from lights and occupants to a downstairs room during a 24 hour period might be 100 MJ. This is to be compared with the 150 MJ of solar energy that might be transmitted through the solar wall into a downstairs room on a perfectly clear December day.

These sources together would in theory, with no ventilation loss, maintain a mean temperature difference between inside and outside of more than 25 K. However, clear days are infrequent, and the heat loss by a ventilation rate of about two changes per hour would equal the conduction loss. A temperature difference of much less than this is therefore met in practice.

In summer, large solar gains are to be expected during extended sunny spells, but the large thermal storage prevents large temperature rises by day, and the structure cools again by cross ventilation during the night.

It is natural to enquire how successful the building is. The question can be approached from different standpoints. The part of the school which was designed by Emslie Morgan (and known as the Annexe), forms the second half of a complex of buildings, the first half of which, the Main School, is conventionally heated. An economic study of the two parts carried out by Mr. J.B. Love of the UKAEA, show that, having made due allowance for the increase in prices between the dates of completion, the Annexe cost more to build but is cheaper to run than the Main School; if capital and running costs are expressed as a total annual cost, the Main School is cheaper than the Annexe comparing the cost of the total floor areas, more expensive comparing the cost per place and approximately the same comparing the costs of the teaching areas.

An investigation has been carried out by Dr. Ann Davies to discover what the staff and pupils think of their environment. Although the two parts of the school are physically very different, they were designed to serve similar functions, each to house 300 pupils. The school is now run as a single unit. Dr. Davies was thus able to obtain a direct comparison between the responses of users to the two environments. In winter, the Annexe is universally recognised as much the more desirable thermal environment. However, to maintain it as such, the windows and ventilators have to be kept shut and teachers complain of a damp, stuffy and odour-laden atmosphere. All in all including a range of non-thermal factors, they prefer the Main School as a teaching environment.

However, in summer the users of the Annexe have complained of overheating. This almost certainly arises because the windows, due to a minor fault in their design, have to be secured closed after 4 p.m., so preventing adequate cooling by night. The situation is aggravated by the fact that, during the day, while the downstairs classroom-to-outside windows and the corridor-to-outside ventilators are kept open, the noise level in the corridor can be sufficient to make staff keep the corridor-to-classroom ventilators closed; this reduces cross ventilation.

The Department of Building Engineering at Liverpool has conducted a survey, as yet unpublished, into the thermal response of the building. By suitable control of lighting and ventilation, it is possible to control of lighting and ventilation, it is possible to control the indoor temperature to a value between about 2K and 17K above ambient, according to need. The highest daily mean temperature observed was 24.5°C. The largest swings of temperature noted during the 19 months of observation were surface temperature 3.2K, air temperature 8.8K. The mean values were 0.9K and 3.7K.

The study as a whole has shown that solar gain can be harnessed in winter by simple architectural means without mechanical services, that temperatures affording satisfactory comfort levels are usually achieved, that the occupants are mostly satisfied with the environment, and also that it is not expensive to run in comparison with other secondary schools. Certain specific defects of the building have been identified, and a number of suggestions for an improved design have emerged.

4.12 The Doug Kelbaugh House

D. Kelbaugh
Princeton, New Jersey
USA

Introduction

A passive approach to solar heating using the Trombe (thermal storage) wall technique is utilized in a house in Princeton, NJ, designed, built, and occupied by Doug Kelbaugh and his family (Figure 1). Princeton (located slightly north of 40 degrees latitude) normally experiences 5100 heating F-days and receives about 55% of the possible sunshine during the winter.

Among the advantages which the owner feels a Trombe wall affords are long life, low operating termperatures, substantial summer cooling, no fans, pumps, and few moving parts. In addition, the vertical collector is self-cleaning and provides a nearly optimal collection angle when there is snow on the ground. Furthermore, the wall may be architecturally integrated with the remainder of the building and permits the use of a variety of backup systems. Finally, perhaps the greatest asset of such a system is the simplicity of its design, maintenance, operation, and understandability.

Design of house and solar heating system

The design of the house was undertaken by the owner, Doug Kelbaugh, an architect and solar consultant. The task he set for himself was to design a house with $45,000 maximum construction cost, a Trombe wall, a large interior space, three bedrooms, a fireplace, a greenhouse, and a usable yard on a small lot.

Fig. 1
Doug Kelbaugh
house

Fig. 2 Ground floor plan

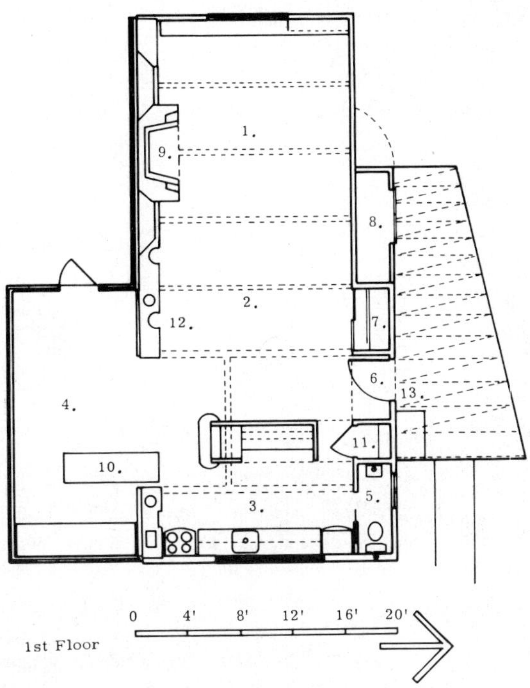

Fig. 3 Second floor plan

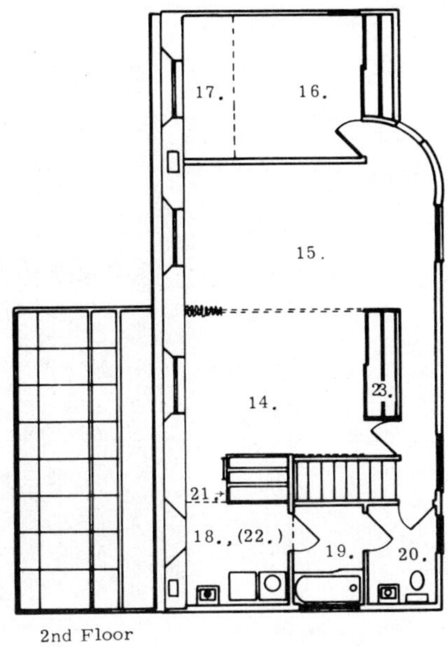

1. living
2. dining
3. kitchen
4. greenhouse
5. w. c.
6. entrance
7. coat closet
8. garden storage
9. circulating fireplace
10. trap door to cellar
11. utility closet
12. candle niche
13. arbor
14. master bedroom
15. study
16. child's bedroom
17. sleeping loft
18. sink, bidet, laundry
19. bath
20. w. c.
21. linen closet
22. storage loft
23. movable closet

The owners found one of the few remaining building lots in town and filed for a zoning variance, which was granted without opposition. This variance permitted the house to be situated on the northern boundary of the 60- by 100-foot lot, thereby providing unshaded access to maximum available sunlight and also allowing the large tree in the center of the lot to be saved. At the same time, this arrangement provided the desired large exterior yard.

Figures 2 and 3 illustrate the floor plans for the first and second floors of the house. The ground floor has been kept simple (one large, continuous space) with the kitchen partially separated from the living and dining area by the stair.

On the second floor, the three bedrooms are strung along the concrete wall with separate rooms for toilet, bath/shower, and utility along the east wall, which has the only plumbing in the house. The European bathroom arrangement allows a minimum number of plumbing fixtures without compromising access and privacy.

A standard Lord and Burnham greenhouse leans against and works sympathetically with the Trombe wall. The interior of the house conveniently adjoins the greenhouse through a wide arch (Figure 4).

There are intentionally no double-height rooms or any other spatial 'tours de force' in order to keep costs down as well as to make air circulation easier to predict and control. A minimum amount of window area is included, particularly on the north wall where windows lose more heat than they collect.

The north, east, and west walls are standard wood frame with rough sawn cedar plywood on the exterior and sheetrock on the interior. The cavities (4.5 inches on the first floor, 3.5 inches on the second floor,

Fig. 4
Greenhouse arch

and 9.5 inches above the ceiling) have been blown full of cellulosic fiber (recycled newspaper) to achieve an R factor of 18 in the walls and 40 in the roof. Styrofoam 1-inch thick has been placed on the outside of the foundation wall around the perimeter to a depth of 2 feet.

Solar heating system

Figure 5 illustrates a simplified diagram of the solar heating system of the house. The basis for this system is a Trombe wall constructed to 15-inch concrete which has been painted with a special black coating (3M Nexel over masonry conditioner). In front of the wall are two sheets of double-strength window glass. The total collection area is roughly 600ft^2 (plus the greenhouse). Vents in the wall at the top and bottom of each floor level are part of a natural convection loop which permits circulation of the air.

Fig. 5
Solar heating system

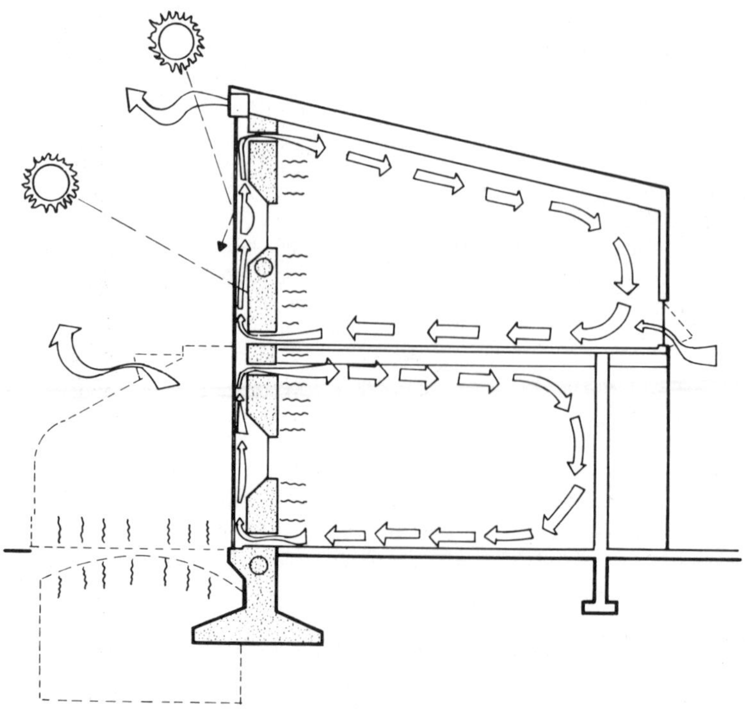

The heat loss for the building is calculated by conventional analysis to be about 65,000 BTU/h at $0^{o}F$ outside temperature. The backup heating system consists of a conventional, gas-fired, hot-air furnace with a net delivery of 58,000 BTU/h. The backup system is separate and distinct from the solar heating system although its ducts are cast in the concrete wall. One branch leaves the wall to supply heat to the bathrooms (which are isolated from the concrete wall). Three 250-watt heaters were also installed in the bathrooms to provide instantaneous supplementary heat.

A lean-to greenhouse which acts sympathetically with the solar wall was added to grown ornamental and edible plants. Its thick, concrete floor is painted black to collect and store heat to heat the cellar as well as the greenhouse.

The design also provides substantial summer cooling. Four fans at the eave exhaust heat buildup at the southern wall and ventilate the entire house by pulling air across the rooms from windows on the north wall. By ventilating the concrete wall at night, the wall is cooled to absorb heat from the rooms on the following day. The greenhouse is shaded in the summer by slat blinds drawn over the glass and also by two large deciduous trees.

Original plans called for a solar water heater to be placed on the greenhouse deck for domestic hot water, but these plans have been postponed since the cost of operating the existing, wrapped, hot-water heater have been extremely low (about $5 per month).

The total system provides heat to an area of $2100ft^2$ and a volume of $19,000ft^3$.

Modifications to the solar design

In response to a number of problems that were discovered during the first winter of occupancy, several modifications have been made to the solar design in order to improve the effectiveness of the solar heating system.

One problem encountered was that of reverse thermosiphoning whereby warm air at the ceiling height would leave the room at night through the upper vents in the concrete wall, wash down along the exterior glass wall, and reenter the room through the lower vents at a lower temperature than it left the room. This problem was solved simply and effectively by installing a passive damper consisting of a screen (on the inside of the concrete wall) covered by a light cloth, which permits air circulation in one direction (inward) only.

Another problem encountered was that of heat migration to the second floor. Because there is an open stairwell and the backup system seldom cycles air, the warmer air has a longer time to rise and collect upstairs. The average winter room temperature upstairs was about $5^{o}F$ higher than downstairs. This problem has been addressed by putting a door at the top of the open stairwell. However, since this was accomplished only in March 1977, no data exist to evaluate its influence on overall system effectiveness.

Problems were also encountered in the greenhouse. During the first winter, excessive heat losses were experienced in the greenhouse. In fact, hourly heat loss in the single-glazed, $200\text{-}ft^2$ greenhouse was about 32,000BTU, which represents a sizable proportion of the 75,000BTU/h total for the entire house. Subsequently, a second layer of glass has been added, which is estimated to cut heat loss by about 15,000BTU/h.

Another problem encountered in the greenhouse was that of excessive temperature fluctuation. The greenhouse temperature often swung from 75°F on a sunny winter afternoon to 50°F that night. Eight five-gallon water drums painted black have subsequently been added in order to flatten the diurnal temperature curve and to double as plant benches.

Curtains and shutters have also been added over several windows. In addition, plans have been formulated to add a curtain between the greenhouse and living area and to provide triple glazing on some windows.

Construction and cost

Design of the house took about one year from conception to groundbreaking, although the actual design stage was much shorter since considerable preliminary effort was expended in researching the Trombe-wall and getting municipal approvals. The basic construction required about one year to complete, but the owners moved in during July 1975, approximately nine months after construction had commenced.

The entire cost of the house was estimated to be $55,000 plus the owner's labor. The additional initial cost of the house directly attributable to the solar heating and cooling system is approximately $8,000 to $10,000. This figure includes a modest allowance for the cost of the wall which the Trombe wall replaces. In addition, several thousand dollars could be saved if solid concrete block was used instead of poured-in-place concrete. It is difficult to estimate this cost since the system is so closely integrated into the design of the house.

Based on varying assumptions regarding general inflation and the escalation of natural gas prices, the owner estimates that the payback period for the system is from 10 to 17 years. Considering the declining availability of fossil fuels and the long life of the system, he expects the Trombe wall to pay for itself many times over in the long run.

Performance

Two winters of operating experience have been obtained since the completion of the house. The first winter (1975-76) was quite mild, about 4500°F-days; the second winter (1976-77) was more severe, about 5400°F-days. However, the modifications made to the house between the first and second winters resulted in similar space-heating requirements for the two winters by significantly reducing heat loss from the house.

Temperature performance

Instrumentation provided by Los Alamos Scientific Laboratory has been installed to monitor the temperature performance of the Trombe wall. Figure 6 depicts some illustrative data taken during the period from March 6 to March 8, 1977.

Temperature fluctuation was tolerably light, although vertical stratification is sometimes pronounced. A 5 to 10°F swing in the indoor temperature during the 24-hour cycle allows the concrete wall to collect and discharge its heat. Daytime settings were between 60 and 64°F and the nighttime setting was 58°F. The seasonal low indoor temperatures were 58°F downstairs and 62°F upstairs, and the corresponding highs were 70°F downstairs and 75°F upstairs. The estimated downstairs average was about 63°F while that for upstairs was 67°F.

Fig. 6
March 6-8, 1977,
Trombe wall
temperatures

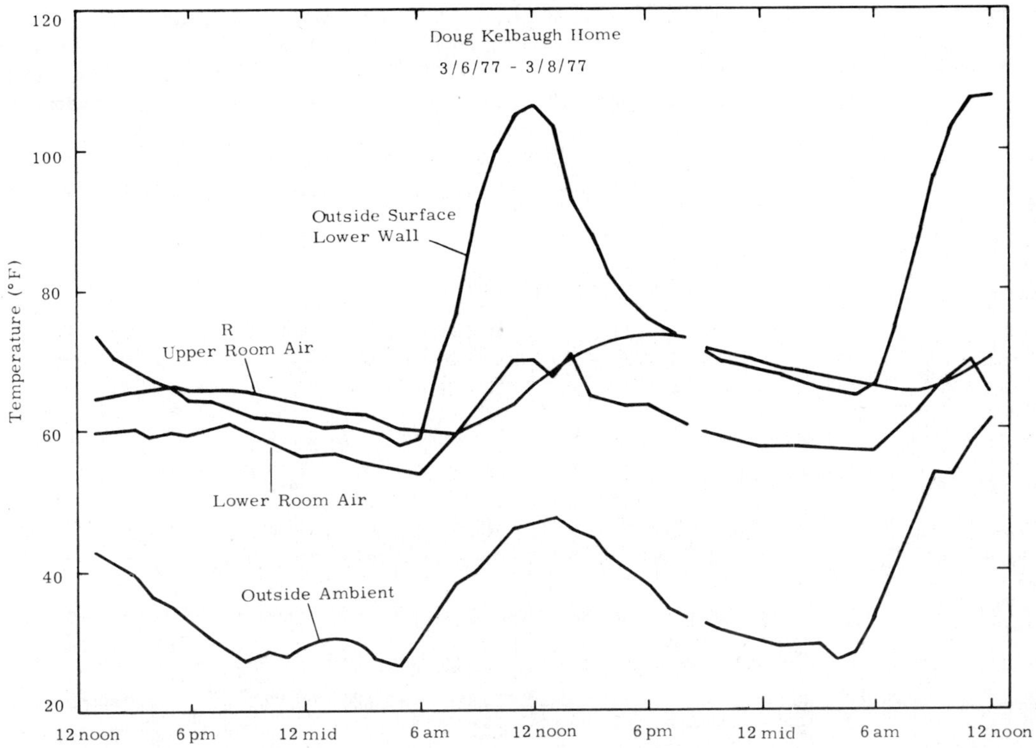

A Heatilator wood fireplace provided additional heat two or three times a week but was not considered in the calculations because it combusts warm room air and pulls in cold outside air, which offsets most of its heat contribution during mild months and all during cold months.

Operational cost

Table I illustrates the relative contributions of the various heating sources and assesses the savings attributable to the solar heating system. Quantities in the table are based on an average room temperature of 65°F in the living area and 58°F in the greenhouse.

For the first winter, the heat loss for the design was calculated to be 27,500 BTU/degree-day. As indicated in the table, the solar heating system reduced the space heating cost 76%, resulting in a $330 saving over the cost of heating by fossil fuels alone.

Table I

Performance summaries for the first two winters

	1975-6 (4500°F-days)	1976-7 (5400°F-days)
Heat loss (BTU)	124,000,000	129,600,000
Miscellaneous heat gain* (BTU)	17,000,000	17,000,000
Space-heating requirements (BTU)	107,000,000	112,600,000
Gas furnace contribution (BTU)	26,000,000	18,000,000
Solar contribution (BTU)	81,000,000	94,600,000
Savings**	$330	$379
Fuel bill (space-heating)	$108	$ 72

* From stove, clothes dryer, lights, hot water, and occupants

** Based on 31¢ per ccf of natural gas or 40¢ per gallon of oil for 1975-76 and based on 32¢ per ccf of natural gas for 1976-77

Modifications made to the system between the first and second winters improved the system so that the calculated heat loss for the design was reduced to 24,000BTU/degree-day. Consequently, the space-heating requirements were approximately the same for the second winter as for the first, despite the fact that the second winter was more severe. The solar contribution for the second winter was greater, reducing by 84% the space-heating bill for a net saving of $379.

Owner observations

The swing in air temperature (which was permitted to achieve high fuel savings) did not prove to be uncomfortable. The owners felt that, in terms of comfort, radiant heat is worth at least a couple of degrees Fahrenheit. In addition, the Heatilator wood fireplace often provided localized comfort (at the expense of increased air filtration), and, in any case, the owners considered the fireplace to be an incalculable aesthetic asset. The infrared heaters which were installed in the bathrooms were seldom needed even though the rooms received no direct radiant heat from the concrete wall.

The owners offer several recommendations based on their experience with this house. The vents in the concrete wall seem to be somewhat undersized. The eave vents are also too small, which hampers proper natural ventilation. Remote manual or automatic controls for the eave vents is desirable during spring and fall, and when cooling needs alternate. A fresh air vent supplied directly to the fireplace for combustion would also be desirable. The final recommendation was that the perimeter styrofoam insulation should be increased to two inches (from the one inch presently provided).

Conclusions

The Trombe-wall solar heating system in this house has performed very successfully. Substantial savings on fuel bills have already been realized and will accumulate at an accelerated rate as fossil fuel costs escalate. When other factors (such as convenience, reliability and cost of maintenance) are considered, the economics of such a design becomes increasingly favorable.

J.B. Wright
Department of Earth Sciences
The Open University

Introduction

The solar heating system for the Anglesey bungalow was designed by S.V. Szokolay, who planned the installation in the Milton Keynes experimental solar house (Szokolay, 1975).

The bungalow (Figure 1) is unique in two respects:

1. It is believed to be the first privately built house in Britain to have an active solar space heating system.

2. The space heating is by underfloor water circulation, which provides background warmth with water temperatures as low as 25°C (77°F). Such temperatures are attainable from solar radiation even in mid-winter provided, of course, that the weather is suitable.

Fig. 1 The solar bungalow, Malltraeth, Anglesey, showing the 16 collector panels under glass on the south facing roof.

The purpose of this article is partly to describe the heating system and confirm that at least 40-50 per cent of domestic energy requirements are provided directly or indirectly by the sun. The principal objective is to show that the energy costs of materials in the installation are appreciably greater than those for a conventional system of comparable efficiency - to the extent that the sun will have to 'pay back' about 25,000 kWht (see footnote, Table 1) before we can expect to break even with conventional systems in energy terms. This will take two to three years of operation.

Heating system

The essential features are summarised in Figure 2 and the associated captions.

Chemical inhibitors and anti-freeze are added to the water to minimise bimetallic corrosion and frost damage.

The system was assembled, installed and tested by T.H. Jones of Holyhead.

To the principal components (1-4) we must add about another 50 metres of assorted copper and steel tubing (diameters range from 15 to 35mm), which complete the various circuits shown in Figure 2.

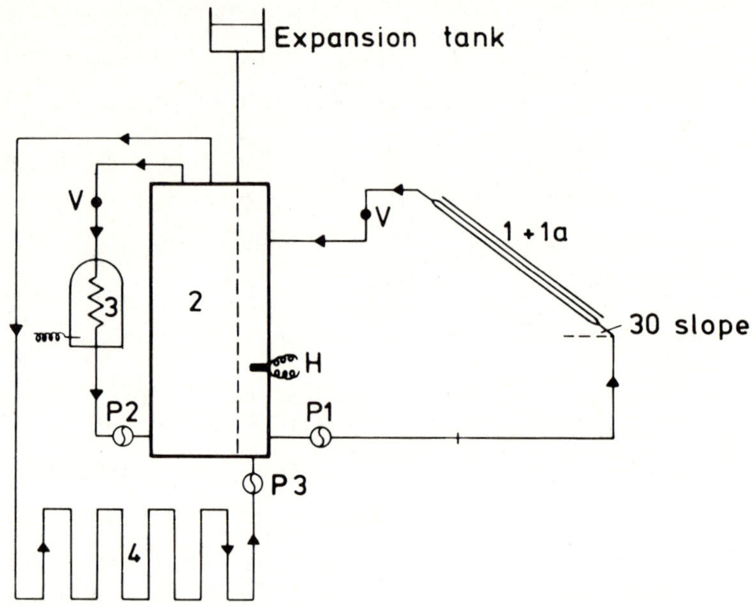

Fig. 2 Diagram of the solar heating circuit, not to scale. For explanation, see text.

Key to Figure 2

1. Sixteen collector panels, each 2.5 x 0.825m in area, capacity 6 litres. Total collection area $33m^2$. (Suppliers: Alcoa, Swansea.)

1a. Patent glazing, one glass pane per panel, with some overlap. Mainly low-Fe St. Gobain glass, 4mm thick. (Suppliers: Ruberoid, Basingstoke.)

2. Storage tank, 3 x 3 x 10 feet, $\frac{1}{4}$ inch mild steel, total volume $90ft^3$ or 550 gallons, i.e. $2.55m^3$.

3. Standard domestic hot water cylinder (30 gal., $0.15m^3$), with heat exchange coil, as well as conventional 2-3kW immersion heater.

4. Underfloor warming pipes, for space heating: 340 metres of 1cm diameter copper tubing in four independent circuits.

P1 Circulation pump between storage tank and collectors.

P2 Circulation pump between storage tank and domestic hot water tank.

P3 Circulation pump for floor circuits.

(All pumps rated at 90 W each. P1 and P2 are controlled by adjustable differential temperature sensors (Suppliers: Satchewell Control Systems, Slough). P3 is controlled by an ordinary room thermostat with an adjustable minimum temperature cut-out set to 25°C - i.e. when the temperature in the top of the storage tank falls below 25°C, the floor circuit pump automatically cuts out.)

H Two 3 kW immersion heaters in the storage tank, linked to the off-peak supply.

V Non-return valves to prevent reverse circulation by convection (thermosyphoning).

The significant constructional features of the house are summarised as follows:

The Volume of the bungalow is roughly 325m^3, of which 270m^3 is heated. All but 15m^2 of the total 95m^2 floor area is warmed by the underfloor circulation system.

The Walls are outer brick/inner block, under plast, cavity-filled.

The Roof has 4 inch polystyrene block under both collectors and tiles.

The Floors are quarry tile on screed on 4 inch concrete slab, with 3-foot polystyrene edge insulation (2 inch thick) along all walls.

The Window area totals approximately 25m^2, of which about 15m^2 is south of the west facing sliding patio-door. All windows are double glazed.

The Doors (additional to above) total approximately 4m in area, made of wood and glass.

Energy costs of the materials

A conventional fossil fuel domestic heating system of comparable type will comprise boiler, radiators, hot water cylinder, and assorted (mainly) copper tubing. The amount of metal involved in such a conventional system is here assumed to be equivalent to component 3 of Figure 2 (the domestic hot water tank), plus half of component 2 (the storage tank), and about 100m of tubing. We need therefore to calculate the energy costs of the additional materials used in the solar installation described here. They amount to:

16 aluminium panels, each weighing 8kg ... 128kg aluminium

37m^2 glass, mostly 4mm thick, some 5mm, assumed density 2800kg m^{-3}, approximately total weight ... 450kg glass

Half the amount of mild steel in the storage tank: 2.97ft^3, or 0.084m^3, assumed density 7800kg m^{-3}, 650kg; total ... 325kg steel

250m piping, mainly copper, estimated average diameter 1.5cm. approx.; total weight ... 60kg copper

The energy costs for these materials are set out in Table 1

Table 1 Energy costs of solar heating system

Material	Weight kg	Extraction & refining kWht kg^{-1}*	Extraction & refining total kWht	+10% (est) fabrication & installation	Total kWht (rounded off)
Aluminium	128	91	11650	1165	12800
Glass	450	6	2700	270	3000
Steel	325	13	4225	423	4700
Copper	60	20	1000	100	1100
					21600

* Data from Chapman (1975). kWht = kilowatt hours thermal. These are the total energy costs needed to obtain the products from raw materials, and allowing for energy conversion factors in processing.

The solar panels and patent glazing replace an equivalent area of tiles, which would have an energy cost (@ 51 kWht m^{-2}, Chapman 1975) of 1900 kWh. When this is subtracted from the total in Table 1, the final figure for the energy costs of additional materials required for the solar heating system is very close to 20,000 kWht. If we allow a further generous 5,000 kWht for pumps, switches, valves, anti-freeze, and other items not mentioned above, the total 'surplus cost comes to about 25,000 kWht.

Solar contribution to domestic energy requirements

Prediction: In his specification, Szokolay (1974) computed the average annual space heating demand at 17,630 kWh, and predicted a solar contribution of 7,400 kWh (42 per cent) to that demand, the balance being made up from auxiliary heating.

The calculations were based on a different collector type and slightly different insulation characteristics, and improvements in the final design might be expected slightly to decrease the demand and increase the relative solar contribution to perhaps 17,000 kWh and 7,600 kWh (45 per cent) respectively.

Performance: The first year of operation was from end-May, 1975 to end-May, 1976. There is no precise way of assessing the performance of the system, because there are too many variables and unknowns, and no systematic monitoring programme was built in, as at Milton Keynes. However, the available information suggests that performance is not significantly below what was predicted, and may be somewhat better.

Immersion heaters in the storage tank (H on Figure 2) were set at 90°F (32°C) and operated from 20 October to 20 April, during which period they consumed 3160 kWhe (kilowatt hours electrical). During periods of nil sunshine in mid-winter, nightly consumption averaged 25-30 kWhe but this was halved after sunny days, when water entering storage from the collectors attained about 80°F (24°C). At the beginning of March, collection temperatures of 110°F (about 45°C) were recorded on sunny days, and nightly consumption by the immersion heaters was in the range of 0-19 kWh, depending upon the amount of solar collection the previous day. By mid-April collection temperatures of 120°F (about 50°C) were regularly attained during sunny days and the immersion heaters were switched off soon afterwards. Temperatures of 130°F to 140°F (about 55-60°C) are quite commonly achieved during summer months.

While underfloor circulation of water from the storage tank for a few hours each day provided background warmth during the colder months, additional space heating was necessary. This was provided almost entirely by a wood-burning stove, which consumed about two tons of fuel during the autumn-spring period - and did most of the cooking as well, thus effecting considerable savings on use of the electric cooker. Given a reasonable estimate of 3.5 kWht kg for the calorific value of the wood, two tons would provide about 7,000 kWht of energy. Unfortunately, we do not know the thermal efficiency of the stove, which must be fairly high because the stove is set back from the wall and the chimney flue is inside the house; nonetheless, about 30 per cent of the heat probably escaped, leaving an estimate 5,000 kWht of useful heat.

A negligible amount of space heating was occasionally provided by a small convector heater on very cold days, but this cannot have used more than 50 kWhe throughout the whole period. The circulating pumps never ran for more than a few hours per day, even in the sunniest weather, and their annual electricity consumption is probably overestimated at 150 kWhe. The internal room thermostat was often used as a manual on-off switch for the circulation pump for the underfloor circuits, but in the colder months was usually set in the 18-20°C (65-70°F) range, so that it cut out the pump when the wood-burning stove took over the main heat load in the evening and on those occasions during the day when the direct solar gain was adequate to maintain the desired temperature.

The auxiliary energy used for space heating therefore works out approximately as follows:

		kWh
storage-tank heaters		3160
wood burning	est.	5000
convector heater	est.	50
pump operation	est.	150
Total approx. (say 8500)		8360

If the total annual space heating demand has been correctly estimated at 17,000 kWh then these provisional figures suggest that some 8,500 kWh of space heating are provided directly by the sun, though not necessarily all of it via the collector system: the house is insulated to above-average standards, and large double-glazed, south-facing patio doors provide a sun-trap effect, especially at low sun angles.

Water heating requirements were not included in the estimate for space heating demand, and there is no certain way of estimating what the energy requirement for hot water would be in a conventional house. It is clear, however, that the system portrayed in Figure 1 has some decided advantages: the water in the domestic tank cannot fall much below 25°C (77°F), because the storage tank temperature is rarely less than that, and the heat exchange system (see item 2 on Figure 2) ensures temperature balance between storage and hot water tanks. Even in the depths of winter therefore, water from the hot tap is at least tepid, and the immersion heater in the hot water tank need only be switched on for short periods to boost the temperature when required: about 20 minutes per day on average, over the six month period October to March. For the remainder (April to September), with solar collection temperatures reaching 50-60°C (120-140°F) on sunny days, there is often an excess of hot water, and the immersion heater in the domestic tank remains switched off for weeks at a time. The annual electricity requirement for hot water is probably about 150-200 kWhe. Without the solar installation, water heating requirements for a comparable establishment (1½ persons and three dogs) could well amount to two or three times that figure, but there is no positive evidence for such an estimate.

Real savings on conventional fuels: the pay-off

About 8,500 kWh of space heating and perhaps some 500 kWh of water heating are provided annually by the sun. If this total of 9,000 kWh had to be provided by electricity at a conversion efficiency of about 25 per cent, then the total primary energy required would be, roughly, 36,000 kWht. Direct use of fossil fuels, in coal, oil, or gas fired heating systems, is a good deal more efficient at between 60 and 80 per cent, which means that the real savings come down to between 11,000 and 15,000 kWht.

The excess energy costs of the additional materials used in this solar installation, estimated at about 25,000 kWht, could thus be paid off in one to three years. In money terms we can be less certain: the annual saving on fuel bills will be in the order of £100. At an installation cost of some £1,300, compared with about £900 for a conventional system in a house of this size, the additional financial outlay to build in the solar heating would also be paid off in about four years; but this assumes trouble-free operation without significant maintenance costs, which cannot yet be guaranteed.

The problem of excess heat

Figure 3 illustrates the chief drawback to the use of solar energy in temperate latitudes: a huge excess in summer months, when space heating requirements are minimal, and a large deficit in winter months, when most heat is required. The 8,500 kWh which we estimate is contributed by the sun towards the space heating requirements corresponds to the shaded areas on Figure 3. Some of the excess produced during summer months provides virtually all the hot water (only during a prolonged cloudy spell would the immersion heater be needed). That is only a small proportion of the total summer input, however, and the rest would remain unused. In heat-waves therefore (cf, June-August, 1975), this excess could lead to over-heating - were it not for the unique design feature of this house: the floor can be used as heat sink and heat store. Even in high summer, with water at up to 60°C (140°F) being circulated through it for a few hours each day, the floor remains relatively cool, because ambient temperatures are already high. Having been warmed throughout the summer, the floor never cools down completely, so that when cooler water (27-32°C in winter months) is subsequently circulated through it, an adequate background warmth is maintained.

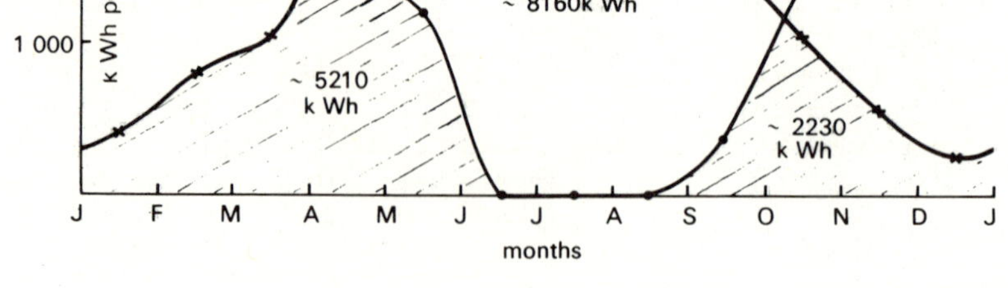

Fig. 3 Curves of estimated average monthly space heating demand (dots) and solar contribution (crosses), based on predictions by S. Szokolay (written communication). The true demand and supply curves are probably a little lower and higher (respectively) than shown because of design improvements since the predictions were made. Shaded areas show the estimated effective solar contribution to space heating requirements in winter months.

In conclusion, then, if trouble-free operation continues, a solar heating system of the kind described here would seem to be a viable proposition.

It is worth noting that Chapman (1975a) has come to very similar conclusions about the energy costs and benefits of solar heating, based on calculations for the Milton Keynes house - some of the figures he uses are open to question, but the 'errors' cancel each other out, and his estimate for the energy pay-off period comes to 2-3 years.

References

Chapman, P.F. (1975). 'The energy costs of materials'. (Energy Budgets, Part 4). Energy Policy, March, 1975, 47-59.

Chapman, P.F. (1975a). 'Fuel's Paradise'. Penguin Special.

Szokolay, S. (1975). 'Solar Energy and Building'. Architectural Press, 1975.

Solar technology for building: Volume 2

Volume 2: Sessions 5 to 9

Solar technology for building

Proceedings of the first International Conference on Solar building technology
25–29 July 1977
At the Royal Institute of British Architects, London
Organised by UNESCO and NELP

Edited by Costis Stambolis

RIBA Publications Limited

Organised by:

United Nations Educational and Scientific Organisation and

North East London Polytechnic, Faculty of Environmental Studies,

Department of General Surveying and Construction

Sponsored by:

International Solar Energy Society

Intermediate Technology Development Group

New Mexico Solar Energy Association

UK Solar Trade Association

ISBN: 0 900630 66 3

Published for UNESCO and NELP

by RIBA Publications Limited, Finsbury Mission,

Moreland Street, London EC1V 8VB

Printed in Great Britain by Unwin Brothers Ltd,

The Gresham Press, Old Woking, Surrey

Contents

Space heating and cooling

5

5.1 Utilisation of solar energy for space cooling

J.C.V. Chinnappa
P.N.G. University of Technology
Lae, Papua
New Guinea

Introduction

In many environments the provision of a comfortable enclosed space sometimes requires heating only, sometimes cooling only, and sometimes both. The designer, while never entirely free from all constraint, has considerable scope to determine the manner in which this may be achieved. Szokolay (1) recognises a polarisation of design approach and describes the extremes:

(a) to make the building envelope an effective barrier between the inside and the outside and to rely on some energy based mechanical equipment to ensure indoor comfort; and
(b) to design the building in such a way that indoor conditions are maintained within (or as near as possible to) the comfort limits by 'natural means'.

Literal interpretation of the title suggests that this address should confine itself to just those means of effecting space cooling, so-called 'active systems', which are operated by solar radiation as a heat source. Such an interpretation may be considered restrictive in the light of Szokolay's remarks and so the coverage has been extended to include examples of 'passive systems' as well.

Many types of active systems using solar energy as heat source for their operation have been proposed and include:
(a) LiBr-water single and multistage absorption systems,
(b) organic working fluid Rankine cycle systems coupled to vapour compression units,
(c) ammonia-water single stage and multistage absorption systems, and
(d) adsorption and absorption dehumidification.

Two of these schemes, namely the LiBr single stage absorption and the Rankine cycle systems, have progressed well beyong the proposal stage and are now in the process of full scale testing. Others are at various stages of development from feasibility studies through the design and testing of components to the testing of experimental prototypes. Examples of these systems will be reviewed.

Passive systems, while using solar energy for heating, use radiation to the night sky and convective heat transfer to the cool night air as the sources of cooling. However they are illustrative of an important principle in environmental design, and a few examples are reviewed.

Aqueous Lithium Bromide systems

The pioneer work on the solar operation of the aqueous LiBr cycle was due at the University of Wisconsin Solar Energy Laboratory (2,3), and the first house to be cooled by a prototype system was the solar house at Brisbane at the University of Queensland (4,5). Since then a considerable amount of work has been done on all aspects - thermodynamic, plant design, and full scale performance testing.

The ideal thermodynamic vapour absorption cycle can be represented on the p-t-x diagram (Figure 1). On the abscissa scale we can read off values for evaporator temperature, condenser temperature, the temperature ranges during absorption and generation, the associated pressures and solution concentrations.

Fig. 1
Simple vapour absorption cycle

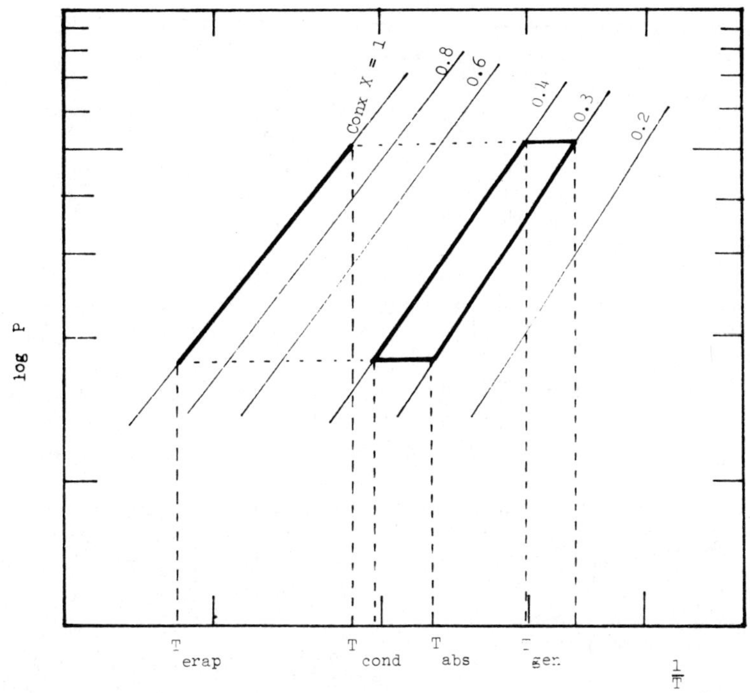

Allen, Morse et al (6,7) have performed a systematic analysis of the thermodynamic cycle. They have investigated the basic cycle, and thereafter the effect of off-design operation on the coefficient of performance, heat exchanger size, solution circulation rate etc. Miller (8) has computed the effect of variations in cooling water and chilled water temperatures on heat supply temperatures.

Multistage LiBr systems have been theoretically investigated in computer studies by Allen, Morse et al (7) and by Wilbur and Mitchell (9). In addition Wilbur and Mitchell have enquired into the effects of a heat rejection buffer and of refrigerant storage. Interesting developments are foreshadowed by these enquiries.

In addition to the theoretical studies just mentioned full scale testing of air-conditioning plant incorporating single-stage aqueous LiBr coolers are in progress in the USA and in Japan. The programme at the University of Colorado is chosen to illustrate these investigations. The objectives and design criteria of the programme are worth quoting in full because they are appropriate and comprehensive and implicitly describe the problem (10):

(1) Establish the practicality of space cooling with solar energy.
(2) Design an effective and economical system for residential heating and cooling and service water heating with solar energy.
(3) Construct a fully automated and instrumented solar heating and cooling system in a new residential building.
(4) Appraise the performance of the complete heating and cooling system and each of its principal components.
(5) Determine which of several operating modes minimizes auxiliary energy requirements.
(6) Appraise the utility of a mathematical model in design and prediction of performance by comparing actual and predicted results.
(7) Modify system design and operation to minimize total annual cost of heating and cooling.

To accomplish these objectives, criteria for system design and performance were established as follows:

(1) Conducting a proof-of-concept experiment and demonstration with an optimized system of predictable satisfactory performance with currently-known technology.
(2) Utilizing a building design which has 'typical' rather than 'extreme' or 'unusual' appearance, size, form, materials, energy requirements, insulation, fenestration and all other characteristics of a modern single-family house.
(3) Providing at all times a comfort level in the building fully equal to that obtainable with the best conventional heating and cooling systems.
(4) Designing the building and the heating and cooling system so that the solar collectors and other components can be easily replaced with completely different units for testing after the present system has been fully evaluated.
(5) Designing a collector for mounting on the roof surface but also suitable for modifying and roof replacement.
(6) Designing a system which will carry about three-quarters of the annual heating load and three-quarters of the annual cooling load with solar energy.
(7) Providing sufficient flexibility and 'redundancy' on the heating and cooling system to permit comparison of several variations in design and operating method.
(8) Employing commercially available controls in a fully automatic system requiring no human intervention other than setting a wall thermostat for the desired house temperature.
(9) Assuming reliability, durability, and safety considerations take priority over least cost incentives in this experimental installation.
(10) Minimizing system cost without jeopardizing performance or durability.
(11) Utilizing only those designs and materials which could provide low cost solar energy with high volume manufacture.

The programme includes the construction and testing of three houses. The first of these has flat plate collectors heating water and a LiBr unit, and will be described more fully later. The second has collectors heating air and a pebble bed storage. When cooling of the house is required, i.e. during the summer the pebble bed is cooled by the circulation of night air. During the day the pebble store cools the air supplied to the house. The third house will be provided with evacuated tubular collectors and an LiBr unit of more advanced design.

The schematic of the plant used in the first house is shown in Figure 2. This conglomerate of systems poses a design problem of some complexity. The size, orientation and design of the collector, the storage capacity, the capacity of the auxiliary heater and boiler, the capacity of the air-conditioner are some of the major variables. The performance has to be assessed over the whole year. This problem, specifically for the CSU house, has been handled by computer modelling of the whole system (11). Several modes of operation are possible and these are described in ref.(10). Clearly there need to be automatic controllers. All this calls for increased sophistication in modelling procedures and these have been forthcoming (12,13,14).

The full details of the plant are given in ref.(10). The major specifications are:
House living area: 140m2
Flat plate collector area: 95.2m2 (4.88m x 19.5m)
Storage tank volume: 4275 litres
Air Conditioner: modified Arkla Servel 3 ton LiBr.

Fig. 2
Flow diagram for
CSU House I.
Solar heating and
cooling system. (10)

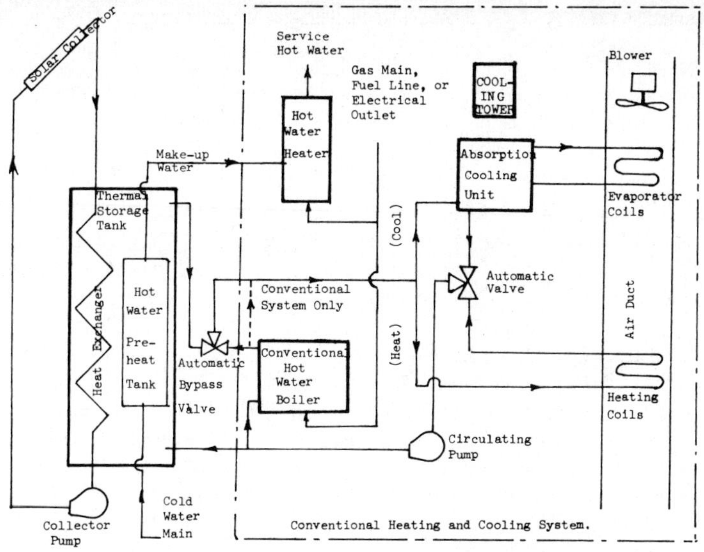

Fig. 3

Preliminary performance of solar heating and cooling system, CSU House I. August 1974 to January 1975. (15)

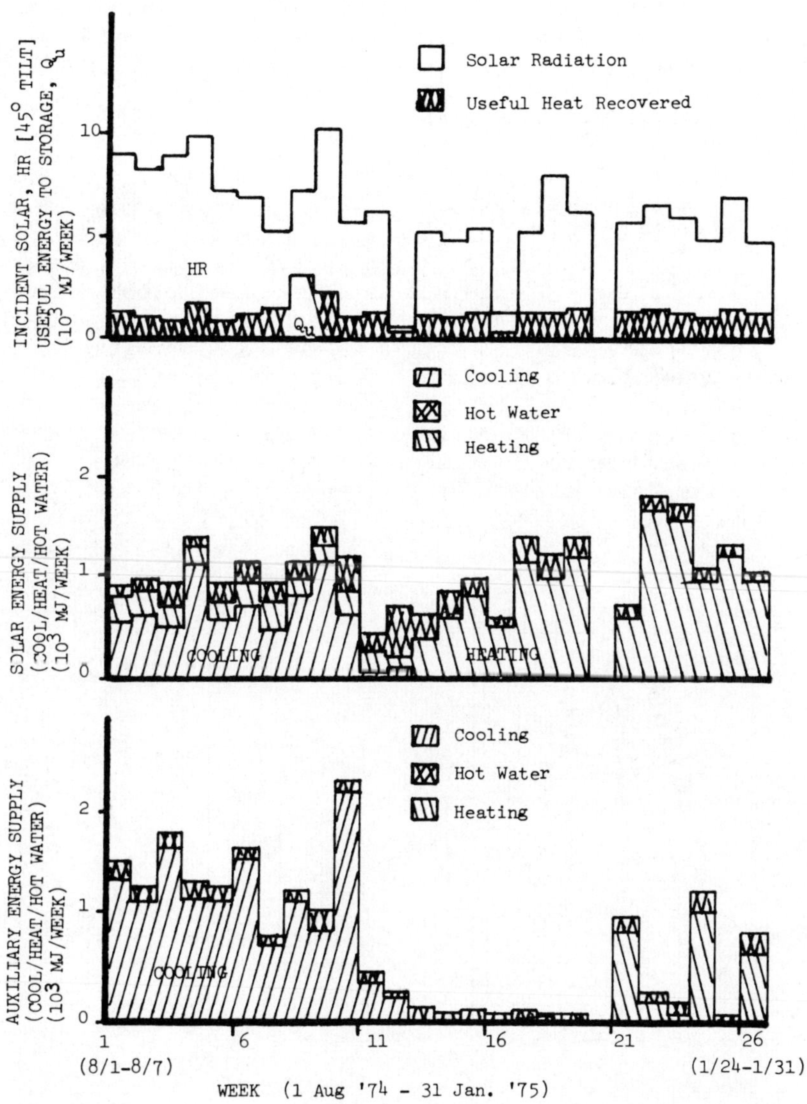

SOURCE	TOTAL ENERGY DELIVERED TO LOAD				COOLING (MJ)
	TOTAL (MJ)	HEATING (MJ)	HOT WATER (MJ)	COOLING UNIT (MJ)	
SOLAR	33,996	20,687	5,248	8,061	4,705
AUXILIARY	22,442	3,374	2,776	14,185	7,457
PER CENT SOLAR	60	86	68	36	39

Data is being collected on the operation of this system and the first results have been reported (Figure 3) (15). The investigation continues and we can look forward to many useful reports covering all aspects of performance.

One example of a house cooling investigation using an LiBr unit has been chosen for purpose of illustration. At the Los Angeles ISES Conference and the Solar Cooling Workshop there were eight descriptions of similar building cooling investigations all recently commenced (16, 17, 18, 19, 20, 21, 22, 55). The largest of these are the plants proposed for the National Security and Resources Centre at Los Alamos with a capacity of 77 refrigeration tons (270kW), and the 100 ton (352kW) plant for the George Towns School in Georgia.

The extent to which the idea has caught on can be gauged from the fact that Arkla Industries have redesigned their gas-fired LiBr units to be operated by hot water (from solar collectors) by modifying the generator and including a solution circulating pump. This unit was expected to be in production by the third quarter of 1976 (23).

Rankine cycle systems

The second method of utilising solar energy to produce cooling is by operation of a vapour cycle the ideal form of which is the Rankine cycle (Figure 4). The arrangement consists essentially of 3 loops. The solar loop consists of solar collectors heating up water which can be stored as hot water. This hot water is used to boil the working fluid in the vapour cycle. The vapour is expanded in a turbine, condensed in a feed pump and returned to the boiler. The turbine drives the compressor of a conventional vapour compression refrigerator.

There are 4 recent reports of notable advance in this area (24, 25, 26, 27). Details from the paper by Eckard and Bond are used to illustrate the principles and performance of this system.

Fig. 4 Rankine cycle system. (26)

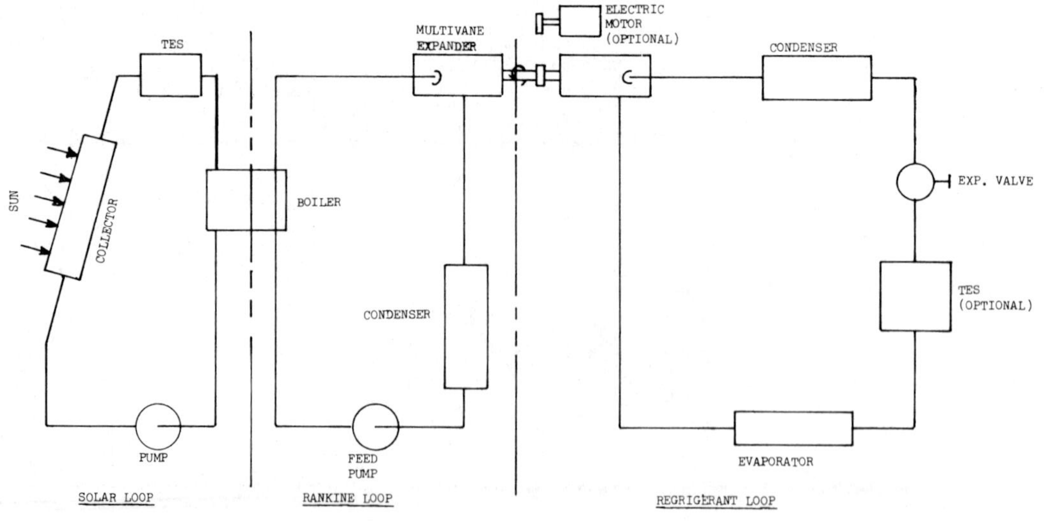

One of several organic fluids may be used in the vapour cycle loop. The Rankine efficiencies of some are shown in Figure 5. The ideal coefficient of performance is the cooling effect produced in the evaporator divided by the heat supplied in the boiler and this is shown in Figure 6 for two different evaporator temperatures against the COP of an absorption system. It will be observed that at lower condensing temperatures the Rankine cycle-vapour compression combination has significantly higher values of coefficient of performance.

The actual performance of the 3 ton prototype is shown in Figures 7, 8 and 9. The unit has a new multivane expander developed by GEC and its performance is shown in Figure 7. Figure 8 shows actual tests compared to Rankine cycle predictions and Figure 9 shows the variation in COP as the heat supply temperature is increased.

The performance described above is characteristic of the other investigations reported. The Barber-Nichols prototype is in the Honeywell mobile demonstration home and the GEC prototype in a similar GEC mobile home (28). A much larger 77 ton (270kW) Rankine cycle unit is being designed by Barber-Nichols for the Los Alamos National Security and Resources Centre (29). This building will therefore be provided with both systems so that they may be compared.

Fig. 5

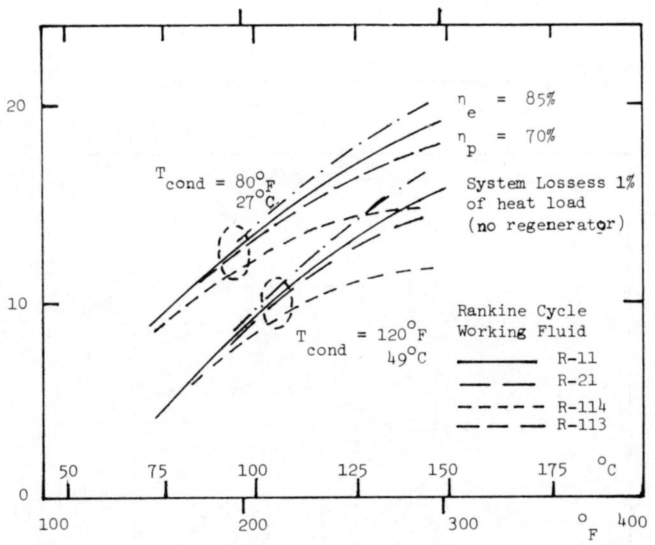

Rankine cycle efficiency for several candidate fluids. (26)

Fig. 6

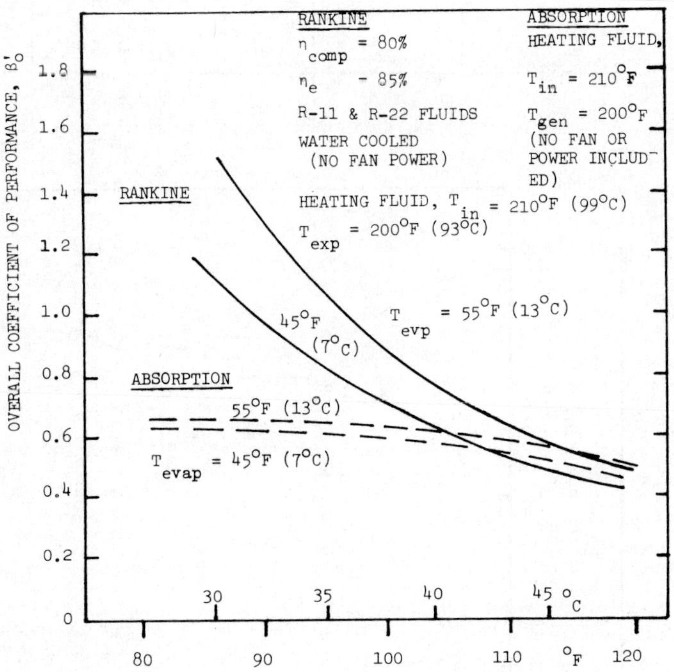

Effect of condensation temperature on the overall COP of a Rankine and an absorption system. (26)

Fig. 7
Efficiency of GEC
Multivane Expander.
(26)

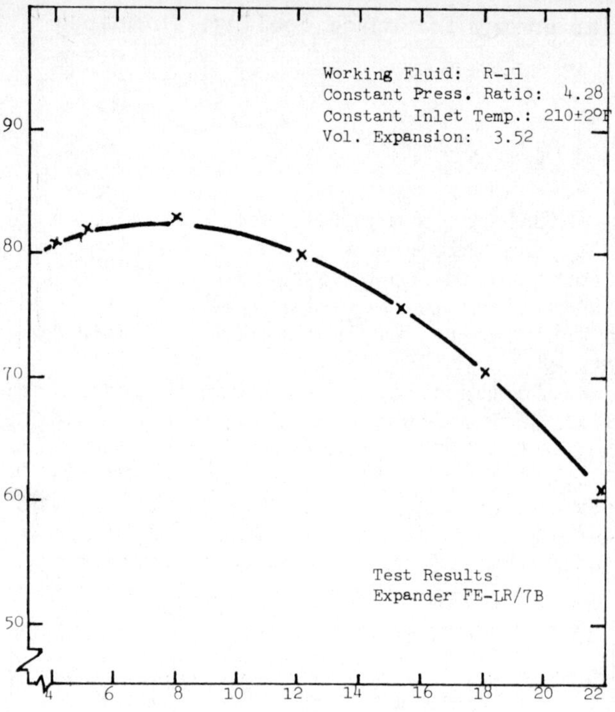

Fig. 8
Comparison of
tests and
predictions:
Rankine cycle
plant. (26)

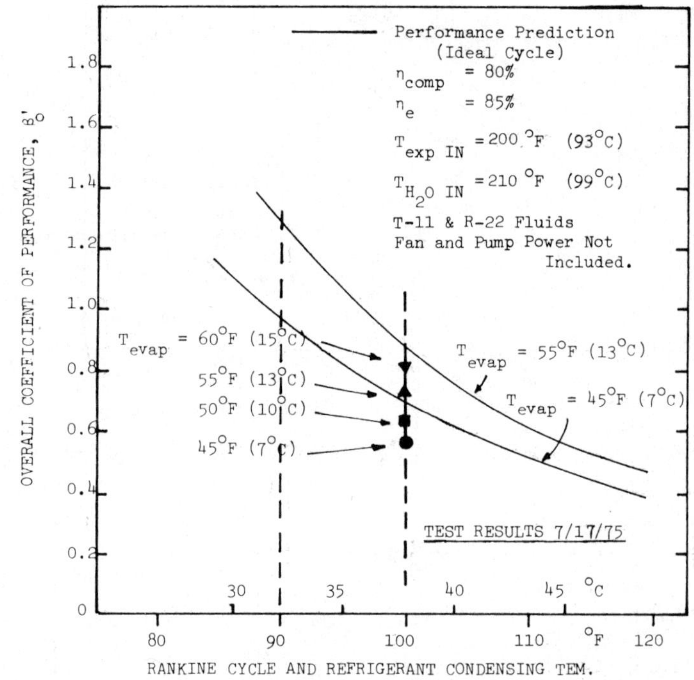

Fig. 9
Test results:
Variation of COP
and cooling effect
with hot water
inlet temperature.
(26)

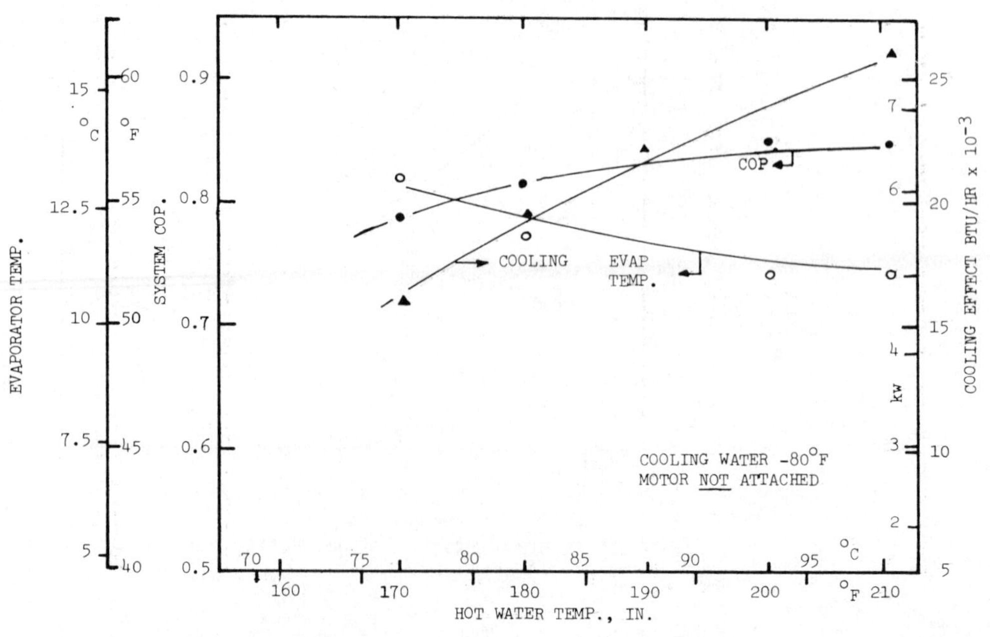

Aqua-Ammonia systems

While the two schemes already described have undergone considerable development aqua-amonia systems using solar energy are still in the initial stages.

The pioneer work was done at the University of Florida by Farber and his associates (30), and a preliminary computed estimate has been made of the performance of the single stage system at Columbo, Sri Lanka (Ceylon) (31).

One of the reasons for the delay in the development is that the commercial aqua-ammonia air conditioning unit operates with generator (or heat supply) temperatures in excess of 300°F (150°C). This temperature is much in excess of temperatures attainable with flat plate collectors and the generator has to be completely redesigned to be heated by hot water at temperatures around 100°C. This work however has now been taken in hand and at University of California, Berkeley, Simmons and his co-workers are modifying an Arkla ACB60 5 ton machine by designing and building a new generator to be heated by hot water (32, 33).

While plant design and experimental work is proceeding studies of the thermodynamics of the cycle and modelling of the solar collector - absorption refrigerator combination are also under way (7).

The intermittent cycle has also been proposed for air conditioning (34) and the performance of an experimental plant working on this cycle reported at the University of Florida by Farber.

Fig. 10
Two-stage
Aqua-amonia
system. (36)

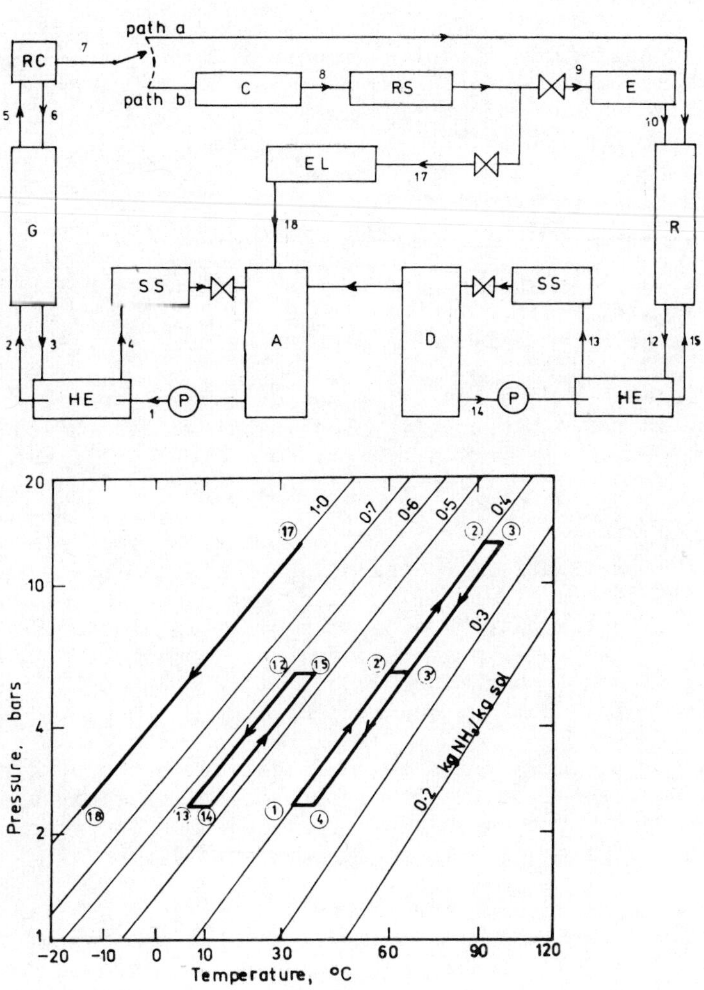

Fig. 11 Computed comparative performance of LiBr and 2-stage NH_3 systems. (36)

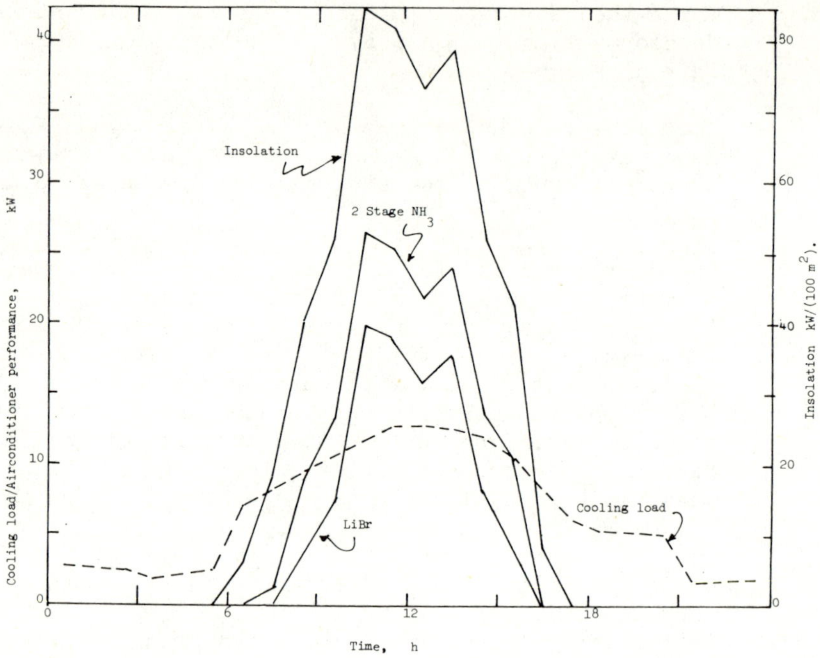

Multistage systems have also been investigated (35, 36). Figure 10 shows a two-stage system which can operate in dual mode and Figure 11 compares the computed performance of an LiBr system with that of the 2 stage ammonia system.

A considerable amount of theoretical and experimental work is still left to be done before the aqua-ammonia absorption system both single and two stage can be adequately assessed. There certainly appear to be quite promising possibilities in this area.

Dehumidification by adsorption and absorption

The earliest proposal regarding the use of solar energy for dehumidification was due to Log in 1955 (37). He proposed the use of tetraethylene glycol as a dehydrating agent.

Since then work has been done on several absorbents both liquid and solid and on adsorbents, the investigations at the C.S.I.R.O. Division of Mechanical Engineering Melbourne by Dunkle and co-workers being notable.

Essentially the procedure for producing comfort conditions using a dehumidification process with a solid desiccant packed into a wheel is as shown in Figure 12. Ambient air which is hot and humid is passed through the matrix of the dehumidifying wheel packed with desiccant (e.g. silicagel or molecular sieve). The air is both heated and dried as it passes through this wheel. It is then cooled and finally humidified to achieve comfort conditions.

Actually the systems are more complex incorporating heat exchangers to reduce the heat inputs and providing for recirculation of room air. The system proposed by Dunkle (38) is shown in Figure 13 and the processes on the psychrometric chart in Figure 14.

Dunkle's proposal for cooling has not yet, as far as the writer is aware, been experimentally tested but testing of components and studies on absorption - desorption have been in progress (39, 40, 41, 42).

Recently Close and Dunkle have proposed a modification of this cycle using silicagel both as heat storage and dehumidifying agent (43).

At the 1975 ERDA Los Angeles Solar Cooling Workshop systems operating on the same principle but differing in detail were described by Rush (44) and Lunde (45). Lunde described a feasibility study using silicagel and Rush a prototype unit in which a gas burner was provided to supply auxiliary heat (when solar heating is inadequate) to regenerate the 50% molecular sieve-impregnated asbestos dehumidifying wheel. The solar collectors were evacuated glass tube collectors and the hot water storage consisted of two 500 gallon tanks. Three of these units, each of approximately 3 ton capacity were under test in 1976.

Tchernev (46) has described a system using zeolite as an adsorbent. He claims that the properties of zeolite are superior to silicagel when solar operation is considered.

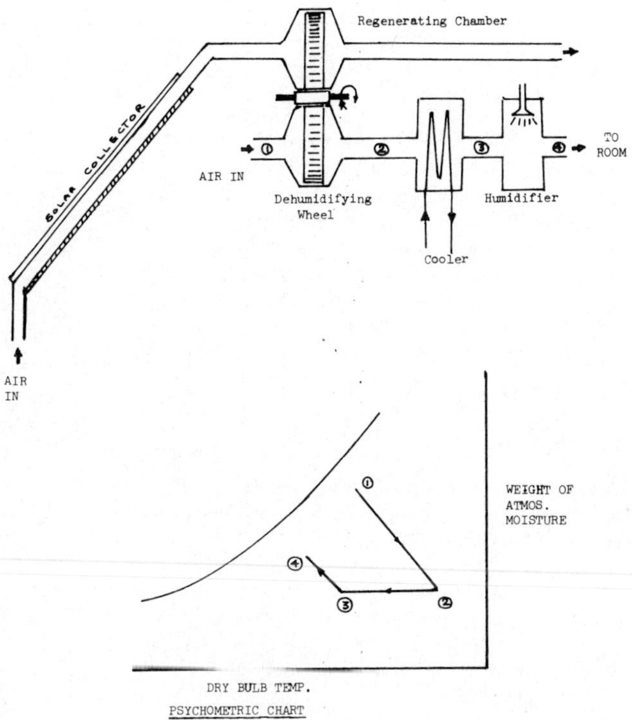

Fig. 12
Principle of air conditioning by dehumidification.

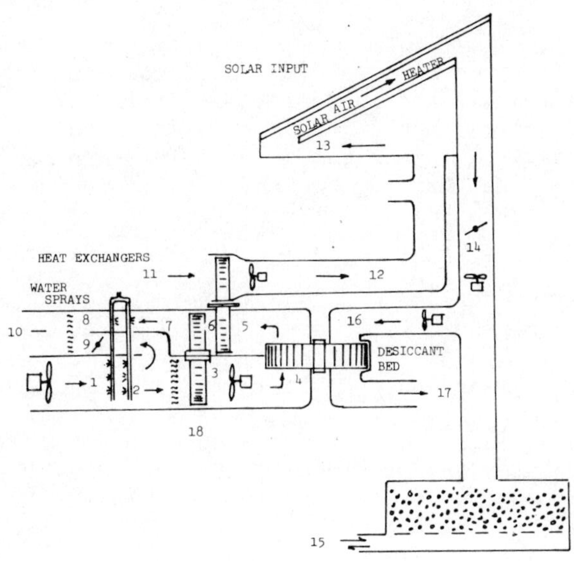

Fig. 13
Solar air conditioning system. (38)

Fig. 14
Cycle on psychrometric chart. (38)

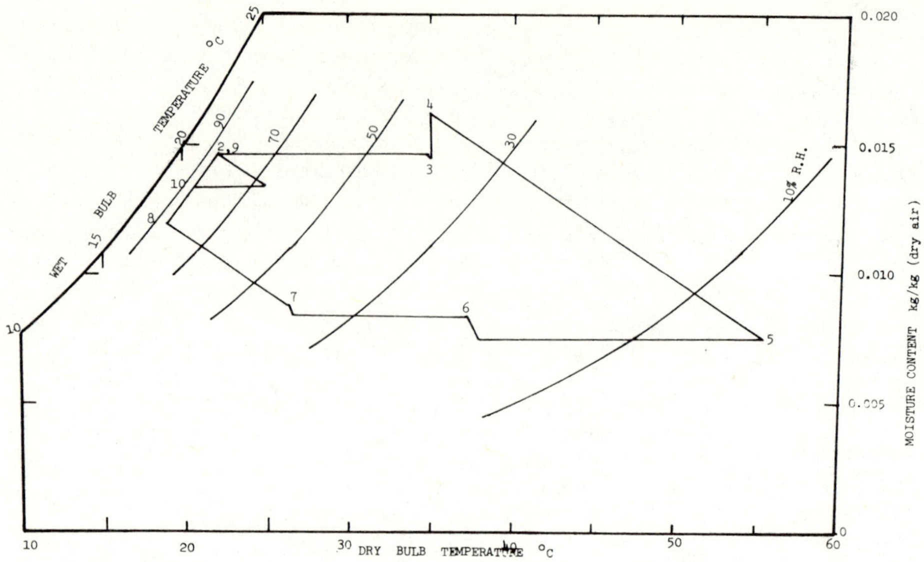

Of interest in these several investigations were various adsorbents proposed and the manner in which solar regeneration was accomplished. Dunkle has suggested silicagel and its regeneration by hot air directly from solar air heaters or indirectly from a rock pile thermal store. Lunde and Rush have proposed solar water heaters, with this hot water being used to heat the air required for regeneration. On the other hand Tchernev has proposed that the adsorbent zeolite be heated directly by the sun.

Meanwhile Mullick and Gupta (47, 48) have systematically investigated CaC1 as an absorbent. The first phase of the investigation has been the design and testing of a free flow solar heater for the refrigeration of the brine with one glass cover, open at top and bottom to the atmosphere. It is expected that this phase will be followed by the design and testing of a complete airconditioning plant. A similar investigation using LiC1 has been reported from the U.S.S.R. (49).

Passive systems

Passive systems in general utilise the diurnal swing of the ambient temperature and radiation to the night sky to produce a low temperature thermal storage which can be used during the warmer times of the day to cool the living spaces.

The system devised by Harold Hay and known as the 'Skytherm' house is well-known and has been described in detail by Stambolis (50). The performance of the house constructed at Atascadero is shown in Figure 15 (51). The equable temperatures maintained are most remarkable and it is clear that in hot dry regions this scheme is quite appropriate. We are informed in the March 1977 issue of the I.S.E.S. News that 5 Skytherm houses are being constructed on the U.S. Federal demonstration programme.

An interesting variation is the thermic diode panel. The valving in these coupled panels (the outer being the radiation sensor and inner water storage) permits heat to be transferred in one direction only, outwards in summer and inwards in winter (Figure 16) (52).

Variations on this theme are obviously possible and while these methods do not use solar energy directly for producing the cooling they are examples of an important valid approach in the creation of comfort conditions, an approach pioneered by the work of Bliss (53).

Air circulation is also a factor in comfort conditions a method using solar heating to create a draught through the house achieves a cleaner, quieter and cooler living space (Figure 17) (54).

Fig. 15

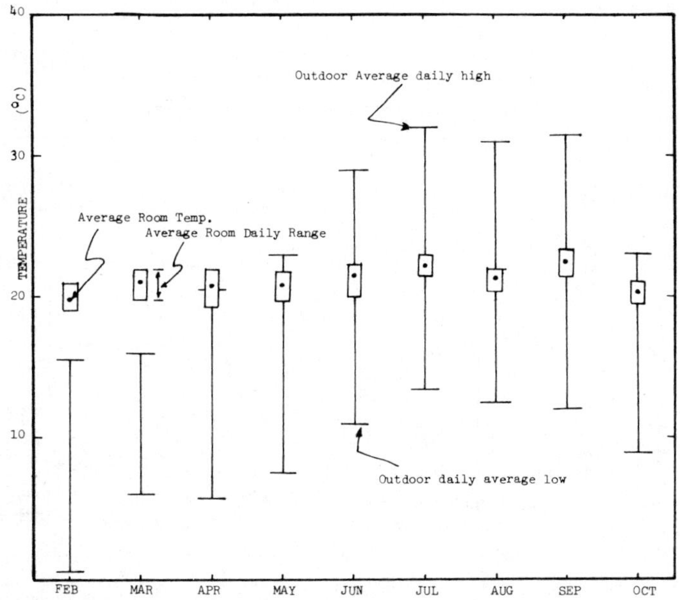

Monthly average temperature, Atascadero 'Skytherm' House. (51)

Fig. 16

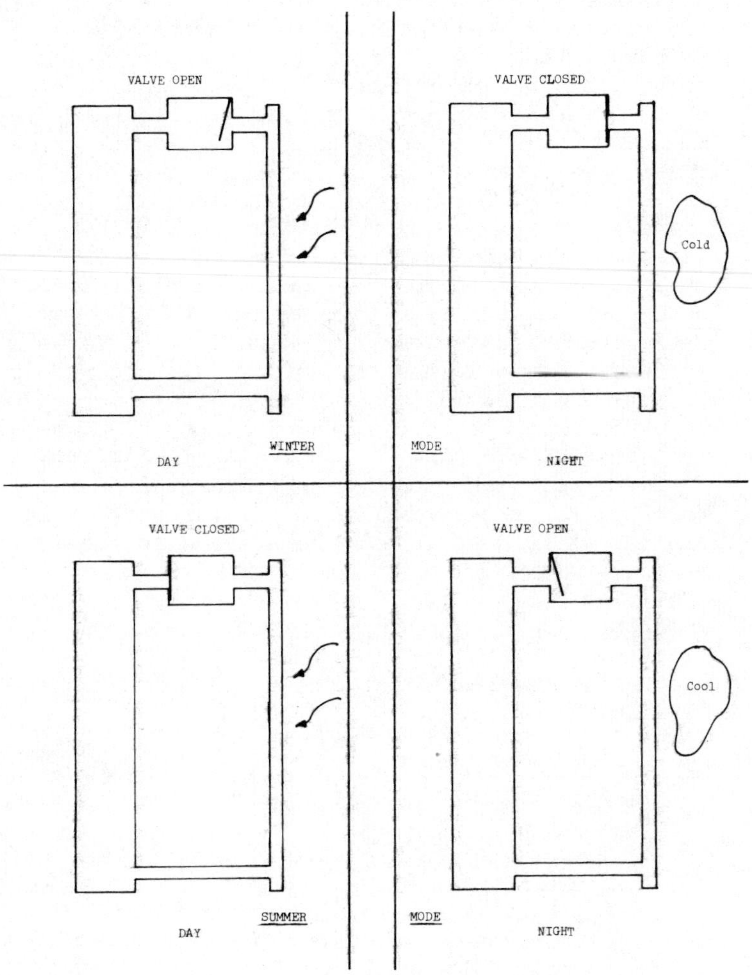

Thermic Diode operation. (52)

Fig. 17 Schematic arrangement for vertical solar induced ventilation.

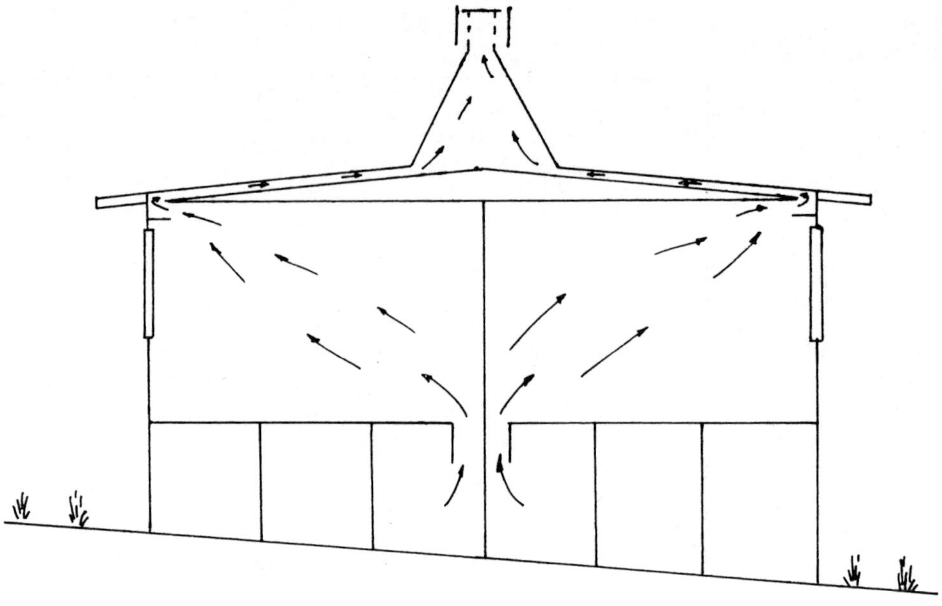

Conclusion

This review has spanned three decades and it can be observed how the ideas which germinated some of them more than twenty years ago are now flourishing investigations in various stages of growth. We can look forward to the fruits of these endeavours in due course. There can be no doubt that the importance attached by the USA to solar heating and cooling and the dramatic increase in Federal government spending (from $\$2 \times 10$ in 1972 to $\$ \times 10$ in 1975) has in great measure assisted in these developments.

There are of course many problems left to be solved. Not the least of these is the necessity for air cooled condensers and absorbers in absorption systems for residential applications. Air cooling implies condenser and absorber temperatures $15^{\circ}C$ or more above temperatures possible with water cooling which in turn necessitates generator temperatures well in excess of $100^{\circ}C$. This raises several difficulties. If water is used in the solar collector loop and heat storage then this loop will have to be designed for pressures above atmospheric. Such temperatures are also at the limit of efficient flat plate collector performance. Higher condenser temperatures also increases the possibility of crystallisation in the LiBr cycle. These and other challenges face us today. There are also promises of an exciting future.

References

(1) S.V. Szokolay, 'Solar energy and environmental design'. Light and Lighting, 68 (2), 70-73 (1975).

(2) R. Chung, J.A. Duffie, G.O.G. Lof, 'A study of a solar air-conditioner'. Mech. Eng. 85, 31-35 (1963).

(3) R. Chung, J.A. Duffie, G.O.G. Lof, 'Solar Space cooling'. Chem. Eng. Progress 55, 74-76, 78 (1959).

(4) N.R. Sheridan & W.H. Carr, 'A solar air-conditioned house in Brisbane', Solar Research Notes No.2 University of Queensland (1967).

(5) N.R. Sheridan, 'Performance of the Brisbane Solar House'. Solar Energy, 13, 395-401 (1972).

(6) R.W. Allen, F.H. Morse, & S.L. Sargent, 'Optimisation Study of Solar absorption air-conditioning systems'. Report No. NSF/RANN/SE/GI - 39117/PR/74/1 University of Maryland (1974).

(7) R.W. Allen, F.H. Morse & S.L. Sargent, 'Optimisation studies of solar absorption air-conditioning systems' (Final progress report). Report No. NSF/RANN/SE/GI - 39117/PR/76/2, University of Maryland (1976).

(8) D. Miller, 'Performance of water-cooled lithium bromide absorption units for solar energy applications'. Procs. Second ERDA workshop use of solar energy for the cooling of buildings (Ed. de Winter and de Winter), N.T.I.S., VA 22161, U.S.A., 186-196 (1976).

(9) P.J. Wilbur & C.E. Mitchell,'Solar absorption air-conditoning alternatives'. Solar Energy, 17 193-199 (1975).

(10) D.S. Ward & G.O.G. Lof,'Design and construction of a residential solar heating and cooling system'. Solar Energy 17, 13-20 (1975).

(11) R.L. Oonk, W.A. Beckman & J.A. Duffie, 'Modeling of the CSU heating/cooling system'. Solar Energy, 17, 21-28 (1975).

(12) G.O.G. Lof & R.A. Tybout, 'The design and cost of optimised systems for residential heating and cooling by solar energy'. Solar Energy, 16, 9-18 (1974).

(13) L.W. Butz, W.A. Beckman & J.A. Duffie,'Simulation of a solar heating and cooling system'. Solar Energy, 16, 129-136 (1974).

(14) S.A. Klein et al, 'A method of simulation of solar processes and its application'. Solar Energy, 17, 29-37 (1975).

(15) D.S. Ward, 'Performance of CSULHouse 1 cooling system. Procs. Second ERDA Workshop, Op.cit 48-54.

(16) L.B. Anderson, C.B. Winn & G.R. Johnson, 'Solar heating and cooling of the Santa Clara Community Centre'. Extended Abstracts. ISES 1975 International Solar Energy Congress, Los Angeles. International Solar Energy Society, 308-309 (1975).

(17) K. Kimura & S. Tanaka, 'Solar heating, cooling and domestic hot water supply at Soka solar house'. Extended Abstracts, Op.cit 310-311.

(18) J.E. Hill & T.E. Richtmyer, 'The design of operation of a retro-fitted residential solar heating and cooling system'. Extended Abstracts, Op.cit 324-325.

(19) Y.B. Safdari, 'Eureka solar house performance data'. Extended Abstracts, Op.cit 348-349.

(20) J.W. Littles, W.R. Humphries & J.C. Cody, 'Design and operation of a solar heating and cooling system for a residential size building'. Extended Abstracts, Op.cit 356-357.

(21) N. Nakahara, Y. Miyakawa & M. Yamamoto, 'Experimental study on house cooling and heating with solar energy using flat plate collector'. Extended Abstracts, Op.cit 358-359.

(22) J.W. Ayres, 'National Security and Resources Study Centre - a Solar cooled building'. Procs. Second ERDA Workshop, Op.cit 55-68.

(23) P.P. Anderson, 'Progress report on solar cooling work at Arkla'. Procs. Second ERDA Workshop, Op.cit 26-29.

(24) D. Prigmore & R. Barber, 'Cooling with the sun's heat: design considerations and test data for a Rankine cycle prototype'. Solar Energy, 17 185-192 (1975).

(25) R.E. Barber, 'Solar air-conditioning systems using Rankine power cycles - design and test results of prototype three ton unit'. Procs. Second ERDA Workshop, Op. cit 91-103.

(26) S.E. Eckard & J.A. Bond, 'Performance characteristics of a 3 ton Rankine powered vapour-compression air-conditioner'. Procs. Second ERDA Workshop, Op.cit, 110-136.

(27) F. Biancardi & M. Meader, 'Demonstration of a 3 ton Rankine cycle powered air-conditioner'. Procs. Second ERDA Workshop, Op.cit 137-152.

(28) S.A. Haas, 'Solar heated and cooled mobile home'. Procs. Second ERDA Workshop, Op.cit 254-260.

(29) R.E. Barber, 'Design and performance prediction of a 77 ton solar air-conditioning system - Rankine cycle predictions'. Procs. Second ERDA Workshop, Op.cit 74-77.

(30) E.A. Farber et al,'Operation and performance of the University of Florida Solar air-conditioning system'. Solar Energy, 10, 91-95 (1966).

(31) J.C.V. Chinnappa, 'Solar operation of a vapour absorption cycle air-conditioner at Colombo'. Trans. Inst.Engrs., Ceylon, 19-32 (1967).

(32) M. Simmons, R. Wolgast & M. Wahlig, 'Ammonia-water absorption air-conditioner'. Extended Abstracts, Op.cit 400-401.

(33) M. Simmons, 'Ammonia-water absorption air-conditioner'. Procs. Second ERDA Workshop, Op.cit 246-253.

(34) R.K. Swartman, 'Intermitten absorption systems for air-conditioning. Procs. Second ERDA Workshop, Op.cit 231-237.

(35) J.C.V. Chinnappa, 'Solar operation of ammonia-water multistage air-conditioning cycles in the tropics'. Solar Energy, 16, 165-170 (1974).

(36) J.C.V. Chinnappa & H.A. Martin, 'Dual-mode operation by solar energy of an ammonia-water two-stage cycle: a comparative study'. Solar Energy 18, 337-342 (1976).

(37) G.O.G. Lof, 'Cooling with solar energy'. Procs. Symp. App. Solar Energy. Phoenix, Ariz. 171-189 (1955).

(38) R.V. Dunkle, 'A method of solar air-conditioning'. Mech. & Chem. Eng. Trans. Inst. Engrs. Australia. MC 1, (1), 73-78 (1965).

(39) K G.T. Hollands, 'The regeneration of lithium chloride brine in a solar still for use in solar air conditioning'. Solar Energy, 7, 39-43 (1963).

(40) I.L. Maclaine-Cross and P.J. Banks, 'Coupled heat and mass transfer in regenerators - prediction using an analogy with heat transfer'. Int. J. Heat Mass Transfer, 15 (6), 1225 (1972).

(41) R.V. Dunkle, P.J. Banks & W.M.J. Ellul, 'Regenerator research, development and applications in Australia. Int. Inst. Refrigeration Meeting, Melbourne, Sept. 1976.

(42) D.J Close and T.L Pryor, 'The behaviour of adsorbent energy storage beds'. Solar Energy, 18, 287-292 (1976).

(43) D.J. Close and R.V. Dunkle, 'Open cycle solar energy cooling systems'. Paper presented to Ann. Conf., Inst.Engrs., Australia, May, 1976.

(44) W.F. Rush, 'Adsorption phenomena as applied to solar cooling'. Procs. Second ERDA Workshop, Op.cit 274-279.

(45) P.J. Lunde, 'Solar desiccant air-conditioning with silica-gel'. Procs. Second ERDA Workshop, Op.cit 280-293.

(46) D.I. Tchernev, 'Solar energy application of natural zeolites'. Procs. Second ERDA Workshop, Op.cit 307-319.

(47) S.C. Mullick and M.C. Gupta, 'Solar desorption of absorbent solutions'. Solar Energy, 16, 19-24 (1974).

(48) S.C. Mullick and M.C. Gupta, 'Solar air-conditioning using absorbents'. Procs. Second ERDA Workshop, Op.cit 336-344.

(49) V.A. Baum et al. 'Utilisation de l'energie solaire dans le conditions particulieres des regions a climat torride et aride pour la climatisation en etc.' Paper No. EH96, International Congress 'The Sun in the Service of Mankind', Paris, July, 1973.

(50) C.N.D. Stambolis, 'Solar houses: a general appraisal of solar building design and technology'. UNESCO/WMO Sclar Energy Symposium, Geneva, Aug/Sept. 1976.

(51) P.W.B. Niles, 'Thermal evaluation of a house using a movable insulation heating and cooling system'. Solar Energy, 18, 413-419 (1976).

(52) S. Buckley, 'Cooling with thermic diode panels'. Procs. Second ERDA Workshop, Op.cit 320-328.

(53) R.W. Bliss, 'The performance of an experimental system using solar energy for heating and night radiation for cooling a building'. Procs. of 1961 UN Conf. on New Sources of Energy, United Nations, N.Y. (1974).

(54) R.K. Hill, 'Utilization of solar energy for an improved environment within housing for the humid tropics'. Report 30. CSIRO Division of Building Research, Melbourne, (1964).

(55) A. Weinstein & C.S. Chen, 'Cooling by solar heat'. Procs. Second ERDA Workshop, Op.cit 197-206.

5.2

Cooling with solar energy: Absorption machine versus expander plus compressor

H. Tabor
Scientific Director
The Scientific Research Foundation,
Jerusalem, Israel

Abstract

Using current technology, the use of a heat engine to drive a compressor - as an alternative to an absorption machine - is shown to call for far higher solar collector temperatures. Only a profound improvement in the COP of compressors could change this result.

The intuitive approach to cooling by means of heat obtained from a solar collector is to use an absorption machine. However, an alternative sometimes considered is to use the solar heat to produce power - by means of an "expander" (i.e. turbine, piston engine or other device for converting heat to mechanical power) which then drives a conventional compressor cooling unit.

This note shows that the use of an expander would call for solar collectors operating at much higher temperatures than those needed for the absorption machine. This implies a higher cost for the total collector, it being either of lower efficiency - because of the higher operating temperature - or of greater complexity in order to provide the higher temperatures needed.

The Absorption System.

It is shown in the Appendix that the coefficient of performance K_a i.e. ratio of "cold" produced to heat input is, for a single stage absorption machine*, always less than unity. Practical units built to date have COP values of about 0.7 for large units and lower values for small units with skimped heat-exchangers.

It is also shown that the temperature "lift" i.e. $(T_i - T_c)$ is, in an ideal absorption system, approximately equal to the temperature "depression" $(T_c - T_e)$ where T_i is the input temperature at the hot end, T_c is the condenser (i.e. heat rejection or "sink" temperature) and T_e is the evaporator or "cold" temperature. In real systems, the lift is always somewhat higher than the ideal value. For example, if $T_c = 40^o$, $T_e = 5^o$, the depression is 35^oC. In theory the lift needed is a little more than 35^oC, say 40^oC, i.e. $T_i - T_c = 40$ i.e. $T_i = 80^oC$. In the real absorption machines commercially available, the input temperature required is in the region of 100^oC; some units have been made to work at about 90^oC (usually with some loss of capacity).

We can thus indicate that, in the present state of the art, an absorption machine to produce 100 units of cold would need an input of $\frac{100}{K_a} \sim 142$ units of heat (taking a good figure of 0.7 for K_a) at a temperature of about 100^oC i.e. say 65-70^oC above the ambient temperature of 35-30^oC.

---

* COP > 1 is possible for double-stage machines: the complication is so great - and a much higher input temperature is needed - that hardly any double-stage units have been made.

The Power Cycle. The efficiency E of any heat engine is given by:

$$E = C_1 \frac{T_i - T_c}{T_i} = C_1 \frac{\Delta T}{T_c + \Delta T} = C_1 \frac{\theta}{1 + \theta}$$

where C_1 = ratio of actual efficiency of the machine to that of an ideal Carnot engine

$$\Delta T = T_i - T_c$$

$$\theta = \frac{\Delta T}{T_c}$$

In small engines or turbines, C_1 can be as low as 0.4 or even less: in a good design, of reasonable size, it can be of the order of 0.7. For the sizes and temperature ranges we are considering, a reasonable value for C_1 would be, say 0.6.

The power produced from an input of heat Q_1 is $W_p = Q_1 C_1 \frac{\theta}{1+\theta}$. This power can be used to drive a compressor unit of the type currently used in cooling systems. Depending upon the conditions, these compressor units have a COP defined as $K_c = Q_e / W_p$ of the order of 3* i.e. they produce 3 units of cooling for one unit of power provided.

Then the cooling produced by a heat input Q_1 when a heat engine is coupled to a compressor unit is:

$$Q_{cooling} = K_c Q_i C_1 \frac{\theta}{1 + \theta}$$

or the overall COP of the system K_o is:

$$K_o = \frac{Q_{cooling}}{Q_i} = K_c C_1 \frac{\theta}{1 + \theta}$$

Comparison

If we wish the heat-engine-compressor system to be more efficient than the absorption system, we require:

$$K_o > K_a$$

i.e. $K_c C_1 \frac{\theta}{1 + \theta} > K_a$ or $\frac{\theta}{1 + \theta} > \frac{K_a}{K_c C_1}$

Taking $K_a \sim 0.7$; $K_c \sim 3$; $C_1 \sim 0.6$

$$\frac{\theta}{1 + \theta} \sim 0.389$$

i.e. $\theta = \frac{\Delta T}{T_c} > 0.636$

i.e. $\Delta T > 199^oC$ for $T_c \sim 313^oK$

* If all the heat exchangers were made larger, a higher K_c value could be obtained but at much higher cost: current designs represent a reasonable compromise arrived at after many years of development, but higher energy costs should move the optimisation over to a somewhat higher K_c.

This result shows that, to get the same overall cooling from a given amount of heat (provided by the solar collector), the collector would have to work at a temperature of 199°C above the sink temperature i.e. close to 240°C whereas the absorption system requires to operate about 60°C above the sink temperature i.e. at about 100°C.

Solar collectors operating at 100°C can be of the conventional flat-plate type: to get 240°C requires a much more sophisticated collector, probably using some form of mirror concentration and has to be designed to suffer very high temperatures should it operate without heat extraction (the "dry" temperature).

We note that the critical term in determining θ is the COP of the compressor, which we have taken as 3 (though ideally it could - for the temperatures chosen - be 8.94, as shown in the Appendix).

We consider the effect of improving K_c.

K_c =	3	3.5	4.0	4.5	5.0
θ =	.636	.500	.412	.350	.340
$(T_i - T_c) = \Delta T$ =	199	166	129	110	95

Conclusions

Using present known technology for absorption machines, heat engines (of any type) and compressor cooling units, the use of heat engine plus compressor requires far higher heat source temperatures than the absorption machine for the same overall COP. Unless high-temperature collectors can be built for the same or lower cost than low-temperature collectors - for the same thermal output - or the cost of an absorption machine is much higher than the combined cost of engine plus compressor - the absorption system is preferable.

If advanced technology could substantially improve the COP of present-day compressor units - and this would be desirable also for electrical operation because of the increasing cost of electricity - then the conclusion given above might need to be reconsidered.

APPENDIX

Ideal Temperature - Entropy Diagrams

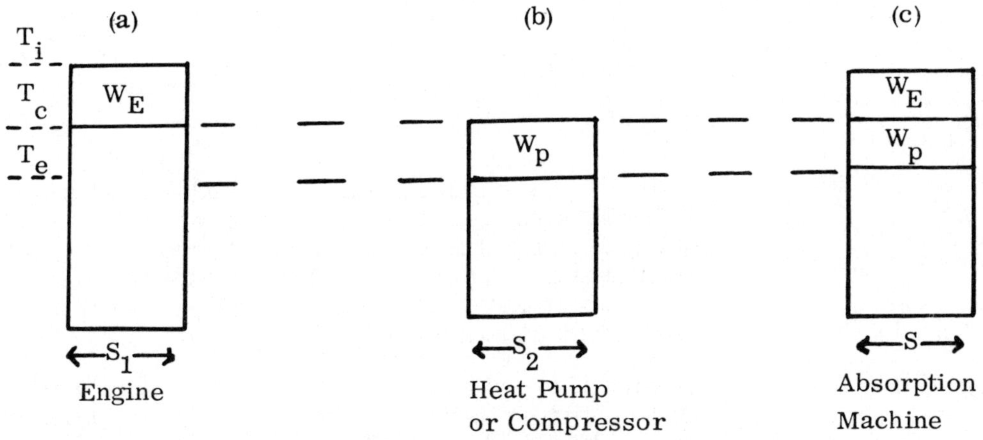

Ideal $W_E = \frac{T_i - T_c}{T_i} Q_i$ $\qquad W_p = \frac{T_c - T_e}{T_e} Q_e$ $\qquad Q_i$ = input heat

Q_e = extracted heat ('cold')

Real $W_E = \frac{T_i - T_c}{T_i} Q_i C_i$ $\qquad W_p = \frac{T_c - T_e}{T_e} \frac{Q_e}{C_2}$

where C_1 and C_2 are the departure coefficients from the ideal (C_1 and C_2 1).

overall COP $K_o = \frac{Q_e}{Q_i} = C_1 C_2 \frac{T_e}{T_i} \frac{T_i - T_c}{T_c - T_e} = C_1 C_2 \frac{T_e}{T_i} \frac{\ell}{d}$

where ℓ = "lift" = $T_i - T_c$

d = "depression" = $T_c - T_e$

An ideal absorption machine must, on thermodynamic grounds, be equivalent to an ideal engine coupled to an ideal heat pump. Because the same unit fluid is used in both, the S value is the same, as shown in figure (c). For $W_E = W_p$ we have once:

$$K_a = Q_e/Q_i = T_e/T_i < 1.$$

Lift $\ell = (T_i - T_c)$ = depression $(T_c - T_e)$ because $W_E = W_p$.

For a real compressor, the compressor COP K_c is

$$K_c = \frac{Q_e}{W_p} = \frac{C_2 T_e}{T_c - T_e}$$

For $T_c = 313^oK$, $T_e = 278^oK$ $\quad K_c = 8.94\ C_2$

Practical units as available "off the shelf" have K_c values of the order of 3 to 4 i.e. $C_2 = 0.34$ to 0.45

5.3 Nocturnal infrared radiative cooling experiments in Mexico

E. Hernández, E. Mayer
Centro de Investigación de Materiales
Universidad Nacional Autónoma de México
Apartado Postal 70-360
México 20, D. F.

ABSTRACT

In order to explore the natural cooling perspectives by means of nocturnal terrestial radiation, several experiments were carried out in México under clear and partially cloudy sky conditions. This study pretends to provide useful informa - tion for natural cooling purposes for instance housing design or preservation of perishable products.

Based on experiments carried out previously by Trombe, Caste-llanos and Fournier a nocturnal cooling equipment was designed and tested. The devices essentially consists of radiators di rected toward the sky and screened only by polyethylene films. Two devices involve a stair-system which consists of a set of aluminium boxes being as radiator steps. The drop of temper-ature experimented in the coolers were of aproximately 15 cel sius degrees with respect to the ambient air temperature.

INTRODUCTION

Most materials composing the surface of the earth radiate near ly as black bodies. Since black body radiation at terrestial or atmospheric temperatures between 200°K and 330°K is practi cally all contained within the limits 2 and 25μ, then at the mean temperature near the earth's surface close to 288°K, Wien's displacement law indicates that the maximum radiation is at a wave lenght close to 10μ. On the other hand, by con-sidering the sun as a black body radiating at a temperature of 6000 °K (Sufficiently close aproximation for practical pur poses), the maximum solar radiation occurs at 0.47μ.

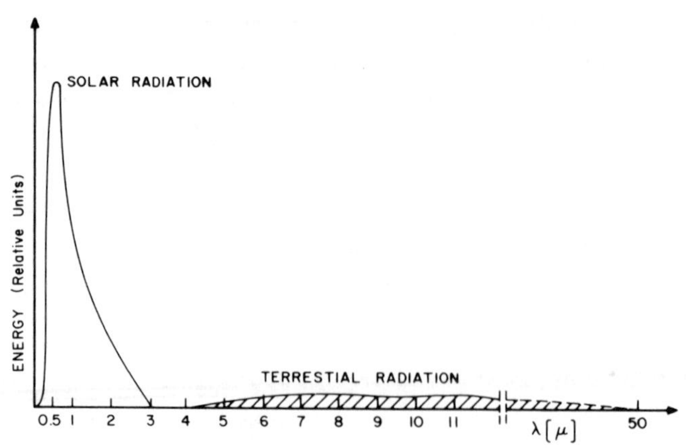

Fig. 1 Energy distribution of solar and terrestial radiation in relative units.

Because of the well defined spectral regions of solar and te-rrestial radiation, they are customary refered as shortwave and longwave regions respectively. Figure (1) shows these re gions as function of the energy emited in relative units.

Since longwave radiation leaving the earth's surface is ab - sorbed by water vapor and carbon dioxide in the troposphere, dissipation of infrared radiation to space is limited mainly by the actual concentration of these atmospheric constituents. The water vapor volumetric proportion in the atmosphere may varie between 0.2 and 3% depending on the relative humidity and air temperature, while carbon dioxide is only about 0.03%.

NOCTURNAL COOLING OF TERRESTIAL SURFACES.

As was first pointed out by Simpson {1}, only the spectral region between 8.5 and 11μ which is known as Simpson's window, constitute a nearly transparent layer for the earth radiation. Simpson regards the atmosphere as being composed of layers of varying thickness with optical depths of 0.039 g. cm^{-2} of CO_2. He concluded that the complex transmission spectrum of thin layers in the atmosphere could be smoothed by reducing the main features of atmospheric absorption of infrared radiation by such a layer characterized as:

- Completely transparent below 4μ and from 8.5 to 11μ.
- Partially transparent from 4.0 to 5.5μ, from 7.0 to 8.5μ and from 11.0 to 14μ.
- Completely opaque from 5.5 to 7.0μ and above 14μ.

Generally speaking in order to reach a maximum nocturnal cooling of terrestial surfaces, the following weather and local conditions are necessary.

- A noticeable change of air temperature between daytime and night.
- A minimum atmospheric content of water vapor, carbon dioxide and aerosols.
- Considerable altitude of the place.

If the first condition holds, nocturnal radiation defined as the net longwave radiation at the ground, i.e. the difference between the upward black body radiation flux at the corresponding ground temperature and the downward radiation of the atmosphere (counter-radiation), will be positive and large for high surface temperatures. The nocturnal dissipation of the energy absorbed during daytime by terrestial bodies which increase their temperature due to absorption of the incoming shortwave radiation will be noticeable whenever the sorrounding air temperature remains lower than that of the emitting surfaces.

Fig. 2

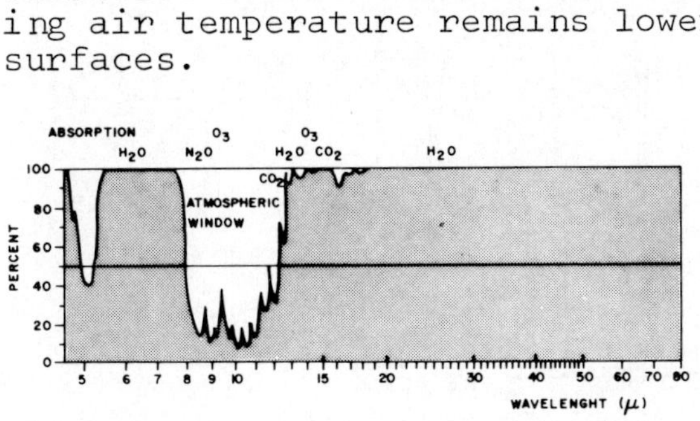

Atmospheric vertical absorption of longwave radiation emitted from the ground and showing the Sympson's window.

From the second condition, the escape of infrared radiation to space through the Simpson's window (Figure 2) known as the negative greenhouse effect could be considerable.

This situation arises under a dry, clean and cloudless sky. Under high humidity conditions, atmospheric water vapor, CO_2 or aerosols will strongly absorb the infrared radiation re-irradiating it to the earth's surface blocking the nocturnal cooling effect.

The third condition arises from the fact that, the longer the effective atmospheric optical depht is, the larger the atmospheric mass will be, hence, absorption of longwave radiation is reduced through a thin atmospheric thickness as in the case of places located at high altitudes.

NOCTURNAL COOLING EXPLOITATION

The nocturnal cooling phenomenon has been exploited long time ago in several regions of the world like in Mid-Orient, Northern Africa and Southern Spain among others, where high temperatures become extremely unconfortable. In earlier experiments carried out by Trombe {2}, {3}, {4}, Castellanos and Fournier {5}, several devices were designed in order to reach a maximum drop of temperature on the emitting surface. One of his best designs is shown in Figure (3), this consists of a set of five emitting surfaces each one covered by a polyethylene film whose global effect is to avoid convection problems with the sorrounding air. Polyethylene has a good transparency to infrared radiation {6} as can be seen from Figure (4).

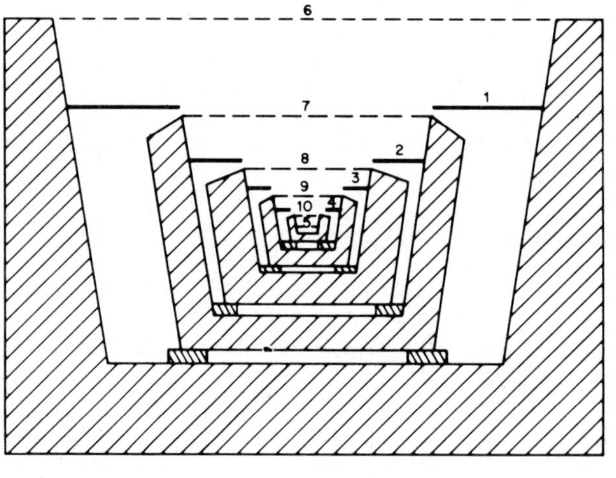

Fig. 3
System designed by Trombe: 1, 2, 3, 4 - radiators, 6, 7, 8, 9, 10 - polyethylene film films.

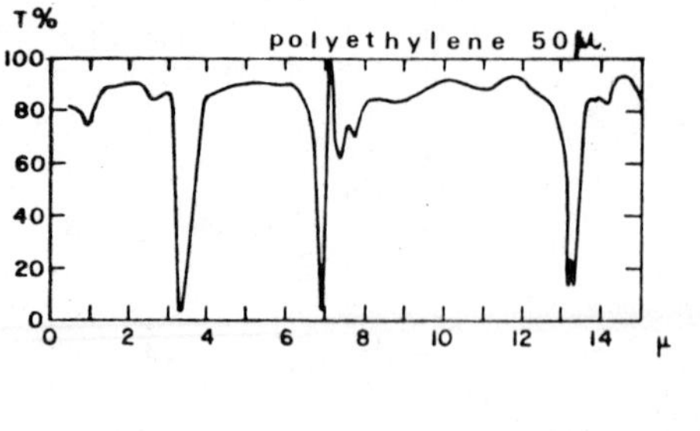

Fig. 4
Polyethylene transmissivity.

EQUIPMENT DESIGN AND RESULTS

One of the devices designed {7}, consist of a wood box containing a set of four black painted aluminium plates acting as radiator steps Figure (5). The objective was to study the drop of temperature in each of the plates and its differences. The way the temperature drops is based on the fact that radiation

Fig. 5
Radiator (four effects).

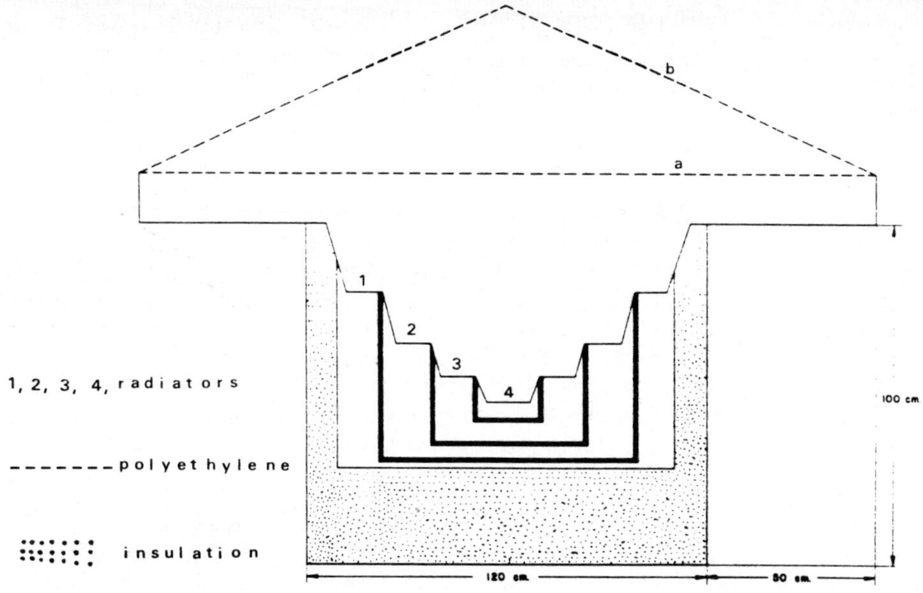

emmited from the first plate produce a continuous cooling of the air mass contained within the box. This air mass also serves as insulator maintaining the walls of the second box at a lower temperature.

By means of this effect the heat flux between the boxes is prevented between the external wood box and the next one, a insulating material was interposed and the external walls of the main box were white painted.

The interior boxes were constructed of pasteboard covered with polyesthyrene. To reflect infrared radiation, the walls at each one of thesteps were covered with aluminized polyethylene and tilted at an angle of 15°. This inclination allows the radiators to increase their solid angle of emission.

The whole radiating system were covered by a polyethylene film. In the (b) form, the cover allowed to insulate a great air mass and it was able to prevent strong fluctuations on the radiator plates when perceptibly winds were blowing. On the contrary, the (a) form caused a noticeable temperature change.

The drop of temperature between the last step at the bottom of the box and the ambient air, was in the (a) case, of about 20°C, while in the (b) case of about 17°C. In Figure (6) corresponding to the (a) case, it can be observed that at 7.00 a.m. (sunrise) the maximum difference was of 19.9°C.

Fig. 6
Drop of temperature at each one of the effects.

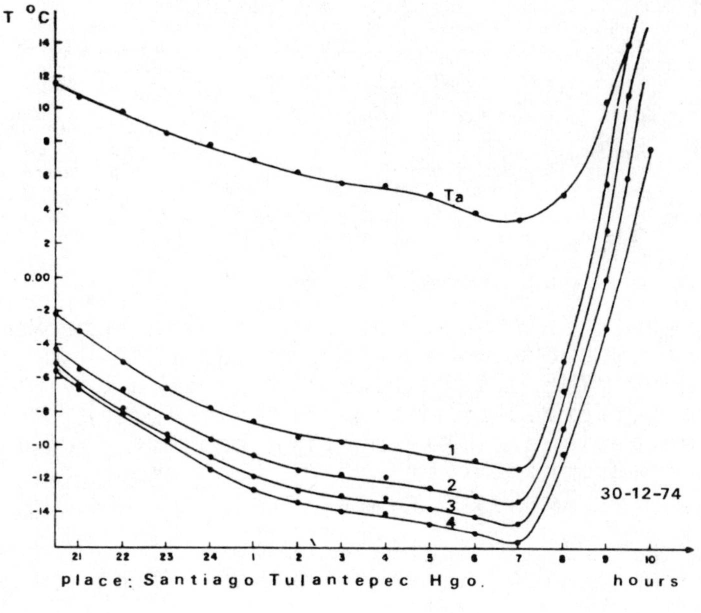

With the purpose of a practical exploitation of this phenomenon, a mass of water of 5.7 Kg. was cooled in a device shown in Figure (7). The water was contained in a galvanized iron box blackened in its upper side with common mat painting. The calculated loss of energy from the emitting surface was of 90 watt/m^2. The registered differences of temperature were of about 9°C. When this device was tested with umblackened boxes the differences found were only of 2 and 3°C This two experiments were performed at the town of Santiago Tulantepec, in the state of Hidalgo.

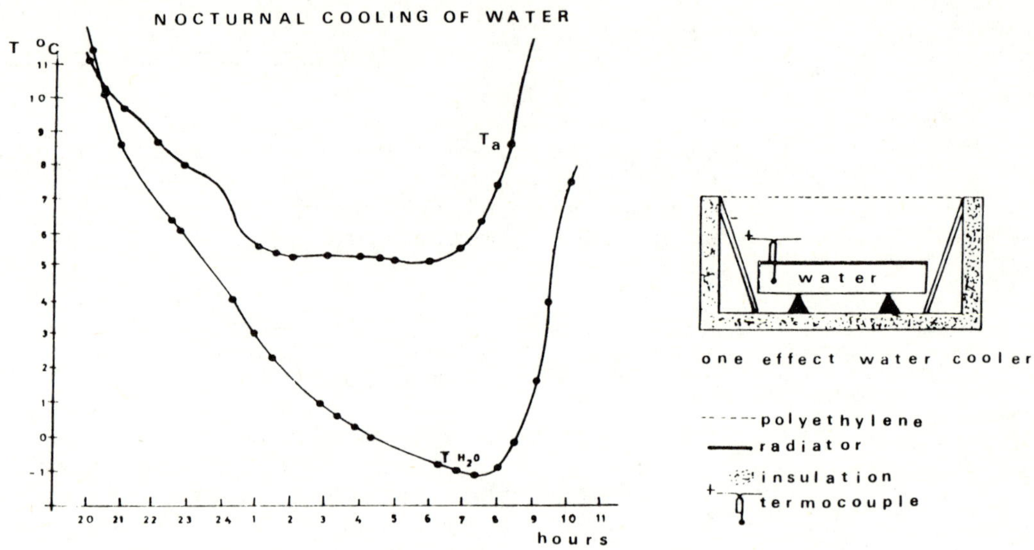

Fig. 7
Drop of temperature in the water cooler and cross section view of the device employed.

On the base of the two devices previously described, another dispositive was designed. A fiberglass structural box with two face to face step form was constructed. To preserve the solid angle of radiation, the walls of the steps were once again tilted at an angle of 15°Figure (8).

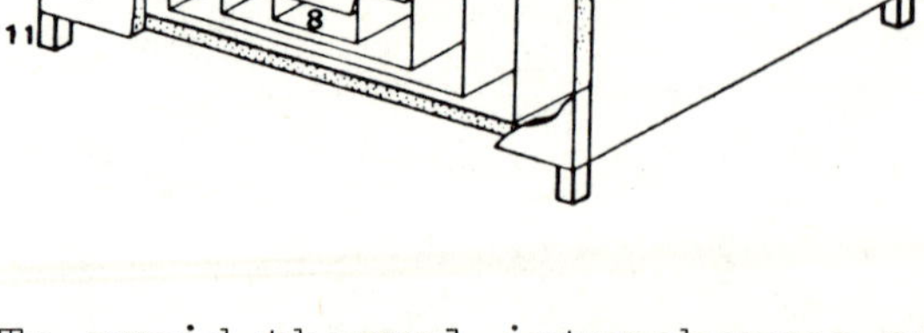

Fig. 8
Natural cooling device.

To avoid thermal interchanges with surroundings, lateral covers of the same fiberglass material were placed under the steps. The dissipation of infrared radiation from the device was performed by using aluminium boxes coated by a cold selective surface. This coating has the property of reflecting shortwave radiation and at the same time actuates as a black body for longwave radiations.

To this kind of surfaces belong most of the white paintings and specially the white oxides. The coating selected was a mat white painting with a high resistance to weathering . The aluminium boxes were placed on each one of the structural steps and connected among them by hoses.

The steps were underside insulated and four air-matress were formed between the two stairs. As insulating material fiber-glass wool was employed.

Finally the device was covered by a polyethylene film (50 μ thickness). Due to the noticeable gradient of temperature between the inside and outside air masses, water vapor condensation may occur on the polyethylene film when the ambient relative humidity is high, consequently this water condensation produce a strong longwave absorption of radiation emitted from the boxes.

To avoid this phenomenon, a second polyethylene film (shed shaped) was installed, so an intermediate air - matress was formed diminishing the condensation and also reducing the inside convection.

The device was tested in the town of Tizayuca in the State of Hidalgo. The first series of measurements were done under critical atmospheric conditions due to the presence of clouds and a high relative humidity caused by a cyclonic perturbation that was developing at that time on the Pacific Coast. The results obtained are shown in Figure (9).

Fig. 9
Drop of temperature at each one of the effects under unfavorable weather conditions.

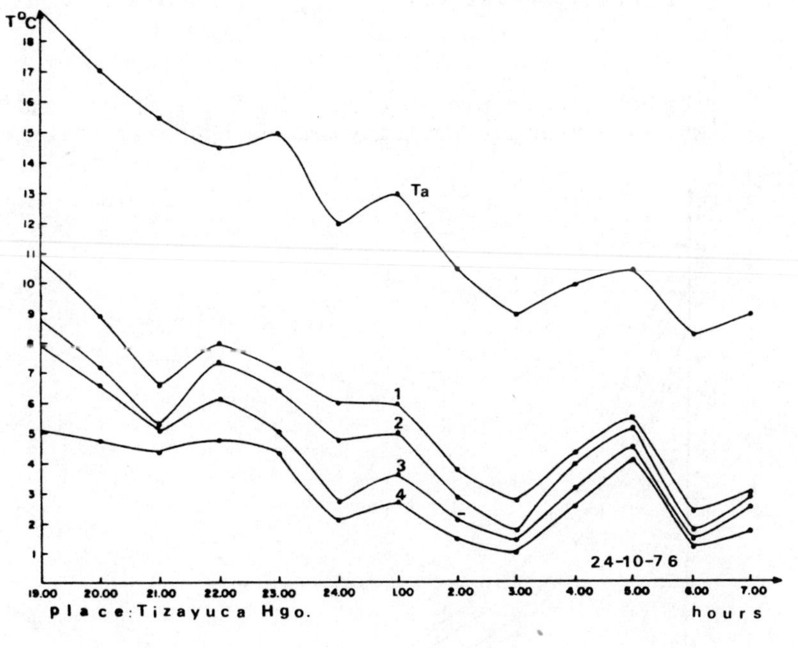

The second series of measurements fortunately were carried out under better weather conditions. In Figure (10). The corresponding temperatures are shown. The sky was cloudless although the relative humidity remained high.

In order to increase the thermal radiative interchange, helicoids were introduce inside the boxes. Figures (10) and (11) includes this effect while Figure (12) does not include helicoids. The results in Figures (11) and (12) were obtained under a partially cloudy sky.

Fig. 10 Drop of temperature at each one of the effects under cloudless conditions high relative humidity and without helicoids.

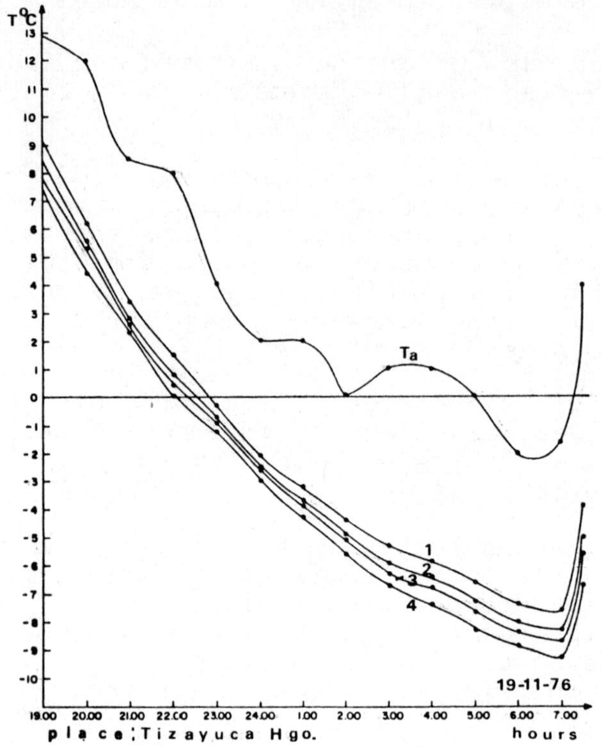

Fig. 11 Drop of temperature at each one of the effects under partially cloudy sky, low relative humidity and using helicoids.

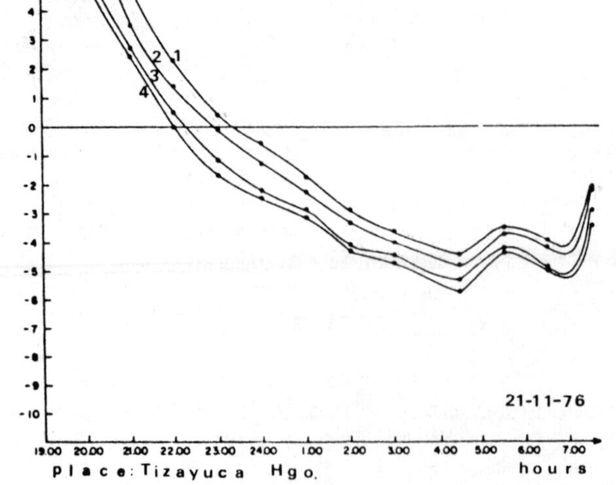

Fig. 12 Drop of temperature at each one of the effects under cloudless conditions high relative humidity and without helicoids.

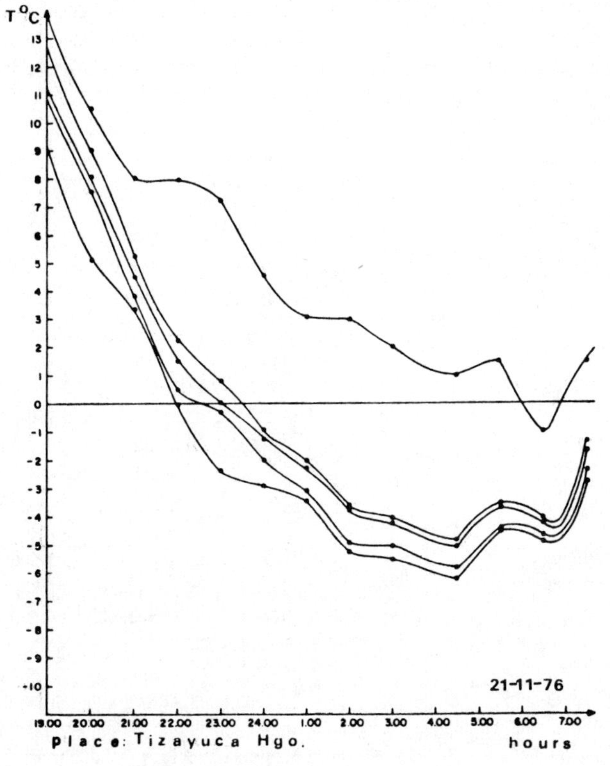

CONCLUSIONS

The drop in temperature reached in the first device with four effects, was about 15°C. This shows that the nocturnal losses of energy from bodies exposed to the sky can be used in a new cooling systems focused to preservation of perishable products and natural cooling of houses and buildings.

Notwithstanding the 10°C the difference of temperature registered, between the radiator and the ambient air, the results obtained, can be considered acceptable if it is kept in mind that the minimum temperature was approximately -10°C. It is convenient to remark that in the first device only radiative plates were used while in the last one, boxes containing an air mass were employed; so the differences found in the drop of temperature are well understandable.

It was found that fluctuations of the cooling process depend directly on the weather conditions specially humidity and cloudiness. The radiative exchange between the cooling boxes and the surroundings was improved and maintained notwithstanding the condensation problems on the polyethylene film. This was solved by means of a double cover film inside which an air-matress was formed.

The results presented here are preliminary. Work is being done in order to determine the amount of water vapor in the atmosphere, in most of the mexican arid area.

ACKNOWLEDGEMENTS

The authors are endebted to all those persons who with their encouragement contributed in one or another form to the fullfilment of this project, specially to Mr. E. Magin, I. Pilatowsky, J. Ambriz and H. Guerrero. Thanks also to Mr. H. Cortez for drawing the figures and to Miss Mercedes Spíndola for the transcription of the manuscript.

REFERENCES

1. Simpson, G. G. "Further Studies in Terrestial Radiation", Royal Meteorological Society, Vol. 3, No. 21 1928.

2. Trombe, M. F. "Sur des Grands abaissements de température obtenus par rayonnement du corps noir sur l´espace", Seance du 14 Janvier 1963, Academie des Sciences.

3. Trombe, M. F. "Nouvelles experiences sur le refroidissement de corps noirs sur l´espace". Academie des Sciences, Sceance du 25 Février 1963.

4. Trombe, M. F. "Some aspects of the utilization of natural radiation, especially in the developing countries" Impact, Vol. 15, No. 4, 1965 .

5. Castellanos, G.S. and Fournier, J. "Refroidissement d´un corp noir irradiant en direction de l´espace dans le deser d´Atacama Chili". Proceedings of the Sun and the Mankind session E20, Paris, 1973.

6. Trombe, M. F. et Moreau, M. "Spectres d´absorption infrarouge de divers matériaux plastiques presentant des caractéristiques utilisables pour les etudes sur le rayonnement terrestre". Publication du Laboratorie des Terres Rares, France, P.P. 553-561.

7. Magin, J. E. "Aprovechamiento de la Emisividad de los cuerpos negros", Professional Thesis, Fac. Quim. U.N.A.M., 1975.

8. Ambriz, J. & Guerrero, H. "Diseño y construcción de un enfriado nocturno aprovechando el fenómeno de radiación terrestre", Professional Thesis, Fac. Química U.N.A.M., 1976.

A methanol – lithium-bromide, zinc-bromide – mixture applied to solar absorption heat pumps

R. Best
Centro de Investigación de Materiales
National University of México
México 20, D.F.

ABSTRACT

This paper describes the use of an absorption system operating as a heat pump. The effect on the coefficient of performance in terms of the temperatures of the different parts of the system, generator, condenser, evaporator and absorber is analysed. A binary system with methanol as the refrigerant and a misture of lithium bromide and zinc bromide as the absorbent is proposed as a better alternative to the two commonly employed mixtures: water-lithium bromide and ammonia-water. A comparison is made on the overall performance of the electric-vapor compressor, the gas fired and the solar operated heat pump systems.

INTRODUCTION

Increasing the efficiency of utilization of energy resources is one of the logical ways of reducing consumption without foregoing any of the benefits associated with its use.

The thermodynamic advantages of heat pumps make them a useful means of using energy with an overall conversion efficiency greater than conventional methods.

Solar energy can be used as a low grade heat source for heat pumps, and in developing countries like Mexico with high levels of solar radiation, (1) a totally solar heat pump could be envisaged if we use an absorption cycle instead of the vapor compressor heat pump used at present.

With the absorption cycle, the utilization of electrical power generated at very low efficiencies (less than 30 per cent) is eliminated, as the pump work required is only 5 to 10 per cent of the power a compressor would consume.

ABSORPTION SYSTEM

The basic absorption system is shown in Figure 1. The working fluid for the system is a mixture of refrigerant and absorbent. Heat is supplied to this solution in the generator, vaporizing the refrigerant and

Fig. 1
Absorption cycle diagram

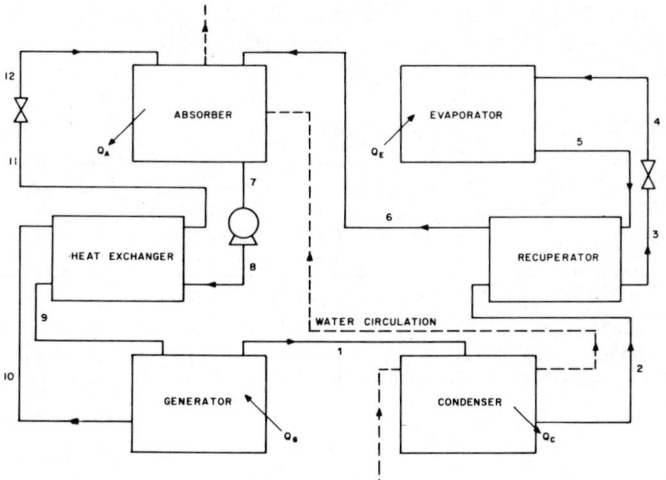

leaving behind a mixture weak in refrigerant. Heat is removed from the vapor and it is liquified in the condenser at high pressure. The liquid refrigerant is passed through an expansion valve into the low pressure evaporator where it is vaporized. The refrigerant vapor is then recombined at the absorber with the weak refrigerant solution from which it was obtained. The solution rich in refrigerant is then pumped back into the generator, closing the cycle.

Heat losses are reduced by using a recuperator between the hot weak solution from the generator and the relatively cold strong solution from the absorber. A second recuperator between the condenser and expansion valve, is sometimes used to cool the liquid refrigerant with the cool vapor entering the absorber, thus raising the thermal efficiency of the system.

The heat rejected at the condenser and absorber is removed by circulating water which could then be utilized for any low temperature purpose.

BINARY SOLUTIONS

Whether or not an absorption cycle operates with a high thermal efficiency and with simple equipment of minimum size depends largely on the properties of the solution at the operating conditions.

Ammonia-water and water-lithium bromide mixtures are the two most employed binary systems. The disadvantages of the NH_3-H_2O system are the high pressure differential necessary between condenser and absorber, the likelihood of outward leaks and the need of rectification due to carry over of water vapor in the generated ammonia vapor, requiring additional equipment and increasing the cost.

H_2O-LiBr systems have the disadvantage that water cannot be used for evaporator temperatures below 0°C, for ice formation occurs; other disadvantages are the extremely low absorber pressures and the resulting high probability of inward leaks. When low refrigerant concentrations are employed, 'salting out occurs'.

Refrigerant selection

There seems to be a gap between these two systems, methanol has intermediate characteristics between ammonia and water: Practical pressure levels required for its condensation; no limitation to evaporator temperatures and comparatively high latent heat of vaporization, as can be seen in Figure 2. Although methanol is inflammable and slightly toxic, the pressures encountered make leakages in and out of the system very improbable.

Fig. 2 Properties of refrigerants

Refrigerant	Vapor pressure Atm. 5°C	65°C	Enthalpy of vaporization K Cal/Kg. 0°C
Water	0.0084	0.288	676
Methanol	0.052	1	284.9
Ammonia	5.0	30	299

Absorbant selection

Hainsworth (2) pointed out eleven different possible absorbants for methanol. Of these, only two, lithium bromide and zinc bromide have been tried experimentally and data is available (3).

Methanol lithium bromide solutions become saturated at relatively high methanol concentrations, the lowest value corresponding to 43 per cent methanol by weight. This is a great limitation when it is necessary to operate at high generator temperatures.

The addition of zinc bromide to the methanol-lithium bromide solution improves the solubility of the refrigerant producing solutions of concentrations as low as 28 per cent methanol by weight. Another advantage of the addition of zinc bromide is that the viscosity is reduced compared with that of the lithium bromide solution.

Information concerning the thermal and physical properties of the mixture is scarce; values were calculated from experimental PTX data (3).

COEFFICIENT OF PERFORMANCE (COP)

The COP for the heat pump is defined as the heat output at stated conditions, divided by the heat supplied to produce the effect:

$$\text{COP} = \frac{\text{Heat output}}{\text{Heat supplied}}$$

$$= \frac{\text{Heat extracted from absorber + condenser}}{\text{Heat supplied to generator}}$$

The greatest possible coefficient of performance that can be obtained is given by a reversible process, (4) based on a Second Law analysis of the system. The ideal COP is given by:

$$\text{IDEAL COP} = \frac{Qw}{Qg} = \frac{Tw}{Tg}\,\frac{Tg - Te}{Tw - Te}$$

Where:

Qw = heat extracted by water from condenser and absorber, $Kcal.h^{-1}$
Qg = heat input into generator: $Kcal.h^{-1}$
Tg = generator temperature: ^{o}K
Te = evaporator temperature: ^{o}K
Tw = water temperature: ^{o}K

A more realistic analysis of the heat pump was obtained by making heat balances on the different parts of the system, based on the thermophysical and thermodynamic properties of the refrigerant absorbent combination. The system of polynomial equations obtained permitted the calculation of the COP starting only the operating temperatures (5).

CALCULATED PERFORMANCE

Effect of generator temperatures

Figure 3 shows the effect of the generating temperature level over the coefficient of performance for various condenser and absorber temperatures and a constant evaporator temperature. As the generator temperature increases, the COP increases rapidly and then tends to a constant value. The rapid increase is due to the fact that as the generator temperature increases, the circulation factor, ie. mass flow rate of strong solution over the mass flow rate of refrigerant vaporized, is greatly reduced. After a certain generator temperature is reached, there is no increase in the COP by raising it further.

Effect of condenser and absorber temperatures

As the condenser and absorber temperatures increase, the COP is reduced and it is necessary to operate at higher generator temperatures if the circulation factor is to remain in a low value (less than 10). The limit for operating generator temperatures is when we reach the saturated solution composition after which crystallization may occur.

Evaporator temperature

As the evaporator temperature rises, the COP increases as indicated by Figure 4. The evaporator temperature will depend on the available heat source used and it should be 5 to 10°C lower.

Comparison with vapor compression system

The COP of a vapor compressor heat pump in terms of the required temperature lift, and using available data (6) is shown in Figure 5. This curve could be compared with Figure 3 for the methanol $LiBr-ZnBr_2$ heat pump.

COP IN TERMS OF PRIMARY HEAT

Electrically operated vapor compression heat pump

It is important to see the effect on the efficiency of the system in terms of the primary heat source (Fig. 6).

Electricity is generated at less than 30 per cent efficiency and this value reduces the overall COP for the vapor compressor system. For a 32°C temperature lift, the COP is reduced from a value of 3.5 to a value of just over unity.

Gas fired absorption heat pump

If a realistic value of 66 per cent efficiency is considered for a gas fired boiler furnace, the overall COP is indicated in Figure 7. For high condenser and absorber temperatures, the overall efficiency of the absorption heat pump is better than that for the vapor compressor, as the efficiency of absorption systems is less dependant on the temperature lift.

Solar operated heat pump

The COP of a heat pump increases as the temperature of the heat source increases. However, the efficiency of flat plate collectors decreases at high temperatures. To improve the efficiency it is necessary to use selective surfaces or arrangements such as honeycombs or evacuated tubes.

Simple flat plate collectors with a single glass cover, operating within temperatures required for absorption systems, (60 to 100°C), have efficiencies in the order of 0.5 to 0.05. If selective surfaces are used, the efficiency rises to values of .65 to .42 for the same temperature range (7).

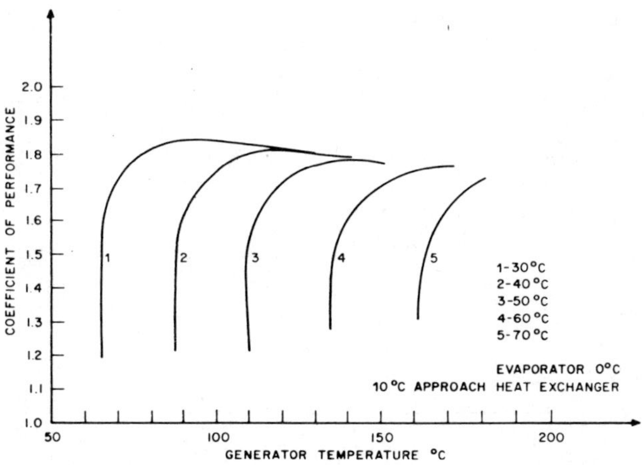

Fig. 3
C.O.P. of LiBr-$ZnBr_2$-Methanol solutions at various condenser and absorber temperatures.

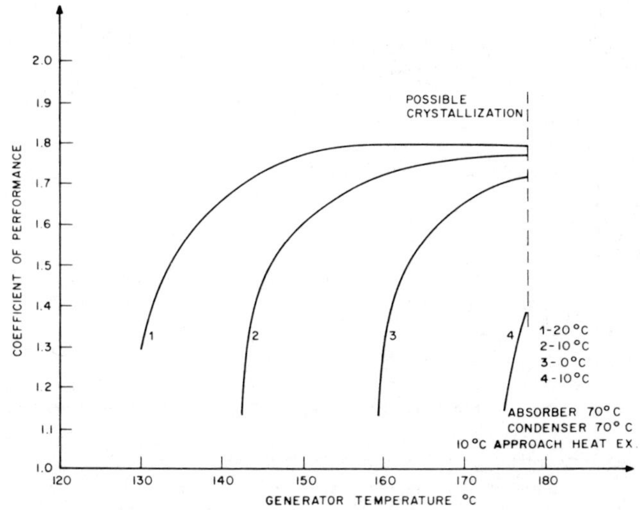

Fig. 4
C.O.P. of LiBr-$ZnBr_2$-Methanol solutions at various evaporator temperatures.

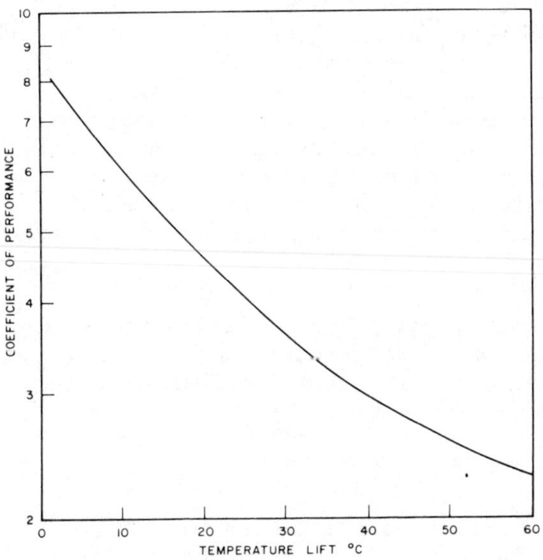

Fig. 5
Realistic coefficient of performance of vapor-compression heat pumps.

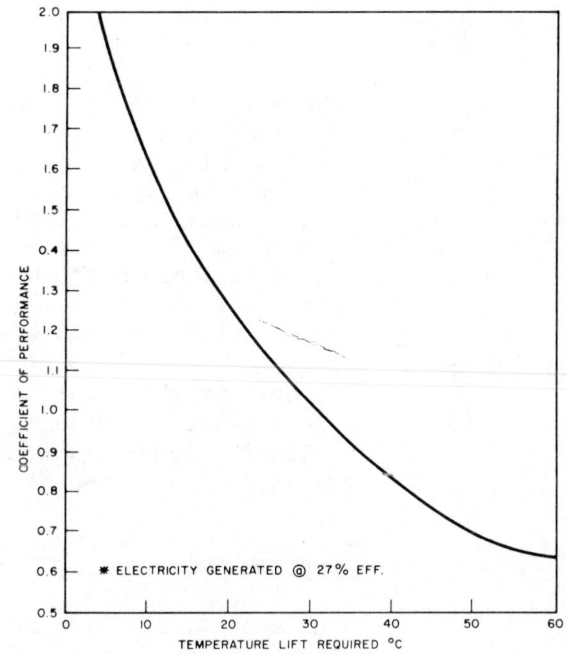

Fig. 6
Coefficient of performance of vapor-compression heat pumps in terms of primary fuel.*

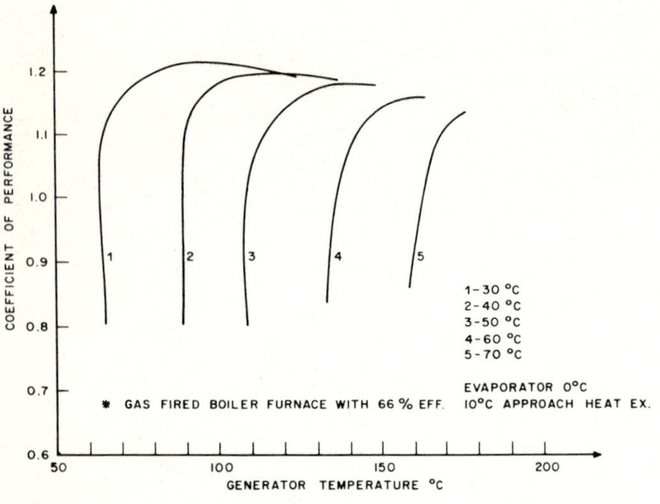

Fig. 7
Absorption heat pump C.O.P. of $LiBr-ZnBr_2$-Methanol solutions. In terms of primary fuel.* At various condenser and absorber temperatures.

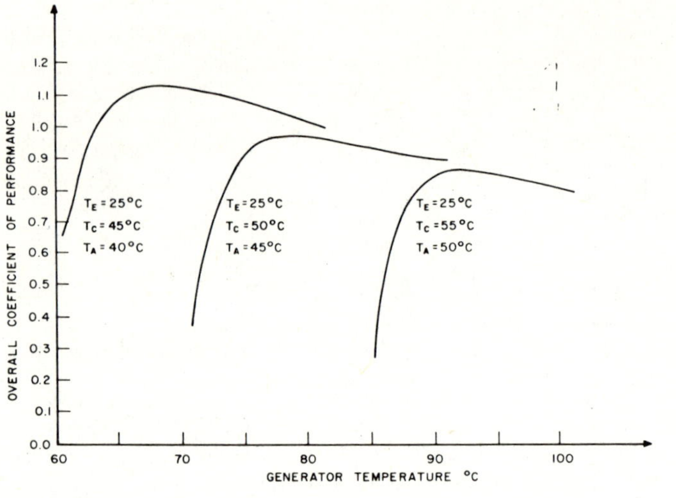

Fig. 8
C.O.P. of $LiBr-ZnBr_2$-Methanol solutions. Operated by a flat plate solar collector.

The overall expected efficiency for the absorption heat pump considered, operated with a flat plate solar collector, with a selective surface and with a solar heated evaporator for conditions encountered in Mexico (8) is shown in Figure 8. Three absorber and condenser temperatures are considered, in all three curves, the absorber is 5°C cooler than the condenser. This form of operating favours a higher equilibrium concentration in the absorber, reducing the circulation factor. In practice, this is achieved by circulating the water which is being heated first through the absorber and afterwards through the condenser.

An optimum generator temperature exists where the COP is highest for each absorber and condenser temperature, which would fix the required collector temperature. At higher generator temperatures the overall efficiency declines due to the solar collector lower performance.

CONCLUSIONS

The operation of the methanol-$LiBr-ZnBr_2$ absorption heat pump seems feasible and with overall efficiencies competitive with the vapor compressor system.

The mixture proposed shows a good performance over the range of operating temperatures considered.

The use of present flat plate solar collectors presents limitations with respect to generator temperatures (around 100°C) and in order to have acceptable COP values it is necessary to work lower absorber and condenser temperatures than with conventional methods.

Future developments on more efficient solar collectors at higher temperatures (in the range of 120°C) would make absorption heat pumps even more competitive.

REFERENCES

(1) Hernández, E. 'The seasonal distribution of the incoming solar radiation in Mexico'. Joint ISES Conference, Winnipeg, Can (1976).

(2) Hainsworth, W.R., 'Refrigerants and absorbents'. Refrigerating Engineering, Vol. 48, 2, August, (1944) pp. 97-100.

(3) Aker, Squires and Albright, 'An evaluation of alcohol-salt mixtures as absorption refrigeration solutions'. ASHRAE Transactions. Vol.71 pp. 14-20 (1965).

(4) Ruhemann, M. 'The ammonia absorption machine' Trans. Instn. Chem. Engrs., 25, 143 (1947).

(5) Best, R., 'Absorption heat pumps'. M.Sc. Thesis, Cranfield Institute of Technology (1976).

(6) E.T.S.U. 1st. Report on heat pumps. Oct. p.33 (1975)

(7) Herrera, R. 'Simulación digital de un sistema alticonvectivo y antirradiativo en colectores solares planos'. B.Sc. Thesis University of México (1977).

(8) Mayer, E.R., Martinez, R., Nissan I., 'Development of a new lost cost flat plate solar collector for tropical countries'. International Solar Building Technology Conference. London (1977).

5.5 Solar cooling and air conditioning at the University of Queensland

S.V. Szokolay
Director, Architectural Science Unit
University of Queensland, St. Lucia, 4067, Australia

Abstract

A brief historical review is followed by a summary of conclusions drawn from the 1965/67 solar house project. The aims of a current project are outlined. The design decision-making process is described in detail under three headings: the building, the collector-roof and the air conditioning system. A simplified optimisation method usable by the designer in practice is given. The paper concludes with an economic assessment, which shows that in areas served by an electrical grid such a system is not competitive yet. However, in remote areas it is already feasible. Future energy cost increases and the mass production potential of the new collector design are likely to make it competitive in all areas of Australia.

Background

After some sporadic individual efforts in the mid-1930's, solar energy research at the University of Queensland began in a systematic way in 1945, with the first grant received for the development of flat plate solar water heaters. Work on solar air conditioning started in 1949 as a co-operative project of the architecture and mechanical engineering departments, under the auspices of the Solar Energy and Tropical Housing Research Committee.

A design for a solar air conditioned house was described at the UN Conference on New Sources of Energy (Sheridan, 1964) in Rome. A revised design was actually built in 1966. This is a single storey, 3 bedroom residence of 130 m^2 floor area. The design included 12 collector panels of 4.8 x 1.2 m each, but only one was constructed with an electrically heated "amplifier". An Arkla-Servel $LiBr/H_2O$ absorption chiller of 10.5 kW (3 ton) capacity was used, with a water-cooled condenser and cross-flow cooling tower. Water hot-storage and rock pile cold-storage was provided, the latter externally to the building: some 30 tonnes of gravel in two cylinders.

The system is fully described by Sheridan and Carr (1967) and a brief resumé is given by Szokolay (1975). The performance monitoring and analysis of results was completed by 1968. Due to lack of funds the SETH Committee was disbanded in 1969.

With the general re-awakening of interest, the Solar Energy Research Committee of this University was established in 1974. Currently work is going on in the departments of agriculture, architecture, chemistry, electrical and mechanical engineering and biochemistry.

Much progress has been made elsewhere in the development of equipment and various applications since 1968. The Japanese Yazaki Corporation is, for example, marketing a $LiBr/H_2O$ absorption chiller, in several sizes, of which one has been obtained by the Architectural Science Unit. However, several of the conclusions of the 1966 project are still valid:

1 the area of the house could be reduced without loss of amenity and service areas need not be air conditioned

2 the glass (window) area should be reduced

3 the planning should be simplified following the rationale of duct layout

4 a combined unit could be built, incorporating all the required machinery, with simple site installation

5 the cold storage should be replaced by condensed refrigerant storage or some other means more efficient than the rock pile

6 the solar collector should be integrated with the roofing system.

Several of these conclusions are incorporated in the current project, the aims of which are:

1 optimisation of the building and its fabric
2 development of an integrated solar collector/roofing system
3 improvement of the air conditioning system itself.

The acute shortage of funds compelled us to link this project with an actual building proposal for which funds were available. The University's veterinary school had a proposal for a small (146 m^2) seminar and study building, at the University Farm, just outside Brisbane. Our project was linked with this proposal.

Optimisation of building and its fabric

A sketch design already existed when this project was first suggested. The introduction of solar air conditioning created a number of mundane but very real architectural problems.

Fig. 1
Site plan

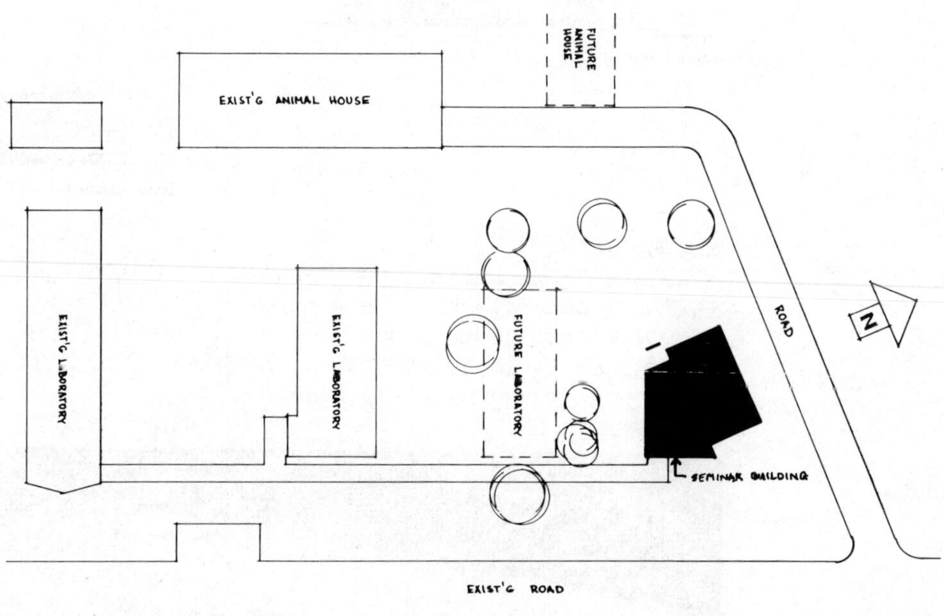

Older buildings on the site form a rectangular grid and the new proposal was to comply with this system. It was to be erected where the "future laboratory" is shown on the site plan (Fig. 1), with its main roof slope facing 24^0 east of north. In that position it would have been severely overshadowed by adjacent mature trees. It was necessary to make three major changes already at this broad, planning stage:

1 to shift the proposed building further north, to avoid the shade of trees, at the edge of a gently sloping open field

2 to replan the building, so that the south and east sides "recognize" the rectangular grid layout, but the northern slope faces due north

3 to change the elongated oblong plan into a more compact, near square, thus thermally more efficient form.

The revised design is shown in Fig. 2 with Fig. 3 giving a model photo.

The intention was to use a brick-veneer construction, i.e., a timber framed, internally plasterboarded wall having a 110 mm brick outer skin. Almost the whole of the north and south walls were to be openable, mostly glass louvres. It is a commonly held belief amongst Brisbane architects that this is a hot-humid climate, where the most suitable solution is the lightweight, fully cross-ventilated building. However, studies such as that by Auliciems (1975) showed that cold discomfort is

Fig. 2
The building

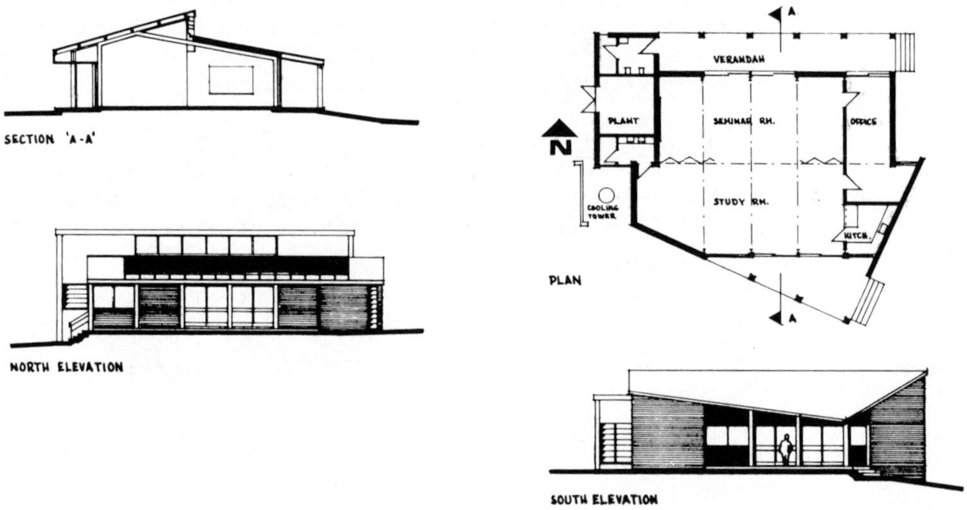

The building
- walls: 110mm brick + 50mm cavity + 100mm block
- floor: concrete slab on ground
- partitions: 100mm block
- roof: galv. iron troughs, 100mm glass fibre, plasterboard

Fig. 3
Model photo

far more frequent than overheating, as the most preferred temperature is 24.5°C. Conditions in a well designed building can be kept comfortable throughout the "winter" without any form of heating. This requires a significant thermal mass. In the summer, daytime conditions are often overheated, but the nights are usually comfortably cool. The typical summertime diurnal variation is between 31° and 20°C DBT and afternoon humidities are rarely more than 50%. There may be 10 - 14 days in a year when night-time conditions are also hot and this would be the only time when a massive construction has no advantage.

The computer program TEMPER, developed by the CSIRO Division of Mechanical Engineering (Wooldridge, 1973) has been used to examine the thermal performance of several alternative constructions. Using the LOAD mode of operation the daily air conditioning load profile was established. Fig. 4 shows a graphic comparison of results for three different constructional systems. In terms of daily total cooling load, the improvement from 'a' to 'b' is almost 36 kWh. The further improvement from 'b' to 'c' is only 14 kWh, whereas the cost increment would be approximately the same in both steps.

Fig. 4
Cooling load
profiles

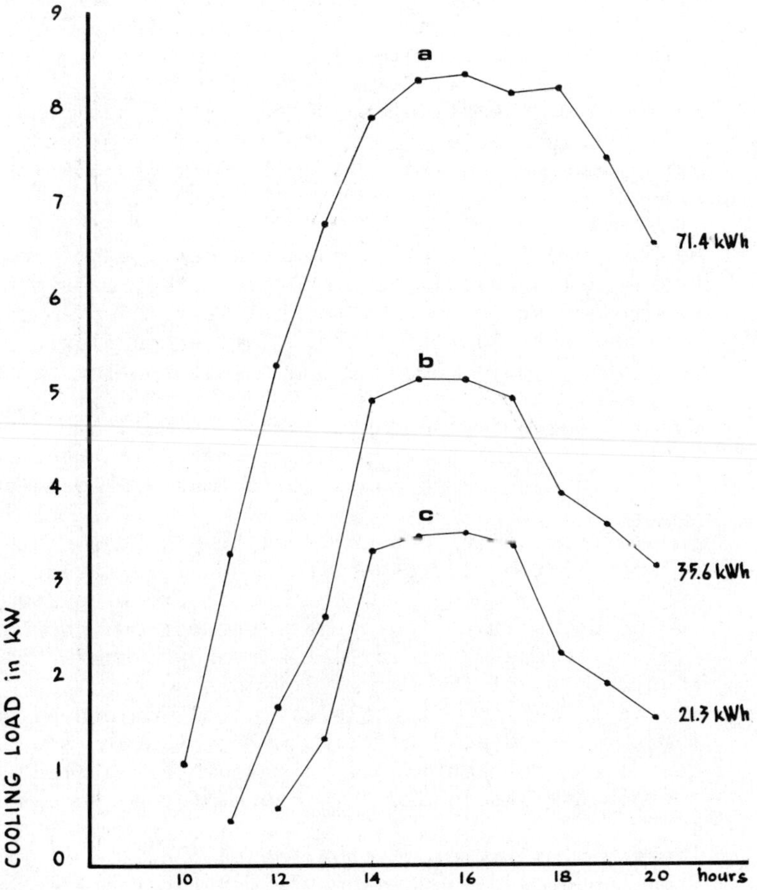

a. brick veneer walls
   timber floor, ventilated under-floor space
b. cavity walls (110 mm brick + 50 mm + 100 mm block)
   concrete slab on ground
c. same, but inner skin 200 mm block and
   clerestorey wall insulation doubled to 100 mm.

Day's total loads given in kWh.

As a result of such simulations, three major design decisions have been made:

1 to use a concrete slab-on-ground floor rather than the elevated timber floor earlier thought of

2 to use a wall construction as massive as feasible within the strict cost limits, i.e., a 110 mm brick outer skin and a 100 mm solid concrete block inner skin - in lieu of the original brick-veneer proposal. (We couldn't afford the 200 mm inner skin and the benefit of this was not proportionate to the extra cost.)

3 all internal walls and partitions to be of 100 mm concrete blocks.

All windows will be fully shaded either by the verandah roof or by external horizontal shading devices.

The collector-roof

All flat plate solar collector panels on the market are factory made complete sealed units. They are usually mounted on top of an otherwise complete roof. In space heating or air conditioning applications, where up to 50% of the roof would be covered by collectors, this is extremely wasteful.

An in situ assembly of a collector-roof has been attempted in the Milton-Keynes solar house project. This proved to be difficult and very labour-intensive. In quality it is inferior to the factory-made product.

We have now attempted to have the best of both worlds: to have a highly reliable, well engineered factory-made product, a collector panel/roofing component, which is assembled in situ. The component includes one sheet of glass. The second glazing can be built over the collectors, using conventional glass roofing technology.

Double glazing will be necessary, as well as selective surfacing because of the high temperature collection requirement (80 - 90°C). Only such high performance collectors are worth considering.

The fabrication of conventional tube-grid + plate absorber panels is a labour-intensive, essentially handicraft type of process, even if assisted by machinery. Sheet metal pressing seems to be the only technology allowing quick mass production. Copper is far too expensive. If closely spaced flat and broad waterways are used, the thermal conductivity of the material is not so critical. Ordinary mild steel may be suitable, but stainless steel would have certain advantages. The chromium black surfacing we use (Cathro and Christie, 1976) requires a first coating of bright nickel by electro-deposition. Using a stainless steel base this stage of the process can be eliminated.

The stainless steel absorber panel consists of two parts: one flat (top) sheet and one pressed, which incorporate the waterways and the headers. These are joined by resistance welding. A 50 mm diameter threaded spigot is welded in at all four corners of the absorber (Fig. 5). The top surface is then coated with a chrome black selective surface (appendix 1). For reasons of convenience a 1 m^2 absorber module has been chosen, which coincides with the international standard A0 paper size: 1189 x 841 mm.

A number of such absorber panels can be assembled in the factory, using short, internally threaded (half left, half right thread) pipe sleeves, joining the spigots of adjacent panels, thus forming continuous top and bottom headers. The large sized header ensures an even

Fig. 5
Absorber panel element

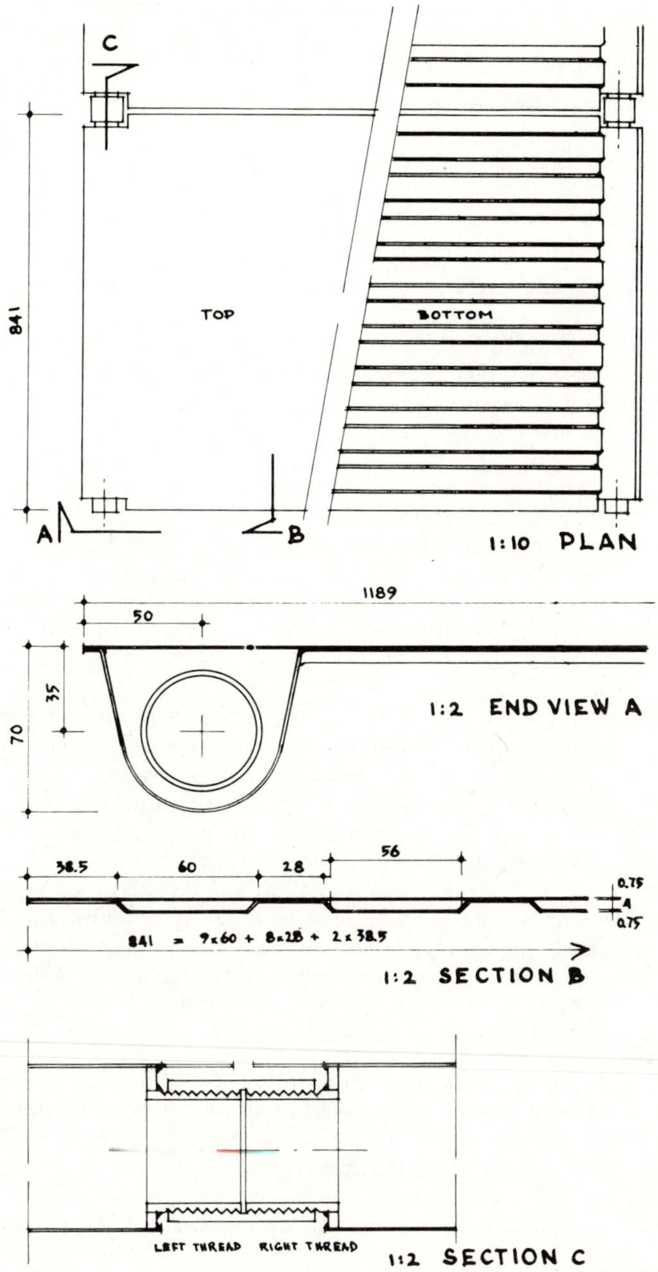

Fig. 6
Collector panel

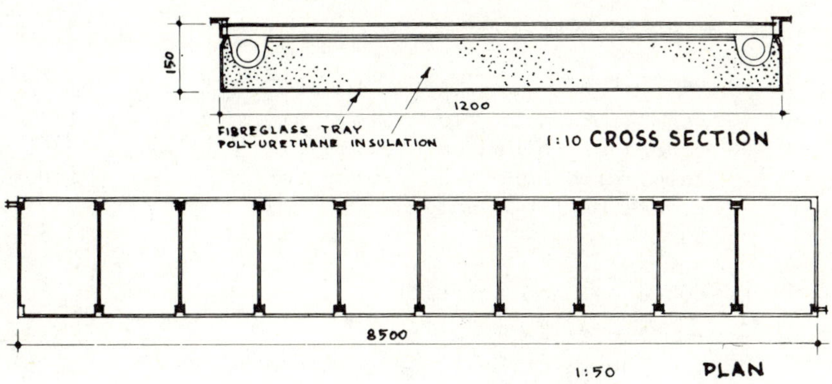

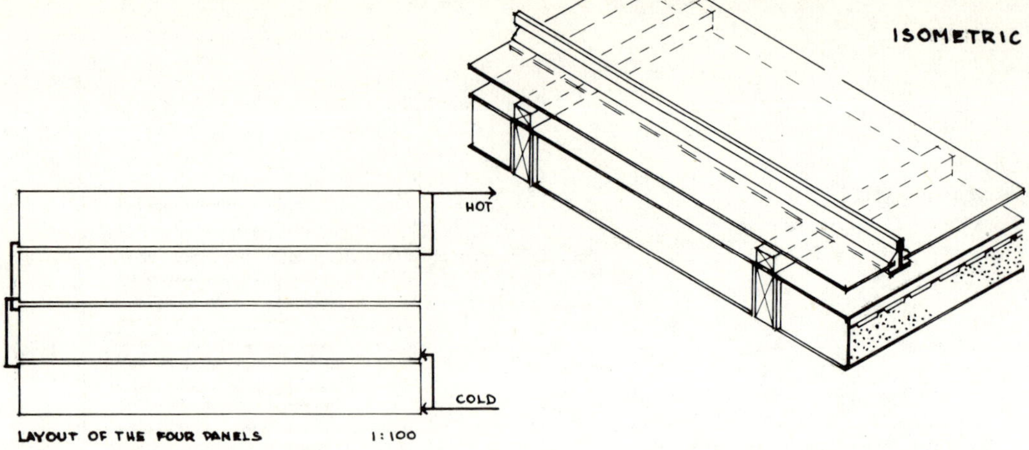

Fig. 7
The collector roof assembly

flow distribution in the waterways, for up to 15 panels, i.e., a continuous 12.75 m long absorber. Any length up to this is possible in 850 mm increments. In our case 10 absorbers will be used in one unit, giving a 8.5 m overall length. A fibreglass tray of the profile shown in Fig. 6 is formed by a continuous process to the desired length. The ends are fixed and the absorber is placed into this tray. The 100 mm space between the back of the absorber and the tray is filled in the factory with a polyurethane foam, in situ expanded. This foam hardens and adds to the rigidity of the collector assembly. A 4 mm low iron glass sheet is fitted, using an extruded rubber strip. The complete collector panel of 1.2 x 8.5 m overall size is delivered and installed.

The main structural members are steel portal frames at 2.5 m centres. Timber purlins are spanning between these, with 1.2 m clear spacing, so that the collector panel fits in between these purlins. The length of the collector equals four spans of purlins (4 x 2.5 m). Four such panels are to be installed, giving a total area of 40 m^2. Aluminium glazing bars, supported on battens over the purlins hold the outer glass cover. The assembly is shown on Fig. 7. The use of a polycarbonate sheet for this outer glazing is now considered, as this would stand up to hail, which can occur about once in three years in the area. The fibreglass tray serves as the finished ceiling.

The air conditioning system

Fig. 8 shows the system diagram. The "heart" of the system is the Yazaki chiller unit, WFC-400S, shown in heavy outline. This is a lithium bromide/water absorption refrigerator, having the following performace characteristics:

Evaporator	
chilled water flow rate	13.3 *l*/min (max 25 *l*/min)
chilled water temperature	9°C (min 8°C)
temp. drop through chiller coil	5 degC
cooling rate	4.6 kW
Generator	
generator water flow rate	18.5 *l*/min
generator inlet temperature	88°C (75 - 100°C)
temp. drop across generator coil	6 degC
generator input rate	7.75 kW
COP	0.6
Condenser and absorber	
cooling water flow rate	35.6 *l*/min
cooling water inlet temperature	29.5°C (24 - 32°C)
temp. increase of cooling water	5 degC
heat rejection rate	12.4 kW

Fig. 8
System diagram

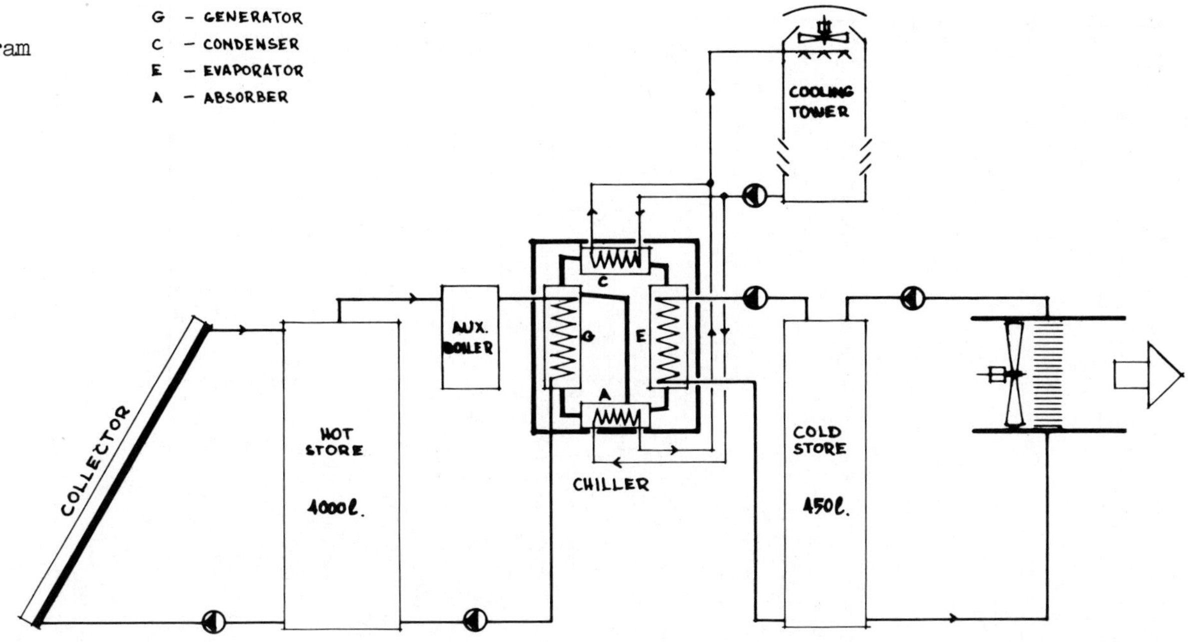

The peak cooling load of the building is some 5.2 kW. The given refrigeration capacity is 4.6 kW. There is a deficiency of 0.6 kW. In energy terms the deficiency over the four hours of its occurrence would total approximately 2 kWh. This must be bridged by some form of cold storage. As the system operates with chilled water, this seems to be the most obvious storage medium. If the operative range is 13 to 8°C, i.e., 5 degC, 1 litre will "store" 5 x 1.16 = 5.8 Wh, thus the required volume would be

$$\frac{2000}{5.8} = 345\ l.$$

If a fan-coil unit is to operate with 5 degC temperature gain across the coil, the 5.2 kW cooling rate can be achieved with a water flow rate of

$\frac{5200}{5 \times 1.16} = 0.25\ l/s$. As the normal flow rate through the chiller is

0.22 l/s, the difference of 0.03 l/s must be supplied from the storage. This would require a volume of 432 l if the peak deficiency existed throughout the four hour period.

The storage of a full day's cooling capacity would require a chilled water tank of over 6 m^3 volume, which is beyond practical feasibility. Any intermediate storage capacity would serve no useful purpose, thus the decision was made to install a 450 l chilled water storage tank. It is more economical to store the required energy at the hot side of the system, where the operative range is at least 20 degC (95 to 75°C), thus each litre could store 20 x 1.16 = 23 Wh of heat.

Having a CoP of 0.6, to provide a full day's cooling 35.6 kWh, the system would require the storage of some 60 kWh of heat. This could be done with a storage volume of 2600 l (2.6 m^3). The decision must however be related also to the available solar heat supply.

The daily average total horizontal insolation in Brisbane in December is 5.6 kWh/m^2 (this is the worst month: the temperatures are as high as in January, but the insolation is less). For midsummer optimisation the collector tilt should be only 4°, but for the three summer months it is 10°. This is also the practical minimum from the point

Fig. 9
Brisbane radiation data

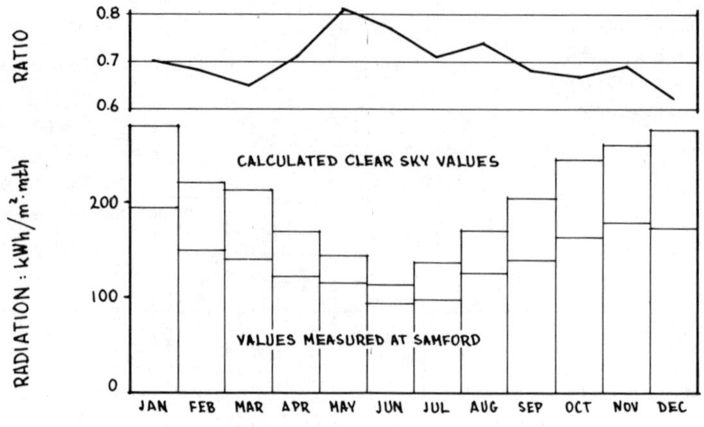

Monthly total horizontal radiation in Brisbane: comparison of calculated clear sky values and 1975 measured data (Samford)

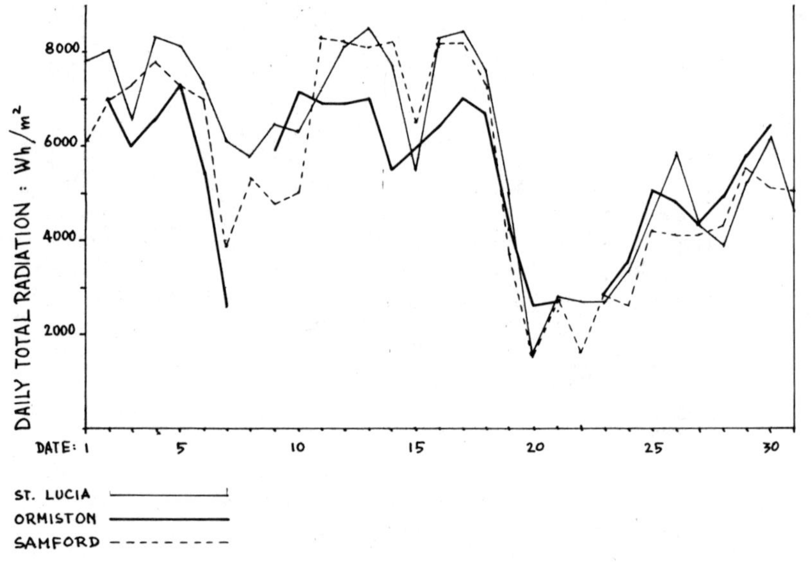

Daily total horizontal radiation in Brisbane: comparison of values measured at three suburban sites during December 1975.

of view of rainwater shedding. Such a surface would receive about 5% more radiation than the horizontal. The collector characteristic curve shows that the average collection efficiency at the required high temperatures will be about 28%. Thus the expected heat production of 1 m^2 collector is 5.6 x 1.05 x 0.28 = 1.65 kWh.

The daily heat requirement has been calculated as 60 kWh. Allowing 10% for system losses, the required collector area is

$$\frac{60 \times 1.1}{1.65} = 40 \text{ m}^2.$$

Thus a decision was made to install 40 m^2 collectors.

Fig. 9 shows a typical month of December radiation climate in Brisbane. Taking the Samford values (nearest to our site) we plotted the daily heat supplied by a 40 m^2 collector in histogram form on Fig. 10. The heat requirement is taken as constant (60 kWh/day). The surplus goes into storage. If we start the month with 30 kWh in the store, the maximum heat available for storing would be 306 kWh, which would require some 13.4 m^3 storage volume. This would be sufficient to supply the collection deficit during the latter part of the month.

Fig. 10 Storage heat content profile during the month of December for a range of storage tank volumes.

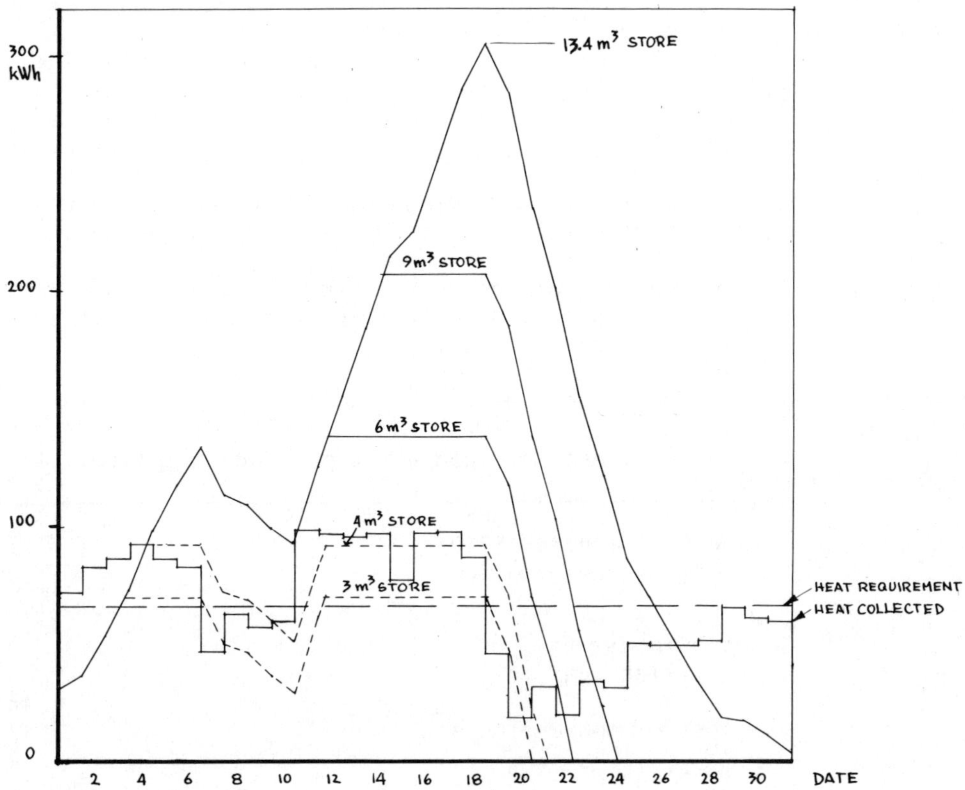

Fig. 11 Storage optimization graph

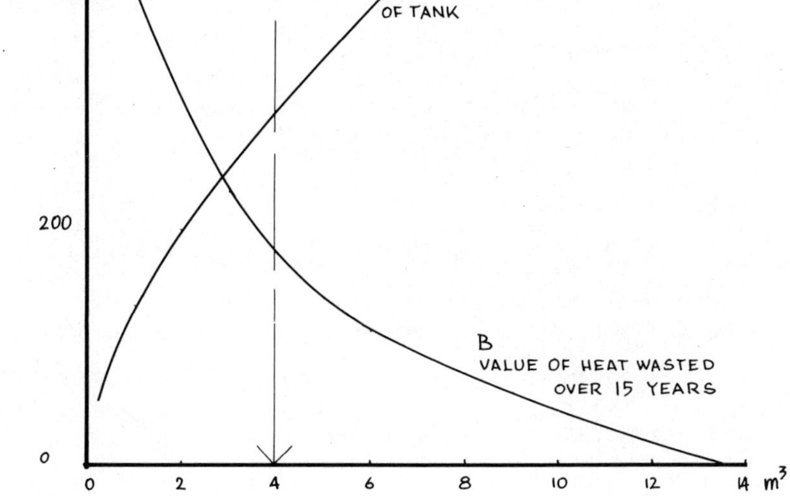

The heat content profile of the store for various tank sizes is also shown in Fig. 10. In all cases (except the full 13.4 m^3) some heat is wasted because of the limited storage capacity. Fig. 11 shows a plot of the value of such wasted heat (which consequently will have to be supplied by an auxiliary source) assuming that the same pattern is repeated for the three summer months and using 15 years as a basis of comparison, with the cost of heat (gas) given as 1.5 ¢/kWh. The cost of tanks is also plotted. The two curves are added graphically and the lowest point determines the optimum tank volume: 4 m^3.

A number of simplifying assumptions have been used, thus it is not an exact optimisation method, but it is a rational estimation.

Hot water from the tank (through the auxiliary boiler) can be supplied to the fan-coil unit for winter operation. This is not shown on Fig.8 for the sake of clarity.

Economic assessment

The cost estimate can be summarised as follows:

40 m^2 collector $120/$m^2$	= $4800		
cost of roof replaced	- 1000	$3800	
chiller unit	$1080		
cooling tower	320		
auxiliary gas boiler	400		
	$1800		
cost of conventional chiller	- 1200	$ 600	
tanks (4 m^3 + 450 *l*)		$ 420	
pumps, piping, valves, controls		$ 380	
net extra capital cost			$5200
for annual total cooling load of 5500 kWh a conventional system (CoP=1.5) would consume 3667 kWh electricity @ 5.6¢		$ 205	
auxiliary (gas) requirement of solar system: 800 kWh @ 1.5¢		- 12	$ 193
annual heating load (elec.) 2000 kWh @ 5.6¢			112
value of annual solar contribution			$ 305

The value of savings would equal the extra capital investment in

$$\frac{5200}{305} = 17 \text{ years.}$$

The criterion for any capital investment can be given as "the net present value should be positive" (Hawkins & Pearce, 1971). This is defined as

NPV = P - C

where P = present worth of all savings (benefits)
C = capital investment.

The analysis was carried out using the following data:

life expectancy	15 years
interest rate	14%
inflation rate	12%
energy cost increase over inflation	5%

Thus we get

$$P = 305 \times \frac{1 + 0.12 + 0.05}{0.14 - 0.12} \times \left(1 - \left(\frac{1 + 0.12}{1 + 0.14}\right)^{15}\right)$$

$$= 305 \times 13.64$$

$$= 4160$$

$$NPV = 4160 - 5200 = -1040$$

The net present value is negative, the investment is not worthwhile.

In spite of this, our project is thought to be worthwhile for several reasons:

1 There are many areas of Australia where the cost of electricity is 12¢ or, in some places, even 17¢, and in such areas the system would already be competitive.

2 We believe that future energy shortages will be reflected in energy costs, and the system will then become competitive. We think we must be ready with the technology.

3 Although a large part of the development cost of the collector system is borne by a manufacturer, its initial cost is high. The unit is capable of being mass-produced, in which case, its cost is expected to be reduced to at least $\$80/m^2$, which would render the system immediately competitive.

References

AULICIEMS, A. (1975) Warmth and comfort in the subtropical winter: a study in Brisbane schools, J. Hyg. Camb., 74, 339.

CATHRO, K.J. & CHRISTIE, E.A. (1976) The application of chrome black coatings to solar panels, Proc. ISES-ANZ Meeting, Melbourne, Nov.

HAWKINS, C.J. & PEARCE, D.W. (1971) Capital investment appraisal, MacMillan, London.

SHERIDAN, N.R. (1964) Prospects for solar air conditioning in Australia, Proc. U.N. Conf. (Rome) on New Sources of Energy, vol.6, pp. 87-96.

SHERIDAN, N.R. & CARR, W.H. (1967) A solar air conditioned house in Brsibane, Solar Research Note. No.2 (Res.Com.on Sol.En.& Trop.Housing), University of Queensland.

SZOKOLAY, S.V. (1975) Solar energy and building, Arch. Press, pp. 86-89.

WOOLDRIDGE, M.J. (1973) Air conditioning load calculation using computer program TEMPER, Aus. Refrig. Air Cond. & Heating, 27, 12, p.26, Dec.

Appendix 1

The chrome black surfacing is deposited on a stainless steel base by the following procedure:

1 degrease in alkali soak cleaner* at 60°C for 5 - 15 min.

2 rinse

3 final degrease by cathodizing at 200 A/m^2 in alkaline cleaning solution at 60°C for 5 min. Steel anodes.

4 rinse

5 deposit chrome black, using the following bath

chromium trioxide	400 g/*l*	
sodium hydroxide	60 g/*l*	
fluosilicic acid	0.5 g/*l*	(calc. as 100% acid)
sucrose	2.5 g/*l*	
barium carbonate	7.5 g/*l*	

using 2000 A/m^2, maintaining bath temperature within 15-25°C (employing cooling coil or immersion heater as required) time 55 to 70 seconds. Lead anode.

6 rinse and dry.

after Cathro and Christie (1976)

* see Metals Handbook, vol.2, p.317, Am. Soc. for Metals, Ohio, 1964

With this process the following surface properties have been obtained:

	absorbance	emittance	a/e
on a new surface	0.923	0.096	9.61
after exposure to 100°C for six months	0.911	0.098	9.3

Solar space cooling with particular reference to the humidification/dehumidification process

5.6

S. Satcunanathan and G.C. So'Brien
University of the West Indies, St. Augustine,
Trinidad, West Indies.

Abstract

The particular attractiveness of solar energy for space cooling lie in the fact that cooling demands in general parallel both in magnitude and time, radiation intensity. Space cooling involves, in general the control of humidity and temperature. Control of humidity is achieved by the removal of moisture either by cooling the air to the dewpoint temperature corresponding to the final required moisture content or by the absorption of moisture in suitable absorbents. In the former case, a refrigeration system either vapour compression or absorption type system driven by solar energy may be used. In the latter case the chemicals used for the removal of the moisture may be regenerated by solar heat.

This paper reviews the various attempts to date on solar space cooling.

Introduction

The particular attractiveness of solar energy for space cooling (both for human comfort and for the preservation of food) lies in the fact that cooling demands in general parallel radiation intensity, both in magnitude and in time. Thus the maximum daily cooling demand is likely to occur just after noon.

Space cooling in general entails the control of the basic air parameters of humidity and temperature and requires in the main removal of moisture and cooling of the air. Removal of the moisture is conventionally achieved by the cooling of the air to a temperature below the dew point, in particular to the dew point temperature corresponding to the final required humidity of the air. Since essentially this is a cooling process, the cooling effect is supplied by a vapour compression or by a vapour absorption refrigeration system.

Alternatively, moisture removal is achieved by the use of suitable absorbing or adsorbing materials, the lowering of temperature being subsequently effected by the partial adiabatic evaporation of water in the dried air stream.

The various means of controlling air parameters are indicated in Fig. 1.

Solar Energy Driven Vapour Compression Refrigeration

Solar space cooling using the conventional vapour compression refrigeration system can, in principle, be achieved by the conversion of solar energy to mechanical power through any of the thermodynamic power cycles and using this power to drive a vapour compression refrigerator. The power cycle most commonly studied in this connection is the Rankine cycle (1, 2, 3, 4, 5) though the Stirling cycle engine has also been considered a possible power unit.

Fig.1
Control of Air Parameters.

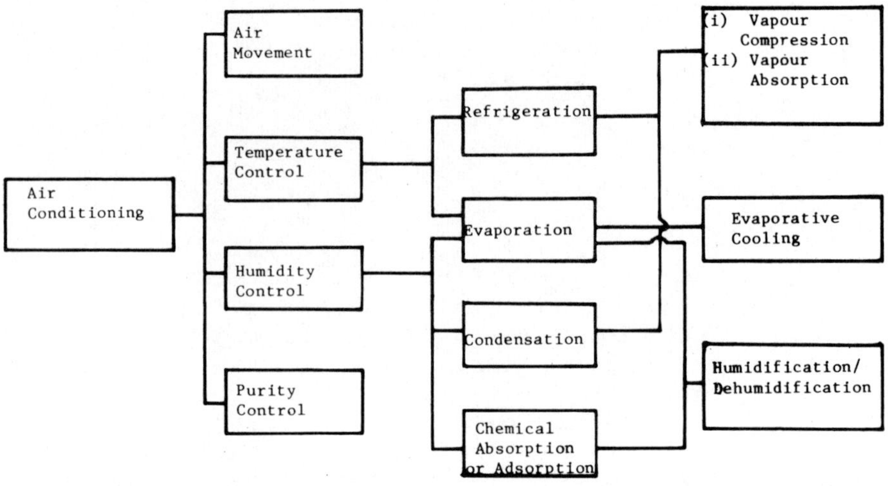

However, the low efficiency of these power cycles, particularly at the low temperatures obtainable with simple flat plate solar collectors - approximately 4% - does not make this method a very viable one at present. In addition to the question of low efficiency, several serious problems have to be solved before such space cooling systems become viable, the most important problem being that of the matching of the power unit to the refrigeration unit at off design conditions. Indications of studies so far are that costs of such systems will be unacceptably high.

Solar Energy Driven Vapour Absorption Refrigeration/Air Conditioning Systems

By far the most serious attention is currently being paid to the vapour absorption refrigeration system. Basically the absorption system operates by driving off a refrigerant vapour from a strong solution of the vapour in a solvent, in a generator, by the addition of heat. The vapour is subsequently condensed and the liquid expanded through a throttle valve to the evaporator pressure. The spent refrigerant leaving the evaporator is then returned to be absorbed in the weak solution returned from the generator, the resulting strong solution then being transferred to the generator. The main advantage of the absorption refrigeration system for solar energy application lies in the fact that generation temperatures are within the temperature range attainable with simple flat plate collectors (6,7). The system can be operated on a continuous basis in which case a mechanical pump is used to transfer the strong solution from the absorber to the generator or alternatively the system may be operated intermittently, the processes of generation and evaporation being alternately carried out and in general coinciding with day and night respectively. A solar driven continuously operated system is shown in Fig. 2.

The feasibility of using solar energy for absorption refrigeration has been demonstrated for a number of refrigerant pairs viz: ammonia-water (8,9,10,11,12,13,14,15) water lithium bromide (16,17,18) ammonia-lithium nitrate (19), ammonia-sodium thiocyanate (15,20,21) and ammonia-calcium chloride (22). Of these the pairs that have been studied most are ammonia-water and water-lithium bromide, the former for

Fig.2
A solar air conditioner. (46)

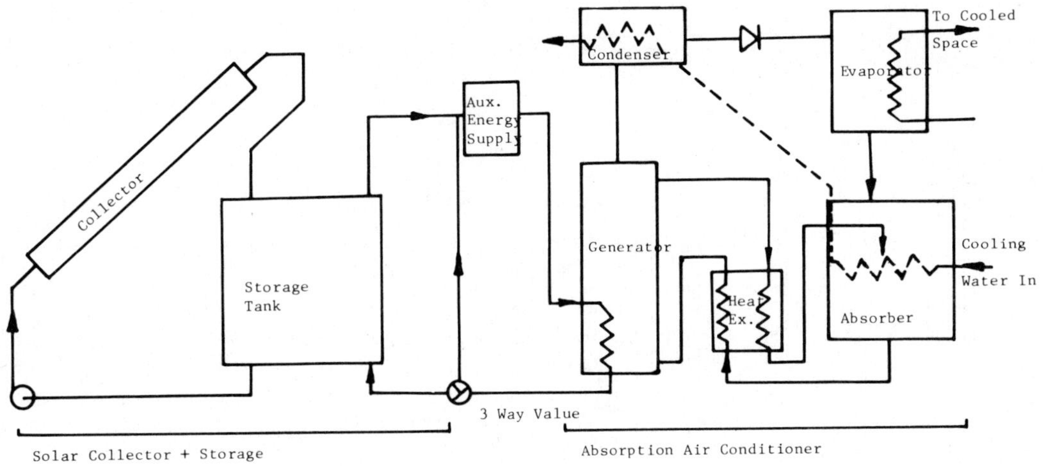

Fig. 3
Comparison of the performance of ammonia-water and water-lithium bromide adsorption cycles. (7)

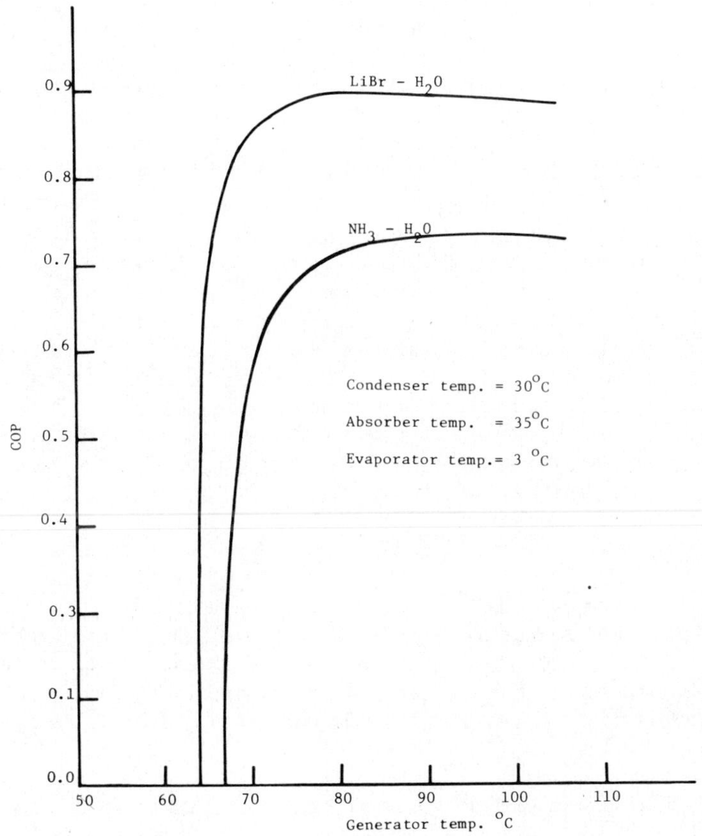

low temperature applications while the latter for temperatures above 0°C (water being the refrigerant). The relative performance of the two pairs may be seen in Fig. 3 (7), the lithium bromide-water pair being significantly superior to the ammonia-water pair for the conditions stipulated and invariably the former is chosen for air conditioning applications.

It should perhaps be noted that to date no commercial solar energy driven units have been installed, all the units currently in operation being experimental or demonstration units. It should also be noted that a few manufacturers are marketing normal absorption chillers modified for use with solar heated hot water.

It may be considered that the technical feasibility of solar energy driven absorption space cooling systems has been established. However, the economic viabilities of such systems are not quite yet established.

Open Cycle Humidification/Dehumidification

Vapour compression and vapour absorption refrigeration systems exchange only heat with the surroundings and may be termed 'closed-cycle' systems. An alternative method of air conditioning, which depends for its operation on water vapour exchange with the atmosphere and which may be modified for solar operation, is classified as 'open cycle' cooling. To date, studies on these cooling cycles have been limited. Experimental work has been reported in the USSR but very little operating data are available. Open-cycle air conditioners have been proposed for some parts of Australia and only recently has development work begun in the U.S.A.

In hot, humid regions the latent heat load, due to the high moisture content of the environment, constitutes an appreciable part of the cooling load. For satisfactory moisture control through refrigeration air conditioning, the air may have to be over cooled and reheated to achieve dry bulb temperatures within the comfort region. The humidification/dehumidification system achieves moisture control by much simplier and probably more economical means. In this system, room air is first dried by contact with a sorbent material. If water is now evaporated into this dry air, extraction of the latent heat of evaporation causes the air to be cooled. At the same time, the moisture content of the air increases. Water evaporation is controlled so that the leaving air stream is at the desired conditioned state. The sorbent material might be either a liquid or a solid. Typical inorganic liquid dessicants are solutions of lithium chloride, lithium bromide, calcium chloride, sulphuric acid, potassium hydroxide and caustic soda. Organic liquids such as glycerol, diethylene glycol and triethylene glycol also serve for dehumidification purposes. For commercial applications lithium chloride and triethylene glycol have the most favourable characteristics.

Typical solid dessicants are lithium chloride, calcium chloride, silica gel, activated alumina or molecular sieves. Of these, silica gel and lithium chloride are generally chosen for commercial units.

A commercial hygroscopic liquid unit is illustrated in Fig. 4. Such units have been manufactured and installed in the U.S.A., U.K. and Holland but they are generally large units intended for industrial and commercial applications. Units such as these have not been manufactured for domestic purposes.

Principle of operation of the HDH System

The principle of operation of the system is best understood with respect to a liquid sorbent system as illustrated in Fig. 4

Fig. 4
A commercial liquid sorbent dehumidifier. (47)

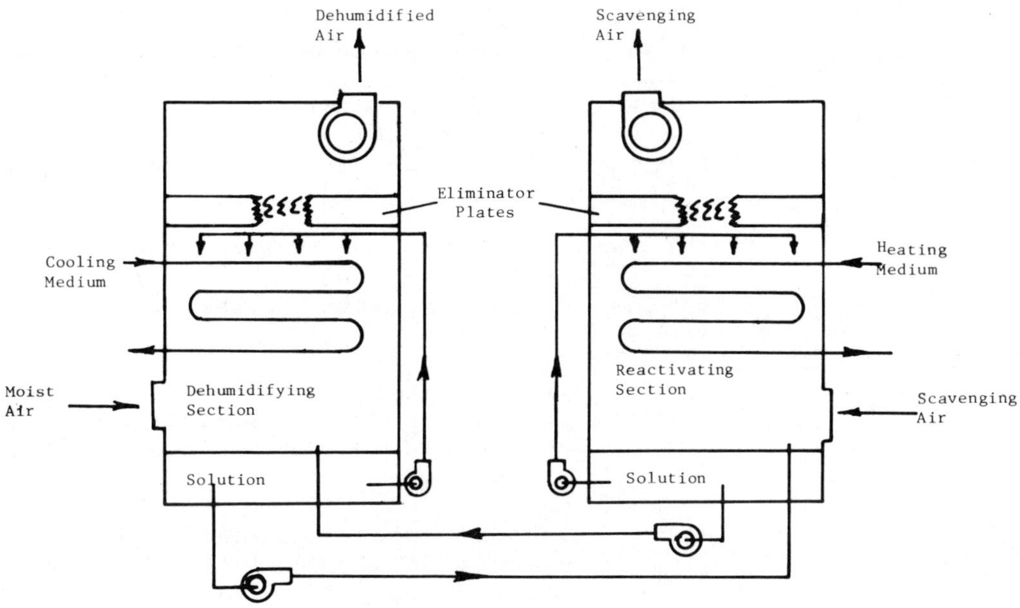

Fig. 5
Equilibrium vapour pressure versus temperature for Lithium Chloride solutions

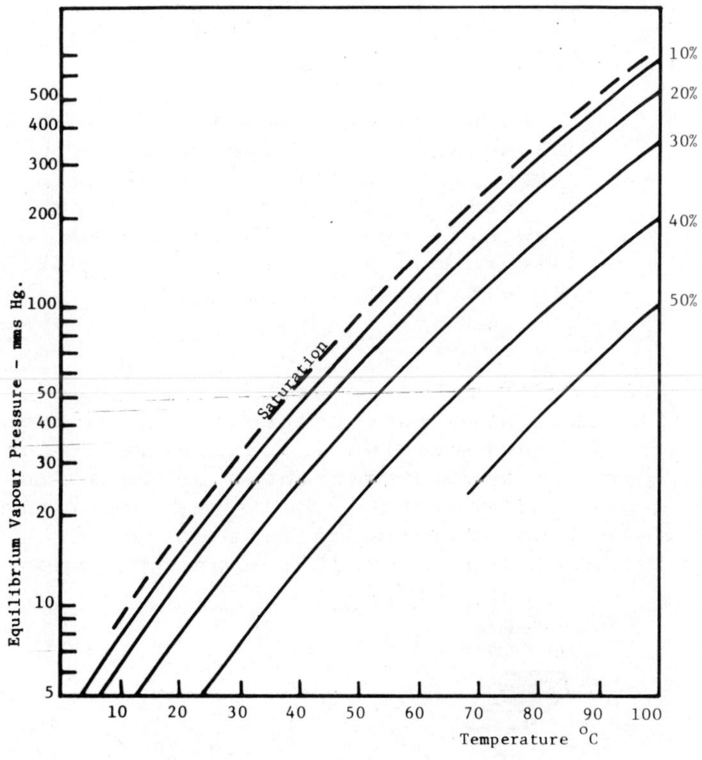

At a given moisture content condition (concentration) of the sorbent material, the equilibrium vapour pressure is a function of temperature (see Fig. 5). For any given concentration and depending on the temperature the equilibrium vapour pressure will be higher or lower than the water vapour pressure of the air to be conditioned. In general, in high humidity areas, at atmospheric temperature, the equilibrium vapour pressure is lower than the atmospheric water vapour pressure. When this humid air is passed over a set of contactor coils over which the strong sorbent solution is sprayed, moisture transfer from the air to the solution takes place. The latent heat loss through moisture absorption and the heat of dilution resulting from solution moisture gain causes an increase in temperature of the solution as well as the air. Cooling water is therefore circulated through the contactor coils to minimise the reduction in the dehumidification capacity due to temperature rise of the sorbent solution and to reduce the sensible heat gain of the air stream.

The temperature of the dehumidified air is generally higher than the inlet air temperature. Air conditions within the comfort region may subsequently be achieved by various means.

(i) Since higher temperatures are tolerable at lower relative humidities, sensible cooling in an air-air heat exchanger may be adequate.

(ii) Sensibly cooled, dehumidified air may be further cooled by adiabatic evaporation of water (evaporative cooling) which also causes an increase in absolute humidity.

(iii) The air may be evaporatively cooled to saturation and mixed with air of different parameters to achieve comfort conditions.

(iv) In some cases, excessive dehumidification may be required to achieve comfort conditions on evaporative cooling. To avoid too much moisture in the conditioned space, outlet room air, evaporatively cooled to saturation, in turn cools the warmer dehumidified air. The air is thus cooled to the wet-bulb temperature without increasing the moisture content.

The dehumidification process causes a decrease in solution concentration so that regeneration is required to maintain constant concentration in the dehumidifier (an increase in moisture content decreases dehumidification capacity at a given temperature). In the regenerator, solution is heated to higher temperatures at which the equilibrium vapour pressure is greater than the atmospheric water

Fig. 6 A commercial rotary dehumidifier. (48)

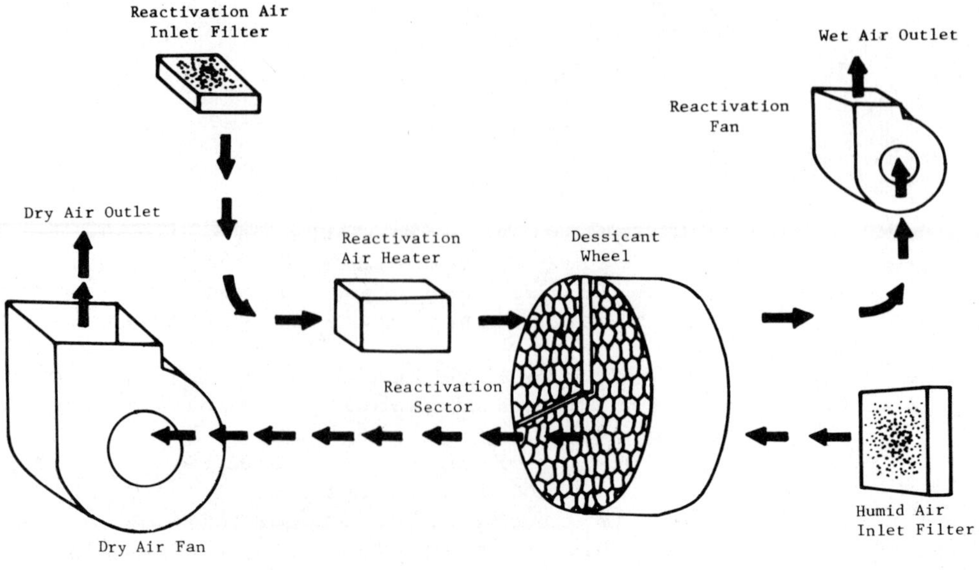

vapour pressure so that moisture transfer to the air takes place. Heat is absorbed in the process.

The principle of operation with solid dessicants is similar except for the constraints imposed by the physical nature of the material. Material transfer from the dehumifier section to the humidifier section is achieved either by packing the material in rotating beds or wheels or by switching the air streams at the appropriate times. A commercial rotary dehumidifier is illustrated in Fig. 6. Perhaps the most serious thermodynamic disadvantage with solid dessicant materials lies in the establishment of temperature and moisture stratification which makes the dehumidification and humidification processes non-isothermal.

Solar Operated Solid Dessicant Systems

The first consideration of a solar operated, open-cycle HDH system for a solar cooled house is credited to Altenkirch (23) Fig. 7 is a schematic diagram of the system subsequently described by Dannies, utilizing solids adsorbents (24). The north-south orientation of the building is such that in the morning when the eastern side of the building becomes heated, the incoming air is dried by passing through cool solid adsorbent on the west wall with subsequent evaporative cooling by passage over a water bath. At the same time, the adsorbent heated on the east wall is regenerated by air flow out of the room. Reversal of the air flow occurs in the evening when the sun heats the west wall. The system is attractive but allows no control over air conditions and cannot, because of its dependence on such variable factors as solar radiation and relative humidity, be reliable. Such a cooling system also places restrictions on building design. Although there are no published reports it is claimed (23) that two houses were constructed in Europe in the 1930's based on the same principle, which were destroyed in the second world war. Cooling of a warehouse in Liberia has been reported (25). During tests the air temperature after passing through the absorbent beds was found to be 32°C rising 3°C above outside dry-bulb temperature. It is claimed that air temperatures as low as 13°C may be achieved through evaporative cooling.

Fig. 7
A solar cooled house.

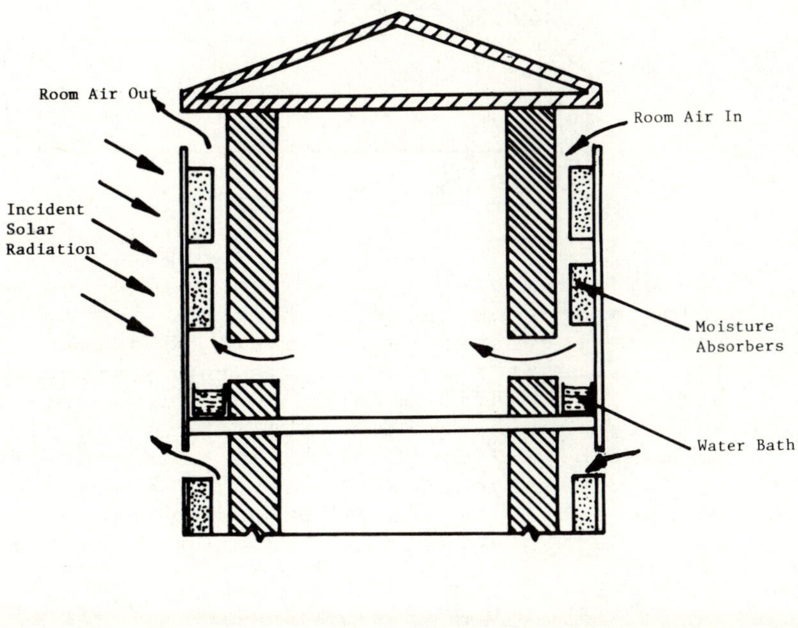

The report on the Altenkirch house (the detail of the wall is shown in Fig. 8), was presented to stimulate suggestions on methods of improvement, the operation being unsuccessful. It was found that natural circulation of room air (chimney effect) was inadequate thereby necessitating the use of a fan. The attractiveness of the system (no moving parts) subsequently reduces. It is also shown that the open-cycle results in low COP and that adequate cooling requires impractical collector areas because of the orientation of the collector surface. The need for better absorbents with lower regeneration temperatures and "roof" collectors is suggested as an approach to improvement. It appears that as a drying chamber (without evaporators) operation is more successful, Dannies having reported on such use for the drying of timber. In this case, the requirement of an isothermal adsorption process is eliminated and the lower moisture content of the outgoing regenerating air results in more effective regeneration.

Fig. 8
Wall detail of the Altenkirch House. (23)

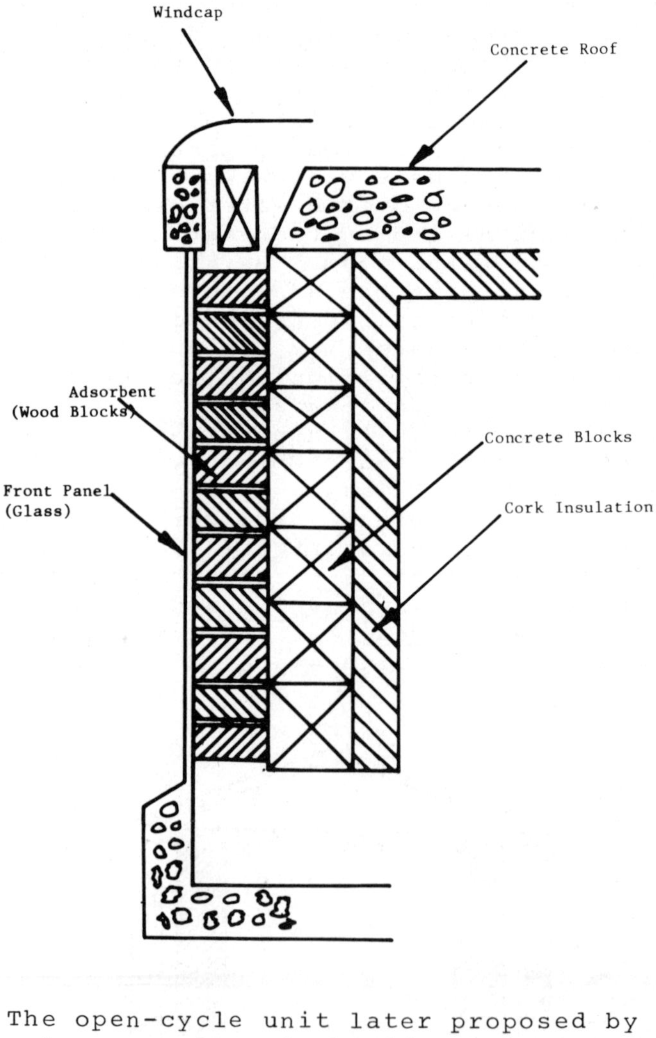

The open-cycle unit later proposed by Dunkle (26) for solar operation in humid tropical and sub-tropical regions incorporates a rotating dessicant bed system. The dessicant silica gel, is regenerated by solar heated air. A schematic diagram indicating rock-pile storage and heat exchangers to maximize cycle effectiveness is presented in Fig. 9. Fig. 10 represents the air cooling cycle on a psychrometric chart, the state point numbers corresponding to the positions on the diagram. Room air is evaporatively cooled, a portion being sensibly

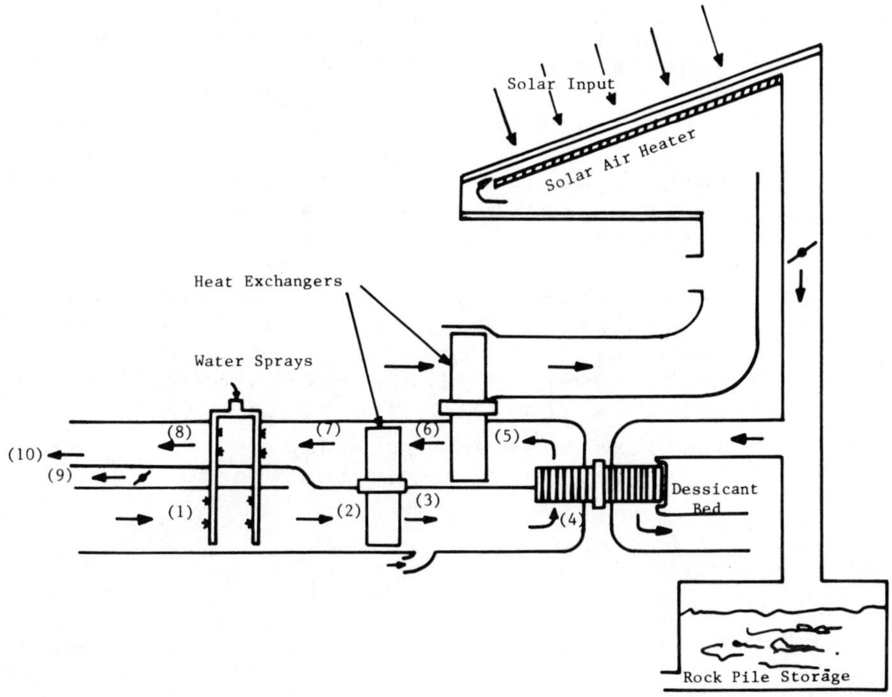

Fig. 9
Dunkle's silica gel system. (26)

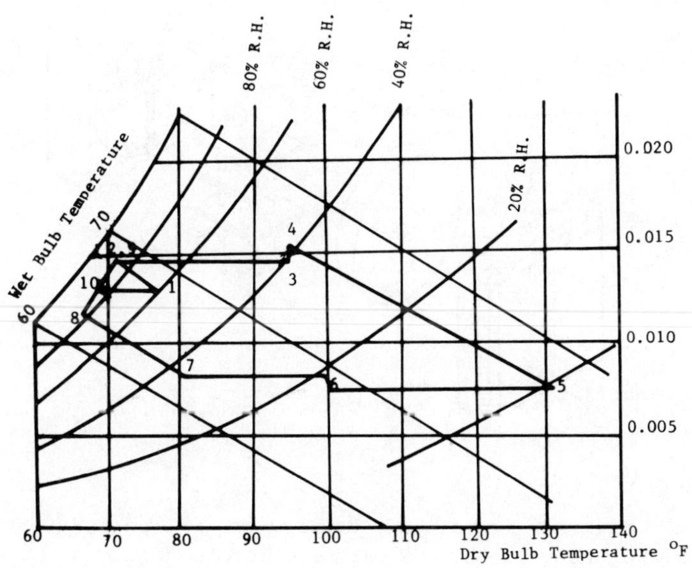

Fig. 10
Air conditioning cycle for Dunkle's system. (26)

heated to state point 3 on passage through a rotary heat exchanger. After mixing with outside air (state point 4), dehumidification and the associated heating effect changes the air state to point 5. Heat exchange with outside air, then with room air and finally evaporative cooling achieve cooled air conditions at the state point 8. Mixing with the diverted portion of the evaporatively cooled incoming air is regulated so that the condition of air delivered to the room is within the comfort region.

Pulverization problems were anticipated with the rotating silica gel system and it was suggested that substitution of a fixed bed, periodically regenerated, might be necessary.

Ahopdhzanyan (27) reported on solar air conditioning with solid adsorbents in Russia. The model of a solar air conditioner, claimed to have been tested for suitability in Southern Russian climates is presented. Regeneration temperatures in the range 70-80°C (attainable with flat-plate collectors) are indicated but no operating data are published.

Rush et al (28) and Lunde (29) have recently investigated solar operated sorption air conditioners similar to that described by Dunkle. Fig. 11 is an exploded view of the Solar-MEC unit described by Rush. It is claimed that dehumdification, sensible cooling and adiabatic humidification processes produce conditioned air corresponding to the quality of air delivered by a conventional compressor unit. Owens-Illinois evacuated glass tube collectors supply solar heated water (88°C) to storage. The total power used as auxiliaries (fans etc) amounts to 1 h.p. which includes 1/16 h.p. for driving the wheels (30).

Fig. 11
Solar-MEC unit.
(29)

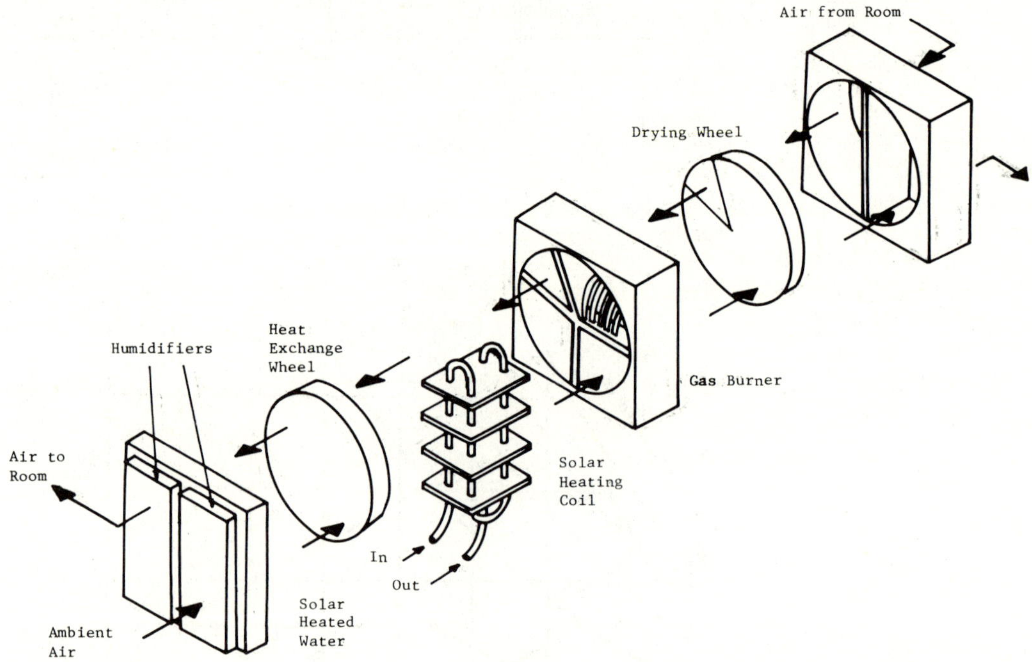

The sorbent material employed in the Solar-MEC unit is a molecular sieve whereas Lunde uses silica gel. The silica gel bed is made up of four rotating discs with four intermediate heat exchangers so that the heat of adsorbtion is removed in four stages. Experimental regeneration temperatures were about 210°F although it is stated that regeneration temperatures as low as 150°F are permissible.

Lunde suggests that since the latent heat load is only a fraction of the total load, a smaller unit may be used for dehumidification if the rest of the cooling load is picked up with another solar air conditioning system (e.g. a lithium-bromide absorption unit). Evaporator temperatures may then be higher allowing air cooling of the absorber and condenser which is a distinct advantage over water cooling.

Problems anticipated with such rotary systems include sealing and pulverization of materials such as silica gel which also has a limited cycling life-time. Reversed air flows for dehumidification and regeneration is effective in eliminating dust deposits.

The theory of rotary regenerators and heat and mass transfer in solid sorbents has been extensively investigated (e.g. 31,32,33,34,35).

The suitability of solid sorbent materials for use in dehumification systems may be gauged from the residual moisture content (a measure of the moisture removal efficiency), the capacity and the speed of drying.

These properties for some selected sorbents are indicated qualitatively in Table 1 (lower numbers indicate better performance). Quantitative data for the residual moisture content of the air in equilibrium is given in Table 2

Table 1

Sorbent material	Efficiency	Capacity	Speed
Phosphorous pentoxide	1	2	1
Sulpheric acid	2	1	2
Calcium chloride	7	3	3
Magnesium perchlorate	3	4	7
Silica Gel	4	5	6
Activated Alumina	5	6	5
Calcium sulphate	6	7	4

Table 2

Material	Residual Moisture Content mg/litre	Equiv. dew-point ^{o}F
Copper sulphate (anhydrous)	2.8	+19.4
Calcium chloride (granular)	1.5	+10.4
Calcium chloride (anhydrous)	1.25	+ 1.4
Zinc chloride (sticks)	0.98	- 2.2
Caustic soda (sticks)	0.80	- 5.8
Potassium hydroxide (sticks)	0.014	-68.8
Calcium sulphate (anhydrous)	0.005	-83.2
Calcium oxide	0.003	-88.6
Sulphuric acid (95% at 35°C)	0.002	-94.0
Barium oxide	0.00065	-106.6
Molecular sieve	0.00065	-107.0

which also gives the equivalent equilibrium dew point temperature while information on the capacity of the more common commercial solid sorbents is presented graphically in Fig. 12.

Solar Operated Liquid Dessicant Systems

Experimentation on the solar operation of a triethylene glycol dehumidification system was reported on by Lof (36). A schematic diagram of this system (Fig. 13) illustrates the use of solar heated air for regeneration. It was found that preheating the glycol solution by heat

Fig. 12
Relative capacities of some commercial solid sorbents. (49)

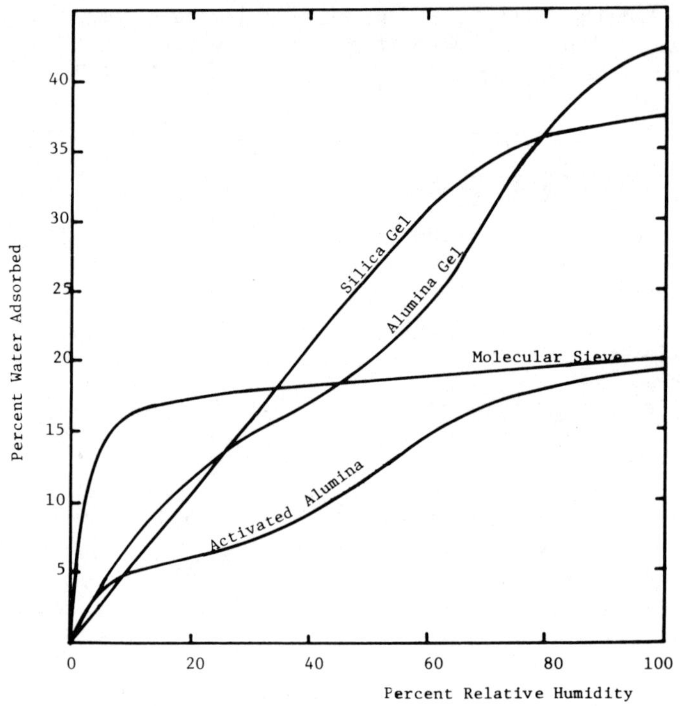

Fig. 13
Lof's Triethylene Glycol dehumidification system. (36)

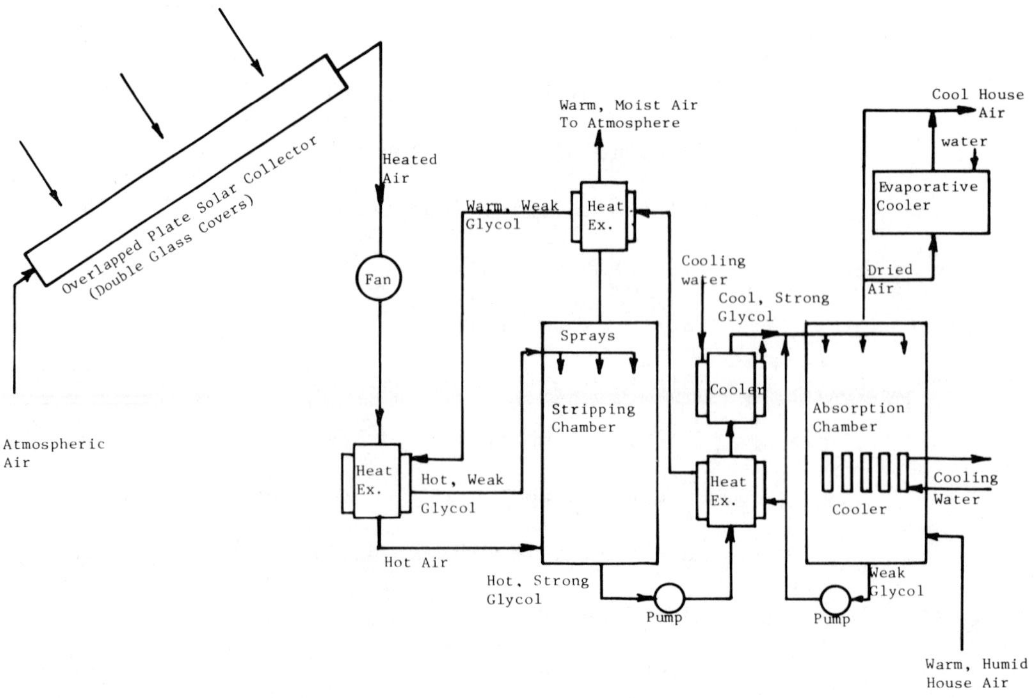

exchange with the solar heated air resulted in improved overall performance even though the regenerating air temperature was reduced. Dehumidification performance was evaluated from glycol solution concentration changes and temperature and humidity changes in the two air streams. In clear weather, the average day-time cooling capacity of a system comprising a 600 ft^2 solar collector has been estimated at four tons of refrigeration.

Hollands (37) attempted to use a solar still for regeneration of lithium chloride brine for air conditioning applications. It was found that the lithium chloride brine could be regenerated in a solar still with a daily efficiency of 5-20% depending on the insolation and concentration of the brine. The efficiency of the still acting as a regenerator is considerably lower than the efficiency of a desalinating unit. It was suggested that increased mass transfer rates, for forced rather than natural convection, might increase regeneration efficiency.

Mullick and Gupta have designed a collector-cum-desorber (Fig. 14) where, rather than by forced convection, air mass flow is achieved by thermosiphon action (38,39).

Fig. 14
Collector-cum-desorber unit. (38)

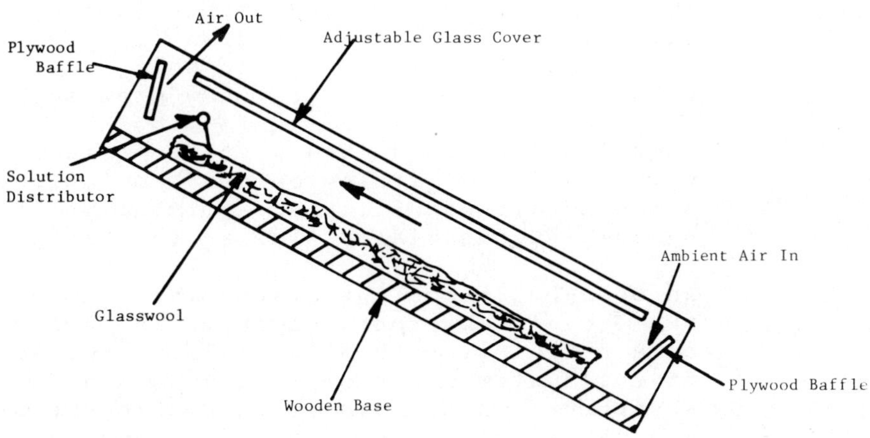

Fig. 15

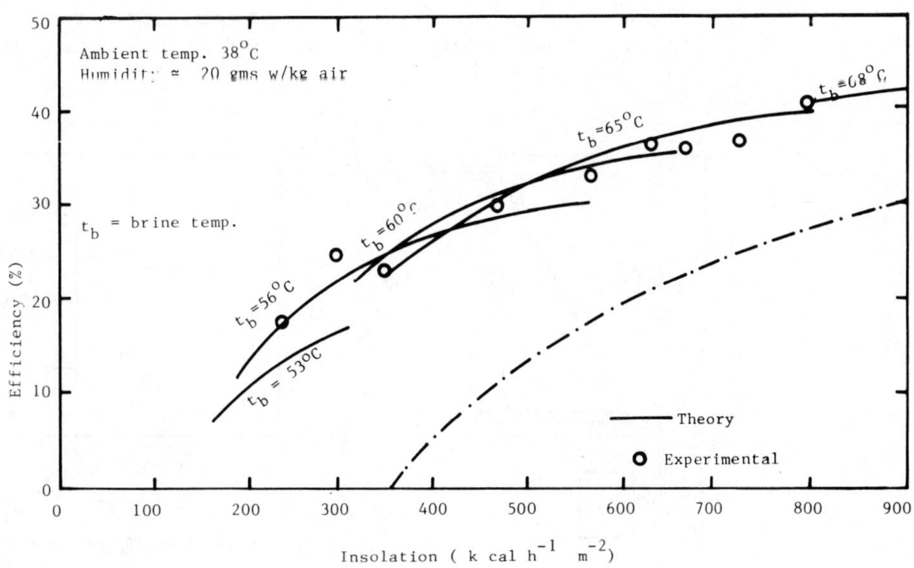

Performance curves for collector-cum-desorber (____.____.____Theoretical efficiency of solar still with 37.5% LICI as given by Hollands)

Performance curves (Fig. 15) show that the system efficiency is greater than that of a solar still acting as a regenerator (Mullick and Gupta use $CaCl_2$ solutions). Operation with much lower irradiation compared to the solar still, extends the operating time. A regenerator such as this is intended for use in climates where winter heating is not required. The unit requires solution pumps and eliminate the use of the solar energy collector for other purposes.

Solar dehumidification of the Citicorp Centre in New York is now being evaluated (40). The system, which uses the liquid dessicant triethylene glycol, was first proposed by Lof (36) (see Fig. 13). A first order feasibility study indicated that the proposed solar dehumidification unit would save more energy than a solar absorption chiller system.

Kapur (41) evaluated the suitability of various solar operated cooling systems for the climatic conditions prevailing in many parts of India with the aim of improving working conditions and hence productivity in industry. He concluded that in the hot, humid regions of India an adsorption or absorption system of dehumidification with sensible and evaporative cooling of air appears to show promise.

Sheridan (42) described a solar operated dehumidification unit utilizing lithium chloride solutions as a potential domestic air conditioning application in Australia. The system is essentially similar to that proposed by Lof.

An open-cycle cooling system more closely related to closed-cycle absorption systems is presently operating on a building in the USSR (43). Fig. 16 is a schematic of the system which employs open regeneration of the absorbent - LiCl solution. The regenerated solution absorbs water vapour from the evaporator thus cooling the water in the evaporator. The cooled water circulates in an air-water heat exchanger, cooling the room air. It has been suggestedthat such a system may be technically

Fig. 16

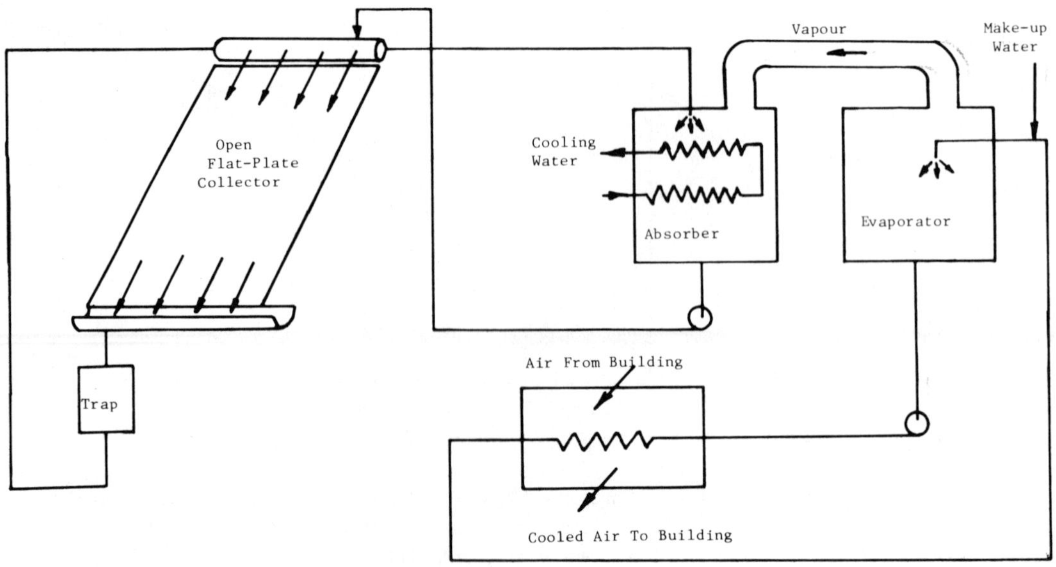

Lithium Chloride-Water air conditioning system with open regeneration. (46)

Fig. 17 Relative capacities of solid and liquid desiccants. (50)

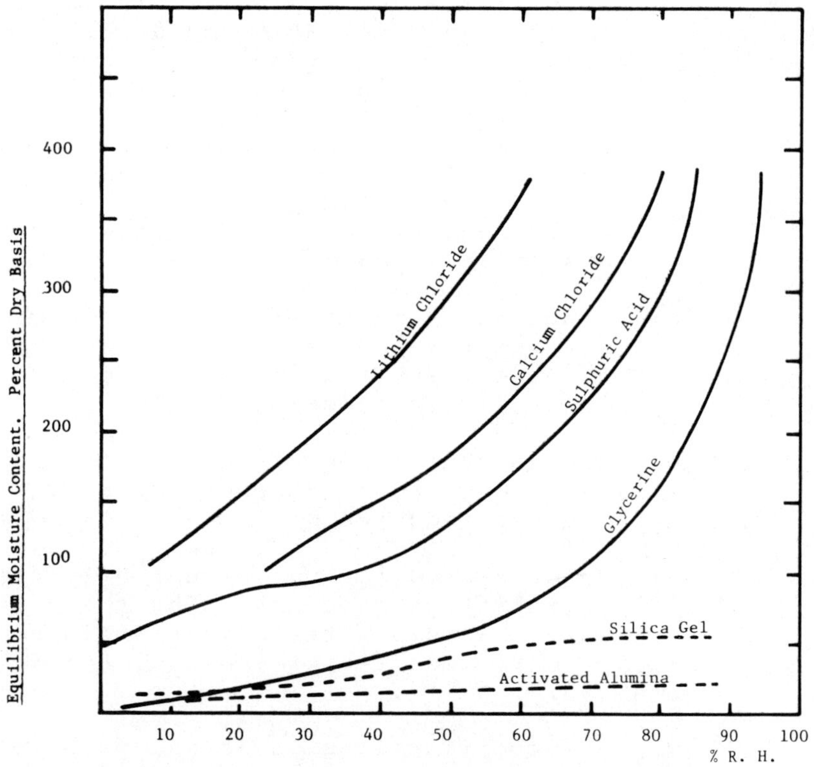

and economically more favourable than other air conditioning units in some dry climates (atmospheric water vapour pressure less than 12-15 mm Hg) (44).

Though the system is reported to have worked well it should be noted that heat losses by convection to the wind (specially in coastal climates with high wind velocities) and radiation to the sky will lower the temperatures and thus lead to unsatisfactory regeneration. The exposure of the solution to the elements during regeneration also leads to contamination by atmospheric dust and dirt necessitating the use of filters to prevent system components becoming clogged.

The moisture removal capacity of typical liquid sorbents is shown in Fig. 17. Curves for two solid sorbents are included for comparison. The equilibrium vapour pressure is a measure of the residual moisture content and has already been indicated in Fig. 5 for lithium chloride solutions of various concentrations over a range of temperatures.

Discussion and Conclusions

Two main features of the HDH system account for its attractiveness for solar energy applications:-

(i) Evaporative cooling is a relatively well known process which is both simple and economical. The HDH system extends the usefulness of evaporative cooling to hot humid tropical and sub-tropical regions.

(ii) Regeneration in general can be achieved with a low temperature heat source and is possible at the fluid temperatures attainable with flat plate collectors. (It may be noted here that the high diffuse fraction in the solar radiation in humid areas makes the flat plate collector a more effective collector).

Additional attractive feactures of the HDH system include:-

(a) since the working fluid is the atmospheric air-water vapour mixture and system operation is at atmospheric pressure, equipment construction is considerably simplified

(b) the ambient temperature is sufficiently low for heat rejection and hence expensive cooling towers are not required

(c) solar heated air may be directly used for sorbent regeneration thus obviating the need for solar water heaters with their associated problems of corrosion, water leakage, boiling, freezing, etc.

(d) the modifications necessary to existing commercial HDH equipment to enable them to be operated off solar heat appears to be much less than is the case with absorption chillers. The indications are also that the changes in operating conditions will have less significant effects on system performance.

Regrettably, sufficient information is not yet available on the regeneration of spent sorbents, particularly at temperatures appropriate to flat plate collectors, though there are indications that it is feasible. In this respect liquid absorbents appear to have an advantage in that regeneration at solar collector temperatures (120°F - 130°F) has been shown to be effective (45). Solid dessicants on the other hand appear in general to require higher temperatures for regeneration (300 - 450°F), though Dunkle (26) has reported obtaining regeneration for silica gel in the range 180-220°F. It is possible however to operate a solid dessicant system, such as the rotating wheel system at partial regeneration conditions leading to a reduction in dehumidification capacity and the possibility of overloading of the system and consequent damage to the wheel core.

The choice and design of appropriate sorbent systems for particular applications is hampered by inadequate information on the performance of these systems at temperatures appropriate to flat plate collectors. Clearly, considerable work needs to be carried out in this area to take advantage of what appears to be a promising and appropriate means of space cooling.

REFERENCES

(1) Prigmore, D. & Barber, R., 'Cooling with the sun's heat. Design considerations and test data for a Rankine Cycle Prototype', Solar Energy, Vol. 17, pp 185-192 (1975).

(2) Barber, R.E., 'Design and performance prediction of a 77-ton solar air conditioning system - Rankine Cycle Predictions'. Proceedings of the second workshop on the use of solar energy for the cooling of buildings, August, 1975, Los Angeles, pp 74-84 (1976).

(3) Freeman, T.L., 'Computer modelling of Solar Rankine Systems for space cooling', Ibid, pp 104-109 (1976).

(4) Eckard, S.E., 'Performance characteristics of a 3-ton Rankine powered vapour-compression air conditioner' Ibid, pp 110-136 (1976).

(5) Biancardi, F., 'Demonstration of a 3-ton Rankine cycle powered air conditioner', Ibid, pp 137-152 (1976).

(6) Satcunanathan, S. and Kochhar, G.S., 'Optimum operating conditions of NH_3-H_2O absorption refrigeration systems for flat plate solar collector temperatures'. Presented at the 1975 International Solar Energy Congress and Exposition, University of California, Los Angeles (1975).

(7) Kochhar, G.S., 'Optimum operating conditions of absorption refrigeration systems for flat-plate solar-collector temperatures', Ph.D. Thesis., Dept. of Mechanical Engineering, U.W.I., Trinidad.

(8) Chinnappa, J.C.V., 'Performance of an intermittent refrigerator operated by a flat plate collector', Solar Energy Vol. 6, No. 4, p 143 (1962).

(9) Chinnappa, J.C.V., 'The solar operation of a vapour absorption cycle air conditioner at Columbo', Trans. Inst. Eng. Ceylon, pp 19-32 (1967).

(10) DeSa, V.G., 'Solar energy utilization at Dacca', Solar Energy Vol. 8, No. 3, (1964).

(11) Eisenstadt, M., Flanigan, F.M. and Farber, E.A., 'Solar air conditioning with an ammonia-water absorption refrigeration system', A.S.M.E. Paper No. 59-A-276(1959).

(12) Farber, E.A., Flanigan, F.M., Lopez, L and Polifka, R.W., 'Operation and performance of the University of Florida solar air conditioning system', Solar Energy, Vol. 10, No. 2, p 91 (1966).

(13) Moore, G.L. and Farber, E.A., 'Combining the collector and generator of a solar refrigeration system, A.S.M.E. Paper No. 67-WA-Sol-4(1967).

(14) Oniga, T., 'Absorption cooling units with fixed conoidal reflector', UN Conference on New Sources of Energy, Rome, 1961, Vol. 6, Paper No. S/70 (1964).

(15) Swartman, R.K., Ha, V.H. and Swaminathan, C., 'Comparison of ammonia water and ammonia-sodium thiocyanate in a solar refrigeration system', International Solar Energy Society Conference, Paper No. E5, Paris (1973).

(16) Chung, R., Lof, G.O.G. and Duffie, J.A., 'Experimental study of a Li Br-H_2O absorption air conditioner for solar operation', A.S.M.E. Paper No. 62-WA-37(1962).

(17) Duffie, J.A. and Sheridan, N.R., 'Lithium bromide-water refrigerators for solar operation', Mech. and Chem. Engr. Trans. Inst. of Engrs. Australia, MC-1, p 79 (1965).

(18) Williams, D.A., Chung, R., Lof, G.O.G., Fester, D.A. and Duffie, J.A., 'Cooling systems based on solar regeneration', Refrigeration Engineering, 66, pp 34-37 and 64-65, Nov. (1958).

(19) Chinnappa, J.C.V., 'Experimental study of intermittent vapour absorption refrigeration cycles employing the refrigerant absorbent systems of ammonia-water and ammonia-lithium nitrate', Solar Energy, Vol. 5, No. 1, p 1 (1961).

(20) Sargent, S.L. and Beckman, W.A., 'Theoretical performance of an ammonia-sodium thiocyanate intermittent absorption refrigeration cycle', Solar Energy, Vol. 12, No.2, pp 137-146 (1968).

(21) Swartman, R.K. and Ha, V.H., 'Performance of a solar refrigeration system using ammonia-sodium thiocyanate', A.S.M.E., Paper No. 72-WA/Sol-3, New York (1972).

(22) Muradov, D. and Shadiev, D., 'Testing of a solar refrigerator', Geliotekhnika (Applied Solar Energy), Vol. 7, No. 3, (1971).

(23) Adler, S., Levite, G., Tabor, H., 'The Altenkirch Solar-Cooled House', Proceedings of the UN Conf. on New Sources of Energy, Vol. 6, pp 60-65, (1964).

(24) Dannies, J.H., 'Solar air conditioning and solar refrigeration', Solar Energy, Vol. 3, No. 1, pp 34-39 (1959).

(25) Avezov, R.R., Lukomskii & Shafeeva, K.A., 'Solar energy utilization for heating and air conditioning', Geliotekhnika, Vol. 8, No. 4, pp 64-74 (1972).

(26) Dunkle, R.V., 'A method of solar air conditioning', Mech. & Chem. Engr. Trans., Inst. of Engrs., Australia, MC 1 (1), pp 73-78 (1965).

(27) Akopdzhanyan, E.S., 'Solar air conditioning with solid sorbents', Geliotekhnika, Vol. 3, No. 6, pp 81-85 (1967).

(28) Rush, W.F. et al, 'A description of the Solar-MEC field test installation', ISES Conf., Los Angeles (1975).

(29) Lunde, P.J., 'Solar dessicant air conditioning systems', Proceedings of the Second Workshop on the use of solar energy for the cooling of buildings, Los Angeles, August 1975, p 280 (1976).

(30) Rush, W.F., 'Absorption phenomena as applied to solar cooling', Ibid, p 274 (1976).

(31) Dunkle, R.V. and Maclaine-Cross, I.L., 'Theory and design of rotary regenerators for air conditioning', Mech. and Chem Engr. Trans Inst. of Engrs., Australia, Vol. MC6, No. 1, pp 1-6 (1970).

(32) Maclaine-Cross, I.L. and Banks, P.J., 'Coupled heat and mass transfer in regenerators - Prediction using an analogy with heat transfer', Int. J. Heat Mass Transfer, Vol. 15, pp 1225-1242, (1972).

(33) Close, D.J. and Banks, P.J., 'Coupled heat and mass convection and diffusion-anaolgies with heat transfer'. Chemeca '70, pp 17-33, Butterworths (1971).

(34) Banks, P.J., 'Coupled equilibrium heat and single adsorbate transfer in fluid flow through a porous medium-1, characteristic potentials and specific capacity ratios', Chemical Engineering Science, Vol. 27, pp 1143-1155, (1972).

(35) Close, D.J. and Banks, P.J., 'Coupled equilibrium heat and single adsorbate transfer in fluid flow through a porous medium-II, predictions for a silica-gel air drier using characteristic charts', Chemical Engineering Science, Vol. 27, pp 1157-1169 (1972).

(36) Lof, G.O.G., 'Cooling with solar energy', Proceedings of the World Symposium on Applied Solar Energy, Arizona (1955).

(37) Hollands, K.G.T., 'The regeneration of lithium chloride brine in a solar still', Solar Energy, Vol. 7, No. 2, p 39 (1963).

(38) Mullick, S.C. and Gupta, M.C., 'Solar desorption of absorbent solutions', Solar Energy, Vol. 16, No. 1, p 19 (1974).

(39) Mullick, S.C. and Gupta, M.C., 'Solar air conditioning using absorbents', Proceedings of the Second Workshop on the use of solar energy for the cooling of buildings', August 1975, Los Angeles, pp 336-344 (1976).

(40) Felske, D. et al, 'Solar dehumidification of the Citicorp Centre', Ibid, p 294 (1976).

(41) Kapur, J.C., 'A report on the utilization of solar energy for refrigeration and air conditioning applications', Solar Energy, Vol. 4, pp 39-47 (1960).

(42) Sheridan, N.R. 'Solar air conditioning', Journal of the Inst. of Engrs., Australia, Jan-Feb. (1961).

(43) Kakabayev, A. et al, 'A large-scale solar air conditioning pilot plant and its text results'. International Chemical Engineering, Vol. 16, No. 1, pp 60-64, Jan (1976).

(44) Baum, V.A. and Kakabayev, A., 'Utilizing solar energy and evaporative cooling methods for air conditioning (Survey)', Geliotekhnika, Vol. 8, No. 2, pp 3-8 (1972).

(45) Niagara Blower Co Literature.

(46) Duffie, J.A. and Beckman, W.A., 'Solar energy thermal processes', John Wiley and Sons (1974).

(47) Threlkeld, J.L., 'Thermal environmental engineering', Prentice-Hall (1962).

(48) Cargo Caire Literature. Bulletin TB 75.

(49) ASHRAE Guide and Data Book. 'Fundamentals and equipment', p 363 (1963).

(50) Williams-Gardner, A., 'Industrial drying', London, L. Hill (1971).

(51) Imperial Chemical Industries Limited. 'Chemical analysis: The working tools', Vol. 1, p 106, Oxford, Clarendon Press (1962).

Design and construction of a pilot solar space heating and air conditioning system using flat plate collectors and an absorption chiller

5.7

Klaus W. Heinemann
Alten Associates, Inc.
Mountain View, California 94043, U.S.A.

Abstract

A solar heating and air conditioning system was designed and installed at a new bank building in Auburn, California. The climate in Auburn is characterized by mild winters and hot summers. The 1,000 m^2 building is equipped with 190 m^2 of double glazed flat plate solar collector panels, augmented with 135 m^2 of reflector area. Collectors and reflectors are mounted at fixed tilt angles of 45° and 28°, respectively. Water is used as heat transfer and storage medium. No heat exchangers are being used. Freeze protection is obtained by hot water recirculation; draindown is provided as back-up protection. The solar heated water drives a 7.5 x 10^7 cal/h absorption chiller for summer cooling. An in-line gas heater is used as back-up. An air handling and distribution system for cooling and heating receives either cold water from the chiller (cooling) or hot water directly from the hot storage tank (heating). Storage capacities are 14,000 l for hot water and 4,000 l for chilled water. The system was financed without public funds. A utility company is monitoring the performance.

I. Introduction

Absorption refrigeration is based on a process by which heat in the form of hot water or steam is used to sustain a vaporization/condensation double sequence of a medium in a closed system in such a way that the heat of evaporation of the second sequence can be extracted from a cool water supply and serve for further chilling of that cool water. Absorption chilling is well known and has been used quite frequently for space conditioning, in particular where excess process heat in the form of low temperature steam (≈150°C) is readily available. During the past three decades, however, absorption chillers have not been cost competitive with compressor driven chillers because of the gas vs. electricity price structure prevailing in the U.S.A.

It is not until very recently that the strongly increasing interest of the American people for the utilization of solar energy has spurred new developments in absorption refrigeration. In 1973 a solar driven absorption system was marketed which operated from 95°C to 115°C hot water. One demonstrational and heavily government funded solar space cooling system that uses this type of absorption chiller was recently completed at the Community Center of the City of Santa Clara, California(1).

The need for 100°C hot water severely restricted the general acceptability of this approach for solar utilization because solar collectors that produce these temperatures at acceptable efficiency levels are unduly expensive. The lowering of the hot water requirements for the chiller to 75°C to 95°C, as was recently introduced by ARKLA Industries, Inc., was a highly welcome improvement in this technology.

This paper describes the design parameters of a solar space conditioning system that uses an ARKLA WFB 360 absorption chiller of 7.5×10^7 cal/h capacity which operates at $80^{\circ}C$ inlet temperature and supplies in excess of 95% of the chilled water needs for air conditioning of 1000 m^2 of the new bank office building.

II. Climatological Data and Building Solar Thermal Analysis

The temperature and sunshine probability data for Auburn, California, are summarized in Table 1. The summers in Auburn are hot and dry, and the winters are mild. The annual percentage of sunshine is relatively high.

The building has two stories and was designed with strict adherence to energy conservation principles. The window areas are samll, and east and west facing windows were avoided where possible. Walls and ceilings were heavily insulated with 140 mm and 220 mm loose fiberglass insulation, respectively. The total heat loss/gain of the building is 1.1×10^6 cal/h per degree Celsius inside to outside temperature difference. The total window gain due to direct solar insolation is 0.5×10^8 cal on a sunny June/July day (adds to the air conditioning load) and 1.2×10^8 cal on a sunny winter day (helps to offset the daytime heating requirements). The lighting level inside the building was kept below 20 W/m^2. The air conditioning system includes an economizer cycle which automatically calls for circulation of fresh air through the building whenever the outside temperature is below $18^{\circ}C$ and the inside thermostats call for air conditioning.

Table 1
Solar system performance

	JAN	FEB	MAR	APR	MAY	JUN	JUL	AUG	SEP	OCT	NOV	DEC
1 ambient temperature at 8:00 ($^{\circ}C$)	3.5	5.5	7.0	9.0	11.5	15.5	19.0	19.0	16.0	10.5	9.5	4.0
2 ambient temperature at 12:00 ($^{\circ}C$)	11.0	14.0	16.5	20.0	23.5	28.5	33.5	32.0	29.5	23.5	20.0	11.5
3 ambient temperature at 2:00 ($^{\circ}C$)	12.0	15.0	18.5	21.0	25.0	30.0	35.0	34.5	31.0	25.0	21.5	12.0
4 probability of sunshine (%)	75	80	90	90	95	95	95	95	95	85	80	75
5 storage tank temp. (1) at 8:00 ($^{\circ}C$)	60	60	60	79.5	79.5	79.5	79.5	79.5	79.5	79.5	60	60
6 storage tank temp. (2) at 17:00 ($^{\circ}C$)	85.5	89	92.5	87.6	87.5	86.9	86.9	87	85.9	91.7	87.2	85
7 daily solar heat coll. ($\times 10^8$ cal) (full sunshine)	4.82	5.45	6.13	6.0	5.78	5.93	6.4	6.76	6.43	5.96	5.11	4.66
8 heat content (3) in storage at 17:00 ($\times 10^8$ cal)	4.80	5.40	6.10	1.53	1.51	1.40	1.40	1.42	1.21	2.31	5.05	4.60
9 heat/cooling demand (4) between 17:00 & 8:00 ($\times 10^8$ cal)	2.29	1.89	1.48	-	.39	.44	.82	.76	.49	.39	-	2.24
10 thermal load for other uses (5) ($\times 10^8$ cal)	.25	.25	.25	.25	.25	.25	.25	.25	.25	.25	.25	.25
11 net surplus (5) ($\times 10^8$ cal/day) (full sunshine)	2.26	3.26	4.37	1.28	.87	.71	.33	.41	.47	1.67	4.8	2.11

III. Description of Building Heating and Air Conditioning System

The system is shown in Fig. 1. An array of 74 ALTEN 200 G solar flat plate collectors with 190 m^2 net absorber area serves as principal heat source. The collectors are facing due south and are mounted at a 45° tilt. The insolation into the collectors during the summer months is augmented by an array of reflectors (135 m^2 area) positioned in front of the collectors at a 28° tilt, facing north. A 1000 W centrifugal pump circulates corrosion inhibited water through the collectors at a flow rate of 250 l/min. The solar heat is stored in two heat storage tanks with 14,000 l combined capacity. The temperature stratification is aided by series connection of the heat storage tanks. No heat exchangers are used between the heat storage and the collector heat transport fluids. The solar heated water is circulated to the generator portion of the absorption chiller as well as to the air handling unit. The building temperature control system operates seven heating and cooling zones and admits flow through the heating coils in the air handling unit and/or through the generator of the chiller upon demand for heating and/or cooling, respectively. An auxiliary furnace of 10^8 cal/h heating capacity is used if the temperature in the solar storage tank falls below a preset value. In order to decrease the on/off cycling frequency of the chiller, a chilled water holding tank of 4000 l capacity is added in the chilled water line between the absorption refrigeration unit and the chilled water coils in the air handling unit.

Fig. 1

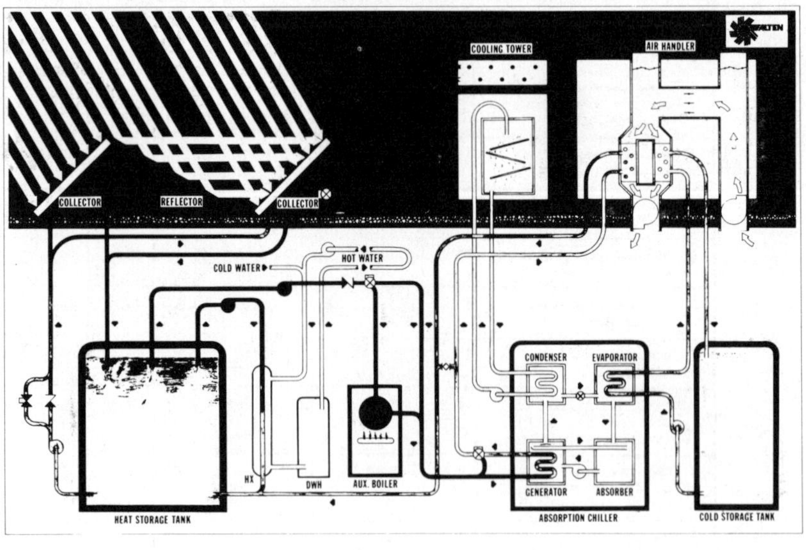

III.1. Collectors

The collectors are of the aluminum-fin/copper-tube construction type with 13 mm I.D. water passages that are spaced 140 mm apart. The fins are manufactured in the aluminum extrusion process and have a thickness of 2.3 mm near the water passages. The thickness of the fins is gradually reduced to 1.5 mm at the outer edges. This assures a maximum temperature difference between any point on the absorber plate and the inside of the nearest fluid passage tube of less than 3°C under highest insolation conditions, resulting in high fin efficiency.

The tubes are press-fitted into the fins. The entire interface area is filled with a silicone-based heat transfer compound. The tubes are dielectrically coated prior to press fitting. The heat transfer compound and the tube coating together virtually eliminate the long term risk of aluminum corrosion in the Al/Cu interface region. The absorber fins are equipped with 20 mm high ribs in horizontal direction, spaced 30 mm apart, which reduces the convective collection losses. Since the system design calls for an average collector temperature of only 80°C during summer operation and 55°C during the winter heating season, a highly absorptive flat black absorber coating was chosen ("Nextel Black Velvet", by 3M Company). The collectors are double glazed. A Teflon film of .025 mm thickness with 97.5% total transmissivity for the visible solar spectrum serves as inner glazing, and a 4.7 mm thick sheet of tempered, low iron glass is used as front glazing. The back side of the collectors is insulated with a 65 mm thick layer of loose fiberglass which has low outgassing rates at temperatures up to 250°C. The fiberglass side insulation is 25 mm thick. Collector sides and backing are made of extruded aluminum and sheet metal, respectively.

III.2. Reflectors

The collectors are mounted in a saw tooth arrangement in 8 banks on the flat rooftop of the building on A-frame mounting trusses. The trusses form a northwards sloping surface which is used to support the reflectors. A metallized, UV inhibited heavy polyester film with adhesive backing (Scotch Cal Series 5400 by 3M Company) is used as reflector material. The reflectivity is above 80% and is expected to remain unaffected by UV exposure. The film is mounted on 1.2m x 2.4m sheets of 3.5mm thick exterior grade masonite which is painted on both surfaces for weatherproofing (see (2) for similar reflector augmentation).

III.3. Heat Storage and Primary System Piping

Two galvanized steel tanks of 2.15 m diameter and 1.8 m and 2.1 m height are used to store 14000 l of solar heated water. The cylindrical tanks are welded to specifications and have heavily reinforced top and bottom surfaces to withstand 2.2 atm working pressure. They are located inside the mechanical room, 10 m below the top of the collectors. They are insulated with 275 mm of loose fiberglass material. The main pipes between storage tanks and solar collector arrays are 2½" steel pipes, the rest of the piping is made of copper. Silicone hose connections are used between the solar collectors and the headers.

III.4. Freeze Protection and Overheating Control

The taller of the heat storage tanks contains a 1000 l air volume for expansion. A pressure of 2.2 atm is maintained with an air compressor which is automatically vented in the event of electricity outage or whenever the collector temperature falls below 1°C in order to allow drainage of the collectors for freeze protection. Upon return of electrical power and/or termination of the freezing temperature condition, the compressor re-establishes the pressure in the expansion pocket and refills the collectors. A temperature of 1°C in the collectors will occur only very occasionally because of a superceding regular freeze protection mechanism which is achieved by short-duration recirculation of solar heated water through the collectors whenever the absorber temperature falls below 2°C. This method is acceptable in climates such as are prevailing in central California. During the coldest night in the winter 1976/77 in a similar location, a solar collector system which was freeze protected with this "hot shot" method was measured to have had pump circulation four times for 30 seconds each.

A heat purging unit capable of dissipating 10^8 cal/h at 100°C is installed to protect against overheating of the storage tanks. Overheating conditions are expected only rarely during the transition months in spring and fall. The purging unit consists of an array of aluminum finned copper tubes and is located above the cooling tower which is part of the absorption refrigeration system. Heat purging occurs automatically by diverting the flow returning from the collectors via a 3-way valve whenever the storage tank temperature exceeds 100°C.

III.5. Absorption Refrigeration System

The most important specifications of the absorption chiller selected for the installation are summarized as follows:

- Delivered capacity at 90°C (80°C) hot water inlet temperature and 7°C leaving chilled water temperature: 7.5×10^7 cal/h (5 x 10^7 cal/h).
- Efficiency between 80° and 90°C inlet temperature: 70%.
- Design hot water flow rate: 350 l /min.
- Permissable range of inlet temperature: 70 - 95°C.
- Design inlet temperature for chilled water: 13°C
- Design outlet temperature for chilled water: 7°C.
- Design chilled water flow rate: 220 l /min.
- Design heat rejection: 1.9×10^8 cal/h.
- Design inlet temperature of condensing water: 30°C.
- Design outlet temperature of condensing water: 39°C.
- Flow rate of condensing water: 350 l /min.
- Electrical power: 115 V, 60 Hz single phase, 1.3 A.

The chiller has dimensions of approximately 3 m x 1 m x 1.5m (LxWxH) and is located in the mechanical room. The chilled water is stored in an insulated 4000 l holding tank and pumped to the chilled water coils in the air handling system which is located above the roof line. Chilled water tank temperature limits are 13° and 7°C.

III.6. Auxiliary Heat Supply

An auxiliary gas furnace of 10^8 cal/h capacity is inserted in the hot water supply line to chiller and air handler and is automatically operated whenever a temperature sensor at the hot storage tank indicates a temperature below a preset limit. This limit is 79.5°C in the summer for air conditioning and 55°C during winter operation. Whenever the auxiliary boiler is not operated, the hot water flow is diverted through a bypass line, thus avoiding unnecessary heat losses in passing through the poorly insulated furnace.

III.7. Domestic Water Heating

The domestic hot water (DWH) heat is supplied from the solar hot storage reservoir through a tube-in-shell heat exchanger. The return from the DWH recirculating loop passes through the shell of the exchanger while a small, 90W centrifugal pump provides the primary hot water flow through the tube. Since the hot storage reservoir temperature remains consistently above 55°C, no auxiliary domestic water heat is required. The boiler is merely used as holding tank, and the heating element is disconnected.

IV. Sizing and Computation of Solar Coverage

IV.1. Reflector Augmentation

The total insolation, I, into solar collectors augmented by reflectors of equal size can be calculated as

$$I = I_o (1 + \varrho r_o \frac{a}{d_o}) \cos \vartheta \qquad (1)$$

I_o is the total (direct plus diffuse) insolation at normal incidence per unit area, ϱ is the fraction of direct radiation, r_o is the reflectivity of the reflector, d_o and a are length parameters as defined in Fig. 2, and ϑ is the angle between the direction of the sun rays and the normal of the collector plane.

Fig. 2

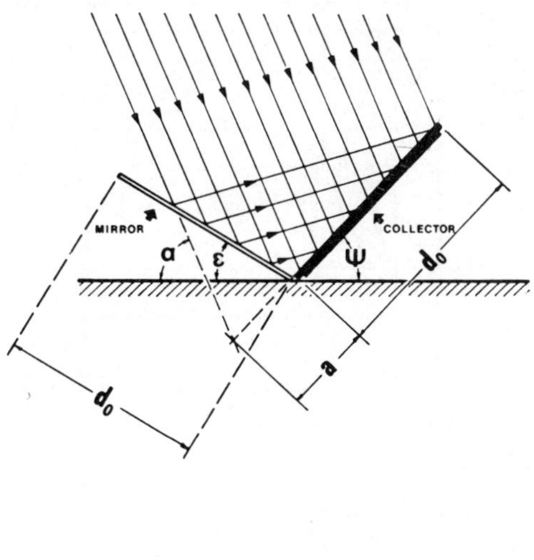

Length parameters and angles for computation of augmented solar heat collection.

The relationship between a/d_o and collector tilt angle ψ, reflector tilt angle ε, and sun altitude angle α is given by

$$\frac{a}{d_o} = \begin{cases} \dfrac{\sin\,(\psi-\alpha-2\varepsilon)}{\sin\,(\psi+\alpha)} & \text{for } \sin\,(\psi-\alpha+2\varepsilon) < \sin\,(\alpha-\varepsilon) \\ \dfrac{\sin\,(\alpha-\varepsilon)}{\sin\,(\psi+\alpha)} & \text{for } \sin\,(\psi-\alpha+2\varepsilon) \geq \sin\,(\alpha-\varepsilon) \end{cases} \qquad (2)$$

Since vertical reflectors are positioned at the ends of the collector/reflector banks, the augmentation is virtually independent of the azimuth angle.

IV.2. Computation of Daily Solar Heat Collection

Fig. 3 shows a block diagram for the computation of the daily solar heat collection. The parameters that are considered include the weather data (ambient temperature profile and sunshine probability), geographical latitude, time and date, collector size and efficiency data including wind effects over the collector surfaces, reflector size and reflectivity, tilt angles of collectors and reflectors, storage tank size, building heat gain/loss values, direct insolation values through windows, internal heat gain by lighting, appliances and people, and heat losses in the piping and storage system. Computed are, at hourly increments, storage tank temperature and accumlative solar heat collection.

Fig. 3 Block diagram for computation of daily solar heat collection.

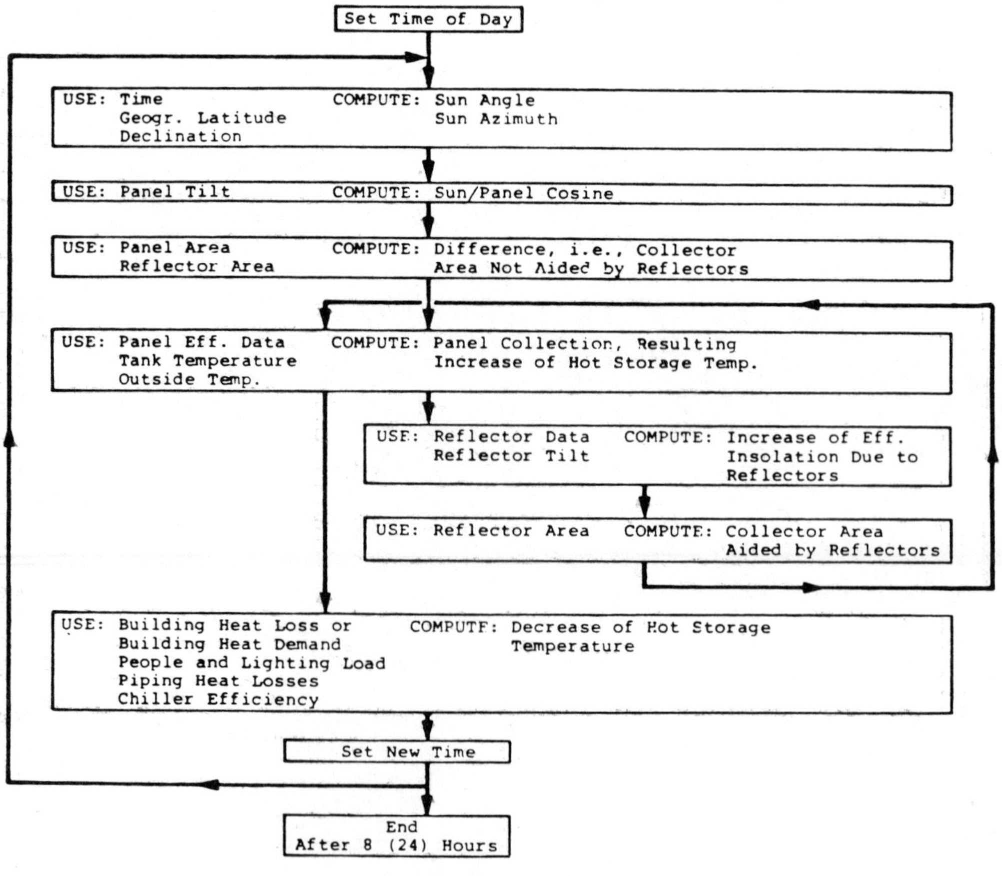

Although this variety of parameters allows a high degree of relative accuracy in the solar system heat collection computation, a particularly large computer is not required for the execution of this calculation. A Hewlett - Packard HP 67/97 pocket calculator with 224 programming steps and 26 permanent storage registers was successfully programmed to execute the calculation within one program. However, the computation time for one day is approximately 6 min, and the solar collection program has to be preceded with a separate program of approximately the same length and complexity to compute the building heat loss/gain values including the direct solar gain through the windows at their particular orientations.

Since the solar collector system is generally operating at moderate to high absorber temperature when compared to the ambient temperature, it can be assumed that during overcast conditions the collection losses exceed the effective insolation values, and the circulation of the heat transfer fluid has to be interrupted in order not to loose heat. This facilitates the computation of projected average monthly and annual heat collection. With good approximation the annual solar heat collection, Qa, can be computed as

$$Q_a = 30 \sum_{i=1}^{12} p_i \, Q_i \qquad (3)$$

Where Q_i is the daily solar system heat collection under full sunshine conditions, p_i is the probability of sunshine and the index i denotates the month.

IV.3. Optimization of Reflector Tilt Angle

The reflector tilt angle is optimized for the month of July during which the solar air conditioning requirement is highest. The optimization was performed with the calculator program described in IV.2., keeping the entire parameter bank constant except for the reflector tilt angle. The result is shown in Fig.4. The

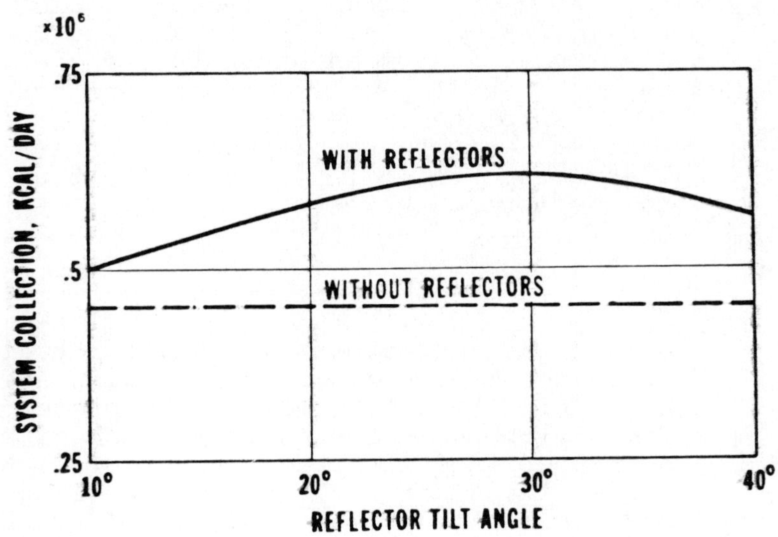

Fig. 4 Optimization of reflector tilt angle for July.

Fig. 5 Effective insolation into 45° tilted collectors with augmentation by 28° tilted reflectors (solid curves) and without reflector augmentation (dashed curves).

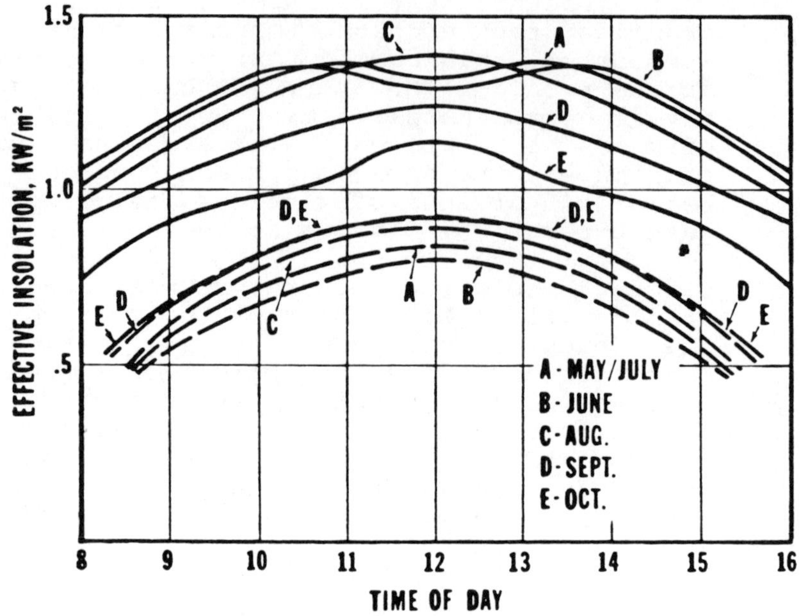

solar heat collection (135 m² reflectors, 190 m² collectors at 45° tilt operating at approximately 85°C in an average 32°C environment) peaks at a reflector tilt angle of 28°. The daily collection is increased by 40%. This increase of solar heat collection due to reflectors is strongly dependent on the general operating parameters of a solar installation as well as on the ratio of reflector to absorber areas, and can therefor not be generalized. However, the result concerning the optim tilt angle is generally applicable for 38°N. latitude and July optimization.

The hourly distribution of the effective insolation I (see equ. (1)) into a 45° tilted solar collector augmented by a 28° tilted reflector is plotted in Fig.5 for different months. It was assumed that the mirror reflectivity is 90% and that 10% of the total insolation of I_o = 930 W/m is diffuse and does not get reflected into the collectors. For comparison, the dashed curves indicate the respective effective insolation values I (w/o refl.) = $I_o \cos \vartheta (t)$ that would prevail if no reflectors were used.

V. System Solar Performance

The main data and expected results pertaining to the performance of the solar installation are listed in Tab.1. Line 5 of Tab.1 gives the preset temperature minimum with which the computation begins at 8:00. Line 6 then shows the resulting storage tank temperature at 17:00 after all heating/cooling requirements have already been satisfied for that daytime period. A substantial amount of heat is still contained in the storage tanks (line 8) when the office closes at 17:00 and can serve for heating/cooling of the building during the off-hour periods. The direct heating and cooling requirement to maintain the building at 25.5°C during the summer evenings and nights and 18.5°C in the winter off-hours is given in line 9. The net surplus (line 11) can be used to further increase or decrease the interior temperature during the winter or summer, respectively, or to heat and cool parts of the building used for off-hour meetings or not normally heated and cooled by the solar energy system.

Acknowledgements

Thanks are due to Mr. Robert McChesney of the Central California Federal Savings and Loan Association and to Prof. Everett D. Howe of the University of California, Berkeley, for stimulating discussions and critical review of the project. The skillful assistance of many ALTEN Corporation employees, in particular Mr. Prentiss Cole, Mr. Robert Foglesong, and Mr. Thomas Moutoux, is also acknowledged.

References

1. L.B. Anderson, "The Santa Clara Community Center Project", AIAA Paper No. 75-608, Solar Energy for Earth Conference, Los Angeles, Ca., 1975.

2. Richard T. Duncan and E. Doering, "The Atlanta Solar Heating and Cooling Experiment", Baltimore Engineer, May 1975.

5.8 Solar assisted heat pumps

T.S. Dean
School of Architecture and Urban Design
University of Kansas, Lawrence, Kansas, U.S.A.

Abstract

The efficiency of a flat plate solar collector declines during severe winter weather when it is necessary to elevate fluid temperature a substantial amount above ambient temperature. By combining the collector system with conventional air-to-air or water-to-air heat pumps the required temperature elevation may be minimized, thereby maintaining utmost collector efficiency. This strategy can increase usable daily energy harvest by 50 percent.

This paper describes systems in which solar heated liquid may be used directly for space heating when adequate storage temperature is available. Otherwise, the solar contribution is used to enhance the performance of conventional heat pumps. Four different schematics for solar assisted heat pump systems are displayed. All of these are in use or under construction. Current research underway at the University of Kansas is described.

Finally, it is emphasized that the method described requires no new technology nor equipment. It is the joining of two well developed technologies. The results obtained are superior to those resulting from either technology used alone.

Introduction

A major barrier to the widespread use of flat plate solar energy systems for space heating has been their apparent high investment cost when compared with systems using energy from conventional sources. Where winter climates are severe or short term radiation is variable, a system using conventional energy is necessary as supplemental or alternate to any solar installation. Thus the additional capital cost is completely attributable to the solar supplement. (The amortization of this capital cost is a function of the amount of captured radiation and escalation rates for the cost of energy from conventional sources.) The amount of radiation which is expected to be captured may be calculated. Escalation rates for fossil fuels are at best only guesses, but the most common guess among knowledgeable energy experts in the United States is 15 percent annually, for a quadrupled price in ten years. This figure ignores effects of possible social or political confrontation or the discovery and successful development of some new source of energy. Using this information, as much as ten to fifteen years may be necessary to amortize the additional cost of a solar energy heating system. Frequently this time frame is unacceptably long for owner or investor.

All of the elements in such a system except the collectors themselves are common to other heating or piping systems and have benefited from long term development and cost stabilization. Substantial technological effort is being invested in attempting to devise more economical flat plat collectors. However, the known materials most suitable for collector fabrication -steel, copper, aluminum, glass fiber and glass - are themselves energy intensive and not likely to decrease in cost. Even in the industrialized nations the manufacture of conventional collectors must be regarded as intermediate rather than high technology and thus not amenable to the economics of mass production. Consequently, it is unclear if this technological effort will yield positive results.

Additional scientific and technological effort is directed toward discovering and developing new solar energy system types, including the direct conversion of solar radiation into electric energy and the use of solar radiation to drive steam generating stations. Although such efforts should and must be continued there is no quarantee that such efforts will yield any greater cost effectiveness than existing systems. When these are large scale systems they require the infusion of public or corporate funds and are thus additionally beset with a host of socio-economic and political considerations. These can delay as well as accelerate a program. Furthermore, the transferal of laboratory discovery to marketplace hardware and public acceptance may require from five to ten years. In many instances this time frame can be unacceptable or even disastrous.

Fortunately, at least one method is available for enhancing the efficiency of a solar heating system. In this instance, a flat plate solar heating system is integrated with the heat pump, a conventional heating source. Not only is the solar portion enhanced, but the two major heat pump short comings are ameliorated. In order to describe this hybrid system, it is first desirable to identify its individual components.

Solar Heating System

A typical flat plate solar system for space heating is shown in Fig. 1. In this system, liquid heated in the collectors is transferred to a storage tank by a coil within the tank or an external heat exchanger. Another coil or other heat exchanger transfers heat from the tank to a finned tube coil mounted in a plenum of the air distribution system. Air is moved past it by a centrifugal fan. Such air distribution systems are desirable since they allow for the inclusion of filters, humidifiers, or another coil for cooling or dehumidification if needed or desired.

Fig. 1

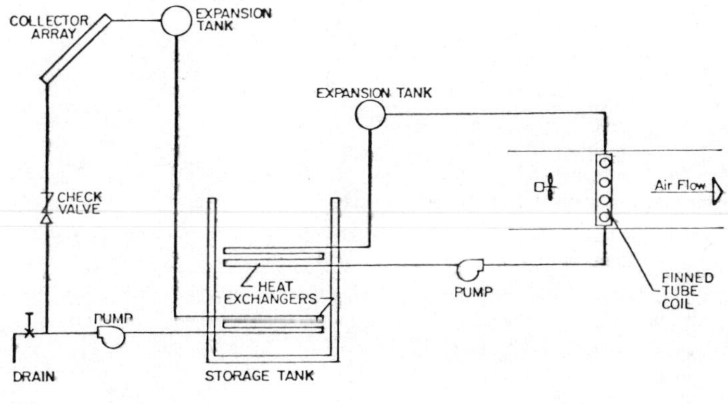

In the design of the usual forced hot air system, air temperature at a terminal (register, diffuser, or grill) is conventionally specified to be about 140°F. However, terminal temperature may be as low as 85°F if supply and return ducts are properly sized and terminal velocity is kept below about 350 fpm. Using this lower terminal temperature and average finned tube coils, liquid in the circuit should be 95°F and tank temperature about 102°F. With carefully controlled approach temperatures, this requires that liquid temperature in the collector loop be above 108°F. Obviously, this would be the lowest temperature at which the absorber could collect usable energy.

The minimum storage tank temperature at the end of a collection period of fruitful energy harvest will probably be 20°F - 30°F higher than starting tank temperature, depending on tank size and daily heating load. Thus, for heating by solar alone, tank temperature should be around 130°F at the end of a collection day.

The actual efficiency of a particular collector can only be determined by test. It is defined as:

$$\% \text{ eff} = \frac{\bar{T}_f - T_a}{q_i} \tag{1}$$

where

$\bar{T}_f$ = average fluid temperature through collector
T_a = ambient air temperature
q_i = solar radiation on cover

The efficiency of a high quality commercially available collector is shown in Fig. 2. This collector utilizes a copper absorber and fluid passages and is glazed with low iron glass.

Fig. 2

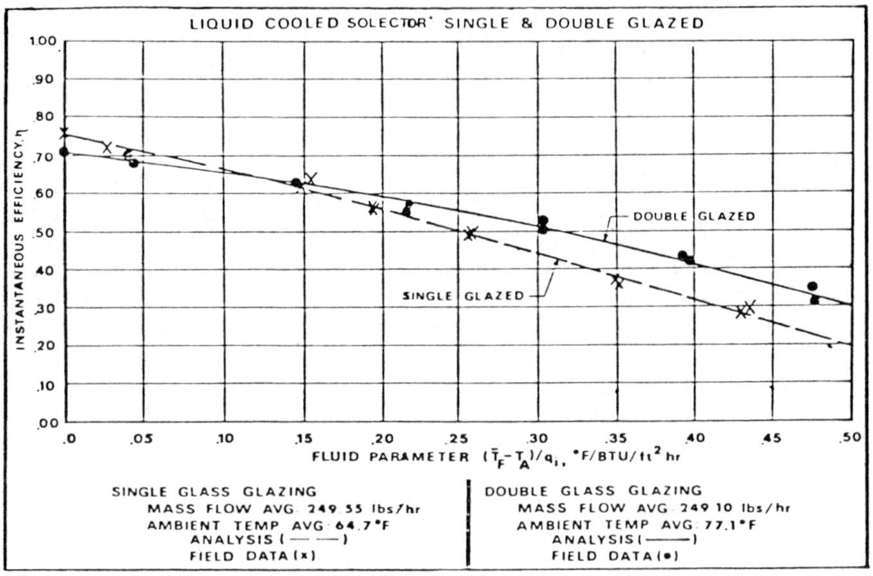

It may be observed that for a given fluid temperature, efficiency is decreased by a lower ambient temperature or less solar flux, or both. This situation may, in fact, shorten the daily collection period. One should also note that for a fluid parameter as high as .25, single glazing has about the same or greater efficiency as double glazing. Since double glazing costs about 2.5 times as much as single glazing this factor has obvious economic implications.

In some parts of the world, all of these factors conspire to reduce the attractiveness of solar systems for space heating. For example, in Lawrence, Kansas, U.S.A., the number of heating degree days (base 65°F) per year is 5161. The most severe heating month is January with an average of 1153 degree days and an average mean temperature of 28°F, although an extreme of -15°F has been officially recorded. In this environment a collector fluid temperature rise of about 100°F is necessary. Assuming clearness and smog factors of 1.0 and dirt factor of .95 and a tilt angle of 50°, the collector efficiency at 9:00AM is 15%, 10:00AM is 40%, 11:00AM is 46%, 12:00 Noon is 47%, 1:00PM is 45%, 2:00PM is 37% and 3:00PM is 10%, for a daily energy harvest of 612 BTU per square foot. These calculations take into account the rise in ambient and tank temperature throughout the day.

Let us now suppose that the system could effectively operate with a collector fluid temperature about 50°F above ambient, although it may be seen that effective heating could not occur with the system described above. In this case, the 9:00AM efficiency is 51%, 10:00AM is 57%, 11:00AM is 59%, 12:00 Noon is 60%, 1:00PM is 59%, 2:00PM is 56%, and 3:00PM is 48% for a daily harvest of 936 BTU per square foot. This collection represents a 53% daily improvement over the required 100°F rise. It remains to demonstrate how this solar heated water which may be at less than a comfort temperature, may, however, be effectively used for space heating. This is accomplished in concert with a heat pump.

The Heat Pump

The heat pump was long considered an engineering novelty. Although first conceived by Lord Kelvin in 1852, its first practical installation was made in Scotland in 1927. The heat pump enjoyed a brief popularity in the U.S. during the 1950's, but was plagued by reliability and control problems. Both reliability and control have been remarkably improved. Today most manufacturers of refrigeration equipment supply a range of heat pumps for the majority of applications. The efficacy of heat pumps in the future is generally recognized by informed persons in the U.S. Except in the coldest weather they are far more efficient than electric resistance heating. The rising cost and possible unavailability of natural gas and fuel oil enhances their cost effectiveness. Although coal reserves are adequate for many years, the delivery of coal to most homes in the U.S. is a formidable task. At this time legislation is in process to encourage users to install heat pumps.

A heat pump operates on the thermodynamic principles of the mechanical refrigeration cycle which is related to the ideal, or Carnot, cycle. Its ideal coefficient of performance (C.O.P.) is given by

$$\text{C.O.P.} = \frac{T_1}{T_1 - T_2} \tag{2}$$

where T_1 is the condensing temperature and T_2 the evaporating temperature, both expressed in degree Rankine. From this expression it is obvious that the most economical operation of a heat pump occurs when the difference in temperature between condenser and evaporator is as small as possible. This fact is central to solar assist.

A commercial heat pump used for cooling operates as any other refrigerating machine. Heat is absorbed from room air into the cold evaporator and rejected to outside air or water from the hot condenser. When the heat pump is used for the heating cycle, heat is absorbed from the outside air or water by an even colder evaporator. This heat along with the heat of compression, is rejected to the air of the conditioned space.

In regions of mild winter climate where a small temperature difference exists between evaporator and condenser, the heat output theoretically may be more than six times the heat equivalent of compressor work. In actual practice the Carnot cycle is not attained, but a heat output of two to four times the compressor work may actually be realized in mild winter climates. For a real heat pump, its actual coefficient of performance may be defined as

$$\text{C.O.P.} = \frac{\text{heat obtainer from the condenser (heating coil)}}{\text{heat equivalent of electric energy input to compressor motor}} \tag{3}$$

Fig. 3 indicates the actual capacity and C.O.P. of a small high quality heat pump manufactured by the Carrier Corporation. An examination of this figure reveals two unique aspects of heat pump systems. First, the heating capacity of a system decreases with decreasing outside temperature. Low outside temperatures generally imply the need for a greater heating capacity. Consequently, when such systems are installed in regions where extremely cold weather persists they are usually fitted with supplemental electric resistance heaters.

It has been noted that the C.O.P. decreases with decreasing outside termperature. With an outside temperature of 50°F, about 3 BTU are supplied to the space for each BTU supplied by the electric utility grid. On the other hand, with an outside temperature of -10°F, the system

Fig. 3

Outside Temperature °F	Capacity 1000 BTUh	Current Draw kw	Coefficient Of Performance
70	46.0	3.8	3.55
60	39.0	3.5	3.27
50	33.0	3.3	2.93
47	31.0	3.3	2.75
40	27.5	3.1	2.60
30	22.0	2.8	2.30
20	18.0	2.6	2.03
17	17.0	2.6	1.92
10	14.5	2.5	1.70
0	11.0	2.3	1.40
-10	7.5	2.0	1.10

performs hardly better than electric resistance heating used alone. Taken together, these two reasons account in part for the favorable acceptance of heat pump systems in locations experiencing mild winters. Heat pump systems are frequently avoided elsewhere.

Water-to-air heat pumps are not subject to such wide variations in capacity and C.O.P., since they are independent of outside air temperature. However, since the water-to-air heat pump requires a source of water in a specified temperature range, it is seldom used in small installations. Its usual application has been in large multi-zone buildings where simultaneous heating and cooling might be called for and the cost of boilers and chillers or cooling tower can be justified.

Recognizing that heat pumps - at least the air-to-air heat pump - is itself a sort of solar device, every manufacturer of this device is heavily committed to research studies aimed at enhancing the unit's performance by means of solar assist. The ACES (Annual Cycle Energy System) proposed by the Oak Ridge Laboratories has been heavily publicized. One manufacturer is circulating a system's refrigerant through flat plate collectors. Another system locates the outside unit of an air-to-air unit in a solar heated attic. Some of these or other schemes will doubtless prove fruitful in the long run. However, they all suffer the multiple curses of advanced technology - extended development and marketing time, lack of compatibility with existing installations, unknown public acceptance, and high cost.

The Solar Assisted Heat Pump

As an intermediate step toward some ultimate solution, the system presented has some advantages. It has been successfully tested in several homes during one of the worst winters on record. It uses components which are readily available. It is relatively inexpensive. It is more efficient than either direct solar heating or a heat pump used alone. A schematic drawing of the system as it has been employed is displayed in Fig. 4.

Although the systems which have been installed have required different size collector arrays, storage tanks, and heat pumps, the elements are common to each system. Heat harvested by the collectors is transferred to a storage tank by means of an antifreeze solution circulating through a copper coil.

Fig. 4

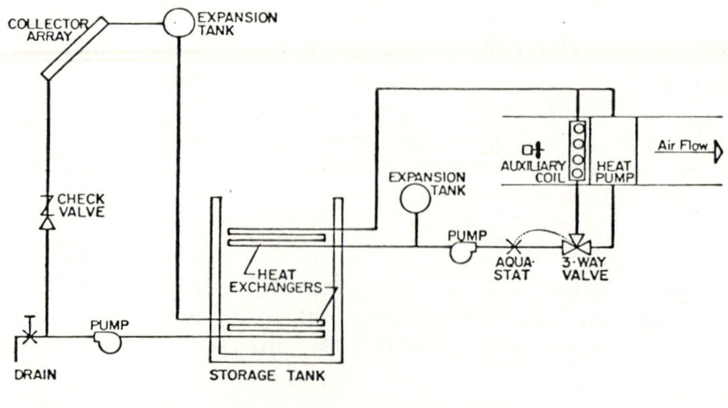

Heat is removed from the tank by an identical coil placed immediately below the water surface. When the tank temperature is sufficient to produce 90°F air off a hot water finned-tube coil located in the return air plenum, heated water circulates through this coil. Otherwise, a three-way valve actuated by a suitably placed aquastat directs the heated water through an air-water heat pump. The heat pump's blower and filter module are active in both modes. This heat pump is designed to be served by water whose entering temperature is between 60°F and 95°F, with leaving temperature about 10°F lower. Its C.O.P. remains almost constant at about 2.95 throughout the entire range of water temperatures. Heating performance for this series of heat pumps is shown in Fig. 5.

Fig. 5

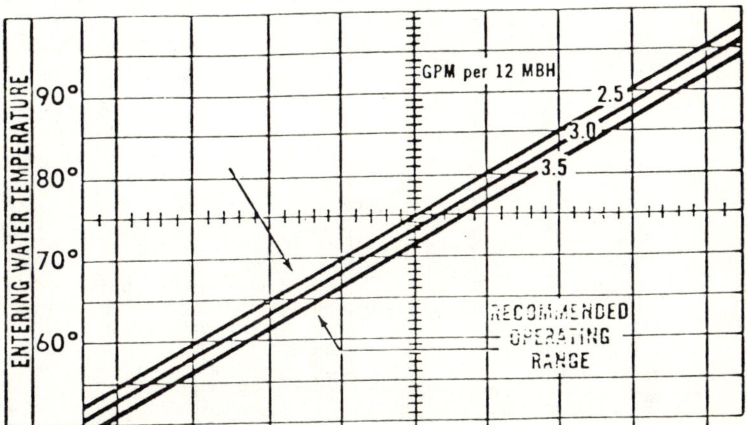

Model	LEAVING WATER TEMP.	ARI Approx. 51	55°	60°	65°	70°	75°	80°	85°
CC 80	TOTAL POWER INPUT, KW	.95	.98	1.02	1.05	1.09	1.14	1.17	1.20
	HEATING CAPACITY BTUH	6290	6940	7680	8320	9060	9800	10450	11200
	HEAT OF ABSORPTION BTUH	3350	3980	4610	5160	5790	6420	7060	7600
CC 100	TOTAL POWER INPUT, KW	1.10	1.13	1.17	1.22	1.27	1.31	1.35	1.39
	HEATING CAPACITY BTUH	8890	9690	10490	11380	12180	12980	13500	14280
	HEAT OF ABSORPTION BTUH	5110	5760	6520	7280	7935	8690	9460	10200
CC & FV 140	TOTAL POWER INPUT, KW	1.50	1.55	1.66	1.82	1.95	2.04	2.17	2.33
	HEATING CAPACITY BTUH	16420	16460	16560	17230	18000	18670	19820	21310
	HEAT OF ABSORPTION BTUH	10520	10550	10740	11160	11480	11800	12250	12800
CC & FV 190	TOTAL POWER INPUT, KW	1.95	2.00	2.15	2.37	2.52	2.63	2.80	2.97
	HEATING CAPACITY BTUH	20300	20900	22850	24000	25100	26050	27300	28600
	HEAT OF ABSORPTION BTUH	14800	14900	15150	15700	16050	16400	16900	17400
CC & FV 260	TOTAL POWER INPUT, KW	2.60	2.66	2.83	3.09	3.24	3.38	3.57	3.77
	HEATING CAPACITY BTUH	26100	26700	28000	30000	31300	32500	34900	35400
	HEAT OF ABSORPTION BTUH	17200	17500	18300	19400	20100	20600	21400	22100
CC & FV 330	TOTAL POWER INPUT, KW	3.20	3.29	3.50	3.76	3.92	4.05	4.21	4.40
	HEATING CAPACITY BTUH	31300	32900	34850	36700	37800	38550	39450	40300
	HEAT OF ABSORPTION BTUH	20200	20750	21800	23100	23800	24400	24200	26100
CC & FV 400	TOTAL POWER INPUT, KW	4.30	4.43	4.65	4.96	5.14	5.29	5.49	5.69
	HEATING CAPACITY BTUH	40000	41200	42800	44600	45800	46800	48700	50200
	HEAT OF ABSORPTION BTUH	25600	26000	26700	27600	28000	28300	28700	30800

The system, therefore, operates in two modes. When sufficient radiation is available, ambient temperatures are sufficiently high for efficient energy harvest, and space loads are moderate, the systems (with exception of pump and blower) is pure solar. Otherwise, with a lower tank temperature, the solar system assists the heat pumps whose C.O.P. remains about 3. One additional benefit during this second mode of operation is the additional thermal storage available from a given tank capacity.

Recently, this basic concept has been adapted to the more common air-to-air systems. There are hundreds of thousands of air-to-air heat pump systems in use in the U.S., most of which could be successfully retrofitted with solar assist. Whether new or retrofit, the system schematic is as shown in Fig. 6.

Fig. 6

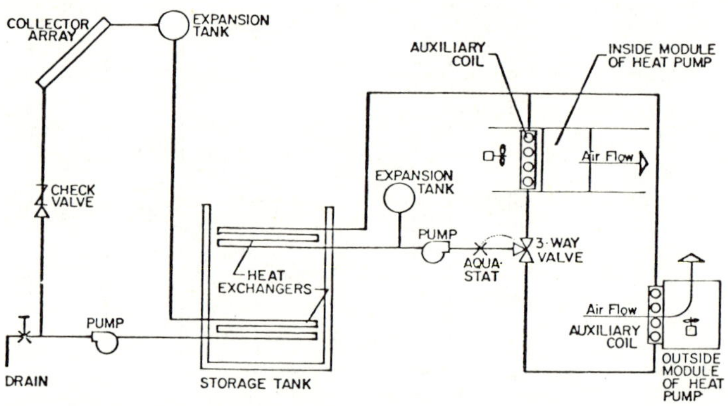

Unless an air-to-air heat pump is grossly oversized for the cooling load, there exists an outside temperature below which the heat pump is unable to meet the building's heat requirements. This is called the balance point and usually occurs between 30°F and 40°F. Below this temperature supplemental electric strip heat (C.O.P. = 1.0) is necessary. Critical unbalance usually occurs at 0°F outdoor temperature at which point the compressor is shut off and system heat is supplied entirely by resistance heaters. By placing a solar heated coil in the evaporator supply air stream the balance point is effectively lowered and high C.O.P. is maintained.

Air-to-air heat pumps require a defrost cycle to prevent coil icing. Defrost may be by a timed cycle during each operating hour or it may be initiated by coil air flow pressure drop or by coil temperature or pressure changes. Most units use reversed cycle operation for defrost. Inclusion of the solar assist coil reduces the number of defrost cycles.

Returning to Fig. 6, it is clear that this configuration differs from the air-water system in that the solar assist circuit, which must now use an antifreeze solution is directed to the outside unit, where it preheats air before it is passed over the outside evaporator coil. This strategy effectively increases both the system capacity and C.O.P.

Three systems of this sort have been installed, but they have not been in use for a full heating season so that no performance data is available. Furthermore, the selection of assist coils was made by extrapolation of known performance data and assumptions regarding operating conditions.

Other Hybrid Systems

Two other applications of these principles are included to indicate the versatility of the method. The first of these examples is a 12,000 square foot building having a 2,200 square foot collector array. It will be occupied by a financial institution in Leavenworth, Kansas. The project is currently in the bid phase with construction scheduled to begin 1 September, 1977 and occupancy on 8 March, 1978.

For this project it was desired to minimize the solar amortization by using the system year round. The basic circuit is similar to the one shown in Fig. 4, except that eight heat pumps are used. Additionally, a 300,000 BTUH solar driven absorption chiller and cooling tower is incorporated in the system. Although the cooling capability of the heat pumps will not normally be used in this design, it is available for use in the event that solar radiation is insufficient to drive the absorption chiller.

The other application is a research house sponsored by the Kansas Power and Light Company and located in Lawrence, Kansas. A drawing of the house is shown in Fig. 7 and a schematic of its heating and cooling system with solar assist is displayed in Fig. 8. Array for space heating contains 481 square feet of single glazed collectors and the domestic water heating circuit contains 111 square feet of double glazed collectors. Both systems are valved to permit use of 33%, 67%, or 100% of the arrays.

Use of electric energy in the U.S. Midwest where extreme temperatures often reach 110°F normally peaks during summer months, usually about 4:00PM. A primary goal of this study is aimed at reducing energy use during the peak load periods by precooling 2,000 gallons of water to 47°F during night time (off peak) hours. The solar contribution to winter heating and year round domestic water heating is considered by the sponsors to be of secondary importance.

Fig. 7

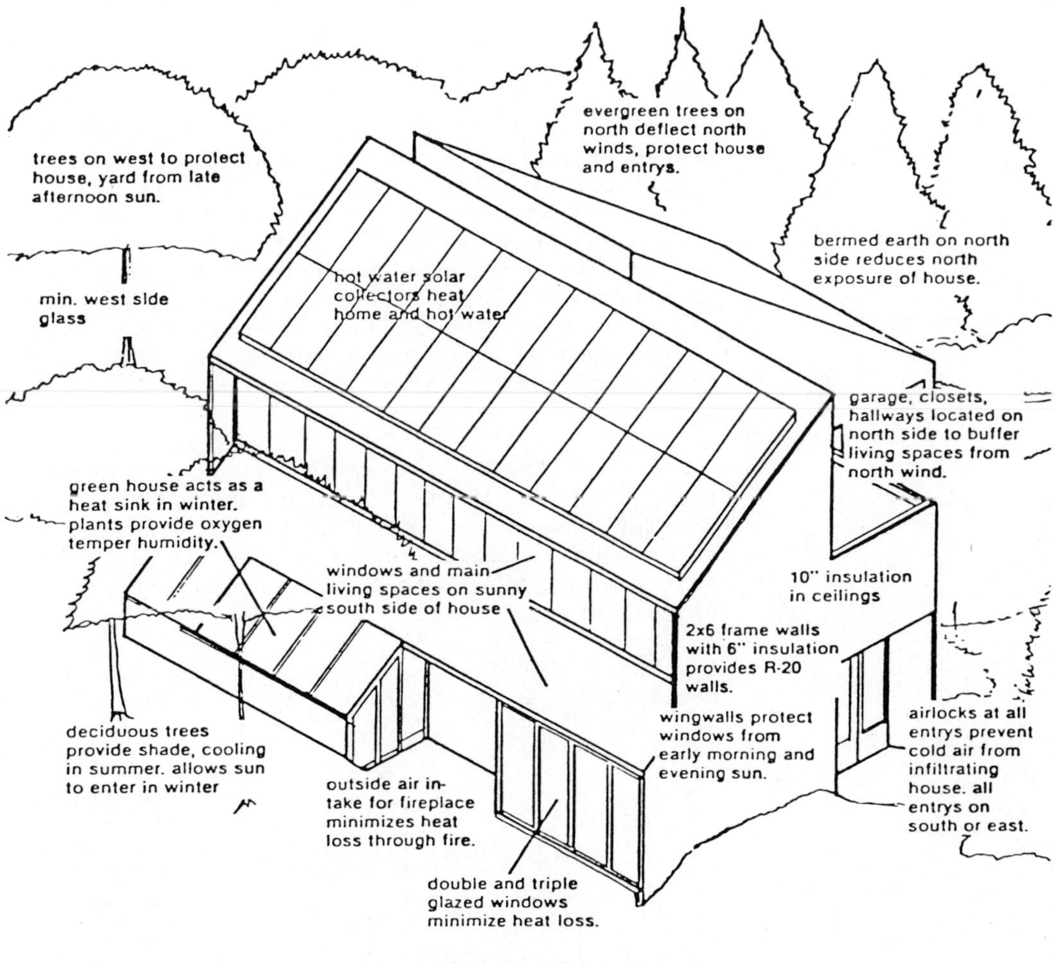

The Sunpower House is designed with many features that help to conserve energy use with passive controls such as optimum insulation, those illustrated above, and others. The Sunpower House will be heated and cooled with a solar heating system and a unique chilling system with the following characteristics.

Cooling: A heat pump / chiller chills a 2000 gallon water tank located below the house during night time, off peak hours. Chilled water is circulated during the day to cool the house.

Heating: A heat pump / chiller using water type solar collector assist. Solar collectors provide heating and hot water needs.

Fig. 8

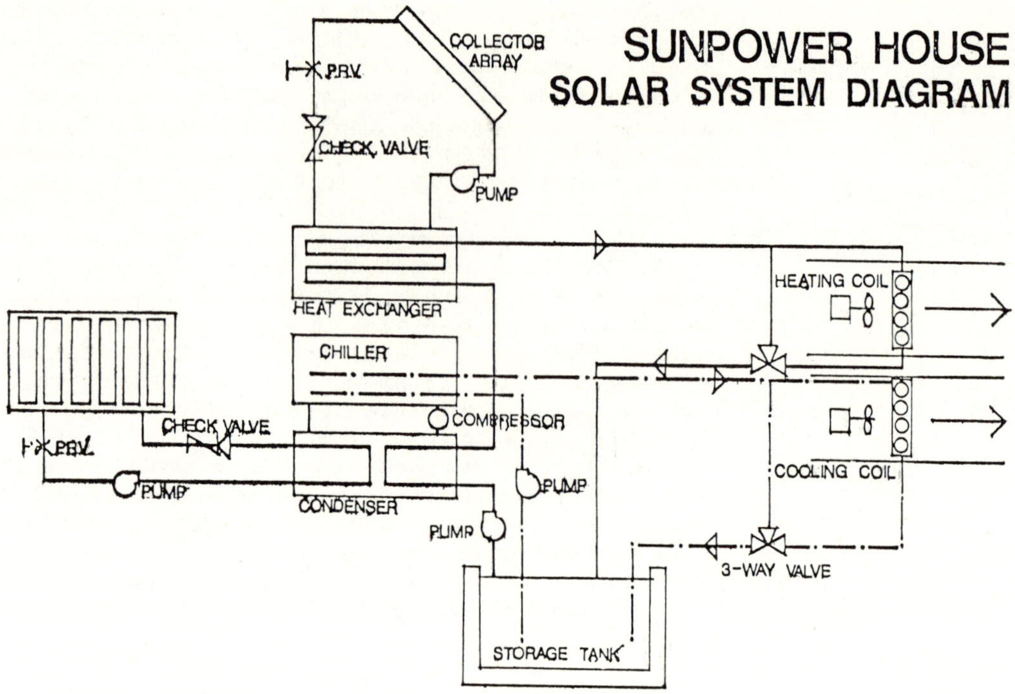

Current Research

It will be recognized that selection of assist coils for the evaporator (outside) unit of air-to-air heat pumps presents problems not ordinarily associated with coil selection. Not only is the air entering the assist coil at a variable temperature, but so also is the solar heated liquid flowing through the coil. Also, since air across the evaporator is customarily moved by a propeller fan, caution must be exercised that the auxiliary coil not create excess static pressure. Since the entire process occurs at temperatures lower than those usually encountered for hot water coils, little published data is available. Finally, the plethora of configurations for the outside units suggests a generalized solution to the problem.

Research at the University of Kansas, sponsored by grants from the University of Kansas, the Lennox Corporation (a manufacturer of heat pumps), and the National Bureau of Standards is directed to quantifying these variables. One doctoral degree candidate, one masters degree candidate, and four fourth-year students are assigned to the project under the direction of the author.

A schematic of the test arrangement is shown in Fig. 9. Simulated solar heated water is provided by a hot water heater. Flow rate and temperature

Fig. 9

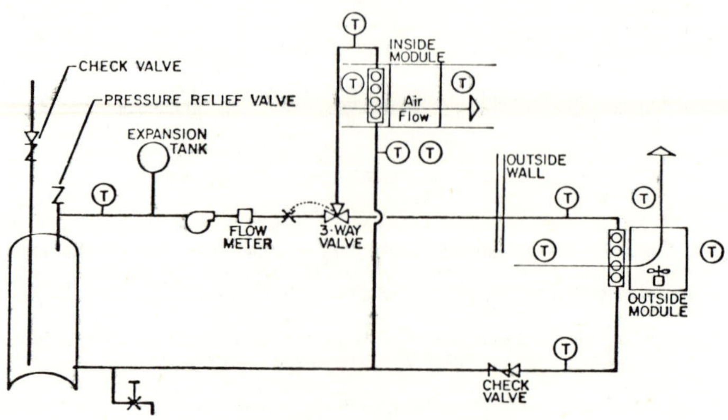

of this water may be varied and recorded. Water may be supplied to an auxiliary coil in the return air plenum of the inside unit or to a preheat coil in the supply air stream of the outside unit. When water is supplied to the inside unit, entering and leaving coil water temperatures as well as the temperature of entering and leaving air through the unit are recorded. Quantity of supply air and external static pressure drop are recorded. Ambient room air temperature may be controlled by a wall mounted air conditioner.

Similiar measurements are made for the outside unit. Additionally, wattage consumed by the compressor, which is located in this unit, is recorded. Use of an Anematherm permits instantaneous determination of velocity, static pressure, and temperature of the air at any point in the system.

The information gained from this study will allow one to optimally match an outside preheat coil with a given heat pump and to predict its performance. This information is particularly useful in a retrofit application. It is anticipated that sufficient data will be available by January 1978 for accurate engineering selection of assist coils. This information will be made available to anyone interested.

Summary

The use of a flat plate solar collection system with a heat pump provides many advantages over either system used alone. Because of the lower tank temperatures needed for heat pump operation, usable energy harvest may be increased at least 50%. With a heat pump whose C.O.P. is 3, it appears that collectors should be sized to supply not over 67% of a building's maximum heat requirements. Frequently, collectors can be single rather than double glazed with attendant investment cost savings. Storage tanks may be smaller. Alternatively, for a fixed size, they provide energy for a substantially longer heating period.

It should be obvious that the system described in this paper required the development of no new technology. Indeed, it simply combined two existing technologies to enhance the overall performance of either used alone. This type of approach - the synergetic joining of existing technologies vis-a-vis the development of new technologies - may in fact contribute significantly to the world's energy and resource needs in the future. If this is true, then among the industrialized nations at least, a dramatic reorientation of scientific-technological-industrial thinking is in order.

The solar assisted heat pump in northern climates

R. Stewart, Associate Director
J. Healey, Research Associate
Atmospheric Sciences Research Center
State University of New York at Albany
1400 Washington Avenue
Albany, New York 12222 U.S.A.

Abstract

The Alumni House-Conference Center (AHCC) at the State University of New York, Albany, New York, incorporates 214 sq.m. of copper solar panels with a water-to-water heat pump, two air-to-water heat pumps, and an electric boiler. Heat storage is maintained in two 30,000 l tanks of water. The solar asystem provides approximately 50% of the annual heating load of this building.

Introduction

The Alumni House-Conference Center (AHCC) on the campus of the State University of New York at Albany is a 743 sq.m. (8,000 sq.ft.) building housing:

(1) 214 sq.m. (2,300 sq.ft.) of Revere copper collectors
(2) a 70kw (20 ton) Carrier water-to-water heat pump
(3) two 42kw (12 ton) air-to-water heat pumps
(4) an 80kw electric boiler
(5) two 30,280 l (8,000 gal.) water storage tanks (heat storage tank and peak loading tank)
(6) 60-channel data acquisition system

Each of the heating systems may operate independent of each other or in consort.

Major components

The solar panels use copper plates and tubes (Figure 1). The copper surface was painted with a non-selective nextel paint (absorptance = emittance = .967) and glazed with two plates (Guardia thermopane, perpendicular transmittance = .78).

Fig. 1
Schematic of Revere copper solar panel.

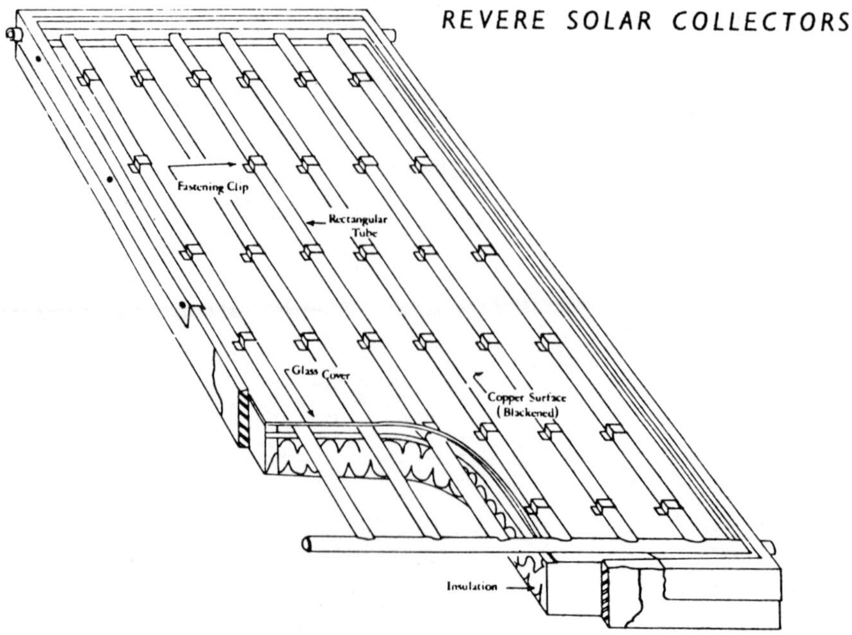

Fig. 2 Performance characteristics of the Carrier water-to-water heat pump.

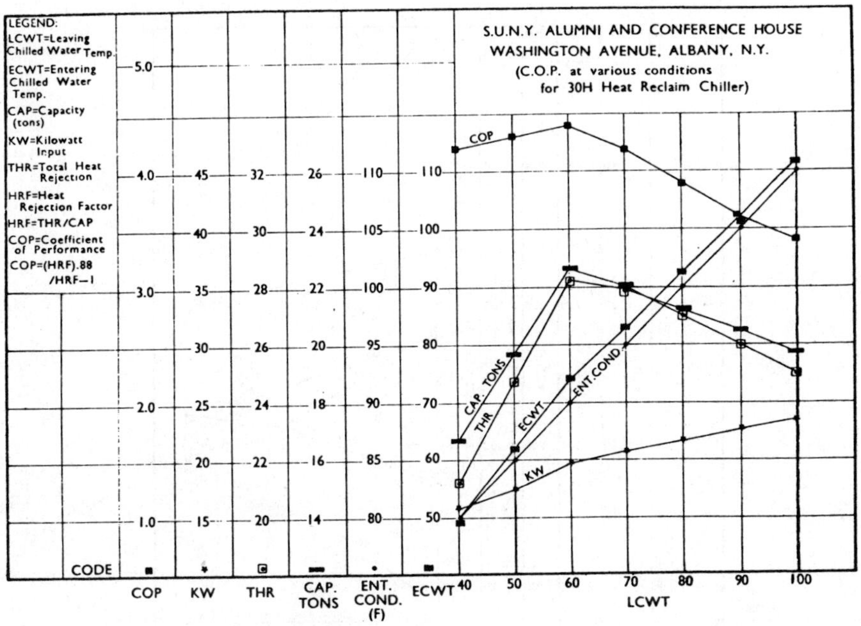

Fig. 3 Performance characteristics of Carrier air-to-air heat pump on heating cycle.

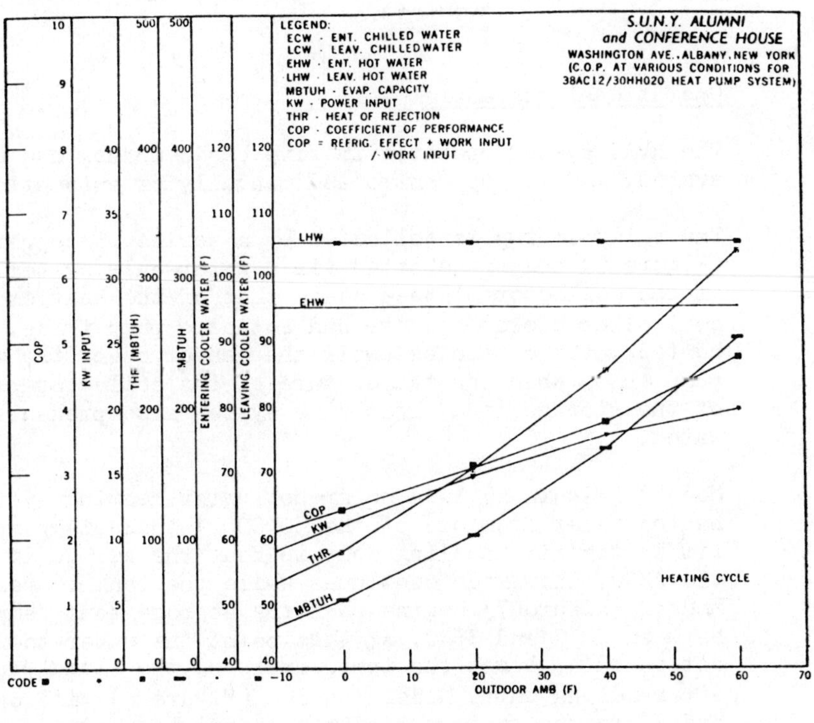

The Carrier water-to-water heat pump may be used for both heating and cooling. The specifications and performance characteristics are presented in Figure 2 as taken from Carrier literature. Two Carrier air source heat pumps which may operate independently provide another option for heating the building. These units are sized to meet the heating requirements of the AHCC with ambient outside air at $-7^{\circ}C$ and full ventilation. The performance specifications are included in Figures 3 and 4.

Fig. 4 Performance characteristics of Carrier air-to-air heat pump on cooling cycle.

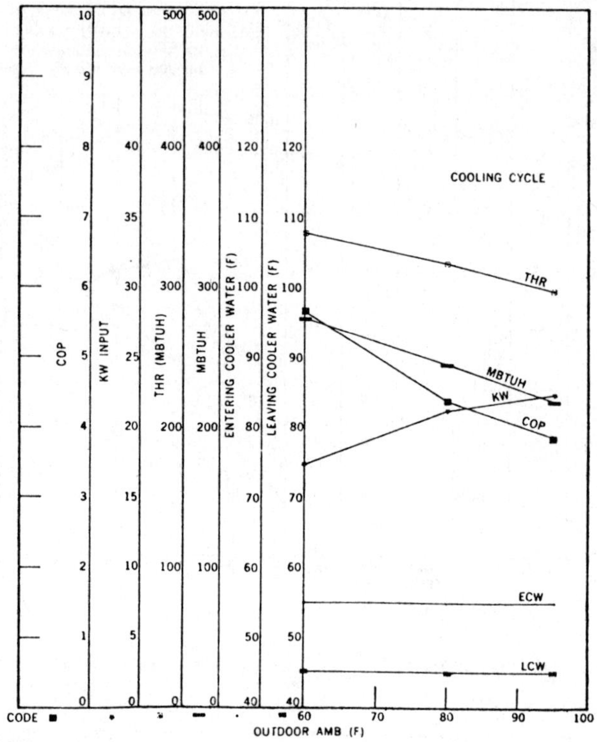

Operational procedure

The HVAC system operates in five basic modes, has a demand limiting system, and may be controlled manually or automatically.

The solar energy is collected by a series of copper solar panels (Figure 5) and transported via a water-only system. When the temperature of the collector plate exceeds that of the heat storage tank, a controller closes a valve and actuates pump three. The collection of heated water continues until the sensor opens the valve and closes down pump three when the temperature of the collector plate approaches that of the storage tank. This same system also preheats the domestic hot water.

Mode 1 (Figure 6) is a solar hot water heating system. The stored solar heated water from one of the 30,000 l tanks is pumped through a coil within the air handling unit to heat the air as it circulates throughout the AHCC. This mode continues while the tank temperatures exceed 38°C. Mode 2 (Figure 7) begins when the storage tank temperature ranges between 10°C and 38°C. At this point the water-to-water heat pump is activated to boost the temperature of the fluid passing through the air handling unit. Modes 3 and 4 (Figure 8) will operate when Modes 1 and 2 are unable to maintain the desired conditions in the AHCC. The electric boiler will raise the temperature of the peak storage tank to 74°C, about enough heating for one winter day. This mode (3) is designed to operate during off-peak hours (midnight to 6 a.m.). Then (mode 4) the AHCC will be heated directly from the peak storage via the air handling units. This process is activated when the temperature in the heat storage tank has dropped below 10°C. Mode 5 presents the additional air-to-water heat pump system. This is an additional heating system which has the capacity of heating the AHCC by itself. Figure 9 presents a simplified version of the total system to illustrate the interconnection of piping, valves and pumps.

Fig. 5 Schematic of solar system.

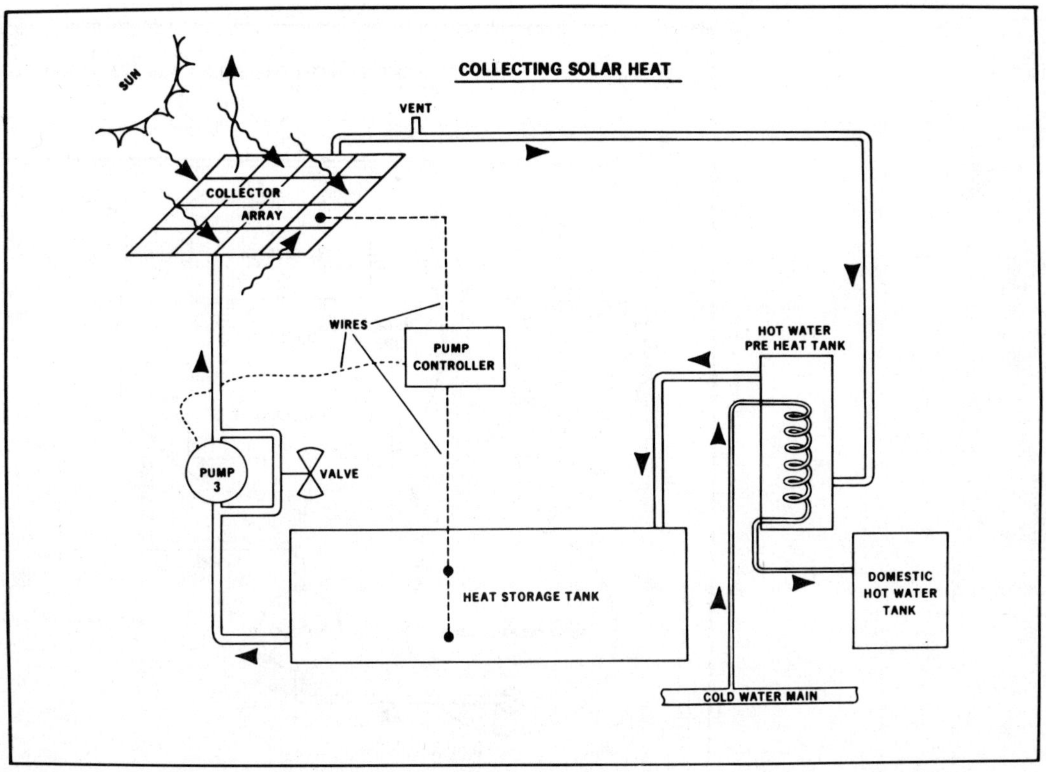

Fig. 6 Mode 1 - heating with stored solar energy.

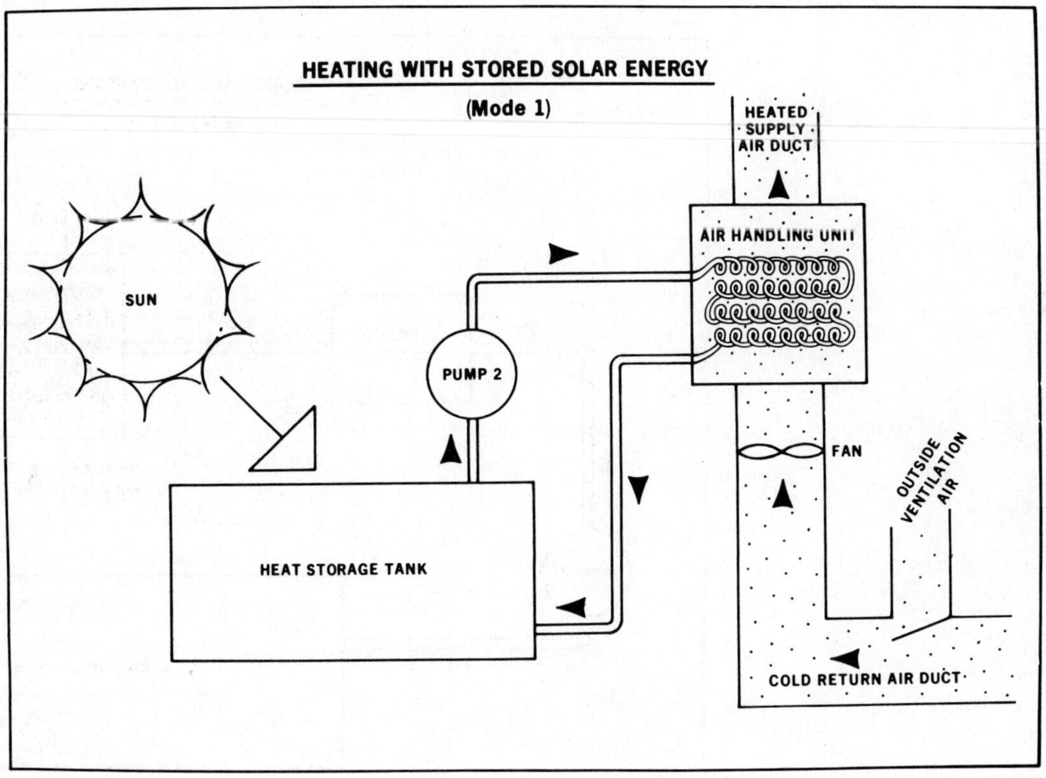

Fig. 7 Mode 2 - heating with the solar assisted heat pump.

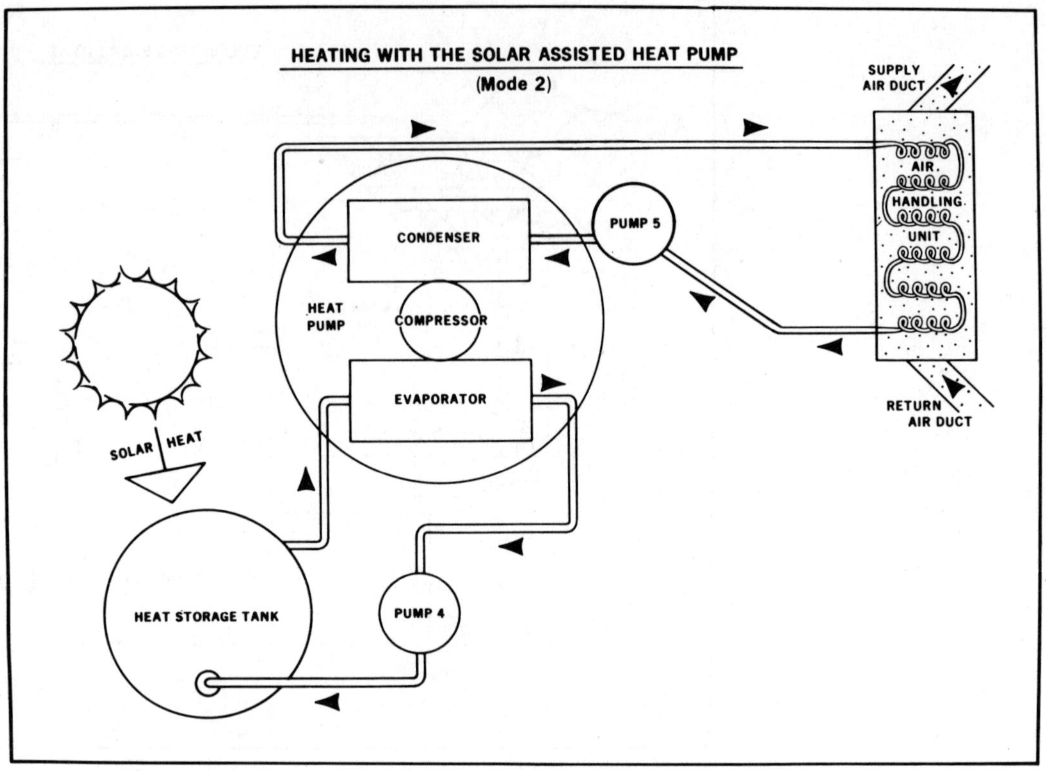

Fig. 8 Modes 3 and 4 - peak tank operation.

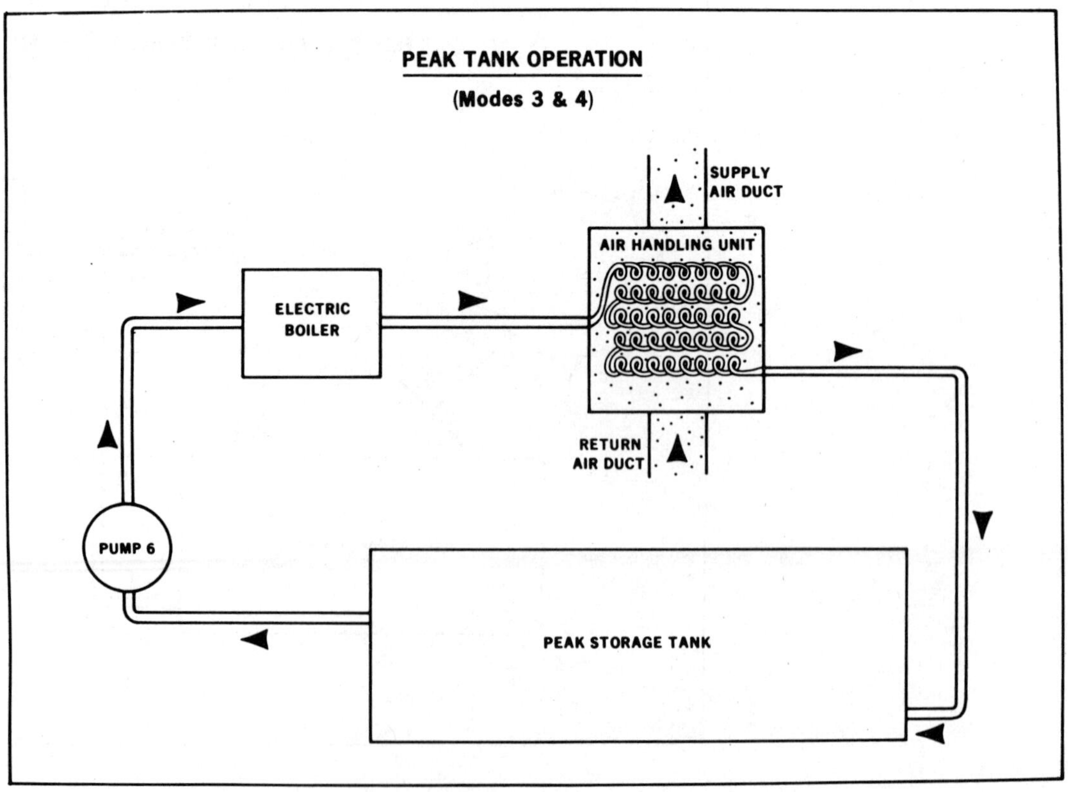

Fig. 9 Simplified AHCC heating system schematic.

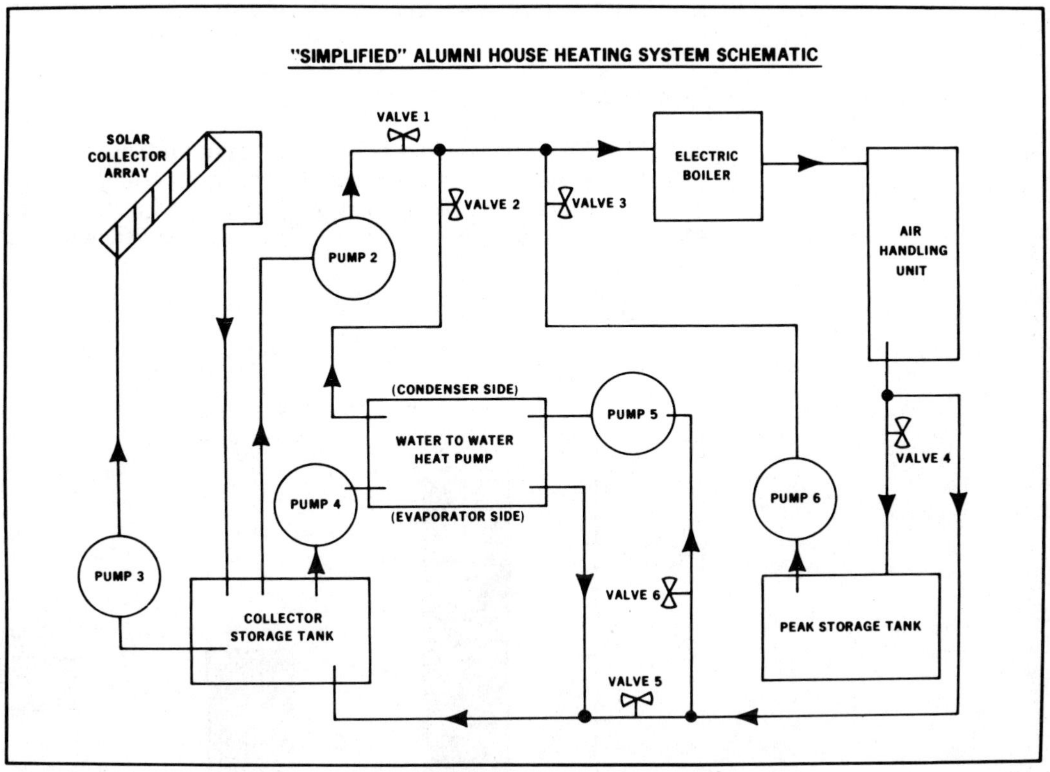

Results

The performance data are presented in Figures 10 and 11. During January the solar collection system provided 33% of the heating demand for the month (January had an average air temperature of -10°C and 5% more sunshine than normal). The heat pump COP (including pump kWh but excluding circulating fan kWh) for this period was 2.95. Ventilation fan kWh amounted to 16% of the total kWh consumed. Neglecting fan kWh the system performance factor (energy delivered to the load/energy consumed on site) for the month was 1.42. Figure 12 provides a heating demand analysis in 15-minute segments, averaged over 29 days. For example, 29 15-minute average readings were calculated for the period 0900-0915. The maximum, minimum and average readings are then plotted. Comparing these data to Figure 13 which contains the power curves for the total building, not just the heating demand, two points are quickly obvious. For this period (1) the total demand follows the heating demand, and (2) the demand limiting system, with a cutoff of 65kw, is effective in holding down the peak demand. The heating curve shows the expected rise between 0700 and 0800 with a sharp decrease between 1600 and 1700 as the building thermostats are reset for the night conditions. For approximately 2-3 hours the building needs very little heating then maintains a night-time temperature.

During February (1977) the heating performance factor rose with readings ranging from 1.39 to 2.40 during the first week.

During late March and early April the solar energy collected exceeds 1.3 x 10 J per day and the HPF exceeds 4. Based on the results of this first winter of operation the solar system will provide over 50% of the total heat demand of the AHCC. Although not all of the systems have been tested and modelled, the computer simulations provide a means of projecting the month by month performance of the systems. Tables 1 and 2 provide the simulated heat budget for the solar assisted heat pump, and for the air-to-water heat pump, respectively. The projected solar energy collected over the heating season is 58% of the total heating load. However, based only on modelling, the air-to-water heat pump provides a more efficient heating service than the solar assisted heat pump.

Conclusion

The use of direct solar heating and the solar assisted heat pump, is a viable means of meeting the heating load of a commercial size building even in severe climates. The possibility of peak load control exists when a storage tank is provided and/or demand limiting is activated.

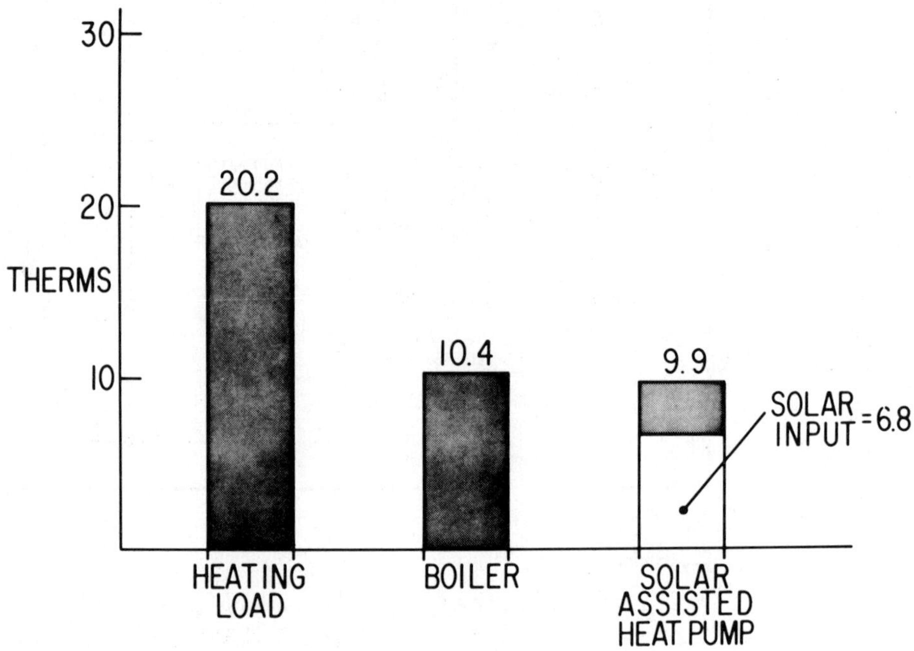

Fig. 10 Average daily heat budget for a 29-day winter period.

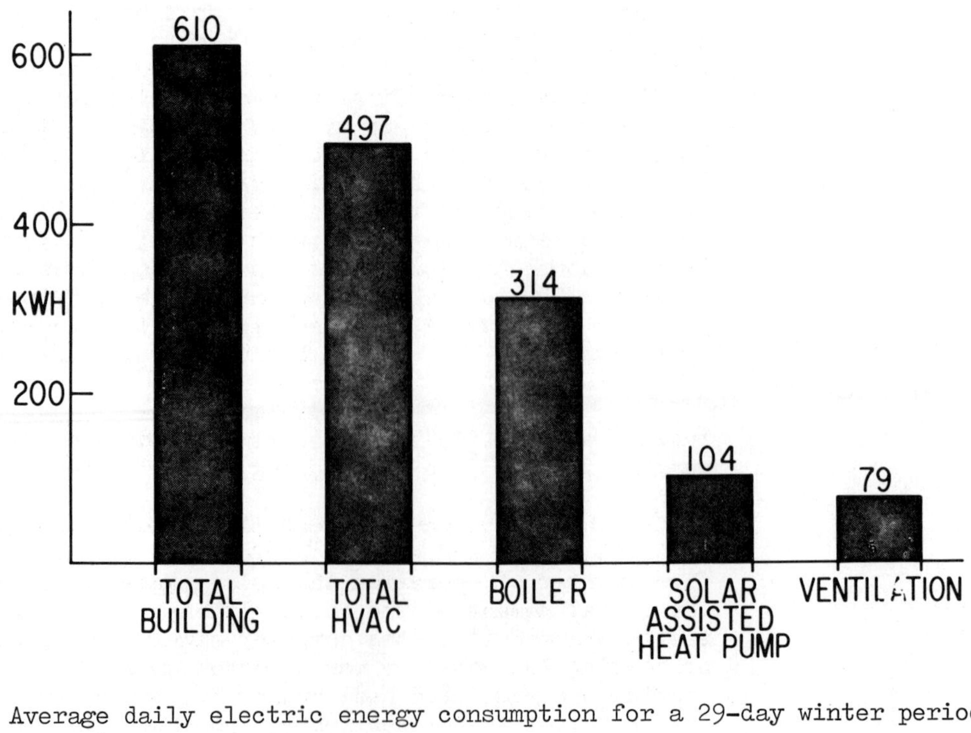

Fig. 11 Average daily electric energy consumption for a 29-day winter period.

Fig. 12 Running fifteen minute analysis of the total power curve.

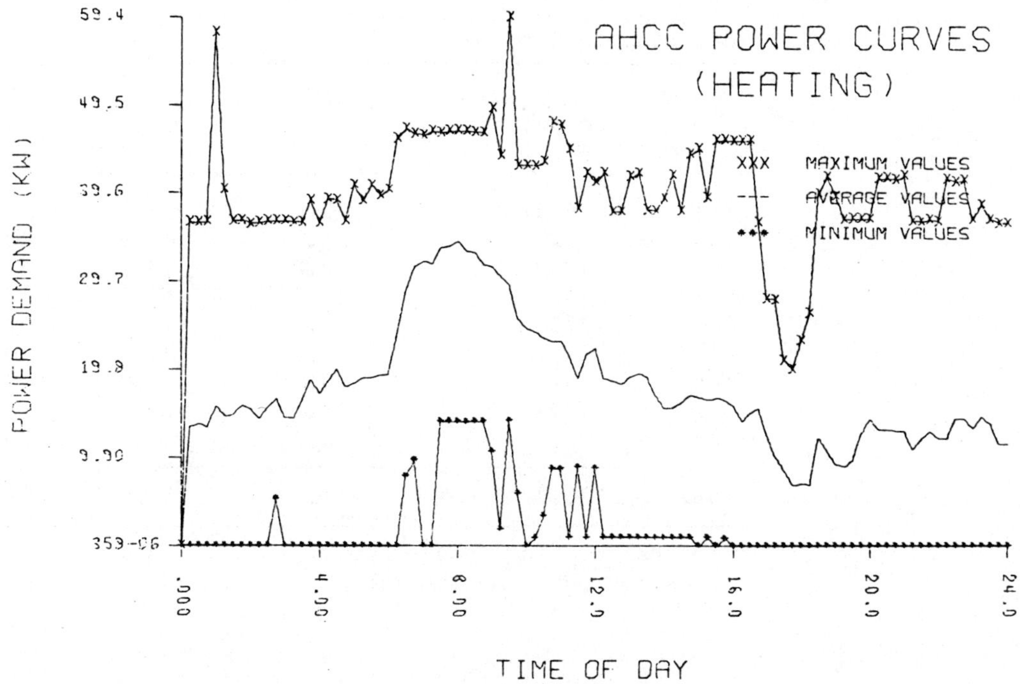

Fig. 13 Running fifteen minute analysis of the total power curve.

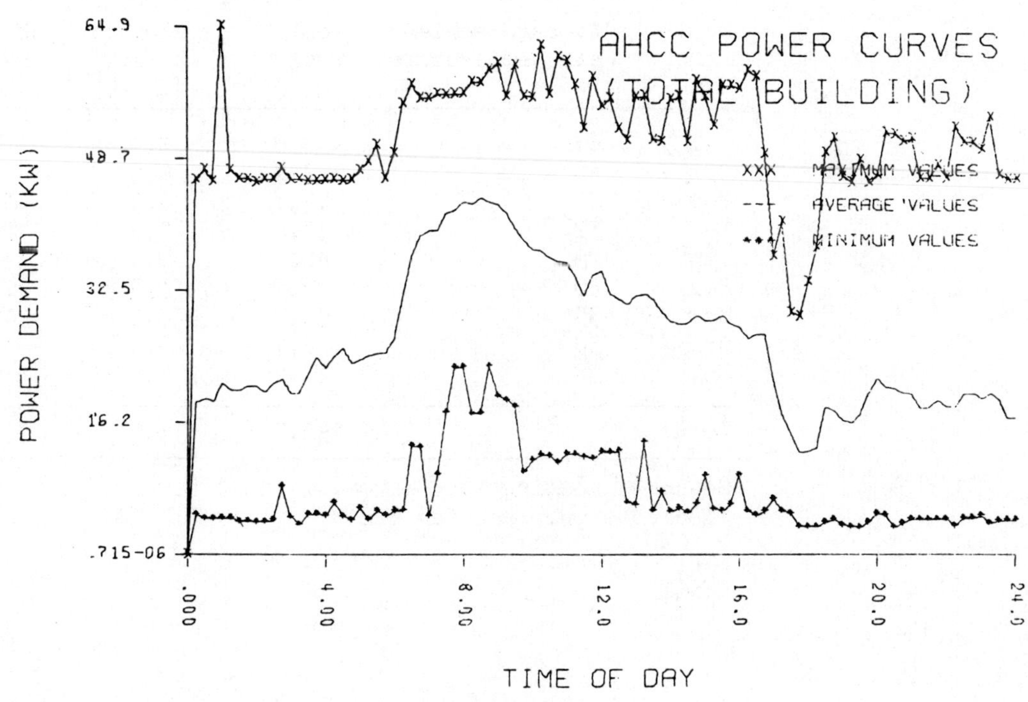

Table 1

Simulated heat budget for the Alumni House Conference Center solar assisted heat pump system projected for a heating season

Heat supplied to load

Month	Solar energy collected 10 J	Direct solar 10 J	Heat pump 10 J	Electric boiler 10 J	Heating system* 10 J
Sep	4.5	0.8	0.0	0.0	7.9
Oct	6.6	4.7	0.0	0.0	8.0
Nov	12.3	2.2	18.7	1.7	15.8
Dec	13.1	0.0	19.0	21.2	37.4
Jan	17.8	0.0	20.7	27.4	44.6
Feb	21.7	0.0	29.4	10.1	28.4
Mar	24.7	0.0	34.0	0.6	20.1
Apr	17.3	2.0	10.0	0.7	12.2
May	8.8	4.4	0.0	0.0	8.1
Total	126.8	14.1	131.8	61.7	182.5

Seasonal heating load = 208. x 10 J
*Includes pump and fan kWh

Table 2

Simulated heat budget for the Alumni House Conference Center air-to-water heat pump system projected for a heating season

Heat supplied to load

Month	Average ambient air temperature °F	Heat pump 10 J	Electric boiler 10 J	Heating system* 10 J
Sep	61	0.8	0.0	7.8
Oct	56	4.7	0.0	9.1
Nov	42	22.7	0.0	14.7
Dec	32	37.4	2.8	26.6
Jan	28	40.0	8.1	39.5
Feb	30	36.5	3.0	25.9
Mar	35	34.5	0.0	19.1
Apr	50	12.7	0.0	11.2
May	58	4.4	0.0	9.0
Total		193.7	13.9	162.9

Seasonal heating load = 208. x 10 J
*Includes pump and fan kWh

Parameters influencing the performance of a solar powered aqua ammonia absorption system used for low temperature refrigeration in Baghdad, Iraq

5.10

J.A. Mynett, Lecturer, I.M. Hussain, Asst. Prof.*
Department of Aeronautical and Mechanical Engineering
The University of Salford
Salford, U.K.

Abstract

This is a study of the feasibility of producing refrigeration effect at low temperatures from a solar powered aqua ammonia absorption system. A triple-glazed flat plate collector with selective absorber coating is used as the energy collecting equipment and the heat transfer media is pressurised water with anti-freeze. It is expected to heat the water to 127°C in the collector. Due to the favourable year round meteorological conditions in Baghdad the problem of storing collected energy is reduced and liquid refrigerant is stored at ambient temperature and high pressure.

The effect of different operating conditions on the performance of the system are investigated, these being: - heat transfer fluid temperature, cooling water temperature, evaporator temperature and aqua ammonia solution concentration. The results are presented in terms of the refrigeration effect produced per unit area of collector surface for a range of operating conditions and using 5 year averaged meteorological data for Baghdad.

Introduction

The purpose of this investigation is to study the feasibility of producing a refrigeration effect for low temperature applications in Baghdad, Iraq, latitude 33° 26'N, using a solar powered aqua ammonia absorption system. This application has great promise in Iraq and similar climatic areas where high ambient temperatures and periods of high insolation prevail almost all year round. The duration of sunshine during the year averages more than eight hours daily. The temperatures during these sunshine days can reach as high as 45°C (113°F) in summer and as low as 10°C (50°F) during winter. The amount of insolation on a horizontal surface reaches a maximum of 700 langleys per day (2500 BTU/ft^2.day) during summer and 300 langleys per day (1125 BTU/ft^2.day) during winter, that is a daily average of 500 langleys per day (1870 BTU/ft^2.day) over the year, Mynett and Hussain [1].

Solar powered refrigeration systems have advantages over other uses of solar energy in that the need for refrigeration is usually in phase with periods of greatest insolation and high ambient temperature. Further an absorption system does not need a great deal of maintenance in comparison with the vapour compression systems currently in use. This makes absorption systems attractive for applications in the developing countries and areas where the average user lacks technical know how. The absorption refrigeration cycle is one of the oldest methods of producing refrigeration effect. It depends upon the utilisation of a small amount of work to raise the pressure of a refrigerant without appreciably changing its volume. This is accomplished by the absorption of vapour refrigerant by a liquid forming a miscible homogeneous mixture. The thermodynamic properties and state of the mixture are established by its pressure, temperature and concentration which is the ratio of the mass of one constituent to the total mass of the mixture. A three cover flat plate collector with selective coating on the absorbing surface is used as the energy collecting equipment. It is expected to heat the heat transfer fluid in the collector to 127°C (260°F)

*on sabbatical leave from University of Baghdad, Engineering College.

Fig. 1 Schematic diagram for continuous solar powered aqua ammonia absorption refrigeration system

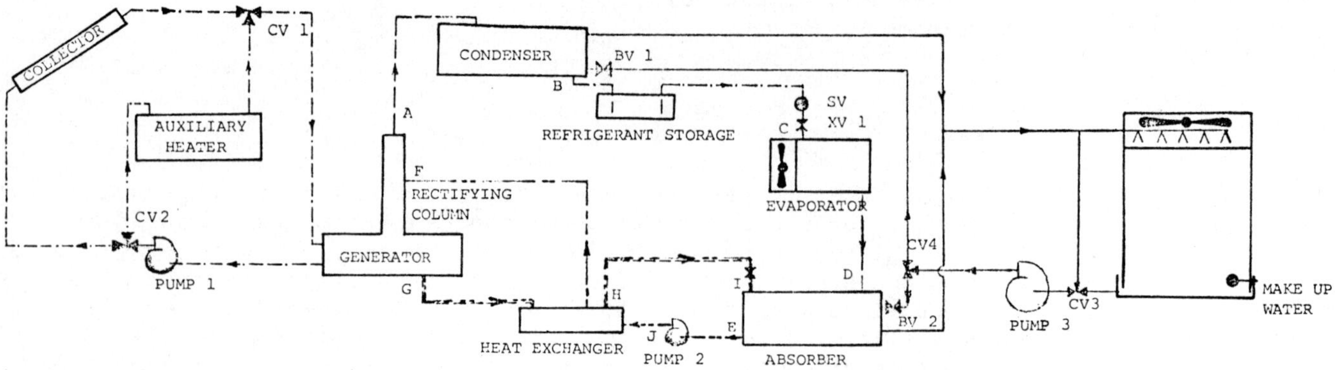

- Fluid Entering Collector
- Fluid Leaving Collector
- Cooling Water Leaving Condenser & Absorber
- Cooling Water Entering Condenser & Absorber
- H.P. Ammonia Vapour Leaving Rectifying Column
- H.P. Ammonia Liquid Leaving Condenser
- Weak Solution Leaving Generator
- L.P. Ammonia Vapor Leaving Evaporator
- Strong Solution Entering Rectifying Column

XV1	EXPANSION VALVE
XV2	EXPANSION VALVE
CV1, CV2,CV3,CV4	THREE-WAY CONTROL VALVES
BV1	COOLING WATER BALANCING VALVE FOR THE CONDENSER
BV2	COOLING WATER BALANCING VALVE FOR THE ABSORBER
P1	COLLECTOR/FLUID PUMP
P2	STRONG AQUA/AMMONIA SOLUTION PUMP
P3	COOLING TOWER PUMP
SV	SOLENOID VALVE

without boiling. The performance of the collector is based on long term average radiation data incident upon a horizontal surface; in this investigation five years data were used. These data and other meteorological data are available from the weather bureau.

System Description

A sketch of the overall solar refrigeration system is shown component by component in Figure 1. The system consists of three separate flow circuits detailed as follows:

1. The Collector Circuit
   The circuit consists of the following items:

a) A flat plate collector - the construction of which is such that the collector is well insulated behind the absorber plate and at the edges in order to reduce the heat losses to a minimum. The collector is tilted at 30 degrees from the horizontal in order to obtain best year round performance. The best tilt was found to vary with the purpose for which the collected energy was used, Mynett and Hussain [1]. The overall coefficient of heat transfer for the solar collector, U_L, is 2.84 $W/m^2\ {}^oC$ (0.5 $BTU/hr.\ ft^2.{}^oF$). The solar energy gained by the collector is transferred to the generator of the absorption system by a fluid that is pressurized water mixed with ethylene glycol, as an anti freeze, and sodium silicate, as a corrosion preventative, allowing it to reach 127°C (260°F) without boiling.

b) A pump to circulate the heat transfer fluid through the solar collector, the auxiliary heater and the generator. The capacity of the pump is specified by the maximum flow rate through the solar collector that takes place during the maximum incident solar radiation per hour upon the surface of the collector. The energy gained by the heat transfer fluid in the solar collector causes a temperature increase of 11°C (20°F) between inlet and outlet i.e.

$$\dot{m} = \frac{Q_{max\ coll}}{C_p(Tout-Tin)} \qquad (1)$$

The outlet of the pump is connected to a three-way control valve that regulates the flow to the solar collector so that the inlet fluid temperature remains constant and does not fall or exceed the specified inlet temperature. The pump is switched on at the presence of solar energy that is sensed by the rise in the absorber plate temperature or when the auxiliary heater is called on. The minimum absorber plate temperature needed to start the pump is set at 2.5°C above the specified inlet fluid temperature to the collector. The pump is turned off in the absence of sunshine or when the outlet fluid temperature is equal to or less than the absorber plate temperature that was previously referred to, provided that the auxiliary heater is not turned on. The operation of the pump is accomplished through a controller.

c) An auxiliary heater is connected in parallel with the solar collector. The purpose of the heater is to supplement the energy gained from solar radiation when the dependence upon it is limited or to supply all of the energy needed by the generator. The inlet and outlet of the heater are connected to three-way control valves (CV1, CV2) as shown in Fig.1. When the outlet fluid temperature from the solar collector falls below the specified temperature, which is 11°C(20°F) above the inlet fluid temperature to the collector, the auxiliary heater is switched on and the outlet control valve (CV1) regulates the flow to maintain a constant fluid temperature entering the generator. It is turned off once the outlet temperature from the collector has reached the specified temperature or when the need for generation has ceased.

2. <u>The Absorption Refrigeration Circuit</u>

In this circuit the energy gained by the solar collector and that supplied by the auxiliary heater are fed directly to the absorption system generator. The absorption cycle has two distinct fluid streams in a completely enclosed system. These are:

a) The vapour ammonia refrigerant stream is produced when the generator temperature is high enough to cause the absorption system to operate. The high pressure, high temperature refrigerant vapour is then condensed and stored as liquid refrigerant in a separate vessel.

Storing energy in the form of liquid ammonia is preferred to storing it in a hot water thermal storage tank as this tends to minimise the problems associated with energy storage. At the operating design conditions ammonia has a capability of producing approximately 11.63 kJ/kg (500 BTU/lb) of refrigeration effect which is equal to its enthalpy of vaporisation, while energy stored in hot water thermal storage is proportional to its specific heat and temperature rise which is approximately 0.465 kJ/kg (20 BTU/lb) at the design storing temperature. Thus by storing liquid refrigerant there is a reduction of mass stored of almost 25 times and hence a reduction of storage tank size by almost 16 times. In addition, liquid ammonia can be stored at high pressure and ambient temperature thereby eliminating the need for insulation of the storage tank.

The flow rate of the liquid refrigerant from the storage vessel is controlled by a normally closed solenoid valve (SV), as shown in Fig.1. When the thermostat in the refrigerated space rises above the set temperature the pilot of the solenoid valve is energised thereby opening the valve and allowing the liquid refrigerant to pass to the expansion valve. The refrigerant is then expanded in the evaporator resulting in a mixture of vapour and liquid ammonia at low pressure and temperature. The energy needed to evaporate the liquid ammonia is obtained from the cold space where the temperature is above the boiling temperature of the ammonia passing through the evaporator. This energy transfer is the *refrigeration effect*. The resulting low pressure, low temperature ammonia vapour then passes to the absorber where it is absorbed in the weak solution producing a strong aqua ammonia solution. The heat released in the absorber due to the mixing of the binary fluid is removed by the cooling water circulating from the cooling tower where the energy is transferred to the atmosphere.

During periods when there is no energy supply to the generator then the refrigeration effect is wholly dependent upon the usage of the stored liquid refrigerant. The stored refrigerant is fed into the evaporator as per cooling demand. The low pressure vapour from the evaporator is absorbed by the circulating weak solution in the absorber. In order to maintain low evaporator/absorber pressure and high chemical affinity between the solution and the low pressure ammonia vapour during periods of no generation it is necessary to increase the amount of cooling water circulating in the absorber. This is accomplished by diverting the cooling water from the condenser to the absorber to increase the heat transfer there. The diversion of the cooling water to the absorber is done by a three-way control valve, CV4, that is operated by a temperature sensing element installed in the outlet of the cooling water from the condenser. When this temperature is less than 2°C above the inlet temperature, the three way valve, CV4, reduces the flow to the condenser and increases it to the absorber. The cooling water flow rate to the condenser is reduced because there is no need for cooling in the condenser when there is no generation.

The absorption of the stored refrigerant after its usage in the evaporator requires an increase in the absorber vessel size by such an amount as is necessary to accommodate the increase in volume of the solution.

When generation starts again the strength of the solution at this instant is greater than that when the system is operating normally, but this increase in concentration would result in an excess ammonia generation that would be stored once more to provide refrigeration effect during periods of reduced or zero insolation.

b) The other stream is the aqua ammonia solution. The weak aqua ammonia solution left behind in the generator is drawn through a heat exchanger into the absorber where it will be mixed with low pressure, low temperature ammonia vapour from the evaporator to form a strong aqua ammonia solution which is pumped back again by pump, P2, into the generator for re-evaporation. The heat exchanger uses the hot weak aqua ammonia solution to preheat the cold strong aqua ammonia solution, thus reducing both the generator energy consumption and absorber cooling water consumption.

3. The Cooling Water Circuit

The heat released from the condenser due to the condensation of the high pressure high temperature refrigerant is removed by cold water supplied from a cooling tower. The temperature of the cooling water depends upon the efficiency of the cooling tower but it is never less than the ambient wet bulb temperature. The cooling tower also supplies the cold water needed in the absorber of the refrigeration system.

Thermodynamic Analysis of the Cycle

The thermodynamic analysis of the aqua ammonia absorption system is done with the following assumptions:

a) The vapour refrigerant leaving the generator is pure i.e. no water vapour is carried over to the condenser.

b) The vapour refrigerant leaves the evaporator dry.

c) The liquid refrigerant leaves the condenser as saturated liquid.

d) The pump work is negligible.

e) Steady state conditions prevail.

Referring to Fig.1. the following mass and heat balances can be written for the system:

a) For the generator - rectifying column.

$M_f = M_a + M_g$ (2)

$M_f X_f = M_a X_a + M_g X_g$ (3)

and according to assumption (a) $X_a = 1$ thus

$M_f X_f = M_a + M_g X_g$ (3a)

also

$M_f h_f + Q_G = M_g h_g + M_a h_a$ (4)

b) For the condenser

$Q_{COND} = M_a (h_a - h_b)$ (5)

c) For the evaporator

$Q_{EVAP} = M_a (h_d - h_c)$ (6)

but $h_c = h_b$ (throttling process) thus

$Q_{EVAP} = M_a (h_d - h_b)$ (7)

d) For the absorber

$Q_{ABS} = (M_a h_d + M_g h_i) - M_f h_e$ (8)

e) For the heat rejected to the cooling tower

$Q_{CT} = Q_{ABS} + Q_{COND}$ (9)

and also

$Q_{CT} = Q_{EVAP} + Q_G + Q_P$ (10)

and according to assumption (d) equation (10) can be rewritten as:

$Q_{CT} = Q_{EVAP} + Q_G$ (10a)

and finally

f) $COP = \frac{Q_{EVAP}}{Q_G}$ (11)

and the overall COP of the system is

$SOLCOP = \eta \times COP$ (12)

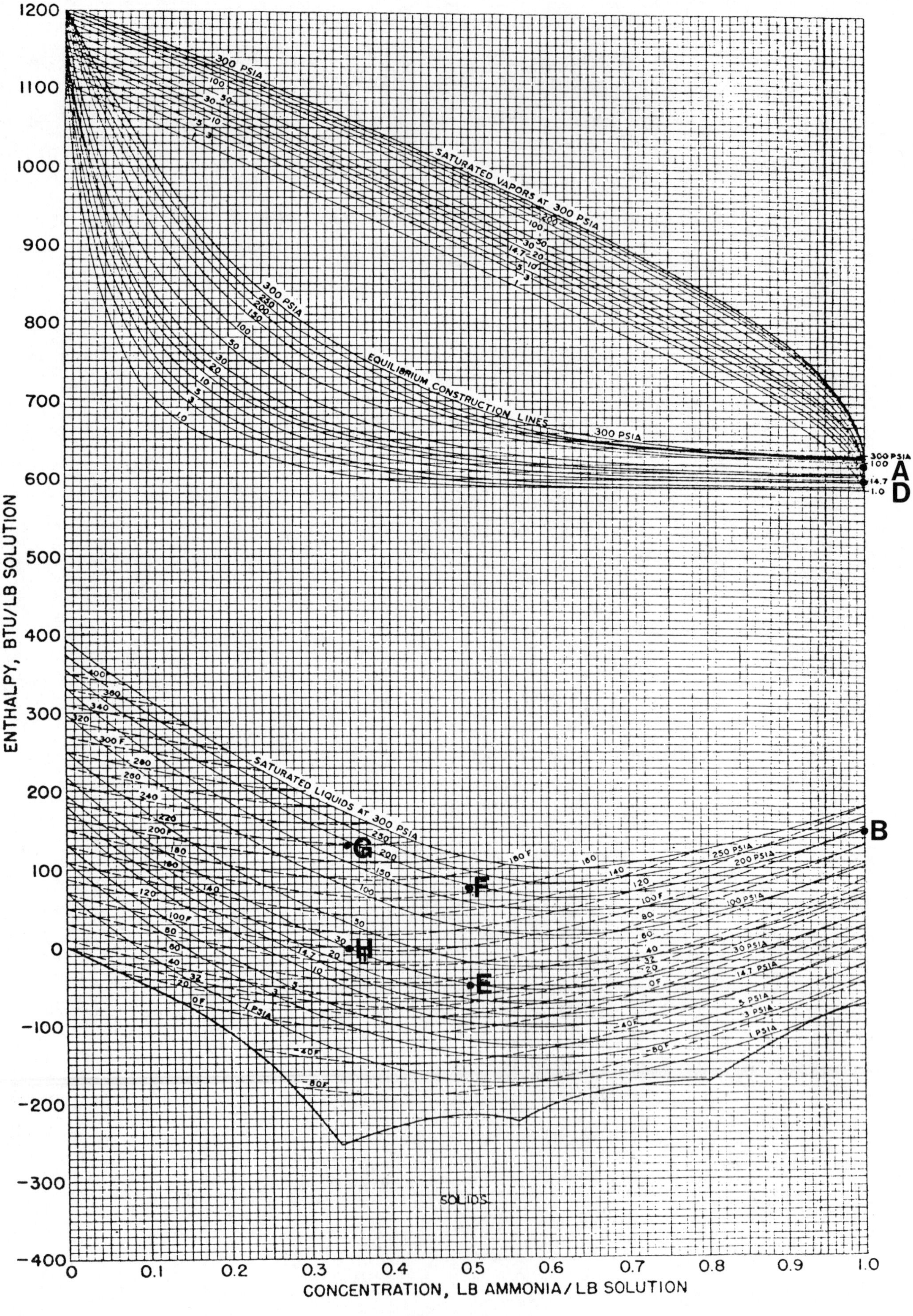
ENTHALPY, BTU/LB SOLUTION
CONCENTRATION, LB AMMONIA/LB SOLUTION
SATURATED VAPORS AT 300 PSIA
EQUILIBRIUM CONSTRUCTION LINES
SATURATED LIQUIDS AT 300 PSIA
300 PSIA
SOLIDS
A
B
D
E
F
G
H

Fig. 2 (opposite)

Discussion of Results

The analysis of the performance of an aqua ammonia absorption system for an average day of each month of an average year is predicted from the enthalpy-concentration relationships shown in Fig.2. A computor program is developed to study the effect of different operating conditions on the performance of the refrigeration system and system parameters.

The operating conditions are:

a) Heat transfer fluid temperature, i.e. generator temperature ranging from 82°C (180°F) to 115°C (240°F) at 11°C (20°F) intervals.

b) Initial concentration of ammonia in aqua ammonia solutions varying from 40 to 70 per cent at 10 per cent interval.

c) Generator/condenser pressures corresponding to condensing temperatures of 35°C to 23.8°C at 2.8°C (5°F) intervals. The generator/condenser pressure is a function of the cooling water temperature. It increases with the increase in the cooling water temperature and decreases with the decrease in the cooling water temperature.

The cooling water temperature depends upon the efficiency of the cooling tower and the ambient wet bulb temperature.

d) Evaporator temperatures of - 17.8°C to - 6.7°C at 5.5°C (10°F) intervals.

The system parameters are: Coefficient of Performance, Solar Coefficient of Performance, weight of ammonia generation and finally refrigeration effect.

The performance is computed for each operating condition and the results are shown in Figures 3-11.

The results indicate that although the amount of energy added to the generator during the year varied from day to day the coefficient of performance remains constant for the same generator temperature, evaporator temperature and initial solution concentration. This shows that the coefficient of performance is a function of the sink and source temperature and not of the amount of heat added to the generator or absorbed at the evaporator.

Figs. 3 and 4 show that at generator temperatures of 82°C to 93°C it is practically impossible to generate any refrigerant at 40 per cent initial solution concentration and high generator/condenser pressure. Figures 5-7 indicate that to operate the refrigeration system at low source temperature, stronger ammonia concentrations have to be used. For a year round performance of the solar refrigeration system with consistent high coefficient of performance, high solar coefficient of performance, high ammonia generation and high refrigeration effect it is recommended to run the refrigeration unit with initial solution concentration of 55 to 60 per cent as seen in Figures 8 and 9. This range of concentration will yield approximately 0.575 coefficient of performance, 0.24 solar coefficient of performance, 5.75 kg of ammonia generated per day per square metre of collector and 6000 kJ of refrigeration effect per day per square metre of collector. This is the equivalent to production of around 16 kg of ice per day per square metre of collector. These values of the system parameters are obtained during the summer months May to October when the generator/condenser pressure is at its highest level of 1350 kPa because of the high ambient wet bulb temperature that

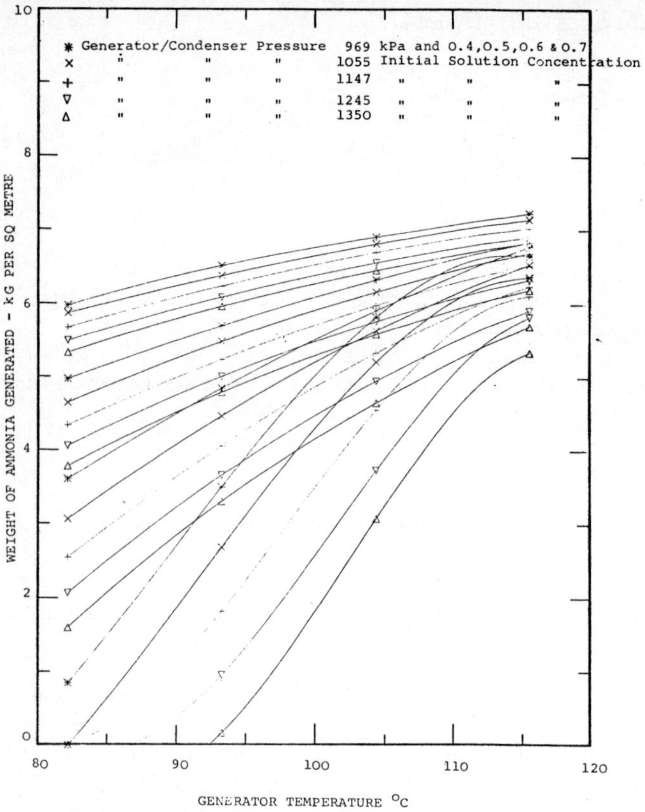

Fig. 3
Weight of ammonia generated at different generator/condenser pressures and solution concentrations for various generator temperatures and -17.8°C evaporator temperature

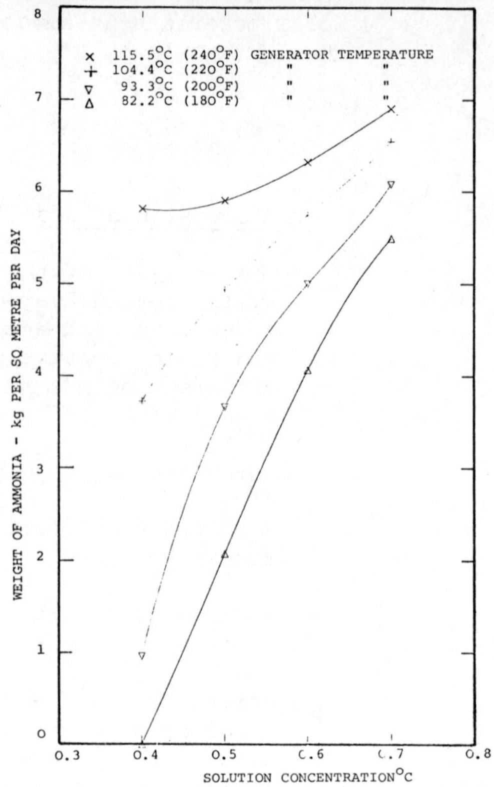

Fig. 4
Ammonia generation as a function of solution concentration for various generator temperatures at 1245 kPa (180 psia) generator/condenser pressure and -17.8°C evaporator temperature

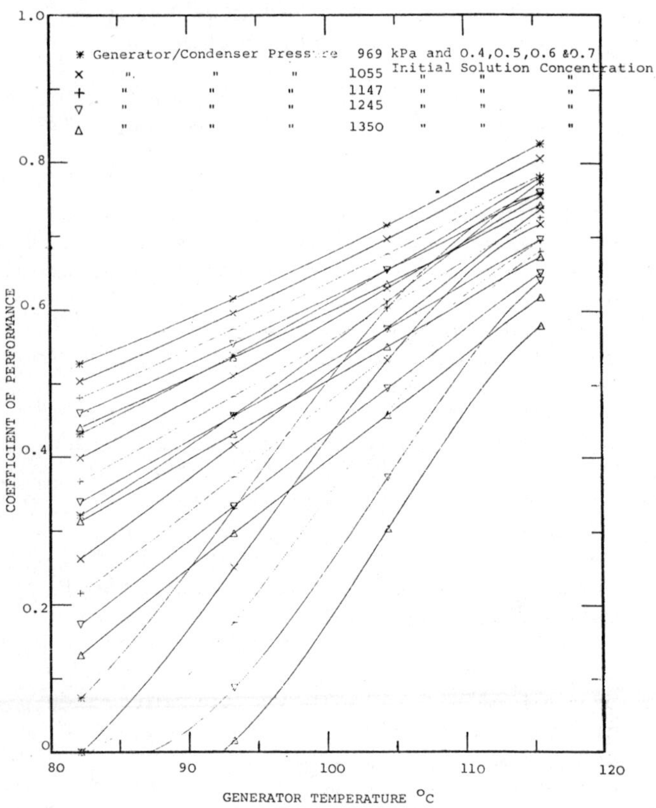

Fig. 5
Effect of solution concentration and generator/condenser pressure on the coefficient of performance at various generator temperatures and -17.8°C evaporator temperature

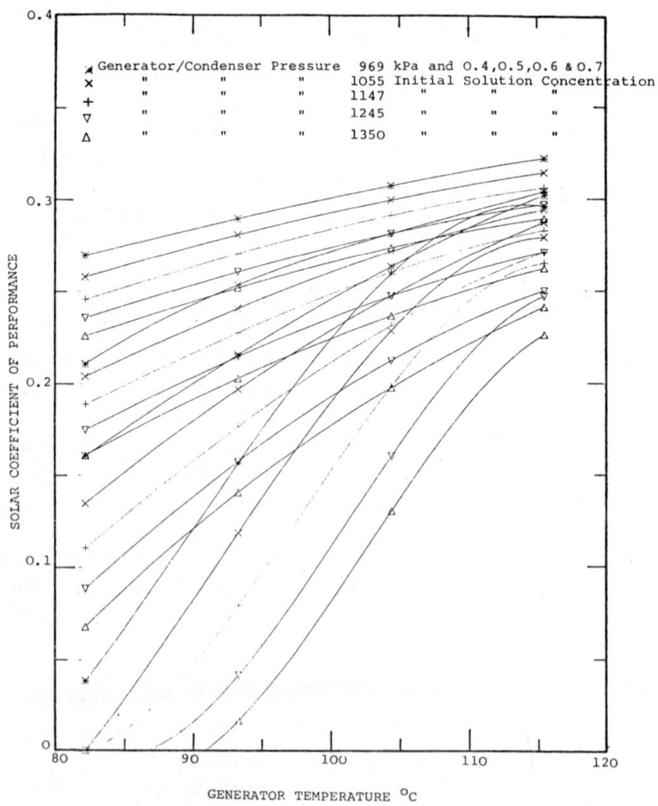

Fig. 6
Solar coefficient of performance variation with generator/condenser pressure and solution concentration at different generator temperatures and -17.8°C evaporator temperature

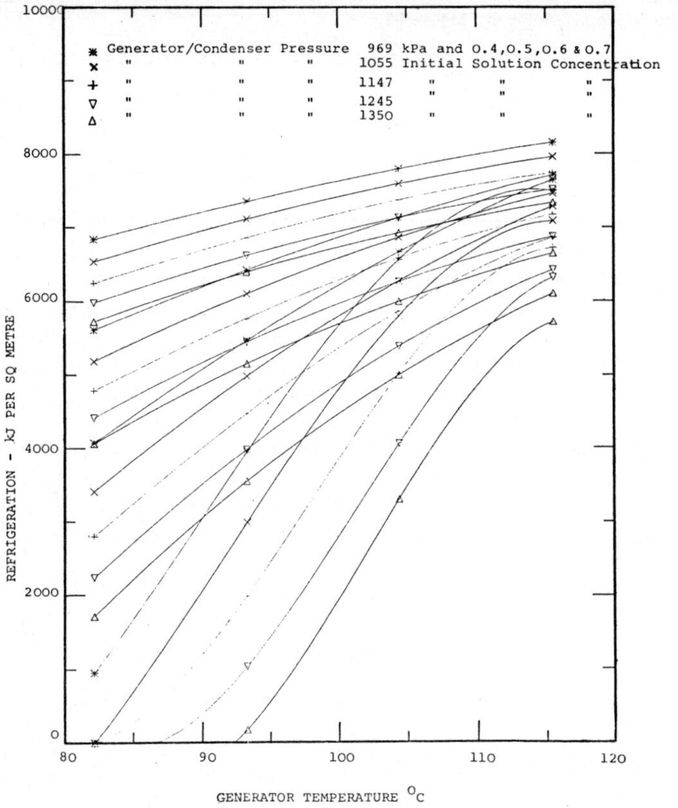

Fig. 7
The effect of generator/condenser pressure and solution concentration on the refrigeration effect at various generator temperatures and -17.8°C evaporator temperature

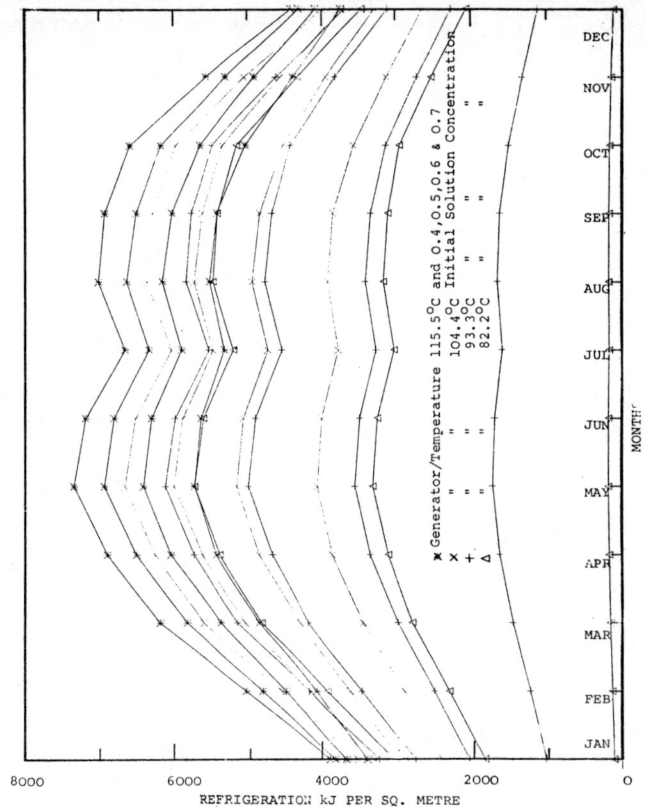

Fig. 8
Refrigeration produced during the year as a function of solution concentration and generator temperatures at -17.8°C evaporator temperature

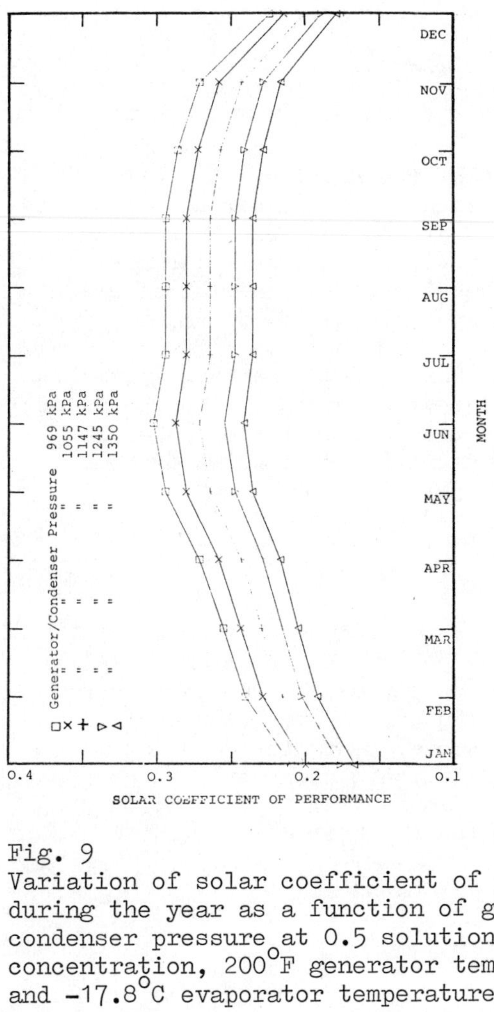

Fig. 9
Variation of solar coefficient of performance during the year as a function of generator/ condenser pressure at 0.5 solution concentration, 200°F generator temperature and -17.8°C evaporator temperature

reaches to 26°C (78°F). The high ambient wet bulb temperature will not allow the cooling water to fall below 29.5°C (85°F) thereby resulting in a high generator/condenser pressure. The average generator temperature is between 110° - 115°C and the solar collector efficiency is at its peak. Also it is seen that the weight of ammonia generated is affected appreciably by the change in the solution concentration at generator temperature of 82°, 92°, 104° and that it levels off at higher temperature of 115°C. Table 1 shows the changes in the coefficient of performance, the solar coefficient of performance, the weight of ammonia generation and the refrigeration effect corresponding to an increase in the generator temperature from 82°C to 115°C and a reduction in the generator/condenser pressure from 1350 kPa to 969 kPa.

Table 1

CHANGE IN GENERATOR TEMPERATURE °C	PER CENT CONCENTRATION	GENERATOR/CONDENSER PRESSURE kPa	CHANGE IN COP		CHANGE IN NH_3 WEIGHT kg/m^2 DAY		CHANGE IN SOLCOP		CHANGE IN REF. Eff. kJ/m^2 DAY	
			FROM	TO	FROM	TO	FROM	TO	FROM	TO
82°C - 115°C	0.4	1350	0.0	0.578	0.0	5.32	0.00	0.226	0.0	5719
		1245	0.0	0.640	0.0	5.81	0.0	0.247	0.0	6328
		1147	0.0	0.680	0.0	6.10	0.0	0.265	0.0	6729
		1055	0.0	0.716	0.0	6.35	0.0	0.279	0.0	7084
		969	0.074	0.758	0.85	6.64	0.038	0.296	961	7500
82°C - 115°C	0.5	1350	0.132	0.617	1.59	5.68	0.067	0.241	1715	6104
		1245	0.174	0.650	2.07	5.90	0.088	0.250	2255	6429
		1147	0.216	0.695	2.55	6.23	0.110	0.271	2810	6869
		1055	0.262	0.736	3.06	6.52	0.134	0.287	3412	7282
		969	0.322	0.774	3.61	6.77	0.160	0.302	4080	7651
82°C - 115°C	0.6	1350	0.313	0.672	3.78	6.18	0.160	0.262	4066	6649
		1245	0.340	0.696	4.06	6.32	0.174	0.271	4422	6880
		1147	0.368	0.726	4.34	6.51	0.188	0.283	4787	7176
		1055	0.399	0.754	4.64	6.68	0.203	0.294	5179	7457
		969	0.432	0.780	4.97	6.83	0.210	0.304	5610	7713
82°C - 115°C	0.7	1350	0.440	0.742	5.32	6.82	0.225	0.289	5721	7335
		1245	0.460	0.760	5.49	6.90	0.235	0.297	5984	7519
		1147	0.481	0.783	5.67	7.02	0.245	0.306	6251	7744
		1055	0.503	0.805	5.86	7.13	0.257	0.314	6535	7959
		969	0.527	0.825	5.97	7.22	0.269	0.322	6845	8159

Figures 8 and 9 show the variation of the refrigeration effect and the solar coefficient of performance during the year as a function of generator temperature and pressure, and the results confirm once more that it is possible to obtain a constant solar coefficient of performance during the year if the initial solution concentration is set at the recommended value of 55 to 60 per cent.

Figures 10 and 11 show that the higher the evaporator temperature the greater is the amount of refrigeration produced and the solar coefficient of performance remains constant over the range of evaporator temperature under consideration provided that the generator temperature remains constant.

Conclusion

The results indicate that it is possible to produce an average of 3500 kJ of refrigeration effect per day per square metre of collector (the equivalent of producing 10 kg of ice per day). The performance of the refrigeration system is controlled by the operating conditions. These conditions are: generator temperature, generator/condenser pressure, evaporator temperature and solution concentration. For best year round performance of the refrigeration system a solution concentration of 55 to 60 per cent will give the best coefficient of performance over the year.

Fig. 10 Refrigeration effect as a function of generator temperature at various evaporator temperatures, 0.5 solution concentration and 1245 kPa (180 psia) generator/condenser pressure

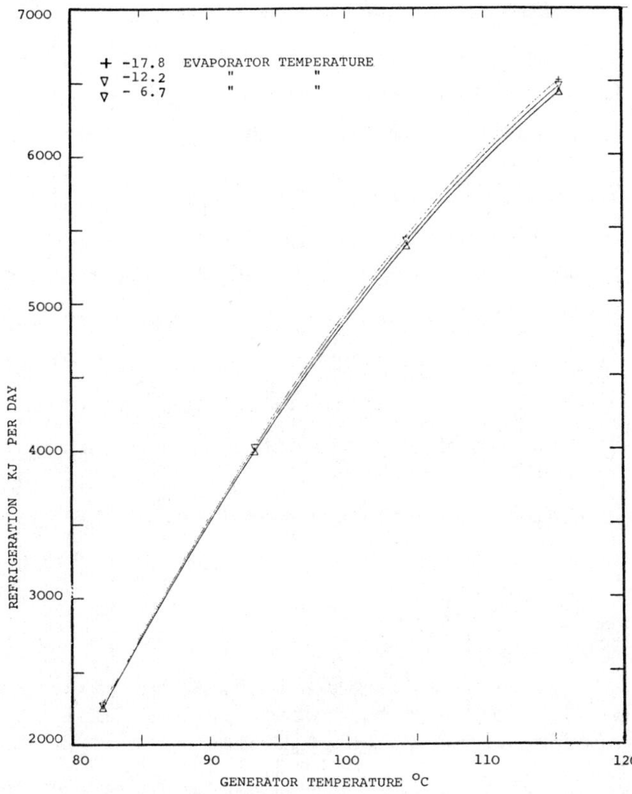

Fig. 11 Variation of coefficient of performance with evaporator temperature as a function of generator temperature for 0.5 solution concentration and 1245 kPa (180 psia) generator/condenser pressure

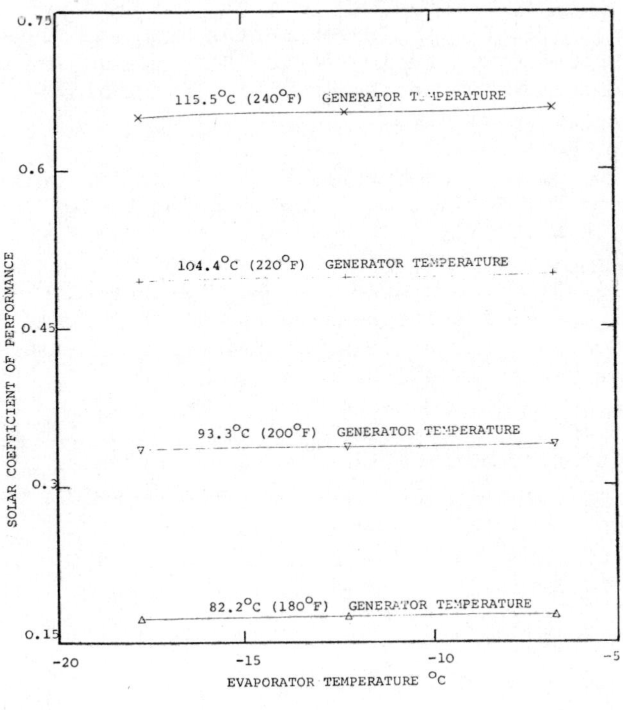

Nomenclature

C_p	Specific heat of the heat transfer fluid
h_a, h_b	enthalpies of saturated vapour and liquid refrigerant at generator/condenser pressure
h_d	enthalpy of saturated vapour refrigerant at evaporator/absorber pressure
h_f, h_g	enthalpies of strong and weak solution at generator/condenser pressure
h_i, he	enthalpies of strong and weak solution at evaporator/absorber pressure
$\dot{m}$	Rate of fluid circulated through the collector
M_a, X_a	Mass and concentration of ammonia refrigerant
M_f, X_f	Mass and concentration of strong solution
M_g, X_g	Mass and concentration of weak solution
$Q_{max\ coll}$	Maximum energy collected by the circulating fluid in the collector per hour, per unit area of collector surface
T_{in}, T_{out}	Inlet and outlet temperature of the fluid into and out from the collector

References

1. Mynett, J. A. and Hussain, I. M. "Availability of Solar Energy and Design Data for the Performance of Flat Plate Collectors at Baghdad, Iraq" to be published.

2. Williams, D. A., Chung, R., Löf, G.O.G., Fester, D. A., and Duffie, J. A. "Intermittent Absorption Cooling Systems with Solar Regeneration" ASME paper No. 57-A-60, 1957.

3. Eisenstadt, M., Flannigan, F. M. and Farber, E. A. "Solar Air Conditioning with NH_3-Water Absorption Refrigeration" Engineering Progress at the University of Florida, Technical Progress Report 14 No.2 pp 22-28, 1960.

4. Chinnapa, J. C. V. "Experimental Study of the Intermittent vapour Absorption Cycle Employing the Refrigerant Absorbent Systems of Ammonia Water and Ammonia Lithium Nitrate" Solar Energy Vol.5 pp. 1-18, 1961.

5. Löf, G. O. G., and Tybout, R. A. "The Design and Cost Optimized Systems for Cooling Dwelling by Solar Energy" Solar Energy Vol. 16 No.9 1974.

6. Little, J. W., Humphries, W. R. and Cady J. C. "Design and Operation of a Solar Heating and Cooling System for a Residential Size Building" NASA, George C. Marshall Space Flight Center EP 45, Alabama, USA.

7. Wilbur, P. J. and Mitchell, C. E. "Solar Absorption Air Conditioning Alternatives" Journal of Solar Energy Science and Technology, Vol. 17, No.3 pp. 193-199, July 1975.

Solar energy utilisation in developing countries

6

The role of solar energy in developing countries

P.D. Dunn,
University of Reading, U.K.

Introduction

It is possible to make several generalisations on energy use in developing countries.

(i) The energy per capita per annum is very much less than for the developed countries. e.g. the current energy use in the USA is 12 t.c.e./capita/annum, compared with 0.15 t.c.e./capita/annum in the Sudan.

(ii) The manner in which the energy is used differs considerably between the developing and developed countries.

Table I

Energy Input			
Sudan (Ref.1)		USA	
Oil	38%	Oil	46%
Firewood/Charcoal	57%	Natural Gas	31%
Bagasse	2.2%	Coal	18%
Hydro	1.8%	Hydro/Nuclear	5%
Muscle	0.6%		

Energy Use				
Direct Heat			Commercial and Domestic	19%
Domestic	61%		Industrial	26%
Industrial	7%		Transport	25%
Total		68%	Electricity Generation	25%
Transport		20%	Non-energy use*	5%
Agricultural tractors/pumping		7%		
Electricity Generation		5%		

*This heading refers to the petrochemical industry.

In the case of the Sudan, with its largely rural agricultural economy and low total energy use, the energy demand is dominated by the domestic heat component, mainly used for cooking, a demand which is met by firewood and charcoal. Electricity generation is both a small fraction of total energy demand and a very low value per capita. Comparing these percentages with those for the USA we see that in the USA electricity generation accounts for 25% of the total energy input, the industrial requirement is a much greater fraction of the total than for the Sudan. The domestic load is an important fraction of the total and of course in absolute terms is much higher per capita than that for the Sudan. The USA dependence on oil and natural gas is brought out by these figures.

(iii) Wood and vegetation contribute considerably to the energy budget of the developing countries. This contribution does not always appear in the published figures, so that the actual energy use per capita may be higher than suggested by these figures. On the other hand, the use of energy in developing countries tends to be very much less efficient than in the developed countries, so the two effects tend to some extent to cancel out.

(iv) Factories, power stations, etc. are required in the developing as in developed countries. It is however the problem of providing energy in the rural areas of the developing countries which is particularly acute.

Use of Energy in Rural Areas of Developing Countries

There are a number of considerations to be taken into account when selecting an energy source for these applications; some of the more important are:

(i) Power level, and whether this is for a continuous or discontinuous supply. For example, if the requirement is for pumping water, then providing that a storage tank can be incorporated in the system one is only interested in pumping the required quantity of water daily, or even over a longer period, and a solar powered pump may be suitable. On the other hand, if a supply of electricity is required to power an educational t.v. set or operate the lights in an emergency operating theatre, then continuity of supply is essential.

(ii) Cost. There are several cost components which must be considered. Initial cost, and its amortisation over the lifetime of the plant. Running cost, including fuel cost, if applicable, operator costs, frequency and cost of maintenance.

(iii) Complexity of operation. This is of course in relation to the local skills available. A simple hand pump may in some circumstances be preferred to an engine driven pump simply because of the lack of suitably skilled operators.

(iv) Maintenance and availability of spares. What has been said under (iii) also applies here. The problem of arranging for adequate maintenance and repair of equipment is perhaps the single most important aspect to be considered in the provision of equipment for rural areas. Development literature contains many references to this problem, for example the estimate by the New Internationalist Magazine (February 1975) quoted with reservations by Ref.2. This source estimated that of 150,000 boreholes in India, 90,000 may be out of action at any one time due to pump breakdown.

(v) Lifetime, under field conditions.

(vi) Suitability for local manufacture.

The principal uses for energy in the rural areas of developing countries are summarised in Table II.

Table II

Agricultural machinery, e.g. the two-wheel tractor
Crop processing, milling, grinding, drying
Pumping water for irrigation and domestic use
Desalination
Powering small industrial equipment, lathes, circular saws, grindstones
Electricity generation for lighting and power, schools, clinics, domestic use
Refrigeration
Domestic; cooking, water heating, heating, cooling
Transport; small vehicles and boats.

Application of Direct Solar Radiation

Solar energy is particularly applicable to the energy needs in the rural areas of developing countries, since it is generally available and the disadvantages of low power density and unreliability are not necessarily too inconvenient for these applications. The principal areas of application for direct solar radiation are:

(i) water heating, using flat plate collectors

(ii) air heating, particularly for crop drying

(iii) small scale stills

(iv) refrigeration and cooling

(v) solar cooking

(vi) pumping, using either flat plate collectors or focussing collectors in association with either conventional heat engines or liquid piston engines

(vii) electricity generation, using engines

(viii) electricity generation, using photo cells. For this application to be of real interest the price of solar cells will require to be reduced; recent advances with amorphous silicon are encouraging.

(ix) building design. There is considerable scope for improvements in building design in order to provide an acceptable environment at low energy cost. This includes the revival of some traditional practices, e.g. Ref. 3.

The above applications are for the use of direct solar radiation, which is the subject of this conference. Indirect solar energy is of course also relevant and includes the use of vegetation, wind power, and water power.

In addition to the small scale applications, solar energy may also have some application to large scale plant, including

(i) desalination

(ii) electricity generation - either in association with other generation equipment or for process electricity.

References

1. Dr. O. M. Abayazid. "Prospects of Fuel and Energy in the Sudan." Report to the National Council for Research, Council for Scientific and Technological Research. January 1975.

2. The Water Panel, Intermediate Technology Development Group Ltd., Edited Arnold Pacey. Pergamon Press 1977. "Water for the Thousand Millions".

3. Mehdi N. Bahadori. "Solar Energy Utilization Research in Iran." UNESCO/World Meteorological Organization Solar Energy Symposium, Geneva, 30 Aug.-3 Sept. 1976.

Solar energy utilisation for developing countries

Mehdi N. Bahadori
Professor of Mechanical Engineering and
Director of Solar Energy Centre
Pahlavi University
Shiraz, Iran

Abstract

Passive solar heating and natural cooling designs which are believed to have more immediate applications in developing countries are briefly described. Other solar energy utilization systems mentioned are solar desalination, solar cooking, solar dehydration, and solar sterilization of medical instruments. It is concluded that with the present and future energy shortages the developing countries need to consider energy conservation and passive systems for their building designs.

Introduction

The utilization of solar energy by man for his dwellings is not anything new. He has built his house with this in mind to collect as much solar energy during the winter days as possible and store this energy within the building materials for the night uses. In fact, there is no building which does not make use of solar energy in a passive way. Most of the buildings are built to reduce the solar radiation gains in summer by proper shading and make use of the radiation to the sky for night cooling.

It was only when liquid fuels became abundant and readily available that man ignored the importance of the passive systems to make his dwelling comfortable and resorted to active devices for this purpose. Developing countries, following the example of the developed and industrialized nations, set aside their own tradition of building technology and imported machinery to keep the buildings built (with no energy conservation in mind) comfortable.

Developing countries need to relearn the dwelling styles and construction methods of their old times and emphasis, more than anything else, on the conservation of energy and the use of passive systems. In this article some of the techniques for solar energy utilization which can employ both indigenous materials and local manpower are discussed, with emphasis placed on passive systems.

Passive solar heating

There are several designs which are briefly described below.

1. Harold Hay Design
Figure 1 shows schematically this system. It consists of a plastic water bag placed on the roof. The bag is made of a transparent cover and a black bottom layer. During the sunny winter days the bag is exposed and solar energy is absorbed by the water. At night the movable insulation covers the bag and heat is transferred to the room through the ceiling by radiation and convection and maintains a rather uniform temperature in the room (1). The height of the water in the bag depends on the energy storage capacity needed for the room.

2. Steve Baer Design
Figure 2 shows the schematics of this design. It consists of a series of barrels placed immediately behind the window of the room. The barrels are painted black on the side receiving solar energy and with a light coloured paint on the side facing the room. The barrels are filled with

Fig. 1
A passive solar heating,
Harold Hay Design

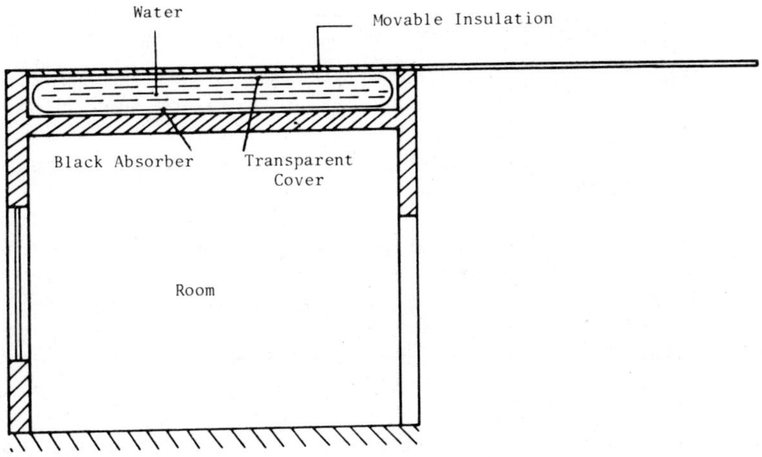

Fig. 2
A passive solar heating,
Steve Baer design

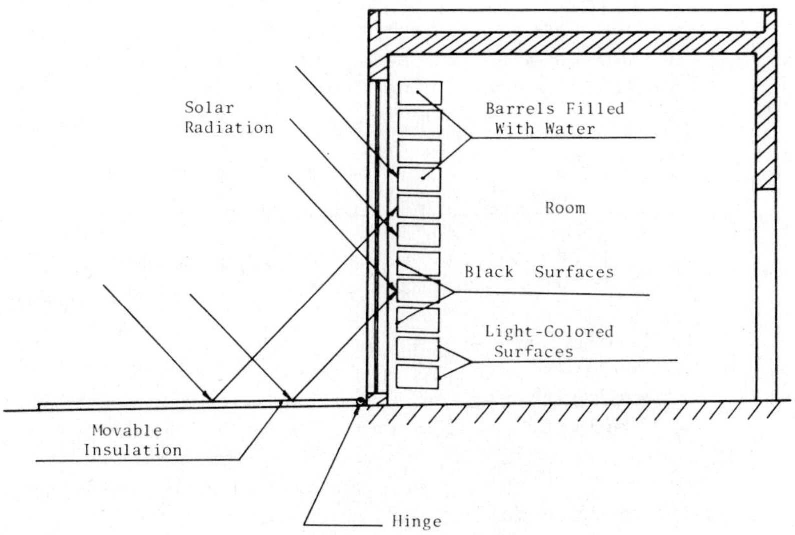

Fig. 3
A passive solar heating,
Odeillo Design

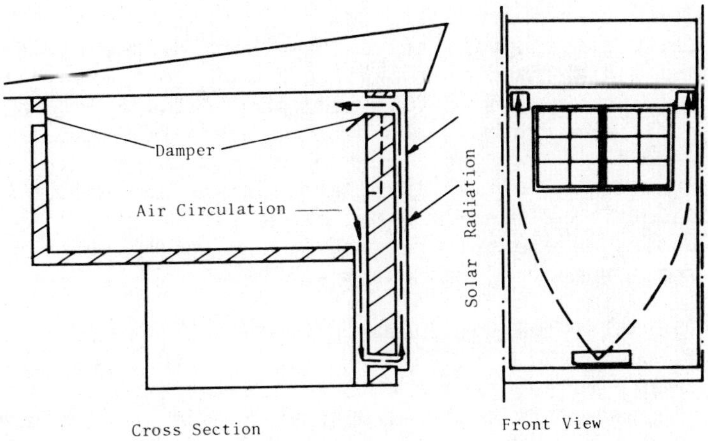

water and spaced at will to allow a desired light to come to the room. A movable insulation covers the window during the winter nights and summer days. The bulk of water provides a good energy storage volume which maintains a uniform temperature in the room (2).

3. Odeillo Design
Figure 3 shows the operation of this system (3). The wall facing south is made of concrete and painted black on the outside. The wall is covered by glass and the concrete wall acts both as a collector and energy storage medium. Heat is transferred to the room by radiation and natural convection.

Active solar heating

The passive solar systems described above, while simple to design and operate, are limited in their applications to rooms with favourable conditions. Other rooms in the house either have to be heated by other means or the utilized solar energy has to be transported to these rooms by some active means.

The active solar heating systems consist generally of black flat plates covered on top by a transparent cover (glass or plastics) and insulated on the back. Air or water can be employed to remove the utilized solar energy and either heat the rooms and/or charge a storage. Uniform temperature can be maintained throughout the building. The storage capacity may be as small as to meet the heating needs for few hours of no insolation or as high as to meet the heating needs for an entire season. Auxiliary sources of heat are employed for the periods of continuous cloudiness or when not enough solar energy has been collected and stored. Selection of the size of the storage and the solar collector depends, among other things, on the cost of fuel which is to be saved, and the cost of the solar installation. It is found that eliminating auxiliary source of heat and meeting all the heating needs of the building by means of solar energy is not economical in Iran (4).

Passive cooling of buildings

The passive cooling of buildings is generally done by making use of low ambient temperature and radiation to the sky during the nights in summer.

1. Harold Hay Design
The movable insulation shown in Figure 1 is removed during the summer nights, exposing the water bag for sky radiation. Further cooling may be provided when the bag is sprayed on or covered with water allowing further cooling by evaporation. The movable insulation covers the water bag during the day and the bag of cold water on the roof keeps the room comfortable during summer days.

2. Wind Towers
The cities surrounding desert in Iran have very little rainfall and very clear atmosphere. They enjoy a high degree of insolation during the day and a high rate of radiation losses to the sky at night. People living in such areas have long made use of these natural phenomena and have built their homes, and adopted a way of life, as to have very little need for winter heating and no need for active summer cooling. For example, in cities like Yazd, Semnan and Bam located at the edge of a desert, one finds the buildings built over fifty years ago having flat roofs, basements, thick walls and roofs, and shades for their relatively small number of doors and windows.

The solar-thermal energy storage capacity of the heavy walls and roofs of these old buildings provided a rather uniform and moderate temperature during the winter season, with only a small need for supplementary heating. During the hot and dry summer months the performances of these buildings are excellent. The building's heavy structure is cooled at night by means of a natural circulation of the outside air through the building and by radiation to the sky.

Natural circulation of the ambient air through the building is provided by one or more wind towers, as shown in Figure 4.(5)

Fig. 4
Operation of a wind tower in summer

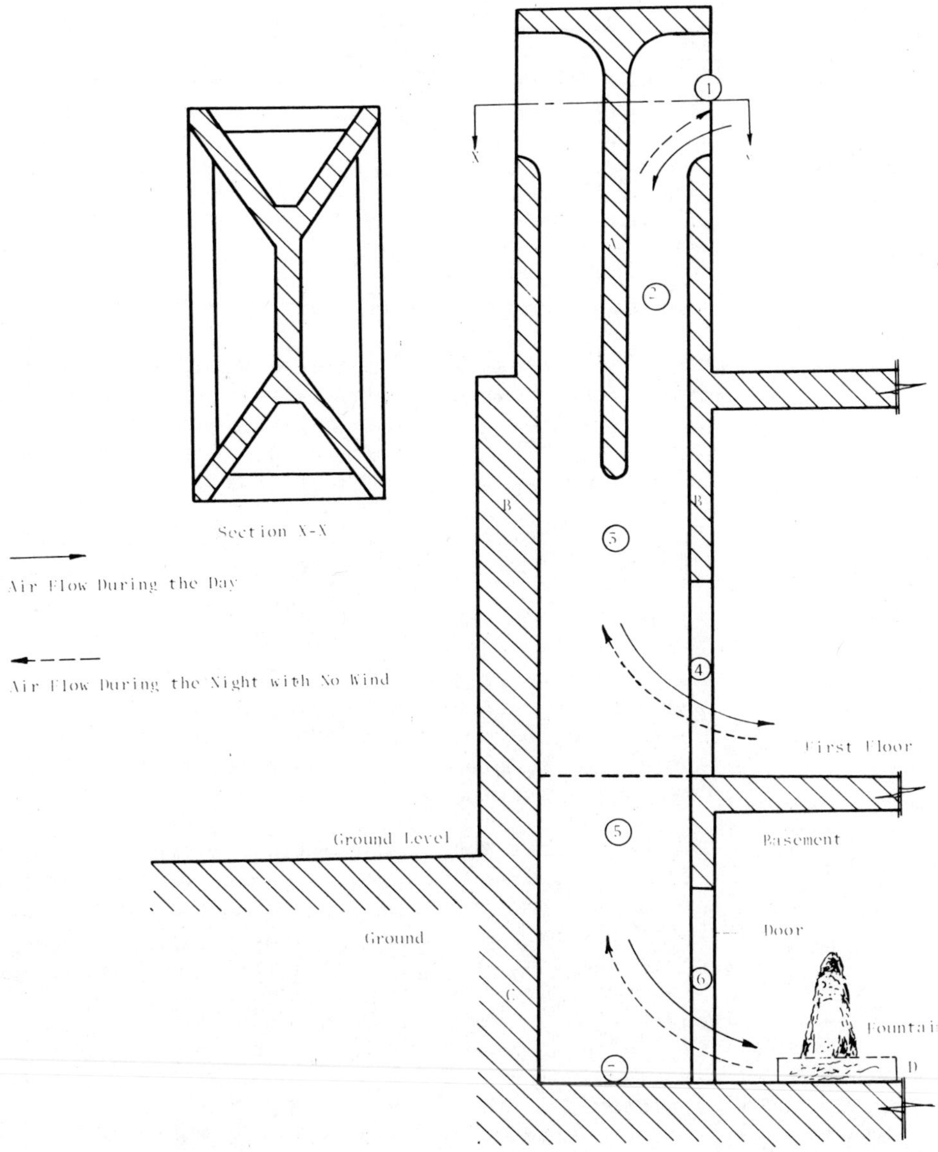

Two air flow patterns are recognized:

(a) No wind is blowing at night.
In this case the air tower acts as a chimney during the night. The cold ambient air is drawn through the rooms and enters the tower at openings (4) and (6), is heated by the warm masonary walls A and B of the tower and is expelled at the opening (1) at the top of the tower. This process continues all during the night, cooling the tower's structure to the lowest daily temperature, which is about 22°C lower than the maximum daily temperature.

If there is no wind blowing during the day the wind tower acts as a reversed chimney. The warm outside air in contact with the cold walls at sections A and B is cooled and is pulled down through the tower. When wind is blowing during the day a higher air circulation is maintained.

(b) There is a wind blowing during the night.
In this case cold night air is forced down through the tower and cools the walls at A, B and C. The operation during the day is the same as that described earlier.

The air is cooled sensibly by about 5-10°C as it passes through portions (2) and (3) of the wind tower and may be let out at the first floor through the door (4) or at the basement through the door (6). In the latter case when going through portion (5) further sensible cooling is accomplished. If the walls at C are moist due to water diffusing through the ground the air is further cooled evaporatively. In some places small pool and fountain D are provided in the basement for further evaporative cooling, especially when the walls at C are not very moist. In some other places the heat and mass transfer area C is increased by placing the wind tower head and the basement away from each other and connecting them by a long horizontal underground shaft.

The wind towers with their effective sensible cooling followed by an evaporative cooling produce air at a temperature a few degrees below the ambient wet bulb. In this way they are more effective, in addition to not using any active energy, than the evaporative coolers which have become very popular in Iran in the past 10 years.

In many of the houses and before refrigerators became popular the area (7) was used as a cold chamber.

3. Natural Ice Makers

Before the advent of mechanical refrigeration in Iran ice was made during the cloudless winter nights and stored in deep underground storages for summer uses. The ambient air temperature during the ice making period was mostly a few degrees above freezing point and the heat losses to the clear sky by means of radiation was mostly responsible for the solidification of water.

Figure 5 shows the cross section of one of the natural ice makers or 'Yakhchal'. It consists of a series of long shallow ponds A, with depths of about 40cm, dug into the ground and protected from the daily solar radiation by tall adobe walls B. A typical unit may consist of 5 or 6 ponds about 600m long in east-west direction with adobe walls about 12m high. The ponds' width was such as to receive no direct solar radiation during the winter days. The ponds were protected from early morning and late afternoon solar radiation by walls C, about 3m high, which were located at the ends of the ponds. The ice made during the night was cut and stored in the storage D which was about 15m deep. The volume of the storage depended on the unit's ice making capacity. (6)

Fig. 5
Natural ice maker, 'Yakhchal'

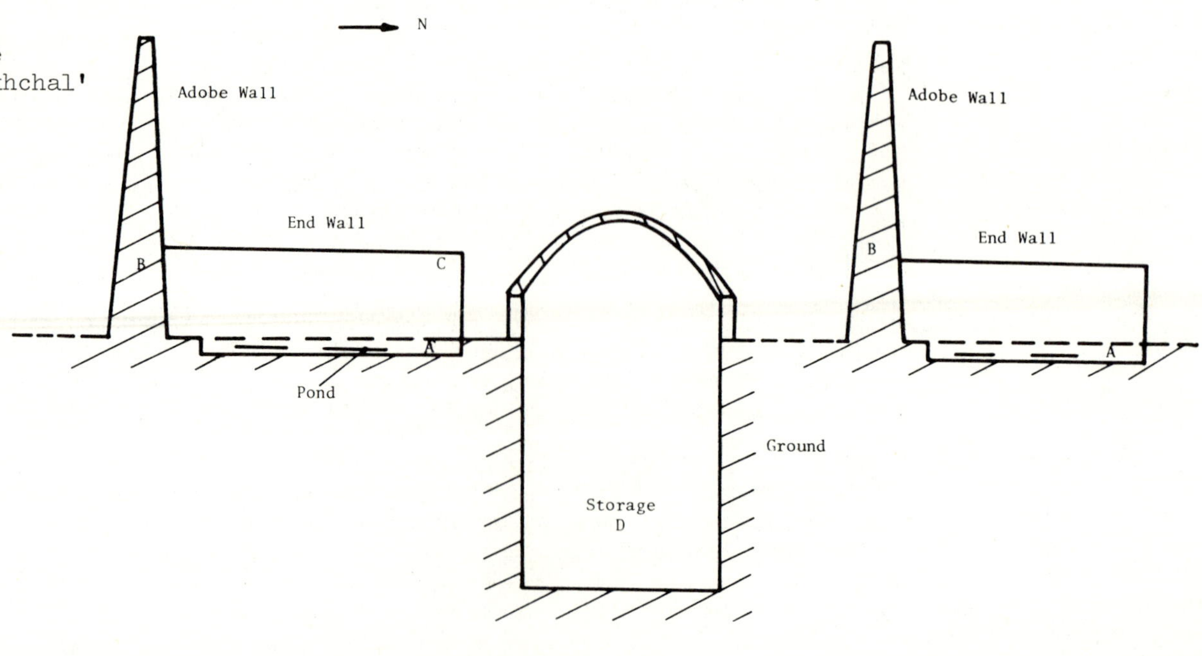

Other applications of solar energy

1. Solar desalination
There is a great deal of knowledge and data available on solar desalination. Most of this data is collected in relatively cold and moderate climates where the fresh water production is limited by the evaporation rate and design improvements are made to increase the evaporation rate (7). In hot climates, however, condensation, and not evaporation, is the limiting factor. A simple solar desalinator consists of a blackened tray which contains sea water and is covered on top by sloped glass or other transparent cover. Such a still is estimated to produce 1m3/m2 a year of fresh water in Shiraz, Iran.

2. Solar cooking
Solar cooking with parabolic concentrators has been tried unsuccessfully in India. The failures of these cookers were due to their lack of social acceptance. The solar cooker shown in Figure 6 consists of a flat plate collector which produces steam at atmospheric condition. The steam may be transported to the kitchen and condensed on the cooking pots. The food is then cooked very slowly at low temperature. This cooker seems to have good chances for social acceptance because it does not require constant focusing (it may only be oriented towards the sun twice daily for better performances), and makes indoor solar cooking possible, where the housewife can tend to her other household chores. A model of such a cooker with a collector area of 1.5m2 and with 3 panes of glass has been used to cook several Iranian dishes in Shiraz (8).

Fig. 6
Solar-steam cooker

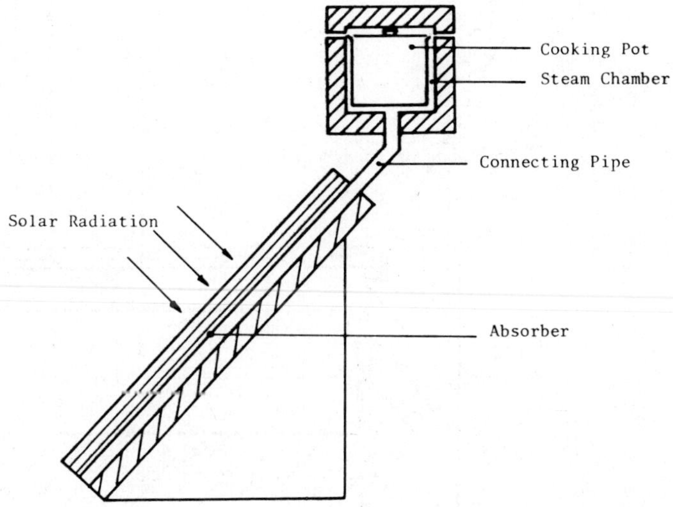

Fig. 7
Solar dehydrator

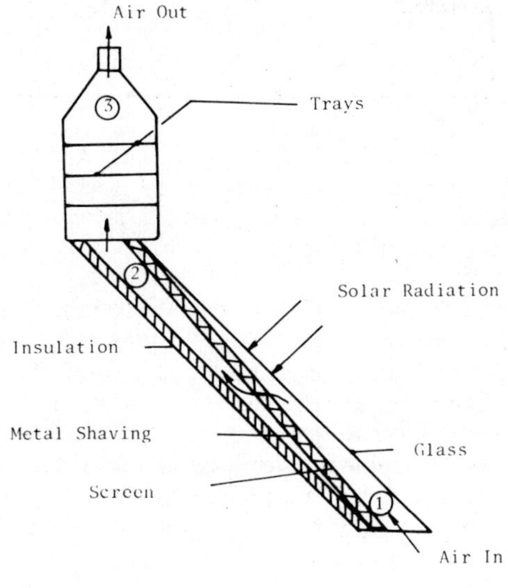

3. Solar dehydration
Fruits, vegetables, and crops are often sundried in developing countries. The products, however, may be contaminated by dust or insects. Figure 7 shows a design of solar dehydrator which has been used for the dehydration of fruits and vegetables (8). It is basically a flat plate collector with one sheet of glass cover. The absorber plate consists of metal shavings produced by a lathe machine which is stacked to a height of about 10cm. The solar energy absorbed by the metal scraps heats the air which flows through the absorber and the trays containing fruits and vegetables. The natural circulation of air through the system was found to be dehydrating more effectively than when a forced air circulation was maintained by a blower. This unit may be produced in different sizes and placed at the disposal of the farmers to produce much cleaner dehydrated agricultural commodities.

4. Solar sterilization of medical instruments
The rural communities in developing countries often lack well-equipped hospitals. The physicians in these locations boil their medical instruments in water for minor surgeries or inoculations. This process is not very effective for killing all the viruses, however.

To produce hot air at 160°C or wet steam at 121°C for complete sterilization of medical instruments solar sterilizers and solar autoclave were designed, as shown in Figures 8, 9 and 10 respectively.

Fig. 8
Solar sterilizer

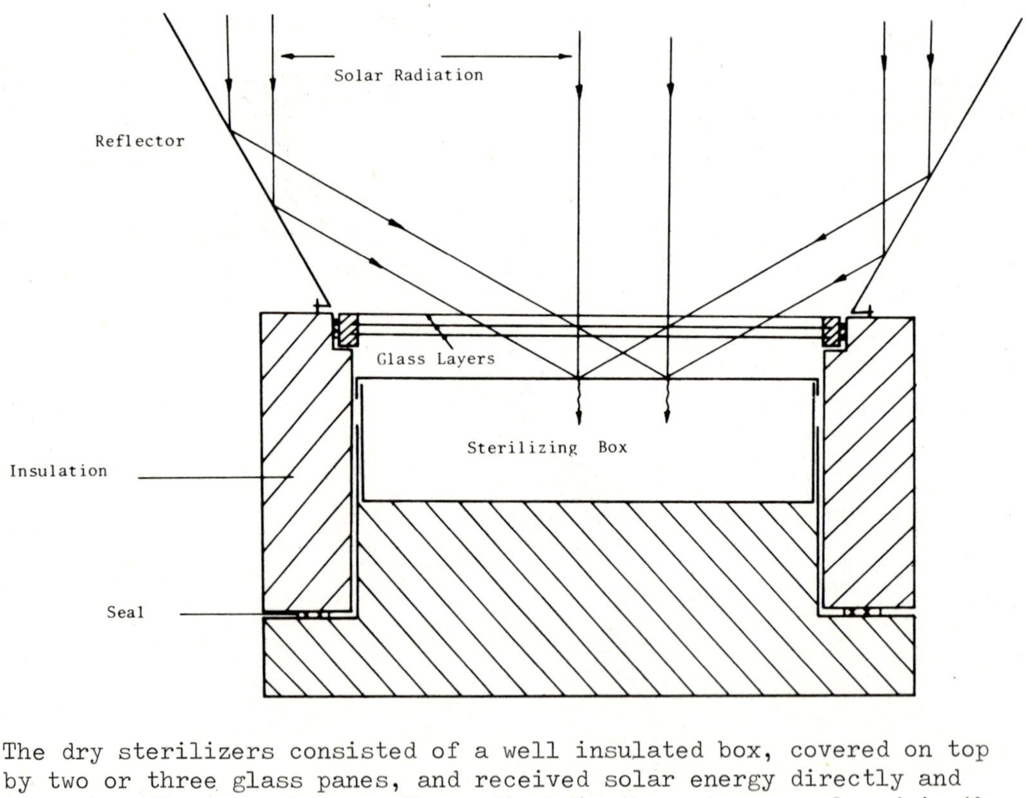

The dry sterilizers consisted of a well insulated box, covered on top by two or three glass panes, and received solar energy directly and through reflecting mirrors. The medical instruments were placed in the box and were held at a temperature of 160°C for at least one hour.(9)

The wet sterilizers utilized wet steam (similar to the autoclaves used in the hospitals), which was produced in the space between two coaxial cylinders and condensed in the inner cylinder. The outer cylinder was blackened and received solar radiation reflected by mirrors covering the inside surface of a truncated cone. The inner cylinder contained the instruments. The cylinders were pressure tight, so that pressures up to 4 atmospheres could be maintained. In this sterilizer it was only necessary to keep the instruments for about 20 minutes at 121°C in order to have a complete sterilization. (10)

Fig. 9 The Solar Autoclave, showing the mirrors, the sterilizing chamber and the wire-mesh container of instruments.

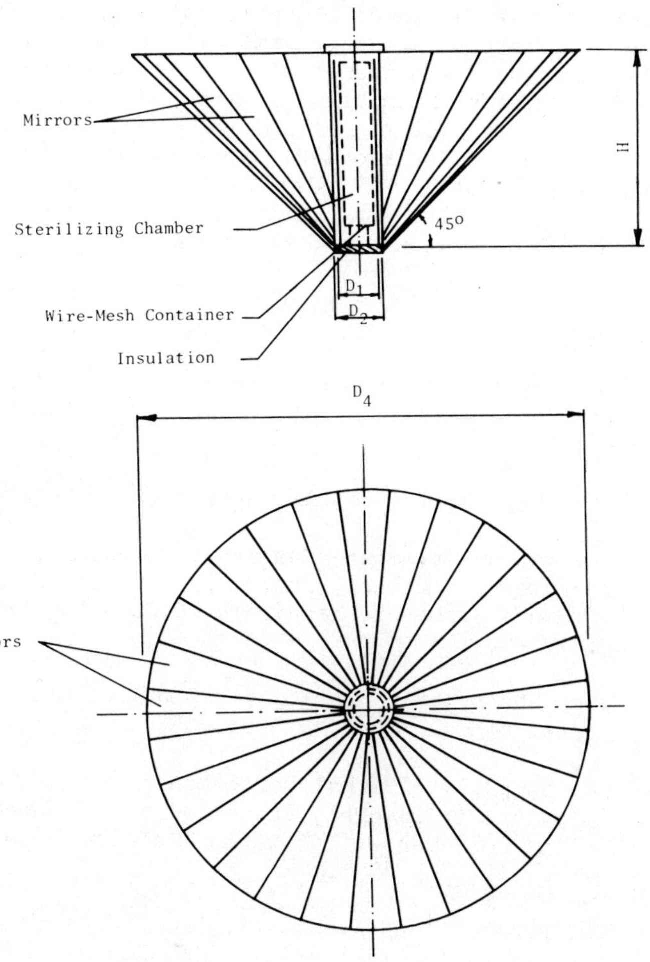

Fig. 10 The Solar Autoclave, showing the safety valve, air vent, thermometer and the focusing device.

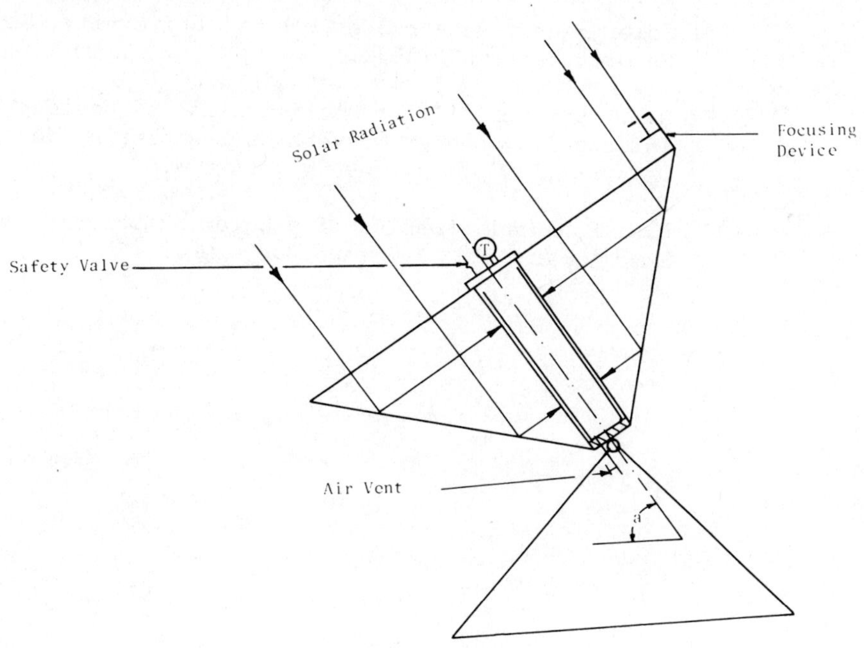

Conclusions

Extensive research and development in solar energy utilization are currently underway in developed countries. The developing countries instead of borrowing the solutions to their energy problems of their dwellings should consider their own building tradition and styles, especially those which are energy conserving. Energy conservation along with the passive solar heating and natural cooling systems should be investigated and implimented. Since the solar energy utilization technology is not highly advanced, developing countries have the opportunity to solve some of their energy problems themselves, mostly independent of the advanced countries.

References

(1) H.R. Hay & J.I. Yellott, 'International Aspects of Air Conditioning with Movable Insulation', Journal of Solar Energy, Vol. 12, 1969 p. 427.

(2) Steve Baer Residence, Albuquerque, New Mexico, USA. Personal visit.

(3) J.D. Walton Jr. Space Heating With Solar Energy at the C.N.R.S. Laboratory, Odeillo, France, Proc. Solar Heating and Cooling for Buildings Workshop, Washington D.C. 1973, Part 1, NSF - RA-N-73-004 pp. 127-139.

(4) M.N. Bahadori, 'A Feasibility Study of Solar Heating in Iran', Journal of Solar Energy, Vol. 15, 1973, pp. 3-26.

(5) M.N. Bahadoi, 'Thermal Performance of Old Buildings in Central Iran Employing Air Towers', Presented at the Passive Solar Heating and Cooling Conference, Los Alamos Scientific Laboratory, Albuquerque, New Mexico, May 1976.

(6) M.N. Bahadori and A. Kosari, 'Performance of the Natural Ice-Makers', to be presented in the 1977 ISES Congress, New Delhi, India.

(7) M.N. Bahadori and F.E. Edlin, 'Improvement of Solar Stills by the Surface Treatment of Glass', Journal of Solar Energy, Vol. 14, 1973, pp. 339-352.

(8) Solar Energy Research at Pahlavi University, Shiraz, Progress Report, September 1976.

(9) M.N. Bahadori, 'Solar Sterilization of Medical Instruments', Tenth Intersociety Energy Conversion Engineering Conference, University of Delaware, 1975, pp. 189-193, Part 2.

(10) M.N. Bahadori, 'Design of a Solar Autoclave' Journal of Solar Energy, Vol. 18, 1976, pp. 489-496.

Solar energy application for buildings in developing countries

6.3

E. Tedeschi
C. de Rosa, R. Vilapriñó
G. Lelio, E. Fernández
Instituto Argentino de Investigaciones
de las Zonas Aridas, Mendoza, Argentina

Abstract

Different cases of insufficient development produce contradictory effects. They can hinder new solar technologies as well as they can offer favorable elements. Other differences among the developing countries, concerning the use of solar energy in buildings, emerge from the physical, geographic and climatic situations. For these reasons solar energy utilisation must take place by taking into consideration the particular problems of each area, as in the arid zones of Argentina where it seems convenient to use passive systems and eventually active systems by air. The performance depends primarily on a correct design integrating the architectural qualities with the needs of solar energy systems. It follows several recommendations for the design of buildings in dry zones. Plans show the design of a middleclass house and a demonstration structure to be built in forthcoming months which will attempt to evaluate different passive systems and a low cost active system.

Introduction

The expression 'developing countries', although frequently used, neither defines an homogenous situation nor has a generally accepted meaning as to its limits; this is well known and has been discussed by specialists, so that it is not necessary to examine it here. In the particular case of the use of solar energy in buildings, the words 'developing countries' include a great variety of cases that must be considered separately in order to avoid unjustified generalizations and mistakes. The variation of cases depends on three factors:
(1) Lack of proper development;
(2) The availability of conventional energy;
(3) The geographic and climatic situation.

Causes and effects

The causes of an insufficient development can be social, political, economical and technological - generally a combination of all of them - and they are conditions for the possibility of using solar energy and the solutions that we can propose. If the cause is principally social and it affects low classes or groups having particular habits, like nomadism, the energetic spending in buildings will be very low, making inconvenient the use of solar energy at present, even if it could be useful for covering low spending uses, for instance to cook foods in the case of nomad people.

If the causes are political as, for instance, in cases when power is concentrated in small groups, it will be very difficult to obtain progress in the use of solar energy - although a great part of the population can be in favor of using it - because the dominant groups would prefer to restrain the development of the low classes way of life in order to avoid competitors.

If the deficitary development depends mainly on shortage of money, although there exists some technological development, it will be attractive to substitute the conventional energy by the solar one, but this will find serious difficulties due to the increased initial investment for the construction and equipment of the buildings, payable in a relatively long period of time before producing substantial savings for the consumer.

Finally, the technological deficit is accompanied commonly by social and political factors and not at any time by the economic ones. It may occur in countries with a relative high income per capita out with unequal distribution of it. It seems easier to acquire a foreign technology rather than developing a solar technology of their own. Therefore, it is probable that the technological solutions introduced from abroad will be accessible to a limited part of the people, while it would be possible to elaborate technologies of lower cost with local resources. Besides, if these developing countries are producers of conventional energetic resources, it is probable that solar energy could be considered as a future possibility with little present interest.

Briefly, a deficitary development can hinder the use of solar energy in buildings specially when the housing needs affect lower social levels without saving capacity. Nevertheless, the situation has the following elements in favour for the use of solar energy:

(1) The bioclimatic comfort standards for the inhabitants are lower than in developed countries. So, it is possible to use less complicated technology.

(2) People have not yet acquired habits in the use of conventional energy and therefore they do not oppose to the introduction of a new type of energy.

(3) In general, people are rural, or just urbanized, with a traditional experience in managing climatic elements to obtain bioclimatic comfort, and practice or know traditional technologies efficient from a thermal point of view, that can be revalorated in the construction of buildings or other types of housing.

(4) In many cases, the existing aim of technological advancement may lead people to accept and collaborate in using non traditional technology.

(5) In general, these countries have a low population density - although there are towns of great concentration - that makes difficult to build great centers of traditional energy production, since the profits required from sales and the high transport cost involved cannot repay the high investment.

(6) Besides, this type of investment requires imported equipment and it is only possible to buy it by international loans.

(7) For these reasons, there seldom exists a developed frame of traditional energy that can be a factor of resistance to the introduction of the solar systems in buildings.

Other differences among the developing countries, concerning the use of solar energy in buildings, emerge from the physical, geographical and climatic situations, since these countries cover about 60% of the populated area of the earth, in all latitudes.

The physical situation determines the magnitudes of the energy needs in buildings, for heating and cooling of spaces, hot water and distillation of saline water. It indicates also the possibility or convenience of satisfying them through the use of solar energy. The way of facing the problems and their solutions is very different and it varies for seashore or mountain areas, warm or cold regions, humid or dry climate.

We need also to know the daily regime of temperatures, the wind directions and speeds and the incident solar radiation.

At the same latitude, but at different altitude, we find the African continental desert zones and the Cordillera Andina: in both cases we notice a great variation of temperature between day and night and in both it would be convenient to store heat, for instance, by the use of pebbles. But in the African desert cold is accumulated during the night and it is used during the day for cooling the inhabited spaces; on the contrary, in the Cordillera, the solar energy is accumulated during the day for heating the inhabited spaces during the night. If the latitude varies more than the height, we notice that a collection system like the vertical Trombe-Michel wall, is efficient in the Pyrenees' latitude but unuseful in Ecuador's mountains, where the sun culminates at Winter noon at 65° on the horizon, instead of the Pyrenees 25°.

These examples make evident the differences in the use of solar energy from the point of view of the physical situation. If the differences of the development are considered at the same time, it is clear that generalizations are deceitful. We must take into account the local and special problems of the area.

Situation in the Argentine arid zones

In our Center* we analyzed particularly the use of solar energy in buildings of the arid zones of Argentina, mainly for low cost housing. As Argentina spans from the 22° to the 56° South latitude (without Antarctic territories) and its arid zones occupy more than 60% of its territory, with great differences in latitude, landscape and climate, we have given priority to the moderate-warm zones of continental climate; that is the West-Center area of the country and the most populated among the arid and semiarid zones, including some important towns. (See Figures 1 and 2 for solar energy distribution in Argentina). We plan to study the moderate-cold and warm zones successively. We have prepared some tentative designs - one of them is presented here - and we

---

*In the first two years of work we have gathered a good quantity of information by means of printed matter, inquires, meetings and assistance to conferences, visits to more advanced research centers - for instance in USA - and also receiving the visit of an expert from the University of New Mexico, Dr. Francis C. Wessling.

Our staff has attended short courses of physics applied to the use of solar energy and of thermal engineering in buildings. On the other hand, we have worked on some researches of local situations. One of our research subjects is the traditional housing of an Argentine arid region, in order to study an actual experience of simple technologies with local materials, fitted to the climatic conditions of the zone. In other work, we have analyzed some types of houses built by the Government in a semiarid zone, evaluating its behaviour from the point of view of their practical uses, thermal needs, economic conditions, etc. We have also studied thermal qualities of the elements of houses built according to the conventional technologies of the zone, and the availability of natural local materials suitable for economic buildings with a good thermal quality.

Besides, we have published information on solar energy applicable to Architecture in newspapers and reviews, and also in a booklet on the matter published by the main magazine of Architecture in Latin America.

We present this information hoping that it may be useful for a new research center.

Fig. 1, 2

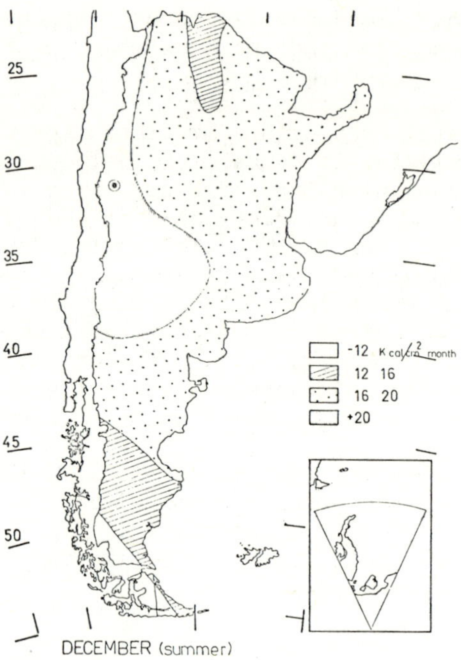

Solar radiation distribution in Argentina for December.

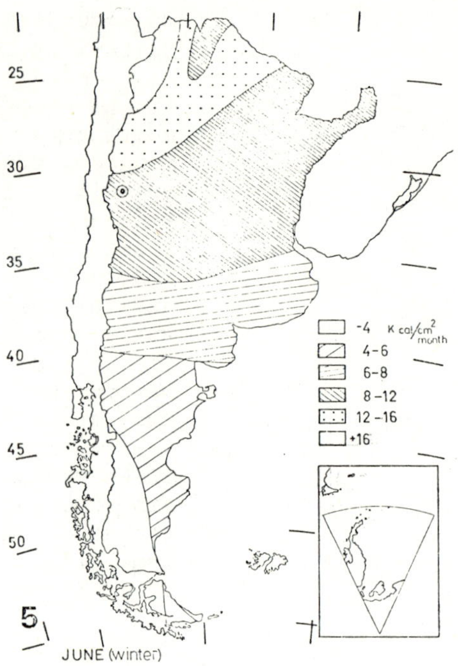

Solar radiation distribution for June.

have also elaborated the project of a demonstration structure. The first step for the design of the demonstration structure has been to choose the technology better fitted to the particular and concrete situation of our work, according to the condition, development factors and physical environment (Figs. 3-13).

In Argentina's arid zones of moderate-warm climate we find many positive factors for the use of solar energy. There is also in Argentina a noticeable technological advance but because of the constant inflation pressure - specially hard in recent years - capitalists look for short term investments and increased profit.

The cost of a house is beyond the means of a great part of the population; they have few saving possibilities and cannot use loans due to their high interest rates. Several factors may affect the use of solar energy in buildings, but the principal is the economic one; if the users are unable to pay the cost of a common house, they can less afford this cost increased by the use of solar energy, in spite of recognizing the long term benefit. The expected cost of an economic house is no more than 150US $ per m2 of construction, but a simple solar heating equipment - using water as medium - adds at least the 20% of the cost, with a figure of merit of approximately 0.7. In consequence the increment of the initial investment is high for a user of low financial capacity, and the sinking period is long. This is due to the fact that the country produces gas and oil, although not as much as to cover the national needs. The price of fuel is high when it is compared to the income of the workers, but not in the same proportion as that of metallic components used in solar equipment, which have to be imported at international prices. A complete unit of heating-cooling air conditioning can only be employed in upper class houses, its price being equivalent to the total cost of an economic house.

Concerning the environmental conditions, we find many favorable factors; dry climate, great variation of temperatures between day and night, scarce cloudiness, atmospheric transparency relatively affected by pollution, winds frequency, plenty of insulating local materials of low cost, specially in the Andean Cordillera, very rich in pumice stone. Besides, there is a type of traditional construction of 'adobe' that has good thermal qualities and is well accepted by the population; but its use is limited to non seismical zones of low cultural development, because in the more developed areas the technology of 'adobe' is disappearing and the common mason does not know how to construct properly.

Fig. 3
Solar house
Sketch

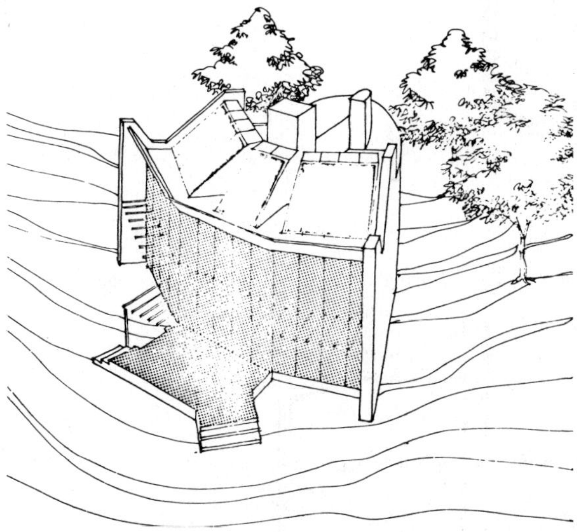

floor area	53m2
bedroom area	40
laundry area	5
basement area	12
north glazed area	56
hot air collector area	22
hot water collector area	5
storage capacity	8m3

Figs. 4, 5
Floor plan and
roof plan

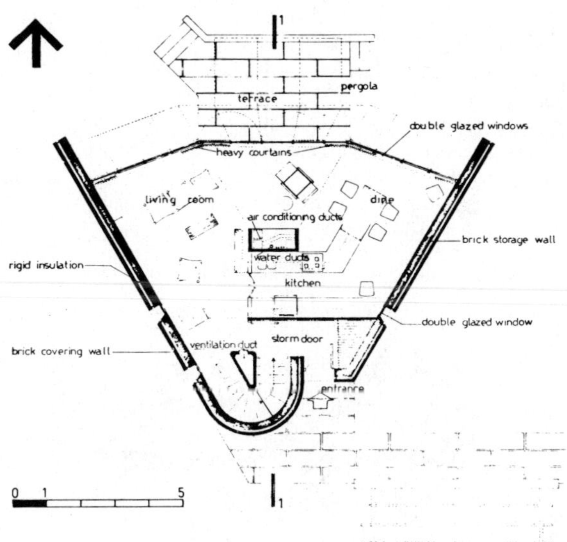

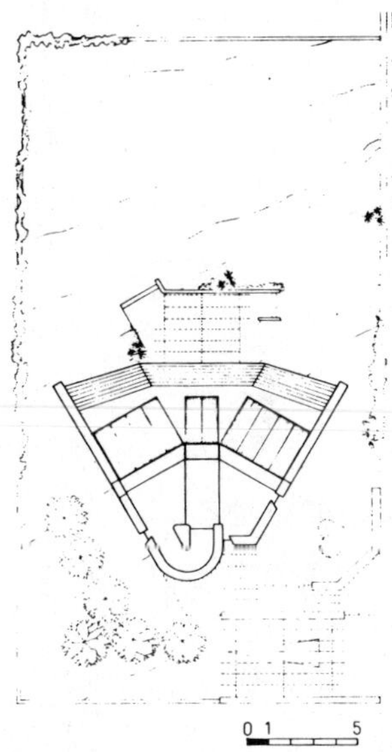

Figs. 6, 7
Bedroom plan and
terrace plan

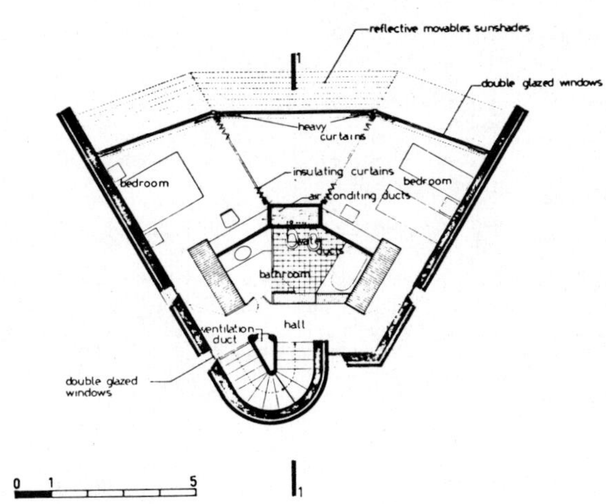

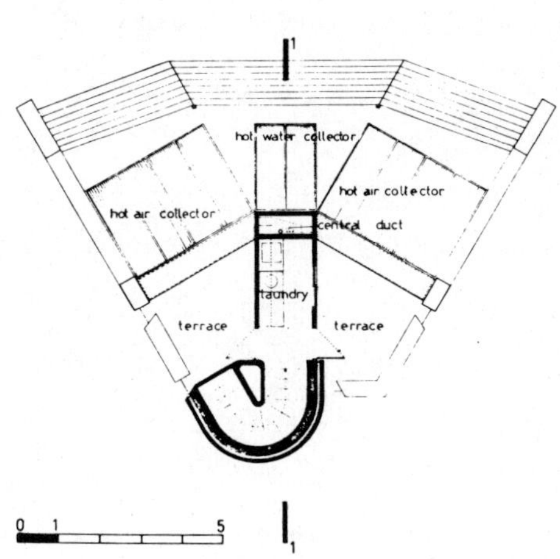

Fig. 8
Basement plan

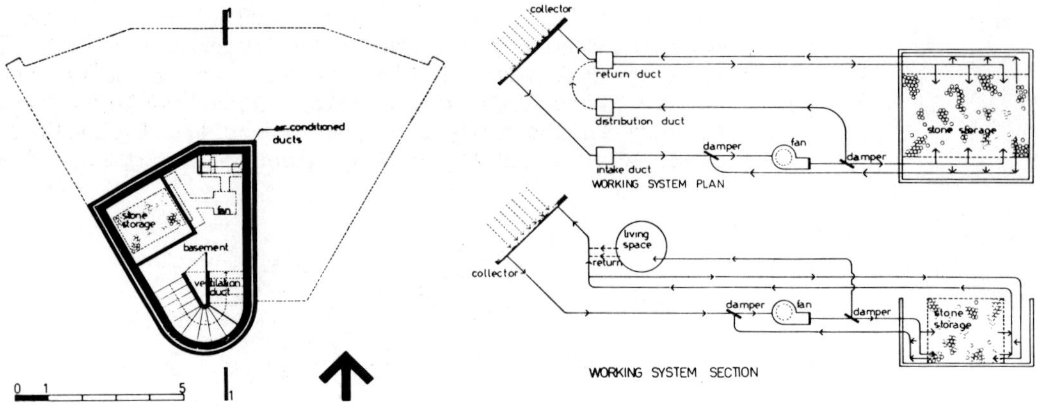

Fig. 9

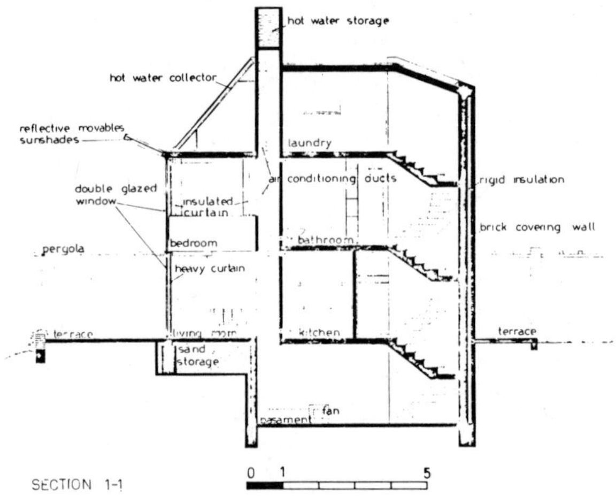

Fig. 10
Experimental solar house

Fig. 11

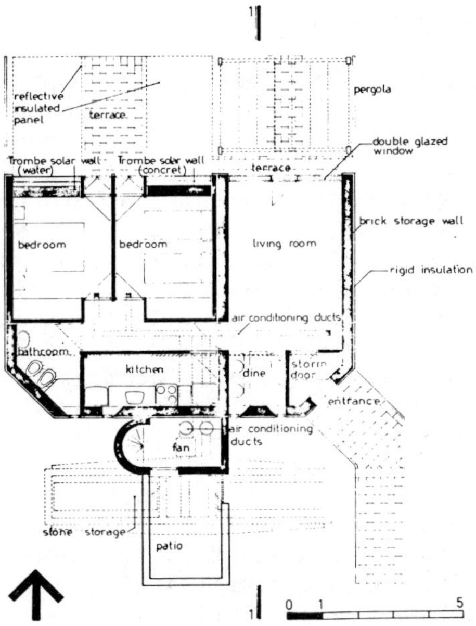

Fig. 12

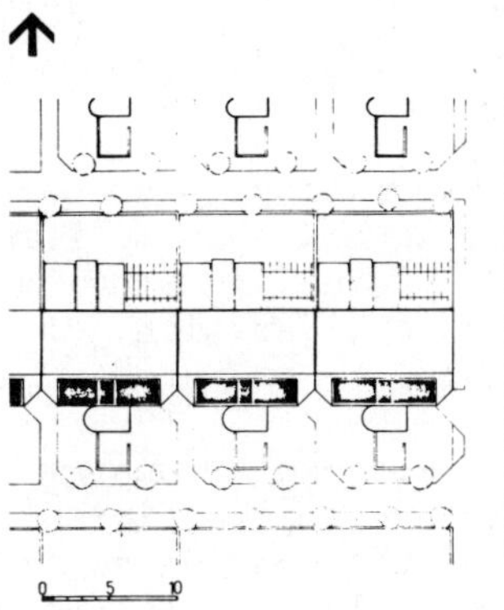

SITE PLAN esc 1:200

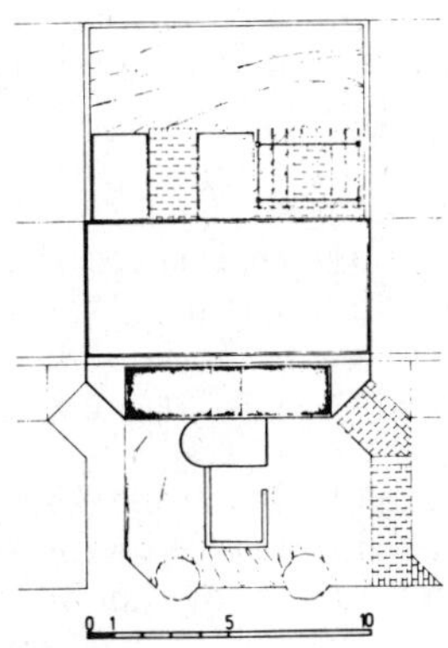

house area	68 m2
air collector area	16
water collector area	3
storage capacity	8 m3
north glazed area	32 m2
Trombe solar wall (concrete)	6.8
Trombe solar wall (water)	6.8
bedroom area	8.4
livingroom area	16
collector inclination	54°N

Fig. 13

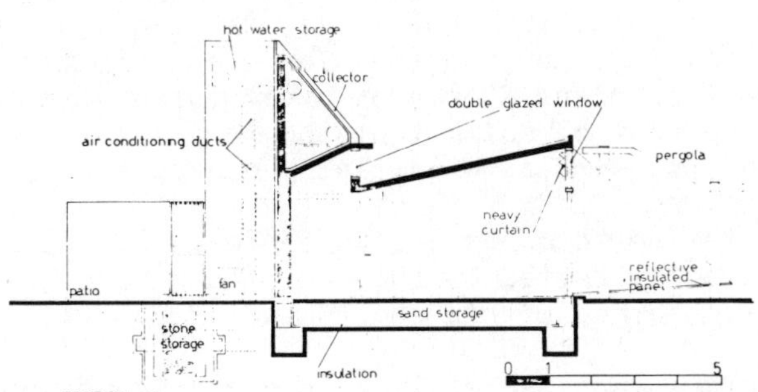

SECTION 1-1

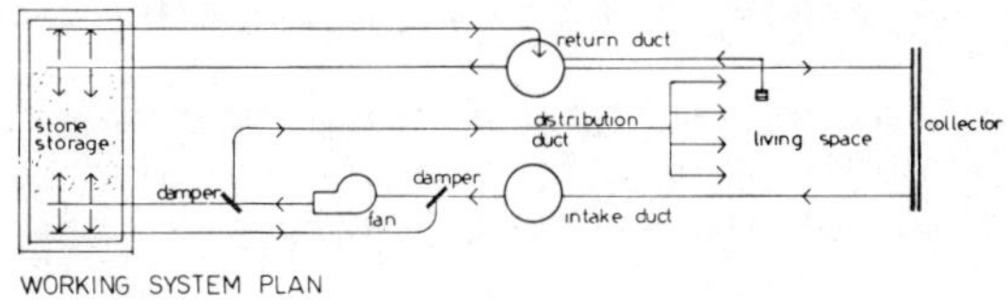

WORKING SYSTEM PLAN

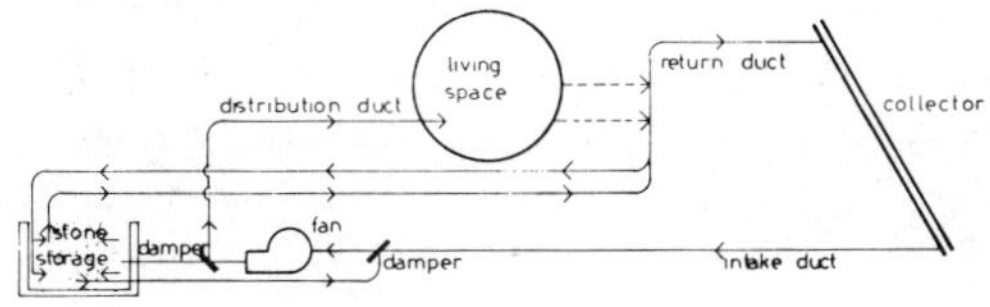

WORKING SYSTEM SECTION

Recommendations and conclusions

In consequence, taking into consideration the present development pattern and the environmental factors, it seems convenient to prefer passive systems, which make good use of solar heating through locally available construction materials. The success of these systems depends on an appropriate architectural design that integrates the functional, constructive and economical qualities with the needs of collection and storage of solar energy, either by direct gain systems, taking advantage of windows, walls, roofs, skylights, or by the addition of intermediate elements, as specially prepared walls. That can satisfy the specific conditions of a passive system, in which the collection and storage of solar energy are contained in only one element eliminating the transference from the collector to the storing which is an active systems' characteristic and the reason for the increment of its cost.

Balcomb & Hedstrom, from Los Alamos Scientific Laboratory have analyzed the performance of passive systems by a simplified model (1). In that work they have arrived at some quantitative basis for incorporation of the basic concepts into architectural design using passive solar heating systems. This study has greatly clarified the empiric criteria used so far by designers, and reinforces their projects. One important result of this research indicates the possibility of obtaining an optimized building design with a passive solar heating system that can achieve good performance, favorably compared to that of the common active systems.

Gathering results from other experiences plus our own research with the conclusions of Balcomb's study we are able to formulate a series of recommendations for the design of buildings using passive solar heating system that we have taken into consideration for our designs. These recommendations are as follows:

- The challenge confronting passive solar energy design is not one of heat collection, since a window is an efficient solar collector, but one of storage and control of the heat so as to maintain suitable comfort standards within the buildings. The paradox is simply put. Sensible heat storage requires a temperature change in the storage medium and yet the objective of building thermal control is to maintain temperature constant. If the storage is to be a part of the living building, how can these be made compatible? One approach is to minimize the inconsistency by brute force - that is, to use such a large mass heat capacity that the temperature variations are tolerable. A second approach is to partially decouple storage from the room (1)

- Much better performance can be achieved if storage is located so that it is directly heated by the sun rather than indirectly heated by sunlight absorbed within the heated space. (1)

- More storage mass is required in a passive solar heating system than in the conventional active system. A heat capacity of at least 30BTU/^{o}F-ft2 of glass (613KJ/^{o}Cm2) should be used. (1)

- Storage should be partially decoupled from the heated space (provided storage is directly heated by the sun as recommended above). In fact, there exists an optimum thermal coupling of roughly 1.5BTU/ft2 - ^{o}F-hr (8.5w/m2^{o}C). (1)

- The low thermal conductivity of masonry results in high daytime outside surface temperatures leading to increased heat losses. A high conductivity, high specific heat material (such as water contained in cans) will have a somewhat better performance. (1)

- For a given set of material properties, there is a thickness of masonry wall which will give a maximum yearly solar energy yield. This thickness is roughly one foot (0.3m) for a concrete wall. Other considerations, however, such as architecture, costs, and comfort will have an influence on the selection of wall thickness. (1)

- Use two sheets of glass for the windows collecting sun by direct gain or placed in front of the storing element and for the windows with East or West orientation.

- Use one glass with a movable thermal protection for the windows located on the opposite side to the collecting window.

- Use storm doors in the entrance or better a hall with storm door.

- The optimum value of U (thermal transmittance) between collection and storage will depend on the chosen storage heat capacity. (1)

- Maintain the allowable temperature swing around 3°C for houses. It may be larger in other buildings. (1)

- Water container as a drum wall, are better collecting-storing devices than masonry or concrete walls.

- If direct gain systems are used, the surface of windows for collection must be, at least, 25% of floor area and the area of the storage wall must be the same as the floor area.

- If the system of storage wall (Trombe-Michel type) is used, the area of collecting glass must vary around 50% of the floor area.

- Insulation on the exterior parts of the house, roofs, walls, floors, etc., keeps the storing mass into the house avoiding heat loss in Winter and heat storage in Summer.

- In passive systems, natural ventilation is very important. In Summer, to accelerate the cooling process at night and in Winter, to remove the eventual overheated air. So we recommend:
  (a) the size of the opening for ingoing air must be the same for outgoing air;
  (b) vertical distance between openings must be as large as possible;
  (c) the area of the openings cannot be inferior to 1% of the area of the floor;
  (d) there cannot be obstructions in circulation area. It may be convenient to accelerate air circulation using chimneys or ventilators. The overheated air can be stored, for instance by stone beds and ventilators.

- Try to place the building front in such a direction as to face the highest sun radiation at noon (South for North hemisphere and North for South hemisphere) with a deviation less than 15°; ensure that solar radiations will not be obstructed; try to protect the house from hard cold winds, to avoid heat loss.

- Choose compact shaped buildings to obtain an economic construction and to avoid heat loss.

- The communication inside the house must be easy in order to obtain a thermal equilibrium at low cost.

- Use thick and compact materials for interior walls to accumulate heat.

- Study carefully the position, form, size and materials of doors and windows.

- Calculate exactly the cost of the insulation and energy economy before the final design. In some cases the insulation could be reduced with a little energy increment which is dependent on extreme temperatures.

- In passive systems for developing regions a figure of merit should be used:

$$FM = \frac{\text{value of energy saved by the solar system in 5 years}}{\text{extra cost of house due to solar passive system}} = 1 \text{ or more}$$

- Heat losses are produced through windows, doors, roofs, walls, and floors. The most important losses are through windows (one glass) and by air infiltration. Windows cause the most important loss per m2 when compared with other surfaces of the house. In many cases this loss reaches the 30% of the total loss and the infiltration losses could reach similar percentage. The total thermal loss reaches 50%. The losses can be reduced by the half or a bit more if the windows are well assembled with insulating components and the infiltration reduced with weather-strips.

- The first consideration in the thermal design of the building should be to minimize the load within the constraints imposed by economics and architecture. Adequate insulation, double glazing, control of infiltration, and passive control of the solar gains are usually more cost effective than active solar heating. After the loads have been minimized by these techniques, then active systems might be considered to satisfy the bulk of the remaining load. (2)

- In passive system the heat storage is the house (Michel). So it is very important to place correctly the storage mass inside a well insulated skin.

- The floor and the soil beneath are useful storages, especially if solar radiation directly reaches the floor. The insulating effects, whether be perimetral - in the foundation - or below the floor are not clearly recognized. Therefore, Wessling et al. computation model indicates that the insulation offers advantages on good transmitter grounds by its nature or by its humidity, but its benefits are not appreciable in common grounds although the perimetral insulation improves the temperature distribution on the floor, which tends to diminish from the centre to the sides of the house without insulation. Apparently, the economy of energy does not compensate the insulation cost.

- Although the materials of major thermal capacity - thick materials - can store more heat than the less dense materials, its use in walls and roof must be accompanied by a good external thermal insulation. In case that thermal insulation does not exist, the loss of internal heat will be more rapid. External insulation is a basic factor for any passive system and can only be achieved by very thick walls and roofs, in this case it is convenient to build them with less dense materials.

- Mass provides natural thermal storage (4)

- Materials of high mass and low thermal diffusivity can delay the arrival of a thermal wave until the heat can be used effectively. (4)

- Close fitting heavy drapes that allow no air circulation near the windows probably have a resistance similar to storm windows.

- The temperature of the storage wall when it is delivering heat to a room, is 2 or 3°C higher than of the air inside the room, if the wall is thick and the house is well insulated.

- The walls that receive sun and store heat must be of dark colors, the other ones of light colors.

- Provide for auxiliary energy.

The fact that we are dealing with developing countries, may lead us to think that the formal aspects of solar houses do not present special difficulties, and that the proof of their economical advantages would be enough for their acceptance. This would also mean an over-simplification.

It is true that in many developing countries there exists a broad strata of population that need any kind of shelter at a minimal cost. When this boundary, formed by a level inferior to that of the possible users of solar houses, is overcome, a psycho-social factor has been verified in Chile, Argentina and other countries. The user sees in his house, even if it has been obtained by a state subsidy, an evidence of advance in his own social status, and wishes the house itself could express it.

It has been observed that many users try to improve the house, for instance by means of exterior decorations trying to imitate the appearance of those of higher classes, even before improving the conditions of use and comfort.

In the case of users of higher socio-economic levels, the meaning of the house as an expression of social status is obviously more evident.

On the other hand, the introduction of the new solar technology in architectural design poses to the designers a problem of changing of expression, in the some way that happened in other instances due to the effect of new technologies and materials.

In the developed countries the new technology shows itself by using active solar energy systems, joining artifacts to the buildings.

In the developing countries the change is referred more directly to the total design. In this case the desire to express the new technological situation by means of non canonic forms, may lead to a conflict with the wishes of the user, of showing-off his social advance, based on the repetition of well known forms, related to ways of life of the privileged social levels.

This situation poses some difficulties to the designers. It is possible that some kind of propaganda may be necessary to overcome this situation in order to assign a value of social status to the solar architecture itself, stressing the fact of its novelty and its non-polluting functioning. In any case the challenge to architects has been put forward.

References

(1) Balcomb, J.D. et al. 'Simulation analysis of passive solar heated buildings - Preliminary results' LA-UR-76?1719. Los Alamos Scientific Laboratory of the University of California. Los Alamos, New Mexico.

(2) Balcomb, J.D. et al. 'Design considerations of air cooled collector/rock-bin storage solar heating systems'. LA-UR-75-1334. Los Alamos Scientific Laboratory of the University of California. Los Alamos, New Mexico.

(3) Wessling, M.E. et al. 'Temperature response of a sunlit floor and its surrounding soil'.

(4) Balcomb, J.D. et al. 'Solar heating handbook for Los Alamos' Los Alamos Scientific Laboratory of the University of California. Los Alamos, New Mexico.

6.4

Feasibility of solar energy utilisation and research for buildings in developing countries

N.I. Ngoka MSc(Eng), Dip BldgSc, MIOB, AMBIM, AIWSc
Environmental and Tropical Design Consultant,
Towning Hill and Partners,
18 Great George Street, Bristol 1 United Kingdom

Abstract

The rapid increase in oil price has highlighted the world's energy crisis. As we need energy in order to maintain and improve our standard of living, there is a need to explore alternative sources of energy. An attempt has been made in this paper to examine the world's energy problems and its effect on the developing countries.

Without neglecting the fundamental thermal design problems of buildings in the developing countries, most of which lie within tropical climate, their possible utilisation of solar energy is discussed. There is a need for further research and co-operation between the developed and developing countries.

Introduction

The rapid increase in oil price has had a great effect on the world's energy crisis. According to Norton Belknap, vice president of planning for Exxon Corporation, the worlds biggest oil company, "The Arabian peninsular is the dumping ground of the worlds energy problems. After 1980 all the growth in OPEC output will come from the peninsular countries, yet the policies of these countries with regard to their oil reserves is impossible to predict. How the conflict between the world's demand for more oil and the Arab's possible reluctance to exhaust their supplies at the rate demanded by the West is resolved is one of the cosmic imponderables of the world today". OPEC - the Organisation of Petroleum Exporting Countries - an anachronism which remained mainly unknown to the world in general for the first 13 years of the organisations life, is now a household word. The true potential of this organisation as a politico-economic power and its strong influence on the world's energy needs became apparent in October 1973. After 13 years during which the price of a barrel of oil had changed from $1.80 (Jan. 1960) to only $2.90 (June 1973), the price was quadrupled to $11.65 a barrel in January 1974. The cathartic aftermath of this quadrupling has had a great impact on the West. Fig. 1 illustrates the magnitude of oil revenue for the main producing countries. As most of these are developing countries, their oil revenue are now being used to fund massive industrialisation plans aimed at producing a sophisicated economy no longer dependant on oil by the time the reserve of crude oil runs out. For example, Iraq has a five-year development plan costing an estimated $30 billion and Nigeria has committed itself to a $20 billions five-year plan. Due to the need for improved living conditions in these countries, substantial funds are usually allocated for housing and buildings.

Although at present we rely heavily on oil for our primary energy, it is evident that we can only do so for a few more decades.[1] Apart from oil, other useful fossil fuels are coal and natural gas. At present over 98% of our energy comes from fossil fuels, but however large, its stock is finite. There are however other sources of energy available to mankind. As this paper relates to the developing countries, it is essential to mention that muscle power represents an important source of energy in most developing parts of the world. Nuclear energy has been developed, but is not universally welcomed as the construction of the plant is energy-intensive and the disposal of its long-lived active wastes still poses problems. Geothermal, solar, tidal, and wave energy are also significant where conditions are suitable.

Fig. 1 Oil revenue of the main producing countries (Source: The Economist Intelligence Unit Ltd)

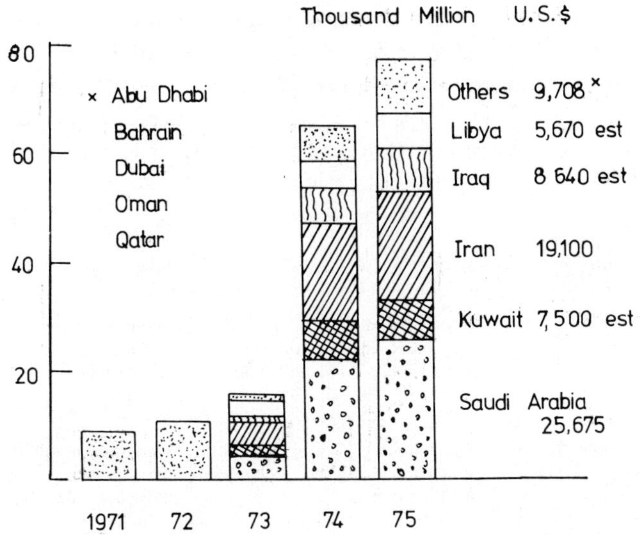

It is a common practice to highlight capital sources when discussing future energy supplies. In assessing a possible energy source, it is essential to ascertain that it is of sufficient magnitude to satisfy our needs and that the means of its exploitation exists or can be developed. The economics of the exploitation and application of the system should be compared with other existing systems.

2.0 World Energy Demand.

It is a generally accepted view that the amount of energy used throughout the world is increasing quite rapidly. When compared with the rest of the world, a high per capita consumption has always been a way of life in America.

In order to assess the efficiency of our energy consumption, let us consider the American situation. About three hundred years ago, it was said that each American family required half an acre of woodland so as to obtain sufficient wood for fuel. Yet its per capita consumption of fuels has increased by only a factor of 2.3 in the last 150 years. This small increase in consumption in a century and a half can be attributed to the efficient conversion of fuel into usable energy. An example of this improved efficiency is found in generating electricity. In 1840 the thermal efficiency of power station was 4%, at present most power stations can attain an efficiency of 40%. By using oil instead of wood for home heating it is possible to obtain an improvement in efficiency from 8% to over 50%. However, it may not be possible to increase efficiencies infinitely as they are at present very close to theoretically established limits. It is therefore correct to assume that future improvements in the standard of living will depend on increased energy consumption. Examination of energy problem in the universe was carried out by Dyson (2). In his findings, he emphasized the preponderance of gravitational energy on a cosmological scale. As the second law of thermodynamics gives us a quantitative measure of disorder through the entropy, Dyson classified the forms of energy in accordance with their entropy per unit energy. This is illustrated in Fig. 2. Low values in this figure denote a high degree of order which will allow entensive abstraction of energy, whereas a high value for the entropy per unit energy signifies a state of disorder with comparatively low potential for useful energy conversion.

Energy consumption has changed in amount and distribution with time. This is illustrated in Figure 3 by comparing the per capita energy used in United States, India, the rest of the world in 1900 and 1970, and the primitive man. For the rest of the world, the per capita energy demand is lower than that of the United States by about a factor of 6, and in the case of India it is lower by a factor of 30. When compared with the primitive man, India uses only about three times per capita energy. This is the case with most developing countries of the world. As energy contributes to our standard of living, it is essential to increase the energy demand of all developing countries.

Fig. 2 Energy forms and their practically achievable utilization efficiencies.

Energy forms	Applications	Practically achievable efficiency, per cent	Fundamental limitation in utilization efficiency
energy released by relocation of material bodies	flywheels to produce translational motion	~100	energy conservation (first law of thermodynamics)
energy released by nuclear reactions	nuclear reactors to generate electricity	35 to 45	Carnot efficiency for the thermal cycles used
energy released by chemical reactions	fossil-fuel power plants to generate electricity	35 to 45	Carnot efficiency for thermal cycles used
	fuel cells to generate electricity	80 to 120 for the definition of efficiency used	energy conservation
solar radiation	photosynthesis	~1	energy conservation
	solar space and water heaters	5 to 20	Carnot efficiency for the thermal cycles used
	wind energy to generate electricity	10 to 60(?)	energy conservation
	hydroelectric power conversion	60 to 80	energy conservation
	photovoltaic cells	1 to 20	energy conservation
fusion energy	the goal is electricity generation	probably 35 to 45	Carnot efficiency for the thermal cycles used

Fig. 3

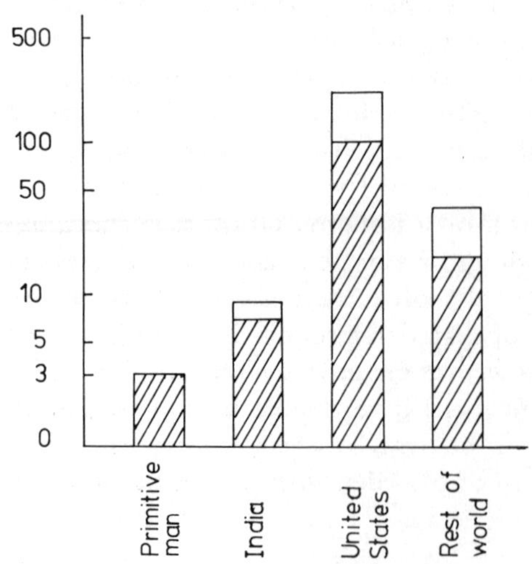

Growth of energy demand. (Shaded area, to 1900; unshaded area 1900-70)

3.0 Fundamental climatic building design for the developing countries.

We have seen earlier in this paper that energy contributes to our standard of living. As far as building design is concerned, it is essential to attain a comfortable living condition and to achieve substantial functional performance. As an illustrative introduction, house's prehistoric precursor could be used to demonstrate mans energy need for comfort. Assuming that we lived in a cave located within a tropical climate, it is likely that the temperature would be stable and no energy may be required for space conditioning. On the other hand, if we are to live in a cave within a temperate climate, we may require heating all the year round as the ground temperature is usually about 13°C.

If we live above ground we are faced with diurnal and yearly temperature variations. Heat input from solar radiation, heat losses to the ground and to the atmosphere, as well as air movement will affect the thermal environment of buildings. Fig. 4 illustrates some internal and external environmental factors which influences thermal comfort in a building.

Fig. 4

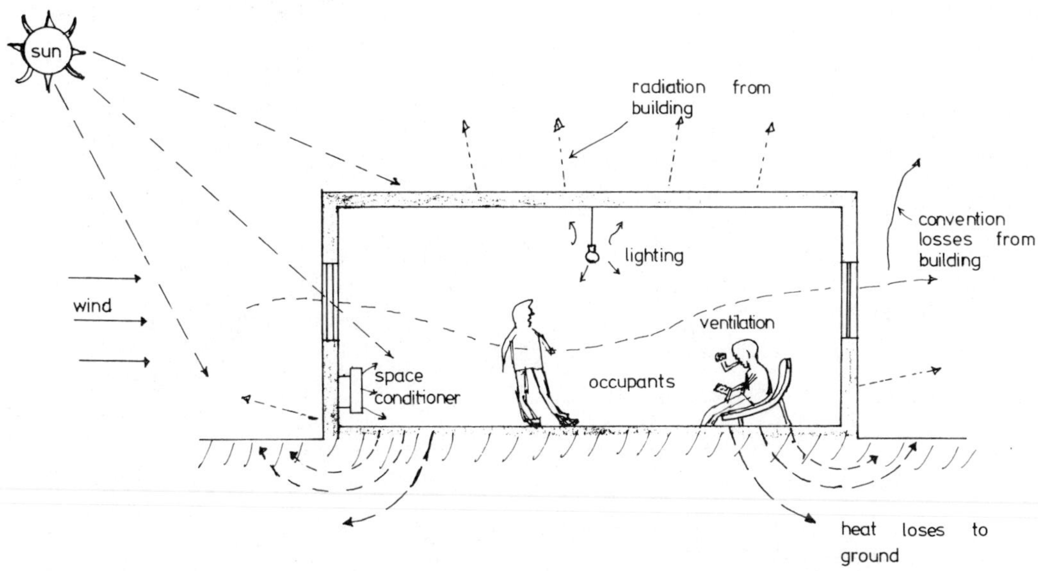

Internal and external influences on heat balance of building.

Generally speaking, almost all developing countries have tropical climates. (3) The design of tropical buildings differ in nature with those in temperate climates. In order to maintain suitable environmental comfort levels, a building envelope should act as a suitable modifier or filter of the external climate. Orientation, shape, form, materials, economic and social conditions are some of the important parameters (4) to be considered by anyone designing for the tropics. An understanding of the thermal performance of buildings in the tropics poses a difficult problem. This is due to the dearth of unco-ordinated research data and the complexity of the problem. Van Straaten (5) has shown that the main factors which determine the thermal response of a building under conditions of large diurnal variations are the thermal resistance and the heat storing potential of its various components. For heavy-weight and light-weight structures exposed to variations of external temperatures and solar radiation, the fundamental difference in their thermal response will depend on their heat storing capacity. For heavy-weight components, a good amount of heat can be absorbed without a significant rise to its temperature. However, light-weight materials will transmit heat almost directly to the interior of the building. The degree of heat gain or loss through light-weight components is therefore mainly dependant on their thermal resistances, but for heavy-weight components, it is dependant on both the influence of heat, storing capacity and thermal resistance. Fig. 5 shows a schematic representation of the differences in thermal response for both structures. From this it will be seen that for hot-humid climates, a light-weight construction has great advantages over heavy-weight one. The converse applies for hot-arid climates. However for hot-humid areas, a combination of light-weight and heavy-weight construction may prove satisfactory in certain cases. For instance, heavy-weight materials may be used in the east and west walls of a house when these walls cannot be shaded, while the shaded north and south walls could be of light-weight construction with sufficient openings for through ventilation. (6)

Fig. 5 Schematic representation of the variation of indoor air temperature for lightweight and heavyweight structures

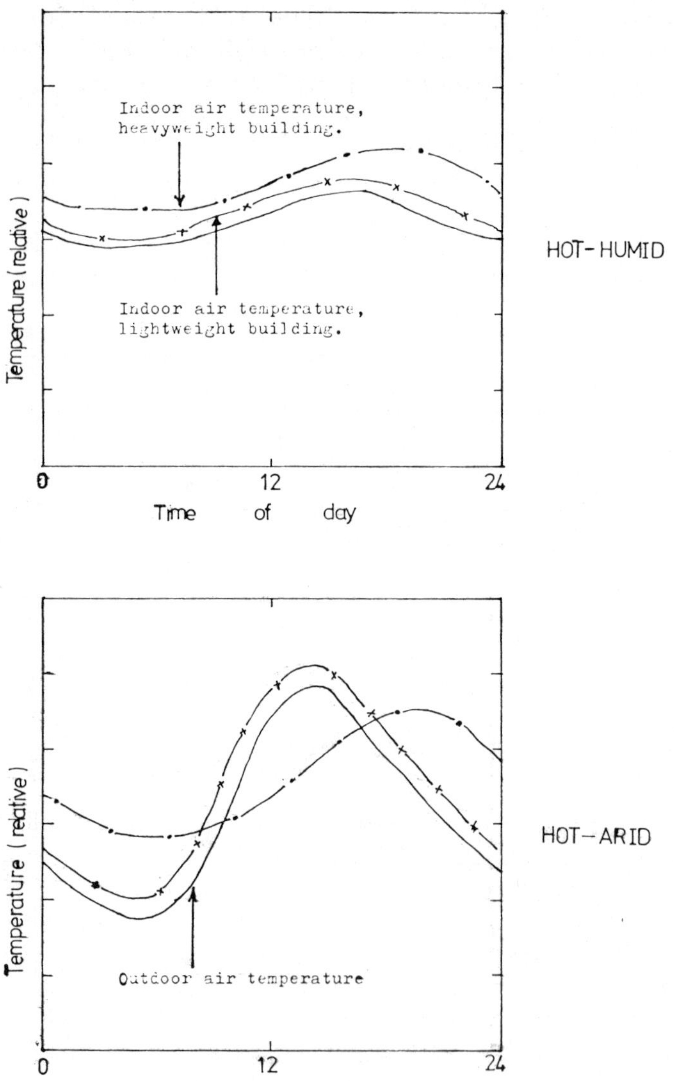

It is practicable to attain a reasonable degree of control of internal room temperatures and air velocities by adjusting the details of design and materials to the prevailing climatic conditions. A summary of possible variations in indoor climate due to the influence of the external environment is given (7) in Fig. 6. The elimination of solar radiation from the interior of a building is the most important design factor. Various types of shading devices are now available for controlling heat transmission through window openings. To achieve a

Fig. 6
Range of variation in indoor climate

The climatic variable	Range of variation
Solar radiation absorbed in the walls	15 - 90% of incident radiation
Solar radiation penetrating through windows	10 - 90% of incident radiation
Indoor air temperature amplitude	10 - 150% of outdoor amplitude
Indoor maximum air temperature	-10 to + 10°C from outdoor maximum
Indoor minimum air temperature	0 to +7°C from outdoor minimum
Indoor surface temperature	-8 to + 30°C from outdoor maximum
Average internal air speed, windows open	15 - 60% of outdoor wind speed
Actual air speeds at any point in room	10 - 120% of outdoor wind speed
Indoor vapour pressure	0 - 7 mmHg above outdoor level

satisfactory design of these devices, accurate solar angle prediction techniques should be used. The orientation and shaping of a building also provides a means of reducing solar heat gains.

In designing buildings for tropical climates it is essential to obtain adequate information on aspects of external climatic conditions which may affect the physical performance of the building and human comfort. Information such as sky conditions, the diurnal and annual temperature ranges, humidity, annual rainfall, radiation heat loss or gain, airflow characteristics, averages and extreme conditions and finally the microclimate of the location of the proposed building, will serve as essential design guide.

4.0 Solar Energy and Collectors.

Before considering the possible use of solar energy for buildings in developing countries, it is necessary to attempt a brief description of this natural energy source and its collection techniques. Although the sun is an enormous energy source, its energy density at the surface of the earth is relatively small. At the highest part of the atmosphere, the energy flux of the solar radiation is about 1.35 kW/m^2, with a spectral distribution resembling that of a black-body radiator at approximately 5800°K. Due to a variety of scattering and absorbing processes in the atmosphere, the maximum flux is reduced to about 1 kW/m^2 at the surface of the earth. A majority of the ultra violet radiation is taken away by absorption processes which is involved in the oxygen-ozone balance above 25 Km. There are also considerable fissures in the energy spectrum in the infra red radiation due to H_2O and CO_2 absorption. Although there is some scatter by atmospheric constituents and pollutants in clear conditions, the principal agencies for scattering are droplets of water in the clouds. Most of the scattered radiation reaches the ground, even in cloudy environments. Radiation is generally received at the surface of the earth in either direct or diffused form. Diffused radiation cannot be brought to a focus on a reflector.

Fig. 7 Some aspects of energy flows (expressed as equivalent continuous power inputs) to and from the earth are shown. The solar power input is seen to be far greater than the tidal or geothermal power.

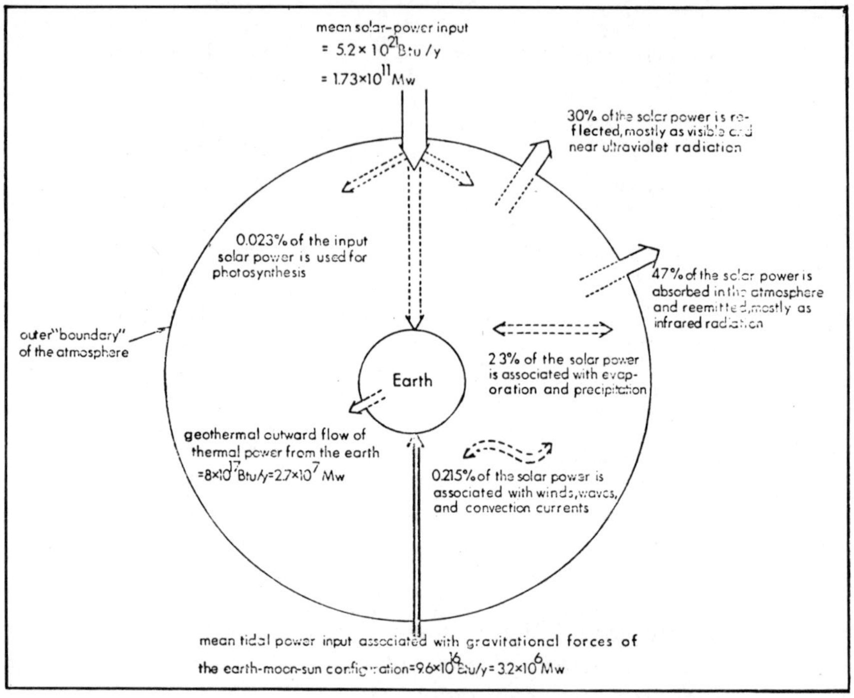

The wavelength of radiation determines its energy content, with shorter wavelengths representing higher grade energy. The intensity of solar radiation is less when the sun is at low altitude angle. Because of the inclination of the earth's axis to the plane of its orbit, the maximum altitude of the sun and the length of the day vary substantially with latitude and season. Fig. 7 shows the distribution (8) of the solar energy received by our planet, giving details of some energy flow to and from the earth.

For building design, it is usually possible to obtain solar radiation data from meteorological stations. However in the absence of this data, it is possible to estimate (9) the daily total radiation if the duration of sunshine is recorded, viz:

$$Q = Q_{sc} \left(0.29 \times \cos \emptyset + 0.52 \frac{n}{N}\right)$$

where Q = daily total radiation on horizontal plane (Wh/m^2 day)
Q_{sc} = solar constant per day.
$\emptyset$ = geographical latitude.
N = possible sunshine hours/day.
n = actual sunshine hours/day.
the value of Q_{sc} can be taken as 9830 Wh/m^2 day.

Devices for collecting solar energy usually consists of glass window and an absorbing surface with fluid underneath to transfer the collected heat. These surfaces could either be plane or concentric cyclinder. The choice of shape depends on whether direct solar radiation or radiation from a concentrating device such as a circular or parabolic cylinderical reflector, is to be collected. The flat-plate type shown in Fig. 8 is usually the simplest active collector. In this system, radiation goes through a transparent cover plate and is then absorbed by the collector panel. From this panel, heat is communicated to a fluid flowing in the

Fig. 8
Flat plate collector and water-heating system

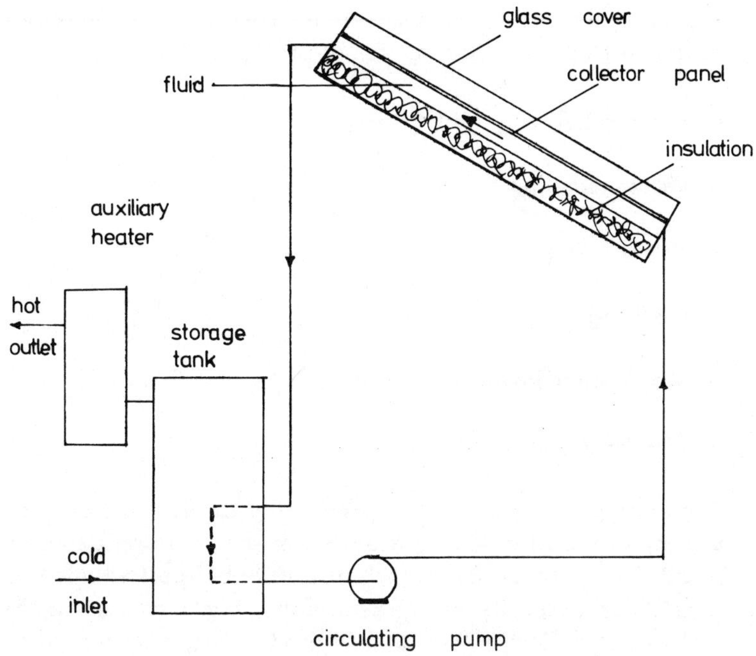

enclosure in contact with the panel thereby energy is extracted from the system through the fluid for use elsewhere. Several factors will affect the efficiency of any device used. It is therefore essential to consider the efficiencies to be expected from solar collectors before chosing a system. Collector efficiency may be computed by the following formula:

$$E = 0.71 - 0.0054 \Delta T ^{o}C$$

where E is the daily collection efficiency and ΔT is the temperature rise above ambient.

5.0 Solar energy in buildings for developing countries.

The fundamental design requirements for buildings in the tropics which covers most of the developing countries have been described in section 3 of this paper. We shall now examine the feasibility of using solar energy for improving the living conditions of people in these countries.

The design of buildings for the tropics calls for adequate protection from solar radiation. So if we have to make use of this 'unwanted' natural energy it is essential that adequate precaution should be taken in order not to defeat our objectives. If we are to harness the sun's energy, effective insulation devices should be incorporated in the basic building design methodology. For example, if a system is to be adopted for room airconditioning, it will have a negative effectiveness if the system gives rise to increased heat gain into the building, with higher initial construction and running cost. Taking this case further, Fry and Drew (10) give examples of how reasonable comfort can be obtained in tropical buildings without having recourse to expensive air-conditioning plants. One of these examples illustrates the use of sunbreakers. With this in mind, it is essential to carry out a thorough cost effectiveness of any solar system to be adopted for tropical buildings. A suitable formula for computing this cost effectiveness is given in Appendix 1.

Assuming that cost effectiveness is satisfied, solar energy could be of practical value in tropical buildings for the following:

- Air conditioning and cooling.
- Water pumping.
- Water heating.
- Cooking.
- Water distillation.
- Electricity production.

The energy requirements for performing above tasks vary a great deal. As solar energy may be limited by the area of the building exposed to the sun and sky, the feasibility of attaining practical functional performance of any device should be considered carefully before adoption. For example, as the energy requirement for distillation is in the form of low grade heat, a system of this nature will be very suitable for application in developing countries in areas where fresh water supply is difficult to obtain. On the other hand, due to a high energy demand for air-conditioning, a room air-conditioned with solar energy needs a collector surface equal in area to the floor space of the room itself. Similarly, a lot of energy will be required to produce electricity.

For communal buildings in the rural areas of developing countries where power supply and other essential services are non-existent, solar energy could be adapted to meet with the villager's basic needs. A good example is the use of solar energy for water pumping in a remote desert school in Mauritania. (11) This project which was based on the work of Professor Masson of the University of Dakar consists of solar mechanical pumping installation using a cheap semi-concentrating aluminium roof trough system as collector. The system diagram of this installation is shown in Fig. 9. This system has a pumping height of 20m with a pumping rate of 8-10 m^3/h and operates 5-6 h/day.

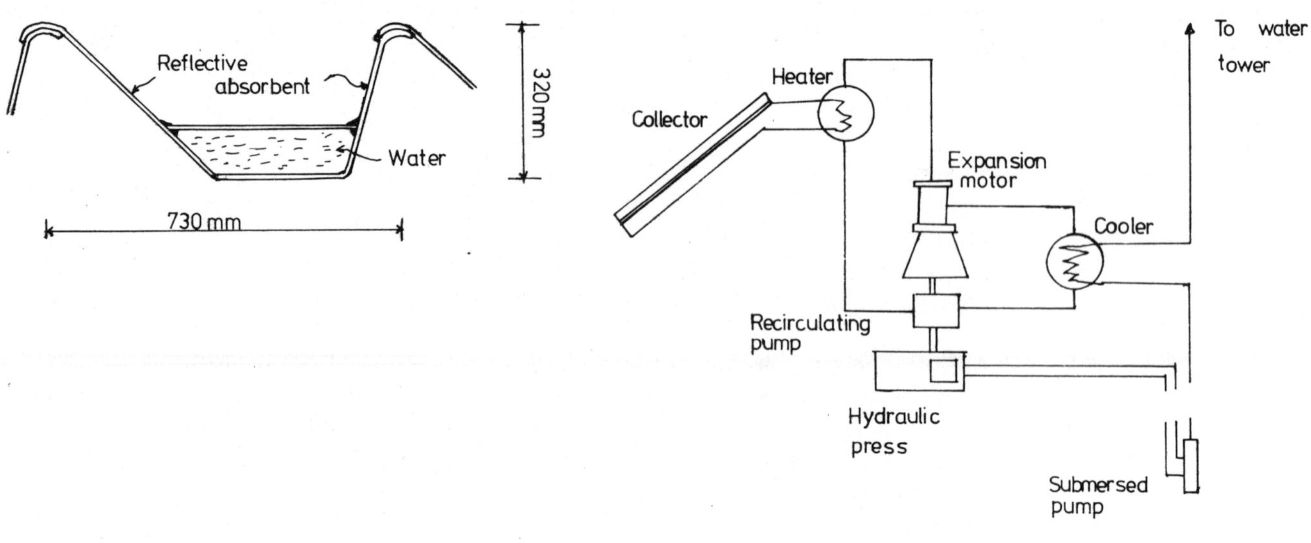

Fig. 9 Collector shape and system diagram for a solar-mechanical-pumping installation in Mauritania.

6.0 Solar energy research and building design.

Generally speaking, the fundamental design principles for buildings in the developing countries are well known by now. The thermal performance of buildings of given design and construction can be predicted with some accuracy by using either mathematical or physical models. Difficulties may however arise in the design and prediction of thermal performance of solar energy buildings in the tropics. This is mainly due to the fact that we now need to attract more solar radiation on buildings of this nature instead of aiming at offering protection against the sun. Further research is therefore needed in the following areas:

- Improving design methods and standards of non-solar energy buildings.

- Improving component insulation in areas of the building where collector surfaces are mounted.

- Cost effectiveness of solar systems.

- System efficiencies.

- Suitable solar energy systems for the developing countries.

7.0 Co-operation between the developed and the developing countries.

There is a need to increase the energy demand in the developing countries if the people in these countries are to achieve improved living conditions. The greater energy use by the developing world have a long term implication; such as increased drain on energy resources. In order to maintain the world's energy resources it has been suggested (12) that the developed countries should reduce their energy consumption per capita. These could be achieved in the following ways:

(1) Improvements in energy use efficiency.

(2) Energy conservation.

(3) Systematic replanning of our life style.

We also need to develop renewable energy sources, such as solar, wave and geothermal to augument the worlds energy needs.

So, apart from encouraging the developed world, such as the United States, to reduce their capita energy consumption, the developed world could assist and co-operate with the developing countries in research and development work in the energy field. The United Nations, UNESCO and other International bodies could establish joint research projects between institutions and research centres in both developed and developing countries. More research work should be carried out in the developing countries. Although it may be possible to monitor the conditions in the developing countries from the developed countries, differing cultural, social, climatic and economic conditions may constitute fundamental problems.

Appendix 1: Formulas for Computing Cost Effectiveness

The formula for computing cost effectiveness of a solar system for buildings is given below.

If an item is purchased and

> the initial capital cost per unit area is, c
> the expected lifetime is L (years),
> the interest rate is r (expressed as a fraction),

the payment per year for equal yearly payments over the lifetime L will be

$$P = \frac{(r+1)^L rc}{[(r+1)^L-1]}$$

If f_0 represents the initial fuel savings/unit area, and the expected average fuel inflation rate is α, then the average fuel savings/year–area, $\bar{f}$, over the lifetime L will be

$$\bar{f} = \frac{(1+\alpha)^L-1}{L\alpha} f_0$$

For the capital improvement to be cost effective the average savings per year must exceed the average payment,

$$\bar{f} \geqslant P,$$

or

$$\frac{(1+\ )^L-1}{L} f_0 > \frac{(r+1)^L rc}{[(r+1)^L-1]}$$

Alternatively, this can be reexpressed such that the ratio of the initial fuel savings/unit area, f_0, to the initial capital costs/unit area must be greater than

$$\frac{f_0}{c} \geqslant \frac{(r+1)^L \alpha r L}{[(1+\alpha)^L -1]\,[(1+r)^L -1]}$$

As an example, take $L = 20$ years, $r = 10\%$, $\alpha = 6\%$; then the ratio of the initial expected fuel savings to the initial capital costs must be greater than 0.064 to be cost effective.

References

1. King, Hubbert M; World energy resources. Proceedings of the 10th Min. and Met. Congress., Canada, 1974.

2. Dyson, F.J; Energy in the Universe, Scientific America 224, 41-59, 1971.

3. Ngoka, N.I; Guidelines on integrated building design for the tropics; Proceedings of the Third International Symposium on Information System for Designers, University of Southampton, March 1977.

4. Ngoka, N.I; The functional aspects of building design in Nigeria with special reference to thermal and ventilation considerations; Proceedings of the International CIB Symposium on Energy Conservation in the Built Environment, Construction Press/CIB, 1976.

5. Van Straaten, J.F; The thermal performance characteristics of curtain wall construction in warm climates; Paper presented at the First Australian Building Research Congress, Melbourne, 1961.

6. Ngoka, N.I; A field study of natural ventilation of buildings; MSc(Eng). thesis, Department of Architecture, University of Bristol, 1974.

7. Givoni, B; Man, Climate and Architecture, Applied Science Publishers, London, 1969.

8. Penner, S.S. and Icerman L; Energy, Vol. 1, Addison-Wesley Inc., Massachusetts, 1974.

9. Glover, J. and McCulloch, J.S.G; The empirical relation between solar radiation and hours of sunshire; Quarterly Journal. Royal Meteorological Society, 84, 1956.

10. Fry, M. and Drew J; Tropical Architecture in the humid zone, Batsford, London, 1956.

11. Alexandroff, Guennec and Givardiev; The use of solar energy for water pumping in arid areas; University of Dakar, 1973.

12. Dunn, P.D, Energy and the developing countries; Aspects of Energy conversion, Pergamon Press, Oxford, 1976.

6.5 An appraisal of solar building prospects in Central Africa

W.K. Kennedy
Engineering Department
University of Malawi
The Polytechnic
Blantyre, Malawi

Abstract

An appraisal of appropriate solar buildings suitable for Central Africa, with particular reference to Malawi, has been carried out. The main purpose of this paper is to consider the possibility of solar building programmes in largely agricultural-based developing countries where the financial resources for such a development is limited.

After a summary of meteorological data for Malawi and the consideration of the country's development priorities, areas where solar buildings would be appropriate are outlined. Building requirements in the following categories are considered:
(a) buildings requiring cooling or heating for the storage or processing of agriculture/fishery produce;
(b) buildings usually requiring comfort cooling.
Buildings appropriate to the economic and technical limitations of both rural and urban areas in Malawi are considered.

The problems associated with the promotion of solar buildings in developing countries are fully discussed. Suggestions are put forward in order to aid the introduction of solar building pilot schemes and a possible solar building project for Malawi is outlined. The implications of such a project from the financial and technical assistance viewpoint is considered and recommendations put forward.

Introduction

In most developing countries consideration of developments in technology such as solar buildings have not previously been taken seriously. This attitude is understandable when one considers that solar buildings are too often interpreted as relating to (usually expensive) solar houses designed largely to a heating mode for temperate climates. There has been little work carried out in this area of solar energy geared to the real needs of developing countries.

The present study is concerned with the possibility of solar buildings being constructed in Malawi which are relevant to the development priorities of the country. These priorities relate primarily to improvements and expansion in agricultural production and the processing of produce.

The economics of solar buildings for developing countries must be the foremost consideration in deciding their design criteria. In this area the appropriate level of technology is an essential factor. In general, the maximum utilisation of the building fabric is necessary together with simple, natural circulation systems for heating/cooling requirements. More complex arrangements may have a limited application in urban areas.

Background information and terms of reference

As the appraisal of solar building prospects in Central Africa is to be limited to that in Malawi, it is from there that background information will be drawn.

Malawi is a small, land-locked country in the central/southern part of Africa (latitude $9\frac{1}{2}^{o}$S to 17°S). The country has considerable annual insolation with average hours of sunshine of 7 to 8.3 hours per day (depending on the location) at an intensity of solar radiation (daily total radiation on a horizontal surface) in the range of 4600 to 7600 Wh/m2 day (estimated) throughout the year. A summary of meteorological and solar data is given in Figure 1.

Economically, Malawi is considered to be one of the poorest nations in the world (on a G.N.P. basis). The country has little mineral resources and the economy is heavily dependent on agriculture. The main cash crop for export is tobacco. Malawi has good hydro-electric potential and this is increasingly being tapped. The major consumer of electricity is industry. The country's industry is largely devoted to the processing of tobacco, tea, sugar and cotton and the production of import substitution items. A small-scale industries policy is shortly to be implemented.

Malawi is the most densely populated but the least urbanised of the Central African countries. The major urban centres are Blantyre in the Southern region with a population of approximately 120,000 and the new capital, Lilongwe in the Central region.

The development priority of the country is geared to improvements and expansion of agricultural production - both export cash crops and local staple food. Various land-development programmes have been implemented by the Government and a national rural development policy has also been drawn up.

The utilisation of alternative energy sources in Malawi has recently been considered by the National Research Council of Malawi (a body set up by Government to co-ordinate and promote research activities in the country and whose Secretariat is based at the Office of the President). The Council approved that the development of solar and other alternative

Figure 1 Malawi meteorological and solar data

Temperature

Hottest month: October/November
Coldest month: June/July
Typical temperature range: (^{o}C)

Location	Period (month) July	October
Nsanje	27.5-32.5	37.5-42.5
Blantyre	20.0-25.0	30.0-35.0
Lilongwe	22.5-27.5	32.5-37.5
Karonga	25.0-30.0	30.0-35.0

Duration of sunshine hours and solar radiation intensity (Wh/m2 day)

Month	Sunshine hours (average)	Solar radiation
February	5.3	5250
October	9.5	7210

Solar altitude angle ( )

Date	Angle
December 21	79.2^{o}S
September/March	76.7^{o}N
June	53.1^{o}N

energy sources in Malawi be actively encouraged and project proposals are being prepared. Work on solar energy has been started at the University of Malawi where at the Polytechnic College low-cost, flat-plate collectors have been constructed. Further work is planned in the area of solar energy utilisation for the preservation of food - i.e. solar driers and refrigeration/air-conditioning units operating on the absorption principle. Plans are also underway to establish a Technology Development Unit at the Polytechnic whenever funds become available. It is proposed that the Unit would develop prototypes of solar equipment and assist in the implementation of pilot projects in rural areas. This work would be directly related to the Government's rural development programme.

Any solar building project for Malawi must have a realistic objective and relate directly to the country's development priorities. It should be relevant to national development policy and to the requirements of industry and commerce. Problems of acceptance and implementation cannot afford to be underestimated. The initial objective should be to construct appropriate solar buildings on a pilot project basis. Such buildings should be of proven design and be appropriate in terms of economy and level of technology.

Solar building prospects in Malawi

Many authors have made the plea that a concerted effort should be made to utilise solar energy in developing countries. The use of solar energy must be related to the real needs of the country. A recent paper has considered the appropriateness of solar energy in rural development programmes of developing countries. (1)

In the case of Malawi it has been seen that agriculture and related processing industries are the major areas of development. In order that solar buildings be seen to be relevant to development it is important that they be identified as an aid to improvements in agriculture. Other areas more associated with solar building applications may find some limited use for comfort cooling but this of necessity must be of low priority. Solar buildings in Malawi have application in the following areas:

(a) buildings which require hot air or steam circulation for drying or curing purposes;
(b) buildings which require air-conditioning for the storage of agricultural or fishery produce;
(c) buildings which require comfort cooling.

In considering designs suitable for the above applications it is imperative that the various limitations present in a developing country such as Malawi are fully considered. It must be emphasised from the outset that the system(s) most likely to find acceptance will be one which makes use of simple principles, be easily constructed and maintained, and be of realistic cost. Sophisticated solutions are certainly not desirable, especially for rural or semi-urban areas. It is realised that there are exceptions to this criterion in the case of some possible applications in urban areas or in large estate farms. However these exceptions should be considered in perspective and they should not be taken as justification that medium/high level solar technology has applicability in the poorer developing countries.

Wherever possible the aim should be to make use of building designs which require the minimum of solar hardware. They should be designed with full consideration of climatic conditions and make use of the building fabric to the full. Where solar units are incorporated these should be of simple, natural circulation principle wherever possible.

If more complex solutions have to be resorted to then these should be based on some degree of appropriateness in terms of level of technology and capital cost.

The order of priority of developments in building design can be stated as follows:
(a) designs which eliminate or minimise the need for cooling by making use of the building fabric (such designs are not 'solar buildings' in the sense that this term is usually defined);
(b) Simple structures making use of natural circulation air systems or flue-type arrangements - especially for the drying of crops, fruit, vegetables and fish;
(c) designs which make use of a more 'static' fluid operating principle - e.g. a hothouse principle in the case of drying/heating and a solar pond (roof mounted) in the case of cooling applications;
(d) designs for heating/cooling buildings which make use of semi-manual/mechanical assisted circulation with an emphasis on realistic capital cost level.

With the aid of surveys of solar buildings throughout the world (2), (3) it has been possible to select some designs which with some modification could meet the objectives/criteria which have previously been outlined as appropriate for Malawi. These will now be briefly considered.

(a) Buildings utilising the building fabric for cooling

Brick dome or vaulted structures have been successfully utilised in the past as a viable means of building technique. Such methods should be increasingly promoted once again as the result is a low-cost structure which is well within the capability of local builders in the rural areas of developing countries given the necessary training. This method can utilise burnt brick or mud brick. With the latter there is a need for suitable water-proofing.

In Malawi, as in many other developing countries where the traditional round house is the norm in rural areas, there is a resistance to the introduction of a domed roof form. There is however a possibility of the acceptance of a vaulted structure.

With such structures it is quite feasible to maintain an interior air temperature within the range of 20° to $25^{\circ}C$ with an outside air temperature in the range 10° to $35^{\circ}C$. Such interior temperatures could certainly not be maintained without air-conditioning with buildings with the (popular) corrugated galvanised-steel roof or with a prefabricated concrete arrangement.

(b) Buildings utilising the Trombe solar wall system

The Trombe solar wall principle is one which has been successfully used in several solar houses which have been built in the northern hemisphere. It is a simple and attractive natural convection arrangement which makes use of a massive wall as the solar collector. The question which is raised is how can such a principle be made use of in the tropics where the sun is directly overhead at times - e.g. the solar altitude angle (throughout the year) varies from $54^{\circ}N$ to $79^{\circ}S$ in Malawi. A horizontal collector would naturally seem to be the solution. For such a system to operate by natural circulation it is necessary to have the collector inclined to some degree. It is possible to use an arrangement whereby the flat-plate collector is inclined on the ground rather than on the roof of the building and thereby provide air circulation without mechanical assistance. Such an arrangement does not require to support a heavy roof and indeed a vaulted roof could be used which would further keep costs to a minimum.

The use of this massive base arrangement in the topics implies that the solar building will only operate as intended for the part of the year when the solar altitude angle is compatable with the North or South inclination of the solar collector. It should be noted that it has already been pointed out that the major application of this type of building in Malawi could be for drying purposes. After harvesting in March/April the solar altitude angle is to the North for a period of about seven months. The solar collector would be operational during this period with a northern inclination.

(c) Buildings utilising the Skytherm principle
The Skytherm principle is one where a solar pond is mounted on the roof of a building with control facilitated by means of moveable panels. This method could have some application in Malawi if it is possible to achieve an arrangement with a satisfactory life at reasonable cost. Apart from the question of maintenance under tropical conditions there is a need to ensure that water does not collect on the horizontal roof during the rainy season and harbour mosquitoes.

(d) Buildings utilising low-cost flat-plate water heaters
It is possible to achieve cost reduction of solar buildings by means of local manufacture of solar water heaters. The collectors are constructed of standard pipe fittings and/or galvanised iron sheet. It is also possible to have an arrangement with the collector and a corrugated-iron roof as an integral unit.

Pilot projects in Malawi

In Malawi as with other developing countries it is imperative that pilot solar building projects be promoted. This can only become a reality if financial and technical assistance is provided by aid agencies. The author has drafted proposals of possible pilot project(s) in Malawi. It was intended that these draft proposals be included in the present paper but this has not been possible because of a number of reasons. The author would welcome correspondence with interested researchers and development workers and agencies in order that co-operation in this area of work be promoted.

Conclusions

It is not possible to formulate meaningful conclusions as the present paper is merely an introductory one. A further paper is planned once the pilot project outlined about is finalised. It will suffice at this stage to conclude that solar building programmes offer considerable possibilities in a country such as Malawi. Acceptance of solar buildings can only become a reality once pilot projects have been shown to be feasible. The utmost co-operation is necessary from international bodies in order to ensure that solar buildings contribute to development policies especially in the area of rural development.

References

(1) W.K. Kennedy, 'The Appropriateness and Implementation of Solar Energy in Rural Development Programmes of Developing Countries', UNESCO/WMO Solar Energy Symposium Geneva, August/September 1976.

(2) S.V. Szokolay, 'Solar Energy and Buildings', Architectural Press, London, 1975.

(3) C.N.D. Stambolis, 'Solar Houses: A General Appraisal of Solar Building Design and Technology', UNESCO/WMO Solar Energy Symposium, Geneva, August/September 1976.

Feasibility study on the development of solar operated cold storages located in rural areas

N.C. Srivastava
Department of Mechanical Engineering
University of Nairobi, Nairobi, Kenya

Abstract

This paper presents a feasibility study carried out for developing cold storages operated by solar energy in rural areas as the solar energy is abundantly and freely available everywhere. Design requirements including size, refrigeration load and insulation for a medium size cold storage of about 200 tons holding capacity located near the crop areas have been established. It is found that the storage for commodities requiring about 10°C storage temperature can be developed for operation with solar energy. Intermittent absorption refrigeration systems have been proposed for this purpose due to simplicity of construction and operation. Building design and economic aspects of the proposal have been studied and it is recommended that the developing countries should take initiative for the indegenous development of such cold storage facilities.

Introduction

It is beyond doubt that most of the developing countries have agriculture based economy. Quite a few of them have capacity to produce more than their usual agricultural crop but they are unable to do so because of lack of cold storage facilities. Often some portion of the crop is destroyed because it is not stored in proper temperature conditions. The necessity of a chain of cold storages has been ever existing in the countries like Kenya but due to shortage of electrical power in rural areas, conventional cold storages cannot be installed. It is believed that such cold storages located right near the crop harvesting point would be of direct use to the farmer and would help in the rural uplift. In view of these facts, this study includes a survey of relevent available literature, establishment of users requirements, building design requirements and the design of refrigeration system suitable for operation with solar energy.

A detailed review of solar refrigeration and air conditioning literature has been carried out by the author (1) which reveals the fact that solar heating of space is economical and nearing the commercial stages. Solar air-conditioning is also a practicability but the equipment involved are much expensive as compared to conventional airconditioning equipment. However due to increasing cost of fuels and other conventional energy resources it is anticipated that solar operated air conditioning equipment will economically compete with the conventional equipment and may gain commercial popularity by 1985 to 1990. Many national programmes, research laboratories and institutions particularly in the developed western countries are actively engaged on the projects for using solar energy for comfort conditioning. Most of these investigations are making use of conventional Lithium-Bromide water absorption refrigeration units modified to operate with solar energy. The investigations for low temperature refrigeration, so far reported, are very few. Most of these investigations i.e. Trombe and Foex (2), Williams et-al (3), Chinnappa (4,5), Chung (6) and Sargent and Beckman (7) have found an intermittent absorption system to be simple and effective. But all these investigators have experimented with very low capacity systems. In the knowledge of the author there is no evidence of an effort for developing low temperature large capacity refrigeration systems operated by solar power particularly for applications like cold storages, milk coolers and ice plants etc. Scope of solar energy application for the above purpose has already been studied by the author (8) and it is found that enough information is now available for the idegenous development of intermittent absorption refrigeration equipment operated by solar energy and suitable for cold storages operating around 10°C. In the following section basic requirements of such cold storages located in rural areas have been established.

Basic Requirements in rural areas

Rural population in most of the African countries basically lives on meat, vegetables, fruits and cereals. Out of these, the items which require the storage temperature of the order of 5-10°C are listed in Table - 1. Undoubtedly the potatoes are produced and consumed most all over the world and hence the potato cold storage is a prime necessity. A large cold storage designed for potatoes can be provided with some extra space for other seasonal vegetables and fruits. The need for cold storages having capacity of the order of 100 tons to 300 tons of potatoes has been indicated by local authorities and the refrigeration firms. However due to the power shortage in rural areas the conventional refrigeration equipment cannot be installed. Other requirements of such cold storages are:

(a) Location

Location of the cold storage should be such that farmers can conveniently transport their produce to the cold storages. Ideal choice of location will be near the main road or highway passing through or near by the fields.

(b) Ownership

Such cold storages should be owned and managed by either the local government or the cooperative societies as it may be beyond the capacity of an individual farmer to own and run the cold storage.

(c) Simplicity, Reliability Durability

The refrigeration equipment should be simple and reliable so that the users may find it easy to train their man for routine maintenance of the plant. The equipment should be rugged, strong and durable.

Table 1
Fruits and vegetable storage requirements*

Item	Storage Temperature °C (°F)	Relative Humidity %	Duration
Avocado	7-13 (45-55)	85-90	4 Weeks
Green Beans	7 (45)	85-90	10 days
Cucumber	7-10 (45-50)	90-95	10-14 days
Eggplant	7-10 (45-50)	85-90	10 days
Lemons	13-15 (55-58)	85-90)	1 month
Mango	10 (50)	85-90	2-3 weeks
Papayas	7 (45)	85-90	2-3 weeks
Pine apples:			
Green	10-16 (50-60)	85-90	3-4 weeks
Ripe	5-7 (40-45)	85-90	2-4 weeks
Potatoes, Early Crop	10-12 (50-55)	85-90	—
Late Crop	3-10 (38-50)	85-90	(at 5°C 6-8 months (at 10°C 2-4 months
Sweet Potatoes	13-16 (55-60)	90-95	4-6 months
Unripened tomatoes	13 (55)	88-95	2-5 weeks

*Extracted from ASHRAE Guide Data Book, Applications p.486, 1962

(d) Energy Supply

During cloudy or no sunshine periods the refrigeration equipment should have the facility for its operation with locally available fuel i.e. wood, charcoal or coke.

(e) Material Handling

The cold storage design should provide for manual handling of the produce as the labour is normally cheap in rural area. This would call for single or at the most two storied buildings with piling heights which would be convenient for an average height man.

Cold storage size and Refrigeration Requirements

In order to establish the requirements of the size and the refrigeration load of a typical cold storage, the holding capacity of 200 tons of potatoes has been considered in this study. The storage should operate at 5 to 10°C holding temperature. It is further considered that the cold storage has daily average turnover of about 20% of the holding capacity i.e. 40 tons of potatoes are brought in for cooling and the same amount is taken out for sale. This will be loaded only during 8 to 10 hrs of the day, with temperature of the produce about 38°C (100°F). Based on the recommended practice (9, 10) the storage space of 20 m long x 12.2 m wide x 5.2 m high (@ 65 ft x 40 ft x 17 ft h) would be required. A slotted mezzanine floor would be used at mid height so that piling height is about 2m maximum. It is further assumed that standard potato bags of size 84 cm x 44 cm x 28 cm (appx. 33 in x 17½ in x 11 in) would be used. Constructional requirements of the storage have be discussed later.

Based on the above general considerations, the refrigeration requirements of the cold storage have been estimated for 100 mm (4 ins.)and 200 mm. (8 ins.) insulation thicknesses and the data is given in table 2. The insulation has thermal conductivity of 3.6 W cm/m^2 °K (appx. 0.25 Btu in/hr ft^2 °F). As shown in the table the estimates have also been made for the precooled products assuming that potatoes or other vegetables can be easily cooled to 27°C either by locally available water or by exposing to night sky. The product after precooling during night can be loaded into the storage space during the convenient morning hours before the solar energy makes it warm. It may be observed from table - 2 that total refrigeration requirements are quite high if the product is not precooled and brought into the storage at 38°C. The precooling of the product to 27°C reduces the refrigeration load by over 25%. Hence the precooling of product is a necessary requirement. In order to evaluate the effect of insulation thickness the cooling load estimates have been made for 100 mm. and 200 mm. insulation thicknesses, both during the pulldown peoriod of loaded product and the holding period i.e. when the all the stored product has reached the required storage temperature. It may be observed that the effect of insulation for the pulldown period on the total refrigeration load is only 5 to 8% but during holding period the cooling load is reduced by 18 to 20% for 8" thick insulation. As the cold storages run more for holding the product at required temperature it may give us appreciable fuel saving particularly during the no sunshine hours when the refrigeration plant is run by locally available fuel.

Table 2
200 Tons potato cold storage refrigeration requirements

Parameter	Insulation Thickness	
	100 mm	200 mm
Holding Temp.	5°C (40°F)	5°C (40°F)
Size of Store	20m x 12.2mx5.2m (65ft x 40ftx17ft)	20m x 12.2m x 5.2m (65ft x 40ftx17ft)
Daily Load at 38°C (100°F)	40 tons	40 tons
Product Load without precooling.	13.6 TR	13.6 TR
Product load after precooling to 27°C (80°F)	9.1 TR	9.1 TR
Total refrigeration Load without precooling.	18.2 TR	17.27 TR
Total refrigeration Load after precooling to 27°C	13.7 TR	12.77 TR
Holding load for clear sky period	4.55 TR	3.61 TR
Holding load for no sunshine period.	3.72 TR	3.14 TR

Refrigeration system design

Based on the data given in Table - 2, it is evident that the precooling of the product is necessary and with the provision of precooling a refrigeration plant of 13 TR (Tons of refrigeration i.e. heat removal rate of 3.517 kJ/s) would be required for a 200 tons potato cold storage. Although one refrigeration unit of this capacity can be designed but for the sake of flexibility in operation it is suggested that the refrigeration load should be shared by two units of 6.5 TR each. This would not only help in maintenance, but also take care of the partial load operation of the cold storage. Solar operated intermittent refrigeration systems which can be designed for the purpose may be of two types, one having direct expansion cooling coils placed into the cold storage and the other producing ice outside the cold space which inturn may be used for cooling the space. Both these systems have been discussed below.

(a) Direct Expansion Intermittent Solar Refrigeration System

Principle of operation of an intermittent solar refrigeration system is shown in Fig. 1. As shown, during the charging mode, solar energy (or heat) is supplied to concentrated ammonia-water solution in vessel-**A** which acts as generator. Ammonia vapours are liberated from the solution and condensed in the vessel-B acting as condenser. After the required amount of ammonia is condensed valve - V is closed. Now the cooling mode starts and solar heat source is removed from vessel - A and vessel - B is placed in the space to be cooled. Valve

Fig. 1 Principle of intermittent solar refrigeration system

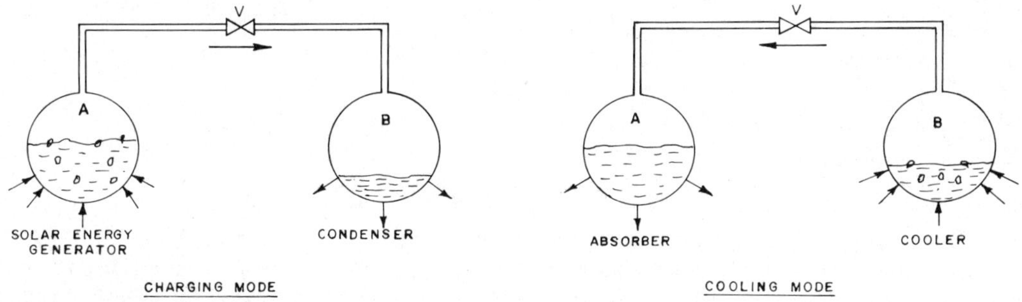

V is opened and heat from the space boils the ammonia liquid in vessel - B at low temperature and pressure. Vessel - A rejects heat to surroundings and ammonia vapours from vessel - B are absorbed in the weak solution of vessel - A.

In the proposed practical system, as shown in Fig. 2, the hot water heated to 80°C by flat plate solar collectors is circulated through generator assembly. The liberated ammonia is condensed into the condenser and collected in the receiver. During the charging or condensing mode valve V_1 remains open. After the receiver is full the valve V_2 is closed. Now during the cooling mode valve V_3 and V_4 are opened. Ammonia flow is regulated by valve V_3. Hot water flow to generator is stopped, and cooling water circuit is opened so that the heat from the absorber is removed. The water circulation for both generation and absorption processes is by thermosyphon action. A detailed design of the cooling equipment based on above principles is beyond the scope of this paper and is being presented by the author elsewhere (11). A smaller machine based on the above principles and suitable for small walk-in coolers is under fabrication stages at the University of Nairobi. Finances required for developing large capacity plant are high and the agencies which can afford and use the system are being contacted.

Fig. 2

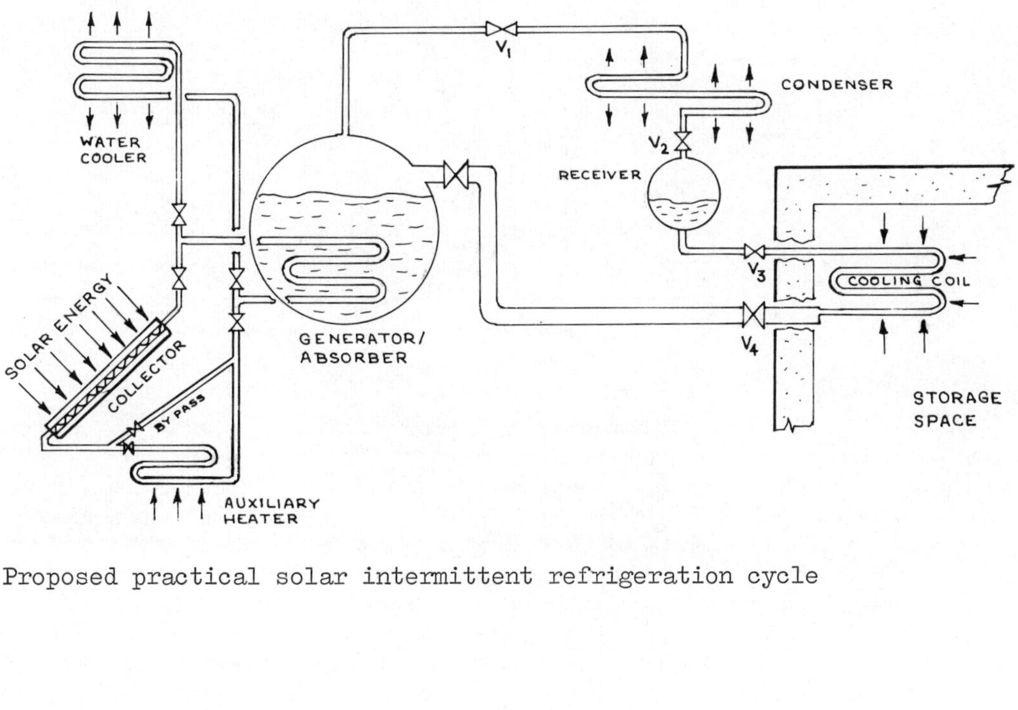

Proposed practical solar intermittent refrigeration cycle

(b) Ice Bank system

This system would make use of ice produced by intermittent solar refrigeration equipment. Cooling units using natural air circulation across the ice bank can be placed in the cold storage space. The ice will be made separately with the help of solar energy and stored in a separate cold room designed for storage of ice at - 2°C (@ 28°F). Ice making can be achieved by use of either continuous or intermittent absorption refrigeration system. Chung (6) has already been successful inproducing about 4 kg (9 lbs) of ice using sodium thiocynate-ammonia system. The system employed about 5.5 kg (12 lbs) of solution having concentration of 0.5 and a 1.2 m (4 ft) diameter parabolic collector. The ammonia was condensed in a smaller vessel immersed in water for about 4 hrs. Based on these results one ton of ice production would need solar concentrator area of about 200 sq. m. This area seems to be two large and perhaps on a larger scale higher efficiency may be expected which would bring down this requirement. The major problems with such a system in developing countries are the parabolic concentrator and the tracking system. In a similar effort Swartman et. al. (12) have reported some success in ice production and the experimental results with the modified equipment are still awaited.

Building Design Considerations

A proposed layout and the building design requirements of the cold storage operated by solar energy are shown in Figs. 3(a) and 3(b) respectively. As shown in Fig. 3(a) the walls, ceiling and floor of

Fig. 3a
Cold storage
layout

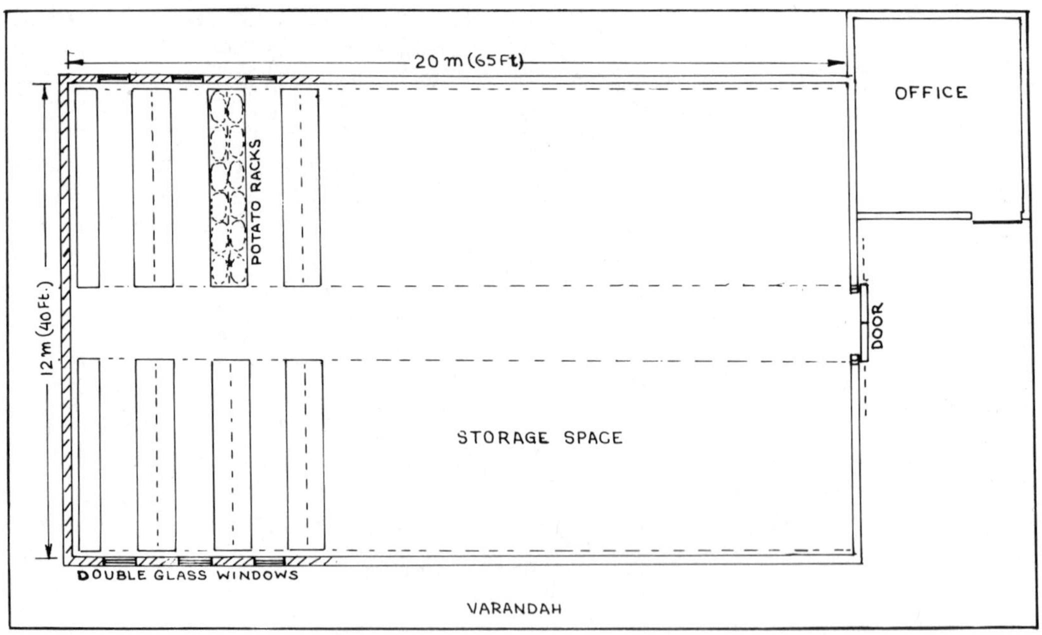

Fig. 3b
Building design
requirements

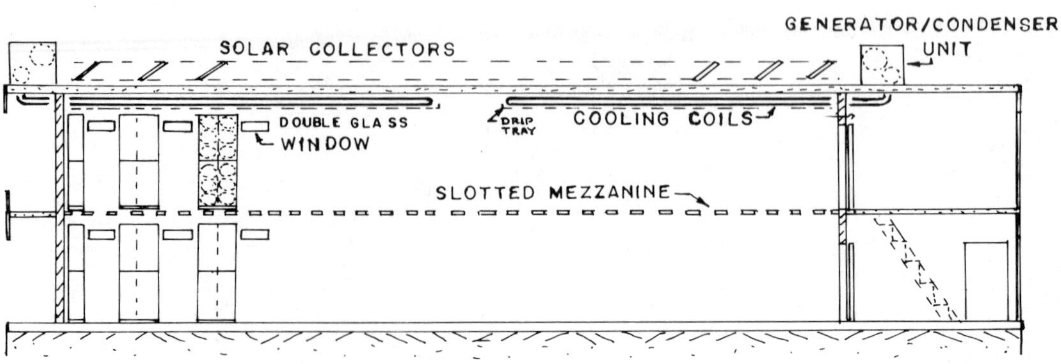

the cold storage has been provided with 200 mm (8 inches) thick insulation. Insulation of 100 mm (4 inches) thickness can also be used inorder to save on initial investment but at the cost of higher running cost. The cooling coils have been located on the two side walls and some portion of the ceiling along with a drip tray to collect the moisture condensed on the coil surface. The solar collectors are located on the roof of the building as shown in Fig. 3(b). The hot water from the collectors is circulated through the generator and condensing units located on both ends of the building, one set on each end. Roofing has to be designed for the static weight of the solar collectors and the refrigeration units when charged with water and refrigerant. As shown in Fig. 3(b) the building has to be provided with verandahs and necessary shadings so as to avoid direct sun rays on the walls and roof particularly during the peak hours of the day. Roofing should be water proof so that any water leakage is avoided. Inside insulation of walls, ceiling and floor should be as per the standard practice recommended by ASHRAE (10).

Economic Aspects

The overall cost of the solar operated cold storage as recommended will depend upon the local material and labour cost, and taxes and duty on the equipment. As per the very general rates prevailing in Kenya it is anticipated that the building cost of the recommended size cold storage located in isolated area will be about KSh. 600000/=, the insulation cost will be about KSh. 400000/= for 4" thick and KSh. 550000/= for 8" thick insulation. The equipment cost along with the solar collectors would be about KSh. 560000/=. As compared to this a conventional refrigeration plant would cost about KSh. 400000/=. These estimates are very approximate and initial development contingencies would be quite high. The difference in initial cost for solar operated plant would be well compensated by its negligible cost of operation since the solar energy is available free. As compared to this the power cost of a conventional refrigeration plant may be about KSh. 40000/= per year.

Conclusions

It is found that a cold storage of about 100 - 300 tons capacity is a practical feasibility and with the prevailing prices of conventional equipment, the proposed solar operated plant compares reasonably well. Such cold storages are necessities as they can be located right near the crop harvesting point where they can be directly used by the farmers on cooperative basis. This would also be a good incentive for producing more, once the farmers know that the produce would not be spoiled. The refrigeration plant is simple and can be made in the developing countries indegenously.

Acknowledgements

Author is very much grateful to the Scientific Committee of the N.E.L.P./U.N.E.S.O. International Solar Building Technology Conference and to the University of Nairobi for providing the necessary funds to participate in the Conference.

References

1. Srivastava, N.C., Present status of solar energy application to refrigeration and air-conditioning. 12th Symposium, Energy Resources in East Africa, East African Academy, 30th Aug. - 4th Sept., 1976, Arusha, Tanzania.
2. Trombe, F., and Foex, M.J., Production of cold by means of solar radiation. Solar Energy, 1(1), 51, 1957.
3. Williams, D.A., Lof, G.O.G., Fester, D.A., and Duffie, J.A., Intermittent absorption cooling systems with solar regeneration. Refrigerating Engineering, 66, 33, 1958.
4. Chinnappa, J.C.V., Experimental study of intermittent vapour absorption refrigeration cycle employing the refrigerant absorbent system of ammonia-water and ammonia-lithium nitrate. Solar Energy, 5, 1, 1961.
5. Chinnappa, J.C.V., Performance of an Intermittent refrigerator operated by a flat plate collector. Solar Energy, 6, 143, 1962.
6. Chung, R., Letter to the Editor. Solar Energy, 7, 187, 1963.
7. Sargent, S.L. and Beckman, W.A., Theoretical performance of an ammonia-sodium thiocynate intermittent absorption refrigeration cycle. Solar Energy, 12, 137, 1968.
8. Srivastava, N.C., Scope of solar energy application to refrigeration and air-conditioning in developing countries. (To be presented at the IFAC-Conference on Systems Approach for Development- Nov. 26-28, 1977, Cairo).
9. International Institute of Refrigeration (IIR), Practical Guide to Refrigerated Storage. Pergamon Press, 1965.
10. ASHRAE Guide and Data Book, Applications, p.486, 1962.
11. Srivastava, N.C., A solar operated system for a 100 tons potato cold store. (To be presented at the International Solar Energy Congress 1977, Nov. 14-19, 1977, New Delhi).
12. Swartman, R.K., Ha, V., and Newton, A.J., Survey of solar powered refrigeration. ASME Publication no. 73-WA/Sol-6, 1973.

P.D. Lessieur
Compagnie Francaise des Petroles
5, rue Michel Ange, 75016, Paris

J.P. Durand
G.I.E. Commissariat a l'Energie Atomique
SOFRETES
Zone Industrielle d'Amilly, B.P.163, 45203, Montargis

ABSTRACT

It is the purpose of this paper to describe the CFP-SOFRETES programme for solar air conditioning in tropical climate. This description will be presented with some experimental data.

This paper will examine the influence of the geometric situation of resistive and capacitive layers on the daily cooling load in air conditioned building. A programme for calculating these characteristics has been set up and applied to the same building as an air conditioning example for hot-wet and hot-dry climate zones.

An analysis is also made of the criteria used for selection of subsystems as cooling tower, automatic and storage tank. Then an experimental programme was carried out with a test loop facility which was constructed to test the absorption machines and subsystems. Finally a special two temperature variable volume storage system, which limits the enthalpy degradation, was designed.

1 INTRODUCTION

Sponsored by the DGRST and the Délégation aux Energies Nouvelles (1) CFP and SOFRETES joined forces to design and test climatic control installations for buildings using solar energy.

The term 'climatic control installation' is not used in this paper in the narrow sense of 'Air conditioning' but in the broader sense, including all available resources which can be used to create comfortable temperature conditions in a building; structure, windows, solar shielding, insulation etc.

Current design work has dealt with buildings located at a latitude 20°N in two extreme hypothetical cases:
- dry climate corresponding to continental regions
- wet climate corresponding to coastal regions

This was done so as to maintain the following conditions:
- temperature (°C) indoor T: $\frac{\text{outdoor T}}{2} + 10$, corresponding to 25°C indoors for 30°C outdoors, and 32°C for 45°C outdoors, these temperatures being perfectly acceptable from the comfort standpoint, in view of humidity control at 50% relative humidity, and at 25°C and 30% at 32°C.

The imposition of more severe indoor conditions or a constant indoor temperature irrespective of the outdoor temperature, in line with the usual tendency in the United States gives rise to very high energy expenditures, with excessive temperature differentials between the air conditioned indoor zone and the outdoors and vice versa (this can also produce a physical shock which is harmful to health).

2 RESEARCH ACCOMPLISHED

Research was undertaken in two different areas:
- Climatic control principle and selection of various components. These studies reveal the specific problems of climatic control with solar energy as compared with conventional air conditioning.

- The influence of the building structure on the heat transfer loads to be taken into account.

The above studies have enabled us to define the ideal structure for climatic control with solar energy, as a function of local climatic conditions.

The following is a brief summary of the main investigations undertaken and the graphs in the appendices illustrate the type of results obtained.

2.1 Studies dealing with the climatic control principle and the selection of various components

2.11 Climatic control system

We have analysed various climatic control systems in detail as follows:
1. Cooling at night by long wavelength reflection with unglazed collectors (radiative cooling);
2. Evaporative cooling;
3. Dehumidification with absorbtion cycle and regeneration by a fluid heated by solar energy.

None of these alternatives proved capable of satisfying the requirements, namely:
- the use, downstream of the cold production unit, of conventional mass-produced systems;
- the suitable "management" of solar energy by a microprocessor type regulation unit with optimization for each type of climate;
- the possibility of use in different climatic conditions without altering the system technology.

The alternative selected provides for use of a hermetic type absorption system producing chilled water, which supplies standard air conditioning units (fan-coil units).

2.12 Selection of the cooling system

The efficiency of an absorption machine is equal to the ratio:

$$\frac{\text{Negative K cals obtained at the evaporator}}{\text{K calories supplied at the generator}}$$

This depends on the temperature of the heat source and the cold source, and on that of the chilled water leaving the evaporator. The temperature of the heat source depends on the area and efficiency of the collectors, and on the maximum allowable temperature in the hot storage tank.

The water temperature of the cold source is determined by the efficiency of the evaporative cooling tower for the cooling water. It is evident that the water leaving the cooling tower will, at best, be at a temperature 3 or 4°C above the outdoor wet bulb temperature (see in Fig. 1 the variation in these temperatures in one day, in the two types of climate considered: hot-dry and hot-wet).

Fig. 1

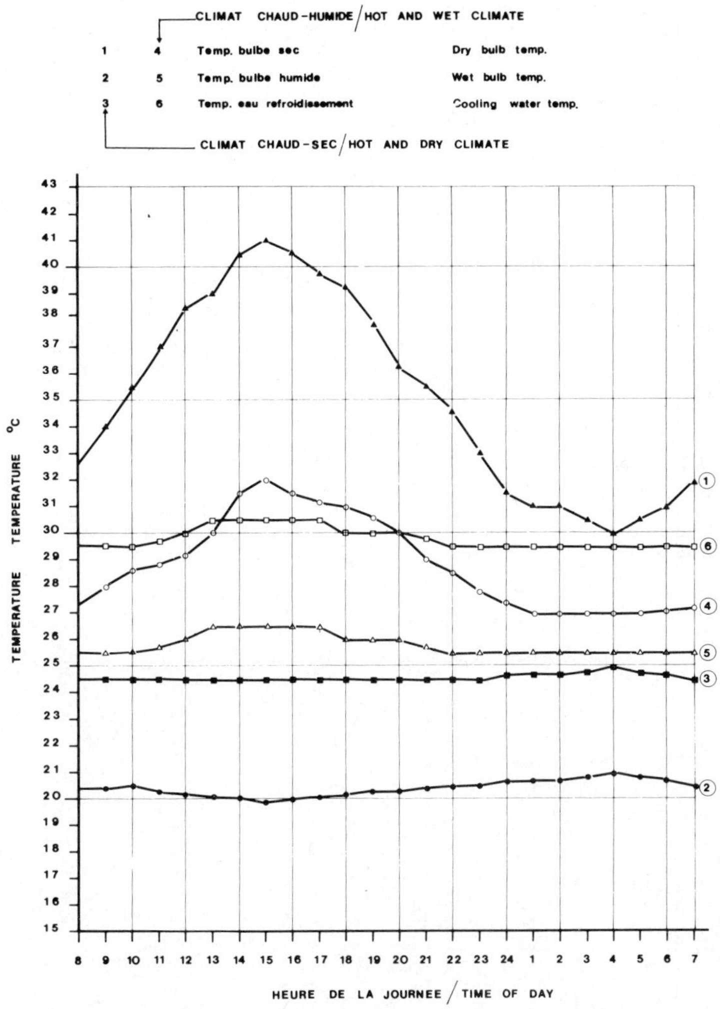

An absorption cooling system is very sensitive to the slightest increase in the condensation water temperature. An increase in this temperature reduces the indispensable difference in pressure between the absorber and the generator, which determines the flowrate of the absorbant through the different heat exchange components of the machine.

To set up again this efficiency, is then necessary to increase heat source temperature. The curve in Fig. 2 shows the increase in hot water temperature as a function of the condensation water temperature for a given refrigeration capacity, together with the curve of the constant quantity of heat energy required by the generator.

As for the chilled water, it is customary in the case of a conventional energy application, to produce cold water at a temperature ranging from 4 to 7°C.

These values are economically and technically unacceptable in the case of a solar energy type absorption cooling plant, because the machine would have to be considerably over-dimensioned in order to attain these temperatures.

We feel that a temperature level lying between 9 and 11°C provides a good basis for comfortable conditions without excessively penalizing the efficiency of the cooling plant.

Fig. 3 and 4 show one of the characteristic curves of an absorption machine optimized for solar energy tested in our experimental loop facility.

Fig. 2

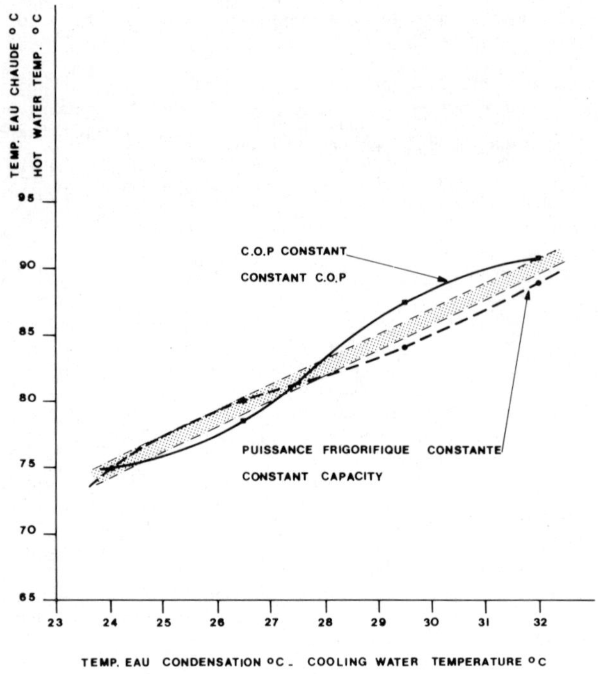

Fig. 3

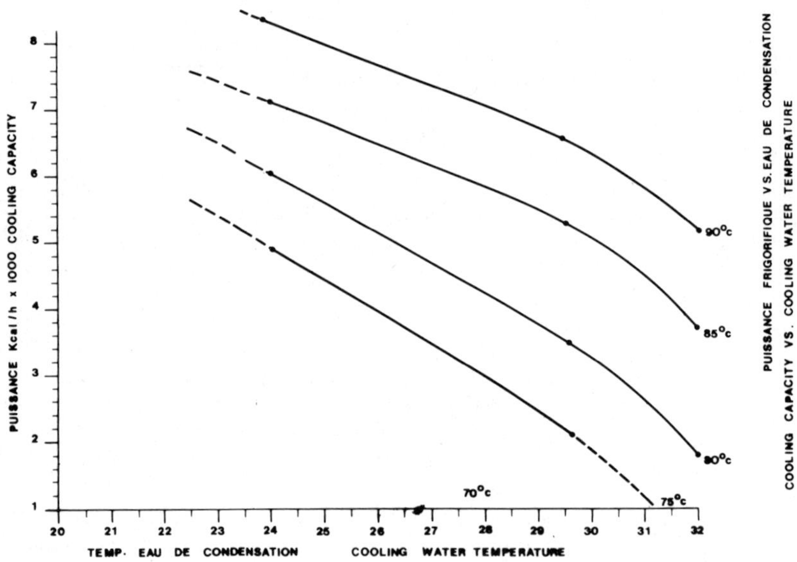

Fig. 4

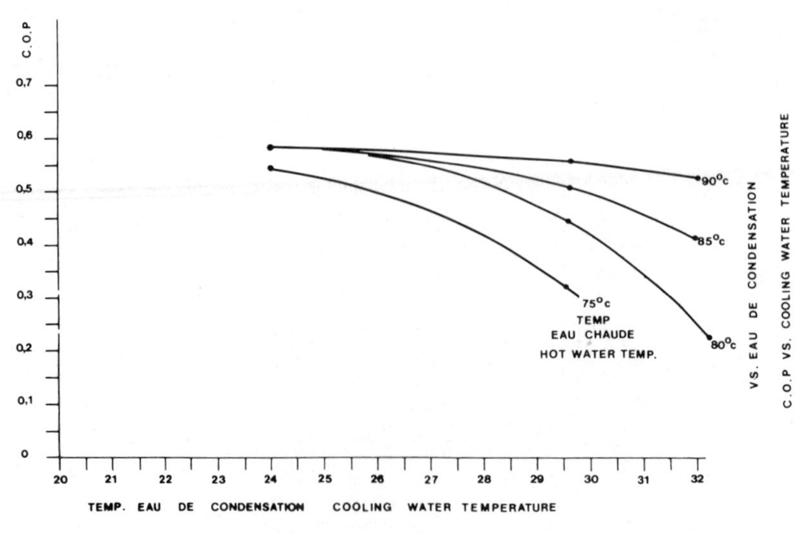

2.13 Scheme of the planned system

Compression and absorption chillers with a capacity ranging 3 000 to 20 000 negative K cal/hour and employing conventional energy sources generally feature 2-position regulation: on and off.

This type of regulation system is unsuitable for an installation using solar energy, except possibly in the case of a system including a cold storage unit.

An analysis of the cyclic operation conditions characterizing an installation with a hot storage reservoir shows that the energy savings in relation to the use of regulation with modulation makes it possible to reduce the collector area, or to improve the total efficiency.

The reduction in efficiency of an absorption machine during modulation is not highly significant. On the other hand, the energy losses involved in the starting and stopping of the refrigerator are very high, because the activation of the operational process requires an interval of 5 to 8 minutes to reach steady state conditions; this interval obviously varies in accordance with the temperature difference between the hot and cold sources. The heat energy thus consumed additionally penalizes the system, because part of the heat is directly transferred to the cold water circuit, and consequently not only are the calories lost, but an artificial load is created by the increase in the cold water temperature.

This is why we feel that the most valid solution consists of the use of a hot storage system, used as a buffer between the collectors and the cooling plant, and a cold water storage system.

The use of a cold storage system is, moreover, recommended for all systems with high thermal peaks.

A diagram of a system of this type is shown in Fig. 5.

Fig. 5

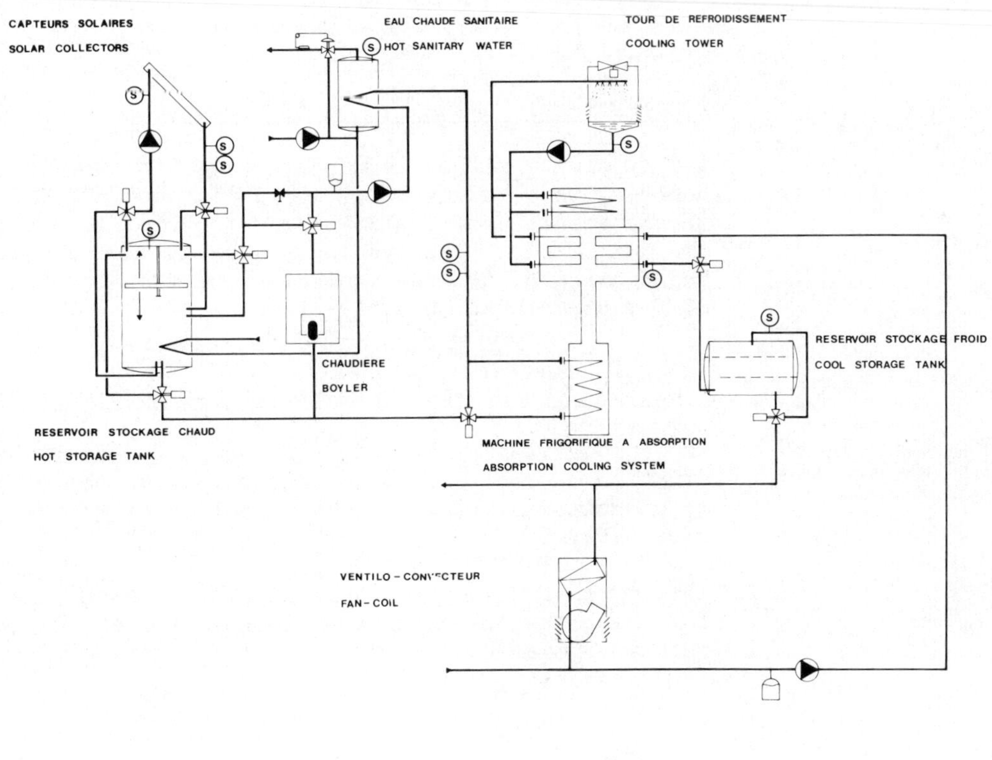

2.14 Electric power supply - an economic approach

For an installation with a compression cooling plant, the overall equipment for cold water production requires about 0,40/0,55 kW to produce 1 000 useful negative K cal.

A well designed solar system, should make it possible to reduce the electric power requirement to 0,090/0,13 kW per 1 000 useful negative K cal produced. This means that a suitably designed solar system can reduce the electrical power needed at the connections to about 18-20% of that required with a compression system.

In order to compare a solar climatic control system with a conventional power installation, one must consider that the investment expenditures are higher in the case of a solar system. This percentage varies from 40 - 70% depending on the installation.

The first design hypotheses provide hope for a solar system which realizes energy savings of 70 - 80% in relation to a conventional power system. It is clear that these values are mainly valid for regions with high insolation. The depreciation period also depends on the time of use of the system. Hence the more the climatic control installation is used, the higher the savings realized by the solar system.

One advantage of the solar system is the relative simplicity of the power production unit, which is undoubtedly an important factor for maintenance and operation.

2.2 Research dealing with the structure of the building

These studies were aimed to determine the influence of building structure on transfer, to the indoors, of a variation in outdoor temperature conditions:
- sudden variation in outdoor temperature;
- sudden variation in insolation;
- sudden variation in available heating or cooling capacity, because in the case of solar climatic control, the power of the cold source is variable.

2.21 Heat/cold accumulation of masonry

A very simple method is used to determine the "response" of the building to a sudden variation in outdoor conditions as a function of the materials constituting the structure. As one example, Fig. 6 shows the summer heating curves caused by an air conditioning shut-down in the heating of two test buildings of identical dimensions, but of different weight: 56 and 94 tons.

2.22 Solar house with high-efficiency double jacket

The calculations give the values of the heat transfer coefficient K as a function of the respective thicknesses of the structure of masses and of the insolation. Fig. 7 shows a section of this type of a double-jacket structure.

This type of structure is highly interesting in the case of a building for which the operating conditions feature wide temperature differences between summer (cooling) and winter (heating).

Fig. 6

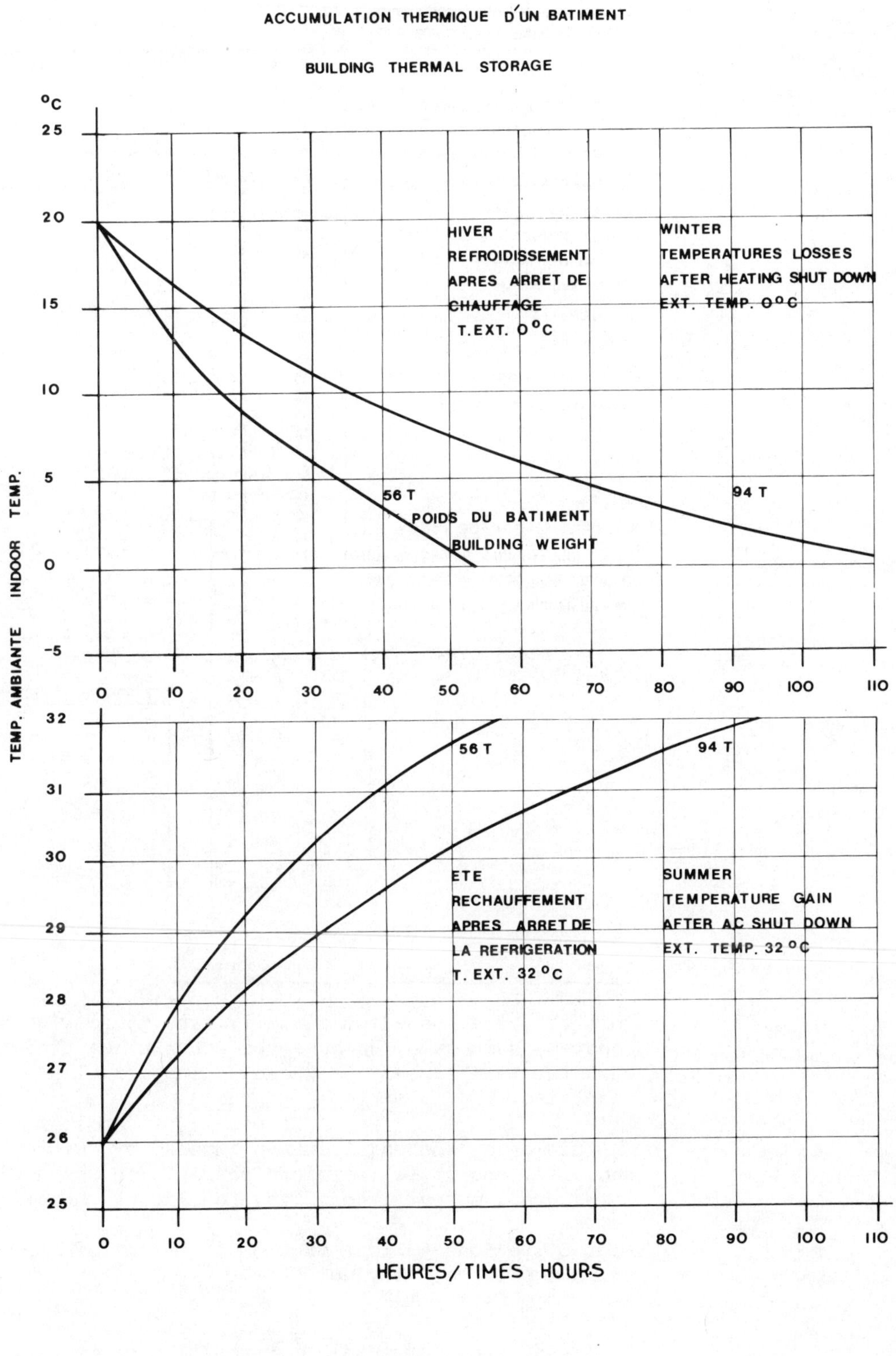
ACCUMULATION THERMIQUE D'UN BATIMENT
BUILDING THERMAL STORAGE
°C
25
20
15
10
5
0
-5
HIVER
REFROIDISSEMENT
APRES ARRET DE
CHAUFFAGE
T.EXT. 0°C
WINTER
TEMPERATURES LOSSES
AFTER HEATING SHUT DOWN
EXT. TEMP. 0°C
56 T
94 T
POIDS DU BATIMENT
BUILDING WEIGHT
TEMP. AMBIANTE INDOOR TEMP.
0 10 20 30 40 50 60 70 80 90 100 110
32
31
30
29
28
27
26
25
56 T
94 T
ETE
RECHAUFFEMENT
APRES ARRET DE
LA REFRIGERATION
T. EXT. 32°C
SUMMER
TEMPERATURE GAIN
AFTER AC SHUT DOWN
EXT. TEMP. 32°C
0 10 20 30 40 50 60 70 80 90 100 110
HEURES/TIMES HOURS

Fig. 7

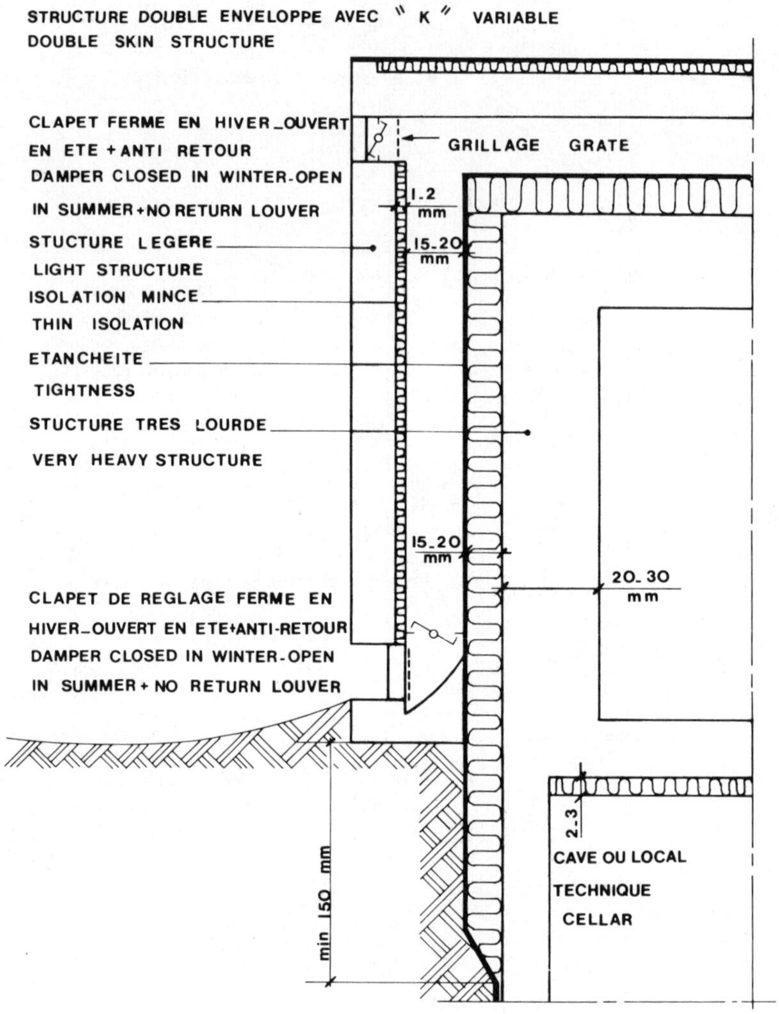

2.23 Low cost single-jacket building

This study was based on a family type house with a pre-fabricated concrete structure, because the pannels required for this type of structure are already produced at industrial scale in many hot countries.

In this case, the requirements imposed by "winter" conditions are not decisive, and it is therefore possible to use a simple, economical structure, without excessively penalizing the efficiency of the system.

Fig. 8 provides an idea of the dimensions of the houses and Fig. 9 shows a perspective view.

2.231 Influence of insulation position on damping of the flux through the walls

For an equivalent heat transfer coefficient, the position of the resistive and capacitive layers in the wall has a considerable effect on the cooling energy consumed during a day, and on the instantaneous cooling energy. Hence, depending on the insulation layout, it is possible to build structures with different thermal responses. As demonstrated some years ago by M. Tavernier, this factor can be exploited to facilitate the use of solar energy, because the characteristics of both the cooling plant and the collectors are thus conditioned by the damping of the heat flux by the walls.

Fig. 8

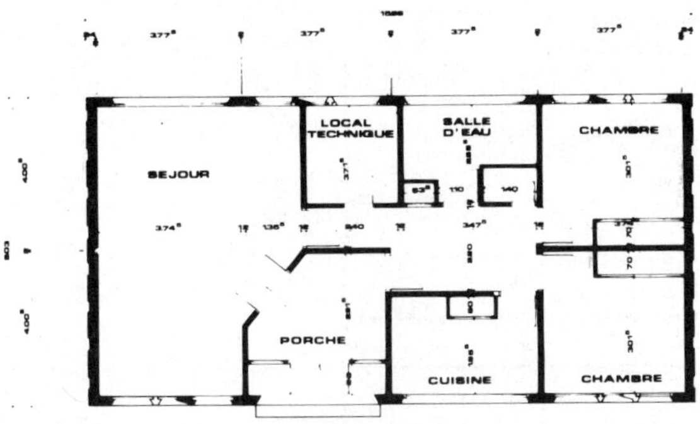

VUE EN PLAN
PLAN VIEW

FACADE LATERALE
SIDE FRONT VIEW

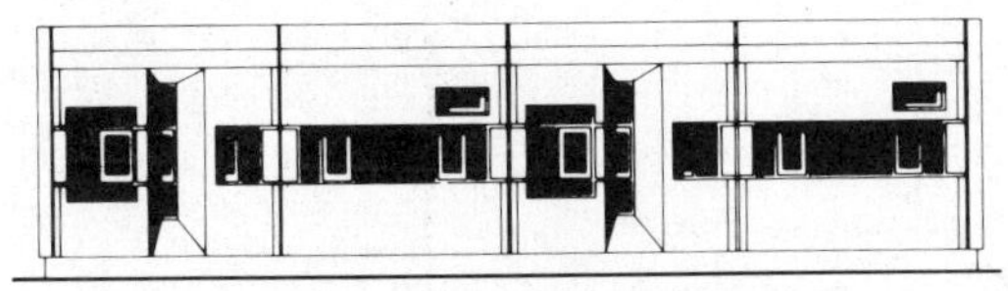

FACADE ARRIERE
REAR FRONT VIEW

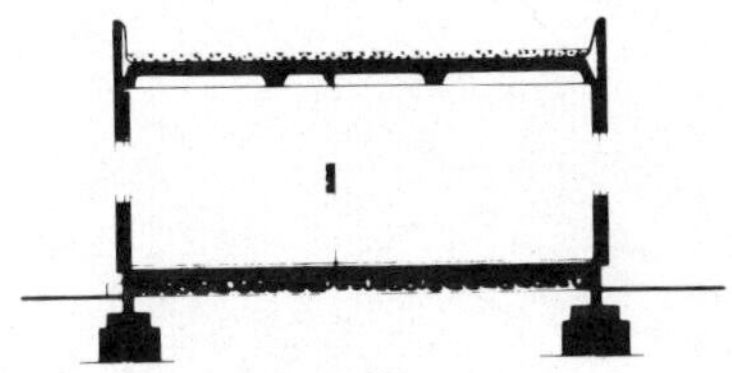

COUPE
SECTION

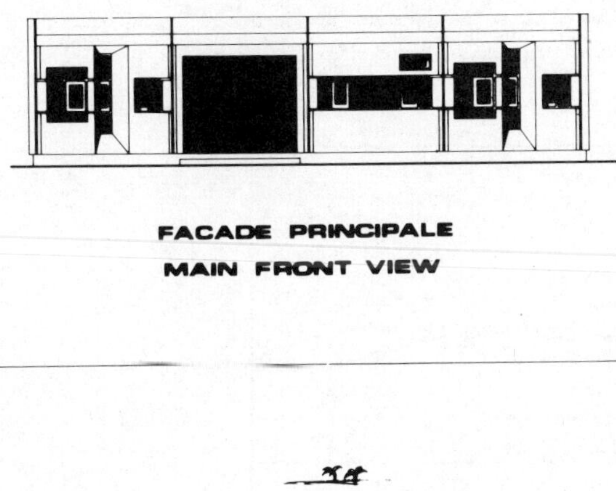

FACADE PRINCIPALE
MAIN FRONT VIEW

Fig. 9

Fig. 10

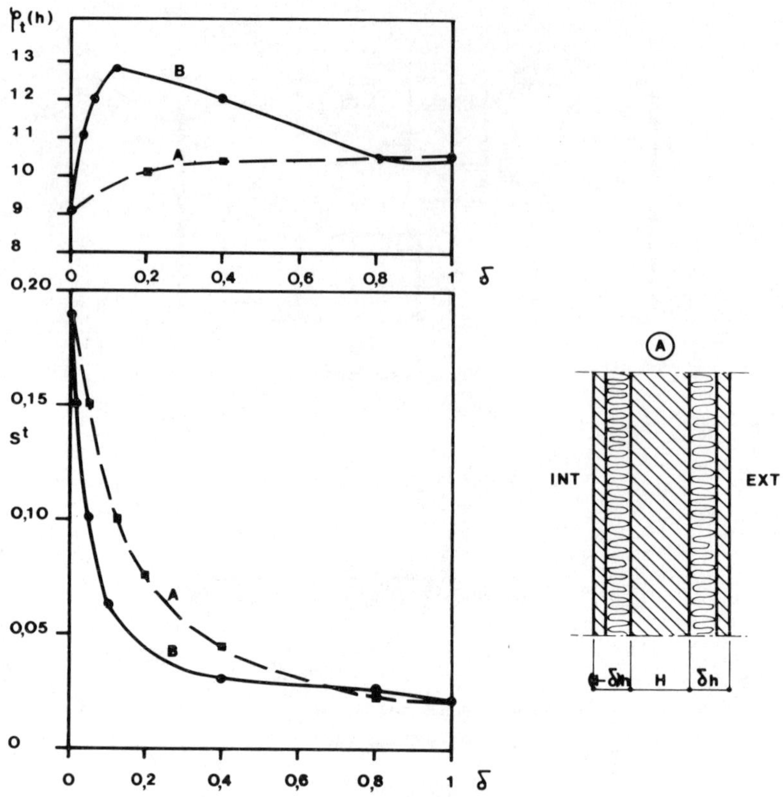

Amortissement de la température des parois
Temperature dumping factors trough walls

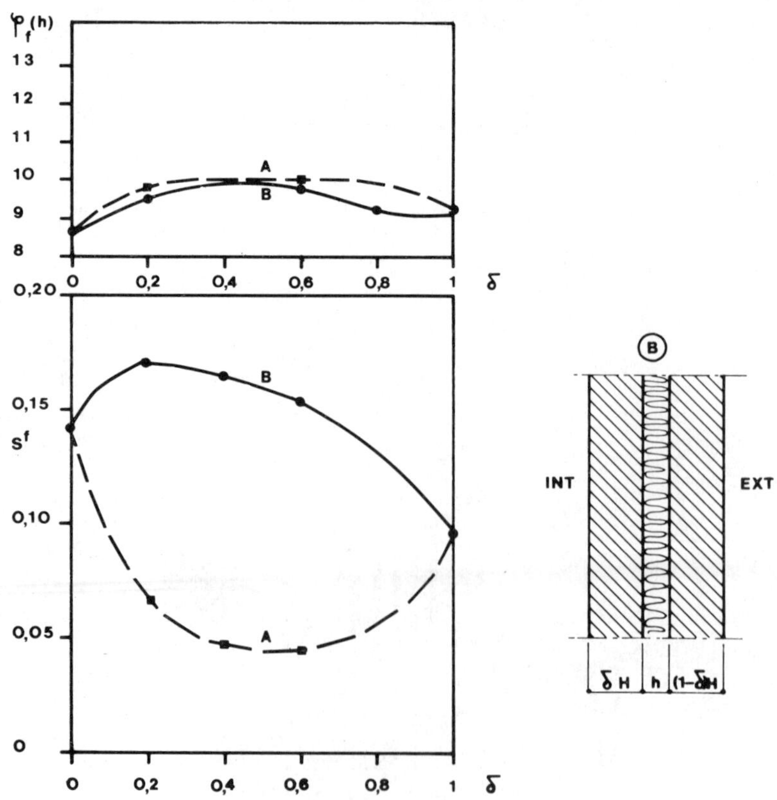

Amortissement du flux des parois
Flux dumping factors trough walls

Fig. 10 shows the diagram relating to damping of the temperature and damping of the heat flux through the walls.

With layout A, the insulation is located at the two ends of the heavy wall, in two parts. The insulation is obviously protected by a very light mechanical protection system. If the variable is caused to vary gradually from 0 to 1, the insulation shifts from the outer surface of the wall to the inner surface.

In the case of solution B, the heavy component of the structure is found on both sides of the insulation.

$\delta = 0$ The insulating surface is entirely on the indoor side of the wall.

$\delta = 1$ The same insulating surface is located on the outdoor side of the wall.

2.232 Comparative analysis

These parameters have been varied in this analysis:

1. Type of glazing: three types of glazing were examined:
- Single glazing with outdoor solar protection
- Thermopane insulating glazing with Parsol type absorbant external surface and internal solar protection
- Thermopane insulating glazing with Soltran type reflecting external surface and internal solar protection

2. The location and thickness of the insulation (see Fig. 11)
- Insulation placed at the outdoor or indoor side of the wall
- Standard or reinforced insulation

In all cases, we prepared diagrams of the cooling load for the two climates considered, hot and dry, hot and humid, giving 24 series of curves.

Fig. 11

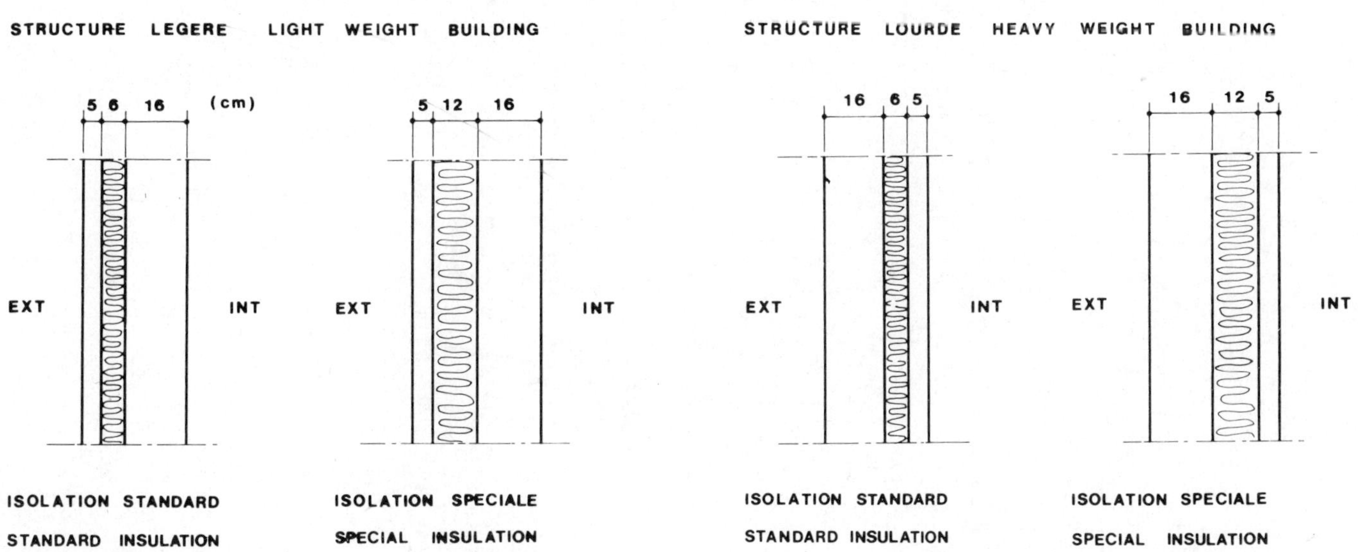

Fig. 12

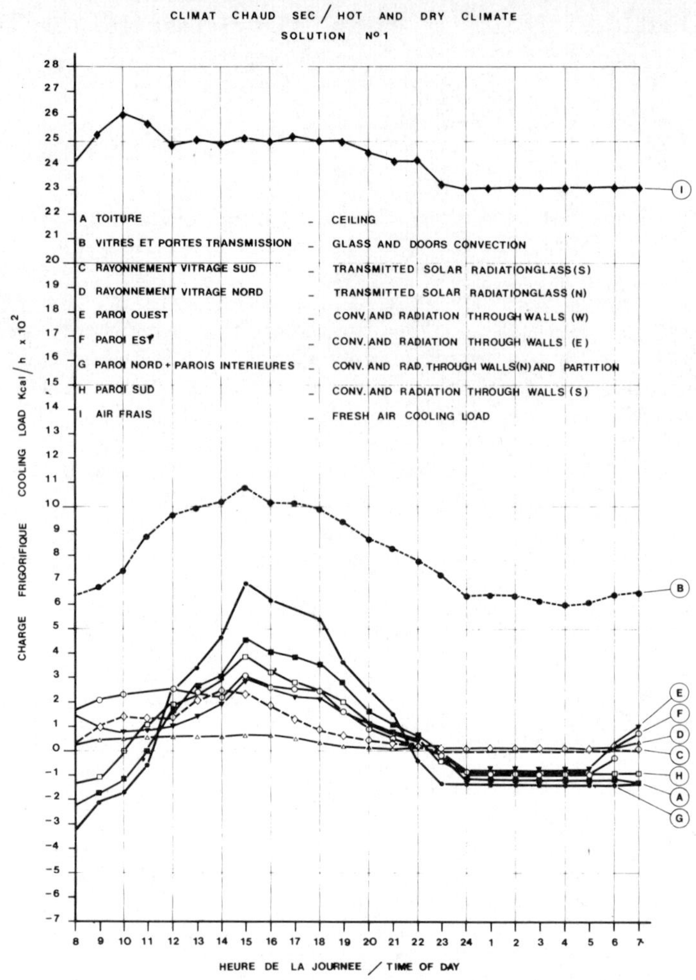

CLIMAT CHAUD SEC / HOT AND DRY CLIMATE
SOLUTION N° 1
A TOITURE _ CEILING
B VITRES ET PORTES TRANSMISSION _ GLASS AND DOORS CONVECTION
C RAYONNEMENT VITRAGE SUD _ TRANSMITTED SOLAR RADIATIONGLASS (S)
D RAYONNEMENT VITRAGE NORD _ TRANSMITTED SOLAR RADIATIONGLASS (N)
E PAROI OUEST _ CONV. AND RADIATION THROUGH WALLS (W)
F PAROI EST _ CONV. AND RADIATION THROUGH WALLS (E)
G PAROI NORD + PAROIS INTERIEURES _ CONV. AND RAD. THROUGH WALLS(N) AND PARTITION
H PAROI SUD _ CONV. AND RADIATION THROUGH WALLS (S)
I AIR FRAIS _ FRESH AIR COOLING LOAD
CHARGE FRIGORIFIQUE COOLING LOAD Kcal/h x 10²
HEURE DE LA JOURNEE / TIME OF DAY

Fig. 13

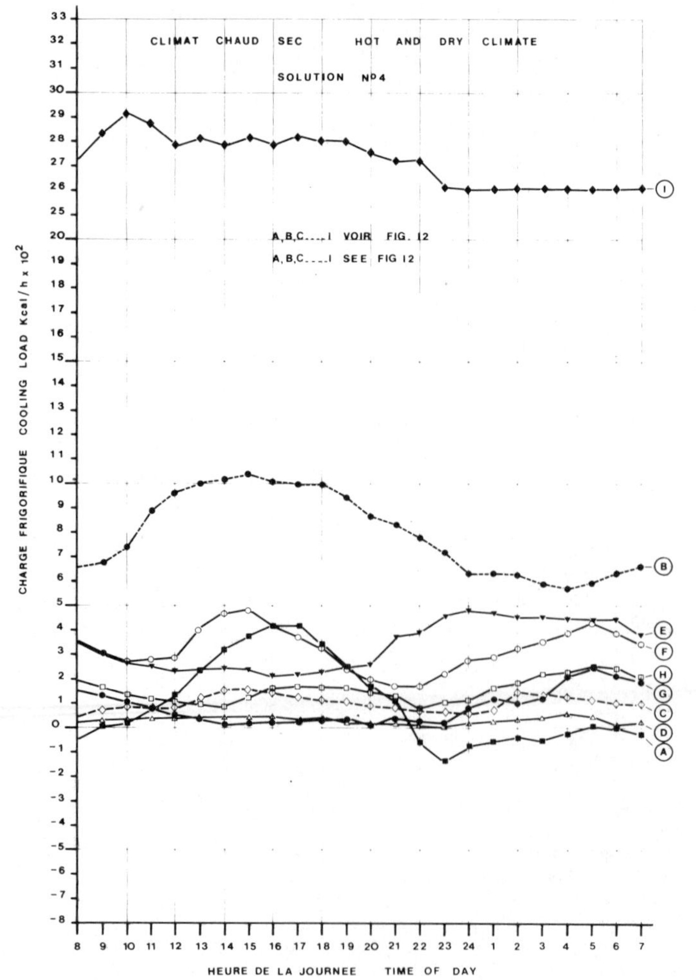

CLIMAT CHAUD SEC HOT AND DRY CLIMATE
SOLUTION N° 4
A, B, C.... I VOIR FIG. 12
A, B, C.... I SEE FIG 12
CHARGE FRIGORIFIQUE COOLING LOAD Kcal/h x 10²
HEURE DE LA JOURNEE TIME OF DAY

The diagrams were plotted for each structural component (glazing, wall, roof, fresh air, etc.)
cf. Fig. 12: hot, dry climate, <u>outdoor</u> standard insulation, single glazing.
cf. Fig. 13: hot, dry climate, <u>indoor</u> standard insulation, single glazing.

It is interesting to note that the weight of each component in the total <u>daily</u> balance differs according to the position of the insulation.

Fig. 14 shows the hourly cooling requirements of the house with the three types of glazing, and with the two insulation positions (standard thickness) in a hot, dry climate.

Fig. 14

Fig. 15 illustrates the same parameters, but for climatic conditions, corresponding to a hot, humid region. It maybe observed that in the case of a hot, dry climate, the positioning of the insulation outdoors imposes a cooling plant with a higher peak power, whereas the positioning of the insulation indoors leads to a significant reduction in the instantaneous cooling requirements.

On the other hand, the total requirements, over a 24 hours period are greater in the latter case.

The same considerations are valid in the case of hot, humid climate, but the influence of load peaks is not truly significant.

Fig. 15

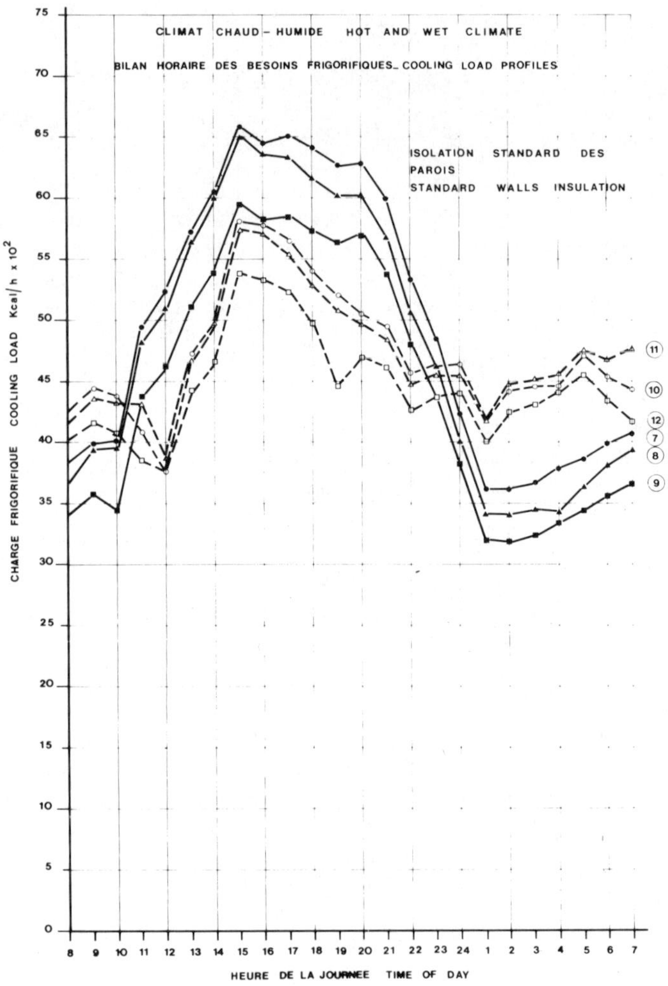

In the case of a conventional power air conditioning system in the climatic conditions in question, it is evident that the best solution is that which corresponds to indoor insulation.

It is then possible to reduce the rated capacity of the cooling plant.

On the other hand, with a solar system, it is necessary to consider not only the maximum instantaneous power, but also the daily cooling requirements.

In effect, the quantity of energy available to energize the absorption system is a finite quantity, determined by the area of the collectors. Any additional energy requirement must be supplied by conventional sources or obtained by considerably penalizing the efficiency of the absorption machine

Fig. 16 summarizes the daily cooling requirements of the 24 variants.

These diagrams provide the possibility of defining the optimum structure of a solar house once the climate of the location is known, in accordance with local technical capabilities.

Fig. 16

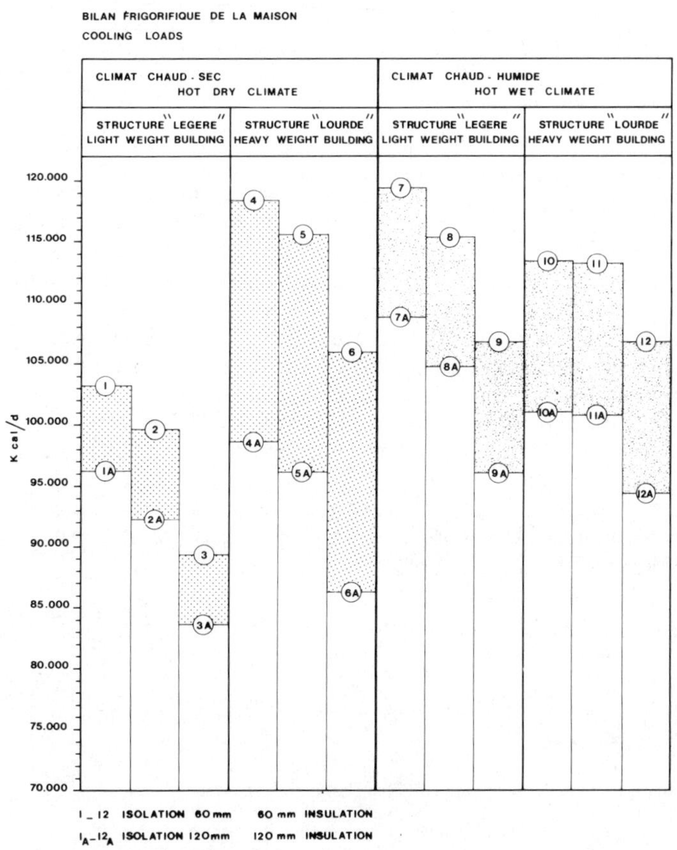

3 TEST LOOP FACILITY

A test loop was designed for analysing the operating conditions of a solar installation in variable conditions.

The cooling requirements of the building, accounting for the inertia, etc. are first calculated theoretically and the low charge obtained enables the determination hour by hour, of the return temperature of the cold water (simulation of air conditioning units).

The temperature of the heat source is variable so as to simulate the heat production of the solar collectors or of the water stored in a tank.

Fig. 17

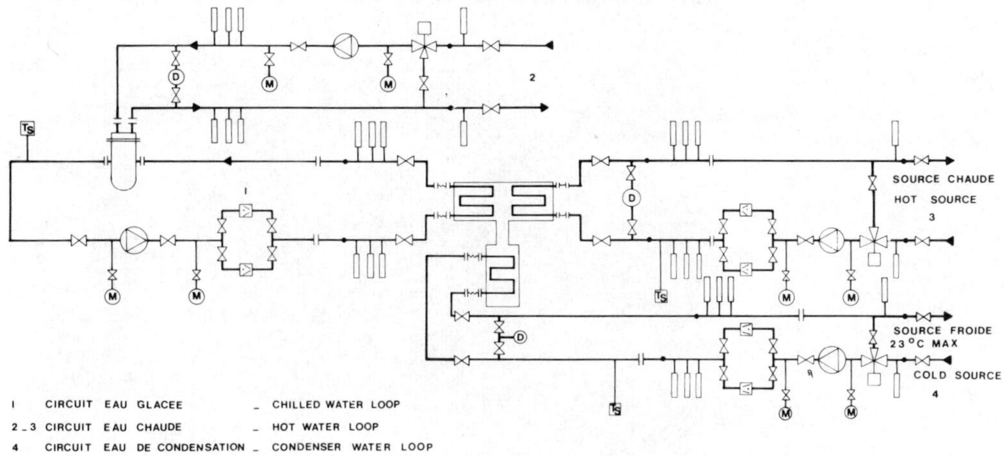

SCHEMA DE PRINCIPE BOUCLE D'ESSAIS POUR MACHINE A ABSORPTION

EXPERIMENTAL TEST LOOP FACILITY

Similarly, the temperature of the cold source can simulate the temperature (variable) of the condensation water leaving the cooling tower, depending upon outdoor conditions.

The tests make it possible to establish, for each system, the following characteristics:

- Minimum and maximum temperature of the heat source.
- Minimum and maximum temperature of the cold source corresponding to the different feed temperatures of the hot source.
- Temperature of the cold water available as a function of the above variables.
- Amount of heat to be rejected to the atmosphere.
- Heating capacity depending on load conditions.
- Corresponding cooling capacity.
- Efficiency of each machine in the different operating conditions.
- Response of the machine to rapid variations in the different parameters.
- Losses due to energization of the absorption machine.
- Total efficiency of a system in variable operating conditions for a 24 hours cycle or longer.

Fig. 17 shows the schematic of the test loop, of which various views are given in Fig. 18, 19 and 20.

An initial series of tests is nearly completed, and the results are being interpreted.

A second series is due to begin shortly.

Fig. 18

Fig. 19

Fig. 20

4 FLUID STORAGE PROCESS BY MEANS OF A RESERVOIR LIMITING THE ENTHALPY DEGRADATION

4.1 Concrete problem to be resolved

In a heating (or climatic control) system (2) with a discontinuous heat source and hot water storage, one must be able to use and store the energy at the highest temperature level. When the installation is started, or at the end of the operating cycle, the temperature of the liquid stored in the reservoir will obviously be at its lowest level, and the efficiency of the heating units is liable to be unacceptable.

For example, in the same operating conditions, an absorption cooling plant has an efficiency of 0,55 if the hot water is at 90°C, and an efficiency of 0,45 if the hot water is at 80°C.

It is also necessary to consider that a degradation in the temperature level in the storage tank will always have two negative consequences:
- reduction of available capacity;
- increase in energy used to produce the same thermodynamic process.

The enthalpy degradation in a conventional storage tank requires an increase in the quantity of heat stored (3).

The storage tank described enables the storage of a fluid assigning priority to quality (temperature level) rather than quantity, unlike conventional tanks which exploit the quantity principle without quality control.

4.2 Description of the proposed process

The solution to the problem involves the use of a system featuring variable ratios of hot to cold masses by means of a floating diaphragm. Another advantage is derived from the fact that the fluid to be heated can be drawn from the low temperature portion of the tank, providing better heat exchange in the heat source.

The heated fluid is then sent to the hot storage tank.

Fig. 21

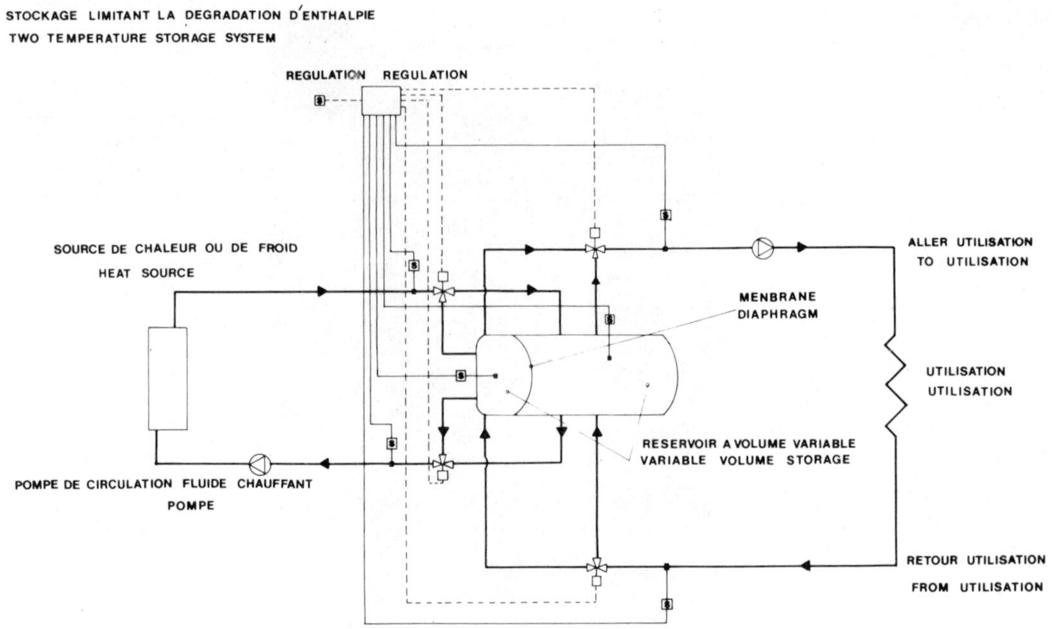

Depending on its temperature, the fluid from the load sections can be introduced into the hot storage tank or the cold storage tank.

The system proposed makes it possible to have, in a single tank, variable hot and cold storage volumes, in accordance with heat requirements and inputs. A tank with practically constant temperature, with variable volume, is used to feed the load units (radiator, convector, absorption chiller, etc.) while the degraded energy receives the return fluids and feeds the heat input source.

The storage tank can be made of metal, plastic or concrete. The choice of a diaphragm for the separation of the liquids depends on the tank volume and on the requirements of specific applications.

A diagram of the system is shown in Fig. 21.

Some details of the storage tank processes are given in Figs. 22, 23, 24, 25, 26.

Fig. 22

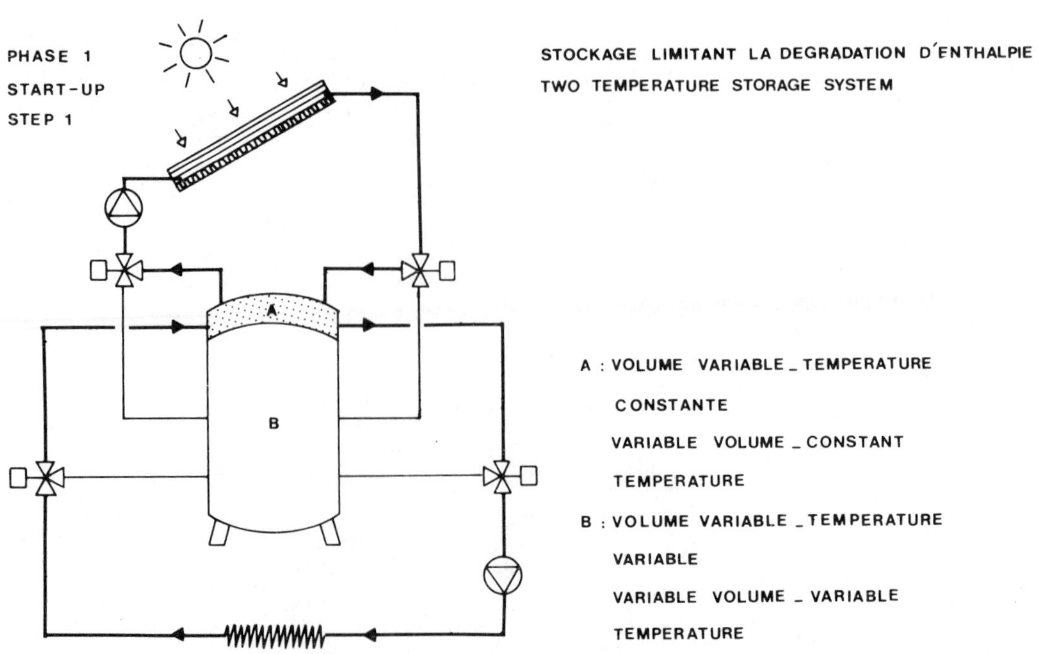

Fig. 23

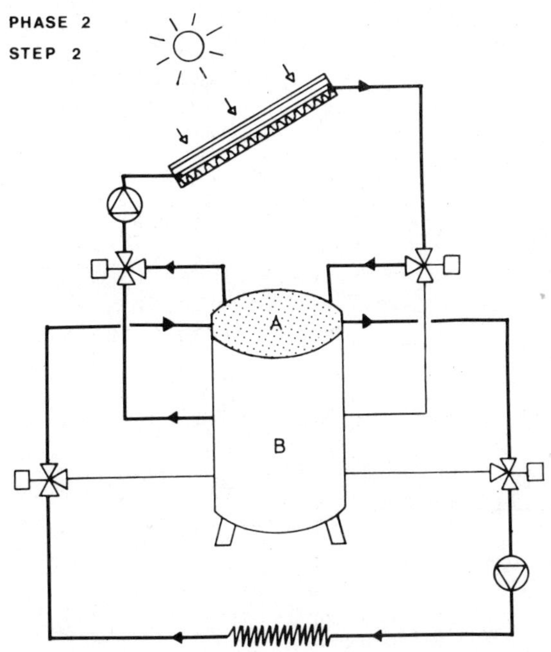

Fig. 24

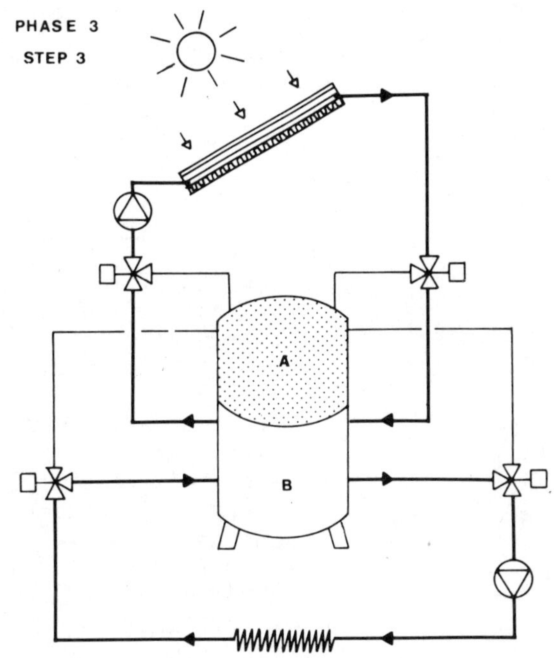

Fig. 25

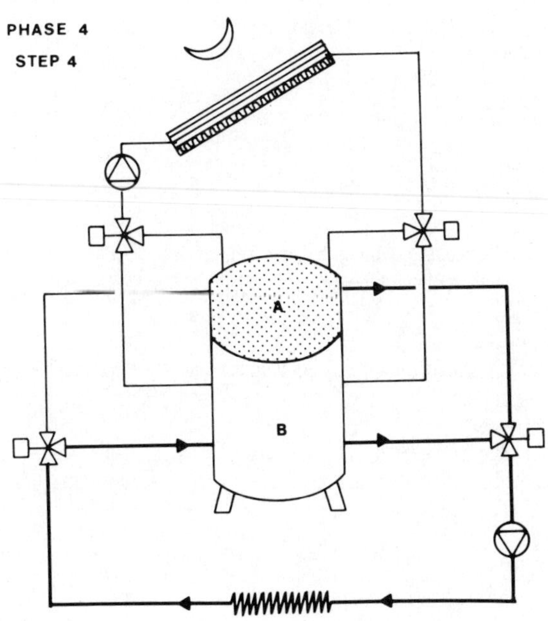

Fig. 26

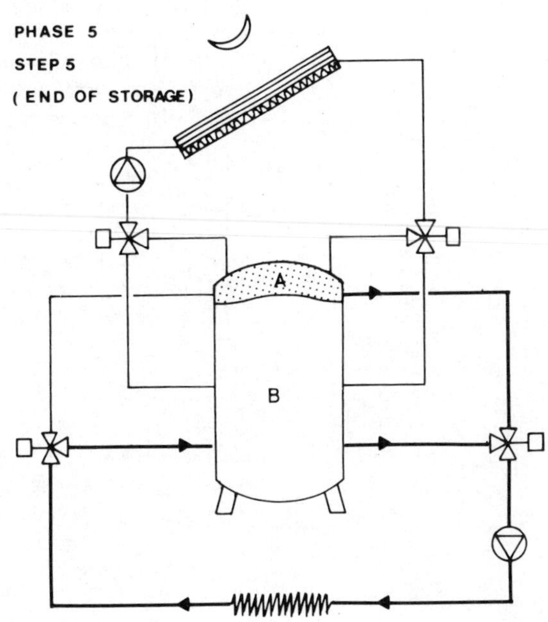

5 RESEARCH PROGRAM

In addition to the supplementary work and the specific investigations already undertaken, two current work trends have been adopted:

1. Development of a regulation system for the overall climatic control installation.
2. Building of experimental houses.

5.1 Regulation system

The objective is to build a number of printed circuits placed in a "black box", which actuates the climatic control system without the intervention of the user, in line with the performance of optimization programs devised.

The composition of the circuit boards depends on the various parameters examined above:
- climatic conditions of the location;
- physical characteristics of the building;
- inertia, heat transfer, etc;
- characteristics of the installation;
- efficiency of the cooling plant;
- efficiency of the solar collectors, etc.

5.2 Experimental houses

The building of one or more experimental climate-controlled houses in the near future will make it possible to:
- confirm the results of the above theoretical studies;
- develop the climatic control system;
- test the various components in actual conditions;
- perfect the regulation of the system.

REFERENCES

1 DGRST (Délégation à la Recherche Scientifique et Technique) and the Delegation aux Energies Nouvelles are committees of the 'Ministère l'Industrie et de la Recherche.'

2 The example considered of heat storage is obviously valid for a cold storage unit.

3 Lamination can partly resolve this problem by the use of a tank featuring large vertical development. This is difficult to accomplish in view of installation constraints.

Solar water heaters in Egypt

Professor Dr. M.S.M. Abou-Hussein
Department of Electric Power and Machines
Faculty of Engineering
University of Cairo
Giza, Egypt

Abstract

This paper gives some statistics concerning the values of the daily total and sky radiation incident upon a south facing surface tilted 30 degrees from the horizontal at Cairo during cloudless days over the year 1976. Also the average air temperature at Cairo during the years 1971-1976 is given.

The results of tests made on a solar water heater of the specifications mentioned in the paper are given including: the water temperature rise function of the discharge at a certain time for a certain heater, water temperature rise for different discharges and for different collector areas at maximum incident radiation, and constant discharge test for different collector areas.

An economical comparison for the total yearly annual charges for the conventional and solar water heating systems is forwarded, and some technical improvements are suggested.

Introduction and Review

Egypt has a limitless amount of sunshine. On the average, only sixty days per annum are cloudy or semi-cloudy around Cairo. The daily total direct and sky radiations on a south facing surface tilted 30 degrees from the horizontal at Cairo (latitude 30 degrees north) during cloudless days varies from a maximum of about 2500 BTU/sq ft day during April to a minimum of about 2000 BTU/sq ft day during January.

A one square meter solar water heater may heat, therefore, about 110 litres of water per day from the April average temperature of 21 degrees centigrade to 55 degrees centigrade or about 80 litres of water per day from the January average temperature of 13.8 degrees centigrade to 55 degrees centigrade. It may also heat 132 litres of water per day during August having a maximum incident radiation of 2300 BTU/sq ft day from the average temperature of about 28.6 degrees centigrade to 55 degrees centigrade. The heater efficiency may be assumed 70%.

In other words the heater of one square meter area may heat annually a quantity of water to 55 degrees centigrade ranging from a minimum of about 80 litres/day to a maximum of about 132 litres/day.

The importance of these figures is apparent when the continuity of the sunshine during at least 10 months per year, and the relatively high cost of fuel (other than solar) used at present for water heating are pronounced.

During the past 10 years, different types of solar water heaters were made and their use was confined to only laboratory experiments. The basin open cycle type, characterised by its low cost, was found suitable for countrymen and the unit was standardized to be of one square meter area and multiples. The metal-in-strip closed cycle type found suitable to heat water for city and country services (e.g.) hospitals, hotels, schools, private and public houses and for country public washing units is standardized to have two square meters area and multiples. For big installations, (e.g.) the university city and the like, any size, other than the standard sizes mentioned, may be made in metal-in-strip. With the two square meter metal-in-strip, heater, the storage tank has a capacity of about 150 litres. The natural circulation (thermo-syphon) system or the forced circulation system with a pump controlled either thermostatically or by radiation may be used. The plastic 'pillow' or 'dish' type of solar water heaters are in their early stage of laboratory testing.

In the following, tests were made on a metal-in-strip heater, 2 square meters area, with forced water circulation. The solar radiation and average ambient yearly temperatures at the place of testing are given below:

Table 1: Daily total direct and sky radiation incident upon a south facing surface tilted 30° from horizontal at Cairo (30° North) during cloudless days.

First day of	Incident radiation BTU/sq ft day
January	1970
February	2100
March	2280
April	2490
May	2400
June	2300
July	2280
August	2350
September	2400
October	2300
November	2200
December	2000

Table 2: Average air temperature °C at Cairo-South during the years 1971-1976.

Year	1971	1972	1973	1974	1975	1976
January	14.3	13.9	13.0	12.4	14.9	14.4
February	15.1	15.5	15.3	14.4	18.0	16.5
March	19.2	17.4	14.7	18.4	18.4	16.0
April	21.4	21.7	21.3	19.9	21.8	21.0
May	26.2	24.9	26.1	26.3	24.6	23.4
June	27.1	26.9	27.7	27.3	28.1	27.7
July	28.1	27.8	29.5	28.6	28.6	28.8
August	28.8	29.5	28.2	28.8	27.4	29.1
September	27.7	28.2	25.6	25.4	26.9	28.0
October	24.9	25.0	23.7	23.5	25.1	22.2
November	19.1	19.6	17.4	19.4	19.6	19.4
December	12.2	16.6	13.7	14.2	15.1	14.3
Average	22.1	22.2	21.4	21.6	22.4	21.8 °C

Performance

The heater is made of one corrugated galvanised iron sheet and a plane sheet. The two sheets are riveted at several points along the lines of contact. The end edges are soldered to square headers made also of galvanised iron of thicker gauge. Tap water is let in at the lower end and flows up the channels (eight per meter) into the upper header. The plane sheet is 1.0 x 2.0 meters and about 0.750mm thick. Two, 3mm thick, glass covers are used with about 2cm spacing. About 5cm space exists

Table 3: Variable discharge test at maximum incident radiation. Heater area about 2 sq m tilt 30° with horizontal latitude 30° North.
Date: 13/4/76 Sky: sunshine Duration of test: 12.30-13.30
Inlet water temperature = 20°C

Discharge Litres/minimum	Temperature rise $^{\circ}$C	Discharge Litres/minimum	Temperature rise $^{\circ}$C
0.2	62	1	25
0.3	50	2	17
0.4	41	3	13
0.5	35	4	12
0.75	30	5	11
		6	10

Similar tests were made with heaters of 4, 6 and 8 square meters and the results are given in Table 4.

Table 4: Variable discharge tests at maximum incident radiation (under conditions of Table 3).

Discharge	4m2	6m2	8m2
	Temperature Rise $^{\circ}$C		
0.2	70	74	79
0.3	60	67	72
0.4	56	60	66
0.5	53	57	61
0.75	42	51	55
1	36	44	51
2	24	33	42
3	22	29	36
4	21	26	33
5	19	24	30
6	17	21	27

The temperature-hour curves of the heaters of the mentioned areas are given in Table 5.

Table 5: Constant discharge test (all conditions of Table 3 except: inlet water temperature: 22.1°C Test duration: 7-17 hours, Date: 27/4/76 Constant discharge of 0.50 litre/minimum).

hour	2m2	4m2	6m2	8m2
	water outlet temperature $^{\circ}$C			
7	28	34	38	41
8	33	41	46	49
9	40	49	55	59
10	45	57	64	70
11	52	64	71	76
12	55	71	76	81
13	57	73	79	84
14	51	71	78	83
15	44	66	74	77
16	37	56	63	68
17	32	45	54	59

between the lower glass cover and the upper iron. The back insulation is glass wool of about 10cm thickness. The iron surface was blackened by a suitable method.

In one test, water at 22°C, was pumped through the heater at different rates obtaining results in Table 3.

Some economic aspects

A 2 sq m solar heater may heat yearly an average of 270 litres with a daily average temperature rise of 25°C.

Thus the useful daily heat energy obtainable from such a heater will be 6750 K cal/day. A yearly figure corresponding to 300 days of sunshine in the Cairo area will be 2,025,000 K cal/annum.

The three sources of fuel mainly in use for water heating are the kerosine, the butane gas and electricity. The fuel weight equivalent to the annual K calories usefully obtained from the heater are respectively 607 kgm of Kerosine, 200 kgms of Butane gas and 2600 Kw an hour.

(The calorific values of kerosine and butane gas are assumed to be 10,000 and 12,000 K cal/kgm respectively. The overall efficiences of the heaters, were taken as 33%, 80% and 92% for the cases of heating by kerosine, butane gas and electricity respectively).

The corresponding price of the fuel saved by using a 2 sq m solar heater for one year will thus amount to the following based on the existing market prices:

14.00 Egyptian pounds or 20.0 US dollars of Kerosine
19.22 Egyptian pounds or 27.46 US dollars of Butane gas
50.40 Egyptian pounds or 27.00 US dollars of Electricity

The manufacturing cost of the solar heater with the tank amounts to 100 L.E. or 140 US ¢ and its sale price is about 122.5 L.E. or 175 US ¢. The average yearly consumer expenses for the existing and solar water heaters are mentioned in Table 6.

Table 6: Average yearly expenses for the different types of heaters.

Type	Kerosine	Butane gas	Electricity	Solar
Sale price	6.0	90	111	175 US ¢
Life	3	10	10	10 years
Fuel cost	20.0	27.46	72.00	0 US ¢/year
Depreciation	2.0	9.00	11.10	17.5 US ¢/year
Total yearly	22.0	36.46	83.10	17.5 US ¢/year

(Neglecting maintenance expenses in each case).

This shows that heating water by solar energy is the cheapest method, and it covers its initial cost in not more than 8 years if it replaces the non efficient Kerosine water heaters, and in 2 years if it replaces the most expensive electric heaters.

Suggested technical improvements

The following proposals are suggested in this respect:

(a) Changing the bulky shape of the heater.

(b) Producing an efficient water pump suitable for the purpose to enable the use of the tank indoors below the heater.

(c) Introducing a selective paint other than using the electroplating method. The paint should be cheap in cost and easy to apply by merely a brush. At the same time, it should not be oxidized nor change its colour by the prolonged exposure to sunlight.

(d) Reducing the cost by introducing new cheap materials which are absorptive to solar radiation and non-emissive to heat energy.

6.9 Development of built-in storage type solar water heaters for large scale use in developing countries

A. Russell-Cowan
Managing Director
Solar Centre Ltd
176 Ifield Road
London SW10

C. Stambolis*
Research Associate
Department of General Surveying and Construction
North East London Polytechnic
Forest Road, London E17 4JB

Abstract

This paper describes the development of low cost solar water heaters which combine the storage tank and collector in one integral unit. Several prototype versions of this solar water heater have been developed by Solar Centre Ltd and have been tested by NELP in Greece. Preliminary tests suggest that significant gains in terms of overall efficiency and maximum water temperatures can be achieved by slight modifications in the design of the above heaters. Results from some of the tests carried out are shown and a discussion follows concerning the possibility of introducing such units on a large scale in a number of developing countries.

Introduction

The principal aim of the development work was to design and manufacture few prototypes of solar water heaters which combined in one unit both the collector and storage capacity. In this way manufacturing and installation costs can be significantly reduced thus enabling the mass production of such units in many poor regions of our world. Extensive work has already been carried out by various researchers on Built-in Storage Type solar water heaters. In particular Chinnappa and Gnanalingam (1), Garg (2), Chauhan and Kadambi (3) and Chinnery (4) have conducted field tests which demonstrate that built-in storage type solar water heaters perform well under variable climatic conditions.

Built-in storage type solar water heaters are generally described as being receptacles of varying shapes such as rectangles, cylindrical circular blackened tanks holding a given amount of water according to size and set in a frame with an air space in front and insulation behind the tank.

The water being in direct contact with the blackened surface absorbing plate. An efficient method of collecting and storing heat. The aim of our work has been to carry out the manufacture and testing of a series of prototypes with variations in order to improve performance and seek a design for mass production.

The idea of integrating the collector plate with the storage in one integral unit is advantageous for the following reasons:
(a) No separate storage tank is required.
(b) The heater can be both gravity and pressure fed.
(c) Production costs can be substantially reduced.

---

*Present address:
Intermediate Technology Development Group, 9 King's Street, London WC2E 8HN

Design of solar water heater prototypes

The first prototype developed showed an improvement in the tank concept by the addition of a bent copper pipe, snake fashion, at the bottom of the tank. By adding the above pipe it was found that the thermo-syphoning process could be better enhanced.

The present models show an improvement in the tank concept by the addition of thermo-syphon apparatus.

In the normal use of thermo-syphon systems the collector is placed so that the top of the collector is a minimum distance below the bottom of a storage tank in order to reduce reverse flow. Usually not less than 300mm.

Experiments have proved that this minimal distance can be reduced to a point where the collector outlet is level with the bottom of the tank without affecting the system's performance to any serious degree. (5)

The flow and return connections are both placed at the bottom of the tank instead of the conventional method of the flow pipe being connected some distance up the side of the tank.

With the flow and return connections being at the bottom of the tank when hot water is drawn off the mixing of the incoming cold water effect is reduced. Resulting in a fairly even temperature throughout the tank.

Preliminary tests indicate a reduction in the amount of reverse circulation at night.

The type of the collector section below the base of the tank can be of various forms such as the snake, ladder grid or whatever shows the most satisfactory results. The collector plate being an extension of the tank's upper surface. Thereby giving a good relationship between the collector area and storage capacity. Tests indicate an optimum depth lies between $2\frac{1}{2}$" to $3\frac{3}{4}$".

Fig. 1 shows a diagrammatic plan of the collector in question and figure 2 shows a photograph of two built-in storage type collectors. The collector on the left hand side has a snake type pipe attached below the main tank while the one on the right hand side is just a plain tank type.

Figs. 1, 2
Tank with integral collector

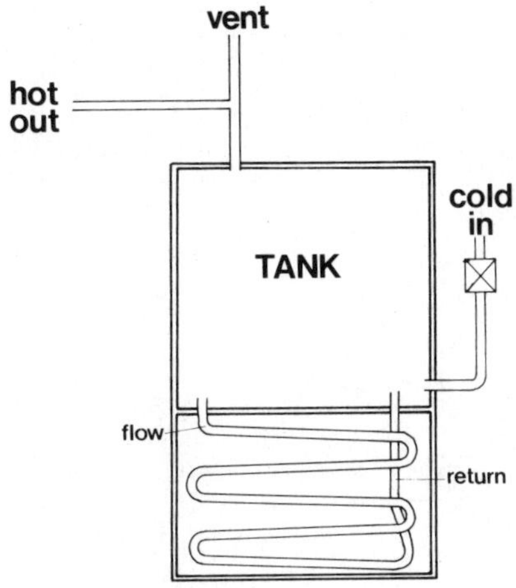

Fig. 3
Internal baffle plate

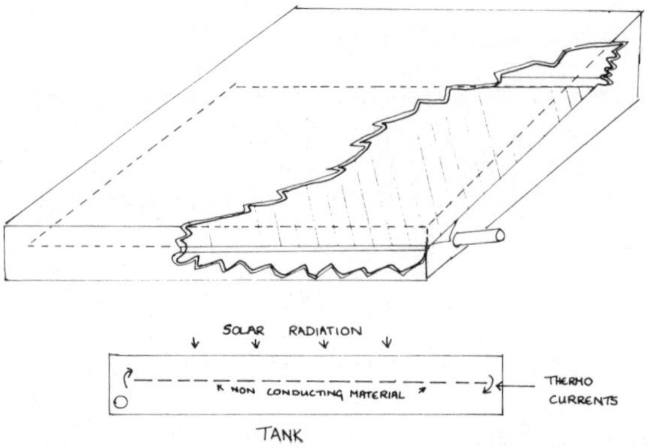

Fig. 4

A further improvement may be made in the collection of heat by incorporating an internal thin baffle plate between the blackened tank surface and the main bulk of water. This thin channel of water nearest to the blackened plate and the absorbed heat is concentrated on a small portion of the water held by the whole tank giving rise to thermo-syphonic currents, and a quick rise in temperature. A thin film of water surface facing the sun. It is important that the inner leaf of the channel should be made of a non conducting material such as a high density plastic in order that there is reduced conduction or heat loss through the back part of the baffle plate nearest to the main bulk of the water.

Fig. 3 shows a cross sectional view of the collector described above and figure 4 shows two prototype units of this type under test. The transparent cover material in this case is 4mm ordinary glass.

In figure 5 four possible methods for operating built-in storage solar water heaters are shown.

Fig. 5

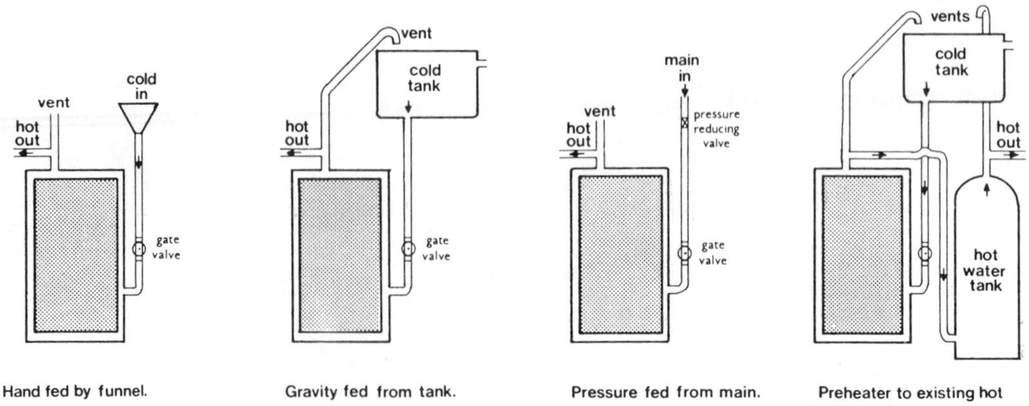

Hand fed by funnel. Gravity fed from tank. Pressure fed from main. Preheater to existing hot water system.

Fig. 6

Performance tests

The first series of performance tests were conducted with two 7.3 lts capacity prototypes, rectangular shaped, tanks which are shown in figure 6. The actual tank was made from copper (15 thou) and the cover material was aluminium with glass fibre used for insulation.

Thermal performance tests were carried out on the 2nd and 3rd of July, 1977 in the town of Rethymnon in Crete (35°N). On the first day the collectors were filled with water at about 11.20am and temperatures were taken at regular intervals with mercury thermometers until 5.00pm. At the same time the total solar radiation was recorded by means of a kipp-zonen pyrheliometer. On the second day three draw-offs were made during the day. The temperatures achieved by the collector which had an added copper pipe enabled three draw-offs while the plain collector achieved satisfactory temperatures for only two draw-offs. Figures 7 and 8 show graphs of the recordings taken during these two days.

On September 21, 1977 thermal performance tests were made on two improved versions of the same type collector, again one using an added

Fig. 7

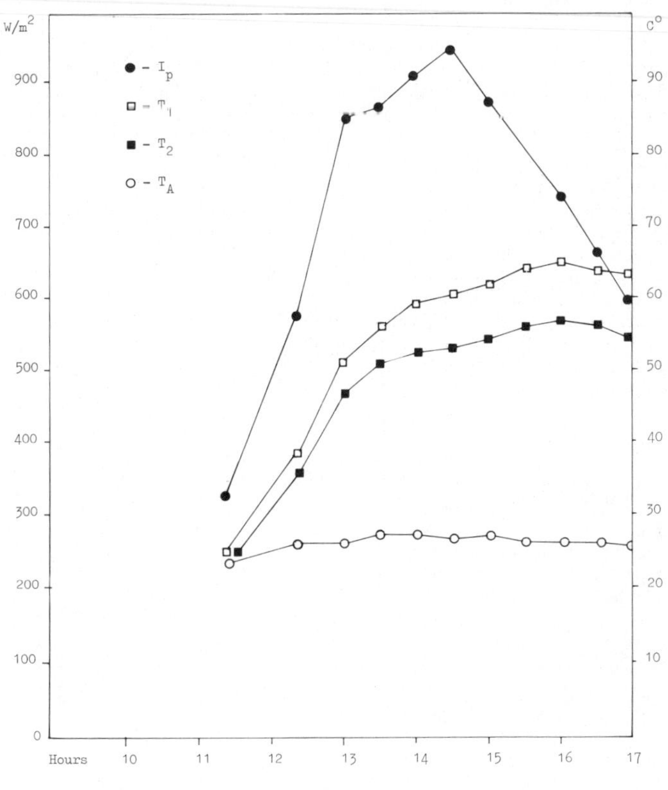

Performance tests: Results for 2.7.77

Fig. 8

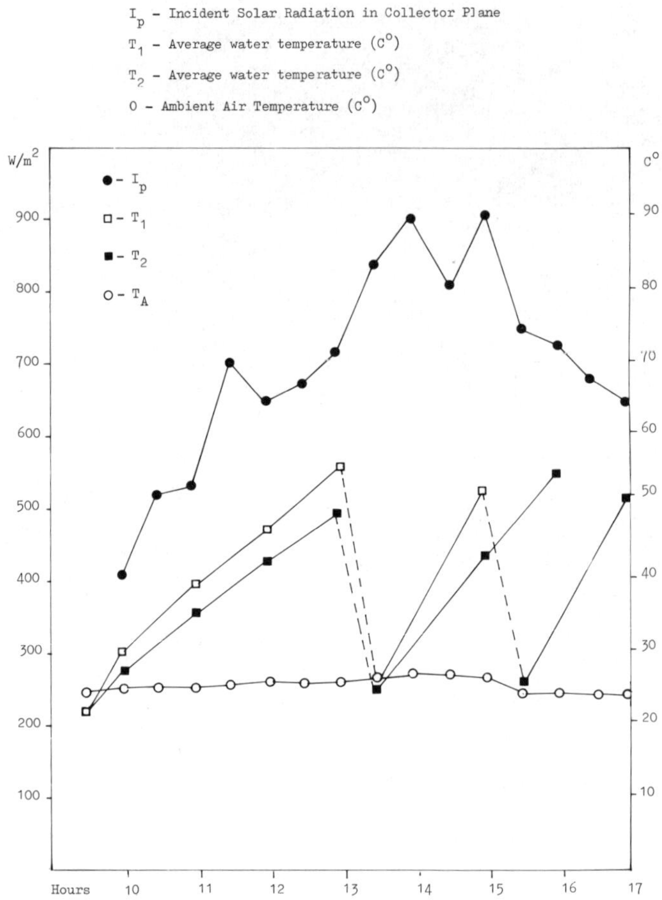

Performance tests: Results for 3.7.77

Fig. 9

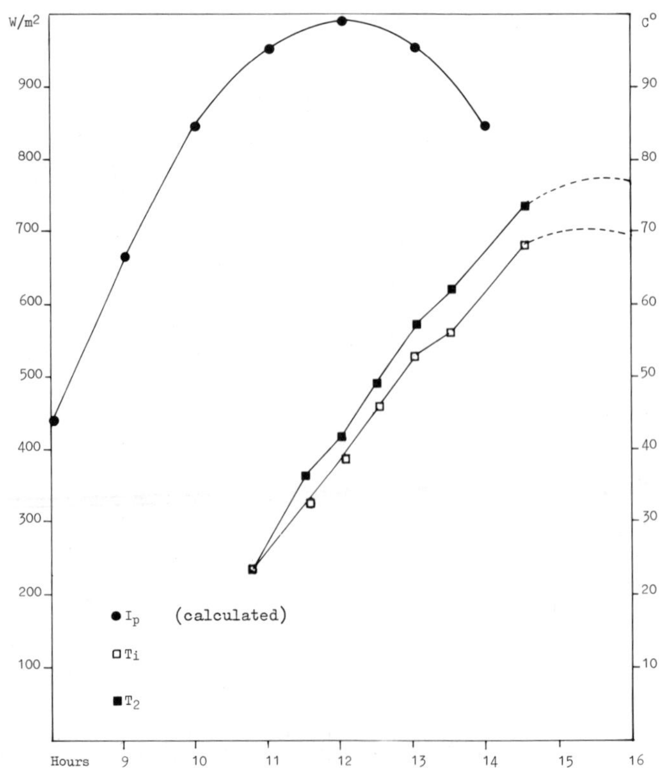

Performance tests: Results for 21.9.77

copper pipe and the other without it. These collectors are shown on figure 2 and the graphs on figure 9 show the temperatures and radiation throughout the test period. The collectors were orientated to the south and were tilted at 42° to the horizontal. The volume of water contained in this improved version was larger than in the previous tests. These collectors could hold 12.8 lts of water. This is not a sufficient quantity for household needs but nevertheless a very good indication of its performance can be obtained at this scale.

Discussion and Conclusions

The 'collector cum storage' concept for solar water heaters could provide a basis for the manufacturing of really low cost water heaters for third world countries. With our present studies we sought to demonstrate that low cost collectors, such as the ones we experimented on, do not necessarily mean low efficiencies. These were found to vary between 35% and 70% for the collectors tested. A more detailed study is underway in order to determine precisely the variation of efficiency with temperature for different radiation levels. The economics of the 'collector cum storage' collectors compare very favourably with those related to classical flat-plate collectors. Following detailed cost analysis it has been concluded that collectors of this type can be constructed on modules 60 to 100 lts for corresponding prices of £20 to £30 per unit. A comprehensive feasibility study is now being planned by the authors for India and the Sudan in order to test different materials and study the socio-economic framework which is necessary for a large scale decentralised production of solar water heaters.

References

(1) J.C.V. Chinnappa and K. Gnanalingam, 'Performance at Colombo, Ceylon, of a pressurised solar water heater of the combined collector and storage type', Solar Energy, Vol. 15, pp 195-204.

(2) H.P. Garg, 'Year round performance studies on a built-in storage type solar water heater at Jodhpur, India', Solar Energy, Vol. 17, pp 167-172.

(3) R.S. Chauhan and V. Kadambi, 'Performance of collector cum storage type solar water heater', Solar Energy, Vol. 18, pp 327-335.

(4) D.N.W. Chinnery, 'CSIR Research Report 237', Pretoria, S. Africa 1967.

(5) J.T. Czarnecki and G.K. Johnson, 'Solar water heater with a horizontal storage cylinder', CSIRO, Internal report no. 128, Division of Mechanical Engineering, Melbourne, November 1973.

6.10 Pipe type solar water heating

S.K. Kaila, R. Ganguly, J.S. Puri & N.K.D. Choudhury
Central Building Research Institute
Rourkee, India

Abstract

A pipe type solar water heater consisting of eight numbers of pipes fabricated from G.I. sheet has been designed and developed at the Central Building Research Institute, Roorkee. The pipes perform the dual function of absorbing heat and storing the heated water. It may be coupled to a separate storage tank which can be installed in the bathroom or kitchen. It can also be coupled to an electric geyser in the house. For day time use the water from the solar heater can be drawn directly. For use in the evening and early morning hours the hot water from the solar heater can be drawn off fully or partially to the storage tank or electric geyser at about 15.00 hrs, when the temperature of water in the heater is maximum.

A prototype of the pipe type solar water heater was tested under different operating conditions. The results show that the designed solar heater supplies 100 litres of water at a maximum temperature of 67-70°C at 15.00 hrs, provided there were no withdrawals of hot water during the day. However, if 10 litres of hot water at hourly intervals are drawn for day time use, starting from 09.00 hrs then the maximum temperature of 10 litres of water drawn at 15.00 hrs is 55°C.

Introduction

Solar water heating probably represents the most widespread engineering use of solar energy at the present time. Solar water heaters are in commercial production in many countries and are competing more or less successfully with fuel operated systems. However, the economic advantages are often marginal due to high investment cost. Research is therefore necessary for the development of efficient solar water heaters. Since the solar energy is available free of cost, the operating cost of any solar device comprises interest on investment, depreciation and maintenance. It would therefore be mostly a matter of opinion only where solar water heaters become uneconomical.

The conventional solar water heaters with a separate flat plate absorber and storage tank are costly for the common man in India. In view of this a cheaper pipe type solar water heater has been developed and its performance tested under different operating conditions. This pipe type solar heater performs the dual function of solar heat collection and storing of heated water.

Hot water is available for day time use directly from the heater. In addition to this it may also be coupled to a separate small storage tank which can be placed in the bathroom or the kitchen in the house. Thus for day time use the hot water from the heater can be drawn directly, whereas for use in the evening and early morning the hot water from the heater can be drawn off partly and stored in the separate storage tank. If the house is provided with an electric geyser then there is no need for the separate storage tank and the solar heater could be coupled with the tank of the electric geyser for the evening and early morning use. It has been observed that the maximum temperature of water in the heater is obtained around 15.00 hrs. It would therefore be desirable to drain off the water to the storage tank at this time of the day.

Fig. 1

Design of the heater

The pipe type solar water heater as shown in Figure 1 consists of eight pipes fabricated from 20 SWG G.I. sheet. It has a storage capacity of 100 litres. Each of the pipes is 100.00cm in length and has an inner diameter of 12.5cm. Each of the two ends of the pipes are closed by two circular discs from the same G.I. sheet and having a central hole of 12.50mm diameter for fixing the heater.

All the eight pipes are placed close together horizontally with a spacing of 2.5cm in between and welded to two G.I. pipe headers at both the ends. The two headers consist of G.I. pipes of size 127.00cm length x 19.00mm diameter. Eight short pieces of G.I. pipes of size 30mm x 125mm diameter are welded to header pipes perpendicularly having a centre to centre distance of 15.0cm. The exposed surface of the absorber is coated by ICI black paint.

The assembly is then placed on a 2.5cm thick layer of insulating material placed in a trough of M.S. or A.C. sheet. The front of the trough is covered with a glass sheet and A1 angles provided at the edges to hold the glass sheet in position. This completed unit is then mounted at an angle of 45° (latitude + 15°) from horizontal and held rigidly in position by means of angle iron frame. The entire unit is oriented due south to collect maximum daily solar radiation.

Performance of the heater

A prototype solar water heater with the above design was fabricated and its performance tested under different operating conditions. It has been found that the maximum temperature of water is attained at about 15.00 hrs. The temperature of water at outlet from the heater at different hours on two different winter days is plotted in Figure 2. It is seen from the curves that the maximum temperature of water at the outlet from the heater at 15.00 hrs during winter is 67 to 70° when no water is withdrawn from the heater for day time use.

To observe its actual performance in domestic situations, 10 litres of water was drawn off hourly from the heater starting from 09.00 hrs in the morning. The average temperature of 10 litres of water drawn off is

Fig. 2

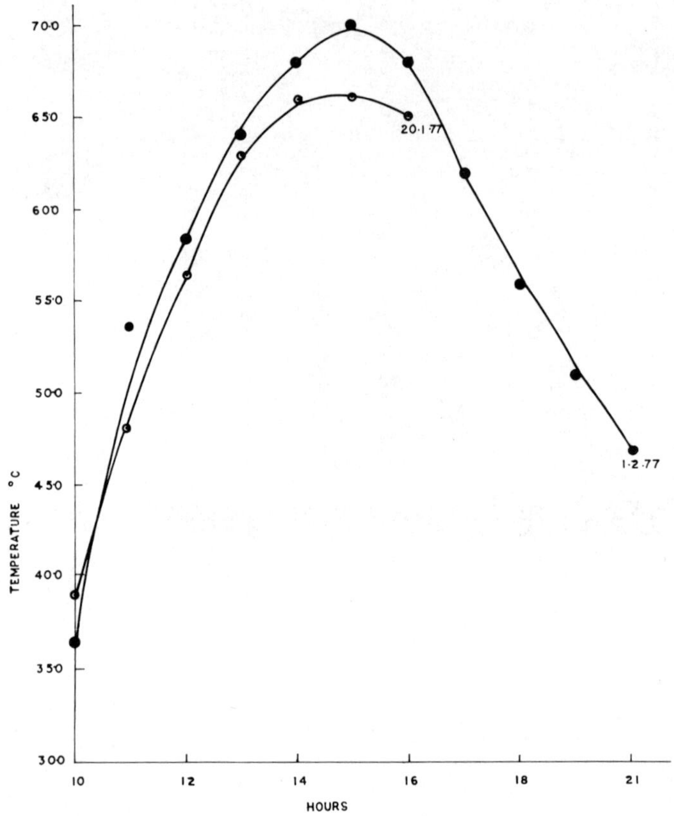

Temperature of water in pipe type solar water heater.

Fig. 3

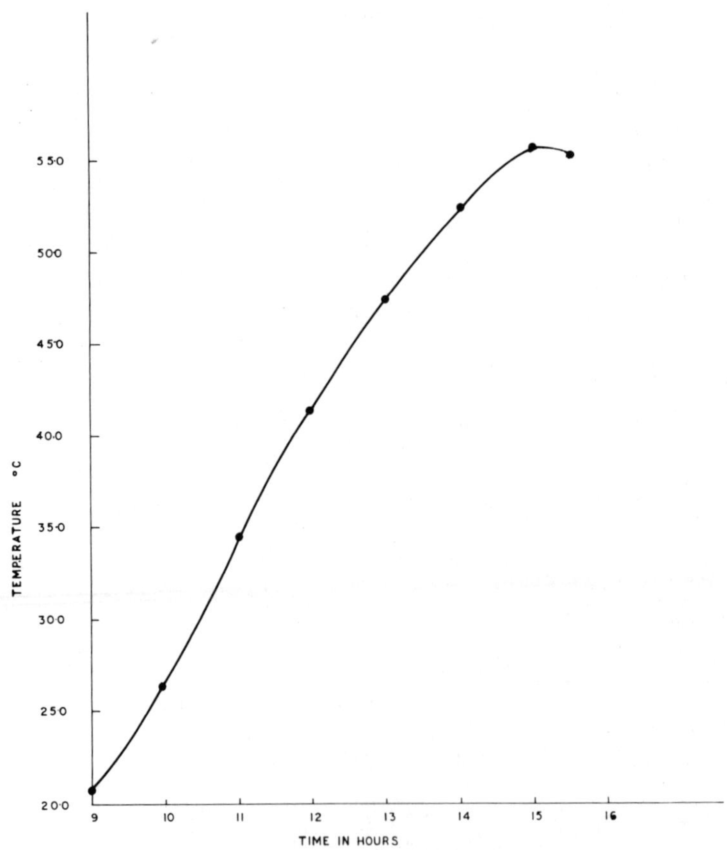

Average temperature of 10 liters of water drained off hourly from pipe type solar water heater on 21.2.77.

plotted against the time of the day in Figure 3. In this case also the maximum temperature of water is attained at 15.00 hrs. The maximum average temperature of 10 litres of water at 15.00 hrs is 55.7°C even when 60 litres of water had already been withdrawn in the earlier hours.

Although the heater has been designed to heat and store the heated water only during the day time, still the temperature of stored hot water in the heater was measured beyond 15.00 hrs on a particular winter day. The results as plotted in Figure 2 show that the temperature of water falls considerably in the evening. Next morning the temperature of water in the heater was found to be as cold as tap water. Thus for obtaining hot water in the evening and for early morning use it is necessary to take off the hot water from the heater around 15.00 hrs and store in a separate tank placed in the bathroom or the kitchen in the house. However, if the house is provided with an electric geyser then this hot water could be stored therein.

An idea of the collection efficiency of this pipe type solar water heater can be had from the following:

Quantity of water heated	=	100 litres
Average temperature of water in the heater at 15.00 hrs	=	55.5°C
Average rise in temperature of water from 11°C at 07.00 hrs to 55.5°C at 15.00 hrs	=	44.5°C
Total useful energy collection from 07.00 to 15.00 hrs	=	100x1x44.5
	=	4450 Kcal
Total solar insolation from 07.00 to 15.00 hrs on the absorber	=	4096 Kcal/m2
Total absorber area (half the total surface area of the pipes)	=	1.6m2
Total insolation on the absorber	=	4096x1.6
	=	6553 Kcals
Collection efficiency of the heater	=	68%

Conclusions

1. The pipe type solar water heater functions both as a solar absorber and hot water storage unit.

2. The temperature of water from the heater is maximum at 15.00 hrs whether the water is withdrawn or not till this time.

3. The temperature of water from the heater is quite high as shown by the performance curve.

4. For use in the evening and early morning hours the hot water from the heater could be drained around 15.00 hrs and stored in a separate small storage tank or the electric geyser.

5. The unit is sturdy enough not having any problem of bulging.

6. The efficiency of the heater calculated for a typical winter day is around 68 per cent.

Acknowledgement

The authors are grateful to Director, CBRI, Roorkee for constant encouragement and interest shown during the progress of the work. The paper is based on the research work done as a part of research programme of CBRI, Roorkee and is presented to the conference with the permission of the Director.

Thanks are due to all those who have rendered assistance in this project.

References

(1) Proceedings of the United Nations Conference on New Sources of Energy, Vol 5, Solar Energy II, Rome 21-31 Aug, 1961.

(2) Stiphen A.Vin-Cze 'A high speed cylindrical solar water heater' Solar Energy, Vol 13, No.3 1971, pp 339.

The use of solar energy in conventional desalination processes

W. P. Teagan
D. R. Cunningham
Arthur D. Little, Inc.
Cambridge, Massachusetts, U.S.A.

Abstract

There are a number of approaches to desalination which are being used, including: multistage flash distillation, reverse osmosis, and vapor compression.

One common trait of all these desalination processes is that large amounts of energy are required. The recent rise in the cost of conventional energy forms has greatly increased the overall cost of desalination--to the point where energy costs are a major cost component. Identifying unconventional energy sources which can be used in desalination processes to lower their cost is, therefore, an important goal of desalination technology.

One source of energy which could be particularly appropriate in desalination technology is solar energy since the need for desalination is often found in arid regions which have high average solar flux levels.

Solar energy can be used in a number of ways to operate conventional desalination processes, including:

- The direct use of solar derived heat to drive multistage distillation systems; and
- The use of solar thermal power units to operate reverse osmosis and vapor compression processes.

The paper discusses the technical/economic performance of the above approaches to solar augmented desalination when using solar collectors and heat engines in a sufficiently advanced state of development to be consistent with near-term implementation.

(A) Introduction

The distilling of sea and brackish water was one of the earliest practical applications of solar energy. A system was built in Chile in the late 1800's and systems are operating today in Greece, Mexico and Australia. There is still a great deal of interest in this subject area as indicated by the numerous articles in the literature. [1,2,3,4]

Most solar distillation units have utilized some form of the "humidification" process. This system arrangement is shown schematically in Figure 1. The operation of this system is the essence of simplicity

Fig. 1
A simple solar still

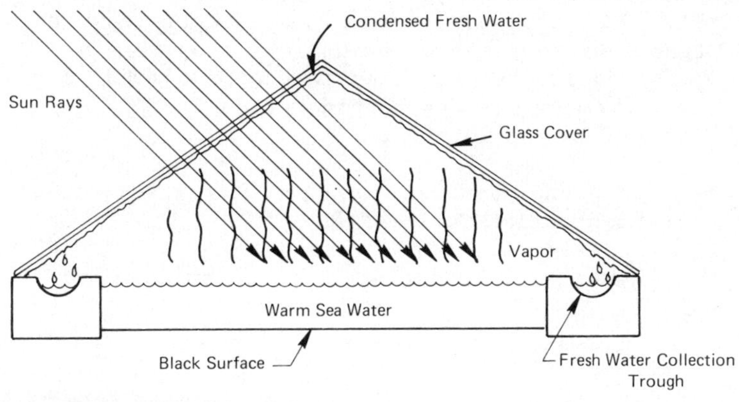

which helps explain its predominant role in solar desalination. Solar energy passes through the glass (or plastic) cover and is absorbed by the water contained in a shallow pond.

The vapor above the solar heated water (typically at a temperature of 110-150°F) condenses on the relatively cool cover plate and dribbles down to a trough and then to a collection point. Much work has been done to lower the cost and improve the efficiency of the solar "humidification" process. The performance of this arrangement is, however, subject to several fundamental limitations.

- Since there is one stage of evaporation, each pound of water requires over 1,000 Btu of energy input for the purification process; and

- The basic heat transfer processes are such that the thermal efficiency must be in the 40-60% range.

As a result, the humidification process requires, at least, 8 ft^2 of collector area for each gallon per day of capacity. Even with the above limitations, desalination with the solar humidification process is a reasonable choice for systems of limited capacity due to its simple construction and operation.

For larger systems, however, it may be preferable to utilize solar energy to drive more energy efficient desalination processes, such as multistage flash distillation, reverse osmosis, and vapor compression units. All these systems require an energy input of only 10-30% that of the simple distillation process previously described.

The multistage flash distillation process requires heat as the primary energy input. This heat could be supplied directly by a solar collector field. The primary energy inputs to the reverse osmosis and vapor compression processes is in the form of mechanical power. This mechanical power can be easily supplied by internal combustion engines or electric motors when conventional energy sources are used to operate the desalination units. With solar energy, the heat from a collector field would have to be converted into mechanical energy in a heat engine (Rankine cycle, Stirling, etc.) in order to operate the desalination units.

This paper discusses the technical/economic performance of various solar collector, heat engine, desalination system options and indicates how the use of solar energy can influence the choice of desalination system approach.

(B) Performance of Solar Heat and Mechanical Energy Sources

There are myriad combinations of solar collector configurations and heat engines which could be utilized to operate conventional desalination units. Collector configurations range from parabolic dishes with a concentratio ratio of 1000/1 to flat plates with no solar concentration at all. Heat engines which have been operated with solar energy include organic Rankine cycle engines which operate at 180°F and Stirling engines which require temperatures in excess of 1000°F for effective operation. The system chosen for consideration in this paper is shown in Figure 2. This system consists of:

- A compound parabolic concentrator (CPC) collector with a concentration ratio of 3X; and

- A Rankine cycle engine using an organic working fluid.

Fig. 2
Solar energy source configurations

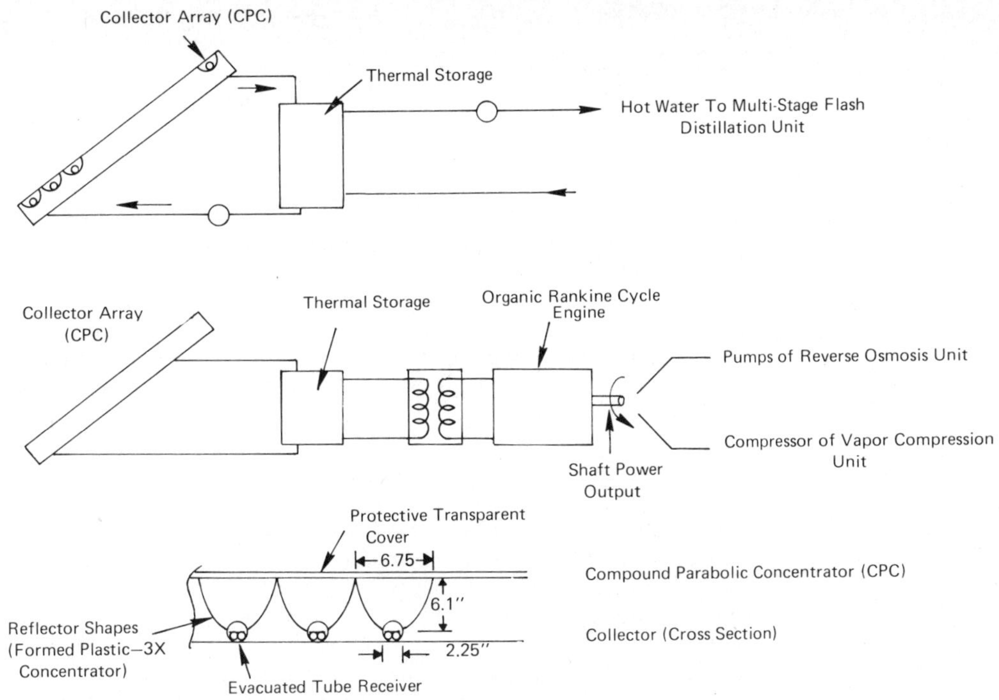

The heat output of the collector array is in the form of a hot fluid (hot water or heat transfer oil). This hot fluid can be used to operate:

- A multistage distillation plant by directly supplying heat to the process; or
- A reverse osmosis or vapor compression unit by supplying heat to the boiler of the organic Rankine cycle (ORC) engine which, in turn, operates the required mechanical equipment.

The compound parabolic concentrator collector configuration indicated uses low levels of solar concentration to achieve the required temperature levels (212°F and higher) and has several advantages for use in remote regions where solar operated desalination units might find their initial application:

- The CPC collector can utilize radiation over a relatively wide acceptance angle (27° for a 3X collector). This eliminates the need for accurate tracking and allows for the utilization of a large fraction of the diffuse radiation. Diffuse radiation represents 10-20% of the total on an annual basis, even in desert areas of above average insolation. At seaside sites the humidity levels tend to be quite high which is further incentive to use a collector which is not particularly sensitive to the spatial distribution of solar radiation.
- Evacuated-tube receivers for solar heating and cooling applications are being developed by a number of large companies which will help ensure their availability at a reasonable and predictable cost for this application.
- The external appearance of the collector is similar to that of a flat plate collector. The highly reflective surfaces are protected from the environment by an outer glass (or clear plastic) cover plate and there are no exposed trough or spherical shapes to collect dust, sand, etc.
- The collector array has a low vertical profile which mimimizes structural requirements. Site preparation will be minimal since accurate leveling and adjustment of the collector support structure is not required (as tends to be the case for highly concentrating systems).

The fact that many different collector arrangements using various levels of solar concentration are being considered indicates that it is still difficult to tell which will prove most viable. However, the collector system chosen for consideration appears to have significant operational advantages and is sufficiently far advanced that cost estimates can be made with some degree of confidence. The performance of this arrangement should be representative of that obtainable from solar thermal heat sources capable of operating conventional desalination processes.

Collector Performance

The annual average thermal performance of a CPC collector array is shown on Figure 3. This performance level is based on computer calculations using weather bureau insolation tapes for Albuquerque, New Mexico. It should be typical of that in many portions of the world where desalination would be a primary concern. For comparative purposes the performance of a flat plate collector is also shown. This flat plate collector assumed uses a selective absorber coating and high transmission glass and has, therefore, particularly good thermal performance for a flat plate unit. At the temperature levels required of multistage distillation systems (212°F and higher), the performance of the CPC collector arrangement is over twice that of the flat plate system. The performance levels indicated are annual averages and take into account variations in climatic conditions and the effects of early morning and late afternoon operation. Midday efficiency levels can be significantly higher than those indicated in Figure 3.

Fig. 3 Variation in collector efficiency with temperature

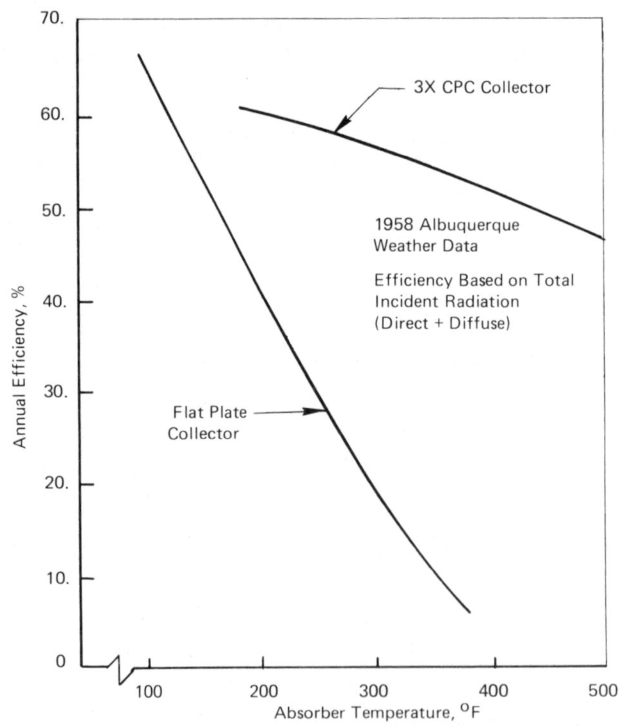

Engine Performance

The most appropriate operating temperature for the collector array when it is used as the heat source for an ORC engine depends on the engine's efficiency/temperature characteristics. The efficiency of these engines as a function of operating temperature is indicated in Figure 4 for a number of commonly used working fluids. As with any heat engine, efficiency increases with temperature. It is important to note, however, that the most significant efficiency gains are made at relatively modest

Fig. 4

Rankine cycle engine efficiency-variation with peak cycle temperature

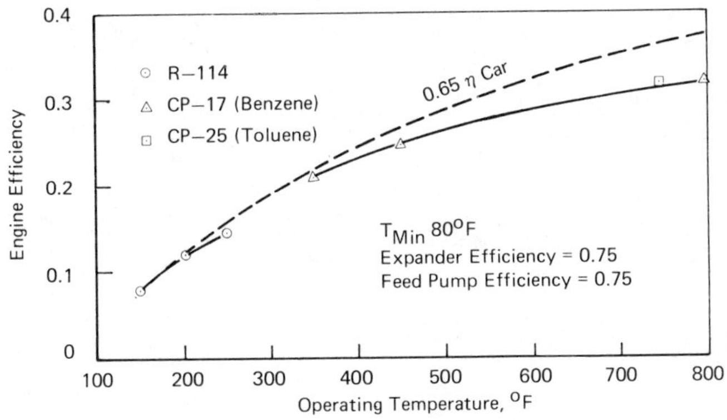

Fig. 5
Overall system efficiency

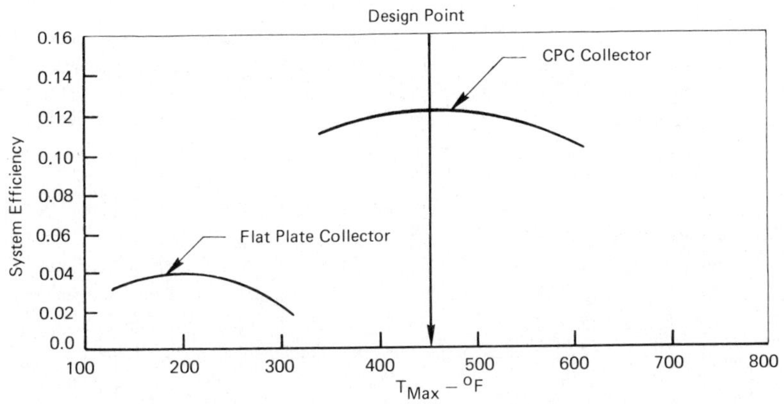

temperature levels (up to about 450°F) which reduces the incentives to attain the higher temperatures associated with high concentration systems.

The collector efficiency curves of Figure 3 and the engine efficiency curves of Figure 4 can be combined to yield overall system efficiency defined as:

$$\eta_s = \eta_c \times \eta_e$$

where

η_s = system efficiency

η_c = collector efficiency,

η_e = engine efficiency

Curves showing the resultant average annual system efficiencies are presented in Figure 5. The efficiency of the system is about 12.5% and occurs at a peak cycle temperature of 450°F. By comparison, a power system using a more conventional flat plate collector as the heat source would have an efficiency of about 4%.

(C) Desalination System Options

There are many different processes for water desalination either in use or in development. Three types of systems which are in general commercial use are:

(a) Multistage Flash Distillation
(b) Vapor Compression
(c) Reverse Osmosis

These three approaches to desalination are described briefly:

Fig. 6
Multistage flash distillation process

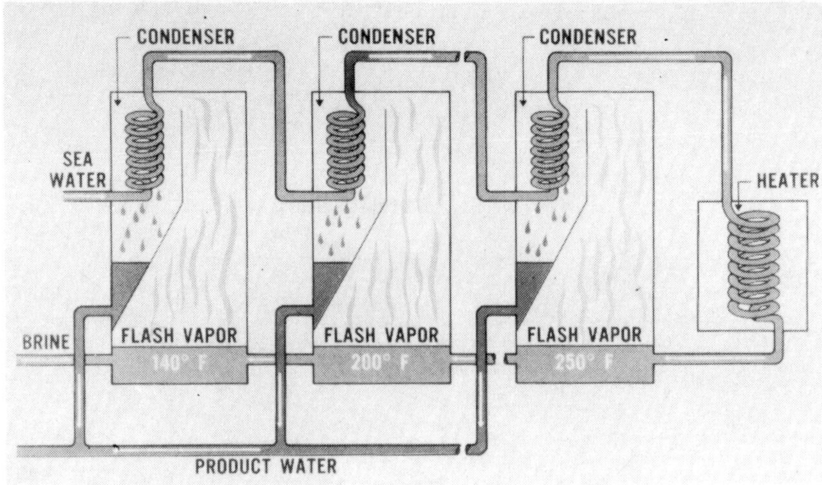

Multistage Flash Distillation

In a multistage flash distillation unit (shown schematically in Figure 6), the incoming seawater is heated and upon introduction into a chamber of sufficiently low pressure, some of the water boils instantly or flashes to a vapor, and in so doing cools the remaining salt solution. This salt solution is then pumped into another chamber of lower pressure and temperature where, again, some of the water flashes to vapor and is drawn off. Several stages are used (typically 10 to 20) at successively lower pressures to permit the needed boiling at lower temperatures. The condensation of the flashed vapor occurs when it is passed through the other side of the preheaters for the incoming ocean water. The primary energy input to this process is in the form of heat although some mechanical energy is required to pull vacuum and circulate water. The heat is usually supplied by an oil or gas fired burner--often with pressurized water or steam as a heat transfer intermediary. Typical heat addition temperatures are from 200-300°F.

Vapor Compression

A vapor compression cycle, Figure 7, is similar to other distillation processes, except in the manner in which heat is added to the brine.

Fig. 7
Schematic of vapor compression process

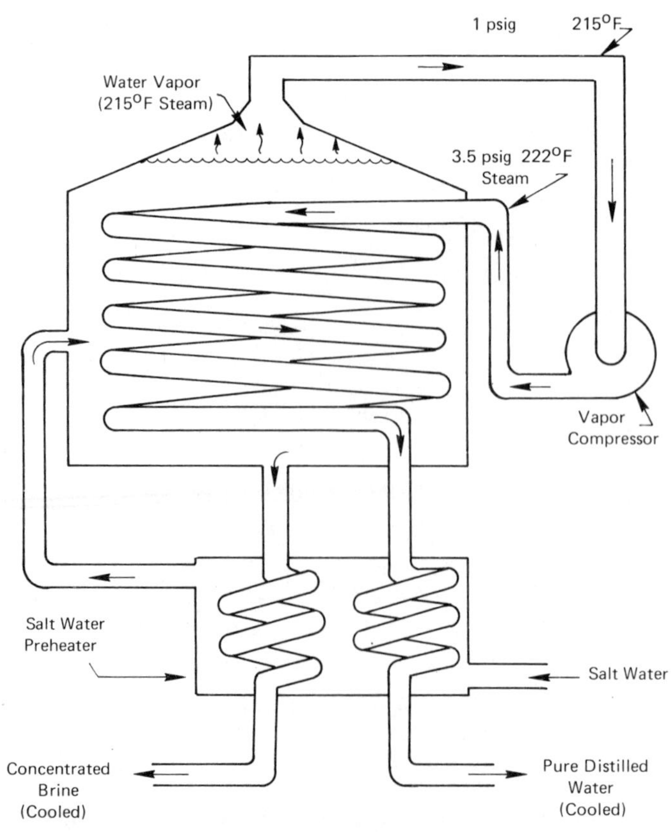

The salt water enters the containment vessel after passing through a preheater. The salt water solution is heated to the point of vaporization by absorbing the heat given off from a condensing coil in which higher pressure, product water vapor is flowing. The vapor created in this evaporation process is drawn off, compressed (to increase the temperature) and is then circulated back through the coil to supply heat to the brine solution. In giving up heat to the brine, the fresh water vapor is condensed. The salt water preheater uses heat from the waste, concentrated brine (which must be removed from the main vessel) and the condensed product water. In a pure vapor-compression process, heat is added only during transient periods (start-up, flow increase, etc.).

Membrane Processes (Reverse Osmosis)

The most common membrane desalination technique is reverse osmosis. In a reverse osmosis procedure, one side of a membrane (typically cellulose acetate) is supplied with saline water at a high pressure. The film absorbs water at a higher rate than suspended or dissolved solids and ions, and as a result water is drawn off the other side of the membrane, as in Figure 8. The capacity of these devices is increased by adding more membrane area. The membranes have a predictable life span which is a function of pressure, feed water impurity, and product water standards, and the equipment only requires the input of energy to run the high pressure pumps (typically 200 to 1000 psi). The pressure difference required across the membranes increases with the degree of water salinity which is one reason why reverse osmosis systems are usually used with brackish water with salt levels well below that of ocean water.

Fig. 8
An illustration of tubular reverse osmosis

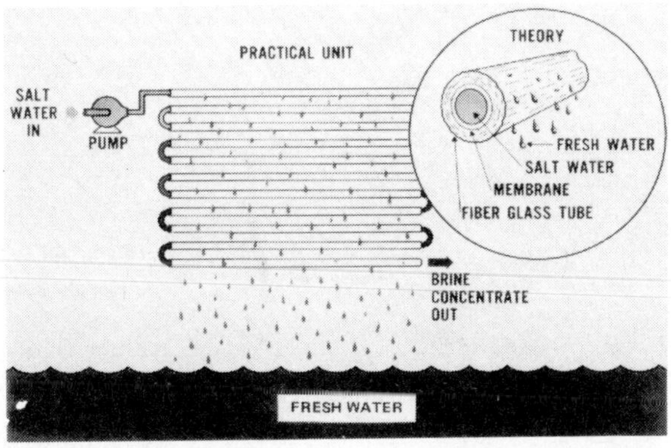

Energy Input Requirements

The single most important operational cost of all desalination systems is now that associated with energy. Therefore, a critical factor in determining the relative viability of different desalination options is the energy input required per pound of water distilled. For fuel fired systems, the impact of this parameter is obvious. When solar energy is utilized for desalination system operation, the size and cost of the collector array will depend strongly on the input energy requirements.

Table 1 provides estimates of the energy inputs for the desalination systems under consideration based on manufacturers' information. The energy input for each process is divided into shaft power requirements and heat addition. The hybrid vapor compression process is a common one which requires both heat and shaft power for its operation.

Estimates were made for the area of collector required to desalinate 1000 gallons of water per day by each of the processes assuming the performance curves of Figures 3 and 5 for the solar equipment.

The collector area requirements to operate modern desalination plants

Table 1

COMPARISON OF ENERGY INPUT FOR DESALINATION PROCESSES

Process	Energy Input Form			Collector Area Requirements (Ft^2 per 1000 gal/day capacity)	
	Shaft (Btu/lb)	Heat (Btu/lb)	Temperature (°F)	CPC Collector	Flat Plate Collector
Hybrid Vapor Compression	17	15	180-220°F	560	1,700
Multistage Flash (18 Stage)	5[1]	150	175-300°F	1,200	2,900
Simple Distillation	--	1,110	120-212°F	N.A.	8,300[2]
Reverse Osmosis	20[3]	--	--	500	1,700

1 Most processes require some shaft input for pumping, vacuums, etc.

2 Area of simple collector used for solar humidification.

3 Additional energy may be required for any water pretreatment. Steps required for reliable operation of a reverse osmosis plant.

are seen to be only 7% - 15% of those required of a simple solar still if the CPC collector configuration is used. If a flat plate collector of good design is utilized, the area requirements increase to 15%-30% those of a simple solar still.

(D) Economic Considerations

The figures of Table 1 indicate that far less collector area is required to operate a modern energy efficient desalination unit than is required by a simple solar still. The cost of CPC solar collectors of the type assumed in this paper is projected to be in the $\$8/ft^2$ - $\$12/ft^2$ range, while solar still costs are one-third to one-half this amount.

The thermal output of the CPC collector array must be utilized by relatively complex thermal/mechanical equipment to perform a desalination operation. The output of the simple solar still is, on the other hand, purified water, i.e., the solar still combines the collector array and the desalination unit in one integrated package. Nevertheless, the very large collector areas required by solar stills will tend to put them at a disadvantage compared to other solar desalination options for large systems where the complexities of a modern desalination unit can be justified.

The potential economic viability of large scale solar desalination can be evaluated by estimating at what cost solar heat and/or mechanical energy sources deliver energy to conventional desalination processes. This cost can then be compared to that from conventional energy sources to determine under what conditions the use of solar energy is economically viable. The cost of heat and mechanical power from the system combinations of Figure 2 were calculated assuming:

- A solar collector cost of $\$10/ft^2$
- An organic Rankine cycle engine cost of \$500/kW.
- An operating temperature of 215°F for the multistage distillation unit.

- Annual capital costs (interest and depreciation) of 12% per year.
- Annual O & M costs of 3% per year.
- Operation in an arid region with a climate similar to the United States southwest.

Based on the above, the installed cost of the solar heat source was projected to be about \$16 per square foot of collector and that of the solar power unit \$2100 per peak kilowatt of capacity. The corresponding costs of the thermal and/or mechanical energy from the solar operated systems are estimated to be \$8 - \$10 per million BTU and 15-25¢/kWh, respectively.

These energy costs are significantly above those produced by conventional energy sources in areas serviced by a utility grid and/or with an established transportation system for oil or gas. However, in many remote regions, the cost of fuel and electricity are very high due to the expenses associated with transportation and/or extending the utility grid. In these areas, the costs of solar derived heat or mechanical energy could become increasingly competitive for operation of desalination systems.

REFERENCES

1. M. N. Bahadari and F. E. Edlin, "Improvement of Solar Stills by the Surface Treatment of the Glass," Solar Energy, Volume 14, No. 3, February 1973.
2. P. I. Cooper, "The Maximum Efficiency of a Single-Effect Solar Still," Solar Energy, Volume 15, No. 3, September 1973.
3. O. St. C. Headly, "Cascade Solar Still for Distilled Water Production," Solar Energy, Volume 15, No. 3, September 1973.
4. S. H. Siliman, "Effect of Wind on Solar Distillation," Solar Energy, Volume 13, No. 4, July 1972.
5. S. G. Talbert, U. A. Erbling, and G. O. Löf, Manual on Solar Distillation of Saline Water, Office of Saline Water, Research and Development Report No. 546, 1970, Washington, D. C., U. S. Department of the Interior.
6. T. A. Lawand, Engineering and Economic Evaluation of Solar Distillation for Small Communities, Technical Report No. MT-6, Brace Research Institute, McGill University.

6.12 The design, construction and operation of solar stills for producing distilled water

O.St.C. Headley
The Department of Chemistry,
The University of the West Indies,
St. Augustine, Trinidad.

J.B. Morris
Central Analytical Laboratory,
Caribbean Agricultural Research & Development Institute,
U.W.I. Post Office, St. Augustine, Trinidad.

Abstract

Solar stills have been designed for producing water for use in teaching and research laboratories. Various designs and configurations have been constructed and tested over the past eight years. The original double surfaced cascade stills (stills A and B) were constructed in aluminium and were rather expensive. Later models termed Series D and E were of galvanised mild steel with stainless steel distillate drains and safety overflow drains. These galvanised steel units suffer from severe corrosion after a few years of operation. The latest design, series F, is built in aluminium with stainless steel distillate drains and automatic feed valves. All of the stills produce very pure distilled water (conductance $\sim 1.5 \times 10^{-6}\ \Omega^{-1}\ cm^{-1}$, pH 6.8-7.4) at a cost which is below that of conventional electricity heated stills.

Introduction

Solar stills have been used for producing distilled water at various laboratories in Trinidad for the past eight years (1). The standard greenhouse still which was originally designed by Carlos Wilson in 1872 was the starting point in our development exercise. Since we were not interested in desalination but in the production of high purity distilled water we modified the still in various ways until we developed the double surfaced cascade still.

Two production models of this still were built and their performance reported in a previous publication (2). The main disadvantage of this design is its high cost. We are therefore looking at ways of reducing the high cost of the cascade still without reducing its efficiency or its service life. In our first modification, we reduced the cover slope from 10° to 5°, reduced the number of cascade bars per side from nine to three, removed the lower condensing surface and changed the construction material from aluminium to galvanised mild steel.

This solar still was called still D and its performance was just as good as stills A and B (3). After three years however, we are now faced with a serious corrosion problem and have come to the conclusion that galvanised mild steel is unsuitable for constructing solar stills, even those fed on fresh water and used for distilled water production.

Design and Construction

In addition to the requirements listed in reference 2, we now add three others; these are:

(1) Minimum maintenance: If possible the still should run unattended for six months to a year.
(2) Good corrosion resistance: The unit should operate for at least five years before any component wears out.
(3) The cost of the unit should be as low as is compatible with the other requirements.

Fig. 1
Comparison of solar still designs

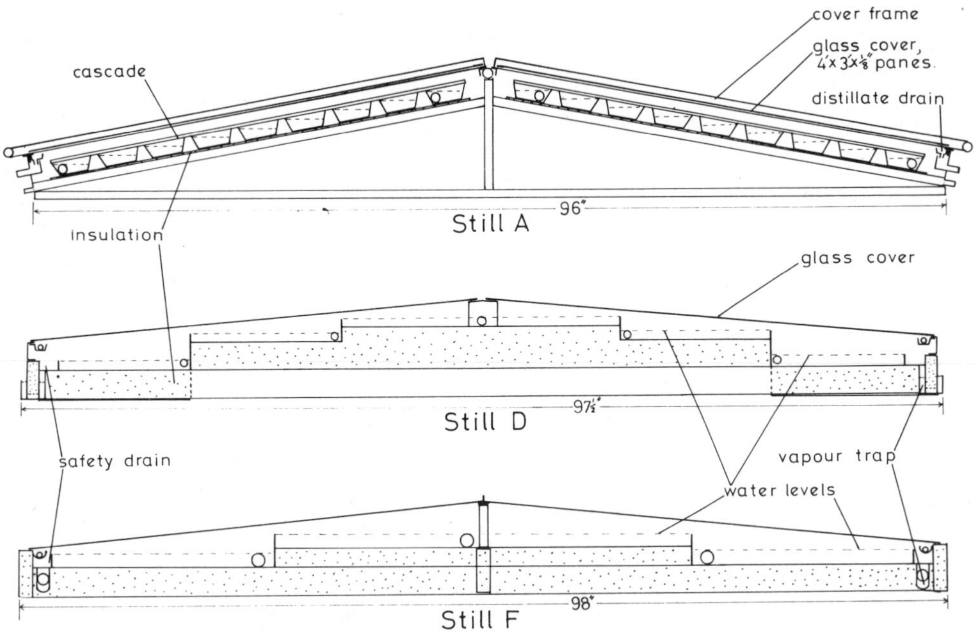

The three designs described in this paper, stills D, E and F attempt to satisfy these three requirements. Stills D and E concentrated almost solely on requirement 3, still F tries to satisfy all three conditions. Fig. 1 shows stills A, D and F in cross section.

To save construction materials all three units are low slope cascade stills. Still D has a cover slope of 5°; still E, 2° and still F, 5°. For low weight and reasonable rigidity, all employ perimeter frames made from box members. These box members are made from 26 gauge galvanised mild steel (Stills D and E) or from 16 or 18 gauge aluminium (Still F). Box members are fabricated from flat sheet by bending and assembled by soldering (galvanised steel) or riveting (aluminium). Since the box members serve as the sides of the still in certain locations, they are filled with insulation and so serve a double purpose.

The steps in the fabrication of a normal parallelsided box member are shown in Figure 2. Since the stills slope away from the centre line, some of the side box members are tapered. One therefore starts with two tapered pieces of sheet and bends them as for a normal box member;

Fig. 2
Fabrication of a standard box member

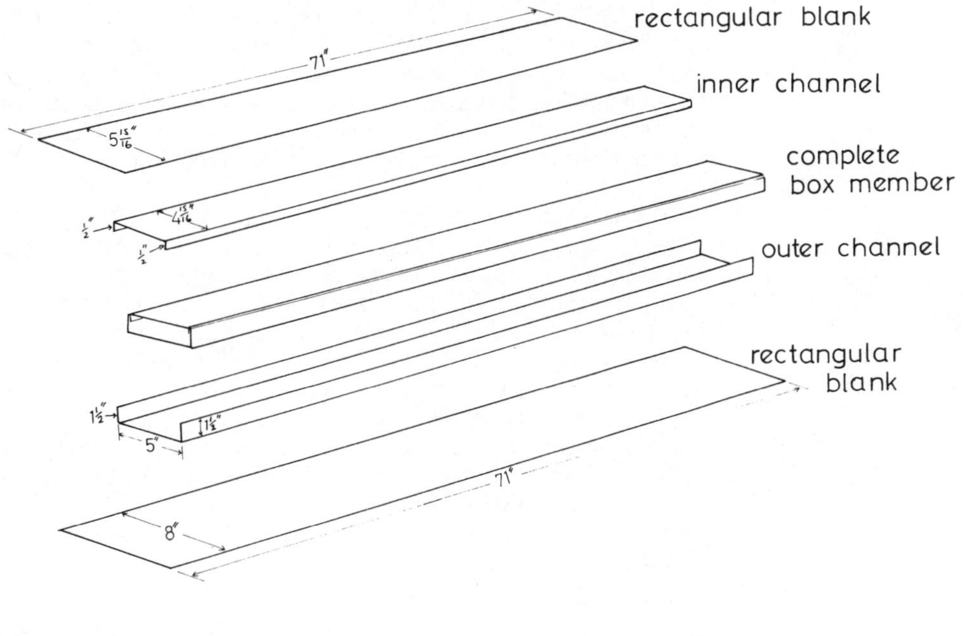

Fig. 3
Fabrication of a tapered box member

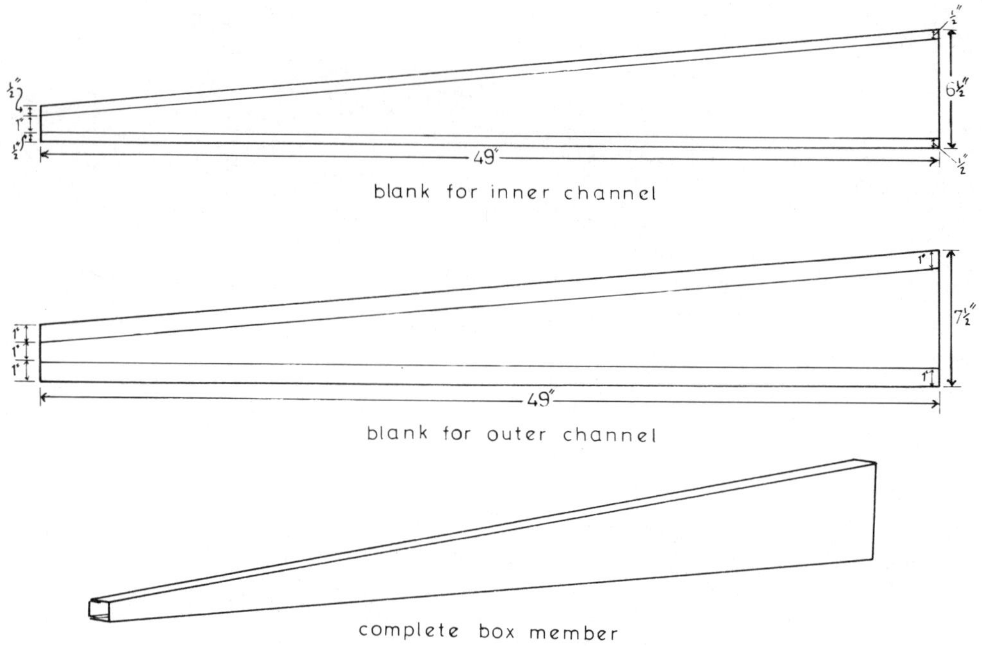

on assembly one has a tapered box member. Figure 3 shows such a box member for still F. Those for still D are identical while those for still E are wider at the narrow end since still E has a smaller roof slope.

These stills are constructed either as a double sided units or as single sided units. In all cases their basic dimensions are based on the size of glass pane available. Glass in Trinidad is available in sizes 48" x 36" x 1/8" or 36" x 30" x 1/8". Stills therefore have nominal sizes of 8' x 6' or 6' x 4' or 6' x 5' or 6' x 2½'. In practice they are somewhat larger since the glass has to fit into a slightly larger frame. Figure 4 shows the basic frame and side views of an 8' x 6' still. The dimensions shown apply to still F but the other two stills are similar. Figure 5 shows the basic frame for a 6' x 5' still. Pieces of angle are used to rivet the box members together and triangular plates in the corners are used to brace the frame. In the case of the 8' x 6' still, the frame starts off with the two 98"x4"x1" longitudinal members and the three 71" x 5" x 1½" transverse members. The two 46 3/4" x 3" x 1½" longitudinal members are then put in and the two 71" x 2" x 1" transverse members put in on top of them. With the base frame assembled, the tapered box members are fitted as shown in Figure 6 which shows their fitting in a 6' x 5' still. They are riveted to the other box members with pieces of angle or flat sheet as is appropriate. With stills D and E, certain joints were soldered since this is possible with galvanised mild steel.

The water trays can be fabricated and fitted next. In a 6' x 5' still one actually has a simple low slope greenhouse still. The only modification is the fitting of a safety drain to prevent salts from creeping into the distillate drain and a vapour trap to allow the still to overflow without loss of vapour. The vapour trap connects the safety drain with the exterior. The water tray is fabricated so that all "wet" seams are vertical. This greatly increases reliability. If the piece of sheet metal is not big enough to make the tray in one piece, it should be made as two separate trays, one with a lip to fit over the other. The outer edge of the tray also has a lip to fit over the edge of the safety drain. In this way all longitudinal seams are above the water line. This greatly reduces the incidence of leaks. Short vertical seams can be welded, brazed, soldered or riveted

Fig. 4
An 8ft x 6ft still

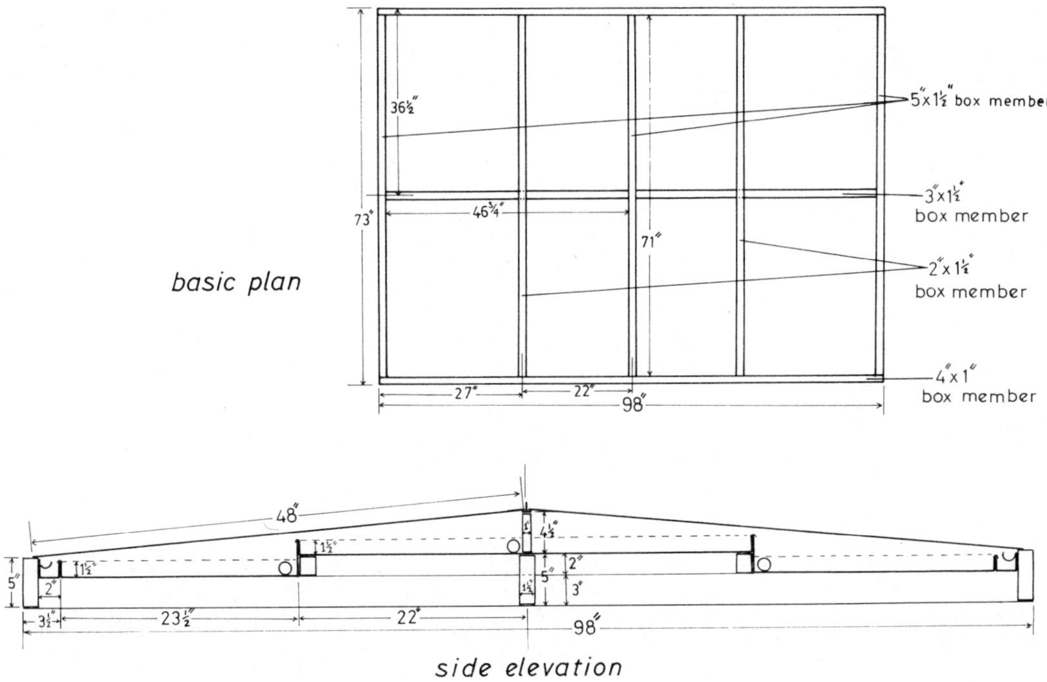

Fig. 5
A solar still frame

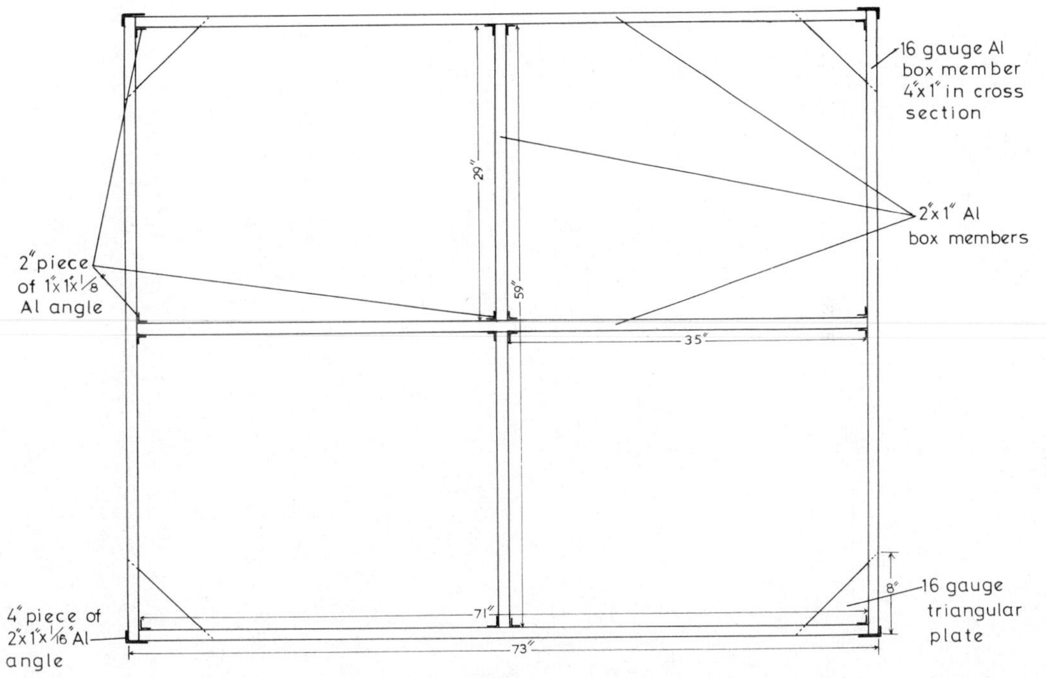

depending on the metal used for construction. Riveted joints have to be sealed e.g. with silicone sealant. Figure 7 shows the water tray in a 6' x 5' still.

In an 8' x 5' still, one can have three (Still D) or two (Stills E and F) cascade bars. Each cascade bar is one water tray which is higher at the back than the front to accept the lip from the tray above it. The uppermost tray is common to both sides. Figure 8 shows the water trays for an 8' x 6' still of the series F design. The order of fitting is, safety drains (which should be drilled before hand for the vapour trap), then the lowest trays, then the intermediate trays (if applicable) then the topmost tray. In the case of the galvanised mild steel trays, the thinness of the metal requires intermediate supports of top-hat section

Fig. 6

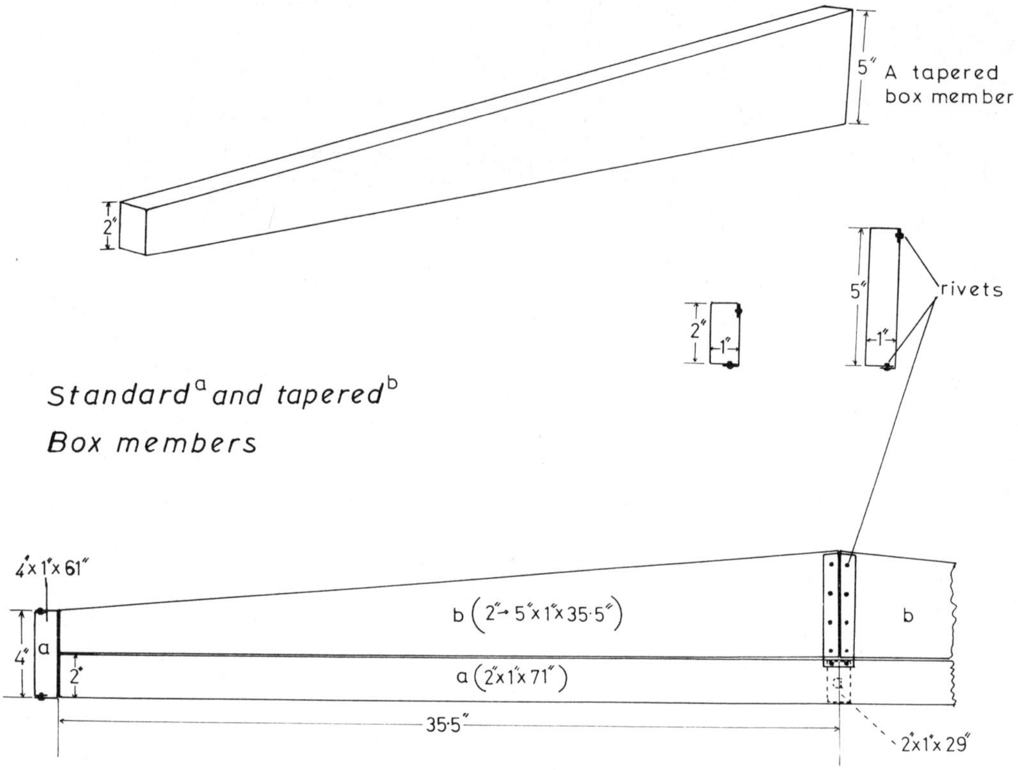

Fig. 7

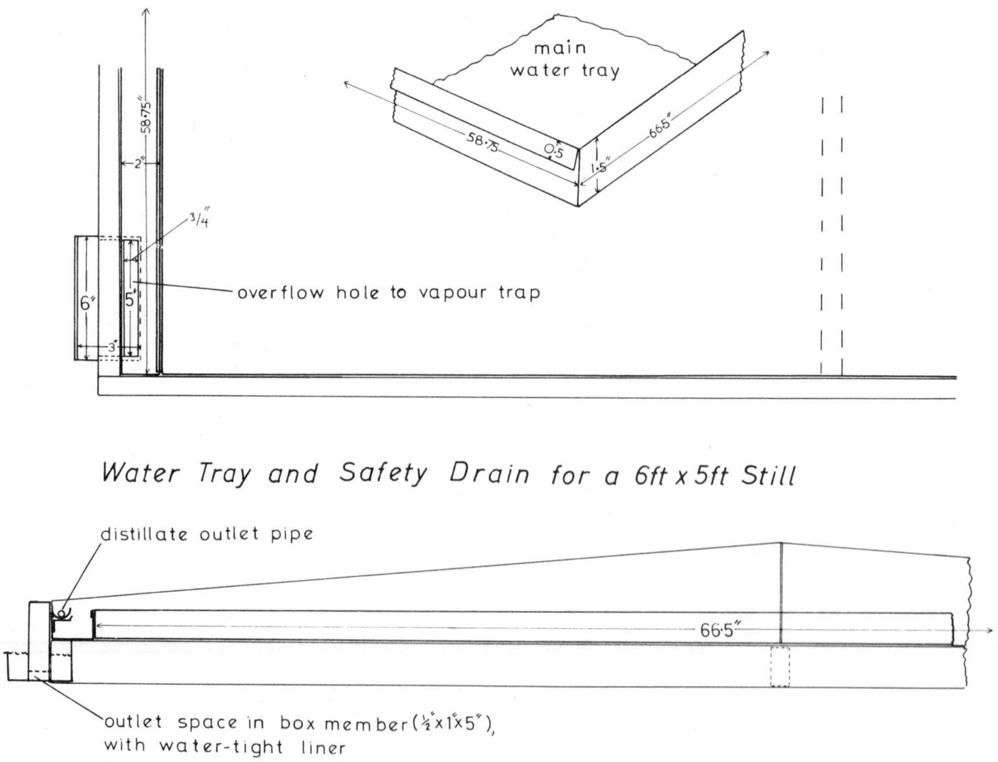

Water Tray and Safety Drain for a 6ft x 5ft Still

Fig. 8

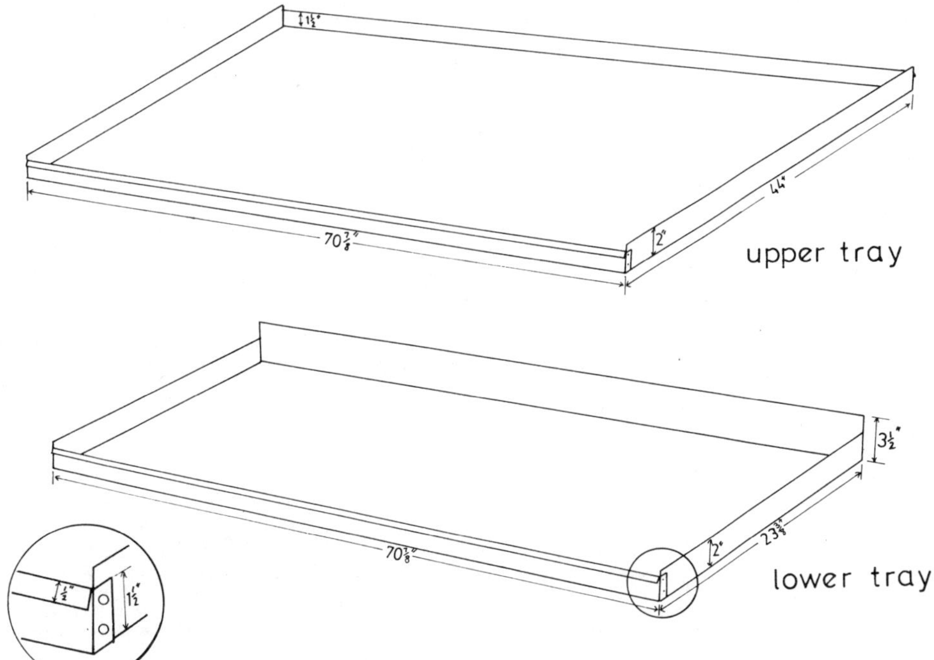

Water trays for still F

at the half distance points, i.e. 18 1/4" and 54 3/4". One should put in as many as are necessary to carry the weight of water in the tray.

With the trays fitted, all seams are either soldered or sealed with silicone sealant. The trays are attached to the side members by soldering or with pop rivets at their upper edges. The central vertical glass support is put in at this point. Two types of vapour traps may be used. The standard plumber's sink trap or one fabricated from sheet metal. The former is much easier to fit. In both cases the hole for the trap should be made in the safety drain before it is fitted. These traps are shown in Figures 7 and 9. The inner part of the still is then painted flat black.

The distillate drain is made of stainless steel sheet or PVC pipe. In both cases it is supported by short brackets riveted to the front box member. The PVC unit is easier to fabricate since one merely cuts a 3/4" PVC pipe in half longitudinally. The advantage of a stainless steel drain is that one can build in a slope much more readily by fabricating it from a tapered piece of sheet as shown in Figure 10. In this way water does not collect in the drain but drains out at once. Re-evaporation of distillate is therefore avoided.

The older stills were batch fed or trickle fed as mentioned in reference 2. With unreliable and variable pressure water supply systems, both of these feed systems lead to problems. In the series D stills with their shallow water trays (3/4" deep) the still could run dry and overheat causing the water tray to warp enough for leaks to be seen when the still was refilled. The series E stills had 1" → 1¼" deep trays while the series F stills have 1½" deep water trays. The series F stills also have automatic float valves which keep the still full whenever water is available. At maximum rates of distillation, 1½" of water lasts for about 4 days. So long as a water tray contains water it does not overheat, hence an intermittent supply is suitable so long as periods of interruption do not exceed 4 days. This greatly improves the still's reliability. One can further improve matters by connecting the stills to a tank which is automatically filled from the normal supply. In this way it is possible to run for periods of two weeks to a month without a regular water supply.

Fig. 9

1. Feed back pipe
2. Upper cascade bar
3. Feed water pipe
4. Valve
5. Still inlet pipe
6. Upper valve chamber
7. Overflow pipe
8. Lower valve chamber
9. Valve float
10. Lower cascade bar
11. Vapour trap
12. Safety drain
13. Distillate drain.

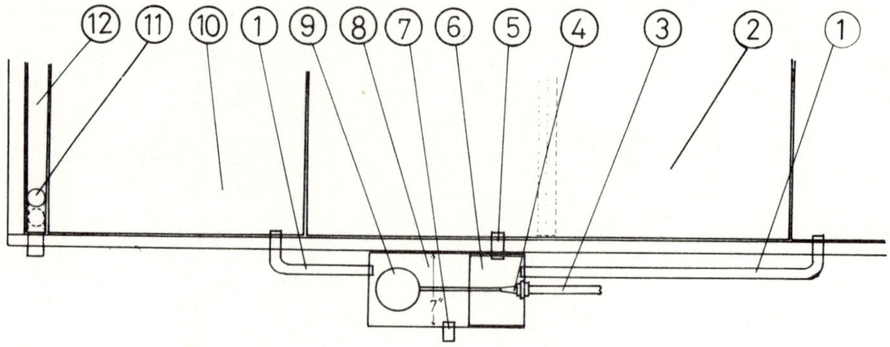

The Water Supply System

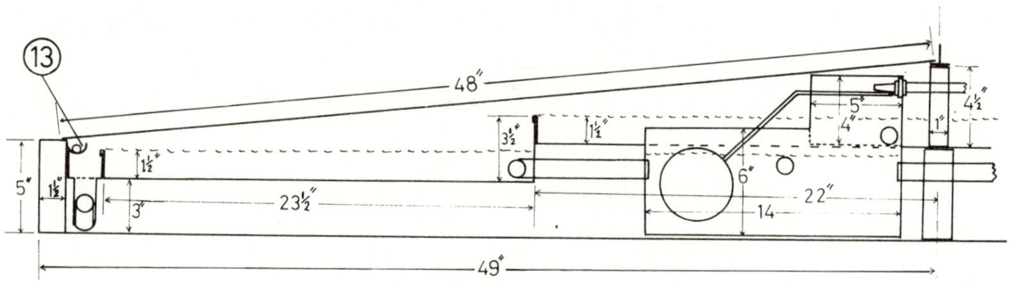

Fig. 10

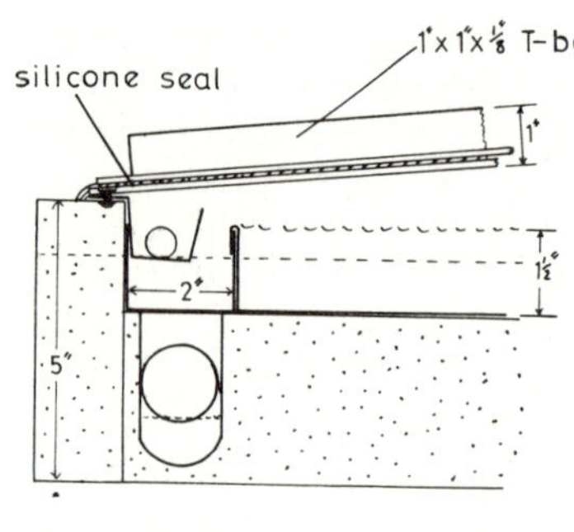

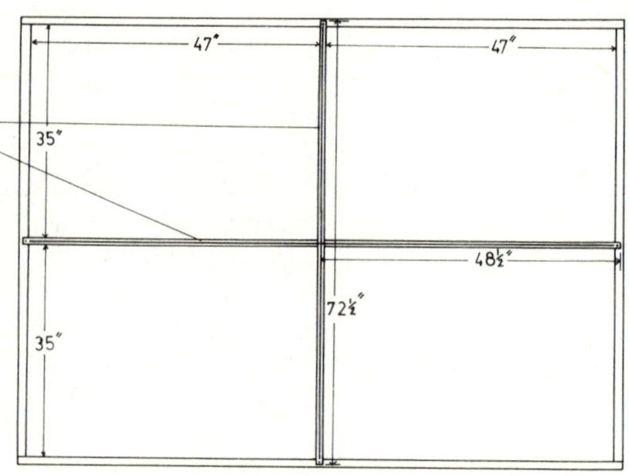

Glass carrying frame

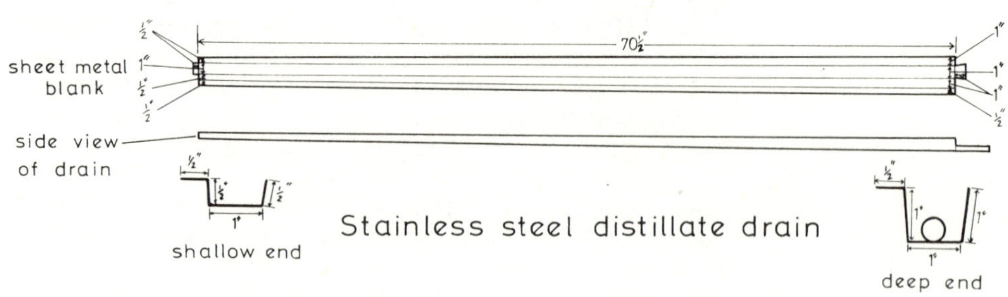

Stainless steel distillate drain

Figures 9 and 11 show the stills connected up to their automatic valves. In a 6' x 5' still the valve merely maintains the water level in the still at its own float chamber level. A pipe below the water line connects the two. The valve itself is a standard toilet unit which is fitted in a chamber fabricated from sheet metal. In an 8' x 6' still with two water levels one needs a double level valve. On opening the valve, water enters the upper valve chamber and from there enters the topmost water tray. When this is full it overflows into the lower trays from which it flows back into the lower valve chamber via the feed back pipes. An overflow pipe with an adjustable outlet is fitted

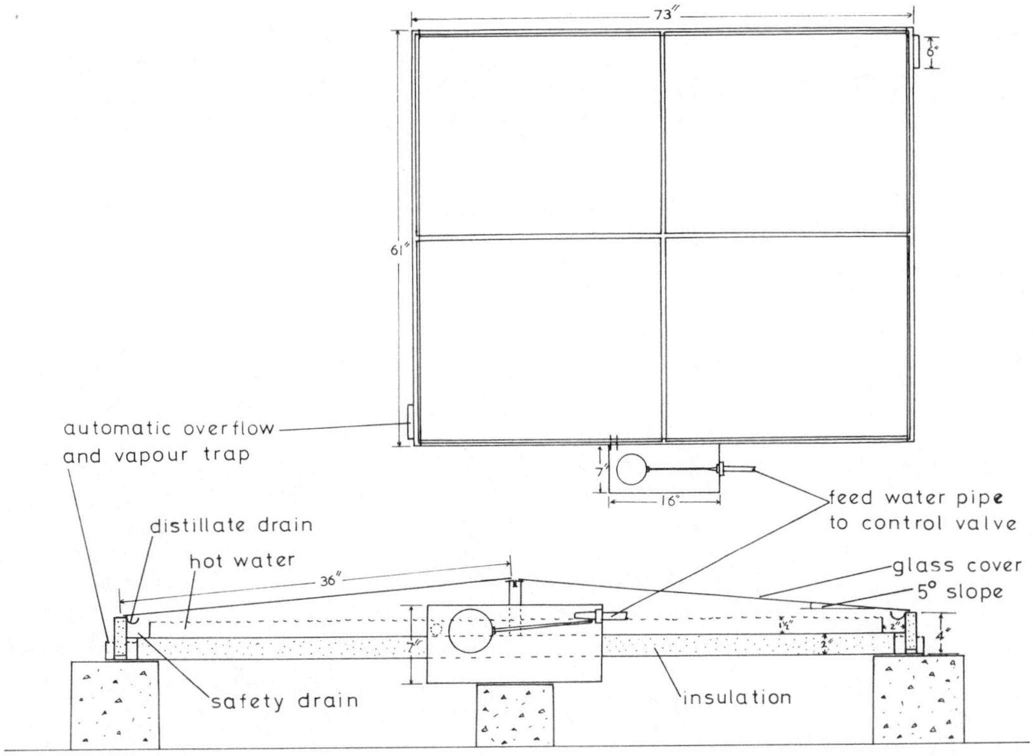

Fig. 11
A simple solar still (with automatic feed)

to the valve float chamber to enable the level in the lower water trays to be adjusted. An important adjustment is salt scale removal. In such a case the still overflows via the safety drains and the vapour traps and any scale present on the surface of the water is removed; this improves evaporation.

Under normal operating conditions the float closes the valve when the feed back pipes fill the float chamber to the predetermined level. Any excess water still in the system drains away via the safety drains and vapour traps or the float chember overflow pipe. One needs two feed-back pipes, one from each lower water tray since if the still is not perfectly level, the upper tray will overflow on one side only. The other side therefore fills via the valve float chamber.

In a 3-bar double-sided cascade still one therefore needs a balance pipe to connect the two intermediate water trays or one will run dry unless the still is absolutely level. This was an early problem in the series D stills.

When the valve is fitted, the still is filled and left to stand at least overnight to check for leaks. If none are found, the float is removed from the valve, the still is inverted and the insulation fitted to the bottom. This can be retained in place with thin sheet metal or polythene sheet and chicken wire. The still is then taken to its place of installation, connected to the mains, filled with water, levelled and glazed.

In the double sided stills, the central, transverse, T-bar, glass support is fitted along the still centre line and riveted to the tapered box members at its ends. The two short glass supports, also aluminium T-bar, are then run from the centre points of the outer edges to meet at the centre over the vertical support. The spaces are checked to ensure that they are square and of the right size; this is shown in Figure 10. The short supports are then riveted to the box members and the glass stuck on with silicone sealant using the procedure described in reference 2. It is absolutely important that the glass be clean and free from finger prints since these lead to breakdown of distillate film and loss of distillate as it drips back into the water tray. With low slope solar stills clean glass is essential to reasonable distillate yields. With a continuous film of distillate on the underside of the pane, one can get good run-off at slope angles as low as $1\frac{1}{2}^{\circ}$.

Performance and Operation

Once all the glass is fitted and there are no vapour leaks the still will fog up within 10 sec. if the glass was initially dry. Since we prefer to fit the glass wet with a water film already in place, the still begins to produce distillate within one or two minutes of completion. Initially the distillate contains solvents from the paint and curing agents from the silicone sealant. These have largely disappeared after about 2 weeks operation. The water is then suitable for general laboratory use. For more accurate work it is necessary to wait until all contaminants and labile materials have been stripped from the glass and the distillate drains. The conductance of the water decreases until after about 6 months it is between 2×10^{-6} and $1.5 \times 10^{-6}\ \Omega^{-1}cm^{-1}$ from a unit with stainless steel drains. At this point the pH is 6.8 to 7.4 .

The high quality of the distillate is the chief advantage of these solar stills. Once they have been leached clean, they maintain their product purity for up to two years when they need dismantling for removal of scale. With regular flushing it should be possible to run for 5 years without removal of the glass cover. This is now being investigated in long term experiments.

The series D and E stills have to be filled every two to four days if they are batch fed and have to be checked every two days if they are trickle fed. The series F stills only have to be checked for sticky valves (algae grow in the valve chambers). This can be done once every two weeks or monthly depending on local conditions. No further maintenance except once a month flushing is required. If one combines the valve check with the flushing operation one has a unit with a very low rate of maintenance.

Except for glass breakage (always due to stones etc., never to thermal stress since the silicone sealant gives a flexible joint) and rare leaks, the unit should run for two to five years before it requires cleaning.

Sixteen series D and E stills have been operated at our own experimental site and at various secondary schools throughout Trinidad for periods of up to five years. Failure is usually due to corrosion or running dry and leaking. At our own experimental site, on the roof of an Analytical Chemical Laboratory, Stills A, B one series D and one series E still have been operating. The distilled water they produce is piped directly into the laboratory below via plastic hoses.

Storage within the laboratory is modular in design the water being stored in a series of plastic drums arranged in cascade with siphon interconnection; the arrangement is shown in Figure 12, Figure 13 summarises the dimension and performance of the stills.

This confirmation was chosen mainly on the basis of safety in case of contamination of the inlet distillate when contaminants are confined to the initial tank which can then be discarded without total loss of the water stored.

The system has proven itself in practice. Massive contamination originating from the flooding of a still during filling - which went uninterrupted for an entire day - was confined to two of the compartments (the conductivity of the water in the second compartment rose to ca $10^{-4}\ \Omega^{-1}\ cm^{-1}$) and losses were confined to 100 litres out of a total of 600 litres.

The stored water is distributed in glass aspirators except to the main work area which is served by polythene tubing which siphons from the main storage. The main distribution valve is a solenoid valve (Alexander Controls Ltd.) replacing glass stopcocks which proved extremely fragile under the stress of 50 - 100 operations a day. The system has been functioning faultlessly for the last four years.

Fig. 12
The distillate storage system

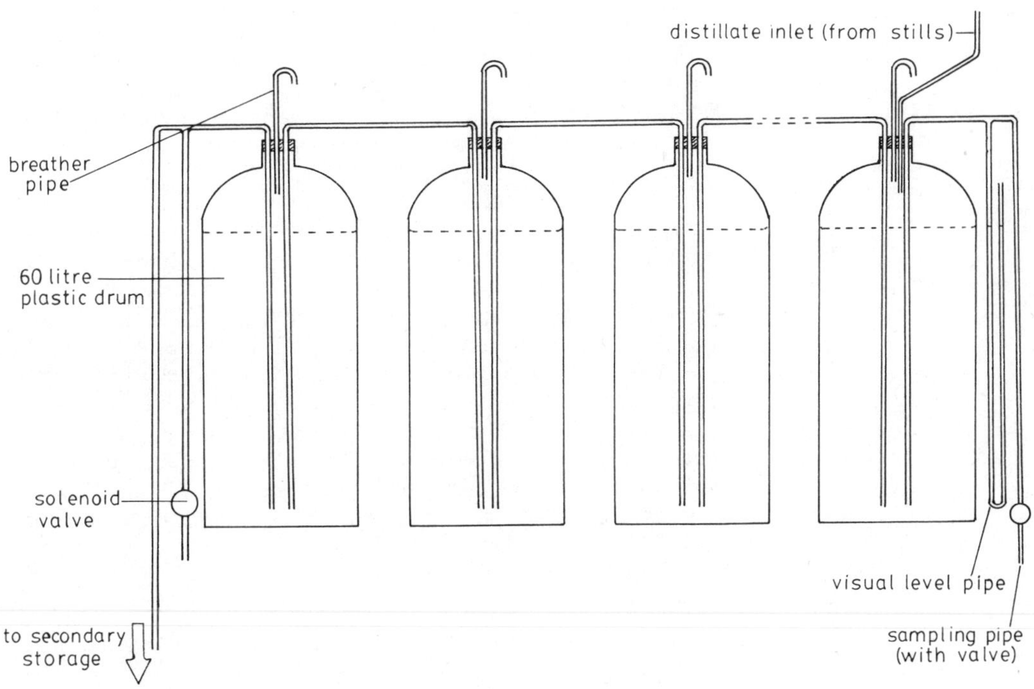

Fig. 13

DIMENSIONS AND PERFORMANCE OF SOLAR STILLS:

	A	B	D	E	F
Unit sizes *	5	6	2,4	1,2,3,4	1,2,3,4
Capital cost,[(a)] ($US/m^2)	117.70	100.54	78.48	78.48	100.90
Average annual distillate (ℓ/m^2)	800	800	800	600	700
Life expectancy (years)	20	20	3-5	3-5	15
Time under field test (years)	6½	6	5	4	<1

(a) Capital cost at the time of construction

* Unit sizes are as follows:

Size number	1	2	3	4	5	6
Dimensions	6' x 2.5'	6' x 4'	6' x 5'	8' x 6'	12' x 8'	12' x 10'
Area (ft^2)	15	24	30	48	96	120
(m^2)	1.39	2.23	2.79	4.46	8.92	11.15

Financial Considerations

Figures 14 and 15 show the cost of producing one litre of very pure distilled water using solar stills or electrically heated stills. The latest design of solar still, still F is cheapest since like all our solar stills it produces water of very high purity in one step while the electrically heated stills' primary product requires secondary distillation for accurate work.

The cost figures for stills D and E are higher than one might expect because these stills have been amortised over 3 years instead of five years. This is to allow for the severe corrosion which can be seen in these units after 3 years. Most of the costs incurred in running solar stills is in their maintenance. If one batch feeds a still it needs about 5 times as much maintenance time as if it is fed automatically.

After 5 years when all capital charges have been paid, aluminium solar stills with automatic feed really become cheap producers of distilled water, the figure for still F being 0.88¢ U.S. per litre compared with 3.86¢ U.S. for an electrically heated unit. The cost figures for electrically heated stills have been helped enormously by the low cost of water and electricity in Trinidad. These are 25¢ U.S. per 1000 gallons and 1.5¢ U.S. per kilowatt hour for large scale consumption. In most countries these figures would be considerably higher.

Purely cost figures do not show the nuisance caused by having to strip and clean an electric still after 40 - 50 hours of operation. In addition one cannot run the electric still in periods of water shortage because of cooling water problems. Solar stills also have the additional benefit of running every day of the year and do not need to be closely watched as do electric stills. If one adds a cost factor for an "electric still watcher," an absolute necessity during the dry season, their running costs would be much higher.

Fig. 14

RUNNING COSTS FOR SOLAR STILLS (IN US $).
ALL FIGURES ARE CALCULATED ON A BASE AREA OF 1 SQ. M
FOR A RUNNING TIME OF ONE YEAR.

	A	B	D	E	F
Capital Cost	117.70	100.54	78.48	78.48	100.90
(a) Capital charges					
(10% interest, 5 yrs. amortisation)	31.05	26.52	31.56*	31.56*	26.62
Maintenance (man hours)	10	10	10	10	2
(b) Maintenance costs @ $ 3.5/hr.	35.00	35.00	35.00	35.00	7.00
Total running costs (a + b)	66.05	61.52	66.56	66.51	33.62
Average distillate production (litres)	800	800	800	600	700
Distillate cost (US ¢ / litre)	8.26	7.69	8.32	11.09	4.80
Distillate cost after 5 years (US ¢ / litre)	4.38	4.38	-	-	0.88

* based on 10% interest, 3 years amortization

Fig. 15

RUNNING COSTS FOR DOUBLE DISTILLATION USING 15 KW ELECTRICALLY HEATED STILLS (IN US $). BASED ON 2000 HOURS OPERATION PER YEAR.

	Primary Distillation	Secondary Distillation
Capital Cost	1167.00	1167.00
(a) Capital charges		
(10% interest, 5 years amortisation)	307.77	307.77
Water production (litres)	40,000	40,000
Electricity consumption KW hrs..	30,000	30,000
(b) Electricity cost (at 1.5¢ per KW hr.)	450.00	450.00
Cleaning (hrs.)	150	--
(c) Cleaning cost ($3.50 per hr.)	525.00	
Cooling water (gallons)	240,000	240,000
(d) Water cost (25¢ US/1000 gals.)	60.00	60.00
Total Cost	1342.77	817.77
	3.36	2.04
Cost per litre of double distilled water (US ¢)		5.40
Distillate cost after 5 years (US ¢ / litre)		3.86

Conclusion

For producing distilled water solar stills are now a reasonable alternative to electrically heated stills. Their main advantages are simplicity of construction and maintenance, lack of dependence on conventional power sources, suitability for operation on intermittent water supplies and, after five years, low cost of product.

These units are perfectly suited to the remote rural areas of the Third World where power and skilled manpower are scarce or non-existent. One must be careful to ensure that materials used in construction are durable since operation of other solar stills (4) has shown that cheap materials prove more expensive in the long run.

References

1. O. Headley and B. Springer, J.Chem.Ed. Vol.48, 49-51 (1971).
2. O. Headley, Solar Energy, Vol.15, pp. 245-258 (1973).
3. O. Headley, Solar Energy, Vol.18, p.75 (1976).
4. E.D. Howe and B.W. Tleimat, Solar Energy, Vol.16 pp. 97 - 105 (1974).
   G. Frick and J. Hirschmann, Solar Energy Vol.14, pp. 405 - 413 (1973).

Solar water heating

7

7.1 Factors affecting the performance of solar water heating systems

B.J. Brinkworth,
Solar Energy Unit,
University College, Cardiff, U.K.

Abstract

A computer model of a solar water pre-heating system is shown to give a predicted performance which agrees fairly well with test results from a real system. The model is then used to indicate the effects of variation in some of the system parameters.

It is shown that a system of this kind is self-regulating to a remarkable degree. For a given hot-water demand, the performance is found to be only weakly dependent on the storage capacity, heat-exchanger effectiveness, collector losses and draw-off pattern. The quantities most significantly affecting the consumption of energy by the auxiliary source are the collector area and the delivery temperature.

Introduction

Although the behaviour of solar heating systems has been under investigation for many years, much remains to be learned about them. Individual components, particularly collectors, have been studied in great detail, but the characteristics of whole systems have received much less attention (1). The designer of a system for a given use in a given location must often resort to a rule-of-thumb to establish even the basic parameters of the system, such as the collector area and storage capacity required. He has almost nothing to guide him on the effects on the system performance of the use of refinements such as selective absorbing surfaces or even the amount of insulation material used behind the collector panel. Accordingly, it is difficult to make rational judgments of the economic value attaching to these, and hence to identify those combinations of parameters which lead to good cost/effectiveness ratios.

These uncertainties are particularly unfortunate in high-latitude countries such as Britain, where at present energy costs, the economics of solar heating is marginal (2). It is recognised that it will have significant place in the future energy economy (3). However, solar heating will not become used sufficiently widely to enable a viable manufacturing industry to develop in time, unless designers can be helped to achieve good performance at low cost.

The identification of optimum system parameters cannot be done effectively by experiment, because there are so many variables that might be thought to be significant, so that such a course would be too costly and would yield results too slowly. Accordingly, it must be done by system modelling. In this paper, some results are presented from a computer modelling exercise intended to identify the optimum system parameters for a domestic hot-water pre-heating system of the type most widely available in Britain at the present time.

The system model

Fig.1 shows a typical water pre-heating system. Fluid heated in the collector gives up energy via a heat exchanger to the water in the storage tank, so that when water is withdrawn from the hot tank, it is replaced by pre-heated water from storage, rather than by cold

Fig. 1
Typical solar water heating system.

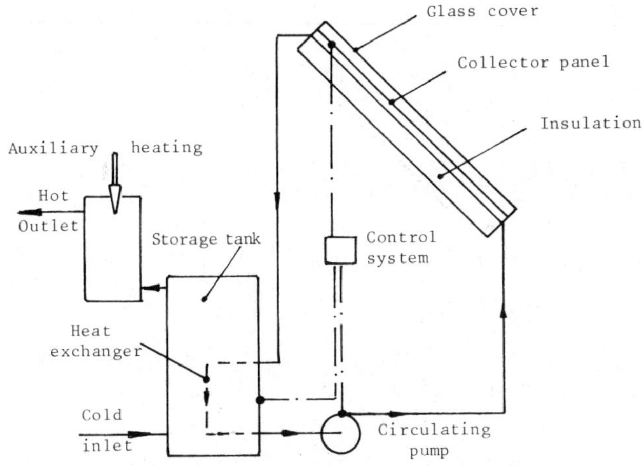

feed water from the supply. An auxiliary heat source, such as an electrical immersion heater, maintains the temperature in the hot tank at the value required at the point of use.

Although the system has few components, its performance depends upon many factors, among which the following might be considered to be important:

Output requirements:

Mean water demand
Draw-off pattern
Water delivery temperature

Input variables:

Water supply temperature
Incident solar energy; direct, diffuse, reflected

Ambient properties:

Outdoor air temperature
Wind direction and velocity
Roof-space temperature
Sky temperature

Component characteristics:

Collector characteristics
Heat-exchanger characteristics
Secondary fluid properties
Secondary fluid flow-rate
Storage tank characteristics
Control system characteristics.

These, in turn, depend upon numerous other factors, and many are variable with time.

It is apparent that a model which reproduced every characteristic of the system as completely as possible would be very elaborate, and would itself become costly to run. The requirement is for the simplest representation which can yield reliable design information.

This is likely to be different from a model which is able to reproduce the performance in fine detail. To give confidence that the information is reliable, however, predictions from the model must at some stage be validated by comparison with results from tests on real systems.

In the model used in the present work, the representation of the system variables, in relation to those listed above, is as follows:

Output requirements:

The mean daily hot water demand is specified in kg/day, and for the most part is kept constant from day to day. A variety of draw-off patterns are imposed, though the draw-off is confined to three periods per day, the first between 7 and 9 a.m., the second around noon and the third after 5 p.m. The final delivery temperature is usually 65°C, though sensitivity to this value is examined.

Input variables:

Water supply temperatures are taken from a 3-year survey made for the purpose, a mean value being taken for each month. Incident solar energy data, global and diffuse, are taken from the Meteorological Office tapes for the recording site at Bracknell. It is considered to be important that the day-to-day variation in the energy input be properly represented. A method has been developed for quantifying the sequential characteristics of the UK insolation, based on the autocorrelation functions for the daily global radiation (4). For use with the present model, a set of data representing one year has been made up, in which the seasonal trend and internal autocorrelation have the same values as the averages of these quantities for an 8-year period at Bracknell. A collector slope of 30°, with a south-facing aspect, is assumed; approximate methods are employed for determining the distribution of diffuse and reflected radiation, an albedo of 0.2 being taken for surfaces within the collector field-of-view.

A major saving in processing time is obtained by simplifying the representation of variation of insolation within the day. In the exercises now reported, the variation is assumed to be sinusoidal, each daily input being specified by three parameters: the daily totals of direct and diffuse radiation and the effective length of the day.

Ambient properties:

The outdoor air temperature is taken from Meteorological Office records, but a single value is used to represent each day. The temperature in the roof-space, which is taken to be the location of the storage tank, is determined from the outdoor air temperature and that of the living space according to assumed heat transfer mechanisms. Wind and sky temperature effects are not explicitly included.

Component characteristics:

Collector characteristics are based on the Hottel-Whillier model, with loss coefficients evaluated by the methods of Duffie and Beckman (1). An overall system effectiveness is determined for the combination of collector and heat-exchanger (5). Characteristics of the heat-exchanger are based on test results with helical coils (6). Fully-mixed conditions are assumed for the storage tank, on the basis of observations of convective motions around heat-exchanger coils. A conductance of 1 W/m^{2}K is assumed for the tank insulation. Secondary fluid properties and flow-rates are chosen so as to allow good control with a small temperature differential between collector plate and storage tank.

In the final stage of heating, an auxiliary source is assumed to raise the delivered water to the required tap temperature whenever the storage tank is below this. Anti-scalding control is assumed, such that flow in the collector circuit is shut down when the storage tank temperature reaches 65°C.

System performance

A solar pre-heating system is used to reduce the requirement for energy from some conventional source. A convenient measure of performance, therefore, is the energy supplied to the auxiliary heater plus that used by the circulating pump. Since different systems are to be compared, the performance index used here is the quantity A, the auxiliary energy consumption per unit mass of hot water delivered. This can, if desired, be compared with the energy required to raise unit mass of water from supply to delivery temperature, giving an indication of the effectiveness of the solar pre-heating component.

It has not been customary for performance data on actual systems to be given in these terms, and few are available which cover a sufficiently wide range of conditions to provide a basis for model validation. One useful set of published results are those given by Sharpley (7), which show the monthly averages of A over a period of three years for a system similar to the one now considered. The 3-year averages and the range of these data are shown in Fig.2, where they are compared with predictions by the present model. In making the comparison, an addition has to be made for the energy losses from the hot tank; these are estimated to be about 0.02 MJ/kg. Having regard to the many approximations made, the results show fair agreement month-by-month, and the annual energy requirements are quite close - 0.173 MJ/kg for the model predictions and 0.169 MJ/kg for the actual system.

The agreement is considered to be sufficient to justify the use of the

Fig. 2
Monthly average of auxiliary energy per unit mass.

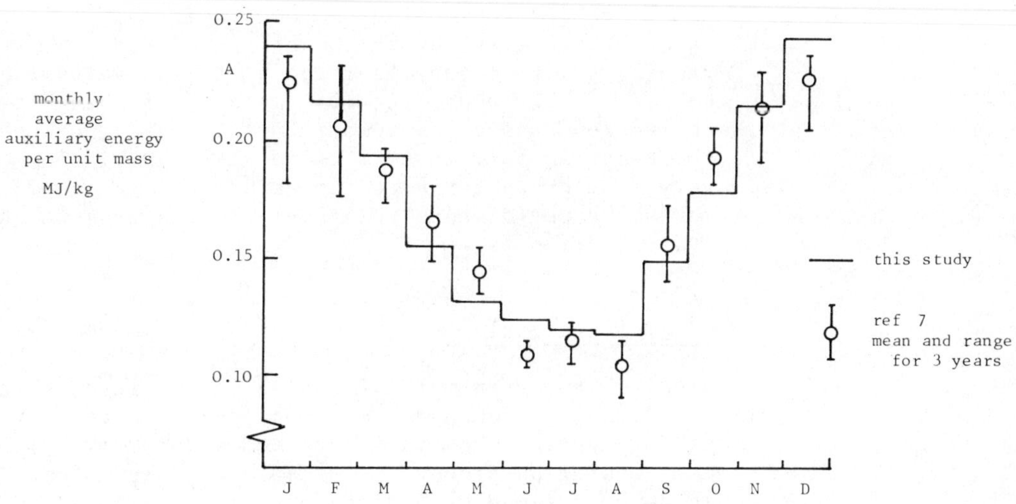

model for its primary purpose, which is to explore the sensitivity of the system performance to changes in the leading parameters, rather than to make predictions which are accurate in absolute terms. Dependence on the major variables of collector size and storage capacity has been shown elsewhere (8). Although some factors, such as pumping power and thermal losses from the storage tank, do not quite scale linearly, it is found that as long as the size does not vary too far from that appropriate to systems for domestic use the energy requirement per unit mass can be conveniently expressed in terms of two

Fig. 3
Annual average auxiliary energy per unit mass.

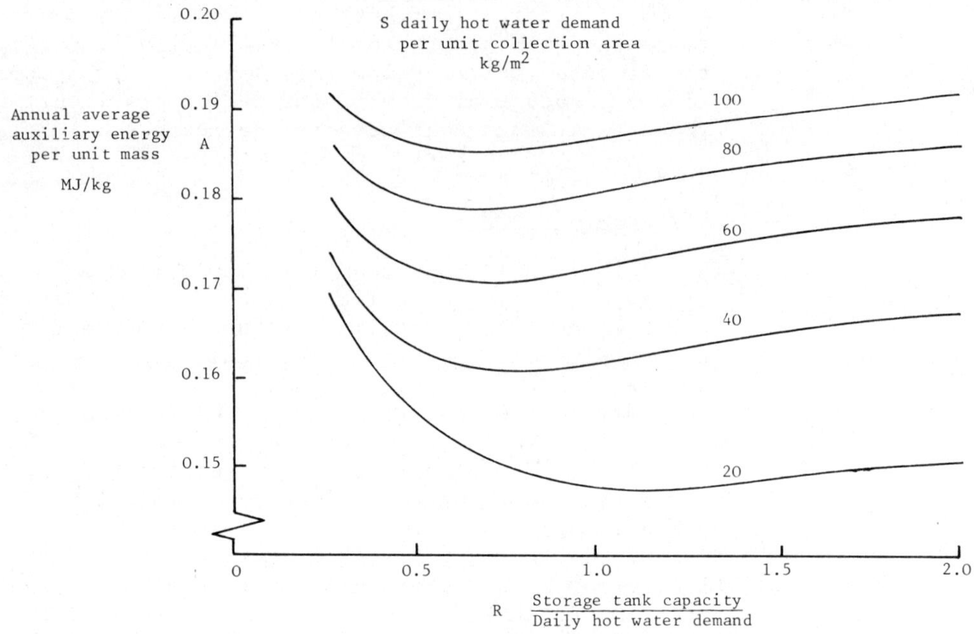

leading parameters: the ratio R of the storage capacity to the daily hot water demand, and the quantity S, the daily hot water demand per unit of collection area. Fig.3 shows the performance curves generalised on this basis, giving the annual average of energy consumption per unit mass of water delivered, for a demand which is invariant throughout the year.

It is seen that the curves pass through shallow minima, occurring where the storage volume is somewhat less than the daily demand. The small variation with R is the first indication of the self-regulating character of the solar pre-heating system, and is readily understandable. If the storage volume is small, the temperature rises quickly leading to shut-down of the system early in the day, and poor overall collection efficiency. A large storage volume, on the other hand, enables the collector to work more efficiently, but results in water being delivered to the hot tank at a relatively low temperature, requiring more energy to raise it to the value required at the point of use. In between, any circumstances leading to an improvement in collection causes the temperature to rise and hence to produce a compensating fall in collection efficiency. This not only results in very flat characteristic curves of the kind shown in Fig.3, but also makes the performance relatively insensitive to changes in other system variables.

Sensitivity analysis

Studies have been made of the variation in auxiliary energy requirement as a result of changes in component characteristics and the mode of use of the system. For this purpose, a datum system is selected, having the leading parameter values of S = 40 kg/m^2 per day and R = 0.75. These correspond with a minimum in the plots of Fig.3, giving an annual average of A = 0.160 MJ/kg.

a) collector area

Sensitivity of the performance to collector area is indicated in Fig.3. From the curves, it is found that in the vicinity of the datum case, a 10% increase in collector area produces a reduction of about 13% in the annual average auxiliary energy requirement, A.

b) Heat-exchanger effectiveness

It has been shown elsewhere that the sensitivity to changes in heat-exchanger effectiveness may be determined analytically (5). Around the datum case, which already has a high effectiveness, an increase of 10% in effectiveness produces a reduction of less than 1% in A.

c) draw-off pattern

The datum figures are for a draw-off pattern consisting of three periods (morning, noon and evening) in which the fractions of the day's demand delivered are respectively 0.2, 0.2 and 0.6. The results obtained with other patterns are shown in Table 1.

It is indicated that the energy consumption is not significantly dependent upon the draw-off pattern.

d) daily draw-off

The datum figures are for a constant daily demand throughout the year. A trial was conducted with the daily demand varying according to the higher demand schedule of BS 1250 (repeated weeks consisting of four days of pattern AA, two of BB and one of CC). The change in annual average A, relative to the datum condition is found to be +2.5%.

It is to be expected that a varying pattern will increase the energy consumption in this case, because the datum state is at a minimum, but the indication is that typical daily variation in demand does not significantly affect the performance.

e) collector front loss

The datum figures are for loss characteristics appropriate to a single-glazed collector with a non-selective absorbing surface. If the characteristics of a single-glazed collector with a selective surface are substituted, the annual average A is changed by - 13%. This is probably typical also of the result to be expected with a non-selective absorber single-glazed with IR reflecting glass.

f) collector back loss

The datum figures are for an overall back-loss coefficient of 0.5 W/m^2K. The effects of changes in this are shown in Table 2.

It is evident from these figures that the performance penalty is small, even if the back loss is allowed to reach a sizable proportion of the front loss. This is a significant conclusion, having regard to the cost of insulation for high temperatures.

g) delivery temperature

The datum figures are for a delivery temperature of 65°C. The effects of reducing this temperature are shown in Table 3.

It is obvious that the auxiliary energy consumption per unit mass will fall with delivery temperature, and over the range considered, the change is found to be almost linear. However, the strategy for determining the optimum delivery temperature is far from clear, since a reduction in temperature is likely to lead to an increase in hot water consumption and thus to changes in collector area and storage capacity.

TABLE 1 EFFECT OF DAILY DRAW-OFF PATTERN

pattern	annual average A MJ/kg	change %
0.2,0.2,0.6	0.160	datum
0.2,0.6,0.2	0.156	-2.5%
0.6,0.2,0.2	0.162	+1.3%
0.333,0.333,0.334	0.162	+1.3%

TABLE 2 EFFECT OF COLLECTOR BACK-LOSS COEFFICIENT

coefficient W/m^2K	annual average A MJ/kg	change %
0	0.157	-1.9%
0.5	0.160	datum
1.0	0.162	+1.3%
1.5	0.165	+3.1%

TABLE 3 EFFECT OF DELIVERY TEMPERATURE

Temp ^oC	annual average A MJ/kg	change %
65	0.160	datum
60	0.140	-12.5
55	0.119	-25.6
50	0.099	-38.1

Development

Results such as those reported have profound implications for optimum system design. Several lines of development are in hand, arising out of the work described. Determination of the sensitivity of overall performance to changes in system variables will be extended to include quantities other than those reported. The exchange rates thus determined will be employed, together with cost coefficients, in the identification of optimum designs, using the techniques of multivariate analysis. Validation exercises will be carried out, using data gathered on the water pre-heating system installed in student accommodation on the university campus at Cardiff. This system, having a collection area of $25m^2$, serves a four-storey block of flats, having five occupants per flat. Arrangements were made, in the design of the system, for the isolation of sections so that tests may be conducted which cover a range of the primary system parameters R and S. It is hoped that through tests of this kind it will also be possible to arrive at the simplest system models which would be acceptable for optimisation work. If these are sufficiently cheap to use, as a matter of routine, optimum designs for given duties can be continually updated as the situation concerning the supply of energy and materials changes. Further, the effects of actual or potential changes in technology can be quickly evaluated, assisting in the rational planning of research and development programmes in this field.

References

1. Duffie, J.A., Beckmann, W.A. — Solar Energy Thermal Processes. Wiley, 1974.

2. Courtney, R.G. — An appraisal of solar water heating in the UK. Building, 224, 1975, 137-139;141.

3. - — Energy R&D in the United Kingdom. Energy Paper No.11, HMSO, 1976.

4. Brinkworth, B.J. — Autocorrelation and stochastic modelling of insolation sequences. Solar Energy, 19, 1977 (in press).

5. Brinkworth, B.J. — Selection of design parameters for closed-circuit, forced-circulation solar heating systems. Solar Energy, 17, 1975, 331-333.

6. Brinkworth, B.J. — Some work of the Cardiff Solar Energy Unit. Proc.Seminar "Solar Energy for Buildings". North East London Polytechnic, Nov.26, 1976.

7. Sharpley, D.E. — Testing of solar collectors and systems in development and production. (in) Testing of Solar Collectors and Systems, ISES (UK), 1977

8. Brinkworth, B.J. — Active collection and use of solar energy. (in) Ambient Energy and Building Design, Construction Press Ltd., 1977.

9. - — BS 1250: 1963. Specification for Domestic Appliances burning Town Gas. Part 3: Water Heating Appliances. Appendix J: Practical Usage Tests.

7.2 Corrosion problems associated with the use of aluminium solar collectors

W.B. Gillett,
Solar Energy Unit, University College,
Newport Road, Cardiff, UK.

Abstract

Experience with the use of inhibited aqueous heat transfer fluids in solar heating systems containing aluminium absorbers is growing, and a number of installations have operated successfully for periods of 2-3 years. It is recommended that precautions be taken before using aluminium in domestic solar water heating systems because of its corrosion potential. The use of a non-aqueous fluid does not necessarily reduce the likelihood of corrosion and micro-organisms can produce additional difficulties.

Introduction

The potential for corrosion in solar heating systems containing aluminium collectors has been discussed by Popplewell (1) and by Butt (2). They have shown that by careful selection of corrosion inhibitors and by the use of pre-treated water, aluminium can be protected even in a multimetal system.

Practical experience suggests however that the ideal conditions in which aluminium will remain protected, frequently do not occur in real systems. Pitting corrosion of aluminium may be encouraged by the presence of very low concentrations of noble metals such as copper in aqueous solution, and also by acid radicals such as chlorides and bicarbonates. With inhibited aqueous solutions, ethylene glycol is often used as an anti-freeze, and this oxidises to form organic acids. Pitting corrosion rates may be correlated with pH in aqueous solutions, and while the reduction of pH associated with the degradation of anti-freeze increases the corrosion of most metals, aluminium is particularly vulnerable.

Commercial pressures appear from one side to encourage the presentation of details of systems where aluminium has been used successfully, while from the other attempts are made to discredit the use of aluminium. In Europe at the present time it is possible to market absorbers made in aluminium more cheaply than those made in copper, and while the evidence presented here suggests that a number of additional precautions should be taken when aluminium is used, a strong effort has been made to present an unbiased account.

Experience with aqueous heat transfer fluids

A system with $10m^2$ of roll-bonded aluminium collectors connected by copper pipes to an indirect copper water cylinder was operated in Llanelli using a 25% solution of ethylene glycol in water with corrosion inhibitors to BS3151 and benzotriazole to inhibit cuprosolvency. A randomly selected panel was found to have no evidence of corrosion after 9 months operation. Similar experience was reported by Hodges in the Solar House at Milton Keynes (3). There is also experience of multimetal systems in motor vehicles where annual replacement of the inhibited fluid is usually recommended, which might suggest that it is not particularly meaningful to judge a system after only 9 months. However, since the degradation of corrosion inhibitors and of antifreeze is both time and temperature dependent, regular monitoring will be required to determine a suitable fluid change interval for solar systems, and this may be expected to vary with geographical location and system design.

One means of detecting degradation in aqueous fluids is to monitor their pH value. A system with $10m^2$ of aluminium absorbers using an inhibited antifreeze solution has been in operation at the National Centre for Alternative Technology since 1974 where the natural water has a pH of 6.5. The inhibitor and antifreeze formulation in this system was designed to raise the pH to 7.5. After 30 months the pH was found to be 7.45 which suggests that the buffers in the solution had contained the degradation of the antifreeze over this period.

Without corrosion inhibitors

It is important to consider corrosion behaviour in the absence of inhibitors because particularly in a "do it youself" domestic installation, inhibitors might be omitted. It may be difficult to ensure that copper filings are excluded from the fluid system because these are most likely to originate from a last minute unofficial "adjustment" to the pipework when something will not fit neatly into place. Similarly, it is common plumbing practice to fill a system to check for leaks before adding inhibitors or antifreeze. Unfortunate timing of this procedure, such as just before the week-end or a holiday period might well permit pitting corrosion to initiate before the system has been properly protected.

A roll bonded aluminium collector was attached by a rubber tube to a cold water supply in Cardiff with a continuous flow of 2.5×10^{-3} $l.s^{-1}$ to waste. After only 55 days a leak was detected and a closer examination revealed a total of six holes and widespread pitting corrosion. Similar experience was reported by Bridgers (4) where an installation in the USA operated for 90 days with untreated water before widespread pitting corrosion destroyed the roll-bond absorbers. This installation was "retro-fitted with serpentine copper tubes rivetted to the back of the aluminium roll-bond absorbers".

Samples of the water supply used in Cardiff were found to contain between 0.2 and 0.4 ppm of copper. Some of this could have been picked up in the mains pipework and some would be due to the natural background level which is acceptable in drinking water at concentrations up to 1 ppm (5). Other examples of pitting corrosion in aluminium from concentrations of copper as low as 0.2 ppm in fresh water are cited by Godard (6). The extent of pitting corrosion in aluminium has been correlated with concentrations for a number of constituents of natural water and can be shown to vary with hardness, copper content, chloride content and pH. Pitting has been found with a copper concentration as low as 0.02 ppm (7) and it has been suggested that in the presence of chlorides and/or bicarbonates that pitting can occur in the absence of copper (8).

The use of biocides

It has been shown by Hill(9) that in concentrations greater than 15% in water, ethylene glycol will act as a biocide, but if the concentration drops below 10% as might well happen in a header tank, then micro-organisms may be actively encouraged. Copper is often considered to act as a biocide, but Booth and Mercer (10) have shown that some micro-organisms are very resistant to it. Biocides should therefore always be used in solar systems and particularly in header tanks.

Experience with non-aqueous heat transfer fluids

A system with $10m^2$ of roll-bonded aluminium collectors was operated using oil fluids. At first a polyaromatic synthetic oil was used but after 3 months this was replaced by a cheaper low viscosity mineral oil containing an anti-oxidant and benzotriazole as a copper passivating agent.

Domestic refrigerators use systems containing copper and aluminium with a non-electrolyte fluid, and the oils were expected, similarly, to provide a corrosion free, long life system which could operate above $100^{\circ}C$ without boiling and below $0^{\circ}C$ without freezing.

After only 9 months, 5 out of 10 panels failed from pitting corrosion. The corrosion products contained potassium and sulphur for which no obvious sources could be identified, although it is known that these are accumulated by micro-organisms. The heat transfer oil also contained some emulsified water, the source of which was unknown. A close examination of the oil and corrosion products revealed only residues from dead micro-organisms but samples of oil from the header tank which had been kept cool were shown to contain live bacteria.

The role of micro-organisms in the corrosion is uncertain because the failures were found during hot weather when the panels were operating at temperatures greater than most micro-organisms can withstand. However, a mechanism similar to that reported by Iverson (11) could have operated within water globules. Micro-organisms would remove oxygen to form an anaerobic region which would become anodic and corrode while the edge of the water globule in contact with more oxygen from the fluid stream would become cathodic (Fig.1). Once a corrosion cell had become established, it might have continued even after the micro-organisms had been killed by the high temperatures.

Fig. 1
Oxygen concentration cell

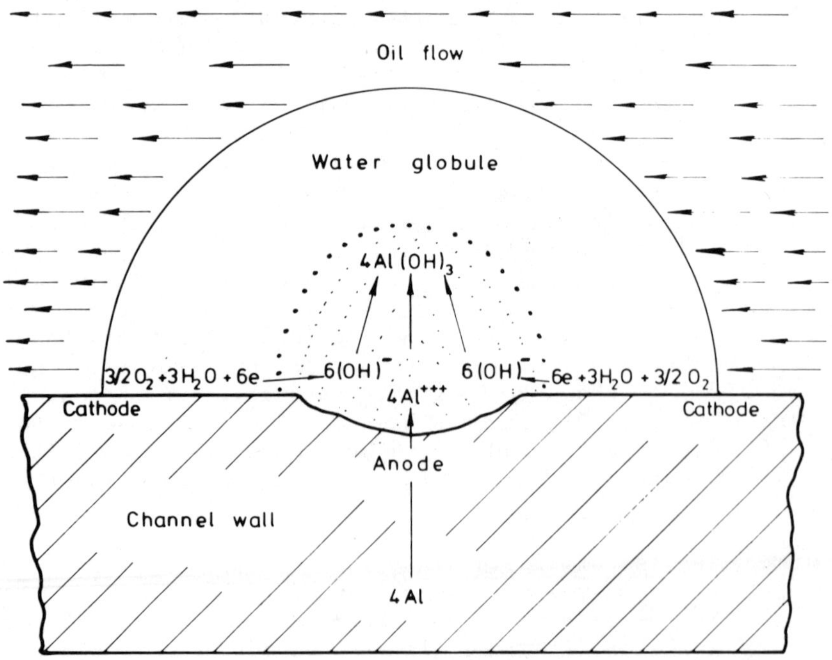

A very faint trace of copper was detected in two samples of corrosion products, but most of the analyses indicated that copper was not present. However, this trace suggests that even in an emulsion with oil containing corrosion inhibitors, water may pick up sufficient copper to cause pitting corrosion on aluminium.

It seems certain that the root cause of the corrosion was the ingress of moisture into the oil and that the resultant emulsion was able both to transfer copper to the aluminium panels and to establish cells suitable for the growth micro-organisms. If heat transfer fluids which are immiscible with water are to be used successfully in solar systems then it would appear that a sealed system should be employed with biocide additives and a desiccating filter.

Discussion

The work described suggests that the risks of corrosion in aluminium solar panels are unreasonably high, especially in domestic applications where the level of technology and the skill of the installers commonly employed may not be sufficient.

Non-aqueous fluids such as oils appear to require additional precautions to ensure freedom from moisture over and above the use of corrosion inhibitors and biocides. These fluids may also require special sealing materials in pumps, valves and pipe joints and their use is likely to incur high installation costs because the technology is more akin to that of air-conditioning or refrigeration equipment than to general plumbing.

Aqueous fluids require corrosion inhibitors, biocides and usually anti-freeze. They must be regularly monitored to ensure that the aluminium remains protected, and from time to time the fluid will need replacement.

The primary reason for these difficulties is the reactive nature of aluminium, and a more noble metal such as copper or steel might be more appropriate for long life domestic solar systems. While it is appreciated that at the present time the initial cost of roll-bonded aluminium absorbers is less than an equivalent in copper or steel, the long term running cost may be expected to be higher if aluminium is used.

One means of reducing the cost of absorbers is to attach copper fluid passageways to aluminium fins or sheets and to prevent external corrosion by the use of a suitable coating. An all copper fluid circuit has the additional benefits that a direct flow design and panel drain-down system may be employed. By using these design features the high initial system cost of copper absorbers may be partially offset by eliminating the need for a heat exchanger and for anti-freeze additives.

References

1. J.M. Popplewell. "Corrosion considerations in the use of aluminium and copper solar energy collectors", paper 30/12. Solar Use Now, ISES conference Los Angeles, 1975.

2. S.H. Butt. "Corrosion protection for aluminium collector plates". Workshop on solar collectors for heating and cooling of buildings, New York, November, 1974.

3. D. Hodges. "Analysis of the performance of the Solar House at Milton Keynes", UK-ISES conference C8 "European Solar Houses" April, 1976.

4. Bridgers, F.H. — Bridgers & Paxton Consulting Engineers Inc., Albuquerque, New Mexico, (Private communications).

5. T. Baumeister & L.S. Marks. — "Standard Handbook for Mechanical Engineers", 7th edition section 6-236, McGraw Hill, 1967.

6. H.P. Godard et al. — "Corrosion of Light Metals", Wiley 1967.

7. - — Institution of Heating and Ventilating Engineers, Guide Book B, Section B7-12 IHVE, 1970.

8. U.R. Evans. — "The Corrosion and Oxidation of Metals". Second supplementary Volume, Edward Arnold, 1976.

9. E.C. Hill — "The control of micro-organisms in aircraft fuel systems", J. Inst.Pet. Vol.56 pages 138-146, 1970.

10. G.H. Booth & S.J. Mercer. — "Resistance capacity of some oxidising and reducing bacteria towards copper", Nature 199 (4893), 622 (1963) Eng.

11. W.P. Iverson. — "A possible role for sulphate reducers in the corrosion of aluminium alloys", Electrochemical Technology, Vol.5 No.3-4, pages 77-79, March-April, 1967.

Editor's note: Following the request of the Aluminium Federation to present their point of view with regard to certain references to aluminium products contained in Mr Gillett's paper we are publishing an edited version of the Federation's comments.

Aluminium Federation statement on the paper presented by W.B. Gillett

We have requested the opportunity of replying to the paper presented by Mr Gillett in view of what we consider as misleading information which it contains relating to aluminium.

We must emphasise that aluminium has, for many years, been successfully used in mixed metal aqueous heat exchange systems and, provided suitable inhibited fluids are used and components made from aluminium isolated from other metals by non metallic connectors, a service life comparable with that of alternative heat exchange materials can be confidently expected.

The success of aluminium in such applications is demonstrated by the number of quality motor cars using aluminium radiators and the number of correctly installed solar heating installations in current use having aluminium collector plates.

Against this background of experience it is difficult to see the basis for Mr Gillett's statement that ... "Practical experience suggests however that the ideal conditions in which aluminium will remain protected frequently do not occur in real systems" ... The reader may think it significant that the only case histories quoted in detail indicate entirely satisfactory performance and, in fact, one of the installations mentioned has now been in operation for over 2½ years.

It is of course true that heat exchange fluids based on ethylene glycol need changing periodically if corrosion is to be avoided but this problem is common to all metals used in solar heating systems and, in fact, the main hazard here relates not to the aluminium collector plate but to corrosion of the secondary heat exchanger where failure will result in contamination of the supply water.

To support his contention that aluminium is unsuitable for use in aqueous systems, the author reports perforation of aluminium tubes through which was passed a raw supply water contaminated with copper. This experiment ignores the simple precautions defined by reputable suppliers of aluminium collector plates.

Another example quoted by Mr Gillett cites a case of corrosion of aluminium in an oil filled system.

Corrosion in oil filled systems may occur from three causes - oxidation of the oil, bacterial action, or water contamination. These factors are recognised by oil manufacturers and due account is taken of them. Oils formulated for solar heating applications generally contain anti oxidants and corrosion inhibitors, and are contained in sealed systems.

If, due to poor design of the system, conditions exist in which aerobic bacteria will grow, agglomerations of organisms can create a differential aeration corrosion cell by removing oxygen locally from an aerated fluid. However, no agglomerations have been reported by Mr Gillett in corroded areas.

It is also true that anaerobic bacteria can permit corrosion to proceed in oxygen free fluids which would not normally support corrosion. This effect however depends on absorbiton by living bacteria of hydrogen produced at cathodic sites in the vicinity of corrosion pits but the author reports the presence of live bacteria only in the header tank. It must be pointed out that in the reference given by Mr Gillett to the work of Iverson, conditions were completely different from those existing in solar heating systems, and the material in question was a high strength aircraft alloy which bears no resemblance whatsoever to the highly corrosion resistant grades of aluminium used for the manufacture of solar panels.

A much more likely explanation for both corrosion and water contamination in the experiments conducted by Mr Gillett is that the system had, at some time, been filled with plain water contaminated with copper from other parts of the system and imperfectly drained. Any deep corrosion pits developing beneath the residual water would probably continue to develop after the system had been topped up with oil.

The optimized performance of a solar regenerator in laminar forced convection

7.3

P. Gandhidasan, V. Sriramulu, and M. C. Gupta
Solar Energy Division
Department of Mechanical Engineering
Indian Institute of Technology
Madras 600036, India

ABSTRACT

Chemical dehumidification of air is an important industrial process in which a strong absorbent solution removes moisture from the air, as a consequence of which the solution becomes diluted. This paper describes a solar regenerator in which the weak absorbent solution is regenerated by using solar energy, and the water evaporating from the liquid surface is removed by a laminar forced air stream. Analysis of the solar regenerator has been made by considering the heat and mass transfer processes occurring in it. The performance characteristics of the regenerator have been obtained for varying insolation and other climatic data. The analysis reveals that there exist optimum flow rates of solution and air which produce maximum rates of desorption.

INTRODUCTION

On account of fast depleting fossil fuels and spiralling costs of conventional energy sources, attempts are being made all over the world to exploit unconventional sources of energy such as wind, tidal, solar, geothermal etc. Among these energy sources solar energy seems to offer a variety of possibilities for its utilization, particularly in tropical countries, which receive sunshine in abundant measure all through the year. A tropical climate can be either hot and dry or hot and humid. In hot and humid climates, the diffuse radiation constitutes a substantial portion of the total radiation received, and hence, flat plate collectors can be conveniently used for relatively low output temperatures. Also, flat plate collectors are simple, cheap, and easy to maintain.

In domestic sectors, solar water heaters are widely used in parts of Australia, Israel and Japan (1). Efforts are being made now to use solar energy in industrial sectors as well. Chemical dehumidification of air is an important process in which a strong absorbent solution removes moisture from the air as a consequence of which the solution becomes diluted. However, the weak solution can be regenerated by making it flow over an absorber surface exposed to solar radiation. A few solar cooling systems incorporating solar regenerators have been proposed (1, 2). In these systems, cooling is achieved by the conventional method, but the weak solution is regenerated in a solar device.

The problem of regeneration of a weak absorbent solution using solar energy has not received much attention so far. Löf (3) has proposed a method in which regeneration is achieved by passing the weak solution through a stripping tower, where it is stripped of water by the solar heated air. Hollands (4) has investigated the regeneration of a solution in a roof type solar still. Kakabaev (5) et al have described a system in which regeneration is accomplished as the solution flows over the surface exposed to the sun. The solar regenerator that is being considered in the present analysis is in many respects similar to this type.

DESCRIPTION OF A SOLAR REGENERATOR

The schematic of a solar regenerator that is being considered is shown in Fig. 1. It consists of a single galvanised iron sheet painted black. An absorbent solution, (such as calcium chloride or lithium bromide or chloride), which is to be regenerated, trickles down the collector as a thin film, and the bottom of the collector is well insulated. In order to reduce radiation losses the collector is covered by a 3 mm thick single glazing with an air gap of about 5 cm. The operating conditions of the regenerator are such that water vapour does not condense on the glass cover, and hence, transmission of solar radiation through the glass cover is not impaired.

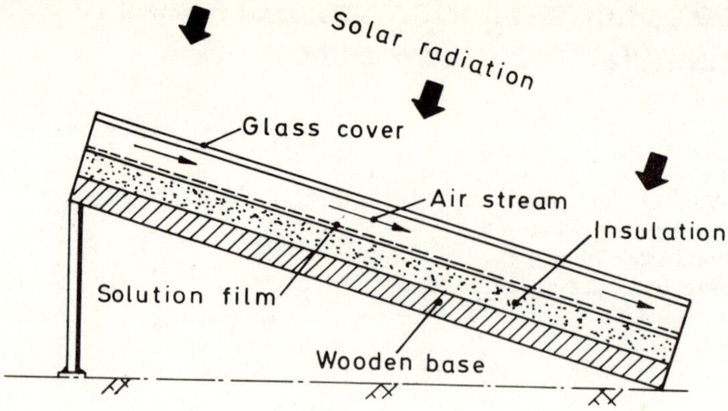

Fig. 1
Solar regenerator

The absorbent solution is heated by solar energy, and consequently water evaporates from the liquid surface. A method (6) has been suggested by which the water vapour leaving the solution surface can be removed. In this method atmospheric wind gusts are made use of. In the case of wind forced flow, the length of the regenerator is limited by the randomness of wind direction. Since long lengths of collectors would create stagnant pockets of air in the regenerator, they are likely to reduce the effectiveness of regeneration. On the other hand by maintaining a forced air flow a continuous flow of air in the unit is ensured. The flow in the unit can be maintained laminar throughout, so that the pressure drop in the unit does not become excessive. It has been shown that (7) the collector area required for laminar forced convection is less than that for wind forced flow. The regenerator employing laminar forced convection performs more efficiently than a wind forced unit for the simple reason that the mass transfer potential in the former is higher, which gives rise to an increased rate of regeneration. But this process lowers the operating temperature of the absorbent solution. When the temperature of the solution is lowered, the potential for mass transfer decreases. Hence, the operation of the unit must always correspond to optimum flow conditions in the regenerator. The present work concerns itself with arriving at these optimal conditions of operation.

ANALYSIS

The present analysis of a solar regenerator takes into account the climatic variables such as insolation, wind velocity, ambient temperature, and humidity in addition to the flow rates of absorbent solution and air. Assuming that steady state conditions exist in the unit, the following energy balance equations can be written:

Heat balance for solution and absorber is given by

$$(1 - e)\, \alpha\, \tau\, G = \dot{q}_k + \dot{q}_c + \dot{q}_r + \dot{q}_d + \dot{q}_a + \dot{q}_s \tag{1}$$

Heat balance for glass cover (8) is given by

$$\alpha_g G + \sigma \left(\frac{1}{\epsilon_s} + \frac{1}{\epsilon_g} - 1\right)^{-1} (T_s^4 - T_g^4) = \sigma\, \epsilon_g (T_g^4 - T_e^4) + h_c (t_g - t_a) \tag{2}$$

$$\dot{q}_k = f \frac{k}{\Delta x} (t_s - t_a) \tag{3}$$

where f is the factor which includes the effect of side losses, and it is taken as 10% of the bottom losses.

$$\dot{q}_c = h\, (t_{o_{av}} - t_g) \tag{4}$$

$$\dot{q}_r = 0.9\, \sigma\, (T_s^4 - T_g^4) \tag{5}$$

$$\dot{q}_d = \beta\, (p_s - p_a)\, \lambda \tag{6}$$

$$\dot{q}_a = \dot{m}_a\, c_a\, (\Delta t)_a \tag{7}$$

$$\dot{q}_s = \dot{m}_s\, c_s\, (\Delta t)_s \tag{8}$$

where $(\Delta t)_a$ and $(\Delta t)_s$ are the temperature rise of air and solution respectively, in the regenerator. The inlet temperature of solution is assumed to remain constant at 30 °C.

Energy balance for air stream:

$$h\left(t_s - \frac{t_{out} + t_a}{2}\right) = \dot{m}_a\, c_a\, (t_{out} - t_a) \quad (9)$$

It has been tacitly assumed that the air always enters the unit at the ambient temperature; h is calculated from the expression given in Ref. (9).

$$h = 1.28 \times 0.17 \times \frac{k}{D_h} \times (Re)^{0.33} \times (Pr)^{0.43} \times (Gr)^{0.1} \quad (10)$$

It is further assumed that the heat transfer coefficients for the solution film and the air stream, and for the air stream and glass cover are equal.

The mass transfer coefficient β is calculated from the measurements of Dropkin, who has propounded the following equation.

$$\frac{h}{\beta} = 0.227 \quad (11)$$

the validity of which has been verified in Ref. (7). The following relation for the vapour pressure of water in calcium chloride solution for a specified temperature and concentration has been used (10):

$$p_s = (-24.4 + \frac{20}{\delta})\, e^{0.05(t_s - 41.5)} \quad (12)$$

RESULTS AND DISCUSSIONS

A computer programme has been developed for predicting the performance characteristics of a solar regenerator. Using this programme the rate of desorption at intervals of one hour has been computed. Any change in the pattern of insolation brings about a change in the performance of the unit. The variation of the temperature of absorbent solution in the regenerator, water desorption per m^2 per hour, and the regeneration capacity of the unit per day have been determined for different flow rates of solution and air. The programmes have been run with the climatic data pertaining to Madras, India for the month of May, the month in which both temperature and humidity are high. The calculations have been performed for a regenerator of size 1 m by 1 m.

For an air flow rate of 36 kg hr^{-1} m^{-2} the variation in regeneration capacity with solution flow rate is shown in Fig. 2. With a 17% $cacl_2$ solution, maximum diurnal regeneration of 2.70 kg of water per m^2 per h is achieved for a solution flow rate of 2 to 2.5 kg. per m^2 per h. However, an increase in the flow rate from 5 to 8 kg

Fig. 2
Effect of solution flow rate on regeneration

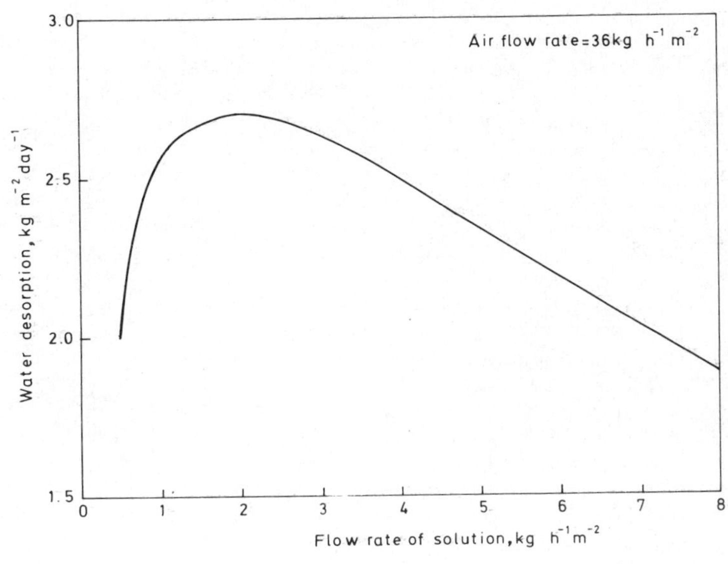

Fig. 3
Effect of air flow rate on regeneration

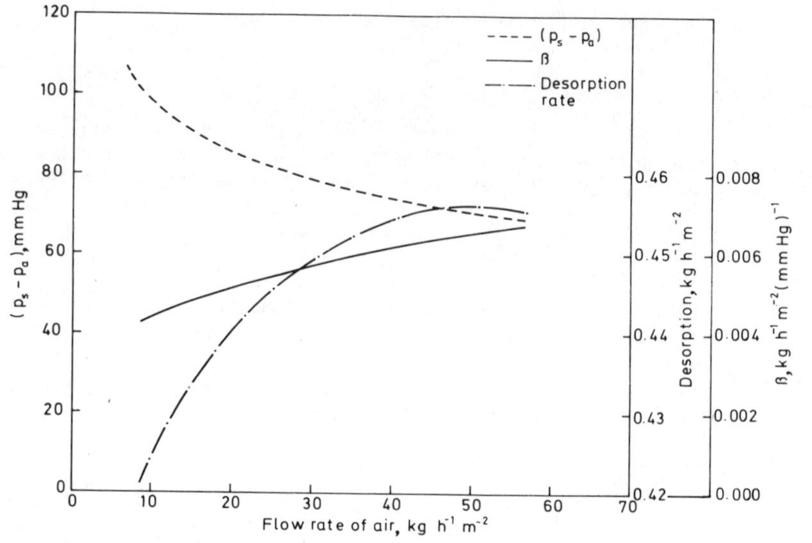

per m^2 per h. affects the performance drastically. Fig. 3 shows the effect of air flow rate on the driving potential for mass transfer (p_s-p_a), and the convective mass transfer coefficient for noon conditions when the solution flow rate is maintained at 2 kg. per m^2 per h. As the flow rate increases, mass transfer coefficient also increases, but the driving potential decreases because of the fall in the solution temperature. Since the rate of desorption is determined by the produce of these two quantities, there exists a flow rate at which the product is maximal. This aspect is clearly shown in Fig. 3. The hourly variation of the rate of water desorption for different concentrations, but for the same air and solution flow rates, is shown in Fig. 4, and the total desorption of water per day for different concentrations is shown in Fig. 5. It is obvious that when the concentration increases, regeneration capacity decreases. The thermal capacitance effect of the collector has been neglected in the analysis since preliminary calculations showed it contributed no more than 1% error.

Fig. 4
Hourly variation of water desorption

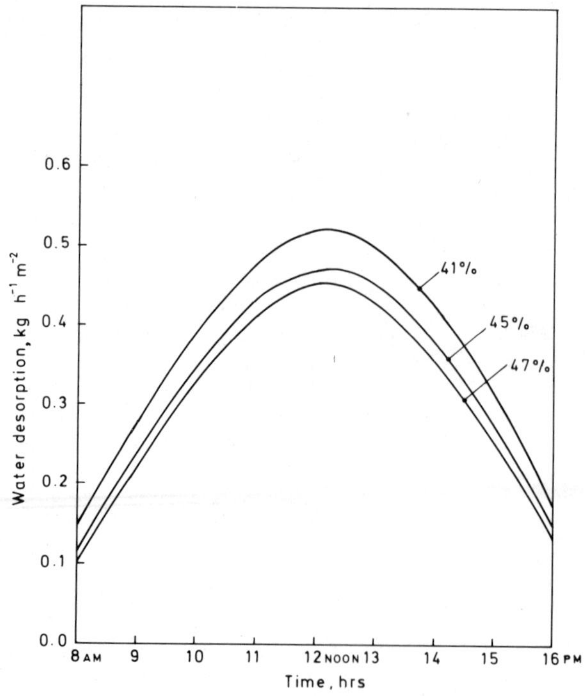

Fig. 5
Effect of concentration on desorption

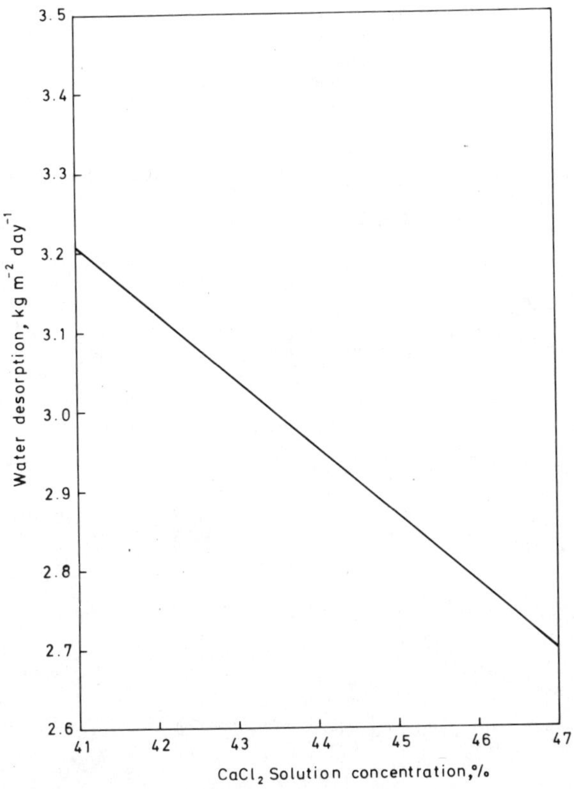

CONCLUSIONS

1. The performance of a solar regenerator for given flow rates of air and solution can be predicted by making use of meteorological data.

2. For a given air flow rate as the solution flow rate increases, the rate of desorption increases to start with and then decreases.

3. For a given solution flow rate, as the air flow rate increases the mass transfer coefficient increases, but the potential available for mass transfer decreases, due to decreased solution temperature. Since the rate of desorption is determined by the product of these two quantities, it attains a maximum at a particular rate of air flow.

NOMENCLATURE

c	specific heat, k cal/kg deg.
D_h	hydraulic diameter, m
G_r	Grashof number
h, h_c	convective heat transfer coefficient k cal/h m^2 deg
k	thermal conductivity, k cal/h m deg
$\dot{m}$	mass flow rate, kg/h
p	water vapour pressure, mm Hg
Pr	Prandtl number
$\dot{q}_k$	Heat conducted through base, k cal/h m^2
$\dot{q}_c, \dot{q}_r,$	heat transferred from solution by convection, radiation, and desorption respectively
$\dot{q}_d$	k cal/h m^2
$\dot{q}_a, \dot{q}_s$	heat transferred by air stream and solution respectively, k cal/h m^2

Re	Reynolds number
t	temperature °C
T	temperature °K
x	thickness, k m
$(\alpha\tau)$	effective absorptance—transmittance product
β	convective mass transfer coefficient, k cal/h m^2 mm Hg
δ	concentration of $cacl_2$ solution
Δ	difference
ϵ	emittance for long wave radiation
λ	heat of desorption, k cal/kg
e	reflectance of solution for solar radiation
σ	Stefan—Boltzmann constant, 4.9×10^{-8} k cal/h m^2 $°K^4$

Subscripts:

a	ambient air
e	sky
g	glass
o	air inside the regenerator
s	solution

REFERENCES

(1) Duffie, J. A., and Beckman, W. A., 'Solar energy thermal processes', Wiley, New York, 1974.

(2) Daniels, F., 'Use of the sun's energy', Yale University Press, New Haven, Conn., 1964.

(3) Löf, G. O. G., 'House heating and cooling with solar energy', Solar Energy Research University of Wisconsin Press, Madison, Wisconsin, 1955, p. 33-46.

(4) Hollands, H. G. T., 'The regeneration of lithium chloride brine in a solar still for use in solar airconditioning', Solar Energy, Vol. 7, No. 2, 1963, p 39-43.

(5) Kakabaev, A., et. al. 'Absorption solar refrigeration unit with open regeneration of solution', Applied Solar Energy, Vol. 5, No. 4, 1969, p 69-72.

(6) Mullick, S. C., and Gupta, M. C., 'Solar desorption of absorbent solutions', Solar Energy, Vol. 16, 1974, p. 19-24.

(7) Gandhidasan, P., and Gupta, M. C., 'A new method of sun powered airconditioning', Paper presented at the conference of International Institute of Refrigeration, Melbourne, Australia, Sept. 6-10, 1976.

(8) Morse, R. N., and Read, W. R. W., 'A rational basis for the engineering development of a solar still', Solar Energy, Vol. 12, 1968, p. 5-17.

(9) Mikheyev, M., 'Fundamentals of heat transfer', Peace Publishers, Moscow.

(10) Mullick, S. C., 'Solar airconditioning: Regeneration of absorbent solution', Ph.D. Thesis, I.I.T., Madras, 1976.

Solar domestic water heating in Malaysia

K.S. Ong
Mechanical Engineering Department,
University of Malaya
Kuala Lumpur, MALAYSIA.

Abstract

Malaysia is situated a few degrees north of the equator. She enjoys a uniform climate throughout the country - hot sunny days with occasional showers at different periods of the year, and warm tropical nights. Daily temperatures range from about 20°C to 35°C (65°F to 95°F). The average daily total solar radiation amounts to about 4.5 kW h m^{-2} (1420 Btu ft^{-2}). Apart from the high humidity and large amount of cloud cover, Malaysian climatic conditions favour the use of solar energy as a supplementary heat source.

Solar water heating has been investigated at the Faculty of Engineering, University of Malaya in Kuala Lumpur since 1971. To date, solar domestic water heating has been found to be the most attractive application of solar energy utilisation in the country. With the impending fuel shortage and spiraling fuel prices the adoption of a solar water heating policy in the country could be forthcoming. Such a policy would help towards alleviating domestic energy consumption. But up to the present moment solar water heaters have not been developed commercially by local manufacturers because of uncertainty of public acceptance. The deterrents are mainly high capital costs with long term recovery rates and the lack of well-proven and reliable demonstration units.

The paper deals mainly with the social and economic factors encountered by the author in his attempt to promote the development and use of solar water heaters in Malaysia. Although this report focuses on Malaysian conditions the discussions presented are generally applicable to other nearby developing Asian countries faced with a similar problem of promoting the use of domestic solar water heaters. Some solar radiation data for the country are included. An effective and realistic method of evaluating the life-cycle cost of solar water heaters compared to conventional heaters is included.

Introduction

We are constantly being reminded that the world is currently facing an energy crises in the form of depleting fossil fuel supplies and consequently, increasing fuel costs. Nett oil-exporting countries or countries which have access to nuclear power would be less affected than others in the near future. Eventually, however, the economic and standard of living of all nations would be affected. Thus there is a global awareness that alternative energy resources would have to be investigated and exploited. Nuclear energy involves high capital investment andpresent environmental as well as health and safety problems. Among the natural and renewable energy resources (solar, geothermal, tidal, wind and wave power), solar energy, at present, lies in the forefront. It provides an abundant and inexhaustible source of enormous amounts of clean energy and is relatively inexpensive to harness.

Solar water heating has been extensively investigated and solar water heaters have been made commercially available since the 1950's in countries like Australia, Israel, Japan and the United States. However to date, apart from these few countries, solar water heaters have yet to make an impact on the international domestic water heater scene. When the savings involved in using solar energy for heating water is marginal compared to conventional means, or where fossil fuel is cheap and easily obtainable, solar water heaters are not so widespread as one would expect. The successful implementation of a solar energy utilisation programme depends not only upon the geographical location of the country but also upon the level of development of that country. It would be more feasible to promote the use of solar energy in countries with an abundant and constant level of insolation throughout the year. Highly developed countries with a sophisticated life-style would hardly consider changing from conventional heating to solar heating unless forced to do so either by a steep rise in fuel cost or in the extreme case, by non-availability of fossil fuels. By and large, the economic status and social habits of the local population would determine public acceptability or rejection of solar water heaters as a standard household item.

Malaysia

A small nation in South East Asia, Malaysia stands today as the world's largest exporter of natural rubber, palm oil, tropical hardwoods and tin. It is one of the few fortunate countries in the world where adequate supplies of oil are available. In 1973, the Gross National Product was estimated at M$16,300 million. Based on 1965 prices, the per capita income was M$1,166 in 1973. Generally speaking therefore, Malaysia could be considered as a developing country with a comparatively high standard of living and where conventional fuel is readily available. Solar energy should then be considered only as a supplementary and not as an alternative main fuel source. It is envisaged that only marginal savings could be obtained using water heaters as compared against electric and gas heaters.

Solar Energy in Malaysia

Malaysia is situated a few degrees north of the Equator. She enjoys a uniform climate throughout the country - hot sunny days with occasional showers at different periods of the year, and warm tropical nights. Daily temperatures range from 20° to 35°C (65° to 95°F). Figures 1 and 2 show the mean daily total solar radiation as recorded over various sites in Malaysia and Singapore by the Meteorology Offices in Kuala Lumpur and Singapore and also by the Harrison and Crossfield Oil Palm Research Station at Banting.. The various instruments used were Kipp and Zonen Solarimeters, Bi-metallic Actinographs, and silicon cells. From the collected data it can be seen that the average mean daily total solar radiation level is about 4.5 to 5.5 kW h m^{-2} (1420 to 1740 Btu ft^{-2}). Instantaneous radiation measurements show that the radiation intensity fluctuated quite rapidly and over a wide range throughout the day due to cloud cover. Rainfall which usually occurs in the late afternoons or evenings does not constitute a major obstacle to the utilisation of solar

Fig. 1

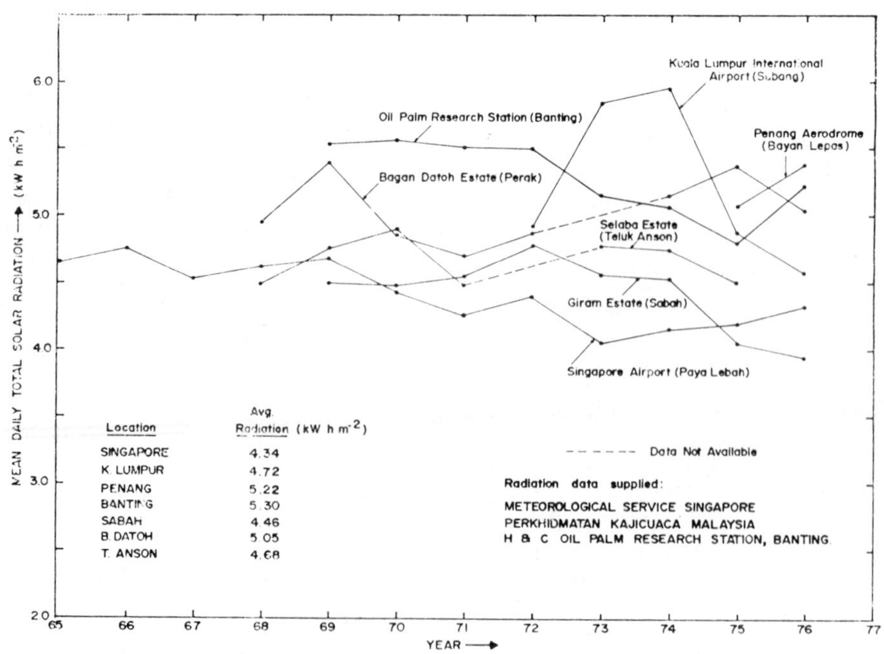

Mean daily total solar radiation for Singapore and Malaysia on yearly basis energy locally. Seldom does rain fall continuosly for long periods. At worst, day-long overcast skies or heavy rainfall occurs for maybe something like one or two weeks in a year. The rest of the year is bright and sunny. Apart from the intermittent nature of radiation intensity and high humidity, local climatic conditions favour the use of solar energy both as a primary or as a supplementary heat-source in conjunction with conventional fuel.

Fig. 2

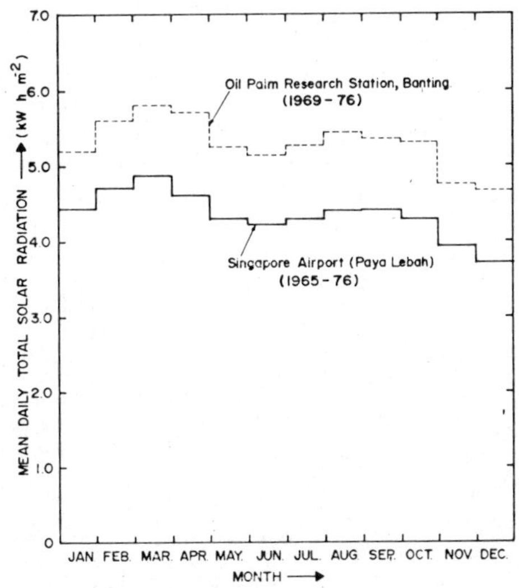

Mean daily total solar radiation for Singapore and Malaysia on monthly basis

Objects of Paper

The performance of solar water heaters in Malaysia have been investigated by Ward [1] , Pillay [2] , Ooi [3], and Ong [4]. But up to this time, it has not been developed commercially because electricity and gas are relatively cheap and easily obtainable. A few units were imported from Japan and Australia in the early 1960"s but they did not catch on with the local population. With the present rising cost of electricity, some local interest is being revived. But the prospects of commercial success seen uncertain. The deterrents are mainly high capital costs with long-term recovery rates and the lack of well-proven and reliable demonstration units.

This paper reports on some of the problems encountered by the author in his attempt to promote the development and use of domestic solar water heaters in Malaysia. Although this report focuses on Malaysian conditions the discussions presented are generally applicable to other developing countries faced with a similar problem of promoting the use of solar energy for domestic water heating.

Economic Aspect

Water heaters are still considered luxurious household items in Malaysia. However, with the rising standard of living, the trend is towards water heaters after air conditioners have been installed, especially in the urban communities. The rural communities and lower income group would be more inclined to use electric kettles or wood as fuel for heating water for bathing and washing clothes. The middle income group would be the most likely potential user of water heaters and especially solar heaters if they prove to be reliable and economical enough. The higher income group would have electric or gas water heaters and would not be very much influenced by the long-term and marginal savings involved by using solar heaters. In general, the major potential users of solar water heaters would be the middle income group of urban dwellers. To a certain but minor extent, technically-minded and imaginative people through out the entire population irrespecitve of their level of income may explore the advantages and long-term economy of solar heaters.

From past sales records, it is estimated that the 1977 demand for domestic water heaters is around 10,000 units. If solar heaters could compete with existing conventional electric and gas heaters, it could capture a third of the market and about 3000 units could be sold in the first year. Subsequent demand would depend upon public acceptability based on economics, aesthetics, reliability, and the availability and cost of conventional fuel. It is difficult to estimate accurately the amount of hot water required daily by an average bousehold. Different people use different amounts and at various temperature levels. From the author's own experience it is recommended that in Malaysia, for an average household of 4 to 6 persons, a 0.14 to 0.16 m^3 (30 to 35 gal) storage tank be used without supplementary heating for economic reasons. An oversized tank is unneccessary because we have good insolation days throughout the year. Very seldom do we have continuous rainfall or cloud cover occuring for more than a few hours a day. Rain usually occurs in the late afternoons or at nights, by which time, we have collected and stored sufficient hot water for our daily requirements. Large tanks with electric booster elements are expensive to manufacture, install and operate. Thermosyphon-flow systems are preferred for obvious economic reasons since they do not require additional pumps for water circulation.

Based on the above, it is estimated that the total annual energy requirement is about 1560 kW h (5.32 x 10^6 Btu) for a typical household for bathing at the end of a working day. Assuming an average available daily insolation level of 4.5 kW h m^{-2} and a 50% conversion rate, this daily energy requirement can be catered for by absorber panels covering 1.9 m^2 (20 ft^2)

A costing formula is presented here for comparing the life-cycle cost of solar water heaters with other alternative systems. The basis of comparison is the accumulated annual cost of each system. The total cost of a solar system (S_t) after the nth year is:-

$$S_t = S_c + \sum_1^n (S_m + S_s) \qquad (1)$$

ans the total cost of an alternative system (A_t) is:

$$A_t = A_c + \sum_1^n \left[A_m + C(1 + a)^{n-1} - i(S_t - A_t)_{n-1} \right] \qquad (2)$$

where:

A_c, S_c = Initial acquisition and installation costs of alternative and solar systems, respectively

A_m, S_m = Annual maintenance and repair costs of alternative and solar systems, respectively

S_s = Cost of supplementary or booster heating in the case of solar system only (= 0 when absent)

C = Annual cost of alternative energy sources based on equivalent annual energy requirement

i = Annual tax-free interest rate on savings

a = Fuel price increase.

In Eg. (2) the last term "$i(S_t - A_t)_{n-1}$" is arrived at by assuming that the difference between the total yearly cost of the solar and alternative systems is invested at the interest rate of i% annually. The solar system is justifiably assumed to cost more initially because of the additional components involved in the absorber panels.

The retail price of a solar system is estimated to be in the region of M\$1000 to M\$1500. The present cost of an electric storage heater with a 0.09 m^3 (20 gal) capacity costs around M\$400. An in-line type of gas heater unit retails for approximately M\$300. The average prices of electricity and gas for domestic consumption are M\$0.10 and M\$0.06 per kW h respectively. The current tax-free interest rate is 6% per annum. For the solar system the maintenance cost is taken as M\$10 annually for cleaning the glass covers on the absorber panels. No supplementary

Fig. 3 Life-cycle costs of solar, electric, and gas water heaters compared for different initial costs for solar system

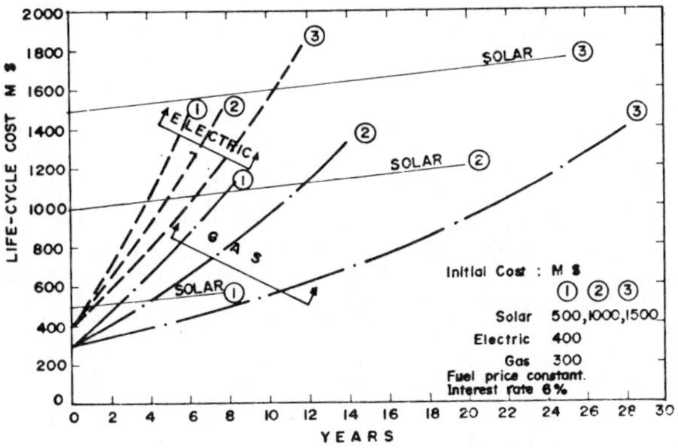

Fig. 4 Life-cycle costs of solar, electric, and gas water heaters compared for different interest rates

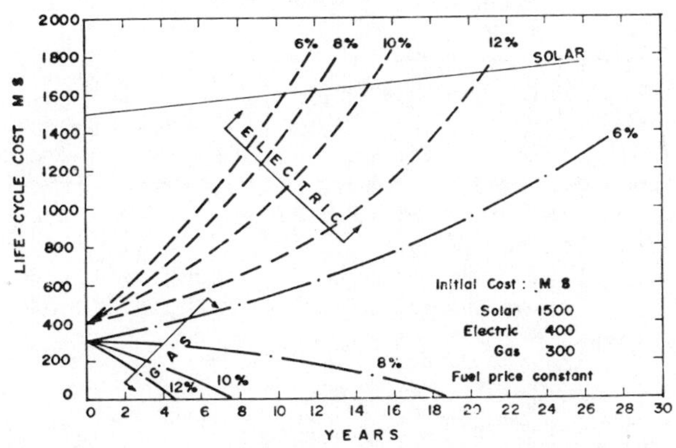

energy source is required, i.e. $S_s = 0$. For the electric and gas heaters the maintenance costs are assumed negligible. With these data inserted into Eqs. (1) and (2), the life-cycle costs of solar, electric and gas systems are obtained and shown in Figs. 3 to 5. Costing formulae different from, or more sophisticated than Eqs. (1) and (2) above can be devised. But, for Malaysia, where consumer goods are usually purchased outright and where hire-purchase is not practised for goods below the M$2000 level, the above analysis is deemed adequate.

Figure 3 compares the life-cycle costs of solar, electric, and gas water heaters for three different initial acquisition costs for the solar system at present day conditions assuming constant fuel price and a tax-free interest rate of 6% per annum. The effect of lower initial solar system cost on the break-even years can be seen here. The higher the initial cost the longer it takes for the solar system to break-even with the other forms of heating. For a break-even figure of five years, the initial acquisition cost of the solar system should therefore not exceed M$1000.

Figure 4 compares the life-cycle cost of solar, electric, and gas systems for different values of interest rates and at a constant fuel price. We have assumed the initial cost of the solar system to be M$1500. When the interest rate increases, the break-even years also increases as expected. The gas system is by far the most economical to operate. At an interest rate of 8%, the savings obtained by investing the amount of money determined by the difference in capital sums involved between the solar and the gas systems more than off-sets the running cost of the latter. Hence for this particular case, the gas system is the most attractive of the three systems compared.

Fig. 5 Life-cycle costs of solar, electric, and gas water heaters compared for different fuel price increases

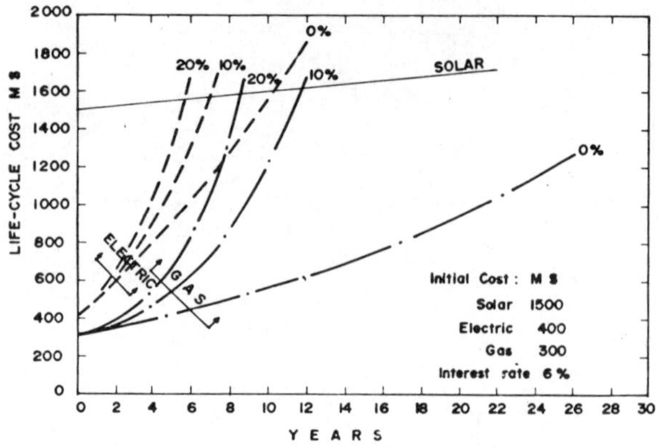

Perhaps the most significant single element affecting the economic. potential of solar heating is the increase in fuel price as shown in Fig. 5. Here it can be seen that the break-even years could be halved by increasing the price of fuel by 20% annually. A more realistic ncrease would be in the region of 5 to 10% annually.

The above results show that:

(i) by increasing the difference between the initial acquisition costs of the solar and the alternative conventional systems, the former becomes less attractive economically because of the increased number of years it takes to break-even with the latter system
(ii) by raising the tax-free interest rates on savings, the break-even period becomes longer
(iii) the solar system would be in a much better position to compete against conventional systems if the price of fuel increases steadily in the future.

Aesthetic consideration

Flat-plate units are recommended as they are more economical to operate than concentrating-cum-tracking types. Also, in the tropics, because of the large proportion of diffuse to direct radiation, concentrator units are generally felt to be less efficient than flat plate units. A typical 0.16 m^3 capacity domestic solar water heater unit would require 1.9 m^2 of collector area as discussed previously. Placed side-by-side, the absorber panels would cover an area od dimensions 1.2 x 1.6 m (approx. 4 x 6 ft). Unless a small area of the building roof was designed where the panels could be aesthetically sited, they could be rather unsightly. A problem associated with thermosyphon units (to keep capital cost down as then a pump is not required) is that the storage tank must be placed higher than the panels. In order to keep the plumbing short, the tank would have to be sited fairly close to the panels.

Figures 6 and 7 show two locally-designed and fabricated solar water heaters installed for demonstration purposes in Kuala Lumpur by the author. The absorber unit is made entirely from aluminium. It consists of a tubular frame fitted with contoured panels and enclosed in a fibre-glass insulated box with single glass glazing as cover. The absorbing surface is painted with ordinary matt-black emulsion paint with the proper corrosion-protection undercoat paints. The storage tank consists of a stainless steel internal tank lined with fiber-glass insulation and protected externally by a galvanised-iron steel jacket. The connecting pipings are ordinary domestic plumbing galvanised iron pipes and fittings. Figure 6 shows two absorber panels (each 1.22 x 0.61 m)placed on an existing roof (15° angle pitch) and connected to a 0.09 m^3 (20 gal) capacity storage tank. The latter is situated inside the bathroom of the house. The unit shown in Fig. 7 is larger and consists of three panels connected to a 0.14 m^3 (30 gal) capacity storage tank. Here, the whole assembly is placed on the existing flat roof and completely hidden from

Fig. 6

A 20 gallon low pressure type solar water heater installed on an existing sloping roof

Fig. 7

A 30 gallon cistern type solar water heater installed on a flat roof

view from the ground. The storage tank here is a cistern-type one and incorporates a float valve in the tank connected to the cold water supply line. Both units have been operating satisfactorily for just under a year. No maintenance has been neccessary so far.

The collector panels, once set up on top of a roof, blending with the roof design or kept out of sight, should operate reliably with a minimum of maintenance for a number of years. In principle they should require no maintenance throughout their lives. But, because of dirt and dust deposition in a dusty environment, the glass covers would have to be cleaned to ensure high collection efficiency. This periodic cleaning operation could be done by a group of service cleaners for a modest fee. However this should be the only form of maintenance required. Problems associated with broken glass caused by vandalism or natural causes are not maintenance problems as such and should be regarded as natural hazards. The unit should last for at least as long as it needs to break-even with the other forms of conventional heating in terms of costs. If it is designed for a longer life, it would be more attractive for potential users to invest in the system.

Legal and environmental aspects

Heat would be radiated, and light reflected, from solar collectors. Legal implications could arise from the additional heat load and glare imposed on buildings situated directly opposite to the panels. These effects are more pronounced in places where the inclination of the collectors is large. The glare from solar collectors installed in houses near airports or along the glide-paths of aircraft would create a navigational hazard. A further problem arises, when after having installed and operated a solar heater for a number of years, a houseowner suddenly discovers that reconstruction work or replanting carried out by his neighbour shade his collectors. Users of solar heaters should, of course, ensure as far as possible that none of the above situations are likely to happen by properly siting their equipment prior to installation.

Nevertheless, some thoughts should be given to resolve these problems if a large scale solar industry were to be set up.

Local building bye-laws should be updated to include the installation and operation of solar water heaters. Public protection from inferior or badly-designed units should be made available to encourage their use. Otherwise, the poor operating performance and experience of some units and users on a new-product market could prove detrimental to the future solar industry.

Discussions and conclusions

In this report we have considered some design factors such as aesthetics, reliability and durability of solar units, and we have briefly dwelt on some legal and environmental aspects. We have analysed the life-cycle costs of solar, electric and gas heaters, but we have not compared the merits and drawbacks of each system, such as power failure, fire-risk, and so on.

Our main conclusion is that, the market for the present design of thermosyphon-flow type of solar heater is promising only if the capital cost is not very much higher than conventional systems. Those who are willing to accept the long-term economy of solar heaters are deterred by the lack of reliably-proven demonstration units. Malaysians have the tendency, common to most humans, to be conservative and would seldom venture into purchasing new products unless and until they have been reliably proven. In Malaysia, the only and most effective way to promote solar heaters is to incorporate them as a standard feature of new houses in future housing estates. As here, the houses could be designed to accomodate the panels and the storage tanks. For aesthetics, the building has to be designed with solar heating in mind. Retro-fitted units could prove unsightly.

The present life-cycle cost analysis should be repeated when more reliable data such as the rate of increase of fuel price and actual retail prices of solar heaters are made available. At present these are only estimates. Without any commercial unit being placed on the market it is very difficult to determine its cost and reliability. In fact, the speculative nature of fuel prices for alternative systems is an important selling point and a key factor in market-capture projections. Public acceptence of solar heaters can be further enhanced by proof-of-concept demonstration units financed by government and industry, by economic incentives to conserve other depletable sources of energy, and by additional research and development. But first and foremost, engineers, manufacturers, and all those involved in the development and promotion of solar water heaters must strive to bring down the initial acquisition and installation cost to that of alternative systems.

References

1. G.T. Ward, Performance of a flat-plate solar heat-collector. Proc. Inst, Mech. Engng. 169, 1091 (1955).

2. R.K.Pillay, Solar energy collector. M. Eng. Sc. Thesis, University of Malaya (1967).

3. H.S. Ooi, Experimental and theoretical evaluation of solar water heaters in Malaysia. M. Eng. Sc. Thesis, University of Malaya (1974).

4. K.S. Ong, An improved computer programme for the thermal performance of a solar water heater. Solar Energy 18, 183 (1976).

Residential solar water heating in Canada 7.5

H. J. McQueen* and S. Lin
Mechanical Engineering
Concordia University
Montreal Quebec H3G 1M8
Canada

Abstract

The possibilities of utilizing solar energy for domestic water heating in existing dwellings are analyzed for the Canadian climate, in particular that of the Montreal-Ottawa region. A system design is proposed to accommodate building design and siting, summer and winter weather, snow cover, storage and interfacing with water heaters.

A solar energy system for heating domestic water, which was designed and fabricated by a team of students in the Engineering Faculty of Concordia University, is described. On a summer day the system operating at 61% efficiency was able to absorb 34 MJ of solar energy heating 50 gallons of water from 21°C to 66°C.

Introduction

The Montreal-Ottawa region is at a latitude of 45°, which is not much different from that of the Riviera. The annual insolation is near the continental average for that latitude, indicating that there are no geographical features that persistently perturb the weather. In the three summer months, the daily high temperatures often exceed 30°C and the rain usually comes as thundershowers. In the winter months of December to February highs of 0°C and lows of -20°C are common. Fortunately the coldest weather is usually associated with clear skies. Furthermore, the radiation on a surface inclined at 60° is almost double the direct solar radiation on a horizontal surface, partly as a result of the snow cover (1, 2).

More than 70% of the dwellings in Montreal have flat roofs with parapets and central drainage. Such a structure ensures an insulating snow cover and a non-freezing drain for the snow as it melts from house heat and sunshine. A common type of home is the two flat 'duplex', frequently attached on one side. Often one flat is owner-occupied, which ensures continual maintenance and improvement. Such flat roofed dwellings are still being constructed although in the suburbs pitched roofs are popular.

The flat roof at a height of some 8m above the ground offers an excellent platform for installation of solar collectors. The collectors can be orientated directly south at the appropriate angle and there is usually space for more than one row. The height avoids shading from trees and building heights in each neighbourhood are fairly uniform because of zoning regulations. Collector supports must be adequate to prevent damage to the roof membrane, to withstand wind forces and to raise them clear of the snow. Since access to the roof is usually provided from a back porch, it would be feasible for interested owners to change the collector angle seasonally. The back porch, which is frequently enclosed and sometimes insulated, can also provide a means for leading the necessary piping to the basement without affecting the interior of the house.

---

*Visiting Professor, Liberal Studies in Science
The University Manchester M13 9PL UK

Heating domestic water has advantages over other applications of solar power because it can be used all year round. A modest collector can supply the full demand for hot water in the summer. In winter it can still be operated at high efficiency since both the inlet and outlet temperatures are lower. An additional incentive is the high cost of electric water heating, since inexpensive off-peak power is not available in Quebec. An effective system can easily be installed in an old house and a satisfactory experience might convince the owner of the feasibility of a larger system for space heating.

The political climate for solar energy has not been very bright in Canada. The price of energy has been very low. However, this is changing as the government gradually raises the prices of all energy to world levels to encourage conservation and extend domestic reserves of gas and oil. The Science Council of Canada recommended a concerted effort to utilize solar energy in 1975 (3). Nevertheless, the government energy strategy up to 1990 projects only a token contribution from renewable resources (excluding hydro) (4, 5). The funding of renewable energy research has been negligible in comparison with that for nuclear power: in 1975-76 $1.6 million compared to $85 million for nuclear and $113 million for all energy (4, 6). On a per capita basis the funding is only one tenth of that of the US even although the heavily populated part of Canada lies within 100 miles of the border. Probably the biggest initiative by the government (and quite aside from its energy policy) arose as a result of hosting the UN Habitat '76 Conference in Vancouver. Some $800,000 was allocated to the construction of eight homes completely or partially powered by the sun (6). This year the funds for renewable energy research have almost been doubled and parliament has granted the charter for a private solar energy foundation. Some hope can be entertained for a program of subsidization or tax relief for the installation of solar equipment since partial subsidization of home insulation has been legislated.

Design objectives

The system for domestic water heating for Canadian conditions was conceived as consisting of several portable collectors mounted on the roof, a heat exchanger mounted in an auxiliary water tank in the

Fig. 1

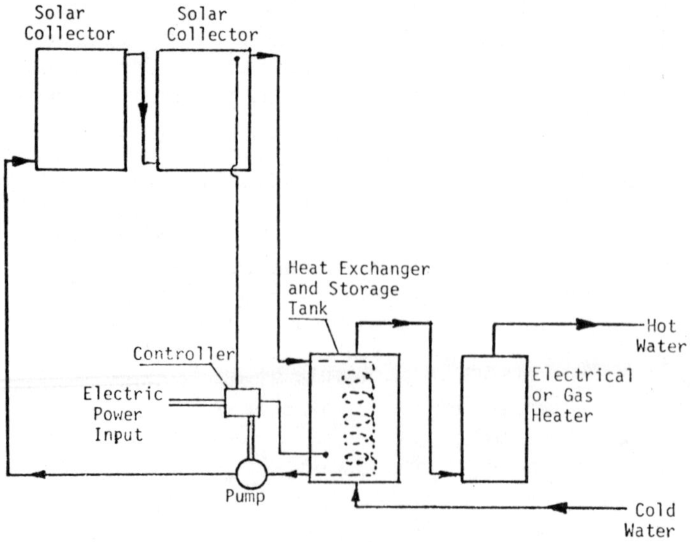

Schematic diagrams of solar water-heating systems showing collectors, circulation controls and storage tanks. (a) Forced convection system with storage in basement and non freezing working-fluid.

basement, a piping and pumping system to circulate the non-freezing transfer fluid and controls which activated the pump only when the collectors were warmer than the reservoir (Fig. 1). The system was to be suitable for installation in existing dwellings which already had a water heater. Its capacity should be such that it could supply almost completely the required hot water in summer but would only preheat in winter. The system should be simple to operate and require little maintenance.

Two flat collectors 3ft (0.91m) wide and 4ft (1.22m) high were considered adequate for providing summer capacity. They were to be constructed to have the highest efficiency, and thus the lowest cut-off radiation level, for winter use. Thus two layers of glass were provided although one would have sufficed for summer. The absorbers were to have a selective coating to reduce emissivity; this makes the use of a third glass sheet counter-productive (7). One other concept was the use of a series of collectors of increasing efficiency. Thus it was planned that a cylindrical parabolic focusing collector 3ft wide and 6ft along the axis would follow the two flat collectors. Since in such a concentrating collector the absorber area is only 5% of the collector area, it would have lower losses to the surroundings. Because it would be difficult to maintain and operate a sun-tracking mechanism under snowy conditions, it was to be used stationary with its focal axis extending east-west with only seasonal adjustment of the inclination of its plane of symmetry.

The flat collectors were to be tilted at 60° to the horizontal even though their normal would be lower than the sun's rays throughout most of the year. However, in winter the high angle makes good use of indirect radiation and assists the snow to slide off.

The antifreeze solution selected was propylene glycol which is less toxic than the ethylene variety by a factor of 30. The hot fluid is pumped to the basement through insulated pipes. There it passes through a heat exchanger in a storage chamber through which the water passes on the way to the conventional water heater. It should be noted that the Canadian climate requires the water supply to be brought in through the basement floor and directly to the heater, without a reservoir in the loft.

Fabrication techniques

Copper was selected for the absorber material for several reasons. Copper is an excellent conductor, can be soldered easily and is corrosion resistant to water or aqueous solutions. Furthermore, copper can be anodized to yield a surface with a differential in absorptivity and emissivity. A black oxide coating of about 10μm thickness has an absorptivity of 97% for the most important solar wavelengths 0.3-3μm but has an emissivity of only 3% for the longer wavelengths characteristic of the plate temperature (8).

The first flat plate collector was of standard construction. The 0.5 in (12.7mm) copper tubes which would run from top to bottom of the collector were silver brazed at 4in (0.1m) intervals into a 0.75in (19.1mm) header with suitable T-joints. The tubes were suitably bent near the header so that the whole piping network would lie flat against the 0.125in (3.1mm) sheet which would be the main absorber. The tubes were then brazed to the plate with soft solder. Since the lead-tin solder has much poorer conductivity than the copper, this decreases the effectiveness of heat transfer from plate to tube.

In order to avoid the problems of the soldered joints and to decrease the weight, and hence the thermal capacity, of the second flat collector, a new design was used which omitted the sheet. The 0.5in (12.7mm) tubes were spaced 1in (25.4mm) apart and brazed into the header tubes (Fig. 2). The tubes were then flattened by stages in unison on a hydraulic press. This was continued until they almost touched each other. Aluminium foil beneath the tubes reflects radiation passing through the gaps onto their under surface. By creating a somewhat irregular cross section and inducing turbulent flow, the flattening also serves to improve the heat transfer.

As there were neither suitably sized tanks nor a suitable power supply at the University, and as no commercial plating house could be found in Canada to perform the task, the selective coating was applied by Enthone, a subsidiary of Chase Brass in New England. Because the thick-soldered joints of the tube-on-sheet absorber had affected the anodizing to cause white streaks, this collector was painted with flat black heat-resistant paint and was used first in the series where its efficiency would not be so critical.

For the cylindrical reflector, parabolic ribs were cut out of plywood, set into the frame and covered by a thin sheet of galvanized steel. The

Fig. 2

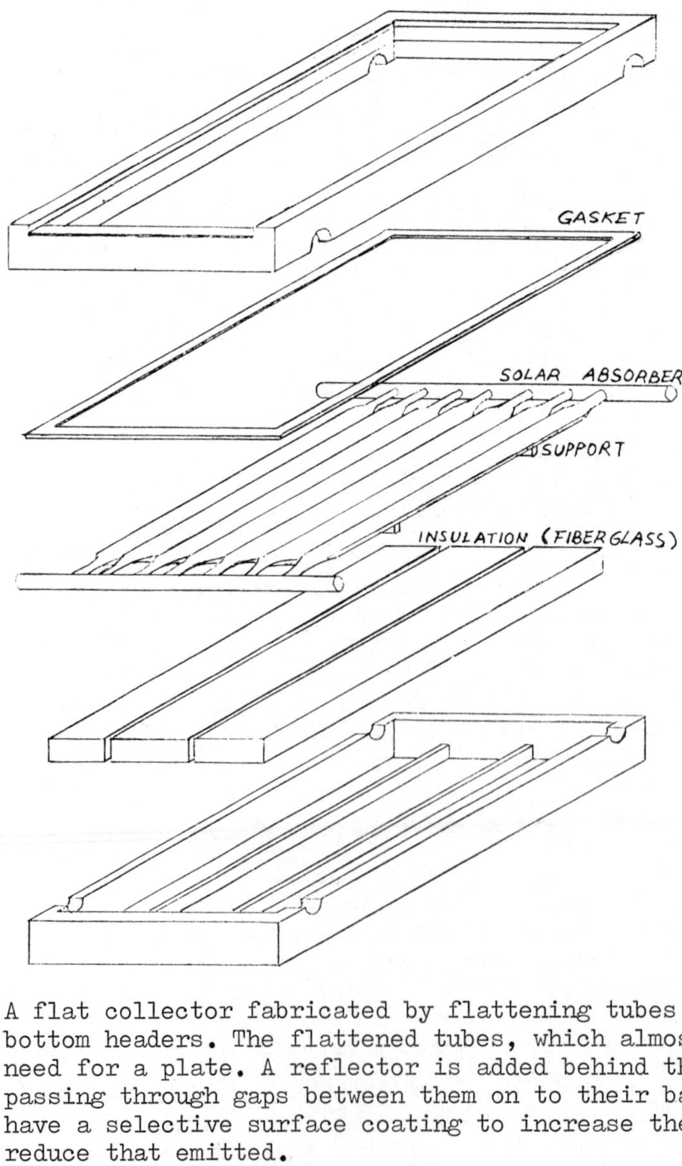

A flat collector fabricated by flattening tubes brazed into top and bottom headers. The flattened tubes, which almost touch, eliminate the need for a plate. A reflector is added behind the tubes to reflect light passing through gaps between them on to their back surface. The tubes have a selective surface coating to increase the radiation absorbed and reduce that emitted.

steel was then covered with a layer of aluminized plastic. Since the parabola was of such a shape that the focal line was inside of the tips of the parabola, a cover glass was added to insulate the absorber and protect the reflector from the weather. The absorber was a flattened copper tube of 1½in diameter which was given the anodizing treatment.

The frames of the collectors were made out of wood and weather proof gasketing was used between the frame and the glass (Fig. 2). Wood is the cheapest material for this purpose when the panels are not mass-produced. The cover glasses were of ordinary window glass 0.125in (3.1mm) thick. This had a transmittance of only 92%, however, no glass of better transmittance could be obtained in reasonably large sheets. The glass was held to the panel by a frame on only three sides. On the lower edge, it overhung the panel so that snow or ice would slide off it more easily. This arrangement has proved effective; storm accumulations disappeared during the following sunny day even when temperatures were -15°C. When the panels were mounted on the flat roof, the lower edge was about 0.6m above the surface so that there was space for the snow and ice which slid off.

Test results

In preliminary trials the collectors were tested individually and in a simple circuit with a double compartment storage tank. The panels were covered with only one glass since this reduced the absorption without a serious loss of heat in summer. In addition, they were mounted at 45° to the horizontal to bring them nearly perpendicular to the sun's rays. The flattened tube collector was found to be superior to the painted tube-on-sheet collector and was placed second in series thereafter.

The storage tank was not a heat exchanger but consisted of two compartments. The lower compartment, which also contained the cold water inlet, was in a loop with the collectors. The upper compartment from which hot water may be drawn was connected to the top of the lower compartment with two 1in (25.4mm) tubes (Fig. 3). Water heated by the collectors circulated into the lower tank without disturbing water at a high temperature in the upper portion. Only when the water from the collector became hotter than that in the upper tank would it begin to enter it. A system such as this would be suitable for a summer cottage where the tank could be mounted in the attic. An immersion heater could be added to the upper tank.

The parabolic reflector did not effectively increase the water temperature above that acquired in the flat collectors, possibly because of the poor reflectivity of the aluminized plastic. Consequently, this collector has been set aside until the reflector can be improved.

Subsequently, the system consisting of the two flat collectors and the storage tank was entered in the Energy Resource Alternatives contest sponsored by Student Competitions on Relevant Engineering for universities in North America. Official tests took place in Albuquerque, New Mexico, in August, 1975. During a day in which the integrated solar flux on the panels was 56 MJ, 34 MJ was collected with efficiency of 61%. Measured in another way the system produced 50 gallons of water heated from 70°F (21°C) to 150°F (66°C). On the basis of these results the system was the best solar water heater and had the highest efficiency of the 40 systems entered in the competition by 33 universities. The flattened-tube collector and the two-compartment storage tank merited third place for innovation.

The collectors have now been installed on a house according to the proposed design. The system will be subjected to testing next winter.

Fig. 3 A multiple compartment reservoir in which the partition partially isolates the hotter water from the cooler volume which may circulate through the collectors under conditions of reduced solar radiation.

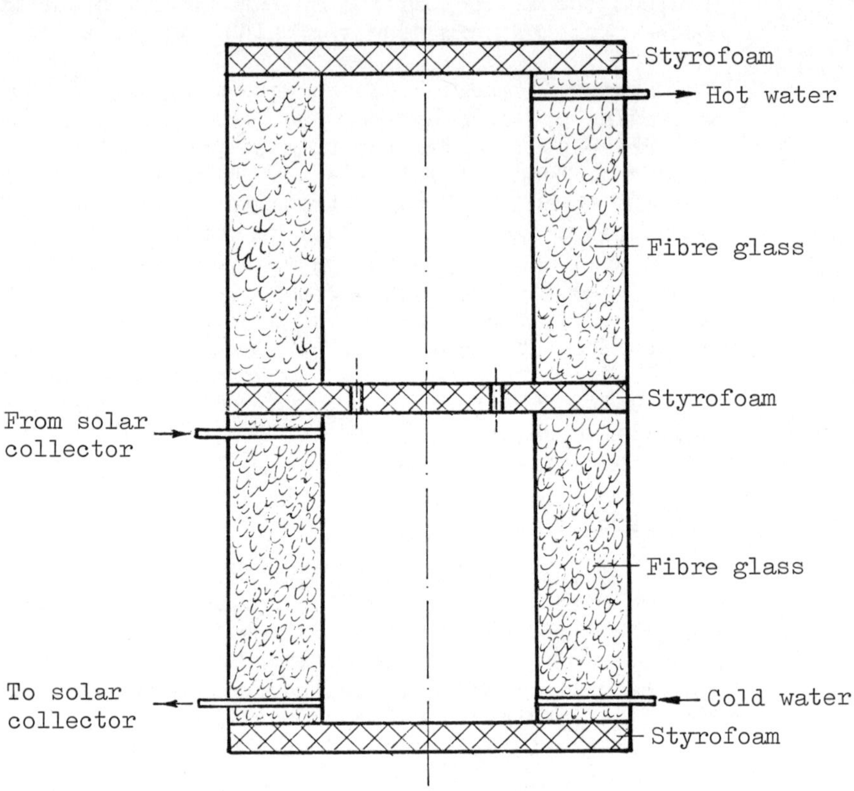

Economic feasibility

Although the system proved technically satisfactory, it is necessary to examine its performance on the basis of economic considerations. This will be done by calculating the annual net energy production and its value in terms of other types of energy saved and comparing these to the annual cost of the capital.

The annual energy production can only be estimated because the efficiency of the collectors in mid winter is not known. Despite what one expects intuitively, the thermal losses will not be much greater than summer because the inlet temperature is lower and the exit temperature substantially so. It is estimated that the increase in cut-off level reduces efficiency to about 50%. Daily values of both measured direct and indirect solar radiation averaged for each month of the year for Ottawa are available (1). Since monthly values vary approximately linearly, the annual collection is calculated from the daily averages at 60% efficiency in July and at 50% efficiency in January (Fig. 4). The amount of electricity or fuel this would save is also listed.

At 1976 prices the annual saving would be $11.80 if electricity were replaced, $11.10 if oil and $8.30 if gas. The cost of these other energy sources are expected to rise; the increase during the next decade could be as much as 100% (6).

The cost of the system which is shown in Fig. 5 is based on the actual costs of components with the exception of the collectors. The amount listed for them was estimated from the prices of commercially available collectors of equivalent area increased by the coating cost ($1.00/ft2). The annual expense of amortization and interest over a 20 year period at 10% interest for the installed cost of $880 is $102. From this analysis it is clear that the system is not economical at current energy costs. If the energy costs doubled, this water-heating system would come closer to being economically feasible. The system could become more feasible if the collector capacity alone were doubled; the investment of $1280 has an annual cost of $149 compared to a doubling of the savings.

If one analyses the problem from the eyes of the common man who does not usually charge himself interest on his investments, it would take about 90 years to recover his investment; however, if he installed it himself it would take about 60 years.

Fig. 4

Annual Energy Collection

Average daily collection	July	20.8 x 0.61	12.7 MJ	
Average daily collection	January	8.5 x 0.5	4.3 MJ	
Annual collection	365 (12.7 + 4.3)/2		3.1 GJ	
Pump consumption (calculated from system friction)			0.1 GJ	
Net energy production			3.0 GJ	
Electrical equivalent[(a)]			840 Kwh	$11.80
Oil equivalent [(b)]			4.3×10^6 BTU	$11.10
Gas equivalent[(c)]			4.3×10^6 BTU	$ 8.30

(a) 100% efficiency, $0.014 Kwh*
(b) 65% efficiency, $0.429/gal, 166×10^3 BTU/gal.
(c) 70% efficiency, $2.05/mcf*, 1000 BTU/cf.

* marginal cost to home owners.

Fig. 5

System Cost in Dollars and Energy

		Dollars	Weight lbs (Kg)	Energy MJ
Collectors :	copper tubing and sheet	$300.00	100 (45)	5.4
	wooden frames	$ 50.00		
	glass	$ 50.00	100 (45)	0.4
Storage tank :	steel with glass lining	$ 80.00	120 (56)	7.3
	copper coil	$ 20.00	(in first line)	
Pump and controls		$ 90.00		
Circulating fluid		$ 90.00		
Plastic tubing		$100.00	90 (41)	6.1
		$680.00		18.2
Installation		$200.00		

Environmental impact

Since the solar collectors would be installed on existing roofs, which generally are not used for any purpose other than shelter, their presence would not have any effect on the local ecology. In their operation they produce no pollutants; however, they do use electricity for the pump and control unit which amounts to about 3% of the energy they trap from the sun.

In examining the overall impact of this system it is necessary to compare its energy production with the energy invested in the manufacture of its components. This has been calculated from the total consumption of energy per unit weight to produce the finished components from raw materials as determined by Roberts (9). These results appear in Fig. 5 and yield a total energy investment of 18 GJ. Since the system would recover this energy in about 6 years, it is a sound project from this viewpoint.

Marketability

The preceding economic analysis revealed that a solar conversion system of this type does not yet compete with other sources in monetary terms. However, when other factors are taken into consideration, its installation appears more favorable.

Since solar energy has no operating costs, it frees the owner from concern for energy price rises due to inflation, shortages or political crises. It also provides security of supply against rationing, strikes, certain natural disasters and resource depletion. The fact that it is pollution free and conserving of non-renewable resources may attract people who are concerned for the environment and limits to growth. It might even become fashionable, and this appears to be one time when keeping up with the neighbours would not lead to useless consumption. Thus, as for many consumer products, acceptance of solar energy will not depend on economic factors alone but on appeal to other desires.

Conclusions

A simple solar heater for producing hot water was designed and constructed. The results were satisfactory with a summer efficiency of 61%. A calculation of heat delivered annually and the amortized capital cost indicates that the system is presently uneconomical. However, there are valid environmental reasons for selecting it.

Acknowledgements

The authors wish to recognize the contributions of five students from the Department of Mechanical Engineering who worked on the project: P.C.C. Kiang, E. Lo, D. Brown, M.D. Champagne and D.Y.P. Lee. The financial support of the SCORE foundation was essential for execution of the project.

References

(1) J.E. Hay, 'Solar Energy Utilization in Canada; a Climatological Perspective.' The Potential of Solar Energy in Canada, Proc. Solar Energy Soc. Canada, Ottawa, 1975, Section I, 23-40.

(2) B.J. Garnier, 'A Note on the Climatology of Solar Energy.' Potential, ibid, Section I, 41-48.

(3) Canada's Energy Opportunities, Science Council Report No.23, 1975. Printing and Publishing, Supply and Services, Ottawa, ($3.30).

(4) An Energy Strategy for Canada: Policies for Self-Reliance, 1976, Department of Energy, Mines and Resources, Printing and Publishing, Supply and Services, Ottawa, ($3.00).

(5) H.J. McQueen, 'Critical Changes in the Canadian Energy Industries', Technology Assessment and the Oceans, 1977, I.P.C. Science and Technology Press, Guildford, 1977; 'Canadian Energy Strategy,' Science, 1977, (in press).

(6) B. McCallum, 'Environmentally Appropriate Technology,' Engineering Journal, 1976, 59 (I), 27-36.

(7) R.E. Chant and D.W. Ruth, 'Solar Heating in Canada,' The Potential of Solar Energy in Canada, Proc. Solar Energy Soc. Canada, Ottawa, 1975, Section III, 1-11.

(8) F. Daniels, Direct Use of the Sun's Energy, Ballantine Books, New York, 1974, 149-160.

(9) F. Roberts, 'Energy Consumption in the Production of Materials,' Metals and Materials, 1974, 8, 167-173.

7.6 The economic feasibility of solar water heating in the United Kingdom

R. Tomkins,
Department of Management Science,
Imperial College, London, UK,

D. Sellmeyer,
formerly of Imperial College.

Abstract

This paper uses various methods of economic evaluation to show that solar water heating in the UK is not currently viable under normal economic criteria. The methods of analysis include a market survey of commercial systems and the calculation of credibility ratios and payback periods for various collector types. Critical payback periods are defined. The sensitivity of uncertain factors, such as future energy price and cost of systems, is examined. Marginal analysis is used to show that selective surfaces are more cost-effective than double glazing. Energy analysis is used to show that a large scale solar energy investment programme would not yield a net energy benefit in the first 10 years. Some conditions, especially future energy price, under which solar energy might become viable are investigated; the energy price was found to be greater than some price forecasts for the year 2000.

Introduction

The question of whether or not solar energy is currently viable in the UK depends not merely on economic considerations, but they form a large part of the case. This paper will argue that under normal economic criteria solar water heating is not viable, but will then examine under what conditions it might become so. We will also point out what we consider to be some misconceptions about the ways of evaluating solar energy.

The analysis looks only at domestic solar water heating in the UK, commonly thought to be the most viable application of solar energy. We are also concerned with evaluating solar energy from the national energy policy point of view and not from the individual user. Thus the relevant question is whether the country's resources will be more efficiently used in solar energy applications than in alternative uses. The answer to this question can be independent of whether the financial conditions for the individual will or will not induce him to purchase a system. In this context, there is confusion in the literature about the treatment of inflation and the fuel price at which to evaluate solar energy savings.

Methods of Economic Evaluation

A necessary condition before any project can be evaluated is that its purpose should be clearly defined. In this context, it is oversimple to say that the purpose of an energy conservation device is to conserve energy. Energy on its own has no value, it is valuable for the ends it can help achieve. Thus the aim of energy conservation devices must be to conserve total resources. An energy conservation method is unlikely to be beneficial if, in saving amounts of energy, it squanders other valuable resources. An evaluation technique should therefore clearly distinguish, in a consistent way, between alternative projects (of which one is solar energy) on the basis of conservation of resources.

Many writers have attempted economic evaluations of solar energy (1,2,3,4,5,6,7); rather more work has been done for the USA than for the UK. We shall distinguish between three different approaches. One method calculates the cash flow for the individual investing in the project (see e.g. (6) pp 38-39, or (7)), including such factors as general inflation and tax rates. While these may accurately reflect cash flows for the individual, they say little about the relative merits of an investment in solar energy as opposed to alternative energy conservation projects for the country as a whole. This method will therefore not be used in this paper.

Another approach uses investment appraisal techniques such as payback method or discounted cash flow. When used with realistic values of the discount rate they enable projects to be effectively appraised in terms of the purpose of energy conservation as defined above. Although the calculations are apparently similar to the first method, the purposes and assumptions are very different.

A third approach is to use energy analysis. This is a valuable descriptive tool (8) but, like the first method, does not allow a judgement to be made on the economic viability as a whole.

Market Survey Evaluation

Consistent data on commercial systems for solar water heating in the UK is not readily available. This prompted us to undertake an extensive market survey during the summer of 1976. Some results of this survey were given in (9). Manufacturers were asked to provide detailed cost and performance information about their solar heating systems. However, some performance claims appeared to be excessive so that it became necessary to calculate the theoretical performance of the different

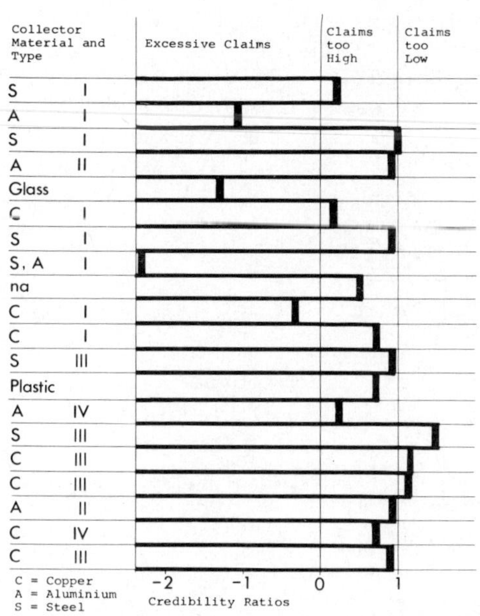

Fig. 1
Credibility of performance claims

systems, based on their technical specification and on a consistent set of assumptions for all systems. The credibility of the manufacturers' claimed energy collection when compared to the calculated performance resulted in the 'credibility ratios' shown in figure 1. The calculation of the credibility ratios has been explained in (9); it is scaled so that a value of 1 or greater indicates that the claimed performance can reasonably be achieved in practise (45% of manufacturers had ratios of 0.9 or higher). A ratio less than 1 indicates that the claimed performance is unlikely to be achieved (35% of claims). A ratio less than 0 is impossible, in the sense that the manufacturer is claiming more energy collected than is likely to be incident on the collector in the UK for a sustained period of time. There were, unfortunately, 20% of claims in this category.

In the absence of reliable manufacturers' or test data, assumptions were made about the likely performance of different categories of collectors in the UK, based on instantaneous efficiencies given by Duffie and Beckman (10) and for two different sizes of solar water heating systems. The assumed overall system efficiencies are shown in Table 1, and these are used throughout the rest of this paper.

TABLE 1 OVERALL SYSTEM EFFICIENCIES

Collector type	Type I single-glazed non-selective	Type II double-glazed non-selective	Type III single-glazed selective	Type IV double-glazed selective
Collector area				
4 m^2	35%	40%	43%	45%
6 m^2	33%	37%	40%	42%

Using these efficiencies we have calculated the payback periods for the various systems based on 1976 prices for gas and electricity and for a system located at Kew (insolation 994 kWh/sm/year). We have used manufacturers' prices for the component costs but assumed comparative installation costs for all systems. The payback period, the number of years to recover the initial investment, is:

$$\text{Payback} = \frac{\text{Capital Cost (C)}}{\text{Value of Annual Savings (A)}}$$

where annual savings are estimated as

$$A = \left(\begin{matrix}\text{price of}\\ \text{useful}\\ \text{heat}\\ \text{substitute}\end{matrix}\right. \times \begin{matrix}\text{Incident}\\ \text{Solar}\\ \text{Energy}\end{matrix} \times \left.\begin{matrix}\text{System}\\ \text{Efficiency}\end{matrix}\right) - \begin{matrix}\text{Running}\\ \text{and}\\ \text{Maintenance}\\ \text{Costs}\end{matrix}$$

Payback is an acceptable method of evaluation for a number of reasons. Since we are dealing with simple cash flows (a single capital outlay and assumed constant annual returns) this method will give the same ranking of alternative investment projects as discounted cash flow methods. It is extremely simple to use and understand. Finally, it avoids having to make assumptions about the lifetime of the equipment and the interest rate to be charged on the capital.

Critical Payback Periods

The viability of an investment project improves as the lifetime of the project increases, the discount rate falls and real fuel prices increase in the future. These three factors can be incorporated into discounted cash flow calculations. If we select appropriate values for the factors, then the critical payback period is the number of years in which the present value of the fuel savings will exactly equal the capital cost. If the actual payback period is greater than the critical value, the project is not viable for the selected discount rate, project lifetime and real fuel price rises. If actual payback is less than the critical value the project is viable. The critical payback period is the maximum period for viability.

We have calculated the critical payback periods for two different fuel price rise forecasts. At present there is a surplus of energy supplies and this situation may well continue for a number of years. Thus one generally acceptable forecast is that fuel prices will not rise relatively until 1985 and will thereafter rise at the real rate of 2% per annum. Let us call this scenario A., then scenario B might be the more pessimistic possibility that fuel prices will rise at 5% per annum.

In between these two would lie the forecast of doubled real prices by the year 2000, a rate of 3.5% per annum. Critical payback periods are shown in figure 2 for these two scenarios and for various project lifetimes, for any selected discount rate.

Under normal commercial criteria, an investor would be seeking a return of 15%-20% over a certain project lifetime of about 10 years, thus the critical payback period would be 4-6 years. In the public sector an acceptable period could be as high as 8-10 years (representing a return of 10% over 15 years, with scenario B). Thus we place the following interpretation on the critical payback period:

Payback period	>10	investment not viable
Payback period in range	4-10	viability possible but uncertain
Payback period	< 4	investment clearly viable

Estimated Payback Periods

The estimated payback periods are shown in figure 3. If solar energy is considered as a substitute for peak rate electricity, then the payback periods range from 12.4 to 28.4 years for 4 sm systems and from 11.5 to 22.7 years for 6 sm systems. On this basis the larger installation is better. Selective surfaces appear to be more cost-effective than non-selective surfaces, but the evidence is not so clear for the use of double glazing. This aspect is examined further below.

However, a great difference emerges when the calculations are based on gas as the substitute fuel; the payback periods range from 44.9 to 111.7 years for a 4 sm system. This is clearly not cost-effective.

The picture would have been roughly similar if the calculations had been based on coal, oil or nuclear power as the substitute fuel. For the country as a whole the resource savings of someone using solar heating can be considered either from the short run or the long run view. In the short run, users will either be saving fossil fuels directly, or if they are saving electricity in the summer the saving will be in the power station running costs only. No generating capacity or other fixed costs will be saved, since the total capacity will still have to be large enough to meet maximum winter demand. Lucas (11) gives the value of these marginal savings in local fuel to be equivalent to 0.8p/kWh, which compares with a price of 0.75p/kWh useful heat for gas used in the

Fig. 2, 3

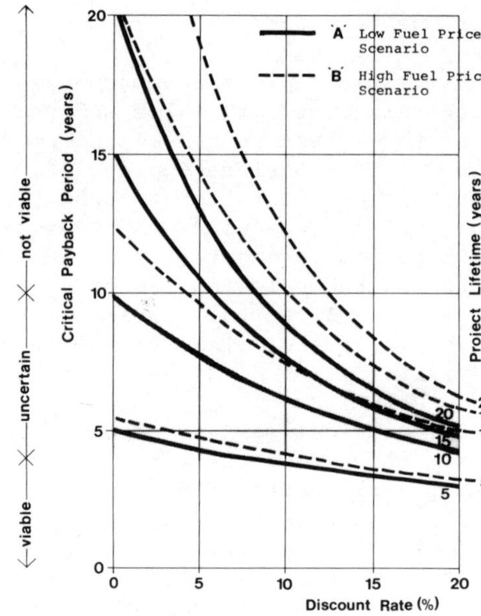

Critical payback periods for 2 fuel price scenarios, various project lifetimes and discount rates

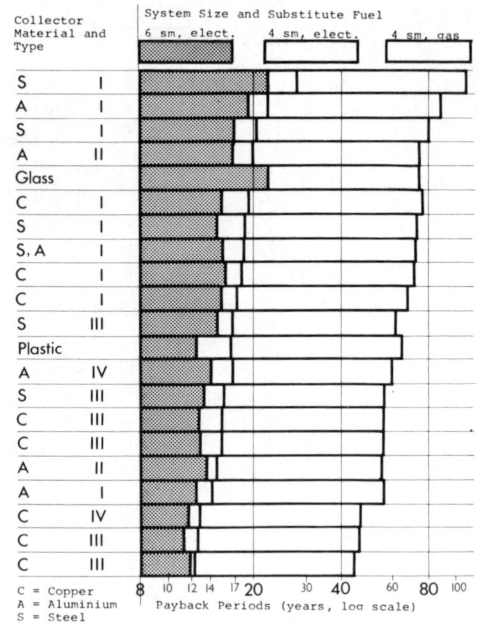

Payback periods for solar water heating systems substituting for electricity or gas

above calculations. In the long run (say beyond the year 2000), the alternative energy sources are likely to be coal or nuclear. The resource value of a marginal unit of nuclear power saved is given as 0.32p/kWh (11). Thus if the resource savings of solar water heating are evaluated at a realistic resource cost, the payback periods are seen to be unacceptably high.

An alternative way of evaluating the system is to calculate the average unit cost of energy provided by the system. Unit costs are given in Table 2 for two sizes of systems and assuming a 10% discount rate and a useful lifetime of the equipment of 15 years to calculate the annual capital charge. Running costs (mainly the

TABLE 2 UNIT ENERGY COSTS IN pence/kWh FOR TYPICAL SOLAR WATER SYSTEMS (1976 prices)

Collector type	Type I single-glazed non-selective	Type II double-glazed non-selective	Type III single-glazed selective	Type IV double-glazed selective
Collector area				
4 m^2	4.5	4.2	3.6	3.6
6 m^2	3.8	3.7	3.2	3.2

electric pump) were assumed to be £4 p.a. and capital costs were the average quoted figures in the market survey.

On this conventional accounting basis the energy costs are considerably more than any substitute fuel.

Sensitivity to Uncertain Factors

All calculations like the ones we have given depend, of course, on the assumptions made. These assumptions are about the very uncertain future trends in key factors. The most important of these factors are:

1. Future price of energy.

2. Capital costs of solar water heaters.

3. Efficiency and useful lifetime of an installation.

Sensitivity analysis is a technique for showing how results are affected by a change in one of the important factors. Below we use the technique to show how these factors affect the assessment of viability of solar water heaters.

Energy Price Effects

The value of solar energy is the cost of the saved substitute fuel. Frequently, solar energy evaluations are done as though the substitute fuel were peak rate electricity (1,6,12). While this may be the case in an individual instance it would be unwise for a country to plan its energy policy as though electricity and solar energy were the rational alternatives for water heating. We have already suggested that in the long run the substitute fuels will be nuclear or coal, both of which are much cheaper than electricity at current prices. Therefore, to evaluate solar water heating for the country as a whole we should calculate the savings at the cost of the real substitute fuels.

Forecasts of the future price of fuels have varied widely. One middle view (8,13) is that prices in real terms will double by the year 2000. This is equivalent to an annual increase of about 3.5%. However, this is just one view; the increase could be considerably more or less (let us not overlook that a successful implementation of a large-scale nuclear programme could even lower real prices!). Figure 4 shows how the payback period of a Type I collector system (6 sm of single glazed non-selective collector surface, situated at Kew) is affected by different energy prices. The shaded area indicates the range of maximum payback periods that might be acceptable. Thus the range of substitute fuel costs (cost of useful heat) above which Type I solar collector systems would become economically viable is about 3-8 p/kWh (1976 prices), or £8.3 - £22/G J, which for coal and nuclear fuels is greater than a lot of forecasts for the year 2000.

General inflation of prices has no bearing on this situation. Inflation redistributes from lenders and savers to spenders and borrowers. General inflation will benefit an individual who has invested in a solar energy system; but so too will an individual who has invested in lumps of copper or tins of food. However, the country as a whole is made neither better nor worse off through general inflation which is a monetary phenomenon and does not affect the supply of real resources. Real price rises only should be taken into account.

Fig. 4

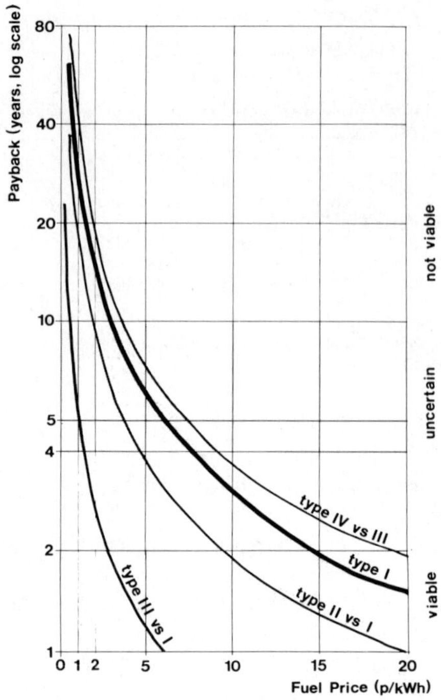

Payback periods for rising substitute fuel prices (useful heat, 1976 prices). Simple system (type I) compared with advanced systems (II-IV)

To see whether the more advanced collectors are more viable than the simple collector we have used marginal analysis. For a given fuel price, the marginal value of the extra fuel savings is compared to the marginal cost (i.e. the difference in cost between the two collectors being compared). The marginal costs were estimated at £4/sm for a selective surface and £7.50/sm for double glazing. The marginal savings were calculated from Table 2. The payback periods for the additional investment in the 3 advanced collector systems are shown in figure 4 and are calculated from:

$$\text{Payback} = \frac{\Delta C}{I * \Delta\eta * p}$$

where ΔC = additional capital cost, £

I = insolation level, kWh/year

$\Delta\eta$ = improvement in efficiency of collector

p = price of useful heat of substitute fuel, £/kWh

The graphs show that a selective surface under these conditions is a better investment than double-glazing. A selective surface is cost-effective for substitute fuel costs greater than about 1p/kWh. Double glazing on its own is cost-effective at fuel prices greater than about 2-4 p/kWh. However, if the better investment of a selective surface is applied first, then double-glazing only becomes cost-effective at fuel prices greater than about 4-12 p/kWh.

Energy Analysis

Energy analysis examines the total use of energy of a system over its lifetime, including the energy used in manufacture, delivery, installation, operating and maintenance. It is a descriptive tool, enabling a comparison to be made between the energy consumed in implementing the system and the energy saved subsequently. The approximate primary energy inputs for a 4 sm Type I system are shown below:

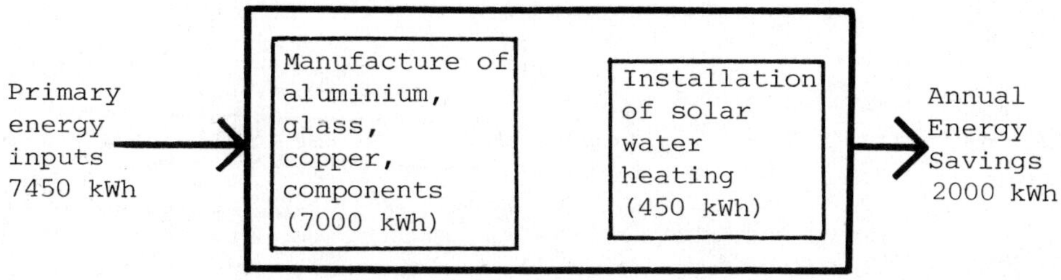

We have included the manufacture of the collector itself, the storage tank, glass cover and electrical components, but ignored components such as the casing, additional pipework, expansion tank, retailers and distribution energy costs. If the solar energy saved substitutes for coal at an overall efficiency of about 70% the savings are equivalent to 2000 kWh of primary energy and the energy payback is 3.7 years.

Energy analysis highlights the fact that solar energy systems use energy intensive materials in their construction. As the price of energy rises, so will the cost of solar water heaters. Thus our previous calculations shown in figure 4 are incomplete; they ignore the rising capital cost and thus overstate the improvement in viability. Appleby (13) has suggested a tripling in the price of industrial goods by the year 2000 due to the higher energy cost. If we make the much less drastic assumption that solar collector prices rise by a sixth of the real energy price rise, then the payback periods for the most viable system, a 6 sm Type III system, will be as in figure 5.

Fig. 5

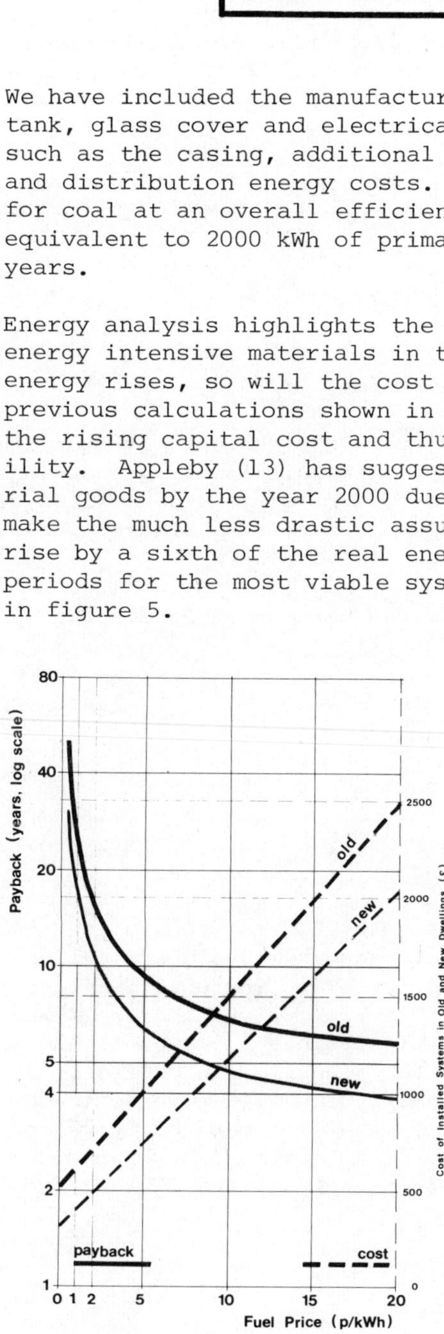

Effect of rising fuel prices on cost of solar heating system and payback period

Cost Reductions

As early as 1966 Sobotka (14) was arguing the possibility of significant cost reductions in the manufacture of solar water heating systems. Later writers (2,4,6) have implied that cost reductions of the order of ½ are possible.

However, current technology uses well established techniques, materials and components and there is little evidence yet that scale economies will have a significant effect. The collector itself is the only component in the system which is not already widely used, but a large reduction in collector cost of, say, 50% would only reduce the total system cost by about 25%. Some reductions may come from more efficient retailing methods (3). Crude averages of various published and manufacturers' figures for the cost of a 4 sm installed system over the last 4 years are:

	Cost of System	Annual Increase
1974	£313	
1975	£373	19%
1976	£456	22%
1977	£550	21%

One factor which will reduce the cost of installed systems is integrating them into new buildings. We have made the following assumptions for the cost reductions associated with installing solar water heaters simultaneously into a large number of new dwellings compared to a one-off conversion:

1. Collector costs can be reduced by 25% for savings in manufacture, distribution and retailing.

2. Standard components (pump, tank etc.) are reduced by 10% for bulk purchasing.

3. Installation costs are reduced by 50%.

4. Where collectors are incorporated into roof structure there is a roof material saving of £12.50 per sm.

The net cost of a 4 sm system is then reduced to about £330 (a reduction of 33%). The corresponding payback periods would also all be reduced by 33%, and these are included in figure 5. Viability could begin to be reached at fuel prices above 3p/kWh. Against the future beneficial effects of possible improved technology and higher energy prices must be set the reduction in collector and system performance as efficiency drops off with age. One estimate (15) gives the drop in system efficiency after 15 years of operation to 72% of the original level, equivalent to about 2% per annum. Such a loss in efficiency would entirely cancel out the 2% real fuel price rises of scenario A.

Long Term Implications

Solar heating, if it ever comes into widespread use, will be a very expensive form of energy by current standards and will have implications for the proportion of a country's resources devoted to energy production and the consequent standard of living which could be supported. One way to examine the energy effects on a country of a large scale conversion to solar energy is to construct an energy account. We have assumed a similar conversion programme to ISES (6); installations growing annually by 10000 p.a. until the year 2000 when there would be 250000 installations p.a.

Figure 6 illustrates the cumulative energy savings from such a conversion programme. Initially there is an energy deficit. After about 10 years the energy savings exceed the energy costs of further installations and thereafter the cumulative savings increase rapidly. The discounted cumulative savings curve shows the energy flows discounted to present value terms at 10%. The implication is that, if it does prove desirable to embark on such a programme in the interests of energy conservation, then the lead time before a net benefit arises can be seen to be about 10 years and this must be taken into account in the planning process.

Fig. 6

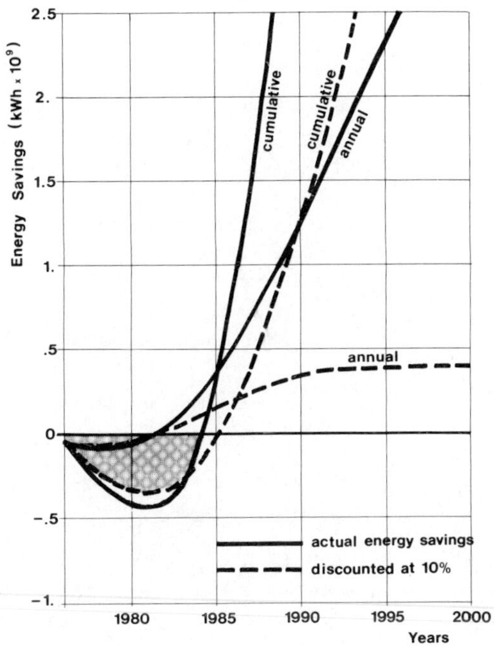

Energy analysis: Cumulative energy savings for a large scale solar energy investment programme

Conclusions

Our assessment of the viability of solar water heating in the UK is not very encouraging. Conventional economic appraisal yields unacceptably high payback periods for current commercial systems and even for an advanced system installed in new dwellings under favourable assumptions about cost reductions and energy price rises to the year 2000, the viability is at best uncertain. Solar energy will be an expensive form of energy. On the other hand, energy analysis yields acceptable paybacks, indicating that if energy was the only resource used in constructing solar water heaters they would be effective energy conservation devices.

The convergence between these two methods of appraisal can be seen, in the economic appraisal graphs, where fuel prices approach extreme values and the cost of energy in a system becomes the major part of the total cost. Economic viability is extremely sensitive to the future price of substitute fuel, about which there is enormous uncertainty, but the sensitivity is reduced when the effect on capital cost is included. The graphs show that certain viability is approached in the fuel price range (cost of useful heat, 1976 prices) of 5-15p/kWh. The lower point of this range is beyond the forecasts of many observers for this century.

References

1.	J.C. McVeigh	Sun Power, Pergamon, Oxford, 1977.
2.	R.G. Courtney	Rays of Hope, Building, November, 1975.
3.	R. Tomkins & M.J.G. Wilson	The Use of Solar Energy in Britain, Engineering, June, 1976.
4.	G.O.G. Löf & R.A. Tybout	The Design and Cost of Optimised Systems for Residential Heating and Cooling by Solar Energy, Solar Energy 16(1), 9-18, 1976.
5.	B.D. Hunn, G.J.E. Willcutt & T.B. Sweeney	Simulation and Cost Optimisation of Solar Heating of Buildings in Adverse Solar Regions, Solar Energy 19(1), 33-44, 1971.
6.	UK-ISES	Solar Energy, a UK Assessment, 1976.
7.	F. Kreith & J.R. Kreider	Preliminary Design and Economic Analysis of Solar Energy Systems for Heating and Cooling of Buildings, Energy (1), 63-76, 1976.
8.	P. Chapman	Fuel's Paradise, Penguin, 1975.
9.	J.C. McVeigh & D. Sellmeyer	The Cost Effectiveness and Performance of Domestic Solar Water Heaters in the United Kingdom, Sun at Work in Britain, Dec. 1976.
10.	J.A. Duffie & W.A. Beckman	Solar Energy Thermal Processes, Wiley, New York, 1974.
11.	N.J.D. Lucas	Space Heating and Energy Supply, Energy World, March, 1976.
12.	P.T. Landsberg	How Much Investment in Conversion Devices?, Conf. on Economic and Commercial Assessment of Solar Energy Conversion, ISES, July, 1977.
13.	A.J. Appleby	Energy Costs and Society, Energy Policy, June, 1976.
14.	R. Sobotka	Economic Aspects of Commercially Produced Solar Water Heaters, Solar Energy 10(1), 9-14, 1966.
15.	J.C. Ward & G.O.G. Löf	Long-term (18 years) Performance of a Residential Solar Heating System, Solar Energy, 18, 301-308, 1976.

Solar water heating for flats: The Harrow project

M.P.K.Keath
London Borough of Harrow
Department of Development and Technical Services

Abstract

This paper describes a solar water heating installation in a block of six, two-person flats currently under construction for the London Borough of Harrow as part of a larger contract. The satisfactory integration of the collector and associated components with the total building fabric was regarded as a high priority design objective.

The reasons for attempting this exercise and the steps taken towards achieving this objective are described. Although such installations for flats appear to be unprecedented the engineering principles involved were not unduly complex. The paper is presented from the point of view of the architect and the scheme is seen as a combined exercise in applied innovative technology and integrated architectural design, rather than as a pure scientific research project.

Introduction

Harrow's investigation into alternative energy sources for housing began early in 1973 when the Architects Division of the Borough's Department of Development and Technical Services identified itself with the wider recognition that an energy crisis was impending and an end to cheap fuel in the longer term was probable. Although it was not possible at the time to deploy resources specifically to problems associated with energy saving in buildings, a growing awareness that changes were coming was evident. It was felt by members of the design team, in this case architects and service engineers, that they had a professional duty (apart from any moral considerations) to consider designing for alternative energy sources, rather than wait until desperation forced it upon them.

Experience was as urgent a requirement for Harrow as for the rest of the country. Despite the novelty of the technology, particularly at that time, the first feasibility studies were undertaken as a part of the normal processes for selecting equipment for heating or any other installation, i.e. just as one might consider the respective merits of one boiler or complete system as against another.

Project Selection

A project had to be identified into which a prototype installation might be introduced. The scheme selected was Phase 4 of the Honeybun Estate redevelopment; it comprised ninety three flats and houses of varying types (contract sum : £1,026,000) and design work was nearing completion.

The Milton Keynes project had already got under way and it was known that a number of "one off" installations in the private housing sector had been completed with varying degrees of sophistication. In view of the feedback likely to be forthcoming from these systems, installed in low rise family dwellings, this dwelling type was rejected for prototype use in order to avoid repetition. Moreover, considering the high proportion of dwellings which are tending to be constructed in multi-storey flat form at low and medium rise, and the growing emphasis on the social need for smaller dwelling units, it was considered that experience in applying solar installations to small blocks of flats would be of greater value.

The Honeybun project contained a plan type for a three-storey block of six flats, a plan to be repeated six times in the layout either singly or in "semi-detached" pairs. One such pair, having a suitable orientation, was selected: one block to be fitted with the prototype installation and one block without.(fig.1)

Fig. 1
Typical floor plan

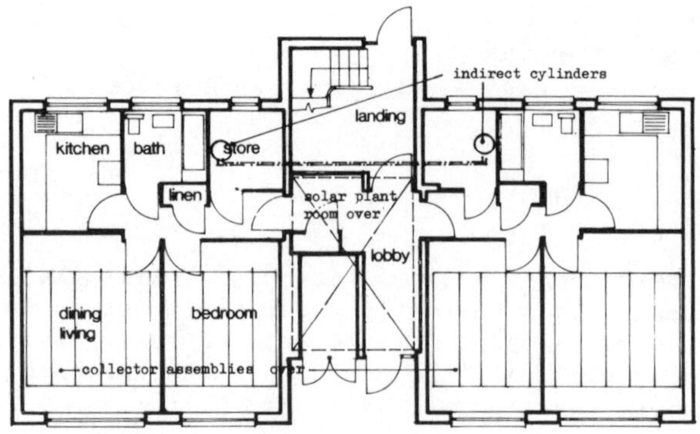

Fig. 2
Axonometric view

Architectural Concept

The architectural concept demanded a collector assembly which could be completely integrated with the roof construction and contained within the overall building form. This was of particular importance in view of the large estimated collector area, up to 30 m^2, under consideration.

It was realised that there might be advantages in utilising absorber panels of the greatest width,or surface area, possible in one piece, e.g. fewer connexions; problems of removal would present themselves if the fixing of the glazing bars for the transparent outer cover was effected, through the panels, to timber supports below. Alternatively, timber joists, or the like, fixed above the collector would cause undesirable overshadowing of its surface. The design of all components to a common module would obviate these objections. The dimension selected was 600mm on account of its suitability for both roof truss and patent glazing centring.

Development of an integrated design

Unfortunately, no suitable panels were known to be available; consequently long, thin central heating radiators (ex catalogue) were considered in the initial design study.

It came to Harrow's notice in September 1974 that Imperial Metal Industries were interested in entering the solar heating field commercially, the company having vast experience appropriate to the manufacture of suitable components. Thus when the first meeting with IMI representatives took place on 17 September 1974, the company was seeking both early customers and a direction for collector panel design development. IMI expressed interest in the integrated roof concept and agreed to consider designing and quoting for an all-copper system incorporating flat plate type collectors accepting the 600mm centres constraint. The company's design and development progress was further stimulated by the Copper Development Association competition for an all-copper system held during March 1975. The jointly conveived integrated collector design was incorporated in a prize-winning scheme entered by IMI in conjunction with the Masimi Franklyn Partnership.

The question of finance was still to be decided. An approach to the Department of Environment was made, but as there was no way of demonstrating that the requirement of a ten year pay back period could be achieved, it was decided to attempt to include the system within the overall cost yardstick for the building scheme. A provisional sum for the supply of solar hardware was included in the bills of quantities; associated builders work and installation costs were covered within the sum allowed for the Mechanical Services installation. As the lowest tender received was within the cost limit no reductions were necessary and, after obtaining formal approval of the Council's Housing Committee, the scheme got under way. By the summer of 1975 a system design had been finalised, builders work detailing was completed and orders placed for the solar hardware.

Thus the system was designed to integrate, with a minimum of compromise, into the building at a stage when the construction common to all similar blocks in the scheme was not yet fully worked out. It was not therefore an 'after the event' modification to a completed building. Conversely, in common with the Milton Keynes, Wandsworth and Croydon installations, the block was not specially designed as a 'solar building'. This approach appears to represent the more typical pattern likely for the foreseeable future.

Objectives defined

The objectives of the pilot study were defined as follows:

1. To provide a block of flats with a pre-heated supply to the domestic hot water system by means of communal solar collector; the water being boosted to the required working temperature by electric immersion heaters in each flat. The amount of energy required for the latter will depend on weather conditions and on individual usage.

2. To achieve an architecturally and visually acceptable physical solution to the problems of roof mounted solar collectors. The Harrow collectors integrated with the normal roof construction in such a way as to show a transparent surface where conventional roof tiles would otherwise have been.

3. To investigate the alternative methods of installing the solar "hardware", taking into account the Contractor's and the Sub-Contractor's operational requirements.

4. To evaluate, by means of simple metering and analysis of all costs involved, the efficiency, use patterns and cost benefit of the installation.

5. To investigate, by means of interview surveys, user attitudes and reactions to solar heating for a pilot scheme set within a larger housing estate.

6. To identify any Housing Management problems which may be encountered, including possible rent adjustments, neighbour envy, misuse or vandalism and maintenance such as cleaning the glass covered collectors.

7. To establish the design parameters of a system suitable for medium rise blocks of local authority flats for wider future application.

The elements of the system and associated architectural construction can now be described in detail. (See Fig.3 System Schematic)

Fig. 3
System schematic

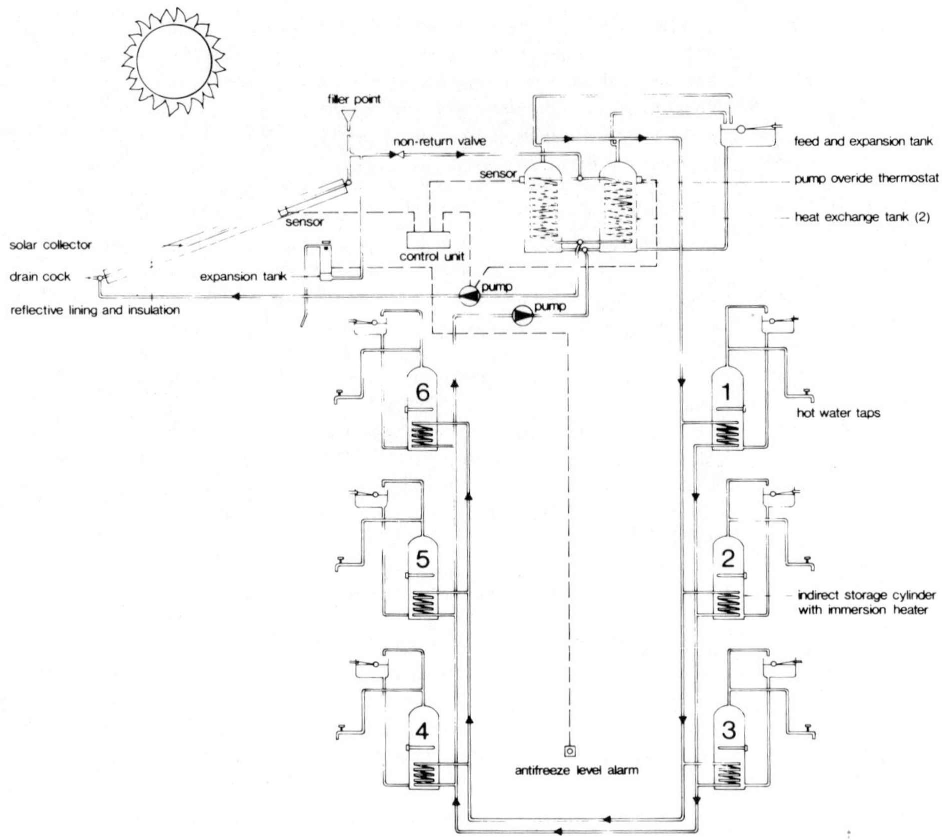

Collector assembly

The collector is in two main sections arranged equally on either side of the stair enclosure. Its constituents are, from the outside in, as follows.

Transparent cover and air space

For the cover medium, glass was selected on account of its transparency to incoming solar radiation and its opacity to the longer wavelength radiation emitted from the absorber plates. It has the well known qualities of being weather and scratch resistant and is not affected by ultra violet radiation. The thickness selected was 6mm clear float glass to take account of selfweight, snow and wind loads for a 600mm (nom) span continuously supported along its long edges and at a tilt of $17\frac{1}{2}^{\circ}$. The air space between the glass cover and absorber surfaces is sealed to avoid the accumulation of dirt on the inaccessible inner surface of the glass by convection currents. With such sealing it was considered that the increase in thermal quality of a double glazed cover is very small and was therefore omitted to reduce costs of both glass and patent glazing bars necessary to hold it.

It was realised after tenders had been received for supplying and fixing the glass covering that some consideration should be given to handling the glass whilst hoisting and fixing to a three storey pitched roof. Likewise the attendant problems of replacement, should breakage occur, or removal for maintenance be required.

Alternatives considered to reduce the length of sheets were (i) stepped lengths of glazing bars and (ii) lead cames; undesirable at low pitch. As the glazing sub-contractor (supplying and fixing glazing bars and glass complete) expressed no qualms about the sheet size (585 x 1200) no changes were made. The glass was fitted in approximately twelve hours without apparent difficulty, with the aid of the main contractor's scaffolding.

Patent Glazing bars

The bars are a simple, standard mill finished aluminium section supplied by The Standard Patent Glazing Co. Ltd. Bedding mastic was used in addition to the rolled-in lead wings.

Roof construction

Standard softwood gang-nailed trusses were supplied by Triad Ltd. The roof finish generally is concrete tiles supplied by Redland Ltd, laid to manufacturers recommendations. The structural engineer designed designed additional bracing and strutting, removable in short lengths for maintenance if necessary, to take account of the lack of continuous tile battening on one side of the double pitched roof.

Absorber plates

The absorber plates are of ribbed sheet copper with close fit, continuously soldered 15mm o/d tubes designed to give maximum thermal conductivity. Plate thickness is influenced by considerations of

i. Economy, in terms of quantity of copper employed: the thinner the cheaper.
ii. Conductivity: the thinner the sheet the closer the spacing of tubes required.
iii. Rigidity, in relation to spans involved in integration with roof construction.

IMI Range Ltd. manufactured the panels, after investigations, with a design thickness of 0.5mm to achieve the optimum balance of these aspects.

The collector assembly incorporates 20 sub-assemblies factory jointed in sets of three for 600mm bay. The long edges are slightly overlapped but unsoldered. (Fig.4)

Selective surface

Absorber surfaces having a high absorbtion coefficient for short wavelength incident radiation and, at the same time, low emissivity for the longer wavelength clearly have advantages over non-selective surfaces. This selectivity is of particular importance in attempting to heat water above the ambient temperature or in seasons and periods of low inulation. The IMI Range plates are treated by a five bath treatment of completed sub-assemblies, converting the surface into a thin, finely divided copper oxide.

In comparing this surface treatment with conventional black paint, IMI's laboratory tests have shown that the ratio of the rate of heat loss for black paint/rate of heat loss of the selective surface at all temperatures above ambient up to 40°C (as required in a domestic hot water situation) is in excess of 2.5.

Reflective lining, Insulation and support panels

Foil-backed glassfibre quilt was selected for simplicity of installation. A total of 125mm of insulation is provided in two layers: 50mm foil-backed superimposed on 75mm ordinary quilt.

Removal of absorber plates from inside the roof space was considered essential to avoid costs of removing glazing from above, as far as possible. In order to facilitate ease of access to the rear of the plates a simple drop-out, toggle-held panel system was constructed on site in hardboard edged with softwood battens on all edges.

Solar Collection Circuit

The elements of this closed circuit, situated entirely within the roof space, are the twin collector assemblies, the solar storage/heat exchange tanks and the control system regulating the flow of the heat transfer fluid through this circuit. Facilities are included for filling, draining down and expansion of fluid. (Fig.4)

Fig. 4
System schematic

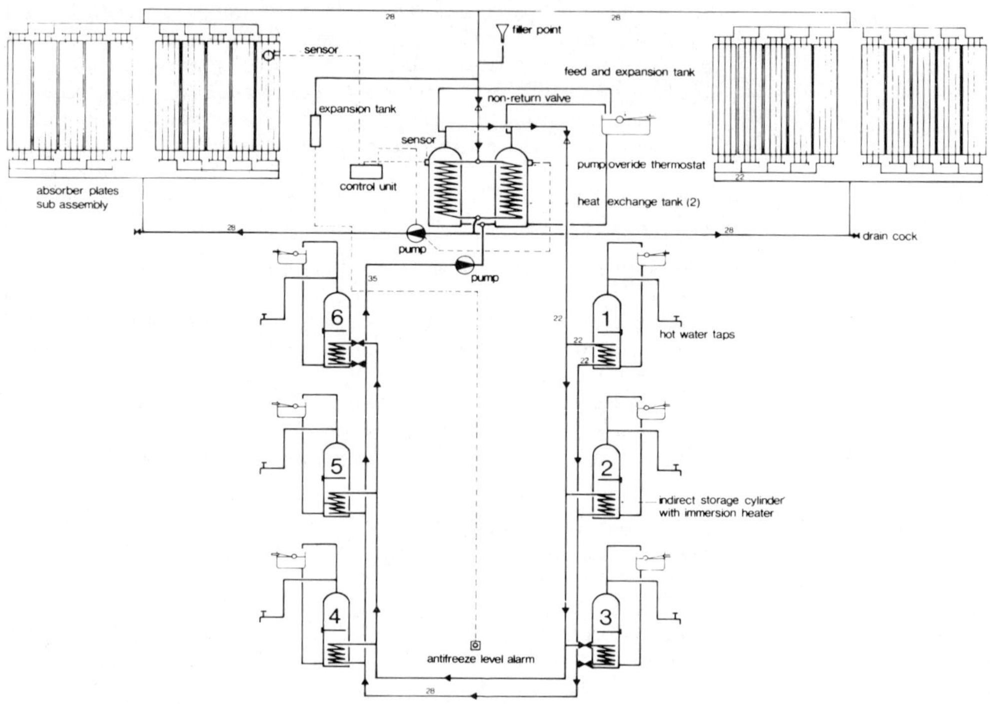

Heat transfer fluid

The fluid is an antifreeze solution, avoiding the need to drain down the collectors when ambient temperature falls below zero. Investigation into ethylene glycol (common car radiator additive) revealed a risk to health should any contamination of consumable water occur by leakage of heat exchangers. Two precautions were introduced to eliminate such risk.

i. A low toxicity glycol was used.
ii. An automatic low level alarm system, warning of any significant drop in glycol level due to either leakage or evaporation.

These, together with the fact that heat exchange occurs at two stages between collection and consumption, make the probability of contamination insignificant.

Propylene glycols are used extensively in the food industry as a heat transfer fluid in refrigeration processes, including immersion freezing, and are considered completely safe.

The anti-freeze in the Harrow collectors is a 33% monopropylene glycol solution containing anti-corrosive, anti-foaming additives and a bacteriacide. Its bulk density at 20°C (g/l) is 1070-1080; its boiling point is 165°C; freezing point (33% solution) is -15°C.

The solution is provided with the collectors and is marketed by IMI Range Ltd. as "Solaric P.33"

Glycol storage and replacement

The delivery containers will be kept clean and stored readily available in the solar plant room, over the stair enclosure, for draining down into and subsequent re-use of the fluid, should this be necessary for maintenance or repairs.

The manufacturers recommend that the fluid be replaced after seven years in service; the life of the circuit components should then equal the life expected for copper plumbing generally.

Circulation pump

The pump employed in an SMC Comodore set initially on No.4 speed. Consideration will be given to turning this down to No.3 for winter conditions if the increased viscosity of the solar fluid will allow this at low temperature.

Expansion tank

Changes in the volume of the transfer fluid resulting from temperature changes are accommodated by the introduction of an expansion tank, on top of which is mounted a pressure relief valve in the form of a car radiator type pressure cap, with a slightly stronger spring than private car practice. The low level alarm, consisting of a wall mounted red light in the entrance hall of the building, relay is wired to this tank.

Solar collection circuit controls

Operation of the solar circuit is controlled in order to take account of weather variations, which might otherwise cause dissipation of heat already in store, and to avoid wastage of pumping power in the circuit. A differential controller is wired to a temperature sensor situated on the collector near the flow outlet; to another at mid height on the storage/heat exchange tanks; and to the fluid circulating pump. The differential controller is pre-set for a 1.5°C temperature drop, but has adjustment of ± 5°C.

The following equipment is employed:

i. Satchwell Solar Heating Controller, type DP266, connected with low voltage twin conductor cable to the following two items.

ii. Satchwell Water Temperature Detector, type DW1204, immersion mounting for tank application.
Detecting element : thermistor mounted on end of stem to give fast response.

Ambient temperature range	-40 to + 70°C
Control temperature range	-5 to + 70°C.

iii. Satchwell Water Temperature Detector, type DWS 1202, pipe surface mounting for collector application.
Detecting element : Negative temperature coefficient thermistor.

Ambient temperature range	-40 to + 70°C
Control temperature range	35 to 120°C

Solar Storage/Heat transfer tanks

Storage is contained in a pair of similar copper cylinders each having a capacity of 190 litres (42 gals.). Coiled heat exchange tube is based on a collector-area to water ratio of one square metre of panel to ten gallons of water. The tanks are insulated by means of a factory applied, sprayed on, urethane foam 25mm thick. The tanks were supplied by IMI Range Ltd., who market similar tanks in their "Solaric" range covered with "Herculag" urethane foam insulation.

The Solar Distribution Circuit

The distribution medium is water, circulating through the coils of solar storage tanks down through 3 flats on one side of the stair enclosure and back up through 3 flats on the other side of the enclosure. A separate feed and expansion tank for this circuit is positioned in the solar plant room. The flow is maintained by a pump similar to that in the collection circuit but fitted with an overide switch operated by a thermostat mounted on the solar collection tanks. The pipework is arranged on the 'reverse return' principle, analogous to a series of central heating radiators, to maintain a balanced distribution to all flats.

Heat transfer, indirect storage Cylinder (6 No.)

These are similar in construction to the solar storage tanks, but have a capacity of 162 litres (35 gals.). Positioned in the storeroom of each flat, adjacent to the stair enclosure, each has its own feed and expansion tank through which the draw-off circuit water flows for ultimate consumption. The fixtures served are bath, wash basin and kitchen sink in that order. Each cylinder is fitted with a 3kw immersion heater manufactured by Redring Electric Ltd. (No 31 321036 complete with thermostat No 91 300514).

Monitoring arrangements

The cost of the system installation, including instrumentation, staff time resources available for recording and processing data, are important factors in an applied technology exercise of this type. The need for utmost economy is obvious. However, unless some meaningful facts are to be derived the value of the system as a prototype is limited. It was therefore felt that the data collected should be essentially practical and direct in type, involving simple, economical instrumentation rather than the more elaborate methods (e.g. data recording machines and computer processing) appropriate to'pure' research projects carried out by institutions endowed with the necessary staff and funds.

In consideration of the above, three types of metering are to be provided initially:

i. A water meter on the cold inlet to each of the six indirect hot water/storage cylinders (one per flat).

ii. A separate electrical meter for each of the six cylinders to measure only the "topping up" consumption of electricity in each flat.

iii. A specially prepared sensitive electrical meter to measure consumption by the control unit/sensors/pumps, i.e. the 'extra' power used in driving and controlling the solar collection circuit and distribution circuit.

Crude comparison with similar flats in the scheme, heated entirely by conventional means will, naturally, be available in addition .

Estimated Performance

IMI Range Limited, the manufacturers of the absorber plates, have submitted the following table of predicted energy gains for a one year period. The water heating requirements have been based on being equivalent to a 48 kw load/day for six flats. (8 kw per flat).

Month	Average Daily Solar Energy Supply *	Electricity Supply	Solar Excess
January	7.2.	40.8	-
February	14.7	33.3.	-
March	29.3	18.7	-
April	43.4	4.6	-
May	58.0	-	10.0
June	61.1	-	13.1
July	56.7	-	8.7
August	48.8	-	0.8
September	39.5	8.5	-
October	23.2	24.8	-
November	9.8	38.2	-
December	5.4.	42.6	-
Annual Total	12,078	6,433	991

Total Solar Contribution = 12,078 - 400 pumps - 991 = 10,687

Percentage Solar Contribution = $\frac{10687}{365 \times 48}$ = 60.9%

* KWH/day

Costs and Conclusions

The main contract has met with delays and for this reason it has not been possible to assess the full costs of the system and its installation. The system is now expected to be commissioned this Autumn and a full analysis of costs will be published as soon as possible thereafter. It is however expected that the total cost of supply and delivery of system components (£±1500), installation (as an extension of the heating sub contract) and associated roof construction, patent glazing etc. may be a little in excess of £4,000. Before conclusions can be drawn it will be necessary to adjust all costs to the same base date, e.g. various component and installation prices will have been received over a two year period. These will then be related to the electricity tarrif at that base date before meaningful calculations can be made.

Of the objectives outlined to date (see p.3) the first three have been substantially achieved. It is expected that all of them will be achieved by Autumn 1978.

Credits

The project was designed in the London Borough of Harrow's Department of Development and Technical Services; Director, G.J.Foxley, FRIBA.

The following staff were directly concerned with the project:

Principal Architect	Michael Keath, B.Arch., RIBA
Principal Engineer	Rodney Gross, MCIBS
Project Architect	Petter Rollo-Smith, BA(Hons)Arch., RIBA
Project Engineer	Alan Northfield, MCIBS
Clerk of Works	David Whitbread

In association with IMI Range Ltd., Stalybridge; consultant engineer, Alan Swinbourne, MA(Cantab),C.Eng., MICE, FIMechE, FIEE.

Contractor : Percy Bilton Limited

Heating Sub Contractor : J.H.Shouksmith & Sons Ltd.

Retrofitting potential in London housing

K. E. McCartney
The Architectural Association
School of Architecture, London, U.K.

Abstract

A case is made for retrofitting solar water heating systems in the U.K.'s existing housing. The constraints on retrofitting are discussed and the methodology and results of a survey of the roof types and orientations of almost 12,000 dwellings are presented. 59% of these dwellings could have solar collectors mounted directly on to their inclined roofs. The use of mounting frames would increase the proportion of dwellings with roofs suitable for collector installation, to 87%. This indicates a greater potential for retrofitting solar systems than has been implied in earlier guess-estimates. Mention is also made of retrofitted system the performance of which is currently being monitored.

Introduction

In studying the potential of the application of a novel energy source, one is faced with many of the uncertainties inherent in any form of futurology. In questions relating to the built environment however, the long life cycle of buildings affords us a firm foundation for making interpolations. Most of the housing we will be living in in the year 2000, has already been built. In the U.K., where the housing stock consists of over 20 million dwellings and the current rate of new building is about 200,000 per annum, more than three quarters of the housing in use at the turn of the century is likely to be pre 1977 stock.

With a housing expansion and replacement rate of only 1%, it is clear that a programme of retrofitting is essential if solar water heating is to make any significant contribution to decreasing the national dependency on fossil and nuclear fuels in the short and medium term future. Existing households consume 6.4% of U.K. primary energy for heating water. (1) About half of this load could be met by solar systems if all houses could be retrofitted. This would represent a saving of 0.28×10^{18}J. each year.

Consideration of retrofitting is particularly timely in Britain, given the changes in the recent years in public housing policy. Local authorities are now encouraged to undertake rehabilitation of older housing as opposed to demolition and redevelopment. Grants of 50% are also available to private house owners who wish to improve their properties to an approved standard. It is foreseeable that such grants could be widened in their scope to include energy saving improvements.

Constraints on a Retrofitting Programme

The scope and rate of expansion of retrofitting programmes will undoubtedly be determined to a large extent by economic considerations. Future prices of fuels, inflation, discount rates, and financial weightings reflecting on evaluation of the benefits of solar energy on environmental grounds and national interest considerations, all lend an element of uncertainty. This paper therefore concentrates upon those physcial constraints which can be observed in the existing housing:

1. Spatial requirements of the solar system.
2. Requirements for additional builders work.
3. Overshadowing.
4. Roof construction and form.
5. Orientation.

The latter three are the major restraints on a retrofitting programme. Considerations of such characteristics has lead an earlier writer to suggest that only between 25% and 50% of households could be retrofitted (2). Whilst detailed studies of the effects of overshadowing are yet to be carried out, surveys of roof types and orientations now indicate that the proportion which could be retrofitted may be considerably higher.

Spatial requirements of solar systems

The solar storage tank is the only item of any bulk to be accommodated. In the conversion of older properties though, space is often at a premium. Against might be balanced the added amenity of an extra drying cupboard. Hot water cylinders are available with two heat exchangers fitted. These can be installed in one tank system, thereby eliminating the need for an additional solar tank. In many cases there will also be a possibility of locating a tank in the attic. Almost three quarters of existing households have attics. (3)

Requirement for additional builders work

In modern roofs, designed within fine structural tolerance, some extra bracing may be required to support the solar collectors, particularly if they are set in to the roof thereby replacing battens which normally serve to brace roof trusses. In older houses however, the roofs were usually over designed and no additional structure is necessary for the collectors. If the storage tank is located in the attic, this may require some additional support.

It is common practice among architects and authorities engaged in rehabilitation work, to strip off and replace roof coverings automatically as experience has shown this to otherwise be a source of high maintenance costs. This obviously reduces the expense and difficulty associated with mounting solar collectors. If they are inset, so as to replace conventional finishes, the savings would represent a discount of about 10% on the installed cost of the solar system.

Overshadowing

Whilst it is possible to accurately calculate the reduction in irradiance that a given obstruction will cause in a particular location, it is difficult to estimate the effect of a variety of obstructions on a variety of locations spread over a wide area.

Observations made during the survey work described below indicate however that the greatest problem lies not so much with neighbouring buildings as with adjacent chimney stacks. Nonetheless the financial loss incurred by a homeowner with a retrofitted solar system which was overshadowed following the construction of a new building could be considerable. New legislation is therefore required to protect the sunlight rights of existing buildings.

The effect of nearby obstructions, such as chimneys can be greatly reduced by careful siting of collectors. The production of charts to guide designers and installers in optimising collector location is an urgent requirement.

Roof construction and form

The roof type determines the method of mounting collectors and their subsequent orientation. The optimum roof in London would be south facing and pitched at about 45 degrees from the horizontal (4) (5). Considerable deviation from the optimum is possible without greatly reducing the amount of energy received.

Flat roofs provide a convenient working platform upon which to mount collectors, tilted on triangulated frames. Single storey extensions in the south side of a house can offer a site for collectors in a thermo-syphoning arrangement. On buildings where no unshaded, south facing roof space is available, it is still sometimes possible to mount collectors on frames spanning above the roof between parapet walls, or even on the walls themselves.

Orientation

In a climate where a large proportion of the available radiation is diffuse, there is a wide range of acceptability in the choice of solar collector orientation. Szokolay has shown that in London, where 54% of the total radiation is diffuse, a collector can be deviated as much as 50 degrees east or west of south, and still receive 90% of the available energy (6). Within 30 degrees of south, the reduction in energy collection will be less than 2%.

Survey of existing dwellings

In order to assess the proportion of dwellings which could be retrofitted, and the installation strategies required to do so, a survey has been carried out in the London Borough of Islington. The survey covers almost 12,000 households. Almost half of these are in the six "Housing Action Areas" - designated areas of need which will receive special subsidies for housing improvements - spread across the north of the borough. The remainder are in a square kilometer surrounding a test house in the south of the borough.

Building types

As in most inner city areas, the vast majority of households surveyed form terraces ranging in height from two to give storeys. Locating the survey in Housing Action Areas has meant that the dwellings covered are predominantly pre-1944 stock. 56% of the national stock belong to this period (3), and these are likely to be the first to be retrofitted in any subsidised improvement programmes.

Methodology

The survey was implemented in three stages:

A. Determination of building orientations.
B. Recording of building characteristics.
C. Classification and analysis of data.

A. Determination of building orientations

Given that most of the buildings are terraces, it is a simple matter to determine their orientations with respect to the street orientation. The majority of dwellings are orientated at right angles to the street angle which can be measured from the grid lines on ordinance maps.

B. Recording of building characteristics

The storey height of the buildings and their roof type were the principal characteristics recorded. Storey height was used as an indicator of the number of households, and to indicate the extent of overshadowing problems.

As already discussed, there is some variety in roof form and the orientation of a pitched roof is not always the same as the house's elevational orientation as measured from maps.

In addition, note was taken of any visually apparent characteristics which would restrict the installation or impair the performance of a solar system.

Initially aerial photographs were used to carry out this stage of the survey. Paired, 225mm. square pictures shot vertically from an altitude of about 450 meters, at a scale of 1:3000 were viewed through a stereoscopic viewer. This allowed a large area to be covered fairly rapidly. Inexperienced operators however find it difficult to maintain stereoscopic vision and as a result it is often difficult to distinguish between horizontal and inclined surfaces. Even with stereoscopic vision it is difficult to determine storey heights of buildings in the centre of the photographs.

Because of these difficulties, stage B of the survey of the Housing Action Areas was carried by field observation. Once familiarity with the house types is gained, most roof configurations can be determined from street level. In some cases the parapet wall on the building facade screens the roof and access to the rear is necessary. In a small number of cases, 0.5% of the sample, it was necessary to deduce the roof type using evidence supplied by drain pipes, rodding eyes and surrounding houses.

C. Classification and analysis of data

Collected data was assembled on duplicated survey sheets classified under the most appropriate of four categories of collector mounting strategy, determined by roof form and orientation, or in a fifth category when collector mounting was not possible under any of the four methods. (See Fig. 1).

The four mounting categories were as follows:

(i) South facing inclined roofs.
(ii) South facing mounting frames spanning between party walls and perpendicular to the orientation of gully type roof slopes.
(iii) Triangulated mounting frames on flat roofs.
(iv) Mounting frames on vertical walls.

N.B. south-facing includes all orientation angles between 130° & 230° measured from north

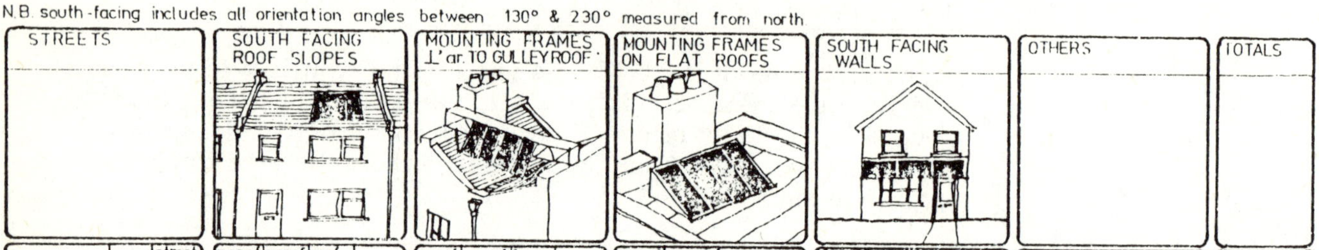

STREETS			SOUTH FACING ROOF SLOPES				MOUNTING FRAMES ⊥'ar. TO GULLEY ROOF				MOUNTING FRAMES ON FLAT ROOFS			SOUTH FACING WALLS				OTHERS				TOTALS	
name	side	street angle	nos of roofs	nos of stories	roof orient	notes	nos of roofs	nos of stories	frame orient	notes	nos of roofs	nos of stories	notes	nos. of houses	nos of stories	stories above gl	notes	nos of roofs	nos of stories	roof orient	notes	houses in street	south mounts
Arthur Rd.	N	60	17	4	150		12	3	150		2	4		31	3/4	2/3						31	31
	S	60	17	3	150		12	3	150		2	4		31	3/4	2/3						31	31
Roden St.	N	58	28	3	148									28	3	2						28	28
	S	58	28	3	148									28	3	2						28	28
Pakenham Street	W	148																5	3	238		5	0
	E	148																10	3	238		10	
Maiton Street	S	52	37	3	148									37	3	2						37	37
Holloway Rd.	E	137	12	5	227	grd flr comr					7	4	grd. floor shops	34	2, 3 4, 5							34	34
			7	4	227						2	2											
			4	3	227						2	3											
Hornsey Road	W	151	24	3	151																	24	24

HOUSING ORIENTATION SURVEY area: Arthur Road Housing Action Area sheet: 02/HAA/A

Fig. 1 Sample survey sheet showing classification of collected data.

Survey results

Detailed results can be seen in table and bar chart form in figs. 2 - 7. It was found that:

(i) 41% of dwellings have an inclined roof which faced within 30 degrees of south.
59% have an inclined roof which faced within 50 degrees of south.

(ii) 19% of dwellings have gully type roofs where neither of the existing slopes faced within 50 degrees of south. These 19% could however have solar collectors mounted on frames perpendicular to the existing pitches and facing within 40 degrees of south.

Fig. 2

Area	90 109	110 129	130 139	140 149	150 159	160 169	170 179	180 189	190 199	200 209	210 219	220 229	231 249	250 270	Flat roofs
								Orientation (measured from north)							
6 HAA	321	0	473	324	448	0	158	639	122	177	0	691	410	341	112
Sq.km.	669	981	18	27	65	382	841	907	298	637	340	0	67	1048	970
Totals	890	981	491	651	513	382	999	1746	420	814	340	691	477	1389	1082
%ages.	7.5	8.2	4.1	5.5	4.3	3.2	8.4	14.7	3.5	6.9	2.9	5.8	4.0	11.7	9.1

Organisational distribution of dwelling roofs for both the 6 Housing Action Areas and the Square Kilometer in the south of the borough

Fig. 3

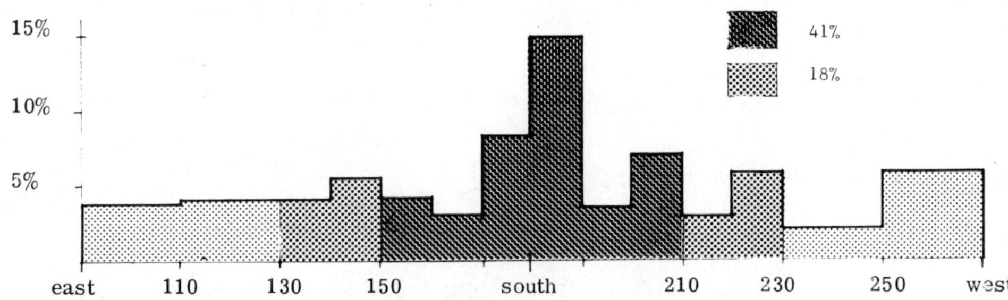

Graph illustrating orientational distribution of inclined roofs

Fig. 4

Orientation (measured from north)

Area	90 109	110 129	130 139	140 149	150 159	160 169	170 179	180 189	190 199	200 209	210 219	220 230	231 249	250 270	Flat roofs	
6 HAA	0	0	0	3	342	54	132	4	4	89	0	0	0	0		
sq.km.	0	0	0	0	24	183	543	100	288	397	129	0	0	0		
Total with frames	0	0	0	3	366	237	675	104	292	486	120	0	0	0		
Other roofs	494	366	491	651	513	382	999	1746	420	814	340	691	108	477	1082	
Grand total	494	366	491	654	879	619	1674	1850	712	1300	469	691	108	477	1082	11866
%ages	4.2	3.1	4.1	5.5	7.4	5.2	14.1	15.6	6.0	11.0	4.0	5.8	0.9	4.0	9.1	100

Number of dwellings whose roofs could be fitted with south facing frames plus other roofs. (NB 'other roofs' are all those shown in Fig. 2 minus those which require and can support south facing frames)

Fig. 5

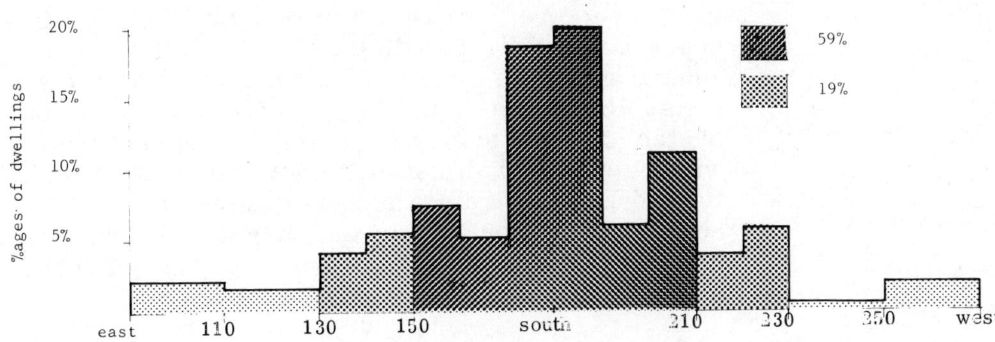

Graph illustrating orientational distribution of all roof mountings i.e. inclined roofs plus frames

Fig. 6

Area	Orientation (measured from north)													
	90 109	110 129	130 139	140 149	150 159	160 169	170 179	180 189	190 199	200 209	210 219	220 230	231 249	250 270
6 HAA	254	0	467	668	695	54	90	939	110	280	0	644	246	205
%ages	5.5	0	10.0	14.4	14.9	1.2	1.9	20.2	2.3	6.0	0	13.8	5.3	4.4
Sq.km.	1599	1421	18	0	160	188	967	114	640	502	257	0	184	1162
%ages	22.0	19.7	0.2	0	2.2	2.6	13.4	1.6	8.9	7.0	3.6	0	2.6	16.1
Total	1853	1421	485	668	855	242	1057	1053	750	782	257	644	430	1367
%ages	15.6	12.0	4.1	5.6	7.5	2.0	8.9	8.9	6.3	6.6	2.2	5.4	3.6	11.0

Orientational distribution of dwellings' front or rear wall orientation

Fig. 7

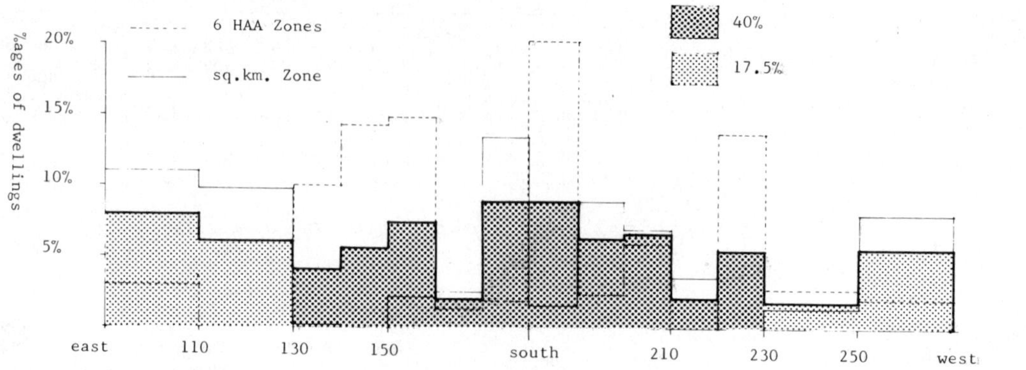

Graph illustrating orientational distribution of front or rear walls

(iii) 9% of dwellings have flat roofs which could support collectors mounted on triangulated frames.

(iv) 40% of dwellings have front or rear elevations facing within 30 degrees of south.
57.5% have front or rear elevations facing within 50 degrees of south.

Discounting wall mounting, one might summarise these results by saying that almost 60% of dwellings have a roof type and orientation which would allow solar collectors to be fitted directly. If simple mounting frames were employed on flat roofs, the proportion of suitable locations rises to almost 70%. If one also adopts larger mounting frames spanning between party walls a total of 87% of the dwellings surveyed have a roof type and orientation appropriate for retrofitting solar collectors.

Comments

It is important to realise that conclusions arising from these results may be wildly inaccurate for specific areas of small size. This is illustrated in fig. 7 where there is a difference of over 40% in the proportion of south facing elevations in the Housing Action Areas compared with the square kilometer area. Nonetheless the sample space is large and of the same order as other surveys from which national trends are interpreted (c.f. the survey basis for Parker-Morris space standards.) A wider geographical distribution would be desirable if conclusions were to be drawn regarding the housing stock as a whole. These results however represent, to the author's knowledge, a first indication of the potential for retrofitting programmes, based upon a study of existing housing.

There is a huge proportion of dwellings which could be retrofitted simply by fitting collectors directly on to an inclined roof. This is the most economic mounting base and these dwellings should be selected for the initial retrofitting programmes.

There is also a significant proportion of dwellings with flat roofs, and triangulated mounting frames for these are already commerically available.

The number of dwellings which could be retrofitted could be greatly increased given the development of a mounting frame which could span between party walls on gully type roofs. This would of course increase the installed cost. There is however scope for mass production as the variation in party wall spacing in the buildings surveyed is limited

Retrofitting field test

In order to test the theoretical predictions of solar water heating system performance, and to demonstrate and gain a working knowledge of retrofitting techniques, a $4m^2$ array of collectors have been mounted on the roof of a terraced masionette recently rehabiliated by the local authority. These are connected to a 200 litre hot water cylinder in a sealed indirect system with forced circulation. The solar system pre-heats the feed water for a conventionally heated hot water cylinder or, via a divertor valve, solar heated water is directed to the point of use upon reaching a selected temperature.

Hot water consumption, fuel consumption in the conventional system and six temperature points throughout the system are being monitored, and a heat meter will be added to the outlet from the solar tank.

Acknowledgements

The work described in this paper forms part of a research programme being carried out at the Architectural Association by R.D. Gray and K.E. McCartney with the collaboration of the London Borough of Islington and the support of the Science Research Council. I wish to thank the following members of the Association who assisted with the survey: A. Ponsi, P. Seamans, M. Shomo, P. Stanley, A. Winlow.

References

1. BRE Working Party
"Energy Conservation"
Building Research Establishment, CP 56/75, June 1975.

2. Ibid.

3. F. Ramig and G. Leach
"Energy Conservation in UK Dwellings: Domestic Sector Survey and Insulation"
International Institute for Environment and Development, London, June 1977.

4. Anon.
"Solar Water Heating"
ISES - UK, London, 1977.

5. SLC Working Party
'Report on Solar Water Heating"
South London Consortium (SLC), London, November, 1975.

6. S. V. Szokolay
'Solar Energy and Building"
Architectural Press, London, 1975.

7.9 Solar system retrofit of row houses

Sidney Shore PhD, PE
Professor and Chairman of Civil and Urban Engineering
University of Pennsylvania
Philadelphia, Penna. 19104, USA

John A. Lepore PhD, PE
Associate Professor of Civil and Urban Engineering
University of Pennsylvania
Philadelphia, Penna. 19104, USA

Noam Lior PhD
Assistant Professor of Mechanical Engineering and Applied Mechanics
University of Pennsylvania
Philadelphia, Penna. 19104, USA

Abstract

On 2 March 1977 the Executive Office of the President of the US published in the Federal Register a request inviting public comment regarding the formulation of a National Energy Policy to be released 20 April 1977. Of the seven areas delineated that must be addressed by such a Policy the number one item was 'a course that places appropriate priority on conservation as a key element in energy policy'.

The work described in this paper emphasizes the energy conservation concept with respect to existing residential structures. The potential for achieving significant energy conservation in this area is obvious on the basis of the number of existing residential structures in the US - 70 million.

In particular, the actual solar system retrofit of a typical row house in Philadelphia for space and hot water heating is presented. The paper describes briefly the retrofit design philosophy, the roof support system used for the solar collection panels, and solar heating system installed. The final sections present the design and installation sequences as well as the levels of effort expended, equipment, and material costs of the system.

Introduction

On 2 March 1977 the Executive Office of the President of the US published in the Federal Register a request inviting public comment regarding the formulation of a National Energy Policy that subsequently was released 20 April 1977. Of the several areas delineated by President Carter's energy policy the item of high priority was energy conservation. In particular the President elaborated on the utilization of solar energy for residential structures and the tax benefits and subsidies that he was going to propose for the period from 1977-1985. The potential for achieving significant energy savings with respect to existing residential structures is obvious, on the basis of the number currently in the US - 70 million.

This paper describes one demonstration retrofit project of a typical row house in Philadelphia for space and hot water heating using a solar energy system. If the retrofit proves successful from both a technical and cost basis, then the potential for using solar energy as an alternate source is excellent, especially in the densely populated cities in eastern US. For example, it has been estimated that 60% of the residential structures in Philadelphia are row houses; other cities such as Boston, New York, Pittsburgh, Baltimore and Washington also have row houses as a high proportion of their residences.

Retrofit design philosophy

Two major constraints were recognized in retrofitting row houses: (a) a relatively high proportion of row houses dwellers were in the low and low-middle income bracket; (b) many of the houses being considered were old structures. Consequently several fundamental specifications were established to recognize these limitations.

To account for the economic level of the row house dwellers, the retrofit consists of utilizing 'off-the-shelf' components. No esoteric or sophisticated controls or hardware were designed with the system. Thus, by purchasing standard and proven parts, the overall reliability and consumer availability of the system was maximized.

Another feature of the retrofit was to minimize the maintenance of the system by installing, for example, steel thermal storage tanks which were Heresite* coated to inhibit corrosion. Further, the fluid medium employed in the closed solar collector panel loop is city water.

On the basis that many of the row houses which are potential candidates for solar retrofit are quite old, it was decided to design the solar system to minimize rehabilitation, renovations, repairs and replacements of the structure and the heating and hot water systems. In fact, it was initially specified that most of the retrofitting should be within the capabilities of anyone who had ability to work with tools. However, as the work progressed, as described below, it appears at this time that although it is feasible for a home owner to retrofit, it is a physically formidable task. However, in the very near future it might be a distinct possibility if, for example, the structural support system of the solar collector panels could be packaged in kits.

Description of the house

The row house being retrofitted is located at 3920 Spruce Street, Philadelphia, PA., adjacent to the campus of the University of Pennsylvania. (Fig. 1) The structure is a conventional three storey building with a full basement and flat roof. The major building materials are brick for the front, side and party walls and a framed bay window in the rear. (Figs. 2 and 3) The interior is typical stud and plaster partitions. The plan dimensions are approximately 70 feet by 15 feet with a floor area, including the basement of 4,000 sq. feet. The long axis of the house is 10° west of south.

The existing space heating system is an oil-fired, forced hot air system; the domestic hot water is heated by a gas-fired water heater.(2)

Roof support system for solar collectors (1)

Five rows of thirty-three solar collectors are oriented at a 55° vertical angle with respect to the horizontal and facing 10° west of south. Although two different types of collectors are used, the same structural support system, consisting essentially of wooden members, was used. On the basis of minimizing structural modifications, the supports spanned the roof and transmitted loads directly by bearing on the side party walls. This scheme obviated strengthening the roof joists which were incapable of carrying the additional loads imposed by the collectors. The loadings used to design the support structure were: dead load, wind load, corresponding to 90 mph and snow load of 20 p.s.f., the latter two on the basis of a 100 year mean recurrence.

---

*Heresite and Chemical Company, Manitowoc, Wisconsin.

Fig. 1
Location of solar retrofit house

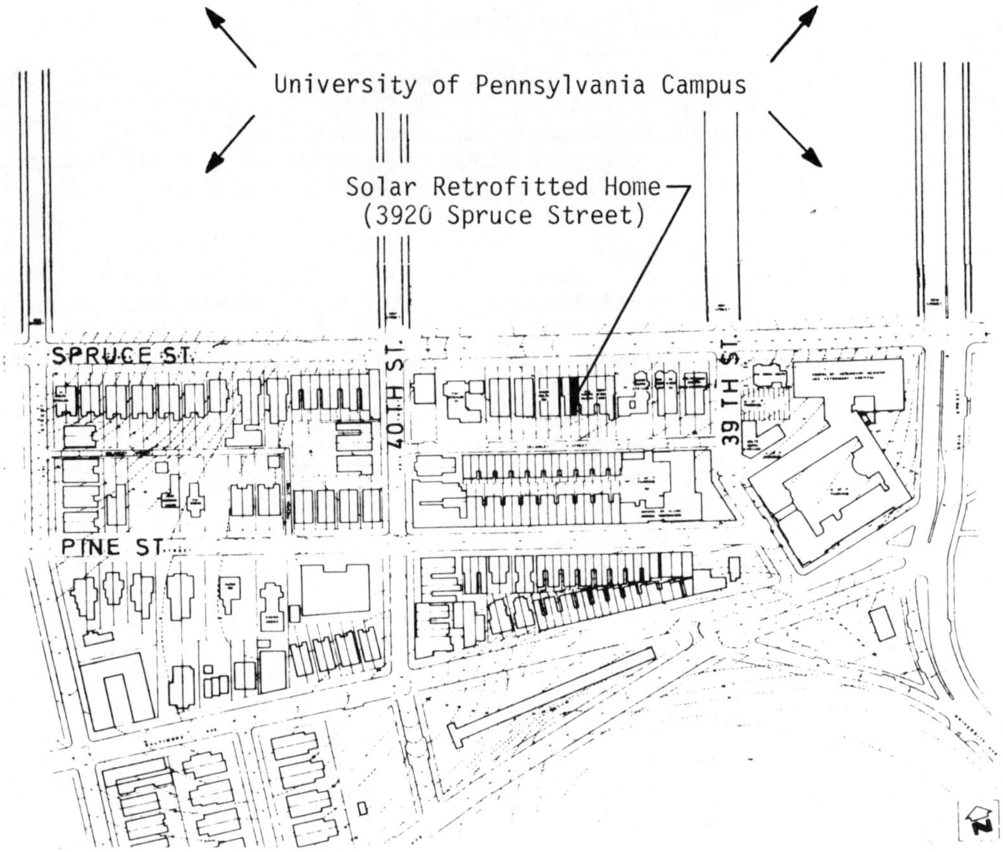

Fig. 2
Solar retrofit house view looking north

Fig. 3
Solar retrofit house view looking south

Fig. 4
Solar collector installation on the roof of the house

Solar heating system (2)

The space and hot water heating system consists of three loops:

(a) the solar energy collection loop which conveys the water from the thermal storage tanks in the basement to the collectors, where it is heated, and returned to the tanks;
(b) the heat demand loop which conveys water from the tanks to a water-to-air heat exchanger (heating the air for distribution through the heating ducts) and returns it to storage;
(c) the warm air loop which conveys and distributes the heated air to the house and returns it to the heater.

The calculated building design heat load, including domestic hot water is 70,000 Btu/hr. It is anticipated that the solar contribution to the annual comfort heating and domestic hot water load will be 40% to 50%.

Retrofit personnel

This retrofit project was planned to utilize University Technicians and students under the supervision of faculty members from the Department of Civil and Urban Engineering and Mechanical Engineering and Applied Mechanics. Consequently, sustained and continuous levels of efforts were difficult to maintain because of varying academic schedules, so that the labor was not as efficient as could be expected in a typical commercial or industrial installation.

Design of retrofit systems

The retrofit project was divided into essentially two major systems: (a) the structural support system; and (b) the solar heating system. Each of these two systems was then subdivided into two tasks: (a) design; and (b) installation. A more detailed breakdown of these systems and tasks are shown in Tables 1 and 2 in which are also tabulated the man-months expended for each sub-task and other explanatory comments.

The design and design drawings of the structural support system were done by graduate and undergraduate students under the supervision of two faculty members in the Department of Civil and Urban Engineering. The design and design drawings of the solar heating system were done by graduate and undergraduate students under the supervision of a faculty member in the Department of Mechanical Engineering and Applied Mechanics.

The actual installation of the structural support and solar heating systems was accomplished by technicians and students under the supervision of faculty members in the two departments of engineering noted above. The total effort expended is therefore greater than if skilled and experienced craftsmen had been employed in the actual installation. The actual man-months required to install the two systems is reported in Table 2.

Retrofit costs

The three direct costs incurred and attributable to the solar retrofit project are labor, material, and equipment. Thus all costs shown are direct costs and the cost of money is not included. The labor component is reported in Tables 1 and 2 in terms of man-months of effort expended since the cost of labor involved (professors and students) is not typical of future solar installation costs that would be performed by contractors and craftsmen; the cost of material and equipment is given in Table 3. A summary of these three components is given in Table 4.

Table 1

Design of Systems - Level of Effort

A.1 Structural Support System		Man Months	Comments
A.1.1	Surveyed residence to determine nature of construction, materials, insulation, plumbing, heating and electrical systems. Precise measurements were made of: the flat roof area available to mount the solar collectors, compass orientation of the long axis of the roof.	0.5	
A.1.2	Access to roof made. Ascertained structural configuration of roof joists and brick bearing side walls by probe holes in roof.	0.5	
A.1.3	Preliminary design and cost analysis of collector supports made considering steel, aluminum and wood as structural materials.	1.0	Wood selected on basis of cost, workability, maintainability and, availability.
A.1.4	Final design of collector support system made and design drawings completed.	2.0	
	SUB-TOTAL 1	4.0	
A.2	Solar Heating System		
A.2.1	Developed computer programs and calculated performance of heating components (e.g., collectors, storage tanks) of total solar system.	5.0	Computer program includes subroutines for determining time dependent position of sun and reading actual weather data from NOAA magnetic tapes.
A.2.2	Developed computer program and determined solar collector area (540 ft^2), inclination with respect to horizonal (55°) and spacing of rows (13.5 ft.) to maximize the total solar energy collected during a heating season. Analysis included trade-off between limited mutual panel shading and maximizing total energy collected.	3.0	Actual hourly weather data were used for the 1973-74 heating season.
A.2.3	Developed computer program and determined: volume of thermal storage tanks required (800-1000 gals. of water) to be consistent with maximization of total energy collected per task A.2.2; thickness of optimal insulation of tanks on basis of minimizing cost/benefit. Used degree-day method to calculate a preliminary building heat load.	1.0	
A.2.4	Designed the flow and piping configuration for the solar heating system.	1.0	
A.2.5	Defined and executed a preliminary design of the control and measurement systems for the flow of water and air in the solar heating system.	0.5	
A.2.7	Developed a self-draining freeze protection scheme for the collector flow loop.	1.0	
	SUB-TOTAL 2	11.5	
	TOTAL 1 + 2	15.5	

There are a number of factors characteristic of this particular retrofit that produced both higher levels of efforts and higher costs of material and equipment than should be expected if a contractor with skilled tradesmen is employed. These factors follow:

(a) This solar system serves both a utilitarian and research function since one of the sponsors requires a monitoring of the solar system performance for a 5 year period. Thus additional costs were incurred to permit the installation of special recording and control devices. However, the monitoring instrumentation and installation costs were not included in the costs reported herein;
(b) The structural support system that was designed and used was extremely conservative to allow construction by the handyman-owner;

Table 2

Installation of Systems - Level of Effort

B.1 Structural Support System		Man Months	Comments
B.1.1	Built and installed saddle supports on roof joists along east wall, lower roof. Sealed roof openings around saddles. Made pockets for 8" x 8" cross beams in upper roof party walls and lower roof west party wall.	1.5	Saddles transmit collector loads to ends of roof joists which then transmit loads by direct bearing on brick wall.
B.1.2	Pre-assembled and weather-proofed members of the structural support including cross beams, diagonals, and collector framing members.	1.0	
B.1.3	Installed: 8"x8" cross-beam; all other framing members to support collectors; safety railing along perimeter of roof.	1.0	
	SUB-TOTAL 1	3.5	
B.2 Solar Heating System			
B.2.1	Installed: One row of 5 PP & G solar collectors (35"x72" nominal size),, and four rows each with 7 International Environment Corp. solar collectors (25"x98" nominal size).	1.0	Total collector area = 475 ft^2; 10° west of south
B.2.2	Pre-assembled and installed manifolds on solar collectors.	1.0	
B.2.3	Modified existing auxiliary oil fired heater by installing water-to-air heating coil. Three steel thermal storage tanks (fabricated and lined by outside contractors) installed on concrete pad and anchored.	0.50	Nominal dimensions of each tank: 36" dia. x 84" long (320 gals.)
B.2.4	Piping installed: from collector manifolds to thermal storage tanks; from tanks to oil burner and hot water heater.	(1.25)*	
B.2.5	Pumps, flow meters, solenoid values and other controls to be installed.	(0.50)	
B.2.6	Insulation of thermal storage tanks and piping to be performed.	(0.50)	
	SUB-TOTAL 2	4.75	
	TOTAL = 1 + 2	8.25	

*Man-months in ( ) are estimated since work has not been completed at the time this paper was written.

Table 3

Cost of Materials and Equipment

	Item	Cost, $
A.	Structural Support System	
	1. Lumber	1,580
	2. Stairs and access to roof	150
	3. Hardware	330
	Subtotal 1	2,060
B.	Solar Heating System	
	1. Solar Collectors	7,210
	2. Thermal Storage Tanks	2,030
	3. Pumps	480
	4. Controls	950
	5. Piping, Fittings & Valves	2,600
	6. Heat Exchanger	610
	7. Insulation	1,500
	Subtotal 2	15,380
	Total 1 + 2	17,440

(c) Several major components in the solar heating system which were originally estimated on the basis of experiences reported by others were found to be significantly more expensive (e.g., thermal storage tanks).

(d) The material and equipment were purchased at retail prices rather than at discounted prices available to bona fide builders purchasing in quantity, therefore, if 'adjustment' factors are applied to the costs shown in Table 4, based on the factors given in the preceding paragraph, the cost of retrofitting a row house would be substantially lower. For example, if the following 'adjustment' factors are assumed-level of effort for installation only, 0.50; material, 0.70; equipment, 0.80 - then the adjusted quantities are labor, 4 man months, materials $4,300, equipment $9,000.

Table 4

Summary of Level of Effort & Cost

Item	Level of Effort, Man-Months			Cost, $
	Design	Installation	Total	
Labor				
1. Structural Support System	4.0	3.5	7.5	--
2. Solar Heating System	11.5	4.75	16.25	--
Material				
1. Structural Support System	--	--	--	2,060*
2. Solar Heating System	--	--	--	4,100**
Equipment				
Solar Heating System	--	--	--	11,280***
Total	15.5	8.25	23.75	17,440

* Table 3, Subtotal 1
** Table 3, Items B5, 7
*** Table 3, Items B1, 2, 3, 4, 6

Current status

The retrofit project is essentially completed at this time and no significant problems were encountered in the design or installation of the solar system. The shake-down of the system should start 1 July, 1977 at which time the solar hot water heating aspects of the system will start in the Fall 1977 at which time students will be living in the house as residents as well as monitors of the entire solar system.

Conclusions

Based on the current status of this project it is concluded that:

1 It is technically possible and feasible to retrofit row houses without substantial or major modifications to the structure or the back-up heating system.

2 It is possible to use 'off-the-shelf' material and equipment.

3 The cost/benefit ratio of retrofitting in today's energy market is too high on a single row house retrofit basis. If, however, a sufficiently large number of row houses are retrofitted simultaneously, the economy of scale will improve the ratio. For example, quantity purchases of material and equipment would reduce unit costs, repetitive operations

on many housing units would increase the efficiency of tradesmen, and delivery charges would decrease the unit costs of equipment and material. Further, as fossil fuel prices increase, and more builders and suppliers enter the retrofit market, the cost-benefit ratio should reach a level at which solar system retrofit of row houses will prove to be a viable conservation method.

4 It does not seem feasible to expect a home owner, however handy he may be, to retrofit his own house. The sizes and weights of too many components of the solar system precludes a one or two person installation. (e.g. the solar collectors, and the thermal storage tanks).

Acknowledgements

The major grant for this project was awarded by the Pennsylvania Science and Engineering Foundation; an equipment grant was provided by the US Department of Housing and Urban Development. Additional support was also given by the University of Pennsylvania and the International Environment Corporation, Gladwyne, Pa. The efforts and assistance of two technicians as well as a number of students in the Department of Civil and Urban Engineering and Mechanical Engineering and Applied Mechanics, University of Pennsylvania, is gratefully acknowledged.

References

(1) Lepore, J.A., Lior, N., and Shore, S., 'Energy conservation in housing construction'. American Society of Civil Engineers Preprint 2781. 27 September - 1 October 1976, Philadelphia, Pa.

(2) Lior, N., Lepore, J.A., and Shore, S., 'Residential solar heating retrofit in the urban environment'. Proceedings of 'Sharing the sun' Joint Conference of the American Section of the International Solar Energy Society and the Solar Energy Society of Canada, Vol. 4, pp 36-52. 15-20 August, Winnipeg, Canada.

7.10 The efficiency of ideal heliothermal devices

M. A. V. Devanathan
Tea Research Institute of Sri Lanka,
St. Coombs, Talawakele, Sri Lanka

Abstract

The efficiency of ideal heliothermal devices has been derived from the principle of radiation balance. The corresponding equations for glazed devices and optically intensified solar radiation devices have also been derived. It is shown that the theoretical maximum temperature attainable, $T_{h.max}$, for a radiation regime provides a simple formula for the efficiency of a heliothermal device operating at a temperature T_h which reads

$$\eta = 1 - \left(\frac{T_h}{T_{h.max}}\right)^4$$

Detailed tables for T_{max}, η, $\bar{T}_{max}$ and $\bar{\eta}$ averaged over the day, average available energies on a cloudless day, and average energies available per month, at various differences of temperature over the ambient temperature, are given for Southampton. The method can be applied to any location. Such data enable the assessment of the efficiency of various designs of devices in any location and the estimation of supplemental energy requirements.

Introduction

Devices which utilize solar radiation to produce thermal effects such as solar water heaters, solar water stills, solar furnaces and solar thermoelectric cells are called heliothermal devices. Unlike photoelectric or photovoltaic cells using solar radiation these operate between two temperatures the ambient sink temperature, and the solar radiation adsorbing panel temperature as the source, in the manner of heat engines.

The theoretical maximum to the efficiency of heat engines is prescribed by Carnot's theorem according to which an ideal heat engine working between the two temperature limits T_2 °K for the source and T_1 °K for the sink is given by

$$\eta = 1 - \frac{T_1}{T_2} \qquad (1)$$

Knowledge of this theoretical maximum enables designers to estimate the operational efficiency of devices as a percentage of this theoretical efficiency. With increasing interest in thermal devices using solar radiation it is useful to ascertain the existence of any such theoretical limits. We shall in particular be concerned with solar water heaters since these appear to be the most likely devices for widespread use in the future.

Radiation balance

The radiation reaching the surface of the earth consists of shortwave radiation from the sun in the wavelength range 0.4μ to 3.0μ, and the downward radiation of the atmosphere of longwave length about 3μ to 30μ. The shortwave radiation consists of both direct solar radiation and diffuse or scattered radiation from the upper atmosphere. The amount of solar radiation reaching the top of the atmosphere varies with the latitude and time of the year and tables of values are available[1]. Expressions for the amount falling on the surface of the earth, consisting of a diffuse component and a direct component which is a function of cloud cover are available, though the statistical nature of cloud cover relations must be recognised.

The downward radiation of the atmosphere arises from emissions from water vapour and to a lesser extent carbon dioxide in the atmosphere. The amount of this radiation is given by the empirical equations of Brunt[2] and of Swinbank[3]. The former expression correlates this to the screen temperature of the atmosphere and the relative humidity, while Swinbank's equation requires only the screen temperature.

Over the whole globe all radiation received by the earth and its atmosphere is re-radiated as terrestrial longwave radiation at an average radiation temperature of 285°K[4] according to Stefan's radiation law, thus maintaining the thermal equilibrium of the earth on a long time scale. This radiation balance does not apply over small areas of the earth surface over short periods of time, so that small areas can be nett adsorbers or nett radiators, depending on the instantaneous balance of radiation.

General expression for thermal efficiency

Let the shortwave solar radiation at any instant be represented by Q_t, the downward atmosphere radiation by $L\downarrow$ and the longwave radiation outwards by $L\uparrow$, then the nett radiation represented by R_n is given by the equation[4]

$$R_n = Q_t + L\downarrow - L\uparrow \tag{2}$$

Since the nett radiation is available for absorption by solar devices and $Q_t + L\downarrow$ is the total radiation input the thermal efficiency is given by

$$\eta = \frac{R_n}{Q_t + L\downarrow} = 1 - \frac{L\uparrow}{Q_t + L\downarrow} \tag{3}$$

Equation (3) would apply to solar water heaters and other thermal devices, but not to photoconverters of solar radiation, since such processes are temperature independent.

For solar water heaters maximisation of η according to (3) is not the sole criterion since a substantial temperature difference is also required. Let us assume that an ideal solar water heater exists in which all heat transfer processes are 100% efficient and that the insulation where required is also perfect, so that all the nett radiation is adsorbed with no loss. Under these ideal conditions the maximum temperature attainable is set by the term $L\uparrow$ which is strongly temperature dependent. Clearly the maximum temperature limit will be that at which the re-radiation is equal to the radiation input. At this limit the nett radiation and therefore the thermal efficiency is zero. Thus maximum efficiency is achieved at zero temperature difference in the water inlet and outlet temperatures and maximum water temperature is obtained at zero water flow rate.

Equation for thermal efficiency of non-glazed devices

It is now necessary to use detailed expressions for Q_t, $L\downarrow$ and $L\uparrow$. The solar shortwave radiation is an approximately sinusoidal function on a cloudless day. The total daily input is given by

$$Q_{sh} = R_A\,(0.29\cos\lambda + 0.52\,\frac{n}{N}) \tag{4}$$

where R_A is the radiation at the top of the atmosphere, N the day length[5] at the location of latitude λ and n is the number of hours of bright sunshine[6]. R_A is given in the Smithsonian Meterological tables[1] for various latitudes and times of the year. The term $\frac{n}{N}$ introduces a statistical element, since it depends on the cloud cover and therefore equation (4) will be applicable only if averaged over at least ten day periods. Although n on a cloudless day will not be exactly equation to N, we may use this as a reasonably good approximation to write for the instantaneous radiation flux[7]

$$Q_t = \frac{\pi}{2N} R_A(0.29\cos\lambda + 0.52) \sin\frac{\pi t}{N} \tag{5}$$

where t is the time of the day after sunrise. For convenience we shall write

$$Q_t = Q_m \sin\frac{\pi t}{N} \tag{6}$$

The two empirical expressions for L↓ are the Brunt equation[2]

$$L\downarrow = \sigma T_a^4(1 - 0.08\sqrt{e}) \tag{7}$$

where e is the relative humidity, and the Swinbank equation[2]

$$L\downarrow = \varepsilon T_a^6(\varepsilon = 5.31 \times 10^{-13} W/m^{-2}\ {}^{o}K^{-6}) \tag{8}$$

which involves only the ambient temperature T_a. The Swinbank formula is believed to have a better theoretical basis and moreover has the advantage of requiring the measurement of screen air temperature only, and will be preferred in this treatment.

The solar water heater absorption panel will be assumed to be a perfect black body at the temperature T_h which is the same as that of the water. Hence the expression for L↑ is given by

$$L\uparrow = \sigma T_h^4 \tag{9}$$

where σ is the Stefan constant $5.67 \times 10^{-8} W/m^{-2}\ {}^{o}K^{-4}$. Equation (3) for the efficiency then becomes

$$\eta = 1 - \frac{\sigma T_h^4}{Q_m \sin\frac{\pi t}{N} + \varepsilon T_a^6} \tag{10}$$

The maximum heater temperature attainable at any time is given by making $\varepsilon = 0$ in (10) when we get

$$T_{h.max} = \left[\frac{Q_m \sin\frac{\pi t}{N} + \varepsilon T_a^6}{\sigma}\right]^{\frac{1}{4}} \tag{11}$$

The amount of heat obtained from the solar water heater, if the flow rate is adjusted automatically to maintain a constant water temperature of T_h with an inlet temperature at the ambient value T_a, is then assuming no heat loss,

$$S\left(T_h - T_a\right)V_t = Q_m \sin\frac{\pi t}{N} + \varepsilon T_a^6 - \sigma T_h^4 \tag{12}$$

where V_t is the flow rate at time t and S the specific heat. The ambient temperature T_a if constant throughout the day will require V_t to be an approximately sinusoidal function so that

$$S\left(T_h - T_a\right)V_m \sin\frac{\pi t}{N} = Q_m \sin\frac{\pi t}{N} + \varepsilon T_a^6 - \sigma T_h^4 \tag{13}$$

where V_m is the maximum flow rate at mid-day. The ambient temperature is a weak approximately sinusoidal function about 1-2 hours behind the solar radiation variation. But for most purposes it can be regarded as constant over the day and even for short periods of a few days. If ΔT represents the increase in temperature of the water then (13) can be written in the integrated form covering a day

$$\Delta T_s \frac{2N}{\pi} V_m = \frac{2N}{\pi} Q_m + N\varepsilon T_a^6 - N\sigma(T_a + \Delta T)^4 \tag{14}$$

This when approximated yields

$$\Delta T_s S V_m \frac{2N}{\pi} = \frac{2N}{\pi} Q_m + N\varepsilon T_a^6 - N\sigma T_a^4 - 4NT_a^3\Delta T \tag{15}$$

Denoting the total water output for the day by V_d and using (4) and (6) we get

$$\Delta T \simeq \frac{Q_{sh} + N\varepsilon T_a^6 - N\sigma T_a^4}{V_d S + 4N\sigma T_a^3} \tag{16}$$

Because of the binomial expansion of $(T_a + \Delta T)^4$ and the retention of the first two terms only, equation (15) is useful only for small temperature differences. Since it is the rapid increase in the $(T_a + \Delta T)^4$ term that sets the limits for the maximum temperature it is preferable to use (14) in the form

$$S\,V_d(T_h - T_a) = Q_{sh} + N\varepsilon T_a^6 - N\sigma T_h^4 \tag{17}$$

While (10) gives the instantaneous efficiency, the average efficiency per day free of cloud cover may be obtained from (17) and reads

$$\bar{\eta} = 1 - \frac{N\sigma T_h^4}{Q_{sh} + N\varepsilon T_a^6} \tag{18}$$

Equations (10) and (18) give the instantaneous and daily average efficiency, and provide a theoretical reference for assessing the operational efficiency of various designs of heliothermal devices.

Efficiency of glazed solar heaters

Glazing of solar water heater panels while protecting the surface of the panel from atmospheric deterioration serves also to reduce heat loss by convection from the heater surface. This heat saving is reduced by absorption of part of the radiation in the glazing material. In order to assess the effects of glazing, let us in the first instance postulate a glazing material wholly transparent to solar shortwave radiation in the region 0.4μ to 3μ and totally opaque longwave radiation. Let us assume that the thickness of the glazing is adequate to absorb completely all longwave radiation and that internal convective heat transfer within the panel and the glazing is absent.

Let us assume that the glazing is in equilibrium at a temperature T_g receiving εT_a^6 from the downward atmospheric radiation and σT_h^4 from the heater panel, then the equilibrium in the absence of convective loss to the atmosphere is described by

$$\varepsilon T_a^6 + \sigma T_h^4 = 2\sigma T_g^4 \tag{19}$$

where the factor 2 in $2\sigma T_g^4$ recognises that the glazing emits radiation from both surfaces[8]. The nett radiation received by the solar panel is then from (12)

$$S(T_h - T_a)\,V_t = Q_m \sin\frac{\pi t}{N} + \sigma T_g^4 - \sigma T_h^4 \tag{20}$$

and using (19) we get the alternative form

$$S(T_h - T_a)\,V_t = Q_m \sin\frac{\pi t}{N} + \frac{\varepsilon T_a^6 - \sigma T_h^4}{2} \tag{21}$$

comparing (21) with (12) we note that the effect of this ideal glazing is to halve the longwave radiation terms in (12). The efficiency of a glazed solar device is then

$$\eta_g = 1 - \frac{\frac{1}{2}\sigma T_h^4}{Q_m \sin\frac{\pi t}{N} + \frac{1}{2}\varepsilon T_a^6} \qquad (22)$$

and the maximum temperature attainable

$$T_{h.max.g} = \left[\frac{2Q_m \sin\frac{\pi t}{N} + \varepsilon T_a^6}{\sigma}\right]^{\frac{1}{4}} \qquad (23)$$

On comparing (10) with (22) we note that within the same temperature limits the glazed device has a higher efficiency. The maximum temperature by (23) is also higher and glazing in effect doubles the intensity of the solar shortwave radiation as far as the efficiency and maximum temperatures attainable are concerned. The extent of the increase in efficiency will depend on the actual values of Q_m and the ambient temperature. Equation (8) is valid up to about 30°C which is about the upper limit for ambient temperatures, and up to this limit the term εT_a^6 is always less than σT_a^4. The analogue of equation (18) for the average efficiency over the day for the ideal glazed solar device will then read

$$\bar{\eta}_g = 1 - \frac{N\sigma T_h^4}{2Q_{sh} + N\varepsilon T_a^6} \qquad (24)$$

and the averaged $T_{h.max.g}$

$$\bar{T}_{h.max.g} = \left[\frac{2Q_{sh} + N\varepsilon T_a^6}{N\sigma}\right]^{\frac{1}{4}} \qquad (25)$$

However it must be noted that although the total input to the glazed device is $Q_{sh} + N\varepsilon T_a^6$ only $Q_{sh} + \frac{N\varepsilon T_a^6}{2}$ is the input to the absorbing panel. Since εT_a^6 is the longwave contribution from the atmosphere, heliothermal devices not only use direct and diffuse solar radiation but also solar radiation adsorbed as shortwave radiation and re-radiated as longwaves by atmospheric gases chiefly water vapour and also carbon dioxide. This contribution is significant especially at low solar radiation intensities.

Attainment of high temperatures

The temperatures which can be attained with natural solar radiation intensities are low and good enough only for producing hot water when the solar radiation intensity is at its maximum in days of long length. If higher temperatures are required for raising steam or for other higher temperature applications, the solar radiation intensities have to be enhanced by optical concentration with mirrors or lenses. Let the intensity multiplication by optical means be m, representing the ratio of the area for solar radiation collection to the area on which the radiation is focused, with corrections for losses in reflection and refraction. Equations (10) and (11) will now read

$$\eta = 1 - \frac{\sigma T_n^4}{mQ_m \sin\frac{\pi t}{N} + \varepsilon T_a^6} \qquad (26)$$

and

$$T_{h.max} = \left[\frac{mQ_m \sin\frac{\pi t}{N} + \varepsilon T_a^6}{\sigma}\right]^{\frac{1}{4}} \qquad (27)$$

When m is large the εT_a^6 term is negligible and glazing will make little difference from the radiation point of view, and (26) and (27) can be written in the simpler forms

$$\eta = 1 - \frac{\sigma T_a^4}{mQ_m \sin \frac{\pi t}{N}} \qquad (28)$$

$$T_{h.max} = \left[\frac{mQ_m \sin \frac{\pi t}{N}}{\sigma} \right]^{\frac{1}{4}} \qquad (29)$$

Simplified formulae for the efficiency

Equations (29) (27) (25) (23) and (11) which define the maximal temperatures enable equations (28) (26) (24) (22) (18) and (10) for the efficiency of heliothermal devices to be expressed in a simple form similar to equation (1) for ideal heat engines, namely

$$\eta = 1 - \left(\frac{T_h}{T_{h.max}} \right)^4 \qquad (30)$$

where $T_{h.max}$ is the maximum temperature and T_h is the working temperature of the solar panel. The concept of the maximum heliothermal temperature as defined by equations (29) (27) (23) and (11) enables the radiative quality of solar radiation in various environments and locations to be tabulated for ready use in equation (30) so that the operational efficiency of any heliothermal device can be assessed in terms of the theoretical efficiency possible in the same way as for heat engines.

Some sample calculations

To illustrate the application of the basic principles of heliothermal devices, we present sample calculations for Southampton. The overall mean temperature for the month has been taken as the mean of the average daily maximum and minimum as prescribed in the Meterological Office data[9]. For the latitude of 50°55'N, values of day length N and R_A have been calculated from the tables of McCulloch[5]. From R_A using equation (4) Q_{sh} is obtained, from which average radiation intensity over the period N and the maximum values have been calculated. These data are given in Table 1 along with values for εT_a^6 the downward longwave radiation. Table 2 gives $T_{h.max}$ as given by equation (27) at mid-day for m = 1,2,5 and 10 and Table 3 the corresponding $T_{h.max}$ averaged over N hours as would follow from equation (18) again with m = 1,2,5 and 10. Table 4 gives the efficiency with T_h equal to the ambient temperature at which efficiency is a maximum, and at increments of 10°C over the ambient. Table 5 gives the results for a glazed device which corresponds to m = 2. Table 6 gives similar data for 30°C increments with m = 5 and Table 7 m = 10 and 50°C increments.

Effect of glazing

These tables show the importance of glazing. The maximum temperatures attainable correspond to an optical intensification of the solar shortwave radiation by a factor of two. This produces a favourable increase in the temperature limits and corresponding increases in efficiency. However the total radiation per unit area is the same

TABLE 1. RADIATION DATA

SOUTHAMPTON 50°55'N 1°24'W 20 m

	Mean T_a °C	N hours	R_A cals/day/cm^2	Q_m W/m^2	Q_{sh}/N W/m^2	εT_a^6 W/m^2	$(Q_{sh} + N\varepsilon T_a^6)$ MJ/m^2
Jan	4.6	8.42	202.9	308.9	196.7	244.1	13.66
Feb	4.8	9.90	341.6	442.6	281.7	245.1	18.78
Mar	6.9	11.79	528.9	575.4	366.4	256.5	26.44
Apr	9.5	13.78	737.6	686.4	436.9	271.1	35.12
May	12.6	15.57	901.4	742.4	472.8	289.5	42.73
Jun	15.7	16.54	979.8	759.7	483.6	308.8	47.18
Jul	17.2	16.11	945.1	752.4	478.9	318.5	46.25
Aug	17.3	14.62	801.4	703.0	447.5	319.2	40.35
Sep	15.0	12.73	603.7	608.2	387.2	304.4	31.69
Oct	11.3	10.75	405.0	483.1	307.5	281.6	22.80
Nov	7.8	8.96	244.1	349.4	222.5	261.5	15.61
Dec	5.7	8.09	169.3	268.3	170.8	250.0	12.26

TABLE 2. MAXIMUM TEMPERATURES °K AT MIDDAY SOUTHAMPTON

	m = 1	m = 2	m = 5	m = 10
Jan	314.3	351.1	421.5	492.4
Feb	331.8	375.3	456.3	535.7
Mar	348.0	396.9	484.8	570.6
Apr	360.5	412.6	505.5	595.6
May	367.3	420.6	515.4	607.3
Jun	370.5	423.8	518.8	611.1
Jul	370.7	423.4	517.9	609.8
Aug	366.4	417.7	510.0	600.0
Sep	356.2	404.7	492.8	579.3
Oct	340.8	385.2	467.0	548.0
Nov	322.2	360.8	433.8	507.3
Dec	309.2	343.2	409.3	476.9

TABLE 3. MAXIMUM TEMPERATURES °K DAILY AVERAGE SOUTHAMPTON

	m = 1	m = 2	m = 5	m = 10
Jan	296.9	325.6	383.6	444.4
Feb	310.5	345.6	413.2	482.1
Mar	323.7	363.4	438.1	512.8
Apr	334.3	377.0	456.2	534.9
May	340.5	384.2	465.1	545.4
Jun	343.8	387.3	468.3	548.8
Jul	344.4	387.3	467.7	547.8
Aug	341.0	382.5	460.8	539.2
Sep	332.3	371.5	445.8	521.0
Oct	319.3	354.6	423.2	493.3
Nov	304.0	334.1	394.5	457.6
Dec	293.5	319.6	373.5	431.1

(11)

TABLE 4. MEAN DAILY EFFICIENCY %

SOUTHAMPTON m = 1

	T_a °K	T_a+10°K	T_a+20°K	T_a+30°K	T_a+40°K	T_a+50°K
Jan	23.4	11.7	0.0	0.0	0.0	0.0
Feb	35.2	26.0	15.2	3.2	0.0	0.0
Mar	43.9	35.5	26.1	15.8	4.4	0.0
Apr	48.9	41.2	32.8	23.4	13.2	1.9
May	50.4	43.0	34.9	26.0	16.2	5.4
Jun	50.1	42.9	34.8	26.0	16.2	5.6
Jul	49.4	42.1	34.0	25.1	15.3	4.6
Aug	47.3	39.7	31.3	22.0	11.8	0.6
Sep	43.4	35.2	26.0	15.9	4.8	0.0
Oct	37.0	27.6	17.3	5.9	0.0	0.0
Nov	27.0	16.0	3.9	0.0	0.0	0.0
Dec	18.5	6.1	0.0	0.0	0.0	0.0

TABLE 5. MEAN DAILY EFFICIENCY %

SOUTHAMPTON m = 2

	T_a °K	T_a+10°K	T_a+20°K	T_a+30°K	T_a+40°K	T_a+50°K	T_a+60°K	T_a+70°K
Jan	47.0	39.0	30.0	20.1	9.2	0.0	0.0	0.0
Feb	58.1	51.8	44.7	36.9	28.3	18.9	8.5	0.0
Mar	64.7	59.4	63.5	47.0	39.8	31.9	23.3	13.9
Apr	68.4	63.7	58.4	52.7	46.3	39.3	31.7	23.4
May	69.4	64.9	59.9	54.4	48.3	41.6	34.4	26.4
Jun	69.0	64.5	59.5	54.0	48.0	41.4	34.1	26.3
Jul	68.4	61.8	58.7	53.2	47.0	40.3	33.0	25.0
Aug	66.7	61.9	56.6	50.7	44.3	37.2	29.5	21.1
Sep	63.7	58.4	52.6	46.1	39.0	31.2	22.7	13.5
Oct	58.6	52.4	45.6	38.1	29.9	20.8	10.9	0.1
Nov	50.0	42.4	34.1	24.9	14.8	3.7	0.0	0.0
Dec	42.0	33.2	23.5	12.7	0.9	0.0	0.0	0.0

TABLE 6. MEAN DAILY EFFICIENCY %

SOUTHAMPTON m = 5

	T_a °K	T_a+30°K	T_a+60°K	T_a+90°K	T_a+120°K	T_a+150°K	T_a+180°K
Jan	72.5	58.5	39.9	15.5	0.0	0.0	0.0
Feb	79.5	69.1	55.2	37.1	13.9	0.0	0.0
Mar	83.3	74.9	63.7	49.1	30.4	7.1	0.0
Apr	85.3	77.9	68.2	55.5	39.3	19.1	0.0
May	85.7	78.7	69.4	57.4	42.0	22.9	0.0
Jun	85.5	78.5	69.2	57.1	41.9	22.8	0.0
Jul	85.1	78.0	68.5	56.2	40.7	21.4	0.0
Aug	84.2	76.6	66.5	53.5	37.0	16.5	0.0
Sep	82.5	74.0	62.8	48.2	29.7	6.6	0.0
Oct	79.6	69.5	56.1	38.7	16.5	0.0	0.0
Nov	74.3	61.4	44.2	21.8	0.0	0.0	0.0
Dec	68.9	53.2	32.2	4.8	0.0	0.0	0.0

TABLE 7. MEAN DAILY EFFICIENCY %

SOUTHAMPTON m = 100

	T_a °K	T_a+50°K	T_a+100°K	T_a+150°K	T_a+200°K	T_a+250°K
Jan	84.7	70.4	47.8	14.1	0.0	0.0
Feb	88.9	78.6	62.2	37.9	3.4	0.0
Mar	91.1	82.8	69.8	50.5	23.2	0.0
Apr	92.2	85.0	73.8	57.2	33.7	1.6
May	92.5	85.6	75.0	59.2	37.1	6.9
Jun	92.3	85.5	74.8	59.1	37.0	7.0
Jul	92.1	85.1	74.2	58.2	35.8	5.3
Aug	91.6	84.1	72.5	55.5	31.5	0.0
Sep	90.6	82.2	69.2	50.0	22.9	0.0
Oct	88.9	78.9	63.1	39.8	6.9	0.0
Nov	85.8	72.6	51.9	21.3	0.0	0.0
Dec	82.5	66.1	40.3	2.0	0.0	0.0

but the glazing reduces the longwave radiations to and from the panel by half. The energy output of the glazed device is therefore $\eta\left(Q_{sh} + \frac{\varepsilon T_a^6}{2}\right)$ whereas that from an optically intensified device will be $\eta\left(2Q_{sh} + \varepsilon T_a^6\right)$. The efficiencies calculated for m = 2,5 and 10 assume that the optical system concentrates only the shortwave radiation and that the longwave downward and upward fluxes are unchanged. This is a fairly good approximation.

In actual practice the reflection at the surface of the glazing will reduce m to below 2, but these inefficiencies can all be collected under the design efficiency factor. The maximum temperatures and efficiencies averaged over a day are valid only for a completely cloudless day.

Effect of cloud cover

The effect of cloud is to reduce the direct shortwave flux represented by $0.52 \frac{n}{N}$ in equation (4) which is applicable when n is averaged over at least 10 days. The equation to be used for instantaneous shortwave flux is equation (5) in which a term $(1-\theta)$ where θ represents the fractional cloud cover of the sky at any instant should multiply the factor 0.52, thus

$$Q_t = \frac{R_A \pi}{2N}\left[0.29\cos\lambda + 0.52\,(1-\theta)\right]\sin\frac{\pi t}{N} \qquad (31)$$

Over a long period of time $(1-\theta)$ is approximately $\frac{n}{N}$ where n is the number of hours of bright sunshine hence the presence of this factor in equation (4)[10]. Cloud cover also affects the longwave flux and the correction is usually written[11]

$$(L\downarrow - L\uparrow)(0.1 + 0.9\,\frac{n}{N}) \qquad (32)$$

for long time averages but for instantaneous cloud cover $(1-\theta)$ can be substituted for $\frac{n}{N}$. However it must be noted that the weighting factor 0.9 depends on the nature of the cloud formation. So this correction becomes rather inconvenient to apply, and the direct measurement by appropriate instruments, of the instantaneous flux of shortwave radiation and downward longwave radiation, will have to be made for exact measurements of the overall efficiency of any heliothermal device.

Maximum available solar energy

In the present state of solar energy technology, the best prospects are for solar water heaters. The norm for water requirements are 15 gals per day per person at 70°C. For a household of four persons this represents an available energy input of about 7.4 MJ to produce a temperature rise of about 65°C. In Table 8 are given figures for the term $(Q_{sh} + \frac{1}{2}\varepsilon T_a^6)$ N and the corresponding available energy to produce a range of temperature rise of 50°, 60° and 70°C. This amount of energy would be available with an ideal solar water heater of one square meter area, on a cloudless day. The average available energy per month is less because of cloud cover. The average available energy per month can be to a first approximation taken as monthly average percentage hours of bright sunshine which are tabulated in the Meteorological Office Tables[9]. The last three

TABLE 8. DAILY AVERAGE ENERGY OUTPUTS (MJm^{-2})
SOUTHAMPTON m = 2

		Cloudless Day			%	Monthly Avergae Day		
	$(Q_{sh}+N\frac{1}{2}\varepsilon T_a^6)$	T_a+50^oK	T_a+60^oK	T_a+70^oK	sunshine	T_a+50^oK	T_a+60^oK	T_a+70^oK
Jan	9.66	0.00	0.00	0.00	21	0.00	0.00	0.00
Feb	14.41	2.72	1.22	0.00	26	0.71	0.32	0.00
Mar	21.00	6.70	4.89	2.92	34	2.28	1.66	0.99
Apr	28.40	11.16	9.00	6.65	42	4.69	3.78	2.79
May	34.61	14.40	11.91	9.14	43	6.19	5.12	3.93
Jun	37.99	15.73	12.95	9.99	44	6.92	5.70	4.40
Jul	37.01	14.92	12.21	9.25	41	6.12	5.01	2.97
Aug	31.95	11.89	9.43	6.74	44	5.23	4.15	2.97
Sep	24.72	7.71	5.61	3.34	39	3.01	3.19	1.30
Oct	17.35	3.61	1.89	0.02	33	1.19	0.62	0.01
Nov	11.40	0.42	0.00	0.00	24	0.10	0.00	0.00
Dec	8.61	0.00	0.00	0.00	20	0.00	0.00	0.00

columns give this estimate of the average available energy. This data will be of use in estimating additional energy requirements for water heating for various months of the year, for solar water heaters of specified dimensions. The inadequacy of solar energy for water heating in the winter months is immediately apparent for the latitude of Southampton. Similar calculations can be carried out for any latitude by the method outlined.

Conclusion

The procedure given above will enable the quantitative assessment of the efficiency of proprietory surface treatments and mechanical designs by proposing theoretical maximal limits and practical limits for performance in various locations.

In conclusion it may be noted that the efficiencies of heliothermal devices are much more sensitive to temperatures because of the dependence on the fourth power of the temperature ratio while heat engine efficiencies are directly dependent on the temperature ratio. Consequently the overall efficiency of an ideal heat engine using the heater panel T_h of an ideal heliothermal heater as source, and the ambient as heat sink, will be the product of the two efficiencies and will be given by the formula

$$\eta_{overall} = \left[1 - \left(\frac{T_h}{T_{max}}\right)^4\right] 1 - \frac{T_a}{T_h} \qquad (33)$$

Acknowledgements

The author thanks Professor G. J. Hills for the facilities provided in the Chemistry Department, University of Southampton and Professor P. T. Landsberg, Mathematics Department, University of Southampton, for his advice on this paper and the stimulus provided by the seminars organised by him on solar energy utilization at the University of Southampton.

The author thanks Lord Garner, Chairman, Commonwealth Educational Awards Commission of the Association of Commonwealth Universities for the award of a Commonwealth Visiting Professorship (spent at the University of Southampton) and the Board of the Tea Research Institute of Ceylon for leave of absence.

References

1. Smithsonian Meterological Tables, 6th Rev. Ed., Washington 1951.

2. D. Brunt, Notes on Radiation in the Atmosphere, Q.J.R.met.Soc., 58, 389, 1932.

3. W. C. Swinbank, Long Wave Radiation from Clear Skies, Q.J.R.met.Soc., 89, 339, 1963.

4. W. D. Sellers, Physical Climatology, University of Chicago Press, Chicago and London, 1965.

5. J. S. G. McCulloch, Tables for the Rapid Computation of the Penman Estimate of Evaporation, East African Agric. & Forestry J., 30, No. 3, 1965.

6. J. Glover and J. S. G. McCulloch, The Empirical Relationship between Solar Radiation and Hours of Sunshine, Q.J.R.met.Soc., 84, 172, 1958.

7. J. L. Monteith, Principles of Environmental Physics, Arnold, London 1973.

8. J. C. Johnson, Physical Meteorology, John Wiley, New York 1954.

9. Tables of Temperature, Relative Humidity, Precipitation and Sunshine for the World, Part III, Meteorological Office Met., 0.856 C, H.M.S.O. London 1972.

10. W. R. Van Wijk, Physics of Plant Environment, North Holland Publishing Company, Amsterdam 1963.

11. H. L. Penman, Natural Evaporation from Open Water, Bare Soil and Grass, Proc. Roy. Soc. A., 193, 120, 1948.

A model to calculate solar radiation absorbed by absorber surface of a flat plate solar collector

H. K. Varma
Department of Mechanical and Industrial Engineering
University of Roorkee, Roorkee, U. P. 247672, India

ABSTRACT

The beam and diffuse solar insolation per unit area per unit time are calculated by method described in ASHRAE Handbook of Fundamentals. The two types of radiation are considered separately and the glass cover properties are determined for the beam radiation on the basis of its angle of incidence and for diffuse radiation on the basis of an average angle of incidence of 60° over the glass surface. The average values of product of transmittance τ of glass and absorptance α_a of absorbing surface, which take into account multiple reflections at glass cover and the absorbing surface are used to compute the rate of energy received by unit area of the absorber surface. The proposed method is illustrated by an example:

It is expected that the proposed model will provide a simple method to compute reliable values of energy received by an absorber surface for localities where sufficient solar data is not available.

INTRODUCTION

The design of a flat plate solar collector involves the calculation of solar radiations being absorbed by the absorber surface. The total energy absorbed by the collector surface is partly transferred to the heat carrying fluid, which forms the useful part, and the remainder is lost to the ambient through the glass cover, and insulation by radiation or convection, as given by the following relation:

rate of energy absorbed by collector surface	=	rate of useful energy	+	rate of heat lost from the collector	+	rate of increase of stored energy of the collector

It is, therefore, important to know the solar insolation data, the angle of incidence of solar radiations and the properties of the glass cover and the absorbing surface. Of these, the incident solar energy on the glass surface is usually calculated by the product of H and R, where

H = rate of incidence of beam and/or diffuse radiation on a unit area of a reference surface (usually horizontal)

R = factor to convert beam and/or diffuse radiation to that on the plane of the collector.

The information regarding the hourly rates of solar insolation is generally not existing for most locations at present. Similarly, the break-up data for total solar insolation into direct and diffuse components is not available. In the absence of local solar data, it is often recommended to use the data from other locations of similar latitude, topography and climate. ASHRAE Handbook of Fundamentals (2) describes a method to calculate the intensity of direct solar radiation, I_{DN}, and the intensity of diffuse radiation, I_{ds}, on any flat surface at any time of 21st day of each month. On the basis of the above calculation procedure, solar insolation data have been computed and reported in British units for south-facing surfaces at various inclinations from 24° to 64° north latitudes (3).

The multiplier factor R and the radiation properties of the glass cover vary significantly for the beam and diffuse radiations and the inclination between the incident radiations and the surface receiving them. For beam radiations, these properties change with the angle of incidence between the incident radiations and the surface receiving them. For diffuse radiations, it is recommended that the mean solar optical properties of glass are close to the properties for direct radiation for the angle of incidence of 60° (2). It is, therefore, important that the beam and diffuse radiations shall be considered separately and appropriate radiation properties be chosen in order to calculate the solar radiations being absorbed by the absorber surface of a flat plate collector.

This paper proposes a simple method to calculate the solar energy collection by a flat plate solar collector. This method is expected to produce reliable data for the rate of energy absorption by a flat plate collector.

CALCULATION PROCEDURE

The solar attitude angle depends on the local latitude, angle L, the solar declination δ and the hour angle H, which is measured from solar noon. The following equation gives the value of solar attitude angle β

$$\text{Sun}\ \beta = \text{Cos L Cos}\ \delta\ \text{H} + \text{Sun L Sun}\ \delta \tag{1}$$

The direct normal solar intensity, I_{DN}, at the surface of the earth on a clear day is well represented by

$$I_{DN} = \frac{A}{\exp(B/\text{Sun}\beta)} \tag{2}$$

where

A = apparent solar radiation for no air

B = atmospheric extinction coefficient

The values of A, B and δ are given in Table 1 (2).

Table 1
Solar data

Date	Declination degree	A W/m^2	B	C
Jan 21	−20	1229	0.142	0.058
Feb 21	−10.8	1214	0.144	0.060
Mar 21	0	1185	0.156	0.071
Apr 21	+11.6	1135	0.180	0.097
May 21	+20.0	1103	0.196	0.121
June 21	+23.45	1088	0.205	0.134
July 21	+20.6	1084	0.207	0.136
Aug 21	+12.3	1107	0.201	0.122
Sept 21	0	1151	0.177	0.092
Oct 21	−10.5	1192	0.160	0.073
Nov 21	−19.8	1220	0.149	0.063
Dec 21	−23.45	1233	0.142	0.057

The angle of incidence of the direct solar radiation over a flat surface inclined at an angle of Σ with respect to the horizontal is given by

$$\text{Cos}\ \theta = \text{Cos}\ \beta\ \text{Cos}\ \gamma\ \text{Sin}\ \Sigma + \beta\ \text{Cos}\ \Sigma \tag{3}$$

where γ is the surface-solar azimuth angle and can be easily determined when the orientation of the surface, expressed as surface azimuth angle ψ, is known and the solar azimuth angle ϕ, has been calculated with the help of eq. (5)

$$\sin \phi = \cos \delta \sin H / \cos \beta \tag{4}$$

Over a flat surface, the intensity of beam solar radiation will be given by

$$I_D = I_{DN} \cos \theta \tag{5}$$

The diffuse radiation incident over the surface will be due to diffuse reflection of solar radiation and forms of three parts, (i) from the sky, I_{ds} (ii) from the ground, I_{dg} and (iii) from the surrounding, I_{da}

The intensity of diffuse solar radiation from a clear sky falling over any surface can be found out by eq. (7)

$$I_{ds} = C\, I_{DN}\, F_{ss} \tag{6}$$

where C is the diffuse radiation factor and F_{ss} is the radiation shape factor between the surface and the sky. The values of C for different parts of the year are given in Table 1. The angle factor F_{ss} is determined by

$$F_{ss} = 1 - F_{sg} \tag{7}$$

where F_{sg} is the shape factor between the surface and the ground. For an inclined flat surface

$$F_{sg} = (1 - \cos \Sigma)/2 \tag{8}$$

Further, the intensity of solar radiations reflected from the ground

$$I_{dg} = \zeta_g\, F_{sg}\, I_{DN}\, (C + \sin \beta) \tag{9}$$

Usually, these radiations are ignored in calculating the total solar insolation for a solar collector. This gives conservative estimates of the values of solar insolation.

The amount of the radiations reaching a collector surface after reflection from the surroundings depends upon the actual location of the surface and the details of the surroundings. This has to be determined for each specific case. In the present analysis, we will neglect such radiations.

It is well known that the properties of glass, viz., transmittance τ, absorptance α_g and reflectance ζ are functions of the angle of incidence, θ. The diffuse radiations are usually considered incident at an average angle of 60° to the glass surface. Since ordinary double strength glass covers are used in flat plate collectors, the values of τ, α_g and ζ for beam and diffuse radiations separately can be taken from Reference 2. Alternatively one can express transmittance τ and absorptance α_g for double strength glass cover as follows:

$$\tau = -0.00885 + 2.71235 \cos \theta - 0.62062 \cos^2\theta - 7.07329 \cos^3\theta + 9.75995 \cos^4\theta - 3.89922 \cos^5\theta \tag{10}$$

$$\alpha = 0.01154 + 0.77674 \cos \theta - 3.94657 \cos^2\theta + 8.57881 \cos^3\theta - 8.38135 \cos^4\theta + 3.01188 \cos^5\theta \tag{11}$$

The above form is suitable for calculation by computer procedures.

Using the properties of glass and absorptance of the absorber surface of the collector, α_a, one can determine the average value of $(\tau\alpha_a)$ to account for multiple reflections at the glass cover and the absorbing surface (1)

$$(\alpha_a \tau) = \frac{\tau \alpha_a}{1 - (1 - \alpha_a)\, \zeta_d} \tag{12}$$

where ζ_d represents reflectance of glass for diffuse radiations.

The energy received by the unit area of the absorber surface per unit time will be given by

I_{DN} Cos θ $(\tau\alpha_a)_b$ + I_{ds} $(\tau\alpha_a)_d$

where subscript 'b' refers to the beam radiation and 'd' to diffuse radiation respectively.

Example

Given that a flat collector surface faces south and is inclined at an angle of 32° to the horizontal. The flat collector is installed in a city at 32° north latitude. Determine the energy received by unit area of the absorber surface per hour at 10 am on 21st May Assume that $\alpha_a = 0.95$

Solution

For 21st May, from Reference 2

$A = 1103\ W/m^2$, $\delta = +20.0$ deg

$B = 0.196$; $C = 0.121$

At 10 a.m., the hour angle $H = 30$ deg.

Therefore, from eq. (2)

$\text{Sin}\ \beta = \text{Cos}(32°)\ \text{Cos}\ (20°)\ \text{Cos}\ (30°) + \text{Sin}(32°)\ \text{Sin}(20°)$

$= 0.690 + 0.1812 = .871$

or $\beta = 60.62$ deg,

and

$$I_{DN} = \frac{1103}{\exp(0.196/0.8714)} = \frac{1103}{1.252} = 881\ \frac{TW}{m^2}$$

with the help of eq. (5), we get

$\text{Sin}\ \phi = \text{Cos}\ (20°)\ \text{Sin}\ (30°)/\text{Cos}\ (60.62°)$

$= \phi.958$

and the solar azimuth angle

$\phi = 73.27°$

For south-facing wall

$\gamma = \phi = 73.27°$

Equation (4) can now be solved for Cos θ

$\text{Cos}\ \theta = \text{Cos}\ (60.26°)\ \text{Cos}\ (73.27°)\ \text{Sin}\ (32°) +$

$\text{Sin}\ (60.26°)\ \text{Cos}\ (32°)$

$= 0.814$

and $\theta = 35.53°$

With the help of equations (8) and (9), when $\Sigma = 32°$

$$F_{ss} = 0.924$$

and $I_{ds} = .121\ (881)\ (0.924)$

$$= 98.5\ W/m^2$$

The transmittance and absorptance are calculated with the help of eqs. (11) and (12). For beam radiation, the angle of incidence is 35.53°, such that

$\tau_b = 0.864, \alpha_b = 0.052; \zeta_b = 0.084$

and by eq. (13)

$$(\alpha_a \tau)_b = \frac{0.864\ (0.95)}{1-\ (0.05)\ (0.148)}\ 0.827$$

For diffuse radiations

$$\tau_\alpha = 0.796, \alpha_d = 0.056, \zeta_d = 0.148$$

and

$$(\alpha_a \tau)_d = \frac{0.796\ (0.95)}{1-\ (0.05)\ (0.148)} = 0.762$$

Hence, the energy received by the unit area of the absorber surface per unit time

$$= (881)\ (0.814)\ (0.827) + (98.5)\ (0.762)$$

$$= 668\ W/m^2$$

It may again be emphasised that the reflected radiations from the ground and the surroundings are neglected. Similar calculations can be made for any time of the day.

REFERENCES

(1) J. A. Duffie, W. A. Beckman, 'Solar energy thermal processes', John Wiley and Sons, New York, 1974

(2) ASHRAE 'Handbook of fundamental, Americal society of heating refrigerating air-conditioning engineers', New York, 1972, Chapter 22

(3) C. A. Morrison, E. A. Farber, 'Development and use of insolation data for south facing surfaces in northern latitudes', ASHRAE Trans., 1974, V. 80, Pt. 2, pp 352-364

7.12 The measurement of solar radiation

D.A. Svendsen
Solar Energy Unit,
Department of Mechanical Engineering and Energy Studies,
University College, Cardiff, U.K.

Abstract

For all aspects of solar energy utilisation some understanding of the quantity and nature of the incident solar radiation is necessary. Measurement is an essential part of gaining this understanding because of the high degree of complexity and variation, both in time and space, of the amount of solar radiation reaching the surface of the earth.

The calibration and changes in calibration of the familiar Kipp and Zonen solarimeter are considered and its use over long and short term periods is discussed. Preliminary results of a programme to establish criteria for appropriate short term sampling rates for the measurement of solar radiation are presented.

The quantity of solar radiation received by a surface of arbitrary orientation is considered and the measurement programme being carried out in Cardiff described. Details are given of the apparatus being used and comparison made between a simple model and measured data.

Introduction

For all aspects of solar energy utilisation some understanding of the quantity and nature of the incident solar radiation is necessary. This understanding is obtained from measurement because of the high degree of complexity and variation, both from time to time and place to place, in the amount of solar energy reaching the surface of the earth. In practice it is not possible to measure all that is required at every locality and so extrapolation of those measurements that are available must be made. The solar energy incident upon a slope of a given orientation is of particular interest especially for the passive utilisation of solar radiation. In this case it is very unlikely that measurements will have been made for the particular slope and so it will be necessary to predict from those measurements that are available. Many varied mathematical models exist for this prediction but there is a lack of extensive measurements from which the models may be checked and compared. For this reason a programme of measurement of solar radiation incident upon slopes of arbitrary orientation has commenced in Cardiff. This paper introduces this programme and also considers some of the aspects of measurement with a pyranometer and in particular the Kipp and Zonen solarimeter.

The pyranometer and measurement

The instruments most commonly used for the measurement of solar radiation fall into two distinct categories. The pyrheliometer or actinometer measures the radiation received from a small solid angle of sky and is most appropriately used in detailed research and calibration where the relative expense and difficulty of use are justified. A simpler instrument and one suitable for long term measurement is the pyranometer where the radiation incident from the whole hemisphere is integrated. The Kipp and Zonen solarimeter is the version of pyranometer which is most widely used in the United Kingdom.

Full details of pyrheliometers and pyranometers may be found in the standard texts (1,2) and so will not be given here. However, some important considerations in the use of a pyranometer will be considered further.

Calibration

The electrical output of the pyranometer is related to the radiative input by the calibration factor which may be obtained either outdoors using the sun or indoors using an artificial light source.

Outdoors the procedure is to measure the direct beam of the sun using a pyrheliometer and at the same time measure the output from the pyranometer both shaded and unshaded. The shade is a small disc which is held so that it completely shades the pyranometer from the sun and also subtends the same solid angle as that of the pyrheliometer. Then knowing the sun altitude the calibration factor may be calculated as is summarised below:

k = calibration factor; V_0,V_1 is the output of the pyranometer unshaded, shaded respectively; I is the value of the solar beam as measured by the pyrheliometer; h = solar altitude. Then

$$k.V_0 = I.\sin(h) + D; \qquad k.V_1 = D$$

where D is the amount of diffuse radiation. Thus

$$k = \frac{I.\sin(h)}{V_0 - V_1}$$

This procedure is repeated many times to obtain an average value of k.

The pyrheliometer itself requires calibration and this is done at regular intervals by comparison with an absolute radiaometer which is a type of pyrheliometer whose calibration factor may be calculated from the geometry of the instrument. The standard pyrheliometer used by the U.K. Meteorological Office is checked at the International Pyrheliometer Comparison meetings held in Davos, Switzerland every five years (3). For indoor calibration which does not depend upon the sun shining, a pyranometer of known calibration is compared against the uncalibrated pyranometer with either a point source or a diffuse source of radiation. The calibration by Kipp and Zonen is using a point source (4) while the U.K. Meteorological Office uses a diffusing chamber. This latter method gives agreement between outdoor and indoor calibration to better than 1% although differing by up to 3% from the manufacturer's value for the Kipp solarimeter(5).

The Kipp solarimeter has a significant temperature coefficient (about -0.2% per ^{o}C (4)) and so it is usual to give the calibration factor at a given temperature. For pyranometers calibrated in the U.K. this temperature is 10°C.

Changes in calibration

There are several circumstances under which the calibration factor can change. The temperature coefficient has already been mentioned and for accurate work the appropriate correction must be made. Imperfections in the glass domes and sensing surface of the pyranometer changes the calibration factor for low angles of incidence of radiation. The change is within ±1% for angles of incidence greater than 75° for a Kipp solarimeter (1).

The pyranometer is usually calibrated with its sensitive surface horizontal but often it is of interest to measure the radiation incident upon a non-horizontal slope. The calibration factor changes with changes in altitude orientation of the solarimeter due to changes in the internal air convection currents. However, the degree of variation is difficult to assess and measurements that have been made differ considerably. For example, the change for a vertical facing Kipp has been quoted as -1% (6), 0% (7), +10% (8) and -7%(9). The large discrepancy is unexplained but it is probable that the latter two values are abnormally high.

Sampling rates

The moment to moment fluctuation of solar radiation due to changes in cloud cover in unsettled weather is sufficiently great and sufficiently random for it to be impractical to sample at the theoretically necessary rate of twice the maximum significant frequency since this would imply sampling every few seconds. In addition there are many days where the solar radiation changes relatively slowly and so it becomes necessary to determine the highest rate of sampling that measures the data sufficiently accurately and yet keeps the cost and complexity of measurement down. Sampling cannot be avoided since it always occurs either explicitly with sampling data loggers or implicitly in the time resolution of chart recorders and analogue or digital integrators.

Fig. 1

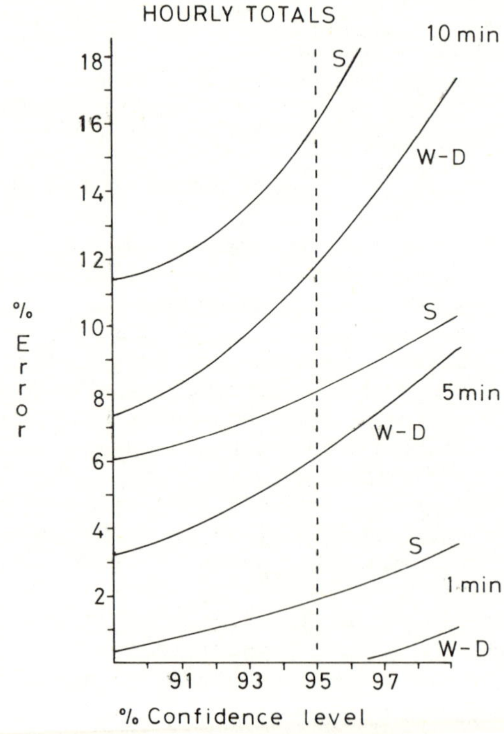

In order to assess the likely accuracies of different sampling rates for measuring solar radiation the total and diffuse radiation on a horizontal surface were measured using a data logging recorder and the high sampling rate of five seconds between samples. The results are still being interpreted but Figure 1. summarises the percentage confidence level that may be placed in the estimate of the hourly mean for any hour being within a given percentage of the true value for one, five and ten minute sampling rates. The true value was taken to be that given by the 5 second sampling rate.

Total radiation was measured during December 1976 (W-D in Figure 1) and March-April 1977 (S in Figure 1). Diffuse radiation was only measured for the Spring period (W-D in Figure 1) but the level of accuracy of the sampling of the diffuse radiation very closely matched that of the total radiation for the winter month. This is because the variability of the winter months was low in that the days tended to be overcast or clear without the type of clear day with scudding clouds that was characteristic of the Spring period. The variability of the diffuse radiation would be expected to be much less than that of the total because of the distributed nature of the diffuse radiation compared with the point source contribution from the sun to the total.

For randomly distributed data the improvement in the stability and accuracy of the result is improved by a factor of approximately $\sqrt{N}$ where N is the ratio of the increased number of samples taken. However, the improvement for going from a ten to five to one minute sampling rate is approximately linear as may be seen at the 95% confidence level shown in Figure 1. This is attributed to the fact that the major variability is due to small clouds briefly obscuring the sun and so the taking of a sample when the sun is obscured or when it briefly breaks through cloud cover could mean that for the whole of the sample period the sample is completely unrepresentative. Halving the time between samples will thus halve the unrepresentative time so giving the approximately linear relationship indicated by the results in Figure 1.

The implication of these results that for regions where unsettled weather is common at least a one minute sampling rate is required for the measurement of total solar radiation if the estimate of the hourly mean is to be better than a few per cent accurate for all conditions. For long term measurements where the average over many hours or days is required then the normal square root rule will operate to improve the accuracies so permitting lower sampling rates.

The pyranometer in use

There are two main reasons for using a pyranometer to measure solar radiation. The first is to obtain long term data for the assessment of the long term performance of solar energy utilisation systems. For this requirement the necessary averaging to reduce the data to manageable amounts means that changes in calibration which are essentially random may be ignored since they will average out. For example the temperature changes in calibration will be small when averaged over many years. However, systematic errors will not be averaged out and will have to be considered. Hence if long term monitoring on a non-horizontal slope was carried out it would still be necessary to consider the change in calibration with angle of orientation of the pyranometer.

The other reason for use is to measure the essentially instantaneous incident solar energy where the amount of radiation as incident for that particular period of time is required. This requirement is becoming common with the extensive testing of thermal solar collector panels. Under these conditions it will be necessary to correct for the temperature coefficient and altitude angle of the pyranometer and ensure high angles of incidence of the radiation if accuracies of better than a few per cent are required. Furthermore, sufficiently high rates of sampling of the data for the particular prevailing conditions must be made as indicated earlier.

Solar Radiation income of slopes

It is of obvious importance to be able to predict the amount of solar radiation incident upon a slope of arbitrary orientation and furthermore it must be possible to do this from a minimum of measured data. Many methods of prediction exist using measured radiation values or associated parameters (10,11). Norris (12) has compared some of these models with measured data and his results show reasonable agreement over monthly periods but there are large discrepancies on a daily basis. In general the amount of measured radiation incident upon non-horizontal slopes is sparse and so the validity of the various prediction models is difficult to verify under a wide range of conditions. A measurement programme is being established in Cardiff to provide such data on a continuous basis with the intention of checking the validity of current prediction models and if necessary creating a new prediction method. The apparatus being used is described in the following sections.

The measurement on slopes of arbitrary orientation

The income of a slope may be measured directly using a pyranometer inclined at the appropriate angle or by measuring the radiation coming from each region of the sky and integrating over the surface. The former approach is the one that has been adopted in Cardiff and since the optimum orientations of slope suitable for characterising the spatial distribution of the solar radiation are not yet known, it was decided to use a moveable solarimeter which could be regularly orientated to particular positions permitting flexibility of use. Since knowledge is required in the long term over the full range of prevailing conditions it is important that the apparatus should be weatherproof and capable of running reliably for many months.

Fig. 2

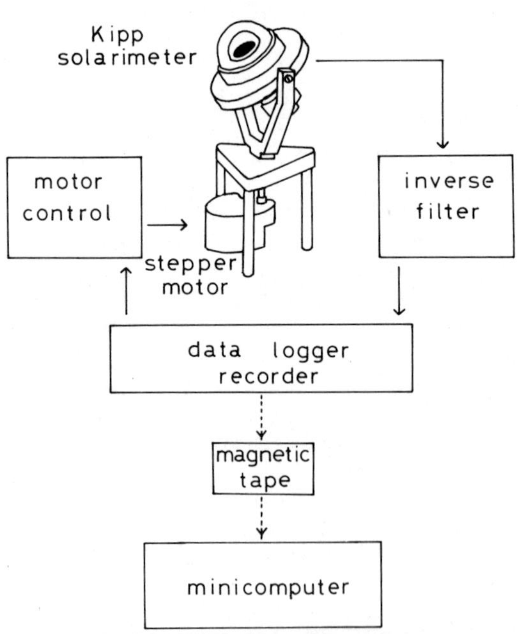

Instrumentation

The prototype system being used at the moment is shown in Figure 2. A Kipp and Zonen solarimeter is fixed at an adjustable angle of inclination in a yoke which may be positioned at any azimuth angle by a miniature stepping motor. The output from the solarimeter is fed through a special inverse filter electronic circuit and the output recorded on magnetic tape by a weatherproof data logging system. The data is thus in a form suitable for off-line processing in a minicomputer. The inverse filter removes the first order lag of the solarimeter hence accelerating its time response by a factor of 10 so that the output settles in less than five seconds after a change of radiation input (13). The data logger is controlled by a crystal clock which initiates the scanning and recording of twelve independent measurement channels at regular preset intervals. On the completion of each scan the solarimeter is moved to the next position of azimuth as determined by the motor drive circuit. This may be programmed to select any azimuth position to a resolution of 1° and automatically changes the direction of rotation of the motor so that the cycle of movement is an oscillatory one from a given position to another given position and back. Between the two extreme positions are a selectable number of positions at equal intervals of azimuth. This prototype is being used to determine the optimum azimuth positions for measurement but lacks positive position registration and is not weatherproof and so cannot be used in bad weather. A further model has been designed to be weatherproof and uses a mechanical detent for each position and this will run continuously.

A simple prediction model

A simple geometric model may be used to calculate the solar radiation income of a slope from a knowledge of the total and diffuse radiation on a horizontal surface and making the assumption that the diffuse radiation comes from an isotropically radiating sky. The total radiation received by the surface G_s is made up of three components, the direct radiation T_s, the diffuse radiation D_s and the reflection from the ground surface facing the slope. T_s is assumed to be given by $F.T_h$ where T_h is the contribution from the direct beam to the total radiation on a horizontal surface. F is the factor which relates the horizontal and sloping surfaces outside the atmosphere and is the integrated value over the period of interest calculated from the geometric position of the sun. With the isotropic assumption for the diffuse radiation D_s and D_h (the diffuse radiation on a horizontal surface) are related by

$$\frac{(1 + \cos(\alpha))}{2}$$

where α is the angle of inclination of the slope, again obtained from simple geometry. If A is the albedo of the surface, that is the ratio of reflected to incident radiation, then the contribution to G_s from the reflection from the ground surface is given by

$$\frac{(1 - \cos(\alpha))}{2} \text{ . A. } G_h$$

where G_h is the total radiation received by a horizontal surface. Thus:

$$G_s = F.T_h + \frac{(1 + \cos(\alpha))}{2} .D_h + \frac{(1 - \cos(\alpha))}{2} .A.G_h$$

but $T_h = G_h - D_h$ and so

$$G_s = (F + \frac{(1 - \cos(\alpha))}{2} .A).G_h + (-F + \frac{(1 + \cos(\alpha))}{2}).D_h$$

This model is critically dependant on the assumption that the diffuse radiation is due to an isotropic sky but with a lack of knowledge of the actual distribution it is the best that can be assumed. The measurement programme being undertaken at Cardiff is expected to indicate how better assumptions can be made to take into account the non-isotropic sky.

Comparison between measurement and model

A typical result of measurements taken with the prototype moving solarimeter is shown in Figure 3. where the full line is the amount predicted by the model described in the previous section. The measurements were taken over four complete cycles at 14.4° intervals of azimuth with the fixed altitude of 45°. With 5 seconds per position the measurement period was thus about 16 minutes. The albedo of the asbestos cement ground surface seen by the solarimeter was measured to be 0.12 but overall the contribution from the reflected radiation was small. Note the systematic discrepancy between the measured and predicted values in Figure 3. which is at its greatest when facing directly away from the sun. This is due to a combination of uncertainty in the exact angle of inclination of the solarimeter and the fact that the direct radiation from the sun was at close to grazing incidence in this position. The errors due to the low incidence angle can be considerable (1) and highlights a difficulty in using the Kipp solarimeter outside the conditions for which it was intended. It may well be necessary in the future to shield the solarimeter from low incidence angle radiation.

Fig. 3
Comparison between measurement and prediction

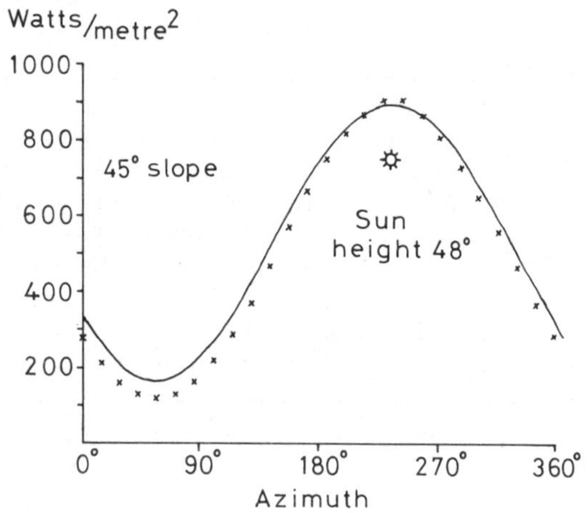

Future programme

The weatherproof stepping solarimeter will be run continuously so as to provide raw data for the purposes of checking mathematical prediction models. At the same time methods of making similar measurements but with a static system so avoiding the problems inherent in the reliability of the motor driven system considered here will be developed. The static system will consist of cheap transducers using a few Kipp and Zonen solarimeters for continual self calibration.

Conclusion

It is perhaps appropriate here to speculate on the future significance of the measurement of solar radiation incident upon surfaces of arbitrary orientation. There is no doubt that as solar energy utilisation becomes better understood and systems become more sophisticated the knowledge of the source of the energy will have to be improved. In particular it will be no longer possible to design on average using measurements of long term averages from one place applied generally. It will be necessary to understand the particular characteristics of the weather of the locality where the solar energy system is to be installed so that it may be tuned for optimum performance. This is particularly important for climates like that of the United Kingdom where the viability of a system could be critically dependant upon the best use of the incident energy for that area. It is unlikely that large centralised collection of solar energy will be made in the United Kingdom but small scale collection for water heating and good design of buildings to use the distributed nature of solar energy could well make a very useful contribution to our energy requirements. For these purposes a thorough understanding of the characteristics of the effect of weather on the incident solar energy on inclined surfaces is necessary, particularly when it is remembered that diffuse radiation contributes 60% of the total radiation on a horizontal surface in the United Kingdom.

The measurement programme being carried out in Cardiff is a contribution to this understanding.

Acknowlegment

The work described in this paper is supported by The Building Research Station, Building Research Establishment, Garston, Herts.

References

1. K.L. Coulson. "Solar and Terrestrial Radiation", Academic Press, 1975.

2. N. Robinson. "Solar Radiation", Elsevier, Amsterdam,1966.

3. - Fourth International Pyrheliometer Comparisons - Results. Working Report No.58, Swiss Meteorological Institute, Zurich, Switzerland, 1976.

4. - Communication from Kipp and Zonen, Delft, Holland.

5. - Communication from U.K. Meteorological Office, Beaufort Park, Bracknell.

6. J.R. Latimer. "Investigation of solar radiation Instruments at the National Atmospheric Centre of the Canadian Meteorological Service", I.S.E.S. International conference, Melbourne, 1970.

7. E. Flowers. "Test and evaluation of the performance of solar radiation sensors at inclination from the horizontal under laboratory and field conditions", NOAA Environmental Research Laboratory, Colorado.

8. D.J. Norris. "Calibration of pyranometers in inclined and inverted positions", Solar Energy, Vol.16, pp.53-55, 1974.

9. J.F. Tricaud. "Contribution à l'estimation des ressources énergétiques solaires ebauche d'un atlas du rayonnement pour la France" These presentee a l'Universite Paris VII, April, 1976.

10. J.K. Page. "The estimation of monthly mean values of daily short wave irradiation on vertical and inclined surfaces from sunshine records for latitudes 60°N to 40°S; Report No.BS32, Department of Building Science, University of Sheffield, July, 1976.

11. B.Y.H. Liu & R.C. Jordan. "Daily insolation on surfaces tilted toward the equator", ASHRAE Journal, Vol.3.pt.10, pp53-59, 1961.

12. D.J. Norris. "Solar radiation on inclined surfaces", Solar Energy, Vol.10, No.2, pp.72-76, 1966.

13. B.J. Brinkworth & T.D.R. Hughes. "A method of obtaining fast response in solar pyranometers", Journal of Physics E: Scientific Instruments, Vol.8, p.902-903, 1975.

General solar energy applications related to building

8

8.1 Solar sustenance project Phase Two: Application of small scale solar greenhouses

William F. Yanda
Executive Director
Solar Sustenance Report, RT1
Box 107AA, Santa Fe, New Mexico 87501

Solar sustenance project phase II

The Solar Sustenance Project, Phase II is a direct action approach to some severe problems facing low-income families in New Mexico. The primary emphasis of the project is to measure heat savings/gain an attached solar greenhouse provides for a home. The twelve experimental units adjoin homes built of various materials in order to record the interaction of each with the greenhouse. In addition, thermally designed window box units (5) were built, tested and available for display.

A benefit derived from the original Solar Sustenance Project is that, because of the demonstration units built, over fifty similar greenhouses were built in north-central New Mexico with owner capital. Phase II expanded this factor by conducting statewide design, construction and operation seminars, utilizing the news media to further its goal. Exposure through media, education and demonstration guarantees major impact for the project.

Another important area of research is to maximize the food and income capabilities of the solar greenhouse. The low-income family has a 'locked in' budget for food that escalates at a greater rate than a 1:1 ratio with rising energy costs. Any vegetable that goes directly from seed to table represents not only immediate family savings, but a measurable energy savings nationally.

Implementation through the workshop approach

The Solar Sustenance Project was the first state supported program in the United States to take a participatory demonstration of solar energy directly into local communities. The emphasis was aimed at the low/fixed income sector, although people of all income and educational levels participated. The twelve demonstration units were assigned to the widely diverse geographical and climatological areas of New Mexico. We have units in areas of low degree days (Carlsbad-2700) up to high degree day locations (Taos-7000). In effect, the only area of the state not within seventy miles of a demonstration solar greenhouse is the extreme northeast (Raton).

The format of the project was as follows:

Interested community organizations (Community Action Programs, local solar energy groups, social service organizations) were contacted by the director and informed of the project and its goals.

A schedule was established.

A package containing organizational plans, materials list, site criteria, media contact information, was sent to the local coordinators. A great deal of the responsibility for the success of the workshop was put into local hands. By doing this, local involvement was stimulated and many of the problems of 'outside' inspired projects were avoided.

At a public meeting, the principles and examples of working greenhouses were explained in an hour and a half lecture/slide show presentation. In five of the communities a representative of the New Mexico Solar Energy

Association also made a presentation. These public meetings were attended by an average of 75 people in the course of the entire project. Besides general education, a primary function of the meetings was to inspire the audience to come out and build the greenhouse the following two days. An unplanned benefit of the public meetings was that in several of the locations (Gallup, Farmington, Portales) the nucleus of a local solar energy association and information exchange forum was established.

Fig. 1
Building on the Navajo Nation Solar Energy Association 'Outreach' project

The greenhouse was built by the participants in the community. During the two day building session many interested non-builders dropped by the site to ask questions and check on the progress of the workshop. As often as possible, the project obtained the names of all attendees. However, many were missed because of the myriad activities taking place simultaneously. It's estimated that an average of 90 persons attended each activity. That would be 1,080 New Mexico citizens directly involved in the program. The end of the two day building phase saw the demonstration units 75-98% complete. The owners of the greenhouses and interested friends completed finishing work, planting layout and planting.

It was not possible to locate a workshop in every community that wanted one. The communities of Roswell, Tierra Amarilla, Ramah, and Raton were not included in the Phase II Project. The director is working with coordinators in these communities for future, New Mexico Solar Energy Association sponsored workshops.

Fig. 2
A clear door under construction, Alamogordo Workshop

A brief survey

One of the primary purposes of the Phase II Project was to determine if a 'multiplier effect' in the private sector can be obtained by government sponsored demonstration solar units being placed in communities. To accumulate data, 100 cards were sent to workshop participants chosen at random. From the return, a proportional sampling of the private building stimulated by the project has been evaluated.

Example of sampling card

Thank you for participating in the Energy Resources Board, Solar Sustenance Project this summer. To determine the impact of our greenhouse workshops we would appreciate your answering these questions.

Yes No

____ ____ I have built an attached greenhouse.

It is____________long and____________deep.

____ ____ I plan to build one this year.

____ ____ I plan to build one when I can.

____ ____ I don't plan to build a greenhouse.

Questions, comments, problems:____________________________

Optional: (Name)________________________Thank you,

(Address)________________________Bill and Susan Yanda

Survey results

Total sample=100 Total participants in project=1,080

Category	of Responses	% of Total Responses	% Sample Factor (10.8)
Have built	13	31	140
Plan to build this year	9	21	97
Plan to build when they can	16	38	173
Don't plan to build	4	10	43
Total responses	42	100	453

Some of the comments by respondees are enlightening:

'I sincerely hope funds are allocated to provide more workshops. I feel this grass roots approach is right on target!' - Steve Meyer, Alamogordo.

'The workshop provided an attempt to realize the practical realities of an attached greenhouse.' - George and Maria Wallace, Sapello.

'Please keep me informed about solar energy. Presently I am renting but I encourage homeowners to build greenhouses.' - Reynaldo Romero, Las Vegas.

'We are interested in solar heating on a scale suited to our modest income. Most projects are unrealistically high for us.' - Ernest E.Shea, Alamogordo. (Plan to build category).

'I am a slave to a garden all summer and decided I did not want to be one all winter.' - No name given, obviously a non-builder.

Regardless of the sampling, it must be assumed that the actual number of privately built units is much higher. Many new greenhouse builders have no contact with the Solar Sustenance Project other than the two books that the director has published. It is estimated that over 300 solar greenhouses have been built in the state in the last year.

Another related effect of the workshops was increased interest by social service organizations in larger scale projects. The director has acted as a volunteer consultant to such groups as the Isleta Pueblo Senior Citizens Center, McKinley Area Services for the Handicapped, and the Canones Community Association. These agencies realize the practicality of combining the multiple use greenhouse with solar heating for their new facilities.

Fig. 3
Greenhouse on mobile home. Built in workshop by Mojave County Solar Energy Commission: Kingman, Arizona.

Many persons, particularly elderly people, cannot build an attached greenhouse by themselves. It was a natural outcome of the workshop format that in several communities (Las Vegas, Alamogordo, Taos) participants who wanted to build for jobs were put in contact with participants who wanted greenhouses built. These small job builders learned greenhouse building techniques at the workshop and have contracted jobs in their own communities as a result of the project. At other workshops (Carlsbad, Shiprock, Albuquerque) C.A.P. winterization crews were trained for their own agencies' future projects.

One final direct result of the Phase II workshop format is that it has served as a model for other states. Nebraska, Idaho, Arizona, and California are using the workshop approach.

Other services and activities

The Solar Sustenance Project with the help of the New Mexico Organic Growers Association constructed a functional, portable solar greenhouse. The 64 ft.2 (5.94 m2) unit is a colorful display that features signs explaining the principles behind the design. It breaks down into five panels which can be easily transported to schools, energy exhibits, and seminars. The greenhouse was first displayed at the New Mexico State Fair where it was examined by over 2,000 people. The visitors were given an explanatory sheet as they asked questions of the attendants. Its next exhibit was at the Energy Seminar at the Albuquerque Convention Center. Here it was shown to over 500 visitors by the project director. At both the State Fair and the energy conference, the unit attracted the attention of Albuquerque TV stations and was given air time in the news segment of their programming. KNME-TV did a 10 minute presentation of the greenhouse in its State Fair special coverage. It was shown at the University of New Mexico Energy Fair in Albuquerque, the Conservation Fair May 14th and 15th in Los Alamos, then at the C.A.P. winterization conference in Santa Fe. The portable greenhouse was in operation at the home of Ray Alfini in Albuquerque all winter, providing that family with free heat and vegetables.

Fig. 4
Portable display unit exhibited at State Fair.

The unit will be used by the New Mexico Solar Energy Association in its Outreach program.

The project director made himself available for free lecturing and consulting services for New Mexico organizations of many varied interests. Examples are: The Los Alamos Garden Club, Society of American Foresters, La Clinica de la Gente, and New Mexico Department of Development. Emphasis was also placed on the Solar Sustenance Project acting as a communications coordinator between organizations and individuals seeking professional help in solar applications and qualified engineers, architects, and builders.

One problem the potential greenhouse builder had was in obtaining greenhouse quality fiberglass, polyethylene and accessories. For that reason, the project established an outlet for these materials. The specialty items sold were not in competition with any existing private business and were sold for a small mark-up to cover the rental of the storage facility and distribution costs. It is interesting to note that since August 1st, 1976, 14,300 ft.2 (1328.5 m2) of fiberglass glazing have been sold. This amount, if the ratio of 1.76:1 (area clear: floor area) is used, would cover 8125 ft.2 (754.8 m2) of greenhouse, or, 51 greenhouses at 160 ft.2 (14.86 m2) each. This is some indication of the amount of building going on in the Santa Fe area alone. There are now new private fiberglass distributors in Las Vegas, Silver City, and Albuquerque.

Produce from the greenhouse

The owners of the eleven attached greenhouses built have all expressed great satisfaction with their greenhouse, though some have harvested much more produce from their unit than others. This has resulted from the fact that some owners took longer to learn the workings and operation of a greenhouse or were not able to spend as much time in it. The owners who had the most success spent an average of five to seven hours a week in the greenhouse. All have commented on the fantastic taste of a vegetable picked fresh in mid winter.

Though one person said her greenhouse 10' x 20' (3.05 x 6.10 m) produced all of the vegetables a family of four plus guests could eat, on the average each project greenhouse produced 40% to 60% of the family's vegetables. In two cases, the greenhouse supplied all of the family's vegetables throughout the winter. The highest monthly average of food production in the project was reported by the Alamogordo unit at a $40.00 saving a month (1976-77 fresh food prices). An average for the total units in the project is $25.00 a month. Though some owners did experiment with more exotic fruits and vegetables, on the whole, everyone in his first year with the greenhouse planted the more common ones: lettuce, spinach, Swiss chard, peas, broccoli, green onions, corn, radishes, beets, carrots, various herbs, and fruiting vegetables such as tomatoes, cucumbers and peppers.

Figs. 5,6

Dwarf corn in the Anton Chico greenhouse, late April.
'I though corn this sweet only grew in Iowa.' Rebecca Chavez.

Albert Martinez displays lettuces from his Taos Project greenhouse.

An important aspect of greenhouse growing, which cannot be overstated, is using the greenhouse throughout the summer period. Much of New Mexico, the Rocky Mountain area, and the northern Midwest experience a short growing season and very cool summer nights. These summer night-time temperatures, often into the low 50's (10°C) in New Mexico, are hardly conducive to high production from fruiting vegetables. The well run summer greenhouse not only avoids the ravages of hail and drought, but puts fruiting vegetables into an environment which can keep an optimum mid 60's (16°C) low temperature range.

The greenhouse must also be looked at in light of its water conservation over field crop conditions. Authorities report water usage for greenhouse crops to be 1/10 to 1/30 of the field crop. (8) Two of the project owners, Tom Rolf of Silver City and the Chavez family of Anton Chico, devised simple systems to trap rain water and snow melt from the roofs of their homes, drain it into tanks in the greenhouse, and gravity feed the water to their plants.

'.... less than $\frac{1}{4}$ of the energy consumed in the U.S. food system is used to actually produce the food. The remaining $\frac{3}{4}$'s are used to transport, process, preserve and distribute it.' (9) There is not only a substantial money savings in growing one's own vegetables year round, which is important for low and fixed income as well as middle income people, there is also a substantial energy savings. It is important to realize that a dollar's worth of December tomatoes or lettuce produced in the home greenhouse garden is an actual 75¢ saving in petrochemical energy.

Window boxes

Five window boxes of various designs were built and displayed at the Albuquerque Energy Fair. The one pictured turned out to be the most successful design. In it, the owner figures she produced 60% of the vegetables she ate; this included a salad every day. Fruiting vegetables were put into the window box in February. The owner notes success with the following crops: Tiny Tim tomatoes, California Wonder bell peppers, Chinese cabbage, chop suey greens, spinach, Black Seeded Simpson lettuce, water cress, sorrell, parsley, rosemary, chives and basil.

The window box units, as they were built in the project, do not substantially provide more heat for the home than do the window they cover. However, they do give the owner space for a mini garden. One problem encountered in testing the units was that it is impossible to equip the hanging window box with enough thermal mass to prevent it from freezing on a cold winter night (without leaving the window to the home open). It is suggested that a clear enclosure could be built under and adjoining a ground floor window box. This small enclosure would have a double clear southern glazing and openings for air circulation to the window box above. Within it could be a black 55 gallon (208 liter) drum. This passive storage container would make the window box capable of sustaining low temperature winter nights without supplementary heat from the home.

Fig. 7
Window box in use
at Santa Fe home

Economics and cost analysis

Many people, particularly senior citizens and families where both adults work, do not have the time resources to build their own greenhouses.

In an attempt to get a grasp on the economic impact of a contract-built, attractive greenhouse, the project had Mr. Michael Coca submit a simplified analysis prepared in conjunction with his economics class at Highlands University. The analysis was done independently of the Solar Sustenance Project, although some of our estimations were used. The data obtained 'fit' our recorded examples in costs and heating potential of the small greenhouse. The vegetable production is slightly lower than we experienced. Note that interest rates are not included in the costs of building the greenhouse, but neither are increases in fuel costs. It is suggested that one might offset the other.

Excerpts from Mr. Coca's report:

'Questions most asked about this type of application are usually related to cost, curiosity about the point in time when accumulated savings in food and heat equal the total cost of the structure, and whether this solar application is a viable alternative to rising food and energy prices.

'A breakdown of cost on a greenhouse this size would look something like this:

Materials - at $2.50/ft.2 ($26.92/m2) x 160 ft.2 (14.86 m2) =	$400.00
Labor - 160 man hours at $4.00/hr.=	640.00
Total cost of greenhouse	$1,040.00

'An analysis of heating fuel cost for the above home indicated an average monthly natural gas bill of $30.00 per month or $360.00 per year. It was estimated that the greenhouse would save approximately 30% or $98.00 of yearly fuel cost.

'Food cost for a family of five runs about $200.00 per month or $2400.00 per year. Of this total, the following vegetables and herbs could be started in the greenhouse:

Description	Quantity Consumed in One Year	Total Yearly Cost
Lettuce	1 head per week at $.35	$ 18.20
Tomatoes	2 lbs. per week at $.60	62.40
Chili (green)	50 lbs. at $10.00	10.00
Chili (red)	3 lbs. at $2.00 per lb.	6.00
Cabbage	$.13 per lb. x 60 lbs.	7.80
Broccoli	$.59 per lb. x 78 lbs.	46.02
Cauliflower	$.59 per lb. x 78 lb.s	46.02
Spinach	$.35 per bunch x 104 bunches	36.40
Misc. herbs		20.00
	Total	$252.84

Chavez Greenhouse: February 3-7, 1977 - No Storage Averages

'In review, a greenhouse could supplement approximately 11% of the total yearly food expense for a family of five.

'A look at annual savings in fuel and food shows the following:

Annual savings in fuel	$ 98.00
Annual savings in food	252.84
Total annual savings	$350.84

Fig. 8

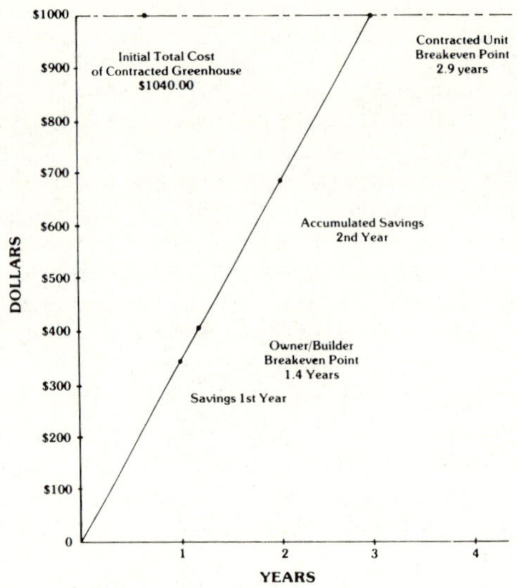

Breakeven point between initial total cost of a solar greenhouse and the yearly accumulated savings.

'Projecting these savings and initial greenhouse cost onto a graph, we can see how many years it takes to break even on the initial investment of a greenhouse.

'Reviewing the above graph, we find that by investing $1,040.00 in an attached greenhouse, a family can accumulate enough savings in 2.9 years to pay for the structure. The breakeven point may be achieved sooner if we assume that food and fuel prices will continue to rise or if a family is willing to do the labor and thus save more than one-half of the cost.'

Solar building impetus has been given to the New Mexico citizen by the State Legislature in the form of a tax credit for solar applications. A passive system workform is given in Appendix 1, for the use of anyone who doesn't know of its existence and those readers in other states looking for help in constructing their own laws.

The actual materials costs of building the 12 demonstration units were compiled by totaling all of the local purchasing receipts and adding on the retail costs of the greenhouse glazings brought to the sites by the project. The total cost was $6,473.32. The total square footage of the 12 project greenhouses was 2,064 (192m2). This gives an average square foot cost of $3.14 ($33.76/m2). The range between low and high costing greenhouses (of the same 160 ft.2 size) was considerable. At the low end we have the Albuquerque unit with a total cost of $394.61, or $2.46/ft.2 ($26.55/m2). At the high end was the Las Vegas greenhouse at $651.74, or $4.07/ft.2 ($43.85/m2). This disparity was a function of: (a) local retail prices; (b) the ability of local coordinators to get bargains and astutely use the suggested materials list to account for individual differences in their site. It should be mentioned that at <u>all</u> sites surplus building materials were left to be used in local projects relating to solar greenhouses. They are included in the average materials costs. Feedback from many owner/builders indicates that $2.50/ft.2 ($26.91/m2) materials cost is an accurate 1977 estimate of the passive, attached lean-to greenhouse. Unlike higher technology, mass produced systems, it is not expected that the cost of this type of application will drop in the future. It will rise reflecting inflationary trends in the building supplies industry. For example, the materials cost has risen 34% since we started building basically the same type units in 1974. The potential owner really has no excuse to wait for further developments.

The Griffin home

The house

A 22 year old frame structure, 1115 ft.2 (103.6 m2), 2 x 4 walls with a flat roof of approximately R-8 insulation. The home has all steel sash windows. It is located in a residential area and has considerable tree and building cover from wind. Heating is by two independent floor furnaces. Owners have reduced thermostat settings by 2° since March 1974. Cooking, heating and domestic hot water are by natural gas.

Add-on greenhouse

Located in a southeast corner of the building. It has an 11' long x 8' high (3.4 x 2.4 m) southerly clear wall. (All of the clear surfaces have outer fiberglass/inner polyethylene glazings.) East greenhouse wall is 12' long x 8' high (3.7 x 2.4 m) and is clear. Clear roof section faces south and is 6' x 11' (1.8 x 3.4 m). Total floor square footage of greenhouse is 120 ft.2 (11.14 m2). A 15' (4.6 m) high apricot tree is 5' (1.5 m) southeast of the southern corner of the greenhouse. The greenhouse covers 3 home windows and a standard size door, for a total of 42.5 ft.2 (4.0 m2) of openings to the home. The unit is operated in a 'open to home during day/closed off from home at night' mode. The owner built the greenhouse entirely by himself for a total cost of $\not{s}$127.55. He is a scrounger, par excellence.

Add-in wood stove

This is a 'Contemporary Franklin' design. It is located in the living room in the western part of the home. The stove has a thermostatically controlled blower that custs on at 150 F. (65.55 C) and blows hot air directly into the living room. It was purchased for $\not{s}$227.36, a wholesale cost.

Some observations about the chart

The first thing that becomes apparent when one examines the chart is the pattern of fuel consumption. Heavy usage begins in January and continues through May. May usage is higher than December usage. This is probably due to several factors: (a) the house, even though it is frame (not a massive structure) is carrying a 'thermal charge' well into the heating

Fig. 9
Fuel consumption in a Santa Fe home before and after add-ons.
(Solar greenhouse and wood stove)

Date	MCF Units	Date	MCF Units	Base (73-74 Ave. MCF)	Date	MCF Units	% Reduction from 73-74 base	Date	MCF Units	% Reduction from 73-74 base
11-73	9	11-74	7	8	11-75	7	12	(11-76)	6	25
12-73	12	12-74	14	13	[12-75]	10	23	12-76	7	46
1-73	19	1-74	26	22.5	1-75	17	24	1-76	13	42
2-74	25	2-75	29	27	2-76	19	29	2-77	13	52 [1]
3-74	24	3-75	19	21.5	3-76	14	35	3-77	9	42
4-74	18	4-75	17	17.5	4-76	11	37	4-77	8	54
5-74	16	5-75	16	16	5.76	10	37	5-77	7	56 [2]
6-74	8	6-75	8	8	6-76	8	0	6-77	8	0
7-74	5	7-75	6	5.5	7-76	6	-9	7-77	6	-9
8-74	4	8-75	4	4	8-76	4	0	8-77	4	0
9-74	4	9-75	3	3.5	9-76	5	-30	9-77	5	-30
10-74	6	10-75	5	5.5	10-76	5	9	10-77	5	9
	150		154	152.0		116	24		91	40

SAVINGS AT MARCH 1977 PRICES:
152-116 = 36 MCF x 1.74 MCF = $62.64 (75-76)
152-91 = 61 MCF x 1.74 per MCF = $106.14 (76-77)

OWNERS—Jerry and Sandy Griffin: data recorded by Sandy Griffin.

▭ Add on Greenhouse
◯ Add in Wood Stove

[1] Stove shut down.
[2] End of recorded data. Assume 75-76 readings.

season. It is a slab on grade and the earth below the home is, perhaps, still warm into January; (b) when the house loses its thermal momentum it takes a long time, into June, to regain it; (c) the high winds of springtime contribute more to the heat loss on this home than the cold temperatures in December. This pattern of fuel consumption should be examined and documented in great detail for the Southwest; it could be a critical factor in home design. We have been looking at degree days as a guide to design, and actual fuel usage and energy consumption may be more important and quite different.

When the heating load becomes heavy (February is consistently the highest month), the greenhouse is going into its own heavy surplus heat-available period, Figure 16. That, of course, becomes greater as we go into March, April and May. It was noted by the owner that no fires were made after 'late February', so we might consider the March through May reductions in 1977 attributable entirely to the greenhouse and a milder spring than previous years. The owner stated that 'in 1975-76 we only opened the greenhouse to the bedroom because we didn't think it would do that much. In 1977 the greenhouse was open through all windows and doors into June.' The results of this operating mode are apparent in the March through May 1977 fuel reductions.

Please note that this greenhouse is not in what would normally be considered an 'optimum' location or geometry, i.e., it is southeast and 'boxy' as opposed to due south and rectangular. To further complicate matters, it is partially shaded until noon in the winter by the apricot tree on its southeast corner. Its energy saving contributions are perhaps due to three factors:

— There is direct gain and continuous daytime airflow to the home.

— It covers 42.5 ft.2 (3.9 m2) of windows and a door of steel sash (and old) construction; if the greenhouse were not there, these surfaces would be contributing to high conductive and infiltrative losses; the greenhouse is also serving as an 'air lock' to the home for 50% of the daily entries and exits.

— As the greenhouse is used as a garden, it is providing the home with high humidity air. The regular home furnaces and the wood stove are dry, forced air systems. The greenhouse is, in effect, providing the other heaters with air which has a greater capacity to store heat. This is an extremely important factor along with the 'buffer effect' of the attached greenhouse. These two complicated effects must be thoroughly examined before anyone will truly understand what an attached greenhouse is doing for a home. There is much more than direct gain going on in these systems.

There are probably subtle lifestyle changes and energy awareness taking place in the family over the last several years. However, they were not given as 'major' or even 'notable' by the operators. Mr. Griffin stated, 'We were energy conscious in 1973 before I ever heard of solar greenhouses.'

Thermal performance

Data used: Eversole greenhouse, Alamogordo, N.M., January and February 1977 -
Chavez greenhouse, Anton Chico, N.M., February 3-7, March 17-21, 1977 -
Roger's heat loss analysis, September 15, 1976

Table 1 summarizes the temperature record of the Eversole greenhouse. Several points of interest emerge. The greenhouse maintains a mean temperature 22°F (12.2°C) above outside air temperature. However, the minimums are only 19°F (10.5°C) higher, while maximums are 24°F (13.3°C) higher. This is due to inadequate storage capacity. The greenhouse tends to overheat during the day, which increases the daytime heat loss; thus less heat is available to offset night-time heat loss. This pattern appears to be typical of many add-on greenhouses, indicating the need for more storage capacity in the cystem.

The amount of temperature variation or temperature range is an important characteristic of the system. The temperature range inside is 26°F (14.4°C) while the outside air temperature range is only 21°F (11.7°C). This is also explained by the low storage characteristics of the system.

Table 1 Eversole Greenhouse: January-February 1977 Averages

Temperature	Maximum	Mean	Minimum	Range
Inside	78.8 (26)	65.6 (18.7)	52.4 (11.3)	26.4 (14.7)
Outside	54.5 (12.5)	43.8 (6.6)	33.2 (0.7)	21.3 (11.8)
Inside-Outside	24.3 (13.5)	21.7 (12.05)	19.2 (10.7)	5.1 (2.8)

Table 2 summarizes the temperature record of the Chavez greenhouse for the period February 3-7, 1977. This structure maintains a mean temperature 29°F (16.1°C) above outside air. Comparison of internal and external maxima and minima reveals a very similar pattern to the Eversole greenhouse. During this period this greenhouse also tends to overheat during the day and underheat at night. The inside temperature range is 16°F (8.9°C), or 162% greater than outdoor air.

Table 2 Chavez Greenhouse: February 3-7, 1977 - No Storage Averages

Temperature	Maximum	Mean	Minimum	Range
Inside	80.4 (26.9)	59 (15)	37.7 (3.2)	42.71 (23.7)
Outside	42.8 (6.0)	29.6 (-1.3)	16.4 (-8.7)	26.41 (14.7)
Inside-Outside	37.6 (20.9)	29.4 (16.3)	21.3 (11.9)	16.3 (9.0)

Table 3 summarizes performance of the Chavez greenhouse during March 17-21, 1977. The pattern here is quite different. The temperature differences between inside and outside are 26°F (14.4°C) for the maxima and 31°F (17.3°C) for the minima, thus reversing the February data. The interior range is also less than the exterior range. This is explained by the addition of eleven 55-gallon (208 liter) drums of water for storage. The performance improves dramatically while maintaining the same temperature elevation.

Unfortunately, no hard data is available on the exact amount of heat transfer from the greenhouse to the house. Opening and closing of doors and vents to the exterior and opening and closing of doors and windows between the house and greenhouse were not recorded. It appears from the temperature records that both the Eversole and Chavez structures were operated with venting to the interior during the day which was closed at night.

Table 3 Chavez Greenhouse: March 17-21, 11-Barrel Storage Averages

Temperature	Maximum	Mean	Minimum	Range
Inside	82.2 (27.9)	66.3 (19.0)	50.3 (10.2)	31.9 (17.7)
Outside	56.7 (13.7)	38.2 (3.4)	19.7 (-6.8)	37.0 (20.6)
Inside-Outside	25.5 (14.2)	28.1 (15.6)	30.6 (17)	-5.1 (-2.9)

Graphs for extracted data are shown in Figures 12, 13, 14 and 15.

Several items become apparent when examing the graphs.

Total solar radiation has increased 30-40% in the March period but the interior daytime high temperatures have not risen accordingly. This is explained by: (a) the increased thermal mass; (b) the geometry of the unit, i.e., the south face is reflecting the higher altitude sun and the partially clear roof is beginning to shade the back part of the greenhouse; (c) increased vegetation and water inside the greenhouse are having an evaporative cooling effect.

The 'thermal wave' through the wall is flatter and more regular in the March reading. The interior channel, 1, has risen in average about 2°F (1°C). The occupants of the home are rarely using any supplemental heat

Fig. 10
Bernardo Chavez and Anton Chico greenhouse/house

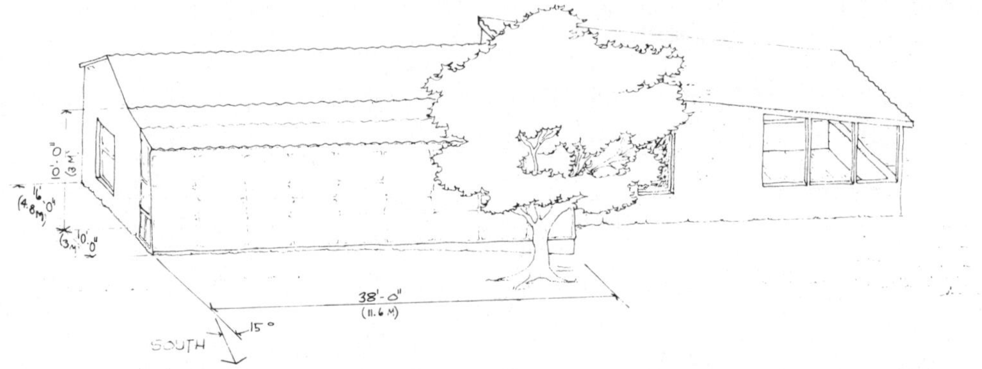

The Chavez greenhouse in Anton Chico, New Mexico, was monitored by Los Alamos Scientific Laboratories Solar Division with a 16 point hourly recording unit. The home is an old adobe structure with 16" (.4 m) thick walls. The greenhouse covers one window and one door. It is 36 ft. (11 m) long. We can look at the structure before and after the addition of eleven 55 gallon (208 liter) water drums.

Fig. 11
Bernardo Chavez greenhouse

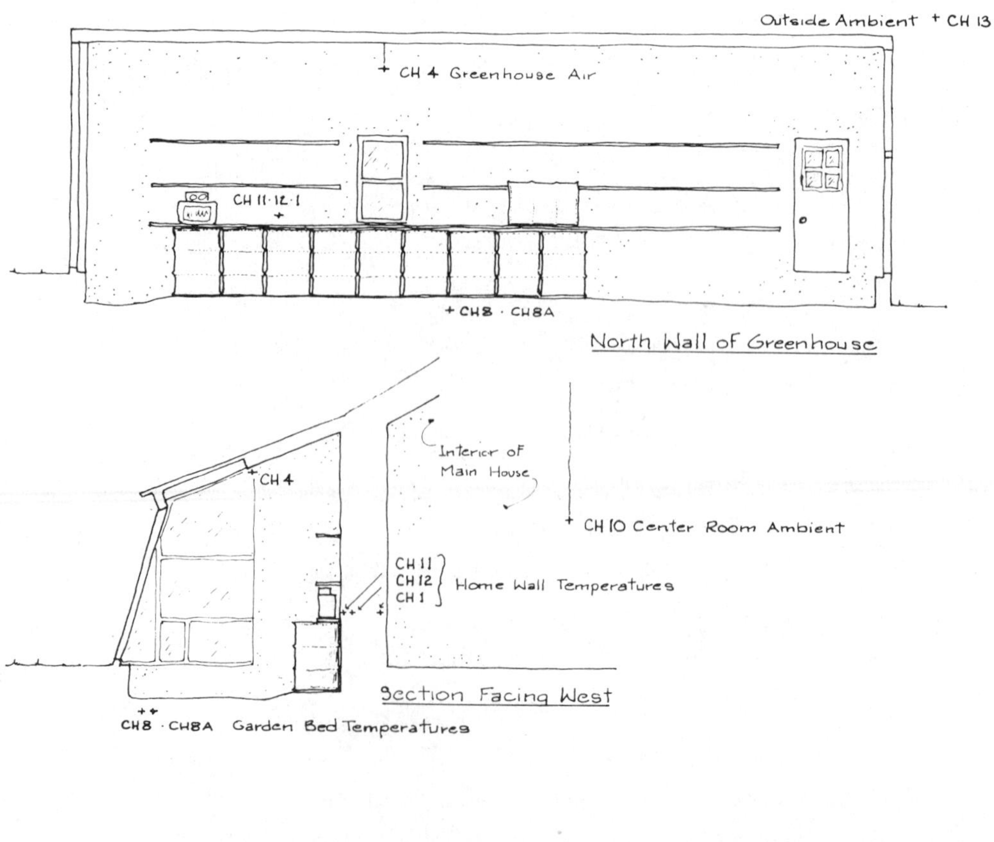

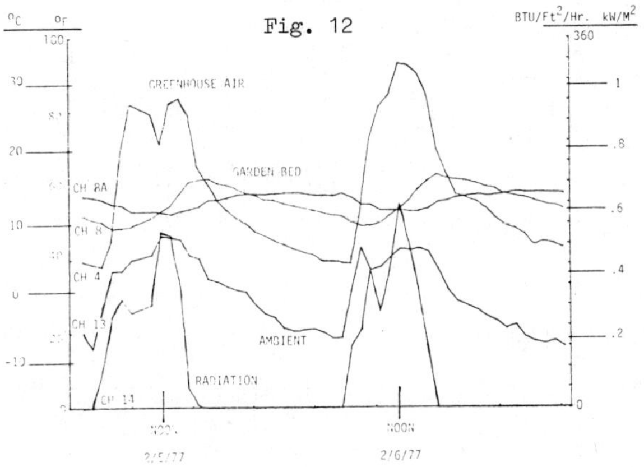

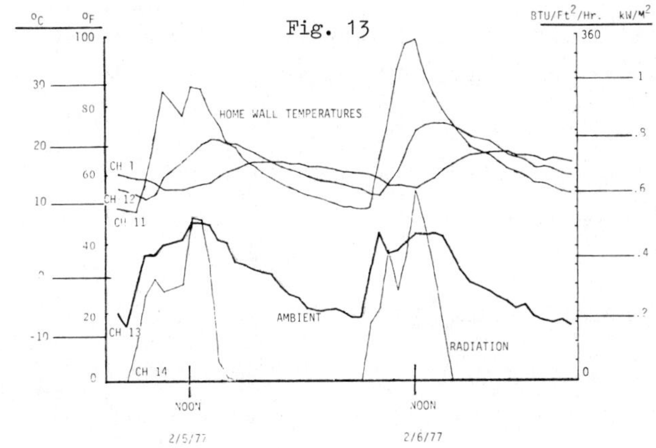

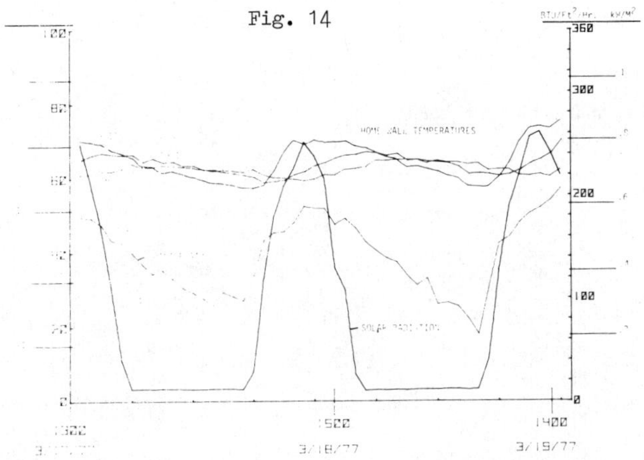

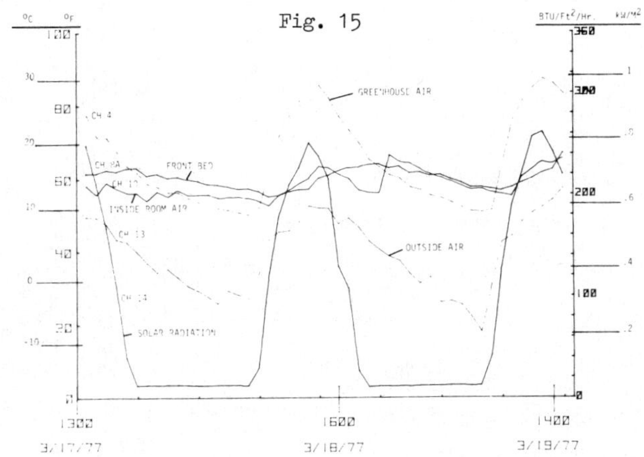

Figs. 12, 13, 14, 15 Bernardo Chavez and Anton Chico greenhouse

within the dwelling. (See inside room air in Figure 14). The greenhouse may be considered the sole heating unit for this part of the home. The massive contiguous wall is a radiant and conductive heater.

In both measured periods the ground beds used for growing constitute thermal storage for the greenhouse. Whenever the interior air temperatures drop below the earth temperatures they help heat the space.

The potential heat available from a greenhouse

The amount of heat which would be available for transfer given optimum storage capacity may be estimated by determining the amount of absorbed solar energy and subtracting from it the heat lost from the greenhouse to the outside air. For a typical add-on greenhouse design using the dimensions 10' (3.04 m) deep by 16' (4.9 m) long and the clear area: floor area ratio of 1.78:1, total heat loss is about 4,200 BTU/Degree Day (7968 kJ/C Day). Data from The New Mexico Solar Resource for Albuquerque gives a heat loss of 126,000 BTU (132,804 kJ) and a heat gain of 277,000 BTU (291,958 kJ) with the greenhouse maintaining a 30°F (16.7°C) temperature elevation and the house maintaining a 33°F (18.3°C) temperature elevation. These are averages for the months of December, January and February. Thus, on the average, some 151,000 BTU (159,154 kJ) is available to the house providing that the heat can be stored when not needed, and released on demand.

Fig. 16

Solar Availability Per Day

	FALL		WINTER		SPRING	
	BTU/Ft.^{2}	kJ/M^2	BTU/Ft.^{2}	kJ/M^2	BTU/Ft.^{2}	kJ/M^2
South 75° tilt:	1806	20,490	1870	21,215	1521	17,256
East & West:	887	10,063	697	7,908	1173	13,308
Roof 15° tilt:	1870	21,215	1521	17,256	2377	26,967

Solar Gains Per Day

	FALL		WINTER		SPRING	
	BTU	kJ	BTU	kJ	BTU	kJ
Total Available:	429,576	452,773	408,036	430,070	530,885	559,553
X .68 transmittance:	292,112	307,886	277,464	292,447	361,002	380,496

Greenhouse Loads Per Day

	FALL		WINTER		SPRING	
	^{o}F. Aver.	^{o}C. Aver.	^{o}F. Aver.	^{o}C. Aver.	^{o}F. Aver.	^{o}C. Aver.
Outdoor Mean	57.6	14.2	37.1	2.8	55.6	13.1
G.H. Mean	67.1	19.5	67.1	19.5	67.1	19.5
ΔT	9.5	5.3	30.0	16.7	11.5	6.4
Total Load	39,900 (BTU)	42,055 (kJ)	126,000 (BTU)	132,804 (kJ)	48,300 (BTU)	50,908 (kJ)

Surplus (For Home) Per Day

FALL		WINTER		SPRING	
BTU	kJ	BTU	kJ	BTU	kJ
252,212	265,831	151,464	159,643	312,702	329,588

Will Heat

421 Ft.^{2} (39.1 M^2) - House load at 12 BTU/^{o}F.D. - Ft.^{2} (245 kJ/^{o}C.D. - M^2)
631 Ft.^{2} (58.6 M^2) - House load at 8 BTU/^{o}F.D. - Ft.^{2} (163 kJ/^{o}C.D. - M^2)

Structure - add on greenhouse/location - Albuquerque

Roof glazing area = 4 x 16 = 64 ft. or 1.2 x 4.9 m = 5.9 m

South glazing area = 16 x 9.3 = 149 ft. or 4.9 x 2.8 m = 13.8 m

East and West area = 46 ft. (4.3 m)

Occlusion Ratio - 90%

Load: Based on A.S.H.R.A.E. Heat loss calculations and 1 air exchange per hour = 4200 BTU/DD (7968 kJ/CD)

Transmissivity - 77%

90% x 77% = 70% effective gain multiplier

An add-on solar greenhouse of 160 ft.2 (14.8 m2) has the potential to heat some 400 to 600 ft.2 (37.2-55.7 m2) of living space on the average.

With light construction such as wood-frame or mobile home this is not possible without the construction of some storage element inside the house itself. For that reason, a greenhouse linked to thermal storage below a mobile home is shown in Appendix 2.

This scheme would also work for a frame house with a crawl space beneath the living area.

There is a great need for measurement of the amount of heat transfer which actually occurs with window and door type openings and the improvement possible with use of fans and better storage devices which interface directly with the house itself.

Fig. 17 Survivors (builders and film crew) of the New Mexico Solar Energy Association Workshop: 1372½ Cerrillos Road, Santa Fe. The solar greenhouse is available for public tours.

Appendix 1

Passive system workform

Introduction

Section 72-15A-11.3, N.M.S.A. 1973 provides for a tax credit of 25% of the equipment cost of solar heating a home. If you have used or are going to use an active system to heat your house, the equipment cost is easily isolated from the cost of the house. Active systems are those which use a heat transfer fluid such as air or water to take heat from a collector and move it to storage. They often use pumps, fans, ductwork, etc. The total cost of the solar heating equipment up to the point where it ties in with a conventional heating system can be counted for the tax credit.

Passive systems are a different matter. They are systems that use the architecture of the building itself to catch and retain the solar heat. Difficulty is often encountered when we try to separate those architectural elements that are particular to the heating system and those which just make the building. In order to evaluate the solar heating system cost in passive house designs, the following procedure has been devised by Mr. M.L. Morton of the State Bureau of Revenue in cooperation with the NMSEA, officials of ERDA, Sandia Labs, and LASL. Each step of the work form has an explanatory paragraph. These paragraphs are grouped together at the end of the form.

Explanatory notes

1. Use floor and ceiling levels because a building with a high floor level above grade or a high parapet would give a misleading outcome on this percentage. We are concerned with the heated volume of the building.

2. This is the open glass area that actually admits energy.

3. This step separates the amount of your South Windows that are especially solar from the amount you would probably have had anyway.

4. You may count the glass, glazing strips, fasteners, sealants, hardware and framing (except for the structural framing of the building).

5. (a) This includes beadwall, styrofoam panels, etc., and their hardware.
   (b) These would be water containers, rock bins, etc.
   (c) Count skylights only if they are provided with movable insulation because without such insulation the skylight will probably lose more heat than it gains. You only count 50% of the cost because you would have to roof the building with something anyway.
   (d) This includes fans, controls, ductwork up to the point that it ties in with a conventional heating plant, etc.

General notes to the designer

Since labor and installation costs may not be counted, it would be best to use as many prefabricated units as possible. Ready made window units for example would be more expensive than glass plus wood strips but may well be cheaper when labor costs and the 25% rebate are considered.

Clerestory windows would be better than skylights for four reasons: You can only credit yourself for 50% of the skylight cost; Vertical clerestories catch more winter sun than low sloping skylights; They are easier to glaze; They are easier to shade from summer sun.

Workform

1. Consider the wall area between ceiling level and floor level of the North, East, and West walls of the house. Determine the percentage of window area within this band. ______ N.E.W. WINDOW

2. Calculate the percentage of window area on the South side. ______ SOUTH WINDOW

3. Subtract N.E.W. percentage of STEP 1 from South percentage of STEP 2.

   ______ % (SOUTH) - ______ % (N.E.W.) = ______ SOLAR WINDOW

4. Multiply this 'Solar Window Percentage' by the materials cost of the South Windows.

   ______ % (SOLAR WINDOW) X $ ______ (MATERIALS COST) $ ______ SOLAR WINDOW COST

5. To this Solar Window Cost, Add:
   (a) Cost of movable insulation for your South Windows. $ ______
   (b) Cost of any architectural elements that are particular to solar heating. $ ______
   (c) 50% of cost of skylights if they have movable insulation. $ ______
   (d) Cost of hardware especially for the solar system. $ ______

   TOTAL OF 4 AND 5 $ ______

This total amount is used in line one of the New Mexico State Bureau of Revenue form PIT-16 'Application for personal income tax credit for solar heating/cooling equipment purchase.'

Appendix 2 Greenhouse: Heated mobile home

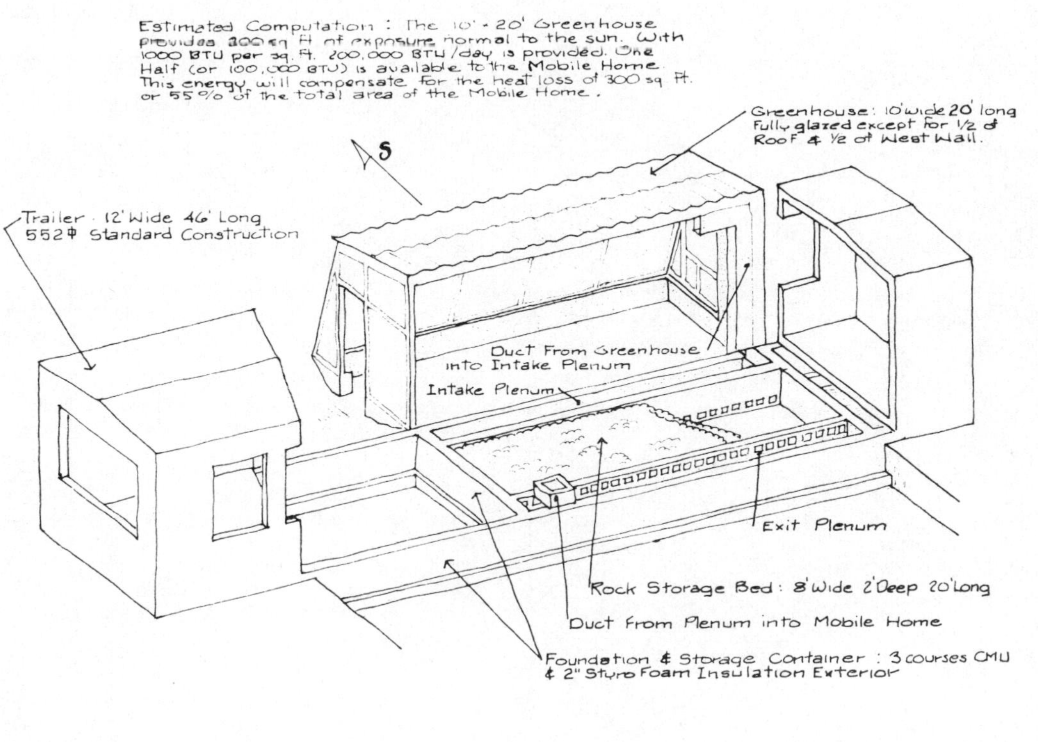

Suggested reading list

1. Abraham, G. and Abraham, K., 'Organic Gardening Under Glass', Rodale Press, 1975

2. Anderson, B., 'The Solar Home Book', Cheshire Books, 1976

3. Boes, E., 'The New Mexico Solar Resource', New Mexico Energy Resources Board, 1977

4. DeKorne, J.B., 'The Survival Greenhouse', The Walden Foundation, 1975

5. Fisher, R. and Yanda, B., 'The Food and Heat Producing Solar Greenhouse: Design Construction and Operation', John Muir Publications, 1976

6. Hoff, E., Jenkins, D. and VanDuyn, J., 'Noti Solar Greenhouse: Performance and Analysis', Center for Environmental Research, University of Oregon, 1977

7. Yanda, S. and Yanda, W.F., 'An Attached Solar Greenhouse', The Lightning Tree, 1976

8. Wartes, Lloyd, 'The Solar Greenhouse: A Means to Increased Food Production', Solar Engineering Magazine, October, 1976

9. Lerza, Catherine, and Jacobson, Michael, editors, 'Food for People not for Profit', Ballantine Books, New York, 1972.

Acknowledgements

Our special thanks go to: New Mexico Solar Energy Association, Los Alamos Scientific Laboratories, Jerry and Sandy Griffin, Larry Palmiter, Bristol Stickney, John Throne, Michael Coca, Jay David, Claude Phipps.

Greenhouses orientated to the sun (for more and early production)

A. Vatansever.
Findikzade, Sair M. Emin Sok.
No. 4/7 Capa - Istanbul,
Turkey.

Abstract

All human beings need free energy in order to prolong their life; and the origin of that energy is the SOLAR ENERGY. The Photosynthesis incident encountered in plants is an incident whereof the irradiation energy of the sun is transformed to chemical energy. The proportion of development and productivity increases as much as a plant receives solar radiation.

Based on the DARWIN THEORY principles, I have improved for Turkey, a type of a greenhouse by taking advantage of the ability of plants to orient to the sun (Photo Tropism).

Fig. 1

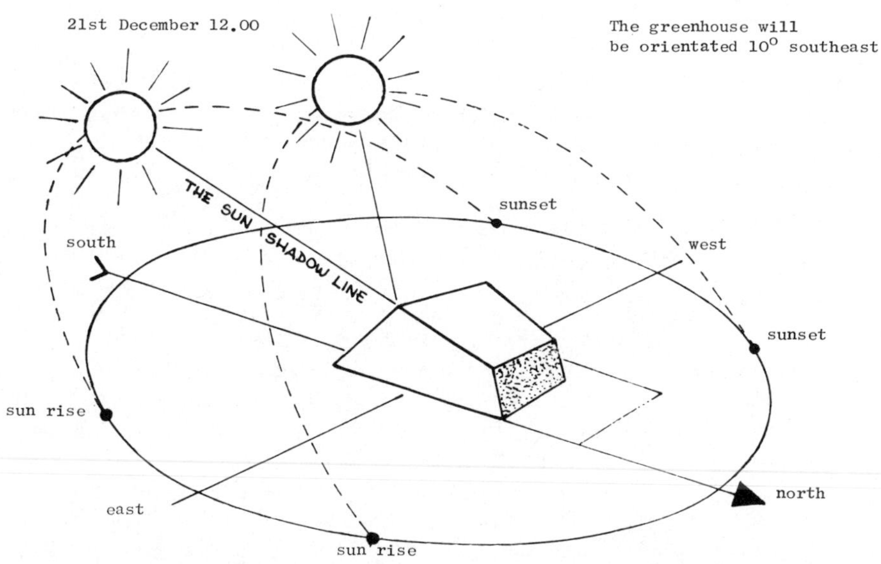

At the Equator, the crown of trees are horizontal. The further from the Equator, i.e. towards the south of the northern hemesphire. The angle of inclination which is 0° at the Equator increases and reaches its maximum value of 23.5^o on the polar circle which is the angle of inclination of the global axis of the sun. The shape of the solar surfaces also changes with the geographical latitude. It is almost flat at the Equator, and convex on the polar circle. The angle of inclination of the crown in Istanbul is 5° (Fig. 2).

Fig. 2

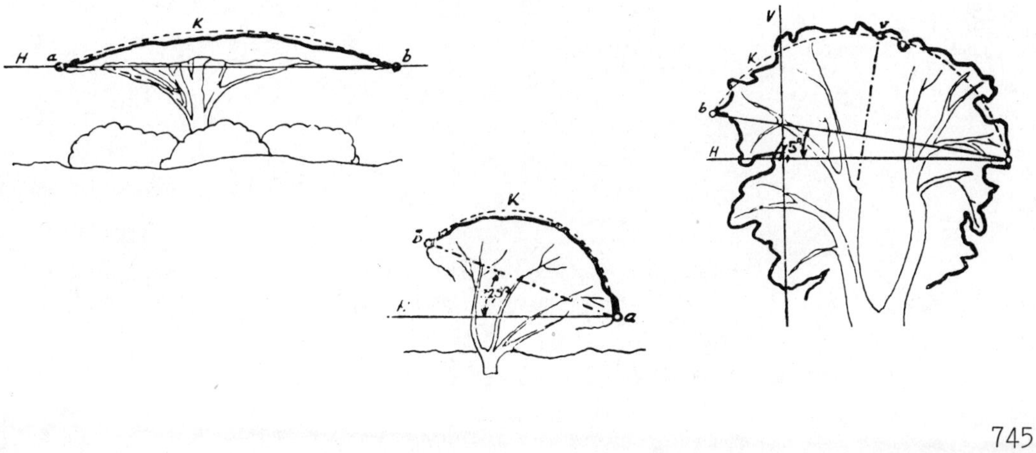

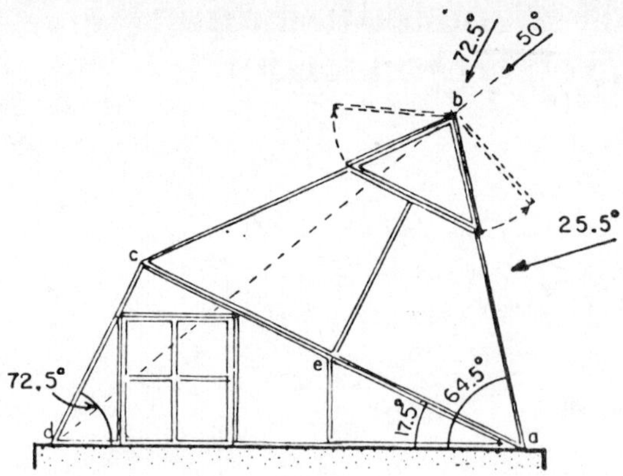

Fig. 3
Greenhouse for
Istanbul.

For the design of greenhouses in Istanbul (41°N), a variation in the elevation of the sun from 50° at the Vernal Equinox to 72.5° in the summer and a crown inclination of 4.5° had to be taken into consideration.

Design of Solar-Orientated Greenhouses

The asymmetric structure of the roof of the greenhouses orientated to the sun allow them to intercept more sunlight than normal greenhouses during the last months of autumn and first months of spring.

Thus, they function at a warmer temperature when auxiliary heating devices are not used. Further more, these asymmetrical roofs orientated to the sun protect the plants from excess solar radiation by reflecting more of the solar rays at noon when the sun is at its maximal elevation. This keeps the greenhouse cool. The shape and inclination of the greenhouse roofs will vary at different latitudes.

The roof of the greenhouse orientated to the sun in Istanbul is perpendicular to the direction of solar rays at noon at the time of the lowest sun elevation, thus transmitting more radiation in winter than the usual greenhouses (Figure 3). The axis diving the upper angle of the roof into two halves is parallel to the direction of solar rays at midday at the time of maximum sun elevation. Therefore, the solar roof reflects in summer more radiation, and protects the plants from excessive heat more efficiently than the roofs of normal greenhouses.

It was found by the comparison of temperature that the greenhouse is warmer in winter and cooler in summer than the usual greenhouses. Long term tests indicate that the average incident radiation transmitted through the asymmetric roof is 75 to 80 percent. Thus, such greenhouse will shorten the maturing time of products and increase productivity about 20 to 30 percent.

The direction of sunlight at noon during equinox determines the width of the greenhouse. The rays at point b of the roof ridge cut the ground plane at point d. The distance from d to the base of the wall orientated to the sun determines the width of the greenhouse.

The design of the roof structure of greenhouses orientated to the sun are a function of:

1. The latitude of the greenhouse site.

2. The angle of inclination of the incident radiation at noon on the 21st of December.

3. The angle of inclination of the incident radiation at noon on the 21st of June.

4. The angle of inclination of the incident radiation at noon on the 21st of March and the 23rd of September.

With this improved type of greenhouse, the danger of frost is prevented, without auxiliary heating.

The other considerations to be taken into account in assembling greenhouses orientated to the sun are:

1. The wall of the greenhouse has to be insulated and to face the southeast by 10°;

2. The wall of the greenhouse has to be insulated with pitch-hay-pitch about 15 cm inside the ground;

3. There should be ventilation windows in the eastern and western sides of the greenhouse;

4. Either Malch (petroleum residue) should be poured along the inside edge of the side facing the south, or better, black plastic should be laid there;

5. The cavity between the two walls facing north should be insulated by filling it with hay;

6. The inside bottom 2/3 of the north facing wall should be painted black and the remainder white;

7. Black plastic sheets should be installed between the lots.

The solar orientated greenhouse is compared to the usual greenhouse in Fig. 4.

Fig. 4 Schematic diagrams comparing a solar orientated greenhouse to a usual greenhouse.

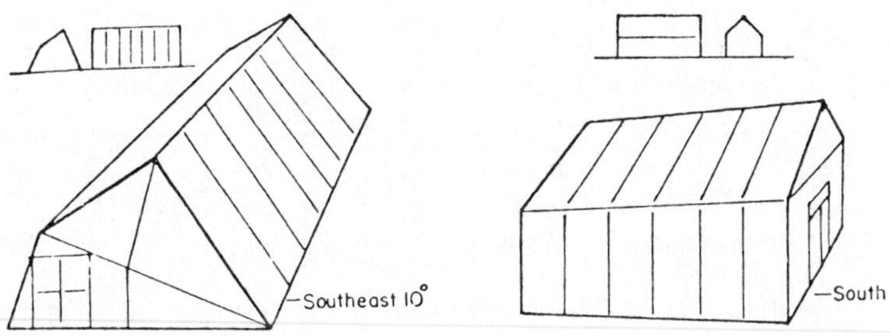

Solar Greenhouse Usual Greenhouse

The difference in collected energy per day between the solar greenhouse and the normal greenhouse is illustrated by the graph below.

Fig. 5

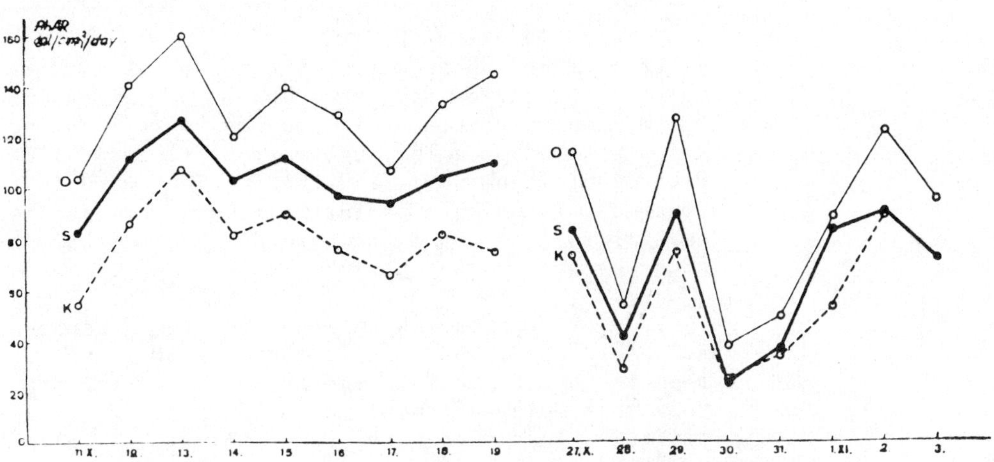

O = Incident energy CAL/cm^2/day falling outside the greenhouses.

K = Incident energy CAL/cm^2/day collected by the normal greenhouse.

S = Incident energy CAL/cm^2/day collected by the solar greenhouse.

Table 1

Comparison of radiation received by two types of greenhouses.

October	CAL/cm^2/day			Percentage of Radiation Collected	
	Solar G.house (S)	Usual G.house (K)	Amount of irradiation outside	Solar Greenhouse	Usual Greenhouse
11.10	82.5	54.5	104.0	79.3	54.4
12.10	111.8	86.1	140.7	79.2	61.0
13.10	126.9	107.5	160.0	79.3	67.1
14.10	103.0	81.4	130.0	79.2	62.6
15.10	111.6	89.5	138.9	80.3	64.4
16.10	99.9	75.3	128.0	78.0	58.9
17.10	93.7	65.0	106.2	88.5	61.2
18.10	103.5	91.2	132.1	78.3	61.5
19.10	109.1	74.2	144.3	75.6	51.4
27.10	82.7	72.5	113.2	73.0	64.0
28.10	40.1	28.2	53.3	75.2	52.9
29.10	89.1	74.1	126.7	70.3	58.5
30.10	23.7	24.0	37.7	62.8	63.6
31.10	36.2	33.3	48.6	74.5	69.7

Table 2

Comparison of plant growth for two types of greenhouses; 50 plants were used in both dates.

Harvest date	8th May		23rd June	
Greenhouse type	Solar (S)	Usual (K)	Solar (S)	Usual (K)
Maximum growth measurement, cm	37	25	260	225
Average growth measurement	30	20	228	199
Number of flowering plants	0	0	30	18
Total leaf area, dm^2	105	53	1873	1354
Total product weight, g	720	326	38320	35800
Total product weight (dry)	53.5	27.5	4616	3482

As it will be observed from the above tables and the combination of comparative curves indicated above, our greenhouse (s) has achieved the maximum improvement and success more than the usual greenhouse. Based on the foregoing information, it is obvious that the greenhouse with cradle type roof (25° inclination) used in the Mediterranean district, is no more efficient as per increased quantity of production, earlier realization of the product and better taste, than the greenhouse orientated to the sun.

Tables 1 & 2 compare a solar-orientated greenhouse with a usual greenhouse at a latitude of 50^oN in Trebon, Czechoslovakia. These results, reported by the Botanic Institute of Central Pruhonice in Czechoslovakia, show that solar-oriented greenhouses are more efficient at these high latitudes than the cradle type roofs (25° inclination) used in the Mediterranean countries.

As a competition two miniature greenhouses of 25 m^2 each are being constructed. One of them will have the asymmetric solar-orientated roof, while the other one will have the conventional greenhouse shape. Plants will be grown simultaneously in both greenhouses in order to compare their yields, speed of growth maturity and taste.

Fig. 6

Production from usual Greenhouse

Production from Solar Greenhouse

Fig. 7

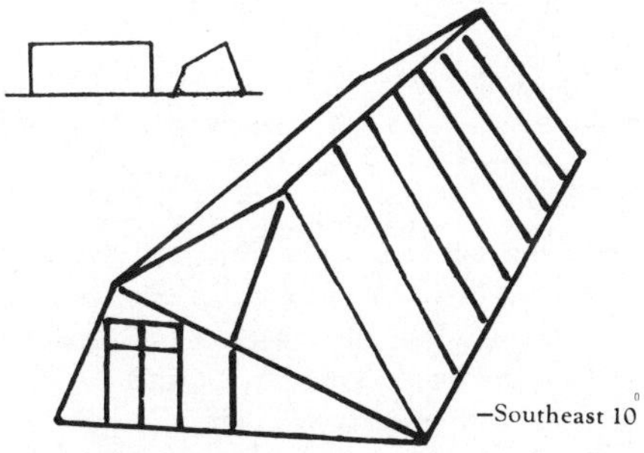

SOLAR GREENHOUSE

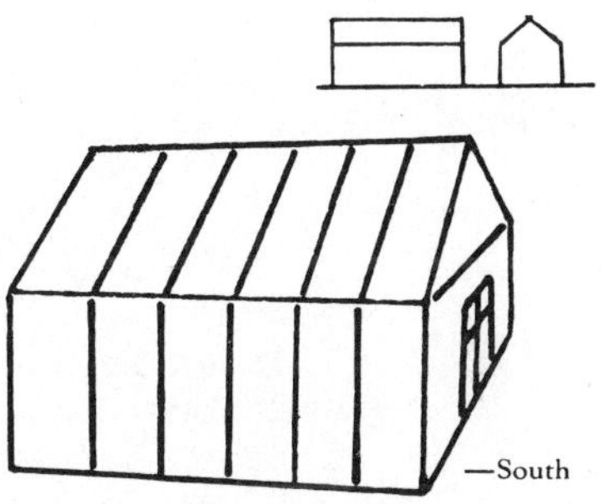

USUAL GREENHOUSE

PRODUCT FROM SOLAR GREENHOUSE

PRODUCT FROM USUAL GREENHOUSE

Solar Air Heater for Greenhouse

The cold air at the base of the greenhouse is heated by the solar air heater located on the southern wall and enters once again into the greenhouse. (Fig. 8).

The solar air heater consists of:

1. exteria glass
2. cavity 10 cm
3. absorbant layer
4. insulation 8-10 cm.

There are air vents located top and bottom at each end of the greenhouse. The cold air which is leaving enters through the lower holes, collects the available energy within the collector, becomes heated and lighter and sequently enters within the greenhouse as warm air. This circulation continues throughout the day during the winter months. In summer the top holes open to the exterior and provide air conditioning. The earth absorbes around 10% of the solar energy.

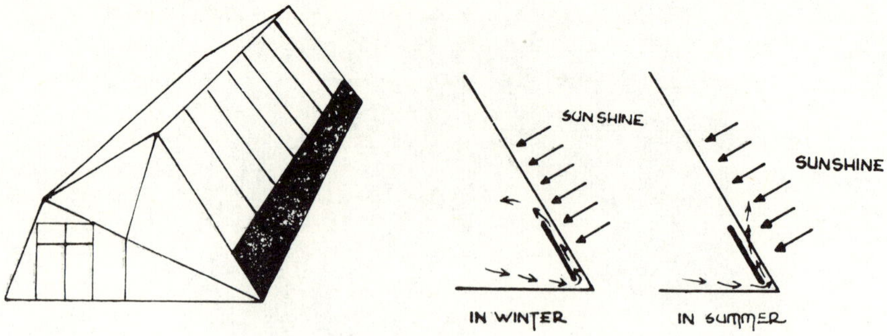

Fig. 8
Solar air heater for greenhouse

The same collector is also useful within the summer months too, in as much as the heated air is directed outside the greenhouse through a discharge channel, and thus makes it possible to keep the inside of the greenhouse cool and ventilated.

At this point I think it important to stress a basic fact concerning greenhouse use. It is essential to produce the maximum crop in the minimum time and as inexpensively as possible.

Another innovation applied to the greenhouses is to arrange the greenhouse walls in such a way that they can precipitate solar radiation. As it is known, the wavelengths of useful solar energy has been determined. Infra-red radiation only elevates the interior temperature of the greenhouse, and this causes an increase in the water requirement of the plants. The energy from the sun rises so much in the summer months that it damages the plants. It is necessary to limit this energy. Moreover, this energy that has been limited can be used in the heating of the greenhouse. This can be realized by using a selective filter at the greenhouse face, which is achieved by circulating a special liquid between the two covers of the greenhouse. In this manner, the plants within the greenhouse receive sufficient light and the extra amount is stored and used for the heating of the greenhouse during the night. Our country and the Mediterranean shores are rather fortunate in this respect and artificial energy input in the form of fuel is not necessary. With the assistance of filtration the water needs of the plants within the greenhouse are reduced by 65%. This fact is of importance in the dry zones.

In heating with solar energy solar radiation can pass through both the glass and plastic covers readily and enter into the greenhouse.

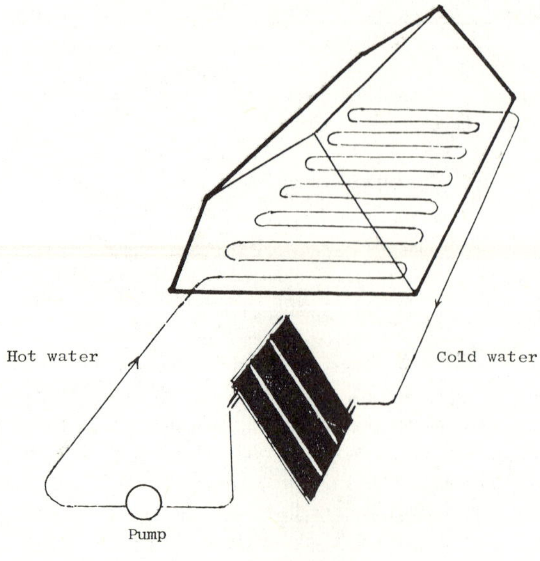

Fig. 9
The heating of the greenhouse ground with solar energy

Solar radiation between 360 milimicrons wavelength and 2600 milimicrons have a great effect in the heating of the greenhouse. This irradiation comprises 40-60% of total solar radiation.

The incident rays are partially reflected by the cooler covering material. Some are absorbed and some retained by the roofing material. One portion of the rays that enter within the greenhouse are absorbed by the plants and the other portion by the earth and thus they are both heated. A part of these rays is also reflected by the ground and plant surfaces. Of these reflected rays, a major part heats the greenhouses air and a small part penetrates through the glass face of the greenhouse.

When a plant is planted it is necessary that the earth must be warm and humid. For this purpose the water heated in the solar energy collector is circulated through plastic pipes which are buried at a depth of 25-30 cms in the greenhouse soil by means of a pump and so the soil of the greenhouse is heated. (Fig. 9).

When the plant starts producing leaves there is need for infra-red and additional ultra-violet rays. When the leaves grow up and mature, there is little need for infra-red and much much more need for ultra-violet rays.

Under these circumstances the heating of the earth by means of solar energy is terminated or cold water is mixed within the cycle. Thereafter, the solar energy collector is covered and taken out of circulation or, if it is used for warm water production, the pump circulates the water only during day time.

Fig. 10

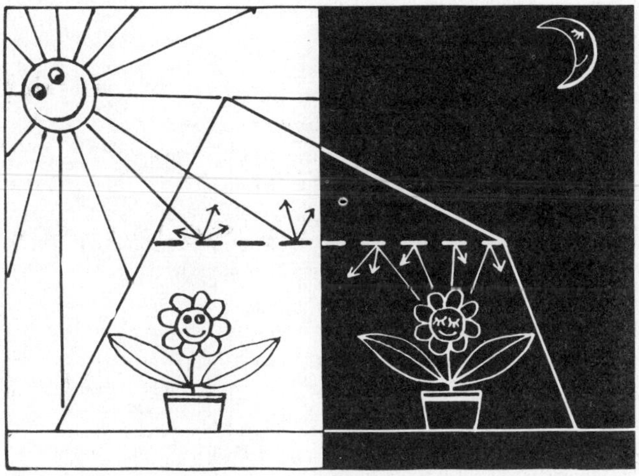

In Summer In the night and Winter

Due to the greenhouse roof cover which can be specially manufactured, the temperature of the greenhouse is controlable summer and winter, night and day. (Fig. 10).

Since this roof cover, which is made of a material of a very low heat transfer coefficient is closed in the evening during winter and cold weather, heat loss through the roof is prevented. In this manner the warm air which remains within the greenhouse prevents the greenhouse from getting cold for a long period and therefore assists it to pass the night without any danger to the plants.

At sunrise the cover is opened again and the greenhouse continues to benefit from the solar energy.

This cover is also closed uner the hot and brilliant sun during summer and with the assistance of its glazed surface nearly 85% of the sun rays are reflected away.

It also acts as shading against 50% of the irradiated sun rays.

In this manner this special greenhouse cover (Fig.6) provides a large fuel economy in winter as well as protecting the greenhouse from the sun's intense heat in summer and this provides the continuation of production.

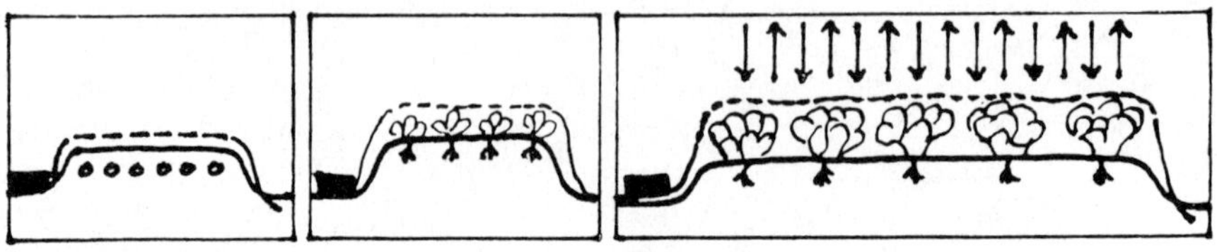

Fig. 11 Use of field greenhouse by depositing the sun energy to the ground by use of special covers (without frame)

Special "Growing" Film (for spring and autumn)

Sow and plant when soil is ready.

Loosely spread special film, for spring and autumn, (see Fig. 11), over the bed and secure sides with planks or stones or by embedding them in the soil.

The film allows moisture to penetrate in from the outside and protects the plants against cold weather and snow. It also prevents excessive evaporation due to the sun or wind. The ideal "micro climate" under the film promotes rapid and safe plant growth.

The film will stretch as plants grow to form a tunnel shaped roof. Water plants through the film, if necessary. The film will stop the soil from hardening and ensure aeration and optimum growth (without airing as is the case with conventional systems). The film prevents rapid evaporation, yet it lets the rain water to seep through. It will therefore give you a carefree holiday.

The film is an efficient barrier against garden parasites such as birds, beetles, insects and snails provided these are not present before you cover your plant with special growing film.

The cover colour also acts as a selective filter for certain rays. The film converts light rays into heat to give your plants all the warmth they need to grow.

The film may also be used successfully on seed beds (as indicated above) in early spring, in the greenhouse, on plant pots and plant containers in the house.

The special "growing" film (for spring and autumn) is particularly suitable for:

Radish, horse-raddish, lettuce, cos lettuce, cabbage lettuce, garden cabbage varieties, fennel, beetroot, spinach, salsify, asparagus, etc.

Cover peas, runner beans and other beans and when the plants have reached a height of 15 to 20 cm. remove the foil.

Remove cover from cucumbers, courgettes, dwarf beans, melons, strawberries, etc., before they begin to flower (to ensure fertilisation) and recover after fruits have formed.

To begin with, the use of this special film protects the seeds against the birds and the wind or rain. It ensures germination by creating uniform moist climatic conditions comparable to those of a greenhouse. However, the film must be assisted in order to perform its function. Although it prevents rapid evaporation it does not altogether obviate the need to water the plants during the hot season. Some additional water will always be required.

Fig. 12
Special plastic cover in greenhouse

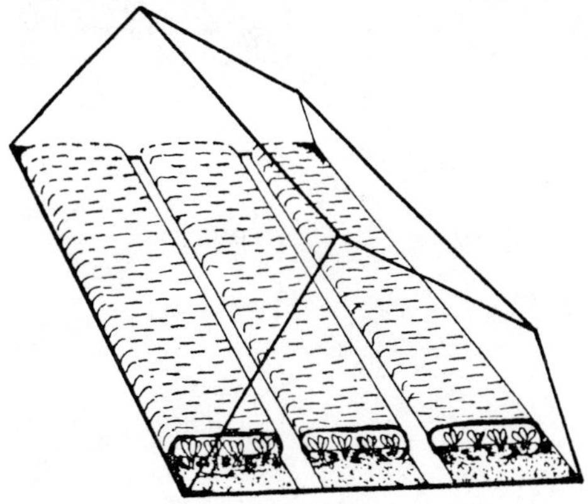

Experience has shown that the film acts as an effective barrier against parasites. Carrot flies and plant lice, for example, will not even attempt to creep in through the small gaps at the sides because the film acts as a camouflage so that they cannot see the colour of the plants underneath the film. Snails cannot enter unless wide gaps are left uncovered. But snails already in the seed beds before sowing and planting must be killed with snail pellets.

Keep bed free from weeds. Make sure all are pulled out before they flower when they are still harmless.

Preparation of the beds

Sowing and planting: Make sure the two outside rows are not further apart from each other more than 90 cms. Gently cover the film immediately after sowing and planting to allow the plants to get used to the pressure and the "micro climate" under the film.

Anti-weed and mulchfoil

Premotes larger, healthier and cleaner garden produce keeps the soil moist warm and biologically active, prevents growth of weeds without the use of chemicals.

Anti-weed foil has thousands of tiny slits which allows penetration of air and water; prevents soil silt-up and crusting, evens out ground temperatures and promotes ripening in the shade, optimum absorption of nitrients through better root development immediately under the surface. Plant two crops without having to dig the soil again, then use a liquid fertiliser. Keeps low growing fruits away from direct contact with the soil. Protects from mildew, decay and dirt. Achieve an extraordinary harvest of strawberries, cucumbers, melon, tomatoes, cabbage, beans, peas, etc.

Fig. 13
Anti-weed plastic cover in the greenhouse

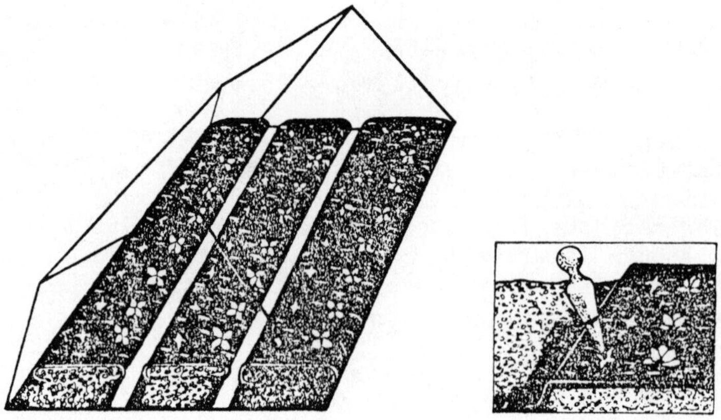

Referances.

1. M. Bitterman. Academia, Praha 1968.

2. Dagmar Dykyjova Trebon Czechoslovakia.

K.J. Wasserman
Solar Age, Port Jervis, N.Y., U.S.A.

Abstract

Greenhouses are enclosed structures which can be integrated into an adjacent building to supply excess heat energy for that adjoining structure. This can be done with modifications of typical greenhouse designs which take into consideration most of the basic concepts of residential solar heating.

In the U.S.A. greenhouses as a source of solar heat have been designed in many areas. A greenhouse should have its own storage area, either water or rock bed. Insulating and shuttering devises are important for evening use.

Growing seasons can be extended and solar energy can be used in an optimum manner, and supply needed humidity in dry areas or seasons.

Where heretofore the greenhouse for the residential building was considered a costly, luxurious hobby, it can now be considered as a solar collector which pays its way in reducing the heating costs and growing food in a symbiotic relationship with a dwelling.

Introduction

All greenhouses, whether commercial or residential, have one thing in common: the growth of off-season plants or crops. There exists in all countries a huge potential for production of family food in residential-related greenhouses. This is of special significance for lower-income families where food and fuel alone can consume an extremely large percentage of an annual income. It also is particularly important in developing countries. In this regard it is the concensus of many people, knowledgeable in the area, that solar heated greenhouses will become a major factor in the utilization of energy and important in helping the quality of life - while greatly improving our environment.

A greenhouse, which is really a controlled environment, when compared to an outdoor field production situation, can show an unbelievably high productivity per acre on a commercial scale. The necessary technology then is here to also bring crop production down to a smaller family and human scale. When utilized on a family scale, even marginal land can be used. For example, a greenhouse can also be built into steeply sloping south orientated hills, thus utilizing land unsuitable for farming or grazing. Furthermore, greenhouses use only a fraction of the water necessary in open field irrigation.

As commonly operated, greenhouses are generally considered to be energy-intensive and they require more heat per unit of area than most buildings existing (1). It has been noticed that in typical greenhouses, fuel accounts for 40% of operating costs and electricity, water systems and heating account for an additional 56% of costs (2). Furthermore, greenhouses can loose about 80% of their collected solar energy when the sun is not available, and especially during the night when conduction losses occur and reradiation and filtration takes place through walls and roof surface (3).

Rarely were greenhouses designed with energy conservation in mind, as formerly cheap fossil fuel or electricity was used generously for heating. A heating system, if installed, was usually independent for free-standing greenhouses, and sometimes came directly from the main residential building for attached greenhouses.

All that has changed. Rising fuel costs have forced the shut-down of many active greenhouses, both commercial and residential. Fortunately, recently developed designs have helped start literally a new industry and technology: solar greenhouses - although the literature will reveal early 19th century designs of greenhouses that amazingly resemble some current solar designs (4).

Since greenhouses are energy intensive, major design improvements are necessary in collecting and storing passively derived solar energy (5, 6, 7, 8, 9).

As in any passive energy system, the collection mode is easy for direct gain heating. You just aim your greenhouse south and try to keep heat-losing north walls well insulated. But also as in passive modes, storage and distribution are greater problems. Dense storage mass, with high heat-holding capacity, such as masonry, adobe, stone, or water must be used to absorb surplus heat from sunlight through the greenhouse glazing. Later at night the energy will radiate back into the greenhouse and hopefully keep plants comfortable, especially in the winter time.

The various designs which we will show consist of free-standing solar greenhouses, attached greenhouses and "live-in" greenhouses. The retrofitting possibilities are good for existing greenhouses since they usually have the proper orientation. In retrofitting a greenhouse to an existing structure we might find orientation problems. However it can be economically attractive (10) and is obviously necessary if greenhouses are to act as solar collectors for buildings.

When designing a solar greenhouse, the same basic rules used to achieve climate control for a solar structure can be employed. A reduction in heat loss takes place. This means plugging up air leaks, insulating surfaces not orientated for sun light, especially on the northern side, as well as a portion of the east and west wall, and part of the roof. Storage for holding heat for cloudy days and night time use is imperative, and only after these basic rules are taken into consideration can a solar greenhouse be a low-energy, cost effective system for plant growth, food production and a source of heat when attached to a building.

There are a growing number of demonstrations of free-standing solar greenhouses as well as attached ones, which supply excess heat to the adjoining building. The question naturally arises as to whether an attached greenhouse is any different in design compared to a free-standing solar greenhouse. It appears that basically the same rules apply in regard to double glazing, insulation at night time, use of reflective surfaces in the proper area if necessary, and ample storage capacity. One of the main differences is the northern wall, which in the case of an attached greenhouse is also a wall of the dwelling. This poses some design consideration. However the principles are the same though the numbers might change if the attached greenhouse were designed to concentrate on the growing of plants or growing plants and supplying some excess energy for an attached building, or used as a solar collector to supply most of its energy to the building.

Greenhouse/Active Collector Hybrid

A well-designed solar greenhouse sometimes is no panacea in replacing a solar collector, if used as a passive greenhouse to get residual heat for transfer to adjoining buildings. A hybrid system could also be a solution (11).

A combined residence/greenhouse solar heated system has been developed and built by the Solar Energy Applications Laboratory at Colorado State University. A solid wall separates the south facing greenhouse from the residential area. There are growing plants on raised benches. In this instance, solar energy is the primary heating source for the entire structure divided up into an active air system and the passive greenhouse system.

Measurements started up January 1977 indicate that the combined system has supplied 87% of the heating requirement of the combined residence/ greenhouse - 85% of which was collector derived and 15% greenhouse derived.

Freestanding Greenhouses

D. Harrison (12) has developed a design for a free-standing school greenhouse with bead wall insulation. The Monte Vista Elementary School greenhouse in Albuquerque, NM was constructed in 1973. This is a system whereby polystyrene beads are blown into a cavity between the 2 glazings to insulate the entire area against heat loss at night or on cloudy days. The cavity is evacuated during the day so that the structure may collect solar energy.

On the northern side behind the exterior wall we have 6 pairs of 55 gallon drums divided equally and serving each bay for storing the polystyrene beads. A vacuum cleaner type motor provides the air power necessary to cycle the beads.

An auxilliary electric heat back-up system has been used whenever necessary. Night time greenhouse temperatures usually lag about 15°F (8°C) below the temperature of the water drums, while day time temperatures usually ended up rising 10-15°F (6 - 8°C) above the temperature in the water drums.

It has been shown by Gillett (13) that by combining thermal mass in the form of water containing storage tubes, double glazing, north side insulation and a circulation fan it was possible to even out the highs and low of the temperature swing in a free-standing solar greenhouse. The temperature in this greenhouse never fell below 55°F (13°C), which is considered acceptable for night time temperatures for plants. The average temperature spread was on the order of 20-30°F (11°-17° C). The only problem occurred when sunny days followed each other in succession and the fan and thermal mass system was unable to keep the temperature below 92°F (34°C)

Unfortunately none of the greenhouses in this experiment were vented, which of course would have solved the problem, but undoubtedly a better solution would be to incorporate this concept into a lean-to, residential situation and store the excess heat in the building.

Attached Greenhouses

Hinsburg, Vermont, in New England is a very cold part of the North-eastern states and in this area a passive solar greenhouse, which can be attached to a home, has been set up and studied.

Taff (9) reaches the following conclusions in his work: that solar heating systems need not be expensive to be significant, that solar retrofits are possible with passive solar greenhouse systems, and that greenhouses do not of necessity have to be energy wasteful but designs must be subject to strict controls in lowering heat costs by maintaining insulation levels. Furthermore modest solar greenhouses can be used to heat homes. Monitoring data show that thermal storage subsystems can stabilize maximum and minimum interior temperatures, even when subjected to variable outdoor temperatures and increasing monthly average temperatures. Also, the heat load coefficient of the house is critical to determine the net usable energy gain from the greenhouse. And finally, reducing the heat loss of the greenhouse via night shutters can have a significant impact on the net yearly solar gain produced by the greenhouse.

The small attached solar greenhouse was built for $800, and acts as greenhouse-solarium, but more importantly, a solar collector. The front surface is made up of double fiber glass. The walls are insulated with 6" (15cm) fiber glass and the room is totally isolated from the home except with the addition of a door and two fan ports.

The greenhouse operates on a differential thermostat such that whenever the temperature of the greenhouse exceeds by 9°F(5°C) the temperature in the house, two small fans moving 200 cubic ft. (5.7 cubic meters) a minute of air become operational. In this way, excess heat produced by the greenhouse is conserved within the home. At the same time, the water storage insures that the greenhouse temperature will never fall below 55°F (13°C) (even in a climate where 25°F below 0 (-32°C) is not uncommon).

The Kelbaugh house (14) in Princeton, NJ has a greenhouse attached to the south-east of the residence with the long side facing south. The greenhouse is double glazed. There are upper and lower vents for good summer ventilation. Wood slat shades provide sun protection in the summer and cut heat loss slightly in winter by keeping wind off exterior glass surfaces. Floor and wall are painted black to absorb heat.

The house is essentially a two-story Trombe inspired design. However the greenhouse area is open to the rear, with no wall separating it from the living quarters. The floor of the greenhouse is a thick masonry slab and it absorbs heat in the winter for reradiation back into the room on cold days, as well as a cellar below it.

During summer months top windows of the greenhouse open to allow an outflow of air - originating from windows on the north side of the house.

A firm, the Solar Room Co. (15) appropriately enough, is in the business of prefabricating greenhouses in northern New Mexico. Typically, in November of last year, a Solar Room maintained a temperature of about 30°F (17°C) above the outside air temperature. The performance of the Solar Room, as far as a 30 ft. (approx. 9 meters) long room is concerned, indicated that there was a net gain of approximately 258,000 BTU's per day and the calculation showed that the gains were essentially through the south sunfacing wall and the roof, while the east and west walls showed net losses.

Doug Balcomb (16) built a solar greenhouse as an add-on to his home in Los Alamos in New Mexico last year. His experience with it was very favorable. As a greenhouse it had somewhat too little sun. However, as a solarium addition to a house, it was just perfect. He indicated that most plants thrived out near the window. Only shade-loving plants were able to grow in the back portion of the room. The room needed no auxilliary heating after the rock bed system became operable in February. This was a good test as there was considerable cold and cloudy weather. The greenhouse was a 400 sq. ft. (37.3 sq. meters) addition to their frame house.

The Balcombs later moved to First Village in Santa Fe into a greenhouse/ residence designed and built by Susan and Dwaine Nichols (17).

The greenhouse has approximately 400 sq. ft.(37.3sq. meters) of thermopane at a 45° angle facing south. It rises in a 2-story home which has the 20 ft. (6 meters) high greenhouse in the middle of the south side. The floor area of the greenhouse is triangular in shape. It contains a circular staircase in the north corner, which has a vent window at the top. This acts as a chimney to cool the space. The side walls are 14"(35.6 cm) thick adobe downstairs and 10" (25 cm.) thick adobe upstairs. Adobe is made out of local mud; bricks are formed and set up with mud as mortar. It is claimed to be a very plastic building material and is indigenous to the New Mexico area. Adobe has the thermal mass necessary for solar storage and many adobe walls exist in solar homes.

Solar collection is a hybrid approach since fans pull the heated air into a storage rock bed. The passive heat gain is totally derived from the greenhouse.

In Northfield, NJ, Don Watson (18) designed a house which uses a "sun room" with greenhouse heat recovery system. Hot air rises and is pulled by a fan through ducts in the ceiling down the wall and into rock storage, buried under the sun room. The sun room and storage will provide 45% of the required space heat. A "sun room" or "solarium", when not called a greenhouse, usually connotates a sunny area devoid of meaningful plants. Accordingly, the area is useful for "hobby horticulture " and the growing of family vegetables, should the owner so desire.

There are solar roof panels to heat water and an oil burner as a back-up heater.

D. Kruschke (19) built a low cost combination solar greenhouse and residence in Wild Rose, Wisconsin. He utilized 28-55 gallon drums filled with water, located between the vegetable plot and his living space. These drums absorb much of the total day's solar energy and help to moderate daytime and night time temperatures within the house. At night time foam insulating panels are put up to reduce heat loss.

This 1,000 sq. ft.(93 sq. meters) combined building maintained inside temperatures up to 88°F (31°C) on cold wintry February days, while outside temperatures were approximately 26°F (-4°C), a difference of 62°F (35°C). However, in the summer the "greenhouse" got too hot and some of the double plastic glazing was replaced with mosquito netting. For cloudy days the back-up system was a woodburning stove.

The 170 sq. ft.(15.7 sq. meters) growing area provides year round lettuce, tomatoes and radishes, and supplies humidity for the winter time. This modest structure should be of interest in many colder areas.

Conclusions

As we have seen, there are situations where, literally and actually, we can have the best of two worlds. Grow your own food and conserve energy, all within the confines of a delightful environment, the solar greenhouse. It may be free-standing, or attached. Solar greenhouses, properly orientated when attached to structures, fulfill a further promise, that of attaining a greater degree of self-sufficiency for many human beings - in living in an environment where food can be grown while simultaneously reducing fuel consumption for the main live-in structure.

A symbiotic solar greenhouse which supplies food, conserves energy, even as it contributes energy, depends on its human occupant for control and guidance. It is a relationship that is worth cultivating in today's world.

References

1. Mears, D., Roberts, W. and Simpkins, J., "New Concepts in Greenhouse Heating," ASAE paper No. NA74-112 (Aug. 1974).

2. Phillon, P. and Kirschling, P., "Profitability of Tomato Production under Plastic Greenhouses," Dept. of Agricultural Economics and Marketing, Rutgers University (1973).

3. Walker, J., "Predicting Temperatures in Ventilated Greenhouses," Transactions of ASAE 8 (3) (1965).

4. Hix, J., "The Glass House," MIT Press (1974).

5. Lawand, T. et al., "The Development and Testing of an Environmentally Designed Greenhouse for Colder Regions," American Sec. International Solar Energy Society, Fort Collins, Col. (Aug. 21-23, 1974).

6. Kusianovich, J., "Solar Heated Greenhouse with a One Year Payout," International Solar Energy Congress, Los Angeles, Ca. (July 28-Aug. 1, 1975).

7. Yanda, W., "Solar Sustenance Project," International Solar Energy Congress, Los Angeles, Ca. (July 28-Aug. 1, 1975).

8. Nash, R.T., et al., "Temperature Stabilization in Greenhouses," International Solar Energy Congress, Los Angeles, Ca. (July 28-Aug. 1, 1975).

9. Taff, D., et al., "Design Considerations, Theoretical Predictions, and Performance of an Attached Solar Greenhouse Used to Heat a Dwelling," Proc. American Sec. International Solar Energy Society, Orlando, Fla. (June 6-10, 1977).

10. Yanda, W., "Considerations for Retrofitting an Attached Solar Greenhouse," presented at "Passive Solar Heating & Cooling Conference, Albuquerque, NM (May 18 and 19, 1976).

11. Smith, C., Annual Meeting "American Society of Agricultural Engineers," North Carolina State University, Raleigh, NC (June 1977).

12. Harrison, D.C., "Review of Monte Vista Elementary School Greenhouse," presented at Passive Solar Heating & Cooling Conference, Albuquerque, NM (May 18-19, 1976).

13. Gillett, D.A., "Solar Powered Greenhouses," presented at New England Solar Energy Association, 1st Annual Conference "Decision Making in Solar Technology (June 24-27, 1976).

14. Kelbaugh, D., "Kelbaugh House", presented at Passive Solar Heating & Cooling Conference, Albuquerque, NM (May 18-19, 1976).

15. Rogers, B.T., "What Is a Solar Greenhouse? A Simplified Approach," Brochure (Spring 1977).

16. Yanda, W., et al., "The Food and Heat Producing Solar Greenhouse - Design, Construction, Operation," John Muir Publications, P.O. Box 613, Santa Fe, NM (1976).

17. Nichols, W., "Visit One, First Village," presented at Passive Solar Heeating & Cooling Conference, Albuquerque, NM (May 18-19, 1976).

18. Watson, D., "Designing and Building a Solar House," GardenWay Publishing, Charlotte, Vermont (1977).

19. Kruschke, D., "A live-in Solar Greenhouse," Mother Earth News No. 39 (May 1976).

The applications of existing solar energy collecting equipment in agriculture, horticulture and fishculture

F. Vogt
Built Environment Research Group
The Polytechnic of Central London
London NW1, England

Abstract

The Built Environment Research Group (BERG) of the Polytechnic of Central London (PCL) has been researching into the use of solar energy for domestic purposes since 1969 (Dr. D. von Hodges, S. Szokolai). In October 1976 research into the practical applications of existing solar energy collecting equipment in agriculture, horticulture and fishculture was added to its work.

Analysis showed that research should be carried out in 4 key fields each of which could in turn be applied to a number of related activities, i.e. the data obtained for pigs could be used for calves, poultry and mushrooms:

1. Economic, partially seasonal heat store.

2. Heating ponds for food producing warm water fish species not requiring high animal protein concentrates.

3. Space heating for pigs post-weaning from 5kg. to 15kg. liveweight.

4. Indirect heating of nutrient in greenhouse crops raised by the nutrient film technique (NFT) or hydroponically and allowing a significant reduction in heating needs because of lower space temperature requirements experienced with this technique.

Project 2 is operational. The others are due or about to be installed.

Introduction

The biological industries, agriculture, horticulture and fishculture are better placed to make use of solar energy than other industries.

The problems of site and siting do not exist as they do in urban locations. It is easier to achieve self-sufficiency in rural, scattered farmsteads than in interlocked conurbations. The new source of energy could be grafted on to existing plants.

Biological industries are bound by fixed cycles and are fundamentally inelastic. Their need for alternative energy is, therefore, greater than that of other industries which can, in any case, adjust more easily to higher costs.

The choice of solar energy among the alternatives is due to the fact that it is backed by a proven and established technology. Data obtained by the Research Group will be of use with other sources of energy such as geothermal et al. Data or information obtained in the course of the programme in this country will be valid for other regions. A computer simulation programme will help here.

The research programme was firmly set out as follows:

1. To design, construct and monitor solar heating systems for biological applications and to evaluate and assess their practical and economic performance.

2. To develop systems which:

   (a) use existing equipment readily available to farmers and growers and which can be installed by non-specialised labour at least in part;

   (b) are reliable and have a long maintenance-free life;

   (c) can be put into practical use without prolonged periods of development;

   (d) are of immediate, widespread and practical use to those branches of the industries concerned with food production.

3. To test and extend the scope of existing computer simulation programmes in relation to biological applications of solar heating systems.

4. Measures to incorporate solar heating systems into the design of new buildings and structures as a means of reducing the overall costs in the next generation of farm buildings et al will be investigated.

   Energy conservation measures for existing and new structures will be considered.

There are indications that it will be essential to integrate biological aspects into the programme.

Description

1. The effective use of solar energy in the biological, or any other, industries depends on an economic method of storing hot water. It is unlikely that any other medium can be considered for at least 20 years, except by trendy architects.

   Based on a farm silo which has been in use for several decades a heat store has been developed at a material cost of 3.75 p./1. (2.63p. after current maximum grant) and U = 0.067 (Vogt 1976).

   The silo consists of a butyl cylinder contained by two concentric rings of weldmesh which are about 55cm to 57cm apart. The insulation is placed between these two rings. The structure then has to be weather-proofed.

   Insulation accounts for between 85% to 90% of the material cost. The most effective method uses a split material system. The inner-most layer of insulation is of polyurethane foam sheeting 2.5cm thick. The remaining 52.5cm to 55cm are filled with fibre glass. The choice of insulating material is critical. Polystyrene is not suitable near the heat store, because computer simulation shows that the temperatures achieved within the butyl cylinder exceeds its tolerance.

2. The project to heat fish ponds is operational, taking 5 months from conception to commissioning. It is unfunded.

   It involves part of a series of 12 identical ponds, four of which are covered by plastic greenhouses or tunnels. The crop is the mirror carp (Cyprinus carpio).

   Results so far seem to confirm that in summer the greenhouse will raise water temperature by 3.5°C and that 50m3 pond water heated by 21m2 of flat solar collector panels (selective black) give a further uplift of not less than 12°C. This heat exchange is via 70m of 42mm diameter submerged copper pipe.

There have been minor mechanical hiccups. Not as many probes are available as is desirable.

An autonomous pond and support system has been designed for replication and to allow the introduction of biological parameters. This artificial pond system will have commercial application. The catfish (Ictalurus spp.) may be introduced later to see how its energy conversion efficiency compares with the carp. Energy here refers to both fuels and foods.

3. The pig project will involve the solar heated silo supplying hot water through radiators in a rearing house. The radiators will also act as pen partitions. The cost of pen partitions and radiators is about equal. Identical houses act as controls.

   The piggery involved is part of a minimal disease (MD) herd. Anyone entering the unit must go through a shower and change to sterile clothing. The showers will also be run off the silo (Vogt 1977).

4. NFT is a method of crop production where plants roots are bathed in a film of water containing the required nutrients within a black plastic trough. It is intended to warm the nutrient evenly throughout an installation by siting the trough in close contact above two large diameter pipes of thin plastic. The hot water flows within the two adjacent pipes in opposite directions.

   The use of filtered off fish detritus to manure the NFT grown plants is considered a possible future field for investigations et al.

8.5 Materials for a solar still greenhouse

D. Tinaut, G. Echaniz, F. Ramos
Instituto de Optica. C.S.I.C., Madrid 6
and
J. Fernandez
J.E.N.-Isótopos, Madrid 3

Abstract

The possibility of associating solar desalination with cultivation in greenhouses opens a perspective for the direct application of solar energy. This will imply work in the photobiological regions of the solar spectrum, needed in the process of photosynthesis, and the calorific bands used in the process of solar distillation of water for irrigation.

With this purpose in mind we are studying a system of a solar still to be incorporated in a greenhouse structure. This system has been designed and will be made out of panels of coloured plastic materials, which will absorb the calorific bands of solar spectrum, for evaporating saline water, and which will also allow a high level of radiation, required for growing vegetables.

In order to establish the characteristics for the selection of the most suitable commercial materials for the construction of distillation devices we have measured the photosynthetic activity for different patterns of materials.

Introduction

The purpose of this work is the study of appropriate materials for the construction of a solar still, which incorporated with a greenhouse of controlled ambient conditions constitutes a solar selective distillation unit.

This unit is foreseen for the cultivation of vegetables in geographic areas which have plentiful saline water supplies and enjoy great insolation.

The first realisations of a solar still incorporated with a greenhouse were carried out by Trombe and Foex in 1957 in France (1), and by Selçuk in Turkey (2).

Selçuk established theoretically the mathematical analysis of the energy balance inside the solar still-greenhouse system.

In the present study we aim at determining the characteristic of the most appropriate materials for use in a solar still working at the interior of a greenhouse.

There are two basic factors which need to be considered at the same time:

(a) The advantage of the calorific effect of the incident solar radiation in a distillation phenomenon, and
(b) The transmission of the necessary bands for a photosynthetic process.

The following methodology was observed in the process of our work:

(a) Selection of commercial materials.
(b) Measurement of the photosynthetic activity when the plants are illuminated by light which is transmitted through the plastic materials.

(c) Evaluation of the output of a solar still, when these materials are used as basic surfaces, as a function of incident radiation levels.

In the next phase of our work we intend to carry out tests, under real life conditions, for verifying some of the laboratory results obtained so far. Fig.1 shows a diagram of the experimental arrangement which we have proposed.

Fig. 1

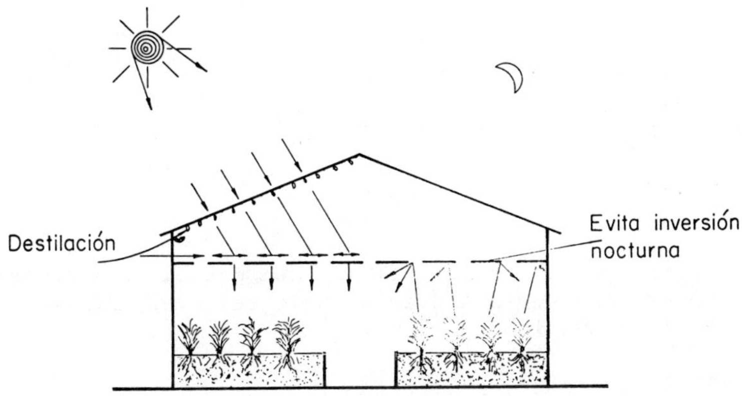

<u>Selection of materials</u>

In order to increase the absorption of incident radiation and consequently the efficiency of the installation generally the solar stills generally have a black surface. We are developing a new solar still which is semitransparent. This solar still allows through the visible solar radiation that is necessary for growing plants by photosynthesis.

The absorption spectrum of chlorophylls (maximum between 640-670 nm and 410-440 nm) and carotenoids (maximum between 400-450 nm) is modified when these pigments are combined with other molecules. Table 2 gives the value of absortion maxima in organic solvents (Hall and Rao, 1972).

Table 1

Pigment	Absorption maxima in organic solvents (nm)
Chlorophylls	
Chl a	420-660
Chl b	453-643
Chl c	445-625
Chl d	450-690
Carotenoids	
α carotene	420, 440, y 490
β carotene	425, 450, y 480
Luteol	425, 445, y 475
Fucoxanthol	425, 450, y 475

The ideal material for the base of a solar still will be one that has a spectral absorption like a magenta filter. Fig.2.

Fig. 2

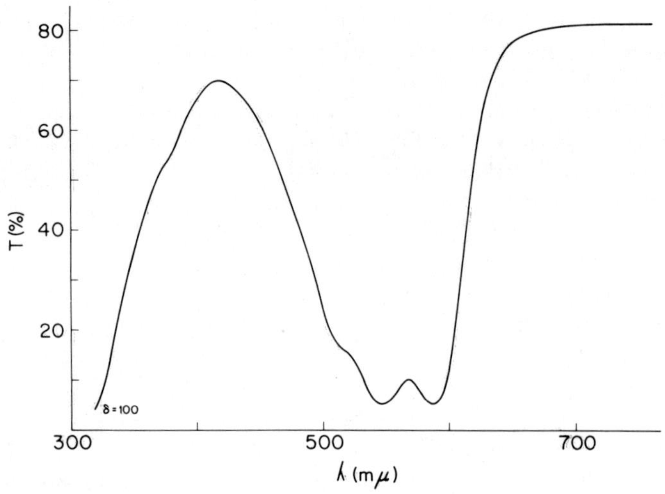

At present there are not any commercial plastic materials with this absorption curve and so we have selected the following ones for experimentation:

1. Red methacrylate rigid
2. Red polyethylene flexible
3. Blue methacrylate rigid
4. Blue polyethylene flexible
5. Blue polyethylene flexible (more opaque than the other one.)

Measurement of photosynthetic activity

The photosynthetic activity is expressed in fixed micromols of carbon dioxide per square centimeter of leaf and per hour of exposure.

This experiment was carried out with barleys which were germinated in our experimental greenhouse twelve days before.

These plants were exposed to an $^{14}CO_2$ atmosphere and later was determinated the ^{14}C incorporated, that is to say the photosynthetic activity.

A microchamber is attached to the leaf and the light comes after passing through the plastic material which is acting like a filter. We used a 14mm diameter area of a barley leaf which was exposed to pulses, having a 15 second frequency of air labeled with $^{14}CO_2$.

Five barley plants were used for testing each material. Five values were founded for each material and then the mean value was calculated.

In this experiment, the percentage of assimilation of ^{14}C was taken as a reference. After these results we can say that all our selected plastic materials are accepted for the growing of the plants when they work as a filter light.

At the end maximum photosynthetic activity was obtained with the red methacrylate and minimum photosynthetic activity corresponded to blue polyethylene filter.

Foresight of still production

An approximate evaluation was made of the volume of distilled water obtained by using each of the selected samples.

We used the spectral distribution of solar radiation and the absorption curves of these plastics to calculate the absorbed radiation for each material.

We obtained hourly values along the year. By integrating we then found the absorbed radiation on a daily basis. This is expressed in $cal.cm^{-2}.day^{-1}$.

Once we knew the values of absorbed radiation for each materials, we used C.S.I.R.O. graphs, that represent output from solar still as a function of radiation level in $cal.cm^{-2}.day^{-1}$., for estimating the solar still production when these materials were used.

Some of these results are given in Table 2:

Table 2

Material number	Photosynthetic activity %	Annual water production litres/100m2
1	102,6	35,050
2	69,5	21,060
3	52,7	23,810
4	44,3	14,350
5	55,7	25,020

References

(1) Trombe & Foex, Brevet Français no. 1168575, Department C.N.R.S., 12 February 1957.

(2) M. Selçuk, 'Greenhouses and Solar Stills Combined', C.O.M.P.L.E.S. Conference Athens 1971, Bull.no. 22. 1972.

(3) D.O. Hall & K.K. Rao, 'Photosynthesis Studies in Biology', 37 Ed. Edward Arnold, London 1972.

8.6 Modelisation of fruits drying by solar energy

R.D. Piacentini
Laboratoire d'Astrophysique
Univ. de Bordeaux I, France and Depto. Física, Fac. Cs. Exactas e Ingenería, Univ. Rosario, Argentina

M. Combarnous
Laboratoire de Mécanique Physique
Univ. de Bordeaux I, 33405 Talence Cedex, France

Abstract

This work deals with a model for fruits drying by solar energy. The model is derived from the equations set proposed by Luikov for heat and mass transfer in porous media, recently developped by Babukha and Shraiber for spherical bodies. Average temperature and water content in the fruit have been computed versus time for different values of the parameters and variables which have been proved to be characteristic of the phenomena. Results have been obtained concerning practical cases of plums drying (prune d'Ente). In particular, the evolution of the drying time versus the internal air temperature in the solar collector as well as the overall efficiency are estimated. Different types of drying systems are analysed.

Introduction

The drying of fruits by direct exposure to the sun radiation is a currently used technique in countries with relatively hot and dry climates. In regions with high values of humidity, it is commonly assumed that solar drying is not suitable. In what follows, it will be shown that, even in this case, simple systems which optimize the collection of solar energy can be used for drying purpose.

The model describes the evolution versus time of both mean temperature $\bar{\tau}$ and water content $\bar{W}$ of the fruits, submitted to given initial and external conditions. The fruits are supposed to be dried in a system, i.e. a kind of container, where the ambient air flows. The solar collector can be either included in the structure of the container, or used as a preheating device for the circulating air. A combination of both possibilities can also be considered.

Mathematical model

The mathematical simulation is based on the analysis of the heat and mass transfer inside the fruit as well as at the interface between the fruit and the air.

Concerning the transfers inside the fruit, they are described with the conservation equations set formulated by Luikov {1} . In the case of homogeneous media of spherical shape and with a spherical symmetry of initial and boundary conditions, the evolutions of the local temperature τ and water content W obey, with the notations listed at the end of the paper

$$\partial(r\tau)/\partial t = a\ \partial^2(r\tau)/\partial r^2 + (\varepsilon L/c)\ \partial(rW)/\partial t \qquad (1)$$

$$\partial(rW)/\partial t = a_m\ \partial^2(rW)/\partial r^2 + a_m\ \delta\ \partial^2(r\tau)/\partial r^2 \qquad (2)$$

The initial conditions and the boundary conditions near the stone ($r = R_o$) and at the fruit-air interface are

- for $t = 0$ $\quad \tau = \tau_i \quad r \in \{0, R\}$ (3)

$W = W_i \quad r \in \{R_o, R\}$ (4)

- for all $t > 0$ $\quad \partial\tau/\partial r = 0 \quad$ in $\quad r = 0$ (5)

$\partial W/\partial r = 0 \quad$ in $\quad r = R_o$ (6)

$$\lambda \frac{\partial \tau}{\partial r} = \alpha\,(\bar{\theta} - \tau) - \alpha_m L\,(1 - \varepsilon)\,\{d_w F(\tau) - f(\bar{\theta})\} + g' \quad (7)$$

$$a_m \rho_s \frac{\partial W}{\partial r} + a_m \rho_s\, \delta \frac{\partial \tau}{\partial r} = - \alpha_m\,\{d_w F(\tau) - f(\bar{\theta})\} \quad (8)$$

where $g' = 0.25\, K_{tr}\, \varepsilon_e\, E$ is due to the direct solar flux.

Considering, as a first approximation, that τ and W can be represented by parabolic functions of the space variable r, the equations (1) and (2) can be simplified in the way detailed by Babukha and Schraiber {2} and generalized by the authors {3} to take into account, in the overall heat balance, the direct radiation flux on the fruit. Hence an equations set is derived whose solutions are the average values in the fruit of both temperature and water content, i.e. respectively $\bar{\tau}$ and $\bar{W}$

$$\frac{d\bar{\tau}}{dt} = \frac{5a}{5\lambda + \alpha R}\left\{ \frac{3\alpha}{R}\,(\bar{\theta} - \bar{\tau}) + L\rho_s \frac{d\bar{W}}{dt} \right\} + \frac{3\, g'}{\rho_s (c_s + \bar{W} c_e)\, R} \quad (9)$$

$$\frac{d\bar{W}}{dt} = - \frac{3\alpha_m}{\rho_s R}\, F(\tau')\,(d_w - H) \quad (10)$$

In deriving (9),(10), by an averaging procedure applied to (1),(2), as the influence of the change of phase has been proved to be negligible, ε has been chosen equal to 0 in equations (1) and (7).

Solar drying of plums (prunes d'Ente)

Equations (9) and (10) have been applied to the particular case of solar drying of plums (prunes d'Ente), produced in a period of the year with a humid and moderately hot weather. The solar radiation energy per unit area and per unit time is schematically described by

$$E = E_{mx}\, \sin(\pi\, t/p) \quad (11)$$

which corresponds to clear or uniformly cloudy days. This implies a mean temperature of the air in the system varying versus time according to

$$\bar{\Theta} = \bar{\Theta}_{mn} + (\bar{\Theta}_{mx} - \bar{\Theta}_{mn})\, \sin(\pi\, t/q) \quad (12)$$

Assuming that during the drying period the water content in the atmosphere outside the system is constant, the relative humidity of the gaseous phase circulating through the system is such that

$$H.F\,(\bar{\tau}') = \{H.F(\bar{\tau}')\}_{t=0} \quad (13)$$

Likewise, if the circulation of the air is due to free convection, the heat transfer coefficient can be expressed as

$$\alpha = \alpha_i + \Delta\alpha \sin(\pi t/q) \tag{14}$$

For the specific case considered in this work, mass transfer and adsorption-desorption coefficients have been chosen such as {4}

$$\alpha_m = \beta \ \bar{W}^{\gamma} \tag{15}$$

$$d_w = \begin{cases} \zeta\bar{W} + \eta & \text{for } \bar{W} < 1.3 \\ 1 & \text{for } \bar{W} \geqslant 1.3 \end{cases} \tag{16}$$

Results and discussions

The coupled differential equations (9) and (10) are solved numerically with the method developped by Bulirsh and Stoer {5} . This high performance method takes about 3 sec. with an IRIS 80 computer to obtain $\bar{\tau}$ and $\bar{W}$ corresponding to a drying period of 12 hours, with a precision better than 10^{-3}.

Before treating a real case, it has been verified that the last term of equation (9), which represents the direct exposure of the fruits to the solar radiation, is small compared with the other terms of the same equation; the predominance of these terms, due to the significant increase of the air temperature and decrease of the air humidity around the fruits when they are placed in the solar system, allow the last term of equation (9) to be neglected for the aim of simplification.

Figure 1 shows the sensibility of the evolutions versus time of both $\bar{\tau}$ and $\bar{W}$ to several characteristic parameters. The reference curves have been obtained with the values given in Table 1, when the weather conditions correspond to a clear day in regions of humid and moderately hot climate.

Results are also given on this figure for dry climate conditions, the humidity of the air being 0.3 instead of 0.8, with the same values for all the other parameters. It must be pointed out that a strong decrease in the humidity of the ambient atmosphere produces only a moderate reduction in the final value of the water content $\bar{W}^*$, confirming the ability of the solar drying even in relatively humid regions.

The other two curves, on Figure 1, are for (a) fruits with a 15% higher initial water content and (b) 25% lower values of both heat and mass transfer coefficients than those considered in the reference situation.

In order to analyse a complete solar drying sequence extended over several days, the evolution versus time of the average water content in the fruit has been computed for different combinations of clear or uniformly cloudy days; intermediate situations can be easily interpolated. It can be seen in Figure 2 that, assuming 0.25 as a convenient final water content for plums conservation, only about 1.7 drying days with a clear sky are necessary with a moderately hot and humid climate. For a sequence of three continuous uniformly cloudy days followed by a clear one, the total number of drying days is about 3.8, which is also quite adequate.

Fig. 1

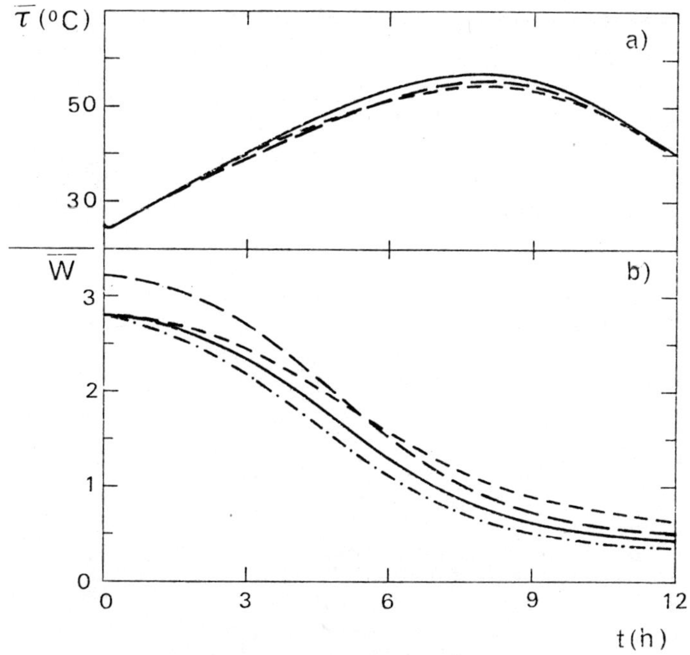

Influence of the parameters on the evolutions versus time of (a) the fruit temperature and (b) the water content.

———— reference curves obtained with the values given in Table 1,

—·—· decrease of the humidity of the ambient air H_i from 0.8 (reference value) to 0.3,

– – – 25% reduction from the reference values of the transfer coefficients α_m and α,

— — – 15% increase of the initial water content.

Fig. 2

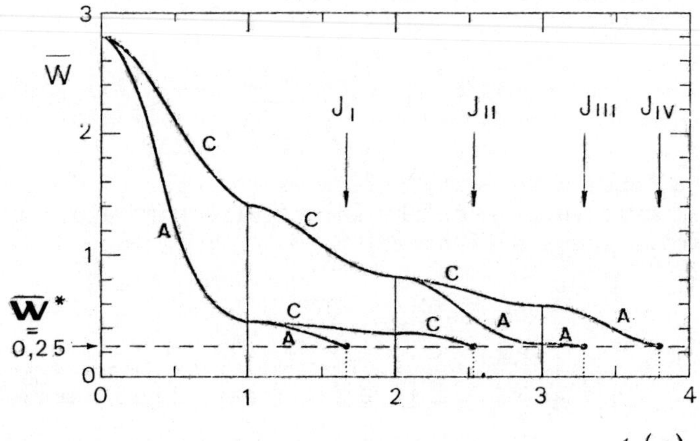

Numbers of days needed for the drying of plums (prunes d'Ente) for different weather conditions with the same solar system.

Each curve A corresponds to a clear day (reference values for the parameters, $\bar{\theta}_{mx} = 60°C$). Each curve C is associated to a uniformly cloudy day ($\bar{\theta}_{mx} = 40°C$ and $\Delta\alpha = 11,6\ W.m^{-2}.°C^{-1}$; the other parameters are chosen according to Table 1).

Four different sequences have been studied:

I . Sequence A-A	(all clear days)	J_I = 1.67 days,
II . Sequence A-C-C		J_{II} = 2.57 days,
III . Sequence C-C-A-A		J_{III} = 3.3 days,
IV . Sequence C-C-C-A		J_{IV} = 3.8 days.

Table 1 Initial and characteristic values for the drying reference curves of Figure 1

$\bar{\tau}_i$	60	°C	E_{mx}	1000	$W.m^{-2}$
$\bar{W}_i$	2.8	(-)	γ	1.16	(-)
a	$16\ 10^{-8}$	$m^2.s^{-1}$	η	0.48	(-)
λ	0.5	$W.m^{-1}.°C^{-1}$	ε_e	0.8	(-)
L	$2.47\ 10^6$	$J.kg^{-1}$	K_{tr}	0.9	(-)
ρ_s	1400	$kg.m^{-3}$	p	12	hours
c_s	$0.32\ 10^3$	$J.kg^{-1}.°C^{-1}$	q	14	hours
α_i	209	$W.m^{-2}.°C^{-1}$	$\bar{\theta}_{mn}$	25	°C
$\Delta\alpha$	29	$W.m^{-2}.°C^{-1}$	$\bar{\theta}_{mx}$	60	°C
β	$3.2\ 10^{-8}$	$kg.m^{-2}.s^{-1}.Pa^{-1}$	H_i	0.8	(-)
ξ	0.4	(-)	(-) dimensionless variable		

Efficiency of a solar drying system

In a general way the efficiency of a solar drying system can be approached through two main characteristic parameters:

- the "surface of the solar collector" per unit mass of fruits to be dried, i.e.

$$S = L \frac{\bar{W}_i - \bar{W}^*}{1 + \bar{W}_i} \Big/ \left[\eta_c \int_0^T E(t)\, dt \right] \qquad (17)$$

- the "drying capacity" given as the mass of fruits to be dried, per unit surface of the solar collector, along the overall collection period of time

$$C = (D/J_N)\ S^{-1} \qquad (18)$$

As an example, the application of these two last equations to the first sequence of Figure 2 (all clear days), assuming that

$$\eta_c = 0.5, \quad E_{mx} = 1000\ W.m^{-2} \quad \text{and} \quad D = 45 \text{ days}$$

yields to: $S \simeq 0.07\ m^2.kg^{-1}$ and $C \simeq 360\ kg.m^{-2}$.

Concerning the types of solar systems to be used, as previously mentionned, they can be:

- a solar collector with a relatively high efficiency {6} working at the same time as the fruits container,
- or composed of two different parts: a solar collector used as an air preheater, the fruits being placed in an other tank connected with the first part,
- or any combination of these two extreme possibilities.{7}

Nomenclature

- a thermal diffusivity of the fruit
- c heat capacity
- d_w fruit adsorption-desorption coefficient
- f^w water vapor partial pressure in the air
- g' direct solar flux on the fruits
- p duration of sunny hours in a day
- q half-period for the evolution vs time of the ambient temperature
- r space variable
- t time variable

- C drying capacity of the solar system
- D overall period of drying, in days
- E solar radiation energy per unit time and unit area
- F saturated water pressure of the air
- H relative humidity, dimensionless
- J_N number of drying days for one fruit
- K_{tr} transmission coefficient through the transparent sheet protecting the fruits from external pollution during drying
- L latent heat of vaporisation
- R external fruit radius
- R_o radius of the fruit stone
- S surface of the solar collector
- T drying time for one fruit, corresponding to J_N days
- W water content in the fruit, dimensionless ratio between the mass of water contained in the fruit and the mass of completely dry material

- α transfer coefficient between the air and the fruit surface (α for heat transfer, α_m for mass transfer)
- β,γ constant values in the definition of α, equation (15)
- δ thermomigration coefficient
- ε dimensionless rate of phase change between liquid and gaseous phases
- ε_e fruit emissivity, including a shape factor
- η constant value in d_w, equation (16)
- η_c solar collector efficiency
- θ temperature of the air in the near vicinity of the fruits
- λ fruit thermal conductivity coefficient
- ξ constant value in d_w, equation (16)
- ρ density
- τ fruit temperature (τ for local temperature and $\bar{\tau}$ for average temperature)
- $\bar{\tau}'$ average surface temperature of the fruit; in the model $\bar{\tau}'$ is taken equal to $(\bar{\theta} + \bar{\tau})/2$

Sub and Superscripts:

e	water	mn	minimum
i	initial	s	dry
m	mass	-	mean value
mx	maximum	*	final

References

{1} Luikov A.V. and Mikhailov Yu. A. - "Theory of energy and mass transfer". Pergamon Press, 1965.

{2} Babukha G.L. and Shraiber A.A. - "Interphase heat transfer in polydisperse gas suspension flows". Communication CT 2.5. 5th International Heat Transfer Conference, Tokyo, September 1974. "Heat Transfer 1974", 5, 69-73, J.S.M.E.-S.C.E.J., Japan, 1974.

{3} Piacentini R.D. and Combarnous M. - "Modélisation du séchage solaire de fruits. Application au séchage de prunes". Revue Internationale d'Héliotechnique de la COMPLES. Under press, 1977.

{4} Bertin R., Pierronne F. and Combarnous M. - "Modélisation et simulation du séchage des fruits en tunnel". Industries Alimentaires et Agricoles, 93, 1431-1436, 1976.

{5} Bullirch R. and Stoer J. - Numerische Math., 8, 1-13, 1966.

{6} Duffie J.A. and Beckman W. - "Solar Thermal Energy Processes", John Willey Interscience, 1974.

{7} Brace Research Institute - "A Survey of Solar Agricultural Dryers" Technical Report T99, 1975.

Assessment of solar applications for transfer of technology

J.K. Parikh
International Institute for Applied Systems Analysis*
Laxenburg, Austria

Abstract

For the large and increasing rural population in the developing countries, the solar pump could be an especially important application of decentralized solar energy. The additional food required to support the growing population in India, for instance, could mean an increase in the number of energized pumpsets from today's 2 million pumps to 20 million in coming decades.

The yields of farm crops depend crucially on the availability of water at certain periods during the season. Therefore, a solar pump would have to be designed to meet the farmer's requirements for irrigation under all possible field conditions. This means that a solar pump has to have a higher capacity to do the same amount of work done by a diesel pump. Our analysis, which takes into account escalation in diesel prices, shows that the break-even costs of a solar pump would be in the range of $250 to $600 per kW. The current price of solar engines are higher by a factor of more than 20. However, the costs can come down if the pumps are manufactured by the developing countries. If the engine is installed to obtain drinking water, then 6 to 8 hours running time per day may be sufficient. The solar pump could then have the same capacity as the diesel engine. In this case, the breakeven costs per kW of the solar engine would be $1200. (Of course, if neither diesel nor electricity are available in some of the remote areas, solar pumps could be the only answer.)

Moreover, even when economic solar pumps are developed, other factors based on climate, geography, locale, social and institutional environment have to be foreseen. Many of these are discussed at length in the text.

The considerations discussed here apply to all the applications where solar application is used for seasonal uses in view of the high capital costs where diesel engines with its low capital cost and also low operational costs (due to occasional seasonal use) will be preferable. The numbers given here are applicable to most of the developing countries and are not relevant for India alone.

1. Introduction

The intensity of solar radiation is low and therefore a considerable amount of land is required for utilizing solar energy. Solar applications therefore would be more suitable for the rural environment where the land is available easily rather than for the urban areas. Conventional centralized energy systems have not yet reached the large population in many developing countries. Solar applications which contribute a decentralized energy system could provide a significant improvement in economic productivity in the rural areas.

---

*Views or opinions expressed in this paper do not necessarily reflect those of the National Member Organizations supporting the International Institute for Applied Systems Analysis, or of the Institute itself.

According to an estimate of the United Nations [1], in 1970 the rural population of the world was 2.26 billion out of which 1.89 billion were in the developing countries. The percentage of persons living in the rural areas of the developing countries is expected to decrease from 75% in 1970 to 59.2% by 2000 A.D. However, their absolute number will still be 2.92 billion, a substantial increase over the present number. Figure 1 shows the region-wise rural population, as projected in Ref. [1].

Fig. 1

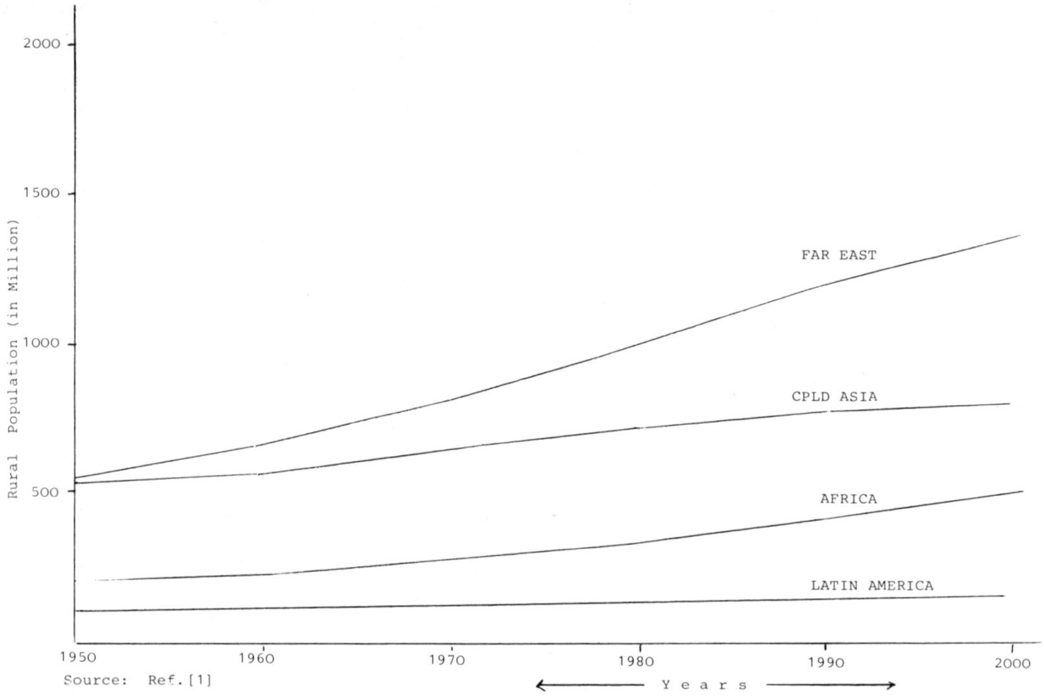

Past and projected rural population of developing regions

The energy requirements of rural people, although extremely low, are largely met at present by locally available non-commercial resources such as firewood, agricultural waste and dung. Yet, energy planners in the governments of most of the developing countries are concerned primarily with the development of large energy systems, appropriate for urban and industrial purposes. Although efforts are being made in most of the developing countries to expand rural electrification, its progress is slow because it is capital intensive, especially when it involves connecting remote villages to the network. Thus, there is a need for developing decentralized energy systems for the rural areas.

Scientists and technologists do come up occasionally with solutions for decentralized energy systems. When these are not adopted they generally complain about the difficulties of technology transfer, resistance from established interests etc. Though these obstacles are not to be underestimated, their claims about the relevance of their research and development are many times not found to be valid for actual adaptations in the field conditions when all the facts are put together. Therefore, a careful appraisal of the difficulties of the transfer of technology is essential.

In this paper we consider first the issues that are important in assessing technology. Then some of the solar technologies are evaluated keeping these issues in mind. Since photovoltaic cells are very expensive at present, we have considered for this analysis only decentralized low thermal devices. In particular solar pumping is evaluated in detail as a case study.

Algebraic expressions can be used for application to any country, although numerical results for the specific case of India are given.

2. Issues Concerning the Transfer of Technology

Here "Technology transfer" would mean the transfer of invention from a laboratory to the field. It has to be recognized that the users cannot run an experimental energy system. Due weight has to be given to the perfection of the invention and the development of institutions which are required, such as establishments which look after the user's problem. The user's viewpoint could be classified into two categories; techno-economic and social or relating to the operating environment in which the technology has to be used. In general, the following points need to be considered.

(i) Private and social benefits.
The benefits, savings, etc., are often calculated on national, state or village levels and not for the consumer who is going to use it. Though benefits at the national level, such as saving of foreign exchange, curbing environmental degradation, overall health effects, etc., are important, they are meaningful only if the new technology is acceptable to the user. If the user does not benefit, yet an invention needs to be promoted for national or social benefits, the individual user has to be compensated if he is to be induced to use it. This requires a national policy action where subsidy, financing facilities, tax rates, etc. have to be introduced to promote better technology.

Thus, the cost-benefit analysis should be done also from the user's point of view along with the analysis from the social viewpoint. One then identifies the loss, if any, that the user has to incur and to what extent the government might subsidize him judging from the indirect costs the society has to bear if the new technology is not promoted.

(ii) Comparison with other alternatives.
The economic benefits to the user should be calculated keeping in mind the best possible alternatives that a user has. For example, if the advantages of bio-gas plants are calculated by taking kerosene or even coal as the alternative, they would look substantial. But actually the comparison has to be made with the cheapest possible alternative, i.e. burning dung and purchasing fertilizer, if at all he needs it. Of course, the fringe benefits and conveniences and nuisances of both the alternatives should be weighed appropriately. Only then one understands why certain innovations are not catching on. In addition, possible future developments in the existing alternatives should also be considered.

(iii) Scale of technology.
Some technologies may turn out to be inappropriate--economically or managementwise--if the proper scale is not chosen. For example, in some situations many small solar pumps may be more expensive than a large pumping station. Yet the small pumps may be preferable when the management problems associated with the different scales are considered. Again, giving an example of bio-gas technology, our earlier analysis [2] shows that a community bio-gas plant may be more economical and socially desirable than family bio-gas plants.

(iv) Introduction of technology.
The manner in which a technology is introduced determines its success. For example, adversely affected or less beneficiary groups may offer resistance. Besides, at the planning stage itself, problems of cooperation, maintenance and repair would have to be foreseen.

(v) Compatibility with the environment:
If an invention requires a change in lifestyle or is in conflict with the surroundings, it will face difficulties in its adoption. In such a case, the strength of the existing establishment of older technology should be carefully assessed and whether or not the society is ready for the change should be considered.

(vi) Acceptance of technology.
An invention has to be appropriate for the kind of use for which it is meant. For example, as will be demonstrated later in this paper, there is a need to consider the manner in which a pump is presently utilized while designing a solar pump for agricultural purposes.

It is therefore necessary that a government with limited resources should evaluate new technologies carefully so that only appropriate ideas are encouraged. The development of inappropriate inventions may waste precious scientific manpower and limited research funds and also cause a loss in the credibility of new technologies in general. Although these may appear to be a matter of common sense, there have been many failures of new technologies due to neglect of simple practical considerations in the past. Therefore we attempt to analyze the difficulties of transfer of technology for one of the applications of solar energy, within the context of the above mentioned criteria. Although the general framework of the analysis is applicable to any country, a case study for India is carried out.

3. Solar Pump

Solar pump for irrigational purposes would be a significant application of solar energy for the developing countries where 40 to 50% of GNP originates from the agricultural sector for which water is an essential input. Table 1 provides some relevant data for energized pump-sets and their electricity consumption in India. About 9% of the total electricity consumption in India is accounted for by energized pump-sets alone [3] inspite of the fact that hardly 20% of the villages were electrified in 1967 as shown in Table 2. The number of pumps required [4] in the next 2 to 3 decades can be more than 10 million. Table 2 also shows [5] that the rate of electrification for small villages of 500 persons is much slower than for large towns. In view of the slow electrification of the rural areas, the importance of the agriculture sector and the necessity of energy for pumping, solar pumps would be an extremely relevant application.

(i) Techno-economic considerations.
The conditions for solar pumps to be acceptable would be that they provide adequate pumped water, are cheaper than the existing alternatives, or alternatively, are more convenient so that farmers would be willing to pay a higher prices.

We shall develop a general framework for such a techno-economic comparison between any two alternatives. The symbols being used in the calculations are explained in Table 3.

Table 1

PUMPSETS AND ELECTRICAL ENERGY CONSUMPTION IN INDIA[a)]

Year	No.of sets in operation	Energy consumption (10^9kWh)	Total electricity consumption 10^9kWh	Consumption per pumpset (kWh)	Consumption in kWh per kW of connected load
(1)	(2)	(3)	(4)	(5)	(6)
1966-67	649182	2.107	33.26	3245	842
1967-68	847357	2.585	36.76	3050	814
1968-69	1088774	3.466	41.46	3183	834
1969-70	1342006	3.770	45.02	2809	738
1970-71	1642006	4.110	48.65	2503	657
on 31.3.74	2444599				
on 31.3.79 (Target)	4022790				
1983	6.5 million)[b)]				
1990	12.0 million)				
2000	20.0 million)				

Source: a) Ref. [3]
b) Ref. [4] p.55.
This projection takes into account the growth of population and the need for additional food production and is compatible with the ground water potential of the country.

Table 2

VILLAGES ELECTRIFIED IN INDIA

Population Range (1961 census)	Total	Number electrified as on				
		31-3-61	31-3-66	31-3-71	31-3-72	31-3-73
Up to 499	351,653	3,986	10,265	31,518	39,730	46,665
500 to 999	119,086	4,306	9,787	26,436	32,602	37,880
1000 to 1999	65,377	5,918	11,567	25,715	27,971	31,586
2000 to 4999	26,565	5,458	9,441	17,036	18,326	19,922
5000 to 9999	3,421	1,319	1,963	2,674	2,753	2,913
10000 to above	776	560	647	702	712	729
Total	566,878	21,547	43,679	104,091	122,094	139,695

Source: Ref. [5]

Table 3

SYMBOL DEFINITIONS FOR TECHNO-ECONOMIC EVALUATION

	solar	electric	diesel
Capital costs	K_s	K_e	K_d
Life time of the pump	l_s	l_e	l_d
number of pumps required for a period T of service	n_s	n_e	n_d
Price of fuel per unit of fuel as a function of distance and time	-	P_e (kWh)	P_d (liter)
Quantity of fuel required per year	Sx365	q_e	q_d
Maintenance costs	m_s	m_e	m_d
Operation costs	O_s	O_e	O_d
Annual work done	W_s	W_e	W_d

In general $O = m + p \cdot q$ for a given option.

The annual discount rate d for all options.

The subscripts are dropped during general discussion of any engine.

There are many ways in which the costs could be worked out, some of which are illustrated below. In the presence of electricity, the solar pump would have to compete with electrical pumps and with the diesel pumps in the absence of electricity.

(ii) Average annual costs per installed kW.
We assume that the annual cost of the loans made to finance the installation would be equal to half the rate of interest (discount) plus the operation costs. Neglecting inflation but considering depreciation, we get the following equation:

$$K_s\left(\frac{d}{2} + \frac{1}{l_s}\right) + m_s = K_d\left(\frac{d}{2} + \frac{1}{l_d}\right) + q_d p_d(s,t) + m_d \qquad (1)$$

Somewhat similar expression has been already derived by A. Takla [6]. However, such an expression does not consider the difference in the work that can be done because of the different availabilities at night of the two alternatives being compared.

Before we consider this problem, we write an exact formula for the discounted costs instead of the approximate form (1),

(iii) The discounted costs for a period of service T.
The discounted cost of installing a capacity at the initial period for any option would be:

$$C = K + \sum_{t=1}^{\ell} \frac{O(t)}{(1+d)^{t-1}} \qquad (2)$$

$$\text{Here } O = m + p \cdot q \qquad (3)$$

In order to compare solar and diesel engines, they must provide service over the same length of time T, since the lifetimes of both the options may be different. We choose the time T such that

$$n_s l_s = n_d l_d = T. \qquad (4)$$

Thus, if the lifetimes of the solar and diesel engines are 20 and 5 years respectively, then one would require 4 diesel engines over these 20 years, requiring new investment every 5 years which have to be discounted to the initial period.

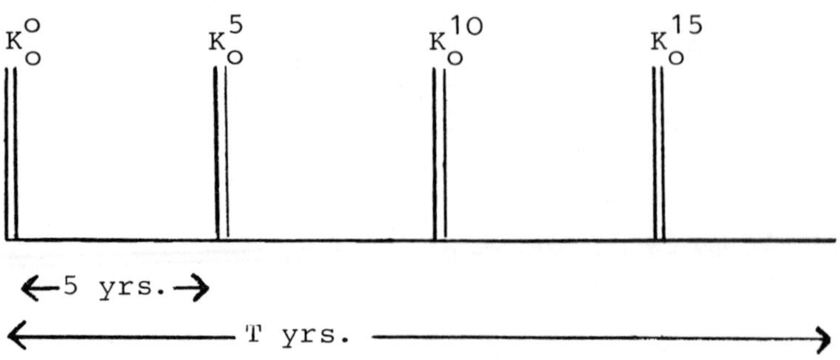

A general formula for discounting this over the period T is given below.

$$c = \sum_{n=1}^{T/\ell} \left(\left(K + \sum_{t=1}^{\ell} \frac{O}{(1+d)^{t-1}}\right) \frac{1}{(1+d)^{(n-1)\ell}} \right) \qquad (5)$$

On the other hand, without a storage capacity the solar pump cannot be operated for the same hours as the diesel pump. In using the formula we have to take into consideration the work that would be done by both these pumps in a day. For this we consider two cases:

a. The rate of pumping:
In some areas the rate at which water recharges may be slow and therefore there may be an effective limit to the rate at which water can be pumped. In this case, we have to compare two pumps of the same capacity. The fact that the solar pump can be operated only for about 6 hours and the diesel pump for 18 to 20 hours make the two pumps non-comparable. In fact, the solar pump without adequate storage may not be considered a feasible option in this case.

b. A comparison of equivalent work done:
A solar pump works with the average efficiency e_s for h_s equivalent hours of full capacity where a solar radiation s kWh/m^2/day is available. The collector area required for the installed capacity c_s is A.

$$\text{Daily work done in kWh } W_s = s\ e_s\ A \tag{6}$$

The diesel pump on the other hand can operate for a much longer time, let us say h_d hours. This may mean higher consumption of the fuel but better utilization of the installed capacity, which is denoted by c_d.

$$\text{Daily work done in kWh } W_d = h_d\ c_d \tag{7}$$

$$\text{We must have } w_s = w_d. \tag{8}$$

$$\text{Therefore } s \cdot e_s \cdot A = h_s c_s = h_d c_d \tag{9}$$

Here $e_s = e_c \cdot e_p$ where e_c and e_p are the efficiencies of the collectors and the pump respectively.

This concept is explained in Fig. 2. Eq. (4) can determine the value of c_s for equivalent work as well as the necessary area of the collectors.

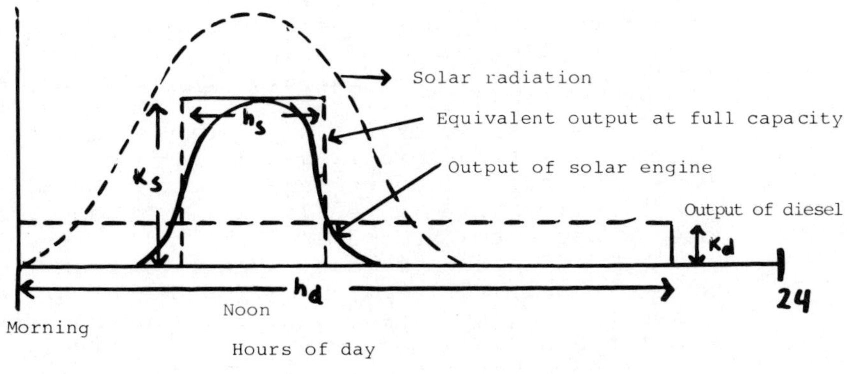

Fig. 2
Daily variation of radiation and output of solar and diesel engines

4. Numerical Comparison Between the Alternatives

Having developed a general framework for a techno-economic comparison, we compare the alternatives for a user. In doing so, one should consider uncertainties in various matters such as future improvements in the efficiencies and costs of solar pumps and escalation of diesel prices etc.

Fig. 3
A schematic design of a solar pump

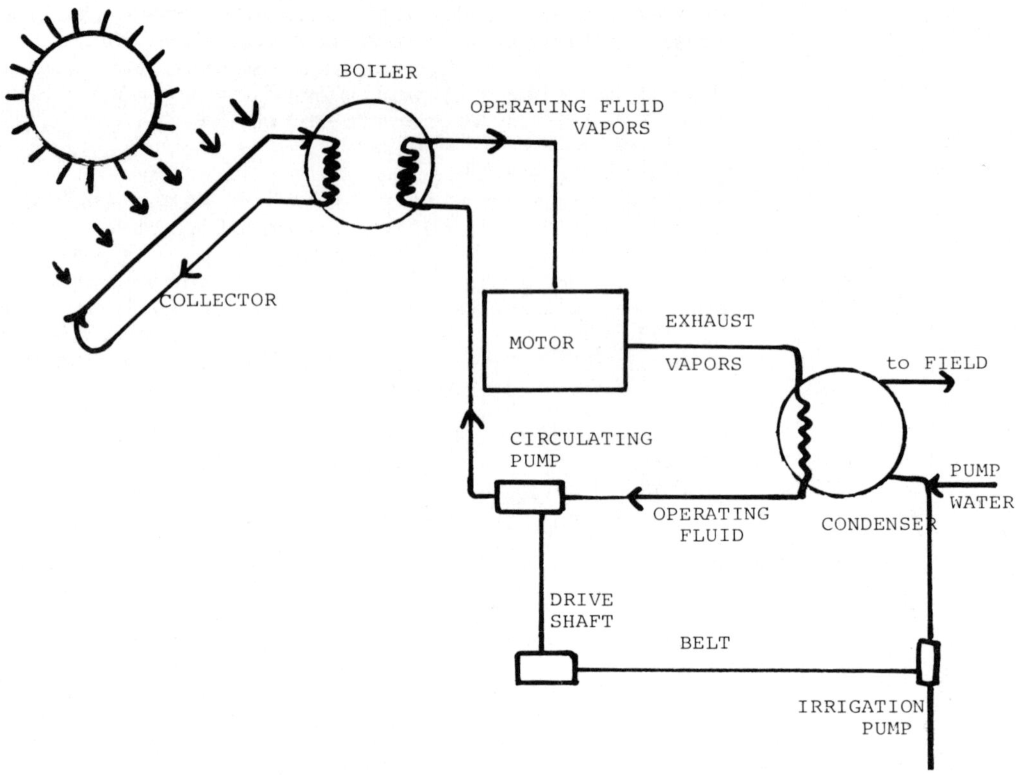

(i) Present design and feasible technical improvements. As we are concerned with a pump for agricultural needs, we do not consider low temperature pump operating only on temperature differences as this technology is not yet developed enough to give the required output but consider an engine driven pump as shown in Fig.3.

If manual tracking system operating with concentrators utilizing Fresnel lenses are developed, which seems to be a realizable goal, then an optimistic figure for the future efficiency could be taken to be 10% along with other improvements in the design of the solar engine and collectors.

(ii) Numerical values.
To be favourable to the solar pump, we assume that electricity is not available in the region and the alternative is to use a diesel pump. Considering the data in Table 1, it seems that on average, farmers' requirements are met by a pump of 4 kilowatt (kw) capacity which runs 1000 kWh/year. We compare a solar engine and a diesel engine to drive such a pump. The lifetimes l_s and l_d are 20 to 5 years respectively; the capital cost today is about Rs 6,000 ($600) for a 4 kW diesel engine (K_d = $150/kW). We consider lubrication costs separately for a diesel engine. Thus, the m_s and m_d would be $50 each per year. The operation cost of a solar engine is only $50 per year. The diesel oil and lubricants consumed by a 4kW diesel pump for running 1000 hours would be 1 ton and 0.028 ton per year respectively. The costs of diesel and lubricant at market prices P_d and P_e, are $150 and $20 respectively. Therefore, using Eq. (3) we get O_d = $220. We consider three scenarios, one with no escalation, and other two with 5% and 10% annual increase in the relative prices of diesel and lubricants. In the discounting procedure, costs are calculated at constant (current) dollars and only that price increase which is over and above inflation is considered.

However, operating conditions in the field are such that pumps have to run 18 to 20 hours, or even 24 hours. A diesel

Table 4 COMPARISON OF COSTS OF SOLAR AND DIESEL ENGINES

	Solar	Diesel Engine (4kW,4000kWhr/yr) Annual increase in relative price of diesel 0%	5%	10%
A. Present Discounted costs for a 20 year service period				
i) Costs of Capital	k_s	1348	1348	1348
ii) Operating costs incl. fuel, lubricants and maintenance	$425	1872	2543	3825
iii) Total	k_s+$425	3220	3891	5173
B. Break-even capital cost of a solar engine for irrigation ($/kW)				
i) Equivalent solar capacity c_s=8kW		349	433	593
ii) Equivalent solar capacity c_s=12kW		233	289	395
C. Break-even capital cost of a solar engine for pumping drinking water ($/kW) Solar engine capacity same as diesel c_s=4kW		698	866	1186

Assumptions:

1. Discount Rate = 10%
2. Life time, solar 20 years; diesel 5 years.
3. Diesel engine is 4kW, runs 1000 hours a year and during peak period is required to run 18 hours.
4. Solar can operate 6 to 9 hours a day
5. Diesel consumption is .2507/kWh, price is $150/ton.
6. Non-fuel operating costs are $50/year for solar as well as diesel pump.

engine can be run round the clock, whereas a solar engine without storage may run for 6 to 9 hours/day. Using Eq. (8) and (9) the capacity of solar engine, c_s, would have to be 2 to 3 times that of the diesel engine depending on whether it runs for 6 hours or 9 hours. Therefore, we consider these two possible capacities with the three scenarios of oil prices. Using Eq. (5) for the discounted costs, we get the results which are summarized in Table 4.

Part A in Table 4 shows the discounted costs for a 4 kW diesel engine and an equivalent solar engine under the various assumptions. The break-even costs of solar engine required to do the same amount of work has been given in Part B. The break-even capital costs, k_s per kW, vary from $233 to $590, depending on whether one requires three times larger solar pump or two times and whether one considers 0%, 5% or 10% increase in oil at constant dollars.

The market price of a solar pump at present is 15 to 20 thousand per kW. Based on our analysis a reduction in price by a factor of 25 to 70 is required before a solar engine would be economically acceptable for driving an irrigation pump.

If the pump is only to be used for obtaining drinking water, then it can be of the same capacity as the diesel pump and may run 4 to 6 hours. This has been also compared in Part C of Table 4. This requires a reduction of price by only a factor of 10 to 20.

(iii) Validity of assumptions.
Most of the assumptions made in the above analysis are quite generous to the solar engine as can be seen from the following:

a. Technical assumptions: A solar engine with a lifetime of 20 years is not yet available. Besides, a solar engine with twice as much capacity of the diesel engine would also require a hydraulic pump--which is driven by the solar engine--of double the capacity as that used by the diesel engine. This additional cost of hydraulic pump for the solar engine is not considered in the calculation.

The present analysis assumes 6 to 8 working hours without storage. So far, the working hours of a solar engine do not exceed 8 hours with storage. The engine designed by the National Physical Laboratory of India works 4 hours per day with storage. Storage requires additional collector area as shown in Eq. (6), the costs of which should also be included.

If adequate storage were to be provided so that the solar engine can be run for 18 hours, the capacity of the solar engine need not be larger than that for the diesel engine. The break-even cost of such a solar engine with the costs of collectors and storage can be as high as $1180 per kW.

b. Economic assumptions: Although an electrical pump provides cheaper alternative, the cost comparison has been made with a diesel pump. Since we are looking into a possibility of nation-wide adoption, the question of unavailability of diesel in the individual remote areas has not been considered. These areas might find solar pump to be useful in the near future, especially for drinking water, as it may be the only feasible technology. However, we do consider an 8-fold increase in the diesel prices (10% annual increase) over 20 years relative to other prices which are kept constant.

However, if the solar pumps are manufactured in the developing countries, they could be cheaper than the current quotations. For example, the pump developed in the laboratory in India [7] has material cost of $1200/kW. However, much progress is to be expected and it has to be seen what the costs of a commercial solar pump would be in the developing countries.

5. Operational Problems of Solar Pumps

Assuming that we have a solar pump which is of a comparable cost to the other alternatives, what are the other factors that need to be worried about? In the above mentioned analysis, only the average solar radiation had been considered for the sake of simplicity.

Moreover, the requirement of water for the crops follows a certain pattern while the demand for water depends upon the crops and cropping pattern, the supply would be determined by the arrival of the rains, the depth of the water table and the rate of recharge. These are the most crucial factors which determine the acceptability of a pump.

(i) Climatic and local variations.
The intensity of solar radiation changes from month to month. The efficiency of utilization depends on the intensity of solar radiation, the temperature, the cloud cover etc. In Table 5, monthly variations of solar radiation, utilization efficiency (Ref. [8]) and utilizable solar

Table 5

MONTHLY VARIATIONS OF SOLAR RADIATION AND TYPICAL EFFICIENCIES IN THE TWO CITIES OF INDIA

		Nagpur			Jodhpur		
Month	Days	Average radiation cal/cm²/day	Utilization efficiency	Monthly Availability cal/cm²	Average radiation cal/cm²/day	Utilization efficiency	Monthly Availability cal/cm²
(1)	(2)	(3)	(4	(5)	(6)	(7)	(8)
January	31	460	88	12549	410	84	10676
February	28	510	83	11852	480	84	11290
March	31	570	76	13429	560	77	13367
April	30	610	72	13176	630	78	14742
May	31	630	70	13671	680	83	17496
June	30	480	40	5760	680	73	14892
July	31	400	23	2852	530	48	7886
August	31	380	23	2709	480	43	6398
September	30	480	51	7344	540	54	8748
October	31	500	69	10695	510	84	13280
November	30	480	87	12528	440	91	12012
December	31	420	84	10937	380	88	10366
Total	365			117502			141155

Source: Columns 3, 4, 6 and 7 are from ref. [8].

energy are given for two places, namely Nagpur, Madhya Pradesh (Central India) and Jodhpur which is in the Western region near Rajastan Desert.

One can see that in Nagpur the utilizable energy drops by a factor of 5 between the months May and August. In fact, these are the months when water is required for cultivation. The reason why the solar radiation drops is that it rains in this period. In case the rains get delayed and it is nonetheless cloudy, then the solar pump installed may not be useful, unless the collectors also collect diffused radiation and desired water is provided. Thus the availability of water at a critical period, which determines the success of a crop is essential.

(ii) Pumping pattern.
In hot regions, some of the farmers may prefer to pump during the evening or night time so as to save loss of water due to evaporation. In such cases, storage may be essential.

(iii) Availability of area in the farm for solar collectors.
The collector area A per kW required for a pump capable to work h_s kWh/day is given by Eq. (9) which is recasted below:

$$A = h_s/s \cdot e$$

This could mean collector area of $100m^2$ for a 4kW pump.

In the developing countries, farms are small in size and an average farmer may not be willing to allocate even a small portion of extraagricultural land for the collectors when it is an excess compared to the areas required for alternative pumps. If these collectors are to be placed such that it obstructs sunlight for the other crop then it may not be a preferred alternative unless the farmer is willing to grow types of vegetables which can be grown under the shade and grow other crops in the remaining land.

These and other factors which may be encountered in the field are summarized in Table 6.

Table 6 FACTORS TO BE CONSIDERED WHILE DEVELOPING A SOLAR PUMP

- Availability of land for the solar collectors
- Compatibility of possible peak load with the quantity of water required, i.e. water pumped in comparison with its requirement over a day
- Maintenance: i) availability of spare parts and necessary services
  ii) availability of skills for repairs
- Adequacy of radiation and its availability especially during cloudy days in the cropping season (seasonal and monthly fluctuations)
- Compatibility of water table with the possible capacity of the presently available pump.
  On the other hand, if the rate of water recharge is small, the pump would have to run at low speed but continuously.

References

1 United Nations: Selected World Demographic Indicators, ESA/P/WP 55, Population Division, Dept. of Economic and Social Affairs, New York (1975)

2 Parikh, J.K. and Parikh, K.S., The Potential of Bio-Gas Plants and How to Realize it, Proceedings of UNITAR Symposium on "Microbial Energy Conversion", Göttingen, F.R.G. (1976)

3 Govt. of India, Ninth Annual Power Committee Report, Ministry of Irrigation and Power, New Delhi (1972)

4 Parikh, K.S., Second India Studies: Energy, McMillan Press, New Delhi (1976)

5 Central Electricity Authority, Government of India, New Delhi (1974)

6 A. Takla, The Development of Solar Energy Utilization in Developing Countries, UNIDO/IOD 59, Vienna (1976)

7 V.G. Bhide, Conversion of Solar Energy into Mechanical and Electrical Energy, report from National Physical Laboratory, New Delhi (1976)

8 G.A. Löf, J.A. Duffie, C.O.Smith, World Distribution of Solar Radiation, Solar Energy Laboratory, The University of Wisconsin, (July 1966)

Giovanni Francia
Instituto di Meccanica Applicata alle Macchine
Universita di Genoa
Consultente dell'Ansaldo
Societa Generale Elettromeccanica S.p.A.

Editor's note

In view of the important implications of solar electricity generation as regards providing power to human settlements this paper was invited from Professor Francia who has been one of the pioneers in this field. In this paper many practical aspects of solar power stations are being discussed which we think will be of interest to people involved with solar building design and construction.

Abstract

In this work solar plants are described and the properties of anti-radiating structures are briefly mentioned. There are then reports on present works and some information about future programs.

In 1964, with the financing of CNR for the University of Genoa, in the proximity of Genoa, the S.Ilario solar station was built for the installation and the study of solar plants: these plants can provide steam at the temperatures and the pressures required by modern steam turbines.

The first plant consisted of a mirror field of 7 x 7m: 121 mirrors, for a total surface of 30 sq.m, reflected the sunlight on a boiler hung at about 3m from the ground. Each mirror was carred and moved by a kinematism which transform the 15°/h uniform rotation into the complex motion necessary to follow the sun. The plant provided, on the average, in summertime, 19 kg/h of steam 100-150 ata, 500°C during six hours per day.

In the winter of 1965-66 the boiler and the mirrors were replaced; the second plant has operated throughout 1966.

The third plant has been designed and built on the basis of all the experimental results: a field of 271 mirrors, for a total surface of 205 sq.m, reflected the sunlight on a boiler hung at 11m from the ground. The plant provided up to 130 kg/h at 150 ata, 600°C.

In the winter of 1971-72 modifications were made; in particular the mirror field has been reduced from 205 sq.m to 135 sq.m.

The third plant is still operating today under these conditions; works are being carried out in order to transform it into a field for testing boilers.

Present programs

The Consortium ANSALDO-MBB has been working since 1974 on the basis of the results of S.Ilario in order to transfer them on a commercial level.

Last April a solar plant, of the type similar to S.Ilario's, for steam generation up to 600°C, 150 atm., was sent by Ansaldo to the Georgia Institute of Technology in Atlanta.

Ansaldo will participate in Consortium with the German Company MBB in the construction in Italy of a demonstrative solar plant having an output of 1 MW/el.

The French Consortium CETHEL will also participate in this project, as well as ENEL which will co-operate with Ansaldo for the Italian part. Studies and plans of solar plants of low output for water pumping and energy production for small communities are being carried out.

1. Anti-radiating structures

The solar plants which we are describing here have all in common the anti-radiating structures which the author submitted for the first time at the United Nations World Conference held in Rome, August 1961 (1), (2).

These consist of honeycomb type structures or of parallel planes perpendicular to the surface to be protected (Figure 1). Both are to be made of a material transparent to solar rays and black to thermic rays (glass, plastic materials, etc.).

Fig. 1
Anti-radiating structure

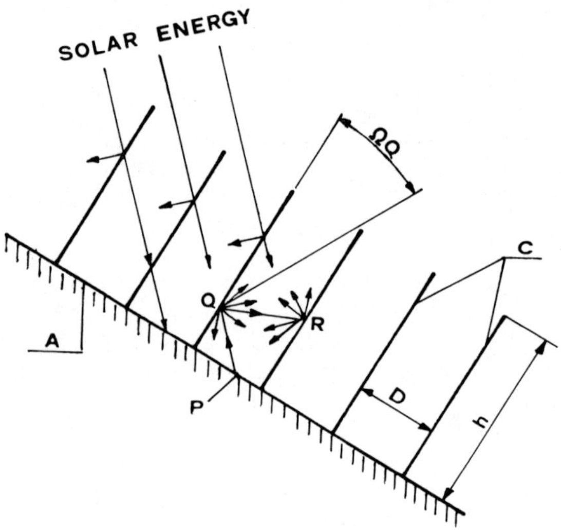

The solar rays which meet with the structure walls partly go through them and partly are reflected but the whole of their energy shall reach the absorber; the anti-radiating structure does not limit the incoming solar energy but for the small part absorbed by the material.

On the contrary, when a thermic ray emitted by a point P of the hot absorber meets with a wall in Q it is absorbed; the wall gets hot and radiates in all directions: part of it is lost in the solid angle which from Q sees the free opening, another part hits the other walls of the cell and repeats the phenomenon. The absorber shall thus recover part of the lost energy which is as great as the diameter of the cell referred to. (3, 4, 5, 6).

The calculation evidences how with the honeycomb type structure the reduction in losses is given by factor 0,7D/H, where D is the mean diameter of the base and H is the height.

The formula is non-dimensional and therefore the height can be reduced as necessary provided D/H remains constant; losses due to conduction shall be taken into account where atmosphère is present: the structure eliminates convective motions; under vacuum height can be reduced down

Fig. 2
Anti-radiating
structure

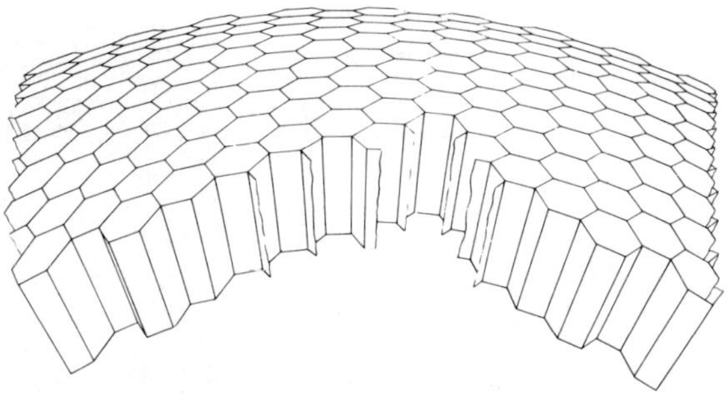

to a few millimeters and the structure can take the form of a very light covering which is transparent to the solar rays and not to the thermic ones (Figure 2).

The rigorous experiments which have been carried out both in air and under vacuum by the Marseilles University (7) and, later on, by Euratom (8) and by the Lockheed (9) confirm this theory.

Figure 3 shows the efficiencies of an absorber, protected by a honeycomb structure, as function of the ratio h = H/D between height and mean diameter of the cells, respectively for 1000 and 20.000 W/m2.

Fig. 3
Graphics of efficiencies with regard to temperatures and ratios h (height/diameter)

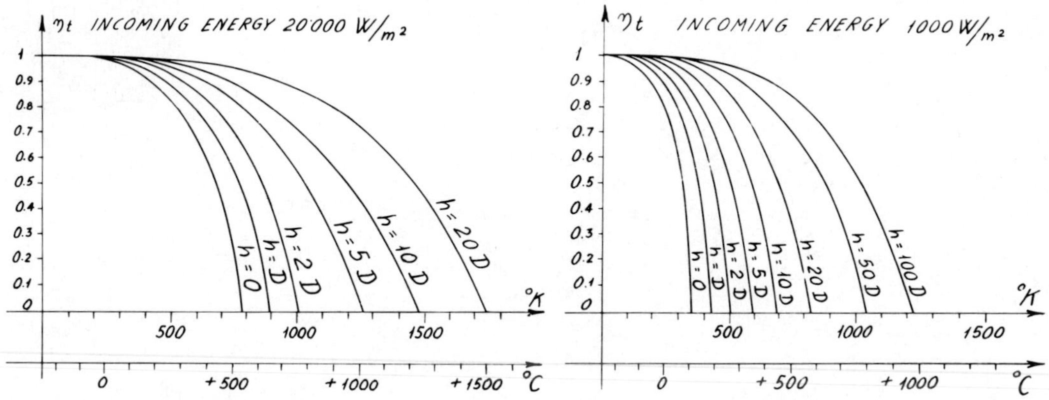

2. Solar boiler at Cesana Torinese

The first solar boiler for high pressure and high temperature superheated steam, utilizing the anti-radiating structure, was built by the author in 1959/60 at Cesana Torinese, 1350m above sea level (2). Three coaxial truncated cones starting from a 3.20m diameter concentrated the solar rays upon an 0.80m diameter boiler with a 16x concentration. The boiler was protected by means of an anti-radiating structure consisting of about 2000 glass tubes, 14mm in diameter, 250mm length and 0,2 ÷ 0,3mm thickness (Figure 4).

The glass tubes were surrounded by a cylinder with reflecting walls. In order to increase absorption the face of the boiler looking towards the sun was spirally grooved like a record. By means of a hand control it was possible to follow the sun: the supply also included thermocouples, volumetric pumps, gauges, etc.

Within the record-shaped boiler water ran in cold from the edge and the superheated steam went out from the center.

Near noontime as much as 7 kg/h of superheated steam at 100 atm, 500-600°C with 70% of efficiency, were obtained. It is to be pointed out that - without this honeycomb structure - efficiency decreases, with the increase in temperature, till it reaches zero at about 400°C.

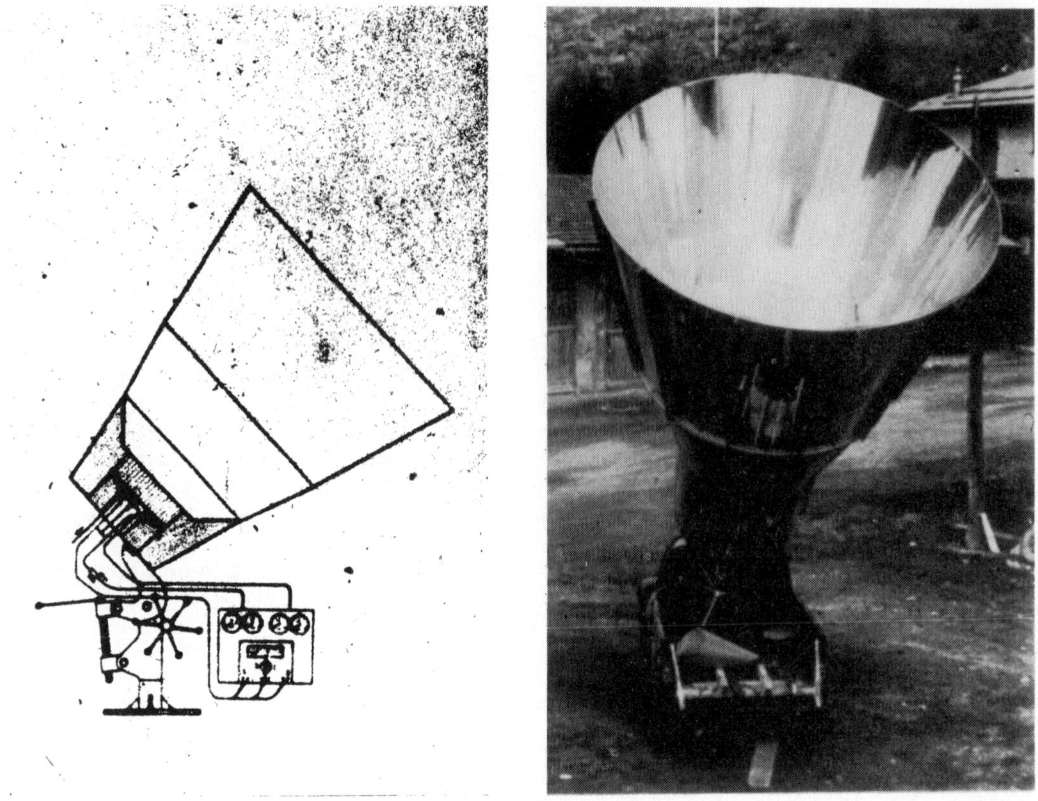

Fig. 4
Solar boiler at Cesana Torinese (1960).

3. Linear solar boiler at Marseilles

The Faculté de Eliotechnique of the Marseilles University, supervised by Professor Perrot, was the first to study the anti-radiating structures and in 1963 with the financial support of DGRST and OTAN it was decided, together with the Genoa University, to build an element of linear solar boiler with the purpose of experiencing the anti-radiating cells on an industrial scale. Basing on a design of the author, a prototype of linear boiler was built in Genoa with the purpose of being utilized for the Marseilles solar plant.

This boiler consists of 7 cylindric-parabolic mirrors, 1 x 8m, arranged in the South-North direction and reflecting the solar rays on the 0,25 x 8m boiler (Figure 5) which is located above them at a height of about 5 meters; for following the motion of the sun all mirrors rotate at the same speed around their longitudinal axis.

The 7 meters of the total width of the mirrors send the solar rays in the 0,25m of the boiler's entrance: thus we have a 28x concentration.

A nest of water tubes is arranged in the upper part of the boiler: the vertical walls going from the boiler's entrance to the nest of tubes are mirror-like in order to reflect on the tubes the energy which arrives in a slanting direction. The anti-radiating structure - arranged under the nest of tubes and consisting of about 1000 pyrex tubes, diameter 40 x H 250mm - cuts down to a minimum the thermic losses due to re-irradiation and ventilation.

A 12-point recorder measures the temperatures; pumps, gauges and flowmeters are also supplied (10).

The boiler supplies as much as 38 kg/h of superheated steam at 100 atm, 450°C with the sun at 1000 W/m2: efficiency exceeds 60%.

When the linear solar boiler at Marseilles was designed we knew little about the solar boilers and the optimal concentration; the recurring troubles caused by temperature peaks seemed to advise us against trying high concentrations.

However, when we set about studying more in detail the solar boilers as such, we realized that these dangerous superheatings were caused by an incorrect geometry of the nest of tubes which utilized only the least part of the very high characteristics of the radiant energy. A correct geometry would have allowed 500x and higher concentrations, with the advantage of reducing of at least a factor 10 the area of the entrance, and the volume, the cost and the weight of the boiler.

Fig. 5
Solar boiler at Marseilles (1963).

4. Point-shaped solar boiler

Concentrations of 500x and higher require point-shaped boilers and a field of hundreds of mirrors for following the motion of the sun and for constantly reflecting its rays into the boiler. In 1964, with the help of CNR (National Research Council) and NATO subventions, and with the patronage and the friendly co-operation of Professor A. Capocaccia, at that time Dean of the Engineering Faculty and Principal of the 'Instituto di Meccanica Applicata alle Macchine' of the Genoa University, the solar plant of S.Ilario near Genoa was built with the purpose of constructing and testing solar plants.

The first point-shaped solar boiler - 'tower-type' as it is called today - was then designed and built.

The mirror motion can be obtained in many ways, as for example:

(a) photoelectric control of each mirror;
(b) central control by means of a computer;
(c) kinematisms which turn a rotation parallel with that of the earth into the motion necessary to the mirror for following the sun.

Technique (c) has been selected and a kinematism - which can be manufactured on an industrial scale - has been designed and built.

5. Kinematisms

The kinematism is based on simple geometrical properties. The normal at the mirror, which is the bisecting line of the angle formed by the direction of the sun and the direction of the boiler, can be materialized as base AB of the isosceles triangle AOB with AO directed towards the boiler and BO directed towards the sun: OA will obviously be stationary and BO will move with the sun.

If BO is connected with axis PO parallel to the polar axis, with $B\hat{O}P = 90^{\circ} - \delta$, a $15^{\circ}/h$ rotation of PO will allow to follow with fairly good approximation the motion of the sun.

Such a scheme cannot however be realized by simply materializing by means of rods the isosceles triangle AOB, in view of the fact that under certain conditions the rod materializing AB side would cut rod PO.

This trouble can be eliminated by shifting, parallel to itself, the rod materializing side AB: this solution has been adopted in the first kinematism (Figure 7).

Figs. 6, 7

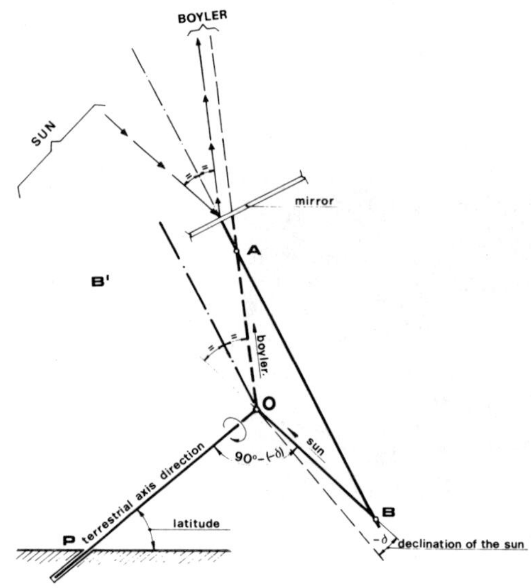

Geometry of kinematism

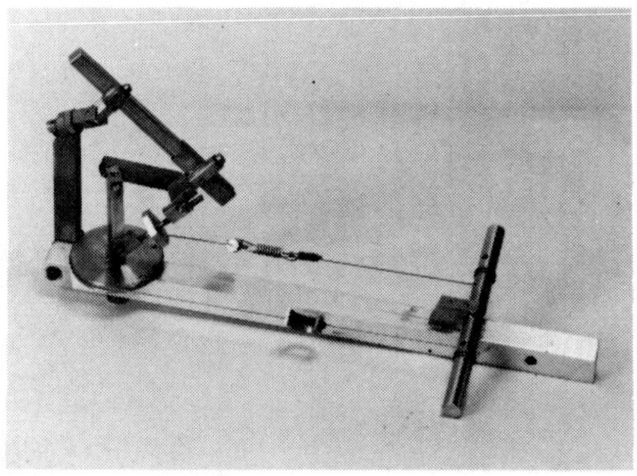

First realization of the mirror-holder kinematism

A steel wire rope rotates the pulley connected with the rod PO parallel with the earth's axis; the rod which materializes side BO of the isosceles triangle follows the sun; apex A of the isosceles triangle is clamped when assembling so that side OA points to the boiler; base AB of the triangle is shifted upward parallel with itself and is materialized by the rod carrying the mirror.

Kinematisms are equal to each other and can therefore be mass-produced; when assembling, allowance must be made for the various mirror-boiler positions.

6. First point-shaped boiler at S.Ilario

The mirror field consisted of 121 kinematisms each carrying a mirror slightly cone-shaped, 58cm diameter, for a total surface of about 30 sq.m and with overall dimensions of 7 x 7m. Figure 8 shows how mirrors and kinematisms are assembled.

Fig. 8
Kinematism and mirrors

Fig. 9
Sketch of motion

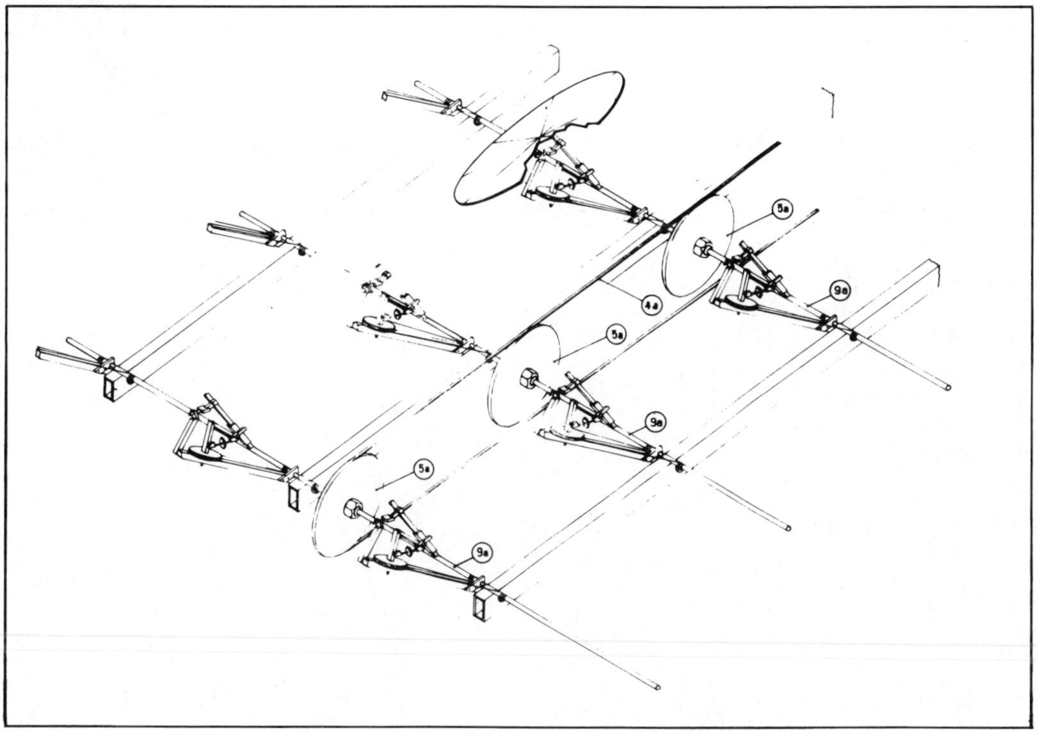

The uniform rotary motion was transmitted to the kinematism axes by a tower clock having oversized weights: the whole required power was in milliwatts. Figure 9 shows schematically the transmission system.

A monotubolar boiler was placed a little more than 3m over the mirror field (Figure 10): this boiler consisted of 60m of steel tube blackened by means of black enamel and wound up partially as a coil and partially as a spiral; the anti-radiating cells consisted of pyrex tubes with a height/diameter ratio increasing from the coldest peripheric area to the hottest central area. The side walls of the cone forming the entrance of the boiler were protected by means of mirrors.

Figure 11 shows the boiler with the pyrex cells and the spiral tube well visible for insufficient absorption. Figure 10 gives a general view of the plant.

The boiler was fed by a centrifugal pump with a high delivery for low pressures and by a piston pump for high pressures.

A 12-point recorder, with full scale at 700°C, gave the temperature curves at various points of the boiler and of the superheated steam: a precision meter measured the water deliveries. Steam-gauges, flowmeters

Figs. 10, 11

First solar boiler at S. Ilario (CNR., 1964-65).

The boiler and the pyrex anti-radiating structure.

Fig. 12
Graph of the temperatures in the boiler: destructive test (> 1000°C).

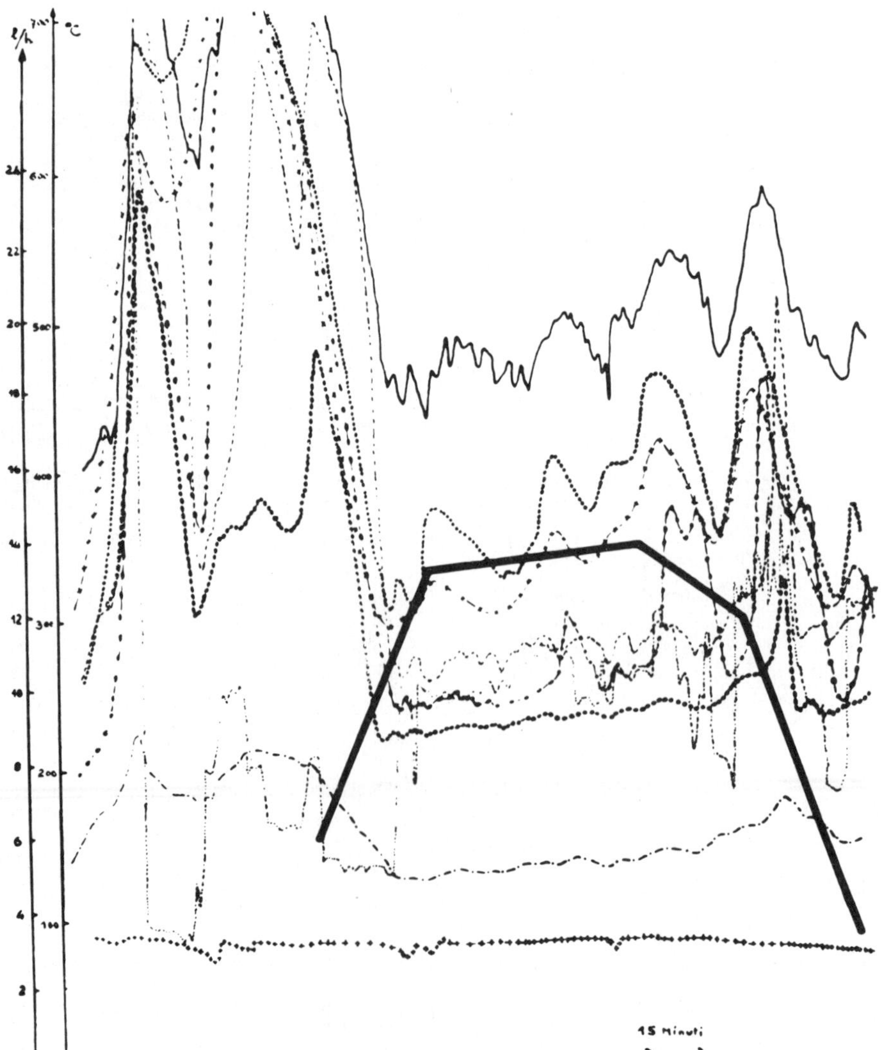

etc. were also supplied. A steam condenser recovered and measured distilled water. (11), (12).

The plant was put in commission in the Spring of 1965. The results bore out the expectations both as regards temperatures and efficiencies.

With the sun at 900 W/sq.m peaks of 21 kg/h of superheated steam at 500°C, 150 ata were obtained and averages of 19 kg/h for 5-6 hours where efficiencies can be considered as being of 70%.

The idle times necessary for reaching rating temperatures varied from 12 to 18 minutes according to the height of the sun (natural obstacles prevented these tests at dawn and sunset).

The limit temperature, too high not only as regards the scale but also as regards the boiler's resistance, has been reached only when it was decided to replace the boiler for the reasons we shall state hereinafter.

Figure 12 shows the recording of temperatures in that test: by reducing the capacities nearly all thermocouples go out of scale until efficiency falls brusquely owing to the melting of the anti-radiating structure.

Temperature has been estimated to be over 1000°C (melting of copper parts) but certainly it would have been increased without the melting of pyrexes.

In synthesis the results of some months of tests can be summarized as follows:

1. Temperatures, deliveries and, consequently, efficiencies were as anticipated;
2. the precision with which the kinematisms followed the sun was quite satisfactory; because of their sturdiness the kinematisms could support larger mirrors;
3. mirrors were too near to each other and, far off noon, efficiency fell owing to reciprocal shadows;
4. the boiler, well protected by the anti-radiating structure from losses due to thermal re-radiation, had an unsufficient absorption of the solar light either owing to the tube colour or to the geometry.

After months of experiments the plants have been disassembled in order to carry out the alterations which had been studied.

7. Second point-shaped boiler of S.Ilario

In the winter of 1965 the second solar boiler was built; the same kinematisms were used but new mirrors, (74cm diameter for a total working area of 52 sq.m) more spaced out so as to limit interferences, and a new boiler was employed.

The most important innovation was the boiler, maybe the first solar boiler to exploit the optical properties of radiating energy for the temperatures and pressures required by modern turbines, 500°C, 150 atm.

The boiler (Figure 13) is of the monotubolar type with finned tubes; the hot part rises with two consecutive cylinders wherein solar energy is bound. (13)

The records of the experiments performed during the winter/spring 1966 prove the possibility of a regular operation also during the periods of lower insolation and at the latitude of 45°. To obviate the relatively high latitude the mirror field was 10° inclined towards South (Figure 14). It is to be emphasized that the kinematisms are adjusted

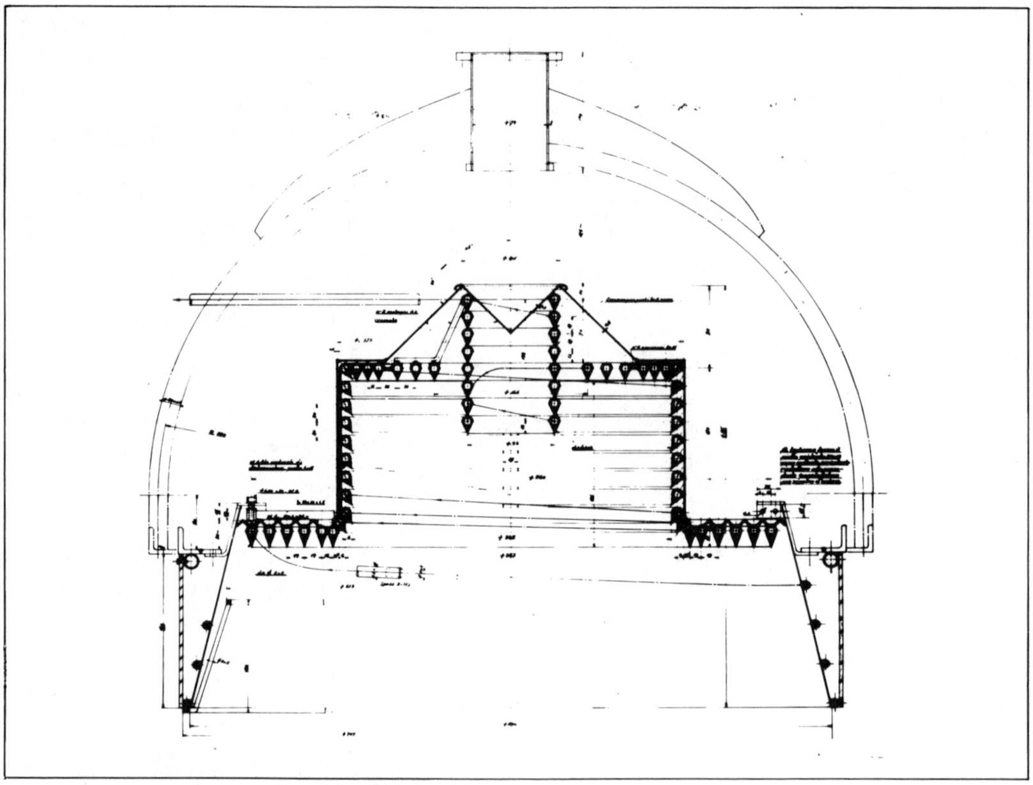

Fig. 13
Section of the
second boiler

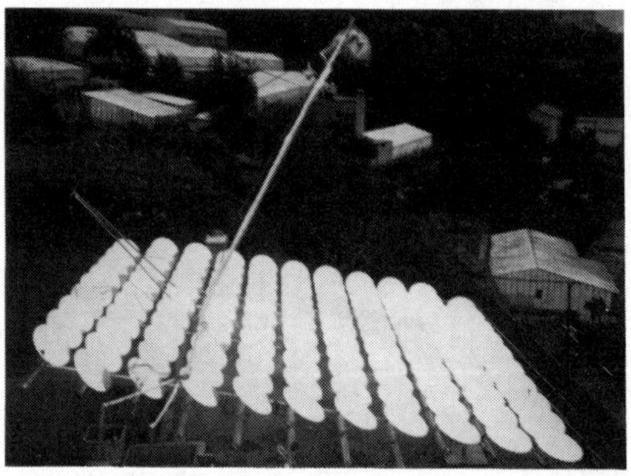

Fig. 14
Second solar
boiler of S. Ilario
(CNR., 1965-66)

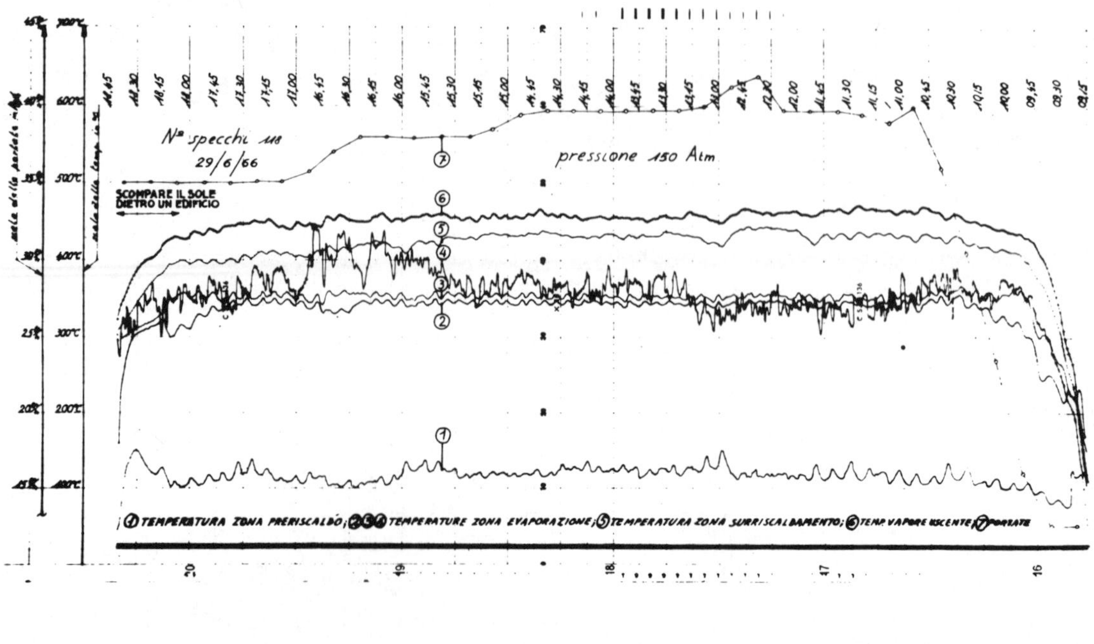

Fig. 15
Graph of a sample
of the second
solar boiler at
S. Ilario: 450°C,
150 atm, 40 kg/h;
29 June 1966.

when assembling so that the reflected ray is in the direction of the boiler whatever is its position: the field therefore needs not to be flat or horizontal.

Up to 38 kg/h of superheated steam, 150 atm, 500°C have been obtained; Figure 15 shows a recording with a temperature higher than 400°C starting from 9.45 am up to 6 pm (before 9 am the sun was hidden by a hill) with an average delivery of 40 kg/h from 11 am to 3 pm and of 38 kg/h from 3 to 6 pm.

After more than one hundred recordings it was decided to build a new larger plant having characteristics as close as possible to those of an industrial plant.

8. Third solar boiler of S.Ilario

The study of the kinematism has been resumed ex novo and during the winter 1966/67 the prototype shown in Figure 16 has been realized. It is always a question of utilizing the properties of an isosceles triangle, but this apex O from which the two equal sides begin does not materially exist and displacements around it are obtained by means of a sector which has that point as its center (Figure 13).

The new kinematism is compact, strong and can be manufactured economically on a large scale.

271 kinematisms like those of the prototype, each supporting a 0.75 sq.m slightly concave mirror, form the mirror field which covers an hexagon 10m in size. The kinematisms are arranged on 19 rows parallel with direction E/W The distance between the rows increases from S to N and the distance between the kinematisms of a row increases from the center to the periphery in order to reduce to a minimum the reciprocal shadows of one mirror on the other; for the same reason the kinematisms of two subsequent rows are offset by a half interval.

All kinematisms of one row are connected by means of steel cables to an axis which, when rotating, transmits the required constant speed of 15 15°/h; all these axes are in their turn connected to each other by means of a chain driven by a clockwork mechanism; the energy required to provide motion to all the 271 mirrors of the field is of about 0,1 Watt.

Figs. 16, 17

The new kinematism (1966-7).

Kinematisms and mirrors of the third installation (CNR., 1967).

The area covered by the mirror field is about 300 sq.m; the total area of the mirrors was 205 sq.m. Figure 17 shows a detail of the assembled mirrors and kinematisms.

The boiler is hung over the mirror field: it has a diameter of 0.9m and it is supported by a steel tubular column containing the thermally insulated pipe which conveys the 500°C superheated steam.

An hydraulic drive lowers the boiler at the level of the mirror field. Eleven thermocouples connected with a recorder make it possible to know in any moment the temperatures of the steam and of various points of the boiler; a twelfth thermocouple measures the solar energy.

The boiler is shown in section in Figure 18.

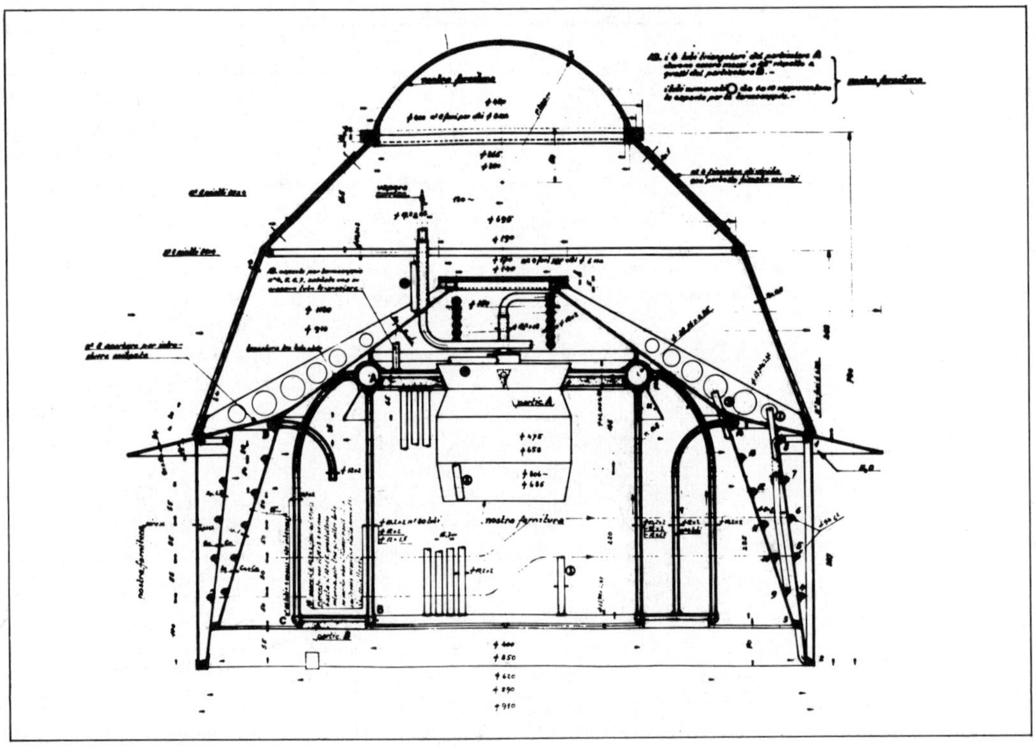

Fig. 18
Section of the third boiler

Cold water goes into a double conic helix preheater which collects the peripheral zone of the beam of rays reflected by the mirrors and goes through to a double collector from which 196 vertical tubes start, leading to a thoroidal collector acting as a condensate separator.

The geometry of the boiler is such that water evaporation takes place in these tubes. Owing to the small dimensions of the evaporation area, evaporation takes place tumultuously and therefore oscillations could originate in the evaporator tubes such as to bring water into the condensate separator and steam into the lower collectors. In order to avoid this special valves have been studies capable of keeping ebullition in the required zone in each tube. Steam, already slightly superheated, goes from the condensate separator into a superheater from which it goes out at the required temperature.

The nest of tubes and the superheater are protected by means of the anti-radiating structure which is formed by pyrex tubes.

The plant is fitted with high-and low-pressure pumps, valves, gauges, flowmeters, water meters, condenser, water distiller etc.

The boiler was started in 1967.

Near noontime and with the sun at 900 W/sq.m up to 130 kg/h of superheated steam 150 atm., 600°C, efficiency 73%, have been obtained.

In an eight-hour period, which was the maximum feasible in the plant owing to the presence of natural obstacles, an average of 115 kg/h of steam at 600°C, 150 atm was reached. The boiler attained the operative temperature within about 6 minutes.

The plant revealed the following defects:
(a) the superheater received energy which had a too high concentration and its inner part was deformed owing to the softening of the material;
(b) too large mirrors, not suitably shaped (square, with rounded angles); in the morning and evening reciprocal shadows were excessive.

In order to find a remedy for these defects the superheater has been replaced with a four-leaved shaped double-helix bitube (Figure 19) and the mirrors have been replaced with other mirrors, round-shaped, with an area of 0,5 sq.m for a total of 135 sq.m of working area.

Figure 20 shows the third S.Ilario plant.

These alterations have been completed in 1972 and a program of tests has been carried on with quite satisfactory results up to the middle of 1973.

Figs. 19, 20

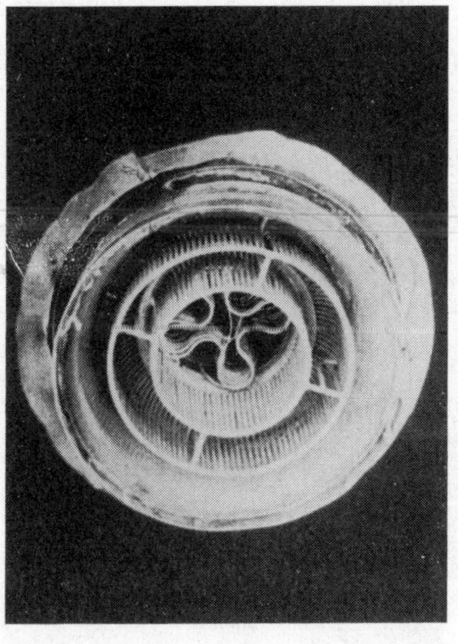

View of the third boiler

Third solar boiler of S. Ilario after the works of 1972 (CNR).

9. New programs

In the second half of 1973 the Italian Company 'ANSALDO Società Generale Elettromeccanica S.p.A.' has started, with the consulting of the author, a program of studies, experiments and industrial realizations in the field of solar energy and especially with regard to the thermodynamic production of mechanical and electrical energy.

The Author has joined in these programs together with ANSALDO specialists and with those of other Companies of the FINMECCANICA Group, and of course with the collaboration of the 'Instituto di Meccanica Applicata alle Macchine' of the Genoa University now supervised by Professor R. Ghigliazza.

These works aim at the realization on an industrial scale of solar plant for the production of electrical and mechanical energy.

The starting point for all these works has been the S.Ilario solar plant on which experiments and tests have been carried out: in particular, samples of the boiler nests of tubes have been considered in order to ascertain the conditions of the material after thousands of thermal shocks, of local overheatings, and of temperature peaks undergone during thousands of tests.

Today, also with the financial support of C.N.R. (National Research Council), the S.Ilario solar plant is in course of transformation in order to use it as a test field for solar plants: the present 271 mirrors 80cm diameter will be replaced with 140 mirrors 110cm diameter, made convergent by means of a simple technique utilizing the elastic deformation of the glass.

The mirrors will be supported by new kinematisms which have been improved even if essentially they are similar to the present ones. A 40-point recorder shall be provided for measuring temperatures, pressures, speed etc., in various points of the boiler.

10. Solar plant of the Georgia Institute of Technology, Atlanta

In May of 1976 the Georgia Institute of Technology, Atlanta, ordered to ANSALDO a plant similar to the S.Ilario's one but having a power four times greater (400 kW/th abt.), provided with a boiler for superheated steam generation up to 600°C, 150 atm. (14).

The plant was completed last January and it was commissioned at Atlanta in May 1977. It consists of 550 kinematisms (Figure 21) which support as many mirrors having a 111cm diameter, focused by means of elastic deformation for sending the rays of sunlight into a boiler (125cm diameter) hung at 23m from the ground.

Figs. 21, 22

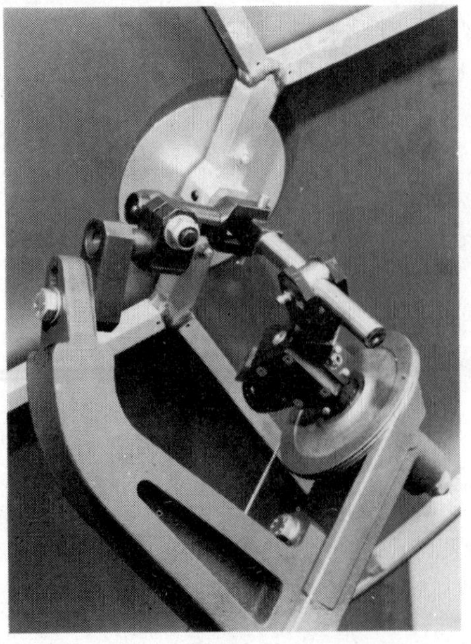

Kinematism of the solar plant of the Georgia Tech. (1976).

Drawing of the Georgia Institute of Technology, Atlanta (1977) solar plant.

The kinematism motion is supplied by a small electric motor; a safety compressed-air system is provided for a rapid defocusing of the mirrors.

The boiler, protected by the anti-radiating pyrex cells, of the type similar to S.Ilario's, is fitted externally with a system of fins in order to avoid damages when focusing of the mirrors takes place with solar power to the full.

This plant will be utilized for testing heat accumulators and solar boiler models for account of the Customer which is one of the most advanced experimental Institutes in this field, sponsored by ERDA, the Agency supervising the national program of the United States of America in the field of energy.

The supply of this plant has offered the opportunity of starting positive relationships for a technical collaboration with research institutes and American industries. ANSALDO and Georgia Institute of Technology intend to continue this collaboration in the future.

Figure 22 shows a drawing of the solar plant.

11. Solar plant of CEE (European Economic Community)

Within the scope of research programs on solar energy planned by CEE, ANSALDO will participate in the construction of a demonstrative solar plant having an output of 1 MW/el., in a consortium with the German Company MBB (Messerschmitt-Bölkow-Blohm). The French Consortium CETHEL and ENEL (Italian State Electricity Board) - which will co-operate with ANSALDO for the Italian part - will also participate in this project.

The Advisory Committee Project Management for solar energy has recommended to the Executive Committee - in the course of its Meeting of the 20th January - to realize this plant located in Italy within a period of three years.

ANSALDO and ENEL will attend to the most important part of the system, that is the boiler and the steam cycle. A power station with a steam boiler of the type studied by the Author has been chosen; this station is based on a system which allows a reduction in costs with a considerably high efficiency.

The design and the realization of this plant will cost 6 ÷ 7 billions of Italian Lire; CEE will contribute about 50%.

This is the first example of the collaboration of three Countries (France, Germany and Italy) on an important community project.

12. Small output plants

The work program is not limited to the realization of large output plants (the 1 MW plant is the first and necessary stage). Some very interesting technical solutions are in the course of study and of design; they concern small output plants (starting from a few kW up to hundreds of kW) for water pumping and electric energy generation for small remote communities. Non-concentration plants and plants with concentration are under study.

The program anticipates having at its disposal within a reasonably short period of time (16-18 months) the industrial prototypes in operating conditions and marketable.

One of the leading policies of this program is the development, especially with regard to collectors, of simple technologies capable of being easily exported so as to permit manufacturing on the spot of some components, also in the case of technologically undeveloped countries.

References

(1) G. Francia, 'Un nouveau collecteur de l'énergie rayonnante solaire: Théorie et vérifications expériemtnales', E/Conf. 35/S/71, O.N.U., Roma, 1961, pp 554-588.

(2) G. Francia, 'A new collector of solar radiant energy - Theory and experimental verification - Calculation of the efficiencies', National Aerospace Engineering and Manufacturing Meeting, Los Angeles, 1962, SAE, 594B, New York.

(3) M. Perrot, J. Oualid, H. Cortes, 'Rémarques théoriques sur la structure cellulaire de G. Francia', IESUA, Alger, Bull. 6, 1962, pp 39-42.

(4) M. Perrot, P. Gallet, J. Gervais, A. Ropké, 'Rémarques théoriques sur la structure cellulaire de G. Francia', C.R. Acad.Sc., Paris, t.256, 1963, pp 1224-1226.

(5) K.G.T. Hollands, 'Honeycomb devices in flat-plate solar collectors', Solar Energy, vol. IX, n.3, 1965, pp 159-164.

(6) G. Francia, 'Le cellule antiirraggianti nell'utlizzazione dell'energia solare', Accademia Nazionale dei Lincei, Convegno Internazionale, Roma, 1975, pp 165-182.

(7) P. Gallet, 'Etudes des propriétés radiatives des structures cellulaires', Thèse Université de Provence, mars 1973, pp 1-154.

(8) A. Bähr, H. Piwecki, L. Rigolini, 'Messungen an einem ebenen thermischen sonnenenergieauffänger mit einer zellstruktur nach Francia', Commission des Communautés Européennes, Euratom, C.C.R., Ispra, 1976, pp 1-27.

(9) K.N. Marshall, P.K. Wedel, R.E. Dammann, 'Development of plastic honeycomb flat-plate solar collectors', Final Report, LMSC/D462879, SAN/1081-76/1, Lockheed Missiles and Space Company, Inc., Palo Alto Research Laboratory, April 1976, pp 1-178.

(10) M. Perrot, G. Peri, 'La chaudière expérimentale solaire du groupe de recherches des Universités de Marseille et de Gênes', Comples, bull. n.6, 1964, pp 93-97.

(11) G. Francia, 'Première série de résultats de l'élement de centrale solaire de S.Ilario de Nervi', Comples, bull. n.9, 1965, pp 42-50.

(12) G. Francia, 'The solar plant of S.Ilario-Genova Nervi', La Ricerca Scientifica, anno 36, n.8, 1966, pp 779-784.

(13) G. Francia, 'Pilot plants of solar steam generating stations', Solar Energy, vol. 12, n.1, 1968, pp 51-64.

(14) J.D. Walton, Jr., S.H. Bomar, Jr., N.E. Poulos, 'Georgia Tech. 400 kWth solar thermal test facility', Engineering Experiment Station, Georgia Institute of Technology, Atlanta, Georgia, June 1976, pp 1-18.

Description of the Polisolar high temperature collector systems and report on integrated systems for cooling, desalination and power production

Ruedi Kriesi
Polisolar SA Ltd
CH-3000 Bern 5

When we evaluated for which temperature we should optimise our collector systems, we worked out the temperature levels of the different possible heat consumers (Fig. 1).

We can put these consumers into three groups:

- The low temperature ones (swimming pool, domestic hot water, heating) their need occurs when there is no sun;
- the demands of the high temperature ones (industrial heat, power production) do not have a distinct correlation with the sun hours;
- and the medium temperature ones (cooling, sea water desalination) their energy need is elevated especially at places and times with much sun.

For the third group, therefore, the storage problem is comparatively small and solar energy must become economic first for these applications. Thus the first interesting market for a solar energy industry will be in desalination and cooling; i.e. especially in southern countries.

Thus our collector systems had to cover primarily this temperature range.

Although storage is more important for the first group (low temperatures) than for the second, the price per useful calory for the first group is lower because of the cheaper collector systems used for this temperature range. Thus our systems also had to cover this temperature range.

For the second group only pilot plants for future applications can be installed at present, because of the higher collector prices for higher temperatures. Thus not much attention was given to temperatures above 200°C.

The various possible systems for thermal conversion of solar energy are shown in Fig. 2.

Figs. 1, 2

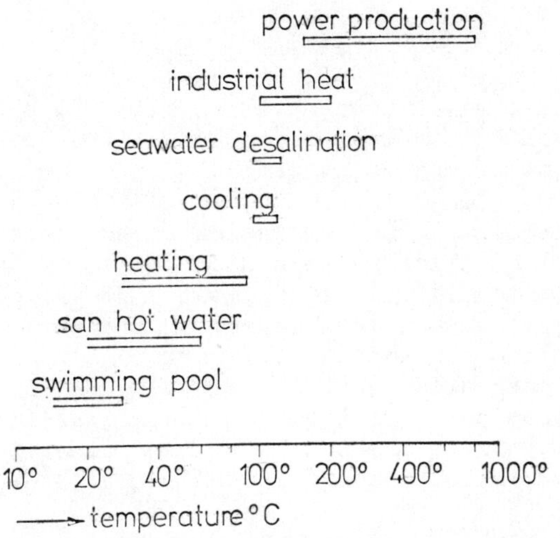

Levels of temperature required for different applications

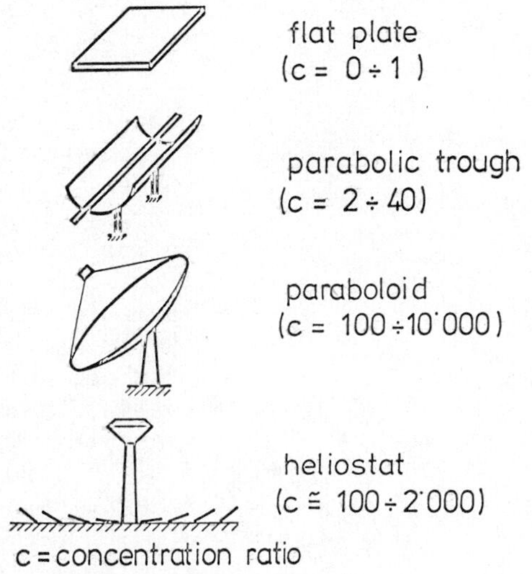

Different collector systems

The concentration ratio C, the parameter used to separate the systems in this diagram, is defined as the collecting surface divided by the total receiver area, which is subjected to thermal losses. For the parabolic trough, for example, C would be the ratio between the mirror width and the tube circumference.

If we use a higher concentration ratio C, higher temperatures can be reached with good efficiencies (Fig. 3), because, at a given absorber temperature, the heat losses become smaller compared to the collected heat, due to the smaller surface of the heated parts.

So a flat plate collector, with C = 1, at present can be satisfactorily used up to about 80°C, or 180°F. A cylindrical parabolic collector has got its theoretical limit of the concentration factor at about 100, but the feasibility under practical field conditions indicate that much lower concentration ratios - in the order of 20 - are viable without excessive cost penalties for sufficient tracking accuracy.

Fig. 3
Efficiency of different systems at a given temperature level

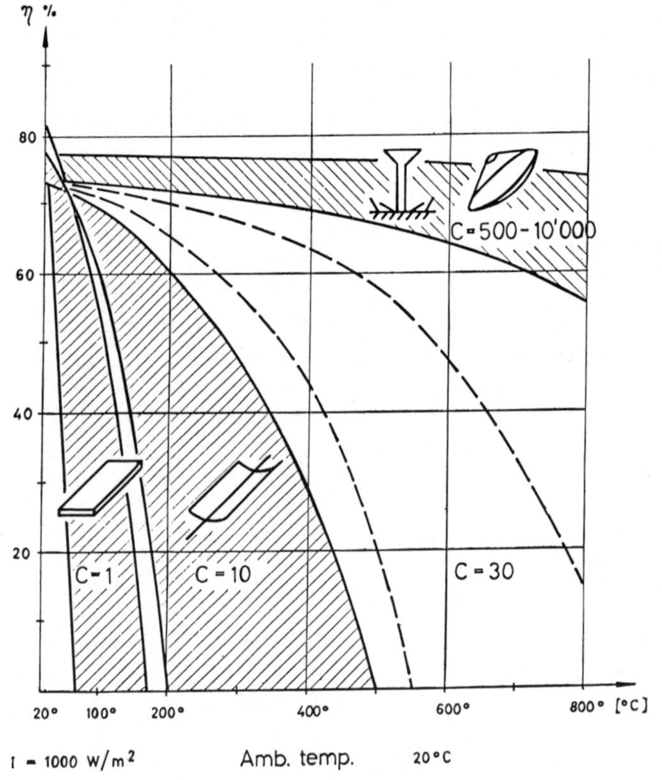

But still, with a given concentration ratio C, a big range of temperatures can be covered, depending on the quality of the absorber element (vacuum, selective coating).

To reach a considerable concentration factor during a longer period of time, some sort of tracking mechanism has to be used. For this, either the absorber or the mirror can be moved. If the mirror is moved, the system becomes more complicated, but more energy per day and mirror area can be collected, because the mirror is positioned perpendicular to the direct insolation during several hours per day (Fig. 4). This means that prices per active mirror area for fixed mirror concentrators must be lower than for the tracked mirror ones, in order to give the same price per collected energy.

With these reflections, we selected our system out of the various possible ones.

Fig. 4

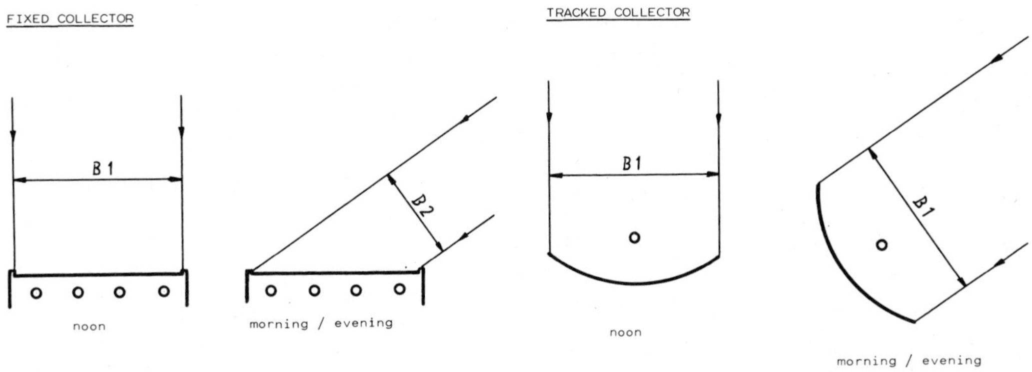

Our first version for outdoor application is a cylindrical parabolic collector with a concentration factor of about 10, north-south oriented axis, silver coated glass mirrors, mounted into a rigid aluminium frame (Fig. 5). Depending on the desired temperature range, only the absorber tube has to be selected, the mirror remains the same.

The low temperature version can be used up to about 100°C, the high temperature one, which has got a highly selective coating, to about 220°C (Fig. 6, 6a).

Our second version, which is presented today for the first time, has been developed in co-operation with a major Swiss company. It is again a cylindrical parabolic concentrating collector, which is put under glass (Fig. 7). Some advantages compared to the field version result.

Under the glass the collector is protected against wind, hail and dust and thus a lighter construction becomes possible. Furthermore, small widths of the cylinders are used in order to have an esthetically acceptable collector, even for residential application. The collector is designed to be an integral part of a roof or facade, saving money on the conventional construction of the building and thus making the use of these elements more economic.

Fig. 5
Hoehenklinik
Heiligenschwendi

Fig. 6

Performance of the high temperature collector-element with selective coating (based on measurements)

absorbertube: ϕ 2,8 cm, glasstube ϕ 4 cm, mirror width: 80 cm

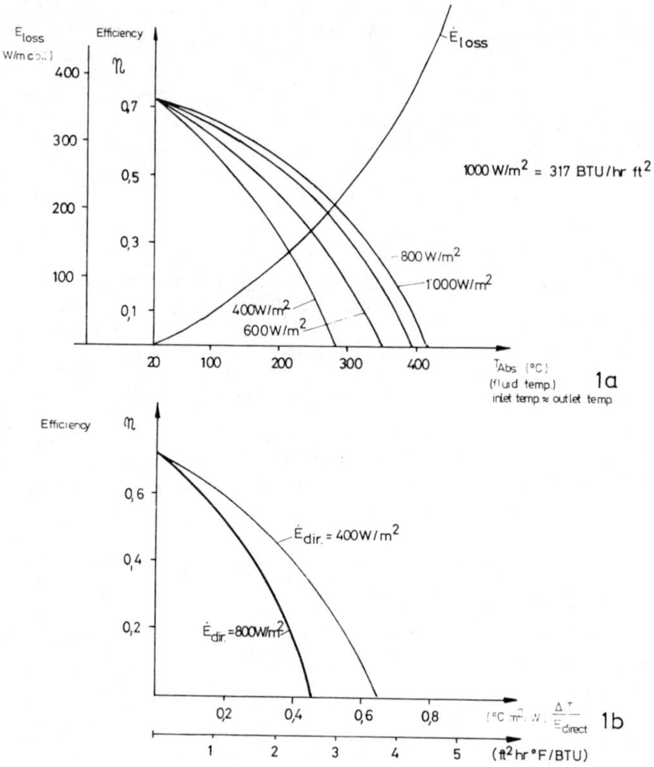

Fig. 6a

Performance of the low temperature collector-element star shaped graphite absorber

absorbertube: ϕ 3.2 cm, glasstube: ϕ 3.2 cm, mirror width: 60cm

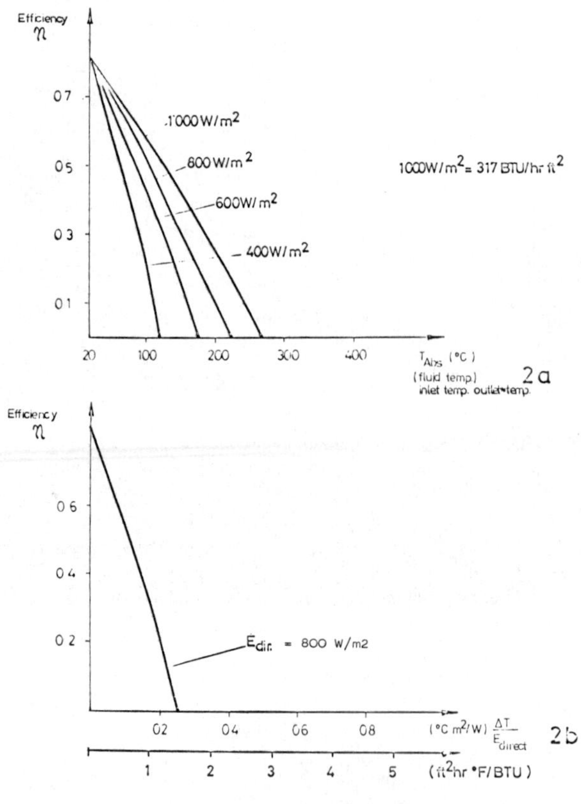

Fig. 7
Example of an energy collecting shading structure for a parking lot

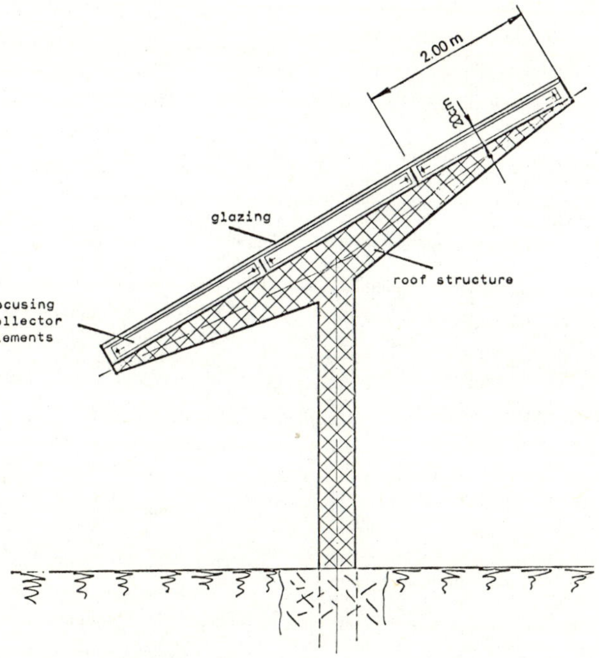

Applications

A very interesting form of the under glass version is the 'Raimann' greenhouse system (Fig. 8). In this greenhouse, the collectors perform as effective and variable shading, they protect the plants from extensive heat and at the same time, they remove this excess heat as highly valuable energy for night heating or other applications.

To distribute the heat during the night, the mirrors can be turned into the upper position (Fig. 8a), the hot water from the storage tank is again pumped through the pipes and the hot pipes radiate the heat into the greenhouse. At the same time, the mirrors provide the roof with a better insulation, lowering the nightly heat demand of the house.

Figs. 8, 8a

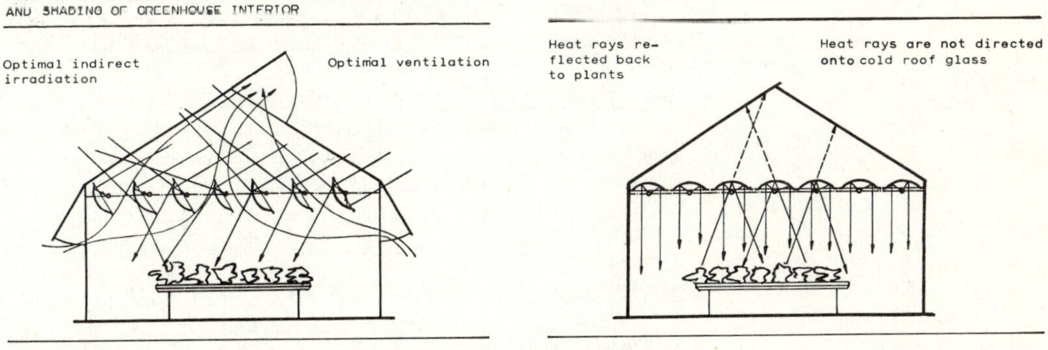

For applications in hot climates, the shading of the greenhouse interior by the collectors is of great importance, because the heat input, which has to be removed by additional cooling, is reduced (Fig. 9). If evaporative cooling is used, a much smaller air exchange rate and water consumption lower the investment and maintenance costs. In addition, the lower air exchange rate favours the CO_2-enrichment, stimulating the growth of the plants.

Fig. 9

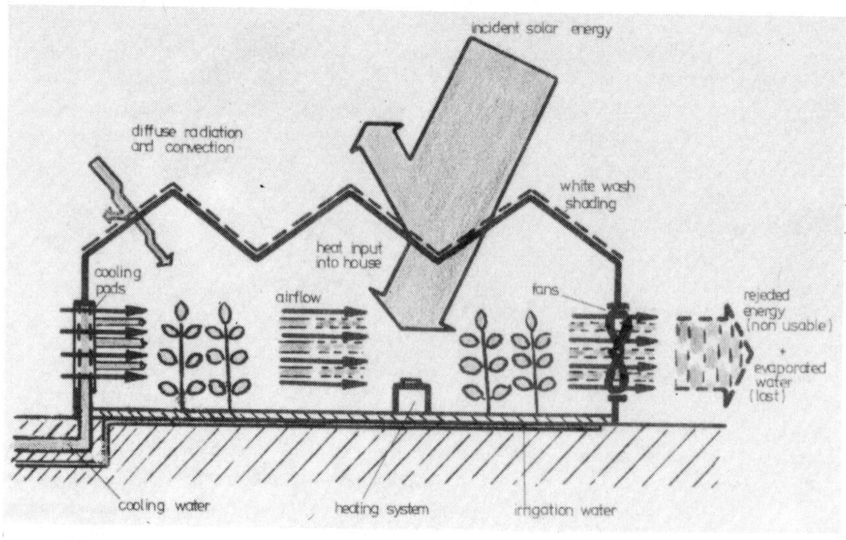

Scheme of the classical greenhouse

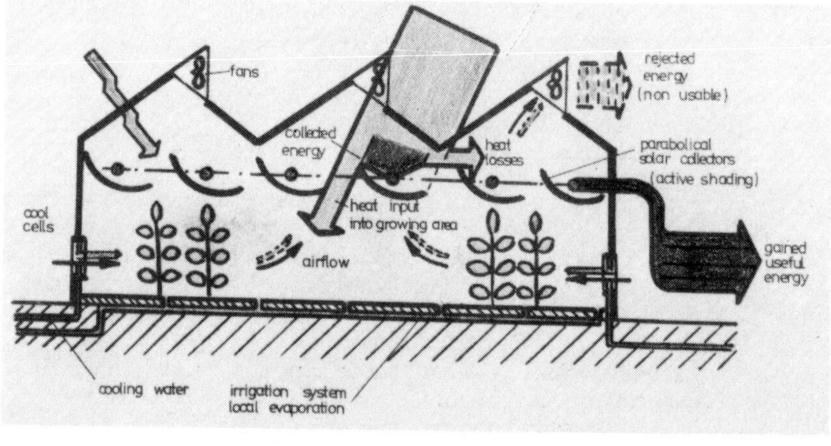

Scheme of the Polisolar greenhouse (System Raimann)

If there is more heat available than is necessary for the night heating, it can be used for the following applications: (Fig. 10, 10a). Obviously the heat for these applications can also be delivered by the outdoor or roof structure collector version:

- Sea water desalination: this application is discussed in the paper of Dr. N. Azhari, Polisolar Ltd., also presented at this conference.
- Absorption cooling: Polisolar Ltd. has pioneered the use of solar powered absorption type air-conditioning units for industry. An initial pilot plant with concentrating collectors has been built near Zurich, having an active collector surface of 50m2 to cool a factory building.
  Using this experience, we are presently installing 2000m2 of field and roof structure type concentrators to run the absorption cooling machine with about 700kW of a modern factory in Rome.
- Power production: the Polisolar collector system is used for a German sponsored 10kW power plant. Other small power stations will also be equipped with this system.
  Economical calculations show that the cost-optimal temperature for a power plant with cylindrical parabolic collectors is well inside the temperature range covered by our high temperature system. Using higher temperatures gives an increase in turbine efficiency, but costs for the collector increase even quicker after a certain limit.

Thus solar collector systems and experience are available today for all heat consumers up to about 220°C.

Fig. 10

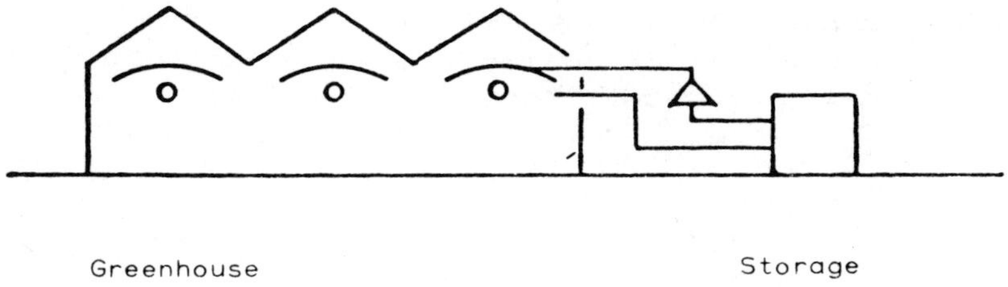

Night heating

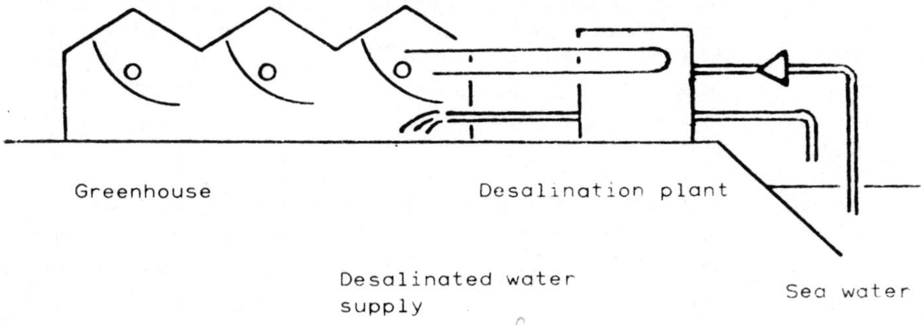

Sea water desalination (or brackish water)

Fig. 10a

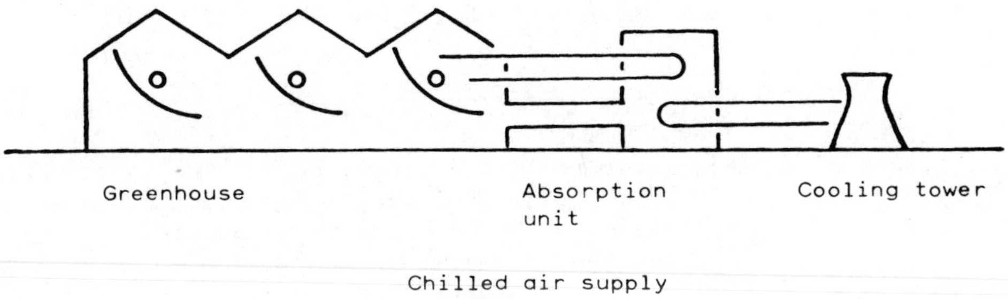

Absorption cooling: General scheme

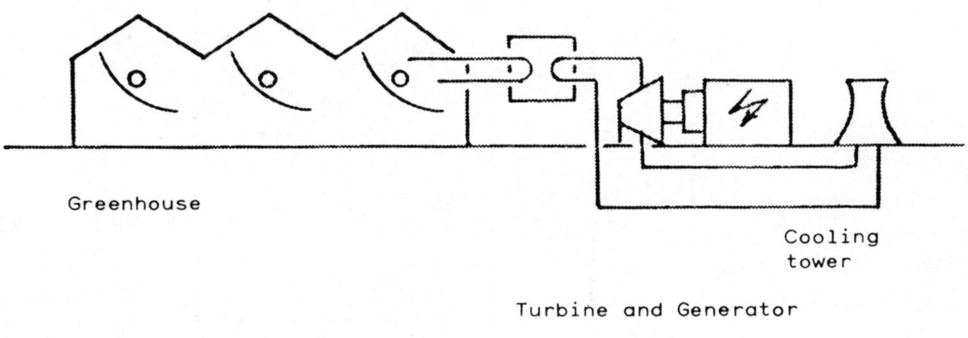

General scheme of a small power plant

Solar MSF sea water desalination plant

Dr. Ing. Naman Azhari
Polisolar Ltd, Güterstrasse 24
P.O. box CH-3000 Berne 5
Switzerland

1 INTRODUCTION

The conventional solar sea water desalination plants work as a result of direct evaporation of sea water by solar energy. Such a plant consists of a shallow basin of water covered with glass or transparent plastic. Solar radiation is transmitted through the cover and heats the water in the basin. Part of this water evaporates and condenses on the inside of the sloping cover. The distillate runs down the cover to a trough and then flows from the still into a tank. The unevaporated sea water is replaced periodically with fresh sea water to prevent crystallization in the basin.

Various forms of basins have been developed to increase the efficiency of the process i.e. the tilted-wick-still form, the 'Church World Service' still, the single-tube expendable type still, the vertical envelope still etc...

Another type of applied solar desalination plant is the heatpipe principle.

The multi stage flash evaporation method has not been applied for commercial sea water desalination plants.

2 SOLAR RADIATION AND PRODUCTIVITY OF SOLAR DESALINATION PLANTS

The solar radiation varies from place to place and from month to month around the year. Budyko (1) has studied the world distribution of solar radiation and has shown this in a number of graphs. Figure 1 shows the average yearly global solar radiation for various places.

Fig. 1
Average yearly global solar radiation

England	927 kWh/m² year
Berlin	1000 kWh/m² year
Zürich	1160 kWh/m² year
Paris	1500 kWh/m² year
Marseille	1860 kWh/m² year
Sahara	2530 kWh/m² year

The productivity of conventional solar desalination plants by a solar radiation of 3,000 kcal/m^2 day (approx 1,000 BtU/(sq. ft. day)) varies between 2 ÷ 3 l/m^2d (approx. 0,04 ÷ 0,065 gal/sqf day.

Figure 2 shows the productivity of various still solar desalination types.

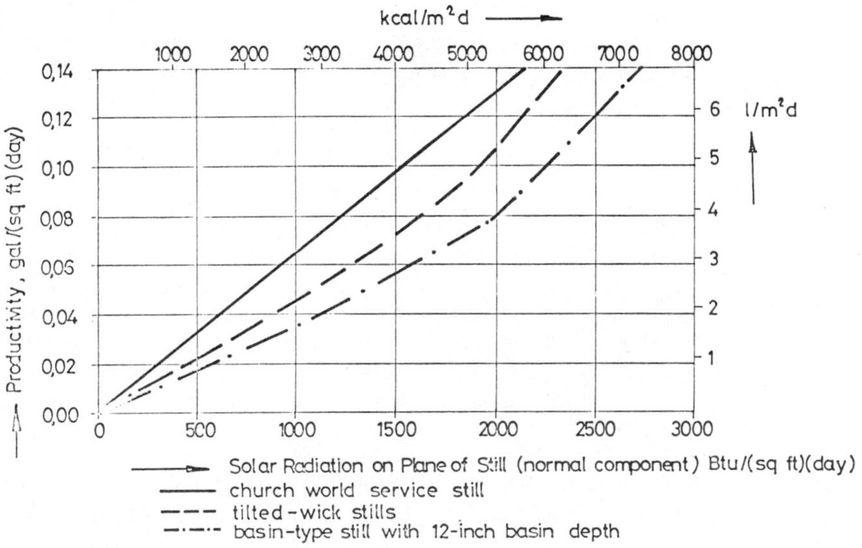

Fig. 2
Productivity of various solar desalination plants

3 SPECIFIC HEAT CONSUMPTION

The average total heat consumption of a still solar desalination plant is approx. 1,000 kcal/kg distillate. This very high heat consumption is due to very low heat efficiency of the system.

More than ninety per cent of all commercial sea water desalination plants are of the multi stage flash type 'MSF! Their specific heat consumption varies between 60 ÷ 110 kcal/kg distillate. Almost all large plants (with capacities over 9,000 m^3/d = 2 MIGD) are combined with steam power stations, where low pressure extraction steam of the turbines is used as heating steam in the brine heater of the desalination plants.

To achieve a low specific heat consumption for desalination plants using solar energy as heating energy, it is necessary to design a collector system with high efficiency and to use the multi stage flash method MSF or the vertical tube evaporator method VTE with high performance ratio.

4 THE POLISOLAR-LIEBI CYLINDRICAL PARABOLICAL COLLECTOR SYSTEM

The Polisolar-Liebi collector system consists of cylindrical parabolical low lead glass mirror segments. In the focal line of the collector, a glass tube with an absorber core is installed. The heated medium, i.e. water, flows through the glass tube.

Figure 3 shows a Liebi collector system.

Fig. 3
Liebi LNC collector system

Fig. 4
Absorber systems

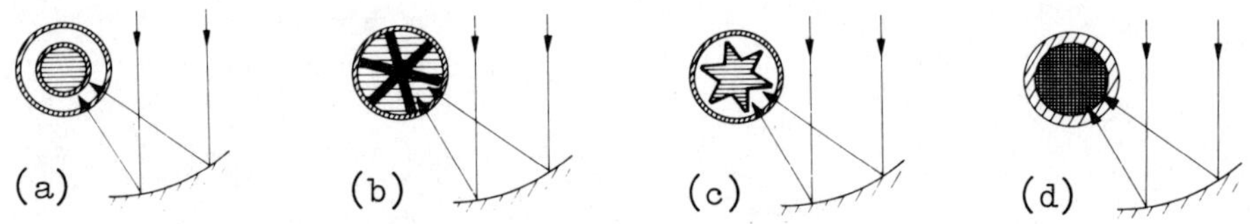

Figure 4 shows 4 types of absorber systems:
type (a) black absorber tube with glass cover
type (b) transparent tube with a starform absorber
type (c) transparent tube with a starform absorber shell
type (d) transparent tube with black absorber fluid.

The mirrors concentrate the sun's radiation on the absorbers and heat the absorber fluid. The collectors are installed slanted in the north-south direction to achieve high solar energy collection per area and day. A sun sensor system drives the collectors for tracking of the sun, so that the sun is always normal to the collector area.

With this system the heated parts only have a small surface compared with flat plate collectors, so that the thermal losses are much lower and higher temperatures can be achieved.

Figure 5 shows the efficiency of the Polisolar-Liebi collector system compared with flat plate collectors.

Fig. 5
Efficiency of collectors

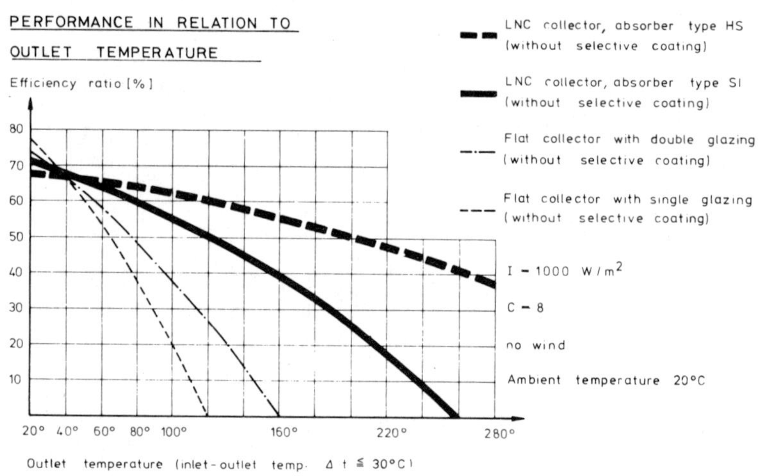

5 THE SOLAR ENERGY MSF-PILOT PLANT

5.1 Design

A pilot plant was specially designed for a large range research program to investigate the operation characteristics of the entire system under different conditions of solar radiation, sea water analyzes, sea water and process top temperatures, various part loads, etc. The plant has a capacity of approx. 350 l/h at a sea water max. temperature of 120°C and the specific heat consumption is approx. 95 kcal/kg.

The plant is designed for a maximum sea water process temperature range between 80 and 120°C. The sea water flow is once through and the flow quantity can be varied between 1 to 3m^3/h.

The evaporator consists of 10 identical stages, the shell material being glass, so that the process can be observed visually.

The solar system of course is the high temperature Polisolar-Liebi collector system.

The collector area is approx. 300m^2 which is sufficient to supply the desalination plant with the energy necessary for 24 hours per day of full load operation. Hot water tanks to store the energy for operation at night are part of the system.

5.2 Process description

The pilot plant operates according to the multi stage flash evaporation once through (without brine recycle) principle.

Figure 6 shows the flow sheet of the plant.

Fig. 6
Function diagram of the pilot plant

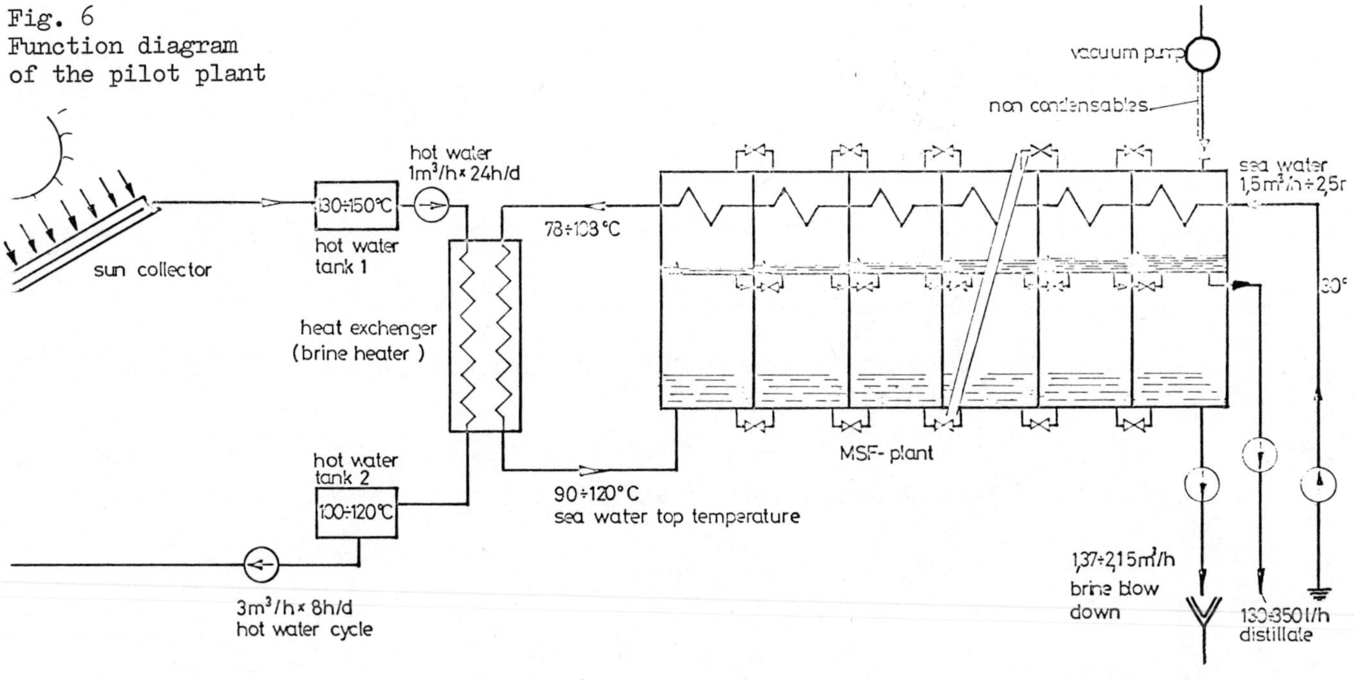

MSF-plant

Sea water is heated in a heat exchanger (brine heater) by means of hot water of the solar system and then flows to the flash chamber of the first evaporator stage. A part of the super-heated sea water is flashed off and the resulting vapour is condensed on the outside of the condenser tubes, which carry the lower temperature sea water to the brine heater. The remaining part of the sea water flows into the subsequent stage, where further flashing occurs and further distillate will be produced. This process is repeated in all stages of the evaporator. In the last stage the lowest temperature and vacuum is achieved. To obtain and maintain the vacuum in the evaporator and to pull out the noncondensable gases of the evaporator, a vacuum pump is provided. The concentrated brine and the distillate are discharged by pumps from the last stage.

Solar system

The hot water is recycled the whole day through the collectors from hot water tank 2 to hot water tank 1 and will be heated by the solar energy. A part of this hot water will be stored in tank 1 for the night hours and the other part flows to the brine heater to heat the seawater to its top temperature. During the night the stored hot water flows from tank 1 through the brine heater to tank 2 and heats the sea water to its top temperature. This hot water circuit ensures a continuous operation of the MSF-plant for 24 hours per day at constant load.

6 OPERATION CHARACTERISTICS OF THE PILOT PLANT

The MSF-plant is designed for self control of all orifices and level control valves for an output range between 60 and 110 per cent i.e. between a production rate between 200 and 350 l/h respectively a sea water max. temperature between 80 and 120°C.

When collected solar energy is constant and hot water outlet temperature of the collectors is constant, the MSF-plant will be operated with constant load. In case of bad weather, when the input of solar energy is low and respectively the hot water outlet temperature of the collectors is low, the sea water top temperature decreases accordingly and the production of the MSF will be reduced automatically. For a long period of bad weather, where the collected solar energy is too low, the MSF plant must be shut down or the auxiliary oil fired steam boiler must be used to keep the MSF-plant on stream.

7 ECONOMIC ASPECTS OF SOLAR MSF-plant and oil fired MSF-plant

To compare a solar and an oil fired MSF-plant it is necessary to compare the energy costs of both systems.

The hereafter described comparison is done under the following conditions:

1	Specific heat consumption of the MSF-plant	65 kcal/kg dis
2	Average global solar radiation	700 kcal/m² h
3	Efficiency of collector	60%
4	Sunshine hours per day	8 h/d
5	Heat value of oil	10,000 kcal/kg
6	Efficiency of steam generator	80%
7	Operation time per year	330 d/year
8	Cost of operation's personnel for the solar system and the steam generation plant are equal.	
9	Costs of maintenance and spare parts for the solar system and the steam generation units are equal.	
10	Interest of capital	8%
11	Depreciation time	15 years
12	Capital cost of the solar system	220 U$/m²
13	Capital cost of the steam generation unit	0.011 U$/kcal

7.1 Energy cost of the solar system per cubic meter of distillate

(1) The productivity of the MSF solar plant can be calculated as follows:

The collected solar heat energy per square meter of collector and day is:

Q = solar radiation x efficiency of collector x sunshine hours per day

= 700 x 0,6 x 8

Q = $\underline{3{,}360\ \text{kcal/m}^2\ \text{d}}$

The productivity for a specific heat consumption of 65 kcal/kg is:

$$\dot{m}_D = \frac{3360}{65} = 51{,}7\ \text{l/m}^2\ \text{d}$$

and for an operation time of 330 d/year is:

$\dot{m}_D$ = $\underline{17{,}061\ \text{m}^3/\text{m}^2\ \text{year}}$

The productivity of the pilot plant is approx. 30 l/m^2 d or approx. 10m^3/m^2 year.

(2) The annuities per one square meter of solar system are:
25,96 U\$/m^2 year

(3) The solar energy cost per one cubic meter distillate production is:

$$C_{solar} = \frac{25{,}96}{17{,}061} = 1{,}522\ \text{U\$/m}^3$$

7.2 Energy cost of the steam generation unit

(1) The investment cost per one kcal of heat energy is

= 0,00118 U\$/year

(2) The annuities per one cubic meter of distillate production at a specific heat consumption of 65 kcal/kg and 330 days per year operation time are:

1 kcal x 24 h/d x 330 d/year = 0,00118 U\$/year
7920 kcal/year ≙ 0,00118 U\$/year

$$\frac{7920\ \text{kcal/year}}{65{,}000\ \text{kcal/m}^3} \triangleq 0{,}00118\ \text{U\$/year}$$

which amounts to specific costs of:

$C_{boiler} \approx 0{,}01$ U\$/m^3

7.3 Equivalent oil cost to the solar system

(1) At break even point the difference in annuities of the solar system and the steam generation unit must equal the oil costs, e.g.:

$$C_{oil} = C_{solar} = C_{boiler}$$

C_{oil} = 1,512 U\$/m^3

(2) The oil consumption per one cubic meter of distillate production at a specific heat consumption of 65 kcal/kg, a heat value of oil of 10,000 kcal/kg and an efficiency of the steam generation unit of 80 per cent is:

$$\dot{m}_{oil} = 8,125 \text{ kg/m}^3$$

(3) The corresponding oil price at break even point is therefore

$$P_{oil} = \frac{1,512}{8,125} = 0.186 \text{ U\$/kg}$$

An oil price like this is nowadays valid for all countries which do not belong to the OPEC group. Since the increase of the oil price is estimated to approx. 10 per cent a year, the water cost of a solar MSF-plant will be accordingly lower compared to an oil fired plant. The development of the solar energy should achieve a lower investment cost to make this system even more attractive.

8 METHODS TO REDUCE THE COSTS OF SOLAR SYSTEMS

Since the collectors require a large installation area, it is advisable to use the collectors as roofs for buildings, garages, halls etc. or use them as shadowing elements for agricultural purposes e.g. for greenhouses.

A multi purpose system for cooling, sea water desalination and greenhouse will result in a lower overall investment cost and a higher overall thermal efficiency of the solar system.

9 SUMMARY

The conventional solar desalination plants have a very low productivity of approx. 6 $l/d\ m^2$ at a global solar radiation of 6,000 $kcal/m^2\ d$. This is equivalent to a specific heat of approx. 1,000 kcal/kg distillate.

More than ninety per cent of the world's installed desalination plants work according to the multi stage evaporation method MSF. Their specific heat consumption varies between 60 and 110 kcal/kg.

A pilot solar MSF-plant was developed, designed and built. Its capacity is approx. 350 l/h and its specific heat consumption is 95 kcal/kg. The plant consists of 10 stage-evaporator and a solar system of 300m^2 collector area.

At break even point, the oil price for an oil fired MSF-plant compared to a solar MSF-plant is approx 0,186 U$/kg or 0,70 U$ per US gallon. This oil price is valid nowaday in all countries except the OPEC group. The increase in oil prices of approx. 10 per cent per year will give the solar MSF-plants even greater advantage in the future.

A combination of solar sea water desalination with solar cooling and greenhouses will reduce the overall investment costs of the solar system plants and will increase the overall efficiency.

REFERENCE

(1) Budyko, M.I. (1958), The heat balance of the earth's surface. English trans. by N. Stepanova, US Department of Commerce Weather Bureau, Washington D.C. 259p.

Solar cookers

M. M. Hoda
Executive Director
Appropriate Technology Development Association,
Post Box 311, Gandhi Bhawan, Lucknow, UP, India

Abstract

Various types of box type solar cookers have been described in this paper. The Appropriate Technology Development Association (India) has tested the various models and found that reflector type solar ovens are very effective, but box type are easier to handle and much cheaper, although they take longer to cook food.

Introduction

Shortage of domestic fuel is assuming serious proportions these days, especially in the rural areas of developing countries. One wonders that even if enough food is produced for the world population how is one going to cook it. At present, fuel wood and cowdung cakes are two most important fuels which help poor people to cook their food. The use of cowdung cakes for cooking is really disastrous for the fertility of the soil because such a high quality manure is being burnt for want of suitable and proper fuel. Wood fuel is also depleting fast because of the indiscrimate felling of trees and denudation of forests. At present, housewives in the villages have to, in some cases, trek for miles to collect wood fuel and spend whole days collecting dried branches of trees fallen by the wayside. It is not possible to supply conventional fossil fuels like coal and cooking gas to village households. It appears that these stupendous problems can be solved only by two technological contrivances which have received attention in recent years. One is the generation of methane gas by anaerobic digestion of animal dung, agriculture waste, etc. In that case, the methane gas produced as by-product due to the digestion of dung and waste can meet a large percentage of the fuel requirement. However, there are still a number of technical imperfections in the technology of biogas generation. It has not yet been designed in such a way as could be within the reach of poor people. Another solution is the harnessing of solar energy for cooking purposes. There is a strong possibility of this technology being successful and making it available to poor people in villages at low cost which can solve part of their cooking problem. The food can be cooked in a solar cooker during the day time and that is also when the sun shines. It has been estimated that in some of the tropical parts of the world, which incidentally is also the home of acute poverty, sunny days vary between 200 - 240 - 300 days in a year. If a solar cooker is properly designed, it is possible to cook day time meals for a family of five for about 250 days in a year.

In all the solar appliances, including solar cookers, the principle is either to concentrate parallel sun's rays to a focus or absorb solar energy radiant on the surface of the earth in a flat plate collector and transfer it to some heat exchanger like fluid or utilise it directly. The sun radiates at an average of 2 calories of heat per square cm per minute. This is also called 'solar constant'. This is the total heat available for utilisation. The main problem in the design of solar appliances is to absorb the maximum percentage of the solar radiation incident and use the surface area in such a way that the total calories absorbed is sufficient to do the work required, that is to boil water or to cook food, to distil water, to dry fruit or to carry out other operations. Since to bring 1 kg of water to a temperature of 100°C from normal room temperature of 30°C it requires 70 kg calories of heat, the surface area to be provided taking into account the efficiency of the absorber should be at least 70 calories per kg. It means that theoretically an area of 9-10 sq cm is required to boil 1 kg of water in two hours, if the total energy incident is completely absorbed and imparted to cooking material without any loss. However, in reality the total overall efficiency is less than 5 per cent and much larger surface area have to be provided.

Principles

The usual principle taught in text books that good absorbers are good radiators is not valid in the case of those solar appliances in which a glass window has been used, because the absorbed and re-radiated radiations are not of the same wave length. The light radiation being of a shorter wave length can enter through glass panes and after being absorbed by the black body are degraded into heat radiation which are of a longer wave length and are unable to pass through sheet glass. This principle has been used very successfully for retaining the thermal radiation obtained through solar radiation after being absorbed by a flat plate.

If the total heat incident on a surface calculated by means of solar constant is = H

And the absorbing efficiency of the collector, taking orientation and other factors = E

Then the total heat absorbed = H x E

If the thermal efficiency of the collector i.e. the total heat radiated after absorption taking into account the loss due to conduction, convection and radiation is = N

Then the total heat absorbed is = H x E x N

This quantity must be more than the heat required by the cooking material for sufficient period of time.

This has been treated by C.L. Gupta (Science Today, April 1975, p.36) more thoroughly in detail which is given below. It has been modified by the author for specific application to the type of cookers described.

To find out the overall efficiency of the solar cooker an experiment can be carried out by keeping 1 kg of water inside the solar cooker and to calculate the heat absorbed by water in a given period of time. Of course, in sophisticated laboratories and workshop where precision instruments for measuring heat and related quantities are available all the factors mentioned in C.L. Gupta's equation could be measured and the various efficiency could be calculated.

To appreciate the relative contributions and interrelationships of solar, geographic and material characteristics and climatic and other operational factors in the mechanism of solar energy conversion, it is helpful to resort to numbers. The primary energy balance equation for any solar energy collection system can be written as:

Useful heat collected = Incident radiation absorbed by the collector - Heat losses due to conduction, convection and radiation.

$$[M\ C_v\ (T_f - T_i] = HRtaX - f_c\ (T_f\ T_i\ V) - f_R$$

$$(T_f\ T_i\ E) - U_c\ (T_f - T_i)$$

Where:

M = Mass of the air inside the box

C_v = Specific heat of air

T = Temperature (^oK)

H = Direct solar radiation intensity on a surface normal to beam radiation

E = Emissivity of absorber

R = Orientation factor (= 1 for sun tracking collectors; depends upon latitude, time of day and year, facing and tilt for flat plate collectors)

t = Transmittance of glass covers

a = Absorptance of absorber

V = Wind velocity

X = Focusing factor (= Concentration factor C*)

X Reflectance of concentrator r*

X Interception factor $_Y$ + * <u>for flat plate collectors</u>

Suffixes: f = final; i = initial.

The left hand side denotes the useful energy removed from the collector. On the right hand side, the first term denotes solar energy absorbed at the absorber, the second term represents the convection losses, the third term represents heat losses due to radiation and the last term is heat loss due to conduction. When the useful heat absorbed is in equibilibrium with the total heat loss, there will be no further rise in the temperature inside the cooker as long as the solar radiation intensity remains the same.

These loss terms can be combined into a single expression similar to conduction without much error if (T_f = T_i) differences are less than 50^oK.

The simplified expression is of the form:

$$M\ C_v\ (T_e - T_i) = HRtaX - U_L\ (T_f - T_i)$$

Where U_L = Overall heat loss coefficient.

However, the collector temperature Tc is usually unknown for a design procedure and one can simplify the expression (following Bliss) to give:

$$M\ C_v\ (T_f - T_i) = F_R\ [HRtaX - U_L\ (T_f - T_i)]$$

and $n = M\ C_v\ (T_f - T_i)\ /\ (HRtaX)$

n = Efficiency of heat extraction

F_R = Heat removal factor

= Flow factor x Heat exchanger factor

The simplified equation thus obtained is the basic equation for the flat plate collector design used for such diverse applications as water heating, air heating, space heating and convective solar drying. By increasing F_R R and T, and decreasing U_L the efficiency or heat gain of the overall system can be increased.

In case the differences (T_f - T_i) are very much higher, as in the case of focusing collectors used for rather high temperature applications (eg solar power generation and solar furnaces), the conduction and convection losses are small enough to be neglected. We now obtain:

$$M\ C_v\ (T_f - T_i) = HR\ Cry - \eta\ ET_c 4$$

where η = Stefan's Constant

In terms of T_c we have:

$$T_c 4 = \eta / E(Cr) \frac{(Y)}{(\eta)} (HR)\ (1 - n)$$

So the temperature of a solar absorber, particularly for high temperature applications, increased if R, (η/E) C, r and y are increased as C is constant; n depends upon end use, H on location and geographical conditions and r and η are maximised as far as possible in the normal course.

Current research in solar energy is geared to maximising efficiency or increasing absorber temperatures in the cheapest and least cumbersome ways.

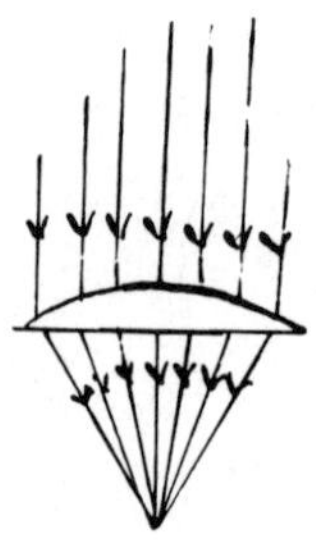

Fig. 3
Plano-convex lens

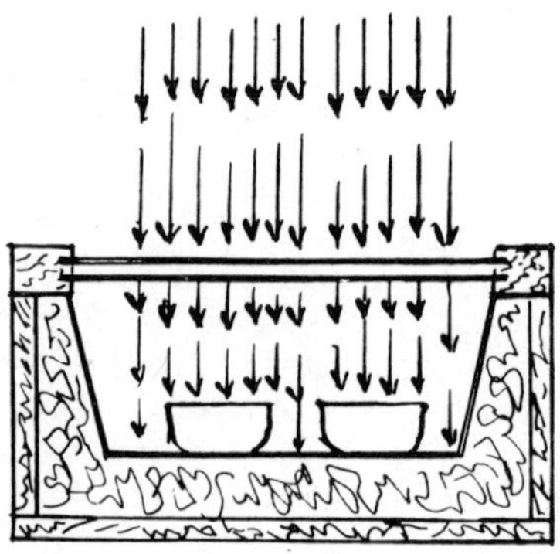

Fig. 1 Box type

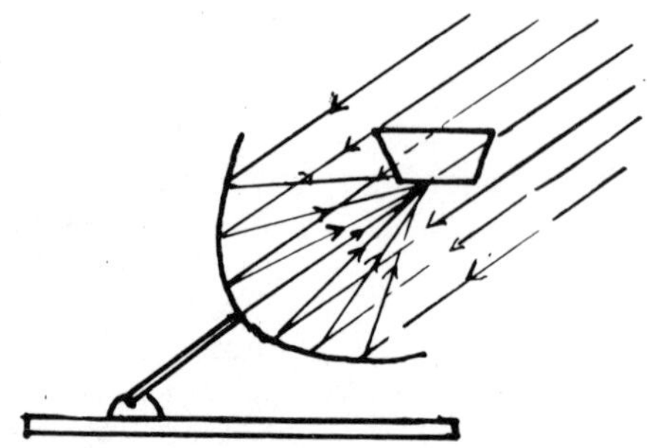

Fig. 2 Parabolic cooker

Design

From the above mentioned principles, solar cookers could be designed in any of the following ways:

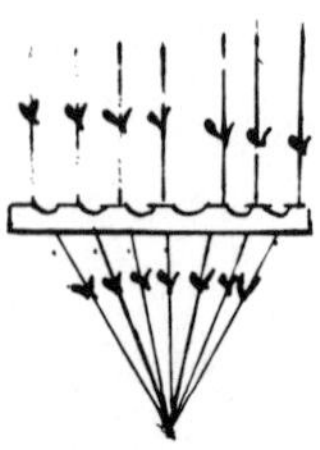

Fig. 4
Fresnel lens

(a) Box-type cooker:
This cooker allows solar radiation to enter through a double walled glass cover, placed inside a blackened box which is well-insulated and made airtight (Fig.1).

(b) Parabolic-type cooker:
In this type, parallel sun rays are made to reflect on a parabolic surface and concentrated on a focus on which the utensils for cooking are placed. The temperature obtained is very high but the loss due to convection and conduction is also considerable (Fig.2).

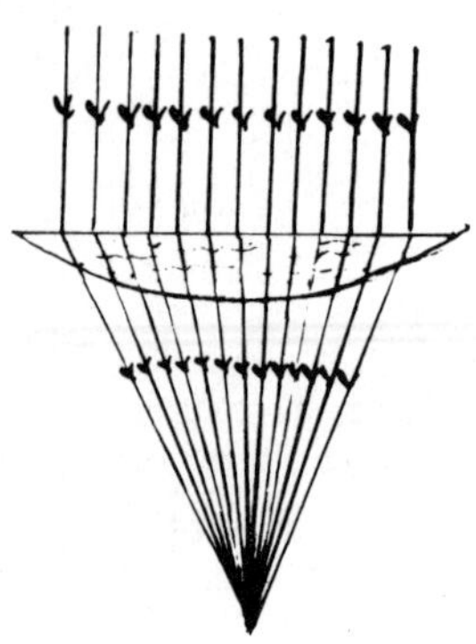

Fig. 5
Water lens

(c) Lens-type cooker:
Sun's rays concentrated through convex lenses and made to converge on to a focus where cooking utensils can be placed (Fig.3). Surprisingly, no design, based on this model has yet been developed. But fresnel lens based on the principles of diffraction gratings have been developed which is reported to be quite effective (Fig.4). Water lens has also been developed by putting water in curved plastic sheets and has given good results in some cases (Fig.5). However, these efforts are still in early experimental stages.

(d) Steam-type cooker:
Steam is raised in a solar water heater outdoors and it is then taken to the kitchen, where the cooking is done in a type of pressure cooker. This enables the housewives to cook indoors in the kitchen and do not have to carry out cooking in the sun.

(e) Box-type cooker:
The Appropriate Technology Development Association (India) are working on perfecting three designs of the box-type cooker. This work is being done in collaboration with the Khadi Gramodyog Prayog Samiti, Ahmedabad, under the guidance of Sri Arvind Pandya, one of the foremost solar appliance designers.

The first box-type solar cooker was designed by Sri M.K. Ghosh of Jamshedpur in 1945, which attracted worldwide attention. Maria Telkes has also done work on this design and called it the Solar Oven.

The box-type solar cooker is based on the following principles. The solar radiation is of short wave length and can therefore enter through glass panes inside a box which is blackened by a dull black paint and kept well insulated from all sides and made airtight. The solar radiation after entering the box through the glass panes is absorbed by the blackened surface and the cooking utensils and transformed into thermal radiation (Fig.1). The wave-lengths of thermal radiation are longer, and it is not possible to pack the wall of the box with insulating material like glass wool or paddy husk ash. This technique can be applied to design a simpler version of this cooker or to make it more efficient by adding a little complication in the design.

After a series of experiments carried out on this cooker, it was found that a temperature of 100°C was reached in winter and 140°C in summer. The cooking time for rice in winter was 3½ hours and in summer 2½ hours. Dal took longest to cook.

Kisan cooker: Kisan cooker is very simple to make and can be made in a any village home by digging a 2' x 2' pit in the ground where it is exposed to sun's rays throughout the day. After packing the sides of the pit with straw or paddy husk ash, a tray as shown in Fig.6, is placed over it. A double glass cover is placed on the tray. The farmer can keep his utensils with food inside the blackened tray, cover the tray with the glass cover, and leave for work. After 3-4 hours, according to weather, the food is ready and it is hot when he comes back.

Fig. 6
Pit type

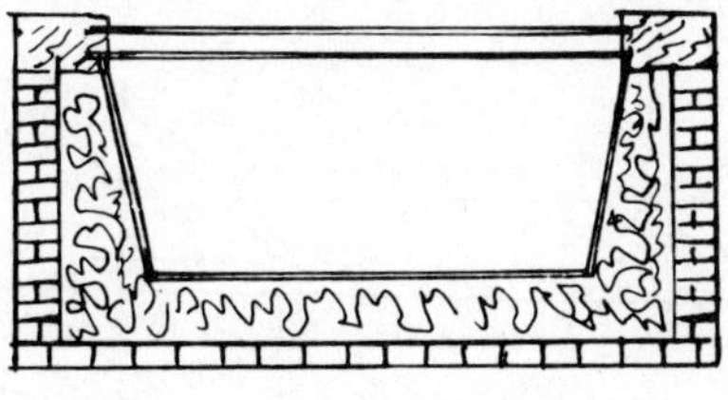

Reflector-type cooker: In the simple box-type cooker, the amount of solar energy incident is rather small for only the direct rays enter the box. An improvement has been made in the design by adding glass reflectors on all sides. This increases the incident energy 4 to 6 times. Although a little more expensive and complicated in construction, this design is highly efficient and brings the cooking time within a range of 45 minutes to 90 minutes, depending upon the weather condition and the types of food to be cooked.

As shown in Fig.7, reflectors are inclined at angles of 60-70° to the glass pane surface and fixed on the top of the box. Geometrically, 60-70° is the most suitable angle for concentrating most of the reflected rays at the centre of the cooking box.

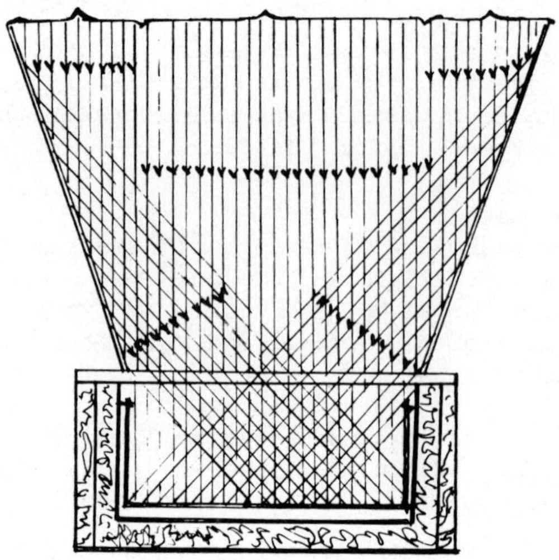

Fig. 7
Reflector type

In addition to the directly incident rays inside the box, reflected rays from all sides of the reflecting surface enter the box, thus multiplying the total solar radiant energy manifold and reducing the cooking time. All the other requirements, such as preserving the heat inside the box by making it airtight and packing the insulating materials on all sides remain the same.

The Association is working on these types of cookers, as its construction and operation are simple and do not require sun tracking.

The parts of the cooker can also be manufactured by the village blacksmith and carpenters.

A working drawing with all the parts and showing the method of assembly is given in Figs. 8 and 9.

Some experiments were carried out on the cooker by the Association and it was found that a temperature of 210°C was reached in summer and 180°C was reached in winter. The cooking time for rice was $1\frac{1}{2}$ hours in winter and 45 minutes in summer. It was also found that the quantity of water added the food to be cooked were also very important as far as cooking time was concerned. Rice needed double the quantity of water, whereas dal needed equal quantity of water. Vegetables and potatoes needed very little water for cooking.

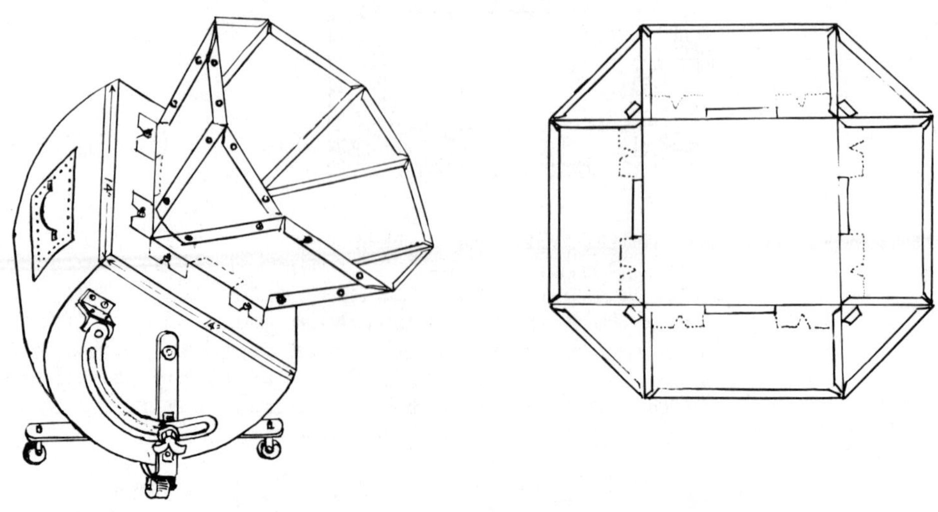

Fig. 8
Solar cooker

Fig. 9
Solar cooker

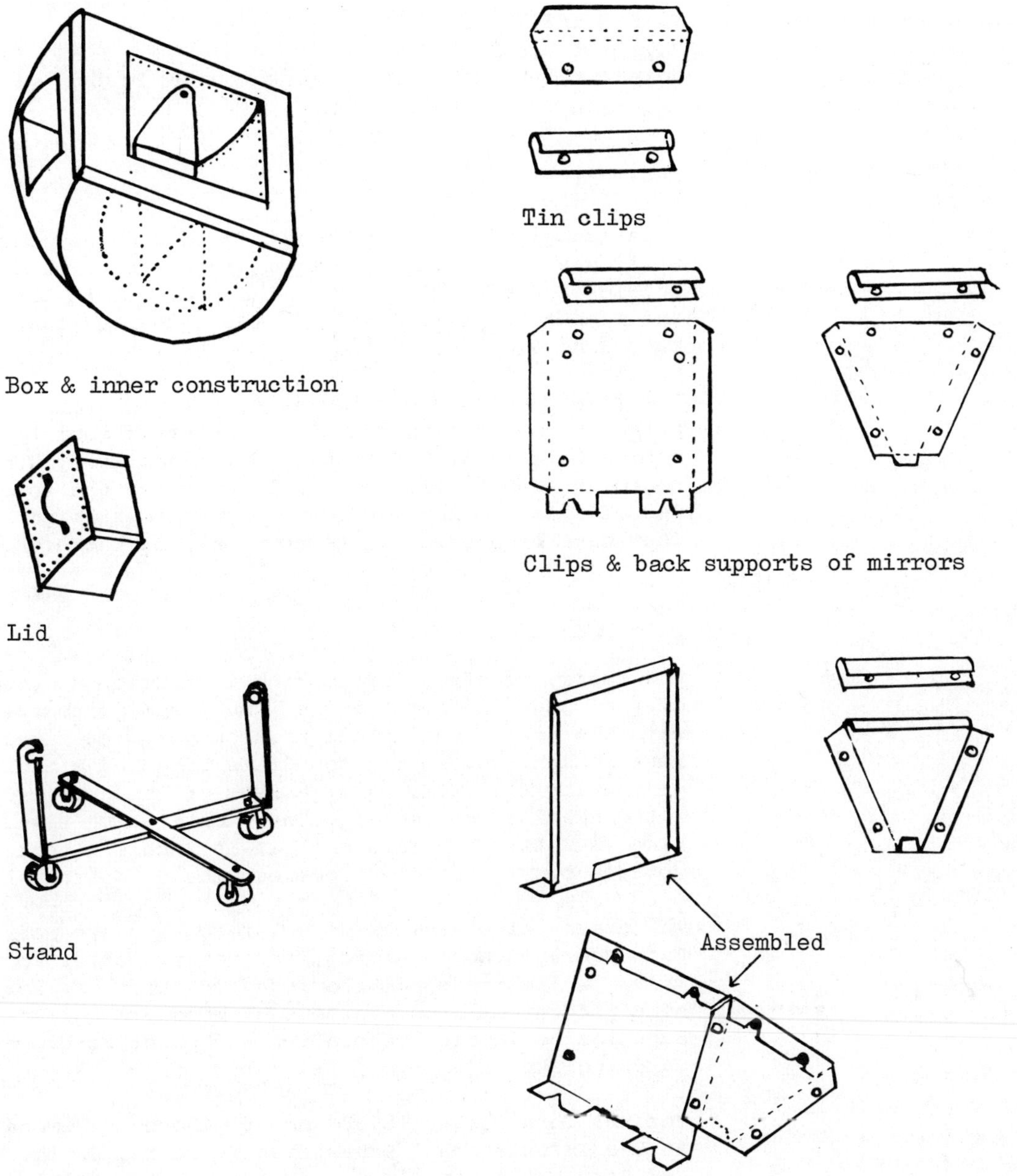

The purpose of the research is to reduce the cost of construction, by using simpler and cheaper materials, and making it more convenient for housewives.

The price of each of the cookers could be as follows if manufactured on a larger scale:

Kissan-type cooker	Rs.50/=	£ 4
Box-type cooker	Rs150/=	£10
Reflector-type cooker	Rs400/=	£30

8.12

Easy-to-build solar cookers

R. Bernard
Physique 1er Cycle
Université Lyon I, 69621-Villeurbanne, France

Abstract

Solar cooking dates back at least to the last century. However it is still used very little, even in those countries whose main resource remains the sun. Perhaps, so far, scientists have bothered more about the technical performances of their devices than about the reactions of those who are to use them.

The five cookers presented here, although less powerful than the usual parabolic devices, have no sophisticated geometry and require only very common materials. They can be built easily by almost anybody. It is hoped that this simplicity will exorcize the foreign and magical character of solar cookers, which may have been responsible for the fact that they have been used, up to now, only by those who don't need them.

Introduction

The energy required for cooking in industrial countries -a few percent of the domestic energy budget- may seem quite trivial in comparison with the space heating demand. However, the situation is quite different in many developing countries because, without any space heating or industrial needs, a large part of the heat produced is required for water heating and cooking. The quantitative significance of solar cooking is better understood if it is recalled that about 80 % of the world's population lives between the 40° parallels [1].

The qualitative aspects of solar cooking are particularly interesting. Each year, in North Africa, the desert border moves forward by a few miles. While trees should be protected there -as fortunately they are in some places- millions of tons of wood are collected yearly out of the Sahelian collapsing vegetation and burnt to prepare meals ... under a parching sun [2].

Yet, as long ago as 1878, Augustin Mouchot demonstrated a solar cooker at the International Exhibition in Paris. In the U.S., Mrs Telkes [3] and Farrington Daniels [4] devised some excellent cookers twenty years ago.

The failure of solar cooking, even in those countries whose main resource remains the sun, teaches us an important lesson: that technical achievements, which might be considered as perfect in a laboratory, must meet some other requirements to be accepted by people in their daily life.

Therefore it appears that a way of overcoming that psychological barrier would be to promote cookers which would be simple enough to be built by the users themselves. In that spirit, five solar cookers were built and tested, and are presented here, together with a few remarks concerning their use, materials and practicality.

Cooker No.1 - Hard-"boiled" eggs without water

It has been reported that, in Africa, some people cook eggs by burying them in the burning sand of the desert. When this big natural collector is not available, however, the cooking can be achieved by means of the following simple device, illustrated in figure 1.

Figs. 1, 2

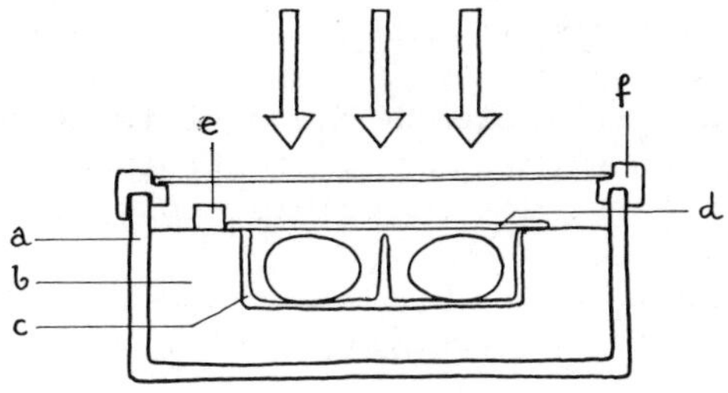

Cooker No. 1

Cooker No. 2 (external lid removed)

This cooker is made from the following materials :
(a) a flat wooden box ;
(b) an insulating material (polystyrene) ;
(c) a cardboard box, coated with a thin aluminium foil and divided into several compartments, one for each egg to be cooked ;
(d) an ordinary glass plate ;
(e) a small wooden stop ;
(f) a lid made of another glass plate in a wooden frame.

The glass plates needed here, as well as in all the other cookers presented, should be as thin as possible (2 mm). They may be purchased at a frame-maker's rather than a seller of window-panes.

The device is put aslant, facing the sun, for about 2 hours. At first sight this duration might be considered to be too long, in comparison with the 10 minutes required for ordinary cooking in boiling water. But it must be emphasized that there is no need to watch the proceedings during all the time, as it is the case when water is placed on a gas burner, and that the time really spent by the user of such a cooker doesn't exceed 3 minutes in total. Due to the very limited possibilities of this little cooker it may be considered mainly as a pedagogical apparatus.

Cooker No.2 - Only carboard and glass

The same principles have been applied to another very simple device in which different dishes can be prepared. As shown in figure 2, it consists of a cardboard box, heavily insulated by means of corrugated cardboard. In a cooker it is not desirable to use black paint which would pollute the food when heated (some glues must be avoided too for the same reason). Owing to its peculiar structure, corrugated cardboard is a good collector of energy when its tubes are directed along the rays of the sun (see Francia's honey comb solar collectors). A central space can receive an aluminium vessel (removable for cleaning) containing the food. Here again, two glass lids are used : the smaller one closing the vessel, the larger one, outside in cardboard frame, closing the box.

With this apparatus it is possible to reach a temperature as high as 90°C (194°F) which is enough to cook such things as tomatoes, mushrooms, eggs, sausages, apples and so on. The desired result is better obtained with the box facing the sun and correctly orientated every 30 minutes or so. But, in very good climatic conditions, successful cooking was possible in such a cooker tied horizontally to a car roof. After traveling during a few sunny hours, the driver found his meal ready for the lunch break.

Cooker No.3 - Two mirrors in a box

It has been possible to increase the efficiency of the preceding cooker by the addition of an elementary concentrating system, consisting of two reflecting surfaces. The materials employed, as illustrated in figure 3 are :

(a) a cardboard box ;
(b) an insulating material (polystyrene) ;
(c) a plywood box ;
(d) two plane reflecting surfaces ;
(e) an aluminium vessel containing the food ;
(f) a plate of ordinary glass covering the vessel ;
(g) a wooden stop to keep this lid from falling when the box is tilted towards the sun ;
(h) the cardboard lid of the box, whose a large central part has been replaced by two plates of glass.

The reflecting surfaces may be two ordinary plane mirrors placed in such a way that the entire cooking vessel may be seen by reflection from a distant point of a line perpendicular to the glass plates. This is the best orientation for maximal use of the sun's rays. This modest concentration (2-3) enabled a temperature of 130°C (266°F) to be obtained after an hour in the sun (without food). In opening (and re-closing) the apparatus as quickly as possible, it is then possible to cook a beef steak or an omelette in 5 to 10 minutes. This cooker may be used also for cooking potatoes, peas, fish, rabbit and so on. It is very convenient for camping owing to its moderate dimensions : a box of 45 x 32 x 24 cm is sufficient for three people (see fig. 4). Its weight (and price) can be considerably reduced in replacing the mirrors by two cardboard plates covered with an aluminium foil. Thus the reflecting surfaces become more fragile but not less efficient.

Figs. 3, 4

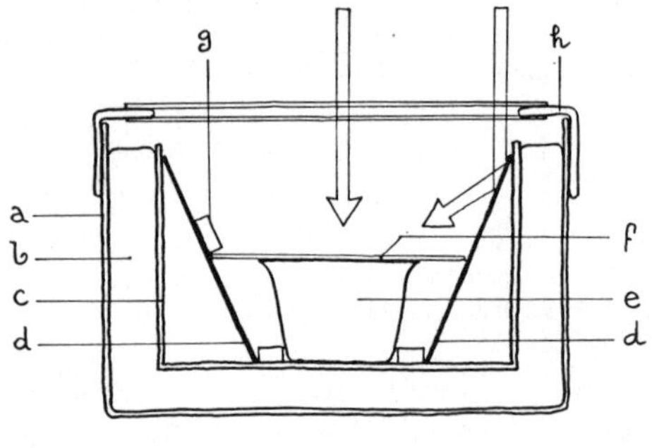

Cooker No. 3 (diagram)

Cooker No. 3 with potatoes and sausages

Cooker No.4 - Hot meals after sunset

We have seen that the preceding devices have to be put aslant to face the sun. Consequently they only are suited for dishes whose cooking requires almost no liquid. For cooking which requires a lot of water, the geometry of the box has been modified so that its bottom may remain horizontal and carry a casserole full of liquid. On figure 5 can be seen :

(a) the usual insulating material (polystyrene) sandwiched between two plywood boxes (b) ;
(c) an aluminium foil ;
(d) a black cast-iron casserole containing the food ;
(e) two thin glass plates held together by a cardboard frame. The plates are placed at an angle to the horizontal plane equal to the latitude of the place, in order to be almost perpendicular to the sun's rays (for instance, in Lyon, this angle is a convenient 45°) ;

(f) an insulating lid carrying the mirror (g). A wooden stick fitted into a small rack is used to orientate the mirror according to the altitude of the sun (see fig. 6). The mirror (g) and the reflecting surfaces (c) provide a concentration of 3-4.

Figs. 5, 6
Cooker No. 4

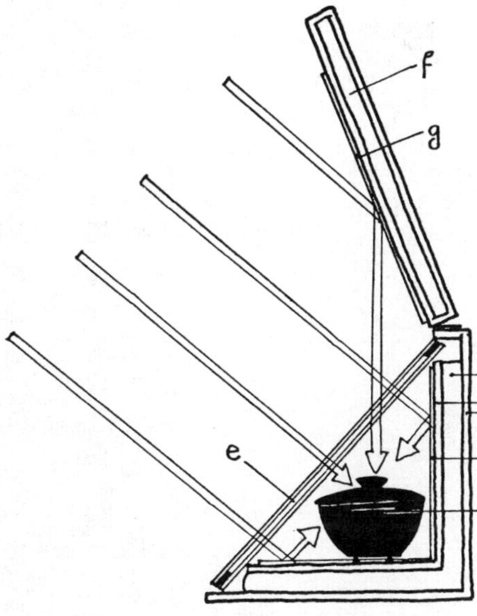

The time required to bring the vessel containing one liter of water to 100°C (212°F) is about 90 minutes. As well as all the foods mentioned previously, all boiled and simmered dishes can be prepared in this cooker, e.g. soups, spaghetti, lentils, french beans, jam and those two basic food for hundreds of millions of poor people : rice and millet.

It has been said that in several developing countries the main meal is eaten after sunset, which makes solar cooking impracticable [5]. However it is possible to cook the food in the sunshine and to keep the meal warm until sunset by shutting down the insulating lid (f). Indeed, it is rather pleasant at night, to find the heat of the sun in one's plate.

Cooker No.5 - More sun in the same space

To be at a convenient height, the preceding cooker must be placed upon a table. It was tempting to use the sun's rays which fell under this table and that result was achieved by the device whose principle is illustrated in fig. 7, with
(a) an insulated box (plywood + polystyrene) ;
(b) an aluminium foil ;
(c) a double glazed window which may be opened to introduce the food container (d) into the box, where it is carried by three steel rods (e) ;
(f) two glass plates ;
(g) four rectangular plane mirrors set in the same wooden frame.
This last device can be described as a roughly cylindro-parabolic mirror whose large focus sweeps the food container. It is sufficient to adjust the angle of the frame every half hour by turning it around its horizontal axis according to the altitude of the sun.

The box and the mirror are carried on a trolley-table which makes the azimuth positioning very easy on the flat ground of a terrace (but, of course, the wheels would be useless on a sandy ground).

Fig. 7
Cooker No. 5

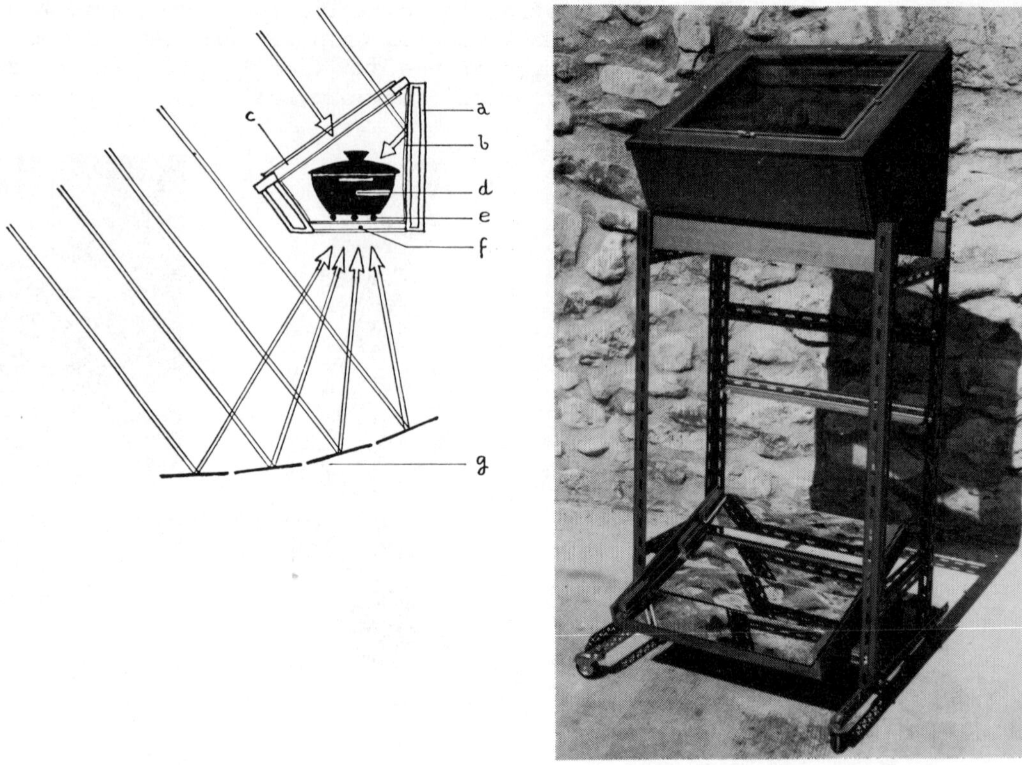

This apparatus can be used to prepare the same dishes as cooker No.4, in a little less time. In fact, this last cooker is probably more convenient to use but more difficult to build than the preceding simple boxes. Generally speaking, there seems to be a constant product of these two factors between which the user may choose the best compromise.

Remarks

The following remarks can be made considering the experience acquired in this field :

(a) Solar cooking with these simple devices requires more time than with traditionnal cookers or parabolic solar cookers ; but for some dishes, as was demonstrated with cooker No.3, it is possible to pre-heat the empty cooker and then the cooking itself may be comparatively quick. The long time taken by the operations requires stable weather for at least two consecutive hours, which will seldom be the case in very changeable climates where, however, solar energy can be valuable for other uses, such as water or space heating. Anyway these slow-cooking devices are much superior to the traditional cookers for the heating of already cooked dishes. When flames or electric elements, at several hundred degrees Celsius, are used for this purpose, it is necessary to stir the food which otherwise would be burnt at the bottom of the vessel. There is no risk of burning in solar cookers where the energy arrives from above (except for cooker No.5) so that solar heating practically requires no attention.

(b) Although the materials used here are very common in Europe, various possibilities must be studied in relation with the local resources of the so-called developing countries such as new insulating materials. J.R. Vaillant [6] reports that in Niamey (Niger) solar cookers have been devised with calabash bark (instead of plywood or cardboard) and peanut shells (instead of polystyrene). Kapok, too, turned out to be an efficient insulating material. The geometry of some devices should also be modified with respect to the latitude where they are to be used.

(c) In order that solar-cooking may be accepted it must meet certain practical requirements that architects should not forget : the kitchen must face the South (in the Northern hemisphere) and be level with a terrace or a balcony. A solar heater may even be built into the wall facing the sun.

Conclusion

The very simple devices presented here will seem almost derisive in comparison with vast programmes of computer simulation or gigantic solar projects. However, it is considered that the general acceptance of solar energy must begin with the faith and interest which can be aroused among the young people,only by the simple devices that they can build and use themselves. Thanks to the Centre Régional de Documentation Pédagogique de Lyon, who diffused our instruction leaflet in the schools, many schoolboys have already enthusiastically built their first cooker and enjoyed their first solar meal. A comparable work is being started in Senegal with our cookers No.3 and No.4 [7] by the General Direction of Scientific and Technical Research of that country.

In this campaign for a sound ecological education of people and for a popular spreading of solar techniques, it is hoped that, where scientists have failed, children might succeed.

References

1 - B.J. BRINKWORTH - Solar Energy for Man (Compton Press, 1972).

2 - P. ELOUARD - Sahel, aujourd'hui et demain (Carnets de l'Enfance, Unicef, Oct. 1974).

3 - M. TELKES - Solar cooking ovens (Solar Energy, Nov. 1959).

4 - F. DANIELS - Colloque C.N.R.S. (Montlouis, 1958).

5 - G.F. WHITE - La Science et l'avenir des terres arides (Unesco, 1963).

6 - J.R. VAILLANT - Utilisation et promesses de l'énergie solaire (Eyrolles, 1976).

7 - GROUPE SOLAIRE LYONNAIS - Cuiseurs solaires (Ecologie, Energie solaire et alimentation, Juil.1976).

Solar powered air structures

G.A. Stevens
Atmospheric Energy Research Organization
18 Albemarle Street, London W1

Abstract

The first part of this presentation at the conference included the film 'Desert Cloud'*(1) which showed a prototype solar air structure operating entirely on atmospheric energy; inflating, erecting and flying in the Arabian desert. The film then explored its use for producing water, food, shelter and transport. This paper takes one particular application, a large span greenhouse, which although using essentially no fossil fuels or mechanical equipment, changes its mode of operation to utilise the various climatic conditions of the world; hot humid, hot arid or temperate. This self-energised system significantly increases the potential crop yield of the main forms of crop production; horticulture, silviculture, algae culture, agriculture or mari-culture, either as monocultures or integrated for symbiosis. The variety of applications of 'Desert Cloud' is illustrated by examples of industrialised buildings, rehabilitated Georgian terrace housing, solar heated swimming pools and fish farming.

'DESERT CLOUD' APPLIED TO BIO-CLIMATIC PLANT HOUSING

Air Structures

Air supported structures as a building technique have been in use since the second world war, (2) and traditionally used for its radiation properties: 'Radomes' 'Echo' Satelites, as well as greenhouses. They can be represented by a simple diagram. (Fig.1) A flexible membrane is

Fig. 1
A standard 'GOUROCK AIRHOUSE' circa 1965 showing main elements and energy exchange.

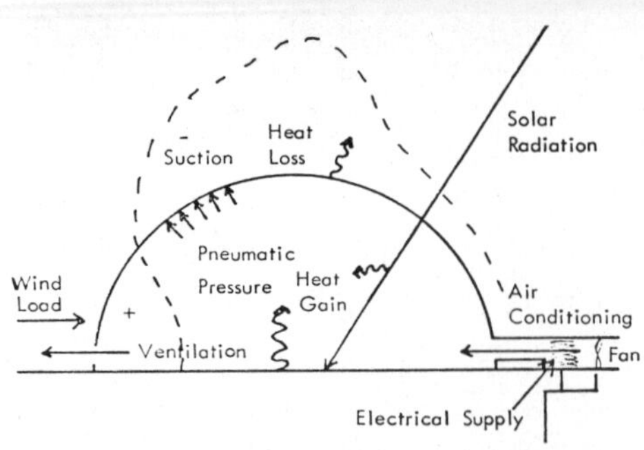

anchored to the ground to form a sealed enclosure, a fan or blower constantly blows air into the enclosure at a greater rate than it can escape through doors, vents, seams, holes etc., resulting in a positive pressure slightly above ambient, which tensions the membrane as a function of internal pressure and radius of curvature. (F)T=(F)P(F)R. This pressure is designed to be high enough to resist any deformation by wind or precipitation. The internal temperature and humidity is controlled by heating appliances inside the structure or air conditioning plant incorporated with the blowers.

The Advantages of this structural system are:

1. Span:
   Pressure being equal throughout the volume there is no moment factor in loading; the tension of the skin is theoretically independent of span, allowing the greatest spans of any known system.

2. Cost:
   The low mass and simplicity of form gives low material and fabrication costs 50p - £50/m^2 installed.

3. The continuity of the membrane and absence of any support frames or walls allows maximum radiation transmission.

Disadvantages:

1. Reliant on continuous supply of fuel and mechanical functioning of blowers.

2. Requires large amounts of fuel and mechanical plant to control temperature since the thin skin allows high rate of thermal losses and gains.

3. Requires specialised entry and exit openings.

The air structure proposed here (Fig.2) has the potential of keeping the advantages and overcoming the disadvantages, and is the result of 12 years experience with air structures (2) (3) (4).

'DESERT CLOUD' ADAPTED TO LARGE SPAN BIO-CLIMATIC PLANT HOUSING

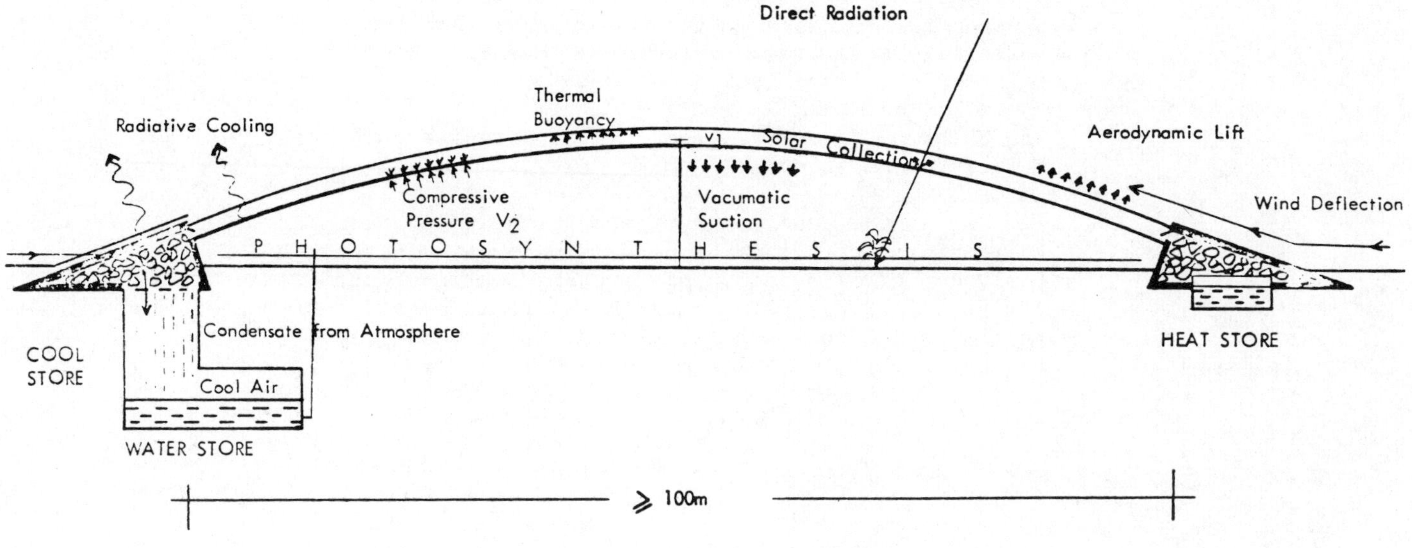

Fig. 2 Schematic representation of the elements, the sizes and configuration changes according to the meteorological conditions of the location.

Principles of the Structure

1. Energy Converter
   The whole structure functions as a collector, store and conditioner of available energy, with plurality of function in designed elements.

2. Structural Energy
   The energy required for structural stability and environmental control is obtained from its ambient environment.

3. No fuel or machinery
   The use of ambient energy obviates the need to import fossil fuels and mechanical plant.

4. Self Operating
   The structure adapts to optimise the use of available forms of energy, radiation, thermal difference, air momentum, etc., of different world climates.

5. Direct and efficient Use of Energy
   Since the energy forms required are the same as the ambient sources their use is by direct transfer and use or by conversion to low intensity dispersed heat for storage, induced flow and designed time cycles, all of which are 'passive', without need of energy conversion and distribution machinery.

6. Optimises Adaptability to function
   For Example: The structure used as a greenhouse is designed for plant requirements, not human; maximised condensation gain, conservation of water, high transmission of photosynthetically active radiation (P.A.R.) etc.

7. Productive Building
   The structure becomes an active part of the productive process it contains, not a passive shelter from the elements.

These principles can be observed in the illustrations from the film 'Desert Cloud' (Figs. 3-7). The operational sequence of the solar membrane can be summarised as follows:

Transparent, partially reflective (silvered) and absorbitive (black) flexible plastic sheet is arranged to form a composite membrane. The membrane is laid on the ground. Short wave solar radiation passes through the transparent and is absorbed into the black. The partially reflective skin prevents radiation passing into the ground and also reflects onto the black.

The black panel re-radiates long wave radiation heating residual air held in the membrane.

The air expands decreasing its density and so becoming buoyant, lifting the top transparent membrane so drawing air in through a flap valve to replace the rising air. This process continues until the composite membrane is fully inflated with the transparent, reflective, absortive panels in the configuration shown in the photograph.

Figs. 3-7 Stills from the film 'DESERT CLOUD' showing inflation and flying sequence in the Arabian Desert.

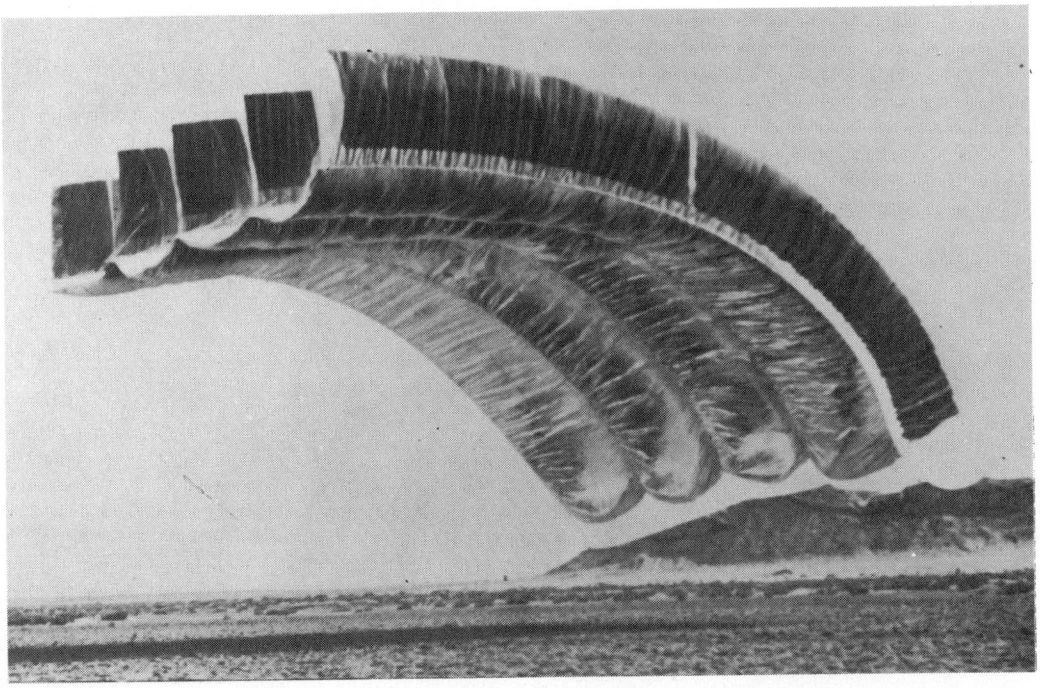

MAIN OPERATIONAL SEQUENCES AND USES OF AMBIENT ENERGY

Structural Energy

Solar/Thermal Buoyancy
Solar radiation heats up V1 providing lift. Excess heat is carried by air or water to the hot rock-store.
Structure erects by raising itself into position.
When increase thermal lift is required, at night or during cold weather, the stored heat re-enters V1 by convection.

Static Air Pressure
Decrease in volume from applied load increases internal pressure to support loading (wind, rain, snow, sand or dead land).
Increase of sealed volume V2 decreases pressure creating suction anchorage.

Wind Energy
Rock-store anchorage deflects wind to flow over low profile of structure creating suction lift forces, lifting membrane.
Elasticity of membrane allows expansion/contraction of V2 to maintain cable tension.

Plant Environmental Control

Heating
The membrane collects available heat.
The rock-store acts as a short term heat storage.
The ground may extend the storage capacity of the structure for inter-seasonal storage.

Cooling
Cold rock-store is insulated from incoming radiation, opened to night re-radiation.
Water may be cycled through membrane for night radiative cooling.

Production of Photosynthesis

The membrane is transparent to P.A.R. reflecting and collecting the rest as heat.

Closed System
The closed system nature of V2 allows control of humidity, CO_2 and microbial content of the microclimate and protects from insects, pests and disease.

Water Production
The cool surfaces of the cold rock-store may be used for atmospheric water condensation.
The membrane panels in contact with atmospheric air may also be used for water condensation, providing a source of water in areas of low rainfall.

Water Collection
The large surface area of the membrane is used for rainfall collection.

Water Conservation
The sealed volume V2 conserves water by eliminating evaporative losses by condensation and recycling.

The system adapts to provide optimum plant environments in the different climates of the world. The system also adapts to specific crop requirements:

Hot Arid
(a) The radiative cooling is used for cooling plants.
(b) The surplus heat may be used for desalination or powering refrigeration cycles.
(c) In excessive heat conditions water may be used for cooling the membrane.

Hot Humid
(a) Tropical climates often have high rainfall for short periods, the surface area may be used to collect water during the rainy season to be stored for the dry season.
(b) The environmental control may be used for creating dry conditions.
(c) The dense humid air increases the effective buoyancy of the dry hot air in the membrane.

Fig. 8 5 acre greenhouse complex using standard air-supported units (centre background). Abu Dhabi Arid Lands Research Centre.

Temperate

(a) The low temperature and dense air increases the effective buoyancy.
(b) Reflective level of lower membrane may be reduced to increase direct heating of air space.
(c) The storage capacity of the rock-store may be extended by using the ground as a heat sink.

Crop Yields

The pioneering work of the Arid Lands Research Centre, Abu Dhabi and the Environmental Research Laboratory, University of Arizona, may be taken as an indication of scale (5 acres see Fig. 8) and crop yields without CO_2 enrichment. (Figs. 9 and 10).

Figs. 9, 10 Comparative and potential crop yields at Abu Dhabi (without CO_2 enrichment).

Comparative yield for a single crop grown in Sadiyat greenhouses or field grown (tons/acre/crop)

Type of vegetable	Sadiyat greenhouse marketable yield	Good yield field grown U.S.*	Al Ain government farm**
Cabbage	31	12	22.0-31.0
Cucumber	102	12	8.8
Eggplant	107	8.25	13.2-22.0
Lettuce	25	10.5	8.8-13.2
Okra	23	5	13.2
Tomato	71	30	6.6-11.6
Turnips	70	10	6.6- 8.8

Potential yield

Type of vegetable	Potential yield (tons/acre/year)
Cabbage	190
Cucumber	450
Eggplant	215
Lettuce	300
Okra	60
Tomato	196
Turnip	470

*From: Knott, James E., 1957, HANDBOOK FOR VEGETABLE GROWERS, John Wiley & Sons, Inc., New York.
** Yields supplied by Mr. Omar Baki of the Al Ain Department of A Agriculture, Abu Dhabi.

The work so far carried out has:

(a) demonstrated self-energised air structures.
(b) and indicates the feasibility of using this type of air structure for plant climate control, ie. Bioclimatic plant housing. (Fig. 7)
(c) Bioclimatic plant houses to be feasible for the main forms of crop production: horticulture, silviculture, algae culture, agriculture or mari-culture, either as monocultures or integrated for symbiosis.

The initial response indicates that links with research stations in the various climatic regions of the works is a desirable way to arrive at the most appropriate design variation for specific conditions and provides the opportunity for joint research participation.

Housing, Industrial and Agricultural Applications

The bioclimatic greenhouse is one example of the application of the principles of 'Desert Cloud' to one of the most current needs, that of food production in previously low yield inhospitable climates; however it is one which requires a large development programme. Other applications, such as desalination, where the improvement on existing technology is mainly cost, require much shorter development times, say three to six months to installation.

The principle may also be applied in combination with the standard air structure, for example, combined cover and water-heating for swimming pools or fish farming; a prototype of which is shown in Figures 11 and 12, carried out as a two week educational exercise at the North London Polytechnic School of Environmental Design.

The applications are not restricted to flexible membrane structures, as shown by the feasibility study rehabilitating Georgian terrace houses (Fig. 13) carried out at the London Borough of Camden. Since one third of the gross national energy consumption in the UK is in domestic space and water heating, adapting existing housing stock offers one of the largest potential energy savings in the UK. The design treats the whole

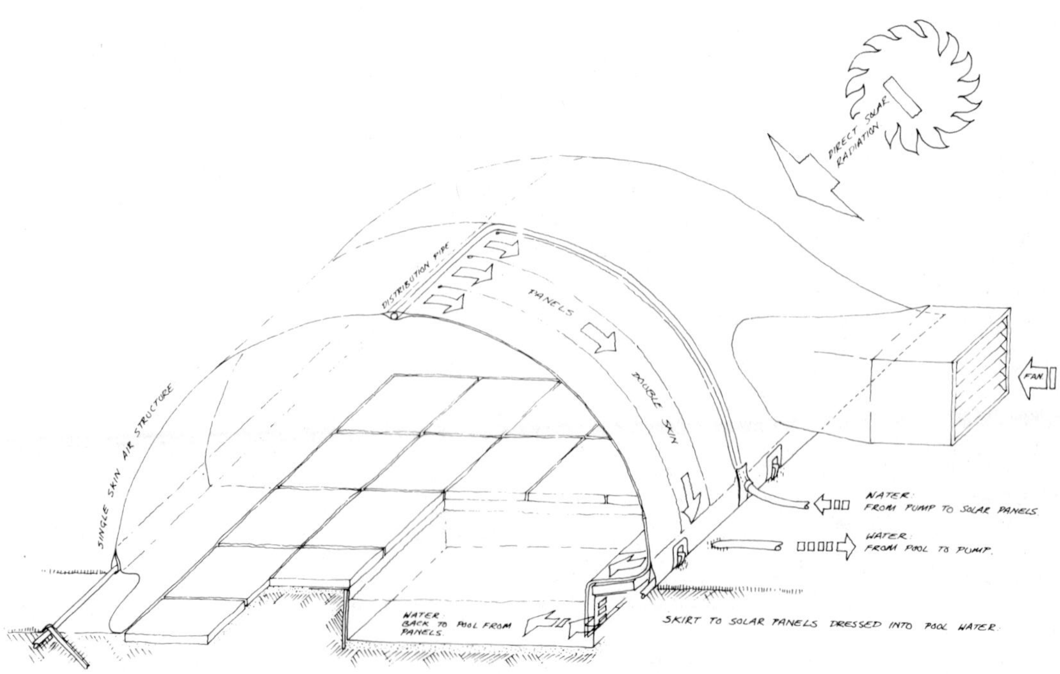

Fig. 11 Combination of 'Desert Cloud' and standard air structure to provide large volume water heating for swimming pools or Fish Farming.

Fig. 12 Working model of Figure 11 showing staggered panels for comparison.

Fig. 13 Cross section of Georgian terrace house adapted to maximise solar energy use.

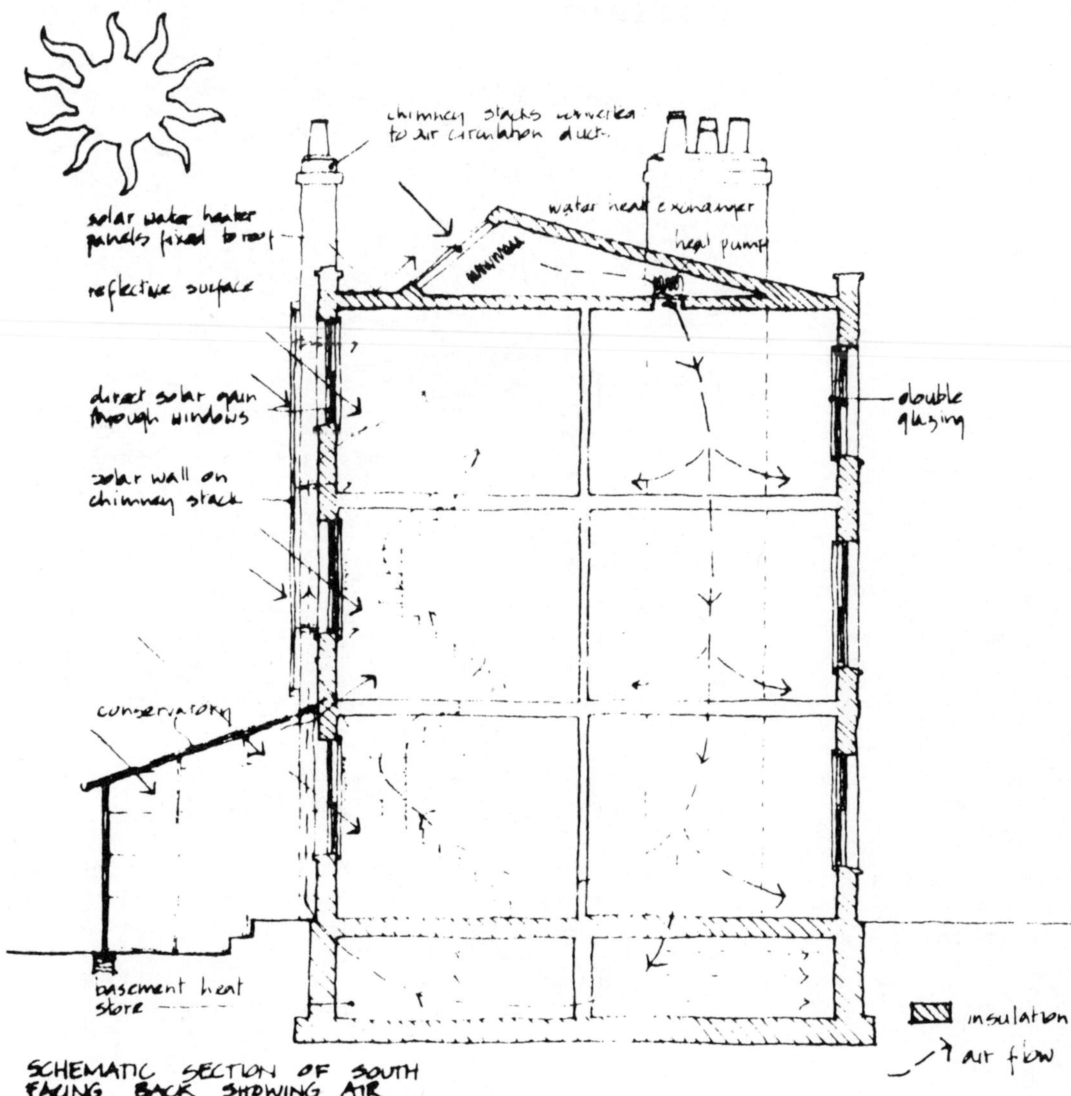

Figs. 14, 15

Example of adaption of 'Desert Cloud' to rigid frame structures for heating and/or cooling.

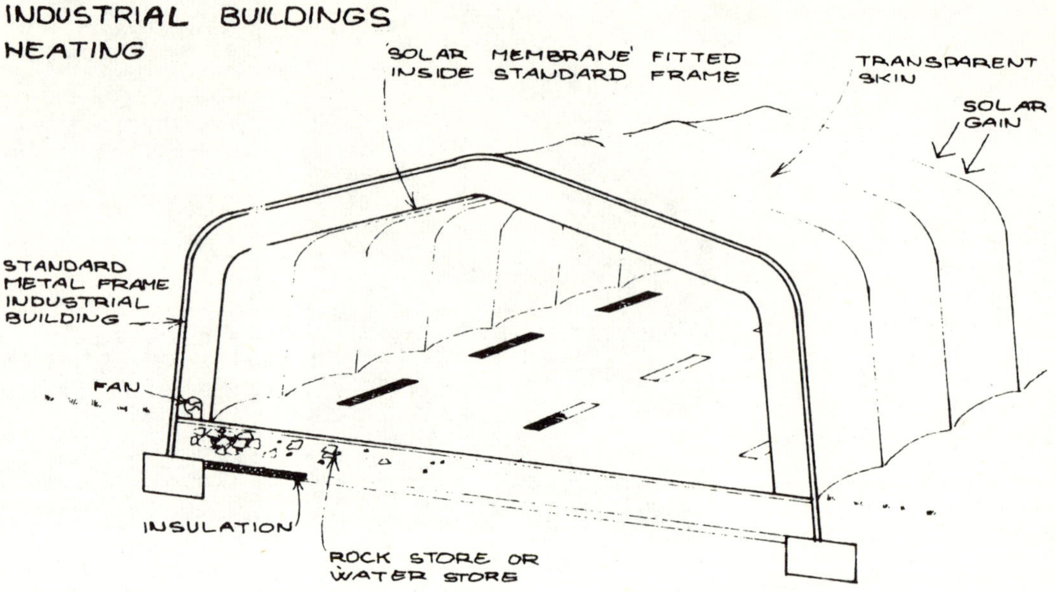

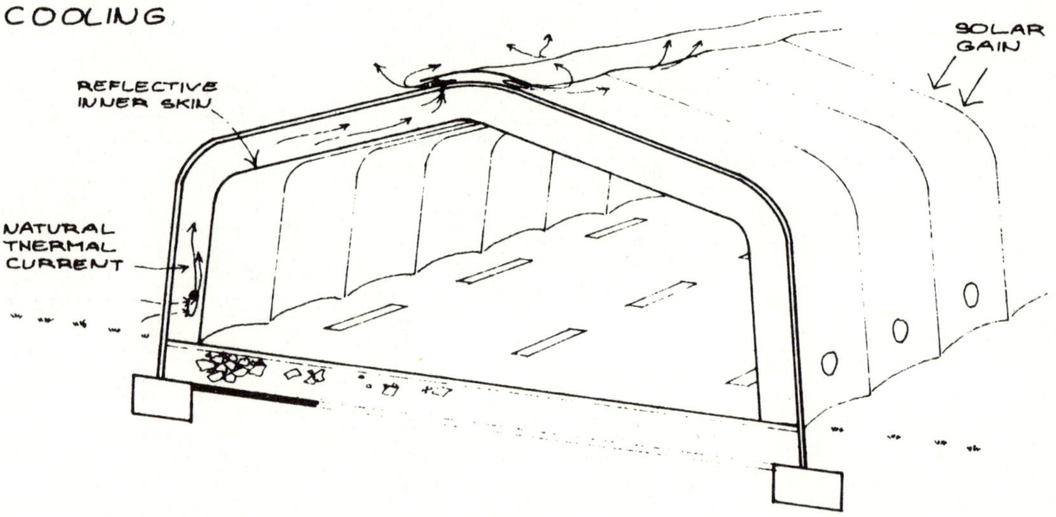

building as an energy collector raising its efficiency as high above the present 7% as physically and economically possible, applying the principles described for the greenhouse and employing a time element for introducing hardware components such as solar water heaters, heat pumps, heat exchangers as and when they become technically and economically feasible. (Fig. 13).

The final example is industrialised buildings (Figs. 14 and 15) which may be used for heating or refrigeration.

The paper has tried to show the process of introducing concepts at a cultural level to be applied and adapted to the specific structural solutions which are most appropriate to the particular physical, economic and social conditions of their location.

References

(1) 'Desert Cloud' and Airworks production for the Arts Council of Great Britain. 16mm colour 18 mins. 1976.

(2) 'Air Structures', a survey by Cedric Price and Frank Newby for the Department of the Environment 1971, HMSO.

(3) 'Pneumatics and Atmospheres'. Architectural Design, March 1972.

(4) International symposium on Pneumatic Structures, IASS, Delft, 1972 (report A.D. January 1973).

(5) First Annual Report of Arid Lands Research Centre, Abu Dhabi, 1972.

(6) 'Bioclimatic Plant Housing' G.A. Stevens and C. Moorcraft, European Seminar on biological solar energy conversion systems. Grenoble, France, May 9-12 1977.

(7) 'Energy Cost Reduction for 6-24 Carol Street'. Initial architects report May 1977.

General

Farrington Daniels 'The Direct Use of Solar Energy'.

B.J. Brinkworth 'Solar Energy for Man'. Compton Press, 1972.

'The Sun is the Service of Mankind'. UNESCO Paris, July 1973.

Conference Proceedings of UK-ISES 1973 onwards.

Conclusions and recommendations

9

9.1 Working groups

During the conference the scientific committee, under the chairmanship of Professor Eric Farber, set up four working groups in the following areas:

1. Educational aspects
2. Passive solar heating and cooling systems
3. Active solar systems
4. Economic, social, legal and political aspects.

The above working groups met during the conference and on the final day, in the afternoon session, they presented their respective conclusions and recommendations to the conference. The exact text of each group's recommendations is reproduced here as follows:

1.1 Working group on educational aspects
Chairman: Peter van Dresser

The sub-committee recognises four categories of activity relevant to solar building developments in the various nations of the world. These are as follows:

1. Cultivation of an awareness of the immediate and future potentials of the solar utilisation in the 'built environment' amongst the general public. This recognition or 'sensitization' might be accomplished through special courses in the various levels of the state school systems: slides, filmstrip and cinematographic presentations made available to various public groups and associations; pamphlets and other literature distributed through such agencies as agricultural extension and home economic services agencies, libraries, mobile exhibits, etc.

2. Information and advisory services to governments and government agencies. This might consist of a special Secretariat of qualified individuals available for consultation with government officials on policies appropriate to the promotion of solar building technology. A strong element of this programme would be the encouragement of the application of solar energy to as wide a range of public buildings as is possible. It should also encourage the adoption of laws and regulations which facilitate and stimulate the widespread construction of solar buildings in the private sectors of national economies.

3. Education and training in solar building design and construction techniques. Several levels of this activity are envisioned. First the establishment of a comprehensive advanced curriculum for professionals in the field of solar building technology. This is envisaged as a new discipline including aspects of engineering, architectural, ecologic and economic analysis. It should produce professionals with a thorough grasp of the full range of solar technology from the most sophisticated to the basic folk or village orientated versions. It should equip the graduate with keen appreciation of the ecological appropriateness of these various versions in differing cultural and physiographic situations and of the vast store of knowledge on climatic design embodied in traditional and vernacular architecture, so as to ensure that recommendations, projects and programmes would be appropriate to the cultural, economic and natural milieu in which they are intended to function.

A more immediate aspect of special training is envisaged as a modification of existing curricula in architecture and engineering to include solar technology and ecological methodologies perhaps, but not necessarily, at the post-graduate level. This may be considered an interim step until the full solar technology curriculum can be developed. An additional level of training would be for artisans and technicians in regional and local situations. This would probably be carried on through such means as field workshops and vocational training courses in local schools. Additionally, travelling grants might be provided for individuals in various regions to visit solar work going on in other regions of the world.

4. The fourth general educational enterprise might be the establishment on a world basis, of a central bureau or clearing house for the filing, retrieval and dissemination of information on all phases of solar technology. If this project appears too complicated or costly, it might be set up as simply a referral centre for agencies, institutions and individuals capable of advising on questions of solar technology, categorized by geographic region and by topic.

1.2 Working group on passive solar heating and cooling systems

Chairman: Dr. Doug Balcomb

The working group has identified four main areas of work where sponsorship by UNESCO, or any other UN agency, would be desirable. Therefore, it recommends that aid is given in:

1. Sponsoring small study grants to document passive solar heating and passive cooling techniques developed prior to the oil age by the indigeneous peoples in various countries and climates (such as the use of air towers for cooling or night radiation for ice making in Iran).

2. Sponsoring a pair of small study grants to achieve the following work objectives:
(a) Conduct a country-by-country survey of the potential applicability of passive solar heating and cooling techniques. How would each technique be expected to work in each country? This should be done separately for heating and for cooling, as applicable. This analysis should be based on weather data for each country and utilize techniques developed (i.e. in France and the US).
(b) Formulate and publish recommendations for the application of passive techniques in various countries. These should be based on the results of the thermal performance expectations (above) and also on considerations of traditional architecture, building materials and techniques and the cultural constraints imposed on building design in the various countries. These should be individually documented for each country in the language of that country (including the use of the local conventional units and terminology).

3. Sponsoring an International Conference on Solar Heating and Cooling. This should emphasize approaches readily applicable to current problems and the use of inexpensive materials easily available within the respective countries.

As there will have been two topical conferences on this subject in the US (May 1976 and March 1978) it would be appropriate to hold this conference outside the US, perhaps during the spring of 1979.

4. Identify the applicability of small greenhouse structures added onto the south side of existing buildings as a solar heating element for the building. (This has been proven in certain areas - e.g. Northern New Mexico, USA - but should have very wide potential). Provide designs suitable for the climates and building types of each country utilizing inexpensive materials available in the country. These designs should be optimized to provide a maximum of heat to the building and food for the occupants.

(Clarification = A passive system is one in which the thermal energy flow is by natural means, i.e. by natural convection, radiation or conduction).

15

1.3 Working group on active solar systems

Chairman: Professor C.V. Chinappa

The working group has identified six main areas for further action and has therefore recommended the following:

1. The nature and scope of meteorological measurements be widened by increasing the number of stations, obtaining insolation data for inclined surfaces and wind data, employing less expensive instrumentation, providing funds for instruments where necessary, making suitable arrangements for the storage and accessibility of data, etc.

2. More small conferences should be planned for specific geographical areas and for specific areas of technical interest, e.g. food processing (refrigeration, drying, etc.), active systems for regions where electricity and/or water is scarce, use of wind power, etc.

3. That funds be provided for interchange of personnel, and travel expenses, wherever projects are undertaken which are of immediate relevance to more than one geographical region, and for sending consultants for training purposes.

4. The dispersal of technical information be assisted particularly to regions where library and information retrieval services are inadequate, e.g. by support of university libraries, etc.

5. Projects appropriate to developing regions, such as crop drying, cooking, waterpumping, refrigeration and desalinisation, be planned and supported in such regions.

6. The provision of solar battery operated lights, radios and television at village level be encouraged as part of rural education programmes.

1.4 Working group on social, economic, legal and political aspects

Chairman: Dr. Cleveland McVeigh

1. All countries are urged to encourage the use of solar energy and other renewable sources by
(a) direct grants for solar demonstration projects, and
(b) appropriate tax incentives.

2. At present the laws of nearly all countries do not protect the individual's rights to sunlight. All countries are urged to adopt legislation guaranteeing these rights.

3. All countries are urged to prepare a national solar energy report such as the UK Department of Energy report, which can establish the potential and appropriate targets for the building sector. Implicit in each report should be a strategy which accepts two principles:
(a) Industrialised countries cannot continue to increase their use of energy per capita and they should examine all means of reducing the present increasing trends.
(b) Other countries should recognise that it is neither necessary nor desirable to follow the same energy patterns of growth adopted in the past by the major industrialised countries.

4. Studies should be commenced for each country on the comparative merits of various solar heating systems including the ones which are already commercially available. Recommendations would then be given for the most appropriate techniques and systems. The use of simple pay-back for comparison is recommended.

5. The long-term implications of the inevitable rise in the cost of fossil fuels on the projected costs of solar energy systems should be studied. As solar collectors are systems which depend on conventional fuels for their manufacture, rises in fuel prices affect collector costs.

Therefore, we urge UNESCO to encourage the preparation of a number of studies on the economic implications of Solar Energy utilisation, particularly item numbers 4 and 5 above, and make the necessary recommendations to the various governments.

9.2 Summary of recommendations

Following the presentation of each working group's proposals and recommendations, by the respective group chairman to the general conference, discussion took place among the conference delegates with further recommendations made and resolutions adopted. After the close of the conference the executive branch of the scientific committee met and considered carefully all the proposals and recommendations made by the four working groups, the proposals made by the various conference delegates and the resolutions adopted by the conference as a whole. Consequently, the following summary recommendations were proposed and submitted to UNESCO for consideration and further action.

2.1 General recommendations

1. UNESCO to co-ordinate the various international programmes of solar energy, undertaken by international organisations in order to reduce duplication of work and allow funds and resources to be used more efficiently and effectively on a world-wide basis. Such programmes for example are already well established by organisations like the EEC and OECD.

2. UNESCO should try to influence the various governments to encourage wider solar energy utilisation through educational programmes - courses, curricula, regional conferences, publications, the introduction of laws and regulations and by any other possible means.

3. UNESCO should try to bring together experts in solar energy with those in architecture, agriculture, engineering, etc., and encourage combined efforts in solving the world's energy problems.

2.2 Specific recommendations

1. UNESCO should take steps to mount a discussion on solar energy at General Assembly level.

2. UNESCO to publish or help publish proceedings of meetings, books, reports and papers and disseminate them widely.

3. UNESCO to help set up groups of experts to pinpoint problems and their solutions on a regional and global basis and disseminate their findings world-wide.

4. UNESCO to provide funds to send expert consultants to needy areas and make it possible for groups to visit centres of solar energy expertise.

5. UNESCO should encourage countries to use solar energy in government buildings and schools to lead the way and demonstrate the viability of solar energy utilisation.

6. UNESCO should try to encourage individual nations to pass laws and regulations beneficial to rapid utilisation of solar energy.

7. UNESCO to sponsor conferences for the education of the general public on a global and regional basis and likewise conferences on very specialised subjects to train technologists in the best methods of solar energy utilisation, such as water heating, building space heating, cooling, etc.